NUCLEAR HYDROGEN PRODUCTION

HANDBOOK

GREEN CHEMISTRY AND CHEMICAL ENGINEERING

Series Editor: Sunggyu Lee
Ohio University, Athens, Ohio, USA

Proton Exchange Membrane Fuel Cells: Contamination and Mitigation Strategies
Hui Li, Shanna Knights, Zheng Shi, John W. Van Zee, and Jiujun Zhang

Proton Exchange Membrane Fuel Cells: Materials Properties and Performance
David P. Wilkinson, Jiujun Zhang, Rob Hui, Jeffrey Fergus, and Xianguo Li

Solid Oxide Fuel Cells: Materials Properties and Performance
Jeffrey Fergus, Rob Hui, Xianguo Li, David P. Wilkinson, and Jiujun Zhang

Efficiency and Sustainability in the Energy and Chemical Industries: Scientific Principles and Case Studies, Second Edition
Krishnan Sankaranarayanan, Jakob de Swaan Arons, and Hedzer van der Kooi

Nuclear Hydrogen Production Handbook
Xing L. Yan and Ryutaro Hino

NUCLEAR HYDROGEN PRODUCTION HANDBOOK

EDITED BY
XING L. YAN
RYUTARO HINO

CRC Press
Taylor & Francis Group
Boca Raton London New York

CRC Press is an imprint of the
Taylor & Francis Group, an **informa** business

CRC Press
Taylor & Francis Group
6000 Broken Sound Parkway NW, Suite 300
Boca Raton, FL 33487-2742

First issued in paperback 2017

ISBN 13: 978-1-138-07468-2 (pbk)
ISBN 13: 978-1-4398-1083-5 (hbk)

Library of Congress Cataloging-in-Publication Data

Nuclear hydrogen production handbook / editors, Xing L. Yan, Ryutaro Hino.
p. cm. -- (Green chemistry and chemical engineering)
Includes bibliographical references and index.
ISBN 978-1-4398-1084-2 (hardcover : alk. paper)
1. Hydrolysis--Handbooks, manuals, etc. 2. Fuel processors--Handbooks, manuals, etc. 3. Hydrogen as fuel--Handbooks, manuals, etc. 4. Nuclear reactors--Industrial applications--Handbooks, manuals, etc. I. Yan, Xing L. II. Hino, Ryutaro. III. Title. IV. Series.

TP156.H82N83 2011
665.8'1--dc23 2011017490

Visit the Taylor & Francis Web site at
http://www.taylorandfrancis.com

and the CRC Press Web site at
http://www.crcpress.com

Contents

Section V Worldwide Research and Development

Section VI Appendices

Foreword

Two enabling technologies for nuclear hydrogen production existed as early as the 1950s. Soon after President Dwight D. Eisenhower of the United States of America spoke about the Atoms for Peace plan to the United Nations General Assembly in 1953, ground was broken for the construction of Shippingport, the first large-scale nuclear power generating plant in the world. The light water reactor went online in 1957, and hundreds more civilian reactors were to follow. At the time electrolysis had been in practice for decades. However, direct combination of the two able to mass produce hydrogen (a manufacturing material in high demand) was not sought after in the market because of plentiful and more affordable oil and natural gas (the hydrocarbon fuels), off which hydrogen can be stripped via a chemical route.

Today, the world demand for the fossil fuels has risen fourfold and the price for them more than doubled. Their proven reserves are estimated to run dry in another 40 and 60 years for oil and natural gas, respectively, at current paces of use. On the day of my writing this foreword, the United Nations Climate Change Conference (COP15) gathered 192 nations in Copenhagen, Denmark for negotiation of an international agreement to limit air-borne emission of climate-altering carbon dioxide gas, a product of fossil fuel consumption. Many came to this meeting with a pledge of deep emission cuts by 2020 including 17% below the 2006 national level in the United States, 25% in Japan and Russia, and 30% in the European Union below the 1990 levels. The threat of climate change is too great to people all around the world and a global accord to mitigate it is imperative.

The Japan Atomic Energy Agency has recently formulated a Nuclear Energy Vision 2100 that proposes how nuclear energy may contribute to a low-carbon society. Relying on a sustainable mix of fast and thermal neutron spectrum fission reactors and future magnetic inertial fusion reactors, our Vision seeks, together with renewable energy and energy efficiency saving, to reduce carbon emission by 25% and 90% below the 1990 level in the coming decade and by the end of the century, respectively, in Japan (our nation is now 16% above that level). In particular, nuclear hydrogen is called upon to replace the majority of fossil fuels used today in the transportation sector through fuel cell engines and in the manufacturing sector through alternative industrial processes such as direct hydrogen reduction of iron ore for steelmaking. In my official capacities in JAEA and AESJ, I am advised by scientists, notably Dr. Xing Yan and Dr. Ryutaro Hino who have over 50 years of collective experience, in the field.

I find that scientists here and abroad have invented more technologies to produce nuclear hydrogen since the dawn of peacetime atomic energy. Besides electrolysis, there are thermochemical, hybrid chemical, thermal reforming, and radiolysis methods combined with several designs of nuclear reactors and systems and with minimal or zero carbon emission. The details of the sciences, engineering, and production applications of these technologies are included in the *Nuclear Hydrogen Production Handbook*. Through development, in which significant public and private interests are currently engaged, these technologies are expected to be put to wide uses, to serve humanity in a low-carbon world.

Dr. Hideaki Yokomizo
Executive Director, Japan Atomic Energy Agency
President, Atomic Energy Society of Japan

Preface

The expansion of the world's population and economy resulted in a 20-fold rise in the use of fossil fuels during the twentieth century. Usage continues to rise and is expected to double the current level by 2050. Neither the trend nor the degree of the present dependence on fossil energy is considered sustainable since the resources of oil, gas, and coal are known to be finite and because intensive use risks grave consequences of climate change. Major alternative fuels are needed on a scale that can keep humanity's development continual in this century and beyond, while steering clear of unwarranted climatic effects. Hydrogen is such an alternative fuel because it can be used in places where fossil fuels are used without emitting the global warming carbon dioxide and is produced in small or large quantities from a variety of resources.

Section I of this handbook introduces the economy-wide roles of hydrogen and the approaches that can be taken to producing it from nuclear energy. The current primary uses of hydrogen in the production of ammonia fertilizers and fuel oils will only grow as agriculture steadily increases with the world's population and as continual volatility in crude oil supply creates incentives to converting widely available and low-priced coal, tar sands, and oil shale into synfuel. It is estimated that increasing the use of hydrogen in hydrocracking by 10-fold from the current 4 million tonnes annually would allow the United States to liquefy enough domestic coal to end oil import. The U.S. manufacture of fuel from coal would be economical if oil is priced at US$35 per barrel. Oil has averaged twice as much in the last 5 years.

Fuel cells are entering markets. Japan calls for 15 million fuel cell vehicles (FCVs) by 2030 and a full replacement of the 75 million strong fleet on its roads today within the century. The National Research Council of the U.S. National Academies sees a more aggressive American deployment scenario of 25 million FCVs by 2030 and 200 million (80% of the light-duty fleet) by 2050, which would need 110 million tonnes of hydrogen fuel annually. Manufacturers will also demand hydrogen to increase sustainability. The present global consensus looks to cut CO_2 emission by 50–80% below the 1990 levels by mid-century. To emerge from the potential applications and policy initiatives is a world economy based on widely available, affordable, and clean hydrogen.

Hydrogen must be produced for it is rarely found alone on the Earth. Half of the current 50 million tonnes annual global production of hydrogen is from natural gas and the rest from oil and coal. Section II describes the basics of the methods with which hydrogen can be produced from nuclear energy with reduced or no fossil feedstock. Incorporating these methods into nuclear reactors to form practical nuclear production systems is discussed in Section III. The resulting systems produce hydrogen through electrolysis of water, nuclear heated steam reformation of hydrocarbons, high-temperature electrolysis, and thermochemical splitting of the water molecule. Section IV reports on applied science and technology and there readers are able to find substantial analyses and data on the present state of the art of nuclear hydrogen production.

The Generation IV International Forum has selected six nuclear reactor concepts for future development that can be licensed, built, and operated to supply economical and reliable electricity, hydrogen, or both while satisfactorily addressing nuclear safety, waste, proliferation, and public acceptance. Section V introduces worldwide up-to-date development and commercialization programs on nuclear reactor systems and associated hydrogen production systems. Section VI presents the properties of the relevant substances.

We would like to thank the large number of the world's leading experts in research institutions, universities, and industries for their contributions that comprise this volume.

Xing L. Yan
Ryutaro Hino

Editors

Xing L. Yan received his PhD from the Massachusetts Institute of Technology in 1990. He participated in the United States Department of Energy's development program on the modular high-temperature gas-cooled reactor and he contributed to the Energy Research Center of the Netherlands' program for small high-temperature reactor cogeneration plant designs. He was a consultant to nuclear reactor vendor industries in the United States, Japan, and France. Since 1998, he has joined the Japan Atomic Energy Agency's design and technology development program for a new generation of GTHTR300C nuclear reactor plants for gas turbine power generation and water-splitting hydrogen production.

Ryutaro Hino received his PhD from the University of Tokyo in 1983. He has since joined the Japan Atomic Energy Agency and currently leads the nuclear hydrogen program on high-temperature reactors. He is the only researcher in the Japan Atomic Energy Agency who has experience in all three leading nuclear hydrogen production methods under worldwide development: steam reforming of methane, high-temperature electrolysis, and thermochemical water splitting. He was awarded the 2007 Prize of the Atomic Energy Society of Japan for his contribution to the successful development of new ceramic heat exchangers used for high-temperature thermochemical hydrogen production.

Contributors

Edward D. Blandford
Department of Nuclear Engineering
University of California
Berkeley, California

Ana E. Bohé
National Atomic Energy Commission
Bariloche, Argentina

Robert Buckingham
General Atomics
San Diego, California

Hongli Chen
Institute of Plasma Physics
Chinese Academy of Sciences
Anhui, China

Charles W. Forsberg
Department of Nuclear Science
and Engineering
Massachusetts Institute of Technology
Cambridge, Massachusetts

Maximilian B. Gorensek
Savannah River National Laboratory
U.S. Department of Energy
Aiken, South Carolina

Jun-ichiro Hayashi
Institute of Materials Chemistry
and Engineering
Kyushu University
Kasuga, Japan

J. Stephen Herring
Idaho National Laboratory
U.S. Department of Energy
Idaho Falls, Idaho

Ryutaro Hino
Japan Atomic Energy Agency
Ibaraki, Japan

Masao Hori
Nuclear Systems Association
Tokyo, Japan

Yoshiyuki Inagaki
Japan Atomic Energy Agency
Ibaraki, Japan

Takamichi Iwamura
Japan Atomic Energy Agency
Ibaraki, Japan

Seiji Kasahara
Japan Atomic Energy Agency
Ibaraki, Japan

Shigeo Kasai
Power Systems Company
Toshiba Corporation
Tokyo, Japan

Satoshi Konishi
Institute of Advanced Energy
Kyoto University
Kyoto, Japan

Shinji Kubo
Japan Atomic Energy Agency
Ibaraki, Japan

Yuta Kumagai
Japan Atomic Energy Agency
Ibaraki, Japan

Won Jae Lee
Korea Atomic Energy Research Institute
Daejeon, Korea

Kazuaki Matsui
The Institute of Applied Energy
Tokyo, Japan

Ryuji Nagaishi
Japan Atomic Energy Agency
Ibaraki, Japan

Toru Nakatsuka
Japan Atomic Energy Agency
Ibaraki, Japan

Horacio E.P. Nassini
National Atomic Energy Commission
Bariloche, Argentina

Tetsuo Nishihara
Japan Atomic Energy Agency
Ibaraki, Japan

James E. O'Brien
Idaho National Laboratory
U.S. Department of Energy
Idaho Falls, Idaho

Hirofumi Ohashi
Japan Atomic Energy Agency
Ibaraki, Japan

Kazutaka Ohashi
Energy and Environmental System Research Center
Fuji Electric Holdings Co., Ltd.
Kawasaki, Japan

Kaoru Onuki
Japan Atomic Energy Agency
Ibaraki, Japan

Per F. Peterson
Department of Nuclear Engineering
University of California
Berkeley, California

Matt Richards
General Atomics
San Diego, California

Nariaki Sakaba
Japan Atomic Energy Agency
Ibaraki, Japan

Hiroyuki Sato
Japan Atomic Energy Agency
Ibaraki, Japan

Carl M. Stoots
Idaho National Laboratory
U.S. Department of Energy
Idaho Falls, Idaho

William A. Summers
Savannah River National Laboratory
U.S. Department of Energy
Aiken, South Carolina

Kazuyuki Takase
Japan Atomic Energy Agency
Ibaraki, Japan

Hiroaki Takegami
Japan Atomic Energy Agency
Ibaraki, Japan

Nobuyuki Tanaka
Japan Atomic Energy Agency
Ibaraki, Japan

Yujiro Tazawa
Japan Atomic Energy Agency
Ibaraki, Japan

Atsuhiko Terada
Japan Atomic Energy Agency
Ibaraki, Japan

Karl Verfondern
Research Center Juelich
Institute for Energy Research
Juelich, Germany

David C. Wade
Argonne National Laboratory
U.S. Department of Energy
Argonne, Illinois

Yican Wu
Institute of Plasma Physics
Chinese Academy of Sciences
Anhui, China

Jingming Xu
Institute of Nuclear and New Energy Technology
Tsinghua University
Beijing, China

Kazuya Yamada
Power Systems Company
Toshiba Corporation
Tokyo, Japan

Xing L. Yan
Japan Atomic Energy Agency
Ibaraki, Japan

Bo Yu
Institute of Nuclear and New Energy Technology
Tsinghua University
Beijing, China

Ping Zhang
Institute of Nuclear and New Energy Technology
Tsinghua University
Beijing, China

Section I

Hydrogen and Its Production from Nuclear Energy

1

The Role of Hydrogen in the World Economy

Ryutaro Hino, Kazuaki Matsui, and Xing L. Yan

CONTENTS

1.1 Introduction

Hydrogen, though the most abundant element in the luminous universe (helium comes in distant second), is rarely found on Earth. The reasons are both that the smallest atom easily diffuses to the outer space and that it is chemically active and readily forms compounds with other elements. Notable compounds containing hydrogen include organic matters and water. In fact, the oceans are the largest terrestrial reservoir of hydrogen.

This handbook concerns the subject of producing hydrogen by splitting it off various chemical compounds including water and describes a range of processes and technologies, from the conventional to the contemporarily researched in the world, designed to convert the primary nuclear energy into chemical energy of product hydrogen.

Hydrogen can be produced from nuclear energy in such manners and quantities that suffice it as a clean and widely available fuel to substitute the fossil fuel uses across the economy, including transportation, stationary and mobile power generation, and energy sources for business, hospital, and home, while meeting substantial demand for hydrogen in sustainable industrial production such as for chemicals and steels.

1.2 Hydrogen Properties

1.2.1 Hydrogen Isotopes

Having atomic number 1, hydrogen is the lightest chemical element. The hydrogen atom is composed of a single electron orbiting a nucleus and can be visualized by the Moon as the electron and the Earth as the nucleus. The electron orbital, which is about a hundred thousand times as large as the size of the nucleus, is formed by the Coulomb interaction between the negatively-charged electron and the positively-charged nucleus.

Hydrogen has three known natural occurring isotopes with the standard atomic weight of 1.00794 u. They include protium ^1H, deuterium ^2H (also represented by D), and tritium ^3H (T). At standard conditions of temperature and pressure, these isotopes naturally form stable diatomic molecular gases, for example, H_2 or HT.

Protium is most common with an abundance of 99.9885% of the natural hydrogen atoms. The isotope, also known as ordinary hydrogen, contains a single proton and no neutron in the nucleus and its atomic mass is 1.007825032 u. The isotope is basically not radioactive. Water primarily is made of molecules of protium with oxygen, namely, H_2O. So are organisms of hydrogen with carbon, for example, methane, CH_4. This handbook is concerned with the production of this isotope and the ordinary hydrogen gas.

Adding a neutron into the nucleus of protium makes what is known as deuterium. Therefore, the latter approximately doubles the atomic mass of the former. Deuterium has a natural abundance of 0.0115% and is nonradioactive. The chemical compound of deuterium and oxygen, D_2O, is known as heavy water. Natural water on the Earth like the oceans contains a small concentrate of deuterium. As a result, heavy water or deuterium can be obtained from water for practical uses. Heavy water is used as a neutron moderator and coolant in some nuclear fission reactors. Deuterium is also useful as a partial fuel for nuclear fusion reactors.

Tritium populates the nucleus with two neutrons and one proton, and weighs about three times as heavy as a protium atom. When combined with oxygen, it forms tritiated water, T_2O and more often HTO. Unlike the other isotopes of hydrogen, tritium is radioactive with a half-life of 12.31 years and decays to ^3He through β decay with release of electron energy (18.61 keV) and emission of an antineutrino. Tritium occurs naturally as a result of the cosmic radiation of atmospheric gases, mainly through fast neutron (>4 MeV) spallation of atmospheric nitrogen ($^{14}N + {}^1n \rightarrow {}^{12}C + {}^3H$). Because of the relatively short half-life, only traces of tritium occurring in this way exist at any moment and accounts approximately 4 per 10^{15} natural hydrogen atoms in the atmosphere. The tritium population is much less concentrated in natural water. However, tritium may be produced in several ways including neutron activation of lithium-6 and neutron capture by deuterium in nuclear reactors. Tritium is considered an indispensable part of fuel for nuclear fusion energy.

Although deuterium and tritium are sought to provide a practical atomic fuel for fusion energy, they are not explicitly required for hydrogen used as chemical fuel and ordinary manufacturing feedstock. This book is thus not concerned with the specific subject of producing the heavier isotopes of hydrogen.

1.2.2 Physical Property

Hydrogen has the second-lowest boiling point (–252.78°C) of all substances, after only helium (–268.92°C), at atmospheric pressure. Pressurization can do little to raise the boiling point of hydrogen. These properties make it difficult, but not impractical, to store hydrogen as liquid. As a result, hydrogen as an automotive fuel has been stored more often as a pressurized gas than a cryogenic liquid in on-board fuel tank. Alternatively, hydrogen may be stored and resupplied via hydrogenation and dehydrogenation of various types of hydrides such as saline (e.g., NaH), covalent (e.g., $NaAlH_4$), and interstitial (e.g., Pd) hydrides.

The density of hydrogen gas (H_2) is 0.08375 kg/m^3 and specific volume is 11.940 m^3/kg at standard conditions of 20°C and 101.325 kPa. To estimate density ρ (and specific volume being inverse of density) in the modest range of temperature and pressure from the standard conditions uses the ideal gas law, $\rho = P/RT$, where the specific gas constant of hydrogen $R = 4124.45$ J/kg K.

At high pressures, hydrogen gas deviates significantly from the thermodynamic behavior of an ideal gas and the density of hydrogen is actually 2.9% less at 5 MPa and 5.7% less at 10 MPa than predicted by the ideal gas law. This is called compressibility factor, which can be measured directly. The equation of state for real (nonideal) hydrogen gas is recently reported [1]. Hydrogen gas is often stored onboard a vehicle as a fuel in a pressure range of 35–70 MPa. At a temperature of 20°C and accounting compressibility factor, the density of hydrogen is 23.651 kg/m^3 and specific volume is 0.042282 m^3/kg at 35 MPa while these values are 39.693 kg/m^3 and 0.025193 m^3/kg at 70 MPa. The density of liquid hydrogen is 71.107 kg/m^3 and the specific volume is 0.014063 m^3/kg at –253°C and 101.325 kPa near the normal boiling point.

Hydrogen gas has the smallest molecular size compared to all other gases, and can diffuse through materials that are impermeable to other gases. Metals or nonmetals exposed constantly to hydrogen may become brittle. Containers of hydrogen gas require deliberate techniques of material and construction, and are an ongoing development issue.

Hydrogen is generally not toxic, but poses a risk of asphyxiation if inhaled. Because hydrogen gas is odorless, tasteless, and invisible to human beings, it is difficult to detect a leak of hydrogen. A leak will not spread but rise quickly due to the highly buoyant nature of hydrogen in atmospheric air. Gaseous hydrogen has a specific gravity of 0.0696 at 20°C and 1 atm and is thus approximately 7% the density of air. Liquid hydrogen has a specific gravity of 0.0708 at the boiling point (–282.78°C) and is about 7% the density of water. A leak of liquid hydrogen, which is 59 times heavier than air, would evaporate and rise quickly in ambient air due to the low boiling point and specific gravity of hydrogen.

1.2.3 Chemical Property

Hydrogen forms a vast array of compounds with carbon. Millions of hydrocarbons are known as organic components. Natural gas and crude petroleum are among them. They originated biologically, and many transformed into being over time.

Hydrogen forms chemical or inorganic components with other elements. Water is the chemical component that hydrogen forms with oxygen. Like hydrogen, pure water is colorless, odorless, and tasteless. It is neither acidic nor basic. Water is the most abundant component on the Earth's surface. Interestingly, water is a renewable source of hydrogen fuel. The use of hydrogen fuel in a combustor or fuel cell forms the same amount of water used to produce it, and no carbon dioxide or pollutants. Hydrogen fuel holds the promise of a clean energy carrier to the future.

Hydrogen can react with organic and chemical components. This property contributes to a broad range of manufacturing activities. Hydrogenation is used to refine or sweeten organic components in petroleum and food processes. Ammonia fertilizers are made by the chemical reaction of hydrogen with the source of nitrogen gas in the air. Hydrogen as an effective reducer is used to remove oxygen (forming H_2O) from metal oxides to produce metals. It is also used to chemically remove unwanted impurities from industrial products. Some major manufacturing applications of hydrogen are reviewed later in this chapter.

Finally, hydrogen can also form compounds with other elements and components through ionic bounding. By taking on a partial positive charge, hydrogen binds to more electronegative elements such as halogens (e.g., F, Cl, Br, and I). Similarly, by taking on a partial negative charge, it forms compounds with more electropositive materials such as metals and metalloids, and these are known as various designs of hydrides, some of which are interesting hydrogen storage media.

1.2.4 Fuel Property

Hydrogen gas is inflammable with a wider range of ignition concentrations in air than other conventional fuels (see Table 1.1). Hydrogen burns in air with a pale blue flame. If burned in pure oxygen, hydrogen flames emit ultraviolet light and are nearly invisible, as observed behind hydrogen–oxygen rocket engines. The hydrogen–oxygen combustion follows the exothermal chemical reaction:

$$2H_2 + O_2 \rightarrow 2H_2O + \Delta H \tag{1.1}$$

The enthalpy, ΔH, of the combustion product is 285.83 kJ/mol (higher heating value or HHV) and 241.82 kJ/mol (lower heating value or LHV) for the conditions of 25°C and 101.325 kPa. The mass-based enthalpy values are given in Table 1.1 for the HHV and LHV as defined therein. Water is produced in combustion as steam, and therefore the LHV represents the amount of energy usable to do work. Hydrogen easily has the highest energy content per mass of not only the fuels in Table 1.1 but all combustion fuels. Multiplying energy content per mass by density, whose values are given in Table 1.1, gives energy density of a fuel. Because its density is so small, hydrogen has the lowest energy density (10.074 MJ/m^3 LHV and 11.915 MJ/m^3 HHV) of all combustion fuels. Methane gas has more than three times the energy content of hydrogen gaseous fuel while gasoline has nearly 3000 times greater energy density than hydrogen.

TABLE 1.1

Comparative Properties of Hydrogen and Conventional Fuels

Properties	Units	Hydrogen [1]	Methane [1]	Propane [1]	Methanol [1]	Ethanol [1]	Gasoline [2]	Notes/ Sources
Chemical formula		H_2	CH_4	C_3H_8	CH_3OH	C_2H_5OH	C_nH_m ($n = 4–12$)	
Molecular weight		2.02	16.04	44.1	32.04	46.07	100–105	[a, b]
Density (NTP)	kg/m^3	0.0838	0.668	1.87	791	789	751	[3, a, c]
Viscosity (NTP)	g/cm s	8.81×10^{-5}	1.10×10^{-4}	8.012×10^{-5}	9.18×10^{-3}	0.0119	0.0037–0.0044	[3, a, b]
Normal boiling point	°C	–253	–162	–42.1	64.5	78.5	27–225	[a, b]
Vapor specific gravity (NTP)	air = 1	0.0696	0.555	1.55	N/A	N/A	3.66	[3, 6, a, d]
Flash point	°C	<–253	–188	–104	11	13	–43	[b, d]
Flammability range in air	Vol%	4.0–75.0	5.0–15.0	2.1–10.1	6.7–36.0	4.3–19	1.4–7.6	[c, b, d]
Auto ignition temperature in air	°C	585	540	490	385	423	230–480	[b, d]
Heat values								
LHV (low heating value)	MJ/kg	120.21	47.141	46.28	20.094	26.952	43.448	[4, e]
HHV (high heating value)	MJ/kg	142.18	52.225	50.22	22.884	29.847	46.536	[5, e]

Sources:
[a] U.S. NIST Chemistry WebBook.
[b] Alternatives to Traditional Transportation Fuels: An Overview. DOE/EIA-0585/O. Energy Information Administration. U.S. Department of Energy (DOE), Washington, DC. June 1994.
[c] *Perry's Chemical Engineers' Handbook* (7th edn), 1997, McGraw-Hill, New York.
[d] Hydrogen Fuel Cell Engines and Related Technologies. Module 1: Hydrogen Properties. U.S. DOE. 2001.
[e] Hydrogen Analysis Resource Center. U.S. DOE. January 2010.

Notes:
[1] Properties of the pure substance.
[2] Properties of a range of commercial grades.
[3] NTP = 20°C and 1 atm.
[4] LHV is defined as the heat released from burning a fuel (initially at 25°C) and cooling combustion products to 150°C, discounting the latent heat of water vapor.
[5] HHV is defined as the heat released from burning a fuel (initially at 25°C) and cooling combustion products to 25°C, counting the latent heat of water vapor.
[6] N/A—Not applicable.

When produced, hydrogen is made into a carrier of the primary energy (such as nuclear, solar, wind, biomass, fossil, and other sources) used to produce it. Most commonly used today are fossil resources of natural gas, oil, and coal. Hydrogen can also be converted from electricity, another energy carrier, in an electrolyzer and vice versa in a fuel cell.

The fuel cell is generally made up of an anode, electrolyte, and cathode, which are sandwiched in a cell unit. The electrolyte is the key element designed to selectively pass ions between the anode and the cathode. As illustrated in the simplified hydrogen fuel cell of Figure 1.1, a catalyst (usually platinum powder) breaks the hydrogen molecules down into positive ions (protonnes) and negative electrons in the anode. The electrolyte passes the ions through it but blocks the electrons, because it is made of a nonelectrical conducting material. Instead, the electrons turn the other way to flow as electrical current in a circuit provided externally to the cathode. In the mean time, the ions pass through the electrolyte and arrive at the cathode, where they meet with the incoming electrons. The ions and

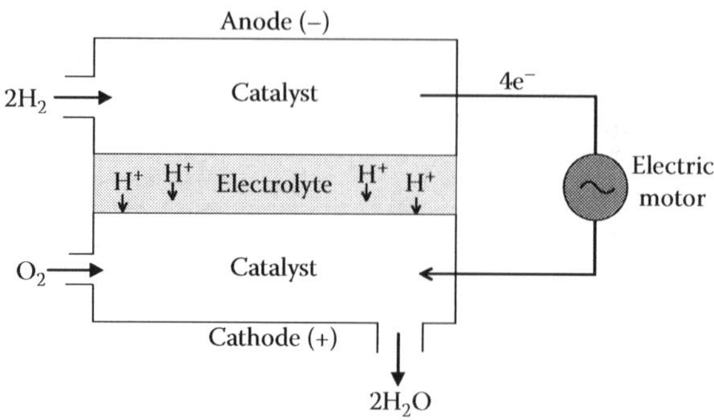

FIGURE 1.1
Working principle of hydrogen fuel cell.

electrons react in the presence of a catalyst (usually nickel) with an oxidant, such as oxygen in the air, to produce the by-product, hot water or vapor. In the process, the chemical energy of hydrogen is converted into electrical energy. The reactions are summarized below.

$$\text{Anode reaction: } 2H_2 \rightarrow 4H^+ + 4e^- \tag{1.2}$$

$$\text{Cathode reaction: } O_2 + 4H^+ + 4e^- \rightarrow 2H_2O \tag{1.3}$$

$$\text{Overall cell reaction: } 2H_2 + O_2 \rightarrow 2H_2O + \Delta H \text{ (electricity and heat)} \tag{1.4}$$

In theory, the maximum efficiency with which a hydrogen fuel cell could convert chemical energy of hydrogen to electricity, via reaction Equation 1.4, is 83%, which is Gibbs free energy to the enthalpy change, ΔH at 25°C and 101.325 kPa. The rest of 17% is ideal heat production.

While the hydrogen ions pass some types of electrolyte (phosphoric acid or polymer membrane) as described above, oxygen can also pick up electrons and travel through other types of electrolyte (molten carbonate and solid ceramic oxide) to the anode, where it combines with hydrogen ions to form water.

Hydrogen fuel and oxygen continuously feed into the fuel cell to generate direct electrical current and by-products of water and heat. Since a single fuel cell due to size limit enables small current and voltage only, multiple fuel cells are stacked to tailor to practical requirements. The sum of the surface area of the cells in parallel stacking determines the total current while the number of the cells in serial stacking determining the total voltage. Multiplying the voltage by the current gives the total of electrical power generated by a stack.

As to be discussed later, hydrogen fuel cells are considered in various important economical applications, particularly for transportation and power generation. Fuel-cell vehicles (FCVs) operate two times more efficiently (fuel tank to wheel) than the most efficient gasoline cars including hybrid cars, because they send out less waste heat, and emit no carbon dioxide in the tailpipe. When burned in an engine, hydrogen combustion needs to control nitrogen oxide as the only major pollutant.

As analyzed earlier, hydrogen is a low volume-dense fuel in comparison with gasoline. As a result, a gas fuel tank of 7 kg compressed (35 MPa) hydrogen sized to travel a range of 700 km requires a net tank volume of about 300 L in a gas tank, comparing to about 70 L of gasoline tank in a conventional gasoline engine car for the range. Although liquid hydrogen tank requires a third of storage volume of gas tank, this approach incurs energy consumption for hydrogen liquefaction and challenges tank insulation design to prevent boil off.

Onboard storage is a key research and development issue for hydrogen as transportation fuel. The current options are to store hydrogen as cryogenic liquid (<–253°C), compressed (35–70 MPa) gas, and hydrides (e.g., LiH and AlH_3). The second option is usually used now at volume density of 25 g/L and the last one is seen most promising. A hybrid compact design combining the compressed gas and hydride options is also proposed [2] that could double the specific mass storage of hydrogen to 5 kg H_2 in a 100-L tank. The hybrid design may include specifications similar to the following:

- Aluminum–carbon fiber-reinforced plastic composite vessel (pressure loading)
- Metal hydride core for hydrogen absorption and heated release
- Tank storage volume density >50 g/L
- Tank gross weight: 40 kg
- Tank fill pressure 70 MPa; burst pressure >200 MPa
- Pressure cycle operating life >13,000 cycles

1.3 Traditional Hydrogen Applications

Hydrogen plays significant roles in the world economy today. The hydrogen consumption worldwide is about 50 million tonnes per year, and in 2008, North America and Asia-Pacific led the world in hydrogen consumption with about 30% each, followed by Western Europe (18%) and other regions (22%).

The main consumers of produced hydrogen are industries. Globally, it is mainly used in the production of ammonia and for petroleum refining, on similar scale in both areas. Other uses are on much smaller scales for semiconductor manufacturing, materials processing such as glass production, food preparation, and chemical production.

1.3.1 Ammonia Production

Ammonia (NH_3) is typically produced by catalyzed chemical reaction of hydrogen, essentially produced from primary energy sources, with nitrogen obtained by processing air to form anhydrous liquid ammonia through the Haber–Bosch process:

$$3H_2 + N_2 \rightarrow 2NH_3 \tag{1.5}$$

Ammonia as fertilizers contributes to the essential nutritional needs of terrestrial organisms. It also contributes to the synthesis of pharmaceuticals. Ammonia solutions are the basis of commercial and household cleaning products. The 2006 worldwide production of ammonia was estimated at 146.5 million tonnes. In 2004, China produced 28.4% of the worldwide output, India 8.6%, Russia 8.4%, and the United States 8.2%. Because, the majority (e.g., 83% in 2003) of the worldwide production of ammonia is used directly or indirectly in fertilizers, the production is expected to steadily increase in future with increase in the world population, which stands to be 6.8 billion as of April 2010 and has consistently added nearly a billion people every 13 years since 1960.

The 2006 global ammonia production consumed 26 million tonnes of hydrogen. In the same year, the U.S. consumption was 2.2 million tonnes of hydrogen. Today, hydrogen required for ammonia production comes mainly from steam reforming of methane (i.e., natural gas) with a large amount of CO_2 emission arises (the mass of CO_2 emitted is 8.8

times the mass of hydrogen produced based on current performance of industrial natural gas reformer plant). Hydrogen can be produced from other sources. In 2002, Iceland produced 2000 tonnes of hydrogen by hydropower electrolysis for the production of ammonia. Future options can include hydrogen produced from water by nuclear, solar, and wind energy sources, thus avoiding CO_2 emission and saving natural gas resource. Nuclear energy can make hydrogen from water by electrolysis and the thermochemical process which will be described in Section 1.5. The introduction of massive nuclear hydrogen would greatly increase agricultural productivity and sustainability by reducing the dependence on hydrocarbon resources.

1.3.2 Petroleum Industry

Hydrogen is substantially consumed in various petroleum refining processes. Hydrogen is used to make petroleum products cleaner, for example, by hydrodesulfurization. Moreover, hydrocracking is typically used in processes of catalytic cracking and hydrogenation, wherein heavy (long-chain hydrocarbons) or difficult (containing excessive sulfur and nitrogen compounds) forms of crude oil are cracked and converted by adding hydrogen to yield synthetic crudes. In 2006, the U.S. consumption of hydrogen for hydrocracking was about 4 million tonnes. In the petroleum industry, hydrocracking is substantially and increasingly performed because of demand for low-sulfur fuel products due to tightened environmental regulations and additionally because rising oil prices justify the cost for the industry to convert low-grade oils such as $(CH)n$ tar sands and $(CH_{1.5})n$ heavy crude into usable $(CH_2)n$ fuels. More than 50% of Canadian oil production is from tar sands. Current practice in hydrocracking consumes hydrogen from steam reforming of natural gas. In 2008, U.S. refiners used 5.3 billion Nm^3 of natural gas as feedstock for hydrogen production which resulted in 11 million tonnes of CO_2 emissions. To use hydrogen produced of water with nuclear energy (or via nuclear-heated reforming) can significantly reduce fossil resources consumption and emissions in the oil sector. Table 1.2 summarizes the results of the case studies to refine a quarter million barrels per day (BPD) of crude oil by a nuclear high-temperature gas reactor (HTGR) via water-splitting hydrogen production at 50% efficiency and by conventional hydrocracking via natural gas reforming at 80% efficiency.

Synfuel can be produced from coal, natural gas, and biomass through numerous processes, of which Fischer–Tropsch and Bergius processes are often encountered. Hydrogen as a process reactant is required and can be obtained from steam reforming of natural gas or by gasification plus water shift reaction of additional solid feedstock. In 2009, the

TABLE 1.2

Estimates of the Energy Consumption and Emissions of Crude Oil Refining Using Natural Gas and Nuclear Hydrogen

	Natural Gas Hydrocracking	Nuclear Hydrocracking
Crude refining capacity[a]	250,000 BPD	250,000 BPD
Refining production process	Hydro-cracking/treating	Hydro-cracking/treating
Process heat and power supply	Fossil energy plant	Nuclear cogeneration plant
Hydrogen supply	Natural gas reforming	Water splitting
	89 t/d H_2	1 HTGR (6OOMWt)
Natural gas consumption	350 t/d	0
$C0_2$ emissions	970 t/d	0

[a] Current world total production = 84 million BPD

worldwide commercial synfuel production capacity was about 0.24 million barrels per day compared with 84 million barrels per day of crude oil production. If hydrogen used could be produced from renewal or nuclear energy sources, synfuel production could expand greatly by the captive use of hydrogen with reduced carbon footprints of synfuel products. It is estimated that 37.7 MMT/year of hydrogen would be sufficient to convert enough domestic coal to liquid fuels to end U.S. oil dependence (57% in 2008) on foreign import (11 million barrels per day in 2008), according to the U.S. Energy Information Administration. This has already been practiced in the South Africa by Sasol, which now runs the world largest synfuel production from coal with a capacity of 150,000 barrels per day and supplies 30% of the country's gasoline and diesel fuel uses.

During the 1980s, a number of demonstration coal-to-liquid (CTL) units were built elsewhere in the world, mainly in Japan and in the US, involving a range of coal types. However, much of the work was stopped in the 1990s because of the top-sided pressure of the then low oil price. With the increase of oil prices in the past decade, CTL units have been reconsidered in these and other countries. Shenhua Group, China's largest as well as technology-driven coal company, commissioned a 1.06 million ton per year direct CTL (Bergius process) demonstration plant in 2008 and has active plans for further capacity expansion. Several indirect CTL (Fischer–Tropsch) plants are also being developed in the country. Since both direct and indirect liquefaction routes are commercialized, current research looks at increasing productivity and improving overall environmental performance, the latter being the largest drawback of synfuel relative to conventional oil.

The hydrogen feedstock to the CTL processes is now produced by steam reforming of natural gas or coal gasification. Either way involves considerable CO_2 emissions, so carbon sequestration is necessary to bring the emission performance of synfuel on par with that of conventional petroleum. The American Clean Coal Fuels company is developing a production facility with an output of 26,000 barrels per day of synthetic diesel and jet fuels with integrated carbon capture and sequestration. The facility will convert 12,000 tons of coal per day from a new mine in Illinois and biomass from waste or agricultural sources via gasification and Fischer–Tropsch conversion. The company aims to start the operation of the facility in 2013.

Use of nuclear hydrogen, heat, and electricity, all of which could be cogenerated by the Generation-IV reactors under current development worldwide, would eliminate nearly all CO_2 emissions from the CTL processes and double the yield of synfuel from a unit of coal. Alternatively, existing nuclear power plants (light and heavy water reactors) can already produce hydrogen by electrolysis, most economically using off-peak electricity. Table 1.3 gives the number of HTGRs required to generate hydrogen feedstock, via thermochemical or steam electrolysis, needed to produce 250,000 BPD of synfuel by direct CTL and compares the estimated performance parameters of such a nuclear production scheme with that using coal-derived hydrogen assuming 60% yield of coal by weight.

1.3.3 Other Applications

Hydrogen finds many other major uses in industrial applications including:

- Food production (butter, margarine, frying oil) from the hydrogenation of vegetable (soybean, sunflower, corn, etc.) oils and some unsaturated animal fats.
- Chemical manufacturing for soaps, plastics, ointments, and so on by hydrogenating nonedible oils; chemical production of methanol (CH_3OH), a common industrial chemical (H_3OCl) (of worldwide demand of about 30 million tonnes a year)

TABLE 1.3

Estimated Coal Feeds and Emissions of Direct CTL Using Coal-Sourced Hydrogen and Nuclear Hydrogen

	Coal-Derived Hydrogen	Nuclear Hydrogen
Synfuel production capacity	250,000 BPD[a]	250,000 BPD[a]
Synfuel production process	Direct coal liquefaction	Direct coal liquefaction
Process heat and power supply	Coal-fired plant	Nuclear cogeneration plant
Coal consumption	84,000 t/d	41,000 t/d
Hydrogen supply	Coal gasification	Water splitting
	7,600 t/d H_2	60 HTGRs (600MWt each)
CO_2 emissions	103,000 t/d	0

[a] Current world total production

made from hydrogen reacting with carbon dioxide or carbon monoxide; and hydrochloric acid (H_3OCl) used in the manufacturing of numerous other chemical products such as vinyl plastic, polyurethane, and food additives.

- Metal work with the use of hydrogen as a protective shielding gas in high-temperature operations, such as manufacturing stainless steel, welding and cutting austenitic steels; and metal production from metal ores with hydrogen used to reduce hot metal oxides such as tungsten oxide (WO_3).
- Aerospace programs to fuel spacecraft and life-support systems. For example, the U.S. National Aeronautics and Space Administration (NASA) uses approximately 5000 tonnes per year of liquid hydrogen for space launches including NASA's space-shuttle flights. Hydrogen is fuel for the shuttle main engines and also for on-bound fuel cells used to power the shuttles' electrical systems, the exhaust of which is only pure water used as drinking water by the crews.
- Semiconductor manufacturing where hydrogen is used as a carrier gas for active trace elements and creates specially controlled atmospheres for etching semiconductor circuits.
- Power generation, where hydrogen is coolant for cooling large-scale high-speed turbine generators taking advantage of hydrogen's high thermal conductivity.

1.4 Developing Hydrogen Applications

The concept of hydrogen-energy economy as depicted in Figure 1.2 is widely discussed. It refers to producing hydrogen economically and environmental-friendly, as energy store and carrier, as industrial material, and of sufficient quantities to replace fossil resources (oil, natural gas, and coal) that are used in today's fossil-energy economy. The concept is being developed because the current practice of fossil energy economy is well understood to be unsustainable. The proven reserves of 1.3 trillion barrels of oil, 185 trillion Nm^3 of natural gas, and 826 billion tonnes of coal would last 42, 60, and 122 years, respectively based on the 2008 world consumption rates [3]. Moreover, the reliance of the world economy on the fossil fuels has accelerated with the consequence of rapidly increasing global emissions of carbon dioxide greenhouse gas into the Earth's atmosphere to 30 billion tonnes in 2006, doubling the amount in 1970 as shown in Figure 1.3.

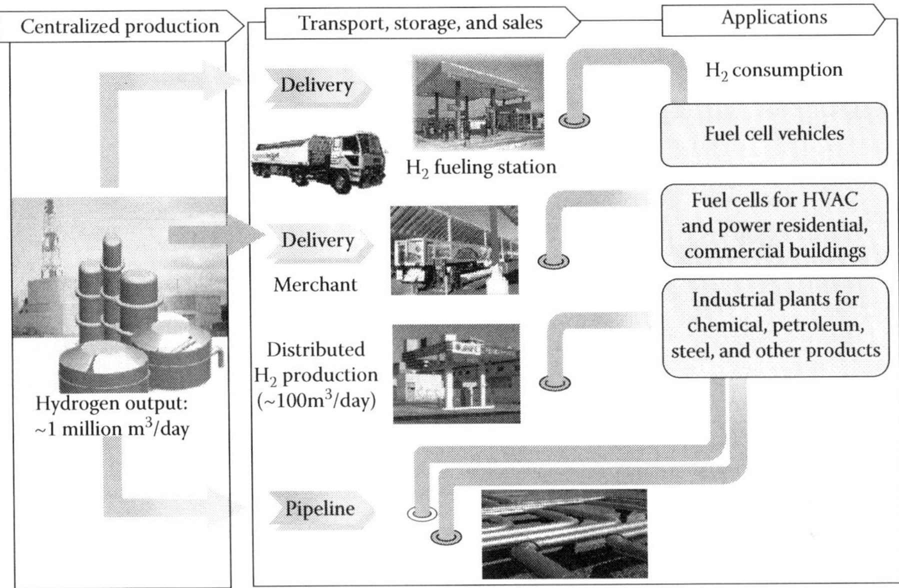

FIGURE 1.2
The elements of developing hydrogen energy economy for the future.

Technologically, hydrogen can be produced from primary energy sources other than fossil resources. Similarly, hydrogen has proven viable in emerging transport and stationary power generation applications based on hydrogen combustion and fuel cells, and in a broad range of advanced commercial and industrial processes. These hydrogen technologies and enabling policies are being developed in many countries and regions for the building of a sustainable hydrogen energy economy.

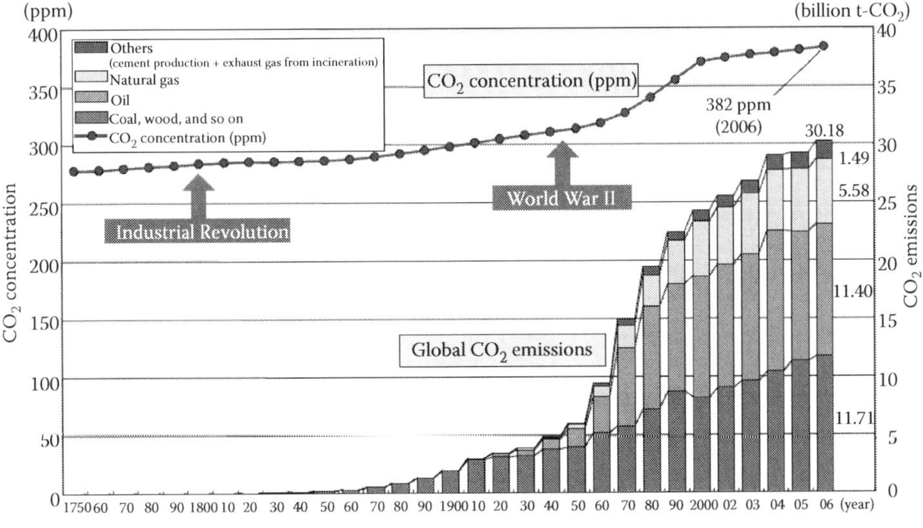

FIGURE 1.3
Global CO_2 emission and atmospheric concentration. (Adapted from carbon dioxide information analysis center at http://cdiac.ornt.gov.)

1.4.1 Development Programs of Applications and Policies

1.4.1.1 The United States

In November 2002, the United States Department of Energy (DOE) issued a National Hydrogen Energy Roadmap for the multiphased development of a U.S. hydrogen economy, as shown in Figure 1.4 [5]. The Roadmap concluded that "Expanded use of hydrogen as an energy carrier for America could help address concerns about energy security, global climate change, and air quality. Hydrogen can be derived from a variety of domestically available primary sources, including fossil fuels, renewables, and nuclear power." In 2004, the National Research Council (NRC) of the U.S. National Academies reported its findings of the technical and policy issues about the hydrogen economy [6]. It found that the United States could have two million hydrogen-powered fuel-cell cars by 2020, which would represent only 1% of all vehicles on roads. After that, the numbers could rise quickly, reaching 60 million by 2035. By 2050, the United States will have the potential of using hydrogen to eliminate essentially all gasoline vehicles and associated CO_2 emissions by overcoming four basic development challenges for practical fuel cells, acceptable onboard hydrogen storage systems, the infrastructure of hydrogen refueling, and the reduced cost and environmental impact of hydrogen production. The petroleum(gasoline and diesel)-based transport in the United States emits 1.8 trillion tonnes of CO_2 in 2009 according to EIA.

The U.S. Energy Policy Act of 2005 authorized the DOE to work with the private sector on technologies related to the production, purification, distribution, storage, and use of hydrogen energy, fuel cells, and related infrastructure. The U.S. Congress has since appropriated funding for the DOE Hydrogen Program, and the fiscal year 2009 funding for the program stood at $269 million.

In a separate 8-year, $180M-budget, industry–DOE cost-shared Advanced Hydrogen Turbine for the FutureGen project, General Electric and Westinghouse have since 2005 been developing the flexibility of conventional gas turbines with minor modifications to operate on pure hydrogen as combustion fuel while maintaining the same performance in terms efficiency and emissions. This project builds on existing gas turbine technology and product developments, and will develop, validate, and prototypically test the necessary

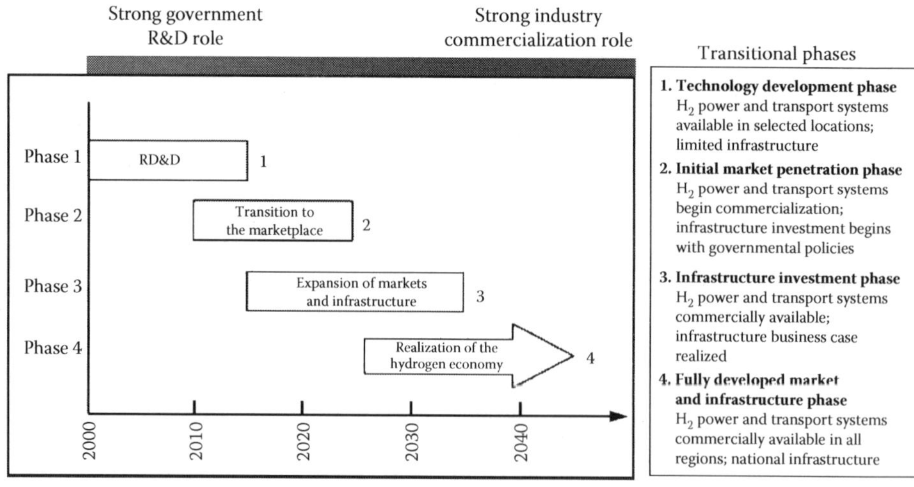

FIGURE 1.4
Phases in the U.S. development of a hydrogen economy.

turbine related technologies and sub-systems needed to demonstrate the ability to meet the DOE turbine program goals. The goal of the project is to develop two hydrogen turbine designs, each by the two gas turbine industry leaders, that could be built and delivered in a 2012 time frame. The prototypes are sought in 2015.

1.4.1.2 Japan

In 2002, the Japan Hydrogen & Fuel Cell Demonstration Project (JHFC) was launched for the demonstration of FCVs and hydrogen refueling infrastructure [7]. The project has the participation of about two dozens of major domestic and foreign automakers (Toyota, Honda, Nissan, GM, Daimler, etc.) and petroleum and gas companies (Shell, ENEOS, Tokyo Gas, etc.), and is subsidized by the Ministry of Economy, Trade and Industry (METI) currently through the New Energy and Industrial Technology Development Organization (NEDO). A total of eight models of hydrogen vehicles including six fuel-cell cars, a fuel-cell bus, and a hydrogen internal combustion engine (ICE) car were demonstrated. The practical road runs were serviced by 11 hydrogen fueling stations in Tokyo and other large cities. The JHFC project was conducted to gather fundamental data on hydrogen supply, by forecourt (on-site) production systems and distribution sources, and vehicle performance including environmental impacts, total energy efficiency, and the safety, all performed under actual road conditions. The data will be used to develop the roadmap of technology, infrastructure (e.g., determining refueling station specifications), regulatory standards (safety, etc.) for the full-scale mass production and widespread use of FCVs in the country.

In 2006 and updated in 2008, Japan follows a national Fuel Cell/Hydrogen Technology Development Roadmap to develop fuel-cell technologies including polymer electrolyte fuel cell (PEFC) and solid oxide fuel cell (SOFC) types, and hydrogen technologies [8,9]. Currently, the development is structured in 11 projects in three development areas including stationary fuel-cell systems (2010 commercial start and 2015 becoming cost competitive), FCVs (2015 commercial start and 2020–2030 gaining commercially mature) and hydrogen infrastructure. The latter includes hydrogen delivery and storage technology, code, and standards necessary to construct a hydrogen society.

For the key onboard storage issue, the Hydrogen Storage Technology Roadmap validated 3–5 kg onboard tank in 2007 based on investigation on various hydrogen storage materials using large-scale facilities including radiation synchrotron and accelerator. The 2008 Roadmap directs the commercial development through compact design and improved hydride storage to a target of 5–7 kg storage tank (necessary to achive a driving range of 500–700 km) during the earlier commercialization period beginning in 2015. The final target will set on 7 kg H_2 fuel tank at affordable cost through mass production for the matured commercial deployment of FCVs in 2020–2030.

Similarly, Japan has been developing fuel-cell stationary energy systems since 2005 and has just begun aggressive introduction of such standardized units of around 1 kWe rating with high thermal efficiency into domestic and oversea markets. As of 2010, several thousands of these units are already operational in the country.

In 2004, the Advisory Panel of Agency for Natural Resource and Energy (ANRE) of the government issued the targets of market introduction of fuel cells for transport vehicles and stationary applications. The plans call for deployment of as many as 2.5 million stationary units totaling 12.5 GWe for power and heat generation, and 15 million FCVs by 2030. On the basis of the targets, the details of official estimates for hydrogen demands are given in Figure 1.5. It is interesting to note that in 2004 ANRE planned a far greater demand for hydrogen by stationary applications than for transportation, which is actually reflected

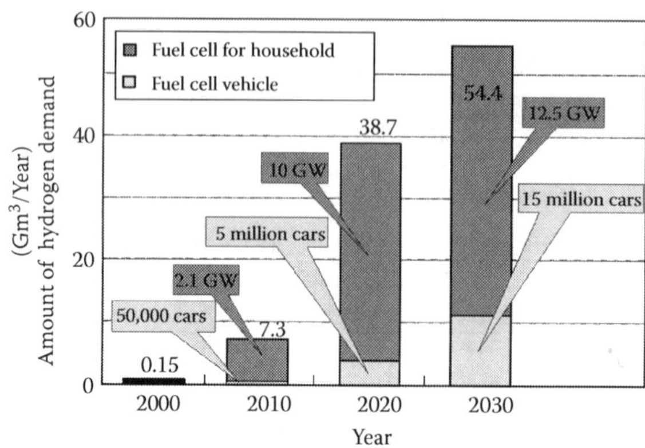

FIGURE 1.5
Hydrogen demands by the officially planned number in Japan of FCVs and household power and heat units.

in the progress of the market in 2010. At this moment, the fuel-cell units are more aggressively introduced into homes annually in large volumes by the joint efforts of government and energy-utility companies.

The code and standards as prerequisite for a hydrogen-economy society have been developed in parallel to those of the above hydrogen technology and product development in the country.

1.4.1.3 Europe

In 2002, an "High Level Group on Hydrogen and Fuel Cells (HLG)" has been established by the European Commission (EC). Its principal task is to initiate strategic discussions for the development of a European consensus on the introduction of hydrogen energy. In 2004, the EC started another policy group, the "European Hydrogen and Fuel Cell Technology Platform" (HFP). The key elements of the European coordinated strategy include a strategic research agenda with performance targets, timelines, lighthouse demonstration projects, and a deployment strategy or roadmap for Europe. The general EU targets by 2020 are a 10–20% supply of the hydrogen energy demand by CO_2-free or lean sources and a 5% hydrogen fuel market share.

HyWays is an integrated project to develop the European Hydrogen Energy Roadmap as a synthesis of national roadmaps from the participating member states [10]. Based on investigation of the technical, socio-economic, and emission challenges and impacts of realistic hydrogen supply paths as well as of the technological and economical needs, the Roadmap details the steps of an action plan necessary to move toward greater use of hydrogen. It projects that an estimated 25 million hydrogen cars will be on the European roads in 2030. The study has also found that introducing hydrogen into the energy system would reduce the total oil consumption by the road transport sector by 40% between today and 2050.

Regarding the large-scale hydrogen production, in the early phase up to 2020, hydrogen production will rely on steam reforming of natural gas, electrolysis, and by-product contributions. On the longer term, by 2050, production will be based on centralized electrolysis and thermochemistry from renewable feedstock and CO_2-free or lean sources (coal and natural gas with carbon capture and sequestration, and nuclear).

In 2005, the HFP adopted a research agenda for accelerating the development and market introduction of fuel cell and hydrogen technologies within the European Community, and called for funding by the EC public and private sectors. In 2006, the agenda was adopted by the European Council. In 2007, the EC adopted two proposals to develop and market hydrogen vehicles. First proposal was designed to simplify the regulatory procedures for hydrogen-powered vehicles, and the second was to establish the "Fuel Cells and Hydrogen Joint Technology Initiative" as called for by the HFP agenda [11].

The EC's second proposal was duly considered by the European Parliament and the Council of Ministers, and in May, 2008 the Council passed a regulation of creating the "Fuel Cells and Hydrogen Joint Undertaking (FCH JU)" that will run from 2008 to 2017, and the energy, nanotechnologies, environment, and transport programs are cofunded by the EC and private sectors for €970 million overall during the period. The FCH JU is a public–private partnership supporting research, technological development, and demonstration in fuel cell and hydrogen energy technologies and aims to accelerate the market introduction of these technologies to the point of launching them commercially by 2020. The application areas of the technologies include hydrogen stationary power generation and combined heat and power systems, hydrogen vehicles of both fuel cell and ICE, refueling infrastructure, hydrogen production and distribution [12].

1.4.1.4 *Worldwide*

Globally, the 2008 Energy Technology Perspectives published by International Energy Agency (IEA) showed the emergence of a considerable hydrogen demand by 2050 [13]. The "Blue Map" scenario aims for a reduction of CO_2 emissions by 50% from current levels by the year 2050, which corresponds to the current consensus of 50–80% emission cuts by many countries and regions, assumes accelerated R&D activities for fuel cells to reduce their manufacturing and operation costs. It is also based on a balanced penetration of both electric and FCVs. The scenario anticipates the annual hydrogen demand by the transportation sector to be about 92 million tonnes, fuelling more than 40% of the transport fleet globally in 2050. A more optimistic "FCV Success" scenario assumed 90% fleet share of FCVs consuming about 200 million tonnes annually for transportation. The annual hydrogen demand globally for stationary hydrogen plants was estimated by the IEA to be 75 million tonnes in 2050.

1.4.2 Transportation

1.4.2.1 *Hydrogen Internal Combustion Engine Vehicles*

One way to boost fuel economy and emission performance with minimal engine modifications to existing vehicles is to add hydrogen to the fuel–air mixture in a conventional gasoline ICE. Since hydrogen can burn alone in a normal ICE, vehicles have also been designed to run on dual fuels of gasoline and hydrogen and they have the potential to provide a transition to FCVs by helping avoid the chicken-or-egg problem of developing hydrogen vehicles and support infrastructure at the same time.

Several major car companies have been developing and road testing the hydrogen-powered ICE vehicles. For instance, BMW has built 100 Hydrogen 7 cars and already collected more than 2 million km in road tests around the world. The company proved that the car is already production ready. Although the fuel efficiency is similar for hydrogen and gasoline fuels in the bifuel engine, the ICE run on hydrogen fuel produces almost no emissions except water vapor. The F-250 Super Chief pickup truck by Ford Motor Company is powered by a hybrid-fuel ICE accepting multiple fuels including hydrogen. In 2007, Mazda rolled out Premacy Hydrogen RE Hybrid that employs the two-rotor Wankel engine on hydrogen or gasoline fuel. The car is equipped of a 110 L hydrogen tank at 35 MPa stores, which contains 2.4 kg of hydrogen gas and another 60 L gasoline tank. The combined range of the dual fuels is 750 km.

1.4.2.2 Hydrogen Fuel Cell Vehicles

Fuel-cell engine is being developed, commercialized, and promises to be widely used as the core of vehicular power train. In the typical layout of such a vehicular power train shown in Figure 1.6, a fuel-cell engine combines hydrogen, retrieved from an on-board hydrogen fuel tank, with oxygen from the air to generate electricity (refer to Section 1.2.4 for the working principle of a fuel cell). A motor drive regulates, according to driver's commands, the electric current sent from the engine to the electric motors that turn the wheels. Water vapor is the only by-product of the engine that is emitted through the tail-pipe. From fuel tank to wheel, the state-of-art FCV is three times more efficient than a conventional gasoline vehicle and about twice as efficient as gasoline-electric hybrid vehicle. The current level of engine performance has allowed for a fuel economy of 100 km/kg-H_2 on a standard size passenger car.

The driving range of FCVs can be extended by increasing fuel tank capacity, which is determined by design trade chiefly among the space, pressure, and cost. The current primary option is compressed hydrogen gas tank of either 35 MPa or 70 MPa. Carbon fiber-reinforced compressed hydrogen gas tanks are under development and some are already used in production vehicles. Typically, the tank is internally lined with a high-molecular-weight polymer that is designed to be hydrogen permeation tight. A carbon fiber composite shell embraces the liner and bears the gas pressure load. Another shell is placed outside for impact protection. Compressed hydrogen gas tank designs have been certified worldwide according to ISO 11439 in Europe and NGV-2 in the United States, and approved by TUV of Germany and KHK of Japan [14].

The successful application of fuel cell as a long-term transportation engine would require not only effective onboard hydrogen fuel tanks, but also for them to be supported by a substantial infrastructure of hydrogen refueling, delivery, and production from nuclear and renewable energy sources. In addition through the hydrogen fuel cell, nuclear energy may also power transportation by generating electricity and recharge electric vehicles. Table 1.4 compares nuclear reactor-to-wheel efficiencies of FCVs and plug-in battery-electric vehicles (BEV) [15]. The two technologies assume supply of nuclear hydrogen and nuclear electricity from the grid. The hydrogen for the FCV is assumed to be produced in a light-water reactor, sodium-cooled fast reactor, and very high-temperature reactor in combination with the most suitable production routes for these reactors.

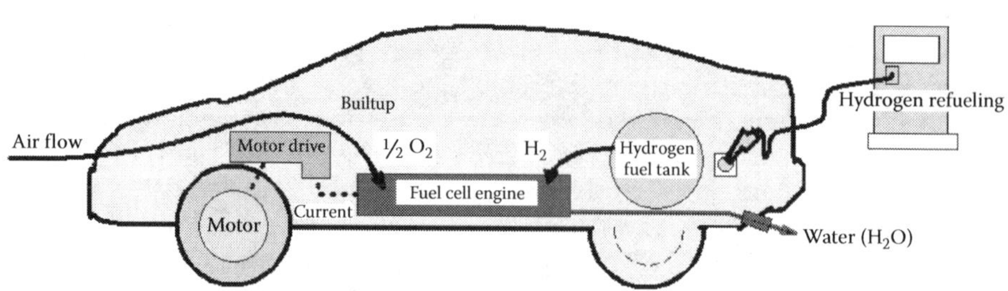

FIGURE 1.6
Automotive power train equipment and operation of FCVs.

TABLE 1.4

Comparison of Nuclear Reactor to Wheels Thermal Efficiency of Plug-in Electric Vehicle (BEV) and FCV

Nuclear Reactor	Electricity/Hydrogen Vehicle Power Train	Efficiency Reactor → Battery/Tank (%)	Efficiency Battery/Tank → Wheel (%)	Overall Efficiency Reactor → Wheel (%)
LWR	Steam turbine BEV	30	70	21
	Electrolysis FCV	23	50–60	12–14
SFR	Steam turbine BEV	30	70	21
	Nuclear-heated steam methane reforming FCV	77[a]	50–60	38–46[a]
VHTR	Gas turbine BEV	45	70	31
	Thermo-chemical FCV	45	50–60	23–27

Notes: Thermal efficiency: LWR (light water reactor) steam turbine 32%, SFR (sodium-cooled fast reactor) steam turbine 41%, VHTR (very high temperature reactor) gas turbine 47%.
Efficiency of H_2 production: Electrolysis 80% from electricity and thermo-chemical, from heat 50%. (LHV) Reforming 85%.
Transmission and distribution loss for electricity: 5%, Compression and transportation loss for H_2: 10%.

[a] Based on the sum of both primary energies.

Figure 1.7 compares life cycle "well-to-wheels" CO_2 emissions of alternative vehicular fuel sources per mile traveled according to the DOE Hydrogen Program. It includes several conventional and advanced vehicles, all of which are based on the projected state of the vehicle technologies in 2020. The hydrogen fuel considers a number of alternative production sources. Although the nuclear hydrogen in Figure 1.7 assumes the production via high-temperature electrolysis, the CO_2 emission performance value shown in Figure 1.7 is representative of all potential nuclear hydrogen production methods from feedstock water because all nuclear methods are characteristically zero emission and the CO_2 emission shown from the nuclear hydrogen fuel cell in the figure is mainly associated with the assumptions in the delivery, storage, and dispensing of hydrogen that would still

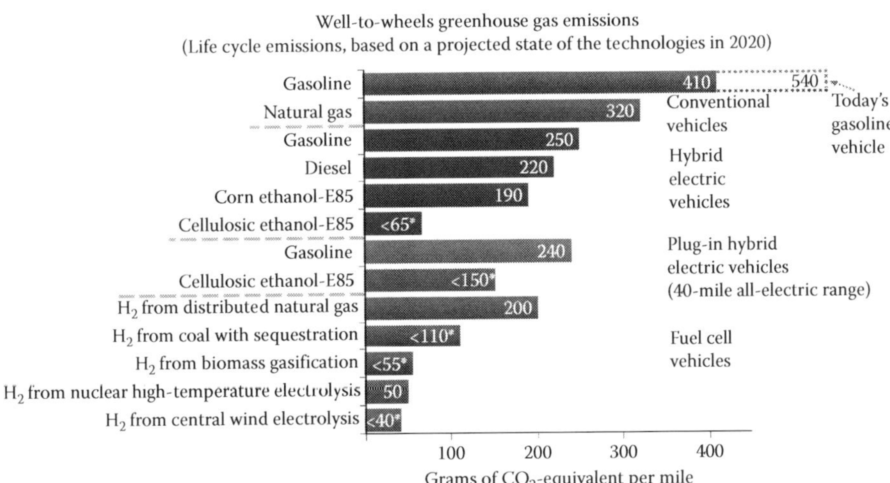

FIGURE 1.7
CO_2 greenhouse emissions of projected vehicle life cycle technologies in 2020.

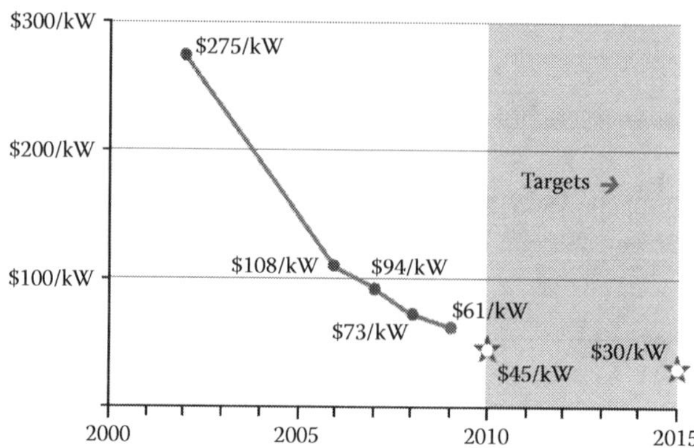

FIGURE 1.8
Fuel-cell cost reduction achieved through the U.S. DOE-funded development with 2009 cost at $61/kW for high-volume production of an 80 kW net power output engine system.

require use of fossil fuel energy in the practical time frame of 2020. It can be seen from Figure 1.7 that fuel-cell cars based on nuclear hydrogen as fuel are among the best emission performers, second only to central wind electrolysis.

As of 2010, FCVs can meet vehicle operation requirements in terms of power output, safety, and functionality. The operating range between refueling is closing in the requirement. The major task at hand now is the cost reduction of making fuel cells to a level that allows FCVs to be affordably priced to dealers. The production cost has rapidly declined and approached to $61/kW for assumed high-volume production based on the 2009 technology status, as seen in Figure 1.8 from the DOE Hydrogen Program's fuel-cell subprogram [16]. The 2015 target for the program is $30/kW, which compares with about $100/kW for a conventional ICE.

Illustrated in Figure 1.9 are some of the current approaches generally taken to cut down the cost of production, including reduced or alternative use as catalyst of platinum (Pt)

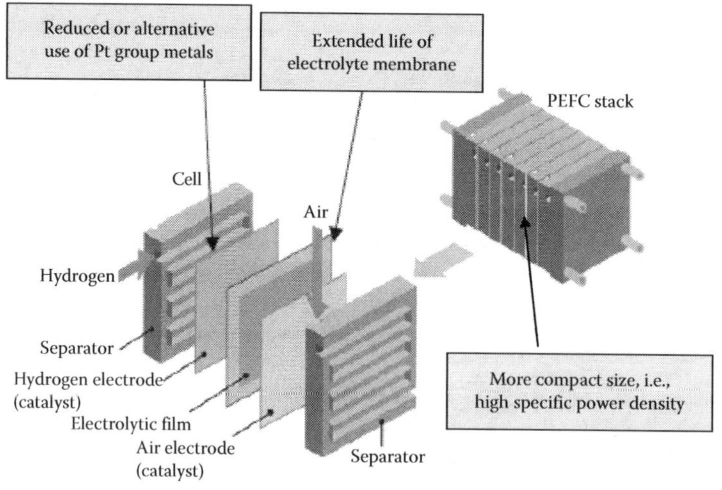

FIGURE 1.9
Current (2010) approaches to product cost reduction of polymer electrolyte fuel cell.

and other precious metals, downsized cell stack, and extended life of the electrolyte, all without compromising performance.

In the United States, Europe, and Japan, the construction of large-sized hydrogen fuel stations has already begun. Since the first U.S. hydrogen refueling station opened in Dearborn, Michigan in 1999, a total of 65 others have been built and operated in 25 states (including 6 more in Michigan), with the largest number (27) located in California. The world total numbers 150 (40 in Europe and 27 in Asia). In March 2010, the London Hydrogen action plan called for the construction of at least six hydrogen fueling stations to service a minimum of 150 hydrogen vehicles expected on the road in London by 2012.

General Motors (GM), Hyundai, Daimler, Toyota, Honda, and Nissan among other automakers plan to market hydrogen FCVs by 2015. GM is testing a production-intent hydrogen fuel cell that can be packaged in the space of a traditional four-cylinder engine and be ready for commercial production in 2015. The system is half the size, 220 pounds lighter, and uses about a third of the platinum of the system in the Chevrolet Equinox FCEVs used in Project Driveway, the world's largest market test and demonstration fleet of 119 FCVs that began in late 2007 and has accumulated nearly 1.3 million miles of daily driving in cities around the world.

Hyundai unveiled its Tucson ix35 Hydrogen FCEV in 2010 that includes several major innovations over the previous generation and which will enable the company to meet its announced goal of beginning mass production of FCEVs by 2012.

As of 2010, Daimler has a fleet of over 100 FCVs including 60 Mercedes-Benz A-Class F-CELL cars, in use worldwide. Some of these vehicles have already covered more than 150,000 km. Since the first model introduced in 1994, the company has logged in more than 4.5 million test km and more than 200,000 h of operating time. Daimler is being joined by seven industrial partners to develop a national hydrogen fuel infrastructure by 2015. The program named H2 Mobility, which has the support of the German national government, consists of Phase 1, which runs until 2011 to assess a nationwide hydrogen network, increase public support, and build a significant number of hydrogen fuelling stations in selected cities such as Berlin and Hamburg, and Phase 2, which develops a full-scale hydrogen fuel infrastructure nationwide and puts 100,000 FCVs on the road by 2015.

The top three Japanese automakers of Toyota, Honda, and Nissan are now demonstrating the FCVs, with Honda leading the pack in commercialization progress. Honda announced the FCX model in 2002 and delivered the first generation of cars to users in Japan and the United States in that same year. In 2008, the next-generation of FCX Clarity in Figure 1.10 became the world's first production hydrogen FCV. Compared to the earlier generation, the power train is over 180 kg lighter, 45% more compact, and 10% point more fuel efficient. Honda has begun leasing a total of 200 cars for a monthly fee of $600 to selected drivers in California where there are more hydrogen filling stations than in other states in the country. The proton exchange membrane fuel cell (PEMFC) engine of FCX Clarity is rated 100 kWe (135 hp) and weighs 67 kg. The driving range is 385 km on a compressed (35 MPa) gaseous fuel tank of 3.92 kg hydrogen. The hydrogen fuel to wheels efficiency of 50–60% is 2–3 times the gasoline cars (15–20%) and the plug-in hybrid cars (30%) based on the same California drive mode, as seen in Figure 1.11.

Since December 2002 when Toyota began testing FCVs in the United States and Japan, Toyota's hydrogen fuel-cell technology has since improved driving range, durability, and efficiency through improvements to fuel-cell stack and the high-pressure hydrogen storage system, while achieving significant cost reductions in materials and manufacturing. In January 2010, Toyota announced to demonstrate an advanced generation of 100 fuel-cell hybrid vehicles on roads in New York and California, and then elsewhere in the United

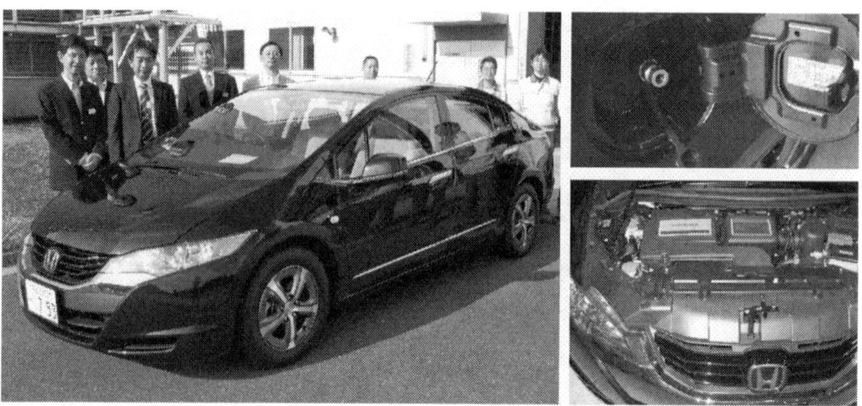

FIGURE 1.10
Honda developers with FCX Clarity fuel-cell production sedan visited the nuclear hydrogen production research and development center of Japan Atomic Energy Agency (JAEA) in June, 2009.

States over the next three years. The two major objectives are expanding the road demo program nationwide in the United States to spur essential hydrogen infrastructure development while proving fuel-cell technologies' reliability and performance prior to 2015, the year Toyota plans to put in showrooms a production hydrogen fuel cell only vehicle that will be reliable, durable, and affordable, with exceptional fuel economy and zero emissions. Consistent with this market goal, the company announced in May 2010 at the annual National Hydrogen Association's Hydrogen Conference and Expo that it had cut the cost of making fuel-cell cars by 90% in about five years by cutting the platinum use in fuel cell

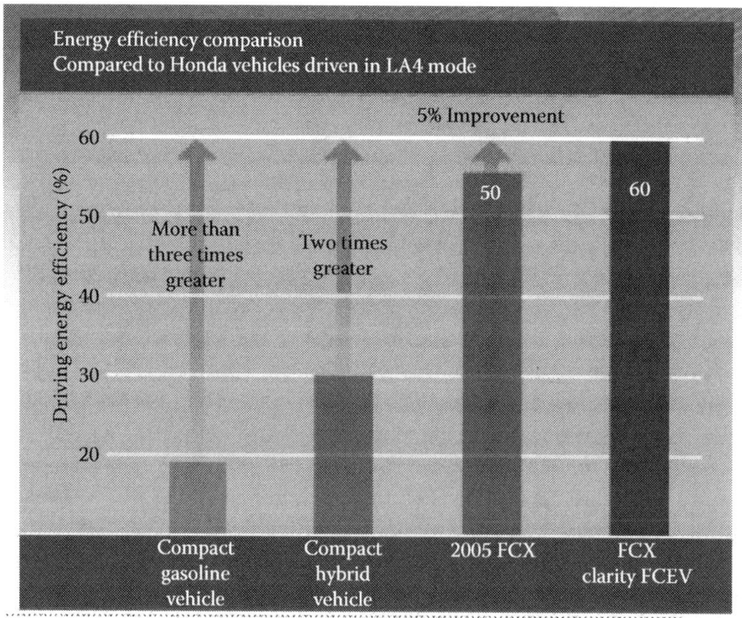

FIGURE 1.11
Comparison of fuel efficiency of Honda made alternative car models. (Data from Honda Motor.)

to 10 g from 30 g and by reducing the cost of the polymer electrolyte membrane used in the cell. As said earlier, GM also succeeded in a similar cut of the platinum use in fuel cell. Toyota said at the conference that it aims to cut the cost further by 50% and retail the first model for a price of US$50,000 by 2015.

One of the chapter's authors got behind the wheel of Nissan's X-TRAIL FCV as photographed in Figure 1.12. Carrying five occupants, the car performed similar to the ICE cousin of X-Trail SUV, except a notable quieter ride. It is equipped with a Nissan-developed fuel cell stack whose size and cost had continually decreased, a compact Lithium-ion battery and a compressed hydrogen gas storage tank. It is capable of speed in excess of 150 km/h with a travel range of up to 500 kilometers. Nissan began developing FCVs in 1996 and has participated in public road tests in Japan and the United States. The Nissan FCVs have logged a total of 500,000 km in the tests and one vehicle has accumulated 150,000 km.

In the domestic development as a whole, Japan's official market introduction goals of FCVs as planned by the ANRE of the Ministry of Economy, Trade and Industry in 2004, which is shown in Figure 1.13, include the start of commercialization for passenger FCVs and hydrogen fueling stations in 2015 (which remains unchanged in 2010) and reaches 5 million in 2020 and 15 million in 2030, of the country's fleet total of 75 million that are expected to remain constant in the foreseeable future. Hydrogen demand by the transportation sector would be 17 billion Nm^3 in 2030, comparing 0.15 billion Nm^3 in 2000. The FCV models already demonstrated on roads today and the future market plans announced by Honda, Toyota, and Nissan as of 2010 are generally consistent with the 2004 ANRE target and schedule.

1.4.3 Power Generation

1.4.3.1 Utility Power Generation

Burning hydrogen fuel in combustion gas turbines has proved technical feasible with limited modifications necessary to conventional turbines burning coal syngas, natural gas, and oil [17–19]. Modifications are primarily required to limit the turbine blade operation temperature, for example, by optimizing fuel with air ratio, and control NO_x emission through fuel dilution and flame conditioning. The large-scale hydrogen gas turbines are being developed through substantial collaborations of government and industry, including

FIGURE 1.12
Test driving Nissan's zero-emission X-TRAIL FCV in June, 2010.

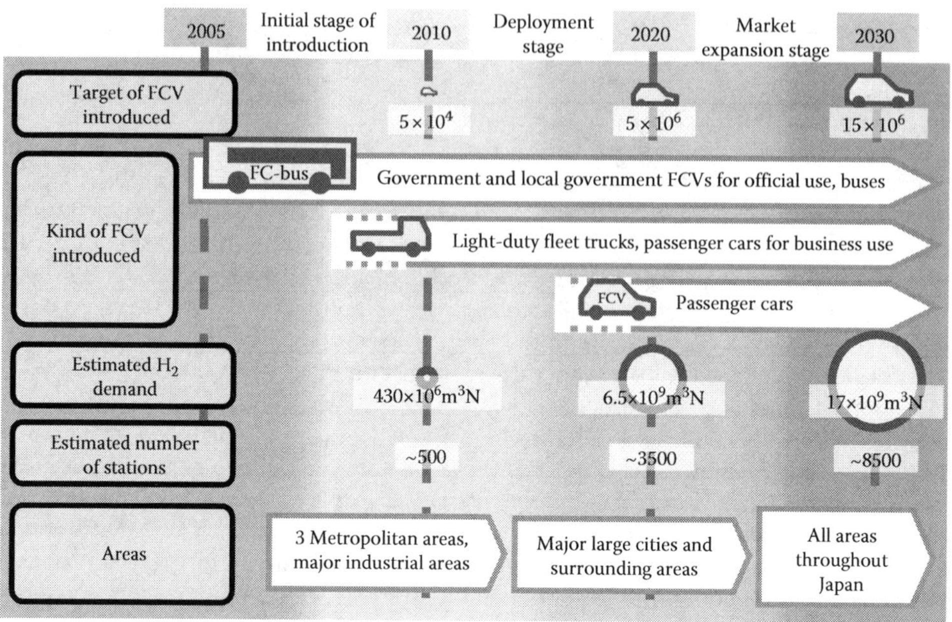

FIGURE 1.13
The planned scenario of FCV market introduction by the ANRE Advisory Panel of the Government of Japan in 2004.

major turbine vendors General Electric and Siemens-Westinhouse Electric, in the U.S. DOE's Hydrogen Turbine for FutureGen program. The turbine systems and components under development include combustor technology, materials research, enhanced cooling technology, and coatings development. These technologies are considered key components to build and operate, by 2012, a turbine fueled with hydrogen fuel, which is produced in an integrated coal gasification plant with CO_2 capture and sequestration. The turbine will demonstrate the fuel flexibility of allowing for turbine operation on 100% hydrogen, in addition to coal syngas, and controlling emissions of NO_x emissions to near zero.

While the hydrogen combustion provides a near-term opportunity to reduce greenhouse emissions of the power industry, another potential application is shaving the peaking power demand. The daily and weekly off-demand (surplus) electricity generated from all-time, base load nuclear power plants would be used to produce captive hydrogen, store it nearby, fuel it in hydrogen combustion gas and steam turbines (or large banks of fuel cells, whose turnkey units rated several MWe are commercially offered as of 2010) to generate electricity at peak demand periods daily and weekly. The daily electricity load variation in mid-summer in Japan is shown in Figure 1.14 [20]. Japan manages the substantial load swing largely with import oil and natural gas thermal power plants, and nuclear energy has played no role in load follow. Therefore, potential exists to reduce fossil fuel imports and CO_2 emission if nuclear plants are configured to enable efficient load follow. A concept of such plant coupling hydrogen production by electrolysis and hydrogen combustion steam turbine to a current nuclear light-water reactor is given in Chapter 9.

Japan Atomic Energy Agency is developing another system concept, combing high-temperature thermochemical hydrogen production from water (reported in Chapter 17) and a Generation IV nuclear high-temperature (950°C) gas reactor (Chapter 10). It aims to commercialize the concept by 2030. Figure 1.15 shows the application system of this plant

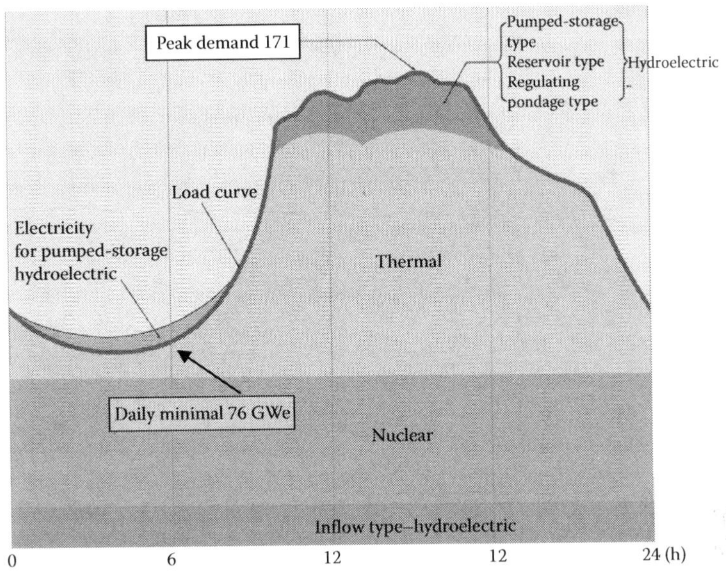

FIGURE 1.14
Combined actual electric power generation profile of all 10 regional electric power utilities of Japan, recorded on August 7, 2009.

concept to load follow. The system consists of the following subsystems and operation characteristics:

1. A nuclear power and heat cogeneration plant based on high-temperature gas reactor (HTGR) that supplies heat of up to 170 MWt at 900°C via an intermediate heat

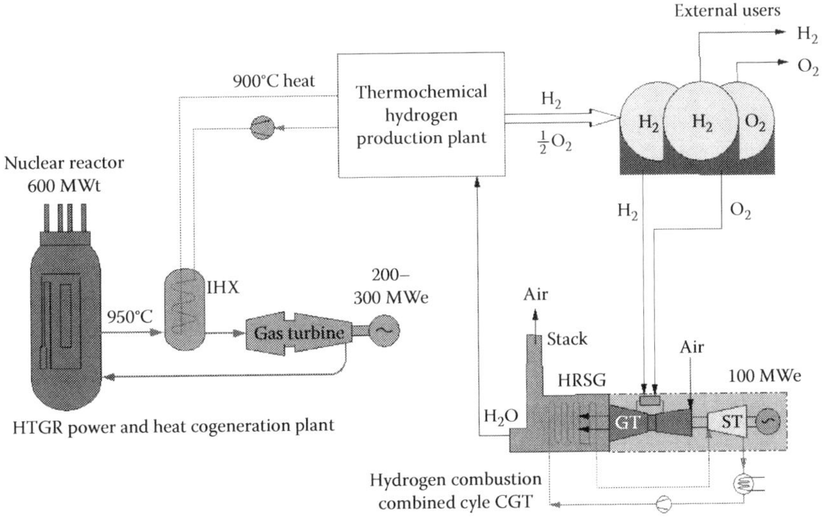

FIGURE 1.15
Combined nuclear reactor (HTGR) hydrogen gas turbine power plant for daily peaking power operations. The plant provide maximum based load 300 MWe and a load follow range of 200–400 MWe while reactor operates steadily in full power.

exchanger (IHX) to hydrogen production while generating power from 300 MWe down to 200 MWt (when heat is cogenerated). The reactor operates at full thermal power of 600 MWt at all time and generates power by a helium gas turbine. The power generation is varied by controlling gas coolant inventory to match load demand of ±5% load/thin between 200 and 300 MWe by the reactor plant alone [21]. The surplus reactor thermal power is transferred via IHX through heat transport lines to the hydrogen production plant.

2. A hydrogen production plant based on a high-temperature thermochemical process (such as the iodine–sulfur process is under development in Japan, United States, and elsewhere in the world). The process produces hydrogen and by-product oxygen from water and from input of major high-temperature heat and minor electricity, both of which are cogenerated by the nuclear reactor plant. The production is performed at a maximum rate of 26,000 Nm^3/h for hydrogen gas and a half of that volume rate for oxygen gas. The thermal production efficiency is about 45%.
3. A combined cycle gas turbine (CCGT) burning on hydrogen and oxygen to provide up to 100 MWe. Air flow is drawn by gas turbine from ambient atmosphere to quench the combustion water vapor to achieve the appropriate gas turbine flow inlet temperature. The gas turbine exhaust heat is recovered in a heat recovery steam generator (HRSG), which feeds a steam turbine. The only emission of the CCGT plant is air, taking in earlier at the gas turbine inlet. The water is condensed and recycled as raw material is fed to the hydrogen production. Alternatively, a fuel cell-based plant may be used in place of the CCGT. In the case of fuel cells, efficiency of power generation would be improved since by-product oxygen, instead of usual air, is used as oxidant for the fuel-cell chemical reaction.
4. In sum, the operation characteristics of the above (1)–(3) subsystems are such that the overall plant enables a peak load of 400 MWe, a continuous base load of 300 MWe, and an ability of rapid load follow in a range of 200 and 400 MWe while the reactor operates steadily in full power and the plant emits zero carbon dioxide.

1.4.3.2 Distributed Power Generation

The hydrogen fuel-cell turnkey units rated up to 1 MWe are now offered by the Canadian-based Ballard and the Dutch-based NedStack in 2010. Both companies use the proton exchange membrane (PEM) fuel cell technology. Such units are applicable for generating electricity and heat for commercial buildings, hospitals, and telecom transmission centers. They are also attractive for remote and mobile applications, and here they offer the advantages of efficiency, reliability, and availability performance over other systems.

Fuel-Cell Energy based in the United States and Canada has hydrogen fuel-cell plants installed at over 50 global locations generating 450 million kWh of accumulated power at 65% efficiency. Hydrogen is presently obtained from on-site reformer of natural gas, propane gas, and coal syngas. Each of these plants contains one or more hydrogen fuel-cell units operating in parallel to provide a range of power from 300 kW to 3 MWe and a balance of plant to convert the DC power generated in the fuel cells to the AC power at the frequency and voltage required by the grid or customer. In April 2010, the company announced an order for 5.6 MWe utility-owned fuel-cell power and by-product heat cogeneration plants with the Pacific Gas and Electric Company and Southern California Edison Company.

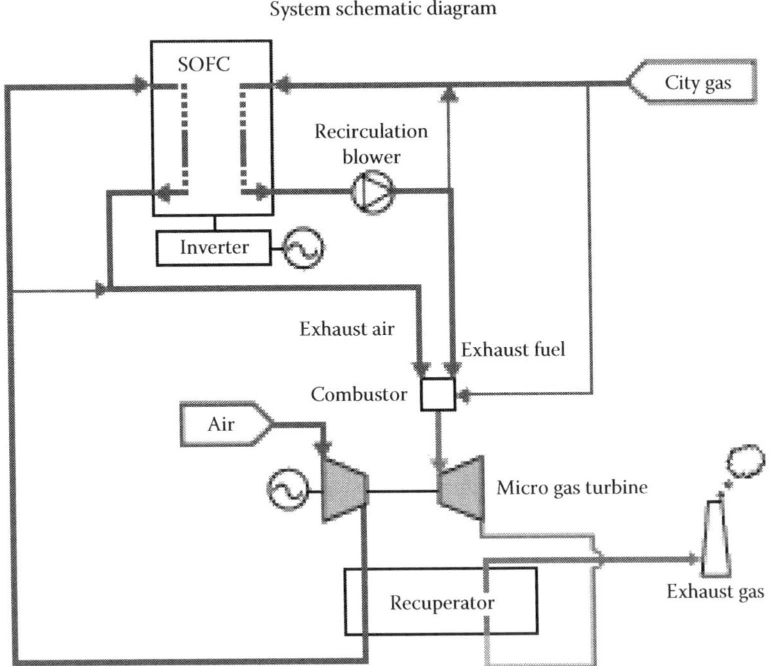

FIGURE 1.16
Process diagram of MHI's 200 kW-class SOFC-MGT combined cycle system for high-efficiency power generation.

While PEM operates in the low-temperature range, the SOFC with power generation efficiency of about 55% operates at a temperature range of 750–1000°C and exhausts high-temperature sensible waste heat. The sensible heat can be captured for increased efficiency in the applications of combined-cycle power generation, heat and power cogeneration, or reformation of hydrocarbons to produce hydrogen from a variety of gaseous fuels including methane and coal syngas. Some of these features are incorporated into a product demonstration unit by Mitsubishi Heavy Industries (MHI) in Japan in a combined cycle 200 kW SOFC and gas turbine (Figure 1.16). The fuel used is city gas (i.e., methane or natural gas). The optimized commercial systems by MHI's estimation can be scaled up to hundreds of megawatts and generate power with an efficiency of 70% or higher with natural gas fuel, and 60% or higher with coal-gasified fuel.

1.4.4 Power and Heat Cogeneration

Fuel cells can be used for stand-alone power and heat cogeneration, because they produce sensible heat by-product from the chemical energy to power conversion and the by-product heat is attractive for such heat applications such as central heating and hot water for homes.

In Japan, verification research by the government, manufacturers, and energy companies started in fiscal 2002, and the government has run large-scale trials on fixed fuel cells since fiscal 2005. Research and development has been conducted since the 1990s by manufacturers and energy companies.

FIGURE 1.17
A residential fuel-cell unit installed in the market in Japan.

Tokyo Gas and five other utilities commenced sales of ENE FARM residential PEM-type hydrogen fuel-cell cogeneration units. In 2005, the first commercial unit was installed in the Prime Minister official residence. In 2009, the particular unit shown in Figure 1.17 had the sales performance specification as follows:

Electric generation capacity	1 kWe (AC)
Electric generation efficiency	>33% (HHV); > 37% (LHV)
Heat cogeneration capacity	1.4 kWt
Heat cogeneration efficiency	>47% (HHV); >52% (LHV)
Heated water storage capacity	200 L at 60°C
Water supply capacity	360 L at 40°C
Unit heat rate	3 kWt
Fuel	Natural gas (city gas) (reformed into H_2)

The fuel cell is commercially marketed to cogenerate electricity and heat for home heating and hot-water supply. The hydrogen fuel needed by the fuel cell is obtained from an auxiliary unit of natural gas reformer on-site. To further reduce the size of carbon footprint of residential sector, home fuel-cell units could switch to hydrogen produced from nuclear or other zero-emission energy sources. As of 2010 over 5000 ENE FARM units have been installed and have logged in over 40,000 h of operation. They now generate electricity and heat for homes with power and heat cogeneration thermal efficiency of greater than 80% (HHV).

The residential units rated up to 1 kW electricity and 900 kW heat and manufactured by Toshiba and ENEOS are priced at about $34,000 each as of 2010, meeting the official cost target (Figure 1.18). Other makers include Panasonic and Ebara Ballard. The official annual production targets 10,000 units in 2010 and 40,000 units in 2015. Overall, the government

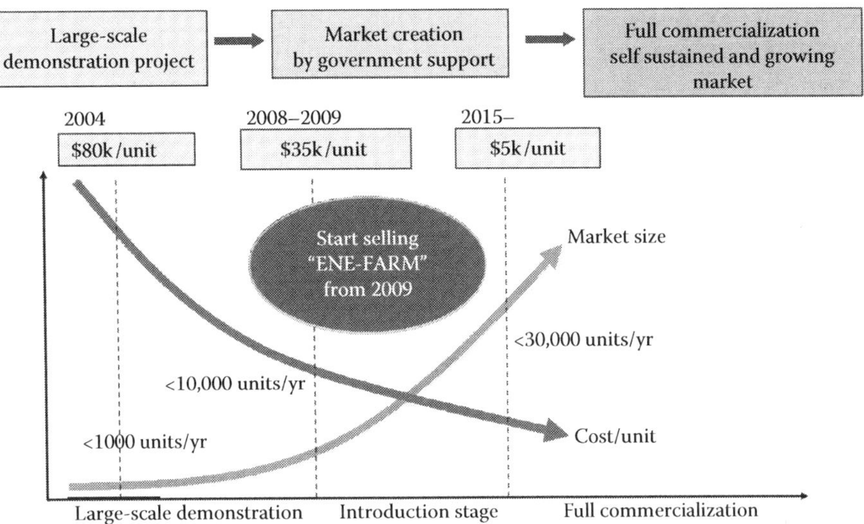

FIGURE 1.18
Japan's official scenario of market deployment for residential fuel-cell units.

targets a total of 2.5 million home fuel-cell cogeneration units, that is, in 5% of the households nationwide, by 2030. To leverage on the economy of scale by accelerating product and market volume, the government institutes a subsidy covering the half of the price of equipment and installation for home users. While ENE FARM is to be considered a standard feature of home in Japan in future, efforts to advance the technology and improve performance, for example, increasing power-generation efficiency by 5% point in 10 years, will continue. This, combined with economies of scale, is expected to bring the unit price down to the government–industry target of $5000 by the time the government cost subsidy ends after 2015.

The U.K. government announced a subsidy support for the adoption of the technology for all fuel-cell owners, starting in April 2010. The U.K. government has estimated that the on-site cogeneration, such as through fuel-cell combined-heat-and-power (CHP) units, has the potential to meet over one-third of the country's electricity needs and thus contributes to meeting its energy and CO_2 emission reduction objectives.

1.4.5 Iron and Steel Making

Global crude steel production stands at 1.3 billion tonnes in 2008, increasing 60% from 10 years earlier. China led the production with (37%), followed by the European Union (15%), Japan (9%), the United States (7%), and Russia (5%). The 2008 CO_2 emissions from the global steel production averaged to 1.9 tonnes for each tonne of steel made. In Japan, the industry as a whole emitted 143 million tonnes CO_2, about 12% of the national total in 2008 [22]. The 2010 global steel demand has recovered to the 2008 level from a decline due to the world financial crisis in 2008–2009 and is forecast by the World Steel Association to grow 5.3% in 2011.

Two conventional routes of producing steel are blast furnace (BF) plus oxygen furnace (about 60%) and electric arc furnace (EAF) (about 35%). A newer process, which accounts for the rest 5% of world steel production and whose share is expected to steadily increase in the future is direct reduction furnace followed by an EAF.

Coke is used for heating and reducing in the BF while natural gas and coal syngas perform these functions in the direct reduction furnace. With the steel industry undergoing sustainability development, near and long-term process technologies enabling replacement of a part or all of the fossil fuels with hydrogen have been studied and pursued for resource saving and emission reduction. These are described in the following two subsections on BF and direct reduction furnace, respectively.

1.4.5.1 Hydrogen-Assisted Blast Furnace

The principal feedstock materials to BF include iron ore (and also pellets, sinter, scrap, and other iron-bearing sources), flux (limestone or dolomite for the purpose of removing sulfur from product iron), and coke (oven-cooked from coal). These materials are charged into the BF where the iron compounds are smelt to release excess oxygen and reduce to liquid iron in a tap-to-tap interval of several hours. The liquid iron (approximately 95% Fe, 4% carbon) tapped from the BF is either sent directly to a basic oxygen furnace for steel making (reducing carbon content and making alloy) or cast into pig iron for later use.

The conventional coke reduction of iron ore in the BF is as follows:

$$\text{Coke reduction: } Fe_2O_3(s) + 3C(s) \rightarrow 2Fe(s) + 3CO(g) - \Delta H \text{ (489.8 kJ/mol)} \quad (1.6)$$

Coke oven gas (COG) can contain about 55% H_2, 30% CH_4, 6% CO, and other gases whereas blast furnace off-gas (BFG) contains about 51% N_2, 23% CO, 22% CO_2, and 4% H_2. The hydrogen and carbon monoxide gases recovered from the COG and BFG may be reused to help reduce coke consumption in an integrated system. Additionally, a reformer can be incorporated in the system to convert the methane in the COG stream to generate more hydrogen. The reducing reactions of iron ore using carbon monoxide and hydrogen gases are below:

$$\text{CO reduction: } Fe_2O_3(s) + 3CO(g) \rightarrow 2Fe(s) + 3CO_2(g) + \Delta H \text{ (27.6 kJ/mol)} \quad (1.7)$$

$$H_2 \text{ reduction: } Fe_2O_3(s) + 3H_2(g) \rightarrow 2Fe(s) + 3H_2O(g) - \Delta H \text{ (95.9 kJ/mol)} \quad (1.8)$$

The processes for separating and reforming the gases from COG and BFG have been investigated by the programs in Europe and Japan. Technical objectives are the reuse of hydrogen and CO to reduce iron ore to save coke use in BF or prereduction of iron ore with hydrogen.

In Japan, six major domestic steelmakers sponsored by the government have been collaborating in the COURSE 50 (CO_2 Ultimate Reduction in Steelmaking Process by Innovative Technology for Cool Earth 50) to develop and deploy the innovative technologies in the BF to help reduce CO_2 emission by approximately 30% from the preprogram level. The focus is on processes that would allow for iron ore to be reduced using hydrogen instead of coke as well as on ways to capture, separate, and recover CO_2 from BF gas. The program seeks to establish these technologies by around 2030 and apply them on an industrial scale by 2050. The budget for the initial five years until 2012 is US$100M. The roadmap for COURSE 50 detailed in Figure 1.19 contains the following technical elements [23]:

1. Develop technologies to control reactions for reducing iron ore with reducing agents such as hydrogen with a view to decreasing coke consumption in the BF.
2. Develop technologies to reform COG for the amplification of hydrogen content in COG reformer by utilizing waste heat of 800°C from coke oven.
3. Develop technologies to produce high-strength and high-reactive coke for reduction with hydrogen in the BF.

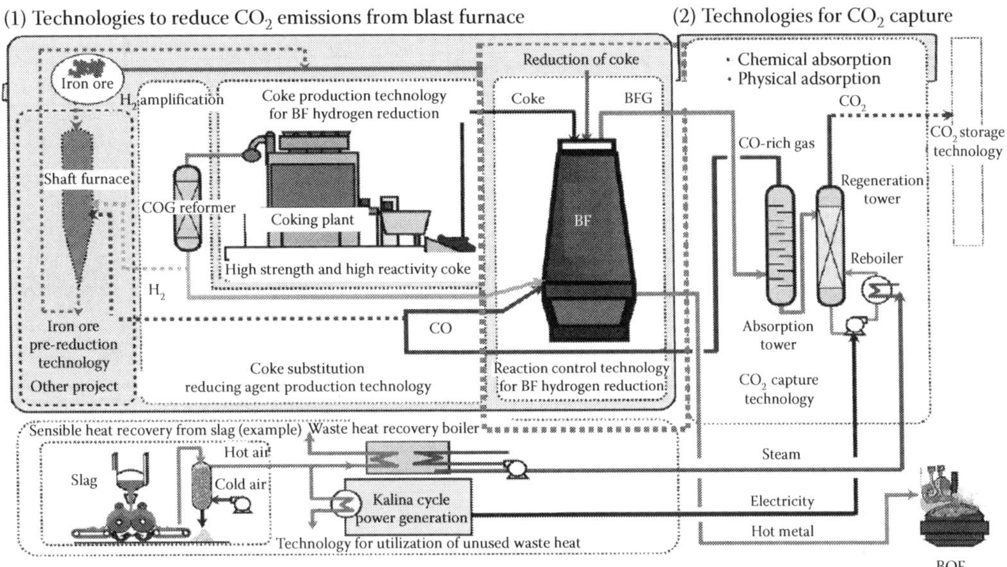

FIGURE 1.19
Japan's COURSE50 to increase use of hydrogen in the iron and steel making.

In Europe, ULCOS (Ultra-Low Carbon Dioxide (CO_2) Steel Making) program is today's largest collaborative program of the world in developing solutions to cut CO_2 emission in steel industry. The program members include all major EU steel makers, energy and engineering partner companies, research institutes, and universities in 15 European countries. The ULCOS aims to halve carbon dioxide emissions from the preprogram industry practice. The budget is 75 million Euro for Phase I over a 6-year period until 2010 with the cost share of 60% by consortium members and 40% by the European Commission. Phase II will be launched for a period of five years until 2015, and an implementation phase will follow.

Besides using hydrogen obtained from coal or natural gas, the integrated BF mill can use additional hydrogen produced from a noncarbon source such as nuclear and biomass to further increase the share of hydrogen use for iron ore reduction in the BF process. In Figure 1.20, the estimated material balance shows that using 100 Nm^3 hydrogen for a tonne of pig iron produced can reduce carbon feed of 50 kg based on the typical production parameters of a modern BF. This means in practice that an HTGR nuclear reactor plant rated at 500 MW thermal power (a typical size) can produce captive hydrogen (1.4 million Nm^3/day), via the route of thermochemical production from water as described later in this chapter, enough to meet the iron ore reduction need of the world's largest BF (13,500 tonnes/day iron), and cut carbon dioxide emission of 2500 tonnes each day (a 8.4% reduction) in the process.

1.4.5.2 Hydrogen Direct Reduction Furnace

Unlike the BF intermittent operation, the iron ore is charged continuously from the top of the direct reduction shaft furnace (SF), reduced to iron while descending in counter to reducing gases (CO and/or H_2) in the furnace, while the product of so-called direct reduction iron (DRI) is removed continuously at the bottom end of the furnace. The solid DRI

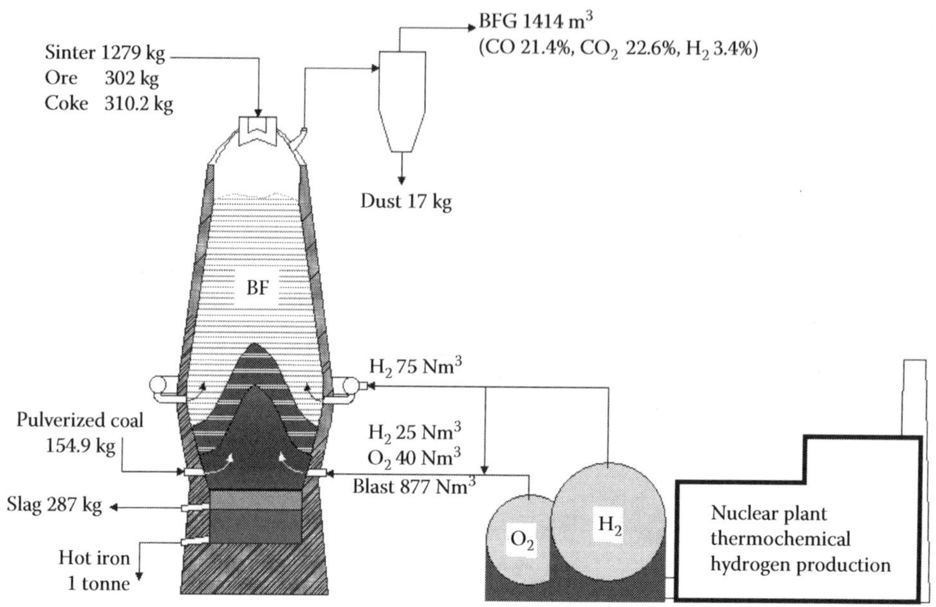

FIGURE 1.20
Process material balance of a nuclear hydrogen-assisted BF.

typically containing 90–94% Fe is either hot briquetted, cooled for reusable DRI, or hot charged into an electric arc steel furnace (EAF), thereby saving energy by as much as 20% in comparison with EAF steel making from pig iron and scrap metals.

The reducing gas is presently obtained in an integrated steel mill from either reforming of natural gas as shown in Figure 1.21 or from gasification of coal. In either way, the reducing gas obtained contains hydrogen and carbon monoxide. A natural gas

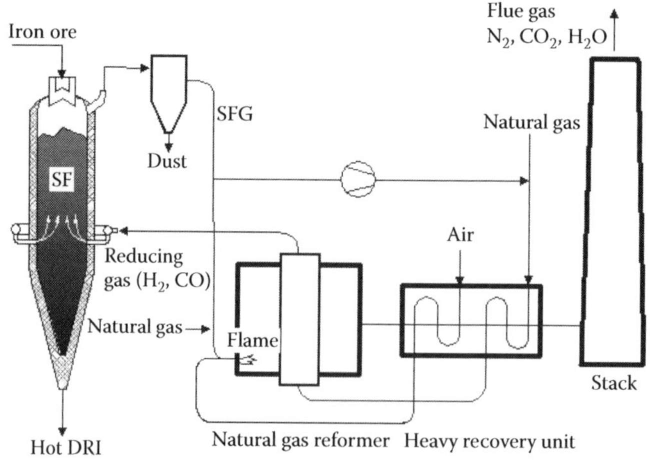

FIGURE 1.21
DRI shaft furnace (SF) combined with natural gas reformer.

reformer integrated in the SFG (shaft furnace gas) stream accomplishes the net reactions below:

$$\text{Steam reforming:} \quad CH_4 + H_2O \rightarrow CO + 3H_2 \tag{1.9}$$

$$CO_2 \text{ reforming:} \quad CH_4 + CO_2 \rightarrow 2CO + 2H_2 \tag{1.10}$$

The hot reformed gas containing about 95% of hydrogen and carbon monoxide is fed into the reduction zone of the furnace to reduce the descending solids of iron ore according to Equations 1.7 and 1.8. At the top of the furnace, some of the SFG (containing about 70% of hydrogen and carbon monoxide) is blended with natural gas feed, preheated to 750°F, and directed into the reformer, and the rest of SFG is burned in the reformer to heat the reformer.

The worldwide DRI production was about 67 million tonnes in 2007, a 50% increase over five years earlier. The largest natural gas-based DRI mill in the world is the Hadeed Steel Works in Saudi Arabia. It began operation in July 2007 with annual production capacity of 1.76 million tonnes of DRI. Saldanha Works in South Africa makes coal-derived syngas as the reducing gas and has a production capacity of 0.7 million tonnes of DRI per year.

Coupling a nuclear reactor plant to a large direct reduction iron process can be attractive because the scale of nuclear splitting of water to produce hydrogen and oxygen suffice the potential gas demands of the largest steel mills. In addition, the nuclear plant can cogenerate electricity and heat to meet in-house consumption of steel mills. Depicted in Figure 1.22, such a nuclear-based steel making facility includes the nuclear power and hydrogen (and oxygen) cogeneration plant and a reduced iron and steel production mill. Since the nuclear plant and the steel plant are linked not thermodynamically but through gas pipe lines (equipped with suitable-sized buffered tanks) and electricity transmission,

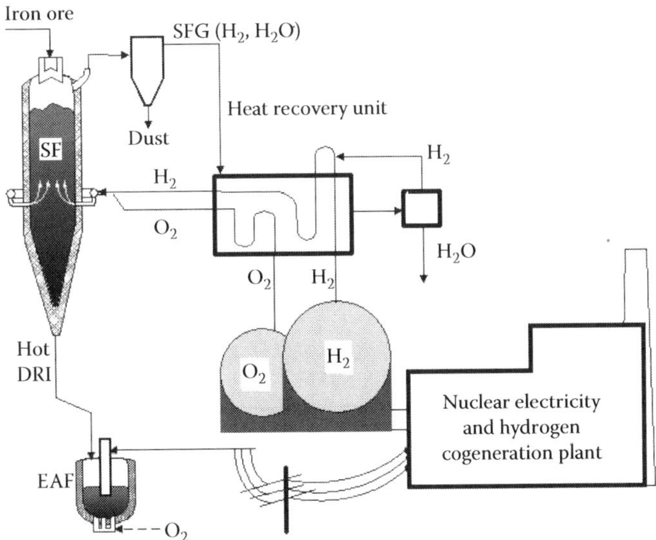

FIGURE 1.22
Process diagram of nuclear hydrogen DRI shaft furnace and EAF steel making system.

the operations and safety protection of the two plants are performed independently of each other. This permits the steel making plant to be built and operated as conventional (nonnuclear grade) industrial plant. As a result, the steel making plant has greatly simpler system and operation than those involved in the traditional BF and SF since the supply of reducing gases and electricity comes from the separated nuclear plant.

Charging 100% hot DRI into EAFs for the production of various steels with low impurities (Si, C, P) is proven. Moreover, the nuclear steelmaking mills would eliminate completely the demand for fossil fuels. As a result, these mills are essentially free of CO_2 greenhouse gas emission. The SFG contains about 70% unused hydrogen and 30% water vapor. The hydrogen and water are separated by water condensation. The hydrogen is reused while the water is returned as the feed water to nuclear hydrogen production.

The simplified flow sheet in Figure 1.23 gives the estimates of energy and material balance for the nuclear steel making process, which consists of hydrogen shaft furnace for the production of DRI from hematite (F_3O_2) and a subsequent EAF for steel making. The nuclear plant consists of a gas turbine electric generator and IHX. The IHX transports heat to an industrial (nonnuclear-class) thermochemical plant for the production of hydrogen and oxygen from water. The majority of electricity generated in the nuclear plant is consumed by the hydrogen plant to power pumps, fluid electrolytic concentrator (electrolyzer), and so on, and the rest is supplied to the steel plant. Combustion of hydrogen in the presence of oxygen meets the temperature and endothermic conditions of hydrogen reduction of iron from iron ore in the SF.

Across Japan, a total of 13 BFs produce 84 million tonnes of steel and emit 99 million tonnes of CO_2 annually. Japan Atomic Energy Agency envisions the deployment of 86 HTGRs to replace these BFs with nuclear hydrogen direct reduction furnaces within this century. This would contribute to 8% reduction in national CO_2 emission from the today level.

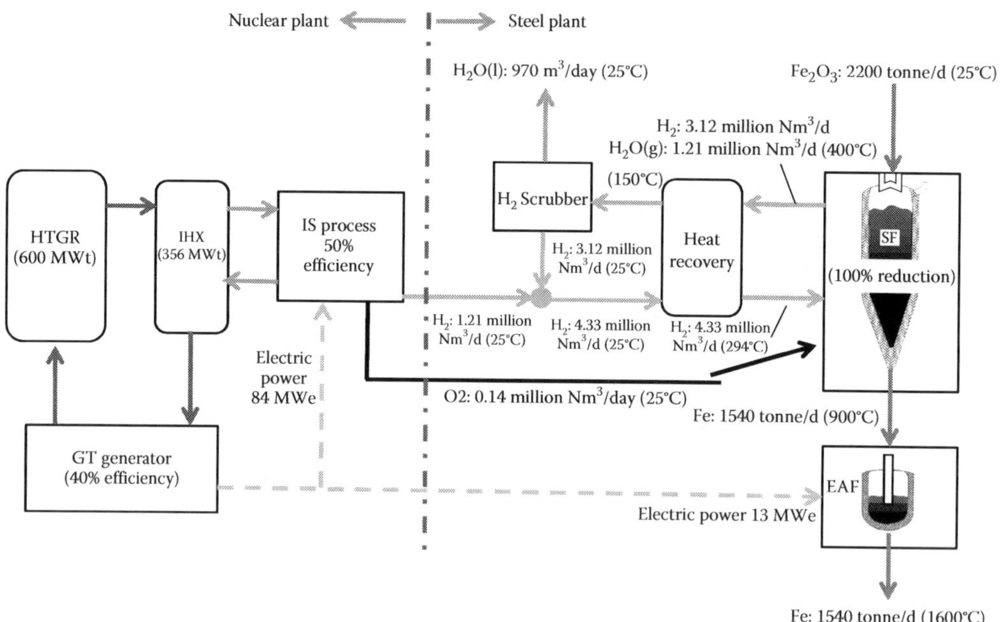

FIGURE 1.23
Energy and material balance for a nuclear reactor DRI + EAF steel making process.

1.5 Hydrogen Production

1.5.1 General Requirements

As introduced in the opening of this chapter, hydrogen is rare in nature and must be produced to meet the demands of world economy as described in the preceding section. To do so, fossil fuels are now used as the material sources of hydrogen produced in the world, including 48% from natural gas, 30% from oil, and 18% from coal. The remaining 4% comes from water. Biofuel or biomass is not significantly used now but interest in it is fast growing in the utilization of this renewable source.

Hydrogen if produced is merely made into an energy carrier, similar to electricity. The input of a primary energy source is thus required. In theory, any form of primary energy sources will be applicable. In practice, some sources are more attractive than others due to such factors as accessibility and cost. And some of the energy sources such as fossil fuels act simultaneously as the hydrogen sources. The primary energy sources may include:

- Fossil energy (natural gas, petroleum, and coal)
- Renewable energy (solar, wind, and hydro electricity)
- Biomass or biofuel sources
- Nuclear energy (atomic fission and fusion energy)

The sources of material and energy as diverse as named above are the bases for hydrogen to be produced anywhere in the world. Furthermore, these sources can be linked to product hydrogen via a variety of potential production routes that have been invented. Figure 1.24 shows only the typical routes of chemical reforming, thermochemical process,

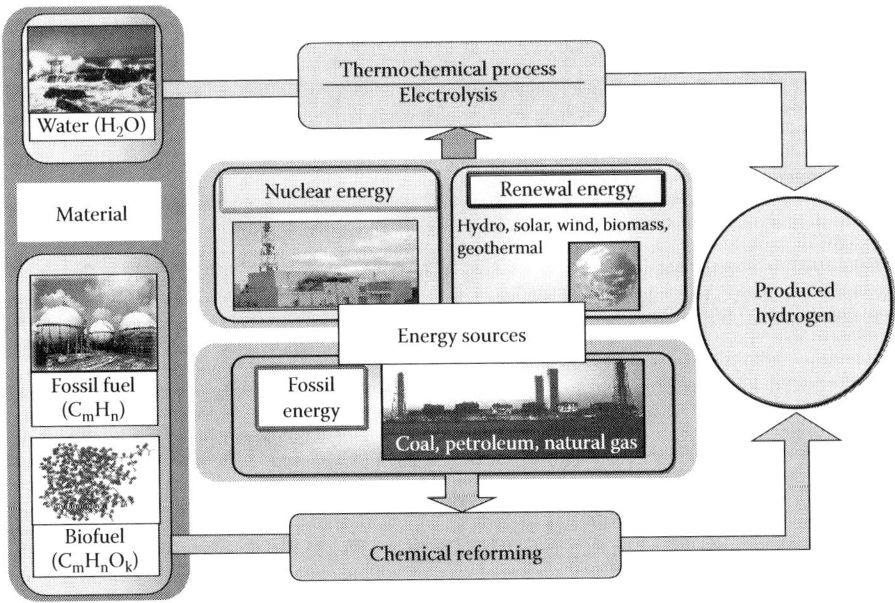

FIGURE 1.24
Potential pathways and requirements of hydrogen production.

and electrolysis, although other routes such as radiolysis and biochemical processes are also studied. Each of these routes starts with the choice of feedstock. The chemical reforming is based on chemical reactions that consume carbon and hydrocarbon sources (coal, natural gas, petroleum, and biomass) as essential reactants. On the other hand, the thermochemical process is based on chemical reactions but using water as the only reactant, and all other chemicals as reagents. These two methods receive heat as the major input of energy. Electrolysis splits water molecules into hydrogen and oxygen using electric power generated from any primary energy source such as solar, wind, and nuclear energy. The remainder of this chapter describes details of these hydrogen production processes.

1.5.2 Chemical Reforming

Chemical reforming for the purpose of hydrogen production almost always implies the use of steam as a reactant, and is commonly referred to as steam reforming. Steam reforming of various carbon and hydrocarbon sources leads to a product gas mixture rich in hydrogen. For example, steam reforming of methane (CH_4) at high temperatures (700–1100°C) over a catalyst (nickel based) yields carbon monoxide and hydrogen as follows:

$$\text{Methane:} \quad CH_4 + H_2O \rightarrow CO + 3H_2 - 206\ \text{kJ/mol} \tag{1.11}$$

Steam reforming of other light hydrocarbon fuels are similar:

$$\text{Ethanol:} \quad C_2H_5OH + H_2O \rightarrow 2CO + 4H_2 - 290\ \text{kJ/mol} \tag{1.12}$$

$$\text{Propane:} \quad C_3H_8 + 3H_2O \rightarrow 3CO + 7H_2 - 497\ \text{kJ/mol} \tag{1.13}$$

$$\text{Butane:} \quad C_4H_{10} + 4H_2O \rightarrow 4CO + 9H_2 - 651\ \text{kJ/mol} \tag{1.14}$$

$$\text{Naphtha:} \quad C_6H_{14} + 6H_2O \rightarrow 6CO + 13H_2 - 985\ \text{kJ/mol} \tag{1.15}$$

$$\text{Kerosene:} \quad C_{12}H_{26} + 12H_2O \rightarrow 12CO + 25H_2 - 1925\ \text{kJ/mol} \tag{1.16}$$

A step of exothermal heterogeneous water–gas shift reaction then follows in lower temperatures (200~350°C) to react water vapor with the product carbon monoxide of the steam reforming. This step, which is common for all hydrocarbons, takes place over a catalyst (noble metals) to produce additional hydrogen according to the reaction below:

$$CO + H_2O \rightarrow H_2 + CO_2 + 41\ \text{kJ/mol} \tag{1.17}$$

Finally, hydrogen is separated from carbon dioxide and any other impurities by pressure swing absorption to yield as high as 99.98% pure hydrogen gas product. Figure 1.25 summarizes the essential steps taken in the steam reforming of a hydrocarbon fuel stock.

Noting the overall (reforming + shift) net reaction of different hydrocarbons, methane has the highest yield of hydrogen molecules per elemental carbon of feedstock and

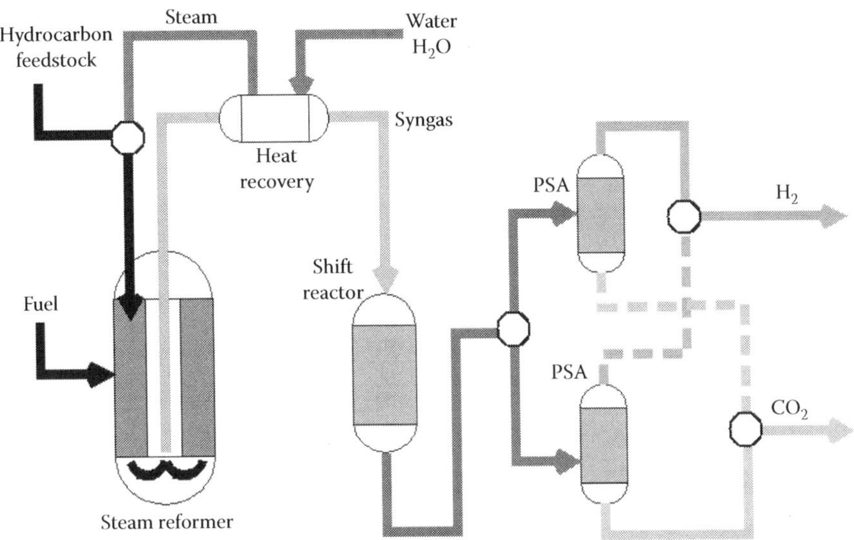

FIGURE 1.25
Steam reforming process of hydrocarbon fuels to produce hydrogen.

thus emits the least CO_2 for the same quantity of hydrogen produced. For these reasons, methane is the most appropriate feedstock of all hydrocarbons and also coal considered for hydrogen production. Each tonne of hydrogen produced consumes 3.5 tonnes of natural gas (mainly methane) and emits about 8.8 tonnes of CO_2 according to current commercial natural gas reformer performance.

The net reaction of chemical reforming of light hydrocarbon fuels is strongly endothermic, as demonstrated by the net heat of Equations 1.11 and 1.17 for methane. The heat requirement may be supplied by the combustion of additional feedstock fuel as in conventional reformer. It can also be met by other heat sources. Nuclear heated allothermal steam reforming processes have been demonstrated up to production scale in Germany and Japan (Chapter 20). It will be shown there that the nuclear heated steam reforming of methane would save the 34% feedstock of natural gas used as combustion fuel of the conventional autothermal process and cut the same percentage of CO_2 emission.

Steam reforming of natural gas offers high efficiency (65% to 75% LHV or 70% to 80% HHV) and generally low cost, and is most widely used process for hydrogen production. In the United States, about 95% of the hydrogen is presently produced this way, and in the world, 48%. A major disadvantage of the process to produce hydrogen is by production of CO_2. Capture and sequestration of CO_2 is feasible and would reduce production efficiency by about 1%.

Practical forecourt reformers with production capacity of 3–5 kg H_2/h have been tested at a number of refueling stations by the Japan hydrogen fuel cell (JHFC) vehicle demonstration project [7]. Various feedstocks used include gasoline, methanol, naphtha, kerosene, liquefied petroleum gas (LPG), and city gas. Overall efficiency of hydrogen production and fueling (from reformer feedstock to compete charging of 35MPa onboard hydrogen fuel tank) was demonstrated from the low end of 54.6% LHV (61.1% HHV) for kerosene to the upper end of 65.0% LHV (68.8% HHV) for methanol.

The steam reforming of coal, often called coal gasification, begins with pyrolysis in medium temperatures (400–600°C) to expel volatile matters (H_2, CH_4) of coal feedstock, and proceeds to steam gasification of the residual char via the reaction below:

$$\text{Char: } C + H_2O \rightarrow H_2 + CO - 118.5 \text{ kJ/mol} \quad (1.18)$$

The CO product gas is further reacted with steam via the water–gas shift reaction, Equation 1.17, to increase hydrogen product yield with CO_2 as the by-product.

The heat required by the reaction coal gasification may be provided internally in an autothermal process by the heating of exothermal chemical reactions of char with limited oxygen to produce carbon monoxide, or externally in an allothermal process using various heat sources. Chapter 20 describes the development in Germany of a nuclear energy-assisted allothermal coal gasification plant, which was tested during many thousands of operating hours under gasification conditions of 750–850°C and 2–4 MPa and with a total of 2400 t of hard coal gasified. The development confirmed the feasibility of an industrial scale plant and resulted in the design of a large commercial nuclear coal gasification plant.

An alternative method of coal gasification in combination with nuclear energy is synthesis of coal with a source of hydrogen feedstock that has been produced in a nuclear vapor electrolysis or thermochemical process. This is achieved in the hydro-gasification exothermic reaction:

$$C + 2H_2 \rightarrow CH_4 + 75 \text{ kJ/mol} \quad (1.19)$$

This is a high-pressure and high-temperature step and produces methane directly and efficiently [24,25]. The product of methane then undergoes the conventional steam reforming process described above, which completes the following next reaction to produce hydrogen:

$$CH_4 + 2H_2O \rightarrow CO_2 + 4H_2 - 165 \text{ kJ/mol} \quad (1.20)$$

As described earlier in the section, the nuclear heated allothermal system has been demonstrated to accomplish the methane to hydrogen reforming-shift process.

Biomass, a renewable organic resource, includes agriculture crop residues, forest residues, dedicated energy plants, organic municipal solid waste, and animal wastes. Gasifying biomass also yields a mixture of CO and H_2. For example, steam reforming of a woody biomass may proceed as follows:

$$CH_{1.47}O_{0.63} + 0.37H_2O \rightarrow CO + 1.11H_2 - 112 \text{ kJ/mol} \quad (1.21)$$

The CO product gas is further reacted with steam via the water–gas shift reaction, Equation 1.17, to increase product hydrogen yield with by-product of CO_2. If emitted, equivalent amount of carbon dioxide is re-captured by photosynthesis to grow the same amount of biomass gasified, resulting in a hydrogen production cycle of effectively zero carbon emission. The steam gasification of biomass is endothermic; in other words, it combines chemical energy of biomass and substantial quantity of heat, provided internally or externally to the reforming process, into chemical energy of hydrogen. Chapters 2 and 7 discussed the integration of nuclear thermal energy sources with the conversion process of biomass to produce hydrogen.

1.5.3 Electrolysis

Electrolysis splits water molecules into hydrogen and oxygen constituents by using electrical power. The basic equipment is the electrolyzer consisting of two electrodes (called anode and cathode) separated by an electrolyte. The process is highly valued for design simplicity and the flexibility of accepting virtually any primary-energy source able to generate electricity.

Electrolysis of liquid water is the earliest hydrogen generation method and was commercialized as early as in 1890s. It remains practiced to date, and is particularly suited to production of very pure hydrogen (99.99%) and oxygen. Research and development efforts have continued to improve the efficiency and cost of the process and equipment. In particular, high-temperature electrolysis of steam is the topic of recent studies for its potential application for nuclear hydrogen production.

The design of electrolyzer is identified with the type of electrolyte used. The conventional design is alkaline water electrolyzer (AWE), in which an alkaline solution of about 30% sodium hydroxide (NaOH) or potassium hydroxide (KOH) concentrate is employed as electrolyte. Electrodes are made of low-carbon steel or nickel-coated low-carbon steel. A porous diaphragm is used to separate product gases and electrodes. The reaction temperatures are in the range of 100–150°C. As shown in Figure 1.26, water reacts with electrons at the cathode to produce H_2 and OH^- ions. The ions pass the electrolyte and through the diaphragm to the anode side, wherein they discharge electrons to the anode while forming oxygen gas and water. The electrons as electric current are sent by the DC power source to cathode via an external circuit. The electrochemical reactions of AWE are summarized below:

$$\text{Cathode reaction:} \quad 2H_2O + 2e^- \rightarrow H_2 + 2OH^- \tag{1.22}$$

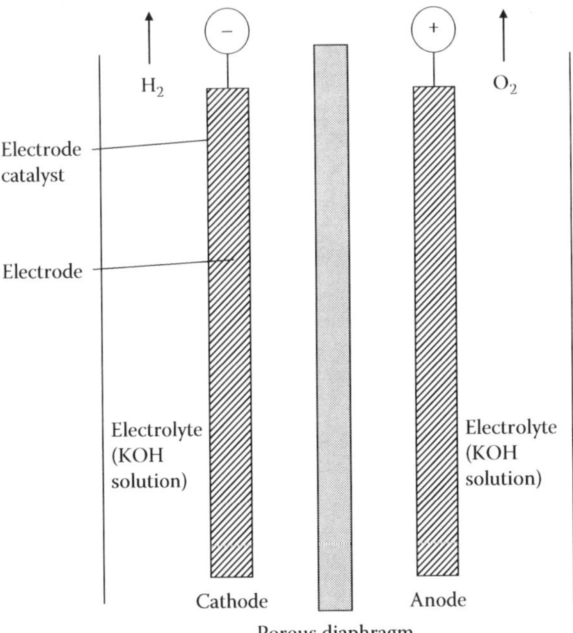

FIGURE 1.26
Design of conventional alkaline water electrolyzer.

$$\text{Anode reaction:} \quad 2OH^- \rightarrow H_2O + 0.5O_2 + 2e^- \tag{1.23}$$

The electricity required to drive the electrolysis to completion is theoretically 237.1 kJ for a mole of hydrogen gas produced at 25°C and 101.325 kPa. In practice, a high-performance electrolyzer consumes roughly 370 kJ (or 0.103 kWh) of electricity for a mole of hydrogen produced, therefore 65% (LHV) electrical efficiency.

A newer design of a polymer electrolyte membrane (PEM) electrolyzer employs an acidic proton-conducting material as electrolyte. Perfluorosulfonic acid polymer membranes such as Nafion® of DuPont are suitable because they are thermally resistant and corrosion resistant. Because electrodes are made to come in contact directly with the PEM in compact design, metal alloy and oxide in the platinum group are used for electrodes to resist the strong acidity of the PEM.

The process operates in low temperatures (80–100°C). Water reacts at the anode to form oxygen, protonnes, and electrons. The electrons are driven by the external power to flow through an external circuit to the cathode while the protonnes pass through the PEM to the cathode. At the cathode, hydrogen ions reunite with electrons to form hydrogen gas. The reactions are summarized below:

$$\text{Anode reaction:} \quad 2H_2O \rightarrow O_2 + 4H^- + 4e^- \tag{1.24}$$

$$\text{Cathode reaction:} \quad 4H^- + 4e^- \rightarrow 2H_2 \tag{1.25}$$

Comparing the AWE design, the PEM has the advantages of operating at high pressure and being free of liquid electrolyte to allow compact and low-maintenance design. The disadvantages are low efficiency (about 50%) and high material cost of membrane and electrodes, which have slowed the commercial progress made since the first PEM was operated by the U.S. General Electric in 1966. While the 2 MWe size AWE electrolyzer units are commercially offered, the largest PEM operated is limited to one-tenth of that size in 2010.

The advanced design of solid oxide electrolyzer (SOE) employs a solid ceramic membrane as the electrolyte that selectively passes oxygen ions. Water at the cathode combines with electrons from external power circuit to form hydrogen gas and oxygen ions. The oxygen ions pass through the membrane and react at the anode to form oxygen gas and give up the electrons to the external circuit. The reactions of SOE are summarized below:

$$\text{Cathode reaction:} \quad 2H_2O + 4e^- \rightarrow 2H_2 + 2O^{2-} \tag{1.26}$$

$$\text{Anode reaction:} \quad 2O^{2-} \rightarrow O_2 + 4e^- \tag{1.27}$$

High temperatures (about 500–850°C) are required for the solid oxide membranes to operate properly. Therefore, high-temperature steam is prepared outside the electrolyzer and using a heat source such as high-temperature nuclear heat available at these high temperatures. As a result, the electrolysis of high-temperature steam consumes about 20% less electricity than the electrolysis of liquid water at about 100°C. This increases overall thermal efficiency and is the major attractive feature of the SOE. Its potential disadvantages are material issues such as thermal strength and membrane performance degradation, which could limit operation life of practical system. The United States currently leads the technology development of the SOE and in 2008 tested the integrated system of three modules (each module 4 × 60 SOECs) continuously for 1080 hours with maximum production rate achieved of 5.65 Nm^3 H_2/h (0.504 kg H_2/h). Chapter 16 presents the details of the U.S. research and development on high-temperature steam electrolysis for nuclear hydrogen production application.

1.5.4 Thermochemical Process

Thermal decomposition, namely thermolysis, of pure water molecules into hydrogen and oxygen requires the heating of water to 2000°C for the decomposition just to begin and until 5000°C for it to complete. Because all materials melt at these temperatures, construction for production plant becomes impractical.

Fortunately, by incorporating cyclic chemical reactions, water can be made to react with chemical reagents and the products of the chemical reactions can then be thermally decomposed to yield hydrogen and oxygen at far lower temperatures. This is called the thermochemical process.

A thermochemical process consists of several chemical reactions with the net products of these reactions to be hydrogen and oxygen. This implies that the water is the sole feedstock to the process and all other reactants can be recovered and reused within the process.

Studies of thermochemical processes for hydrogen production began in the 1960s [26,27] and peaked during the period of 1975–1985 when the interest was heightened in finding alternative fuels in response to the 1973–1974 oil embargo. More than 100 processes were proposed. The recent 10 years have seen the second peak of studies of thermochemical processes in the quest to slow global fossil fuel uses and associated CO_2 emissions. The studies are focused on making some of the more promising proposals practical for economical and large scale hydrogen production with the use of nuclear and solar primary energy sources.

The list in Figure 1.27 contain the processes of recent interest, in which the reaction temperatures are generally lower than 1000°C and in some cases less than 600°C. These temperatures can be satisfied by solar heaters and nuclear reactors as heat source. Also at these temperatures construction materials for chemical plant are available. In addition to pure thermochemical process where only heat as energy input is essential, some of the processes proposed to date also incorporate electrochemical reactions and these are known as hybrid processes, to which electricity is also supplied as minor energy input.

The processes whose names are underlined in Figure 1.27 select thermal decomposition reaction of sulfuric acid as the highest-temperature endothermic reaction as follows:

$$H_2SO_4 \rightarrow H_2O + SO_2 + 0.5O_2 - 325.4 \text{ kJ/mol} \tag{1.28}$$

This reaction actually proceeds in two steps below.

$$H_2SO_4(aq) \rightarrow H_2O(g) + SO_3(g) \quad 300\text{–}500°C \tag{1.29}$$

$$SO_3(g) \rightarrow SO_2(g) + 0.5O_2(g) \quad 800\text{–}900°C \tag{1.30}$$

These steps can proceed smoothly without side reactions and at high conversion ratio at the temperature ranges indicated. Overall, the reaction is intensively endothermic at high temperature. These characteristics are matched by the heat capability of several high-temperature nuclear reactors including salt-cooled advanced high temperature reactor (AHTR), high temperature gas reactor (HTGR), gas-cooled fast reactor (GFR), and fusion reactor. These are presented in corresponding chapters in Section III of this handbook.

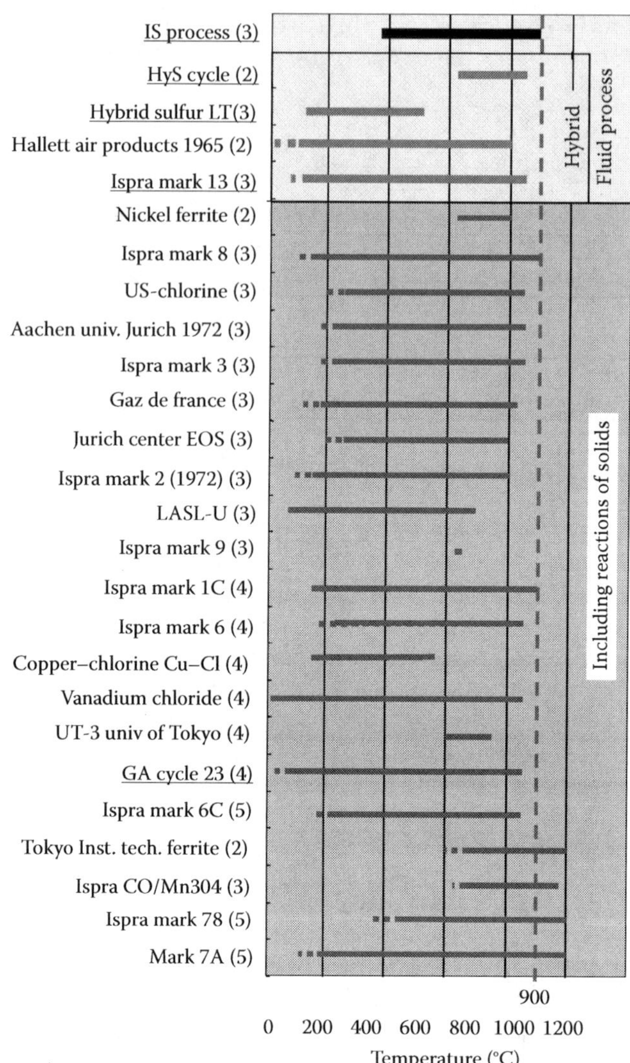

FIGURE 1.27
Selected major thermochemical processes of hydrogen production.

The IS (iodine–sulfur) process, which originated and still develops with General Atomics of the United States, utilizes additional two chemical reactions to close the process. The reactions of the IS process are as follows:

$$\text{Production of two acids:}\quad SO_2 + I_2 + 2H_2O \rightarrow 2HI + H_2SO_4 \quad 100°C \tag{1.31}$$

$$\text{Decomposition of sulfuric acid:}\quad H_2SO_4 \rightarrow H_2O + SO_2 + 0.5O_2 \quad 300\text{–}900°C \tag{1.32}$$

$$\text{Decomposition of hydriodic acid:} \quad 2HI \rightarrow H_2 + I_2 \quad 450°C \tag{1.33}$$

$$\text{Net reaction:} \quad H_2O = H_2 + ½O_2 \tag{1.34}$$

Reaction 1.31 is exothermic, in which the feedstock water is reacted with reagents iodine and sulfur dioxide with the spontaneous production of hydriodic acid and sulfuric acid at a relatively low temperature. The two other reactions 1.32 and 1.33 follow to thermally decompose these acids, producing oxygen and hydrogen, respectively, and all other products of these two reactions are recycled back to reaction 1.31 to resume production of the acids. A closed process is completed with the net effect of reaction 1.34. The IS process has a relatively small number of reactions and can be a pure thermochemical process with all fluid materials. These features simplify scale-up and operation in the practical plant. Much interest of research and development worldwide has thus been paid to this process. The status of the process technology, including engineering development and closed process operations, is reviewed in Chapter 17.

The HyS hybrid, originated with Westinghouse of the United States, is another heavily investigated process and also utilizes the thermochemical decomposition of sulfuric acid. The products sulfur dioxide and water of reaction 1.28 are taken over to a second reaction, adding feed water, to produce hydrogen by electrolysis as below:

$$\text{Electrolysis:} \quad SO_2 + 2H_2O \rightarrow H_2SO_4 + H_2 \quad 80\text{–}120°C. \tag{1.35}$$

The HyS process is one of the few thermochemical processes that only have two reactions and thus minimize the number of chemicals, which are all fluids, and operation complexity in process plant. The development status of this process is presented in Chapter 18.

While the HyS requires high-temperature heat source, a modified process proposed by Japan Atomic Energy Agency, adds a second electrolysis step of SO_3 to replace the high-temperature partial endothermic reaction (Equation 1.30).

$$\text{Electrolysis:} \quad SO_3(g) \rightarrow SO_2(g) + 0.5O_2(g) \quad 500°C \tag{1.36}$$

The result is a hybrid decomposition process of sulfuric acid, instead of pure thermochemical one, and caps the process temperature to 500°C. This temperature can be supported by wider choices of nuclear reactors such as sodium fast reactor and supercritical water reactor. Other medium-temperature processes among those in Figure 1.27 include UT-3 (Cr–Br, <700°C) and chlorine (Cu–Cl, <550°C) which have also been heavily investigated. The simplified chemistry in the copper–chlorine process can be represented by the following three reactions:

$$\text{Hydrolysis:} \quad CuCl_2 + H_2O \rightarrow Cu_2OCl_2 + 2HCl(g) \quad 375°C \tag{1.37}$$

$$\text{Decomposition:} \quad Cu_2OCl_2 \rightarrow 2CuCl \quad 450\text{–}550°C \tag{1.38}$$

$$\text{Electrolysis:} \quad 2CuCl + 2HCl \rightarrow CuCl_2 + H_2 \quad 100°C \tag{1.39}$$

Canada and the United States have been studying the copper–chlorine process [28].

The advantages of thermochemical processes are generally considered to be the potential of thermal efficiencies with the projected range of 25–55% and attractive scaling characteristics for large-scale nuclear hydrogen production applications. The technology is at an early stage with significant global interest and development at the moment. Although currently uncertainties exist about their ultimate performance and economics, several thermochemical cycles have been operated and demonstrated at scales that indicate the risks to be acceptable against the promises they hold to help transform the future energy economy.

1.6 Summary and Conclusions

Today, about half of the 50 million tonnes of hydrogen globally produced each year is consumed by fertilizer producers, and the remainder by oil refineries, chemical plants, and other industries. This hydrogen is 96% produced via chemical reforming of natural gas, oil and coal without carbon dioxide capture and sequestration.

Both the use and production mixes of hydrogen will change considerably, because the world economy has stepped up the speed of sustainable transformation by reducing fossil fuel uses and cutting CO_2 greenhouse gas emissions by 50–80% in 50 years.

This chapter analyzes the anticipated new demands for hydrogen in the vital economical sectors of transport, power generation, and industrial production. The new use of hydrogen for 50% U.S. transport fuel would require some 100 million tonnes of hydrogen a year. Similar use in Japan would require 50 million tonnes of hydrogen annually. The more populous China and India would each demand several times greater for hydrogen transport fuel when these national economies are developed in time.

TABLE 1.5

Synergy of Current and Future Hydrogen Production Options from Fossil, Nuclear, and Renewable Energy Sources

Production Status	Production Routes	Feedstock	Energy Source	Emissions
Current practice	Chemical reforming	Natural gas	Fossil fuels	CO_2 capture and sequestration (CCS) is not practiced
	Petroleum and coke by production	Oil, coal	Fossil fuels	
	Electrolysis	Water	Fossil and hydro electricity	CO_2 with fossil electricity
Future practice	Thermochemical process	Water	Nuclear energy, solar energy	CO_2 free
	Chemical reforming	Biomass	Biomass, nuclear energy	CO_2 free
		Fossil fuels	Fossil fuels nuclear energy	• CO_2 free by CCS • Limited practice
	Electrolysis	Water (Steam)	Solar, wind, hydro, nuclear electricity	CO_2 free

Similarly, the new use of hydrogen for 50% global steel output would need 35 million tonnes of hydrogen yearly. The traditional use of hydrogen for agriculture will increase with the anticipated growth of world population. These are plausible scenarios for the twenty-first century.

The primary energy sources that do not generate greenhouse gases and have the potential to produce hydrogen at scales large enough to replace substantial uses of fossil resources in the economy have to be essential components of the long-term global energy supply. Nuclear, biomass, solar, wind, and hydro are all possible as seen in Table 1.5. In a shorter term, hydrogen can be produced economically by electrolysis of water using power generated from any of these sources. In the mid-term, a new potential path, that is being actively researched and developed, is via the combination of nuclear fission energy with thermochemical processes or electrolysis of high-temperature steam. The feedstock requires water only and the products are hydrogen and oxygen. Before the twenty-first century is out, nuclear fusion energy will be developed to offer additional opportunity of producing hydrogen from water and biomass.

References

1. Lemmon, E.W. et al. Revised standardized equation for hydrogen gas densities for fuel consumption applications, *Journal of Research of the National Institute of Standards and Technology*, **113**, 341–350, 2008.
2. Takeichi, N. et al. "Hybrid hydrogen storage vessel," a novel high-pressure hydrogen storage vessel combined with hydrogen storage material, *International Journal of Hydrogen Energy* **28**, 1121–1129, 2003.
3. BP Statistical Review of World Energy June 2009.
4. Carbon dioxide information analysis center at http://cdiac.ornl.gov.
5. National Hydrogen Energy Roadmap, U.S. DOE November 2002.
6. The Hydrogen Economy: Opportunities, costs, barriers, and R&D needs, Committee on Alternatives and Strategies for Future Hydrogen Production and Use, National Research Council, National Academy of Engineering, 2004.
7. Japan Hydrogen & Fuel Cell Demonstration Project, http://www.jhfc.jp.
8. Fuel Cell/Hydrogen Technology Development Roadmap, NEDO, 2006.
9. Development of fuel cell and hydrogen technologies, NEDO, 2009.
10. Hyways, the European Hydrogen Roadmap, 6th Framework Programme, 2007.
11. Fuel Cells and Hydrogen Joint Technology Initiative, http://ec.europa.eu/research/fch/index_en.cfm.
12. FCH JU Multi-Annual Implementation Plan 2008–2013, Document FCH JU 2009.001, Final Version, May 2009.
13. Energy Technology Perspectives 2008—Scenarios and Strategies to 2050, International Energy Agency, 2008.
14. Hydrogen Storage, U.S. Department of Energy–Energy Efficiency and Renewable Energy Fuel Cell Technologies Program http://www1.eere.energy.gov/hydrogenandfuelcells/storage/hydrogen_storage_testing.html.
15. Hori, M. Nuclear energy for transportation: Paths through electricity, hydrogen, and liquid fuels, *Progress in Nuclear Energy* **50**, 411–416, 2008.
16. Annual Progress Report, U.S. DOE Hydrogen Program, 2009.
17. Chiesa, P. et al. Using hydrogen as gas turbine fuel, *Journal of Engineering for Gas Turbines and Power*, **127**, 73–80, 2005.

18. Wu, J. et al. Advanced gas turbine combustion system development for high hydrogen fuels, GT2007–28337, Proceedings of ASME Turbo Expo 2007: Power for Land, Sea, and Air, May 14–17, 2007, Montreal, Canada.
19. Todd, D. and Battista, R. Demonstrated applicability of hydrogen fuel for gas turbines, GE power system, http://www.netl.doe.gov/technologies/coalpower/turbines/index.html.
20. Nuclear Energy 2009, Japan Agency for Natural Resources and Energy, 2009.
21. Yan, X. et al., HTGR for flexible energy production, *Proceeding of Japan Society of Mechanical Engineering Annual Meeting*, Nagoya, September 5–8, 2010.
22. National greenhouse gas inventory report of Japan, Ministry of the Environment, Japan, April 2010.
23. Takashi, M. and Haruji O. Technology Development for Environmentally Harmonized Steelmaking Process, presentation in the joint Workshop on Delivering Green Growth—Seizing New Opportunities for Industries by OECD and Korean Ministry of Knowledge Economy, Seoul, Korea, 4–5 March 2010.
24. Substitute Natural Gas from Coal—Japanese Market Demand, The Australian Coal Review November 1997.
25. Naoki, T. and Takeshi, A. Potential cases of coal-derived substitute natural gas to future Japanese market, IGRC2008, Paris, 2008.
26. Funk, J.E. and Reinstrom, R.M. System study of hydrogen generation by thermal energy, TID 20441; EDR 3714, Vol. II, Suppl. A, Allison Division of General Motors Corporation, 1964.
27. De Beni, G. and Marchetti, C. Hydrogen, key to the energy market, *Eurospectra*, **9**, 46–50, 1970.
28. Lewis, M. et al. R&D status for the Cu–Cl thermochemical cycle, Section II.G.1, FY2009 Annual Report of DOE Hydrogen Program, 2009.

2

Nuclear Hydrogen Production: An Overview

Xing L. Yan, Satoshi Konishi, Masao Hori, and Ryutaro Hino

CONTENTS

2.1 Elements of the Approach

When atoms split or join, a bit of the original atomic mass gets uncounted for. The mass lost (m) shows up as energy (E) of an enormous quantity of $E = mc^2$, where c is the speed of light, stated Albert Einstein in 1905. Nearly all of this *atomic energy* may be harnessed in *nuclear reactor* that sends it out to *processes* of conversion in to one or more forms of *energy carrier* known as electricity, hydrogen, and heat.

The atoms that release atomic energy by fission or fusion are fuel of the nuclear reactor. They are naturally abundant and also can be synthesized from other abundant sources. In theory, enough resources are available to fueling a large number of nuclear reactors the world would need for centuries and longer.

Currently 438 commercial nuclear reactors are operating around the world. Another 54 units are under construction and hundreds more are proposed to be built. Together, the nuclear reactor capacity could double in 2030. These nuclear reactors built and proposed, with a few exceptions, are the third and earlier generations of reactor systems. They invariably connect to turbine power conversion process to generate electricity. If desired, the electricity produced may be carried to a further process to produce hydrogen. That process likely takes the conventional route of water electrolysis and, for economical reasons, produces hydrogen by using off-peak-demand electricity.

The few exceptions are the proposed fourth generation of nuclear reactors, better known as Generation IV reactors, which are being developed to improve safety and economics, minimize nuclear waste, and enhance nonproliferation, to meet future user acceptance and needs. Some of these reactors have been constructed and operated in smaller scales. The earliest commercial deployment of these can occur in the next 10 years. While design improvements over previous generations are varied, all Generation IV reactor proposals specify a higher operation temperature, with 950°C already demonstrated in operations, than the 325°C maximum achievable with the previous three generations mentioned earlier. While the higher temperature supports higher thermal efficiency of power conversion process, it also creates the possibility of nuclear reactors for direct heat applications, such as hydrogen production via high-temperature steam and thermochemical processes.

All four generations of nuclear reactors and proposals above are based on the principle of nuclear fission, which splits heavy atoms to release atomic energy. Going 50 years into the future, a new breed of nuclear reactors is anticipated to be commercially ready, according to the development plans and efforts currently under way in the world. They are based on the science of nuclear fusion that powers the Sun and farther stars. Fusion joins very light atoms to release atomic energy in fusion reactor. From there, converting fusion energy into electricity and hydrogen follows processes similar to nuclear fission.

Nuclear reactors are now the largest scale energy sources that do not emit significant carbon dioxide greenhouse gas. The 400 plus operating nuclear reactors are the source to 2.56 trillion kWh (about 15%) of the electricity produced globally in 2009, equivalent to meeting the annual electricity needs of 1 billion people or more than 300 million homes on the world average. They avoid emission of 2.1 billion tons of CO_2 compared to the electricity generation from fossil fuels based on the world generation mix of 42% from coal, 21% from natural gas, and 6% from oil [1]. With these qualities in production, nuclear reactors can continue to be a significant source of sustainable energy in the long term.

The world economy will demand substantially more hydrogen in future than the annual 50 million tons consumed today, because not only thus more hydrogen to ammonia fertilizers are needed to increase food production as the world population steadily grows and expects to add two more Indias by 2050; but also new demands in potentially larger scales have begun to come from transport and industrial sectors, as detailed in Chapter 1.

One of the key attributes of hydrogen for potentially economy-wide uses is that it can be produced from flexible feedstock options ranging from fossil resources such as natural gas and coal to renewable biomass and water. Similarly, primary-energy sources required are flexible, including nuclear, renewable, and fossil energy. As a result, hydrogen production follows more processes or pathways than listed below:

- Chemical reforming of fossil fuels and biomass, using nuclear heat.
- Electrolysis of water, using nuclear power.
- Electrolysis of steam, using major nuclear power and minor nuclear heat.
- Thermochemical process, using nuclear heat and often minor nuclear power.

The last three pathways essentially need nuclear cogeneration of electricity and hydrogen, as shown in Figure 2.1. Also seen in the figure is that the two energy carriers are convertible, one to the other via the routes of water electrolysis and fuel cell.

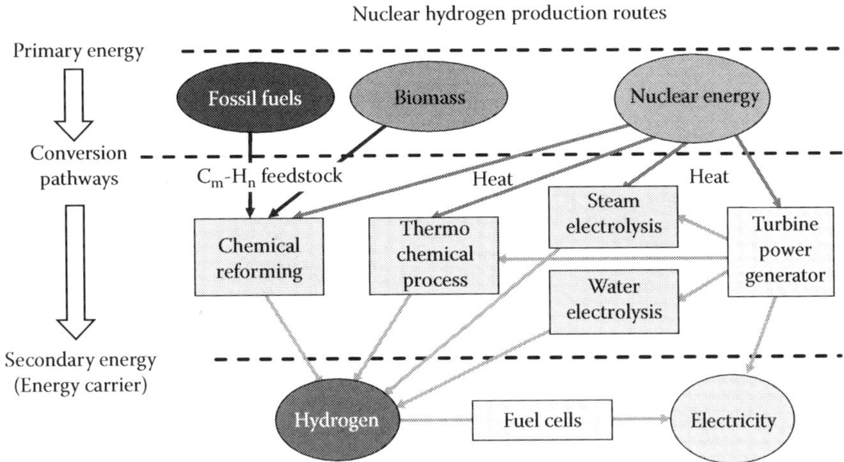

FIGURE 2.1
Potential pathways of hydrogen production from nuclear energy.

The chemical reforming relies on net endothermic chemical reactions of carbon or hydrocarbon feedstock with steam, in which nuclear energy meets the heat requirement. Conventional electrolysis splits water molecules into hydrogen and oxygen by consuming nuclear power. Advanced electrolysis raises water to high-temperature steam by nuclear heat prior to the electrolysis step, resulting in an improvement of thermal efficiency for hydrogen production. The thermochemical process consists of a series of chemical reactions using water as the only reactant and all other chemicals as reagents. Nuclear energy, in the form of heat and often additional nuclear power, drives a pure or hybrid thermochemical process. All four pathways use renewable sources of water or biomass as feedstock and are consistent with the sustainability goal of nuclear energy.

The next section introduces details of hydrogen production by nuclear fission reactors and associated pathways. Section 2.3 gives the details of hydrogen production for nuclear fusion energy.

2.2 Hydrogen Production from Nuclear Fission Energy

2.2.1 Nuclear Fission Energy

Nuclear reaction involves nuclei or particles reacting to produce new ones with the redistribution of nucleons. In contrast, chemical reaction involves molecules reacting to yield new molecules through rearrangement of electrons while keeping nuclei intact.

Fission is a subset of nuclear reaction. Another subset is fusion, the subject of Section 2.3. Fission splits a large nucleus into fragments termed fission products. It happens in a nuclear reactor designed to split nuclei with striking neutrons or in a particle accelerator designed to beam speedy particles to strike the target of a nucleus and split it into parts of fission products. Fission can occur naturally, either spontaneously or interactively. The former is decay of any radioactive isotopes of varying half lives. An example is radioactive β-decay of carbon-14 at a half-life of 5730 years (accuracy ±1%) to nitrogen-14. This natural phenomenon is what archaeologists rely on to date organic samples. An example of interactive nuclear reaction in nature is the continuous production of beryllium isotopes including ^{10}Be (half-life of 1.52 million years) and ^{7}Be (half-life of 53.28 days), by cosmic high-energy proton spallation of nitrogen and oxygen nuclei in the stratosphere. The isotopes are subsequently precipitated and found on the Earth's surface.

Like chemical reactions, nuclear fission reaction can be either exothermic or endothermic. In theory, the line is drawn at elemental iron, Fe, which has the highest binding energy and is the most stable of all elements. Splitting elements heavier than iron produces energy whereas splitting those lighter than iron consumes energy. Obviously, energy production needs the heavier elements.

Though the choices seem many, only a few very heavy ones including uranium, plutonium, and thorium are practical. And not all isotopes of the three are usable, at least directly. To make matters simpler, a usable isotope is one whose nucleus can absorb a striking neutron and produce more than one neutron. The neutrons so produced are absorbed by other nuclei of the same isotope to produce even more free neutrons. When this process repeats again and again, the result is a chain reaction. Such isotope that enables a chain fission reaction is termed fissile and is directly useable as nuclear fuel. Of the four known fissile isotopes, only one, ^{235}U, is found naturally in the Earth's crust and seawater.

Uranium-235 can absorb the low-energy neutrons (~0.1 eV) that approaches it at low speeds (on the order of 10^4 m/s). The low-energy neutrons are named thermal neutrons because they have lost most of their original kinetic energy at fission birth to heat. The neutrons are born in fission energetic (~2 MeV) and fast at about 7% of the speed of light. These so-called fast neutrons are too energetic to be caught by ^{235}U, and fission becomes improbable. In nuclear reactors, a moderator, which can be water or graphite, is used to slow the fast neutrons down to thermal neutrons to cause fission.

Although uranium-238, which is 99.3% naturally abundant of uranium (the balance is ^{235}U) on Earth, is not fissile, it is able to absorb a fast neutron and breed the transuranic isotope of ^{239}Pu, which is fissile. For this reason, ^{238}U is said to be fertile and a source of fission energy, albeit indirectly. Similarly, thorium-232 is fertile. ^{232}Th, which is estimated to be 3–4 times as naturally abundant as uranium, can absorb a thermal neutron to breed the fissile ^{233}U.

By exploiting the abundant fertile materials, the source of nuclear fission energy can be increased by hundreds of times as using just naturally fissile uranium. In fact, this is in part what happens in a typical thermal-neutron spectrum reactor, where ^{238}U of as much as 96% of the enriched fuel is loaded with ^{235}U. Uranium-238 is bred to ^{239}Pu with free neutrons from fission of ^{235}U in the reactor. As a result, some one-third of the generated power actually comes from the fission of ^{239}Pu, which is not supplied as a fuel to the reactor, but rather, produced from ^{238}U *in situ*.

Breeding is more efficient with the presence of ^{238}U in fast-neutron spectrum. A fast reactor uses no or limited neutron moderation. Since fission of ^{239}Pu by fast neutrons produces more neutrons, it becomes possible to breed more fissile fuel from fertile source than spent in the same core. Hence, fast reactor may extract many times more energy from the same quantity of uranium than thermal-neutron reactor. However, out-of-pile reprocessing is necessary to recover the extra-fissile fuel from the spent fuel (SF).

Fission reactions vary greatly. Generally, a fissile material fissions to a pair of fragments, which are two new nuclei with atomic masses close to 95 and 135 u, some neutrons of typically two or three, and additional γ-ray photons. They are collectively called fission products. The precise makeup of fission products is governed by statistical probability and cannot be predicted for any given event of fission. For example, the many pairs of fragments from the fission of ^{235}U may include Kr and Ba, Zr and Te, Cs, and Rb. Two specific reactions of ^{235}U fission are expressed below:

$$^{235}_{92}U + ^{1}_{0}n \rightarrow 2^{1}_{0}n + ^{100}_{38}Sr + ^{134}_{54}Xe + \text{energy} \quad (2.1)$$

$$^{235}_{92}U + ^{1}_{0}n \rightarrow 3^{1}_{0}n + ^{89}_{36}Kr + ^{144}_{56}Ba + \text{energy} \quad (2.2)$$

Fission reaction (Equation 2.1) is depicted in Figure 2.2. A fission reaction as above observes a set of conservation conditions including:

1. The mass number (total nucleons) and the charge (protons) must balance on each side of the reaction.
2. Linear momentum before and after the reaction must be equal.
3. Angular momentum, parity, and spin of the nuclear levels must balance according to the quantum rules.
4. The total energy must be kept equal before and after the reaction.

The total energy includes the energy equivalent of the particles' rest masses plus kinetic energy. It turns out that the total rest mass of all fission products is less (about 0.1% less) than that of the original particles before fission. The difference in rest mass, m, makes fission energy, E, which is exacted by Albert Einstein's formula $E = mc^2$, where c is the speed of light in vacuum. Because the speed of light is a large value (approximating 3×10^8 m/s) and, with this value squared, the small amount of mass difference is released into an enormous sum of net energy products by fission.

The energy released from each event of ^{235}U fission is approximately 212 MeV, of which about 169 MeV is the kinetic energy of the pair of fragments, which fly apart at about 3% of the speed of light, due to Coulomb repulsion, and emit about 7 MeV γ-rays photons from recoil. Also, prompt and nonfission neutrons carry a sum of about 14 MeV. The total prompt fission energy amounts to about 190 MeV, or ~90% of the fission energy. The remaining ~10% is released from β^- decays of the fragments to daughter isotopes and delayed γ-rays from the decays, and finally as antineutrinos. Since the makeup of fission products is not

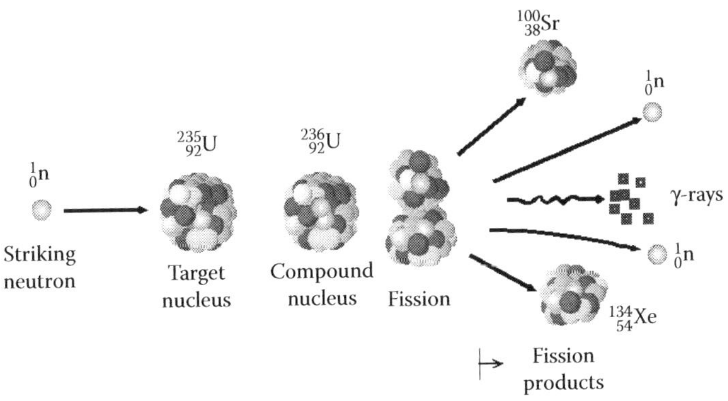

FIGURE 2.2
Uranium-235 fission with a sample set of fission products.

TABLE 2.1

Average Energy Release per Fission (MeV)

	^{235}U	^{239}Pu	^{233}U
Prompt Energy			
Kinetic energy of fission fragments	169	176	168
Radiation of prompt γ-rays	7	8	8
Kinetic energy of prompt neutrons	5	6	5
Energy of nonfission neutrons	9	12	9
Delayed (Decay) Energy			
Energy of β-decay	7	5	5
Radiation of γ-rays from decay	9	7	7
Energy of antineutrinos	6	5	5
Total	212	219	207

fixed, the statistically averaged values for three fissile isotopes are given in Table 2.1 from data source [2].

With the exception of the energy associated with antineutrinos, all others may be recovered as heat in a practically designed nuclear reactor. Thus, the total useable energy per fission event amounts to a little over 200 MeV (1 MeV = 1.602×10^{-13} J) for any of the three fissile materials. To put this value in perspective, fission of a gram of ^{235}U, which contains 2.56×10^{21} atoms and thus permits the same number of fission reactions, releases 84.1 billion Joules, equivalent to the energy released by combusting 3.5 tons of coal, 2200 L of oil, or 2100 m^3 of natural gas. Fission of a kilogram of ^{235}U can generate electricity for use by half a million homes for a day.

2.2.2 Nuclear Fission Energy Reactor

A nuclear reactor is a system devised to start, stop, and regulate a nuclear fission chain reaction, achieving a controlled chain reaction, and remove the fission energy as heat from the reactor. The components that are generally present in nuclear reactors include fuel, core, control rod, coolant, and reactor vessel.

The fuel is usually uranium oxide (UO_2 with fissile uranium enrichment of a few percent to less than 20%) loaded in a fuel assembly. A number of fuel assemblies are arranged to form a core of critical assembly needed to enable a chain fission reaction. When the event of chain fission reaction actually occurs, the core is said to be critical.

To gain the criticality, a neutron source needs to be present in the core. Usually this is an isotope of one of the low atomic weight elements such as beryllium mixed with an alpha emitter such as polonium or radium. The alpha particles emitted in decay cause beryllium to release neutrons. Restarting a reactor, called a re-criticality, with used fuel may not require an external neutron source, as neutrons from the decay of used fuel can be used by maneuvering control rods.

The control rods are effective neutron absorbers containing material of cadmium, hafnium, or boron. They are inserted into or withdrawn from the core to decrease or increase the population of free neutrons that cause fission. In other words, they are used to regulate the number of fission reaction events that can go on at a time, which in effect controls the power level of a reactor. Inserting the control rods deep in the core can halt all chain fission reactions and shutdown the reactor, because neutrons in the core are absorbed, down effectively to zero.

However, when the control rods are pulled out of the core, reactor can start up again because free neutrons are permitted to move around the core to cause fission.

The coolant comes in a variety of fluids including water, gas, liquid metal, or molten fluoride salt. There are further details of each fluid, like light and heavy water, or helium or carbon dioxide gas. The coolant circulates through the core, removes the heat from the fuel by heat transfer, and carries the heat away from the core. The heat removed by the coolant is given primarily by the kinetic energy of the fission-product particles in the fuel. The heat carried away from the core is transported to the balance of a reactor plant for the conversion of heat to electricity and hydrogen. The coolant is looped back into the core to continue the cyclic process of fission-heat production, transfer, and transport.

The reactor vessel usually made of steel contains all the components mentioned above, although a well-known exception to this is the Canadian deuterium uranium (CANDU) reactor that has hundreds of pressure tubes to contain fuel and coolant only.

2.2.3 Nuclear Hydrogen Production

2.2.3.1 Hydrogen Production Processes

2.2.3.1.1 *Low-Temperature Electrolysis*

Electrolysis splits water molecules at low temperature (~100°C) and usually at atmospheric pressure into hydrogen and oxygen constituents by using electrical power. The basic equipment is the electrolyzer consisting essentially of a catalytic electrolyte and a pair of electrodes. The process is highly valued for design simplicity and the flexibility of accepting virtually any nuclear reactors that generate electricity. The nuclear reactor couples to the process simply via electric transmission, meaning that there is no fluid–thermodynamic connection, such that distributed production of hydrogen close to end-user of hydrogen is possible with electricity grids. Megawatt electrolyzer units are commercially available off-the-shelf. Overall efficiency of nuclear reactor to hydrogen is in the 20–40% range. Combining one of the advanced reactors and an efficient electrolyzer would yield in the upper range of the efficiency. Chapter 3 introduces the design types, industrial development, and applications of the low-temperature water-electrolysis technologies.

2.2.3.1.2 *High-Temperature Electrolysis*

High-temperature electrolysis (HTE), also known as steam electrolysis, performs electrolysis of hot water vapor, instead of liquid water. This method promises higher efficiencies than the water electrolysis mentioned above because some of the energy is supplied as heat without prior power conversion and because the electrolysis reaction consumes less power at higher temperatures. Beside the efficiency, the heat supply directly boosts the economics of hydrogen because the heat is generally supplied cheaper than is electricity. When realistic electrolysis component efficiencies, steam generation performance, and operating conditions are considered, overall hydrogen production efficiencies in excess of 50%, compared to 40% with the water electrolysis, are predicted with reactor outlet temperatures above 850°C, which can be reached in several types of nuclear reactors to be introduced later.

HTE is not yet a commercial technology. It attracts great interest in research and development at the moment. In 2008, the U.S. Idaho National Laboratory (INL) operated an integrated laboratory-scale (ILS) HTE unit for 45 days, with the use of solid oxide electrolysis cells (SOEC). This demonstration achieved a peak output of 5650 L/h hydrogen. Despite the important milestone, hydrogen production by means of HTE faces challenges, particularly to sustain high performance operation of electrolyzers. Values for practical SOEC require the area-specific resistance less than 0.40 Ω cm^2 in stacked cell unit and degradation rate of less than 1% per 1000 h. The target Ohm value has been observed at INL in single-cell tests, but not yet in a stack configuration, while the degradation target has not been met. The current research in INL is focused on minimizing SOEC degradation. Although research and development on the HTE have been carried out elsewhere, the results obtained by INL to date have been the most extensive in the world and are reviewed in Chapter 16.

2.2.3.1.3 Thermochemical Process

Thermochemical process is a series of chemical reactions that split water to hydrogen and by-product oxygen. Many combinations of chemical reactions and reaction catalysts are possible. Higher the temperature, the better it is for process-thermal efficiency. In theory, the process is purely heat driven. But practical processes often include electrolytic processes, to concentrate process fluids for example, or are hybrid by incorporating electrochemical steps. The hybrid process is separately discussed in Section 2.2.3.1.4.

Since the first studies on thermochemical water-splitting cycles were reported in 1960s, a large number of thermochemical cycles have been investigated (see Chapter 5). These cycles assume the utilization of energy sources that can supply heat at temperatures up to 1000°C. Largely handling thermal and chemical fluids, these processes offer the potential for high efficiency and economics of scale. The current status of the technology for a leading iodine–sulfur process, known as IS-process, has advanced to the stage of a detailed pilot-plant design.

From 2005 to 2009, General Atomics (GA), which proposed the original IS process in 1970s, collaborated with the U.S. Sandia National Laboratory and the French Commissariat L'Energie Atomique (CEA) for an ILS experimental demonstration of the process. All process components were fabricated of engineering materials, in particular, heat and corrosion-resistant materials of SiC and glass-lined steels. The integrated process was continuously operated with buffer storage of process chemicals between its three key reaction sections. Control system can operate each section in standalone and integrated operation modes. Because of the relatively small scale of the loop, the major difficulties encountered were with pumping and level control of iodine. This limited hydrogen production rate to 75 L/h, below the design target of 100–200 L/h. All key thermochemical reaction sections have been operated numerous times in standalone or partially integrated modes. A novel SiC bayonet sulfuric acid decomposer was built and tested. International collaborations are being sought to support further development on process and component improvements, and scale-up of component designs. The status of the GA investigation until 2009 completing a part of the U.S. nuclear hydrogen initiative program is introduced in Chapter 31.

The Japan Atomic Energy Agency (JAEA) has continued studies on the IS process for some decades and accumulated comprehensive material and process know-how. JAEA has successfully demonstrated laboratory-scale and one-week steady bench-scale hydrogen production at up to 30 L/h. In 2010, JAEA has begun the work of scaling the production up to 200 L/h in a new integrated process test facility. Meanwhile, a pilot-plant demo project producing hydrogen at 30,000 L/h from helium heater of 400 kW is underway to validate the engineering of the IS process on basis of the bench-scale experience. More specifically, the pilot-plant demonstration would allow validation of process construction materials identified, automated operation techniques, process efficiency, and plant cost. If successful, JAEA plans the construction of an IS plant producing 1 million L/h (2 tons/day) of hydrogen and connected it to the nuclear high temperature engineering test reactor

(HTTP) nuclear reactor to demonstrate the performance of an integrated nuclear-hydrogen-production system around 2020. Chapter 17 reviews JAEA's research and development leading up to the success of the bench-scale production operation. Chapter 29 discusses the development of the HTTR IS plant demonstration program.

2.2.3.1.4 *Hybrid Process*

Despite often categorized just as thermochemical process, hybrid process incorporates electrochemical step or steps into a thermochemical process. While the motivations to do so are widely varied, it is seen to minimize the number of the chemical reactions in an overall process. The most studied process, called hybrid sulfur (HyS) cycle, is a hybrid variant of the thermochemical IS-process described earlier.

The HyS cycle comprises two chemical reactions, in contrast to the three steps with the IS process. Like the IS, the HyS involves all fluid reactants and products. The endothermic thermochemical reaction requires high-temperature (>800°C) heat to decompose sulfuric acid while the electrochemical step is the SO_2-depolarized electrolysis (SDE) of water at low temperature (~100°C). The HyS cycle was proposed and studied by Westinghouse Electric Corporation, during 1970s and 1980s. Simultaneous development also took place in Europe during the same period. Currently, the Savannah River National Laboratory (SRNL) is leading the HyS development in the United States. Research works in the EU, France, the Republic of Korea, China, and South Africa are ongoing. In 2009, SRNL developed new operating methods that demonstrated PEM-type SDE operation without the previously encountered limitations caused by sulfur buildup. The future challenges in the SDE include higher-temperature operation and catalyst development, development of high-temperature membranes with low SO_2 crossover, minimal performance degradation, and scale-up to larger cell sizes and multicell modules. Chapter 18 reports the current status of the research and development for the HyS thermochemical process at SRNL.

Alternative hybrid cycles under development intend to lower the maximum endothermic-reaction temperature from the greater than 800°C required by both the IS and the HyS processes, because this temperature is too high to be supported by some nuclear reactors. Two of the alternative cycles now under investigation in Japan, the United States, Canada, and France have the objective of driving the completion of the cycles with the medium-temperature (500–700°C) heat sources. The first cycle introduced in Japan is a modified HyS, which divides the sulfur-acid decomposition task into a step of endothermic sulfur trioxide production from sulfuric acid, then followed by a step of medium-temperature electrochemical decomposition of the sulfur trioxide at 500°C. This increases the total steps of chemical reaction by one from the original two-step HyS. The research on the modified HyS is reviewed in Chapter 11 in the description of the development for a nuclear hydrogen production plant based on sodium fast reactor.

The second medium temperature hybrid cycle investigated is a Cu-Cl cycle that includes a hydrolysis step, a thermochemical reaction step, and an electrochemical step. The maximum temperature of 550°C occurs at the thermochemical step which decomposes copper oxychloride. A major technical challenge is in the hydrolysis step which involves reaction and production of solid and gas. The difficulty is about the design of compact reactor equipment with high yield of copper oxychloride product and separation of excess steam from the product gas that must be used subsequently in the electrolysis step. The research and development work on the Cu-Cl process has been led by the U.S. Argonne National Laboratory and joined by Atomic Energy of Canada Ltd (AECL) and the CEA in France, and universities in the United States and Canada [3].

2.2.3.1.5 Chemical Reforming

Steam reforming of carbon and hydrocarbon sources including fossil fuels and biomass usually requires heat of high temperatures over 800°C to reform the feedstock with steam to yield hydrogen and carbon monoxide. A number of Generation VI reactors, to be described later, support the high temperature heat requirement. The carbon monoxide is then reacted with water to yield further hydrogen product and carbon dioxide waste gas. Reforming processes of coal and methane are commercially practiced with use of fossil combustion heat source. The allothermal processes of coal and methane reforming with nuclear heating have been demonstrated in large scales in Germany and Japan, which are reviewed in Chapters 19 and 20. It has been shown that the nuclear-assisted steam reforming of methane is marginally cost competitive compared with the conventional combustion-reforming process. The more important advantage with nuclear chemical reforming is that it eliminates the use of fossil fuel from the conventional process and reduces emission of carbon dioxide, about 30% less in the case of methane reforming. Recently, a novel membrane reformer of methane has been reported, which lowers the temperature of methane reforming to the 500–550°C range [4,5]. At this temperature, a wider spectrum of nuclear reactors can be considered as heat source for the membrane reformer process.

2.2.3.2 Nuclear Hydrogen Production Systems

The overall process efficiency of nuclear heat to hydrogen product moves from about 25% with today's light-water reactors (33% for reactor power generation × 75% for water electrolysis) to 38% with more efficient reactors and to 45–50% for high-temperature reactors in combination with steam electrolysis or a thermochemical or hybrid process. These nuclear hydrogen production pathways use water as the only material feedstock and the production of hydrogen is completely carbon emission free. On the other hand, chemical reforming with a nuclear reactor heat source reduces the consumption of fossil fuel or biomass feedstock and cuts CO_2 emission by replacing fossil fuel combustion.

Nuclear reactor types and design parameters for hydrogen production are summarized in Table 2.2. More details are discussed below for each type of the reactors.

2.2.3.2.1 Light-Water Reactors

Light-water reactor (LWR) is usually thermal neutron reactor and the light water serves as moderator and coolant. Nuclear fission energy heats up the coolant in the reactor core, after which there are two design approaches, direct cycle and indirect cycle, with respect to how to use the heated coolant. The direct cycle as adopted by boiling water reactor (BWR) boils the coolant to saturated steam (99.9% dry) at about 7.2 MPa and 286°C at the core exit and guides the steam directly to a Rankine cycle turbine to turn a generator to produce electricity. The steam then goes to condenser and the resulting water is returned to the reactor core, completing the coolant circulation. The indirect cycle approach taken by pressurized water reactor (PWR) heats the coolant without boiling it to about 325°C. A high pressure of about 15.5 MPa is necessary to prevent coolant vapor. The hot water is routed to a steam generator to transfer heat to produce secondary cycle steam at about 277°C and 6.1 MPa, which drives a turbine generator to produce electricity. The primary coolant exits the steam generator and returns to the reactor core, completing the coolant circulation.

The BWRs and PWRs account for the majority of operating nuclear power reactors in the world. The recent proposals of advanced LWR incorporate passive safety features.

TABLE 2.2

Summary of Nuclear Reactor Parameters for Hydrogen Production

	Neutron Spectrum	Fuel Cycles being Used or Studied	Coolant	Reactor Outlet Coolant Temperature (°C)	Typical Sizes MWt/MWe	Electricity Generation Route	Hydrogen Production Route
Light-water reactors: PWR, AP, EPR; BWR ABWR	Thermal	UO_2; U–Pu MOX	Light water	280–325	2000–4800 MWt 600–1700 MWe	Steam turbine	Water electrolysis
Heavy-water reactors: CANDU, ACR	Thermal	Natural UO_2; low enriched UO_2;	Heavy water	310–319	2000–3200 MWt; 700–1100 MWe	Steam turbine	Water electrolysis
Supercritical water reactor: S–LWR, CANDU SCWR, SF–LWR	Thermal or fast	UO_2; Th/U; U–Pu MOX	Light water	430–625	1600–2540 MWt 700–1150 MWe	Steam turbine	Water electrolysis; M–T thermochemical
Liquid metal fast reactors: SFR, LFR	Fast	U–Pu MOX; U–Pu nitrides; MOX w/TRU	Sodium; lead; lead bismuth	SFR: 550 Pb or Pb–Bi: 500–800	45–3000 MWt 20–1100 MWe	Steam turbine; SCO_2 turbine	Water electrolysis; M–T thermochemical; M–T methane reform
Molten salt reactors: AHTR	Thermal	UO_2; Th/U	Salts: $^7Li_2BeF_4$ NaF–ZrF_4 and so on.	750–1000	900–2400 MWt 400–1200 MWe	Steam turbine; SCO_2 turbine	Water electrolysis; steam electrolysis; thermochemical; chemical reforming
Gas-fast reactor: GFR	Fast	U–Pu MOX; U–TRU	Helium	850	600–2400 MWt 280–1100 MWe	Steam turbine; gas turbine	Water electrolysis; steam electrolysis; thermochemical; chemical reforming
High-temperature gas reactors: HTGR, VHTR	Thermal	UO_2; Th/U; PuO_2; U–Pu MOX; U–TRU	Helium	750–950	100–600 MWt 45–300 MWe	Steam turbine; gas turbine	Water electrolysis; steam electrolysis; thermochemical; chemical reforming

The advanced BWRs are Generation III(+) systems such as ABWR and ESBWR rated up to about 1600 MWe. Several ABWRs are either operating or under construction in Japan and Taiwan and a few others are planned in the United States. The advanced PWRs are also Generation III(+) including AP1000, US-APWR, and EPR (European pressurized reactor) rated up to 1700 MWe. The construction for the advanced PWRs is either underway or being planned.

Because LWRs are commercial technologies, they offer a ready mean of producing hydrogen from nuclear energy through electrolysis of water. Large electrolyzer units rated at 2 MWe and 485 Nm^3/h hydrogen are available off-the-shelf. A number of these units can add quickly up to result in a large production capacity. The production cost of hydrogen depends primarily upon the economics of electricity. Chapter 9 exploits a number of economical incentives to use lower-cost off-peak electricity for hydrogen production from light-water reactors and to store nuclear power in the form of hydrogen for the shaving of peak electricity demands with fuel cells and combustion turbine electric generator. This expands the application of LWRs beyond the traditional base-load power generation.

2.2.3.2.2 *Heavy Water Reactors*

These are known as CANDU reactors. The thermal neutron reactors use heavy water as both moderator and coolant. Unlike the LWRs, which need fuel enriched to between 3% and 5% of fissile uranium with the remainder being fertile uranium, the CANDUs use fuel of natural uranium containing less than 1% of fissile uranium, thanks to more efficient neutron moderation of heavy water than the light water. Another unique design feature of CANDU is the hundreds of pressure tubes used to contain fuel and coolant, instead of the usual large reactor vessel. This feature facilitates on-line refueling in the individual tubes. The reactor coolant is heated to up to 313°C and 10.6 MPa. Like in a PWR, a steam generator transfers primary heat to secondary loop to produce steam to 263°C and 4.9 MPa, which then powers a steam turbine to generate electricity. More than two dozens of CANDUs are operating in the world. Advanced designs including CANDU-6 and Generation III+ ACR-1000, rated at 700–1100 MWe, are being commercially offered.

Because heavy water reactors are also existing technology with similar reactor coolant conditions, they are considered suited to hydrogen production through electrolysis of water similar to the LWRs described above.

2.2.3.2.3 *Supercritical-Water Reactors*

These are Generation IV of LWR designs and use supercritical light water as the working fluid. The low density of supercritical water requires additional moderator for a successful thermal-neutron spectrum. The SCWR connects to a direct cycle steam turbine much like the BWR but runs the high pressure single-phase coolant similar to the PWR. Although new in nuclear reactor, supercritical water already enters the fossil-fired power plants. The SCWR minimizes equipment count and commands greater operating temperature and pressure (up to 625°C and about 25 MPa, well above the thermodynamic critical point of 374°C and 22.1 MPa for water) than the Generation III+ LWRs (refer to Section 2.2.3.2.1). These features are expected to lower construction cost and raise thermal efficiency to estimated 45%, in comparison to about 33% for other LWRs. In addition to thermal-neutron spectrum, the SCWR due to the low density of the supercritical coolant may accept a fast-neutron spectrum core, which would demand further material research and development. The reactor systems such as Super-LWR or CAND-SCRW are being developed in Japan and Canada, respectively, as described in Chapter 9.

In addition to hydrogen produced by water electrolysis, the SCRW's upper coolant temperature matches the heat requirement for medium-temperature thermochemical processes such as the Cu-Cl described earlier.

2.2.3.2.4 Liquid Metal Fast Reactors

The objective of fast reactors is improved fuel sustainability through efficient irradiation of fertile uranium in the fast-neutron spectrum. To do so, highly enriched fissile uranium or plutonium is used as fuel and the reactor core must be sized more compact than that of a LWR. Liquid metals (as well as helium) become the choices of reactor coolant since they slow little of the neutrons (being neutron transparent) and are superior heat-transport media well matched to the fission-heat generation intensity of a small-size fast-neutron spectrum core. Certain liquid metals are satisfactory as coolant, which requires low melting points (sodium at 97°C, lead–bismuth eutectic alloy at 124°C) and high boiling points (sodium at 883°C and lead–bismuth at 1670°C) at the atmospheric pressure. The low pressure has the safety advantage with respect to leak proof of primary circuit.

Sodium-cooled fast reactors (SFRs) have been designed in the range of medium size (150–500 MWe) to large size (above 1000 MWe). Fuel forms are either mixed uranium–plutonium fuel alone or with minor actinides for waste destruction. The fuel cycle requires reprocessing of SF. The reactor coolant outlet conditions are approximately 550°C and 0.5 MPa. Sodium is compatible with structural steels and many fuel forms. Its disadvantage is its violent chemical reaction with air and water with severe consequences to nuclear reactors. To minimize the risks, an intermediate sodium loop is used to prevent air, water, and other reactants from coming in contact with the primary sodium. JAEA has been operating and up-rating Joyo experimental SFR (now at 140 MWt) since 1977. Developed also by JAEA is the 280 MWe Monju prototype fast breeder reactor (FBR), which was stopped in 1995 because of a leak in the secondary sodium loop and restarted in 2010. Japan plans the full commercial FBR in 2050. As presented in Chapter 11, JAEA is currently performing research and development for a hybrid thermochemical hydrogen production process whose heat requirement is especially tailored to the coolant-temperature range of the SFR. The description of a hydrogen production plant design based on the hybrid process and the SFR is also given in the chapter.

Beside sodium, lead and lead–bismuth coolants are also considered. The reactor design options include small battery type of systems thermally rated at 120–400 MWt with long refueling interval (10–30 year core life), larger modular system rated at 750–1000 MWt, and the largest plant at 3000 MWt. The fuel may be based on metal or nitride, mixed with fertile uranium and transuranics (TRU). The battery type of systems is intended for reactors to be factory-fabricated and transported to the plant site, and sealed for the long operation period. The reactor operation and safety as a battery is enhanced with near atmospheric coolant pressure and natural convective coolant circulation through the core. The reactor outlet coolant temperatures range from 550 to 800°C, depending on the selection and feasibility of structural materials and fuels. The broad coolant-temperature range opens more routes for hydrogen production than with the sodium reactor. The various sizes of the designs that have been shown to date will enable distributed and central production of hydrogen. Chapter 14 presents STAR-H2, a Pb-cooled reactor that utilizes an intermediate heat-transport loop to transfer the primary lead-coolant heat to a cascade energy-cogeneration plant in a third loop, which contains a thermochemical process operating above 700°C for hydrogen production, a further downstream medium-temperature supercritical CO_2 Brayton cycle generator to cogenerate electricity for in-house production and operation needs, and finally a desalinization plant driven by the sensible waste heat of the Brayton cycle. The cascade system makes very high thermal utilization of the nuclear energy.

2.2.3.2.5 Molten Salt Reactors

Two types of these salt-cooled reactor proposals are characterized by fuel forms. The earlier proposal known simply as MSR (molten salt reactor) features fuel-coolant liquid mixture of sodium, zirconium, and uranium fluorides. The homogenous liquid fuel allows various mixing of fuel and actinides for waste destruction. The flexibility with liquid fuel eliminates the need for fuel fabrication. The molten-salt fuel flows through graphite core channels, producing an epithermal to thermal spectrum.

The more recent design, called AHTR (advanced high-temperature reactor) uses coated fuel particles with low-enriched uranium kernel, and a molten salt serves as coolant only. The fuel particles are dispersed in graphite matrix. The core is of thermal spectrum. The particle fuel benefits from the similar well-developed fuel form used in another type of HTGR reactor to be introduced in Section 2.2.3.2.7.

The AHTR's molten-salt coolant is a mixture of fluoride salts with melting point near 400°C and atmospheric boiling point around 1400°C. Various salts have been evaluated such as $^7Li_2BeF_4$ and $NaF–ZrF_4$. The reactor operates at low pressure (<0.5 MPa) with the core outlet coolant temperatures in a range of 700–1000°C. The reactor rating falls in the range of 900–2400 MWt. Heat is transferred from the primary coolant through a compact secondary molten-salt coolant to a third energy conversion loop for generation of electricity or directly hydrogen. If electricity is produced, a multireheat nitrogen or helium-Brayton power cycle is preferred because of the nature of high-average coolant temperature condition of the AHTR. For direct hydrogen production, the high-temperature potential of the reactor outlet coolant temperatures should favor high-temperature steam electrolysis or thermochemical process for efficient hydrogen production. The efficiency at the 1000°C reactor temperature could be 10–20% higher than at the 700°C reactor temperature depending on the hydrogen production process selected. In Chapter 13, the researchers at the University of California, Berkeley discuss about the design and development for AHTR-based hydrogen production.

2.2.3.2.6 Gas Fast Reactors

These are Generation IV reactor designs under study. The designs have been proposed for modular and larger systems rated between 600 and 2400 MWt. They use helium as coolant, which is neutron transparent as is preferable with fast-neutron spectrum, with reactor outlet temperature of 850°C. Helium is inert and benign with structural materials. Several fuel forms are being considered for their potential to operate at the high temperatures and to ensure excellent retention of fission products: composite ceramic fuel, advanced fuel particles, or ceramic clad elements of actinide compounds. Core configurations are being considered based on pin- or plate-based fuel assemblies or prismatic blocks. A commercial feasibility study by JAEA identified the coated particle fuel with mixed nitride kernel and TiN coating to be optimum [6]. The major development activities for the GFR are concentrated in France, Japan, and more recently in the United States.

In the United States, a helium-cooled fast reactor proposal, called the energy multiplier module (EM2), uses as its fuel the SF from existing reactors without prior reprocessing [7]. Many of the technology requirements such as helium coolant and direct cycle gas turbine balance of plant are shared with another long-running thermal-neutron reactor design, the GT-MHR. EM2 is rated at 500 MWt with coolant outlet temperature at 850°C. The core life span of 30 years is based on spent LWR fuel and depleted uranium (tails of uranium enrichment) without the need for refueling. If realized, EM2 would offer a unique advantage of nuclear-fuel recycle without reprocessing.

A key design advantage for the GFRs is the high-temperature nuclear heat capability as a result of selecting helium as coolant. At 850°C, it can drive a direct Brayton-cycle gas turbine for high-efficiency power production and open the full range of possible routes for hydrogen production as indicated in Table 2.2. In particular, the HTE and thermochemical process can be supported by this reactor for highly efficient hydrogen production.

2.2.3.2.7 *High-Temperature Gas Reactor*

The high-temperature gas reactor (HTGR), also known as VHTR (very high-temperature reactor) is helium-cooled and graphite moderated and uses thermal-neutron spectrum. The fuel is in the form of strong ceramic (SiC or ZrC) coated particle with maximum design limit of 1600°C. The designs usually select once-through uranium fuel cycle, but many fuel cycle options have been investigated and confirmed in test and commercial reactors [8,9]. It has been shown that TRU might be consumed in a "Deep Burn" mode where a burn up of 60% fission of initial metal atoms (FIMA) or more can be achieved in single irradiation loading in the core [10,11] . In sum, the HTGR allows multiple fuel cycles due to the strong ceramic fuel design with high burnup ability.

The HTGR outlet coolant temperature is proven to be 950°C. At this temperature, hydrogen production can be performed with conventional electrolysis with overall thermal efficiency of 35–40% (reactor power generation efficiency of 50% times electrolyzer efficiency of 70–80%). It can also be carried out by high-temperature (900°C) steam reforming of coal and natural gas, as has been demonstrated in Japan and Germany. Steam reforming of biomass can be similarly performed. Hydrogen production of HTGR in combination with high-temperature steam electrolysis and thermochemical and hybrid processes is being studied in the United States, South Africa, Japan, Korea, China, and Europe. In general, the passively safe reactor is rated from 200 to 600 MWt dependent on core configurations. Large reactors up to 2400 MWt have been designed and the largest operated is the 842 MWt FSV (Fort St. Vrain) in the United States. These larger systems were the design choices before the emphasis was shifted to fully passively-safe modular reactors of today's designs of under 600 MWt per reactor. In the case of the modular designs, multiple reactor modules can be added for incremental capacity increase or built jointly on the site. A 600 MWt HTGR could produce more than 150 tons of hydrogen fuel daily, equivalent to 1.2 million liters gasoline based on road distance traveled, while cutting 2800 tons of CO_2 emission.

In a 50-day test in 2010, JAEA operated the HTTR (a 30 MWt HTGR test reactor) at high temperatures (>930°C) and full power. It plans to connect a thermochemical process to the HTTR to demonstrate nuclear hydrogen production around 2020. JAEA has been developing commercial designs for domestic and overseas markets using direct cycle gas turbine for electricity production and the IS process for hydrogen production, deploying the first units after 2020. Chapter 10 reviews the HTGR design and development for hydrogen production and includes extensive results obtained in JAEA.

The U.S. Department of Energy and the U.S. industries are cosharing the conceptual designs for the next generation nuclear plant (a 500–600 MWt HTGR) that includes a mission to demonstrate a 50 MWt scale hydrogen production based on the HTE or a thermochemical process. The plant is scheduled to enter the detailed design phase in 2011 and achieve startup in 2021. Chapter 31 describes details of the U.S. DOE programs related to nuclear hydrogen production. Additional chapters in Section IV introduce the works elsewhere in the world.

2.2.4 Economics and Sustainability

2.2.4.1 Economics of Nuclear Hydrogen Production

An assessment on the economics of hydrogen production from nuclear energy and via other alternative routes was reported recently [12]. The technology assessed for nuclear hydrogen production among the many possible combinations, as discussed above, of nuclear reactors and hydrogen production processes is the centralized hydrogen production plant based on the HTGR combined with a thermochemical production process. This specific technology is believed to represent one of the most economical arrangements for nuclear reactor-based hydrogen production for several key reasons. First, the technology requires mostly nuclear heat as energy driver, which is generally cheaper than electricity. Secondly, the reactor-to-hydrogen thermal efficiency of the technology is among the highest of the nuclear hydrogen production options because both the HTGR and the thermochemical process operate in high temperature. Thirdly, the thermochemical process is essentially a fluid process and can be easily scaled up to become comparable with a large nuclear-reactor energy source for centralized hydrogen production. Finally, the HTGR is the most developed of Generation IV reactor systems under current international development for near-to-mid-term deployment.

The reported economics of hydrogen production considered the underlying efficiency of the technology employed, the current state of its development (i.e., conceptual, developmental, commercial), the scale of the plant, its annual utilization, and the cost of its feedstock. Other considerations included the physical distance and availability of potential feedstocks from potential end-use markets for hydrogen gas, and whether to use centralized production in order to take advantage of economies of scale in production and incorporate hydrogen transmission and distribution systems from the plant gate, or rely on distributed hydrogen production, where the feedstocks are transported over a greater distances and the hydrogen gas transmission and distribution infrastructure is minimized.

The economics of the hydrogen production technologies were assessed from a review of literature, adjusted to the U.S. annual average prices in 2007. The results are summarized in Table 2.3. For most of the production technologies considered, plant capital costs are relatively a large portion of the production costs. The capital costs for the distributed wind (electrolysis) and central nuclear thermochemical technologies were obtained from a 2004 study by the National Academies of Sciences, National Research Council of the U.S. National Academies [13], while the other production costs were estimated by the National Renewable Energy Laboratory in 2005 [14], with an exception of central coal gasification with carbon capture and sequestration (CCS), whose costs were estimated in 2008.

Separately, GA evaluated the economics of nuclear hydrogen production based on the IS process and concluded the production cost of hydrogen to be US\$1.53/kg-H_2 based on a 2400 MWt HTGR operating at 850°C, with 42% overall efficiency, and US\$1.42/kg-H_2 at 950°C and 52% efficiency (both 10.5% discount rate). The GA evaluation reports that on the same cost basis, conventional plant steam reforming of natural gas yields hydrogen at US\$1.40/kg-H_2, and sequestration of the CO_2 would push this to US\$1.60/kg-H_2.

JAEA has validated the nuclear steam methane reforming in full scale reformer equipment and system operation. The technical results of this program are given in Chapter 20. The program developed nuclear commercial plant process flow sheets and sized equipment designs at a plant heat rate of 380 MWt and for hydrogen production rate of 2.17×10^5 tons per year [15]. The designs considered two purity grades of product hydrogen with the higher purity incurring additional investment and operation and maintenance (O&M) costs in the multistage PSA (pressure swing absorption) equipment for hydrogen separation

TABLE 2.3
Estimated Hydrogen Production Costs

	Overnight Capacity Cost				Hydrogen Production Cost (dollars per kilogram)			
Technology and Fuel	**Capacity MGPD**	**Million Dollars**	**Dollars per MGPD**	**Capacity Factor (%)**	**Capital[a]**	**Feed-stock**	**O&M**	**Total**
Central SMR of natural gas[b]	379,387	181	477	90	0.18	1.15	0.14	1.47
Distributed SMR of natural gas[c]	1500	1.14	760	70	0.40	1.72	0.51	2.63
Central coal gasification with CCS[d]	307,673	691	2246	90	0.83	0.56	0.43	1.82
Central coal gasification without CCS[d]	283,830	436	1536	90	0.57	0.56	0.09	1.21
Biomass gastification[e]	155,236	155	998	90	0.37	0.52	0.55	1.44
Distributed electrolysis[f]	1500	2.74	1827	70	0.96	5.06	0.73	6.75
Central wind (electrolysis)[g]	124,474	500	4017	90	1.48	1.69	0.65	3.82
Distributed wind (electrolysis)[h]	480	2.75	5729	70	3.00	3.51	0.74	7.26
Central nuclear thermochemical[i]	1,200,000	2,468	2057	90	0.76	0.20	0.43	1.39

SMR = Steam methane reforming; CCS = carbon capture and sequestration; MGPD = thousand kilograms per day; O&M = operations and maintenance.

Note: Table excludes transportation and delivery costs and efficiency losses associated with compression or transportation.

[a] For all cases a 12% discount rate is used. Economic life of 20 years assumed for distributed technologies and 40 years for all other technologies. Average U.S. prices for 2007 are used where practicable.

[b] Assumes industrial natural gas price of $7.4 per million Btu and industrial electric price of 6.4 cents per kilowatthour.

[c] Assumes commercial natural gas price of $11 per million Btu and commercial electric price of 9.5 cents per kilowatthour.

[d] Assumes coal price of $2.5 per million Btu.

[e] Assumes biomass price of $2.2 per million Btu ($37.8 per tonne).

[f] Assumes commercial electric price of 9.5 cents per kilowatthour.

[g] Excludes opportunity cost of wind power produced.

[h] Assumes grid supplies 70% of power at 9.5 cents per kilowatthour and remainder at zero cost.

[i] Includes estimated nuclear fuel cost and co-product credit as net feedstock cost, decommissioning costs included in O&M.

from the reformer off gases. The parameters used to evaluate the economics of the plant include cost of natural gas at US$0.214/Nm^3, discount rate of 10% and return on investment of 8%. The load capacity factor is 90%. The results of the economical evaluation based on the engineering and construction experience obtained are summarized in Table 2.4 [15]. The data there indicate that nuclear SMR for hydrogen production is less than the conventional SMR and the CO_2 emission from the nuclear process is 32% less than from the conventional process.

2.2.4.2 *Hydrogen Production Scenarios on CO_2 Emission*

While using a fossil fuel feedstock, nuclear production helps reduce the CO_2 emission associated with the final product hydrogen, as just said of the case with nuclear-heated

TABLE 2.4

SMR Hydrogen Production Cost Estimates (2004 Cost)

Plant heat rate (MWt)	380	380
Hydrogen product purity (vol %)	97	99.99
Hydrogen production rate (tonne/year)	217,270	204,059
Conventional SMR hydrogen cost (US$/kg-$H_2$)	1.38	1.60
Nuclear (HTGR) SMR hydrogen cost (US$/kg-$H_2$)	1.29	1.37
Conventional SMR CO_2 emission (million tonne/year)	2.39	N/A
Nuclear (HTGR) SMR CO_2 emission (million tonne/year)	1.62	1.66

reforming of methane. Hydrogen production from water with nuclear energy is the most attractive route because it can produce large-scale hydrogen and leave near zero carbon footprints.

An examination is first made with the electrolysis routes using electricity from various fuel sources. Life cycle CO_2 emission intensity for electricity production, in gram CO_2 per kWh electricity sent out from plant, is compared with other energy sources in Figure 2.3 [16]. The life cycle considers emissions associated with the construction of plants, operation and maintenance for the plants, and decommissioning the plants. The emission from combustion of fossil fuels constitutes a larger share of the total emission for fossil fuels. The life cycle emission for the nuclear plant, in this case the current LWRs, considers uranium mining, conversion, enrichment, fuel fabrication, SF storage, and low-level waste disposal, whose tasks are assumed to be performed as typically with fossil energy uses. The CO_2 emission for nuclear plant life cycle is less than 3% of the coal-fired plant and one of the lowest of all sources.

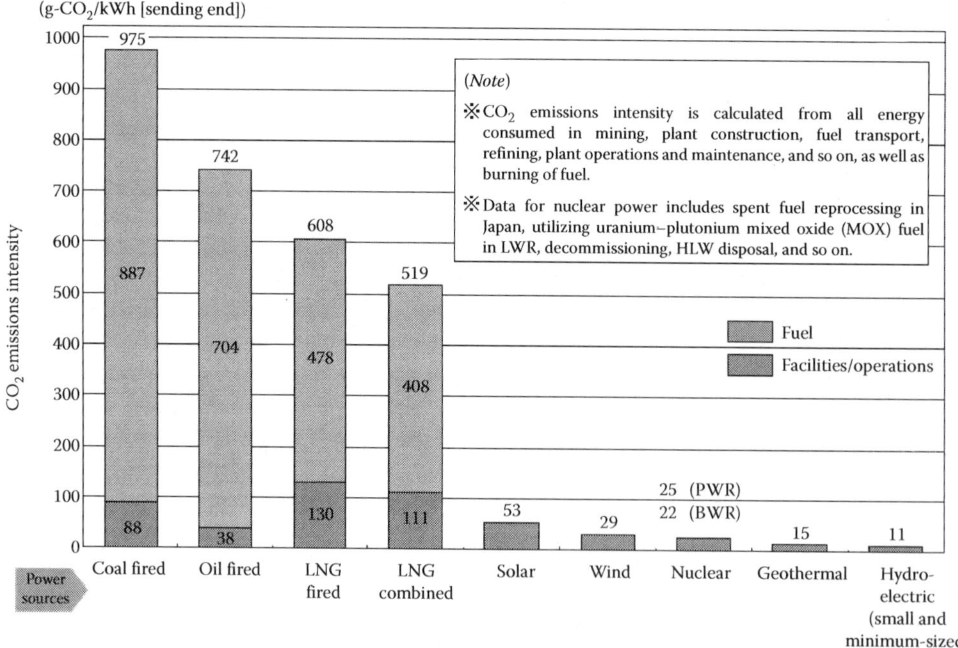

FIGURE 2.3
Life cycle CO_2 emission intensity for electricity generation by energy sources.

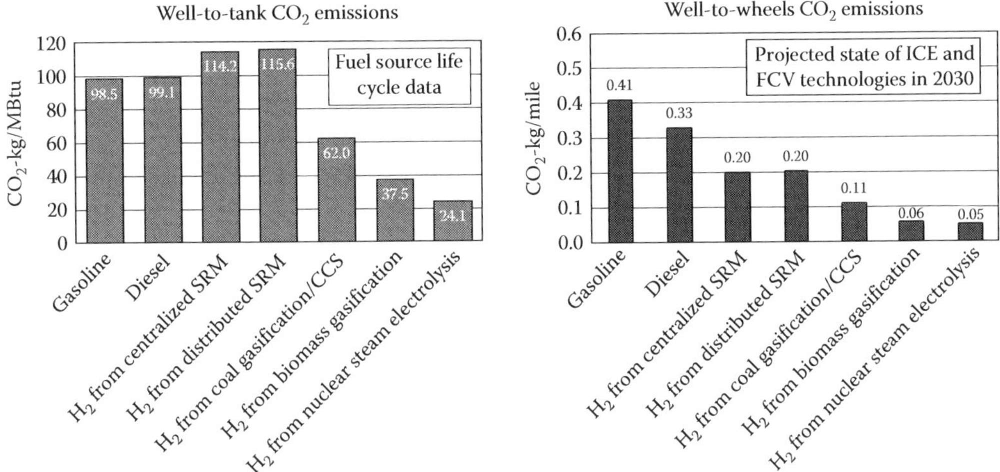

FIGURE 2.4
Vehicular CO_2 emissions by hydrogen production sources.

The second examination is mode on the potential impacts on full fuel cycle CO_2 emissions associated with the transportation market transition to hydrogen FCVs from gasoline and diesel powered vehicles. Five sources of hydrogen production were considered, including (1) centralized natural gas steam reforming of methane (SRM), (2) distributed SRM, (3) centralized integrated coal gasification CCS, (4) centralized integrated gasification–biomass gasification, and (5) centralized nuclear-steam electrolysis.

The hydrogen production methods chosen are not intended to provide an exhaustive list of possibilities but were selected to demonstrate a range of outcomes, given current expectations of CO_2 emissions for the fuel delivered to the vehicle. The CO_2 emissions associated with each of the life cycle fuel sources of production are provided in Figure 2.4, from the data taken in literature [3,12]. The "wells to tank" emissions are per fuel energy equivalent while the "well to wheels" values are per the equivalent mile traveled. The latter accounts the vehicle life cycle emission and relates to vehicle performance assumptions. The vehicles of internal gasoline and diesel combustion engines and hydrogen fuel cells are based on the projected state of these technologies in 2020, when fuel cells are expected to be in earlier stage of market penetration. Although nuclear hydrogen assumes the production of HTE, and the CO_2 emission is representative of all other nuclear hydrogen production methods from feedstock water because all nuclear methods are characteristically near zero emission and the CO_2 emission shown from the nuclear hydrogen fuel cell in the figure is mainly associated with the assumptions in the delivery, storage, and dispensing of hydrogen that would still require use of fossil fuel energy in the practical time frame of 2020. It can be seen from Figure 2.4 that fuel-cell cars based on nuclear hydrogen as fuel are among the best emission performers.

2.2.4.3 Nuclear Fuel Demand and Supply

2.2.4.3.1 Near- and Mid-Term Fuel Strategy

About 5.47 million tons of naturally occurring uranium are known to be recoverable from the mines [17]. Australia leads with a share of 23% of the global known reserve, followed by Kazakhstan's 15%. Another 14% is in North America. Russia has 10%, followed by South Africa for 8%.

The world consumes 67,000 tons of mined uranium annually. At this rate, about 80 years of uranium supply remain from the world's known reserve. At the end of 2006, the secondary uranium sources drawn from government and commercial inventories including disarmed military stockpiles and recycled uranium, supplied about 40% of the world's actual commercial reactor fuel. This secondary uranium is estimated to extend the world's known uranium supply to more than 100 years.

In the mean time, the world nuclear electricity capacity is posed to increase due to the significant number of the reactors that are now being constructed and proposed. It is expected that a significant number of nuclear reactors additionally for hydrogen production and process heat applications will be deployed during this period, likely beginning in 2020–2030, to fuel a then-rapidly expanding fleet of hydrogen fuel-cell vehicles.

A strategy in the near-to-mid term can be increasing investment in uranium exploration on the top of the world's current expenditures to add new deposits to the uranium reserve bank at comparable rates with actual nuclear capacity growth.

Two evidences support the viability of this strategy. Speculative resources, that is, uranium deposits that can be expected to be found based on the geological characteristics of already discovered resources, have been estimated to be 10.5 million tons. And the known uranium resources have increased in synchronization with the expenditure on exploration for decades as far back as the statistics were available. In terms of recent figures, the combined worldwide exploration expenditures during 2006–2007 were US$ 1.5 billion, and more than the total of the preceding 10 years (1995–2005), inspired by the signs of a nuclear renaissance. This added 15% more uranium to the world known reserve during that two-year period.

The said near-to-mid strategy will aim at maintaining an "insurance margin" of several decades at minimum between demand and the known reserves and have all the effect to assure the energy market, despite the forecast global nuclear capacity growth, to be real or false

2.2.4.3.2 Long-Term Fuel Strategies

Continued global reliance on nuclear energy for electricity, hydrogen, and other heat applications in the long term could be assured of fuel supply by the availability of the four options presented below.

The first three options involve the alternative fuel cycles to the currently mainstream uranium cycle and their successful implementations will need to be approached through the long-term development of both technologies and international policies. In the fourth option, uranium of practically unlimited quantities at a currently proven cost, which is in the range of recent uranium prices, could be extracted from seawater.

2.2.4.3.2.1 SF and Depleted Uranium The SF from the current LWRs contains about 1% of residual fissile uranium and 1% of plutonium generated from fertile uranium by neutron capture during reactor operation. The fissile uranium and plutonium can be recovered by processing of SF. In particular, more easily recovered plutonium, which is 65% fissile, can be mixed with depleted uranium left over from normal enrichment plants to produce fuel of mixed oxides, that is, MOX ($UO_2 + PuO_2$), which is enriched to 7% of recovered plutonium to be operation equivalent to the 4.5% of uranium enrichment used in the current reactors.

Only a sum of 2000 tons of MOX fuel has been made in this way and used in 35 commercial reactor plants, mostly in Europe, which loaded 180 tons in reactors in 2006. Japan began using MOX in 2009. The reactors originally designed for 100% uranium fuel can accept up to 50% MOX for the core without significant redesign or change of operation behavior. APWRs such as the EPR or AP1000 have been designed to accept complete fuel of MOX if necessary. One new ABWR reactor in Japan is expected to enter service to allow 100% MOX in 2014.

A single recycle of uranium and plutonium from SF increases energy output from mined uranium by about 22%. The percentage can be increased with higher burn up core. The HTGR design can reach not only a burn up of 120 GWd/t, compared to 45 GWd/t in current LWRs, but allow 100% MOX, as evaluated by JAEA [18].

The nuclear reactors have been able to use just about 1% of the potential energy from mined uranium. The rest has accumulated in the forms of SF at the reactor plant sites and depleted uranium at the enrichment plants. For example, the United States has an estimated 63,500 quadrillions of energy in the inventory of spent nuclear fuel, equivalent to about 9 trillion barrels of oil sufficient for about 1500 years at the current consumption of 16 million barrels per day. Advanced fuel-cycle technologies such as MOX are being researched and introduced to practically recycle and reduce these legacy "wastes" for energy production.

2.2.4.3.2.2 Fast Reactors In these reactors a moderator is not used and fission occurs with fast neutrons. Fissile plutonium is the fuel instead of fissile uranium, because ^{239}Pu has higher probability to fission in fast-neutron spectrum and its fission releases more neutrons than uranium. The surplus neutrons beyond fission energy production are utilized by a deliberate fast reactor design feature, known as blanket, in which fertile ^{238}U from a depleted uranium source is placed for breeding by irradiation of fast neutrons into fissile plutonium. It is possible to breed more plutonium in the blanket than burned up in the fission core. Technically various breeding ratios (BR) are feasible. The core with BR = 1 means it produces fissile plutonium at the same rate as used in the core. At BR > 1 the reactor produces more fuel than consumed. In this sense, fast reactors are intended as a breeder to produce fuel.

Several Generation IV designs are fast-neutron spectrum reactors. Fuel reprocessing of SF, especially the blanket, is integral to the fast-breeder strategy. The reprocessed plutonium is returned to the core or fabricated into MOX fuel for thermal reactors. As an example of the latter, a deployment scenario of a fleet of FBR and HTGRs is shown able to meet a significant portion of future energy demands in Japan. The FBRs operating with BR = 1.2 generates 58 GWe electricity, the forecast nuclear electricity capacity in 2030 in comparison to the present 48 GWe total of Japan, while producing MOX fuel enough for use by 34 HTGRs of 600 MWt each. The HTGRs produce hydrogen at the output sufficient to supply 30 million hydrogen fuel-cell vehicles, or 40% of the cars on the future roads in Japan.

If successful, a fleet of fast reactors is believed to extend the global known uranium resources to more than 2000 years based on the current nuclear capacity.

2.2.4.3.2.3 Thorium Fuel Thorium is said to be three to four times the natural deposits of uranium. Although not fissile, thorium-232 will absorb thermal neutrons to become fissile uranium-233. U-233 is better fissile than U-235 and Pu-239 because its fission releases more neutrons than either of them. As a result, it is possible to breed fissile U-233 in thermal neutron core. Using neutron starter of another fissile material (U-233, U-235, or Pu-239) to initiate neutrons, an efficient breeding core can be realized to form U-233 from irradiation of thorium at the same or faster rate than the fissile materials are consumed in the core. In theory, this could lead to most of the thorium mined being useable in thermal reactors, compared to the 0.7% of mined uranium. However, there are disadvantages with thorium-based fuel, including higher radioactivity of U-233 and cost of fuel fabrication [19].

Significant experience with thorium has been gained and is continually accumulated in LWRs, heavy water reactors, and HTGRs. The thorium base fuel was loaded in several HTGRs including 300 MWe THTR (thorium high-temperature reactor) in Germany and 330 MWe FSV in the United States. Canada has had over 50 years of history experimenting thorium fuel up to burn up of 47 GWd/t and is studying thorium fuel for heavy water reactors CANDU-6 and ACR-1000. In 2009, AECL agreed with partners in China to develop and demonstrate thorium fuel in its CANDU reactors in China. In the same year, AREVA and Thorium Power agreed to study thorium fuel for AREAV's EPR. India, which has many times more natural thorium than uranium, is developing thorium fuel for both its heavy water reactors and fast reactors.

2.2.4.3.2.4 From Seawater The uranium in seawater—around 3.3 parts per billion (3.3 mg per ton of seawater)—is estimated to be about 4.5 billion tons, close to 1000 times the known deposits on land. Collecting each one-tenth of this sum would carry the current global nuclear capacity once-through fuel cycle for seven millenniums.

Although studies on "mining" uranium from seawater were started more than 40 years ago in England [20], it is the advanced research stage, currently being carried out in Japan that has proved the technical and economic feasibility of the technology [21]. The Quantum Beam Science Directorate of JAEA reported in 2009 that it had synthesized a fabric of irradiated graft polyethylene, an amidoxime polymer, which has shown a high affinity for uranium [22]. The JAEA anchored three cases of the adsorbent fabric (weighing 350 kg) 25 m deep in the seawater at 7 km offshore. The experiment collected a total of 1.1 kg uranium. Following that, the collection was optimized through the development of a braid adsorbent, which stands free on ocean floor, eliminating much of the anchored structure used earlier, and enables more efficient contact with seawater. The results cut 40% cost of the collection system and confirmed by a marine experiment of the braid adsorbent a performance value of 1.5 g U per kg of adsorbent per month, which doubles the performance of the former adsorbent system. The economics of the collection is further augmented by the confirmed 8 repetitions of using the same adsorbent in the experiment. The feasibility of as many as 18 repetitions is anticipated. The cost of uranium was assessed by including costs of the braid adsorbent equipment, collection operations, and post uranium purification for an annual production scale of 1200 t U. The assessed equivalent cost of $96 per pound of U_3O_8, the usual form of mined uranium trade containing 85% uranium, falls in the middle of the world uranium spot price of US$41/lb-$U_3O_8$ in May 2010 (today) and the record high of US$136/lb-$U_3O_8$ in July 2007. JAEA also considers simultaneously collecting some of the other 76 minerals such as vanadium contained in seawater to practically reduce cost of uranium.

The results of the advanced research above suggest that uranium from seawater at any price reasonably different from JAEA's assessed value is expected to have only marginal effect on the cost of nuclear energy production, because uranium purchase counts only a 10% share of nuclear reactor fuel cost and the fuel contributes to only one-third cost of electricity production.

Given the sound economical fundamentals above, the effort should be directed to accelerate sea-bound collection of uranium to a production scale. It is estimated that the future success in collecting just 0.2% of the 520,000 tons of uranium naturally carried each year by

"Kuroshio" or Black Current up along Japan East Coast would meet the country's annual need of 8000 tons. This can be similarly carried out with some 16 major surface ocean currents around the globe, Gulf Stream in the Atlantic, Agulhas in the Indian Ocean, East Australian Current in the Pacific, to cite a few.

The proven cost and guaranteed availability of uranium in the sovereign and international waters to all countries facilitate development of widely deployable nuclear energy reactors optimized for such features as modular economics, enhanced or passive safety, coproduction of multiple energy products, and proliferation-resistant uranium fuel.

2.2.5 Future Development and Outlook

The heat applications, especially large-scale hydrogen production, represent new missions and corresponding challenges for nuclear energy. The new missions would require a larger fleet of reactors than is presently used in the traditional mission of electricity production. The new challenges include but not limited to the following:

- Research and commercial development for efficient, large-scale hydrogen production processes suitable for use with the Generation-IV nuclear reactors. The promising hydrogen production processes are those that utilize nuclear mid-to-high temperature heat directly and/or highly efficient nuclear electricity to produce hydrogen from renewable feedstocks including water and biomass. Some of these production methods have been demonstrated at the laboratory scale and are continually advanced in many countries, as described in Sections IV and V of this handbook.
- Advanced nuclear systems must be developed to provide the suitable qualities of heat and electricity to these processes. The Generation IV International Forum is addressing the development of six advanced reactor systems that not only meet the energy requirement for one or more processes but also improve safety and cost to encourage worldwide deployment.

In performing the new and traditional missions, nuclear plant will remain capital-intensive. Recent studies point to the costs of nuclear electricity production ranging from US$30 to $57/MWh and of nuclear hydrogen production between US$1.3 and $3.0/kg-$H_2$, depending on specific economic and plant parameters. For example, the technology learning for new reactors or unfavorable economic factors would lead nuclear electricity and hydrogen to the upper range of the costs mentioned above. The electricity and hydrogen at these costs are comparable with those produced from other alternative primary-energy sources. Based on the consistent cost parameters and today's commercial technologies, the estimated costs of electricity generation are reported by Japan's electricity industry to be US$53/MWh for nuclear, US$57/MWh for coal-fired, $62/MWh gas-fired, US$107/MWh for oil-fired, and US$119/MWh for hydroelectric plants.

And with the CO_2 emission of nuclear electricity and hydrogen plants being from near zero to substantially less than that of alternative large-scale energy plants, the nuclear plants will be competitive in the future energy economy that expects the broad implementation of price and trade on greenhouse gas emissions.

Based on the conditions above, the nuclear share is posed to expand in future over today's. The near- to long-term nuclear fuel strategies presented earlier in this chapter are diverse and adequate to support this expansion for the century and beyond. However, the reality of nuclear energy's expansion will also depend on how successfully the additional technology and policy issues, including nuclear waste and proliferation that have limited nuclear energy's role to perform its traditional mission around the world and on which the chapter has earlier discussed some solutions of reactor and fuel, are continually addressed in connection to its new missions, particularly its contribution to the foundation of a hydrogen energy economy.

2.3 Hydrogen Production from Nuclear Fusion Energy

2.3.1 Introduction to the Science and Technology

Nuclear fusion is a reaction between relatively light nuclides that release energy by combination to form heavier atom that is more stable. Proton of atomic weight 1.007825 and neutron 1.008665 make helium nucleus of 4.0026 that is lighter than the sum of the reactant 4.03298. This difference in mass can be converted into energy according to the famous equation $E = mc^2$. As shown in Figure 2.5, energy release from the mass difference is particularly large between hydrogen and helium. The product particles of the fusion reaction have MeV orders of energy, and to utilize them, adequate conversion technology is required. The reaction between deuteron and triton (heavy isotopes of hydrogen) results in two product particles by

$$d + t = \mathrm{He}(2.3\,\mathrm{MeV}) + n(14.3\,\mathrm{MeV}) \tag{2.3}$$

in which kinetic energy of the alpha and neutron particles carries the released energy.

Such a fusion reaction occurs in high-temperature plasma over 100 million °C (10 keV). To generate and control hot plasma is technically very difficult, but several decades of effort to produce and control fusion plasma for peaceful purpose, study of plasma physics, and research and development of plasma confinement devices finally came to the phase to actually generate burning plasma for prolonged time and generate fusion energy in engineering scale. Fusion research is now in the phase to actually convert its energy for practical use.

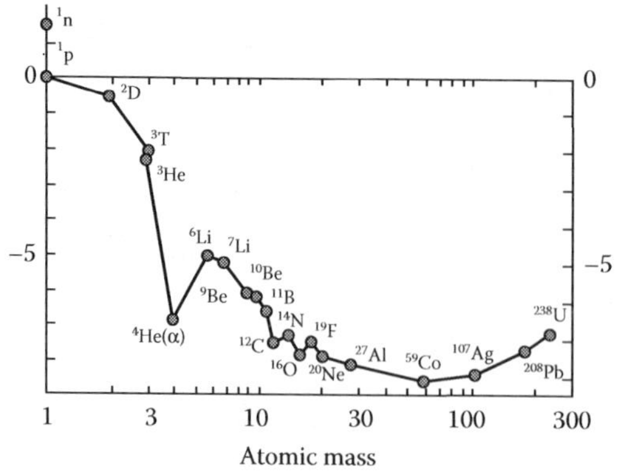

FIGURE 2.5
Formation energy and mass of nuclei.

One of the features of fusion energy is, unlike in fission reactors that have fixed temperature of heat as primary-energy product depending on specific reactor concept, various temperatures can be considered to be extracted from the same type of plasma device. In the case of fission, high-temperature reactors, typically HTGR or VHTR are being developed for the utilization of high-temperature heat. To utilize fusion energy, high-energy neutron must be converted into the forms adequate for the use by consumers. Because the energetic neutron generated by fusion reaction has no electrical charge and does not interact with electromagnetic field applied for the confinement of burning plasma, it penetrates through materials consisting of reactor device facing burning plasma. Equipped "Blanket" surrounding the plasma slows down these fast neutrons, and the energy of the neutron is converted into heat. Thus the location of fusion reaction in plasma and the device to extract its energy is geologically separated, and blanket concept has some independence from plasma confinement device. Development of blanket has been independently implemented from fusion-plasma study, and the International Thermonuclear Experimental Reactor (ITER) being built under an international endeavor (see Chapter 32) will be the first device that will test the harnessing and conversion of fusion energy. In the ITER, six different types of test blanket modules are planned to be installed by different parties to compare the different energy conversion concepts.

A blanket is composed of vessels of several 10 cm thickness and filled with lithium containing materials, and heat transfer media circulate in it. Neutrons slow down in the process of collisions in the blanket while heating the fillings of the blanket. Finally, neutrons are absorbed by lithium and generate tritium nuclei to compensate the consumption, by the reaction, $^{6}\mathrm{Li} + n = t + \alpha$. The blanket temperature increases, and the heat is carried away by the heat-transfer media (coolant) such as water, liquid metal or gas. Most of the current blanket designs consider the temperature from 300 to 900°C, but because original energy of neutrons is far higher, this temperature is mainly limited by the materials and technology to handle it. Some of the blanket designs utilize liquid metal or helium to transfer fusion neutron energy to the form of heat at high temperature. Energy utilization technology of fusion therefore, depends on the heat-transfer media and the blanket concept to generate it. Because water [23,24], liquid metal or high temperature helium [25] are considered as blanket coolant, energy utilization technology planned for water-cooled reactors, liquid metal fast reactor and HTGR can be applied to fusion reactor plants. Technology developed for the utilization of fission reactors is thus applicable for fusion reactors with adequate modification.

Fusion is an advanced energy that requires decades to be commercially introduced into the market. In the selection of energy conversion, consideration on the social, economical, and environmental features are as important as technical issues. Blanket selection and utilization of the product energy must meet the requirement of the market decades later, and hydrogen is regarded as one of the attractive energy products, potentially larger than the electricity market. Hydrogen production processes such as electrolysis including vapor electrolysis, IS process, and biomass conversion can all be regarded as possible applications for fusion [26,27], if adequate blanket that provides heat at required temperature is developed.

Among these hydrogen production processes, waste biomass conversion proposed by the author attracts particular attention because of its superior potential efficiency [28]. Figure 2.6 schematically shows the process, where biomass mixed with steam forms hydrogen and carbon dioxide,

$$(C_6H_{10}O_5)_n + H_2O + 830\ \mathrm{kJ} = 6CO + 6H_2, \tag{2.4}$$

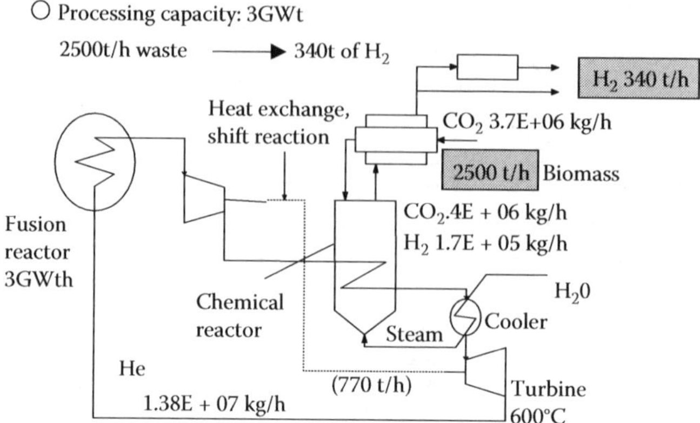

FIGURE 2.6
Hydrogen production from biomass using fusion heat.

with heat from a fusion reactor. Here, cellulose, the chain of sugar, is considered as representing compound of biomass, because almost all plant fibers are made of this compounds. This biomass decomposition is an endothermic reaction that occurs at the temperature above 700°C and almost complete at around 900°C. The primary product of this reaction is a mixture of hydrogen and carbon monoxide. When hydrogen is the intended product of the process, carbon monoxide mixed with steam can further be used for hydrogen production by the well-known shift reaction,

$$CO + H_2O = CO_2 + H_2, \tag{2.5}$$

Total reaction is $m(CH_2O) + nH_2O = mCO_2 + (m + n)H_2$, requiring 120 kJ/mol to completely convert cellulose into 1 mol hydrogen and carbon dioxide mixture. It should be noted that the primary product, $CO + H_2$ can also be converted into synthetic oil by Fischer–Tropsch reaction.

Table 2.5 compares the theoretical energy efficiency of hydrogen production processes applicable for a 3 GW thermal fusion reactor as a heat source. Conventional electrolysis using alkaline solution is possible, but generation efficiency for its electricity at 300°C will be limited to ca. 30%. With a 500°C reactor blanket, the electricity production can use supercritical water technology at 40% efficiency. Also with polymer electrolyte electrolysis technique, efficiency of water decomposition improves somewhat, because of the reduction of electrical resistance.

TABLE 2.5

Energy Efficiency of Hydrogen from Fusion

3 GW Heat	Generation Efficiency (%)	Electricity (GW)	Energy Consumption (kJ/mol)	Hydrogen Production (t/h)
300°C-conventional electrolysis	33	1	286	25
900°C-SPE[a] electrolysis	50	1.5	231	44
900°C-vapor electrolysis	50	1.5	181	56
900°C-biomass	–	–	60	340

[a] Solid polymer electrolyte-based electrolysis process.

With vapor electrolysis that is operated above 900°C energy efficiency will drastically improve. It can supply energy for gasification of reactant liquid water by heat of the $T\Delta S$ term energy, whose fraction of the total energy increases with temperature (which is desired as heat is about a half of the cost of electricity), while it provides the ΔG term energy only by electricity to split water molecules.

However, all these technologies have inherent limit of efficiency due to the thermal cycle for electricity generation that discards waste heat at low temperature. Thermochemical water decomposition is a chemical process that may not look to be limited by Carnot's efficiency, but in fact, it is a combination of endothermic and exothermic reactions. It must also discard low-temperature heat resulting from exothermic reaction while high-temperature reaction is endothermic, and thus is regarded as a kind of thermal cycle.

Hydrogen production from biomass [28] proposed by the author is, far more efficient because it is a once-through reaction like steam reforming. The chemical process involves a single reaction to gasify cellulose with superheated steam at ca. 900°C. This reaction utilizes almost all the energy of nuclear heat and does not discard any low temperature heat, because this chemical process does not involve a heat cycle.

Furthermore, this process also utilizes stored chemical energy of biomass that originally was converted from solar energy without significant loss. In other biomass based technologies such as biomass combustion electricity generation or biochemical ethanol production cannot effectively use the conserved chemical energy and thus goes waste as heat or in chemical reactions. Biomass itself is a low-quality fuel that does not generate high-temperature heat and produces various by-products such as tar, smoke, or char. In the biochemical conversion process such as fermentation, significant fraction of original chemical energy is used by bacteria or exothermic reactions. On the other hand, high-temperature reaction of biomass proposed here converts its original chemical energy effectively into hydrogen because the energy required for this endothermic reaction is provided by external source, nuclear heat. As a result, when we look at the energy efficiency from nuclear heat to the hydrogen product, its apparent efficiency exceeds unity and approaches 3. This biomass conversion may be regarded as one of the attractive potential candidate for the use of nuclear heat, and among hydrogen production processes. It can also be used for HTGR or VHTR (see Chapter 7) or other high-temperature sources if available.

However, due to the above-mentioned energy multiplication effect, biomass conversion has a distinguished feature specific for combination with fusion. In the design of fusion power plant, consideration on overall power balance and its conversion is inevitable in order to sustain a burning plasma and operate it as well as to generate its final product—commercial energy that is ready to use. Fusion reactor requires large fractions of product energy to be recycled to heat its plasma to sustain the temperature and reaction because plasma performance is not sufficiently high to sustain its spontaneous reaction. Current plasma is still premature for practical electricity generation due to the poor energy confinement in the plasma and requires large external heating. Based on the assumption of some 30% conversion efficiency to electricity to be used for plasma heating, without further improvement of plasma physics fusion energy production for net power output is impossible. For instance, ITER, the largest and probably only attempt of the burning fusion plasma experiment in the world, in the next two decades cannot generate net power output even if it successfully achieves its target. Because plasma performance exceeding it cannot be expected before ITER, fusion energy production can only be planned after the burning plasma experiment by ITER. Similarly, by the concept of fission–fusion hybrid reactor, energy output by fission reaction must be added to the fusion device.

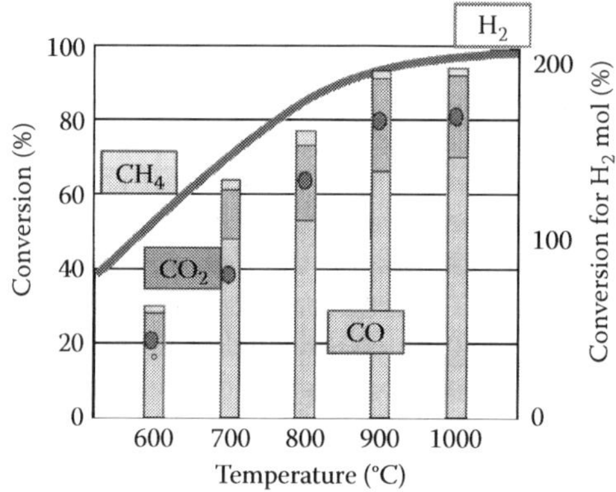

FIGURE 2.8
Gasification of cellulose.

CO, CO_2, and CH_4 to generate corresponding amount of hydrogen, suggests loss of hydrogen or error in analysis. At the lower temperature region where conversion of carbon of the cellulose does not look like high is inconsistent with the equilibrium calculation that suggests insufficient gasification of hydrocarbon. Corresponding loss of carbon was suspected to become tar that was sometimes observed at the downstream of the experiment.

The amount of absorbed heat by this pulse mode reaction was measured by the integration of the temperature decrease in the packed column section, calibrated with known endothermic reaction of the decomposition of strontium carbonate. The results showed good agreement, indicating the conversion of heat to the chemical energy occurred as expected with this reaction.

2.3.4 Conclusions of Nuclear Fusion

Energy conversion technology particularly for hydrogen production with fusion is introduced in this chapter. Although fusion may be an exotic advanced energy, it may be

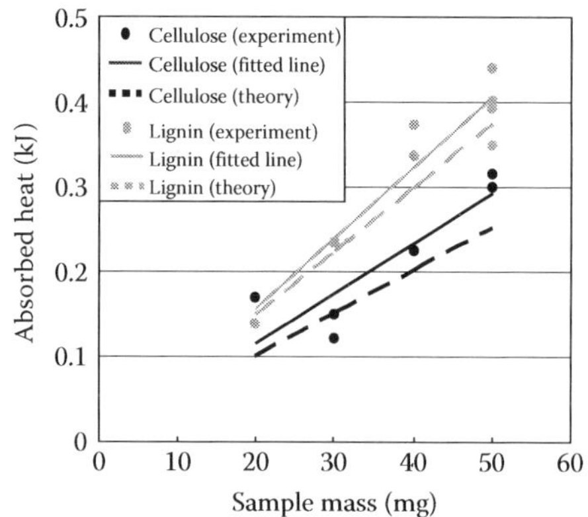

FIGURE 2.9
Heat adsorption by the endothermic reaction of cellulose gasification.

understood that its essential technology is based on the previous and current knowledge of nuclear technology. Developments and challenges are needed to establish a fusion energy conversion to be a viable energy supplier in the future, and coordination of the improvements in physics, chemistry, materials, chemical, mechanical, and nuclear engineering, will be essential. However, in this chapter, the author emphasizes and expects readers that not only the technical knowledge and consideration, but also socioeconomic study and a common sense as a consumer are inevitable, because even if an energy conversion technology is successfully developed, market selection and social acceptance are the most important measures.

References

1. *Energy Balances of OECD Countries* (2009 edn), International Energy Agency, Paris, France.
2. *Kaye & Laby Tables of Physical & Chemical Constants*, 16th edn. National Physical Laboratory, 1995, www.kayelaby.npl.co.uk.
3. FY2009 Annual Report of the U.S. Department of Energy Hydrogen Program. 2009.
4. Hori, M., Matsui, K., Tashimo, M., and Yasuda, I., Synergistic hydrogen production by nuclear-heated steam reforming of fossil fuels, *Progr. Nucl. Energy*, 47(1–4), 519–526, 2005.
5. Tashimo, M., Advanced design of fast reactor-membrane reformer (FR-MR), *OECD/NEA Second Information Exchange Meeting on Nuclear Production of Hydrogen*, Argonne, 2003.
6. Final Report of Feasibility Study on Commercialized Fast Reactor Cycle Systems-Executive Summary, Japan Atomic Energy Agency and The Japan Atomic Power Company, March, 2006.
7. Energy Modular Multiplier, General Atomics, http://www.ga.com/energy/em2/.
8. Nickel, H., Nabieleka, H., Potta, G., and Mehner, A. W., Long time experience with the development of HTR fuel elements in Germany, *Nucl. Eng. Des.*, 217, 141–151, 2002.
9. Greneche, D. and Szymczak, W., The AREVA HTR fuel cycle: An analysis of technical issues and potential industrial solutions, *Nucl. Eng. Des.*, 236, 635–642, 2006.
10. Rodriguez, C., Baxtera, A., McEacherna, D., Fikania, M., and Venneri, F., Deep-Burn: Making nuclear waste transmutation practical, *Nucl. Eng. Des.*, 222, 299–317, 2003.
11. Besmann, T.M. Thermochemical assessment of oxygen gettering by SiC or ZrC in PuO_2x TRISO fuel, *J. Nucl. Mater.*, 397, 69–73, 2010.
12. The impact of increased use of hydrogen on petroleum consumption and carbon dioxide emissions, SR/OIAF-CNEAF/2008-04, August 2008, Energy Information Administration.
13. *The Hydrogen Economy: Opportunity, Costs, Barriers, and R&D Needs*, The National Research Council, the U.S. National Academies. The National Academies Press, Washington DC, February 2004.
14. DOE H2A Analysis, U.S. Department of Energy Hydrogen Program, 2005, www.hydrogen.energy.gov/h2a.analysis.html.
15. Shiina, Y. and Nishihara, T., Cost estimation of hydrogen and DME produced by nuclear heat utilization system, Phase II, JAERI-Tech 2004–057, Japan Atomic Energy Agency, 2004.
16. Hiroki Hondo, Life cycle GHG emission analysis of power generation systems: Japanese case, *Energy*, 30, 2042–2056, 2005.
17. *Uranium 2007: Resources, Production and Demand*, A Joint Report by the OECD Nuclear Energy Agency and the International Atomic Energy Agency, OECD, Paris, 2008. ISBN 978–92-64-04766-2.
18. Mouri, T., Nishihara, T., and Kunitomi, K., Nuclear and thermal design for high temperature gas cooled reactor (GTHTR300C) using MOX fuel, *AESJ Trans.*, 6(3), 253–261, 2007.
19. *Thorium Fuel Cycle—Potential Benefits and Challenges*, IAEA-TECDOC-1450, International Atomic Energy Agency.

20. Davies, R.V., Kennedy, J., McIlroy, R.W., Spence, R., and Hill, K.M., Extraction of uranium from seawater, *Nature*, 203, 1110–1115, 1964.
21. Tamada, M., Current status of adsorbent development for recovery of uranium from seawater (in Japanese), *Genshiryoku Eye*, 54(11), 13–16, 2008.
22. Technology of uranium collection from seawater, JAEA presentation to Japan Atomic Energy Commission, June 2009.
23. Konishi, S., Nishio, S., Tobita, K. and the DEMO design team, DEMO plant design beyond ITER, *Fusion Eng., Des.*, 63–64, 11–17, 2002.
24. Enoeda, M., Design and technology development of solid breeder blanket cooled by supercritical water in Japan, *Proc. 19th IAEA Fusion Energy Conference*, Lyon, France, Oct. 14–19, IAEA-CN-FT/P1-8, 2002.
25. Ueda, S., Nishio, S., Yamada, R., Seki, Y., Kurihara, R., Adachi, J., Yamazaki, S., Dream design team, "maintenance and material aspects of DREAM reactor," *Fusion Engineering and Design*, 48 (3–4), 521–526, 2000.
26. Sheffield, J., Brown, W., Garrett, G., Hilley, J., McCloud, D., Ogden, J., Shields, T., and Waganer, L., A study of options for the deployment of large fusion power plants, Technical Report: JIEE 2000-06, June 2000, Joint Institute For Energy & Environment, Knoxville, TN, USA, 2000.
27. Waganer, L.M. and ARIES team, Assessment of markets and customers for fusion applications, *Proc., 17th IEEE SOFE*, San Diego, Oct. 1997.
28. Konishi, S. Use of fusion energy as a heat for various applications, *Fusion Eng. Des.*, 58–59, 1103–1107, 2001.

Section II

Hydrogen Production Methods

3

Water Electrolysis

Seiji Kasahara

CONTENTS

3.1 Introduction

Electrolysis of water is a method of producing hydrogen, and by-product oxygen, by the direct decomposition of water molecules using electric energy. Water electrolysis was already commercially practiced in 1890s. Some installations were operating around the beginning of the twentieth century. In the 1920s and 1930s, several plants in over 10 MWe size were constructed [1]. Industrial research and development to improve the economical performance of the method have continued to this very date. The newly developed electrolysis cells include the high-pressure designs and state-of-the-art membrane electrolyte designs.

Electrolysis of water supplies only a few percentage of world hydrogen used today. The method is preferred for the production of high purity hydrogen and oxygen. It is used more often in places where hydropower is abundantly produced, for example, Iceland and Norway. Its wider use in industrial applications has been limited mainly because the cost of electricity remains high. In contrast, hydrogen produced by reforming fossil-fuel resources, chiefly methane or natural gas, has been developed and made economical and supplies the remainder of the world hydrogen demand now [2].

This chapter discusses electrolysis of liquid water, that is, alkaline water electrolysis and polymer electrolyte water electrolysis. The discussion is based on review of the literature [3–6].

3.2 Principle

Heat and work are added to and taken from a reaction to maintain enthalpy and entropy balance of the reaction in accordance with the equation below.

$$\Delta H = \Delta G + T\Delta S \tag{3.1}$$

When reaction temperature (and pressure, to be accurate) are decided, heat and work requirement are fixed. In the case of water decomposition, the relation of heat and work is schematically illustrated in Figure 3.1. A certain work input is needed to obtain products when the temperature is below T_d. Note that the product gases of H_2 and O_2 are the same as the initial pressure. H_2 and O_2 of lower pressure are made by thermal equilibrium of water decomposition at lower temperature. To increase the pressure of these gases to the initial pressure requires a certain kind of work. Electrolysis of water can be regarded as a reaction of which the work requirement is provided as electricity. Electricity demand depends on temperature. The requirement is smaller at higher temperature. Electrolysis at several hundred to 1000°C is called high-temperature steam electrolysis. This type is explained in Chapter 4 because the technology is different from electrolysis at ambient temperature. Here, electrolysis methods applied to liquid water are discussed.

Electrolysis of water is a combination of two half-reactions as shown below. The equations are different by electrolyte type.

Acid electrolyte:

$$\text{Anode: } H_2O \rightarrow 2H^+ + 0.5O_2 + 2e^- \tag{3.2}$$

$$\text{Cathode: } 2H^+ + 2e^- \rightarrow H_2 \tag{3.3}$$

Alkaline electrolyte:

$$\text{Anode: } 2OH^- \rightarrow H_2O + 0.5O_2 + 2e^- \tag{3.4}$$

$$\text{Cathode: } 2H_2O + 2e^- \rightarrow H_2 + 2OH^- \tag{3.5}$$

Theoretical voltage of electrolysis is described as

$$\Delta G_e = nFE \tag{3.6}$$

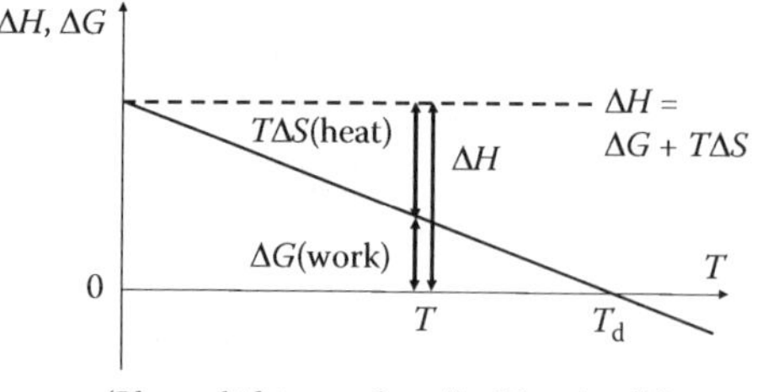

FIGURE 3.1
G–T diagram of decomposition of water.

ΔG_e depends on temperature, composition of electrolyte, and pressure of gas as in Equation 3.7. From Equations 3.6 and 3.7, theoretical voltage is described as in Equation 3.8.

$$\Delta G_e = \Delta G_e^0 + RT \cdot \ln\left[\frac{\left(\frac{f_{H_2}}{p^0}\right)\cdot\left(\frac{f_{O_2}}{p^0}\right)^{0.5}}{a_{H_2O}}\right] \tag{3.7}$$

$$E = E^0 + \left(\frac{RT}{nF}\right)\cdot \ln\left[\frac{\left(\frac{f_{H_2}}{p^0}\right)\cdot\left(\frac{f_{O_2}}{p^0}\right)^{0.5}}{a_{H_2O}}\right] \tag{3.8}$$

Actual cell voltage is greater than theoretical voltage because of over potential of electrodes and ohmic resistance of cell components as in Equation 3.9.

$$E_{\text{cell}} = E + E_{\text{ov.pot.A}} + E_{\text{ov.pot.C}} + E_{\text{ohm}} \tag{3.9}$$

Breakdown of actual cell voltage is illustrated in Figure 3.2. Over potential of electrodes means excess voltage to theoretical cell voltage in order to progress cell reactions at practical rate. Over potential is made from the composition difference in between bulk electrolyte and around electrodes. The approximate value of over potentials is described by Tafel Equation 3.10.

$$E_{\text{ov.pot.}} = C_1 + C_2 \ln i \tag{3.10}$$

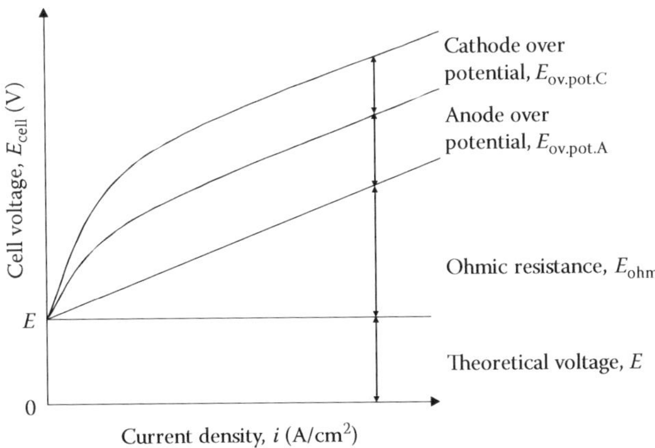

FIGURE 3.2
Breakdown of cell voltage.

Ohmic resistance is made from electric resistance of components: electrolyte, separator, gas bubble between electrodes, outer electric circuit and electrodes. Ohmic resistance is approximately linear to current density. The total cell voltage is high in large current density. When current density is large, operation cost is higher because greater electric power is required. However, cell size can be made small and cell cost can be low. Designing the total cell system and optimization of cell operation are required for lower total cost.

Efficiency of a water electrolysis cell is defined in Equation 3.11 as the ratio of reaction enthalpy to the electric energy supplied to the cell. Not only work of ΔG but heat of $T\Delta S$ should be supplied to operate the cell. This is the reason to use ΔH, not ΔG as numerator. This efficiency can be defined as ratio of thermoneutral voltage, E_H defined by Equation 3.12 to actual cell voltage in Equation 3.13. Theoretically, heat production by Joule loss is the same as heat requirement and no heat supply is required at thermoneutral voltage.

$$\eta_{el.} = \frac{\Delta H}{W} = \frac{E_H}{E_{cell}} \tag{3.11}$$

$$E_H = \frac{\Delta H}{nF} \tag{3.12}$$

$$E_{cell} = \frac{W}{nF} \tag{3.13}$$

It is noted that the efficiency of water electrolysis is a different concept from other efficiency, such as that of thermochemical water splitting (see Chapter, Section 5.1). While the calculation in Equation 3.11 uses electric work, the efficiency of thermochemical water splitting uses heat. When efficiency of electrolysis is compared with other methods, the same definition in those methods has to be used.

3.3 Alkaline Water Electrolysis

3.3.1 Outline

Figure 3.3 is a schematic of an alkaline-water electrolysis (AWE) cell. Ions in the electrolyte solution work as transfer agents of electricity. When the electrolyte contains cations which are reduced more easily than H^+ or anions which are oxidized more easily than OH^-, water decomposition reaction cannot progress. Therefore, a strong acid or strong alkali is used so as to decompose only water. Alkaline electrolyte is usually chosen in order to avoid corrosion of cell materials. High concentration KOH solution of 25–30 wt% is often used. Low over potential, large contact area with electrolyte, and good detachment of product bubbles are desired for electrodes. Low-carbon steel mesh or nickel coated low-carbon-steel mesh is used as cathode in normal cells. Alkali- and oxidation-resistant materials like nickel-coated low carbon steel or nickel series metal are applied for anode. Electrode catalysts on which reaction occurs more easily such as Pt are sometimes used together. A porous diaphragm works for preventing mixture of products gases and direct contact of electrodes. Though asbestos was used first, alternative materials

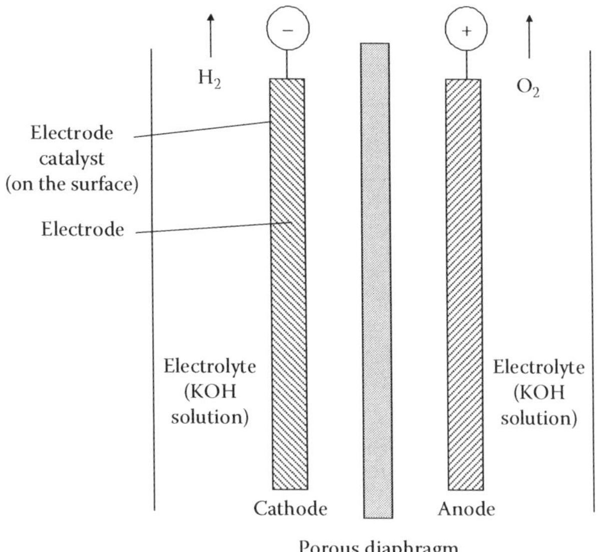

FIGURE 3.3
Schematic of an alkaline-water electrolysis cell.

such as potassium titanide, polyantimonic acid, oxide-coated metallic materials, polytetrafluoroethylene (PTFE), and their composites were tried considering health-related issues and low-temperature resistance of asbestos [7].

AWE cells are grouped into unipolar and bipolar electrolysis cells. A schematic diagram of a unipolar cell is shown in Figure 3.4. Several pairs of anode and cathode are in one tank of electrolyte. Electrolyte can transfer through porous diaphragms. Total voltage in one electrolyzer is the same as a pair of electrodes because all electrodes are parallel. Instead, electric current is large. Several cells connect with each other in serial in large plants. This type is advantageous in simple structure and low leakage current. However, current density is usually smaller compared with bipolar cell. This type requires space among cells and large plant area is necessary. Figure 3.5 illustrates a schematic of a bipolar electrolysis cell. Electrolyte circulates through the cell to release the heat made in electrolysis. The electrolyte is fed from the bottom and mixture of the electrolyte and product gas flows out from the top. The gas and the electrolyte are separated in drums at the top. One side of an electrode is the anode, and the other side the cathode. Electrons generated by reaction Equation 3.2 or 3.4 at cathode in a cell unit compartment are transferred to the anode of the neighboring unit, and are used in reaction 3.3 or 3.5. Electrodes work as also flow separators. The total electrolyzer is equivalent to a serial connection of many cells. Total voltage is large instead of small electric current. This type is commonly used because the floor area is smaller than that of unipolar type and mass production of cell components is possible. However, this type has a disadvantage that entire stack should be stopped in case of repair.

3.3.2 Research and Development

Basic technology of conventional AWE cell is very old. The early industrial water electrolysis cells as mentioned in the introduction of this chapter were based on AWE. Research on

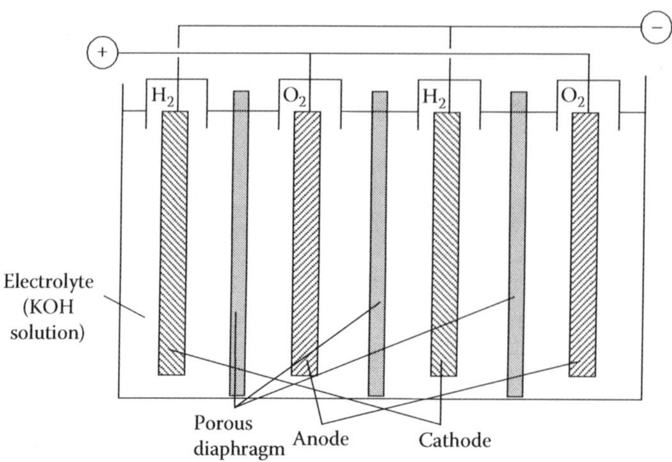

FIGURE 3.4
Schematic diagram of a unipolar AWE cell.

"advanced alkaline water electrolysis" had been carried out from the latter half of 1970s to the first half of 1980s all over the world. The study aimed at high performance by increasing the operation temperature and pressure. When the temperature of electrolysis is high, theoretical energy of water electrolysis ΔG decreases considering Equation 3.1. Work input to the cell decreases by supplying $T\Delta S$ as heat because temperature dependence of ΔH is not so large (see Figure 3.1). And ohmic resistance of electrolyte and over potential of electrodes are expected small at high temperature. High pressure is necessary to increase the temperature for keeping the electrolyte as liquid. High pressure is also advantageous in that volume of bubbles of the product gas that causes electric resistance can be reduced. An important issue was the materials that resisted high temperature and high concentration alkaline electrolyte. Low-carbon steel could be used for construction material with confidence only below 80°C. There were few alternatives besides PTFE-based materials for

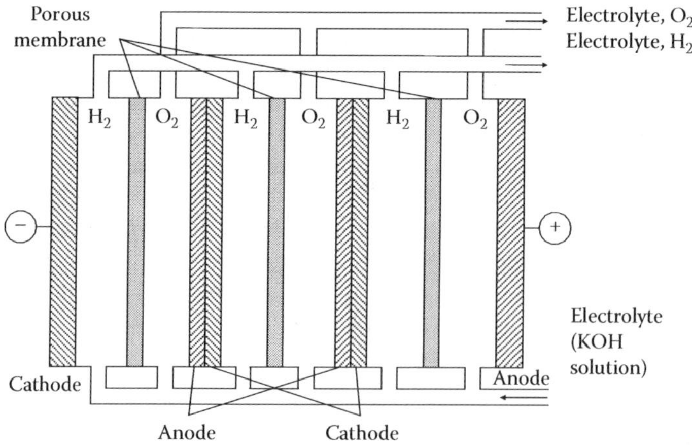

FIGURE 3.5
Schematic diagram of a bipolar AWE cell.

use as gasket and insulating materials at temperatures higher than 100°C. Potassium titanate and polyantimonic acid were proposed as novel diaphragm materials substituting asbestos [2]. And reduction of electric resistance by improvement of electrodes and porous diaphragm was a target of the research. New materials were studied and cell designs of small distance between an electrode and a diaphragm was investigated.

Teledyne Energy Systems, USA, had been involved in development of advanced AWE since 1976 supported by U.S. Department of Energy (DOE). Operation at high temperature over 125°C, improvement of catalyst/electrode structure including alternative electrode materials and a long life electrode separator were focused upon. Material screening, design, and cost evaluation were investigated for the objective. Brookhaven National Laboratory (BNL), USA, conducted trials mainly on electrodes with some universities for advanced AWE [8].

In Japan, study on AWE began in 1975 as a plan of Sunshine Project, which had aimed at the development of alternative energy sources. The study was conducted mainly at Osaka National Research Institute (ONRI) (now Kansai Center, National Institute of Advanced Industrial Science and Technology) and a consortium of several public and private laboratories. Their theme was improvement of operating current density and energy efficiency of the electrolyzer under 120°C and 20 kg/cm^2 [9]. A pilot plant of 20 Nm^3/h of hydrogen production capacity was operated under 120°C and 20 kg/cm^2 by mid-1980s. Risk of explosion by mixing hydrogen into oxygen was a problem in high-pressure operation [3].

European Community (EC) also had R&D programs in 1975–1979 [10] and 1979–1984 [11] within the 1st and 2nd Framework Research Programmes. Many contractors participated in the R&D on components, constituent materials, nature of electrolytes, operating conditions, and total cell concept in order to lower total cell voltage and to increase current density. Operating temperature was set mainly on 120–200°C though some researches were on around 90°C. Target of current density was 1 A/cm^2. Three moderate temperature (120°C) pilot cells of 3–10 kW scale were developed (e.g., [12]) and economic analysis was conducted based on the data from these cells. Electrolysis was estimated as most economic among gas reforming and coal gasification in small plant of 15 Nm^3/h hydrogen at 1984 though coal gasification was most economic in the case of ca. 100,000 Nm^3/h hydrogen [11]. Electricité de France and Gaz de France also had been investigating materials for structure, diaphragm, electrode materials, and electrocatalysts since 1975. A bench-scaled test cell (20–30 kW) design was also carried out from 1978 [13]. Noranda Research Center, Canada, worked from 1979. The laboratory studied on promising separator and electrocatalyst materials and researched for construction of a long-term test cell operating at 150°C [14].

High cost of alkali resistant materials increased hydrogen production cost. Activity of research on advanced AWE reduced in the mid-1980s since low price hydrogen by reforming of hydrocarbon was available. Reduction of cell voltage and increase of current density at below 100°C by improvement of electrodes became the main targets in the period [5].

3.3.3 Industrialization

Industrialization of large-scale AWE is very old as explained in the introduction of this chapter. Table 3.1 shows examples of industrial mass hydrogen-production plants. Sites were selected where low-cost hydroelectric generation and large amounts of pure water were available.

Smaller size AWE systems were commercialized afterwards. Table 3.2 shows examples of small-scale commercial AWE systems as of August, 2009. It is noted that data is not

TABLE 3.1

Example of Large-Scale Electrolysis Hydrogen Production Plants

Manufacturer (At the Establishment of the Plant)	Nation of Manufacturer	Site	Nation of Site	Type	Establishment Year	Hydrogen Production (Nm^3/h)	Objective
Brown Boveri[a]	Switzerland	Aswan	Egypt		1977	21600	Ammonia synthesis
Cominco	Canada	Trail	Canada	Tank	1939	17000	Manufacture of fertilizer
Demag	Germany	Aswan	Egypt	Filter press	1960	41000	Ammonia synthesis
De Nora	Italy	Nangal	India	Filter press	1958	26000	Ammonia synthesis
Lurgi GmbH[b]	Germany	Cuzco	Peru	Filter press	1958	5000	
Norsk Hydro[c]	Norway	Kristiansand	Norway			1050	Nickel refining
Norsk Hydro[c]	Norway	Reykjavik	Iceland			2600	Ammonia synthesis
Norsk Hydro[c]	Norway	Fredrikstad	Norway			1800	Ammonia synthesis
Norsk Hydro[c]	Norway	Glomfjord	Norway		1950	6800	Ammonia synthesis
Norsk Hydro[c]	Norway	Rjukan	Norway	Filter press	1927[d]	27900	Ammonia synthesis

Manufacturer (At the Establishment of the Plant)	Reference	Note
Brown Boveri[a]	[15]	Replacement of the Demag plant
Cominco	[15]	Type: Ref. [17], Objective: Ref. [18]
Demag	[15]	Type: Ref. [17], Objective: Ref. [16]
De Nora	[15]	Type: Ref. [17], Objective: Ref. [1]
Lurgi GmbH[b]	[15]	Operation at high pressure (30 (kg/cm^2))
Norsk Hydro[c]	[16]	
Norsk Hydro[c]	[16]	
Norsk Hydro[c]	[16]	
Norsk Hydro[c]	[16]	Establishment year (time of start operation): Ref. [1]
Norsk Hydro[c]	[15]	Type: Ref. [17], Hydrogen production: Ref. [5], Objective: Ref. [1]

These plants are not always at work as of August 2009.

[a] Brown Boveri and Asea AB merged into ASEA Brown Boveri.

[b] At present, marketing and realization of new installation and the maintenance of existing sites was transferred from Lurgi.

[c] Statoil and oil & gas division of Norsk Hydro merged into StatoilHydro.

[d] Began in 1927. Upgraded through 1965.

TABLE 3.2

Example of Specification of Small Scale Commercial Alkaline Water Electrolysis (AWE) Cells

Company	Nation	Model	Type	Hydrogen Production (Nm^3/h)	Maximum Temperature (°C)	Maximum Pressure (atm)	Hydrogen Purity (%)	Electricity Consumption (kWh/Nm^3)	Reference
Avalence		Hydrofiller	Unipolar[e]	0.4–4.6		650 psi[i]	99.7[e]	5.1[e, k, l], 5.3[e, k, m], 5.4[e, k, n]	[19]
Hydrogenics[a]	Canada	IMET 300	Bipolar[e]	1–3		25[i]	99.9	4.2[o], 4.9[k]	[20]
Hydrogenics[a]	Canada	IMET 1000	Bipolar[e]	4–60		10[i, j]	99.9	4.2[o], 4.8[k]	[20]
Industrie Haute Technologie[b]	Switzerland		Bipolar[f]	110–760	90[f]	32	99.8–99.9	4.3–4.6	[21]
Industrie Haute Technologie[c]	Switzerland		Bipolar[f]	3–330	80[f]	Atmospheric	99.8–99.9	3.90–4.22[p], 4.20–4.54[q]	[21]
StatoilHydro[d]	Norway		Bipolar	10–377[g], 10–485[h]	80	Atmospheric	99.9 ± 0.1	4.1 ± 0.1[g], 4.3 ± 0.1[h]	[22]
Teledyne Energy Systems	USA	TITAN HM	Bipolar[e]	2.8–11.2		10[i]	99.9998	5.6–6.4[r]	[23]
Teledyne Energy Systems	USA	TITAN EC	Bipolar[e]	28–56		4.2–8.1[i]	99.9998	5.6[r]	[23]

[a] Though IMET was dealt by Stuart Energy, Hydrogenics has acquired Stuart Energy.

[b] At present, marketing and realization of new istallation and the maintenance of existing sites was transferred from Lurgi.

[c] At present, marketing and realization of new istallation and the maintenance of existing sites was transferred from Lurgi, which had sold Bamag cell.

[d] Statoil and oil & gas division of Norsk Hydro merged into StatoilHydro.

[e] Ref. [24].

[f] Estimation by the description of Ref. [3].

[g] At 4000 Amp DC.

[h] At 5150 Amp DC.

[i] Cell is at atmospheric pressure and product gas is pressurized at delivery.

[j] 25 atm is available only up to 30 Nm^3/h and power consumption of 4.9 kWh/Nm^3 h.

[k] Including rectifier and auxiliaries.

[l] Of Hydrofiller 15.

[m] Of Hydrofiller 50.

[n] Of Hydrofiller 175.

[o] Electrolysis only.

[p] At 20°C, 1013 mbar, wet.

[q] At 0°C, 1013 mbar, day.

[r] Estimation by the description of Ref. [16].

necessarily on uniform standard. Pressurized electrolysis is advantageous as shown in Section 3.3.2. Lurgi Umwelt und Chemotechnik GmbH, Germany, manufactured a module operating at over 3.0 MPa (now available from Industrie Haute Technologie, Switzerland). However, the combination of an atmospheric pressure electrolysis cell and a compressor for storage of hydrogen is preferable to pressurized electrolyzer for safety and economy [3]. Lurgi system is the only commercial large-scale pressurized cell. Though both unipolar and bipolar cells were used in 1970s, the latter operating at temperatures lower than 100°C are commonly used now [25]. Technology of conventional AWE has matured and property has already reached the limit. For example, the efficiency of the cell alone defined in Equation 3.11 is 83% and that of total system including rectifier and auxiliaries is 73% in Hydrogenics IMET® 1000 [24].

3.4 Polymer Electrolyte Water Electrolysis

3.4.1 Outline

Polymer electrolyte water electrolysis (PEWE) uses a polymer electrolyte membrane as a medium of ion transfer instead of solution electrolyte in AWE. This method is often called polymer electrolyte membrane or proton exchange membrane (PEM) water electrolysis, too. Figure 3.6 shows a schematic of a cell. The reactions using a cation-exchange membrane are shown in Equations 3.2 and 3.3. Reactant water is fed into channels of the anode side only of the bipolar plate. The water flows from the plate to the anode through the current collector, and reacts to make protons. Current collectors are porous conductors that allow electrons to transfer from electrode to outer circuit and allow reactant gas from bipolar plate to electrode. The protons are transported through the PEM to cathode side, and hydrogen is generated at the cathode. The PEM also works as a separator of product gases. Perfluorosulfonic acid polymer membranes, such as Nafion® of DuPont, USA, are typically used because of its excellent thermal resistance and oxidation resistance. Figure 3.7 shows

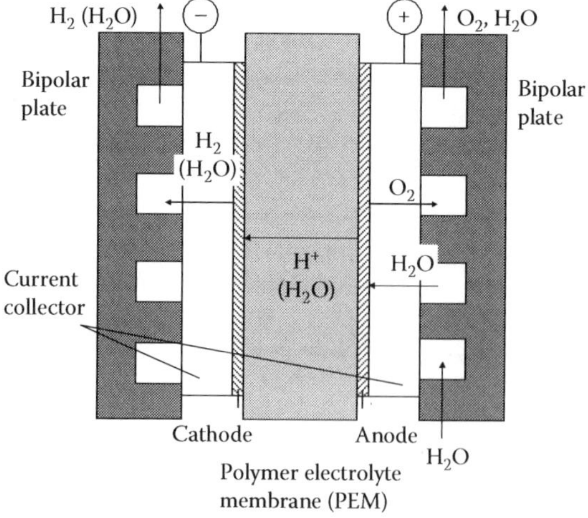

FIGURE 3.6
Schematic diagram of a PEWE cell.

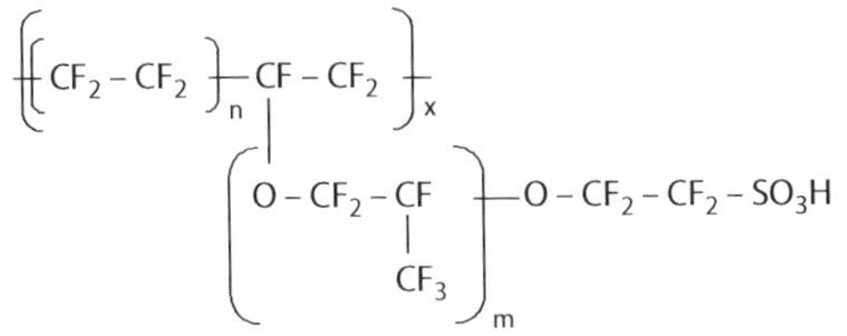

$$\left[\left(CF_2-CF_2\right)_n-\underset{\left(O-CF_2-\underset{CF_3}{CF}\right)_m-O-CF_2-CF_2-SO_3H}{CF}-CF_2\right]_x$$

FIGURE 3.7
Chemical structure of Nafion.

the chemical structure of Nafion. The membrane works as strong acid due to the sulfonate groups (–SO_3H) at the end of side chains. Electrodes come in contact directly with the PEM to avoid interface electric resistance. Corrosion resistance to strong acidity of the PEM is required for electrodes in addition to the requirements for AWE electrodes. Platinum group metal, alloy and oxide of Pt are used for electrodes. Cathode over potential is the main source of the total cell over potential and it is influenced by selection of material. Oxides of Ir and Ru are used. These materials are often mixed with inert components for structural stability. Metallic Pt is commonly used as cathode [26].

Advantages and disadvantages of this method compared with AWE are as follows [3].

Advantages:

1. Corrosive liquid electrolyte is not required. Therefore, range of structural material selection is wide and easy to maintain.
2. Construction of facility is easy because pressure difference between anode side and cathode side is allowed.
3. No electric resistance by gas bubbles between electrodes can be made.
4. Purity of product gas is high with no droplets of solution electrolyte and good product gas separation by the PEM.

Disadvantages:

1. Components which come in contact with the PEM (electrodes, current collectors) should be corrosion resistant due to strong acidity of the PEM.
2. Uniform contact between the PEM and the electrodes should be achieved to reduce electric resistance.
3. Electric current loss resulting from backward permeation of gas within the PEM (around 0–5%).
4. Cost of the PEM, electrodes and current collectors is high.

3.4.2 Research and Development

The first trial of PEWE was carried out by General Electric (GE), USA, in 1966 using the technology of polymer electrolyte fuel cell in the Gemini space program. The DOE effort with GE began in 1975 and resulted in the operation of a 200 kW system consisting of 60 electrolysis cells. Application of thinner PEMs and manufacturing technology lowering the current collector resistance were developed. The major technological problems were cell sealing, water purity, and cell impedance. The costly PEM, electrodes and current collectors prevented the cost goal to be achieved for hydrogen production at the time. The funding support to GE from the DOE stopped in 1981. BNL conducted works on PEM/electrocatalyst interface [8]. In Japan, a research was conducted by ONRI in Sunshine Project (see Chapter 3.3.2). Reduction of noble metal loading for electrode catalysts was achieved by special plating methods [9]. From 1976 to 1989 Brown Boveri, Switzerland (now Asea Brown Boveri) also researched on a commercial cell. A pilot plant with 3 m^3/h hydrogen-production capacity operated for more than 6000 h [27,28]. And two 100 kW scale demonstration plants using this type cell were operated in 1990s [29].

Some laboratories in France investigated on the technology in 1990s. Pt- and Ir-based efficient electrode–membrane–electrode (EME) composite was developed. A laboratory-scale electrolysis test was performed for 25,000 h continuously. A medium size cell of 0.5 kW input power was designed and tested based on these achievements [30]. In Japan, World Energy Network (WE-NET) project, which was a successor of Sunshine project (1973–1991) and New Sunshine project (1992–2000), had been administrated by New Energy and Industrial Technology Development Organization (NEDO), Japan, from 1993 to 2002 (end year was moved up from 2003 in the original schedule). Operation of a cell of electrode area of 50 cm^2 was performed for over 6000 h at 80°C and current density of 1 A/cm^2. A cell stack test equipment with electrode area of 2500 cm^2 was constructed, and high-energy conversion efficiency was attained at 120°C, 0.5 MPa and 1 A/cm^2 current density [31,32]. After WE-NET project, NEDO continued a research aiming at reduction of mass of Pt electrode and at improvement of durability [33].

Even though small size cells are commercialized, R&D is still underway at present. Recently, Générateur d'Hydrogène par électrolyse de l'eau PEM (GenHyPEM) program was conducted as a project of the 6th Framework Research Programme of European Commission from 2005 to 2008. The main objective of the program was to develop efficient PEWE in the scale of several m^3/h of hydrogen production at 1 A/cm^2 current density and 50 bar operating pressure. Direct deposition of electrocatalysts on the surface of the PEM and non-noble metal electrocatalysts were investigated in order to reduce cost of catalysts [34]. Russian Research Center Kurchatov Institute has studied PEWE for over 20

TABLE 3.3

Example of Specification of Commercial Polymer Electrolyte Water Electrolytes (PEWE) Cells

Company	Nation	Model	Hydrogen Production (Nm^3/h)	Maximum Pressure[a] (atm)	Hydrogen Purity (%)	Electricity Consumption (kWh/N m^3)	Reference
h-tec	Germany	EL 30	2.4	30			[36]
Kobelco Eco-Solutions	Japan	HHOG	1–60	4–9[b]	99.99993	6.5	[37]
Proton Energy Systems	USA	HOGEN® H	2–6	15[c]	99.9995	7.3[d], 7.0[e], 6.8[f]	[38]
Proton Energy Systems	USA	HOGEN S	0.265–1.05	13.8	99.9995	6.7	[38]

[a] Cell is at atmospheric pressure and product gas is pressurized at delivery.
[b] Maximum 30 atm is optional.
[c] 30 atm is optional.
[d] At 2 Nm^3/h.
[e] At 4 Nm^3/h.
[f] At 6 Nm^3/h.

years [35]. The institute participated in GenHyPEM project. Now some projects aiming at high temperature PEWE are under way in the 7th Framework Research Programme from 2008 to 2010.

3.4.3 Industrialization

The first commercial scale PEWE plant was installed in 1987 for metallurgical process. The designed hydrogen production was 20 Nm^3/h. However, the plant was stopped in 1990 after about 15,000 hours of operation because of safety problem of hydrogen mixture in the production oxygen. By the time, drastic voltage drop was made by short circuit in the cells [29]. Table 3.3 summarizes examples of commercial PEWE plants as of August, 2009. Production scale of PEWE is smaller than that of AWE. Plant of several hundred m^3/h scale is not commercialized yet. Larger size PEM, long-term stability, and low-cost components are the key to wide use [39].

Nomenclature

a	Activity of liquid component (dimensionless)
C_1, C_2	Constants depending on temperature, surface state, and electrode material
E	Theoretical voltage (V)
E_{cell}	Cell voltage (V)
E_H	Thermoneutral voltage (V)
E^0	Standard theoretical voltage (= 1.229 V at standard state (298.15 K and 10^5 Pa))
f	Fugacity of gas component (can be taken as partial pressure at low pressure environment) (Pa)
F	Faraday constant (= 96,485 C/mol)
i	Current density (A/cm^2)
n	Molar ratio of transferred electron to decomposed water (=2) (dimensionless)
p^0	Standard pressure (= 10^5 Pa)
T	Reaction temperature (K)
T_d	The temperature of $\Delta H = 0$ (= 4310 K in the case of decomposition of water)
W	Electric energy consumption (kJ)
ΔG	Reaction Gibbs energy (= work requirement) (kJ)
ΔG_e	Gibbs energy of electrolysis (kJ/mol)
ΔH	Reaction enthalpy (kJ)
ΔS	Reaction entropy (kJ/K)
$\eta_{el.}$	Efficiency of water electrolysis (dimensionless)

Superscripts and Subscripts

A	Anode
C	Cathode
o	Standard state (298 K and 10^5 Pa)
ohm	Ohmic resistance
ov.pot.	Over potential

References

1. Aureille, R., Deux siecles de production d'hydrogene. *Proc. 5th World Hydrogen Energy Conf.*, 21–43, Toronto, Canada, July 15–20, 1984.
2. LeRoy, R. L., Industrial water electrolysis: Present and future. *Int. J. Hydrogen Energy*, 8, 401–417, 1983.
3. Abe, I., Arukari mizu denkaihou. In *Suiso Enerugi Saisentan Gijyutu*, ed. T. Ohta, 37–58, 1995, Tokyo, NTS.
4. Takenaka, K., Kotai koubunsigata mizu denkaihou. In *Suiso Enerugi Saisentan Gijyutu*, ed. T. Ohta, 74–84, 1995, Tokyo, NTS.
5. Takenaka, K., Mizu denkai. In *Denki Kagaku Binran*, 5th edition, Electrochemical Society of Japan, 371–380, 2000, Tokyo, Maruzen.
6. Takenaka, K., Mizu denkai ni yoru suiso seizou gijyutu. In *Suiso Riyou Gijyutu Syuusei*, Vol. 3, 344–355, 2007, Tokyo, NTS.
7. Renaud, R. and R. L. LeRoy, Separator materials for use in alkaline water electrolysers. *Int. J. Hydrogen Energy*, 7, 155–166, 1982.
8. Bonner, M., T. Botts, J. McBreen, A. Mezzina, F. Salzano, and C. Yang, Status of advanced electrolytic hydrogen production in the United States and abroad. *Int. J. Hydrogen Energy*, 9, 269–275, 1984.
9. Ohta, T. and M. V. C. Sastri, Hydrogen energy research programs in Japan. *Int. J. Hydrogen Energy*, 4, 489–498, 1979.
10. Imarisio, G., Progress in water electrolysis at the conclusion of the first hydrogen programme of the European Communities. *Int. J. Hydrogen Energy*, 6, 153–158, 1981.
11. Wendt, H. and G. Imarisio, Nine years of research and development on advanced water electrolysis. A review of the research programme of the Commission of the European Communities. *J. Appl. Electrochem.*, 18, 1–14, 1988.
12. Vandenborre, H., R. Leysen, H. Nackaerts, and Ph. van Asbroeck, A survey of five year intensive R&D work in Belgium on advanced alkaline water electrolysis. *Int. J. Hydrogen Energy*, 9, 277–284, 1984.
13. Bailleux, C., A. Damien, and A. Montet, Alkaline electrolysis of water-EGF activity in electrochemical engineering from 1975 to 1982. *Int. J. Hydrogen Energy*, 8, 529–538, 1983.
14. Bowen, C. T., H. J. Davis, B. F. Henshaw, R. Lachance, R. L. LeRoy, and R. Renaud, Developments in advanced alkaline water electrolysis. *Int. J. Hydrogen Energy*, 9, 59–66, 1984.
15. Fickett, A. P. and F. R. Kalhammer, Water electrolysis. In *Hydrogen: Its Technology and Implications*, Vol. 1, eds. Cox, K. E., and K. D. Williamson, 3–41, 1977, Cleveland, CRC Press.
16. Bello, B. and M. Junker., Large scale electrolysers. *Proc. 16th World Hydrogen Energy Conf.* II-215, Lyon, France, June 13–16, 2006.
17. Funk, J. E., Thermochemical and electrolytic production of hydrogen from water. In *Introduction to Hydrogen Energy*, ed., T. N. Veziroglu, 19–49, 1975, Coral Gables, International Association for Hydrogen Energy.
18. Laskin, J. B. and R. D. Feldwick, Recent development of large electrolytic hydrogen generators. *Proc. 1st World Hydrogen Energy Conf.* 6B-3–6B19, Miami Beach, U.S., March 1–3, 1976.
19. Avalence. http://www.avalence.com/ (accessed August 20, 2009).
20. Hydrogenics. http://www.hydrogenics.com/default.asp (accessed August 20, 2009).
21. Industrie Haute Technologie. http://www.iht.ch/technologie/electrolysis/industry/tailormade installations.html (accessed August 20, 2009).
22. StatoilHydro. http://www.electrolysers.com/ (accessed August 20, 2009).
23. Teledyne Energy Systems. http://www.teledynees.com/index.asp (accessed August 20, 2009).
24. Ivy, J., Summary of electrolytic hydrogen production milestone completion report. *NREL/MP-560-36734*, 2004.

25. Welboren, D. J. S., Expectations of hydrogen production technologies since the 1970s, Master Thesis, Technical Univ. Eindhoven, 2006.
26. Rasten, E., G. Hagen, and R. Tunold, Electrocatalysis in water electrolysis with solid polymer electrolyte. *Electrochim. Acta*, 48, 3945–3952, 2003.
27. Stucki, S. and R. Müller, Evaluation of materials for a water electrolyzer of the membrane type. *Proc. 3rd World Hydrogen Energy Conf.* 1799–1808, Tokyo, Japan, June 23–26, 1980.
28. Oberlin, R. and M. Fischer, Status of the MEMBREL® process for water electrolysis. *Proc. 6th World Hydrogen Energy Conf.* 333–340, Vienna, Austria, July 20–24, 1986.
29. Stucki, S., G. G. Scherer, S. Schlagowski, and E. Fischer, PEM water electrolysers: Evidence for membrane failure in 100 kW demonstration plants. *J. Appl. Electrochem.*, 28, 1041–1049, 1998.
30. Millet, P., F. Andolfatto, and R. Durand, Design and performance of a solid polymer electrolyte water electrolyzer. *Int. J. Hydrogen Energy*, 21, 87–93, 1996.
31. Hijikata, T., Research and development of international clean energy network using hydrogen energy (WE-NET). *Int. J. Hydrogen Energy*, 27, 115–129, 2002.
32. New Energy and Industrial Technology Development Organization, *NEDO WE-NET-0208*, 2003.
33. New Energy and Industrial Technology Development Organization, *NEDO 06990091-0-1*, 2008.
34. Millet, P., D. Dragoe, S. Grigoriev, V. Fateev, and C. Etievant, GenHyPEM: A research program on PEM water electrolysis supported by the European Commission. *Int. J. Hydrogen Energy*, 34, 4974–4982, 2009.
35. Grigoriev. S. A., V. I. Porembsky, and V. N. Fateev, Pure hydrogen production by PEM electrolysis for hydrogen energy. *Int. J. Hydrogen Energy*, 31, 171–175, 2006.
36. h-tec. http://www.h-tec.com/html/web/industrial/english/index.asp (accessed August 20, 2009).
37. Kobelco Eco-Solutions. http://www.kobelco-eco.co.jp/product/suisohassei/hhog_seihin.html (accessed August 20, 2009).
38. Proton Energy Systems. http://www.protonenergy.com/ (accessed August 20, 2009).
39. New Energy and Industrial Technology Development Organization, *NEDO WE-NET-0201*, 2003.

4

Steam Electrolysis

Ryutaro Hino, Kazuya Yamada, and Shigeo Kasai

CONTENTS

4.1 Reaction Scheme

The process of the high-temperature electrolysis (HTE) of steam is a reverse reaction of the solid-oxide fuel cell (SOFC): an oxygen ionic conductor is usually used as a solid-oxide electrolyte as shown in Figure 4.1. Steam (H_2O) is dissociated at the cathode with hydrogen (H_2) molecules forming on the cathode surface: $H_2O(g) + 2e^- \rightarrow H_2(g) + O^{2-}$ while oxygen ions migrate simultaneously through oxygen vacancies in the lattice of the electrolyte material. Oxygen (O_2) molecules form on the anode surface with the release of electrons: $O^{2-} \rightarrow \frac{1}{2}O_2(g) + 2e^-$. The products, H_2 and O_2, are separated by the gas-tight electrolyte. Reactions on the two electrodes are summed up as,

$$H_2O(g) \rightarrow H_2(g) + \frac{1}{2}O_2(g) \tag{4.1}$$

In the reaction, theoretical energy demand (ΔH) for water and steam decomposition is the sum of the Gibbs energy (ΔG) and the heat energy ($T\Delta S$). Figure 4.2 shows an energy demand for water and steam electrolysis. The electrical energy demand, ΔG, decreases with increasing temperature as shown in the figure; the ratio of ΔG to ΔH is about 93% at 100°C and about 70% at 1000°C. Thus, the HTE demands less electricity to produce hydrogen than does the conventional water electrolysis. This reaction can be expressed as follows:

$$\Delta G = \Delta G_o + RT \ln\left(a_{H_2}\, a_{O_2}{}^{1/2}/a_{H_2O}\right) \tag{4.2}$$

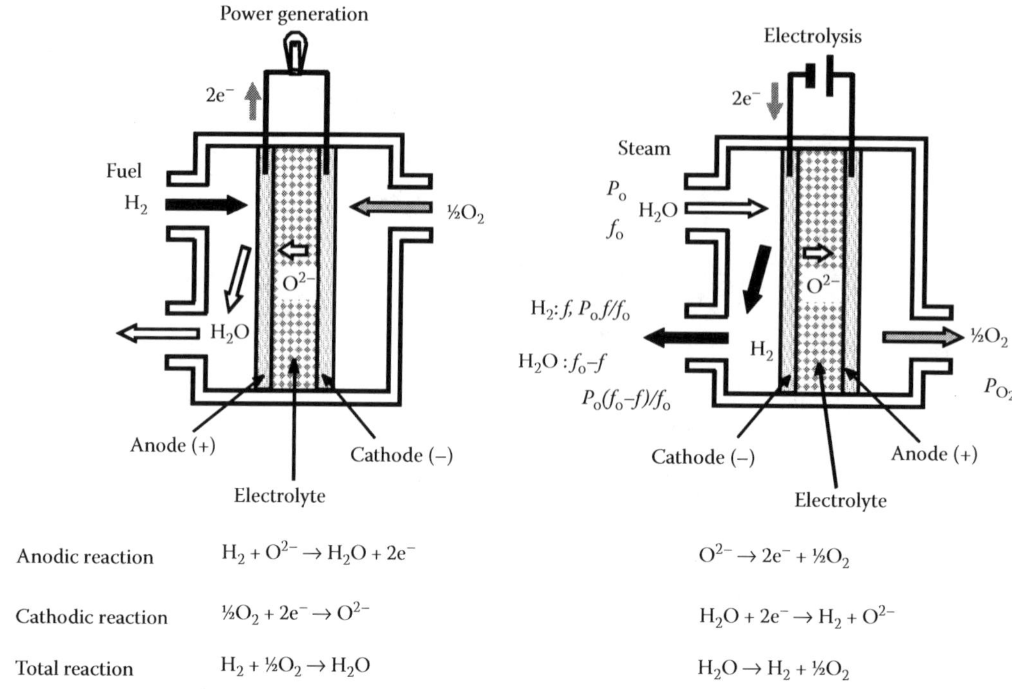

FIGURE 4.1
Principle of high-temperature electrolysis of steam. (Reverse reaction of solid oxide fuel cell.)

where ΔG_o is the standard Gibbs free energy change (per mole) for the reaction 4.1 at a temperature of T, R the gas constant, and a_{H_2}, a_{O_2}, and a_{H_2O} the activities of H_2, O_2, and H_2O in the cell. Equation 4.2 can be rewritten using relations of $E = \Delta G/2F$ and $E_o = \Delta G_o/2F$ as

$$E = E_o + (RT/2\text{F})\ \ln(a_{H_2}\ a_{O_2}{}^{1/2}/a_{H_2O}) \tag{4.3}$$

where F is the Faraday constant, and E_o (= $\Delta G_o/2F$) the standard electromotive force (EMF) of the following reaction in the quasi-static state: $H_2(g) + ½O_2(g) \rightarrow H_2O(g)$.

Equation 4.3 is equivalent to the Nernst equation for EMF of SOFC. Activities of reactants and products can be expressed as partial pressures in the cell, because the reaction proceeds in the gas phase regardless of the electrolyte. The applied voltage increases with polarization in the operating cell. The overvoltage, η, is caused mainly by shortage of steam-concentration at the cathode (concentration overvoltage) and by electrical resistance including electrical leads, interconnections and others (resistance overvoltage). Equation 4.3 can be rewritten as:

$$V = E_o + (RT/2F)\ \ln(P_{H_2}\ \ P_{O_2}{}^{1/2}/P_{H_2O}) + \eta \tag{4.4}$$

where P_{H_2}, P_{O_2}, and P_{H_2O} are the partial pressures of H_2, O_2, and H_2O, respectively.

In the model cell shown in Figure 4.1, steam of molar flow rate, f_o, and pressure, P_o, is reduced to hydrogen gas of flow rate, $f_o - f$, and partial pressure, $P_o(f_o - f)/f_o$, in the cathode

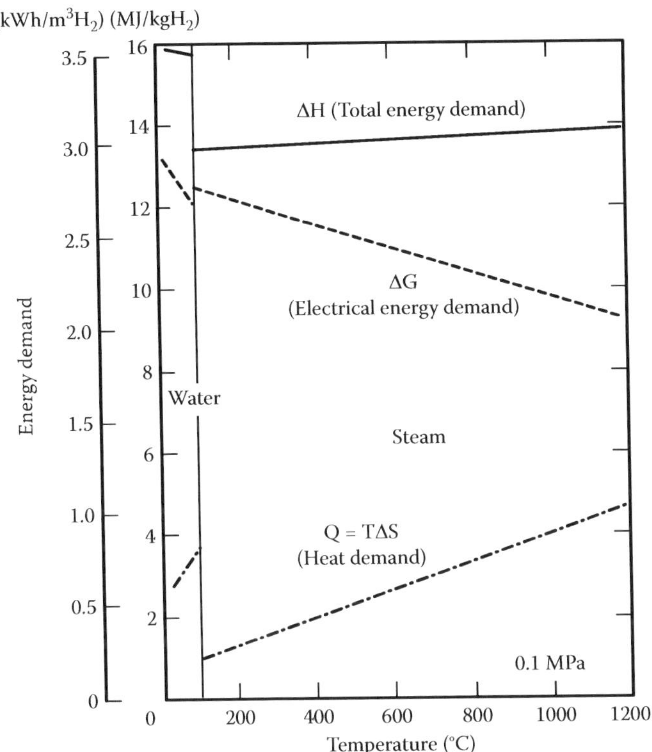

FIGURE 4.2
Energy demand for water and steam electrolysis.

compartment: f/f_o is a conversion ratio from steam (H_2O) to H_2 (steam conversion ratio). The partial pressure of O_2, P_{O_2}, is assumed to be unity. Using the steam conversion ratio, $x = f/f_o$, Equation 4.4 can be written as follows:

$$V - \eta = E_o + (RT/2F)\ \ln(x/(1 - x)) \tag{4.5}$$

The value of $V - \eta$ is called the open-circuit voltage of the cell, which is related to the composition of the product. Note that the steam conversion ratio, x, depends on the open-circuit voltage, and is not affected by the pressure or flow rate of the reactant. Also, the open-circuit voltage decreases with increasing temperature, because of the endothermic nature of the reaction. However, due to the temperature dependence of the logarithmic term in Equation 4.5, this effect decreases with the value of x.

Performance parameters of the electrolysis include the applied voltage, E (V), the applied current, I (A), and the hydrogen production rate, Q (NL/h) at reference condition of 0.1 MPa (1 bar) and 273 K (0°C). The Faraday efficiency, ε_f, expressed as Equation 4.6 below, is the ratio of ΔG to the applied power, $I*E$, that is, the ratio of the theoretical electric power needed for the electrolysis to the actually applied power to the cell. So, the Faraday efficiency is one useful measurement to judge electrolysis performance.

$$\varepsilon_f = \Delta G/(I*E) = 2F*(E - \eta)*Q/(I*E) \tag{4.6}$$

The energy efficiency, ε_e, is the ratio of the produced hydrogen (H_2) energy and the applied power. Then, the produced H_2 energy is estimated with the combustion heat of produced H_2, by using the lower heating value, ΔH (kJ/mol), provided along the reaction (H_2 (g) + ½O_2 (g) → H_2O (g)). The energy efficiency is useful for comparing performances of hydrogen production systems.

$$\varepsilon_e = \Delta H^*Q/22.4/(I^*E) \quad (4.7)$$

where $Q/22.4$ is molar rate of produced hydrogen.

4.2 Research and Development Overview

Several experimental studies have been carried out using practical tubular electrolysis cells, a representative of which was the HOTELLY (high operating temperature electrolysis) program, which included R&D on solid-oxide cells and their assembly technologies as well as demonstration tests using a pilot plant. In the program, self-supporting electrolysis tubes were fabricated by connecting solid-oxide cells in series. Basic hydrogen production data on HTE was obtained with a multicell electrolysis tube [1–3]. The cell consisted of a gas-tight cylindrical electrolyte and porous electrodes. The electrodes (cathode and the anode) were porous thin layers formed on the inner and outer surfaces of the electrolyte cylinder. Ten cells were connected in series by interconnections. The cell specification is included in Table 4.1.

Using the 10-cell electrolysis tube, hydrogen was produced at the maximum rate of 6.78 NL/h of 997°C of electrolysis temperature, 370 mA/cm^2 of current density and a power of 21.7 W. Based on the HTE experimental results and operational experience, an HTE module consisting of 10 electrolysis tubes was integrated, and thereafter 10 modules (including 1000 electrolysis cells) were assembled into a stack, in which hydrogen was produced at the maximum rate of 0.6 Nm^3/h [4]. However, the stability in a long-term operation and the reliability of the electrolysis tube and the module against thermal cycles were not published.

Maskalick [5] had also carried out the HTE experiment using a single-cell electrolysis tube supported by a ceramic tube, which was developed for SOFC at Westinghouse Electric Co. The electrolyte and electrodes layers were formed on a closed-one-end porous support tube made of calcia-stabilized ZrO_2 (CSZ), whose porosity was around 35%. Dimensions of the support tube were 12–13 mm in inner diameter, 1–1.5 mm in thickness and 1000 mm in

TABLE 4.1

Design Specification of the Tested 10-Cell Steam Electrolysis Tube

Active cell length		10 mm
Electrolyte outer diameter		14 mm
Thickness	Electrolyte	0.3 mm
	Anode	<0.25 mm
	Cathode	<0.1 mm
Material	Electrolyte	Yttria (Y_2O_3)-stabilized Zirconia (ZrO_2) (YSZ, Y_2O_3 containment 8–12 mol%)
	Cathode (air electrode)	Nickel containing YSZ (Ni cermet)
	Anode (fuel electrode)	Strontium-doped LSM

length. The support tube was overlaid with a porous strontium-doped $LaMnO_3$ layer, 1.4 mm thick, working as the anode. A gas-tight electrolyte layer made of yttrium-stabilized zirconia (YSZ) (containing 10 mol% of Y_2O_3), 0.04 mm thick, covered the anode along the cell's active length. The Ni cermet layer working as the cathode, 0.1 mm thick, covered the entire electrolyte surface. By using this type of cell, hydrogen was produced at a maximum rate of 17.6 NL/h at a current density of 400 mA/cm^2, an applied power of 39.3 W, and 1000°C of electrolysis temperature [5].

However, the HTE data obtained below 950°C in the 1990s were not sufficient to support the data need for the engineering design of the HTGR (high temperature gas-cooled reactor) heat utilization system. In addition, the HTE operational issues were not clarified. To improve the situation on HTE, the JAEA carried out the laboratory-scale experiments under HTGR operation temperatures from 850 to 950°C, by using a practical electrolysis tube with 12 solid-oxide cells connected in series, and a self-supporting planar cell of a practical size [6,7]. The details of the JAEA experiments are discussed in Section 4.3.

Toshiba as a private company is performing HTE engineering experiment and system design to develop an HTGR cogeneration system [8]. The results from the Toshiba's works are detailed in Section 4.4.

Since 2003, the Idaho National Laboratory (INL) has been conducting extensive research and engineering development aiming at enabling HTE for nuclear hydrogen production. They have fabricated refined self-supporting planar cells composed of YSZ (containing 8 mol% of Y_2O_3), Ni cermet and Strontium-doped $LaMnO_3$ (LSM), and demonstrated 1000-h continuous hydrogen production at a rate of about 160 NL/h (H_2) by using a 25-planar cell stack [9]. Chapter 20 reports the up-to-date activities and the extensive results of the INL research and development to advance the HTE technology.

4.3 Experimental Studies in JAEA

4.3.1 Experimental Results with Tubular Cell

Figure 4.3 illustrates the structural drawing of an electrolysis tube with 12 cells, and Photo 4.1 shows the view of the tube. This type of electrolysis tube had been originally developed for SOFC [10]. The electrolysis tube was composed of 12 electrolysis cells of 19 mm in length. The length was determined so as to keep current density uniform in the electrolysis region. These cells were connected in series. The electrolyte layer made of YSZ (containing 8 mol% of Y_2O_3) was sandwiched between the porous cathode and anode layers: Ni cermet (Ni + YSZ) for the cathode and $LaCoO_3$ for the anode. At the ends of the electrolysis tube, platinum (Pt) wires working as electric leads were welded on copper coating layers. The copper coating layers were connected to thin layers of electric conductors. The outside of the electrolysis tube, except the cells and the copper coatings, was coated by thin alumina (Al_2O_3) gas-tight layers and YSZ protective layers. These layers were formed on a porous calcia-stabilized ZrO_2 (CSZ) tube (support tube) of 22 mm in outer diameter, 3 mm in thickness, and around 38% of porosity, by using the plasma spraying. The thickness of each layer was in the range 0.1–0.25 mm. The total length of the electrolysis tube was 710 mm.

Experiments were carried out at the electrolysis temperatures of 850, 900, and 950°C under the pressure of 0.11 MPa (abs). Steam (H_2O) was supplied with argon carrier gas at a

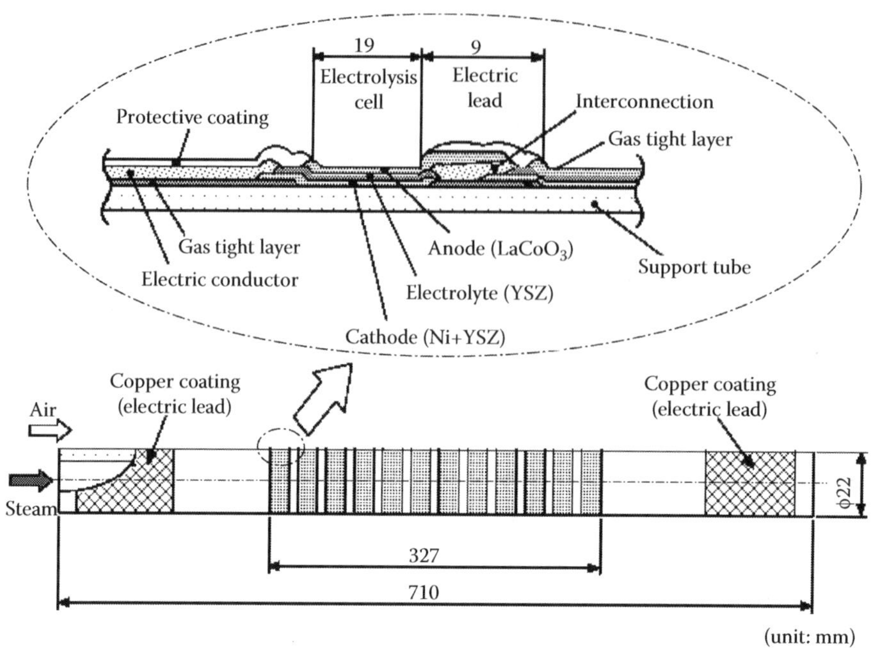

FIGURE 4.3
Structural drawing of electrolysis tube with 12 cells (tubular type cell).

rate of about 0.32 g/min. The heating and cooling rates were set to be less than 20°C/h during start up and shut down of the test section so as to avoid generating high interfacial stress from differential-thermal expansion among the electrolysis tube components otherwise occurring under rapid heating or cooling. Before applying electrolysis voltage, the cathode material, Ni cermet, was chemically reduced with hydrogen mixed with argon carrier gas at least for 1 h.

Figure 4.4 shows the relation between the H_2 production rate and the applied voltage. H_2 production rate increased with the applied voltage and the electrolysis temperature.

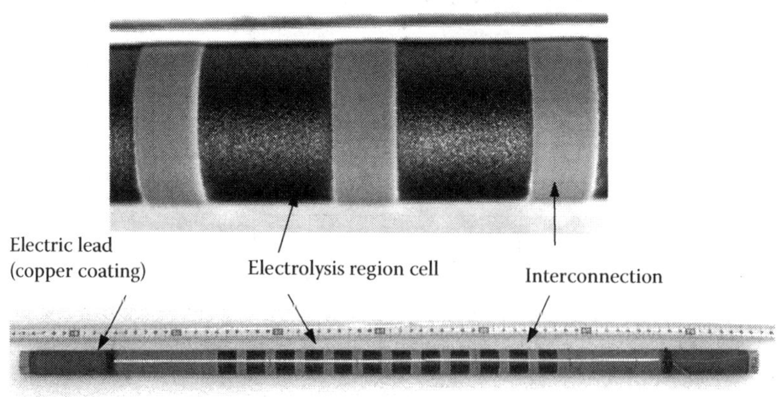

PHOTO 4.1
Outer view of electrolysis tube with 12 cells.

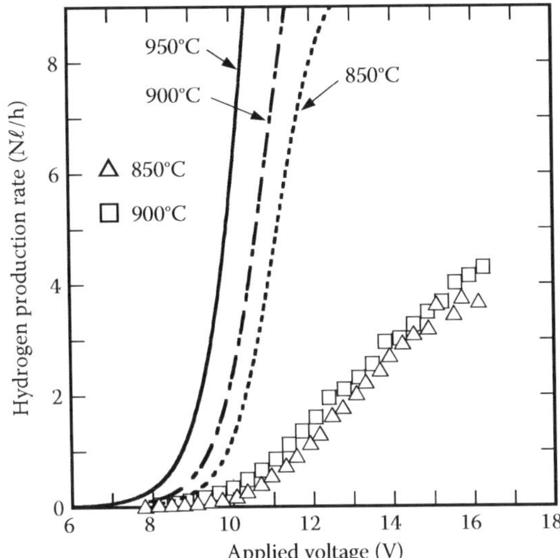

FIGURE 4.4
Applied voltage versus hydrogen production rate.

The maximum production rate was 3.8 NL/h at 15.7 V, 1.42 A, and 850°C, 4.3 NL/h at 16.3 V, 1.38 A, and 900°C, and 6.9 NL/h at 15.6 V, 1.72 A, and 950°C, respectively. These H_2 production rates were quoted to the reference condition of 273 K (0°C) and 0.1 MPa (1 bar) as mentioned above. Lines in the figure express the hydrogen production rate calculated with Equation 4.5. Differences between experimental values of the applied voltage and the corresponding theoretical lines indicate the extent of overvoltage, η. As seen in the figure, the overvoltage increases with the hydrogen production rate. The Faraday efficiency at 950°C ranged from 0.74 to 0.85 V in the applied power of more than 6 W (corresponding to the current density more than 45 mA/cm^2), which was derived from Equation 4.6, was ranged 0.27–0.56. The experimental result obtained in the HOTELLY program, 6.78 NL/h of the hydrogen production rate under the condition of the electrolysis temperature of 997°C and the applied power of 21.7 W [2], showed the Faraday efficiency of 0.75. Then the open-circuit voltage, E–η, was 1.006 V at 997°C. This low Faraday efficiency could have been caused mainly by lower oxygen ion conductivity than expected for 997°C. Also, other possible causes included rather high ohmic losses at interconnections and electric lead layers, and an increase of the concentration overvoltage, since steam could not be supplied effectively to the cathode through the support tube and because the hydrogen produced at the cathode was not fully discharged into the main flow through the support tube.

In these experiments, the steam-conversion ratio (steam-utilization ratio) was less than 40%. The lower than expected steam-conversion ratio might be caused by low-permeation rate of steam through the support tube to the cathode, even though the support tube had rather high porosity of around 38%. If steam could reach the cathode sufficiently, the hydrogen production rate would have increased while the concentration overvoltage would have decreased. Turbulent flow generated by twist tapes to be inserted inside of the tube might be effective. Otherwise, from the viewpoint of steam supplying performance to the cathode, the self-supporting cell structure would be better in HTE than in the multicell structure supported by the porous ceramic tube.

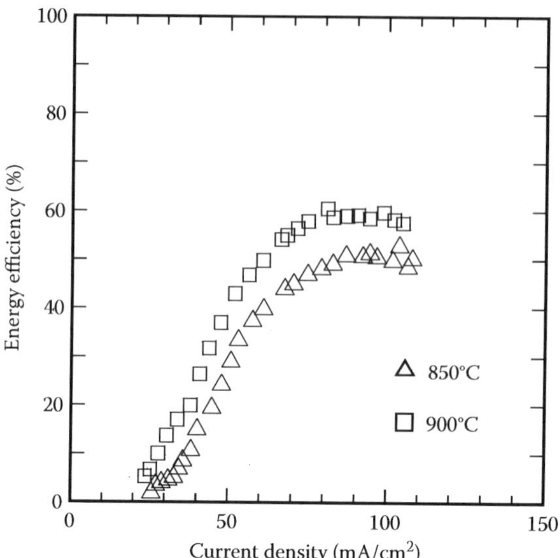

FIGURE 4.5
Current density versus energy efficiency.

Figure 4.5 shows the relation between the current density and the energy efficiency. The energy efficiency is a quotient of combustion heat of generated hydrogen and applied power, which could be derived from Equation 4.7 using the low-heating value. As seen in the figure, the energy efficiency has a peak in the range of current density from 80 to 100 mA/cm^2. The maximum energy efficiency below the electrolysis temperature of 950°C was 86% at a current density of 98 mA/cm^2. This would be mainly due to rather low-oxygen ion conductivity at 950°C and shortage of steam supply at the interface between electrolysis and electrode as mentioned above as the reason of increase of the concentration overvoltage.

When connected to the HTGR, HTE operating temperature will be up to 900°C. It is necessary to use other solid-oxide electrolytes working below 900°C: and having considerably higher oxygen ion conductivity than that of YSZ. One candidate is ytterbia (Yb_2O_3)-stabilized ZrO_2 (YbSZ). It is also necessary to enhance hydraulics of steam supply to the interface between electrolysis and electrode, which might be realized by decreasing the fuel-electrode layer thickness and by increasing its porosity.

After the experiments, it was observed that large parts of the anode (air electrode) layers had been separated from the electrolyte layers, though the electrolysis tube had served for one thermal cycle only. This indicates that the durability of the cell against thermal cycles, especially the durability of the anode layer, is one of the key issues of HTE. It could be met by raising the bonding force and by keeping high residual compression stress of the anode layer to the electrolyte layer against large differential thermal expansion of these layers in consideration of relevant service conditions.

4.3.2 Experimental Results with Planar Cell

Since the tubular cells shown in Figure 4.3 are rather bulky and costly, preliminary HTE experiments were conducted with a planar cell, especially a self-supporting planar cell.

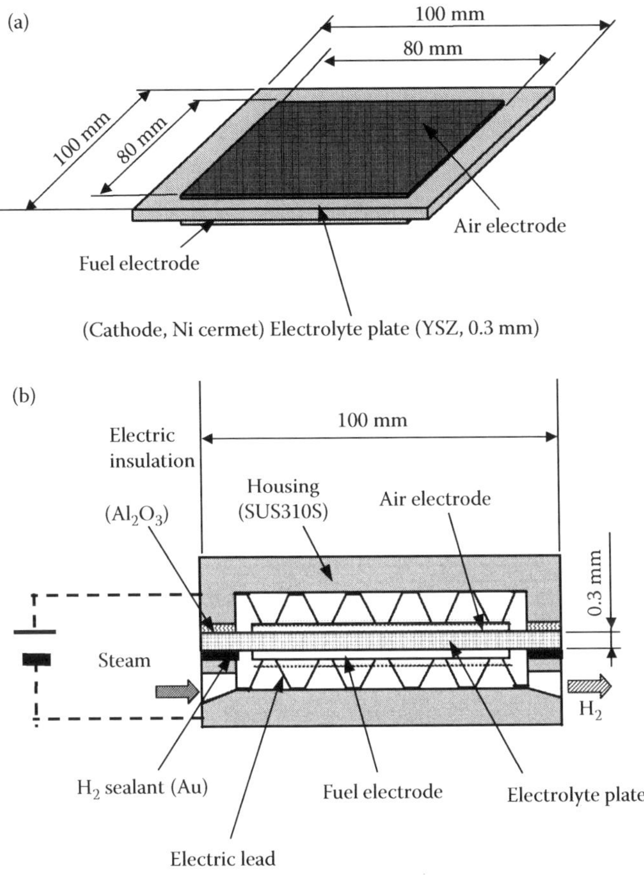

FIGURE 4.6
Schematic drawing of (a) self-supporting planar cell and (b) electrolysis chamber.

Figure 4.6 shows a schematic drawing of the self-supporting planar cell and an electrolysis chamber [6,7]. The cell consisted of the electrolyte plate of YSZ and porous electrodes. The YSZ plate was a 100 mm^2 plate with a thickness of 0.3 mm, and electrodes were baked on the area of 80×80 mm^2 with a thickness of less than 0.03 mm. Material of the cathode was Ni cermet (Ni + YSZ), and that of the anode strontium-doped $LaMnO_3$. The cell was sandwiched with metal housings made of SUS-310S. Each housing had a metal rod for the electric lead, an inlet and an outlet piping for H_2 and H_2O, and Pt sheet of 0.1 mm thick was welded on inner surface of each housing opposite to the electrodes of the cell. A corrugated electrical lead plate (0.1 mm thick) was installed in each electrode compartment to work as the electric lead. Then, the flat Pt mesh sheet was inserted between the corrugated electrical lead plate and the electrode to realize a uniform current density. DC power was passed through the metal rods, housings, Pt sheets, and corrugated electrical lead plates to the cell. Compression seals were installed at the edge of the cell plate; the compression load was up to 20 kg. An alumina sheet of 0.3 mm thick was then inserted between the cell plate and the housing in the anode side (air supplying side) to prevent current leakage through

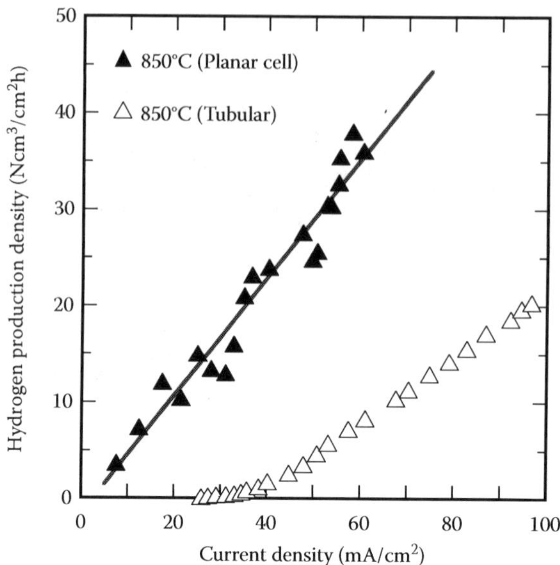

FIGURE 4.7
Relation between current and hydrogen production densities.

the cell-plate edge, and gold (Au) sheet of 0.1 mm thick was inserted in the cathode side (steam supplying side) to prevent any H_2 leak.

Figure 4.7 shows the relation between current and H_2 production densities obtained at 850°C of electrolysis temperature. The maximum H_2 production density was 38 N cm³/cm² h, which was higher than that of the electrolysis tube obtained at 950°C. The maximum H_2 production rate was 2.4 NL/h at the applied power of 10 W: the applied voltage and current were 2.68 V and 3.72 A, respectively. The open-circuit voltage then was 0.847 V. Hence, the Faraday efficiency and the energy efficiencies were 0.5 and around 0.73, respectively, which were almost the same values as those of the electrolysis tube obtained at 950°C.

After the experiment, a crack crossing the cell was observed, which could have been caused by the thermal expansion due to the Joule heating. In designing a module consisting of stacked planar cells, special relief structures against the thermal expansion should be installed, which are also seem to be necessary for providing both seal and electric insulation performances. Recently, Herring et al., have been improving the self-supporting planar cell by using update materials such as NbB_2–ScSZ, Strontium-doped $LaCrNiO_3$, and so on, and its stack structure based on numerical analysis [9].

4.4 Experimental and Analytical Studies in Toshiba

4.4.1 Experimental Single Electrolysis Cell

Toshiba selected tubular cells because they have high leak-tightness. Figure 4.8 shows the structure of the tubular-electrolysis cell. The manufactured electrolysis cell has a hydrogen-electrode-supported configuration, with YSZ electrolyte, nickel-YSZ cermet

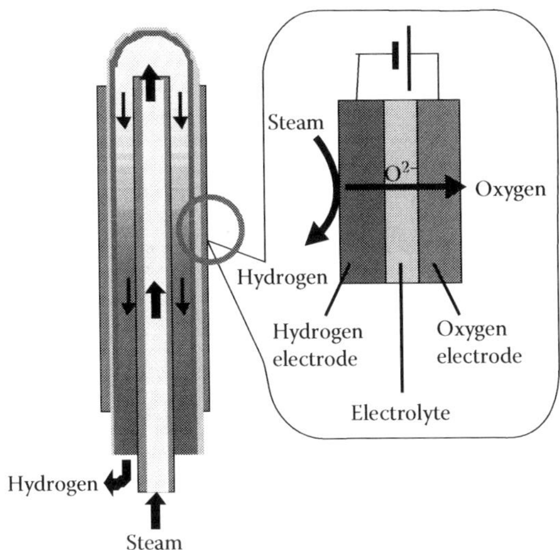

FIGURE 4.8
Structure of tubular electrolysis cell.

steam/hydrogen electrodes, that is, a cathode and lanthanum, strontium and cobalt mixed-oxide oxygen electrodes, that is, an anode. Steam/hydrogen mixed gas is supplied to the hydrogen electrodes via a steam induction pipe provided inside the electrolysis cell, and oxygen/nitrogen mixed gas is supplied to the oxygen electrodes. The supplied steam is decomposed into hydrogen and oxygen ion on the cathode, and then the oxygen ion is conducted via the electrolyte to the anode, where it releases an electron to become oxygen.

Figure 4.9 shows a prototype single tubular cell for constructed for laboratory-scale tests. The outer diameter of the cell was 12 mm, the electrolyte was 13 μm thick, the oxygen electrode was 25 μm thick, and the hydrogen electrode was 7 μm thick.

Figure 4.10 shows the current–voltage characteristic of a 15 cm^2 cell. It was estimated that a current density of more than 0.45 A/cm^2 was obtained at a thermal–neutral voltage of 1.3 V at 800°C, since the current density was 0.45 A/cm^2 when the operating voltage was 1.23 V. The measured open-cell potential (E_{OCV}), which is the operating voltage for a current density of zero, of the 15 cm^2 cell was 0.94 V at 800°C and 0.91 V at 900°C. The E_{OCV} can also be predicted from the test conditions using Equation 4.5. The theoretical E_{OCV} values predicted from Equation 4.5 are 0.941 V at 800°C and 0.908 V at 900°C. The measured values were mostly in agreement with the theoretical values predicted from Equation 4.5. Therefore, it was judged that the cross leakage between the hydrogen side and the oxygen side did not occur. It was thus confirmed that the electrolysis cell remained leak-tight [11].

4.4.2 Experimental Multielectrolysis-Cell Assembly

An assembly unit consisting of 15 cells, whose hydrogen production rate design value is 100 NL/h, was developed for a first-stage demonstration of large-scale H_2 production. Figure 4.11 shows the internal structure of the unit in which 15 tubular electrolysis cells, each 12 mm in outer diameter and 75 cm^2 in active area were installed. Each cell was set on a pedestal with glass interposed between them to serve as a seal between the hydrogen side and the oxygen side, as well as for electrical isolation. These were installed in the

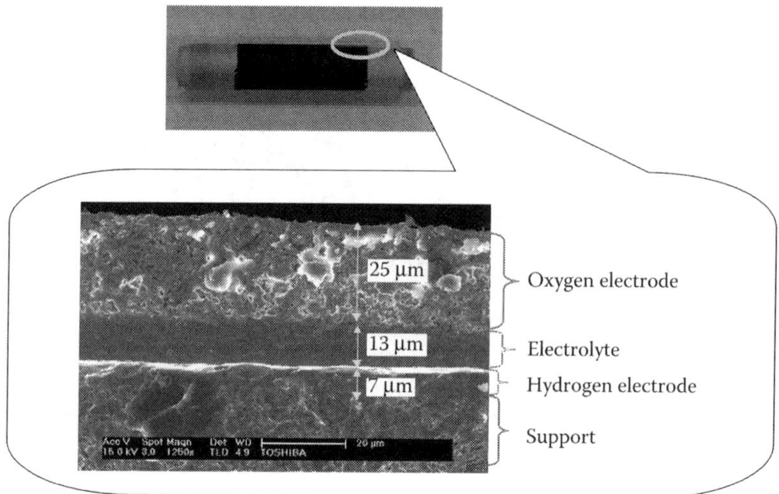

FIGURE 4.9
Prototype tubular single cell.

casing when tested. Fifteen cells were installed in the casing and divided into three blocks of five each for supplying electricity individually to each block, and all cells were connected in parallel. Electricity was supplied to the hydrogen electrode from both ends of the cell and was supplied to the oxygen electrode via silver expanded metal which covered the oxygen electrode. The electricity supply lines to the oxygen electrode were attached at three locations on the silver expanded metal: the upper part, the central part, and the lower part. The steam–hydrogen gas mixture entered the inlet manifold at the bottom, (in the left side view photograph of Figure 4.11), flowed into the steam induction pipe provided inside the electrolysis cell, then exited through the outlet hydrogen manifold, also at the bottom (in the left side view photograph of Figure 4.11). The air flow entered the

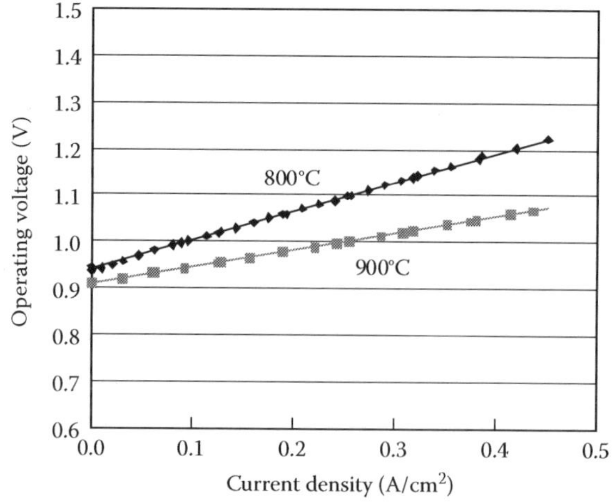

FIGURE 4.10
Current–voltage characteristics of 15 cm^2 cell.

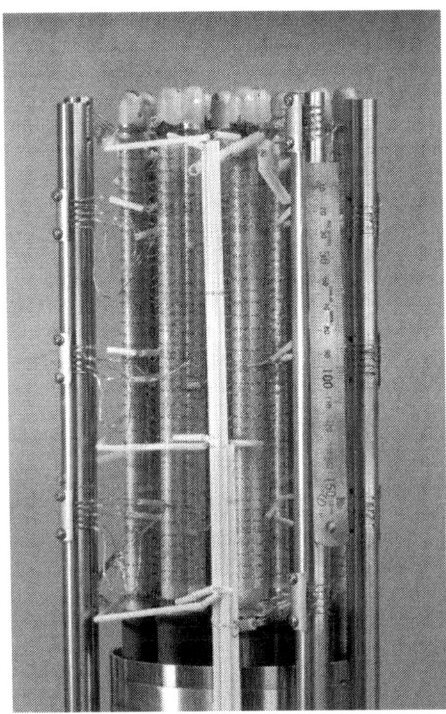

FIGURE 4.11
Internal structures of 15 cells assembly unit.

space between the internal structure shown in Figure 4.11 and the casing, and exited from another pipe installed in the space. Unit operating voltages were measured using wires attached to the hydrogen and the oxygen electrodes of one of the five cells in each of the three blocks.

Figure 4.12 shows hydrogen production test results. It was confirmed that the observed hydrogen production rate was mostly in agreement with the theoretical value, which was predicted from the current density. The hydrogen production rate of the unit was 130 NL/h at an applied voltage of 1.4 V. From the results, it was confirmed that the unit, which consisted of cells, seals, electrical isolation, electricity supplies and so on, worked well [11].

4.4.3 Simulation Method for Thermo-Electrochemical Coupled Phenomena

An analytical model to predict the thermal and electrochemical characteristics of the electrolysis cell was constructed. An electrode reaction model and an electrical conduction model were added to a computational fluid dynamics (CFD) model. The electrode reaction is described by the empirical formula of the SOFC that was the reverse reaction of the high-temperature steam electrolysis. For general incompressible and compressible fluids, the mass, momentum, enthalpy, and species conservation equations are solved using commercial CFD code, STAR-CD®. With this approach, the profiles of the temperature and the current density in the electrolysis cell could be analyzed. There was no extreme distribution in these physical values as a result of the calculation, and they were comparatively uniform. Figure 4.13 shows the temperature profiles at each applied voltage.

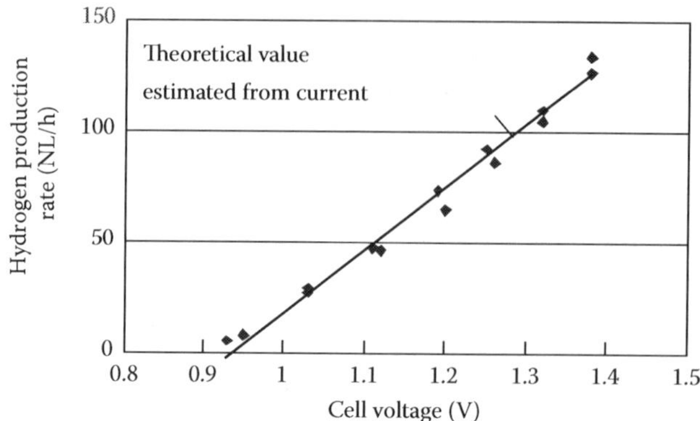

FIGURE 4.12
Hydrogen production test result of 15–75 cm^2 cells assembly unit.

In addition, the computational model was enhanced to a multicell electroanalysis model for predicting the performance of the multielectrolysis-cell assembly unit. Figure 4.14 shows a comparison of the hydrogen production rates between experiment and simulation. The solid line shows the linear approximation of the experimental data. The simulation results give a hydrogen production rate of about 101 NL/h at a voltage of 1.28 V. This is in very good agreement with the experimental results, and the error is smaller than 5% [12].

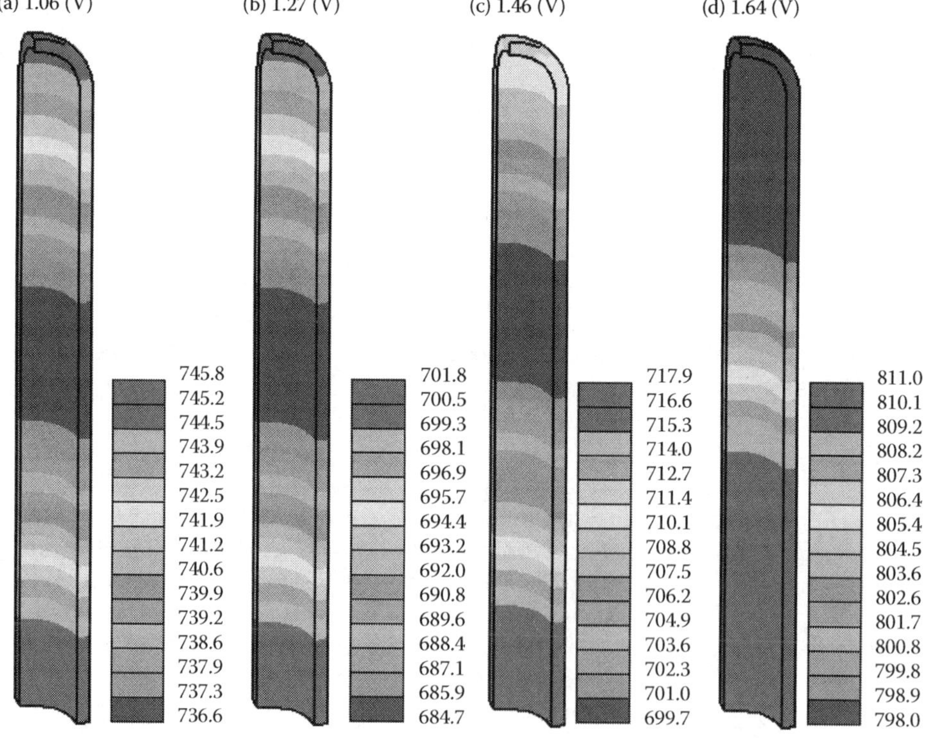

FIGURE 4.13
The temperature profiles (°C) at each applied voltage.

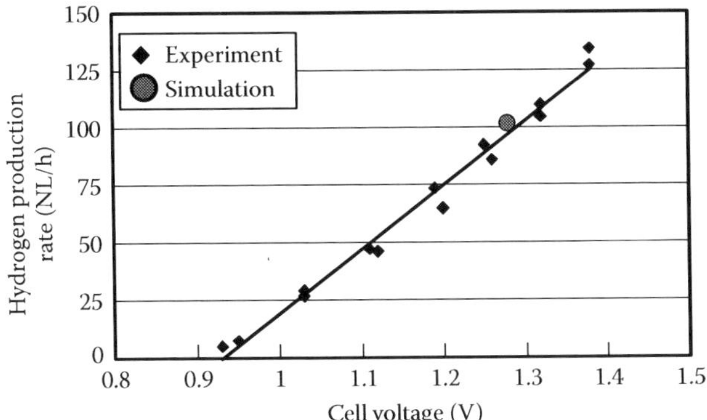

FIGURE 4.14
Comparison of hydrogen production rate between experiment and simulation.

4.4.4 Design Analysis of the HTE Plant System

Figure 4.15 shows a conceptual design combining an HTE plant with a nuclear reactor. The required thermal energy and electrical energy in the HTE reaction is supplied from the nuclear plant. A high-temperature gas cooled reactor was selected as the nuclear reactor in Figure 4.15. Make-up water to produce hydrogen and oxygen is supplied to an H_2 separator where hydrogen product gas is also supplied together with undecomposed water and mixed with the make-up water. Mixed water goes down to a water circulator, and hydrogen gas goes up to an H_2 circulator. Mixed water is supplied to an economizer and

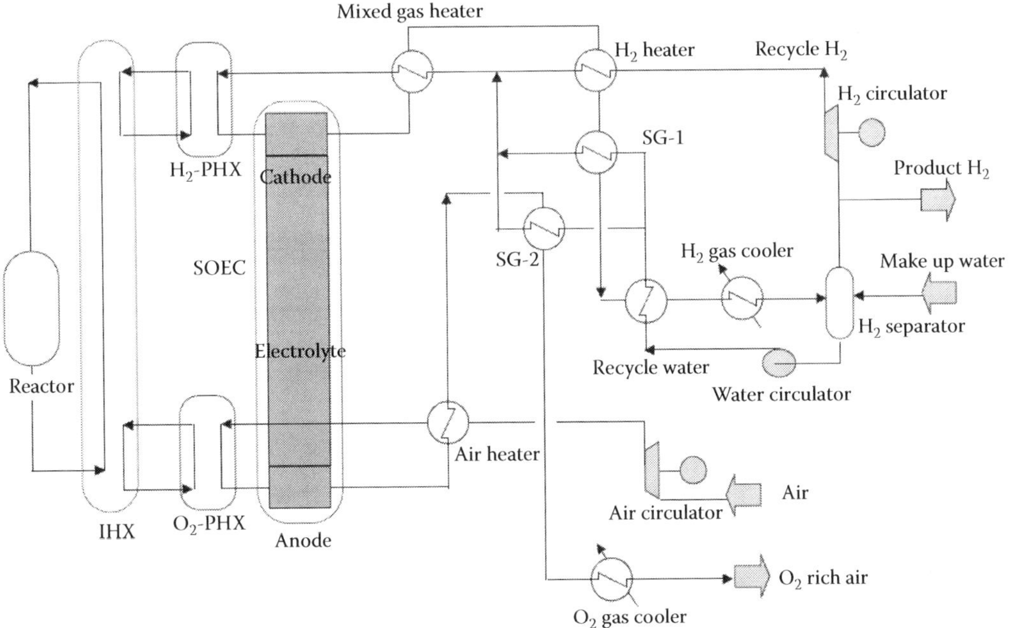

FIGURE 4.15
Plant system coupling with high-temperature gas-cooled reactor.

TABLE 4.2
Standard Design Conditions

Parameter	Value
Electrical generation efficiency	53% [13]
Operating pressure	0.4 MPa
Operating temperature	900°C
Steam utilization	90%
Recycle ratio (recycle H_2/product H_2)	1
Air to product O_2 ratio	1
Heat loss	0
Pressure drop	0

steam generator (SG), where the heat of the hydrogen or oxygen product gas warms up the mixed water to produce steam. Hydrogen from the H_2 separator is removed as hydrogen product before the H_2 circulator, and the rest of the hydrogen is recycled in order to maintain the reductive conditions for the cathode. The heat of the hydrogen product gas warms up the recycled hydrogen in an H_2 heater. Recycled hydrogen from the H_2 heater is mixed with steam from the SG before entering a mixed gas heater. In this heater, mixed gas is warmed up to the H_2 process heat exchanger (PHX) entrance temperature by the hydrogen product gas. In the H_2 PHX, mixed gas is warmed up to the HTE reaction temperature in a solid oxide electrolysis cell (SOEC). In the SOEC, water is decomposed to hydrogen and oxygen. Hydrogen product gas is cooled down before entering an H_2 separator by heating make-up water and recycled H_2. At the anode side, air is used as a carrier gas of the oxygen product. Before entering the SOEC, the air is heated through an air heater and an O_2 PHX, the same as at the cathode side. After exiting the SOEC, the oxygen product gas is cooled down by adding its own heat to an air heater and steam generator (SG-2). Finally, product gas is taken out as oxygen-rich air.

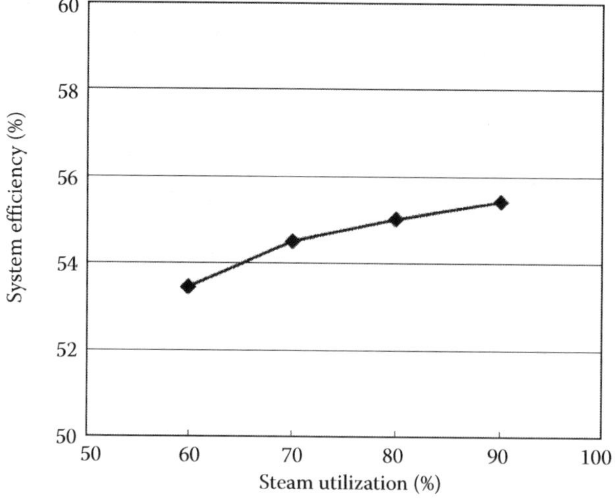

FIGURE 4.16
System efficiency as a function of steam utilization.

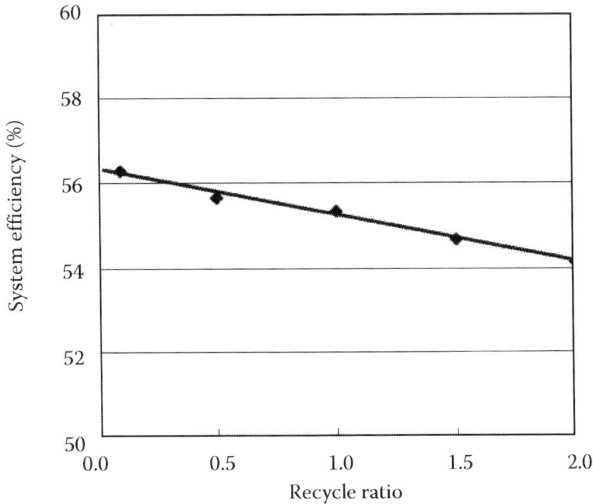

FIGURE 4.17
System efficiency as a function of recycle ratio.

In order to evaluate the system efficiency, we used the following equation:

$$\eta = HHV/(W/\phi + Q) \tag{4.8}$$

where η is the system efficiency, and HHV is the higher heating value of hydrogen, W is electrical energy input to produce hydrogen equal to the HHV amount, ϕ is electrical generation efficiency, which is 53% in a high-temperature gas cooled reactor [13], and Q is thermal energy input to produce hydrogen equal to the HHV amount.

The standard design conditions are shown in Table 4.2. In this evaluation, the SOEC is operated at the thermoneutral point where the inlet gas temperature of the SOEC is considered to be equal to the outlet one. The driving power for the air and H_2 circulators is neglected. Figure 4.16 shows the system efficiency as a function of the steam utilization in the SOEC. The system efficiency increases from 53.5 to 55.4% as the steam utilization efficiency increases from 60 to 90%. This is because the increase of the steam utilization efficiency by decreasing the steam flow into the SOEC cathode results in a decrease of the thermal input power in the PHXs. Figure 4.17 shows the system efficiency as a function of the hydrogen recycle ratio from the SOEC cathode outlet. The system efficiency decreases linearly from 56.3% to 54.3% as the hydrogen-recycle ratio from the SOEC cathode increases from 0.1 to 2. This is because the increase of the hydrogen recycle ratio by increasing the hydrogen flow into the SOEC cathode results in an increase of the thermal-input power in the PHXs [14].

References

1. Doenitz, W. and Schmidberger, R., Concepts and design for descaling up high-temperature water vapor electrolysis, *Int. J. Hydrogen Energy*, 7, 321, 1982.
2. Doenitz, W. and Eedle, E., High-temperature electrolysis of water vapor status of development and perspective for application, *Int. J. Hydrogen Energy*, 10, 291, 1985.

3. Doenitz, W. et al., Electrochemical high temperature technology for hydrogen production or direct electricity generation, *Int. J. Hydrogen Energy*, 13, 283, 1988.
4. Doenitz, W. et al., Recent Advances in the development of high-temperature electrolysis technology in Germany, in *Proc. 7th World Hydrogen Energy Conf.*, Moscow, 1988, 65.
5. Maskalick, N.J., High temperature electrolysis cell performance characterization, *Int. J. Hydrogen Energy*, 11, 563, 1986.
6. Hino, R. et al., Present status of R&D on hydrogen production by high temperature electrolysis of steam, *JAERI-Research* 95–057, 1995.
7. Hino, R. et al., R&D on hydrogen production by high temperature electrolysis of steam, *Nucl. Eng. Des.*, 233, 363, 2004.
8. Matsunaga, K. et al., Hydrogen production system with high temperature electrolysis for nuclear power plant, *Paper 6282 in Proc. ICAPP'06*, Reno, USA, June 4–8, 2006.
9. Herring, S. et al., Laboratory-scale high temperature electrolysis system, 2006 Annual Merit Review Proc., Hydrogen production and delivery, D. Nuclear Energy Initiative, http://www.hydrogen.energy.gov/annual_review06_delivery.html.
10. Nagata, S. et al., Fabrication of high temperature solid electrolyte fuel cell and power generation test (in Japanese), *J. High Temperature Soc.*, 7, 217, 1981.
11. Yamada, K. et al., High temperature electrolysis for hydrogen production using tubular electrolysis cell assembly unit, *Proceedings of the 2006 AIChE Annual Meeting*, San Francisco, USA, November 12–17, 2006.
12. Hoashi, E. et al., Development of simulation method for thermo-fluid-electrochemical coupled phenomena related to hydrogen production technology using nuclear energy, *Proceedings of the ANS 2007 Annual Meeting*; Embedded Topical Meeting, Safety and Technology of Nuclear Hydrogen Production, Control, and Management, Boston, USA, June 24–28, 2007.
13. LaBar, M. P. et al., The gas Turbine—Modular Helium Reactor, *Nuclear News*, October 2003.
14. Ogawa, T. et al., Hydrogen production by high temperature electrolysis with nuclear reactor. *Proceedings of the GLOBAL 2007*, Boise, USA, September 9–13, 2007.

5

Thermochemical Decomposition of Water

Seiji Kasahara and Kaoru Onuki

CONTENTS

5.1 Introduction

5.1.1 Principle of Thermochemical Processes

As described in Section 3.2, not only heat but also work input is required to obtain hydrogen when the heat of lower than T_d (4310 K) is input to the direct decomposition of water at 1 atm. However, incorporation of chemical reactions to the decomposition process can lower the temperature threshold of the heat to be required. Figures 5.1 through 5.3 explain the concept. The reactions in Figure 5.1 satisfy these enthalpy/entropy balances.

$$\Delta H_{WS} = \sum_i \Delta H_i \tag{5.1}$$

$$\Delta S_{WS} = \sum_i \Delta S_i \tag{5.2}$$

$$\Delta G_i = \Delta H_i - T_i \Delta S_i \tag{5.3}$$

Reaction i proceeds without work input at the temperature of $\Delta G_i \leq 0$. When all the reactions are operated at T_i (= $\Delta H_i/\Delta S_i$), ΔG_i of all the reactions are 0. If ΔH_i is plus and ΔS_i is minus, such T_i is impossible (case (1) in Figure 5.2). This is also the same in the case of ΔH_i is minus and ΔS_i is plus (case (4)). And when both ΔH_i and ΔS_i are plus in all the reactions, T_i is higher than T_d at least one reaction shown as line (1) in the upper diagram of Figure 5.3. Therefore, ΔH_i must be plus and ΔS_i must be minus in at least one reaction when all T_i are lower than T_d, as line (1) of lower diagram of Figure 5.3. Figure 5.1 is an example of a 2-reaction process. The two reactions proceed by heat input of ΔH_1 at T_1 and heat output of ΔH_2 at T_2 without any work input/output. Sensible heat to change temperature of reactants is not taken into consideration in the discussion above. Generally, heat input and heat output can be considered to cancel each other if ideal heat exchange is applied.

This principle is not necessarily kept in actual processes. Reactions of cases (1) and (4) in Figure 5.2 are used when absolute value of ΔG_i is small (e.g., 10 kJ/mol-H_2) like HI decomposition reaction described as Equation 5.4 in iodine–sulfur (IS) process. Though IS

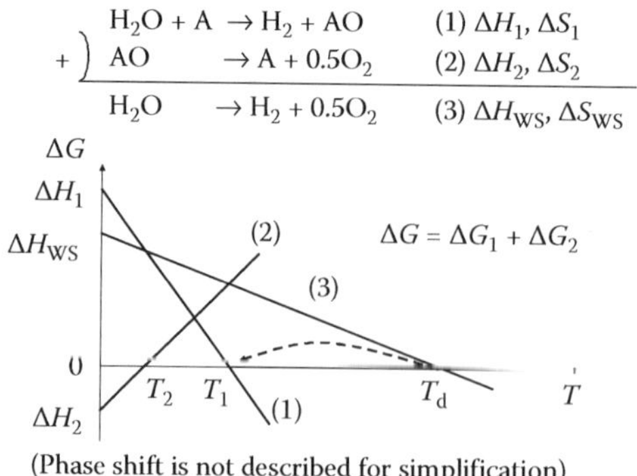

FIGURE 5.1
Concept of thermochemical water splitting.

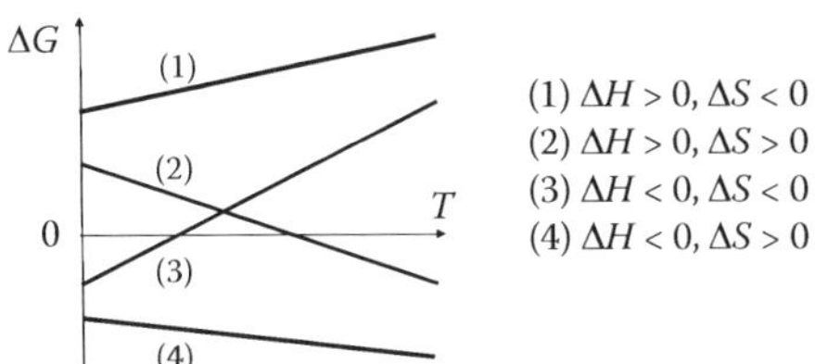

FIGURE 5.2
Relation of reaction enthalpy, entropy and Gibbs energy.

process is called by other names such as sulfur–iodine (S–I) process, the name IS, is used in this chapter.

$$2HI \rightarrow H_2 + I_2 \tag{5.4}$$

Sometimes different temperatures from T_i is selected considering reaction rate or stability of component material even in the cases (2) and (3) in Figure 5.2. More heat than ideal is required in these conditions. Hybrid cycles allow work input and of course do not obey the principle.

5.1.2 Principle of Hybrid Processes

Hybrid processes are a kind of thermochemical cycles that utilize at least one electrochemical reaction. Many hybrid cycles have been proposed in the search of proper thermochemical cycles explained in Section 5.2. Westinghouse (Ispra Mark 11, Hybrid Sulfur (HyS)), Ispra Mark 13, Cu–Cl, Hybrid Ca–Br, and Hybrid hydrogen process in lower temperature (HHLT) processes are examples of such cycles. The reactions comprising these cycles are shown in Table 5.1. And details of these processes are explained in Section 5.3. The Westinghouse process is introduced in detail in Chapter 18 because this

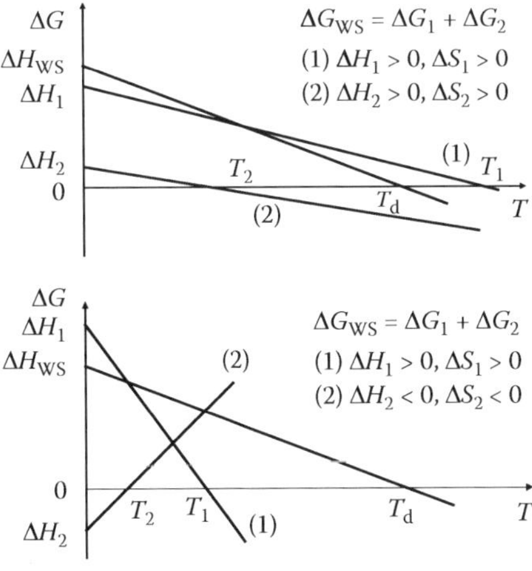

FIGURE 5.3
G–T diagram of thermochemical water splitting.

TABLE 5.1
List of Thermochemical Cycles Mentioned in this Chapter

Name	Reaction	Typical Temperature	Thermochemical (T) or Electrochemical (E)	Section
HgO/Hg	$HgO(g) \rightarrow Hg(g) + 0.5O_2(g)$		T	5.2.1
	$Hg(g) + H_2O(g) \rightarrow HgO(g) + H_2(g)$	360°C	T	
Ta–Cl	$H_2O(g) + Cl_2(g) \rightarrow 2HCl(g) + 0.5O_2(g)$	1000 K	T	5.2.1
	$2TaCl_2(s) + 2HCl(g) \rightarrow 2TaCl_3(s) + H_2(g)$	298 K	T	
	$2TaCl_3(s) \rightarrow 2TaCl_2(s) + Cl_2(g)$	1366 K	T	
V–Cl	$H_2O(g) + Cl_2(g) \rightarrow 2HCl(g) + 0.5O_2(g)$	727°C	T	5.2.1
	$2VCl_2(s) + 2HCl(g) \rightarrow 2VCl_3(s) + H_2(g)$	25°C	T	
	$4VCl_3(s) \rightarrow 2VCl_4(g) + 2VCl_2(s)$	727°C	T	
	$2VCl_4(l) \rightarrow 2VCl_3(s) + Cl_2(g)$	25°C	T	
Mg–S–I	$SO_2(aq) + I_2(aq) + 2H_2O(l) \rightarrow 2HI(aq) + H_2SO_4(aq)$	70°C	T	5.2.2.5
	$2MgO(s) + H_2SO_4(aq) + 2HI(aq) \rightarrow MgSO_4(aq) + MgI_2(aq) + 2H_2O(l)$	70°C	T	
	$MgI_2 \cdot H_2O(aq) \rightarrow MgO(s) + 2HI(g) + nH_2O(g)$	400°C	T	
	$MgSO_4(s) \rightarrow MgO(s) + SO_2(g) + 0.5O_2(g)$	995°C	T	
	$2HI(g) \rightarrow H_2(g) + I_2(g)$	995°C	T	
NIS	$SO_2(aq) + I_2(aq) + 2H_2O(l) \rightarrow 2HI(aq) + H_2SO_4(aq)$	40°C	T	5.2.2.6
	$2HI(aq) + H_2SO_4(aq) + 2Ni(s) \rightarrow NiI_2(aq) + NiSO_4(aq) + 2H_2(g)$	60°C	T	
	$NiI_2(s) \rightarrow Ni(s) + I_2(g)$	700°C	T	
	$NiSO_4(s) \rightarrow NiO(s) + SO_2(g) + 0.5O_2(g)$	880°C	T	
	$NiO(s) + H_2(g) \rightarrow Ni(s) + H_2O(g)$	600°C	T	
CIS	$SO_2(aq) + I_2(aq) + 2H_2O(l) \rightarrow 2HI(aq) + H_2SO_4(aq)$	60°C	T	5.2.2.6
	$HI(aq) + CH_3OH(aq) \rightarrow CH_3I(g) + H_2O(l)$	70°C	T	
	$HI(g) + CH_3I(g) \rightarrow CH_4(g) + I_2(g)$	400°C	T	
	$CH_4(g) + H_2O(g) \rightarrow CO(g) + 3H_2(g)$	730°C	T	
	$CO(g) + 2H_2(g) \rightarrow CH_3OH(g)$	330°C	T	
	$H_2SO_4(g) \rightarrow H_2O(g) + SO_2(g) + 0.5O_2(g)$	840°C	T	
Fe_3O_4/FeO	$Fe_3O_4(s) \rightarrow 3FeO(s) + 0.5O_2(g)$	2200°C	T	5.2.3.4
	$3FeO(s) + H_2O(g) \rightarrow Fe_3O_4(s) + H_2(g)$	400°C	T	

ZnO/Zn	$ZnO(s) \rightarrow Zn(s) + 0.5O_2(g)$	2000°C	T	5.2.3.4
	$Zn(s) + H_2O(g) \rightarrow ZnO(s) + H_2(g)$	1100°C	T	
UT-3	$CaBr_2(s) + H_2O(g) \rightarrow CaO(s) + 2HBr(g)$	700–750°C	T	5.3.1.1
	$3FeBr_2(s) + 4H_2O(g) \rightarrow Fe_3O_4(s) + 6HBr(g) + H_2(g)$	550–600°C	T	
	$Fe_3O_4(s) + 8HBr(g) \rightarrow 3FeBr_2(s) + 4H_2O(g) + Br_2(g)$	200–300°C	T	
	$CaO(s) + Br_2(g) \rightarrow CaBr_2(s) + 0.5O_2(g)$	500–600°C	T	
Hybrid Ca–Br	$CaBr_2(s) + H_2O(g) \rightarrow CaO(s) + 2HBr(g)$	750°C	T	5.3.1.2
	$CaO(s) + Br_2(g) \rightarrow CaBr_2(s) + 0.5O_2(g)$	580°C	T	
	$2HBr(g) \rightarrow H_2(g) + Br_2(g)$	100°C	E	
Fe–Cl	$3FeCl_2(s) + 3H_2(g) \rightarrow 6HCl(g) + 3Fe(s)$	1150 K	T	5.3.2.1
(RWTH Aachen)	$3Fe(s) + 4H_2O(g) \rightarrow Fe_3O_4(s)\ 4H_2(s)$	600 K	T	
	$H_2O(g) + Cl_2(g) \rightarrow 2HCl(g) + 0.5O_2(g)$	1300 K	T	
	$Fe_3O_4(s) + 8HCl(g) + 0.5Cl_2(g) \rightarrow 1.5Fe_2Cl_6(s) + 4H_2O(g)$	700 K	T	
	$1.5Fe_2Cl_6(s) \rightarrow 3FeCl_2(g) + 1.5Cl_2(g)$	600 K	T	
Cu–Cl (five step)	$2Cu(s) + 2HCl(g) \rightarrow 2CuCl(l) + H_2(g)$	450°C	T	5.3.2.2
	$4CuCl(s) \rightarrow 4CuCl(aq) \rightarrow 2CuCl_2(aq) + 2Cu(s)$	30°C	E	
	$2CuCl_2(aq) \rightarrow 2CuCl_2(s)$	100°C	T	
	$2CuCl_2(s) + H_2O(g) \rightarrow CuO \cdot CuCl_2(s) + 2HCl(g)$	400°C	T	
	$CuO \cdot CuCl_2(s) \rightarrow 2CuCl(l) + 0.5O_2(g)$	500°C	T	
Cu–Cl (four step)	$2CuCl(aq) + 2HCl(aq) \rightarrow 2CuCl_2(aq) + H_2(g)$	70°C	E	5.3.2.2
	$2CuCl_2(aq) \rightarrow 2CuCl_2(s)$	70°C	T	
	$2CuCl_2(s) + H_2O(g) \rightarrow CuO \cdot CuCl_2(s) + 2HCl(g)$	400°C	T	
	$CuO \cdot CuCl_2(s) \rightarrow 2CuCl(l) + 0.5O_2(g)$	500°C	T	
HHLT	$SO_2(aq) + 2H_2O(l) \rightarrow H_2SO_4(aq) + H_2(g)$	<100°C	E	5.3.3
	$H_2SO_4(l) \rightarrow H_2O(g) + SO_3(g)$	400°C	T	
	$SO_3(g) \rightarrow SO_2(g) + 0.5O_2(g)$	>500°C	E	
Mg–I	$1.2MgO(s) + 1.2I_2(s) \rightarrow 0.2Mg(IO_3)_2(s) + MgI_2(aq)$	150°C	T	5.3.4
	$0.2Mg(IO_3)_2(s) \rightarrow 0.2MgO(s) + 0.2I_2(g) + 0.5O_2(g)$	600°C	T	
	$MgI_2 \cdot 6H_2O(s) \rightarrow MgO(s) + 2HI(g) + 5H_2O(g)$	400°C	T	
	$2HI(g) \rightarrow H_2(g) + I_2(g)$	400°C	T	

process is most actively researched. The division of pure thermochemical cycle and hybrid cycle is not clear. For example, application of electrochemical reaction to Bunsen reaction (Equation 5.5) in IS process, which is considered as a pure thermochemical process, was proposed [1].

$$SO_2 + I_2 + 2H_2O \rightarrow H_2SO_4 + 2HI \tag{5.5}$$

The process applied the electrochemical Bunsen reaction can be taken as a kind of hybrid process. Research of pure thermochemical cycles and hybrid cycles has been carried out in parallel. Thus, both of them are simultaneously discussed later. Here, specific characteristics of hybrid processes are briefly commented.

G–T diagram of a certain electrochemical reaction is shown in Figure 5.4. Reactions which are not thermodynamically advantageous (great positive reaction Gibbs energy) can be used by allowing direct work input in the form of electric energy (upper diagram). When electrochemical reaction is applied to a reaction preceded by heat input, reaction temperature can be reduced for the same reaction (lower diagram). Generally, reaction voltage is lower than theoretical standard voltage 1.23 V in liquid–water electrolysis. For example, theoretical standard voltage of SO_2 electrolysis (Equation 5.6) for the Westinghouse process is 0.17 V and HBr electrolysis (Equation 5.7) for Ispra Mark 13 or hybrid Ca–Br process is 0.58 V [2].

$$SO_2 + 2H_2O \rightarrow H_2SO_4 + H_2 \tag{5.6}$$

$$2HBr \rightarrow H_2 + Br_2 \tag{5.7}$$

Therefore, thermal efficiency of hybrid processes can be greater than existing electrolysis cell if heat supply to the processes is small.

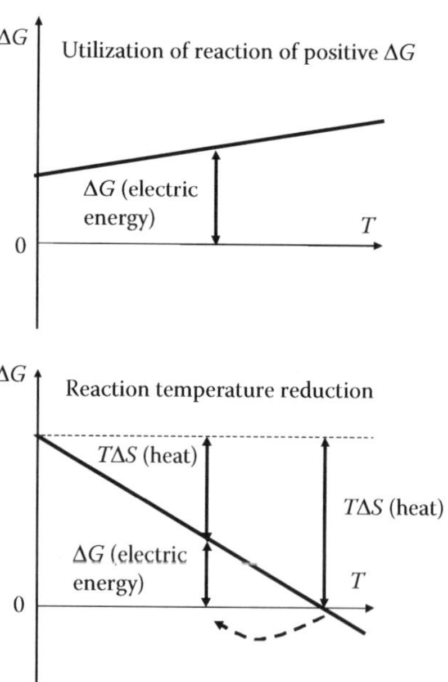

FIGURE 5.4
G–T diagram for electrochemical reactions in hybrid process.

Electrochemical reaction takes place in a cell similar to a PEM water electrolysis cell (see Section 3.4). Carbon cloth electrodes, rare metal electrode catalysts, and poly[perfluorosulfonic] acid membranes are typically used (e.g., [2]). Though voltage is lower at higher temperature, operation temperature is a little below 100°C mainly because of stability of poly[perfluorosulfonic] acid membranes. There are advantages in hybrid processes compared with pure thermochemical processes. High conversion ratio can be expected in electrochemical reactions because the ratio is not limited by thermochemical equilibrium attained by the reaction. Therefore, lower process flow rate can be achieved. Separation step can be omitted theoretically by the electrochemical reaction because products are made separately in the different sides of the membrane of the cell.

However, there are some disadvantages, too. Solid products or by-products may deposit on the electrodes and the membrane to prevent the reaction. For example, when small amount of SO_2 transfers through the membrane from anode side to cathode side in the cell of reaction Equation 5.6, the SO_2 reacts at cathode as in Equation 5.8 [3].

$$SO_2 + 4H^+ + 4e^- \rightarrow S + 2H_2O \tag{5.8}$$

The produced solid sulfur would poison electrode catalyst and block flow. Mass balance in the total process would be lost because the total amount of sulfur element contained in flow decreases with the production of solid sulfur. Membranes that can prevent SO_2 permeation have to be developed [4]. Improvement of cell property is still desired for high thermal efficiency. The goal of voltage of the cell in the Westinghouse process is set at 0.6 V at 0.5 A/cm^2 current density and 50 wt% H_2SO_4. Though thermal efficiency of Higher Heat Value (HHV) base as in Equation 5.9 was calculated 42% with that cell [5], such property has not been achieved yet [2]. Cost estimation showed the ratio of the cost of cell units to total capital cost of the Westinghouse process plant was 56% [6]. And the ratio for the Cu–Cl process was 37% [7]. Large number of electrolysis cell and high price of cell components such as a membrane and electrode catalysts would impact on plant cost. Large size cell component, higher current density, and membranes of lower cost are required for cheaper hydrogen production.

5.1.3 Process Thermal Efficiency

Thermal efficiency is an important performance indicator of the water splitting processes. Although many definitions of efficiency exist [8], the most desirable definition is based on thermodynamics as follows:

$$\eta_{WS} = \frac{\Delta H_{HHV}}{\sum_i Q_{WS,i} + \sum_i W_{WS,i}/\eta_{WG}} \tag{5.9}$$

This definition is applicable to any type of methods including pure thermochemical cycles, hybrid cycles, and electrolysis because it takes electric energy generation into account. Some analysts apply "lower heat value (LHV)," which is defined as enthalpy change of the reaction $H_2O(g) \rightarrow H_2(g) + 0.5O_2(g)$ at 25°C (= 241.83 kJ/mol-H_2), instead of ΔH_{HHV}. HHV is more appropriate because liquid water is fed to the water-splitting processes and H_2O is liquid at 25°C. Gibbs energy of the reaction $H_2O(l) \rightarrow H_2(g) + 0.5O_2(g)$ at 25°C (= 237.14 kJ/mol-H_2) adapted at numerator was discussed. Using hydrogen for electricity generation by fuel cell is assumed in the definition [9]. Amount of waste heat recovery of a process is added to the denominator in some papers (e.g., [10]). Heat recovery should be integrated in the process and calculate thermal efficiency by Equation 5.9 based on the modified flow sheet.

Thermal efficiency of a general water-splitting process including pure thermochemical, hybrid, and hydrolysis is discussed. Figure 5.5 is the total system. In the investigation, any procedure can be applied within water-splitting part, such as chemical reaction, heating/cooling, pressure change, phase shift, heat exchange, and so on and the complex of such procedures. Temperature, pressure, composition within the part can be varied. Work requirement of the part is supplied by a Carnot power generation cycle. This cycle also generates work for Carnot heat pumps. Each heat pump takes up heat from the heat sink and raises temperature to the requirements in the water-splitting part. When the part emits heat, the heat is used for work generation by another Carnot cycle. There is no entropy loss in the Carnot work-generation cycles and the Carnot heat pumps because Carnot cycle is defined to work without loss. The enthalpy/entropy balance as in Equations 5.10 through 5.13 are made in the system.

Enthalpy balance within the water splitting part (all work demands of the part can be treated as summed up (W_{WS}), Q_{WSi}, is heat input or output),

$$\sum_i Q_{WS,i} - \Delta H_{WS} + W_{WS} = 0$$

$$Q_{WS} - \Delta H_{WS} + W_{WS} = 0 \quad \left(Q_{WS} = \sum_i Q_{WS,i} \right) \tag{5.10}$$

Work generation by a Carnot cycle,

$$\frac{T_s - T_0}{T_s} \cdot Q_{CC,in} = W_{WS} + \sum_i W_{HP,i}$$

$$\frac{T_s - T_0}{T_s} \cdot Q_{CC,in} = W_{WS} + W_{HP} \quad \left(W_{HP} = \sum_i W_{HP,i} \right) \tag{5.11}$$

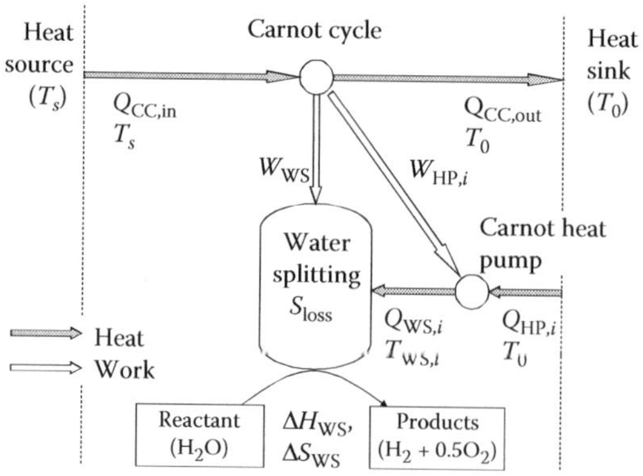

FIGURE 5.5
Heat/work balance of a water splitting process system.

Carnot heat pumps and Carnot work generation form waste heat,

$$\frac{T_{WS,i} - T_0}{T_{WS,i}} \cdot Q_{WS,i} = W_{HP,i}$$

$$\sum_i \frac{T_{WS,i} - T_0}{T_{WS,i}} \cdot Q_{WS,i} = \sum_i W_{HP,i} \tag{5.12}$$

$$Q_{WS} - T_0 \sum_i \frac{Q_{WS,i}}{T_{WS,i}} = W_{HP}$$

When temperature of the procedure within the water splitting part is changed by the heat input/output, $T_{WS,i}$ is the logarithmic average of temperatures before and after the change. Entropy balance of the part,

$$\sum_i \frac{Q_{WS,i}}{T_{WS,i}} - \Delta S_{WS} + S_{loss} = 0 \tag{5.13}$$

Substituting Equation 5.12 into Equation 5.13,

$$\frac{1}{T_0} \cdot \left(Q_{WS} - W_{HP}\right) - \Delta S_{WS} + S_{loss} = 0$$

Substituting Equations 5.10 and 5.11 into equation

$$\frac{1}{T_0} \cdot \left\{\left(\Delta H_{WS} - W_{WS}\right) - \left(\frac{T_s - T_0}{T_s} \cdot Q_{CC,in} - W_{WS}\right)\right\} - \Delta S_{WS} + S_{loss} = 0$$

$$Q_{CC,in} = \frac{T_s}{T_s - T_0} \cdot \left(\Delta H_{WS} - T_0 \Delta S_{WS} + T_0 S_{loss}\right) \tag{5.14}$$

η_{WS}, the thermal efficiency of the total system, which consists of the water splitting part, Carnot work generation and waste heat recovery cycles and Carnot heat pumps is defined as Equation 5.15. The heat from the heat sink $Q_{HP,i}$ is not included because it has no effect on thermal efficiency. This efficiency means just the same as the efficiency η in Equation 5.9.

$$\eta_{WS} = \frac{\Delta H_{WS}}{Q_{CC,in}} = \frac{\Delta H_{WS}}{\frac{T_s}{T_s - T_0} \cdot \left(\Delta H_{WS} - T_0 \Delta S_{WS} + T_0 S_{loss}\right)}$$

$$= \frac{T_s - T_0}{T_s} \cdot \frac{\Delta H_{WS}}{\Delta H_{WS} - T_0 \Delta S_{WS} + T_0 S_{loss}} \tag{5.15}$$

Exergy is a value often used to analyze loss of a procedure. Exergy change of the water-splitting reaction is defined as Equation 5.16. Then heat demand and thermal efficiency of the total system are shown as Equations 5.17 and 5.18 by using the exergy change.

$$\Delta E_{WS} = \Delta H_{WS} - T_0 \Delta S_{WS} \tag{5.16}$$

$$\begin{aligned} Q_{CC,in} &= \frac{T_s}{T_s - T_0} \cdot \left(\Delta H_{WS} - T_0 \Delta S_{WS} + T_0 S_{loss}\right) \\ &= \frac{T_s}{T_s - T_0} \cdot \left(\Delta E_{WS} + T_0 S_{loss}\right) \end{aligned} \tag{5.17}$$

$$\begin{aligned} \eta_{WS} &= \frac{T_s - T_0}{T_s} \cdot \frac{\Delta H_{WS}}{\Delta H_{WS} - T_0 \Delta S_{WS} + T_0 S_{loss}} \\ &= \frac{T_s - T_0}{T_s} \cdot \frac{\Delta H_{WS}}{\Delta E_{WS} + T_0 S_{loss}} \end{aligned} \tag{5.18}$$

If the water splitting part is operated reversibly, S_{loss} is equal to 0. This is the ideal case. And the thermal efficiency in Equation 5.19 is maximum because $S_{loss} \geq 0$. When S_{loss} is 0, Equation 5.20 is obtained form Equation 5.17. Exergy of the water splitting reaction means the same value as the energy generated by Carnot cycle using minimum heat demand of the total system in case of no irreversibility loss within the water splitting part.

$$\eta_{WS,ideal} = \frac{T_s - T_0}{T_s} \cdot \frac{\Delta H_{WS}}{\Delta E_{WS}} \tag{5.19}$$

$$\Delta E_{WS,ideal} = \frac{T_s - T_0}{T_s} \cdot Q_{CC,in} \tag{5.20}$$

Figure 5.6 is the relation of the ideal thermal efficiency to temperature of heat source at 298.15 K of the heat sink. The value of $\eta_{WS,ideal}$ exceeds 1 over 1750 K. This apparent paradox is made because the efficiency is the ratio of HHV of hydrogen (the same value as ΔH_{WS}) to

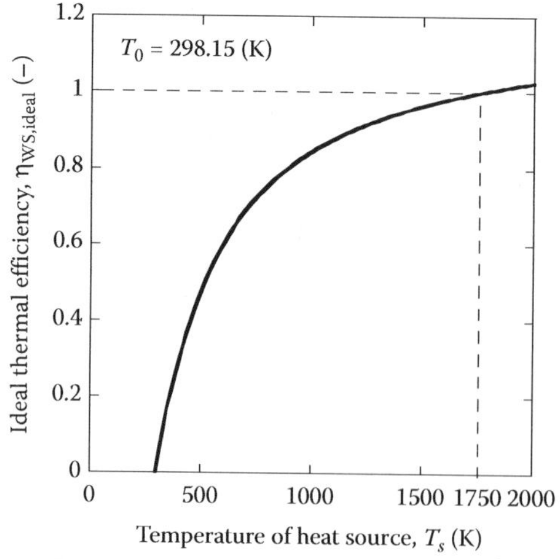

FIGURE 5.6
Ideal thermal efficiency of total water splitting system.

the heat input. When electricity is generated by an ideal hydrogen fuel cell, electricity of ΔG_{WS} is produced from 1 mol of hydrogen. The ideal efficiency of work generation corresponding to HHV of hydrogen at T_0, $\eta_{WG,ideal}$, is

$$\eta_{WG,ideal} = \frac{\Delta G_{WS}}{\Delta H_{WS}} = \frac{\Delta H_{WS} - T_0 \Delta S_{WS}}{\Delta H_{WS}} = \frac{\Delta E_{WS}}{\Delta H_{WS}} \tag{5.21}$$

From Equations 5.19 and 5.21, the total efficiency of the process heat source → hydrogen → electricity is the same as Carnot efficiency of work generation from heat.

$$\eta_{total.ideal} = \eta_{WS,ideal} \cdot \eta_{WG,ideal} = \left(\frac{T_s - T_0}{T_s} \cdot \frac{\Delta H_{WS}}{\Delta E_{WS}} \right) \cdot \left(\frac{\Delta E_{WS}}{\Delta H_{WS}} \right) = \frac{T_s - T_0}{T_s} \tag{5.22}$$

Considering Equation 5.19, ideal hydrogen production thermal efficiency is determined by only temperatures of the heat source (T_s) and the heat sink (T_0). This discussion is applicable in any processes of thermochemical cycles, hybrid cycles and electrolysis. This means that ideal thermal efficiency is the same in any ways. In actual system, such factors as entropy loss within the water splitting part, loss at heat exchange from heat source to the part, irreversibility of electric energy generation, and waste heat which is not recovered make thermal efficiency difference among processes. This is the reason why not only water splitting part itself but electric energy generation and waste heat recovery should be paid attention to in efficiency analysis.

5.2 Research and Development

5.2.1 Proposition of Thermochemical Water Splitting

The concept of thermochemical water splitting is not new. A two-step HgO/Hg cycle was patented as early as in 1925 [11]. However, R&D of the concept was activated in the early 1960s by the theoretical proposition based on thermodynamics in the Energy Depot project of U.S. Army and others [12]. Effective hydrogen production processes with two steps of thermal reaction were not found in the temperature range below 1100°C. Instead, a three-step Ta–Cl cycle and a four-step V–Cl cycle were proposed as examples [13]. Table 5.1 shows the reactions of the cycles discussed above. Hydrogen energy was paid attention to at that time because of the shortage of petroleum. The concept of thermochemical water splitting was so attractive that search of effective processes that consist of more than three reactions or that include electrochemical reaction became active all over the world. Table 5.1 lists the cycles proposed from that time to present which are introduced in this section. The Hydrogen Economy Miami Energy (THEME) conference, the first conference of hydrogen energy was held in 1972. Proposition of a particular process (e.g., [14,15]), review of several processes [16] and calculation of thermal efficiency by reaction enthalpy [17,18] were main themes of this period. In 1975, the symposium of Hydrogen Energy Fundamentals followed the conference (e.g., [19]).

In the period, lists of cycles were reviewed in literatures [9,20,21]. Some researchers tried computer-aided systematic search of high thermal efficiency processes because number of possible cycle was too much [22,23]. IS process [23] and UT-3 process [24], which has been researched in detail, were found promising by such search. Many kinds of calculation methods of thermal efficiency were used as discussed in Section 5.1.3 [8]. After the period, theoretical discussion based on thermodynamics was done. International Energy Agency adopted the definition as Equation 5.9 [9].

5.2.2 Early Research Activities

Publication on thermochemical water splitting increased rapidly during 1972 and 1976, and the great number of publications was kept until 1984 [25]. This period was the first peak of the research. As a succession of the conferences shown above, World Hydrogen Energy Conference (WHEC) started in 1976 against the background of the active study of hydrogen energy. The conference continues since then and the 18th conference was held in 2010. Detailed experimental investigation on selected promising cycles was carried out in this time. There are so much research by many institutes and universities that some examples are introduced here. Many reviews are available to know the detail of the early activities [8,9,26–29].

5.2.2.1 Joint Research Center Ispra

The Joint Research Center (JRC) Ispra, Italy, of the European Commission carried out a series of extensive research. The research was reviewed [27,30]. Following the performance of preliminary study, the first discussion was held at the International Round Table on Direct Production of Hydrogen with Nuclear Heat in 1969. The Mark 1 cycle was proposed based on thermochemical evaluation [31]. A program of hydrogen production from water was approved by the Council of Ministers of the European Communities. The research was carried out in 1973–1983 in three continuous programs, which proposed many cycles. Some promising cycles were selected, though the selection procedure was not systematically done. Table 5.2 lists the main cycles selected.

The research at JRC Ispra proceeded in three phases. In the first phase, verification of each reaction of the Mark 1 cycle by experiment, measurement of thermodynamic property of $CaBr_2$, corrosion test of materials, flow sheet and cost evaluation were performed [32]. However, this process was abandoned because of use of Hg. Alternative cycles, Mark 1C, 2, 3, and 6 were proposed, where Hg was not included. However, results of experiments of these cycles were not always encouraging. The phase was ended as a new series of cycles was found.

Fe–Cl cycles (Mark 7, 7A, 7B, 9, 14, and 15) were investigated in the second phase. The research was carried out in 1973–1975. Some Fe–Cl cycles were proposed and intensively researched outside of JRC at that time. JRC also focused on the series because the chemical components are abundant and chemical property of them is well known. The detail of the study is described in Section 5.3.2.1. The research was concluded by the assessment of the thermal efficiency, investment costs, and corrosion in 1975 [33].

The study in the third phase was on cycles of sulfur series (mainly Mark 11, 13, and 16). Thermal decomposition of H_2SO_4 was common in these cycles. Separation of H_2SO_4 and HI produced by Bunsen reaction (Equation 5.5) was a major problem of Mark 16 process. A variety of extracting agents was tried to separate them [34]. Thermal efficiency of these cycles was calculated by a simulation code OPTIMO, which was developed in this research [35]. Mark 16 cycle was abandoned because of experimental difficulties and less promising economical evaluation. Nevertheless, this cycle was researched for long time after the JRC study as S–I or IS process. This detail is explained in Section 5.2.3. JRC

TABLE 5.2

List of Main Cycles Investigated by JRC Ispra

Name	Reaction	Typical Temperature (°C)	Thermochemical (T) or Electrochemical (E)
Mark 1	$CaBr_2(s) + 2H_2O(g) \rightarrow Ca(OH)_2(s) + 2HBr(g)$	730	T
	$2HBr(aq) + Hg(l) \rightarrow HgBr_2(aq) + H_2(g)$	200	T
	$HgBr_2(s) + Ca(OH)_2(s) \rightarrow CaBr_2(s) + HgO(s) + H_2O(g)$	200	T
	$HgO(s) \rightarrow Hg(g) + 0.50_2(g)$	600	T
Mark 1B	$CaBr_2(s) + 2H_2O(g) \rightarrow Ca(OH)_2(s) + 2HBr(g)$	730	T
	$2HBr(aq) + Hg_2Br_2(s) \rightarrow 2HgBr_2(s) + H_2(g)$	120	T
	$HgBr_2(s) + Hg(l) \rightarrow Hg_2Br_2(s)$	120	T
	$HgBr_2(s) + Ca(OH)_2(s) \rightarrow CaBr_2(s) + HgO(s) + H_2O(g)$	200	T
	$HgO(s) \rightarrow Hg(g) + 0.5O_2(g)$	600	T
Mark 1C	$2CaBr_2(s) + 4H_2O(g) \rightarrow 2Ca(OH)_2(s) + 4HBr(g)$	730	T
	$4HBr(aq) + Cu_2O(s) \rightarrow 2CuBr_2(aq) + H_2O(l) + H_2(g)$	100	T
	$2CuBr_2(aq) + 2Ca(OH)_2(aq) \rightarrow 2CuO(s) + 2CaBr_2(aq) + 2H_2O(l)$	100	T
	$2CuO(s) \rightarrow Cu_2O(s) + 0.5O_2(g)$	900	T
Mark 1S	$SrBr_2(s) + H_2O(g) \rightarrow SrO(s) + 2HBr(g)$	800	T
	$2HBr(g) + Hg(l) \rightarrow HgBr_2(s) + H_2(g)$	200	T
	$SrO(s) + HgBr_2(s) \rightarrow SrBr_2(s) + Hg(g) + 0.5O_2(g)$	500	T
Mark 2	$Mn_2O_3(s) + 4NaOH(l) \rightarrow 2Na_2O \cdot MnO_2(s) + H_2O(g) + H_2(g)$	800	T
	$2Na_2O \cdot MnO_2(s) + 2H_2O(g) \rightarrow 4NaOH(aq) + 2MnO_2(s)$	100	T
	$2MnO_2(s) \rightarrow Mn_2O_3(s) + 0.5O_2(g)$	600	T
Mark 2C	$Mn_2O_3(s) + 2Na_2CO_3(s) \rightarrow 2Na_2O \cdot MnO_2(s) + CO_2(g) + CO(g)$	850	T
	$CO(g) + H_2O(g) \rightarrow H_2(g) + CO_2(g)$	500	T
	$2Na_2O \cdot MnO_2(s) + nH_2O(l)$ $+2CO_2(g) \rightarrow 2Na_2CO_2 \cdot nH_2O(aq) + 2MnO_2(s)$	100	T
	$2MnO_2(s) \rightarrow Mn_2O_3(s) + 0.5O_2(g)$	600	T
Mark 3	$Cl_2(g) + H_2O(g) \rightarrow 2HCl(g) + 0.5O_2(g)$	800	T
	$2HCl(g) + 2VOCl(s) \rightarrow 2VOCl_2(s) + H_2(g)$	170	T
	$4VOCl_2(s) \rightarrow 2VOCl(s) + 2VOCl_3(g)$	600	T
	$2VOCl_3(g) \rightarrow 2VOCl_2(s) + Cl_2(g)$	200	T

continued

TABLE 5.2 (continued)
List of Main Cycles Investigated by JRC Ispra

Name	Reaction	Typical Temperature (°C)	Thermochemical (T) or Electrochemical (E)
Mark 4	$Cl_2(g) + H_2O(g) \rightarrow 2HCl(g) + 0.5O_2(g)$	800	T
	$2HCl(g) + S(s) + 2FeCl_2(s) \rightarrow H_2S(g) + 2FeCl_3(s)$	100	T
	$H_2S(g) \rightarrow H_2(g) + 0.5S_2(g)$	800	T
	$2FeCl_3(g) \rightarrow 2FeCl_2(s) + Cl_2(g)$	420	T
Mark 5	$CaBr_2(s) + H_2O(g) + CO_2(g) \rightarrow CaCO_3(s) + 2HBr(g)$	600	T
	$CaCO_3(s) \rightarrow CaO(s) + CO_2(g)$	900	T
	$2HBr(g) + Hg(l) \rightarrow HgBr_2(s) + H_2(g)$	200	T
	$HgBr_2(s) + CaO(s) + nH_2O(l) \rightarrow CaBr_2{\cdot}nH_2O(aq) + HgO(s)$	200	T
	$HgO(s) \rightarrow Hg(g) + 0.5O_2(g)$	600	T
Mark 6	$Cl_2(g) + H_2O(g) \rightarrow 2HCl(g) + 0.5O_2(g)$	800	T
	$2HCl(g) + 2CrCl_2(s) \rightarrow 2CrCl_3(s) + H_2(g)$	170	T
	$2CrCl_3(s) + 2FeCl_2(s) \rightarrow 2CrCl_2(s) + 2FeCl_3(g)$	700	T
	$2FeCl_3(g) \rightarrow 2FeCl_2(s) + Cl_2(g)$	350	T
Mark 6C	$Cl_2(g) + H_2O(g) \rightarrow 2HCl(g) + 0.5O_2(g)$	800	T
	$2HCl(g) + 2CrCl_2(s) \rightarrow 2CrCl_3(s) + H_2(g)$	170	T
	$2CrCl_3(s) + 2FeCl_2(s) \rightarrow 2CrCl_2(s) + 2FeCl_3(g)$	700	T
	$2FeCl_3(s) + 2CuCl(s) \rightarrow 2FeCl_2(s) + 2CuCl_2(s)$	150	T
	$2CuCl_2(s) \rightarrow 2CuCl(l) + Cl_2(g)$	500	T
Mark 7	$3FeCl_2(s) + 4H_2O(g) \rightarrow Fe_3O_4(s) + 6HCl(g) + H_2(g)$	650	T
	$Fe_3O_4(s) + 0.25O_2(g) \rightarrow 1.5Fe_2O_3(s)$	350	T
	$1.5Fe_2O_3(s) + 9HCl(g) \rightarrow 3FeCl_3(s) + 4.5H_2O(g)$	150	T
	$3FeCl_3(g) \rightarrow 3FeCl_2(s) + 1.5Cl_2(g)$	420	T
	$1.5Cl_2(g) + 1.5H_2O(g) \rightarrow 3HCl(g) + 0.75O_2(g)$	800	T
Mark 7A	$3FeCl_2(s) + 4H_2O(g) \rightarrow Fe_3O_4(s) + 6HCl(g) + H_2(g)$	650	T
	$Fe_3O_4(s) + 0.25O_2(g) \rightarrow 1.5Fe_2O_3(s)$	350	T
	$Fe_2O_3(s) + 6HCl(g) \rightarrow 2FeCl_3(s) + 3H_2O(g)$	150	T

	$0.5Fe_2O_3(s) + 1.5Cl_2(g) \rightarrow FeCl_3(g) + 0.75O_2(g)$	1000	T
	$3FeCl_3(g) \rightarrow 3FeCl_2(s) + 1.5Cl_2(g)$	420	T
Mark 7B	$3FeCl_2(s) + 4H_2O(g) \rightarrow Fe_3O_4(s) + 6HCl(g) + H_2(g)$	650	T
	$Fe_3O_4(s) + 0.25O_2(g) \rightarrow 1.5Fe_2O_3(s)$	350	T
	$1.5Fe_2O_3(s) + 4.5Cl_2(g) \rightarrow 3FeCl_3(s) + 2.25O_2(g)$	1000	T
	$3FeCl_3(g) \rightarrow 3FeCl_2(s) + 1.5Cl_2(g)$	420	T
	$6HCl(g) + 1.5O_2(g) \rightarrow 3Cl_2(g) + 3H_2O(g)$	400	T
Mark 8	$3MnCl_2(s) + 4H_2O(g) \rightarrow Mn_3O_4(s) + 6HCl(g) + H_2(g)$	700	T
	$1.5Mn_3O_4(s) + 6HCl(g) \rightarrow 3\ MnCl_2(aq) + 1.5MnO_2(s) + 3H_2O(l)$	100	T
	$1.5MnO_2(s) \rightarrow 0.5Mn_3O_4(s) + 0.5O_2(g)$	900	T
Mark 9	$3FeCl_2(s) + 4H_2O(g) \rightarrow Fe_3O_4(s) + 6HCl(g) + H_2(g)$	650	T
	$Fe_3O_4(s) + 1.5Cl_2(g) + 6HCl(g) \rightarrow 3FeCl_3(g) + 3H_2O(g) + 0.5O_2(g)$	150–200	T
	$3FeCl_3(g) \rightarrow 3FeCl_2(s) + 1.5Cl_2(g)$	420	T
Mark 10	$2NH_4I(s) \rightarrow 2NH_3(g) + H_2(g) + I_2(g)$	500–700	T
	$I_2(s) + SO_2(g) + 2H_2O(l) + 4NH_3(g) \rightarrow 2NH_4I(s) + (NH_4)_2SO_4(s)$	0–50	T
	$(NH_4)_2SO_4(s) + Na_2SO_4(s) \rightarrow 2NaHSO_4(s) + 2NH_3(g)$	400	T
	$2NaHSO_4(s) \rightarrow Na_2S_2O_7(s) + H_2O(g)$	400	T
	$Na_2S_2O_7(s) \rightarrow Na_2SO_4(s) + SO_3(g)$	450	T
	$SO_3(g) \rightarrow SO_2(g) + 0.5O_2(g)$	800–850	T
Mark 11 (Westinghouse, Hybrid Sulfur (HyS))	$SO_2(g) + 2H_2O(l) \rightarrow H_2SO_4(aq) + H_2(g)$	25	E
	$H_2SO_4(g) \rightarrow H_2O(g) + SO_2(g) + 0.5O_2(g)$	800–900	T
Mark 12	$2NH_4I(s) \rightarrow 2NH_3(g) + H_2(g) + I_2(g)$	500–700	T
	$I_2(s) + SO_2(g) + 2H_2O(l) + 4NH_3(g) \rightarrow 2NH_4I(s) + (NH_4)_2SO_4(s)$	0–50	T
	$(NH_4)_2SO_4(s) + ZnO(s) \rightarrow ZnSO_4(s) + 2NH_3(g) + H_2O(g)$	500	T
	$ZnSO_4(s) \rightarrow ZnO(s) + SO_2(g) + 0.5O_2(g)$	850	T
Mark 13	$SO_2(g) + Br_2(g) + 2H_2O(l) \rightarrow H_2SO_4(aq) + 2HBr(aq)$	50–100	T
	$H_2SO_4(g) \rightarrow H_2O(g) + SO_2(g) + 0.5O_2(g)$	800–900	T
	$2HBr(aq) \rightarrow Br_2(g) + H_2(g)$	25	E
Mark 14	$3FeCl_2(s) + 4H_2O(g) \rightarrow Fe_3O_4(s) + 6HCl(g) + H_2(g)$	650	T

continued

TABLE 5.2 (continued)
List of Main Cycles Investigated by JRC Ispra

Name	Reaction	Typical Temperature (°C)	Thermochemical (T) or Electrochemical (E)
	$Fe_3O_4(s) + 0.5Cl_2(g) \rightarrow 1/3FeCl_3(g) + 4/3Fe_2O_3(s)$		T
	$4/3Fe_2O_3(s) + 8HCl(g) \rightarrow 8/3\ FeCl_3(s) + 4H_2O(g)$	160	T
	$3FeCl_3(s) \rightarrow 3FeCl_2(s) + 1.5Cl_2(g)$	280	T
	$Cl_2(g) + H_2O(g) \rightarrow 2HCl(g) + 0.5O_2(g)$	730	T
Mark 15	$3FeCl_2(s) + 4H_2O(g) \rightarrow Fe_3O_4(s) + 6HCl(g) + H_2(g)$	650–700	T
	$Fe_3O_4(s) + 8HCl(g) \rightarrow FeCl_2(s) + 2FeCl_3(s) + 4H_2O(g)$	200–300	T
	$2FeCl_3(s) \rightarrow 2FeCl_2(s) + Cl_2(g)$	300–400	T
	$Cl_2(g) + H_2O(g) \rightarrow 2HCl(g) + 0.5O_2(g)$	600–700	T
Mark 16 (GA, IS, S-I)	$SO_2(g) + I_2(s) + 2H_2O(l) \rightarrow H_2SO_4(aq) + 2HI(aq)$	25–100	T
	$H_2SO_4(g) \rightarrow H_2O(g) + SO_2(g) + 0.5O_2(g)$	800–900	T
	$2HI(g) \rightarrow I_2(g) + H_2(g)$	400–500	T

concentrated on Mark 11 and 13 processes. Mark 13 process was proposed almost simultaneously in JRC itself [36] and Los Alamos Scientific Laboratory (LASL, presently Los Alamos National Laboratory (LANL)), USA [37]. A bench-scale plant of the cycle, which produced hydrogen with rate of about 100 L/h, was constructed and operated [38]. The continuous operation time is implied for as long as 30 h from the description of the literature. Blowing heated gas directly to process flow in the H_2SO_4 decomposition part was considered for the industrial pilot-plant to avoid heat exchange between corrosive process fluids. The process was called CRISTINA and a Mark 11 plant of the concept was designed. H_2SO_4 decomposition was made in the following two steps in an actual process. First, H_2SO_4 was decomposed into SO_3 and H_2O through vaporization. Then SO_3 gas was decomposed into SO_2 and O_2.

$$H_2SO_4 \rightarrow SO_3 + H_2O \tag{5.23}$$

$$SO_3 \rightarrow SO_2 + 0.5O_2 \tag{5.24}$$

CHRIS, the test facility of the SO_3 decomposition with air blowing was operated [39]. Corrosion test of materials was also carried out on halogens, hydrogen halides [40], and sulfur compounds [41].

5.2.2.2 Rheinisch-Westfälische Technische Hochschule Aachen

Rheinisch-Westfälische Technische Hochschule (RWTH) Aachen, Germany, researched under Euratom, the same as JRC Ispra. The group researched mainly on Fe–Cl cycles earlier and on IS process later. The group proposed a cycle of Fe–Cl series that use metal Fe [19]. The research on the cycle is summarized in Section 5.3.2.1. For IS process, HI distillation from HI–H_2O–I_2 solution was the main theme. Vapor–liquid equilibrium (VLE) data of the solution was measured and estimation was made by Non-Random Two Liquid (NRTL) equation [42]. HI decomposition (Equation 5.4) in gas phase in VLE environment of HI–H_2O–I_2 solution was suggested [43] and a concept of HI reactive distillation obtaining H_2 directly from the top of the distillation column was proposed [44]. Flow sheet of H_2SO_4 decomposition [45] and thermal HI decomposition in HI–H_2O–I_2 solution were also studied [46]. Most of the research was completed around the end of 1980s.

5.2.2.3 General Atomics

R&D of General Atomics (GA), USA, began in 1972. And IS process, the same process as Ispra Mark 16 cycle, was selected as a promising candidate by a systematic computer search [23]. The research continued to mid-1980s [47,48]. The separation problem, found in JRC, between HI and H_2SO_4 made by Bunsen reaction was solved with addition of excess amount of I_2. Distillation of almost pure HI from the HI–H_2O–I_2 mixture requires very high temperature (probably above 300°C) and pressure (order of some MPa). GA proposed the moderation of condition by the separation of HI–H_2O and I_2 before distillation using H_3PO_4 extractive agent to overcome the problem. Screening of catalysts of SO_3 gas-phase decomposition reaction and HI decomposition in liquid phase was also carried out [47]. The closed-loop cycle demonstrator (CLCD) was constructed and operated with hydrogen production rate of 1 L/h scale. The operation of bench-scale subunit facilities corresponding to each reaction followed. Thermal efficiency and hydrogen cost calculation was made in the flow sheet with extensive heat recovery [48].

5.2.2.4 Westinghouse Electric

Westinghouse Electric, USA, has researched on the Westinghouse process, the same process as Ispra Mark 11 cycle, or Hybrid Sulfur process (HyS), sponsored by National Aeronautics and Space Administration (NASA), USA. Though LASL, USA, also proposed the idea of this process around the same time [49], the process was called by that name by the long and intensive activity of the research group. Research on thermochemical hydrogen production started in 1973. The objective of the total program was a demonstration of the process demonstration unit (PDU) by 1983 [50]. The development program on four tasks was planned: SO_2 electrolysis, SO_3 reduction, process studies and integrated testing. Evaluation of membranes, electrodes development, cell design, and demonstration were done for SO_2 electrolysis (Equation 5.6) [51]. Screening of SO_3 decomposition catalysts, examination of materials for H_2SO_4 concentrator and evaporator, and scale up of reactors were done for SO_3 reduction. Thermal efficiency and hydrogen cost were calculated by flow sheet analysis [52,53]. A pressurized test unit assuming solar heat as heat source was designed at the end of the research [54]. Integrated testing was comprised of three sequential steps: the laboratory model, the PDU, and the pilot-scale plant. The laboratory model which was sized to produce 2 L/min of hydrogen was successfully operated in 1978 [52]. The research work at Westinghouse Electric continued until 1983 and ended then [55]. The research is summarized in reports and articles [50,52,56–58]. Details of the process are explained in Chapter 18.

5.2.2.5 National Chemical Laboratory for Industry

National Chemical Laboratory for Industry (NCLI), Japan, presently known as Advanced Industrial Science and Technology (AIST), proposed Mg–I cycle [59]. The institute had studied mainly this process. The research seemed to stop in about 1984 considering publication on the process. The detail of the study is explained in Section 5.3.4. The group also studied Mg–S–I process at the same time [60]. First, intermediate compounds MgI_2 and $MgSO_4$ were processed without separation. Batch system demonstration was carried out. Then the two compounds were separated to replace the complicated reactor where two decomposition reactions in Equations 5.25 and 5.26 occurred.

$$MgI_2(aq.) \rightarrow MgO + 2HI + nH_2O \tag{5.25}$$

$$MgSO_4 \rightarrow MgO + SO_3 \tag{5.26}$$

Demonstration of the process applying the improvement was performed with hydrogen production of 0.5 L/h for 33 h. In addition, electrochemical Bunsen reaction (Equation 5.5) was investigated for IS process [61].

5.2.2.6 Conclusion of the Early Research

The attention to thermochemical water-splitting processes gradually declined during the latter half of 1980s as the attraction of hydrogen energy was decreased by low-cost fossil fuel. Regarding systematic research, only a few were carried out during 1990s. Study on the UT-3 process in Japan, explained in Section 5.3.1.1, was the most active. Japan Atomic Energy Research Institute (JAERI), now known as Japan Atomic Energy Agency (JAEA) researched NIS (Nickel–Iodine–Sulfur) process [62] and CIS (Carbon–Iodine–Sulfur) process [63] during 1980s. These two processes are described in Table 5.1. The target process was changed to IS process, the same as Ispra Mark 16 or GA process [64] in the period. Continuous operation of the laboratory scale facility (1 L/h production of hydrogen) was achieved for 48 h [65,66]. Besides Japan, investigation of metal oxide processes by using solar heat also continued in this period [67].

5.2.3 Current Research Activities

Hydrogen energy was paid attention again around the end of the twentieth century because of different reasons. The risk of global warming by CO_2 emission from fossil fuel has been discussed and energy resource without emitting CO_2 was required. Thermochemical water splitting coupled with nonfossil-fuel heat source like nuclear heat and solar heat can produce hydrogen with little CO_2 emission. Therefore, research activity has gained momentum again around research groups of nuclear and solar energy [68].

5.2.3.1 Research Programs of the United States

Nuclear Energy Research Initiative (NERI) was established by United States Department of Energy (DOE) in 2001. Under the program, General Atomics (GA), Sandia National Laboratories (SNL), USA, and University of Kentucky (UK) collaborated from 1999 to 2002 [69]. The first phase of the study was the screening of processes based on the research before. Twenty-five processes were selected from 10 scoring criteria and essential criteria shown in Table 5.3. The viability of each selected process as a whole was checked by the three institutes separately. Adiabatic version of the UT-3 cycle and IS cycle were selected as the result. The second and third phases were practically a continuous research. A flow sheet of IS process was created and coupled with a nuclear reactor. And sizing of process components were carried out and hydrogen-production cost was estimated. NERI was succeeded by International Nuclear Energy Research Initiative (I-NERI) program. GA, SNL and Commissariat à l'Énergie Atomique (CEA), France, participated in the project on thermochemical hydrogen production. For IS process, GA, SNL and CEA covered mainly the R&D of HI part, H_2SO_4 part, and Bunsen part, respectively. Construction of bench-scale equipments (100–200 L/h-H_2) of each part was carried out and closed-loop operation tests

TABLE 5.3

Screening Criteria in NERI Study

Scoring criteria	Low number of chemical reactions
	Low number of separation steps
	Low number of elements
	Abundant elements
	Low corrosivity
	Continuous flow in the process
	Maximum temperature matching high temperature gas-cooled reactors (HTGRs) (750–850°C)
	Many publishment of the research
	Experience of demonstration of the cycle
	Good efficiency and cost data
Essential criteria	Not using mercury
	Maximum temperature was below 1600°C
	No reaction of large positive Gibbs energy

by connecting these equipments were done from 2008 [70]. The bench-scale test was completed in April 2009 [71].

Nuclear Hydrogen Initiative (NHI), sponsored by I-NERI Program, set a goal of demonstration of economic, commercial scale production of hydrogen using nuclear energy. For thermochemical cycles, GA and SNL have worked on IS process, Savannah River National Laboratory (SRNL), USA, covered the Westinghouse process, and Argonne National Laboratory (ANL), USA and many universities investigated on other cycles, respectively [68]. The R&D is summarized in papers [4,5,72,73]. Detail of the NHI programs is mentioned in Chapter 31.

5.2.3.2 European Works

HYTHEC (Hydrogen THErmochemical Cycles) project, a project of European Community Framework 6 program "Innovative Routes for High Temperature Hydrogen Production (INNOHYP)," was carried out in collaboration with several European institutes, universities, and companies from 2004 to 2007 [74]. The project focused mainly on IS process and on the Westinghouse process as a reference. CEA, the coordinator of the project, made flow sheets of IS and the Westinghouse processes and measured vapor-liquid equilibrium of HIx mixture (mixture of $HI\text{-}H_2O\text{-}I_2$). University of Sheffield, U.K., investigated on HIx membrane distillation. Università degli studi, Roma Tre took charge of component sizing and techno-economical evaluation, and solar H_2SO_4 decomposition flow modeling. Deutsches Zentrum für Luft und Raumfahrt, Germany, conducted H_2SO_4 decomposition in a solar furnace and coupling the Westinghouse process to solar and/or nuclear heat source. Empresarios Agrupados, Spain, took charge of coupling to the reactor, safety evaluation and thermo-structural analysis of the solar test reactor. Prosim SA, France, implemented a simulation code for IS process flow sheet model [75,76]. A successive HYCYCLES project was carried out focusing on materials and components of H_2SO_4 decomposition from 2008 to 2010 [77].

CEA had another research strategy to develop for massive hydrogen production including thermochemical cycles from 2001. IS process was selected as main target and some other cycles such as the Westinghouse process, UT-3 cycle as alternatives [78]. An integrated program to choose the most promising way of hydrogen production was launched in 2008 and 2009 [79]. HYPRO (HYdrogen PROduction) project proceeded simultaneously. The main objective of the project was to demonstrate feasibility of pilot-scale hydrogen production (20–80 m^3/h at standard state) by IS process coupled with a high-temperature helium loop. Preliminary design of a small SO_3 decomposer and technological development of a compact-plate heat exchanger in silicon carbide were done. The project has been integrated in HYCYCLES project since 2008 [77].

Ente per le Nuove tecnologie, l'Energia e l'Ambiente (ENEA), Italy, worked on solar hydrogen production using IS process within the framework of the Italian National Program TEPSI (New Systems for Energy Production and Management—Hydrogen). The project started in 2005 and was scheduled for three years [80–83].

Detail of these research projects is mentioned in Chapter 28.

5.2.3.3 Asian Researches

In Japan, the research on IS process by JAEA continued. [84–86]. 175 h continuous operation of a bench-scaled facility of 31 L/h hydrogen production was achieved [87]. Chapter 17 includes the recent JAEA's research on the process. HHLT process, a modified process of the Westinghouse process, was newly proposed. The detail is expressed in Section 5.3.3.

South Korea and China also have researched IS process. In South Korea, Korean Atomic Energy Research Institute (KAERI) established a plan in 2003 aiming massive hydrogen production by early 2020s using a Very-High Temperature Reactor (VHTR) [88]. KAERI, Korea Institute of Energy Research (KIER), and Korea Institute of Science and Technology (KIST) have developed technology for IS process [89–92]. Chapter 30 mentions the research in South Korea. Tsinghua University, China, started R&D on the hydrogen production through IS process in 2005 considering HTR-10, a High Temperature Gas-cooled Reactor as heat source [93]. Details are in Chapter 27.

5.2.3.4 Solar Hydrogen Production

Higher temperature than that of nuclear reactors is expected by solar heat (ca. 1500°C) [94]. Therefore, two-step pure thermochemical cycles which require such high temperature are possible (see Section 5.1.1). Metal oxide methods are typical and many cycles have been proposed. 30 advantageous cycles were screened considering the criteria: appropriate maximum temperature, small number of reactions, small number of elements in the cycle, no electric reaction, moderate corrosivity compounds in the cycle, easy separation of gases, no separation of solids, and no rare elements in the cycle [95]. Fe_3O_4/FeO process and ZnO/Zn process were considered as promising candidates [94] and these processes has been well researched. Application of solar heat to IS, Westinghouse, and UT-3 processes have been also studied [94] even thermodynamical temperature requirement (temperature of $\Delta G_i = 0$) are lower than the potential of solar heat. Many projects on solar heat application including thermochemical processes are now under way. Present status is summarized in literature [96].

5.3 Technical Description of Processes

5.3.1 Iodine–Sulfur (IS) and Hybrid Sulfur (HyS) Processes

Of all known thermochemical and hybrid water splitting processes, the IS process and the HyS process are most intensively researched and developed at the moment and have attracted the most interest worldwide in the application of thermochemical and hybrid methods, respectively, to nuclear hydrogen production. Chapters 17 and 18 of this handbook are devoted to detailed presentations on these processes, which are thus not discussed here.

5.3.2 Calcium–Bromine Series

5.3.2.1 UT-3 Process

UT-3 process was first proposed by a research group in University of Tokyo, Japan, around 1976 [24]. The name of the process was named after the university and the third proposition order. The reactions of the process are shown in Table 5.1. Conceptual flow sheet of the process is illustrated in Figure 5.7. All reactions are thermal solid–gas reaction.

This process was made based on the screening of the set of reactions whose equilibrium tended to products considering reaction Gibbs energy [97]. This process has been studied mainly in Japan since the proposition. CEA, France, has selected and has researched the

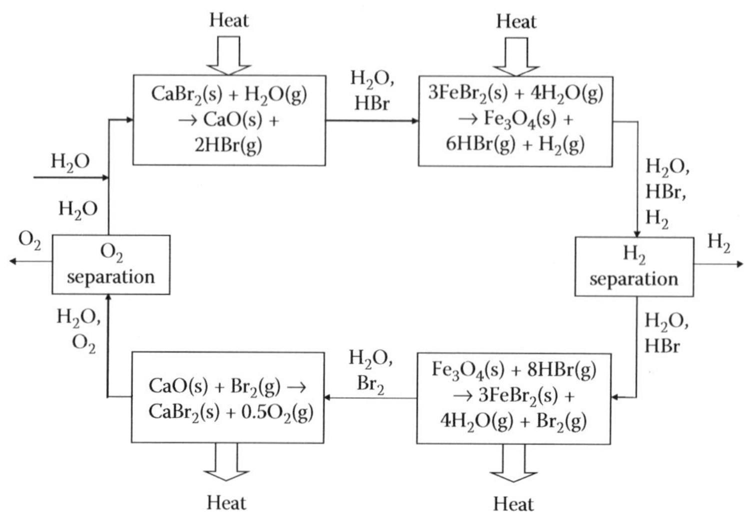

FIGURE 5.7
Conceptual flow sheet of UT-3 process.

process since 2001 as an alternative method of high-temperature steam electrolysis and IS process [78,98].

There were some problems on operation of reactions because all reactions were solid–gas reaction where solid reactants were used cyclically. Reaction rate of $CaBr_2$ hydrolysis (Equation 5.27) was slowest and rates of other reactions were limited by the rate.

$$CaBr_2 + H_2O \rightarrow CaO + 2HBr \tag{5.27}$$

Reactant pellets were degraded quickly in cyclic reaction due to the difference of molar volume between reactants and products. Preparation of pellets of the Ca compound [99,100], and the Fe compound [101,102] was investigated to overcome these problems. Modeling of reactors was also carried out [103–105]. Membrane separation was considered for separation of H_2 and O_2 from flow gas as an alternative of separation by condensation. Oxygen ion-conductive membranes were expected for O_2–H_2O separation. An amorphous silica-based membrane was studied for H_2 separation from H_2–H_2O–HBr mixture. Permeance and selectivity of a membrane were investigated experimentally [106] and exergy analysis of the process was done in various separator configurations [107].

A bench-scale test apparatus of 3 L/h hydrogen production, "MASCOT" (Model Apparatus for Studying Cyclic Operation of Tokyo) was constructed and demonstrated. Rough flow sheet of the original apparatus is shown in Figure 5.8. H_2O flow was separated into reactors containing $CaBr_2$ and $FeBr_2$. H_2 separation step was set at the outlet flow of the reactor containing $FeBr_2$. When reaction of solid reactants was completed, flow direction was changed to recover the original reactant in each reactor. Hydrogen production continued through the cycle procedure. The first cycle of the process was then completed. This method had an advantage of no solid transport. Continuous operation time was less than 2 h and hydrogen production rate was not stable in the original apparatus [108]. Process flow was modified to be simpler. The flow was the same as that in Figure 5.7 [103].

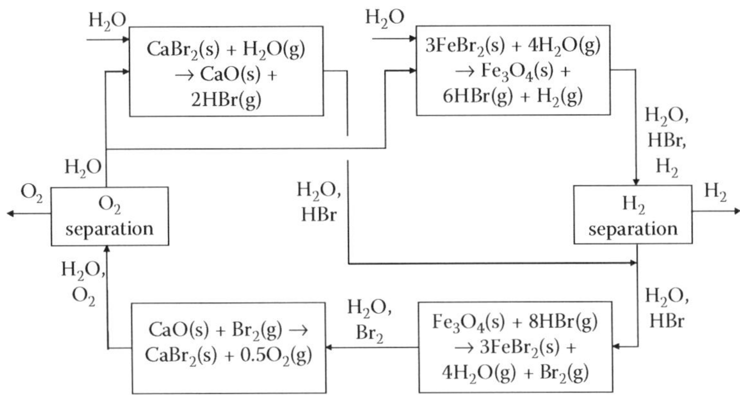

FIGURE 5.8
Flow sheet of original "MASCOT" of UT-3 process.

A continuous operation for 11 cycles with constant hydrogen production rate was successful in this modified facility [109]. Dynamic simulation of the total process was also conducted to confirm the stable operation method [110].

Thermal efficiency of the process was calculated. Influence of conversion ratio of Equations 5.27 and 5.28 and efficiency of heat exchange within the process were investigated [97].

$$3FeBr_2 + 4H_2O \rightarrow Fe_3O_4 + 6HBr + H_2 \tag{5.28}$$

Effectiveness of high-power generation efficiency and application of a hydrogen separation membrane with high recovery rate on higher thermal efficiency was shown [111]. Thermal efficiency by using membrane separation was compared with that using separation by condensation. Membrane separation was much advantageous in case of excess amount of H_2O over reaction stoichiometric ratio [98]. Hydrogen cost from a commercial UT-3 plant producing 20,000 Nm^3/h of hydrogen was evaluated based on standard chemical engineering cost estimation [10,111]. More economical hydrogen production may be possible by separation membranes with lower cost [111].

For reactions of Equations 5.27 and 5.28 with considerable conversion ratio, excess amount of H_2O (e.g., 40 times of stoichoimetric amount for $CaBr_2$ and 10 times of stoichoimetric amount for $FeBr_2$) and low pressure (below 5 mbar) were required due to overcome the positive reaction Gibbs energy over 100 kJ/mol [98]. This great amount of H_2O causes significant heat loss in heat exchange. When membrane separation of gas is applied, higher permeability of the membranes is required for the commercial size plant. Difference of reaction rates and switching of flow direction are problems for stable operation of the process. Stability of solid reactant particles against volume change with reaction, sintering and volatilization is important for long time operation.

5.3.2.2 Hybrid Ca–Br Cycle

A modified version of UT-3 process was proposed in the Secure Transportable Autonomous Reactor Hydrogen (STAR-H2) project, a part of NERI (see Chapter 14) [112]. The reactions are shown in Table 5.1. Direct decomposition of HBr (Equation 5.7) was applied instead of reactions of Fe compounds. Thermal decomposition of HBr is difficult considering reaction Gibbs energy. Therefore, other methods were proposed. Though HBr decomposition by plasma reaction at 100°C was assumed originally, this was replaced by electrolysis. Therefore, this modified process can be called hybrid Ca-Br process. Gas phase electrolysis was investigated because current density was low and separation of HBr-Br_2 was difficult in solution electrolysis [113]. Electrolysis by a cell using a poly[perfluorosulfonic] acid membrane was examined. Cell voltage was 0.66 V at 0.2 A/cm^2 of current density, which was close to equilibrium voltage of 0.58 V [2]. This suggested that cell resistance of the reaction was small. Thermal efficiency was calculated for the process applying plasma decomposition of HBr [114]. Two simultaneous reactions of $CaBr_2$ and H_2O, CaO and Br_2 (Equations 5.27 and 5.29) in one molten salt-based reactor were proposed to overcome the problem of disintegration of reactant particles.

$$CaO + Br_2 \rightarrow CaBr_2 + 0.5O_2 \tag{5.29}$$

However, small solubility of CaO in $CaBr_2$ and deposition of CaO at the inlet of the reactor made achievement of the idea difficult [115].

5.3.3 Chlorine Series

5.3.3.1 Fe–Cl Cycles

Several research groups found cycles using compounds of Fe and Cl as main reactant in 1970s. These cycles were considered as advantageous because components were not expensive and because knowhow of reactions was well known. The study on the series is summarized [9]. RWTH Aachen carried out a systematic screening of cycles [116]. Eighteen reactions of compounds of Fe, H, O, and Cl were selected which were thermodynamically feasible and suitable with respect to chemical engineering. 35 processes were constructed by using these reactions. Experimental confirmation of reactions was the main research theme for the cycles [117]. Chlorification of Fe_3O_4 with Cl_2 and/or HCl to iron chlorides ($FeCl_2$, $FeCl_3$, etc.), reverse Deacon reaction (Equation 5.30) and reduction of $FeCl_2$ with H_2 (Equation 5.31) were taken as feasible considering chemical engineering.

$$Cl_2 + H_2O \rightarrow 2HCl + 0.5O_2 \tag{5.30}$$

$$FeCl_2 + H_2 \rightarrow Fe + 2HCl \tag{5.31}$$

Hydrolysis of $FeCl_2$ was performed with a fixed-bed reactor charging pellets of $FeCl_2$. Following two-step reaction was recommended for the hydrolysis considering separation of gas products.

$$3FeCl_2 + 3H_2O \rightarrow 3FeO + 6HCl \tag{5.32}$$

$$3FeO + H_2O \rightarrow Fe_3O_4 + H_2 \tag{5.33}$$

Low-equilibrium decomposition ratio of $FeCl_3$ (exist as Fe_2Cl_6 dimer actually) as Equation 5.34 was considered as the greatest problem.

$$Fe_2Cl_6 \rightarrow 2FeCl_2 + Cl_2 \tag{5.34}$$

Condensation of produced $FeCl_2$ vapor and removal by scraping could increase the conversion of $FeCl_3$ to 70% in laboratory experiment. Recovery of Cl_2 from HCl without using $FeCl_3$ as an intermediate chemical was desired to avoid the problem, if possible. Though an alternative reaction,

$$Fe_3O_4 + 8HCl \rightarrow 3FeCl_2 + Cl_2 + 4H_2O \tag{5.35}$$

was proposed [118], thermodynamical difficulty was pointed out [33]. Decomposition of $FeCl_3$ with a two-step reaction by using CO (Equations 5.36 and 5.37) was proposed [119].

$$2FeCl_3 + CO \rightarrow 2FeCl_2 + COCl_2 \tag{5.36}$$

$$COCl_2 \rightarrow CO + Cl_2 \tag{5.37}$$

Flow sheet calculation of Ispra Mark 15 cycle (see Table 5.2) using the two-step $FeCl_3$ decomposition was carried out [119].

JRC's study [30] began with Ispra Mark 9 cycle. Experimental test, especially, of $FeCl_2$ hydrolysis reaction with a continuous moving bed reactor was conducted. Mark 15 cycle was made based on the consideration of problems of the cycle series. The results and problems they found through experiments were similar to those in RWTH Aachen. Moving bed reactors for $FeCl_2$ hydrolysis were not suitable for scale-up because of gas velocity. Low conversion ratio of $FeCl_3$ decomposition reactions was also a problem. These problems were the reasons of low thermal efficiency and high hydrogen production cost compared with electrolysis of water. Though the problem of $FeCl_2$ hydrolysis was expected to be solved by introduction of the two-step hydrolysis, Equations 5.32 and 5.33, it is concluded that this idea was not enough for the improvement of efficiency [33].

5.3.3.2 Cu–Cl Cycles

A series of hybrid processes using Cu and Cl has been researched [120] mainly in Canada in collaboration among University of Ontario Institute of Technology (UOIT), Canada, Atomic Energy of Canada Limited (AECL), ANL, and partner institutes in the framework of Generation IV International Forum (GIF) [121]. This process was based on H-6 process in the selection by Gas Research Institute (GRI, present Gas Technology Institute (GTI)), USA [122].

Several processes with minor difference have been proposed [123]. Some of these processes are described in Table 5.1. Conceptual flow sheet of a five-step cycle is shown in Figure 5.9. Solid Cu reacts with HCl gas and molten CuCl is produced with H_2. The CuCl is cooled to form a solid and fed into an electrolysis cell. The CuCl solution is electrolyzed into $CuCl_2$ solution and solid Cu. Cu is recycled to the first reaction. $CuCl_2$ solution or slurry is dried to powder by a spray dryer. The powder reacts with H_2O gas and solid $CuO \cdot CuCl_2$ is made. The $CuO \cdot CuCl_2$ is thermally decomposed and molten CuCl and O_2 are produced. The temperature is suitable for the heat source of Generation IV Super-Critical Water Cooled Reactor (SCWR) [121]. An alternative process (four-step Cu–Cl cycle) was proposed applying electrolysis of CuCl and HCl solution in one cell instead of the electrolysis reaction of CuCl to $CuCl_2$ and Cu. And this process is selected as a promising one considering viability of reactions, engineering feasibility and expected thermal efficiency [73].

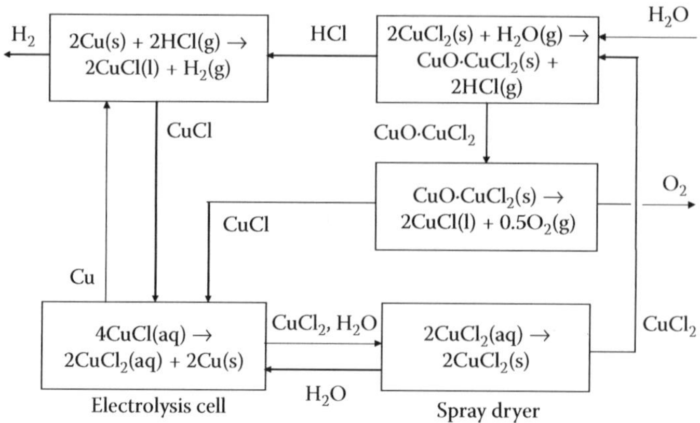

FIGURE 5.9
Conceptual flow sheet of Cu–Cl process (five-step cycle).

Experimental investigation of reactions has been carried out because various types of reaction are included in the process. Electrolysis of both the five-step and the four-step cycles, drying of $CuCl_2$ solution/slurry, hydrolysis of $CuCl_2$ and decomposition of $CuO \cdot CuCl_2$ have been examined [7,121,124]. Viability of these reactions was verified chemically by these studies. Thermal efficiency was calculated by flow-sheet analysis for both the processes. Effect of reaction temperature of Cu and HCl reaction [125] and application of heat pump [126] were evaluated for the five-step cycle. Cost analysis based on the flow sheet and a pilot-plant design was also carried out. A sensitivity analysis of hydrogen production rate (from 1 to 10 t-H_2/day) and capacity factor of the scaling method on the cost was carried out [127]. Thermal efficiency and hydrogen cost calculation was done for the four-step process [7].

Transport of Cu components is a remaining problem. Chocking or clogging of the feeder of the $CuO \cdot CuCl_2$ decomposition reactor may occur by reactant particles. Cation exchange membranes without permeation of Cu species are required for electrolysis in the four-step cycle [121]. Higher performance of electrolytic cell should be attained to achieve high-thermal efficiency [73].

5.3.4 Hybrid Hydrogen Process in Lower Temperature

A thermochemical and electrical hybrid hydrogen process in lower temperature (HHLT) has been proposed and researched by Japan Nuclear Cycle Development Institute (JNC), presently known as JAEA after merger with JAERI [128]. This process is shown as in Table 5.1.

This process is a modification of the Westinghouse process. Decomposition of SO_3 (Equation 5.24) is a thermal reaction around 900°C in the Westinghouse process. Electrolysis of SO_3 similar to solid oxide fuel cells (SOFC) was applied to decrease the temperature. Heat source of lower temperature around 500°C, such as in the Fast Breeder Reactor (FBR), can be used for the process. Components in this process are the same as the Westinghouse process except SO_3 electrolysis. Results of the research on the Westinghouse process shown in Chapter 18 can be applicable to common components of HHLT process. Specific parts of this process are explained here. Electric resistance of the SO_3 electrolysis was large, below 500°C, though lower temperature electrolysis is required considering the temperature of heat source. Therefore, investigation of the detailed reaction system is important to achieve the lower temperature electrolysis. Electrolysis of SO_3 was conducted with an oxygen conductive membrane like 8 mol% yttria-stabilized zirconia (YSZ). Electrolysis experiments showed resistances at anode and cathode, reactions were the main reason of cell resistance. Rate-determining process in the electrolysis was not transport of O^{2-} through the membrane but electrode reactions. Identification of the rate-determining step within the cathode reaction as shown in Figure 5.10 was investigated through detailed modeling of the electrolysis [129]. First-principle calculation of dissociation of a SO_3 molecule on the surface of the Pt electrode was carried out [130]. Operations of a laboratory scale test apparatus of the total process were conducted. Two demonstrations were performed for total of 56 h with stable oxygen production of about 5 mL/h. However, hydrogen production was not constant. A process apparatus of 1 L/h hydrogen production scale was constructed and test operated [131]. Thermal efficiency of the process was roughly estimated. High concentration of H_2SO_4 by SO_2 electrolysis (Equation 5.6) and efficient heat recovery within the process were necessary to obtain high thermal efficiency [132]. A plant design with specification of major components was carried out [133].

5.3.5 Mg–I Process

Mg–I process was proposed by NCLI (now, AIST), Japan [59]. This process had been studied by the institute in the earlier half of 1980s. This process is shown in Table 5.1.

The research is summarized in literatures [59,134]. Figure 5.11 is the schematic flow sheet. The reaction of MgO and I_2 proceeded when excess amount of I_2 was introduced to the reactor. A by-product, expressed as $a\text{Mg(OH)}_2{\cdot}\text{MgI}_2{\cdot}b\text{Mg(IO}_3)_2{\cdot}c\text{I}_2$ (a, b, and c are constants), was made in the reaction. The by-product could be changed into $Mg(IO_3)_2$ by the reaction with dilute MgI_2–I_2 solution fed from a later reaction. $Mg(IO_3)_2$ was thermally decomposed easily into MgO, I_2, and O_2. The former two compounds returned into the first reaction. There were problems in separation of H_2O and I_2 by evaporation from the concentrated MgI_2–I_2 solution obtained from the first reaction: a side reaction of MgI_2 hydrolysis and requirement of large amount of heat. The first problem was overcome by adding MgO to form stable compound $3Mg(OH)_2{\cdot}MgI_2$. For the latter problem, recovery of latent heat in

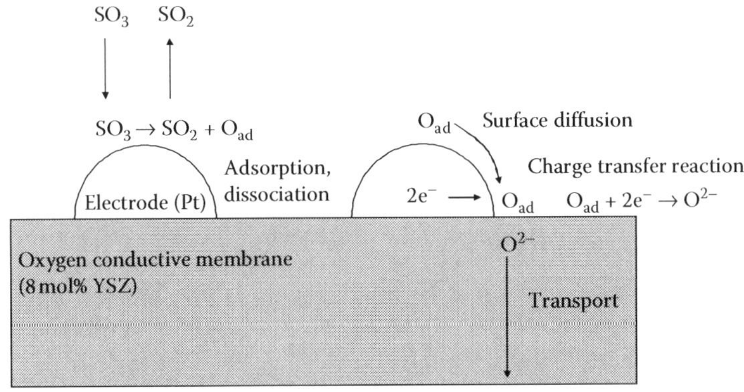

FIGURE 5.10
Cathodic reaction mechanism in SO_3 electrolysis.

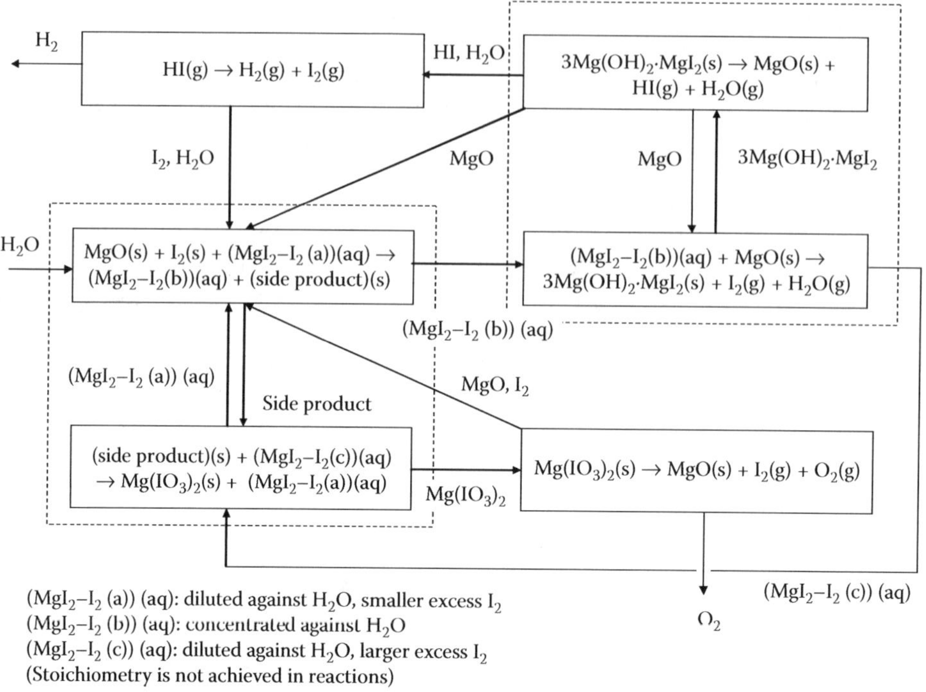

FIGURE 5.11
Conceptual flow sheet of Mg–I process.

the evaporation by a steam turbine was proposed. Screening of catalysts [59,135] and measurement of reaction kinetics [136,137] were carried out for the HI decomposition reaction. However, gas phase electrolysis of HI was proposed later to eliminate the problem of product separation. Heat/mass balance and thermal efficiency evaluation of the process was carried out based on the experimental result and a flow-sheet study [134].

The challenge of the process seems to be the handling of solid reactants because decomposition reactions of Mg compounds proceed in several steps. Chemical and thermodynamic properties of the compounds are not yet known well. Therefore, detailed analysis of the reactions is required.

5.4 Future Technical Challenges

Thermochemical processes proved to be challenging. While many challenges appeared through the research have been overcome, others as described below remain to be resolved before the processes become methods of industrial production of hydrogen.

5.4.1 Chemical Reaction

Some reactions have low equilibrium conversion ratio because of their positive Gibbs energy, such as HI decomposition in IS process and Mg–I process (Equation 5.4) and $FeCl_3$

decomposition in Fe–Cl processes (Equation 5.34). This generates great amount of recycling flow and low thermal efficiency. The processes become complicated if such reactions are replaced by several reactions in order to avoid the difficulty. Nonequilibrium reaction using membrane reactors and absorption of product is effective to improve reaction ratio. However, the technique is not established yet. Reaction rate of some reactions, especially solid reactions, are low like hydrolysis of $CaBr_2$ in UT-3 process (Equation 5.27) and $FeCl_2$ hydrolysis in Fe–Cl processes (Equations 5.32 and 5.33). Particles of solid reactants which do not limit reaction rate should be prepared.

5.4.2 Construction Material

Many thermochemical processes include severe environment for component materials. Materials that resist high temperature (maximum 900°C for nuclear-heat source and higher for solar-heat source) and corrosion (e.g., halogen, hydrogen halide, sulfuric acid) are required. Though rare metals, ceramics and lined materials [40,41,57,138–142] have been proposed and tested, applicability is not yet fully verified. For example, thermal shock and hydrogen embrittlement should also be considered for process reactor and equipment based on candidate materials. Refer to Chapter 21 for detailed study and technical resolutions of the material challenges.

5.4.3 Chemical Engineering

Separation of objective product, by-products, and nonreacted reactants has some difficulties, especially in the mixture of the same phase. When separation of objective product is not enough, recycling of the flow is required and flow rate becomes larger. Complete separation is not possible by phase separation in many cases because of the similar volatility of compounds. Purification to remove impurities is required because the impurities may cause unexpected corrosion and stacking of flow. However, research on the purification method has not progressed yet. Leakage of cycling compounds with H_2 and O_2 is a problem, too. Even a small leakage requires great amount of makeup in the long run.

5.4.4 Process Thermal Efficiency

Table 5.4 summarizes thermal efficiency by studying the flow sheets. Special care is needed in comparing the values because supposition of each evaluation is different especially in electric-energy recovery from waste heat. Thermal efficiencies based on practical assumptions made today reach around 35–45%. Future development may increase the value, since the ideal thermal efficiency of the total water-splitting system is as high as 90% assuming a heat source of 1123.15 K (= 900°C) as shown in Figure 5.6. The ideal efficiency is achieved only in a reversible process which takes an infinite time. Some part of the difference between the ideal value and the calculation values cannot be reduced to attain reasonable hydrogen-production rate (e.g., efficiency of electric-energy generation, temperature difference at heat exchanger, reflux ratio at distillation columns). Nevertheless, some part of the difference can be reduced by the optimization of the process flow sheet and improved process component designs. The substantial results of the work that has been performed in these areas are presented in Section IV of this book. However, when high thermal efficiency is aimed, special components like membrane separation, components for high temperature and pressure, and many and/or large heat exchangers are generally required. Such high-efficiency processes tend to lead

TABLE 5.4

Thermal Efficiency of Thermochemical Cycles Based on Flow Sheet Analysis

Cycle	Thermal Efficiency (%)	Note	Author	Reference
IS (GA process, Ispra Mark 16, S-I)	41	42% in original paper. However, calculation method is not clear. 689.253 kJ/mol-H_2 of heat and 5.35 kJ/mol-H_2 of power are demanded. The value in this table is recalculated by Equation 5.9 applying 40% electric energy generation efficiency.	Brown et al. (2003)	[69]
	57	No temperature difference in heat exchangers. No electricity loss in an electro-electrodialysis cell. Electricity recovery from waste heat with efficiency of 15%.	Kasahara et al. (2003)	[143]
	36	The same H_2SO_4 procedure as [70]. Heat pump is applied in HI reactive distillation column.	Goldstein et al. (2005)	[144]
	34	Temperature difference in heat exchangers and electricity loss in an electro-electrolyalysis cell are considered. No electricity recovery from waste heat.	Kasahara et al. (2007)	[85]
Westinghouse (Hybrid Sulfur, Ispra Mark 11)	45	Electricity recovery from waste heat above 500 K with efficiency of 30%. Work for electrolyzer and SO_2/O_2 separator is twice of the ideal Gibbs energy.	Carty et al. (1977)	[145]
	42	SO_2 electrolysis cell potential is –0.6 V.	Gorensek and Summers (2009)	[5]
UT-3	45	Almost all work in the process can be supplied from electricity recovery from waste heat. Membrane gas separator is applied.	Tadokoro et al. (1997)	[111]
	22	Electricity recovery from waste heat with efficiency of 35%. Membrane gas separator and reaction of H_2O and Br_2 by UV are applied.	Lemort et al. (2006)	[98]
Fe–Cl (Ispra Mark 15)	35	Two-step $FeCl_3$ decomposition is applied. Heat exchange and separation were modified by pinch analysis.	Cremer et al. (1980)	[119]
Cu–Cl	51	43% in LHV base (see Section 5.1.3) in original paper. The value in the table is recalculated by Equation 5.9. Conversion ratio of reactions is assumed 100%. Efficiency of electric energy generation is 50%.	Lewis et al. (2005)	[124]
Mg–I	25	All work in the process can be supplied from electricity recovery from waste heat.	Fujii et al. (1982)	[134]

to high hydrogen-production cost. Tradeoff of thermal efficiency and cost of process construction should be considered in practice.

Nomenclature

Q	Heat input (kJ/mol-H_2)
T	Temperature (K)
T_d	The temperature of $\Delta H = 0$ (= 4310 K in the case of decomposition of water)
S_{loss}	Entropy loss within a water splitting part (kJ/(K · mol-H_2))
W	Work (kJ/mol-H_2)
ΔE	Reaction exergy ($\Delta E = \Delta H - T_0\Delta S$) (kJ/mol-$H_2$)
ΔG	Reaction Gibbs energy (kJ/mol-H_2)
ΔH_{HHV}	"Higher Heat Value" enthalpy change of the reaction $H_2O(l) \rightarrow H_2(g) + 0.5O_2(g)$ at 298.15 (K) (= 285.83 (kJ/mol-H_2))
ΔH	Reaction enthalpy (kJ/mol-H_2)
ΔS	Reaction entropy (kJ/(K · mol-H_2))
η	Thermal efficiency (–)

Subscripts

0	Heat sink (at 298.15 (K))
ad	Adsorbed on the surface of electrode
CC	Carnot cycle
elec.	Electricity generation from heat
HP	Heat pump
i	Reaction number
ideal	Ideal case, that is $S_{loss} = 0$
in	Input from heat source
out	Output to heat sink
s	Heat source
total	Total process of heat source → hydrogen → electricity
WG	Work generation from hydrogen
WS	Water splitting

References

1. Dokiya, M., K. Fukuda, T. Kameyama, Y. Kotera, and S. Asakura. 1977. The study of thermochemical hydrogen preparation (II) electrochemical hybrid cycle using sulfur–iodine system. *Denki Kagaku* 45:139–43.
2. Sivasubramanian, P., R. P. Ramasamy, F. J. Freire, C. E. Holland, and J. W. Weidner. 2007. Electrochemical hydrogen production from thermochemical cycles using a proton exchange membrane electrolyzer. *Int. J. Hydrogen Energy* 32:463–8.
3. Steimke, J. L. and T. J. Steeper. 2007. Generation of hydrogen using electrolyzer with sulfur dioxide depolarized anode. *AIChE 2007 Annual Meeting* 224a, Salt Lake City, USA, November 4–9.

4. Summers, W. A., J. L. Steimke, D. T. Hobbs, H. R. Colon-Mercado, and M. B. Gorensek. 2008. Development of a sulfur dioxide depolarized electrolyzer for hydrogen production using the hybrid sulfur thermochemical process. *Proc. 4th International Topical Meeting High Temperature Reactor Technology*. HTR2008–58196, Washington, DC, USA, September 28–October 1.
5. Gorensek, M. B. and W. A. Summers. 2009. Hybrid sulfur flow sheets using PEM electrolysis and a Bayonet decomposition reactor. *Int. J. Hydrogen Energy* 34:4097–114.
6. McLaughlin, D. F., S. A. Paletta, E. J. Lahoda, and W. Kriel. 2006. Revised capital and operating HyS hydrogen production costs. *Proc. 2006 Int. Congress Advances Nucl. Power Plants* 2263–9, Reno, USA, June 4–8.
7. Ferrandon, M., M. Lewis, D. Tatterson, and A. Zdunek. 2008. Status of the development effort for the thermochemical Cu–Cl cycle. *AIChE 2008 Annual Meeting*. 328d, Philadelphia, USA, November 16–21.
8. Sharer, J. C. and J. B. Pangborn. 1975. Hydrogen production by thermochemical methods. In *Survey of Hydrogen Production and Utilization Methods Volume 2. Discussion*, 108–144. Institute of Gas Technology, *N7615591*.
9. Sato, S. 1979. Thermochemical hydrogen production. In *Solar-Hydrogen Energy Systems*, ed. T. Ohta, 81–114. Pergamon Press, Oxford.
10. Aochi, A., T. Tadokoro, K. Yoshida, M. Nobue, and T. Yamaguchi. 1989. Economical and technical evaluation of UT-3 thermochemical hydrogen production process for an industrial scale plant. *Int. J. Hydrogen Energy* 14:421–9.
11. Collett, E. 1925. Improved process for the production of hydrogen. *Patent GB232431(A)*. April 23.
12. Funk, J. E. and R. M. Reinstrom. 1966. Energy requirements in the production of hydrogen from water. *I&EC Process Design and Development* 5:336–42.
13. Funk, J. E. and R. M. Reinstrom. 1964. System study of hydrogen generation by thermal energy: Final report. *GM Report TID20441* 2(A).
14. de Beni, G. 1974. Consideration on iron-chlorine-oxygen reactions in reaction to thermochemical water splitting. *Proc. Hydrogen Economy Miami Energy (THEME) Conf.* S11-13 – S11-22, Miami Beach, USA, March 18–20.
15. Hickman, R. G., O. H. Krikorian, and W. J. Ramsey. 1974. Thermochemical hydrogen production research at Lawrence Livermore Laboratory. *Proc. Hydrogen Economy Miami Energy (THEME) Conf.* S11-23–S11-33, Miami Beach, USA, March 18–20.
16. Dorner, S. and C. Keller. 1974. Hydrogen production from decomposition of water by means of nuclear reactor heat. *Proc. Hydrogen Economy Miami Energy (THEME) Conf.* S3-37–S3-47, Miami Beach, USA, March 18–20.
17. Funk, J. E., W. L. Cogner, and R. H. Carty. 1974. Evaluation of multi-step thermochemical processes for the production of hydrogen from water. *Proc. Hydrogen Economy Miami Energy (THEME) Conf.* S11-1–S11-11, Miami Beach, USA, March 18–20.
18. Pangborn, J. B. and J. C. Sharer. 1974. Analysis of thermochemical water-splitting cycles. *Proc. Hydrogen Economy Miami Energy (THEME) Conf.* S11-35–S11-47, Miami Beach, USA, March 18–20.
19. Knoche, K. F., H. Cremer, and G. Steinborn. 1975. Thermochemical process for the production of hydrogen. *Proc. Hydrogen Energy Fundamentals A Symposium-course*. S2-29, Miami Beach, USA, March 3–5.
20. Bamberger, C. E. 1978. Hydrogen production from water by thermochemical cycles; a 1977 update. *Cryogenics* 8:170–83.
21. Yalçin, S. 1989. A review of nuclear hydrogen production. *Int J Hydrogen Energy* 14:551–61.
22. Yoshida, K., H. Kameyama, and K. Toguchi. 1976. A computer-aided search procedure for thermochemical water-decomposition processes. *Int. J. Hydrogen Energy* 1:123–7.
23. Russell, J. L., Jr., K. H. McCorkle, J. H. Norman, et al. 1976. Water splitting—A progress report. *Proc. 1st World Hydrogen Energy Conf.* 1A105-1A124, Miami Beach, USA, March 1–3.
24. Kameyama, H., K. Yoshida, and D. Kunii. 1976. Hydrogen production by thermochemical water decomposition method, *Proc. 10th Autumn Meeting of the Soc. of Chem. Eng.* Nagoya, Japan, F-311.

25. Funk, J. E. 2001. Thermochemical hydrogen production: past and future. *Int. J. Hydrogen Energy* 26:185–90.
26. Ohta, T. and M. V. C. Sastri. 1979. Hydrogen energy research programs in Japan. *Int. J. Hydrogen Energy* 4:489–98.
27. Beghi, G. E. 1980. Recent review of thermochemical hydrogen production. *Proc. 3rd World Hydrogen Energy Conf.* 1731–48, Tokyo, Japan, June 23–26.
28. Weirich, W., K. F. Knoche, F. Behr, and H. Barnert. 1984. Thermochemical processes for water splitting—status and outlook. *Nucl. Eng. Des.* 78:285–91.
29. Gillis, J. C., H. C. Maru, and J. C. Sharer. 1975. Survey of hydrogen production and utilization methods Volume 3. Appendixes, Institute of Gas Technology, *N7615592*, Appendix B.
30. Beghi, G. E. 1986. A decade of research on thermochemical hydrogen at the Joint Research Center, Ispra. *Int. J. Hydrogen Energy* 11:761–71.
31. de Beni, G. 1969. Mark 1 as an example of a "direct process" for producing hydrogen. In *Round Table on Direct Production of Hydrogen with Nuclear Heat*, pp. 31–62. Joint Research Center Ispra. *EUR/C-IS/1062/1/69e.*
32. Joint Nuclear Research Center Ispra Establishment. 1972. Hydrogen production from water using nuclear heat Progress report No. 1, ending December 1970. *EUR-4776e.*
33. van Velzen, D. and H. Langenkamp. 1978. Problems around Fe–Cl Cycles, *Int. J. Hydrogen Energy* 3:419–29.
34. de Beni, G., G. Pierini, and B. Spelta. 1980. The reaction of sulphur dioxide with water and a halogen. The case of iodine: Reaction in presence of organic solvents, *Int. J. Hydrogen Energy* 5:141–9.
35. Broggi, A., R. Joels, M. Morbello, and B. Spelta. 1977. OPTIMO—A method for process evaluation applied to the thermochemical decomposition of water. *Int. J. Hydrogen Energy* 2:23–30.
36. Schütz. G. and P. Fiebelmann. 1974. *EUR-COM 3234.*
37. Onstott, E. I. and M. G. Bowman. 1974. Hydrogen production by low voltage electrolysis in combined thermochemical and electrochemical cycles. *Proc. 146th Meeting of Electrochem. Soc.*, New York City, USA, October 13–17.
38. van Velzen, D. and H. Langenkamp. 1980. Status report on the operation of the bench-scale plant for hydrogen production by the Mark-13 process. *Proc. 3rd World Hydrogen Energy Conf.*, pp. 423–38, Tokyo, Japan, June 23–26.
39. Mertel, G., H. Dworschak, A. Broggi, and G. Vassallo. 1986. The thermal decomposition of sulphuric acid by the CRISTINA process. *Proc. 6th World Hydrogen Energy Conf.* 673–87, Vienna, Austria, July 20–24.
40. Porsini, F. C. and G. Imarisio. 1976. The compatibility of containment materials for thermochemical hydrogen production. *Proc. 1st World Hydrogen Energy Conf.* 7A-3–7A57, Miami Beach, USA, March 1–3.
41. Porsini, F. C. 1982. Long term corrosion tests of materials for thermal decomposition of sulphuric acid. *Proc. 4th World Hydrogen Energy Conf.* 2:471–82, California, USA, June 13–17.
42. Neumann, D. 1987. Phasengleichgewichte von HJ/H2O/J2-Lösungen, PhD diss., RWTH Aachen.
43. Engels, H. and K. F. Knoche. 1986. Vapor pressures of the system $HI/H_2O/I_2$ and H_2, *Int. J. Hydrogen Energy* 11:703–7
44. Roth, M. and K. F. Knoche. 1989. Thermochemical water splitting through direct HI-decomposition from $H_2O/HI/I_2$ solutions. *Int. J. Hydrogen Energy* 14:545–9.
45. Knoche, K. F., H. Schepers, and K. Hesselmann. 1984. Second law and cost analysis of the oxygen generation step of the general atomic sulfur–iodine cycle. *Proc. 5th World Hydrogen Energy Conf.* 487–502, Toronto, Canada, July 15–20.
46. Berndhäuser, C. and K. F. Knoche. 1994. Experimental investigations of thermal HI decomposition from H_2O–HI–I_2 solutions. *Int. J. Hydrogen Energy* 19:239–44.
47. Norman, J. H., G. E. Besenbruch, and D. R. O'Keefe. 1981. Thermochemical water-splitting for hydrogen production, final report (January 1975–December 1980). *GRI-80/0105.*

48. Norman, J. H., G. E. Besenbruch, L. C. Brown, D. R. O'Keefe, and C. L. Allen. 1982. Thermochemical water-splitting cycle, bench-scale investigations, and process engineering, final report for the period February 1977 through December 31, 1981. *DOE/ET/26225-1.*
49. Booth, L. A. and J. D. Balcomb. 1973. Nuclear heat and hydrogen in future energy utilization. *LA-5456-MS.*
50. Farbman, G. H. 1979. Hydrogen production by the Westinghouse sulfur cycle process: program status. *Int. J. Hydrogen Energy* 4:111–22.
51. Lu, P. W. T. and R. L. Ammon. 1980. Development status of electrolysis technology for the sulfur cycle hydrogen production process. *Proc. 3rd World Hydrogen Energy Conf.* 1:439–61, Tokyo, Japan, June 23–26.
52. Farbman, G. H. and G. H. Parker. 1979. The sulfur-cycle hydrogen production process. *Proc. Hydrogen: Production and Marketing*, 359–89, Honolulu, USA, April 2–6.
53. Carty, R. H. and W. L. Conger. 1980. A heat penalty and economic analysis of the hybrid sulfuric acid process. *Int. J. Hydrogen Energy* 5:7–20.
54. Lin, S. S. and R. Flaherty. 1982. Design studies of the sulfur trioxide decomposition reactor for the sulfur cycle hydrogen production process. *Proc. 4th World Hydrogen Energy Conf.* 2:599–610, California, USA, June 13–17.
55. Summers, W. A., M. B. Gorensek, and M. R. Buckner. 2005. The hybrid sulfur cycle for nuclear hydrogen production. *Proc. GLOBAL 2005*, paper097, Tsukuba, Japan, October 9–13.
56. Brecher, L. E., S. Spewock, and C. J. Warde. 1976. The Westinghouse sulfur cycle for the thermochemical decomposition of water. *Proc. 1st World Hydrogen Energy Conf.* 9A1–9A-13, Miami Beach, USA, March 1–3.
57. Farbman, G. H. and V. Koump. 1977. Hydrogen generation process, final report, *FE-2262-15.*
58. Westinghouse Electric Corporation. 1983. Solar thermal hydrogen production process, final report, January, 1978–December, 1982. *DOE/ET/20608-1.*
59. Kondo, W., S. Mizuta, T. Kumagai, Y. Oosawa, Y. Takemori, and K. Fujii. 1979. The magnesium–iodine cycle for the thermochemical decomposition of water. *Proc. 2nd World Hydrogen Energy Conf.* 2:909–21, Zurich, Switzerland, August 21–24.
60. Mizuta, S. and T. Kumagai. 1986. Progress report on the Mg–S–I thermochemical water-splitting cycle—continuous flow demonstration. *Proc. 6th World Hydrogen Energy Conf.* 696–705, Vienna, Austria, July 20–24.
61. Dokiya, M., T. Kameyama, and K. Fukuda. 1979. Thermochemical hydrogen preparation–Part V. A feasibility study of the sulfur iodine cycle. *Int. J. Hydrogen Energy* 4:267–77.
62. Sato, S., S. Shimizu, H. Nakajima, K. Onuki, and Y. Ikezoe. 1984. Studies on the nickel–iodine–sulfur process for hydrogen production. III. *Proc. 5th World Hydrogen Energy Conf.* 457–65, Toronto, Canada, July 15–20.
63. Onuki, K., S. Shimizu, H. Nakajima, Y. Ikezoe, and S. Sato. 1986. Studies on the methanol–iodine–sulfur process. *Proc. 6th World Hydrogen Energy Conf.* 723–31, Vienna, Austria, July 20–24.
64. Onuki, K., S. Shimizu, H. Nakajima, et al. 1990. Studies on an iodine–sulfur process for thermochemical hydrogen production. *Proc. 8th World Hydrogen Energy Conf.* 547–56, Honolulu and Waikoloa, USA, July 22–27.
65. Onuki, K., H. Nakajima, I. Ioka, M. Futakawa, and S. Shimizu. 1994. IS Process for thermochemical hydrogen production. *JAERI-Review* 94-006.
66. Nakajima, H., M. Sakurai, K. Ikenoya, G.-J. Hwang, K. Onuki, and S. Shimizu. 1999. A study on a closed-cycle hydrogen production by thermochemical water-splitting IS process. *Proc. 7th Int. Conf. Nucl. Eng.* ICONE-7104, Tokyo, Japan, April 19–23.
67. Steinfeld, A., P. Kuhn, A. Reller, R. Palumbo, J. Murray, and Y. Tamaura. 1998. Solar-processed metals as clean energy carriers and water-splitters. *Int. J. Hydrogen Energy* 23:767–74.
68. Elder, R. and R. Allen. 2009. Nuclear heat for hydrogen production: Coupling a very high/high temperature reactor to a hydrogen production plant. *Prog. Nucl. Energy* 51:500–25.
69. Brown, L. C., G. E. Besenbruch, R. D. Lentsch, et al. 2003. High efficiency generation of hydrogen fuels using nuclear power, final technical report for the period August 1, 1999 through September 30, 2002. *GA-A24285.*

70. Moore, R., E. Parma, B. Russ, et al. 2008. An integrated laboratory-scale experiment on the sulfur—iodine thermochemical cycle for hydrogen production. *Proc. 4th International Topical Meeting High Temperature Reactor Technology*. HTR2008–58225, Washington, DC, USA, September 28-October 1.
71. Russ, B. and P. Pickard. 2009. 2009 DOE Hydrogen program review sulfur–iodine thermochemical cycle. *U. S. Department of Energy Hydrogen Program 2009 Annual Merit Review Proc.*, Arlington, USA, May 18–22, http://hydrogendoedev.nrel.gov/annual_review09_production.html (accessed in August 12, 2009).
72. Schultz, K., C. Sink, P. Pickard, et al. 2007. Status of the U.S. nuclear hydrogen initiative. *Proc. 2007 Int. Congress Advances Nucl. Power Plants (ICAPP' 2007)*. Paper 7530, Nice, France, May 13–18.
73. Lewis, M. A. and J. G. Masin. 2009. The evaluation of alternative thermochemical cycles—Part II: The down-selection process. *Int. J. Hydrogen Energy* 34:4125–35.
74. le Duigou, A., J.-M. Borgard, B. Larousse, et al. 2005. Hythec: A search for a long term massive hydrogen production route. *Proc. Int. Hydrogen Energy Congress and Exhibition 2005*. 1.3 HPT218, Istanbul, Turkey, July 13–15.
75. le Duigou, A., J.-M. Borgard, B. Larousse, et al. 2007. HYTHEC: An EC funded search for a long term massive hydrogen production route using solar and nuclear technologies. *Int. J. Hydrogen Energy* 32:1516–29.
76. le Duigou, A., J.-M. Borgard, B. Larousse, et al. 2007. Recent progress in EC funded project HYTHEC on massive scale hydrogen production via thermochemical cycles. *Proc. AIChE 2007 Annual Meeting*. 89a, Salt Lake City, USA, November 4–9.
77. Poitou, S., G. Rodriguez, N. Haquet, et al. 2008. Development program of the SO_3 decomposer as a key component of the sulphur based thermochemical cycles: News steps towards feasibility demonstration. *Proc. 17th World Hydrogen Energy Conf.* paper 345, Brisbane, Australia, June 15–19.
78. le Duigou, A., X. Vitart, P. Anzieu, et al. 2003. The CEA program for massive hydrogen production from nuclear. *Proc. GLOBAL 2003*. 1470–1477, New Orleans, USA, November 16–20.
79. Rodriguez, G., J. C. Robin, P. Billot, et al. 2006. Development program of a key component of the Iodine-Sulfur thermochemical cycle: The SO_3 decomposer. *Proc. 16th World Hydrogen Energy Conf.* S04–300, Lyon, France, June 13–16.
80. Giaconia, A., G. Caputo, A. Ceroli, et al. 2007. Experimental study of two phase separation in the Bunsen section of the sulfur–iodine thermochemical cycle. *Int. J. Hydrogen Energy* 32:531–6.
81. Caputo, G., C. Felici, P. Tarquini, A. Giaconia, and S. Sau. 2007. Membrane distillation of HI/H_2O and H_2SO_4/H_2O mixtures for the sulfur–iodine thermochemical process. *Int. J. Hydrogen Energy* 32:4736–43.
82. Barbarossa, V., S. Brutti, M. Diamanti, S. Sau, and G. de Maria. 2006. Catalytic thermal decomposition of sulphuric acid in sulphur–iodine cycle for hydrogen production. *Int. J. Hydrogen Energy* 31:883–90.
83. Lanchi, M., A. Ceroli, R. Liberatore, et al. 2009. S–I thermochemical cycle: A thermodynamic analysis of the HI–H_2O–I_2 system and design of the HIx decomposition section. *Int. J. Hydrogen Energy* 34:2121–32.
84. Onuki, K., H. Nakajima, S. Kubo, et al. 2002. Thermochemical hydrogen production by iodine-sulfur cycle. *Proc. 14th World Hydrogen Energy Conf.* Montreal, Canada, June 9–13.
85. Kasahara, S., S. Kubo, R. Hino, K. Onuki, M. Nomura, and S. Nakao. 2007. Flow sheet study of the thermochemical water-splitting iodine–sulfur process for effective hydrogen production. *Int. J. Hydrogen Energy* 32:489–96.
86. Onuki, K., S. Kubo, A. Terada, N. Sakaba, and R. Hino. 2009. Thermochemical water-splitting cycle using iodine and sulfur. *Energy Environ. Sci.* 2:491–7.
87. Kubo, S., H. Nakajima, S. Shimizu, K. Onuki, and R. Hino. 2005. A bench scale hydrogen production test by the thermochemical water-splitting iodine-sulfur process. *Proc. GLOBAL 2005*. paper 474, Tsukuba, Japan, October 9–13.

88. Chang, J. and W.-J. Lee. 2008. Status of nuclear hydrogen development and demonstration program in Korea. *Proc. 16th Pacific Basin Nuclear Conf.* P16P1248, Aomori, Japan, October 13–18.
89. Bae, K.-K., C.-S. Park, C.-H. Kim, et al. 2006. Hydrogen production by thermochemical water-splitting IS process. *Proc. 16th World Hydrogen Energy Conf.* S04–280, Lyon, France, June 13–16.
90. Hong, S.-D., J.-K. Kim, B.-K. Kim, S.-I. Choi, K.-K. Bae, and G.-J. Hwang. 2007. Evaluation on the electro-electrodialysis to concentrate HI from HIx solution by using two types of the electrode. *Int. J. Hydrogen Energy* 32:2005–9.
91. Lee, B. J., H. C. No, H. J. Yoon, S. J. Kim, and E. S. Kim. 2008. An optimal operating window for the Bunsen process in the I–S thermochemical cycle. *Int. J. Hydrogen Energy* 33:2200–10.
92. Lee, B. J., H. C. No, H. J. Yoon, H. G. Jin, Y. S. Kim, and J. I. Lee. 2009. Development of a flow sheet for iodine-sulfur thermochemical cycle based on optimized Bunsen reaction. *Int. J. Hydrogen Energy* 34:2133–43.
93. Zhang, P., S. Chen, L. Wang, and J. Xu. 2008. Overview of hydrogen production through iodine-sulfur process at INET. *Proc. 16th Pacific Basin Nuclear Conf.* P16P1303, Aomori, Japan, October 13–18.
94. Kodama, T. and N. Gokon. 2007. Thermochemical cycles for high-temperature solar hydrogen production. *Chem. Rev.* 107:4048–77.
95. Abanades, S., P. Charvin, G. Flamant, and P. Neveu. 2006. Screening of water-splitting thermochemical cycles potentially attractive for hydrogen production by concentrated solar energy. *Energy* 31:2805–22.
96. Pregger, T., D. Graf, W. Krewitt, C. Sattler, M. Roeb, and S. Möller. 2009. Prospects of solar thermal hydrogen production processes. *Int. J. Hydrogen Energy* 34:4256–67.
97. Kameyama, H. 1978. PhD diss., University of Tokyo.
98. Lemort, F., C. Lafon, R. Dedryvère, and D. Gonbeau. 2006. Physicochemical and thermodynamics investigation of the UT-3 hydrogen production cycle: A new technological assessment. *Int. J. Hydrogen Energy* 31:906–18.
99. Aihara, M., H. Umida, A. Tsutsumi, and K. Yoshida. 1990. Kinetic study of UT-3 thermochemical hydrogen production process. *Int. J. Hydrogen Energy* 15:7–11.
100. Sakurai, M., A. Tsutsumi, and K. Yoshida. 1995. Improvement of Ca-pellet reactivity in UT-3 thermochemical hydrogen production cycle. *Int. J. Hydrogen Energy* 20:297–301.
101. Amir. R., S. Shiizaki, K. Yamamoto, T. Kabe, and H. Kameyama. 1993. Design development of iron solid reactants in the UT-3 water decomposition cycle based on ceramic support materials. *Int. J. Hydrogen Energy* 18:283–6.
102. Sakurai, M., J. Ogiwara, and H. Kameyama. 2006. Reactivity improvement of Fe-compounds for the UT-3 thermochemical hydrogen production process, *J. Chem. Eng. Jpn.* 39:553–8.
103. Kameyama, H., Y. Tomino, T. Sato, et al. 1989. Process simulation of "MASCOT" plant using the UT-3 thermochemical cycle for hydrogen production. *Int. J. Hydrogen Energy* 14:323–30.
104. Yoshida, K., H. Kameyama, T. Aochi, et al. 1990. A simulation study of the UT-3 thermochemical hydrogen production process. *Int. J. Hydrogen Energy* 15:171–8.
105. Sakurai, M., N. Miyake, A. Tsutsumi, and K. Yoshida. 1996. Analysis of a reaction mechanism in the UT-3 thermochemical hydrogen production cycle, *Int. J. Hydrogen Energy* 21:871–5.
106. Sea, B. K., E. Soewito, M. Watanabe, K. Kusakabe, S. Morooka, and S. S. Kim. 1998. Hydrogen recovery from a H_2–H_2O–HBr mixture utilizing silica-based membranes at elevated temperatures. 1. Preparation of H_2O- and H_2-selective membranes. *Ind. Eng. Chem. Res.* 37:2502–8.
107. Sea, B. K., K. Kusakabe, and S. Morooka. 1998. Hydrogen recovery from a H_2–H_2O–HBr mixture utilizing silica-based membranes at elevated temperatures. 2. Calculation of exergy losses in H_2 separation using inorganic membranes. *Ind. Eng. Chem. Res.* 37:2509–15.
108. Nakayama, T., H. Yoshida, H. Furutani, H. Kameyama, and K. Yoshida. 1984. MASCOT—A bench-scale plant for producing hydrogen by the UT-3 thermochemical decomposition cycle. *Int. J. Hydrogen Energy* 9:187–90.
109. Sakurai, M., M. Aihara, N. Miyake, A. Tsutsumi, and K. Yoshida. 1992. Test of one-loop flow scheme for the UT-3 thermochemical hydrogen production process. *Int. J. Hydrogen Energy* 17:587–92.

110. Sakurai, M., K. Moriyama, and H. Kameyama. 2002. Dynamic simulation on the thermochemical hydrogen production process. *J. Hydrogen Energy Systems Soc. Japan* 27:11–16.
111. Tadokoro, Y., T. Kajiyama, T. Yamaguchi, N. Sakai, H. Kameyama, and K. Yoshida. 1997. Technical evaluation of UT-3 thermochemical hydrogen production process for an industrial scale plant. *Int. J. Hydrogen Energy* 22:49–56.
112. Doctor, R. D., D. C. Wade, and M. H. Mendelsohn. 2002. STAR-H2: A calcium–bromine hydrogen cycle using nuclear heat. *Proc. AIChE 2002 Spring National Meeting*, 139c, New Orleans, USA, March 10–14.
113. Kondo, W., S. Mizuta, Y. Oosawa, T. Kumagai, and K. Fujii. 1983. Decomposition of hydrogen bromide or iodide by gas phase electrolysis. *Bull. Chem. Soc. Jpn.* 56:2504–8.
114. Moisseytsev, A. and D. T. Matonis. 2003. Integrated heat balance of STAR-H2 system for hydrogen production. *Proc. 2nd Information Exchange Meeting on Nuclear Production of Hydrogen*, 277–88, Argonne, USA, October 2–3.
115. Simpson, M. F., V. Utgikar, P. Sachdev, and C. McGrady. 2007. A novel method for producing hydrogen based on the Ca–Br cycle. *Int. J. Hydrogen Energy* 32:505–9.
116. Knoche, K. F., H. Cremer, and G. Steinborn. 1976. A thermochemical process for hydrogen production. *Int. J. Hydrogen Energy* 1:23–32.
117. Knoche, K. F., H. Cremer, G. Steinborn, and W. Schneider. 1977. Feasibility studies of chemical reactions for thermochemical water splitting cycles of the iron–chlorine, iron–sulfur and manganese–sulfur families. *Int. J. Hydrogen Energy* 2:269–89.
118. Knoche, K. F., H. Cremer, D. Breywisch, S. Hegels, G. Steinborn, and G. Wüster. 1976. Experience of a laboratory scale cycle for thermochemical hydrogen production and proposal for technical implementation. *Proc. 1st World Hydrogen Energy Conf.* 7A-83–7A-98, Miami Beach, USA, March 1–3.
119. Cremer, H., S. Hegels, K. F. Knoche, et al. 1980. Status report on thermochemical iron/chlorine cycles: A chemical engineering analysis of one process. *Int. J. Hydrogen Energy* 5:231–52.
120. Lewis, M. A., M. Serban, and J. K. Basco. 2003. Hydrogen production at <550°C using a low temperature thermochemical cycle. *Proc. GLOBAL 2003*. 1492–8, New Orleans, USA, November 16–20.
121. Naterer, G., S. Suppiah, M. Lewis, et al. 2009. Recent Canadian advances in nuclear-based hydrogen production and the thermochemical Cu–Cl cycle. *Int. J. Hydrogen Energy* 34:2901–17.
122. Carty, R. H., M. M. Mazumder, J. D. Schriber, and J. B. Pangborn. 1981. Thermochemical hydrogen production, *GRI-80-0023*.
123. Wang, Z. L., G. F. Naterer, K. S. Gabriel, R. Gravelsins, and V. N. Daggupati. 2009. Comparison of different copper–chlorine thermochemical cycles for hydrogen production. *Int. J. Hydrogen Energy* 34:3267–76.
124. Lewis, M. A., J. G. Masin, R. B. Vilim, and M. Serban. 2005. Development of the low temperature Cu–Cl thermochemical cycle. *Proc. 2005 Int. Congress Advances Nucl. Power Plants (ICAPP'05)*. Paper 5425, Seoul, Korea, May 15–19.
125. Orhan, M. F., I. Dincer, and M. A. Rosen. 2008. Energy and exergy assessments of the hydrogen production step of a copper–chlorine thermochemical water splitting cycle driven by nuclear-based heat. *Int. J. Hydrogen Energy* 33:6456–66.
126. Naterer, G. F. 2008. Second law viability of upgrading waste heat for thermochemical hydrogen production. *Int. J. Hydrogen Energy* 33:6037–45.
127. Orhan, M. F., I. Dincer, and G. F. Naterer. 2008. Cost analysis of a thermochemical Cu–Cl pilot plant for nuclear-based hydrogen production. *Int. J. Hydrogen Energy* 33:6006–20.
128. Nakagiri, T. 2002. Research on hydrogen production system. *JNC-TN9420-2002-002*.
129. Suzuki, C., T. Nakagiri, and K. Aoto. 2007. The refinement of the rate determining process in sulfur trioxide electrolysis using the electrolysis cell. *Int. J. Hydrogen Energy* 32:1771–81.
130. Suzuki, C. and T. Nakagiri, 2008, Investigation of SO_3 adsorption and dissociation on Pt electrode, *Solid State Ionics* 179:855–61.
131. Nakagiri, T., S. Kato, and K. Aoto. 2006. Hydrogen production experiment by thermo-chemical and electrolytic hybrid hydrogen production process. *Proc. 16th World Hydrogen Energy Conf.* S04–170, Lyon, France, June 13–16.

132. Nakagiri, T., T. Kase, S. Kato, and K. Aoto. 2006. Development of a new thermochemical and electrolytic hybrid hydrogen production system for sodium cooled FBR. *JSME Int. J. B* 49:302–8.
133. Chikazawa, Y., T. Nakagiri, M. Konomura, S. Uchida, and Y. Tsuchiyama. 2006. A system design study of a Fast Breeder Reactor hydrogen production plant using thermochemical and electrolytic hybrid process. *Nucl. Technol.* 155:340–9.
134. Fujii, K., W. Kondo, S. Mizuta, Y. Oosawa, and T. Kumagai. 1982. Hydrogen production by the magnesium-iodine process, A progress report. *Proc. 4th World Hydrogen Energy Conf.* 2:553–66, California, USA, June 13–17.
135. Oosawa, Y., Y. Takemori, and K. Fujii. 1980. Catalytic decomposition of hydrogen iodide in the magnesium–iodine thermochemical cycle—catalyst search. *Nippon Kagaku Kaishi* 7:1081–7.
136. Oosawa, Y., T. Kumagai, S. Mizuta, W. Kondo, Y. Takemori, and K. Fujii. 1981. Kinetics of the catalytic decomposition of hydrogen iodide in the magnesium–iodine thermochemical cycle. *Bull. Chem. Soc. Jpn.* 54:742–8.
137. Shindo. Y., N. Ito, K. Haraya, T. Hakuta, and H. Yoshitome. 1984. Kinetics of the catalytic decomposition of hydrogen iodide in the thermochemical hydrogen production. *Int. J. Hydrogen Energy* 9:695–700.
138. Onuki, K., H. Nakajima, S. Shimizu, S. Sato, and I. Tayama. 1993. Materials of construction for the thermochemical IS process, (I). *J. Hydrogen Energy Systems Soc. Jpn.* 18:49–56.
139. Onuki, K., I. Ioka, M. Futakawa, et al. 1994. Materials of construction for the thermochemical IS process, (II). *J. Hydrogen Energy Systems Soc. Jpn.* 19:10–16.
140. Wong, B., R. T. Buckingham, L. C. Brown, et al. 2007. Construction materials development in sulfur-iodine thermochemical water-splitting process for hydrogen production. *Int. J. Hydrogen Energy* 32:497–504.
141. Kim, H. P., D.-J. Kim, H. C. Kwon, J. Y. Park, and Y. W. Kim. 2008. Corrosion of the materials in sulfuric acid. *Proc. 4th International Topical Meeting High Temperature Reactor Technology*, HTR2008–58007, Washington, DC, USA, September 28–October 1.
142. Tanaka, N., J. Iwatsuki, S. Kubo, A. Terada, and K. Onuki. 2009. Research and development on IS process components for hydrogen production (II) corrosion resistance of glass lining in high temperature sulfuric acid. *Proc. 2009 Int. Congress Advances Nucl. Power Plants (ICAPP'09)*. Paper 9291, Tokyo, Japan, May 10–14.
143. Kasahara, S., G.-J. Hwang, H. Nakajima, H.-S. Choi, K. Onuki, and M. Nomura. 2003. Effects of process parameters of the IS process on total thermal efficiency to produce hydrogen from water. *J. Chem. Eng. Jpn.* 36:887–99.
144. Goldstein, S., J.-M. Borgard, and X. Vitart. 2005. Upper bound and best estimate of the efficiency of the iodine sulphur cycle. *Int. J. Hydrogen Energy* 30:619–26.
145. Carty, R., K. Cox, J. Funk, et al. 1977. Process sensitivity studies of the Westinghouse sulfur cycle for hydrogen generation. *Int. J. Hydrogen Energy* 2:17–22.

6

Conversion of Hydrocarbons

Karl Verfondern and Yoshiyuki Inagaki

CONTENTS

6.1 Introduction

The principal conversion product of hydrocarbons including major fossil fuels and biomass is synthesis gas, the mixture of hydrogen and carbon monoxide, mainly resulting from the processes of steam reforming, partial oxidation (POX), or autothermal reforming (ATR). If the hydrogen product is to be maximized, the conversion processes are followed by the water–gas shift reaction. The final processing step is the separation and purification of the hydrogen.

6.2 Steam Reforming of Methane

The steam reforming process is the catalytic decomposition of light hydrocarbons (e.g., methane, natural gas, naphtha) to react with superheated steam resulting in a hydrogen-rich gas mixture. A processing scheme for methane reforming is given in Figure 6.1. The reforming reactions are endothermic running at high temperatures of >500°C. Steam methane reforming (SMR) takes place at typically 850°C, and at pressures >2.5–5 MPa:

$$CH_4 + H_2O \leftrightarrow 3H_2 + CO - 206\ \text{kJ/mol} \tag{6.1}$$

$$CH_4 + 2H_2O \leftrightarrow 4H_2 + CO_2 - 165\ \text{kJ/mol} \tag{6.2}$$

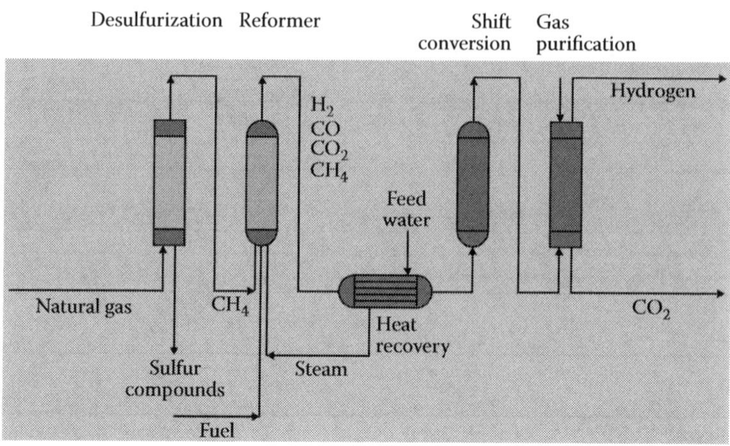

FIGURE 6.1
Processing scheme of steam methane reforming.

Generally, a nickel catalyst is used for the reaction, loaded to an alumina base material at 10–15 wt%. Besides nickel, platinum and ruthenium are also used as catalysts. Because of their excellent hydrophilic properties, carbon deposition on these metals is minimized. The shapes of the catalyst can be different (cylinder, sphere, pellet, etc.) and are chosen according to the shape of the reaction tube.

In order to increase the output of hydrogen and to avoid carbon deposition due to the Boudouard reaction, the CO is catalytically converted in the slightly exothermic water–gas shift reaction with steam according to

$$CO + H_2O \leftrightarrow H_2 + CO_2 + 41\ \text{kJ/mol} \tag{6.3}$$

The result is more H_2 and a lower CO concentration down to 0.5–2% of the dry gas. Catalysts for this reaction consist mainly of noble metals.

In the conventional reforming process, the reformer tubes in the furnace are heated from the outside by burning a part of the natural gas. The main processes of heat transfer are radiation and convection of the flue gas with temperatures above 1300°C. The average equilibrium composition of the dry reformer gas, that is, without steam, is 75% H_2 (about half of which is from the shift reaction), 13% CO, 10% CO_2, and 2% still unreformed CH_4. Strongly depending on the fuel characteristics, the steam-to-carbon ratio, outlet temperature, and pressure are chosen according to the desired products. The gas produced mainly contains a mixture of hydrogen and carbon monoxide, which is called synthesis gas. It is typically used as an intermediate product for the generation of substitute natural gas (SNG), ammonia, or methanol.

High reforming temperatures, low pressures, and high steam-to-methane ratio favor a high methane conversion. If excess steam is injected, typically 300% away from the stoichiometric mixture, the equilibrium at temperatures of 300–400°C is shifted toward more CO_2 and increasing the H_2 yield. The hydrogen gas needs to pass further purification steps to achieve a purity of >99% before being used, for example, in fuel cells. The unwanted constituents CO_2 and others are removed from the gas mixture by pressure swing adsorption (PSA).

If the steam is completely or partially replaced by CO_2, the composition of the synthesis gas is shifted toward a larger CO fraction. The CO_2 can be either imported or taken from the reformer outlet. The catalytic reforming of methane with CO_2 offers an environmental advantage, because two greenhouse gases are combined resulting in a product gas mixture which might be more favorable for certain applications like the synthesis of oxygenated chemicals. Major drawbacks are the rapid deactivation of conventional catalysts and the relatively high soot formation (methane cracking).

Steam reforming of natural gas is a technically and commercially well-established technology on industrial scale and currently the most economical route accounting for almost half of the hydrogen produced worldwide [1]. Reforming technology is mainly used in the petrochemical and fertilizer industries for the production of the so-called "on-purpose" hydrogen. Large steam reformer units with up to about 1000 splitting tubes have a production capacity of around 130,000 Nm^3/h. Commercial large-scale SMRs produce hydrogen at an efficiency of about 75%. The CO_2 intensity is about 9.5 kg for each kilogram of hydrogen produced [2].

If light hydrocarbons are used as fuel, sometimes a pre-reformation is helpful to operate the tubes under the same conditions with methane as feed gas. Steam reforming of heavier hydrocarbons is possible, but requires more complex process equipment and is therefore only less applied.

6.3 Partial Oxidation and Autothermal Reforming

6.3.1 Partial Oxidation

POX of carbonaceous feedstock in the presence of water is also a conversion process at high pressures and high temperatures (950–1100°C) which produces synthesis gas and maximizes H_2 yield, if followed by the water–gas shift reaction. It can be noted for alkanes:

$$C_nH_m + \frac{1}{2}nO_2 \leftrightarrow \frac{1}{2}mH_2 + nCO \tag{6.4}$$

$$C_nH_m + nO_2 \leftrightarrow \frac{1}{2}mH_2 + nCO_2 \tag{6.5}$$

The oxygen required is typically provided by an air-separation plant. POX can easily be performed without the presence of a catalyst. High temperatures of 1200–1450°C and pressures of 3–7.5 MPa (Texaco process) are needed to ensure high conversion rates. The catalytic partial oxidation (CPO) reaction, however, can take place at lower temperatures and may lead to a significantly enhanced H_2 yield from the fuel.

The POX process has the advantage of accepting all kinds of heavy hydrocarbon feed such as oil, residues, coal, or biomass. In comparison to steam reforming, the hydrogen yield is smaller, but the resulting synthesis gas with a H_2/CO ratio of ~2 makes methanol synthesis an ideal follow-on process. POX allows for compact equipment and easy maintenance, since there is no need for external heating or steam supply. Efficiencies of about 70% are somewhat less compared to SMR (80–85%) because of the higher temperatures involved and problems with the heat recovery. POX can be scaled down to 10 kWe units.

CPO of heavy oil and other hydrocarbons is a commercially applied, large-scale H_2 production method, for example, in refineries, where synthesis gas is generated from residual heavy-oil fractions, coal, or coke. The lower feedstock prices of heavy residues are at the expense of higher capital cost and the more demanding operating conditions. Large-scale plants also usually include air decomposition with unit sizes which may reach about 100,000 Nm^3/h. Small-sized units of POX reforming for mobile applications are currently in the R&D and demonstration phase.

Tandem reforming is the combination of a gas-heated reformer and an oxygen-fired autothermal reformer (Figure 6.2). If neither oxygen nor steam is available, the facility needs air separation and steam generator units.

6.3.2 Autothermal Reforming

The combination of the POX process with endothermic steam reforming may lead to internally heat-balanced reactions in a single fixed-bed reactor without heat input from the outside. This method is called ATR. It can be described with the following equation:

$$C_nH_m + xO_2 + yH_2O \leftrightarrow (2n - 2x + 0.5m)\,H_2 + nCO_2 + (y + 2x - 2n)H_2O \tag{6.6}$$

where x is the molar oxygen-to-carbon and y the molar steam-to-carbon ratio in the input mixture. Typical ratios in the reactions are 0.2–0.6 for the oxygen-to-carbon and 1.0–3.0 for the steam-to-carbon ratios [3]. The reaction heat is ideally zero (thermoneutral) to achieve maximum fuel reforming efficiency. In practice, however, side reactions such as reverse shift, methanation, or incomplete conversion result in additional compounds in the reformate.

ATR technology was developed in the late 1970s with the goal to have the reforming step in a single adiabatic reactor. It consists of a combustion zone where the heat-up reaction gas mixture is directly transferred into a fixed-bed catalytic steam reforming zone which

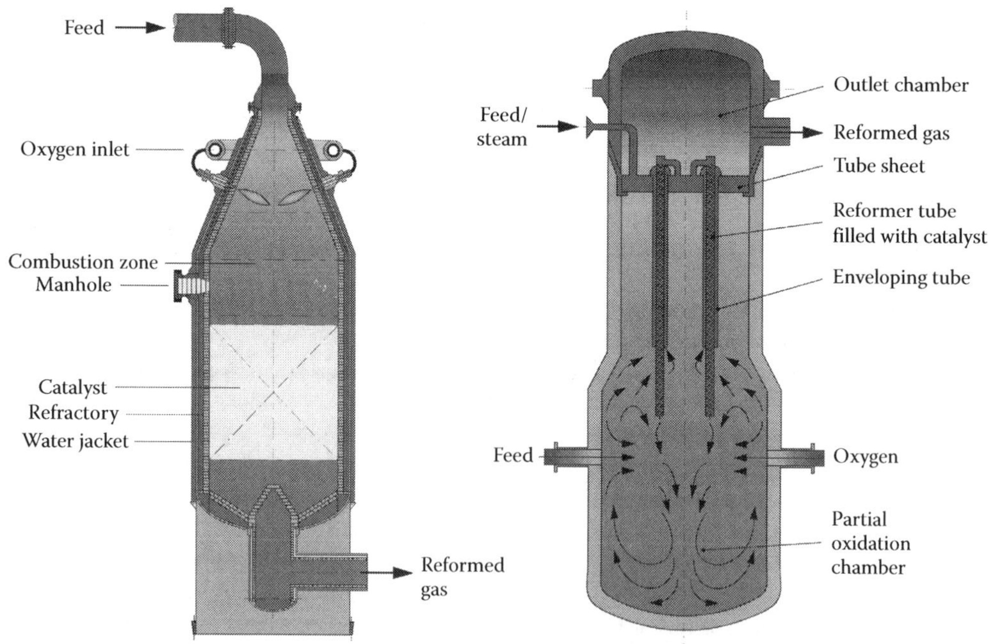

FIGURE 6.2
Autothermal reformer (ATR) (left) and combined ATR (right). (Courtesy of Udhe.)

is implemented in the lower part of the same reactor vessel. Downstream processing is needed for cooling, purification, and separation of the hydrogen from the reformate. When atmospheric air is used, the nitrogen must be removed from the product stream. The hydrogen contents in the reformate can be as high as 50–55%.

ATR is mainly used in large-scale plants for gas-to-liquid (GTL) applications with an important cost factor being the oxygen. Typical capacities of combined autothermal reformers, however, are between 4000 and 35,000 Nm^3/h. But there are also smaller units for local H_2 production with a capacity around 150 Nm^3/h. Increasing interest is given in using ATR for automotive applications to convert hydrocarbons such as gasoline, diesel, bioethanol, or methanol into a hydrogen-rich gas. For onboard reforming, multifuel processors in the 50 kW range have been developed. Catalytic ATR is ideal for fuel-cell systems due to its simple design, low operation temperatures, flexible load, and high efficiency.

In plate-type reformers, plates are arranged in a stack with one side being coated with a catalyst and supplied with the reactants. These reformers are more compact, show a faster startup, and a better heat transfer and therefore higher conversion efficiency. For the high-temperature range, inorganic membranes (ceramics, metals) are under development. They allow new concepts which may make the stages of air separation, POX, or PSA obsolete. So-called "ion transport membranes" (ITM) with their stable oxygen defect crystal structure are operated at >700°C and allows only oxygen ions to move through the membrane, which is gas-tight for all other gases. Conceptual designs promise cost reduction in the generation of high-pressure hydrogen [4].

6.4 Coal Gasification

6.4.1 Coal Conversion Processes

Because of its abundant resources on earth, the conversion of coal to gaseous or liquid fuels has been applied commercially worldwide. The coke furnace process was already in use more than 100 years ago for the production of low-BTU gas, synthesis gas, town gas, or SNG [5]. Today coal gasification accounts for ~18% of the world's hydrogen generation [1].

With the first oil crisis at the beginning of the 1970s, the coal resources were to play a central role and a revival of coal conversion programs were started. Extensive experimental and theoretical studies included coal gasification, liquefaction, and advanced combustion systems aiming at improved methods for the generation of SNG, liquid hydrocarbons, and other raw materials for the chemical industries [6]. Interest in coal refinement faded away again with cheap oil prices since the 1980s. Today, coal gasification is primarily used for ammonia synthesis in the fertilizer industry and for synthesis-gas production to be used in the synthesis of methanol and other hydrocarbons.

If expressed in carbon and hydrogen, coal can be described with the formula of ~$(CH_{0.8})_n$. For the production of higher grade hydrocarbons, either the carbon must be reduced or hydrogen must be added. The conversion of coal into gas is realized by means of a gasification agent which reacts with the coal at temperatures >800°C. Gasification agents can be steam, oxygen, air, hydrogen, carbon dioxide, or a mixture of these. The gasification agent steam (steam-coal gasification) belongs to the most important reactions of commercial interest. If air or oxygen is injected into the gasifier, a part of the coal is directly burnt allowing for an autothermal reaction. The processes have in common that high pressures are needed to achieve a high methane yield, whereas for an optimal synthesis-gas output, high temperatures and low pressures are required.

6.4.2 Steam–Coal Gasification

In the conventional steam–coal gasification process, a part of the coal is partially oxidized, before the residual organic solids are converted to synthesis gas. The first step is the pyrolysis reaction during the heating phase (400–600°C) where all volatile constituents of the coal are rapidly expelled. The gasification reaction with the agent "steam" is given by the heterogeneous water–gas reaction and the homogeneous water–gas (shift) reaction with a further increase of the H_2 fraction:

$$C + H_2O \leftrightarrow H_2 + CO - 118.5 \text{ kJ/mol C} \quad (6.7)$$

$$CO + H_2O \leftrightarrow H_2 + CO_2 + 43.3 \text{ kJ/mol C} \quad (6.8)$$

It is followed by a methanation step if the desired end product is SNG. Heat must be quickly withdrawn to avoid reverse chemical reactions.

$$CO + 3H_2 \leftrightarrow CH_4 + H_2O + 206.0 \text{ kJ/mol C} \quad (6.9)$$

Gasification processes are classified according to the type of reactor. The principal lines used today are those by Lurgi (since 1931), Winkler (since 1922), Koppers–Totzek (since 1941). They all were developed in Germany and exist on a large scale (Figure 6.3). Modified process variants have been developed aiming at an adjustment to the feedstock quality, optimization of the product gas composition, and, of course, efficiency improvement. Commercial-scale plants typically run in an autothermal mode. Depending on the customers' requirements, respective downstream processing allows the optimized generation of either hydrogen or methane or synthesis gas. Coal conversion is estimated to be around 95% and the total efficiency (based on higher heating value) to be ~70%. Table 6.1 lists some of the major characteristic features of the different types of steam–coal gasification processes [5].

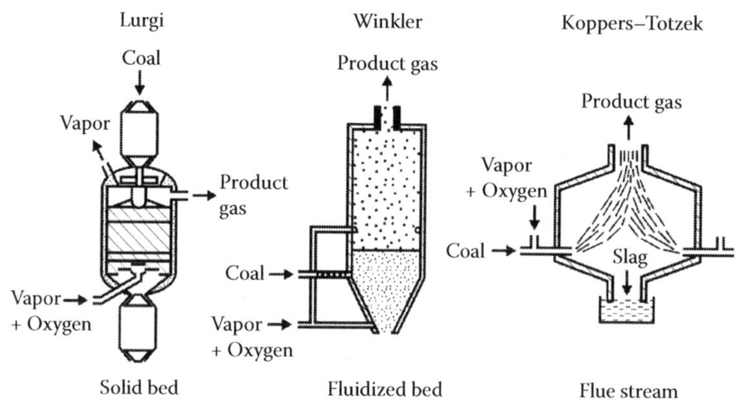

FIGURE 6.3
Three principal lines of steam coal gasification, all developed in Germany and today applied at a large scale.

TABLE 6.1
Characteristic Features of Different Steam–Coal Gasification Processes

	Lurgi	Winkler	Koppers–Totzek
Reactor	Solid bed	Fluidized bed	Flue stream
Grain size (mm)	10–30	1–10	<0.1
Steam-to-oxygen ratio	9–5	2.5–1	0.5–0.02
Movement of reactants, products	Counter-current flow	Vortex co-current flow	Co-current flow
Residence time of fuel (min)	60–90	15–60	<0.02
Requirements to fuel	Must not cake or decay	Highly reactive, must not decay	Melting point of ash <1450°C
Maximum gas outlet temperature (°C)	370–600	800–950	1400–1600
Pressure (MPa)	2–3	0.1	0.1
Composition of product gas (vol%)	62	84	60 + 29
$CO + H_2$	12	2	0.1
CH_4			
By-products	Tar, oil, phenols, gasoline, waste water	None	None

In the Lurgi pressure gasification, a solid bed of coal moving from top to bottom is gasified by adding steam and oxygen from the bottom at a pressure of 1.5–3 MPa forming different reaction zones at different temperatures. The coal should be in smaller pieces (but not too small) and must not cake to allow for a sufficient permeability of the gases. The counter-current flow arrangement leads to higher conversion rates and thermal efficiencies. A drawback is that before exiting the product gas passes through a zone of fresh coal, where it receives a significant load of tar and higher hydrocarbons, and therefore necessitates extensive purification. Sasol company in South Africa became and still is the world's largest commercial user of coal conversion technology operating a total of 97 units. The plants in Secunda and Sasolburg convert more than 30 million tons per year of bituminous coal to yield about 5.1 million Nm^3/h of synthesis gas, which corresponds to almost 30% of the world's production [7].

The high-temperature Winkler (HTW) process takes place in a fluidized bed where fine-grain brown coal is reacted with oxygen and steam that are fed in at the bottom at high speed. The fluidized bed has no reaction zones, but rather forms a homogeneous distribution of solids. Operation temperatures must be below the ash melting point to prevent a softening and agglomeration of the ash, which would lead to a collapse of the fluidized bed. The product gas composition changes with height and contains almost no higher hydrocarbons at the exit. HTW gasification is characterized by simple coal pretreatment, low oxygen consumption, and good performance over a broad load range. It was proven successful for highly reactive coal grades. Several large-scale plants were constructed in Germany and other countries with coal throughputs of up to 35 t/h. The industrial scale is at ~60,000 Nm^3/h of synthesis gas.

In the flue stream gasifier, dry coal dust is mixed with steam and oxygen/air and gasified at atmospheric pressure. The Koppers–Totzek process runs at very high temperatures above the ash melting point with the reaction zone being limited to the flame area. It has the advantage that tar formation is suppressed and other organic substances are destroyed. The conversion rate is at almost 100% with a methane content in the product gas of <0.1%. Industrial plant capacities are in the order of 50,000 Nm^3/h. The Shell process applies the

Koppers–Totzek principle under pressures up to 4 MPa with gasification temperatures of up to 2000°C. In the Texaco gasification process, fine-grained coal is mixed with water to form a suspension. The conversion rate is about 99%. The synthesis gas typically contains 34% of H_2 and 48% of CO.

6.4.3 Hydro-Gasification

In the hydro-gasification process of lignite, dried and milled coal is given to a gas generator which is blown by pure hydrogen. Coal is converted—in an exothermic reaction—into a methane-rich raw gas, ideal for the production of SNG. The hydrogen can be provided either by taking the coke left from the hydro-gasification and convert it further with oxygen and steam in an HTW process, or by taking a part of the methane for steam reforming. The gasification reaction with the agent "hydrogen" and the main product methane is:

$$C + 2H_2 \leftrightarrow CH_4 + 87.5 \text{ kJ/mol C} \tag{6.10}$$

The kinetics of the process are more complex compared to steam gasification. The above reaction runs in several steps, a pyrolytic step, where primary methane is formed plus volatile hydrocarbons, which also react with H_2 to methane. Other chemical reactions taking place are the endothermic steam–methane reforming and again the water–gas shift reaction, both of which serve the purpose to provide the gasification agent.

A high gasification degree can be obtained already with relatively short residence times of 9–80 min. In order to obtain a high conversion rate of coal, the methane fraction should not be more than 5%. The advantage of hydro-gasification compared with steam–coal gasification is its 200 K lower preheating temperature which reduces potential corrosive attack. A major drawback, however, is the low conversion rate, that is, the large amount of residual coke of up to 40%. In contrast to steam gasification, the hydro-gasification process still needs to be demonstrated at a larger commercial scale.

6.5 Biomass Gasification

Steam gasification of biomass is generally conducted in two steps. First, a thermochemical decomposition of the biomass takes place with cracking and reforming of volatiles, the production of tar, char, and gasification of the char. In the second step, hydrocarbon gases and carbon in the biomass react with CO, CO_2, H_2, and H_2O leading to lighter gases. Dry gas yield, carbon conversion efficiency, and also the hydrogen fraction in the gas increase with increasing temperatures. The conversion of biomass such as peat, wood, agricultural residues on the one hand or dedicated bioenergy crops on the other hand in a thermal process leads to a hydrogen containing gas mixture. Its H_2 content is dependent on the fuel/feedstock, the availability of steam and oxygen, and the process temperatures. Processes for decomposition of the organic substances are gasification or pyrolysis with subsequent steam reforming, autothermal or allothermal (outside heat source). The gasifiers are usually indirectly heated or oxygen blown to avoid nitrogen in the product gas, and are operated at low pressures. Although the conversion rate of biomass is high, H_2

production is highly inefficient due to the relatively low specific energy content of the biomass; only 0.2–0.4% of the total solar energy is converted into hydrogen.

The gasification of biomass or the microbial H_2 production by converting organic wastes is limited to mid-sized plants for decentralized applications. Reasons are the distributed nature of biomass connected with high transportation cost, where the economy of scale does not apply. Facilities for wood treatment are on the verge of getting commercial [9]. Demonstration of biomass gasification in pilot plants is being performed in various countries and has reached power sizes in the range of several tens of MW [10]. The technology still needs further improvements in feedstock preparation and raw gas handling and cleanup.

References

1. Scholz, W.H., Verfahren zur großtechnischen Erzeugung von Wasserstoff und ihre Umweltproblematik, *Linde Berichte aus Technik und Wissenschaft* 67:13–21, 1992.
2. European hydrogen and fuel cell technology platform, Strategic Research Agenda, July 2005, (https://www.hfpeurope.org/uploads/873/HFP-SRA004_V9-2004_SRA-report-final_22JUL2005.pdf), 2005.
3. Liu, D.-J., Kaun, T.D., Liao, H.-K., and Ahmed, S., Characterization of kilowatt-scale autothermal reformer for production of hydrogen from heavy hydrocarbons, *International Journal of Hydrogen Energy* 29:1035–1046, 2004.
4. Ranke, H. and Schödel, N., Hydrogen production technology—status and new developments, *Oil Gas European Magazine* 2: 78–84, 2004.
5. Franck, H.-G. and Knop, A., Kohleveredlung an der Schwelle der 80er Jahre, *Die Naturwissenschaften* 67:421–430, 1980.
6. Teggers, H. and Juntgen, H., Stand der Kohlevergasung zur Erzeugung von Brenngas und Synthesegas, *Erdöl und Kohle – Erdgas* 37:163–174, 1984.
7. Van Dyk, J.C., Keyser, M.J., and Coertzen, M., Syngas production from South African coal using Sasol-Lurgi gasifiers, *International Journal of Coal Geology* 65:243–253, 2006.
8. Uhde GmbH, Hydrogen. (http://www.uhde.biz/informationen/broschueren.en.html), 2003.
9. Jurascık, M., Sues, A., and Ptasinski, K.J., Exergetic evaluation and improvement of biomass-to-synthetic natural gas conversion, *Energy Environ. Sci.* 2:791–801, 2009.
10. Patel, J. and Salo, K., *CARBONA Biomass Gasification Technology*. Presentation at the International Conference on Renewable Energy TAPPI 2007, Atlanta, USA, 2007.

7

Biomass Method

Jun-ichiro Hayashi

CONTENTS

7.1 Introduction

A potential method of recuperating nuclear heat into chemical energy of hydrogen using biomass as a carbonaceous reducer of water, that is, steam gasification of biomass is described in detail. The description presents the thermodynamic significance of the method and its basic chemical reaction scheme. A type of nuclear-heat-driven process with a particular reactor configuration is examined in detail toward enabling effective conversion of the heat into product H_2 or H_2/CO.

7.2 Significance and Function of Hydrogen Production from Carbonaceous Resource

Increasing attention has been paid to biomass expecting its dual role as future carbon/energy resources. Such a role will further be intensified if a variety of biomass resources are integrated into syngas consisting mainly of H_2 and CO, which is a most important platform common to energy and carbon-based chemicals. From a viewpoint of thermodynamics, H_2 and CO are nearly equivalent with each other since the former is easily derived from the latter through slightly exothermic water-gas shift reaction. Olefins such as C_2H_4 and C_3H_6 are most important chemical platforms, and these can be derived from H_2 and CO by Fischer–Tropsch synthesis.

Biomass as well as fossil fuels can be converted into H_2 and CO via steam reforming or gasification. The terminologies of steam reforming and gasification are referred to as conversion with steam of vaporous fuel into syngas and that of solid fuel, respectively. Taking a type of woody biomass as an example, stoichiometry of its conversion into H_2 and/or CO lies between the following extreme expressions:

$$CH_{1.47}O_{0.63} + 0.37H_2O = CO + 1.11H_2 \Delta H^\circ = +112 \text{ kJ/mol-C (no } CO_2 \text{ formation)} \quad (7.1)$$

$$CH_{1.47}O_{0.63} + 1.37H_2O = CO_2 + 2.11H_2 \Delta H^\circ = +71 \text{ kJ/mol-C (full CO conversion)} \quad (7.2)$$

Total chemical energy (lower calorific value) of the products is 116–126% of that of the original biomass (lower calorific value = 438 kJ/mol-C). Thus, regardless of H_2/CO ratio, the steam gasification of biomass is largely endothermic, in other words, it integrates chemical energy of biomass and substantial quantity of heat (thermal energy) into chemical energy of H_2 and/or H_2/CO.

Here is considered steam gasification of biomass with steam at 750°C without consideration of its practical feasibility. Quality of energy is often evaluated by using exergy rate (ε) that is defined as the fraction of energy transformable into mechanical or electric energy in principle. Exergy rate of thermal energy is given by the following equation.

$$\varepsilon = \frac{(H - H_0) - T_0(S - S_0)}{H - H_0} \quad (7.3)$$

H, enthalpy; *S*, entropy; *T*, temperature; subscript 0, standard state (25°C, 1 atm).

Steam at 750°C and atmospheric pressure has ε = 0.50 (see Figure 7.1). On the other hand, ε of fuel (chemical energy) is given roughly by $\Delta G^\circ/\Delta H^\circ$ of combustion. The same woody biomass as above and H_2 have ε = 0.97 and 0.83, respectively. The steam gasification transforms heat with a low exergy rate into H_2 with a much higher one, and thus has a function of chemical heat pumping. It is in fact expected H_2 can be burned at temperature as high as 1700°C (ε = 0.68) in a future gas turbine combustion system.

High-temperature gas-cooled reactors (HTGR) can produce gas at temperature as high as 950°C [1], but its exergy rate of about 0.55, is still much lower than that of H_2. Integration of a biomass gasification process with an HTGR is a potential way to improve the quality of nuclear heat. Thermochemical water-splitting iodine–sulfur process (IS-process) [2] has

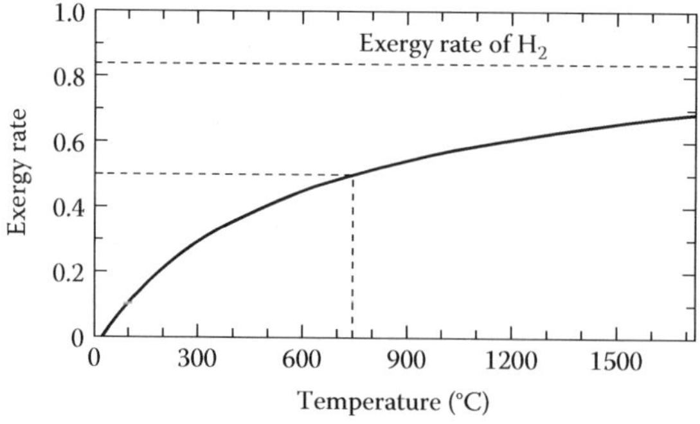

FIGURE 7.1
Exergy rate of steam at atmospheric pressure as a function of temperature.

the same significance as the steam gasification of biomass, since both processes can transform heat into chemical energy of H_2. It may be said that steam gasification can produce more H_2 than IS process with equivalent input of heat. On the other hand, IS process needs no carbonaceous reducer of water and it also produces O_2. The gas fast reactors (GFR) have the potential similar to the HTGR as a heat source to be integrated with the biomass gasification process. The sodium fast reactors (SFR) as a heat source cannot be applied for the biomass gasification process because the reactor outlet temperature is too low to produce the hot steam at above 750°C, which is required for biomass gasification.

7.3 Chemical Reaction Scheme of Biomass Gasification

General biomass resources originating from plants, that is, woody biomass and herbaceous one, consist of the three major chemical constituents. These are cellulose, hemicellulose, and lignin. Chemical structures of cellulose and a typical type of lignin are illustrated in Figure 7.2. Cellulose is a homopolymer consisting of D-glucose units that are connected in a linear manner. Hemicellulose is a heteropolymer comprising of some or many types of sugar monomer units such as xylose and mannose. The lignin consists of not linear but cross-linked macromolecules, and more importantly, it is aromatic in nature [3].

Reaction scheme of biomass pyrolysis is shown in Figure 7.3. The pyrolysis is the primary step of biomass gasification, and it involves degradation, the polymers evolving their fragments generally termed volatiles and also repolymerization/carbonization forming carbonized solid termed char [4]. Reactive gases such as steam and oxygen (O_2) are minimally involved in the primary pyrolysis since it takes place rapidly inside the solid matrix. Rather, such gases react with the pyrolysis products, that is, the volatiles and char. Yields of the primary pyrolysis products are influenced by nature of feedstock and also operating variables such as heating rate, peak temperature, pressure and particle size, and so on. Table 7.1 shows combined effects of particle size and heating rate on the char yield from the pyrolysis of woody biomass [5,6]. As seen in Table 7.2, the volatiles consist of inorganic gases, light hydrocarbon gases and condensable organic compounds that are generally termed tar. Tar is the most abundant among the constituents of the primary volatiles and its molecular mass distributes over a range from about 100 up to 1000 or even greater.

The volatiles released into the vapor phase undergo thermal cracking (secondary pyrolysis) and reactions with steam (steam reforming) and/or O_2 (partial or full combustion). The thermal cracking produces not only light gases such as H_2, CH_4 and CO but also refractory monoaromatic compounds (e.g., benzene and phenol) and polyaromatic hydrocarbons (PAH) in parallel. Reactivities of such aromatics with O_2 are in fact much lower than those of light gases. Decomposition of tar, particularly that in the gasifier, has long been a most important technical subject in biomass gasification because the tar brings about troubles and problems in downstream processes. Substantial efforts have therefore been made on developing catalysts and catalytic reactors [7,8]. Meanwhile, the volatiles undergo such reactions as above, the char experiences partial/full combustion or steam gasification reacting with O_2 and steam, respectively. The steam gasification of char is in general the slowest reaction among those involving the pyrolysis products, and hence it often determines overall rate of the biomass conversion into gases. Review papers on steam gasification of char from biomass are available [9].

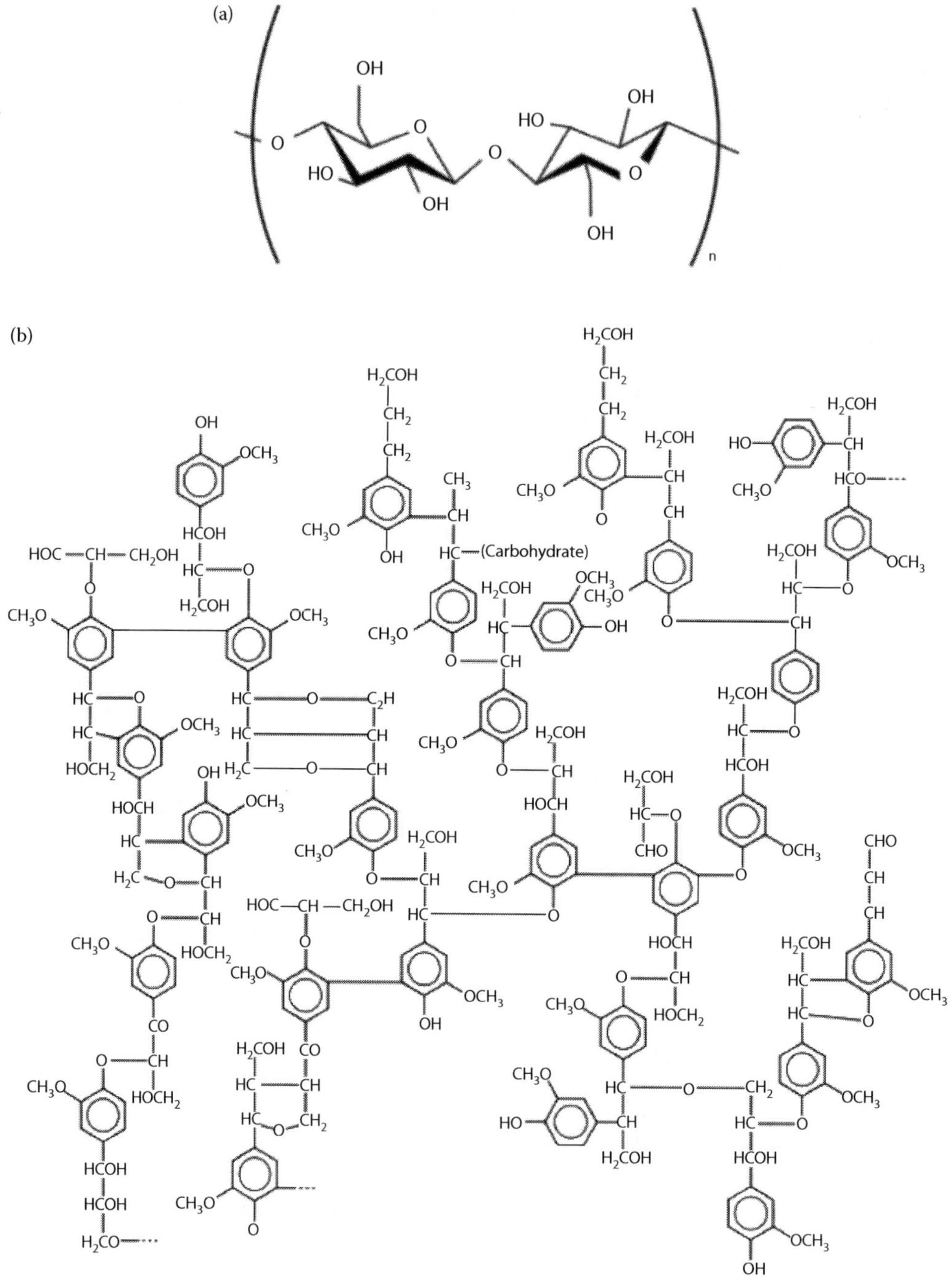

(a)

(b)

FIGURE 7.2
Chemical structures of (a) cellulose and (b) lignin. (Adapted from A.N. Glazer and H. Nikaido, *Microbial Biotechnology*, W.H. Freeman, New York, NY, p. 340, 1995.)

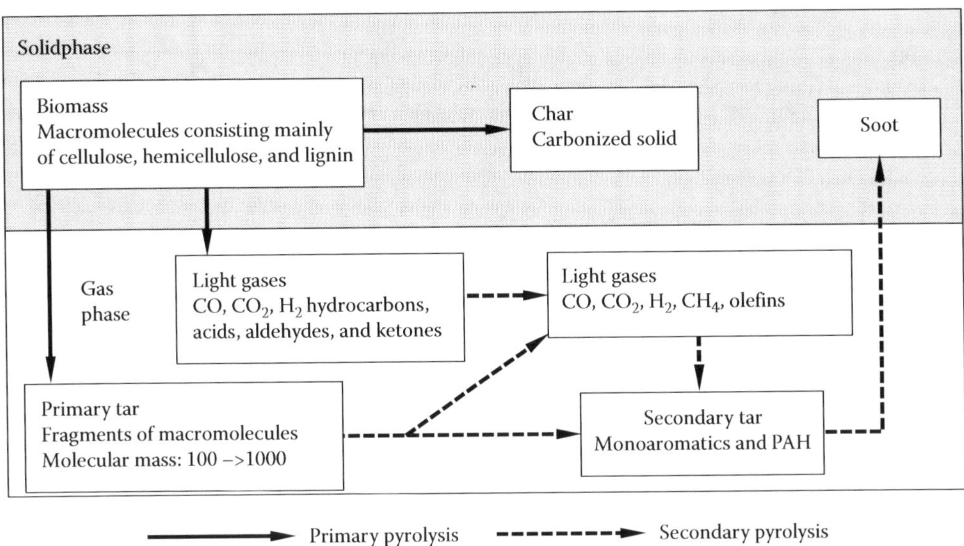

FIGURE 7.3
Reaction scheme of biomass pyrolysis.

7.4 Steam Gasification of Biomass with Nuclear Heat Supply

Nuclear heat, at the temperatures required for biomass gasification, may be supplied by several types of advanced nuclear reactor technologies under international development. In this section, examination is made on effectiveness of supplying heat from HTGR to biomass steam gasification assuming that hot steam at temperature as high as 750°C is available. The examination leads to understanding of necessity of a type of gasification process with a particular reactor configuration and sequence of biomass conversion steps.

7.4.1 Case I: Gasification with Hot Steam but without O_2

Ideal transformation of heat to chemical energy is to convert biomass into syngas with only steam and heat, in other words, to realize a stoichiometry as indicated previously. Most biomass resources, except secondary fuels such as ethanol and biodiesel oils, are originally solid, and this nature limits variety in the way of supplying heat to a reactor or reactors of a gasification process. Different from the case of catalytic steam reforming of vaporous fuel such as natural gas and naphtha, heating the gasifier externally is not a reasonable way, but internal heating by feeding hot steam together with biomass, or otherwise, by employing solid heat carrier that circulates between the reactor and heat receiver, would be necessary.

There have so far been a number of technical reports on gasification processes with air, air-steam, O_2-steam, [5,7,10] but none of them were operated at temperature lower than 750°C. This is clearly due to difficulty of both steam reforming of volatiles and steam gasification of char at such low temperature. Generation of steam at temperature as high as 800°C by using heat from a HTGR may be possible, but even so, such temperature is too low to drive largely endothermic steam gasification with some thermal margins. Feeding

TABLE 7.1

Combined Effects of Heating Rate and Particle Size on the Char Yield from the Pyrolysis of Woody Biomass under Atmospheric Pressure

Biomass	Particle Size (mm)	Heating Rate (°C/s)	Peak Temp.[a] (°C)	Char Yield[b] (%)	Tar Yield[b] (%)
Pine sawdust	0.13–0.21	1000	600	15	70
Pine sawdust	0.13–0.21	1	600	30	55
Cedar sawdust	0.21–0.35	>3000	700	13	58
Cedar chip	10 × 10 × 2.0	5.5	550	36	38

Source: Data from J.-i. Hayashi, S. Hosokai, and N. Sonoyama, *Process Safety and Environmental Protection*, 84(B6), 409–419, 2006; T. Okuno et al., *Energy and Fuels*, 19, 2164–2171, 2005.

[a] Holding time at peak temperature was 0 s.

[b] On a carbon basis.

O_2 together with steam is hence inevitable even with heat supply from a HTGR. As mentioned in another chapter, processes for thermal decomposition of water such as IS process [2] have been under development. Assuming simultaneous production of H_2 and O_2 by such a process, use of O_2 in a neighbor gasification process would be reasonable.

7.4.2 Case II: Gasification with Simultaneous Use of O_2 and Hot Steam in a Single Reactor

A process assumed in this case study is schematically illustrated in Figure 7.4. The pyrolysis of biomass, reforming of volatiles and gasification of char are operated in a single reactor. The volatiles and char undergo thermal cracking, steam reforming/gasification and also partial or full combustion. Feeding of more oxygen (O_2) enables more feeding of

TABLE 7.2

An Example of Composition of Pyrolysis Products from Cedar Chip

Product	Yield, % on C Basis
Water[a]	18.4
H_2[a]	0.95
CO	11.7
CO_2	6.6
CH_4	3.6
C_2H_4	0.9
C_2H_6	0.7
C_3H_6	0.6
C_3H_8	0.2
C_4	0.4
Methanol	0.1
Acetaldehyde	0.2
Tar ($CH_{1.76}O_{0.37}$)	38.8
Char ($CH_{0.57}O_{0.12}$)	36.1

Note: Peak temperature, 550°C; heating rate, 5.5°C/s.

[a] Yield is indicated in a unit of mol/100-mol-C.

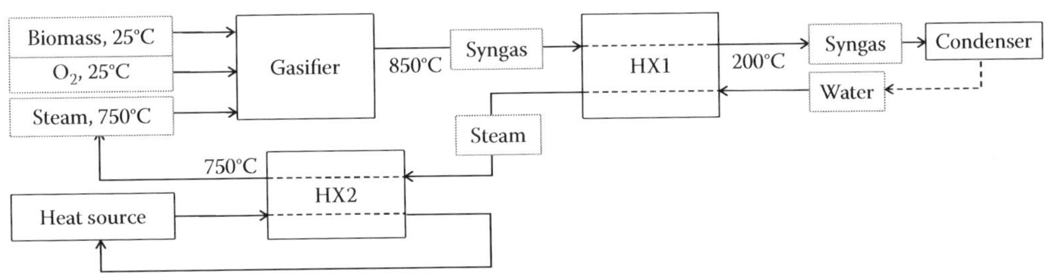

FIGURE 7.4
Diagram of single-gas-flow process.

biomass and steam at a given reactor temperature, or otherwise, higher reactor temperature at a given steam/O_2 or steam/biomass ratio, while more loss of chemical energy of the original fuel is inevitable. A most suitable type of reactor may be a fluidized bed reactor.

A single reactor is assumed, into which a typical woody biomass (cedar), oxygen (O_2), and hot steam at 750°C are fed, and complete conversion of the biomass into syngas takes place at 850°C. A process diagram is depicted in Figure 7.4. Other main assumptions are as follows: First, the syngas is cooled down to 200°C in a series of ideal heat exchangers (HX1) in which steam is generated and heated to temperature as high as possible but no higher than 750°C. Second, if the steam temperature is lower than 750°C at HX1 exit, it is raised up to 750°C in the second series of heat exchanges (HX2) with the aid of heat from HTGR. Other assumptions are listed below:

1. The biomass is a type of cedar wood that contains no moisture, and it has the following properties. Elemental composition: C, 50.9; H, 6.30; O, 42.8 wt%. Calorific value: HHV, 20.0 MJ/kg, LHV, 18.6 MJ/kg. Heat of formation ($\Delta H_f°$): –133.2 kJ/mol-C.
2. Temperatures of the biomass and O_2 are both 25°C at the gasifier inlet.
3. The syngas consists of CO, CO_2, and H_2 (i.e., no hydrocarbon gases).
4. The composition of the syngas including steam at the gasifier exit is governed by chemical equilibrium.
5. Loss of heat from the reactors and that from the heat exchangers to the environment is negligibly small.

Figure 7.5 shows results from the simulation that was performed by solving equations of heat balance, material balance, and chemical equilibrium simultaneously. The cold gas efficiency (CGE) is an expression of recovery of chemical energy, and defined using the following equation.

$$\text{CGE} = \frac{\text{Chemical energy of syngas on an LHV basis}}{\text{Chemical energy of the original fuel on an LHV basis}} \tag{7.4}$$

CGE seems to decrease very slightly with increasing steam-to-biomass (S/B) ratio and this is due to that a higher S/B ratio results in more extensive progress of an exothermic reaction, that is, water-gas shift reaction: $CO + H_2O = CO_2 + H_2$.

Feeding of O_2 together with steam inevitably results in CGE of around $0.88 < 1$. It is also noted that this process requires heat supply from HTGR of at most 10% of chemical energy

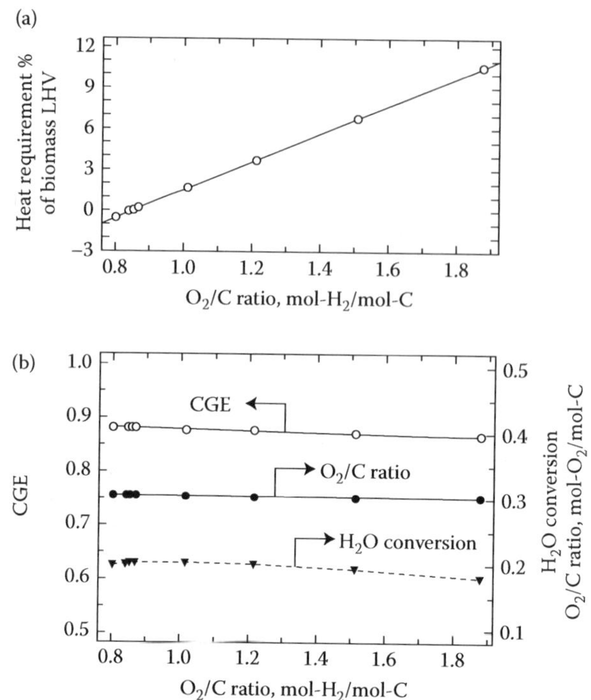

FIGURE 7.5
Results from simulation of single-gas-flow process: (a) heat requirement as % of the biomass LHV and (b) process performance indicators.

of the biomass, and more importantly, there is no heat demand if S/B < ca. 0.85. In other words, Case II process with a lower S/B ratio can be stand-alone thermally. No existing biomass gasification processes have so far reached CGE as high as 0.88, and Case II process is thus a very efficient process. However, this type of process, even if realized by developing practical high-performance catalysts, could not recuperate heat from HTGR into chemical energy of H_2/CO effectively.

7.4.3 Case III: Dual-Gas-Flow Gasification

Here is considered a type of process involving three different reactors: biomass pyrolyzer to generate volatiles and char, reformer to convert the volatiles into syngas with steam, and gasifier to convert the char into syngas with steam and O_2. There are two streams of syngas from the volatiles and that from the char. A schematic diagram of this process is shown in Figure 7.6. A particular feature of this process is that the reformer temperature is maintained by external heating while the volatiles are fed into the reformer together with preheated steam but without O_2. The volatiles are gaseous and this enables such external heating. A reactor consisting of a fixed catalyst bed or fixed beds in parallel is applicable to the steam reforming, in the same way as that of naphtha. Another type of reactor consisting of a bed of fluidizing catalyst particles may be applicable if heat exchanger tubes are inserted into the fluidized bed. Char formed by the pyrolysis is isolated from the volatiles and sent to the gasifier together with O_2 and hot steam. The use of O_2 results in reduction of CGE due to the same reason as in Case II process. However, chemical energy retained in

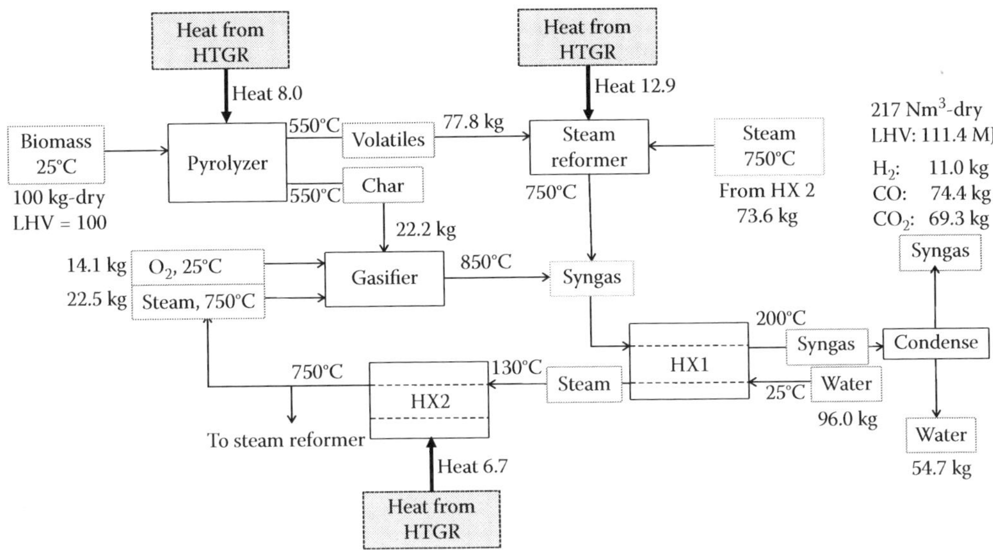

FIGURE 7.6
Diagram of dual-gas-flow process.

the char is as small as 10–40% that of the original biomass, as mentioned previously. In other words, a major part of the chemical energy of the original fuel is carried by the volatiles that undergo fully endothermic steam reforming in the reformer.

Another important feature of Case III process is that external heating is also applied to the pyrolysis. The pyrolysis is an endothermic process and it can therefore recuperate externally supplied heat to chemical energy of the volatiles and char. External heating of the pyrolyzer is not as critical as to the steam reforming since the heat of reaction per reactant mass of the pyrolysis is smaller than that of the steam reforming as well as steam gasification. The yield of the volatiles generally increases by raising the pyrolysis temperature, but a temperature as high as that for the steam reforming is not necessary for evolution of tar and light hydrocarbon gases. Previous studies on the biomass pyrolysis showed that temperature and time of as low as 500°C and as short as 10 s, respectively, are enough to obtain maximum tar and hydrocarbon gas yields [5,6]. A conventional type of reactor for the pyrolysis is a rotary kiln. Combination of the pyrolyzer with the gasifier may also be effective. For example, in a dual fluidized bed system, biomass is pyrolyzed in a fluidized bed reactor that feeds the char to the other fluidized bed for char gasification together with the fluidizing medium such as silica sand, which circulates fast between the fluidized beds playing a role of heat carrier. Heating with such a heat carrier is effective for increasing heating rate of the biomass and thereby increasing the volatile yield.

The assumptions made in the simulation of Case III process are summarized below:

1. The same type of biomass as in Case II is the original fuel.
2. The biomass at 25°C is fed to the pyrolyzer. The pyrolysis is performed at 550°C producing volatiles and char with yields/composition shown in Table 7.2.
3. Syngas from the reformer is mixed with that from the gasifier and then cooled down to 200°C in a series of HX1 in which water at 25°C is converted into steam.

4. Steam generated in HX1 is further heated up to 750°C in another series of HX2 where heat from HTGR is transferred to the steam.
5. A portion of the steam from HX2 is supplied to the gasifier while the other to the reformer.
6. The steam-to-volatiles ratio (mol-steam/mol-C-volatiles) is 1.8, which is high enough to avoid significant carbon deposition from the volatiles onto the catalyst.
7. Conversion of steam by the char gasification is below 70%.

In addition to the above, the same assumptions as those of Case II (3–5) are made. The seventh assumption is based on author's experiences of steam and steam–O_2 gasification of biomass-derived chars in fixed-bed reactors.

The product composition listed in Table 7.2 was used for the simulation. Figure 7.6 also shows temperatures and mass flows at inlet/exit of the components working in Case III process. These parameters were optimized through repeated trial and error. The pyrolysis converts 64% of the biomass into the volatiles and 36% to the char on the carbon basis under supply of a quantity of heat accounting for 8% of LHV of the biomass. The reformer requires more heat (12.8%) while the gasifier needs less (6.7%). Resulting from net transformation of heat from HTGR into chemical energy of the syngas, Case III process can attain a CGE as high as 1.11. This CGE is 1.26 times that of Case II process that is an ideal stand-alone process.

Performance of Case III process may be evaluated in terms of efficiency of external heat utilization in two different criteria. On the first criterion, chemical energy of the syngas is compared with total input (chemical energy plus external heat), and the efficiency is defined by

$$\eta_1 = \frac{\text{Chemical energy of syngas (LHV)}}{\text{Chemical energy of biomass (LHV)} + \text{External heat received by Case III process}} \tag{7.5}$$

The η_1 value of Case III process is 0.87. From a viewpoint of exergy rate, definition of η_1 without distinguishing chemical energy from thermal energy may not make a significant sense. However, η_1 is a measure of utilization of heat when an important function of endothermic reaction is taken into consideration. Latent and sensible heats of steam involved in the syngas are released to the environment without use. This loss is equivalent with 7.0% of chemical energy of the original fuel, and is even more than the heat exchanged in HX2. Minimization of steam generation and/or maximization of steam conversion are hence important for improving the efficiency.

The second way of evaluation takes another definition of efficiency as follow:

$$\eta_2 = \frac{\text{Chemical energy of syngas from Case III process} - \text{Chemical energy of syngas from Case II process}}{\text{External heat accepted by Case III}} \tag{7.6}$$

In other words, η_2 shows how effective is the external heat supply to biomass gasification in comparison with stand-alone gasification with a highest CGE. η_2 is 0.85, and this indicates that input of 100 kJ heat contributes to increase in the resulting chemical energy by 85 kJ. Taking difference in the exergy rate between heat and H_2/CO into account, it is clear that Case III process has a function of chemical heat pumping.

Though the process simulations assume complete conversion of biomass into syngas, such conversion is in practice very difficult unless the gasification temperature is well above 1000°C. It is therefore believed that use of a catalyst is indispensable for reforming of volatiles from the pyrolysis of biomass, in particular, tar, at 750°C or lower temperature. Incomplete reforming allowing a portion of tar to escape from the reformer results in installation of expensive and energy-consuming tar removal processes at downstream of the reformer. Catalytic processes for H_2 purification or production of liquid such as methanol or dimethylether (DME) are not tolerant to syngas, if contaminated with aromatic compounds. It is estimated that such catalytic processes can accept only syngas with concentration well below 1 mg/Nm^3-gas, which corresponds to a residual tar yield of 10^{-5}%. Development of catalysts and catalytic reforming processes is an essential and challenging task, since the following requirements must be satisfied simultaneously in the tar reforming:

- No or little poisoning of catalyst by species containing Cl, S, and/or volatile alkali and alkali earth metallic species.
- Little coke accumulation over catalyst or otherwise easy regeneration of catalyst by decoking.
- Elimination of particulate matter such as fines of char and ash minerals from syngas without cooling it between the pyrolyzer and reformer.

Previous studies on the catalytic reforming of tars from biomass were reviewed comprehensively by Devi et al. [7]. Very recent studies showed possibility of manufacturing of catalysts with not only high activity but also high tolerance to coking [11,12].

Complete gasification of char with steam and O_2 is promising but efforts are needed for developing efficient processes. It is known that biomass resources contain less or more amounts of alkali and alkaline earth metallic species (i.e., Na, K, and Ca) that play catalytic roles in the steam gasification [13,14]. The gasifier should be designed so that overall activity of such inherent catalysts is maximized while their volatilization [6,15] is minimized for avoiding erosion/corrosion problems in the gasifier and downstream facilities.

7.5 Conclusions and Future Perspective

Biomass gasification with an optimized sequence of pyrolysis, steam reforming of volatiles and steam-O_2 gasification of char can potentially accept nuclear heat supply from the HTGR and recuperate it into chemical energy of syngas or H_2 effectively. However, further efforts are needed for developing novel processes of catalytic steam reforming and steam–O_2 gasification as well as for generating high-temperature steam.

References

1. S. Saito, T. Tanaka, Y. Sudo, et al., Design of high temperature engineering test reactor (HTTR), JAERI 1332, Japan Atomic Energy Research Institute 1994.

2. S. Kubo, H. Nakajima, S. Kasahara, et al., A demonstration study on a closed-cycle hydrogen production by the thermochemical water-splitting iodine-sulfur process, *Nucl. Eng. Des.*, 233[1–3], 347–354, 2004.
3. A.N. Glazer and H. Nikaido, *Microbial Biotechnology*, W.H. Freeman, New York, NY, p. 340, 1995.
4. T.A. Milne, R.J. Evans, and N. Abatzoglou, Biomass gasifier tars: Their nature, formation and conversion, NREL/TP-570–25357, National Renewable Energy Laboratory, U.S. Department of Energy, 1998.
5. J.-i. Hayashi, S. Hosokai, and N. Sonoyama, Gasification of low-rank solid fuels with thermochemical energy recuperation for hydrogen production and power generation, *Process Safety and Environmental Protection*, 84(B6), 409–419, 2006.
6. T. Okuno, N. Sonoyama, C. Sathe, C.-Z. Li, J.-i. Hayashi, and T. Chiba, Primary release of alkali and alkaline earth metallic species during pyrolysis of pulverized biomass, *Energy and Fuels*, 19, 2164–2171, 2005.
7. L. Devi, K.J. Ptasinski, and F.J.J.G. Janssen, A review of the primary measures for tar elimination in biomass gasification processes, *Biomass and Bioenergy*, 24, 125–140, 2003.
8. Z.Y. Abu El-Rub, E.A. Bramer, and G. Brem, Review of catalysts for tar elimination in biomass gasification processes, *I&EC Res*, 43, 6911–6919, 2004.
9. W. Klose and M. Wölki, On the intrinsic reaction rate of biomass char gasification with carbon dioxide and steam, *Fuel*, 84, 885–892, 2005.
10. S.P. Babu, Biomass gasification for hydrogen production—Process description and research needs, IEA Bioenergy (available online), 2002.
11. J. Nishikawa, K. Nakamura, M. Asadullah, T. Miyazawa, K. Kunimori, and K. Tomishige, Catalytic performance of $Ni/CeO_2/Al_2O_3$ modified with noble metals in steam gasification of biomass, *Catalyst Today*, 131, 146–155, 2008.
12. M. Sugawa, S. Hosokai, K. Norinaga, C.-Z. Li, and J.-i. Hayashi, Activity of mesoporous alumina particles for biomass steam reforming in a fluidized-bed reactor and its application to a dual-gas-flow two stage reactor system, *Ind. Eng. Chem. Res.*, 47, 5346–5352, 2008.
13. D.M. Keown, J.-i. Hayashi, and C.-Z. Li, Drastic changes in biomass char structure and reactivity upon contact with steam, *Fuel*, 87, 1127–1132, 2008.
14. D.M. Keown, G. Favas, J.-i. Hayashi, and C.-Z. Li, Volatilization of alkali and alkaline earth metallic species during the pyrolysis of biomass: Differences between sugarcane bagasse and cane trash, *Bioresource Technology*, 96, 1570–1577, 2005.
15. N. Sonoyama, T. Okuno, O. Mašek, S. Hosokai, C.-Z. Li, and J.-i. Hayashi, interparticle desorption and re-adsorption of alkali and alkaline earth metallic species within a bed of pyrolyzing char from pulverized woody biomass, *Energy and Fuels*, 20, 1294–1297, 2006.

8

Radiolysis of Water

Ryuji Nagaishi and Yuta Kumagai

CONTENTS

8.1 Theoretical Studies

8.1.1 Water Decomposition by Ionizing Radiations

When aqueous solutions are irradiated by ionizing radiations, water molecules are decomposed to form radiolysis products, leading to the chemical reactions of products with solutes dissolved in aqueous solutions. Figure 8.1 shows the schematic diagram of water radiolysis as a function of time after the radiation energy is deposited into water molecules [1,2].

Just after the energy deposition, H_2O is mainly ionized to form $H_2O^+ + e^-$, and partially excited to be decomposed into H + OH (or H_2 + O) (this stage is called the "physicochemical stage"). These take place in localized regions around positions at which the energy was deposited. The region is called a "spur" or "track." The H_2O^+ cation is ready to react with a surrounding H_2O to form a hydroxyl radical, OH, and the ejected e^- is solvated by H_2O to form a hydrated electron, e^-_{aq}. Sequentially the radiolysis products of e^-_{aq}, H, OH, H_3O^+, H_2 are formed 1 ps (10^{-12} s) after the energy deposition, where the radiation-chemical yields of products are called "initial" yields.

Since the radiolysis products are distributed inhomogeneously at the physicochemical stage, the products are ready to diffuse to be distributed homogeneously, and simultaneously react with each other ("spur or track reaction") because of their high reactivity and local concentrations. Sequentially the e^-_{aq}, H, OH, H_3O^+, H_2, H_2O_2 escaping from or formed in the spur reactions, are distributed homogeneously 0.1 μs (10^{-7} s) after the energy deposition, where the radiation-chemical yields of products are called "primary" yields. Finally, the products are involved in homogeneous reactions (this stage is called "chemical stage").

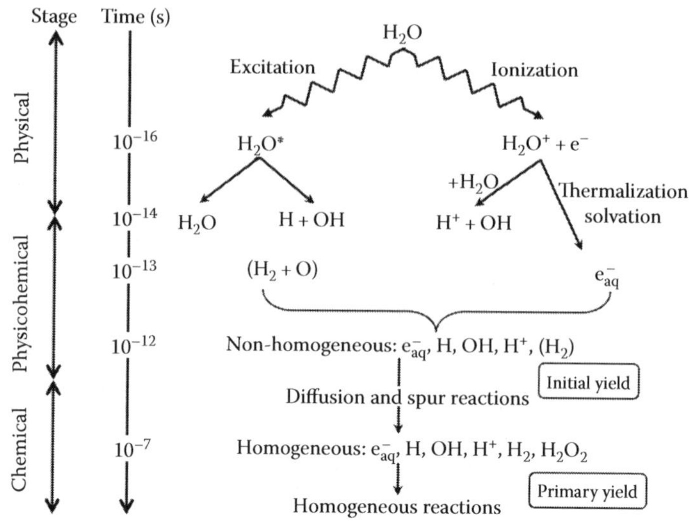

FIGURE 8.1
Schematic diagram of water radiolysis.

Therefore, the behavior and yield of H_2 production in water radiolysis are dependent on the physicochemical stage (energy deposition of radiation and water decomposition) and chemical stage (spur and homogeneous reactions). The two types of energy deposition in media at the physicochemical stage are illustrated in Figure 8.2.

In the case of radiations like 1 MeV electron beams or Co-60 γ-rays with low linear energy transfer (LET), which means the deposited energy per unit length through the passing way of radiations, spherical spurs (a) are formed at more than 100 nm separations. On the other hand, in the case of high LET radiations like α-rays or charged particle ions, a series of the spurs are connected to form a cylindrical track (b). Primary yield of H_2 as a molecular product for high LET radiations is consequently more than that for low LET radiations.

More quantitative and qualitative discussions about the radiation-chemical yields are described in the next section.

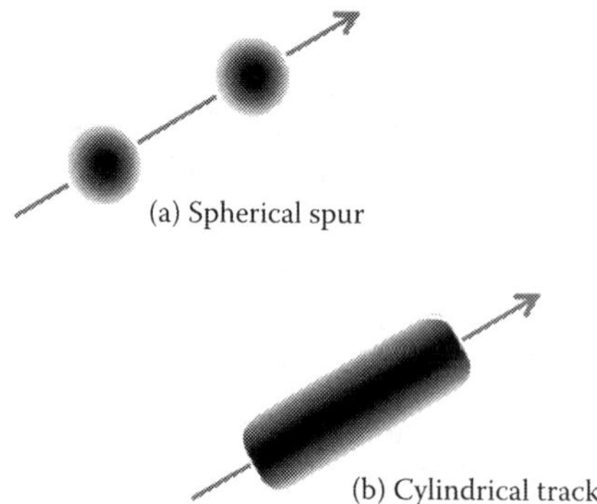

FIGURE 8.2
Deposition of radiation energy to water.

8.1.2 Radiation-Chemical Yields of Radiolysis Products

In the spur reactions, e^-_{aq}, H, and OH are formed as "radical products," while H_2 and H_2O_2 as "molecular products." The radiation-chemical yields of products ("G-values") are separated into two types of yields for convenience sake: "initial" and "primary" yields as shown in Figure 8.1. The original unit of these yields was "molecules $(100\ eV)^{-1}$," which is still used. The unit "μmol/J" in the SI system of units is also used in recent literature. The relationship between two units is 1 μmol/J = 9.649 molecules $(100\ eV)^{-1}$. The primary yields involved in homogeneous reactions in water radiolysis are discussed hereafter.

The relationship among primary yields (expressed in g) of products can be written from the law of conservation of mass as

$$g(-H_2O) = 2g(H_2) + g(H) + g(e^-_{aq}) = 2g(H_2O_2) + g(OH), \tag{8.1}$$

where the $g(-H_2O)$ denotes the primary yield of water decomposition. From the principle of conservation of charge, the relation of $g(e^-_{aq}) = g(H^+)$ is also satisfied. When a smaller yield of HO_2 is introduced to the relationship, it can be rewritten as

$$g(-H_2O) = 2g(H_2) + g(H) + g(e^-_{aq}) - g(HO_2) = 2g(H_2O_2) + g(OH) + 2g(HO_2). \tag{8.2}$$

These relationships are independent of time, kinds of radiations, conditions of solutions, and so on.

Since the spur reactions of radiolysis products are, on the other hand, affected by LET of radiations, pH, and temperature of solution, the primary yields of products depend on those. Table 8.1 illustrates the primary yields (molecules $(100\ eV)^{-1}$) of water radiolysis for different LET radiations [3]. When the LET of radiation increases, the yields of radical products (e^-_{aq}, H, and OH) decreases while those of molecular products (H_2 and H_2O_2 including H_2O) increased. It is because at higher LET, the concentrations of radicals in the track become higher and then the combinations between radicals (e.g., $H + H \rightarrow H_2$, $OH + OH \rightarrow H_2O_2$, $H + OH \rightarrow H_2O$) are promoted.

The primary yields are almost independent of pH of the solution [4], irrespective of acid–base equilibria of products ($OH = O^- + H^+$, $HO_2 = O_2^- + H^+$, $H_2O_2 = HO_2^- + H^+$). Since there are the temperature effects of spur reactions in terms of the activation energies of reactions and of the diffusions of products, the primary yields depend strongly on temperature. For low LET radiations at temperature up to 300°C, the primary yields were expressed as a function of temperature (°C) [5]:

$$g(e^-_{aq}) = 2.56 + 3.40 \times 10^{-3}\ T, \quad g(OH) = 2.64 + 7.17 \times 10^{-3}\ T, \quad g(H_2) = 0.43 + 0.69 \times 10^{-3}\ T,$$

TABLE 8.1

Primary Yields (Molecules $(100\ eV)^{-1}$) of Water Radiolysis for Different LET Radiations

	G-Values						
LET/eV nm^{-1}	$-H_2O$	e^-_{aq}	OH	H	H_2	H_2O_2	HO_2
γ-rays/0.23	4.08	2.63	2.72	0.55	0.45	0.68	0.008
H^+/12.3	3.46	1.48	1.78	0.62	0.68	0.84	—
D^+/61	3.01	0.91	0.91	0.42	0.96	1.00	0.05
He^{2+} /108	2.84	0.54	0.54	0.27	1.11	1.08	0.07

$$g(H_2) + g(H) = 0.97 + 1.98 \times 10^{-3}\,T, \quad g(H_2O_2) = 0.72 - 1.49 \times 10^{-3}\,T \tag{8.3}$$

$$g(H) = 0.54 + 1.28 \times 10^{-3}\,T, \quad g(-H_2O) = 3.96 + 6.06 \times 10^{-3}\,T. \tag{8.4}$$

These show that the yields of products excluding H_2O_2 increase with increasing temperature.

8.1.3 Radiation-Induced Reactions

After the spur or track reactions, the reactions of radiolysis products in approximately homogeneous distribution start with their primary yields to form final products, which are measured as observed yields by steady-state analysis. In dilute solutions, the radiolysis products react with other solutes and also with each other. In Table 8.2, the typical reactions of radiolysis products are listed with their reaction rate constants [6,7].

These radiation-induced reactions are characterized by reactions of OH radical, e^-_{aq} and H atom, and by their simultaneous proceeding because of the high reactivity of products [1].

TABLE 8.2

Selected Reactions of Radiolysis Products, Oxygen Molecule, and Bromide Ion

No.	Reactants		Products	Rate Constant (10^{10} M^{-1} s^{-1})
1	$e^-_{aq} + e^-_{aq} (+2H_2O)$	→	$H_2 + 2OH^-$	0.644
2	$e^-_{aq} + H^\bullet (+H_2O)$	→	$H_2 + OH^-$	2.64
3	$e^-_{aq} + {}^\bullet OH$	→	OH^-	3.02
4	$e^-_{aq} + H_2O_2$	→	${}^\bullet OH + OH^-$	1.41
5	$e^-_{aq} + H^+$	→	$H^\bullet$	2.25
6	$H^\bullet + H^\bullet$	→	H_2	0.543
7	$H^\bullet + {}^\bullet OH$	→	H_2O	0.153
8	$H^\bullet + H_2O_2$	→	${}^\bullet OH + H_2O$	0.00516
9	${}^\bullet OH + {}^\bullet OH$	→	H_2O_2	0.474
10	${}^\bullet OH + H_2$	→	$H^\bullet + H_2O$	0.00415
11	${}^\bullet OH + H_2O_2$	→	$HO_2^\bullet + H_2O$	0.00287
12	$e^-_{aq} + O_2$	→	$O_2^{\bullet-}$	1.79
13	$H^\bullet + O_2$	→	$HO_2^\bullet$	1.32
14	$Br^- + {}^\bullet OH$	→	$BrOH^{\bullet-}$	0.1
15	$BrOH^{\bullet-}$	→	$Br^\bullet + OH^-$	0.12
16	$Br^\bullet + Br^-$	→	$Br_2^{\bullet-}$	0.12
17	$Be_2^\bullet + H^\bullet$	→	$H^+ + 2Br^-$	1.4
18	$Br_2^{\bullet-} + e^-_{aq}$	→	$2Br^-$	1.1

Source: Adapted from Ohno, S., *CRC Handbook of Radiation Chemistry*, Chapter IV, CRC Press, Boca Raton, 1991 [6]; Elliot, A. J., *Rate Constants and G-Values for the Simulation of the Radiolysis of Light Water over the Range 0–300°C*, Report AECL-11073, Atomic Energy of Canada Ltd., Chalk River, Ontario, 1994 [7].

The OH radical is an oxidizing agent. It abstracts electron or hydrogen from many ions and molecules. In contrast, the e^-_{aq} and H atom are reducing agents. The e^-_{aq} reacts with many kinds of solutes by electron attachment. In addition to the attachment reaction, the H atom can also abstract hydrogen especially from saturated organic molecules to give H_2. Then the term "scavenge" is used in connection with them, and means to convert the highly reactive species to others, ordinarily lower one.

A sequence of these reactions of radiolysis products with primary yields results in final products with observed yields. In Figure 8.3, results of numerical calculation on the production of H_2 and H_2O_2 in two solutions, deaerated pure water, and 1 mM (= mol/dm^3) potassium bromide (KBr) solution are shown to be compared with each other.

For the calculation in pure water, input data on the yields and reactions (1–13) of radiolysis products in Table 8.2 are used, and for that in 1 mM KBr solution, reactions (14–18) are further added to the reaction set. In pure water, the H_2 production comes to equilibrium at absorbed doses of ca. 10 Gy (= J/kg), and then no more H_2 is produced. In contrast, H_2 is produced at the same rate indicating the primary yield of H_2 in the 1 mM KBr solution. This is because OH radical, H atom, and e^-_{aq} are scavenged in reactions (14–18) including bromo-containing ions. Therefore, they are not involved in ether production or decomposition of H_2.

As can be expected from Figure 8.3, the observed yields of final products are affected even by the addition of small amount of solutes. Because the experimental researches on the radiation-induced H_2 production are mainly based on analysis of final products, appropriate care on the chemical stage in water radiolysis is necessary for discussion on the experimental results.

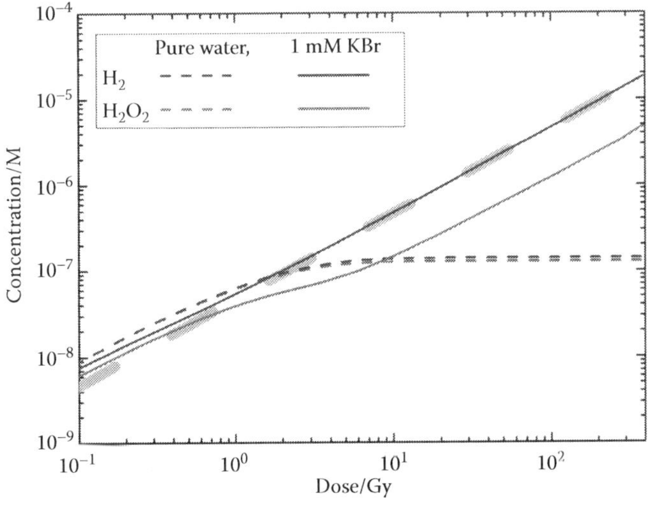

FIGURE 8.3
Numerical calculation of H_2 and H_2O_2 concentrations under γ-ray irradiation as a function of absorbed dose at the dose rate of 1 kGy/h, with a line (gray, broken) corresponding to H_2 production at the rate of the primary yield.

8.1.4 Formation of Molecular Hydrogen

After the sequence of many reactions of radiolysis products, radical species are consumed to give molecular products. H_2 and H_2O_2 are the main products in radiolysis of pure water [1]. As expected from the foregoing, H_2 is also involved in radiation-induced reactions, and the yield observed as a final product changes depending on the solution conditions. At the beginning of chemical stage, H_2 is already produced as a primary radiolysis product. In chemical stage, it is produced in reactions (1), (2), and (6) in Table 8.2, while it is attacked by OH radical and consumed in reaction (10).

When a solution contains a reducing agent like organic compounds, increase in the observed yield of H_2 is expected. Hydrogen abstraction from the organic compounds by H atom increases the yield of H_2 especially in acidic solution, where e^-_{aq} is converted to H atom by reaction (5). In addition, OH radical would be scavenged out by reactions with reducing agents. The scavenging reactions contribute to inhibition of H_2 decomposition, leading to the increase in H_2 production.

On the contrary, in a solution containing an oxidizing agent like O_2, little H_2 production is expected. The e^-_{aq} and H atom are scavenged by oxidizing agents, and the primary radiolysis product of H_2 is oxidized by the OH radical. Moreover, in the case of continuous irradiation, even if H_2 were produced at the beginning of irradiation, the oxidative counterpart such as H_2O_2 would be accumulated in the solution to inhibit the H_2 production.

As seen in the Figure 8.3, the radiolysis of pure water by γ-rays scarcely produces H_2 mainly due to the reactions of OH radical and accumulated H_2O_2. Therefore, additional efforts are necessary for extracting H_2 from water radiolysis.

8.2 Experimental Studies

8.2.1 Concept of Effective Hydrogen Production

As discussed previously in Section 8.1, H_2 is finally produced through spur and the subsequent homogeneous reactions in water radiolysis, where H_2 is formed mainly by radical combinations ($e^-_{aq}(H) + e^-_{aq}(H)$) and by hydrogen abstraction from solutes like alcohols ($H + RH \rightarrow H_2 + R$) while it is consumed by its reaction with the OH radical ($H_2 + OH \rightarrow H + H_2O$). Therefore, the yield, $G(H_2)$, of H_2 observed as a final product is much smaller than the primary yield, $g(H_2)$ in general.

On the other hand, it is well known in radiation chemistry of water that the H_2 production can be effectively realized by scavenging OH radical to highly convert the radiolysis products to H_2 without consuming H_2 [1]. In aqueous alcoholic solutions as a typical example, alcohol (RH) molecules are ready to react with OH radical ($\rightarrow H_2O + R$) and with H atom ($\rightarrow H_2 + R$) so that the observable $G(H_2)$ can be 10 times larger than the primary $g(H_2)$. In this case, the RH works as a precursor of H_2 production as well as an OH scavenger.

Recently, it has been reported by several groups [8–14] that adding ceramic oxides to aqueous solutions enhanced the H_2 production. The ceramic oxides seem to enhance the yield of radiolysis products of reducing agents and/or to interact with the products of oxidizing agents. In next sections, the measurement procedure of H_2 produced from water radiolysis (Section 8.2.2) will be described, and then latest experimental results obtained by adding ceramic oxides to water (Section 8.2.3) will also be discussed.

8.2.2 Measurement of Molecular Hydrogen from Water Radiolysis

Production of H_2 in water radiolysis has been studied mainly by steady-state radiolysis experiments [14,15]. In the experiments, samples are continuously irradiated and then final products formed by radiation-induced reactions are analyzed after irradiation. A lot of ionizing radiation sources have been used: for example, γ-, β-, and α-rays from radioisotopes, and electron, proton, and ion beams from accelerators [1,16].

Concerning the H_2 measurement, the experimental setup in steady-state radiolysis can be roughly divided into batch-type and combined-type experiments. The batch type is suitable for exploring the H_2 production under many different sample conditions [14]. Small amount of samples are irradiated in sealed vials, and gases finally produced after irradiation are analyzed sample by sample. In the combined type, detailed information on the production rate according to accumulated dose can be obtained [15]. Samples are irradiated in gas-tight vessels with flow or loop line for gas sampling, and produced gases varying from hour to hour are analyzed online. Figure 8.4 shows the schematic diagrams of combined-type irradiation experiment, where the produced gases can be measured by commonly used gas chromatography.

Accurate measurement of absorbed dose of sample is always important in irradiation experiments. There are many kinds of chemical dosimeters for low LET radiations, and some of them are listed in Table 8.3 with references [17–20].

Suitable dosimeter should be chosen in terms of total dose and dose rate. It is to be noted that the response characteristics of the dosimeter can be different depending on LETs and types of radiations. For ion beams, more complicated protocols are necessary because the LETs of beams widely change depending on ions and their energies [21].

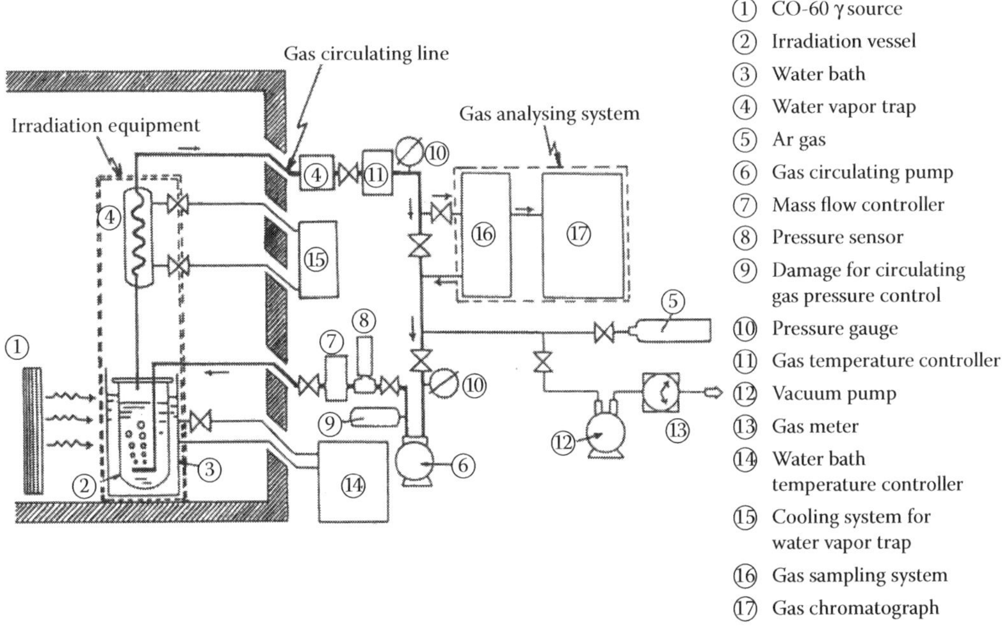

FIGURE 8.4
Schematic diagrams of experimental setup in combined system [15]. (From Nakagiri, N. and Miyata, T., *J. Atom. Energy Soc. Japan*, 36, 744, 1994, (Japanese article). With permission.)

TABLE 8.3

Some of Dosimeters and their Applicable Dose

Dosimeter	Dose and Dose Rate Range
Fricke [17]	40–400 Gy, < 10 kGy/h
Dichromate [18]	1–50 kGy, < 10 kGy/h
Alanine [19]	10–100 kGy, < 1 MGy/h
CTA film [20]	5–300 kGy, 10 k–10 MGy/h

Validity of experiment can be checked by measuring the H_2 production in solutions, where the observed yield of H_2 is established. Deaerated KBr solution is one of suitable solutions as mentioned in Section 8.1.3, and the observed yield of H_2 in the solution is 0.45 molecules $(100 eV)^{-1}$ equal to the primary yield for low LET radiations [6].

Check on the impurity effect on H_2 production is also recommended, because contamination of organic solutes can seriously affect the results of H_2 measurement in steady-state radiolysis as previously mentioned. This can be carried out by measuring oxidative counterpart produced in radiation-induced reactions such as H_2O_2 in the case of pure water and by checking the mass conservation.

8.2.3 Experimental Results Obtained by Adding Ceramic Oxides to Water

Enhancement in H_2 production is observed in water radiolysis in the presence of ceramic oxides [8–14], which heterogeneously exist without being dissolved in aqueous solutions. In this subsection, the effect of ceramic oxides will be introduced as one possibility of effective usage of water radiolysis for the H_2 production.

There are many experimental studies on the effect of oxides, especially silicon dioxide(SiO_2)-, titanium dioxide(TiO_2)-, and zirconium dioxide(ZrO_2)-based materials [8–14]. The common finding in these studies is a phenomenological result that the observed yield of H_2 in the presence of oxides tends to increase comparing with that in the absence of oxides, though the extent of increment depends on oxide composition, structure, size, surface area, weight fraction, and so on. As an example, Figure 8.5 shows the H_2 production in γ-radiolysis of aerated 0.4 M sulfuric acid (H_2SO_4) solutions containing different kinds of oxide particles at the same weight fraction of 33 wt.% [14].

The H_2 production in H_2SO_4 solution was observed larger than that in pure water, although the observed yields of H_2 in the aerated solutions were lower than the primary yield. Then the addition of oxides to the H_2SO_4 solution resulted in the further increase in H_2 production, the extent of which was different depending on the kinds of oxides.

Two mechanisms can be proposed for the enhancement in H_2 production on the basis of experimental results under different conditions as schematically summarized in Figure 8.6. One is some kind of energy transfer from oxides to water, leading to water decomposition on oxides surface. H_2 production in radiolysis of surface water on SiO_2, ZrO_2, and other oxides has been reported to be more efficient than that of bulk water [8–10]. The other is involvement of oxides in radiation-induced reactions. There are reports on increase in the observed yield of H_2 in radiolysis of aqueous solutions containing relatively small amount of oxides [11,12] and also reports on reactions of nano-sized TiO_2 particles with radical products of water radiolysis [13]. Because these mechanisms are proposed at different stages in radiation-induced reactions, these are not conflicting but both have possibilities to increase the yield of H_2.

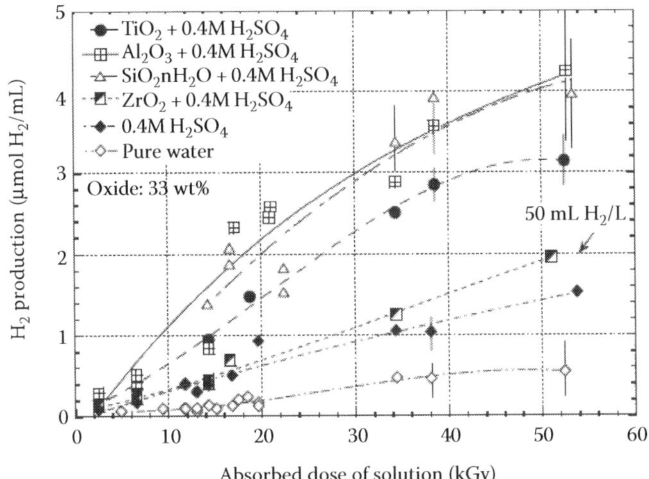

FIGURE 8.5
H_2 production in steady-state Co-60 γ-radiolysis of aqueous 0.4 M H_2SO_4 solution containing oxide particles at 33 wt.% [14]. The amounts of H_2 production were plotted as the number of H_2 (mol) produced in the unit volume (mL) of solution (µmol H_2/mL). (From Yamada, R. et al., *Int. J. Hydrogen Energy*, 33, 929, 2008. With permission.)

In the former case, the energy of radiation is deposited mainly on oxides. The increase in the yield of H_2 is often defined as increase by comparing a small amount of water on a large amount of oxides (sample 1) with the small amount of water only (sample 2) to illustrate the energy transfer from oxide to water, although there is a significant difference in whole absorbed energy of radiation between two samples. In the latter case, the reactions

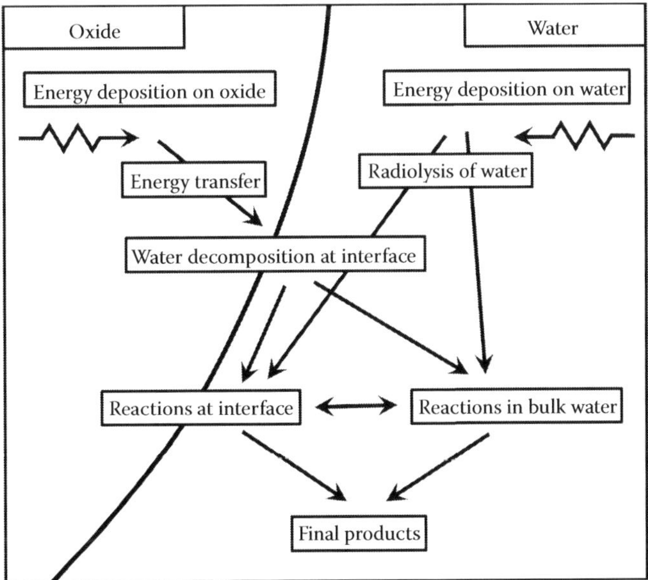

FIGURE 8.6
Schematic diagram of the effect of ceramic oxides from energy deposition to final products.

of radical products of water radiolysis on the oxide surface in a similar way to catalytic reactions are expected to contribute to the increase of H_2 production. However, kinetic scheme leading to enhancement in the H_2 production is still an open question.

References

1. Spinks, J. W. T. and Woods, R. J., *An Introduction to Radiation Chemistry* (3rd edn), John-Wiley & Sons, Inc., New York, 1990.
2. Mozumdar, A. and Hatano Y., *Charged Particle and Photon Interactions with Matter*, Marcel Dekker, Inc., New York, 2004.
3. Buxton, G. V., Radiation chemistry of the liquid state: (1) Water and homogeneous aqueous solutions, in Farhataziz and M. A. J. Rodgers (Eds.), *Radiation Chemistry. Principle and Practice*, pp. 321–349, CVH Publishers, New York, 1987.
4. Draganic, I. G., Nenadovic, M. T., and Draganic, Z. D., Radiolysis of HCOOH + O_2 at pH 1.3–13 and the yields of primary products in γ radiolysis of water, *J. Phys. Chem.*, 73, 2564, 1969.
5. Elliot, A. J., Chenier, M. P., and Ouellette, D. C., Temperature dependence of g values for H_2O and D_2O irradiated with low linear energy transfer radiation, *J. Chem. Soc., Faraday Trans.*, 89, 1193, 1993.
6. Ohno, S., Radiation chemistry of aqueous systems, in Y. Tabata, Y. Ito, and S. Tagawa (Eds.), *CRC Handbook of Radiation Chemistry*, Chapter IV, CRC Press, Boca Raton, 1991.
7. Elliot, A. J., *Rate Constants and G-Values for the Simulation of the Radiolysis of Light Water over the Range 0–300°C*, Report AECL-11073, Atomic Energy of Canada Ltd., Chalk River, Ontario, 1994.
8. LaVerne, J. A. and Tandon, L., H_2 production in the radiolysis of water on CeO_2 and ZrO_2, *J. Phys. Chem. B*, 106, 380, 2002.
9. Petrik, N. G., Alexandrov, A. B., and Vall, A. I., Interfacial energy transfer during gamma radiolysis of water on the surface of ZrO_2 and some other oxides, *J. Phys. Chem. B*, 105, 5935, 2001.
10. Rotureau, P., Renault, J. P., Lebeau, B., Patarin, J., and Mialocq, J.-C., Radiolysis of confined water: Molecular hydrogen formation, *Chem. Phys. Chem.*, 6, 1316, 2005.
11. LaVerne, J. A. and Tonnies, S. E., H_2 production in the radiolysis of aqueous SiO_2 suspensions and slurries, *J. Phys. Chem. B*, 107, 7277, 2003.
12. Seino, S., Yamamoto, T. A., Fujimoto, R., Hashimoto, K., Katsura, M., Okuda, S., and Okitsu, K., Enhancement of hydrogen evolution yield from water dispersing nanoparticles irradiated with gamma-ray, *J. Nucl. Sci. Tech.*, 38, 633, 2001.
13. Gao, R., Safrany, A., and Rabani, J., Fundamental reactions in TiO_2 nanocrystallite aqueous solutions studied by pulse radiolysis, *Radiat. Phys. Chem.*, 65, 599, 2002.
14. Yamada, R., Nagaishi, R., Hatano, Y., and Yoshida, Z., Hydrogen production in the γ-radiolysis of aqueous sulfuric acid solutions containing Al_2O_3, SiO_2, TiO_2 or ZrO_2 fine particles, *Int. J. Hydrogen Energy*, 33, 929, 2008.
15. Nakagiri, N. and Miyata, T., Evaluation of value for hydrogen release from high-level liquid waste, (I) Gamma-ray radiolysis of aqueous nitric acid solutions, *J. Atom. Energy Soc. Japan*, 36, 744, 1994 (Japanese article).
16. Tabata, Y., Radiation sources and dosimetry, in Y. Tabata, Y. Ito, and S. Tagawa (Eds.), *CRC Handbook of Radiation Chemistry*, Chapter II, CRC Press, Boca Raton, 1991.
17. Mathews, R. W., Aqueous chemical dosimetry, *Int. J. Appl. Radiat. Isot.*, 33, 1159, 1982.
18. Sharpe, P. H. G., Barrett, J. H., and Berkley, A. M., Acidic aqueous dichromate solutions as reference dosimeters in the 10-40 kGy, *Int. J. Appl. Radiat. Isot.*, 36, 647, 1985.

19. Regulla, D. F. and Deffner, U., Dosimetry by ESR spectroscopy of alanine, *Int. J. Appl. Radiat. Isot.*, 33, 1101, 1982.
20. Tamura, N., Tanaka, R., Mitomo, S., Matsuda, K., and Nagai, S., Properties of cellulose triacetate dose meter, *Radiat. Phys. Chem.*, 18, 947, 1981.
21. Andreo, P. *et al.*, Absorbed dose determination in external beam radiotherapy, Technical Reports Series 398, Vienna: IAEA, 2000.

Section III

Nuclear Hydrogen Production Systems

9

Water Reactor

Charles W. Forsberg, Kazuyuki Takase, and Toru Nakatsuka

CONTENTS

9.1 Introduction

All of the world's existing commercial nuclear power plants and those under construction are water-cooled reactors. These plants were expected to operate for 30–40 years. Today it is expected that many plants will operate for 60 years. There are operating demonstration fast reactors and high-temperature reactors; however, it will be several decades before significant numbers of these reactors could be built. If any significant quantities of nuclear hydrogen are produced in the next several decades, it will be from water-cooled reactors.

This chapter provides a brief description of different water-cooled reactor technologies and how these nuclear reactors may be used for the generation of hydrogen. While it is feasible to build dedicated water-cooled reactors for hydrogen generation, this is not the most likely deployment scenario. The only hydrogen production technologies that are coupled to water-cooled reactors are low- and high-temperature electrolysis (HTE). In both cases, electricity is the primary cost of hydrogen production. The demand and cost of electricity varies daily, weekly, and seasonally. As a consequence, there are large incentives to produce hydrogen at times of low electricity cost. Large-scale hydrogen production using water-cooled reactors would significantly impact electricity markets; thus, Section 9.4 also addresses the integration of hydrogen production with electricity production—including the production of peak electricity using hydrogen.

9.2 Light Water Reactors

The light water reactor (LWR) is a conventional type of nuclear-fission reactor and the technology basis for the majority of the world's nearly 500 commercial nuclear power stations operational or under construction to date. As so named, light water (H_2O) is used as both coolant and neutron moderator in the reactor. In contrast, heavy water (D_2O) is used exclusively or in part in another type of water reactor to be described in Section 9.3.

The LWR produces heat from controlled fission chain reaction in a reactor core. The reactor coolant removes the fission heat from the core to the balance of plant for energy conversion and delivery. The reactor core contains mainly of nuclear fuel assembles and reactivity-control rods. A steel pressure vessel structurally supports and encloses the reactor core while withstanding all thermal and seismic loadings and irradiation exposure over the lifespan of the reactor operation.

A fuel element is generally a cylindrical rod that includes many small fuel pellets inside a metallic tube. The fuel pellet is made typically of low enriched uranium dioxide ceramic. As many as 200 fuel rods may be bundled to form a fuel assembly. Equally many fuel assemblies may be loaded in a large size reactor core. A part of these fuel assemblies need to be replaced with fresh ones normally in less than two years of fuel burn up in the reactor.

A control rod is made of various elements which easily capture neutrons. A number of control rods are arranged for distributed insertion in the core. Inserting the control rods deeper into the core causes more neutrons to be captured and consequently the chain reaction to be slowed or stopped. Pulling the control rods out of the core results in more neutrons absorbed in the core and thus the chain reaction is accelerated. The active movement of the control rods is carried out by control rod drive mechanism.

The three general design variants of the LWR include pressurized water reactor (PWR), boiling water reactor (BWR), and supercritical water reactor. They are further described in the separate sections below.

9.2.1 Pressurized Water Reactor

Since detailed technical reference are available [1], the description given here is an introduction to the design and development of the PWR. A PWR has characteristically two main cooling systems, primary and secondary, while other LWR design variants have only one. In a PWR, water under high pressure flowing through the reactor core, is heated, flows to a steam generator, which transfers that heat to a secondary coolant, and is returned to the reactor core. The water pressure is sufficient to prevent boiling; thus, these reactors are called PWRs. Steam is produced in the secondary system to turn a turbine power generator.

The history of PWR began with the development of a research nuclear reactor by the collaboration of Argonne National Laboratory and Westinghouse in the United States. Westinghouse followed this with construction of the world's first PWR power generating plant (60 MWe) operated during 1958–1982 in Shippingport, Pennsylvania. Although producing power, the reactor was built also to facilitate research and development, and its design features such as seed blanket reactor core, highly enriched (93%) uranium-235, control rods made of hafnium, and so on, which differed significantly from the commercial reactors that would follow.

The newer PWR designs employ Zircaloy-4 (a zirconium alloy resistant to heat and corrosion) clad fuel, control rod cluster, and vertical steam generators. By 1970, these design features were used in the construction of the Beznau power plant in Switzerland and the Ginna power plant in the United States. Today, the PWR is used in a majority of commercial LWRs.

Figure 9.1 shows the design outline of a PWR power plant. In the gigawatt electric output class plant, the primary water coolant is pressurized to 15.7 MPa so that the water is always a liquid as it exits the reactor core at a temperature of about 320°C before flowing to the steam generator under the circulation of the primary coolant pump. In the steam generator, the primary coolant transfers heat to generate steam of about 277°C and 6.1 MPa in the secondary coolant system. The steam expands in the turbine for electric power conversion, is condensed to a liquid in the condenser, and pumped back to the steam generator. Typically sea water is used to remove the thermal waste heat of the secondary system power conversion in the condenser.

Figure 9.2 illustrates the structural design of the reactor steel pressure vessel and internals. The reactor core contains the fuel assemblies and the control rods. The control rod cluster guide tubes are located above the core, and the control rod driving mechanism, one for each cluster of control rods, is externally placed on the top of the vessel head.

In typical designs, a fuel assembly consists of 14 × 14 to 17 × 17 square array of tubes including 12 or more control rod guide channels, instrument guide tube and the rest of fuel rods. The array of tubes are spaced apart and bundled by several support lattice plates in the fuel assembly of roughly 4 m long and 20 cm square cross-section. The fuel rod contains fuel pellets, usually made of several percent enriched UO_2 ceramic, in the Zircaloy-4 cladding tube. The pressurized helium gas is filled in the fuel rod to enhance the heat conduction from the fuel to the cladding tube. Both ends of the fuel rod tube are welded to seal helium inside the reactor core.

The control rods are bundled in a cluster inserted into each fuel assembly. The cluster of control rods is driven by the control rod drive mechanism into or out of the fuel assembly to

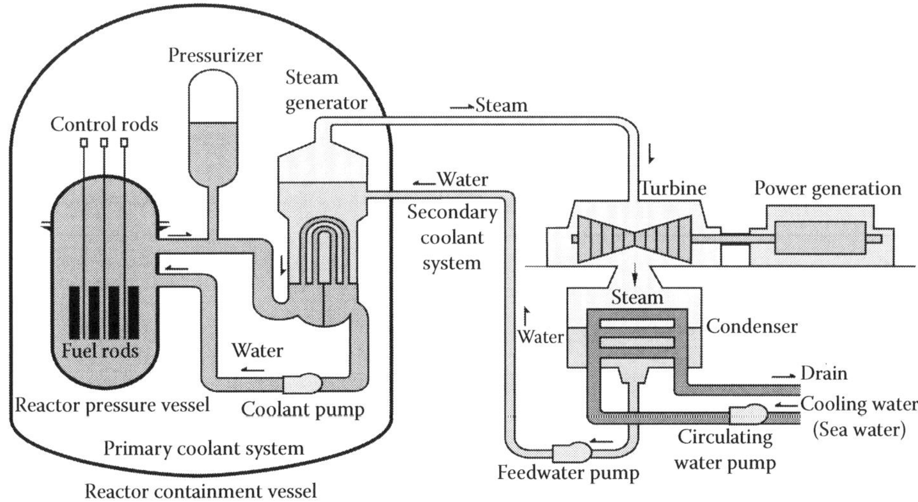

FIGURE 9.1
Outline of a PWR power plant. (From Japan Atomic Energy Relations Organization, Drawing collection of nuclear power and energy in 2008, No. 5-05, 2009. With permission. [2])

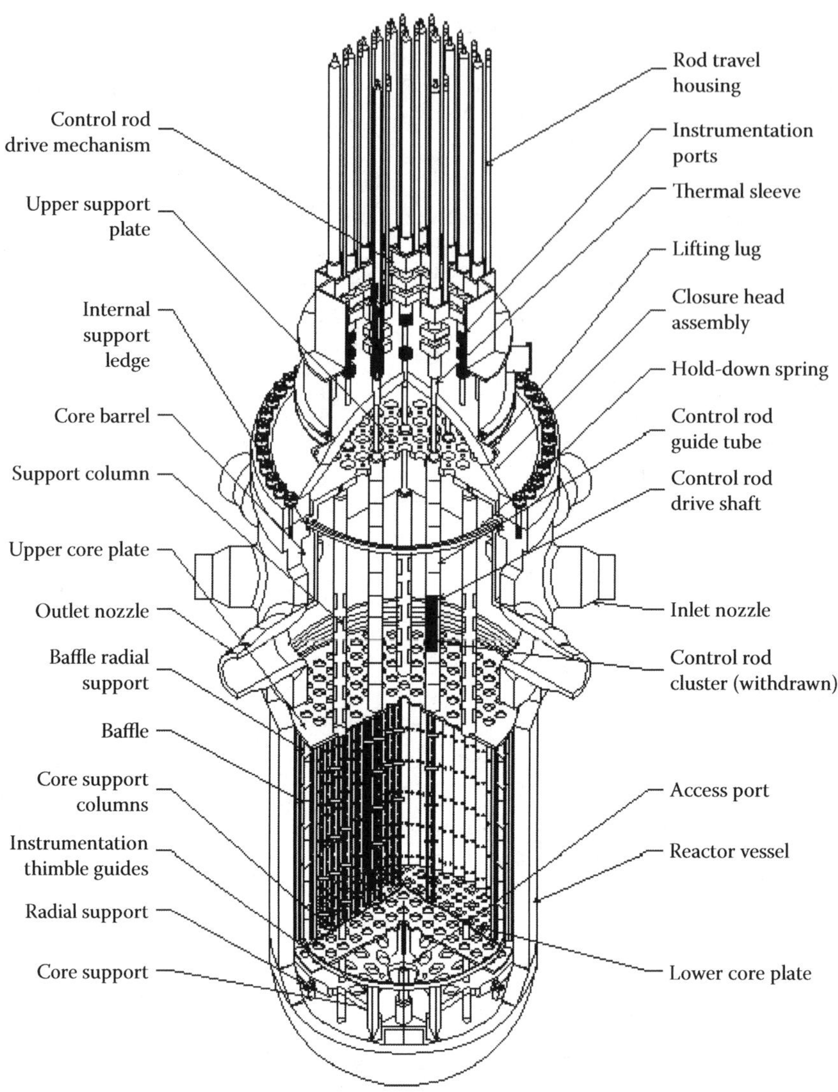

FIGURE 9.2
Schematic of a PWR pressure vessel and internals. (From http://www.eia.doe.gov/cneaf/nuclear/page/nuc_reactors/pwr.html, Pressurized Water Reactor and Reactor Vessel, Energy Information Administration, USA. With permission. [3])

perform reactor startup and shutdown and to change reactor power. A degree of inherent following of reactor power to turbine load demand is achieved by the negative temperature feedback of reactivity with the water moderator. The turbine load change causes the steam flow in the steam generator to change accordingly. This results in the primary water temperature to increase or decrease. The negative temperature coefficient of reactivity causes the reactor to fission less and more, thereby generating less or more power, due to the negative response of the water density to water temperature. Adjusting the inserted position of control rods in the core or altering the concentration of boric acid, which readily absorbs neutrons, solution in the coolant can then be used to return temperature to the design point.

In case of abnormal event or emergency, all control rods are dropped down into the reactor core simultaneously to shut the reactor down. In case of such an event as a loss of coolant accident, a reactor emergency core cooling system (ECCS) is further activated to inject water through multiple and independent flow circuits to remove the decay heat continuously generated in the fuel rods after the reactor shutdown.

The advanced PWR designs including Generation III+ AP1000 (Advanced Passive 1000 by Westinghouse) and EPR (European pressurized reactor by AREVA) have been developed. The AP1000 improved the previous PWR designs by relying upon passive reactor safety and employing modular plant construction in a 1117 MWe standardized reactor design. The AP1000 is estimated by the Westinghouse to have an overall core damage probability of 2.41×10^{-7} per plant per year [4]. The Nuclear Regulatory Commission (NRC) issued the design certification of AP1000 in December 2005. Several reactor units of AP1000 have been ordered by China.

The EPR rated at 1650 MWe was designed with enhanced active as well as passive safety features, achieving an estimated overall core damage probability of 6.1×10^{-7} per plant per year [5], and with improved economic competitiveness, for example, through larger scale, over previous PWR designs. As of 2007, two EPR units were under construction, one each in France and Finland.

9.2.2 Boiling Water Reactor

In a BWR, liquid water coolant is heated, boiled, and converted to steam in the reactor core; thus, the name BWR. BWRs are the second most common type of power reactor after PWRs. The BWR plants operational today roughly include two types, the conventional BWR and the advanced boiling water reactor (ABWR). The development of the BWR was started in the 1950s by General Electric (GE) and several U.S. national laboratories. The first BWR power plant in the world is Dresden No. 1 (200 MWe), which was built in Illinois, United States. It was operated between 1960 and 1978.

What follows is a brief introduction to the BWR general design. A detailed design description is available elsewhere [6]. In contrast to the PWR, a BWR has a single main cooling flow system as illustrated in Figure 9.3. The plant consists of a steel reactor pressure vessel, recirculation pumps, and a pressure suppression pool inside a reactor containment vessel. The balance of the plant is a turbine power generator system installed outside of the containment vessel. The BWR heats water to boil to generate steam in the reactor pressure vessel. The steam is used to drive the turbine power generator. The turbine exhaust steam is condensed to liquid water by external cooling of typically seawater. Water is pressurized and returned to the reactor by the feed water pump.

The steel reactor vessel structurally supports and encloses the reactor core. The reactor core contains mainly of fuel assemblies and control rods. Moreover, a steam-water separator and a steam dryer is situated above the reactor core.

A fuel assembly for BWR consists usually of a square array (e.g., 8×8) of fuel rods. A fuel rod consists of a Zircaloy cladding tube containing UO_2 ceramic pellets. Both ends of the cladding tube containing high-pressure helium gas are welded to seal the gas in the tube.

A control rod contains neutron-absorbing materials such as boron carbide (B_4C), hafnium (Hf), or their combinations. Because the steam conditioning equipment is placed above the core, the control rods driven by the control rod driving mechanism attached to the bottom head of the reactor vessel are inserted from below the core. The control rod drive mechanism includes hydraulic pressure drive and electric motor drive. An additional device is designed to limit the accidental fall of a control rod.

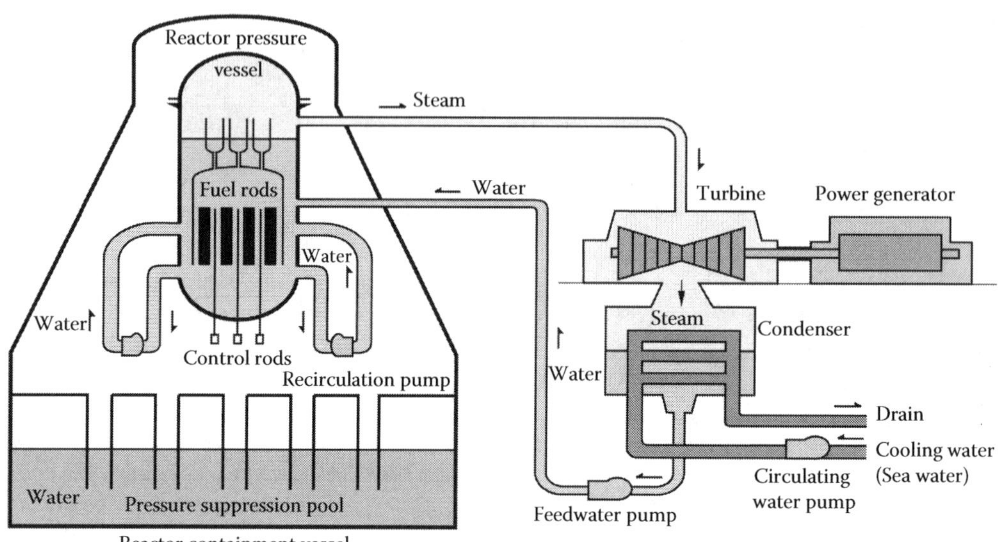

FIGURE 9.3
Outline of a BWR power plant. (From Japan Atomic Energy Relations Organization, Drawing collection of nuclear power and energy in 2008, No. 5-02, 2009. With permission. [7])

The coolant water maintained at about 7.5 MPa is boiled in the reactor core to generate steam at 285°C. Like in the PWR, control rods are inserted into or pulled out of the core to modulate the core reactivity and power-generation rate. In addition, the BWR may control power generation by adjusting the coolant-recirculation flow through the reactor core by the recirculation pumps. As the flow of the water is adjusted, the presence of the steam voids or bubbles is decreased or increased inversely responding to the recirculation flow rate, which causes the fuel to fission more or less, releasing more or less fission heat. In the effect of this control, the reactor power may be adjusted by controlling the speed of the recirculation pumps.

In case of any abnormal event or emergency, all control rods are rapidly inserted into the reactor core by pneumatically driven nitrogen gas stored in tanks. In case of such an event as an accidental loss of reactor coolant, a reactor ECCS is further activated to inject water through multiple and independent flow circuits to remove the decay heat continuously being generated in the fuel rods. The ECCS consists of the high- and low-pressure reactor core spray systems, automatic depressurization system, and the low-pressure injection system.

The reactor containment vessel is provided to prevent the external release of radioactive material because of a combination of fuel failure and loss of reactor coolant. The pressure suppression pool is designed to suppress the high pressure by condensing the reactor coolant steam entering the containment vessel. In addition, a containment water spray system is used to limit the rise of temperature and pressure in the containment subjected to a protracted period of decay heat generation after the reactor fission has been stopped.

The design of the ABWR that was incorporated with improved design features and increased rating (1350 MWe) was developed between the late 1980s and early 1990s. Use of internal recirculation pumps inside of the reactor pressure vessel is a major improvement over the BWR. The internal pumps are powered by wet-rotor motors. The pump housings are integrated into the lower vessel, eliminating large diameter external recirculation pipes. The ABWR was certified by the U.S. NRC as a standardized design in the early

1990s. The Kariwa nuclear power plant in Japan, the first ABWR build in the world, began its operation in 1996. Additional units have since been built and operated in Japan.

GE also proposed the concept of the 600 MWe simplified boiling water reactor (SBWR) that first introduced the passive safety design to the LWRs. The latest proposal by GE is the Generation III+ design of economic SBWR, known as ESBWR that combines the large-scale design features of the ABWR and the passive safety features of the SBWR. One thing characteristic of the 1600 MWe ESBWR is natural recirculation of the coolant in the reactor pressure vessel without any pumps. In fact, pumps are not required to operate any safety systems, simplifying safety design and reducing plant cost. The overall core damage probability is estimated to be 6.16×10^{-8} per reactor year [8]. The design is presently under review for Final Design Approval by the U.S. NRC. Figure 9.4 illustrates the general design layout of the ESBWR plant.

9.2.3 Supercritical Water Reactor

The supercritical water-cooled reactor (SCWR) is an advanced high-temperature, high-pressure water reactor designed to operate above the thermodynamic critical point of water (374°C, 22.1 MPa). Above the critical pressure, water does not exhibit a change in phase. Although SCWR was proposed during the early years of the LWR development [9], it had not attracted significant development interest until new SCWR concepts were emerged in the 1990s in Japan [10], Russia [11], and Canada [12]. The SCWR is now one of the Generation IV nuclear energy systems and enjoys worldwide development efforts [13–15]. No SCWR reactor has yet been built.

The new concepts of SCWR are intended to improve plant economics by targeting high thermal efficiency and low construction cost. Figure 9.5 depicts the schematic of a Generation IV SCWR. Similar to the BWR, it has one main cooling loop. However, the water is superheated without boiling, like the PWR. The power cycle is that adopted from commercial supercritical water (SCW) fossil-fuel power plants (FPPs). The SCWR coolant flow rate is about one-eighth of equivalent BWR's. Therefore, the design is more compact

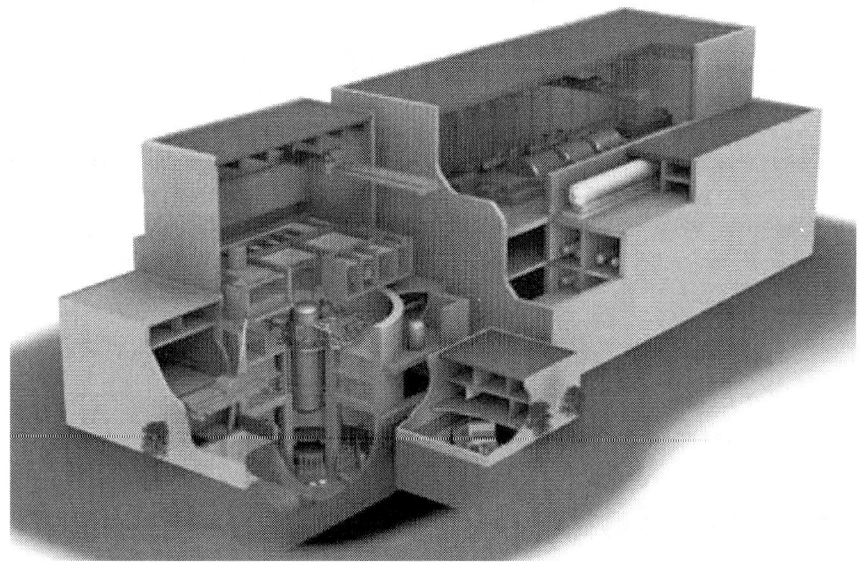

FIGURE 9.4
Design layout of the ESBWR plant. (Courtesy of U.S. NRC.)

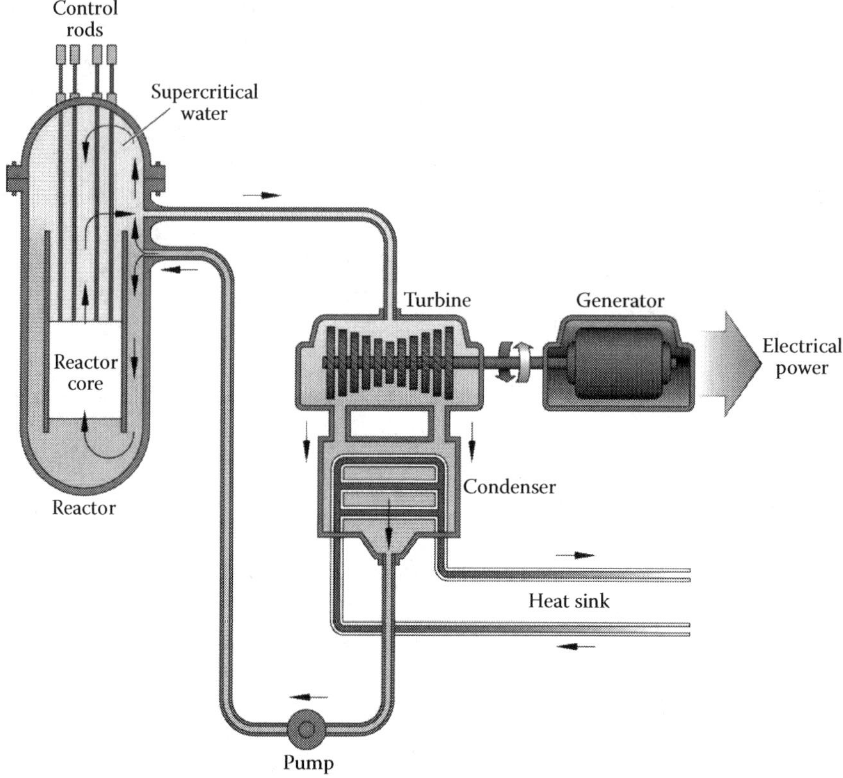

FIGURE 9.5
Schematic of Generation IV SCWR. (From Generation IV International Forum, A Technology Roadmap for Generation IV Nuclear Energy Systems, GIF-002-00, December 2002. With permission.)

than the conventional LWRs. The coolant outlet temperature is also significantly higher such that the thermal efficiency of the SCWR is projected to increase to about 45% from 33% in the conventional LWRs.

Figure 9.6 shows a simplified schematic of a direct cycle SCWR system in comparison with a BWR and a SCW-FPP. The development of the SCWR has the benefit of the industrial experience gained in the conventional water reactors. Both the pressure vessel and pressure tube designs have been proposed. The former evolves based on the conventional LWRs and the latter on the Canada deuterium uranium (CANDU) reactors described in Section 9.2.

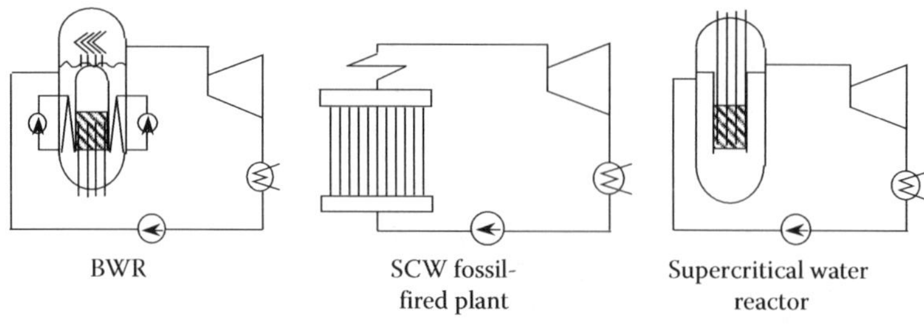

FIGURE 9.6
Evolution of supercritical water reactor from conventional systems.

Furthermore, the balance of the plant design of the SCWR takes advantage of existing SCW-FPP systems, which operate in conditions similar to the 25 MPa and 625°C design values at the SCWR core outlet.

In the case of the pressure vessel design, although an indirect cycle system like a PWR had been designed in the past research [16], nowadays once-through, direct cycle like that of a SCW-FPP receives the focus of development. The direct cycle design eliminates the need for steam generator, steam separator and dryer, pressurizer, and recirculation system in steady-state operation.

Once-through fuel cycle based on traditional ceramic UO_2 or MOX fuel pellets has been used. The concept of a fast reactor has also been studied with the same plant system [17]. The application of materials in the LWR and SCW-FPP serves as a helpful guidance for the selection of structural materials in the SCWR. Stainless steel may be adopted as a fuel cladding tube. Nickel-based alloy may also be considered in the design of the high-temperature core.

The major advantages of SCWR are as follows:

- SCW-FPPs are well established in the electricity production industry. Turbine development is not required for outlet temperatures < 625°C, since SCW-FPPs have been operating at these conditions.
- The SCWR is an evolution from existing water reactors (i.e., PWR, BWR, and PHWRs), and can thus benefit from today's relevant expertise.

The SCWR share many of common equipment, systems, and operation methods with the conventional LWR and FPP plants.

For startup, the SCWR may follow similar procedures to SCW-FPP. Although separators, dryers, and recirculation systems are not required for the steady-state operation, some of these components may still be required for plant startup operation. Two types of the SCW-FPP designs exist: those that operate at constant pressure through wide range of the load and those that change operating pressure with the load.

Two types of startup sequences were estimated based on the SCW-FPP [18–20]. One is constant pressure startup with which the reactor starts at a supercritical pressure and a flash tank and pressure-reducing valves are the required components to lead steam to the turbine. The other is sliding pressure startup with which the reactor starts at a subcritical pressure and a steam–water separator and a drain tank are necessary for two-phase flow. Heat can be recovered from the drain tank of the water separator by using either a recirculation pump or an additional heater.

During SCWR startup at subcritical pressure, the moisture content in the steam becomes a limiting factor as the turbine inlet steam is saturated steam. The moisture content in the steam is limited not to be larger than 0.1% so that turbine blade damage can be avoided. This criterion is consistent with that of the current BWRs.

9.3 Heavy Water-Moderated Reactors

9.3.1 CANDU Reactor

CANDU is a heavy water reactor (D_2O) and the world's third most common type of commercial water-cooled reactor after PWR and BWR. The reactor core comprises hundreds of small-diameter, horizontal channels called pressure tubes, which contains nuclear fuel. The fuel channels are individually accessible and refueled online, eliminating reactor refueling outages. Separate sources of heavy water (D_2O) are used as coolant in high temperature and as moderator in low temperature and pressure. The coolant is circulated through the fuel channels to units of steam generators in the primary circuit and transfer the fission-produced heat to the secondary light water circuit to produce steam to drive the turbine-power generator. The reactivity control devices are located in the moderator. The use of the heavy water moderator, and heavy water coolant, creates an efficient thermal–neutron spectrum core so that the natural uranium (containing <1% fissile uranium) or other similarly low fissile material can be used as fuel. Besides eliminating a need for fuel enrichment, the design lowers core excess reactivity and permits relatively long lifespan of prompt neutrons in the core, both of which improve reactor safety.

An extended summary of design features is prepared by Atomic Energy of Canada Limited (AECL), the developer, and vendor of the CANDU reactors [22]. As of today, some 50 units of CANDU reactors are being operated or planned to be constructed worldwide.

9.3.2 Advanced Designs

The ACR-1000 or Advanced CANDU Reactor is a Gen III+ , heavy-water-moderated and light-water-cooled pressure-tube reactor. Figure 9.7 shows the schematic of ACR-1000 [23]. It evolves from the earlier CANDU and retains many notable design features such as the heavy-water moderator and horizontally oriented reactor core, individual pressure tube coolant channels, and online refueling. The new design features includes the light-water coolant and low enriched uranium fuel, intended to reduce reactor core size and spent fuel volume. In addition, the coolant temperature and pressure are increased to improve power generation thermal efficiency to about 36% from 35% in the conventional CANDU. A study showed the application of ACR-1000 for competitive hydrogen production through conventional electrolysis or HTE of steam [24]. An overall thermal-to-hydrogen production efficiency is estimated to be around 33% in an ACR-1000 coupled HTE plant.

The CANDU-SCWR is the latest proposed CANDU type of Generation-IV SCWR. Like the ACR-1000, heavy-water moderator and light-water coolant are the key design features of the CANDU-SCWR. Unlike the earlier generation, the coolant is highly pressurized supercritical water and operates in single phase. Thus, the design shares similar design features with its supercritical LWR counterpart described in Section 9.2.3.

Rated at 1220 MWe, the CANDU-SCWR power generation efficiency of about 45% is projected at the design point of the reactor core outlet water conditions of 625°C and 25 MPa [25]. At this high thermal efficiency, hydrogen production by electrolysis of water reaches 35%. If superheated steam (~800°C) can be produced with a supplementary heat source, the reactor could produce hydrogen more efficiently from steam electrolysis. Several thermochemical hydrogen production processes described in Chapter 5, such as Cu–Cl cycle, may be supplied by the high-temperature heat of the CANDU-SCWR. The reactor vendor AECL has been collaborating with the U.S. Argonne National Laboratory and the Commissariat à l'Énergie Atomique (CEA) in France in the research and development of the Cu–Cl thermochemical hydrogen-production technology for future application with the CANDU-SCWR.

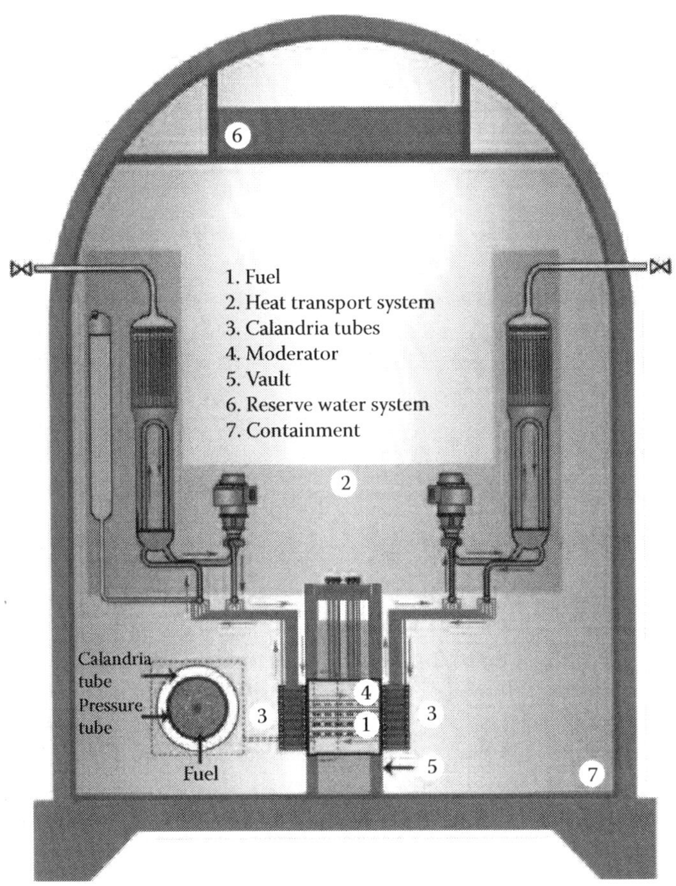

FIGURE 9.7
Schematic of ACR-1000, advanced CANDU reactor. (From Miller, A. I. and Duffey, R. B., Meeting the near-term demand for hydrogen using nuclear energy in competitive power markets. *Second OECD-NEA Information Exchange Meeting on the Nuclear Production of Hydrogen*, Argonne, IL, October 2–3, 2003. With permission.)

9.4 Coproduction of Electricity and Hydrogen

Most of the world's commercial reactors are water-cooled reactors. If hydrogen is to be produced in large quantities using nuclear energy in the next several decades, the nuclear technology will be the water-cooled reactor. These nuclear reactors have several characteristics: peak steam temperatures are between 260 and 320°C, power outputs typically in excess of a 1000 MW(e), and high capital costs with low operating costs.

Water-cooled reactor characteristics limit the applicable hydrogen-production technologies to low-temperature and HTE. The primary cost of hydrogen production for both technologies is the cost of the electricity; thus, hydrogen economics depend upon electricity costs. Electricity costs vary with time; thus, when electricity is used for hydrogen production it has a large impact on hydrogen costs. Furthermore, hydrogen can be used to support electricity production by stabilization of the electricity grid and production of peak electricity. For nuclear hydrogen production using water-cooled reactors, the interconnections between electricity generation and hydrogen may become central to the large scale use of hydrogen.

9.4.1 Electricity System Characteristics

To understand nuclear hydrogen production using water-cooled reactors, the economics and challenges of electricity production must be understood. Electricity demand varies daily, weekly, and seasonally. In higher latitudes, there is also an approximately 3-day cycle associated with weather patterns. A representative example of this variation in electricity demand is shown in Figure 9.8. This example shows the electricity demand over a week for each of three representative weeks in the winter, spring, and summer in the state of Illinois in the United States. While no two parts of the world are the same, all electrical grids show large changes in the demand for electricity with time.

To meet variable electricity demands, utilities operate different types of power plants. Nuclear and renewable power systems with high capital costs and low operating costs are operated at full capacity. Renewable power outputs depend upon local solar or wind conditions. FPPs are the primary technology used to match nuclear and renewable electricity production with fluctuating electricity demand. Fossil fuels are used for intermediate and peak electricity production because (1) they are inexpensive to store until needed (coal piles, oil tanks, and underground natural-gas storage) and (2) the technologies for conversion of fossil fuels to electricity have relatively low capital costs.

The production costs for electricity are higher at times of peak electricity demand. Power plants must be built to meet peak electricity demand; but, these plants operate only for a limited number of hours per year. Consequently, the wholesale prices of electricity vary dramatically. For example, Figure 9.9 shows the number of hours each year that electricity

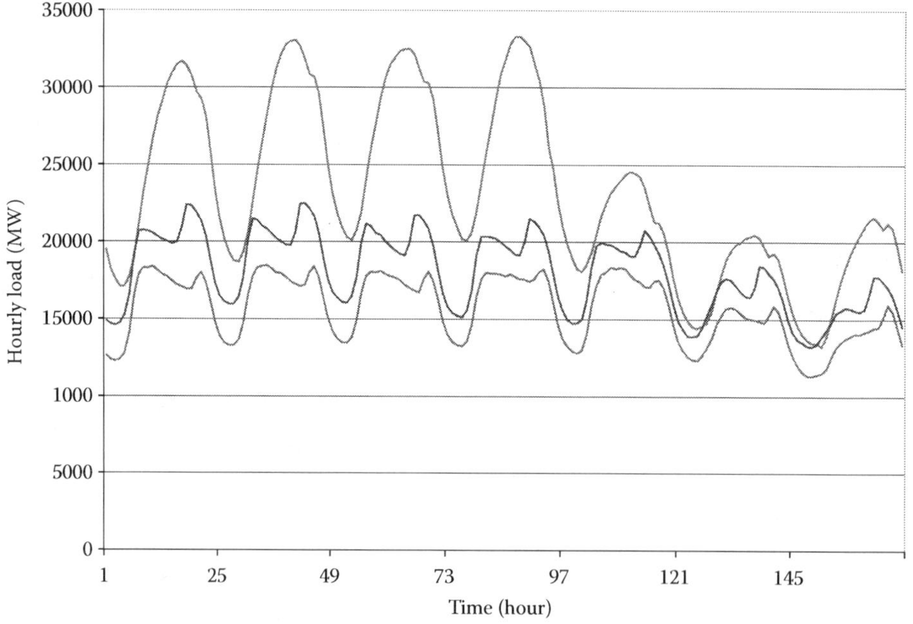

FIGURE 9.8
Weekly variation in Illinois electricity demand in summer (top curve), winter (middle curve), and spring (bottom curve).

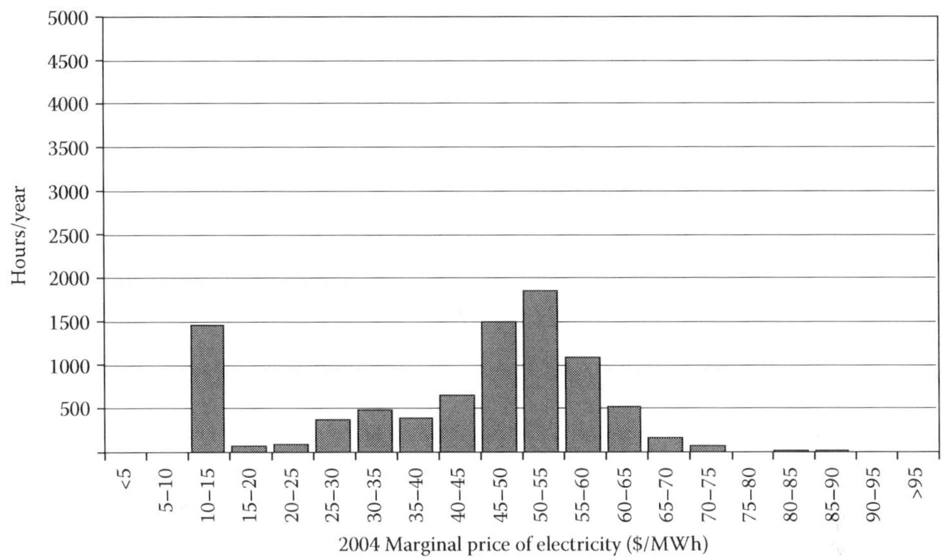

FIGURE 9.9
Marginal prices of electricity in Los Angeles.

can be bought at various prices in Los Angeles [26]. There are low electricity prices [$10–15/MW(e)-h] for almost 1500 h per year but high-priced electricity for over 5000 h a year. This double-hump price curve reflects the low nighttime demand and the high daytime demand for electricity in Los Angeles. Each electrical grid has its own patterns depending upon the types of local industries, weather, and living patterns.

The use of fossil fuels to meet variable electrical demands may be limited in the future because of concerns about the price of natural gas and climate change. While carbon dioxide from fossil power plants may be sequestered underground, such FPPs are likely to be uneconomic for the production of intermediate and peak electricity because of their high capital costs [27] and the difficulties in operating such plants with variable output. Renewable forms of electricity compound the challenges of intermediate and peak electricity production because production does not always match the demand for electricity. This implies for the future (1) larger differences in the cost of electricity between times of low electricity demand and high electricity demand and (2) large incentives to develop technologies for storing electricity.

9.4.2 Nuclear Hydrogen Production

The commercial technology for hydrogen production using electricity is low-temperature electrolysis for the conversion of water into hydrogen and oxygen. Such systems can be sited anywhere where there are the appropriate connections to the electricity grid. If hydrogen storage is available, there are large incentives to produce hydrogen at times of low electricity demand and low electricity prices [28]. Limiting production to times of low electrical prices is a common practice in other industries that are large electricity consumers.

HTE is a new technology (Chapter 17) under development that uses heat to convert water to steam and then uses electricity to convert the steam to oxygen and hydrogen. Because the cost of steam from a water-cooled reactor is about a third the cost of electricity, HTE has the potential to produce hydrogen at significantly lower costs than traditional electrolysis. Although high-temperature electrolyzers operate at ~800°C, water-cooled reactors can be used to provide most of the required heat. Most of the heat input into a HTE system is used for the conversion of water to steam to avoid using electricity for conversion of a liquid (water) into gases (hydrogen and oxygen). The electricity in the HTE process is primarily used to break the chemical bonds.

In an HTE system, the water-cooled reactor steam temperature (~300°C) can be increased by using a counter-current heat exchanger where the steam temperature is increased while the hydrogen and oxygen from the electrolyzer is cooled. The heat from inefficiencies in the electrolyzer can provide the additional temperature increase for the steam to electrolyzer temperatures.

The HTE steam requirements imply that an HTE system must be located within several kilometers of a nuclear power plant so the nuclear plant can supply the required steam. Steam that comes directly from the reactor core, such as in a BWR cannot be used because it contains small amounts of tritium (a radioactive form of hydrogen) that would ultimately be in the end-product—hydrogen. This is avoided by using the steam from the nuclear reactor to provide heat to a secondary steam generation system that produces clean steam. In a number of countries (Switzerland, Canada, Russia, etc.) nuclear power plants sell steam for district heating and industrial customers; thus, the technical and commercial basis to provide clean steam for nonelectric purposes is well established.

A recent study [29] compared hydrogen production by low-temperature electrolysis and HTE using steam from a water-cooled reactor. The water-cooled reactor they selected had a thermal efficiency for electricity production of 36%. With low-temperature electrolysis, the overall thermal efficiency for hydrogen production was 25.7%. With HTE, thermal efficiency of hydrogen production was 33–34%. The efficiency of HTE hydrogen production is almost equal to the efficiency of electricity production. There are large incentives to commercialize HTE for hydrogen production at water-cooled reactors and thus siting hydrogen production facilities and reactors together.

With the HTE process, at times of peak electricity demand, the reactor would only produce electricity for the electrical grid. At times of low electricity demand and available low-cost electricity, the reactor would produce steam and electricity for hydrogen production. If there was a large demand for hydrogen, the reactor can produce just steam for HTE with electricity supplied from the grid.

The viability of variable hydrogen production using low-cost off-peak electricity depends upon the capability to store hydrogen. Unlike electricity, hydrogen can be stored inexpensively [30] for days, weeks, or months in large underground facilities with the same technology used to store natural gas. In the United States, natural gas is produced in states such as Texas and Oklahoma and transported through pipeline to approximately 400 underground facilities that store at high pressure a quarter of a year's production of natural gas in the fall, before the winter heating season. It would not be practical to build production and pipeline systems to meet peak demands. Similar storage facilities exist in Europe to store natural gas from Russia, Norway, and North Africa. This is a low-cost technology, with market prices for storage typically 10% of the value of the natural gas.

A limited number of hydrogen storage facilities exist in Europe and the United States to support the chemical and refining industries. Equally important, measurements of helium concentrations in different geologies from radioactive decay and the long-term existence of natural gas deposits show that many geological environments have the low gas permeability required for hydrogen storage. However, storage and handling economics demand large facilities [31]. Several factors that are almost independent of facility capacity drive the facility size: siting and site development costs. In addition, gas storage requires compression of the gases, typically to pressures of ~7 MPa. Gas equipment efficiencies and costs are strong functions of the size of the equipment. For bulk hydrogen storage, the capital costs are estimated to be \$0.80–\$1.60/kg which is lower than the total production costs for hydrogen.

Low-cost hydrogen storage makes it viable to use low-cost off-peak electricity for hydrogen production independent of when the customer requires hydrogen to be delivered. This also has major implications for the electricity grid. If there are failures on the electrical grid and hydrogen is being produced, its production can be stopped immediately with that electricity diverted to other customers. If the production of nuclear and renewable power in an electrical grid exceeds demand, hydrogen production is a way to efficiently use the excess electricity that is available from energy sources with low operating costs.

Oxygen could be stored using the same technology; however, bulk high-pressure underground storage has not been demonstrated and there remain some uncertainties. There are many oxygen markets.

9.4.3 Using Hydrogen for Peak Electricity Production

The commercial technology for low-cost, large-scale bulk storage of gaseous hydrogen, combined with advancing hydrogen technologies such as HTE, creates the option to use hydrogen to meet daily, weekly, and seasonal peak electricity demands. This is potentially the second-largest market for hydrogen after transportation. One configuration [32] of such a system is shown in Figure 9.10. There are four major components.

- *Steam production.* Water-cooled reactors produce steam. The power plant may operate as a conventional nuclear plant where all the steam is used to produce electricity. Alternatively, electricity and some of the steam may be diverted to hydrogen production.

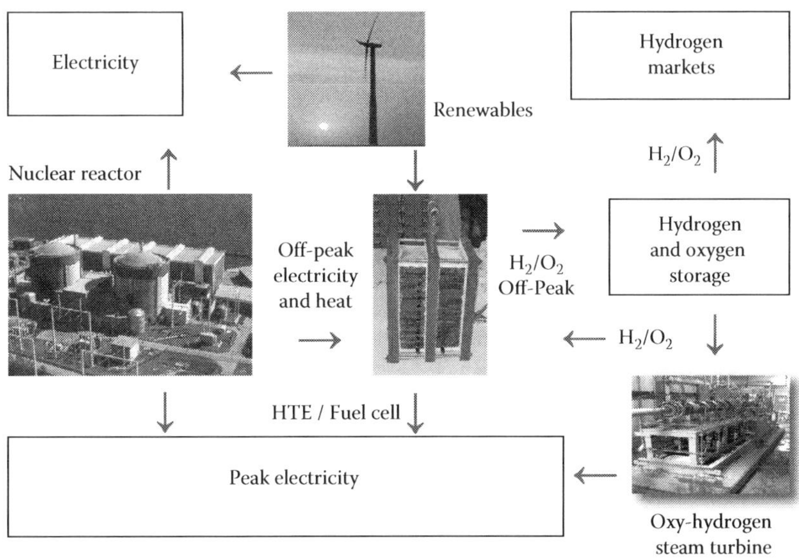

FIGURE 9.10
Peak-power production system using HTE. (Figure of oxy-hydrogen steam turbine courtesy of Clean Energy Systems.)

- *Hydrogen production.* Hydrogen is produced at times of low-power demand using low-temperature electrolysis or HTE.
- *Hydrogen and oxygen storage.* Underground storage facilities are used for the low-cost storage of hydrogen and oxygen on a daily, weekly, and seasonal basis.
- *Peak electricity production.* Variable electricity production to match demand is achieved by two methods: (1) variable production of hydrogen and oxygen to storage when electricity is not needed for the electrical gird and (2) using the hydrogen and oxygen for electricity production to meet peak load demand. The near-term technology for peak electricity production is the gas turbine. The longer-term options include HTE operating in reverse as a fuel cell or use of an oxygen–hydrogen steam turbine.

The economics depends upon (1) the efficiency of the electricity to hydrogen to electricity conversion processes, (2) full use of capital-intensive equipment, and (3) the large price difference in electricity between times of peak electricity demand and low electricity demand. The high-capital-cost nuclear reactor operates at full power. At times of high electricity demand the electricity goes to the grid but at times of low electricity demand, electricity and steam are used to produce hydrogen. The storage costs for hydrogen are low and thus enabling seasonal storage of hydrogen to address seasonal variations in electricity demand. The hydrogen storage facilities enable supplying hydrogen based on demand—independent of the instantaneous production rate.

The requirements for peak-power production are for an efficient method to convert hydrogen into electricity and low capital costs. The low capital costs are necessary to minimize electricity costs, because peak-power equipment is used for a limited number of hours per year. In this context, HTE has unique advantages. It can convert steam and electricity to hydrogen and oxygen at times of low electric demand and be operated in reverse as a high-temperature fuel cell to convert hydrogen and oxygen back into electricity at times of high electricity demand. This allows the same equipment to operate with high capacity factors and thus minimize the economic impacts of its capital costs. There are several special characteristics of this system.

- *Oxygen.* The power output of a given fuel cell is controlled by the oxygen electrode. If air is used in the fuel cell, the nitrogen in the air creates a mass diffusion barrier for oxygen reaching the surface of the oxygen electrode. If oxygen from electrolysis is stored and used for peak electricity production, fuel cell output is increased by a factor of 2 to 4—depending upon various design details. There are significant incentives with fuel cells to use oxygen as the oxidizer.
- *Fuel cell/gas turbine.* Siemens has been developing a hybrid fuel cell and gas turbine where the gas turbine recovers energy from a fuel cell that effectively replaces the combustor within the engine. With hydrogen and air as an oxidizer, the efficiency for large systems is estimated ~70%—substantially higher than other such technologies.

The storage of oxygen and hydrogen at pressure provides a second method of peak electricity production—the oxy-hydrogen steam turbine (Figure 9.11). Hydrogen, oxygen, and water are fed directly to a burner to produce high-pressure, very high-temperature steam. The resultant steam is fed directly to a very high-temperature turbine that drives an electric generator. Through the use of advancing gas-turbine technology with actively cooled blades,

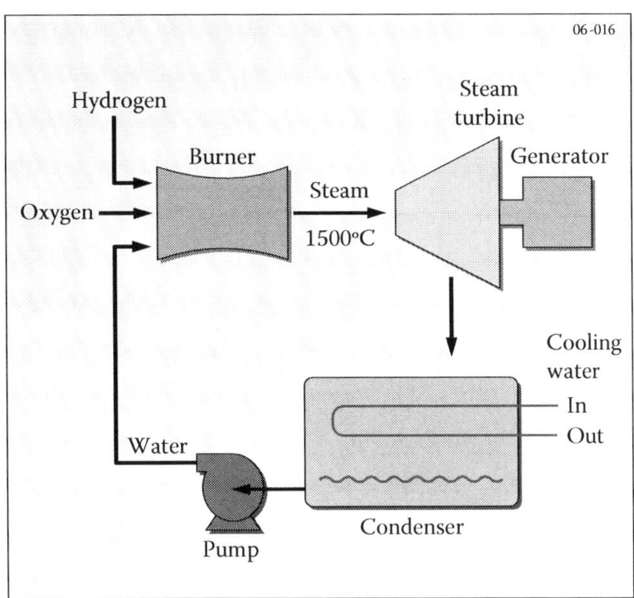

FIGURE 9.11
Oxygen–hydrogen–water steam cycle.

it is expected that peak steam temperatures at the inlet of the first turbines can approach 1500°C. The projected heat-to-electricity efficiency for advanced turbines approaches ~70%, starting with compressed oxygen and hydrogen from the storage facilities.

The technology is based on ongoing development of an advanced natural-gas electric plant that uses oxygen rather than air [33]. Combustors with outputs of ~20 MW(t) are being tested by Clean Energy Systems. With a feed of natural gas and oxygen, a mixture of steam and carbon dioxide is created. After it passes through the turbine to the condenser, the steam is condensed and the carbon dioxide is available for (1) injection into oil fields to increase the recovery of oil and/or (2) sequestration. The higher heat-to-electricity efficiency and the production of a clean carbon dioxide gas stream for long-term sequestration of the carbon dioxide greenhouse gases has created strong incentives to develop oxy-fuel combustors for burning of fossil fuels.

The capital costs [34] for the oxy-hydrogen steam turbine system are significantly less than those for any existing method to convert a fuel to electricity. Today, the low capital-cost peak-power production technology is the natural-gas-fired combined-cycle plant with a high-temperature gas turbine producing electricity and the hot gas turbine exhaust being sent to a steam boiler with the steam used to produce additional electricity. For the oxy-hydrogen system, the turbine remains but the need to compress air as an oxidizer is eliminated, as well as the gas flow of nitrogen (80% of air) through the system. The high-surface-area steam boiler is also eliminated. These changes increase efficiency (55–70%) and lower the capital costs.

The HTE/fuel cell and oxy-hydrogen steam cycle are complementary technologies. The high-temperature electrolyzer as a peak-power fuel cell is the likely economic peak-power option up to the capacity of the HTE/fuel cells required for hydrogen production because the same equipment is used for both hydrogen and electricity production. For added peak-power capacity, the oxy-hydrogen steam turbine is expected to have significantly lower capital costs but it cannot generate hydrogen from electricity. Its low cost and capability to rapidly vary electrical output may make it a critical technology for an electrical grid with significant renewable electricity production and the need for very large quantities of backup power when local weather conditions shut down wind or solar systems.

If such a peak-power system were to be built today, low-temperature electrolysis would be used for hydrogen production, and a combined-cycle (gas turbine with steam bottoming cycle) plant would be used to convert hydrogen to electricity. The round-trip efficiency of electricity to hydrogen to electricity would be between 40% and 50%. With near-term improvements in electrolyzers and the use of oxy-hydrogen steam cycles, the round-trip efficiency would increase to between 50% and 60%. HTE may further increase round-trip cycle efficiency to exceed 60%. Such conversion efficiencies are potentially competitive for peak-power production. In many electrical grids there is a factor of 2 to 4 in price difference between base-load and peak electricity. Equally relevant, today this system is the only alternative to large-capacity hydroelectricity for seasonal storage of electricity.

9.5 Conclusions

Today's commercial nuclear hydrogen production option is low-temperature electrolysis. The economics are primarily dependent upon the cost of electricity; thus, the incentive to use lower-cost off-peak electricity for hydrogen production. Economics are grid specific. Analysis and laboratory prototypes of HTE systems show the potential of HTE to significantly reduce hydrogen production costs by partly substituting heat for electricity and thus the incentive to develop this hydrogen production technology for coupling with water-cooled reactors.

All existing and proposed hydrogen production technologies that use water-cooled reactors are electrolysis processes. Consequently, the distinctive characteristic of these nuclear hydrogen options is that they couple hydrogen production with electricity production. This creates the option of using hydrogen as a method to store electricity to meet daily, weekly, or seasonal peak energy demands. The technology for using hydrogen to meet peak electricity demands exists today; however, it is not competitive with low-cost fossil fuels. Fossil fuel price increases, restrictions on greenhouse gas emissions, or improving hydrogen technologies such as HTE could rapidly change these economics making hydrogen peak-electricity systems producers and users of hydrogen.

References

1. Tong, L. S. and Weisman, J., *Thermal Analysis of Pressurized Water Reactors* (3rd Edition), American Nuclear Society, USA, 1996.
2. Japan Atomic Energy Relations Organization, Drawing collection of nuclear power and energy in 2008, No. 5-05, 2009.
3. Http://www.eia.doe.gov/cneaf/nuclear/page/nuc_reactors/pwr.html, Pressurized Water Reactor and Reactor Vessel, Energy Information Administration, USA.

4. Chapter 19: Probabilistic Risk Assessment, AP1000 Design Control Document rev. 16, 2007-05-26, U.S. Nuclear Regulatory Commission.
5. UK-EPR Chapter R: Probabilistic Safety Analysis, Fundamental Safety Overview, Vol. 2.
6. Lahey, R. T. and Moody, F. J., *The Thermal-Hydraulics of a Boiling Water Nuclear Reactor* (2nd Edition), American Nuclear Society, USA, 1993.
7. Japan Atomic Energy Relations Organization, Drawing collection of nuclear power and energy in 2008, No. 5-02, 2009.
8. ESBWR Design Control Document. Rev 3. February 2007.
9. Y. Oka, *Review of High Temperature Water and Steam Cooled Reactor Concepts*, Proc. SCR-2000, November Tokyo, Japan, 104, 6–8, 2000.
10. Oka, Y. and Koshizuka, S., Supercritical-pressure, once-through cycle light water cooled reactor concept, *J. Nucl. Sci. Technol.*, 38(12), 1081–1089, 2001.
11. Silin, V. A., Voznesensky, V. A. and Afrov, A. M., The light water integral reactor with natural circulation of the coolant at supercritical pressure B-500 SKDI, *Nucl. Eng. Des.* 144, 327–336, 1993.
12. Bushby, S. J., Dimmick, G. R., Duffey, R. B., et al., *Conceptual Designs for Advanced, High-Temperature CANDU Reactors, Proc. 8th International Conference on Nuclear Engineering (ICONE-8),* Baltimore, USA, ICONE-8470, April 2–6, 2000.
13. Generation IV International Forum, A Technology Roadmap for Generation IV Nuclear Energy Systems, GIF-002-00, December 2002.
14. Chow, C. K. and Khartabil, H. F., Conceptual fuel channel designs for CANDU-SCWR, *Nucl. Eng. Technol.*, 40(2), 139–146, 2008.
15. Squarer, D., Schulenberg, T., Struwe, D., et al., High performance light water reactor, *Nucl. Eng. Des.*, 221, 167–180, 2003.
16. Cheng, X., Liu, X. J., and Yang, Y. H., A mixed core for supercritical water-cooled reactors, *Nucl. Eng. Technol.*, 40(1), 1–10, 2008.
17. Jevremovic, T., Oka, Y., and Koshizuka, S., Conceptual design of an indirect-cycle, supercritical-steam-cooled fast breeder reactor with negative coolant void reactivity characteristics, *Ann. Nucl. Energy*, 20(5), 305–313, 1993.
18. Cao, L., Oka, Y., Ishiwatari, Y., et al., Three-dimensional core analysis on a super fast reactor with negative local void reactivity, *Nucl. Eng. Des.*, 239(2), 408–417, 2009.
19. Nakatsuka, T., Oka, Y., and Koshizuka, S., Startup thermal considerations for supercritical-pressure light water-cooled reactors, *Nucl. Technol.*, 134, 221, 2000.
20. Ji, S., Shirahama, H., Koshizuka, S., and Oka, Y., Stability analysis of supercritical-pressure light water-cooled reactor in constant pressure operation, *Proc. of 9th Int. Conf. Nucl. Eng.* (ICONE-9), Nice, France, ICONE-9306, April 8–12, 2001.
21. Yi, T. T., Ishiwatari, Y., Liu, J., Koshizuka, S., and Oka, Y., Thermal and stability considerations of super LWR during sliding pressure startup, *J. Nucl. Sci. Technol.* 42, 537–548, 2005.
22. CANDU 6 Technical Summary, CANDU 6 Program Team, June 2005.
23. ACR-1000 Technical Summary, Atomic Energy of Canada Limited, August 2007.
24. Miller, A. I., An update to Canadian activities on hydrogen, *Proc. of OECD/NEA 3rd Information Exchange Meeting on Nuclear Production of Hydrogen*, Oarai, Japan, October 5–7, 2005.
25. Chow, C. K. and Khartabil, H. F., Conceptual fuel channel designs for CANDU—SCWR, *Nucl. Eng. Technol.*, 40(2), 139–146, 2008.
26. Federal Energy Regulatory Commission. *Form 714—Annual Electric Control and Planning Area Report.* Washington, DC, www.ferc.gov/docs-filing/eforms/form-714/data.asp#skipnavsub., 2004.
27. Massachusetts Institute of Technology, *The Future of Coal: Options for a Carbon-Constrained World.* Cambridge, MA, 2007.
28. Miller, A. I. and Duffey, R. B., Meeting the near-term demand for hydrogen using nuclear energy in competitive power markets. *Second OECD-NEA Information Exchange Meeting on the Nuclear Production of Hydrogen*, Argonne, IL, October 2–3, 2003.

29. O'Brien, J. E., McKellar, M. G., Stoots, C. M., Hawkes, G. L., and Herring, J. S., *Analysis of Commercial-Scale Implementation of HTE to Oil Sands Recovery*, DOE Milestone Report, NHI Program, Idaho National Laboratory, Idaho Falls, September 15, 2006.
30. Forsberg, C. W., *Nuclear Hydrogen for Peak Electricity Production and Spinning Reserve*, ORNL/TM-2004/194, Oak Ridge National Laboratory, Oak Ridge, TN, 2005.
31. Forsberg, C. W., Is hydrogen the future of nuclear energy? *Nucl. Technol.*, 166, 3–10, 2009.
32. Forsberg, C. W. and Kazimi, M.S., Nuclear hydrogen using high-temperature electrolysis and light-water reactors for peak electricity production, *Proc. Fourth OECD-NEA Information Exchange Meeting on the Nuclear Production of Hydrogen*, Oak Brook, IL, April 14–16, 2009, http://mit.edu/canes/pdfs/nes-10.pdf.
33. Anderson, R. E., Doyle, S.E., and Pronske, K. L., Demonstration and commercialization of zero-emission power plants, *Proc. 29th International Technical Conference on Coal Utilization and Fuel Systems*, Clearwater, FL, April 18–22, 2004.
34. Forsberg, C. W., Economics of meeting peak electricity demand using hydrogen and oxygen from base-load nuclear or off-peak electricity, *Nucl. Technol.*, 166, 18–26.

10

High-Temperature Gas Reactor

Xing L. Yan, Ryutaro Hino, and Kazutaka Ohashi

CONTENTS

10.1 Introduction

10.1.1 High-Temperature Gas Reactor Development History

High-temperature gas reactor (HTGR) development has continued over half a century during which the reactors listed in Table 10.1 have been built and operated. The development at the moment is focused on the next-generation commercial designs, alternately called very high-temperature reactor (VHTR), for a range of economic applications, of which one that promises to exploit the highest potential of the HTGR is the production of hydrogen.

Of the reactors constructed, the Dragon pioneered the tri-isotropic(TRISO)-coated particle fuel, still in use today. The Arbeitsgemeinschaft Versuchsreaktor (AVR) tested more fuel variants and gained know-how about fuel performance through long-term operational measurement and examination. The prototypical FSV (Fort St. Vrain) validated the prismatic core physics design with high burnup (90 GWd/t) on thorium fuel and proved easy load follow and 39% net thermal efficiency with steam-turbine power generator. The FSV, however, suffered from coolant circulator and related technical problems that forced excess plant outage and undermined the plant economics. The other large commercial thorium high-temperature reactor 300 (THTR-300) of pebble-bed core prototypical plant experienced technical problems after a short period of operation and ensuing extensive public debate, which prevented early restart. The FSV and THTR-300 plants were prematurely taken off service largely on merit of business judgment.

Asia is home to the newest builds, high-temperature engineering test reactor (HTTR) and high-temperature reactor 10 MWt (HTR-10), which are prismatic and pebble-bed designs, respectively, and are operational today. The HTTR rated at 30MWt has a reactor outlet coolant temperature of 950°C and allows for 863°C process-heat output. Such high-temperature capability is compatible with modern process technologies and widens market roles of the reactor, as shown in the recent commercial designs.

Sponsored mainly by the U.S. Department of Energy (DOE), a research and industrial group proposed the gas turbine modular helium reactor (GT-MHR) in 1994 [1]. The design is based on a 600 MWt and 850°C prismatic-core reactor that is passively safe and employs gas-turbine power conversion at thermal efficiency approaching to 50%. The cost of electricity generation was shown competitive to other generation options [2]. General Atomics (GA) has since continued the GT-MHR development in cooperation with partners in the Russian Federation. The Japan Atomic Energy Agency (JAEA) has offered the GTHTR300C, a gas turbine high-temperature reactor of 300MWe for cogeneration of variable rates of electricity and hydrogen [3]. The GTHTR300C uses a 600 MWt prismatic-core reactor with outlet coolant temperature of 950°C to power a direct-cycle gas turbine for electricity generation and a thermochemical process for hydrogen production. South Africa has been developing the 400 MWt pebble-bed modular reactor (PBMR) with reactor outlet coolant temperature at 900°C for the production of electricity and hydrogen and for other process-heat applications [4].

In 2001, Generation IV International Forum (GIF) of ten member countries endorsed six nuclear systems that can be licensed, constructed, and operated by the year 2030 and which will deliver affordable energy products while satisfactorily addressing the issues of nuclear safety, waste, and proliferation [5]. Recognizing the VHTR to be deployable in the near future and exceptionally suitable not only for electricity generation but also for hydrogen production and industrial heat applications, the U.S. DOE has placed the Generation-IV priority on the VHTR. This led the U.S. to the creation of the next generation nuclear plant

TABLE 10.1
The HTGRs Constructed in the World

	Test HTGRs				Prototype HTGRs		
	Dragon	AVR	HTTR	HTR-10	Peach Bottom	FSV	THTR-300
Country	U.K. (OECD)	Germany	Japan	China	USA	USA	Germany
Period of operation	1963–1976	1967–1988	1998–present	2000–Present	1967–1974	1976–1989	1986–1989
Reactor type	Tube	pebble	prismatic	Pebble	Tube	Prismatic	Pebble
Thermal power, MWt	21.5	46	30	10	115	842	750
He coolant outlet temperature,,°C	750	950	950	700	725	775	750
Coolant pressure, MPa	2	1.1	4.0	3.0	2.25	4.8	3.9
Electrical output, MW	—	13	—	2.5	40	330	300
Process heat output, MW	—	—	10	—	—	—	—
Process heat temperature,°C	—	—	860	—	—	—	—
Core power density, W/cc	14	2.6	2.5	2	8.3	6.3	6.0
Kernel fuel particle coating	UO_2 TRISO	(Th/U, U)O_2, C_2 BISO & TRISO	UO_2 TRISO	UO_2 TRISO	ThC_2 BISO	(Th/U, Th)C_2 TRISO	(Th/U)O_2 BISO

(NGNP) program to demonstrate commercial high-efficiency generation of electricity and hydrogen (see Chapter 31).

10.1.2 HTGR Technology Characteristics

10.1.2.1 Energy Production

In 2007 nuclear energy was the source of 78% electricity in France, 24% in Japan, 19% in the United States, and 14% average worldwide [6]. Practically all the nuclear electricity is generated by water reactors with the Rankine cycle steam turbine. For decades, fossil-fired power plants have relied on the evolution of Brayton cycle gas turbines to generate power, just as aircraft engines have done. The gas turbine inlet temperature has seen steady increase to well above 1000°C and the gas turbine plants have recorded the thermal efficiency near 40% in simple cycle (e.g., MHI's M501G), 44–50% in recuperated and intercooled cycle (e.g., GE's 100 MWe LMS 100), and 60% in combined cycle (e.g., GE's latest H System). However, temperature of the water nuclear reactors at 325°C maximum is too low to take any advantage of the evolution.

The HTGR promises to change that. The modular reactor is rated in the same range of hundreds of megawatt capacity as gas turbine, and the reactor coolant temperature of 950°C is high enough to approach 50% generation efficiency by regenerative Brayton cycle gas turbine (Figure 10.1). While the Rankine cycle undergoes phase changes of water, the Brayton cycle remains in gas phase such that the cycle temperature and pressure can be independently selected, often on hardware considerations. For instance, the highest temperature allowable for turbine blade is usually selected to maximize cycle-thermal efficiency while the pressure may be set to optimize the bulk of the plant.

Figure 10.2 shows the potential for cost of electricity improvement with power conversion cycle options, according to a U.S. evaluation based on a common U.S. utility costing method [2] and to a Japanese study for the HTGR direct-cycle gas turbine plant and the light water reactor (LWR) based on the Japanese utility cost evaluation method [7].

The promise of the HTGR goes beyond power generation. As shown in Figure 10.3, the process heat supplied by the HTGR satisfies the temperature requirements for broad

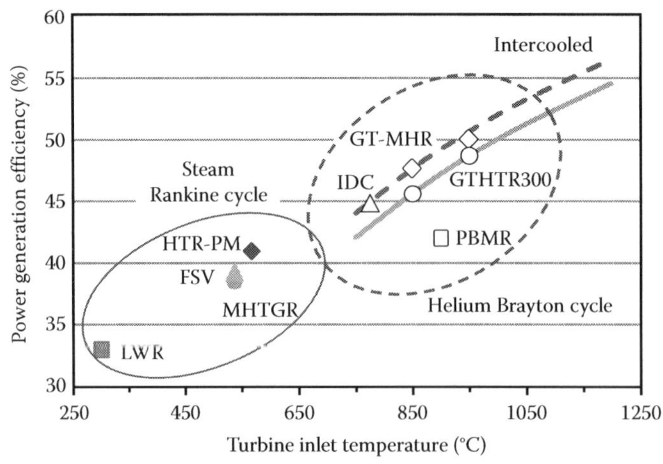

FIGURE 10.1
Nuclear plant efficiency versus turbine inlet temperature.

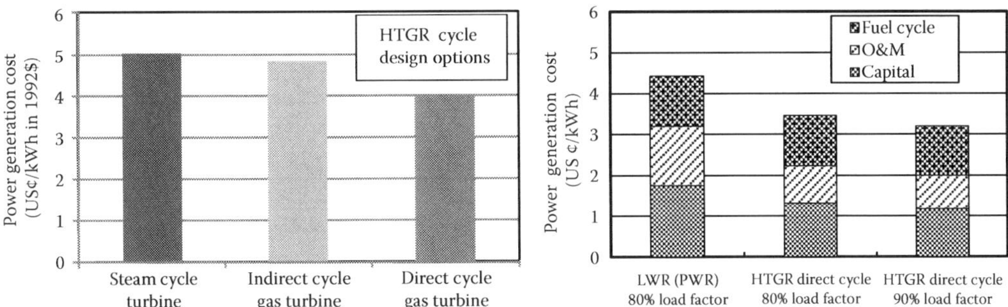

FIGURE 10.2
Cost of electricity for nuclear power generating plant options from the U.S. study (left graph) and Japanese study (right graph).

industrial applications, of which hydrogen production processes with reduced or zero CO_2 emission will be discussed in this chapter. A 600 MWt HTGR at 45% thermal efficiency could produce more than 150 tons of hydrogen daily, which would displace over 1.2 million gasoline liters based on distance traveled while eliminating 2800 tons of CO_2 emission each day from road transport.

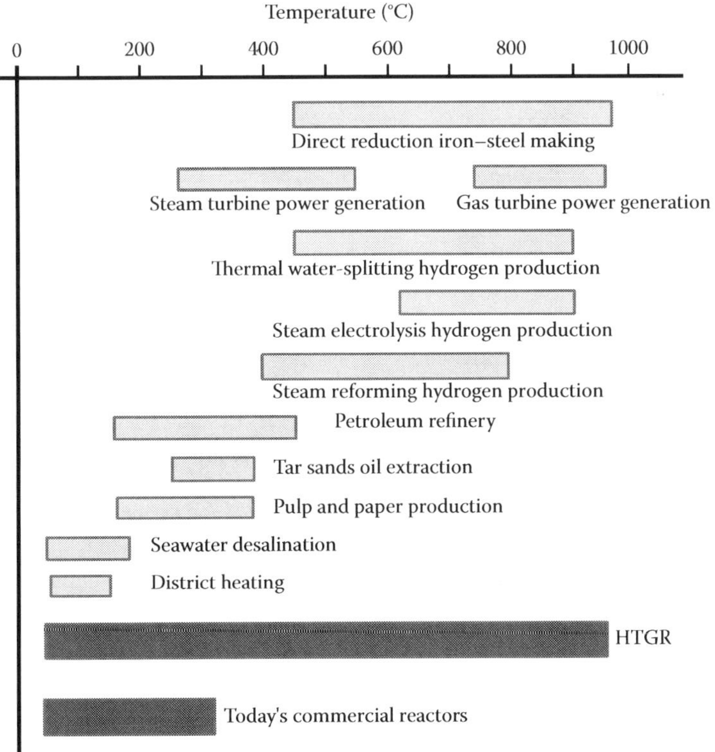

FIGURE 10.3
Temperature ranges of reactor heat supply and industrial heat demand.

10.1.2.2 Passive Safety

Using graphite and ceramics for core construction makes HTGR capable of withstanding extreme temperatures in accident. Moreover, the large quantity of these materials used and the low power density of the core design are intended to limit reactor temperature excursion in any accident. Even in this limit, should an abnormal rise in temperature take place, the strong negative temperature coefficient of reactivity of the core would shut the reactor down. Decay heat is removed from the core by thermal conduction only. The coolant helium contributes a handful of inherent properties to enhance nuclear passive safety, being strong in cooling, always single phase, practically subsonic, neutronically transparent, noncorrosive (inert) to any material that comes into contact with it, and so on. Some of those that directly mitigate consequences due to loss of coolant in an accident, include no direct reactivity excursion, small spontaneous heat release, and prevention of core-graphite oxidation. Together, these inherent and passive safety design features provide proof of a core melt and of significant radioactivity release in case of an accident.

10.1.2.3 Fuel Cycle

Low enriched (<20%) uranium fuel cycle is in use and remains appropriate in the near future. Thorium fuel cycle [8] may be interesting regionally or in the longer term since world thorium reserve is several times more abundant than uranium. Although not fissile, Th-232 is fertile and breeds fissile U-233 by absorbing neutrons produced, for example, by fission of initial uranium. The reactors in Table 10.1 already employed this type of fuel.

More fuel options exist for HTGR [9]. The fuel cycles that can effectively destroy weapons-grade plutonium and transmute minor actinides while engaging in energy production are being studied [10,11]. The HTGR particle fuel has demonstrated greater than 700 GWt-d/t burnup, an important asset in the proposed deep-burn trans-uranium (TRU) fuel cycle since the high burnup reduces mass flows and thus cost of fuel reprocessing [12]. Figure 10.4 summarizes the TRISO-coated particle fuel options for the HTGR. Waste management strategy includes recycle and direct repository disposal of spent fuel [13]. Mechanical separation of spent-fuel compacts from bulk graphite block, pulsed currents to free TRISO particles from the compact, and subsequent removal of ceramic coating layers by high-temperature oxidation or by carbochlorination to access spent kernels of the particle have been studied [14]. In case of direct disposal of spent fuel, separation and reduction of waste streams could be made prior to disposal. Separated graphite blocks may be treated and reused. Separated fission products and actinides can be confined in stable matrices such as glasses based on existing industrial practice.

The HTGR fuel is proliferation resistant since the fuel structural design adds difficulty to illicit operation of accessing the isotopes in spent fuel kernels and more importantly because the high burnup in commercial systems will leave little and poor isotopes in spent fuel. The low contents of nuclear substance in spent fuel would impede diversion of large nuclear material quantities that pose a weapon's risk.

10.2 Technical Description of the Reactor

The typical prismatic and pebble-bed reactor designs differ in how common fuel particles are packaged and loaded in the reactor core. Figure 10.5 compares the two designs

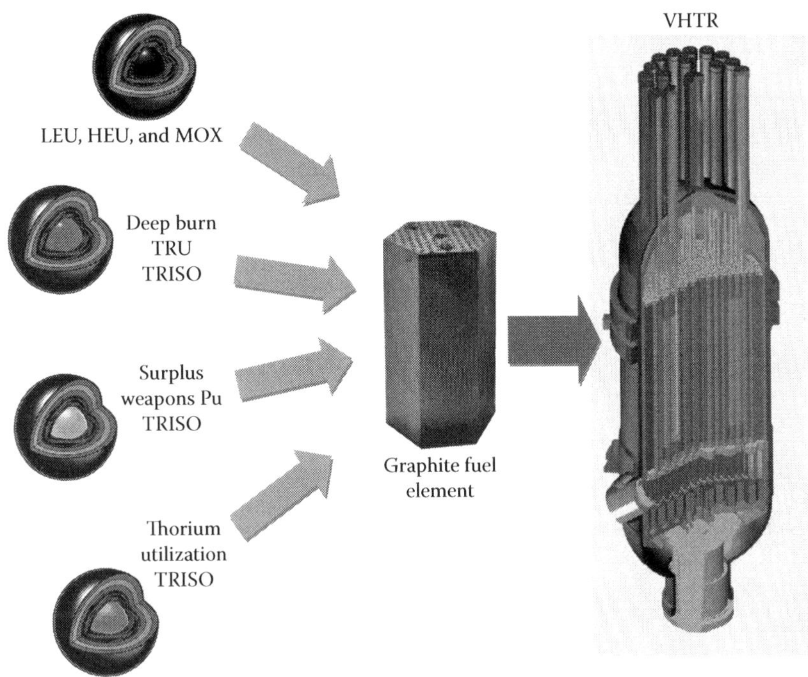

FIGURE 10.4
Options of TRISO-coated particle fuel for the HTGR.

adopted in the HTR-10 [15] and HTTR [16], respectively. A fuel particle measuring 1 mm in diameter consists of an inner nuclear kernel and outer successive layers of carbon and ceramic coating. Thousands of the particles are packed in graphite matrix forming a spherical pebble of roughly a tennis-ball size or a cylindrical compact about the size of man's thumb. A great number of the pebbles are piled up in the pebble-bed core and helium coolant flows in the void space between the pebbles. Large hexagonal graphite elements embedded with the fuel compacts and coolant flow channels are stacked in columns forming a prismatic core. Both cores are enclosed in steel reactor pressure vessel (RPV). Reactivity control rods are inserted from above RPV.

10.2.1 Fuel Technology

10.2.1.1 TRISO-Coated Particle Fuel

The TRISO-particle fuel is standard today. Innermost part of the particle fuel (Figure 10.6) is a low-enriched fuel kernel typically of uranium dioxide (UO_2) and less often uranium oxycarbide (UOC). The kernel is coated with a layer of porous carbon buffer and then by TRISO layers including a high-density inner-layer pyrolytic carbon (IPyC), a layer of silicon carbide (SiC), and a high-density outer-layer pyrolytic carbon (OPyC). These layers are intended for the specific functions explained in Table 10.2. The ceramic layers perform the functions to retain fission products as designed at any temperature that can be reached in operation or accident. New coat of zirconium carbide (ZrC) is being developed in JAEA, which offers better heat resistance by about 200°C compared with SiC coat and is considered the next generation fuel for future reactors [17].

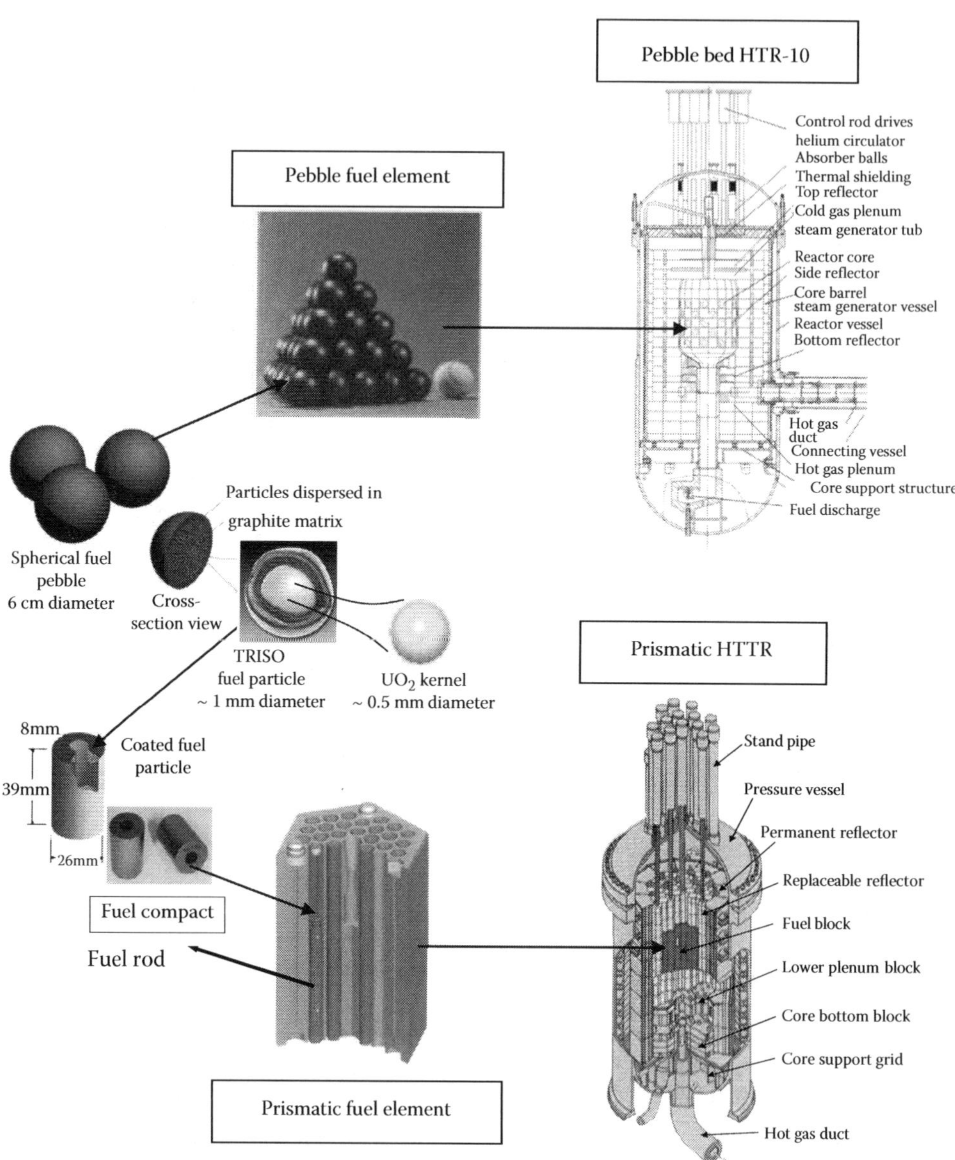

FIGURE 10.5
Packaging particle fuel into the pebble-bed and prismatic reactors.

The technology of TRISO-particle fuels has been well established for the UO_2 kernel type in Germany, China, and Japan, and that for the UOC kernel type is continually investigated in the United States, currently within the NGNP program. South Africa, Korea, and France have recently begun developing fuel technology including fuel irradiation.

10.2.1.2 Fuel Fabrication

USA, Germany, Japan, and China have fabricated fuel on industrial scale. The HTTR fuel is fabricated by Nuclear Fuel Industries (NFI) in Japan following the process in Figure 10.7.

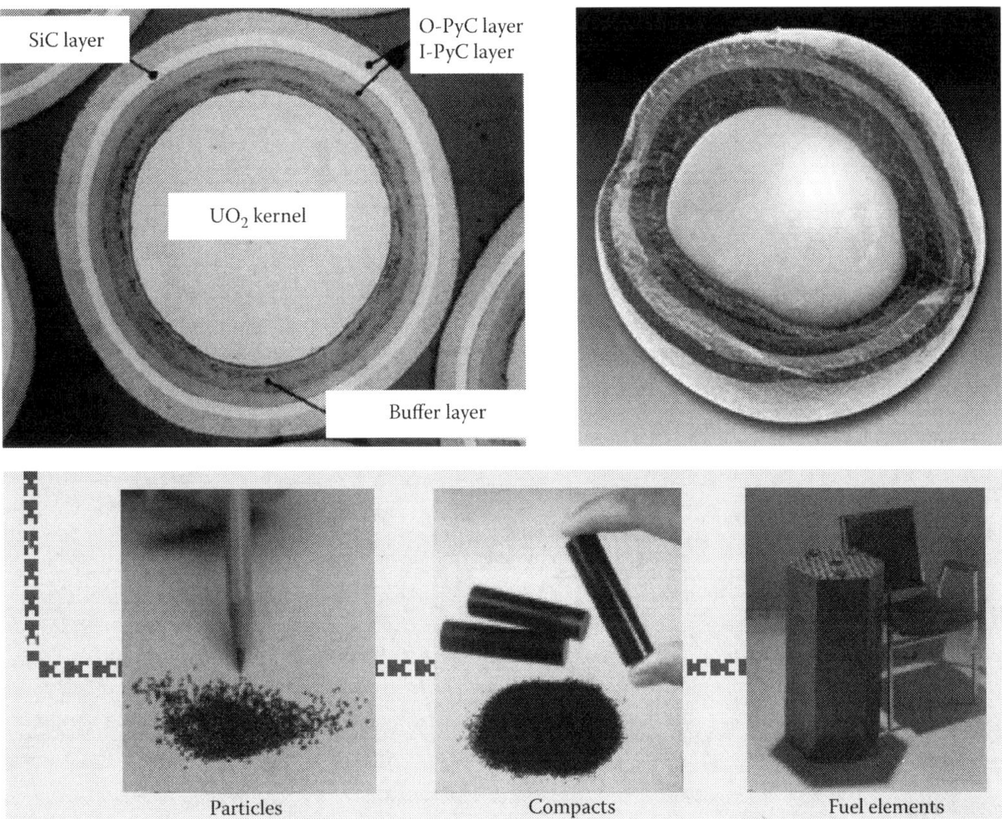

FIGURE 10.6
From the TRISO-coated particle (top left) to the prismatic fuel element assembly on the bottom right.

The steps on the left column form the UO_2 kernels from gel-precipitation to hardening by sintering. The steps in the middle coat the kernel with porous buffer and consecutive TRISO layers derived from the chemical vapors as indicated in a heated fluidized-bed coater. The operation conditions and procedures are controlled according to proven experience to ensure consistent production quality of the particles. The fuel particles are then coated and hot pressed in a mixture of graphite and binder powders into individual fuel compacts. The compacting process controls the temperature and speed of the pressing to avoid the direct contact of overcoat particles in the final compact. Each compact contains

TABLE 10.2
Performing Functions of Fuel Particle's TRISO Coating Layers

Coating Layer	Performing Function
Buffer layer	Plenum container for such kernel produced gases as gaseous fission products and CO
IPyC layer	Retaining product gases protecting UO_2 kernel from material gas during coating application of outer layer
SiC or ZrC layer	Fuel particle structural and pressure containment vessel barrier for metallic FPs
OPyC layer	Cushion for SiC or ZrC layer

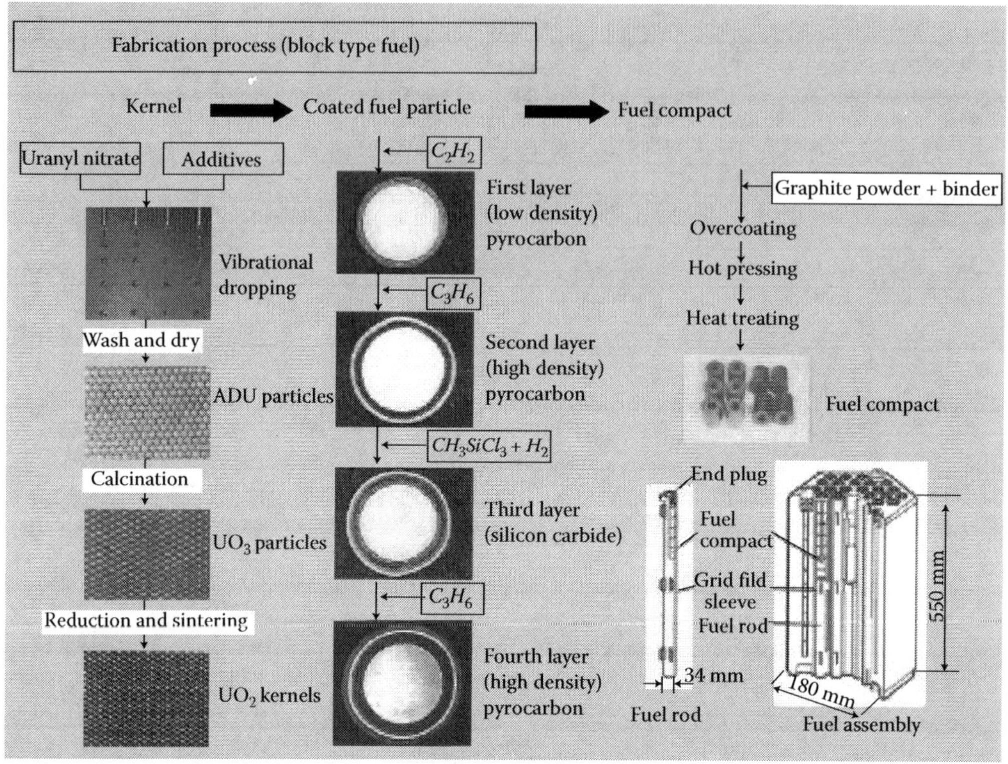

FIGURE 10.7
The HTTR reactor TRISO UO_2 fuel fabrication process.

about 10,000 particles. The compact is carbonized at 800°C under nitrogen atmosphere and then sintered at 1800°C in vacuum to get the final product.

The HTTR employs a LWR-type containment vessel and is thus relatively forgiving to the fabricated fuel quality. The safety requirement specifies a failure fraction of 1%, the same value for LWRs. The actual fuel production is aimed at a higher quality than this in order to prepare fuel for future commercial HTGRs, which do not expect the use of a containment. As an acceptance criterion for the HTTR production fuel, the fuel particle that has through-coating defect is limited to 1.5 per 10,000 particles or 0.015% as fabricated. The operation of the HTTR first loading of fuel has confirmed that the actual as-fabricated defect fraction is about two orders of magnitude less than the acceptance criterion.

The HTR-10 reactor uses spherical pebble fuel element. Each element is sized 60 mm in diameter. The element consists of an inner spherical core of 50 mm diameter in which about 8000 coated fuel particles are dispersed in the graphite matrix. The inner core is wrapped in 5 mm thick graphite shell that contains no fuel. Similarly a fraction of through-coating defect is limited to 0.08% during fabrication and irradiation [15]. The source-term safety analysis assumes that the gaseous fission products be fully released from the defective particles into the coolant. An on-line helium purification system removes the fission products and keep the concentration of fission products in the coolant at a low level such that in the case of a loss of coolant by accident, the fission products that are released with the primary coolant to the atmosphere—because the HTR-10 is housed in vented reactor building—would meet regulatory limits set for the HTR-10.

10.2.1.3 Fuel Performance

Table 10.3 lists three sets of JAEA's fuel designs to enable progressively higher burnup against the major failure modes including migration of the kernel through the inner coating, corrosion of the SiC layer by fission product palladium, and pressure rupture by the fission gases produced in the kernel. The fuel could also fail if exposed to the β to α transition temperature of SiC from 1600 to 2200°C, which is generally not allowed in design. In JAEA, the integrity of the HTTR fuel has been investigated and validated in irradiation tests. The fuel continues to perform well in the HTTR reactor operations up to date. As shown in Figure 10.8, during the average burnup of 11 GWd/t-U and fast neutron fluence of about 1×10^{25} n/m^2 reached in the HTTR first-core fuel, the measured fractional releases remain nearly constant in each of the operation modes performed, which is 2×10^{-9} at 60% of the rated reactor power, and increases to 7×10^{-9} at the full

TABLE 10.3

JAEA's Fuel Design Specifications

	HTTR	Extended Burnup	High Burnup
Technology Status	Reactor operated	Particle irradiated	Design specified
Burnup (GWd/t, %FIMA)	33	95, 5–10	120, 12–15
Fuel rod			
Rod design	Sleeved		Cladded
Length (mm)	546		1000
Diameter (mm)	34		26
Fuel compact			
Length (mm)	39		83
I.D./O.D. (mm)	10/26		9/24
Cladded thickness (mm)	0		1
Particle packing fraction (vol%)	30–35		21–29
Coated fuel particle			
Coating type	TRISO	TRISO	TRISO
Diameter (μm)	920	970	1010
Fuelkernel			
Material	UO_2	UO_2	UO_2
Enrichment (wt. % average)			14
Diameter (μm)	600	600	550
Density (g/cm^3)	10.80	10.80	10.80
Buffer layer			
Thickness (μm)	60	95	140
Density (g/cm^3)	1.15	1.15	1.15
IPyC layer			
Thickness (μm)	30	25	25
Density (g/cm^3)	1.85	1.85	1.85
SiC layer			
Thickness (μm)	25	40	40
Density (g/cm^3)	3.20	3.20	3.20
OPyC layer			
Thickness (μm)	45	25	25
Density (g/cm^3)	1.85	1.85	1.85

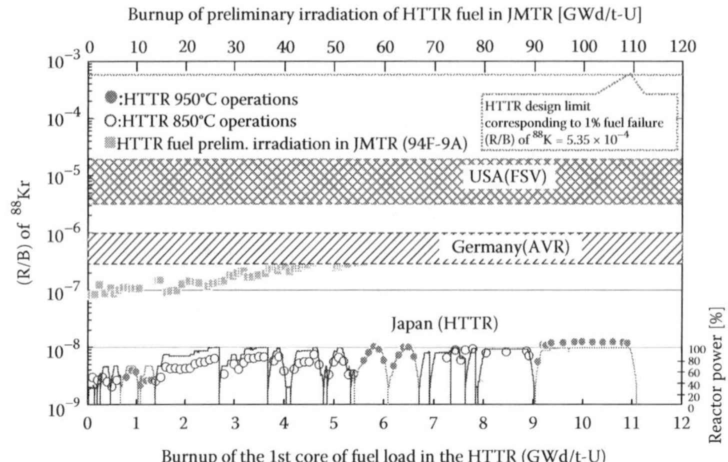

FIGURE 10.8
Measured release rate of ^{88}Kr, indicative of fuel coating failure fraction.

power and 850°C core outlet coolant temperature and to 1.2×10^{-8} at the full power and 950°C core outlet coolant temperature [18]. The maximum temperature of the fuel compact in normal reactor operations was estimated to be 1428°C by pre-operation analysis, but has since been reliably updated from the relevant measurements of the reactor operation at 950°C core outlet coolant temperature to 1435°C, which is below the 1495°C design operation allowance. The data confirm to a failure fraction of SiC-layer of 8×10^{-5} and as-fabricated all-through coating layers of 2×10^{-6} about two orders of magnitude lower than the HTTR licensed limit.

The first-core of the HTTR will reach full burnup in 2014, which is to be followed by post-irradiation examination (PIE) on the spent fuel including graphite blocks and coated particles. Burnup and FP inventory of the fuel kernels will be measured and the integrity of the particles will be examined by x-ray radiography to detect failed particles and by means of ceramography and electron-probe microanalysis to evaluate failures such as kernel migration and Pd corrosion in the SiC layer.

The extended burnup fuel in Table 10.3 is designed to permit greater burnup and fast neutron fluence than those of the HTTR fuel by increasing the thickness of the buffer and SiC layers [19]. Its fuel compacts were irradiated at the high-flux isotope reactor (ORNL in the United States) up to 7% FIMA and at the Japan material test reactor (JMTR) (in JAEA) over 9% FIMA [20]. Based on the irradiated performance data obtained in the HTTR fuel and the extended burnup fuel, the design of high burnup commercial fuel (120 GWd/t average) was specified and evaluated to be adequate against the major failure modes mentioned above [21,22].

Advanced gas reactor (AGR) fuel development program in the United States was initiated in 2002 and consists of eight steps to accomplish the final objective of fully developing and qualifying the fuel production process for the U.S. DOE's NGNP plant. The program includes irradiation tests followed by safety tests and PIE for each test, and the data obtained will complete validation of performance and fission product transport models and codes.

In the first step called AGR-1, the TRISO-coated fuel particles of 780 μm diameter containing 19.8% low enriched uranium (LEU) UCO kernel in 350 μm kernel diameter were fabricated [23]. The particles were formed into cylindrical fuel compacts sized 12.4 mm diameter by 25.4 mm length. Each compact has about 20% particle-packing fraction and contains 4150 particles and 0.9 gram uranium. The kernels were produced in the pilot-scale equipment by Babcock and Wilcox and the particles and fuel compacts are fabricated in 5 cm fluidized coater and other laboratory-scale equipment in ORNL.

Between December 2006 through November 2009, the ARG-1 completed the first irradiation test of the fuel compacts in the Idaho National Laboratory's (INL) advanced test reactor (ATR) in 13 irradiation cycles equivalent to 620 effective full-power days in reactor with the particles' burnup range of 11.5–19.6% FIMA (108–184 GWd/tU). The upper burnup reached was more than twice the previous record set by the particle fuel experiments in Germany in the 1980s, and more than three times that achieved by existing light-water reactor fuel. The irradiated fuel temperatures were typically 1163–1242°C and fast fluencies (>0.18 MeV) were between 2.2 and 4.4×10^{25} n/m^2. The result of the AGR-1 irradiation test confirmed the performance of the particles as design intended and has the initial indication that none of the 300,000 particles contained in the irradiated compacts failed during the tests. The irradiated particles are currently in PIE [24].

10.2.2 Prismatic Reactor

10.2.2.1 Test Reactor—HTTR

JAEA developed the 30 MWt HTTR that is currently the world's highest temperature (at 950°C) operating reactor. Since its first criticality in November 1998, all development-scheduled tests including long-term reactor operations and safety demonstration tests have successfully been performed. The reactor has accumulated more than 14,000 h at operation and generated 11,000 MWD (megawatt day) thermal power as of March 2010.

The reactor, as so named, is developed and now operated for the purpose of high-temperature engineering validation for technologies, including operation and maintenance know-how that are essential for the development of a commercial prototype HTGR as the next step. The key engineering technologies incorporated in the HTTR are many and more notable ones include the production-scale fuel as already described above, prismatic-core physics design techniques (e.g., computational design codes), high-temperature materials (alloys and graphite), and structural features which JAEA had developed especially to build the HTGR design bases. The details of the core physics is described below while the description on the materials and structures will be given in Section 10.3.3, focusing on the intermediate heat exchanger (IHX), the most important reactor equipment for a reactor-coupled hydrogen-production system.

Figure 10.9 shows the HTTR reactor layout design and includes three-dimensional mapped core enrichment zoning of fuel and burnable poisons. Table 10.4 summarizes the HTTR key design parameters. The RPV is made of low-alloy steel 2¼Cr–1 Mo and sized to 5.5 m diameter and 13.2 m height. The cylindrical core is made up of columns of removable graphite blocks containing fuel in various uranium enrichments, control rod guide channels, and additional irradiation channels. The permanent graphite reflector blocks, which are affixed by the lateral-restraint mechanism to the RPV, embrace a ring of replaceable side-reflector blocks. The boron carbide (B_4C) control rods atop of the RPV are moved in and out of the core for adjustment and shutdown of core power in addition to compensating for reactivity due to changes in core temperature, fuel burnup, and concentration of fission products such as ^{149}Sm and ^{135}Xe with large neutron-absorption cross-sections. Another reserved system backs up reactor shutdown by releasing B_4C pellets into the control rod guide channels.

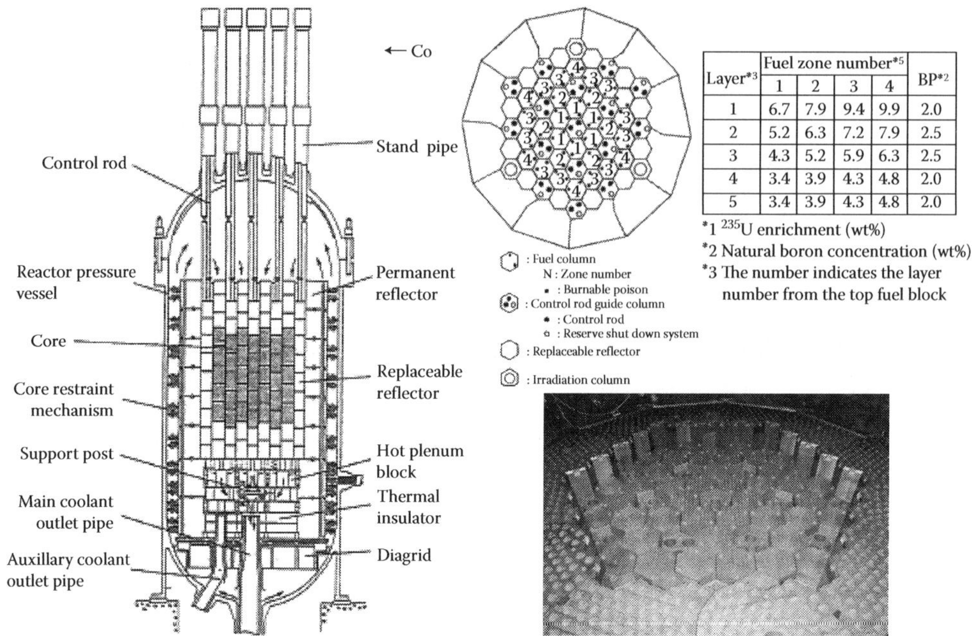

Layer*3	Fuel zone number*5				BP*2
	1	2	3	4	
1	6.7	7.9	9.4	9.9	2.0
2	5.2	6.3	7.2	7.9	2.5
3	4.3	5.2	5.9	6.3	2.5
4	3.4	3.9	4.3	4.8	2.0
5	3.4	3.9	4.3	4.8	2.0

FIGURE 10.9
The HTTR test reactor design (photo is top view of fueled reactor core).

TABLE 10.4
The HTTR Design Parameters

Thermal power (MW)	30
Outlet coolant temperature (°C)	950
Inlet coolant temperature (°C)	395
Primary coolant pressure (MPa)	4
Core structure	Graphite
Equivalent core diameter (m)	2.3
Effective core height (m)	2.9
Average power density (W/cm^3)	2.5
Fuel	UO_2
Uranium enrichment (wt.%)	3–10
Type of fuel	Pin-in-block
Burnup period (days)	660
Coolant	
Material	Helium gas
Flow in core	Downward
Reflector thickness	
Top (m)	1.16
Side (m)	0.99
Bottom (m)	1.16
Number of fuel blocks	150
Number of fuel columns	30
Number of pairs of CRs	
In core	7
In reflector	9

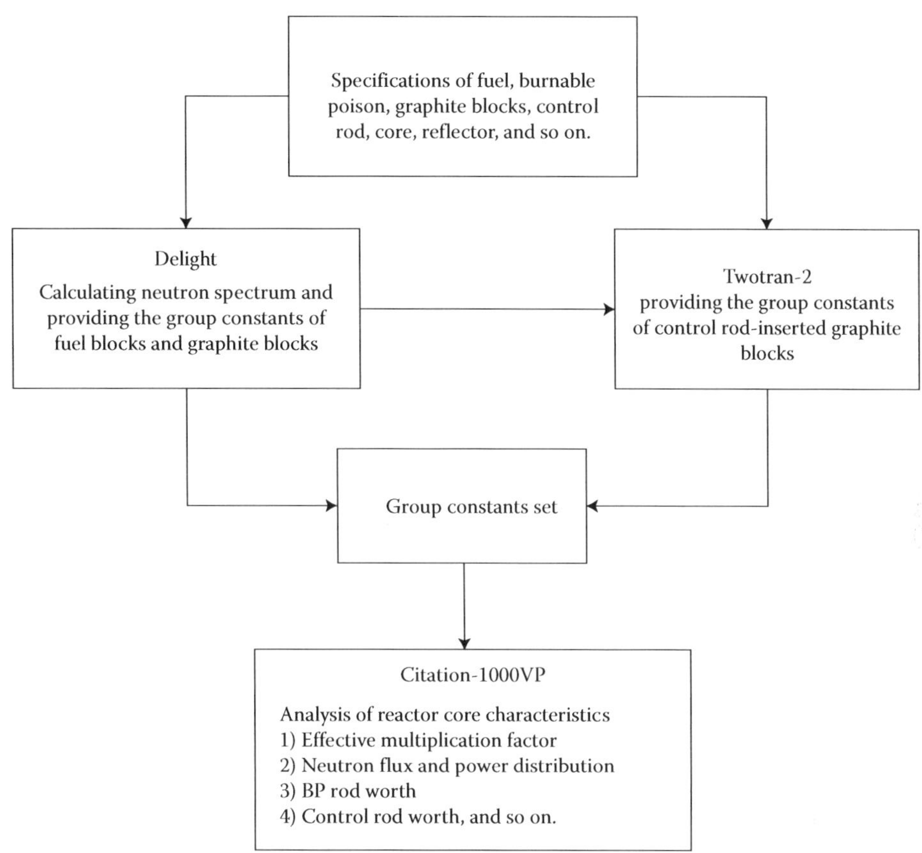

FIGURE 10.10
Nuclear physics design code system used in designing the HTTR prismatic reactor.

The nuclear design-code system in Figure 10.10 has been used and validated with the design and commissioning of the HTTR. The system is continually improved for accuracy and computation efficiency in the ongoing test operations of the HTTR such that it can be reliably applied to the commercial reactor design to be described later. The DELIGHT is a one-dimensional lattice burnup cell calculation code developed for the HTTR nuclear design of the HTTR [25]. It computes the multigroup neutron spectrum of the cylindrical fuel cell and burnable poison cell models (see Figure 10.11) and generates the group constants required in subsequent core calculation. The dimensions of the cell models are sized to conserve the quantity of the materials in a graphite block. The TWOTRAN-2 transport code [26] provides the average group constants of the control rod and control absorber blocks using the mesh model shown in Figure 10.12. The CITATION-1000VP code, improved from the original CITATION [27], calculates nuclear physics characteristics of the HTTR whole core using the mesh model in Figure 10.13, in which a fuel block is divided into six meshes in plane. The CITATION results in Figure 10.14 shows that the initial excess reactivity with fresh fuel can be compensated adequately (small and flat curve of K_{eff} factor) by the in-fuel-block placement of burnable poisons alone over the fuel burnup period from the beginning of life (BOL) to end of life (EOL). The reactivity control, power distribution, reactivity coefficients, and so on, are then considered in the design to add in active and inherent shutdown margins of the core reactivity. It is found that there is little need to

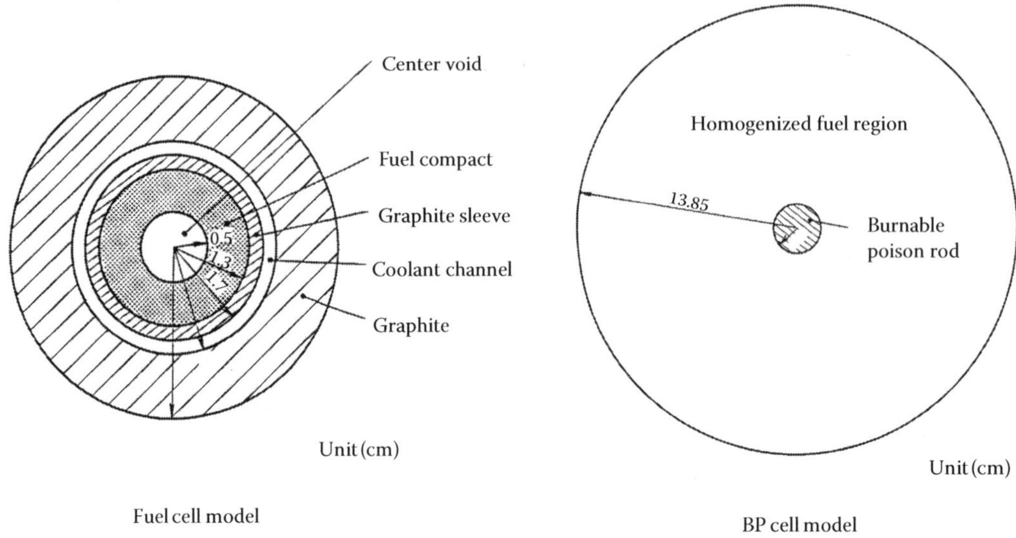

FIGURE 10.11
Fuel and BP cell models used in the DELIGHT code.

change the control rod positions for normal power operation in the entire burnup period and that the distribution of the core power (Figure 10.15) as well as of the fuel temperature remains steady. The calculation models have been improved through evaluation of the nuclear physics experiments on the HTTR. For instance, the BP cell model in Figure 10.11 is revised from one- to two-dimensional to consider the physical configuration of the BP rod, and the number of the triangular meshes in the CITATION model shown in Figure 10.13 has been increased to 24 from six cells in the plane of each fuel and graphite block. The detailed models can be executed efficiently in terms of computer run time. The power distribution of the core predicated by the improved code system is found to agree with the

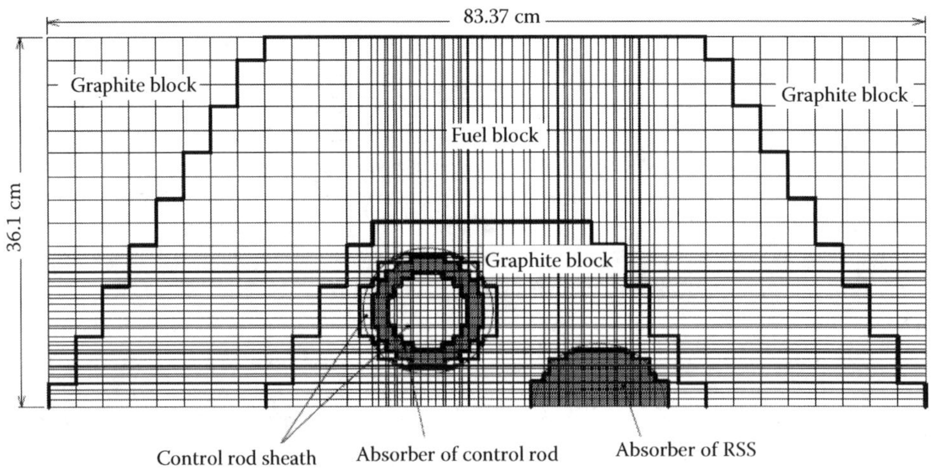

FIGURE 10.12
The control rod cell mesh model used in TWOTRAN-2 code.

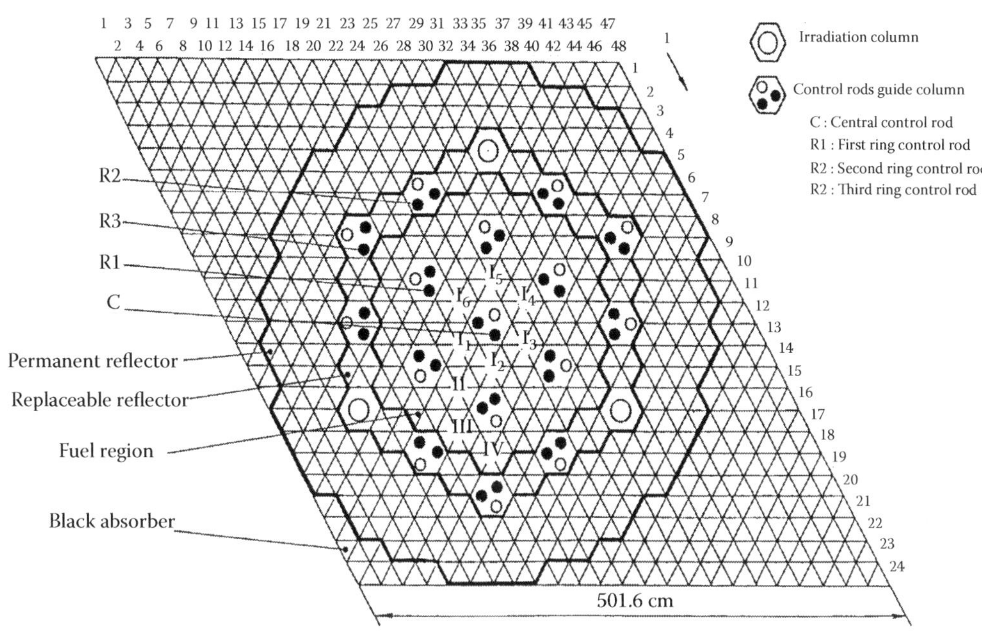

FIGURE 10.13
The 3D whole core calculation model used in CITATION-1000VP.

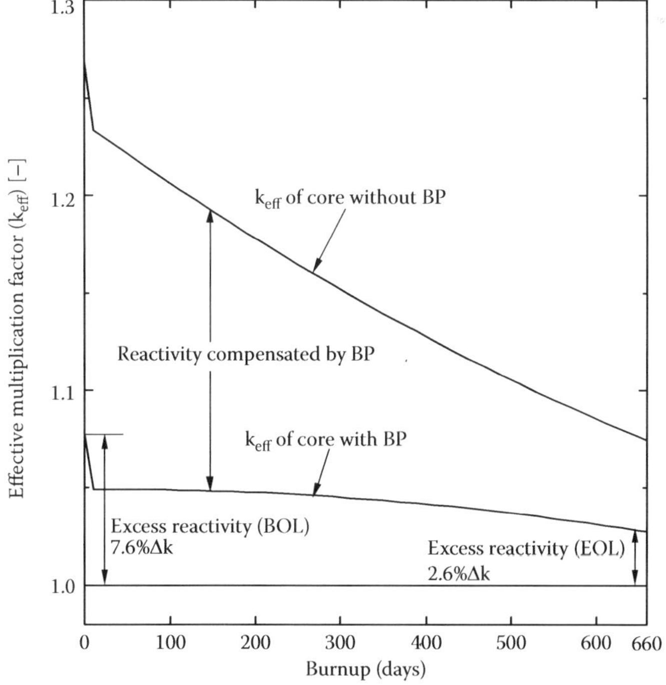

FIGURE 10.14
Excessive reactivity and burnup design of the HTTR.

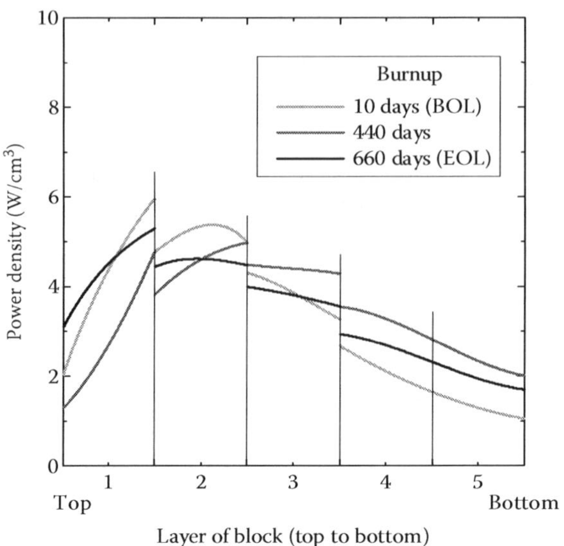

FIGURE 10.15
Power distribution design of the HTTR through the burnup period.

HTTR nuclear physics measurements and with the predication, in maximum discrepancy of 3%, of the Monte Carlo MVP code [28].

Similarly, a thermal design-code system including FLOWNET and TAC-2D for the HTTR was developed in the HTTR design and construction mock-up [16]. The FLOWNET calculates the core coolant-flow distribution while the TAC-2D of general purpose heat transfer code calculates fuel temperature in operation and accident conditions. Large- or full-scale experimental reactor models were built to recreate the thermal hydraulic conditions in the core and pressure-vessel structures. Important data were generated to validate the design models computing the coolant flow distribution, and heat transport in the core structure and through the reactor vessel. Presently, the HTTR operation data continue to be used to benchmark the thermal design-code system.

10.2.2.2 Commercial Reactor Designs

JAEA applied the same nuclear and thermal design-code systems used in the design and operations of the HTTR to the design of commercial reactor. Table 10.5 includes four sets of the design parameters for the GTHTR300 600 MWt reactor. The sets are varied only as needed to meet their specific application mission requirements.

The baseline design is applied to the direct-cycle gas-turbine power generation. The characteristics of the design include 850°C reactor coolant outlet temperature, 600 MWt core thermal power with passive ability of reactor shutdown and decay heat removal in accident, conventional steel construction of the RPV, and a refueling interval of two years. Unlike the HTTR cylindrical core, the greater thermal power of the GTHTR300 commercial design as shown in Figure 10.16 requires annular configuration of the active core in cross-section to passively conduct decay heat from and accept the reactivity control in all fuel regions of the large core. In general, the thermal power of an annular core can be raised with increase in diameter until maximum size of practical RPV construction is reached.

TABLE 10.5
GTHTR300 Reactor Design Results

	GTHTR300 - Power Plant - Baseline Design 850°C	GTHTR300+ -Power Plant- Growth Design 950°C	GTHTR300C -H_2 Cogen. Plant- Pin-in-Block 950°C	GTHTR300C -H_2 Cogen. Plant- Multihole 950°C
Reactor rating	600 MWt		< — same	
Core design	Prismatic annular core		< — same	
Fuel particle	TRISO UO_2		< — same	
Fuel block design	pin-in-block		< — same	Multihole
RPV steel	SA533/508		< — same	
Reference fuel cycle	LEU		< — same	
Refueling method	2-batch axial shuffling		< — same	
Core Physics Design			< — same	
Fuel columns	90		< — same	
Inner reflector columns	73		< — same	
Outer reflector columns	48		< — same	
Core height (m)	8.4		< — same	
Average power density (W/cm^3)	5.4		< — same	
Average burnup (GWd/t)	120		< — same	
Fuel block height/across flat (mm)	1050/410		< — same	
Fuel rods per block	57		< — same	2.62
Fuel rod diameter (cm)	2.6		< — same	1.25
Core enrichment count	8	7	< — same	4
Average enrichment (%)	14.3	14.5	< — same	13.7
Burnable poison (BP) count	6	5	< — same	6
BP pin diameter (mm)	4.8	3.6	< — same	12.5
Refueling interval (months)	24	18	< — same	18
Maximum fuel power peak factor	1.16	1.12	< — same	1.10
Core Thermal Design				
Coolant (helium) flow rate (kg/s)	439	403	322	318
Coolant inlet temperature (°C)	587	663	594	587
Coolant outlet temperature (°C)	850	950	950	950
Coolant inlet pressure (MPa)	7.0	6.4	5.1	7
Core coolant pressure drop (kPa)	58	50	35	30
Minimum fuel coolant flow fraction	0.82	0.82	0.82	0.88
Maximum fuel temperature (nominal) (°C)	1108	1206	1244	1150
Maximum fuel temperature (LOCA) (°C)	1546	1564	1535	1569

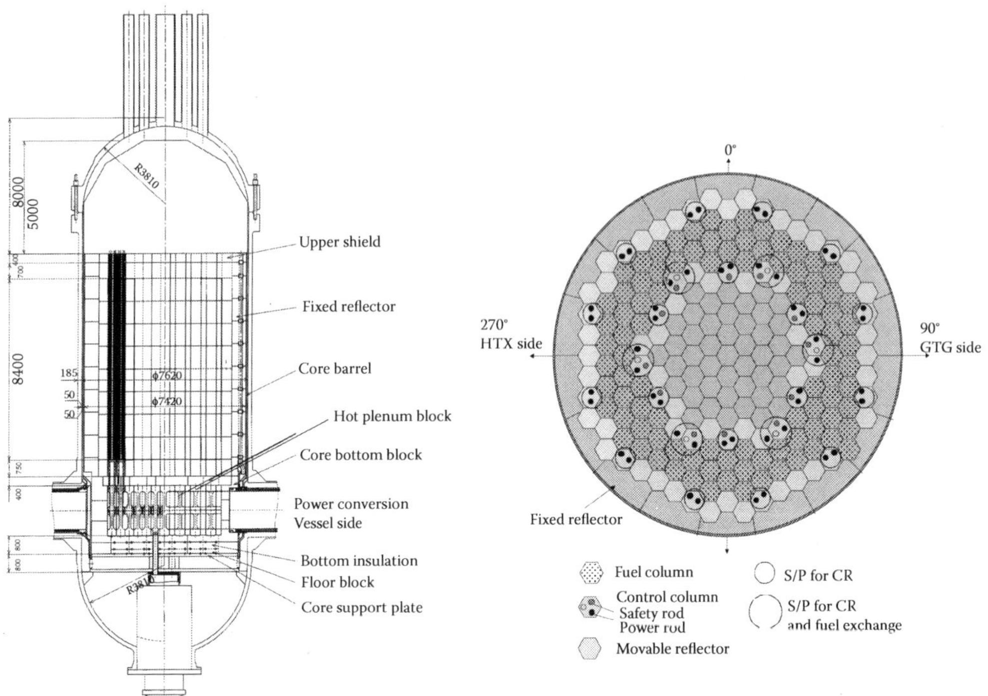

FIGURE 10.16
The cross-section views of the GTHTR300's prismatic core HTGR design.

The reactor core consists of an annular ring of fuel columns surrounded by inner and outer graphite reflector columns that partly contain control rod insertion channels. Each fuel column is stacked of eight hexagonal fuel blocks high and capped at top and bottom with reflector blocks. Dowels are used to align fuel blocks in a column. The core is enclosed by a steel core barrel, which is in turn housed in the steel RPV. The coolant enters the reactor via the inner pipe of the horizontal coaxial duct on the left and travels upwards through the flow channels embedded in the side reflector, turns in the top core plenum to flow downward in the active core, and exits through the inner pipe of the horizontal coaxial duct on the right. The particular coolant flow path through the pair of coaxial ducts on the lower main body of the RPV is a unique feature of this reactor design that provides intrinsic cooling to the RPV [29]. As a result, the operating temperature of the RPV is kept below the 371°C limit of Mo–Mn steel (SA533/508), a conventional and economical material of nuclear RPV construction.

Alternate steel for vessel construction is 2-¼ Cr-1Mo-V used in the HTTR RPV and qualifies for design temperature up to 425°C. It has lower strength than the Mo–Mn steel and results in thick wall section of larger-diameter vessel. Another steel, 9Cr–1Mo, has better strength at high temperature and has been selected in recent designs [1,30], but it has not yet been qualified for nuclear construction.

The pin-in-block fuel design as shown in Figure 10.17 is similar to the HTTR fuel of the HTTR. The fuel rod comprises annular fuel compacts around a central graphite rod. The fuel rods are placed in the coolant holes of large graphite block. Burnable poisons are stored in three holes with a dowel cap. Fuel enrichment is varied between several core regions to distribute core power that minimizes peak fuel temperature through the fuel

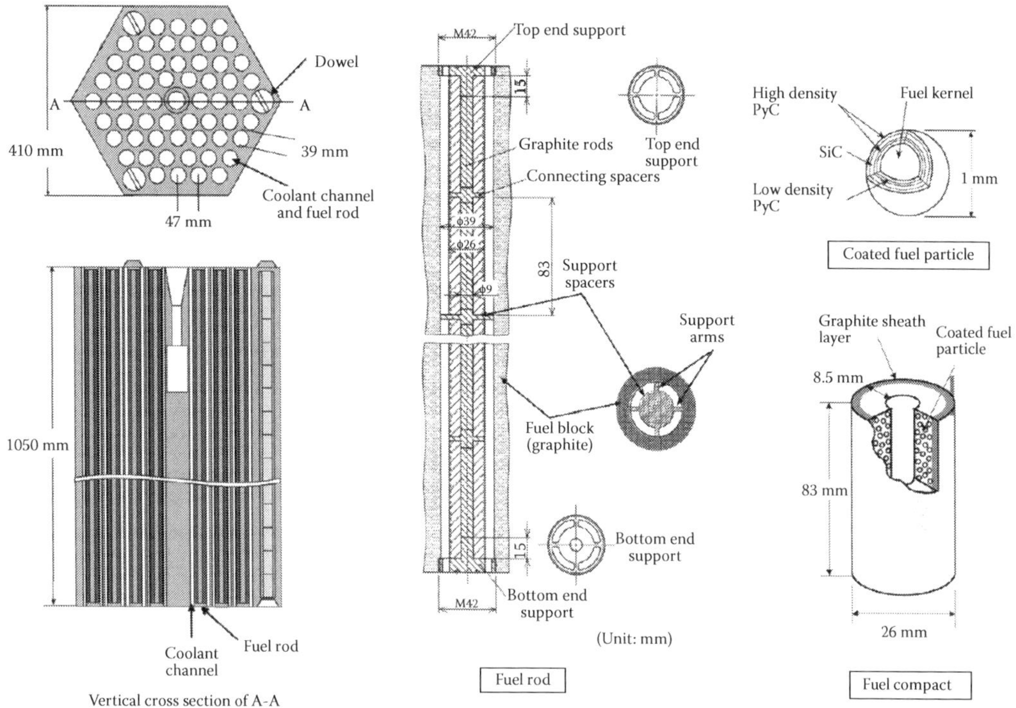

FIGURE 10.17
Prismatic pin-in-block fuel design.

burnup period. The through-core coolant flow rate is raised by increasing core inlet temperature to augment the core heat convection. As a result of these design measures, the estimated fuel operating temperature is limited to 1108°C, in which reliable fuel performance is expected. In the safety design against a loss of coolant accident (LOCA), decay heat is assumed to be conducted from the core to the RPV and then removed by thermal radiation from the RPV to reactor cavity cooling panels. The maximum temperatures reached in this as well as other bounding safety events satisfy the design limits of the fuel and the pressure vessel.

The other three reactor designs in Table 10.5 specify a 950°C core outlet coolant. At this temperature, GTHTR300+ growth design boosts power generation efficiency to 50%. The GTHTR300C is designed to provide high-temperature heat for thermal chemical process hydrogen production. As intended, the 950°C reactor designs share a great deal of common design features with the baseline design. The major differences are the corresponding higher fuel-operating temperature and a shortened length of the refueling interval.

The GTHTR300C multihole design (Table 10.5) employs the multihole fuel element, which appeared originally in Fort St Vrain (FSV) commercial reactor in the United States (Table 10.1). Two features of the fuel design differ from the pin-in-block fuel. The first is that the fuel compacts and the coolant flow occupy separate holes as depicted in Figure 10.18, and the compacts are embedded in the graphite web of the block. The second is the smaller diameter and thus denser population of the fuel compacts in the block cross-sectional area. The overall core configuration is similar to that shown in the Figure 10.16 by replacing the pin-in-block fuel block with the multihole fuel element in the identical

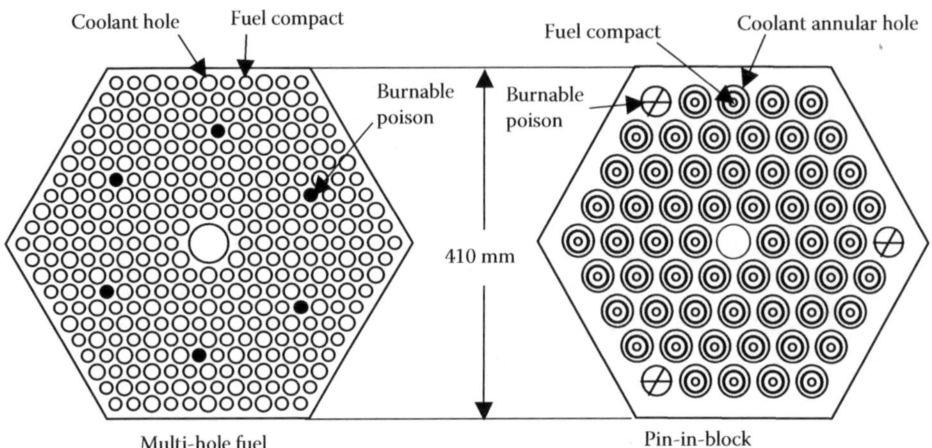

FIGURE 10.18
Multihole (left) and pin-in-block prismatic fuel block designs.

graphite block size and core volume. The major benefits of the multihole design are reduced fuel operating temperature and coolant radioactivity.

In the multihole reactor design, additional fuel temperature reduction is made by the use of a sandwich shuffled refueling scheme, in which alternate fuel blocks in each fuel column are changed in the refueling period, and by the core thermal and structural design measures that increase effective coolant flow in the fueled blocks. As shown in Figure 10.19, the nominal or best estimate of maximum fuel temperature throughout the 550-day burnup period was basically below 1150°C, whereas the conservative estimate of the maximum fuel temperature during the period was less than 1300°C taking into account the maximum analytical model uncertainties.

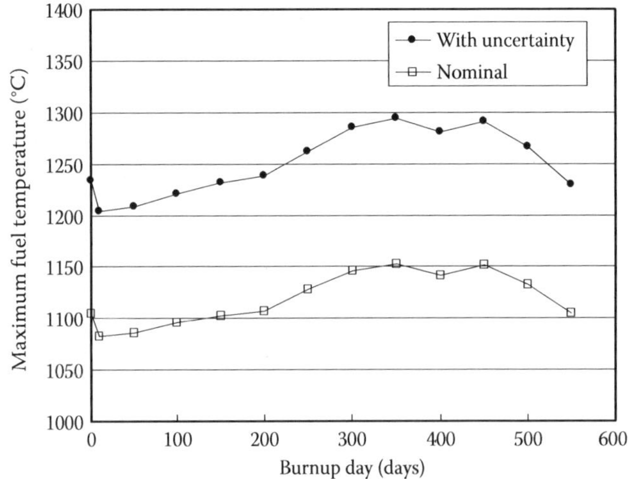

FIGURE 10.19
Fuel temperature during burnup with uncertainty.

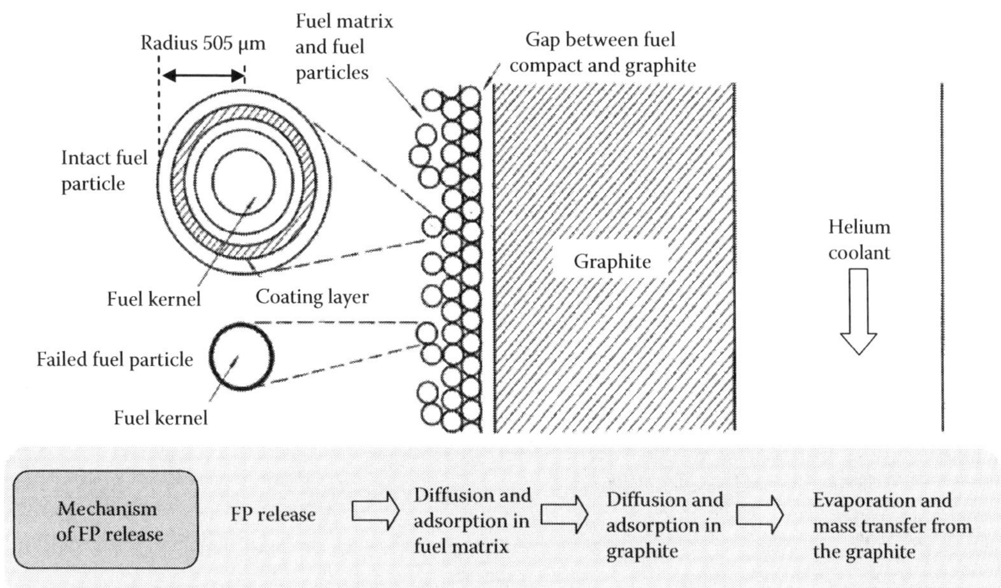

FIGURE 10.20
Analytical model of the metallic FP transport from the fuel.

The fractional release of fission product is the fraction of cumulative release to cumulative generation of a specific FP nuclide and is calculated with the model shown in Figure 10.20. Incorporating the HTTR fuel performance data as described in Section 10.2.1.3, the fractional release rates of the typical metallic FPs, ^{137}Cs, and ^{110m}Ag for the GTHTR300C multihole fuel during a burnup period (1100 days) of a fuel element in core are obtained as shown in Figure 10.21.

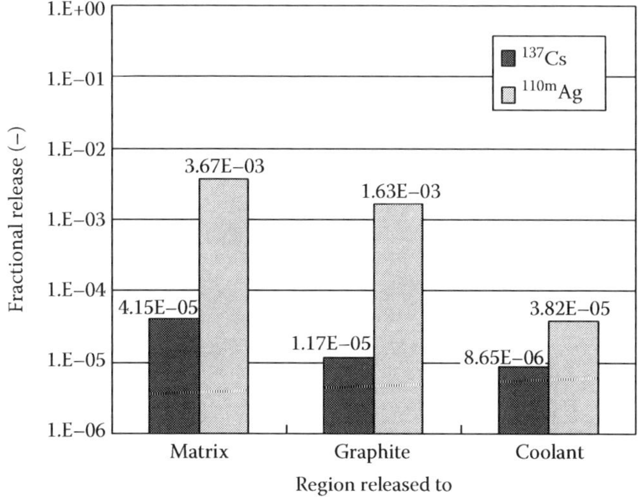

FIGURE 10.21
Fractional release of ^{137}Cs and ^{110m}Ag.

TABLE 10.6

Comparison of the Maintenance Dose Rates

	(Unit: μSv/h)			
	Gas Turbine		Compressor	
	At Surface of the Casing	At Surface of the PCV	At Surface of the Casing	At Surface of the PCV
Multihole fuel	653	18	752	21
Pin-in-block fuel	2220	63.6	5480	154

The maintenance exposure dose rates are estimated for the 600 MWt direct-cycle gas-turbine design, GTHTR300, with the multihole fuel and significantly reduced from the estimates made of the same plant with the pin-in-block fuel in Table 10.6 [31]. This reduction in dose rates is mainly attributed to the FP retention effect of the relatively thick graphite web separating the fuel compact holes from the coolant holes in the multihole fuel design. The data in Table 10.6 are at the end of the 50th reactor operating year when the gas-turbine equipment is to be accessed for the last scheduled overhaul, thus in the severest contamination in 60-year plant life with respect to the maintenance concerns. The estimated levels of exposure with the multifuel would allow personnel access and hands-on maintenance on the gas-turbine equipment for a period of several hours.

Figure 10.22 shows the conservative estimates of transient temperatures in a LOCA passive conduction cool down in the GTHTR300C 950°C reactor design (Table 10.5). The conservative estimated peak fuel temperatures in the same event are listed for the other GTHTR300 reactor designs in Table 10.5. The maximum fuel temperatures remain below the design limit of 1600°C for the fuel and 538°C for the vessel.

In the case of a primary pipe-rupture accident, air ingress into the core of the HTGR can take place by natural circulation (NC) and damage fuel and core structure by means of graphite oxidation in the high-temperature condition present in the core. A leak-tight containment vessel is traditionally used, like in the HTTR, to limit the quantity of the air

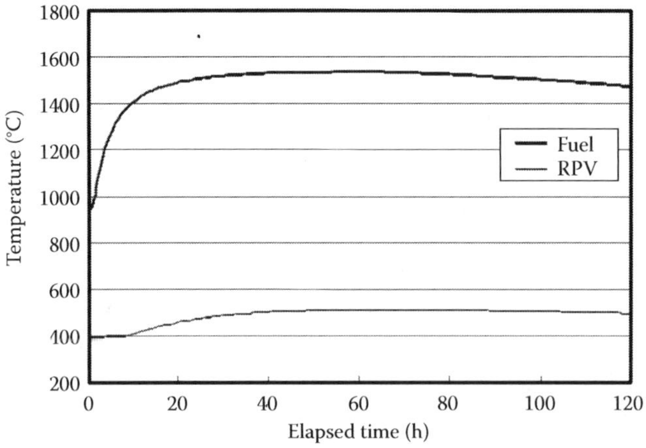

FIGURE 10.22
Reactor transient temperatures during the LOCA passive conduction cool down.

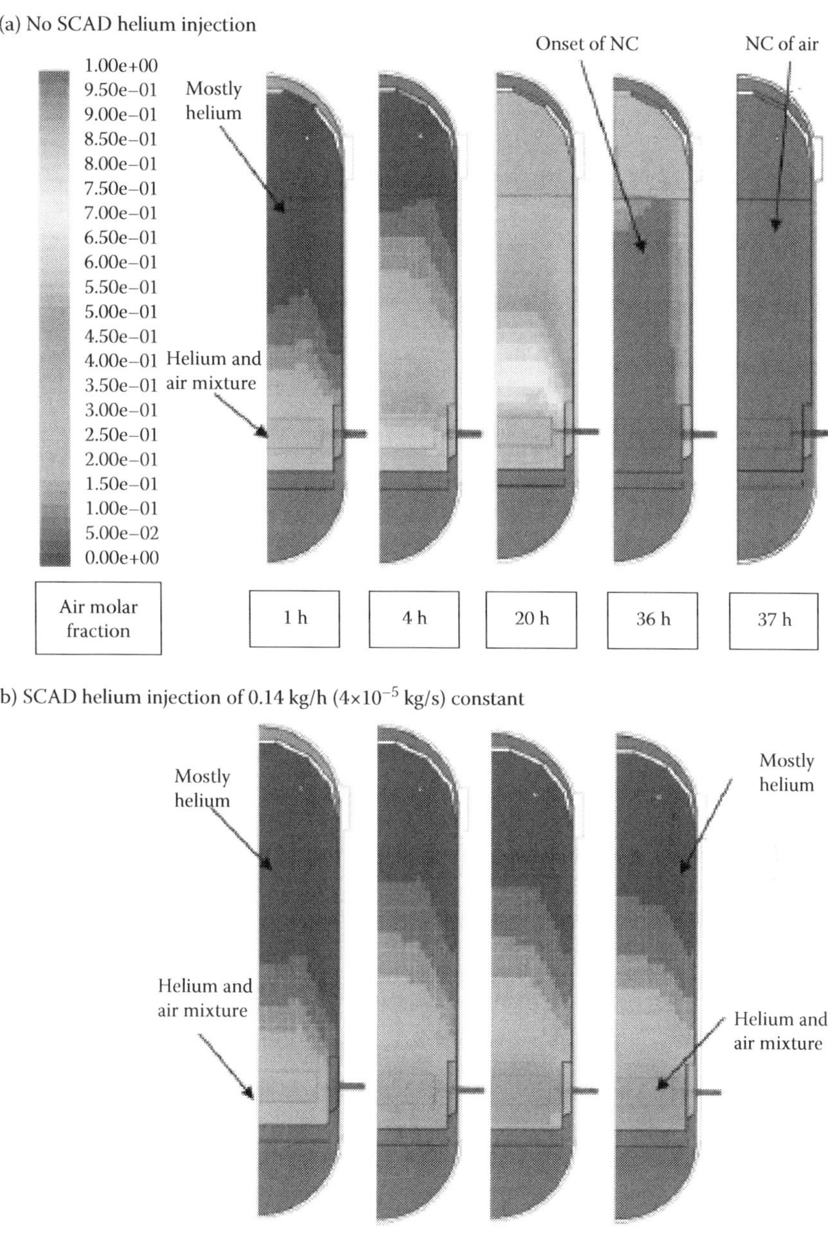

FIGURE 10.23
CFD results of SCAD air ingress and its prevention mechanism.

around the reactor. The use of a vented confinement for the reactor is instead considered in the current commercial reactor designs.

A mechanism to mitigate significant air ingress has been shown effective. Figure 10.23 shows the computational fluid dynamics (CFD) results of an event postulating guillotine breaks of the main inlet and outlet coolant pipes of a 600 MWt HTGR [32]. The external

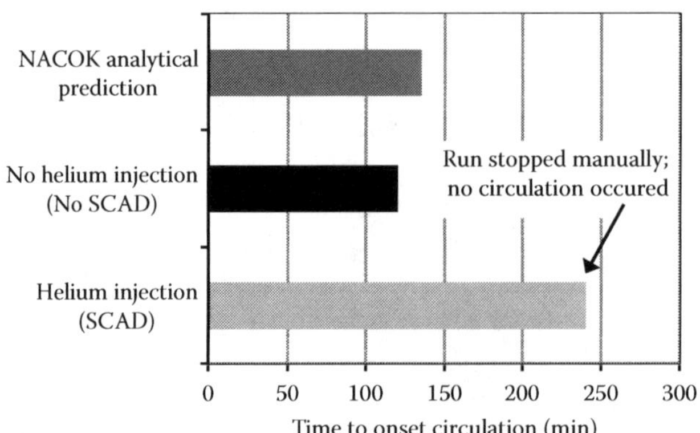

FIGURE 10.24
MIT reactor model experimental results of SCAD air ingress prevention.

ambient is assumed constant at 100% air at 20°C and 0.103 MPa. The reactor is initially filled with helium at pressure equal to the ambient. Figure 10.23a shows a time sequence of air concentrations in the reactor after the pipe breaks, followed by passive core conduction cool down and air ingress without use of mitigation. Air enters the reactor by diffusion and natural convection as seen with the gradual increase in air concentrations over time. The onset of NC is observed at about the 36th hour. The velocity of NC is 0.2 m/s in the core and the rate of air ingress into the reactor is 320 kg/h. Using a simple device, a sustained counter-air diffusion (SCAD) mechanism begins injection of helium at a minute rate of 0.14 kg/h upon pipe rupture to the area above the core. As seen in Figure 10.23b, the SCAD mechanism is effective to establish stable equilibrium of helium–air concentration and prevents advent of NC of air from the external to the internal of the reactor. The equilibrium continues as long as the helium is injected at that rate.

Figure 10.24 shows the data of an Massachusetts Institute of Technology (MIT) experimental investigation on a simplified reactor mockup [33]. In the case that SCAD is not used, NC of air ingress is shown to take place in about two hours following the simulated pipe rupture in an experimental reactor model apparatus. The experimental onset time of NC without SCAD is agreed by the analytical prediction based on the model obtained in the NACOK air ingress experiment in Germany. In the experimental run marked as helium injection (SCAD), a minute flow of helium is injected into the upper region of the model and as seen NC of air is prevented in twice the time long till the experimental run is manually ended. The injection flow in this case is less than 1% of the experimental mockup internal volume per hour.

The SCAD installation in practical reactor can be simple and reliable because it essentially requires two stationary equipment, a helium store and an orifice to begin injecting helium passively upon reactor depressurization. In a 600 MWt HTGR, 300 kg of helium is injected during a period of three months until the reactor cools below the temperature condition of graphite oxidation.

10.2.3 Pebble Bed Reactor

10.2.3.1 Test Reactor: HTR-10

The HTR-10 pebble-bed reactor developed by the Institute of Nuclear and New Energy Technology (INET) of China is presented in Figure 10.25. The photo in Figure 10.26 shows

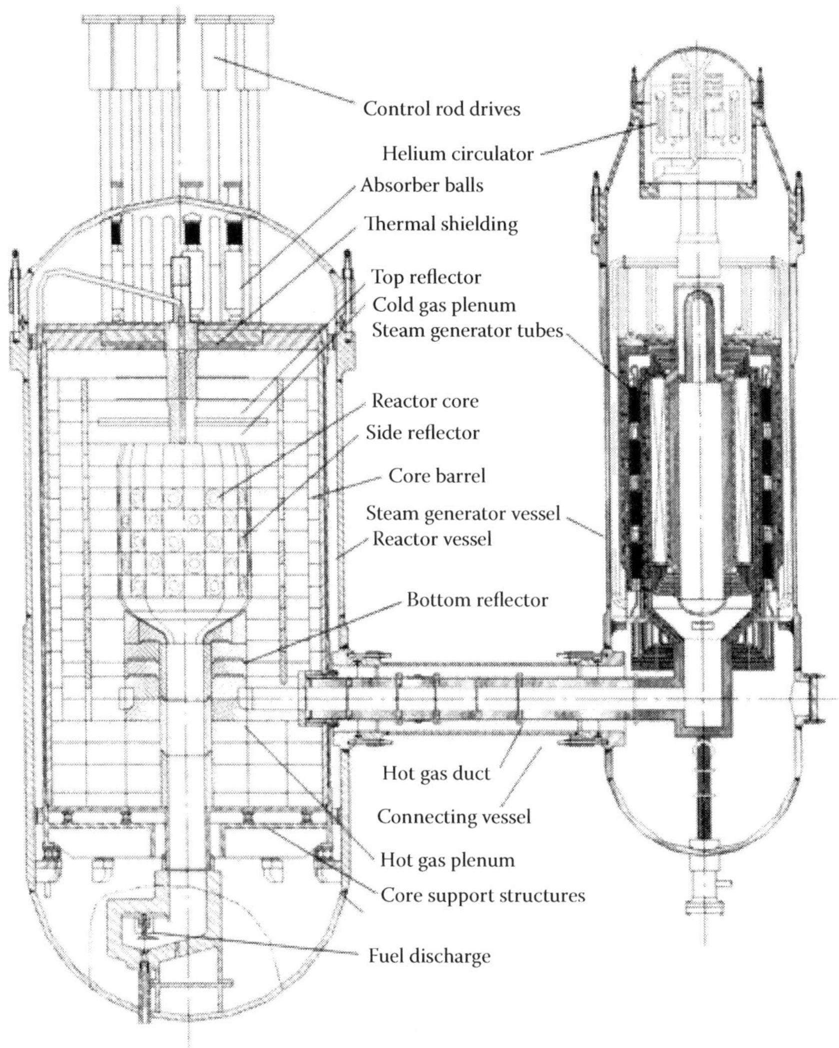

FIGURE 10.25
INET's HTR-10 test reactor design layout.

the view overlooking the constructed core structure. The cylindrical pebble bed core of 1.8 m diameter and 2.0 m effective eight is formed by graphite reflector wall. A large fuel discharge tube is visible in the center of the core bottom. Around this tube is web of small holes from where the heated helium coolant goes to exit the core. The larger boreholes in the inner peripheral of the side reflector are control rod insertion channels and the smaller boreholes in the outer peripheral of the reflector are core inlet coolant channels, in which the coolant flows upward to the top to enter the central core. The core contains 27,000 fuel spheres which slowly move downward in the core. Fuel pebbles are reloaded continuously online into the core from a top charging tube, by gravity move downward, and removed from the core through the bottom discharge tube. The removed fuel pebbles are checked for defect and burnup. While defective and spent fuel pebbles are taken to storage, partially burned and fresh pebbles are transported pneumatically

FIGURE 10.26
A top view of the HTR-10 core with pebble fuels in the central fuel discharge tube.

to the top to reload in the core. Each fuel pebble is expected to pass through the core multiple times before fully spent. It is a deliberate design intent to pass fuel pebbles in the core multiple times, which keeps the core power and temperature distribution as uniform as possible so that excessive reactivity is not present in the core. This is a major nuclear core-design advantage over the prismatic core, which loads fuel once-through and unloads the fuel off-line.

10.2.3.2 Commercial Reactor Designs

The 250 MWt high-temperature reactor pebble-bed module (HTR-PM) whose construction is under way in China is shown in Figure 10.27. The design parameters are in Table 10.7. The design is based on the HTR-10 technology and resembles the HTR-MODUL in Germany [34]. The cylindrical core is sized to 3 m diameter by 11 m height and filled with 420,000 fuel pebbles. Relative to the HTR-10 parameters in Table 10.8, the core outlet coolant temperature and the core power density are increased with more heavy-metal loading. The fuel is spent after making 15 passes through the core with average burnup of 90 GWd/tU. The reactor internals and the pressure vessel are structurally designed following their counterparts in the HTR-10. In operation, the annular area between the RPV and the core barrel is filled with helium of 250°C to limit the pressure vessel operating temperature.

Annular type of core was originally proposed to increase the pebble bed reactor thermal output [35]. It introduces a central column of nonfuel graphite pebbles and thus moves the active core region radially outward with the provision of an efficient decay heat-conduction path such that the fuel temperature will not exceed the design limit of 1600°C in the event of a depressurized conduction cool down. Although the designers of the HTR-PM had initially selected the annular core in a 458 MWt per reactor unit, they have now opted for the reduced 250 MWt cylindrical core (Figure 10.28) due to following considerations [15,36]:

1. There are greater technical uncertainties in the annular core about the controlling of fuel movement in the core and in the prediction of fuel and coolant temperatures. There is additional difficulty for reactivity control in the case of a dynamic

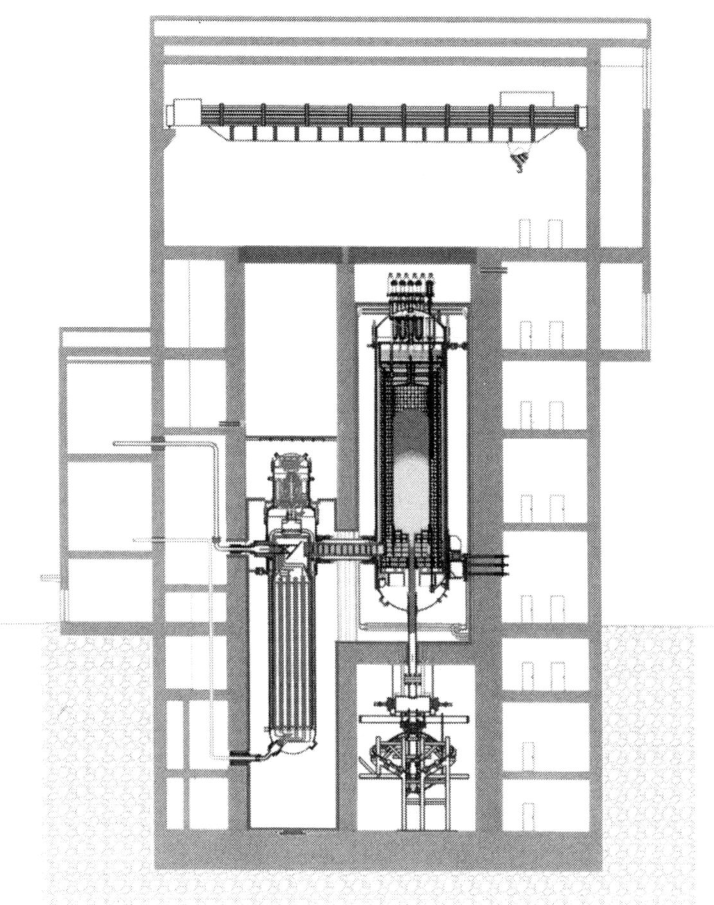

FIGURE 10.27
The HTR-PM commercial reactor design.

(moving graphite pebbles) central core and for graphite replacement in the case of a solid central core. On the other hand, the cylindrical core is one-zone pebble-bed reactor similar to the HTR-10 as well as to the HTR-MODUL design developed in Germany. The experiences gained in these systems will reduce technical uncertainties.

2. There is only 5% capital cost advantage for the 458 MWt annular core reactor plant than a plant employing two units (2 × 250 MWt) of the cylindrical core reactors. The main cost differential items include 14% more RPV steel and 20% more graphite material used in the 2 × 250 MWt plant and more refueling trains needed for the annular plant. The reactor building is found to be identical size in the two plant options.
3. Overall, the 2 × 250 MWt reactor plant is judged to be more attractive because it excludes many technical risks associated with the larger annular core.

The PBMR design in Figure 10.28 selects the annular core in the 400 MWt unit reactor thermal power NGNP. The graphite reflector blocks surrounding the active annular core

TABLE 10.7

Main Design Parameters of the HTR-PM

Parameter	Unit	Value
Rated electrical power	MWe	210
Reactor total thermal power	MWt	2 × 250
Designed life time	A	40
Average core power density	MW/m^3	3.22
Electrical efficiency	%	42
Primary helium pressure	MPa	7
Helium temperature at reactor inlet/outlet	°C	250/750
Fuel type		TRISO (UO_2)
Heavy metal loading per fuel element	g	7
Enrichment of fresh fuel element	%	8.9
Active core diameter	m	3
Equivalent active core height	m	11
Number of fuel elements in one reactor core		420,000
Average burnup	GWd/tU	90
Type of steam generator		Once through helical coil
Main steam pressure	MPa	13.24
Main steam temperature	°C	566
Main feed-water temperature	°C	205
Main steam flow rate at the inlet of turbine	t/h	673
Type of steam turbine		Super high-pressure condensing bleeder turbine

provide moderation for neutrons and act as a transient heat sink. They include a replaceable central reflector, an inner ring of replaceable side reflector, and an outer ring of permanent side reflector. There are additional bottom and top graphite support structures for the core. A total of 24 control rods are located in the side reflector. Half of the rods are used for control and the other half for shutdown. The shutdown rods are longer and run the length of the reflector blocks, while the control rods only run in the upper half of the reflector blocks. The neutron absorber rods make use of boron carbide (B_4C). A reserved shutdown system is comprised of small B_4C pellets, which are dropped into channels in the central reflector if required for additional negative reactivity margin in accident.

The core barrel consists of metallic structures made of mainly stainless steel 316 H and maintains the core geometry. The RPV consists of a main cylindrical section with hemispherical upper and lower heads. The upper head is bolted to the main section and incorporates penetrations for control rods and instrumentation. An opening is provided in the upper head to allow access to the core for reflector replacement. The lower head is welded to the main section and incorporates penetrations for fuel discharge. The RPV has a maximum external diameter of approximately 6.8 m, and its total length is about 30 m. The RPV is manufactured from conventional carbon steels SA533/SA508. A separate stream of

TABLE 10.8

Main Design Parameters of the HTR-10

Item	Unit	Value
Thermal power	MW	10
Reactor core diameter	cm	180
Average core height	cm	197
Primary helium pressure	MPa	3.0
Average helium temperature at reactor inlet/outlet	°C	250/700
Helium mass flow rate at full power	kg s^{-1}	4.3
Average core power density	MW m^{-3}	2
Power peaking factor	—	1.54
Number of control rods in side reflector		10
Number of absorber ball units in side reflector	—	7
Nuclear fuel		UO_2
Heavy metal loading per fuel element	g	5
Enrichment of fresh fuel element	%	17
Number of fuel elements in core		27 000
Fuel management	Multipass	
Average residence time of one fuel element in core	EEPD	1080
Maximum power rating of fuel element	kW	0.57
Maximum fuel temperature (normal operation)	°C	919
Maximum. burnup	MWd tHM^{-1}	87 072
Average burnup	MWd tHM^{-1}	80 000
Maximum thermal flux in core ($E > 1.86$ eV)	n cm^{-2} s^{-1}	3.43×10^{13}
Maximum fast flux in core ($E > 1$ MeV)	n cm^{-2} s^{-1}	2.77×10^{13}

helium actively cools the RPV to within the vessel operating temperature below the design limit of 371°C.

10.3 Balance of the Reactor Plant

10.3.1 Power Conversion System

Electricity that is a significant input to a number of hydrogen production processes can be more efficiently generated or cogenerated by the HTGR than by other nuclear reactors. Figure 10.29 shows the typical options of power conversion system. In general, system simplification and efficiency are gained by moving from the steam turbine cycle to the direct gas-turbine cycle. The system is simplified through reduction of equipment count, particularly eliminating the often troublesome steam cycle equipment including steam generator, moisture separator, steam turbine, condenser, feedwater heater, condensate and feedwater pump, and steam chemistry conditioning [2]. Being able to take the full advantage of the high temperature of the reactor heat source, the gas-turbine power conversion system

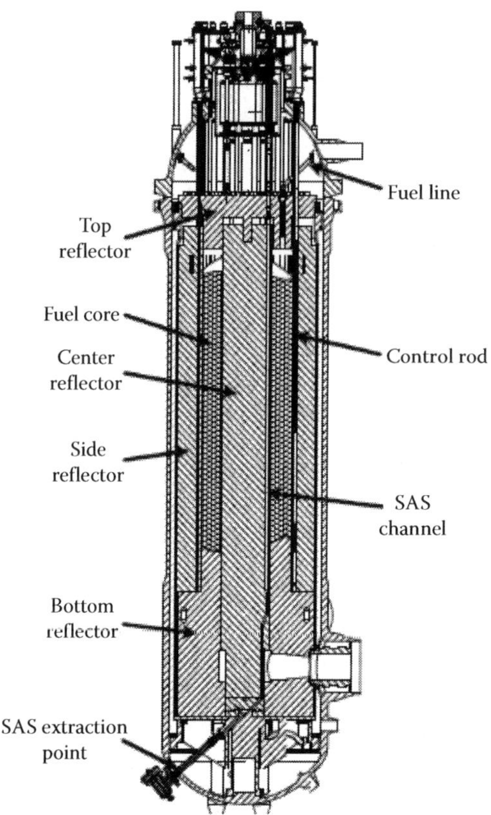

FIGURE 10.28
The annular pebble bed core 400 MWt PBMR design in a RPV of 30 m height and 6.8 m diameter.

options increase thermal efficiency (refer to Figure 10.1). As a result, the gas turbine options yield lower generating costs of electricity (Figure 10.2).

In practice, the gas-turbine power conversion system is tailored with design features to meet specific project requirements for development and deployment. This is reflected in the wide range of the design approaches taken by the design programs listed in Table 10.9 [37]. The PBMR selects a horizontal direct cycle (DC) gas-turbine unit rotating an external electric generator on a single shaft, although its earlier design opted for a multishaft and vertical unit [38]. The turbomachinery shaft seals are used that allows suspension of the rotors on conventional oil bearings and also use of the external generator. These are dry gas seals, which are themselves development items and whose operation reliability is a must because they seal high-power turbine shaft and are located on the primary pressure boundary. The GT-MHR prefers a single-shaft, asynchronous vertical direct-cycle gas-turbine unit. Magnetic bearings are used to avoid the use of bearing lubricants, whose accidental leak risks contamination of reactor primary circuit. The design intends to minimize construction cost by selecting a compact integral vessel system and small building lot. The vertical turbomachinery facilitates compact system and building footprint but the orientation is not widely practiced in the industry and involves complex rotor support and machinery seal interface with other components in the vessel. The GTHTR300 employs a DC gas-turbine unit that features a single shaft, a nonintercooled compressor, and a direct synchronous drive of electric generator. The design features are selected to maximize compliance to the

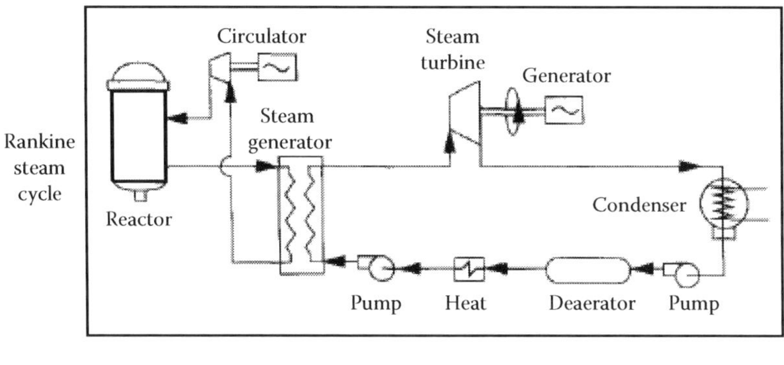

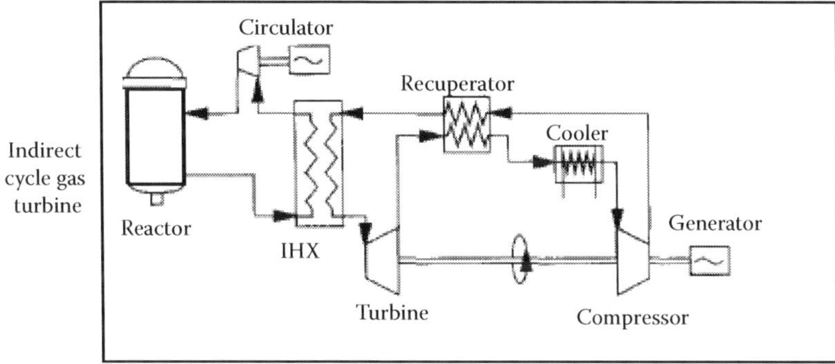

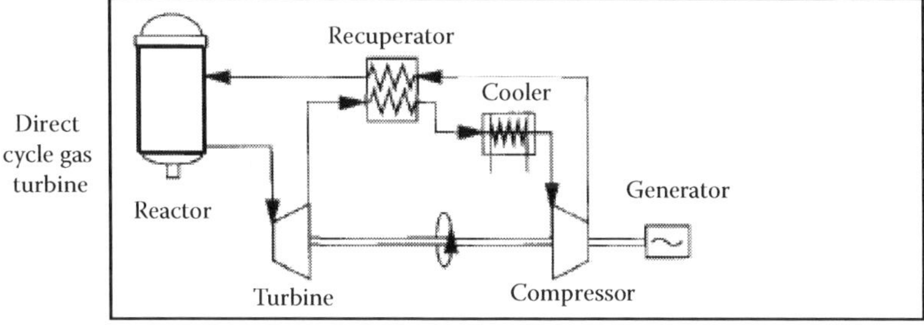

FIGURE 10.29
The evolution of simplified power conversion system options.

prevailing industry practice in heavy duty gas turbine in order to minimize development risks [39]. Although the distributed system layout increases vessel system volume, it facilitates modular construction and maintenance access to turbomachinery and heat exchangers by placing them in the separate pressure vessels. Framatome, now AREVA, settles on an indirect combined cycle using a mixture of nitrogen and helium gas to relate to the more familiar combustion gas turbine conditions [30]. Indirect cycle requires IHX and primary circulator and results in efficiency reduction, due to temperature drop over the IHX, and more complex operation control requirements. The advantage of the indirect cycle includes flexible selection of the turbine working fluid, in this case a nitrogen–helium mixture, and easier maintenance on the power conversion unit (PCU) because PCU is placed outside of the primary circuit.

Note that a direct-cycle helium-gas turbine is employed in all but one of the above plant designs, mainly because the DC maximizes plant generation efficiency while simplifying

TABLE 10.9

Summary of gas Turbine PCU Design Features of Representative HTGR Systems

	PBMR, Ltd. PBMR (Horizontal)	General Atomics GT-MHR	JAEA GTHTR300	Framatome (now AREVA) ANTARES Indirect
Reactor thermal power level	400 MWt	600 MWt	600 MWt	600 MWt
PCU working fluid	Helium	Helium	Helium	Nitrogen/helium mixture
Direct versus indirect cycle	Direct	Direct	Direct	Indirect
Recuperated versus combined cycle	Recuperated	Recuperated	Recuperated	Combined
Intercooled versus nonintercooled	Intercooled	Intercooled	Nonintercolled	Intercooled
Integrated versus distributed PCU	Distributed	Integrated	Distributed	Distributed
Single versus multiple TM shafts	Single (formerly multiple)	Single	Single	Single
Synchronous versus asynchronous	Reduction geat to synchronous	Asynchronous	Synchronous	Synchronous
Vertical versus horizontal TM	Horizontal (formerly vertical)	Vertical	Horizontal	Horizontal
Submerged versus external generator	External (shaft seal)	Submerged	Submerged	External (shaft seal)

the plant design and construction. However, the development of such a direct-cycle helium-gas turbine to meet the requirements of aerodynamic performance, reliability, and maintainability for nuclear service presents technical challenges, the extent of which depends heavily on the design choices. JAEA has been developing, through subscale component test, several component technologies related mainly to the helium working fluid for the GTHTR300 helium gas turbine unit in Figure 10.30. As seen, the helium compressor tends to have narrow and numerous-stage flow path as a result of inlet pressurization and working in helium, whose specific heat capacity is five times that of air. The difficulty with helium compressor aerodynamics was evident in a prior German installation [40,41].

The JAEA design unit rules out compressor intercooling to create large volume flow and low-pressure ratio with a given unit power rating. This key design decision led to the single bladed flow path having a sufficient flow area and an acceptable stage count, which is necessary to rein in boundary layer growth. A high-reaction blade airfoil is used to balance the dual design goals of efficiency and surge margin. The blades are incorporated with the controlled diffusion airfoil and three-dimensional blade stacking of camber to mitigate end wall boundary layers and flow separation. Designing compressor to rotate at grid-synchronous speed eliminates the need of gear box or frequency converter for electric generator drive, which avoids penalty to overall machinery efficiency.

A performance test program of a ⅓-scale compressor generated the correlation of the compressor efficiency with Reynolds number in Figure 10.31. The steep change in efficiency data with Reynolds number is attributed to the viscous effect of prevailing turbulent attached flow on smooth surfaces until the ⅓-scale critical Reynolds number $R_{e,cr} = 4 \times 10^5$ is reached. Above this Reynolds number the surfaces become hydraulically rough and

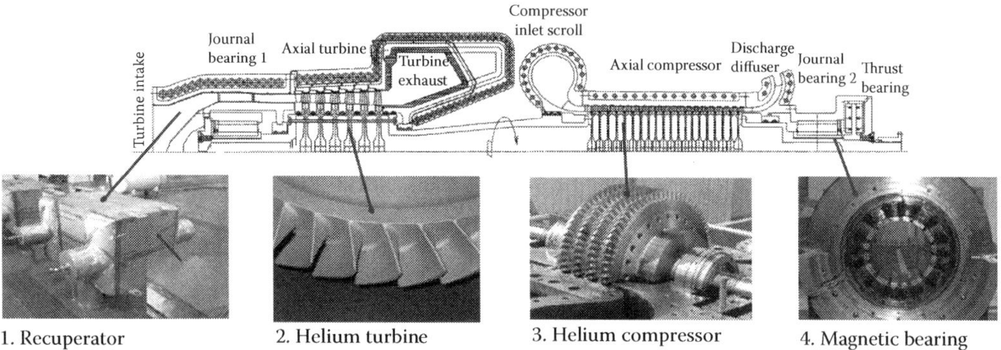

FIGURE 10.30
The helium gas turbine design in the GTHTR300.

efficiency becomes independent of Reynolds number. More specifically, the data correlate polytropic efficiency (η_p) with Reynolds number (R_e.) for the helium compressor as follows:

$$1-\eta_p \sim R_e^{-n} \text{ where } n = \begin{cases} 0.35 & R_e < R_{e,cr} \\ 0 & R_e \geq R_{e,cr} \end{cases} \tag{10.1}$$

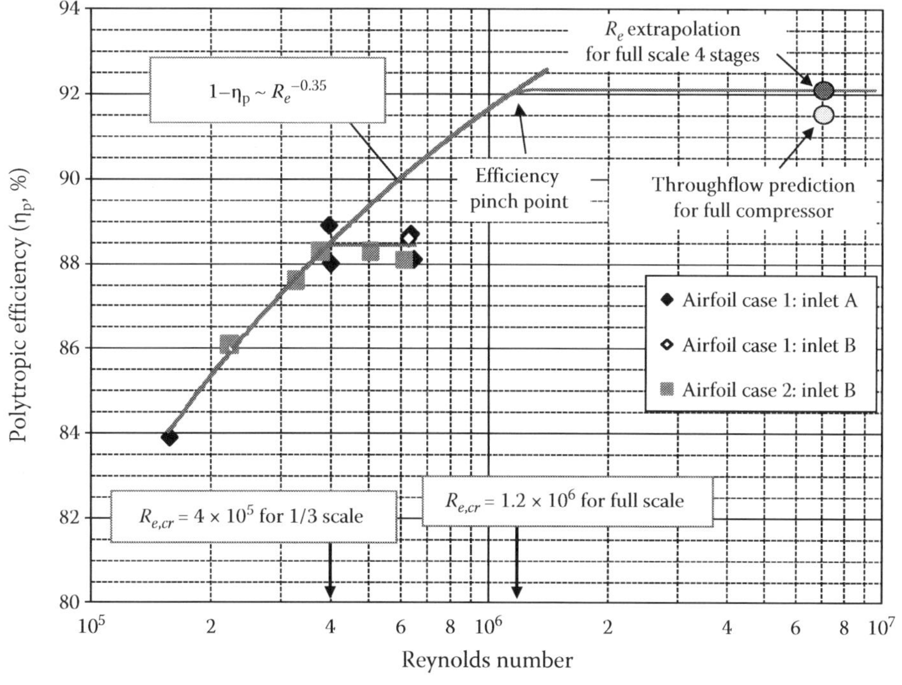

FIGURE 10.31
Correlation of helium compressor efficiency data with Reynolds number.

where critical Reynolds number, $R_{e,cr}$, is determined by the reported procedure [42]. The power law index of $n = 0.35$ in Equation 10.1 suggests strong viscous loss contribution to the aerodynamic inefficiency of the three-dimensional, low-subsonic flow field in the helium compressor. The validated design can be geometrically scaled to compressors of varied ratings based on the principle that one can increase or decrease inlet pressure and/or rotor diameter while holding speed constant to maintain aerodynamic and mechanical performance similarity in machines of larger or smaller unit capacity. Based on this principle of scale design, aerodynamic stage pressure rise and efficiency are not changed; neither are the stresses in blades and discs. The critical Reynolds number is determined to be $R_{e,cr} = 1.2 \times 10^6$ for the full-scale compressor in Figure 10.11. According to Equation 10.1, the polytropic efficiency is projected to be about 92% for the full-scale compressor unit. The efficiency projection agrees well with the prediction of a calibrated through flow code as marked in Figure 10.31.

10.3.2 Helium Circulator

Helium circulator is required in an HTGR devoid of DC gas-turbine for circulation of the reactor coolant. Additionally it is often used in a secondary helium loop for nuclear heat transport to the hydrogen production plant. Table 10.10 includes several sets of circulator design parameters [1,16]. If installed in the primary circuit, gas film or magnetic bearings are required to eliminate the risk of fluid–lubricant contamination to the circuit.

The HTTR uses as many as four identical units of helium circulator, whose design is shown on the left of Figure 10.32. The design consists of a single-stage centrifugal impeller, inlet gas filter, motor drive, and gas film bearings. The circulator casing is cooled by a water jacket and kept helium-leak tight. Flow rate is regulated by varying the motor drive speed with frequency converter. The filter is intended to protect the rotating assembly, particularly the gas film bearings, by filtration of any dust that could circulate in the primary circuit. It has a capture efficiency of 95% for dust larger in size than 5 μm. The filter consists of cylindrical filter elements made of 316 stainless steel. The inspection of the filter in the course of the HTTR operation reveals that the trace of graphite dust captured in the filter has been minimal in the prismatic HTTR reactor.

TABLE 10.10

Selected Helium Circulator Design Data

	HTTR 10MWt IHX	GTHTR300C 170MWt IHX	MHTGR 450MWt SG	MHTGR 450MWt IHX
Flow rate, Kg/sec	4.14	80	212	240
Inlet temperature,°C	430	475	288	490
Inlet pressure, Mpa	4.7	5.0	7.07	7.17
Pressure rise, kPa	79.4	200	91	145
Circulator power, MWe	0.26	8	5.0	13.2
Impeller type	Centrifugal	1 stage centrifugal	2 stage axial	2 stage axial
Nominal rotational speed, rpm	3000–12000	7900	4234	5800
Induction motor drive	Variable speed	Variable speed	Variable speed	Variable speed
Bearing type	Gas bearings	Magnetic bearings	Magnetic bearings	Magnetic bearings

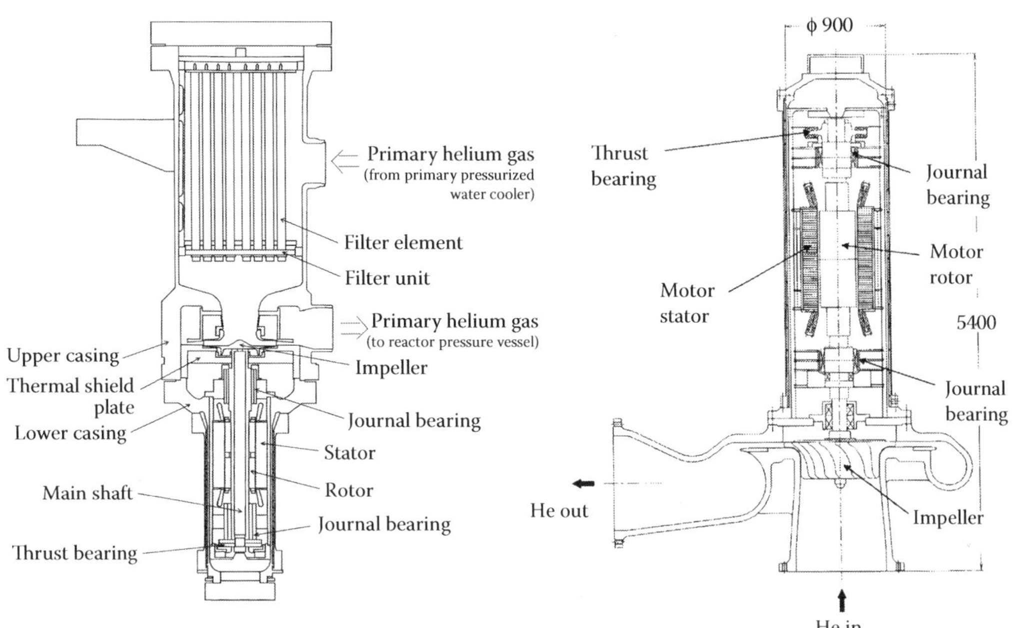

FIGURE 10.32
Helium circulator designs.

Because of generally-limited load duty of gas bearing, the large commercial circulator designs in Table 10.10 select magnetic bearings. The magnetic bearing unit shown in Figure 10.30 has been successfully tested in operation above the first bending of a 5-ton rotor. The right side of Figure 10.32 depicts the GTHTR300C's secondary-loop circulator design, whose design parameters are included in Table 10.10, based on centrifugal impeller and heavier-duty magnetic bearings.

The MHTGR designs in Table 10.10 employ two-stage axial-flow circulator based on the conceptual design shown in Figure 10.33. A single unit is mounted atop the IHX vessel as shown in Figure 10.34 [2]. The axial flow impeller can offer more efficient and larger flow circulation duty than centrifugal impeller. In general, one to two axial-flow stages are adequate to deliver the pressure rise required in process HTGR systems. The design database obtained in the development and test of the axial-flow helium compressor (Figure 10.31) suffices the aerodynamic design need of a helium circulator with the flow capacity of 300 kg/s at inlet conditions of 5 MPa and 350°C. Multiple circulator units operating in parallel will offer even greater flow capacity for a commercial reactor. This mode of operation is used in the HTTR.

10.3.3 Intermediate Heat Exchanger

IHX in the HTGR high-temperature (850–950°C) service condition performs both heat exchange and safeguards against material exchange between the primary system and a heat-utilization system such as a hydrogen production process. Successful safeguard against material exchange protects reactor from chemical ingress while allowing the heat-utilization system to be designed and operated as conventional plant, which saves cost of

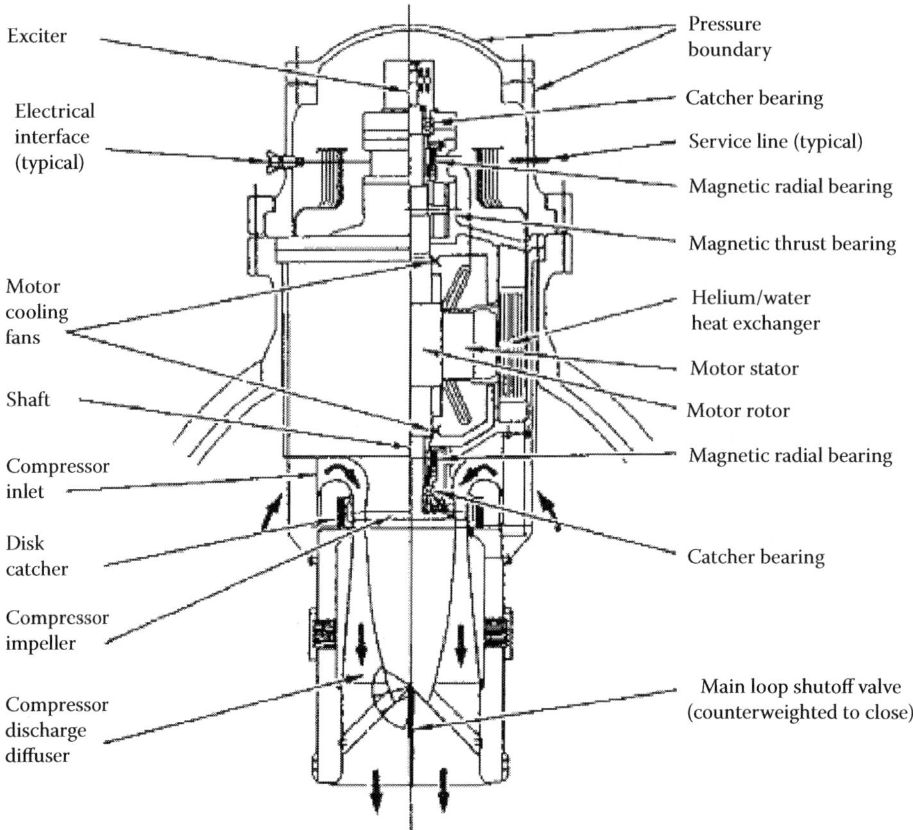

FIGURE 10.33
Main helium circulator design concept for the MHTGR.

the process plant. Figure 10.34 shows the typical installation arrangement of an IHX with the reactor.

The particular design shown in the figure employs a tube bundle and was proposed during the U.S. DOE sponsored GT-MHR design evaluation [1]. Industries have vast experience on tubular heat exchangers and have developed them specifically for the HTGR applications. The development has included the selection and new development of heat-resistant alloys according to the service requirements for creep properties, structural stability, and workability. An important focus of the development was qualification and mitigation on the generally significant effect of HTGR helium impurities on the long-term creep behavior. The results of the material development works mainly done prior to 1984 in Germany, the United States, and Japan were reported in the three special issues of a journal [43]. The material applications, design code qualifications, and successful IHX tubular designs were discussed for the German HTR program [44] and for the Japanese HTTR program [45]. The successfully developed tubular IHX designs for the HTGR application include a 10 MWt mock-up (950°C He/He) tested in the KVK facility in Germany and a 10 MWt IHX (950°C He/He) operated in the HTTR reactor in Japan. The limit of scale up based on the know-how of these test units was estimated to be around 300–400 MWt. Greater thermal power could be attained by multiple parallel units coupled to the same reactor.

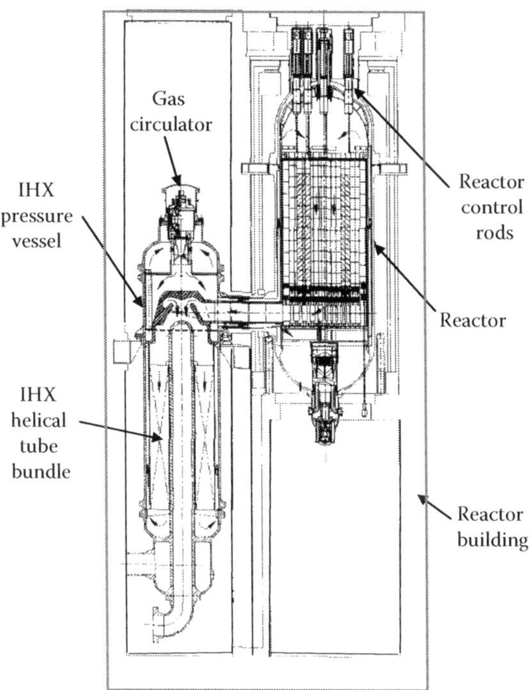

FIGURE 10.34
The MHTGR with IHX in side-by-side installation arrangement.

Depending on cost performance and development requirements, alternative plate-type IHXs have also been proposed in many system designs. The plate-type IHXs can achieve specific thermal power of an order of magnitude higher than tubular type and are thus compact and cost attractive. The plate type IHXs have not been used in high-temperature conditions in nuclear reactors. Similar to the efforts done on tubular design development and qualification, the unique structural design features of the plate IHXs to withstand the pressure and thermal loads as well as the effect of the HTGR impure helium on long-term mechanical behavior must be validated. Several designs of plate IHX will be discussed.

10.3.3.1 Tubular Type

Table 10.11 lists several tubular IHX designs [1,46,47]. The two associated U-tube IHX units were investigated in Germany. Figure 10.35 shows the component construction of the KVK unit. The other larger unit was similarly designed for the prototypical process-heat nuclear reactor, HTR-MODUL [48]. The investigation concluded that the prototypical IHX can be operated for 100,000 h on the basis of the KVK unit's experience. Especially the extensive wall material testing for high-temperature alloys at a temperature of 950°C showed that this component can be applied in conjunction with the process heat HTGR.

Shown in Figure 10.36 is the IHX operational in the HTTR. It consists of a helium-to-helium helical tube bundle with 96 tubes coiled in six radial layers. The tube bundle is supported in suspension on a center pipe. The suspension support is structurally designed

TABLE 10.11
IHX-Tubular Type Designs

	KVK	HTR- MODUL	HTTR	GTHTR300C Smooth Tube	GTHTR300C Finned Tube	MHTGR Indirect Cycle
Design type	U-tube	U-tube	Helical tube	Helical tube	Helical tube	Helical tube
Design status	Tested	Design	Reactor tested	Design	Design	Design
Thermal rate, MWt	10	170	10	170	390	454
LMTD, °C	61	69	113	157	122	67
Heat transfer area, m^2	350	4700	215	1448	3653	7572
Primary helium in shell side						
Flow rate, kg/s	3.0	50.3	3.4	323.8	325.1	239.4
Temperature (inlet/outlet), °C	950/293	950/292	950/389	950/850	950/719	850/485
Inlet pressure, MPa	3.99	4.050	4.060	5.020	4.980	7
Pressure loss, kPa	60	60	4	31	30	41
Secondary helium in tubes						
Flow rate, kg/s	2.9	47.3	3.0	81.0	176.7	250.0
Temperature (inlet/outlet), °C	220/900	200/900	237/869	491/900	475/900	425/775
Inlet pressure, MPa	4.350	4.350	4.210	5.150	5.130	7.460
Pressure loss, kPa	100	160	45	58	70	205
Tubing						
Type	Smooth tube	Smooth tube	Smooth tube	Smooth tube	Finned tube	Smooth tube
Tube material	Nicrofer 5520	Inconel 617	Hastelloy-XR	Hastelloy-XR	Hastelloy-XR	Inconel 617
Number of tubes	180	2470	96	724	3685	1120
Tube sizing (OD × t), mm	20 × 2	20 × 2	31.75 × 3.5	45 × 5	31.75 × 4	44.45 × 5.156
Tube effective length, m	—	30	22	14	10	48
Tube bundle						
Bundle diameter (ID × OD), m	2.0 OD	1.5 OD	0.84 × 1.31	1.84 × 4.57	1.84 × 5.92	1.94 × 4.84
Effective height, m	—	—	4.87	2.94	2.25	11
Number of tube columns	—	—	6	22	52	27
Tube pitch (traverse/axial), mm	—	—	47/47	65/65	40/40	55.6/59.7
Effective bundle weight, ton	—	69	5	50	125	284
Pressure vessel						
Vessel material	WB 36	WB 36	2¼ Cr-1 Mo	SA 533 (Mn-Mo)	SA533 (Mn-Mo)	9Cr-1 Mo
Outer diameter, m	2.4	4.89	1.90	5.40	7.04	5.86
Lifetime, h		140,000	120,000	157,000	157,000	300,000

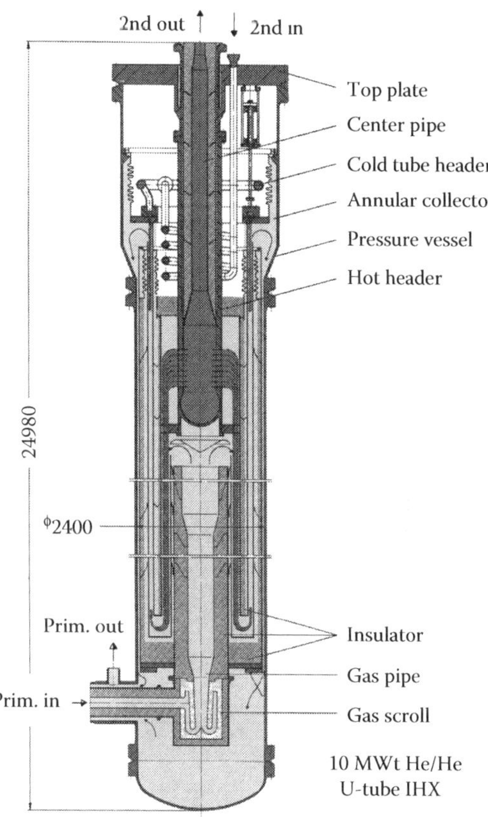

FIGURE 10.35
The KVK U-tube IHX component test unit.

to allow axial and radial thermal movement of the tubes while landing the bundle weight on the cold vessel head and transmitting lateral seismic load to the vessel shell through stoppers. The primary helium flows upward in the shell side and the secondary helium downward in the tubes. Research and development was carried to examine the design issues of (1) tube creep buckling, (2) thermal induced creep fatigue of the tube header, and (3) augmented thermal behavior of the tube bundle. In addition, the seismic design and in-service inspection method were also studied [46].

The IHX heat transfer tubing material is a nickel-based superalloy, Hastelloy XR-II, which was developed jointly by JAEA and Mitsubishi Materials over a period of two decades. In 1970s, material selection tests of available heat-resistant alloys identified Cr–Mo–Fe superalloy Hastelloy X, on which vast experience had been gained through its use in jet engines, to be a superior candidate to meet the service conditions of the IHX heat-transfer tubes and other high-temperature components in the reactor circuit. Hastelloy XR was then developed by modifying the alloy composition of Hastelloy X in order to make it more compatible with the helium and irradiation conditions of the HTGR in high temperature. The manganese and silicon contents were optimized to encourage the *in situ* formation of protective films of $MnCr_2O_4$ spinel and SiO_2 to increase heat and corrosion resistance; aluminum and titanium contents were decreased to minimize internal oxidation and inter-granular attack; and cobalt was decreased to reduce activation in the primary cooling

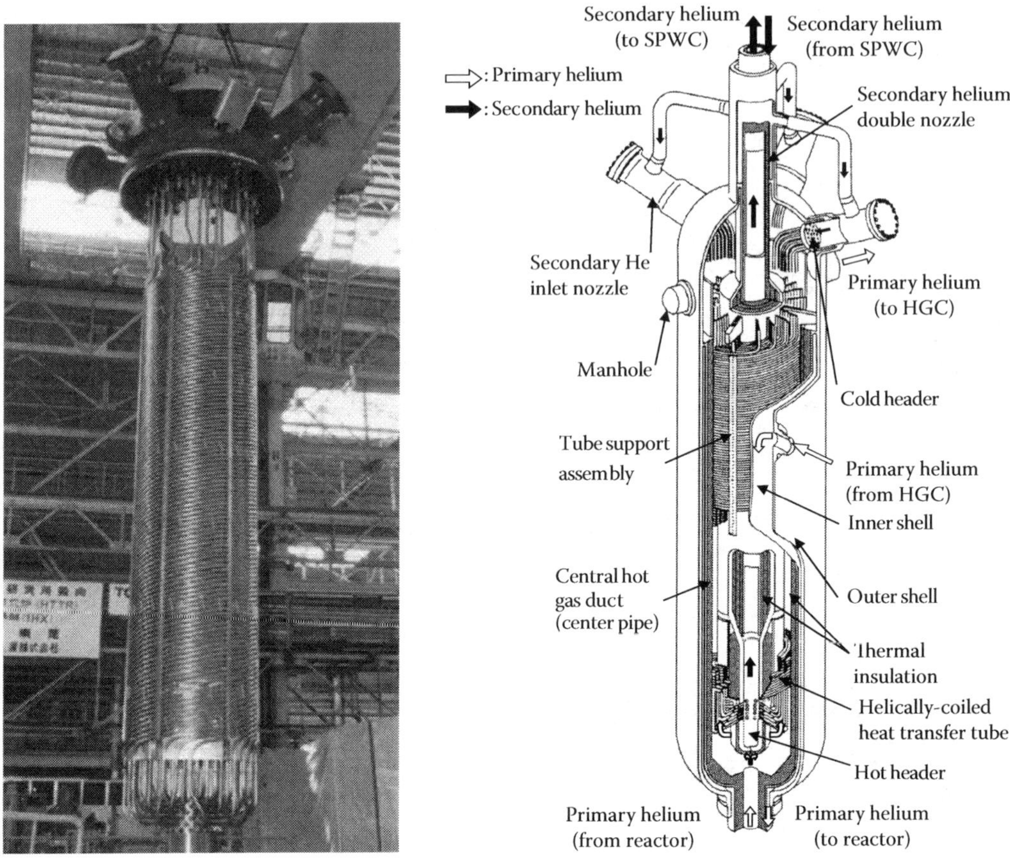

FIGURE 10.36
10 MWt Helical tube IHX installed and tested in the HTTR reactor.

system. In addition, a suitable range of boron content was added to yield intragranular precipitation to reduce creep rate of the base metal and weld. The alloy is named Hastelloy XR-II in deference to the boron addition. Figure 10.37 indicates the clear effect of boron addition on increasing creep-rupture strength of Hastelloy XR as tested in an impure helium environment in high temperatures. More tests on Hastelloy XR were followed to construct engineering database as required for the HTGR design. Based on the engineering database, JAEA by 1990 developed high-temperature structural-design guideline including design allowable limits on Hastelloy XR. The heat transfer tubes and hot header of the IHX in the HTTR is made of Hastelloy XR-II.

To achieve the design life of 100,000 h, the pressure load on the tube wall is limited to 0.15 MPa in normal operations. A higher pressure load can be allowed in the design base accidents. The worst condition occurs when the secondary side of the IHX depressurizes rapidly in an accident, which can expose the outer tubes to a maximum pressure load of 4 MPa in 950°C helium gas. Figure 10.38 shows experimentally obtained correlation between the buckling time and thickness of a tube sized to 31.8 mm outer diameter. The wall thickness is important as expected. Given wall thickness, the roundness of a tube cross-section area proves important. The time before buckling can decrease by several factors with the deviation from perfect roundness of up to 6%. The deviation value is defined

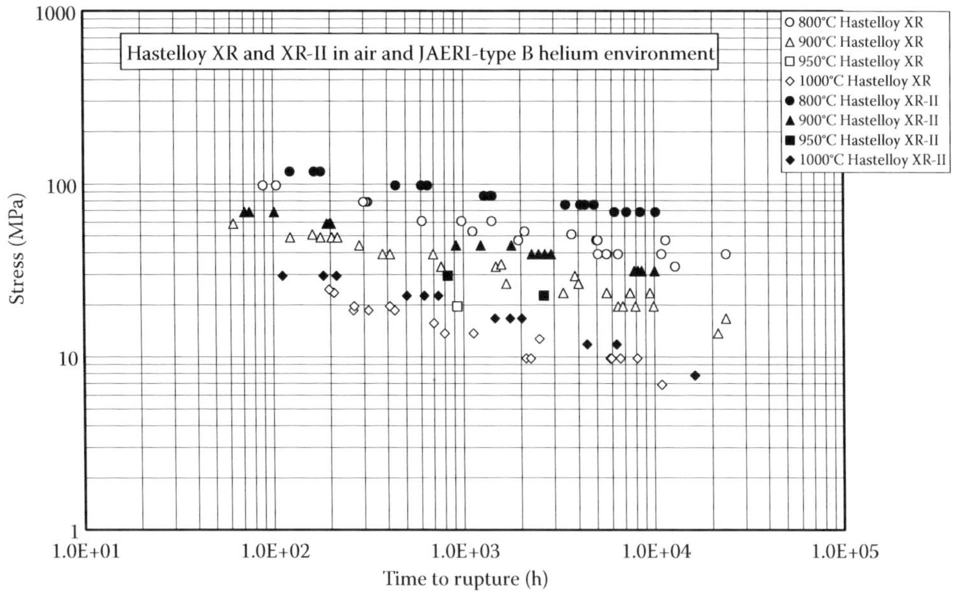

FIGURE 10.37
Comparative creep rupture strength of Hastelloy-XR and XR-II in the HTGR helium service environment.

by the difference between the maximum and minimum of the mean diameter. Although the post-inspection of a number of buckled tubes found many micro cracks on the tube outer surface, through-wall cracks were not observed. The deepest crack measured was 0.6 mm, relative to the nominal wall thickness of 3.5 mm. This establishes that leak tightness is maintained in a buckled tube in the high-temperature design conditions.

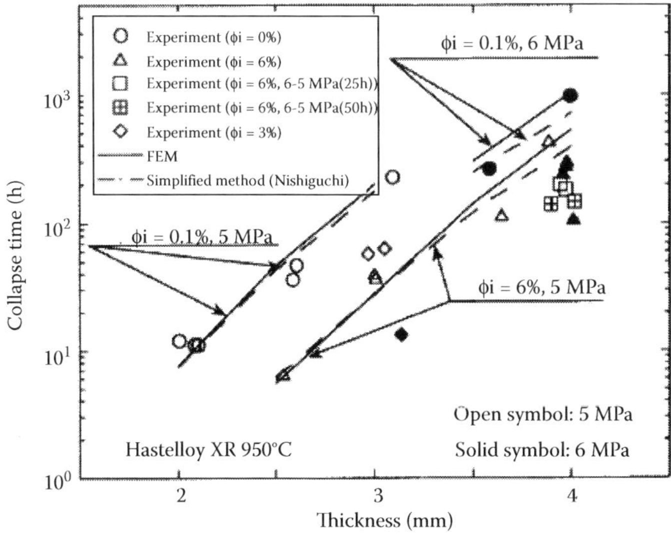

FIGURE 10.38
Elapsed time of buckling versus tube wall thickness.

The tubes are bent and welded in conjunction to the lower header of the central pipe. The connection of the tubes and header has a complex geometry and can be subjected to extreme thermal stress due to differential thermal movement. The material stress limit can be challenged in the high-temperature service. The full-scale mock-up shown in Figure 10.39 is used to test various structural configurations of the header, the connecting helical parts of the tubes from the header tube nozzle to the first tube support fastener. The central pipe was vertically moved in maximum displacement of 50 mm for more than 10,000 cycles in the 950°C helium to yield severer than the expected thermal stress conditions. The test provided the data used to select the optimum design configuration and establish the practical safety margin that is expected of the elastic creep design using ABAQUS of FEM analysis code.

Flow-induced vibration and heat-transfer augmentation for the helical tube bundle were investigated with air flow in full-scale component models. One model shown in Figure 10.40 consists of 54 helically coiled tubes in three radial layers. The plates are inserted between the layers. The vibration was tested in air flowing at the room temperature with the shell Reynolds number up to 1.54×10^4. The vibration resonated in natural frequencies with the amplitude increasing with the shell-side air velocity. The plate insertion proved effective to mitigate flow-induced vibration due to flow-vortex shedding. No abnormal fluid–elastic vibration problems can be identified and the maximum amplitude of vibration is limited to a ratio of $\sim 2.5 \times 10^{-3}$ amplitude to tube outer diameter in the test conditions.

Figure 10.41 plots the measurements and analytical prediction of convective heat transfer coefficient outside the helical-coiled tube bundles in a range of 200 to 300°C air inlet temperatures. The data over the inclined tubes are correlated by

$$Nuc = 0.78\ Re^{0.51}\ Pr^{0.3} \tag{10.2}$$

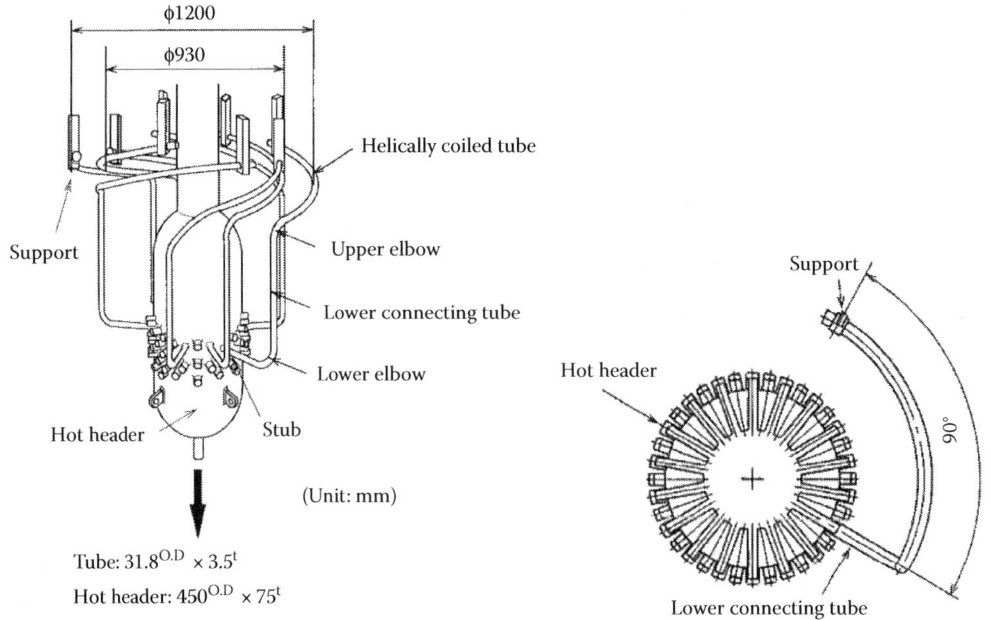

FIGURE 10.39
Fabricated and mechanically tested full-scale mockup of the IHX tube header.

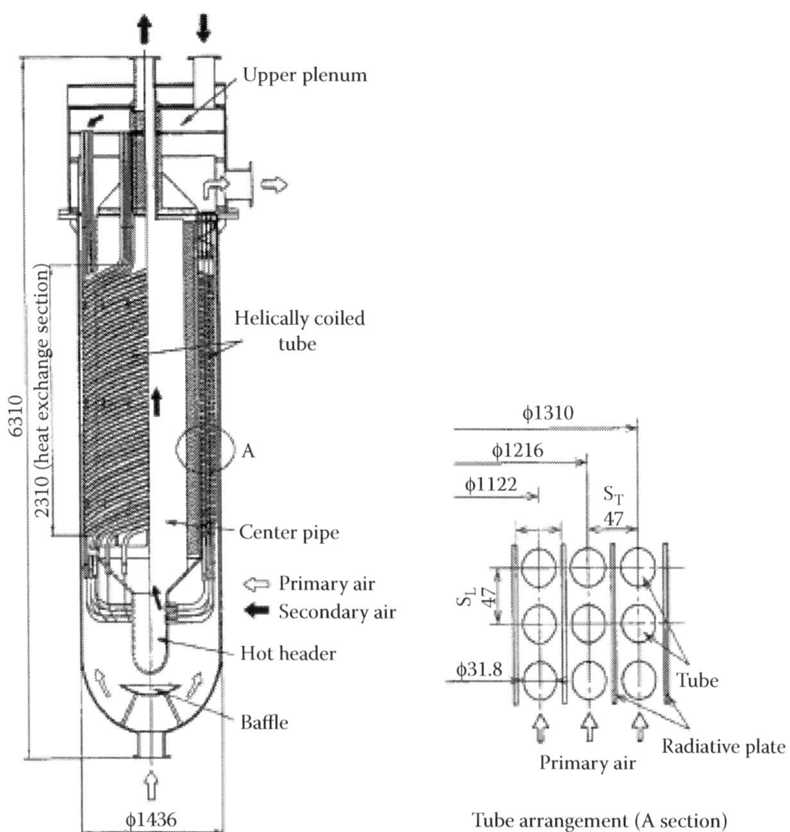

FIGURE 10.40
Thermal hydraulic test mockup of the IHX tube header.

where Nuc, Re, and Pr are Nusselt number, Reynolds number, and Prandtl number, respectively. The data are generally below the predications for perfect cross flow around in-line tubes by Zukauskas (1972) and Fishenden and Saunders (1950).

The plates inserted between the tube layers are found to additionally augment shell-side heat transfer, more effectively in high temperature. It does so by adding a mechanism that allows the heat to be transferred first from hot gas to the plates by convection and subsequently from plate wall to tube outer surface through thermal radiation. The overall shell-side heat transfer coefficient is increased and has been included in the measurements in Figure 10.41. The plate heat transfer augmentation was found to reduce about 10% percent bundle volume in the high-temperature conditions.

In addition, rolled-finned tube can significantly augment the thermal performance of the tube bundle through promoting flow turbulence and extend heat transfer area [49,50]. This kind of tubes is used in the design in Figure 10.42 and contributes to the result of a very compact design. Although the tubular design has been demonstrated to be well-suited to the HTGR process-heat applications including hydrogen production, development of new IHX designs that are considerably more thermally effective and physically compact can significantly improve large commercial plant economics. Two promising design types as described in the following sections have been under development.

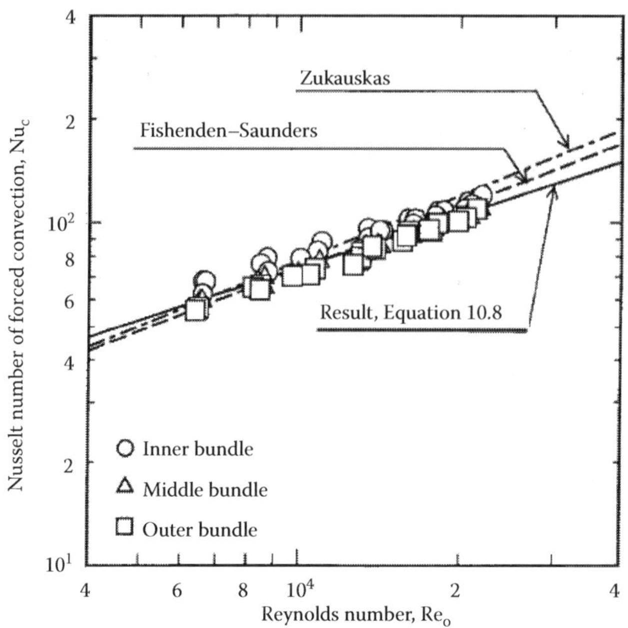

FIGURE 10.41
Forced convective heat transfer characteristics of the shell side of the helical tube bundle.

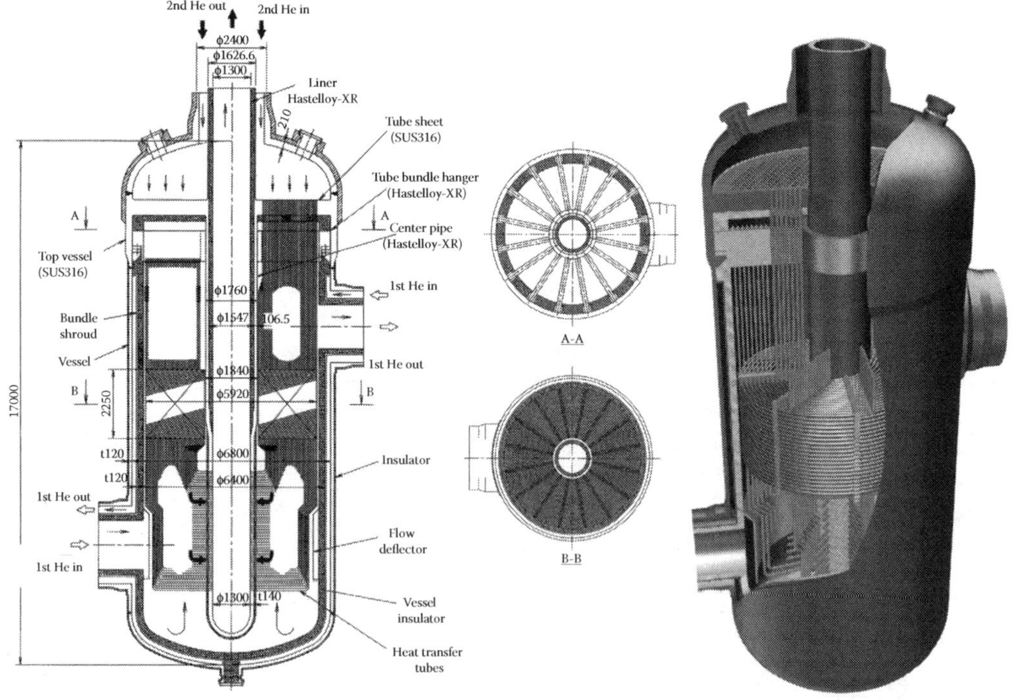

FIGURE 10.42
The 390 MWt IHX industrial design concept for the GTHTR300C.

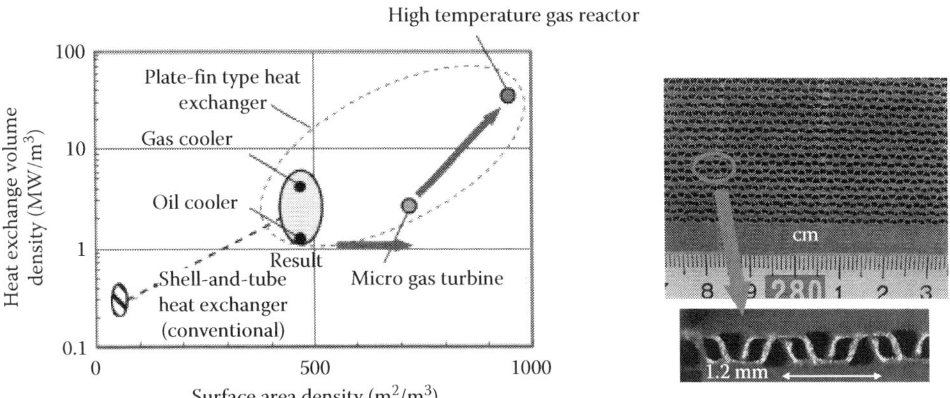

FIGURE 10.43
Thermal duty volume density of heat exchangers left.

10.3.3.2 Plate Type

An industrial development for compact helium plate-fin heat exchanger was carried out by Mitsubishi Heavy Industries, Ltd. of Japan [51,52]. The first objective targets the thermal duty of a practical heat exchanger with a fine surface of 1.2 mm fin-pitch by 1.00 fin-height (plate-spacing) as shown in Figure 10.43, whose thermal duty potential is estimated in excess of 20 MW/m^3 or 100 times that of a conventional shell-and-tube type.

Such heat transfer surface core was built and the offset fins are brazed to the plates to form the large units complete with inlet and outlet headers. The heat transfer and pressure loss characteristic factors were obtained and compared with prediction from the Wieting

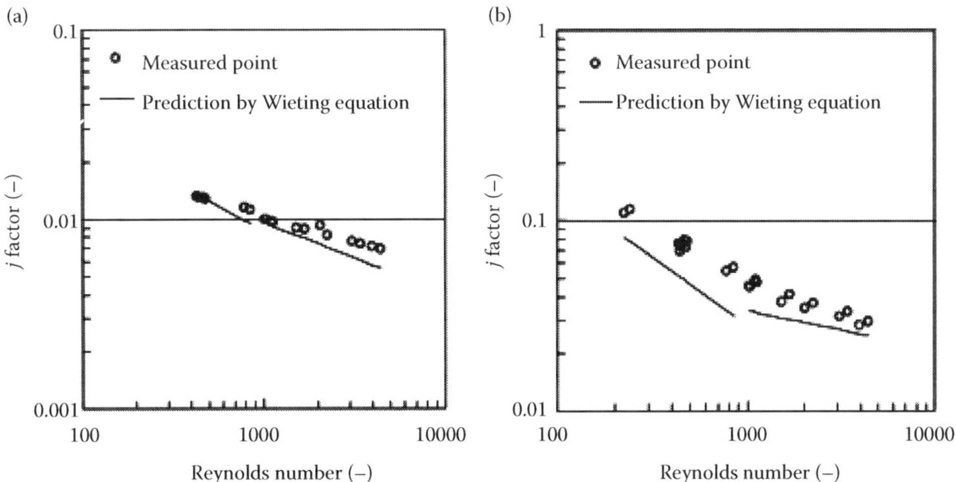

FIGURE 10.44
Heat transfer performance test results of a fine plate-fin surface IHX: (a) Reynolds number vs. *j* factor, where *j* factor $= St \cdot Pr^{2/3}$ and (b) Reynolds number vs. *f* factor, where *f* factor = friction factor.

correlations [53] in Figure 10.44. The data give the Colburn modulus heat transfer coefficient factor j $(=St{\cdot}Pr^{2/3})$ and the pressure loss friction factor f as follows:

$$j = 0.07059 \times Re^{0.27780} \quad \text{for } 450 < Re < 4000 \tag{10.3}$$

$$f = 2.14204 \times Re^{0.54679} \quad \text{for } 450 < Re < 1000 \tag{10.4}$$

$$f = 0.46838 \times Re^{0.33283} \quad \text{for } 1000 < Re < 4000 \tag{10.5}$$

The second objective of the industrial development deals with fabrication difficulty and high cost of fine surface topology and with specific problems related to nuclear service in high temperature and pressure. To improve manufacturing, strength tests of blazed specimens of plate-fin heat exchanger core structure were performed to optimize brazing material and brazing condition. The fine fin topology was confirmed with trial fabrication and strength and reliability tests of full size core. Previously, the smallest fin pitch had been limited to about 3 mm and typically fabricated of stainless steel. The fine pitch is even more difficult to fabricate of Hastelloy-X steel, the selected alloy to be discussed later. Improving press die shapes and performing double pressings of the steel sheet, with heat treatment between the pressings, led to the success of several fabricated fin surface geometries in the range of 1.2–1.3 mm fin pitch, 1.1–1.2 mm fin height and 0.2 mm fin thickness of Hastelloy-X. These fins and 0.5 mm-thick Hastelloy-X plates are brazed to form a heat-transfer core.

The establishment of high-temperature structural design method includes strength analysis method and design criteria to be applied to the IHX under the HTGR service conditions of high temperature 950°C and pressure of 7 MPa. In addition, a new ISI method based on a leak detection mechanism in place of the conventional ISI method has been studied and the safety design scenarios have been specified in the method.

A practical stress analysis method has been developed in an equivalent homogenized solid approach to the heat transfer core. A database including the fundamental deformation of the ultrafine plate offset-fin structure has been developed by tests of tensile strength, creep rupture, fatigue, and creep fatigue. High-temperature design criteria are established taking into account of the particular plate-fin structure. The primary stress limit for tensile and creep rupture and the secondary stress limit for progressive deformation and the creep-fatigue limit for crack initiation have been specified as the design criteria.

Specific material and fabrication requirements for IHXs operating at high temperature (800–950°C) include creep property, weldability, workability, and steady supply. Although Hastelloy-X is not ranked best in creep strength, it is selected because evaluation ranks it high in meeting all other requirements among the Ni-based alloys in Table 10.12.

The industrial program was concluded with fabrication of an IHX unit. The fabrication process is explained in Figure 10.45. The unit being a half of the full-scale IHX module product specified for a commercial reactor is measured in 493 mmW × 493 mmD × 191 mmH. It contains heat transfer surface of 1.2 mm fin pitch × 1.2 mm plate spacing with a specific surface thermal duty of 70 MW/m^3. The structural design and thermal performance of the unit have been validated by tests.

10.3.3.3 Printed Circuit Type

The printed circuit heat exchanger (PCHE) design shown in Figure 10.46 and developed by Heatric Corporation consists of stacked steel plates on which flow channels are either chemically etched or pressed. The plates are diffusion bonded with possibly restored

TABLE 10.12

Evaluation of Ni-Based Heat-Resistant Materials for Fabricating Plate-fin IHX

	Creep Rupture Strength	Weldability	Workability	Supply Availability	Overall Score
Hastelloy X	B	B	B	B	B
Inconel 617	A	B	C	C	C
Nimonic PK33 alloy	B	D	D	C	D
Haynes 263 alloy	B	C	D	C	D
Ni-Cr-W superalloy	A	C	C	D	D

Note: A, excellent; B, good; C, acceptable; D, not acceptable.

properties of base metal to form a strong and compact heat exchange core. Small cores may be welded together to form a larger one. Headers are welded to the core to guide fluids through appropriate flow channels, which may be in counterflow, cross flow, or a hybrid of both. The company claims that alloy 617 and alloy 230 would be the most suitable materials for the VHTR IHX because they have the appropriate combination of mechanical, physical, and corrosion-resistance properties and are commercially available in appropriate product forms of sheet and fin.

The PCHE concept allows for simultaneous high-temperature and high-pressure operation with relatively thin wall thicknesses between primary and secondary coolants. The PCHE may reduce to one-fifth in size of a conventional tubular heat exchanger of equivalent heat duty, and yields greater thermal effectiveness. Several HTGR plant designs have been based on the PCHE type of IHX.

FIGURE 10.45

Fabrication of a 1/2 scale fine topological plate fin IHX module in counterclockwise from top-left: Stacking of fin and plate sheets, fully stacked unit, bound prior to brazing, and the final brazed unit.

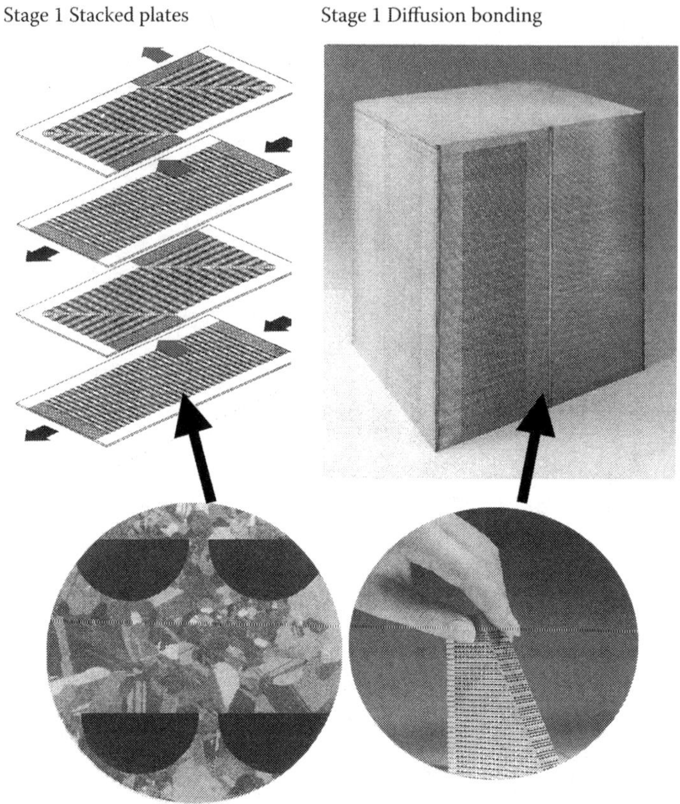

FIGURE 10.46
Heatric concept PCHE.

AREVA's NGNP preconceptual plant design (Figure 10.47) selected a steel PCHE IHX-1 to transport the 900°C heat to the H_2 plant because it was thought to have a reasonable chance of being developed, tested, and ASME-certified in the NGNP development time frame, and may offer breakthroughs for better economics for commercial-scale VHTRs. On the other hand, the tubular IHX-2 is selected to heat the indirect combined cycle power conversion system because of their mature design, greater known reliability, and operational experience.

GA's NGNP preconceptual design (Chapter 31) includes a PCHE IHX option. The IHX transfers a nominal 65 MW of thermal energy to the secondary heat transport system (SHTS), which transports the heat energy to hydrogen production facilities. The PCHE IHX arrangement in pressure vessel was designed by Toshiba as shown in Figure 10.48 and the design data are presented in Table 10.13. The alloy 617 material volume before etching flow channels is 8.05 m^3 and weighs about 67 tons. The large mass of the PCHE is due to the design specified small logarithmic mean temperature difference (LMTD). Eight PCHE modules are assembled by tungsten inert gas (TIG) welding to form a PCHE core unit. The plenums are attached to both sides of the PCHE modules by TIG welding to form the flow passage for the secondary helium coolant. The material used for the plenum is also alloy 617. Six parallel PCHE units are installed in the pressure vessel.

The primary coolant flows from the reactor through the inner cross duct and turns upward to enter the PCHE units, in which heat is transferred from the primary coolant to

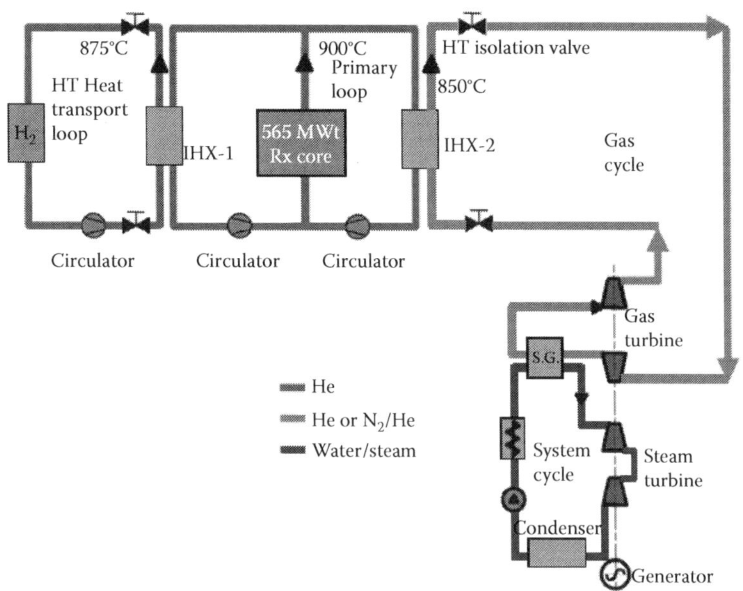

FIGURE 10.47
AREVA NGNP pre-conceptual hydrogen cogeneration plant design.

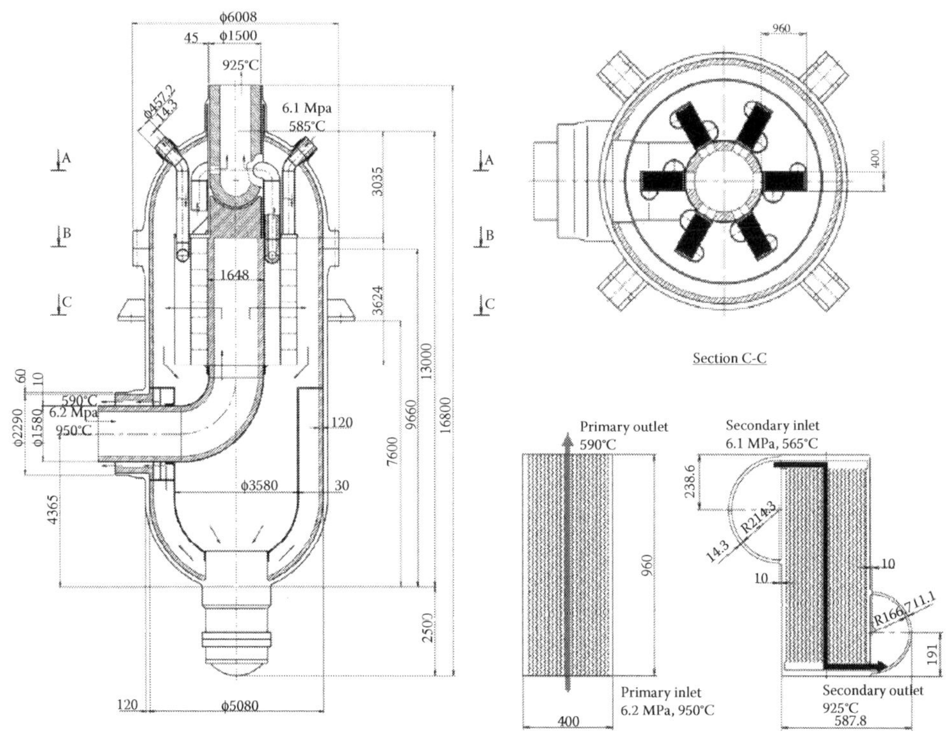

FIGURE 10.48
The PCHE IHX design by Toshiba in GA's NGNP Pre-conceptual Design.

TABLE 10.13

Toshiba PCHE Type IHX Design Conditions

Design Conditions	Design Value
Heat rate	65 MWt
LMTD	25°C
Primary side flow rate	34.7 kg/s
Primary side inlet/outlet temperature	950/590°C
Primary side inlet/outlet pressure	6.20/6.15 MPa
Secondary side flow rate	34.7 kg/s
Secondary side inlet/outlet temperature	565/925°C
Secondary side inlet/outlet pressure	6.10/6.05 MPa
Design Sizing	
Number of PCHE modules	48
Module height/width/length	453/400/960 mm
Helium channel hydraulic radius	1.5 mm
Helium channel flow area pitch	3.9 mm
Helium channel offset pitch	12.7 mm
Offset height	2.286 mm
Plate thickness	2.4 mm

the secondary HTS helium. The exiting primary coolant dropping in temperature descends in the IHX shell to reach the lower-end helium circulator. The primary coolant pressurized in the helium circulator moves upward between the pressure vessel and shell and through the annulus of the cross duct to the reactor. The secondary HTS helium enters the IHX through the nozzles on the pressure vessel top spherical head and distributes through the inlet plenums to the PCHE units. The helium flows through the PCHE units, where it is heated, and collects in the outlet plenums that are connected to the central secondary outlet header on the top of the pressure vessel. The heated secondary helium is transported in the secondary circulation loop to hydrogen production plants.

The GA's NGNP preconceptual design identified a number of design issues that would require more detailed analyses. Some issues are

1. Installation of thermal insulation into small pipe
2. Maintenance considerations with respect to working space and methods, and so on
3. Confirmation of secondary piping and PCHE support design feasibility
4. IHX lifetime design whereas the vessel is designed for 60 years of the plant life, but the IHX heat exchanger core may have to be made replaceable in shorter intervals
5. Implications of ISI on the PCHE as required by ASME Code. Because of this, the primary coolant pressure boundary might be considered to extend to the isolation valves in the SHTS
6. Methodology for monitoring the IHX heat exchanger surface for leakage
7. Determination of the precise pressure drop through the PCHE by experiment
8. Structural analysis is needed to confirm the feasibility of the design selections
9. Many slide joints used to enable maintenance of the IHX need to be evaluated of leak rates and effect on performance degradation

10.4 HTGR Hydrogen Production Plant Systems

In addition to having the general performance characteristics of a Generation-IV system, the HTGR offers the following exceptional attributes as described in the earlier sections of this chapter:

1. High-temperature coolant
2. Small-to-medium modular size
3. Fully passive reactor safety
4. Near-term deployment

The HTTR successfully demonstrated 930–950°C in long-term operations in 2010. The high temperature enables high thermal efficiency for electricity. The high-temperature heat is also a right match to the heat requirement for a number of major candidate processes of nuclear hydrogen generation. The reactor is economically rated in a small to medium thermal range of up to 600 MWt, which is compatible with the scale of practical hydrogen production and consumption. The passive reactor safety ensures that the reactor can be safely sited near a point of hydrogen demand, it being a hydrogen fuel hub in populated area or an industry installation. Since the engineering scale reactors are already in operation, the major step in commercial deployment is prototype development which is being pursued in several projects in the world.

Although many processes introduced in Section II of this book could be coupled to an HTGR, only selected process-reactor combined systems are discussed in the following.

10.4.1 Water Electrolysis

As presented in Chapter 3, electrolysis of water is a commercial method of hydrogen production using electrical energy. About 4% of the world's hydrogen is produced via this route today. Emission performance of the overall process depends upon source of electricity, and approaches to zero CO_2 emission with use of nuclear power like the HTGR. Electrolysis production hubs supplied by existing power transmission grid would lower infrastructure development requirement for hydrogen distribution.

Although high performance atmospheric electrolyzers are commercially available, vendors continue to develop new models for cost reduction. Hydrogen Technologies, a leading supplier, offers the largest commercial units of the kind with the state-of-the-art specification shown in Table 10.14. Efficiency is more than 82% based on hydrogen higher heating value (HHV), which is enabled by the proprietary catalytic coat applied on the electrodes that effectively lowers cell voltage. The range of production capacity is adjustable with the number of electrolytic cells stud-bolted on a skid shown in the left side of Figure 10.49. The largest unit is rated at 2 MWe for 485 Nm^3/h H_2 (40 kg/h H_2 or 50 kWh/kg-H_2) and 242.5 Nm^3/h by-product O_2 and installed on the standard skid frame in a floor area of 4 × 14 m inclusive of maintenance area. The range of production capacity is extended by combining multiple units working in parallel. Egypt has the current largest installation with 35,000 Nm^3/h hydrogen output.

The new generation of high-pressure units as shown in the right side of Figure 10.49 and the polymer electrolyte membrane (refer to Chapter 3) are being developed to improve economical performance of large-scale production. GE reported the progress of developing

TABLE 10.14

Technical Data of Hydrogen Technologies' Atmospheric Electrolyzer

Capacity	
Capacity range (Nm^3 H_2/h)	10 – 485
Maximum Nm^3 H_2 per cell	2.11
Energy	
Power consumption at 4000 Amp DC (kWh/Nm^3 H_2)	4.1 ± 0.1
Power consumption at 5150 Amp DC (kWh/Nm^3 H_2)	4.3 ± 0.1
Purity	
H_2 purity (%)	99.9 ± 0.1
O_2 purity (%)	99.5 ± 0.1
H_2 purity after purification (%)	99.9998% (2 ppm)
Pressure	
H_2 outlet pressure after electrolyzer	200–500 mm WG
Maximum H_2 outlet pressure after compressor	440 bar g
Operation	
Operating temperature	80°C
Operation	Automatic, 20–100% of max capacity
Electrolyte	25% KOH aqueous solution
Feed water consumption	0.9 L/Nm^3 H_2

high pressure plastic stack electrolyzer technology. The cell low cost is approached by the design of thin sheet nickel-coated electrodes and plastic-molded electrolyte with polysulfone material options of Udel and Radel. The design reduces the usual steps of machining. Successful design, construction, and efficiency benchmark testing in 2008 of the full-scale, 10 × 2500 cm^2 cell prototype at 15 bars provides confidence that low capital cost plastic electrolyzer can meet all necessary performance targets. The 2009 technology performance and cost parameters were reported to include cell efficiency of 68% lower heating value (LHV) or 80% HHV and a capital cost of US$150k for a 1 MWe (20 kg/h H_2 capacity)

FIGURE 10.49
Standardized atmospheric electrolyzer and advanced pressurized unit.

production electrolysis module. The balance of plant (BOP) cost based on the state-of-art similar scale system is about \$300/kWe. This gives total plant capital cost of \$450/kW. The production-scale cost of hydrogen is estimated to be US\$2.6–3.1/kg-H_2 in the US\$0.04–0.05/kWh cost of electricity range and with a stack cell replacement life of 10 years [54]. Market research and discussions with electrolyzer manufacturers shows that the GE electrolyzer can be successfully commercialized in the near-term.

An HTGR of 100–300 MWe can thus power 50–150 units of the largest electrolyzer for combined capacity of 24,250–73,750 Nm^3/h high-purity hydrogen plus 12,125 to 36,375 Nm^3/h of oxygen. The capacity is extendable by multiplying nuclear reactor units. The power generating efficiency of the HTGR is about 50% based on 950°C reactor coolant temperature and DC gas turbine (Figure 10.2). Assuming additional 1% loss of converting AC into DC power, overall efficiency of the HTGR hydrogen production by water electrolysis is close to 40% HHV. The power generation cost from the HTGR is estimated to be 3.5–4.0 ¢/kWh [1,7]. The nuclear plant has 40–60 years of life. Based on these cost parameters, the at-gate hydrogen cost from the HTGR water electrolysis plant is estimated to be US\$2.3–2.6/kg-H_2.

As is for water reactors in Chapter 9, there are strong incentives to produce hydrogen using off-peak electricity of daily, weekly, or annual demand since the primary cost of electrolysis is electricity. The off-peak electricity can be priced several times lower than the peak demand, thus hydrogen generation can cost significantly less. In addition to more affordable hydrogen pricing, the hydrogen produced in this way can be stored for later use to meet peak electricity demand with power generation provided by hydrogen fuel cells and hydrogen combustion turbines. The production may be scheduled to satisfy peak demand that exceeds the capacity of HTGR.

Since large electrolyzer units are commercially offered, future technical development challenge is thus prototype development of highly efficient (~50%) Generation IV VHTR power generating plants.

10.4.2 Steam Electrolysis

Electrolysis of water or steam can be thought to undergo an isothermal thermodynamic process, of which the work requirement is an input of electricity. Electricity demand then decreases with increasing process temperature, and at the same time the heat is added to increase the process temperature approximately at the same rate with which electricity demand decreases. In the extreme case of process temperatures exceeding 2000°C, water molecules begin to decompose to hydrogen and oxygen through thermolysis and electricity demand approaches to zero. Therefore, electrolysis of steam at temperatures as high as practically achievable can be more economically efficient compared with the low-temperature electrolysis of water, because electricity demand is reduced and replaced with less expensive heat input and less waste heat is removed from the process because of the reduced power conversion duty. Additional gain in efficiency can be expected from increase in both electrochemical reactivity and electrolyzer conductivity with temperature.

The present electrolysis cell designs envision operation at 800–900°C. At the high temperature, the efficiency can potentially approach 50%, compared with 40% maximum for water electrolysis as discussed above. However, maintaining performance and durable life of the steam electrolysis stack at the temperatures of interest proved challenging. Although HTE is not a commercial reality, significant development has been made in recent years toward the goal of demonstrating a nuclear hydrogen production system

(HPS). The status of research and development on the tubular and planar cells performed in JAEA and Toshiba Corporation are reported in Chapter 4. The most extensive engineering development to date in the world has been undertaken in the INL. In 2008, the INL completed the 1000-h test of an integrated laboratory scale (ILS) technology demonstration experiment. The experiment achieved a hydrogen production peak rate of 5.7 Nm^3/h with a power consumption of 18 kW.

In 2009, INL operated a stack unit of 10 planar cells for 2500 h. However, the operation was interrupted numerous times mostly due to gas control and water chemistry problems. A major cell-related problem of performance degradation was reduced to 8% per thousand hours from 21% three years earlier. This is still far from the near-term target of 3% per thousand hours. There still appeared a lack of understanding on several types of degradation phenomena and the current investigation is focused on the fundamental mechanism of electrode and electrolyte degradation through simulation on atomic level of the electrolyte and its interface with electrodes. Given the overall progress made to date and the relative simplicity of the HTE system, the high-temperature electrolysis (HTE) has been identified as the leading candidate for hydrogen production in the U.S. DOE's NGNP demonstration plant program [55].

There are several plant options integrating a reactor with hydrogen production process. The plant can dedicate a reactor for steam supply to the electrolyzer while drawing electricity from another nuclear power reactor or from grid including off-peak electricity. Alternatively, a standalone plant can cogenerate steam and electricity from the same reactor and supplies both to the electrolyzer as shown in Figure 10.50. The cogeneration system shown in the figure can be designed to potentially accommodate flexible production modes, including electricity or hydrogen generation and cogenerating both simultaneously.

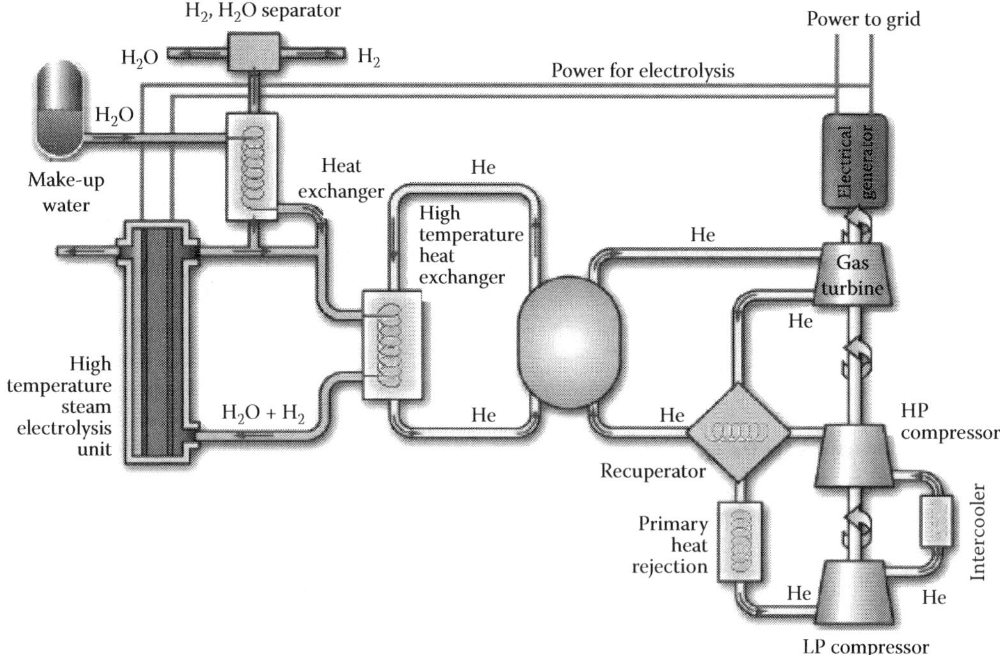

FIGURE 10.50
A conceptual HTGR system for hydrogen cogeneration by HTE of steam.

Furthermore, the system may be designed to follow load in the range of 200–400 MWe with a 300 MWe rated reactor power plant that produces hydrogen to store electricity during low demand periods.

GA estimated the performance of an HTE hydrogen cogeneration plant preconceptual design [56]. The electrolyzer design is a tubular solid oxide cell featuring an Yttria-Stabilized Zirconia (YSZ) electrolyte, a Nickel-YSZ cathode (hydrogen electrodes), and a LSM (Strontium-doped Lanthanum Manganite) anode (oxygen electrodes). The cell operates at 5 MPa and 900°C maximum in an expected life of 5–10 years. A standardized electrolyzer unit is rated at 1.9 MWe and sized in a steel pressure vessel of 2.6 m inner diameter and 4.1 m height. The unit contains 18,000 cells with a production rate of 600 Nm^3/h (54 kg/h) of hydrogen. The plant employs 10 units producing 21.3 MWt thermal energy of total hydrogen based on HHV of hydrogen. The plant process flowsheet was evaluated using the commercial process computational code HYSYS. The cell operating parameters are 1.304 V and at a current density of 0.6 A/cm^2 based on the cell-performance projection. The heat losses associated with process equipment and piping are neglected, losses associated with AC to DC conversion are also neglected, and pressure losses associated with components were assumed to be 1% of the inlet pressures of the components. Electricity is cogenerated at thermal efficiency of 50.5% by DC helium gas turbine in similar configuration of Figure 10.50. The 10 HTE units require a total of 18.7 MWe, of which 0.41 MWe can be met through process heat recovery. A total of 3.59 MWt of hot steam is supplied to the process from the reactor. The overall efficiency of the process is estimated to be

$$\eta_{HTE} = \frac{21.3}{3.59 + (18.7 - 0.41) / 0.505} \times 100 = 53.5\% \text{ (HHV)} \tag{10.6}$$

Flowsheet sensitivity studies predicted that efficiency was only about 1% lower if the cell operating pressure is reduced by 0.5 MPa. Increasing cell inlet temperature from 750°C to 850°C increases the heat duty from 3.42 to 3.67 MWt and simultaneously decreases electric power consumption from 18.81 to 18.56 MWe. This raises overall plant efficiency by about 0.3% over the temperature range. These studies assume that electricity generation efficiency remains constant.

GA foresees that a large commercial plant would consist of four 600 MWt HTGRs in cogeneration of steam and electricity to supply the HTE-based hydrogen production plant having 292 units each consisting of eight modules of planar solid oxide electrolysis cells (SOECs). Assuming such a plant gains maturity of technology, GA estimates that the overall reactor hydrogen production cost would be US\$2.22/kg H_2 [56].

10.4.3 Thermochemical Process

Of the significant number of thermochemical cycles introduced in Chapter 5, thermochemical iodine–sulfur (IS) process has attracted the most interest of development for nuclear hydrogen production. The process splits water molecules into hydrogen and oxygen via three cyclic chemical reactions that are closed in that the process chemicals, sulfur and iodine, are fully recycled. Water goes in the process as the only material feed, and high-temperature heat is theoretically the only energy requirement for the process. In practice, however, electricity that may account for a substantial fraction, say 20%, of total energy input to the process is required to power process equipment such as pumps and gas compressors, and electrolytic fluid processors.

Since 1970s, the United States, Japan, France, Korea, China, Italy, and India have invested in the development of the process. These efforts have resulted in the integrated cycle operation of a variety of process scales as reviewed in Chapter 17. Figure 10.51 shows the milestones accomplished and the next steps in the process development toward the goal of demonstrating nuclear hydrogen production using the HTTR by JAEA.

JAEA combines the HTGR and IS-process as the basis of commercial nuclear hydrogen production in the consideration of the following:

1. The Generation IV HTGR has the design features generally desirable by industrial thermochemical process, including (1) the passive reactor safety that will simplify collocation of the nuclear heat source with a conventional (nonnuclear-grade) industrial process plant, which is a key to affordable product; (2) the thermal rate of the reactor can economically match the typical scale of the industrial applications including transport fuel production, oil refining, fertilizer manufacturing and steelmaking; and (3) the high-temperature heat quality of the reactor supports the hot and heat-intensive endothermic sulfuric acid decomposition reaction of the IS process.
2. The IS process consists simply of three essential chemical reactions, and the high temperature of the process is the basis for high thermal efficiency, which is estimated in the range of 40–50%.
3. The IS process is essentially a thermal fluid process that permits continuous operation and which scales with volume rather than areas, offering the incentive of economy of scale for large-scale nuclear hydrogen production.

Figure 10.52 shows the process schematic of the JAEA's GTHTR300C commercial hydrogen cogeneration plant design. The DC gas turbine generates electricity and circulates reactor coolant, performing both tasks most efficiently of all possible forms of plant arrangement. Hydrogen cogeneration is enabled by adding an IHX in serial between the reactor and the gas turbine. The particular serial arrangement offers several important design benefits. There is only one main coolant loop circulated by gas turbine, eliminating need of separate circulator and motor drive. The cogeneration ratio of hydrogen and power

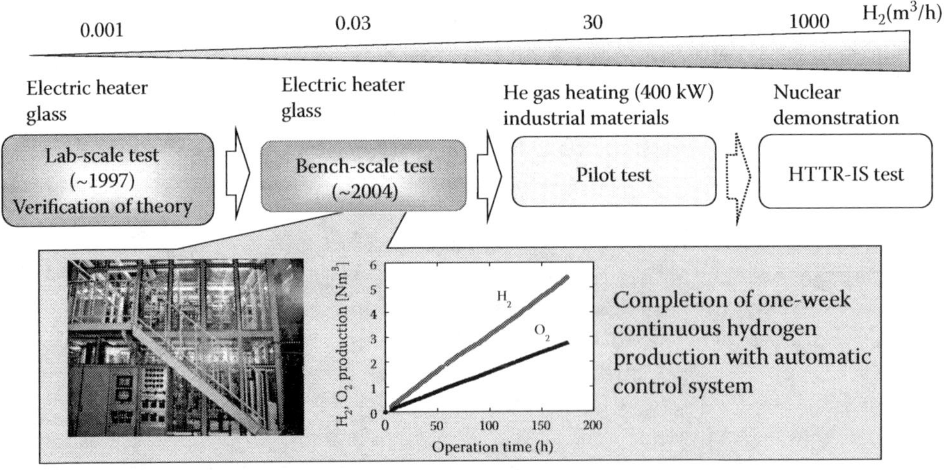

FIGURE 10.51
JAEA's development steps for nuclear hydrogen production based on thermochemical IS process.

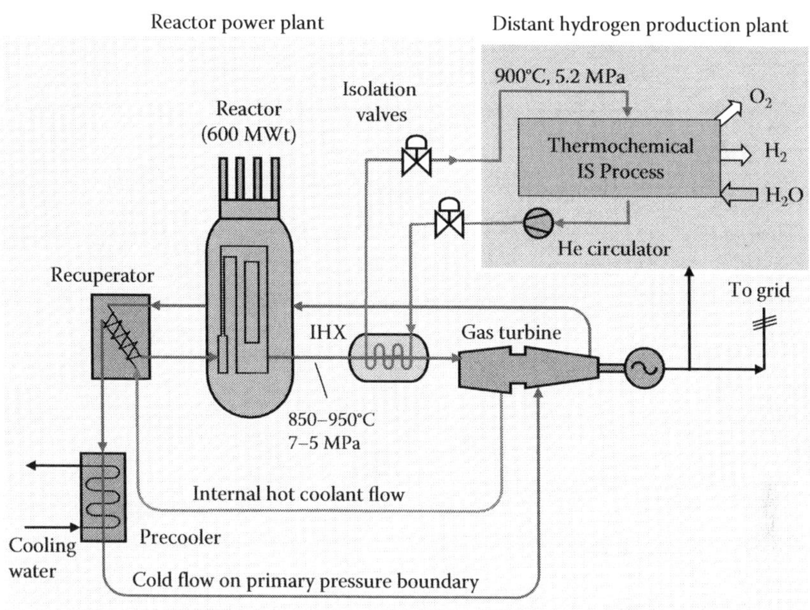

FIGURE 10.52
The HTGR IS thermochemical hydrogen cogeneration process employed in the GTHTR300C.

can be varied with several operational approaches that have been investigated by JAEA, in which power generation efficiency is not compromised up to full thermal power of the reactor. The LMTD is large between the primary and secondary fluids, creating the design condition to yield a small size IHX, which is possible even for a tubular type. The design results of the IHXs for GTHTR300C at two different heat duties are given in Table 10.11.

A secondary helium loop delivers hot helium from the IHX to the hydrogen plant over a sufficient distance. The secondary loop includes the design measures such as the isolation valves to provide for safe physical and material separation between the nuclear plant and the conventional-grade hydrogen plant. Chapter 23 discusses about the physical separation design in the thermal fluid coupling of the reactor to the hydrogen plant. Chapter 24 discusses on the operational control of the power reactor and the hydrogen plant and describes the approach to limiting tritium migration from the reactor primary loop to the hydrogen product.

The "distant hydrogen production plant" included in Figure 10.52 is based on the IS-process, whose detailed flow sheet is shown in Figure 10.53. The flow sheet involves three inter-cyclic thermochemical reactions to dissociate water molecules into hydrogen and oxygen gas products with major heat and minor electricity as energy input and with water as the only material feed. All process materials other than water are reagents.

The exothermic Bunsen reaction produces two aqueous solutions of sulfuric acid and hydriodic acid from material feeds of water, sulfur dioxide and iodine. The reaction favors presence of excess water and iodine to make it spontaneous and with iodine rich hydriodic acid (HI_x) formed to facilitate subsequent phase separation. The excess of water and iodide, however, imposes heavy process stream loads upon subsequent reactions, particularly so in the HI reaction steps. Though not yet reflected in the present flowsheet, improved reaction conditions are being studied with the goal of significantly reducing excessive reactants in order to simplify overall process and production cost.

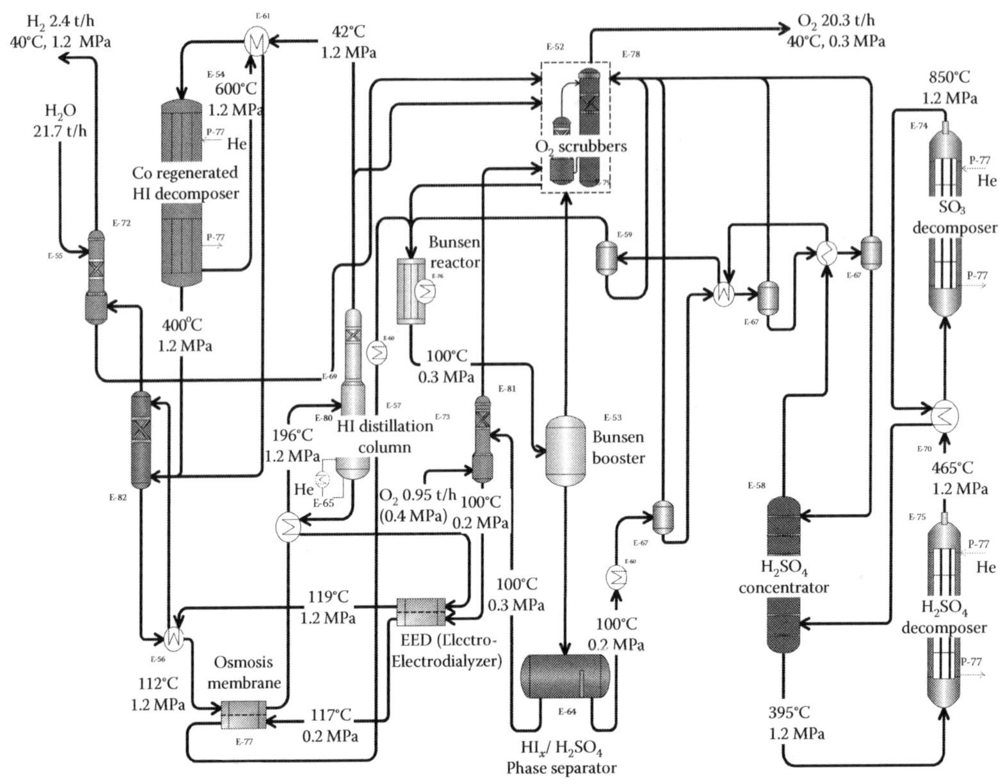

FIGURE 10.53
The initial IS process flowsheet design for the GTHTR300C.

In the endothermic sulfur section on the right side of Figure 10.53, sulfuric acid H_2SO_4 from Bunsen reaction is purified and concentrated before being decomposed in steps into H_2O and SO_3 and then to SO_2 and by-product oxygen gas, involving heat temperatures up to 850°C. The sulfur reaction is relatively well established and the main technical issues are concerned with having decomposers that are sufficiently heat and corrosion resistant. These practical problems are being tackled in industrial trial fabrication of the key component elements complete with strength and performance evaluation.

In the endothermic HI section on the left side of Figure 10.53, hydriodic acid HI_x from Bunsen reaction is concentrated in a number of steps and the resulting hydrogen iodide concentrate is decomposed into reagent iodine and product hydrogen gas. The HI reaction steps appear to have the largest room for process improvement, for which several innovative process techniques have been incorporated in the present flowsheet. The HI concentration steps combine electro-electrodialysis cell and carbonized osmosis membrane to reduce excess iodine and water prior to final distillation. Toshiba Corporation proposed an iodine absorber to be included in the HI decomposer to undergo the following coregenerated process:

$$(1)\ 2HI \rightarrow H_2 + I_2\ (400°C) \tag{10.7}$$

$$(2)\ Co + I_2 \rightarrow CoI_2\ (400°C) \tag{10.8}$$

$$(3)\ CoI_2 \rightarrow Co + I_2\ (600°C) \tag{10.9}$$

$$(4)\ 2HI \rightarrow H_2 + I_2 \tag{10.10}$$

By absorbing product I_2 from reaction (1) in the presence of reaction (2), as high as 80% once-through HI decomposition ratio is achievable in net reaction (4) as has experimentally been observed. The cobalt and iodine are regenerated in endothermic reaction (3).

Figure 10.54 details the heat and mass balance of the flow sheet shown in Figure 10.53, based on the process database best known to the JAEA program. The thermal rate transported by the secondary loop hot helium from the nuclear reactor to the hydrogen plant is used to heat process heat exchangers and decomposers. The electricity power totaling 21.7 MWe is used to power the process HI purification electrolyzers (13 MWe), the secondary helium circulator, the hydrogen process gas circulators and pumps, and other utilities. The process yields a product rate of hydrogen of 26,829 Nm^3/h or 2.4 ton/h, and the by-product rate of oxygen of 13,515 Nm^3/h.

Table 10.15 gives the efficiency values estimated from the heat and mass balance in Figure 10.54. An overall process efficiency is defined as high heating value of total hydrogen produced against total energy consumed. The total energy consumption includes the heat input and the thermal equivalent of electricity input. The net efficiency is seen sensitive to how efficiently the electricity input to the process is generated. The GTHTR300C cogenerates this in-house electricity by DC gas turbine at 47% gross efficiency. The net hydrogen production efficiency by the GTHTR300C is thus 44% based on the flow sheet design in Figure 10.53.

JAEA has continued the extensive engineering studies for the IS process plant design. The existing industrial materials for use in the construction of the high-temperature corrosive acid reactors have been screened with the goal of identifying both durable and

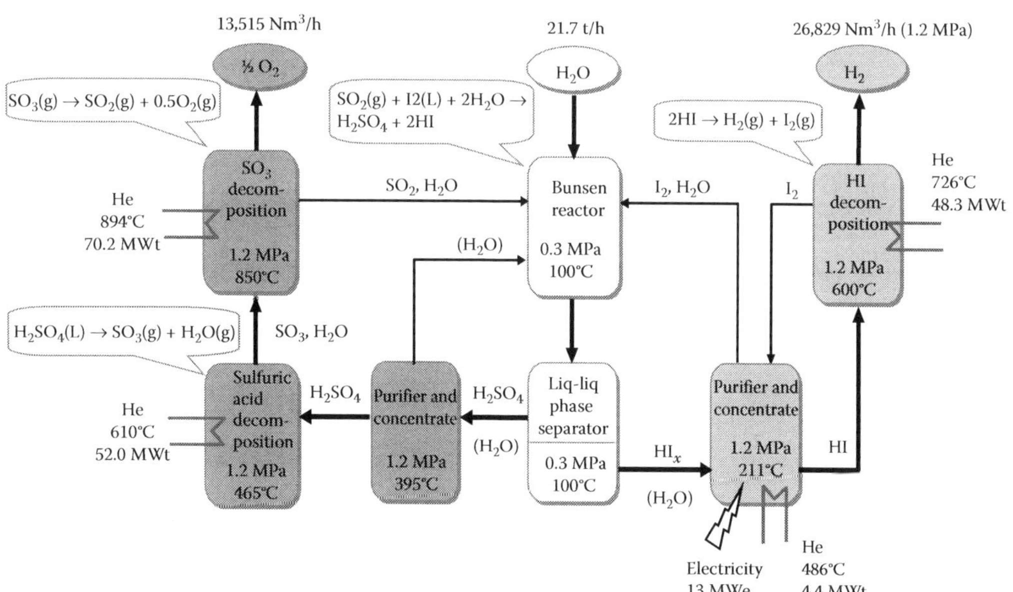

FIGURE 10.54
IS process heat and mass balance.

TABLE 10.15

Thermal Efficiency Estimates for GTHTR300C IS Process Plant

Thermal Heat (MWt)	Electricity Consumed (MWe)	Electricity Cogeneration Efficiency (%)	IS Plant Net Efficiency (%) (HHV)
170.0	21.7	40	42.4
		45	43.6
		50	44.6

low-cost options. Sealing materials and catalysts are also examined. The screening exposes candidate materials to actual plant working conditions and material effects such as stress corrosion cracking and hydrogen embitterment. The development for lining techniques proved essential to obtaining reasonable material cost by separating materials' structural support function from corrosion resistance performance, the latter requires costly materials such as tantalum and ceramics.

The equipment designs have been continually optimized with the objective of substantial simplification and cost reduction to be made to the initial process flow sheet in Figure 10.53. A design proposal intended to improve the Bunsen section by combining the functional equipment including the Bunsen reactor, booster, phase separator, pipes, and pumps into an integral unit shown in Figure 10.55. The production of sulfuric and hydriodic acids is separated into two reaction chambers. The high pressure SO_2 gas is fed in the first chamber to generate turbulent contact in the production of sulfurous acid (H_2SO_3) while a mechanical agitator is used in the second chamber to stir, flow, and contact to finalize production of H_2SO_4 and HI acid. The mixture of acid products will flow on gravity and settle at low velocity into density-differential phases in the tank. The dimensions of the unit indicated in Figure 10.55 are sized for the design conditions of 136 m^3/hr flowrate of the acid products at 100°C and 2 MPa for one-minute duration in each reactor chamber and five-minute duration in the phase separator. The industrial conceptual design for the

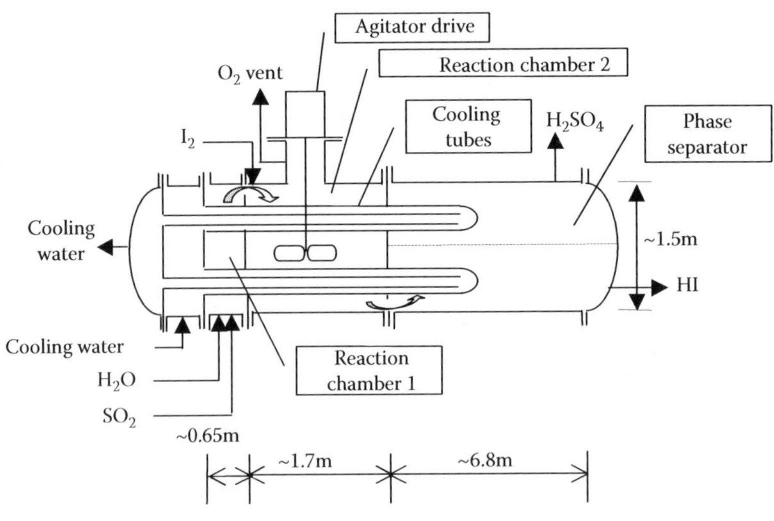

FIGURE 10.55
Schematic of Bunsen reactor and phase separator integrated unit.

integrated Bunsen section unit has been completed, which employs a steel vessel and the tubes that are glass-lined on any surface that contacts corrosive process fluids.

Another practical design concept simplifies the sulfuric acid decomposition section by reducing the functional equipment of the H_2SO_4 decomposer, heater, and SO_3 decomposer, pipes and pumps into an integrated unit shown in Figure 10.56. A significant number of piping and seals are eliminated, and the total use of costly heat and corrosion resistant alloys and ceramic materials used to construct the sulfuric acid decomposition section is reduced. The hot helium enters to flow up in the inner part of the concentric tube while heating and decomposing the downward flow of SO_3 in the annulus of the tube. Helium at a reduced temperature turns to flow down in the tubes of the upper heat exchanger wherein the SO_3 is preheated. Helium leaves the heat exchanger to exit the unit at mid-level. H_2SO_4 liquid enters the shell side of the lower heat exchanger and is decomposed with the heating of hot SO_2 in the tube side. A 5.7 MWt industrial conceptual equipment unit of the integrated sulfuric acid decomposition section is designed as shown in Figure 10.57. Figure 10.58 includes the thermal conditions of the unit. The lower and upper heat exchangers contain tube heat transfer surface area of 195 and 1603 m^2, respectively. The material for the pressure vessel and tubes is alloy 800 H and is lined with SiC to protect it from the corrosive process fluids.

The improvement to the HI section has been made by the conceptual decomposer design shown in Figure 10.59. The design data for the unit are listed in Table 10.16. The novel unit consists of internally sequential stages of the HI decomposing reaction. The process flow scheme is configured to facilitate the inter-stage removal of the reaction product iodine by the scrubbing of the iodine-lean aqueous HIx solution stream that flows counter to the overall HI decomposition solution stream. Figure 10.60 shows that the effect of the number

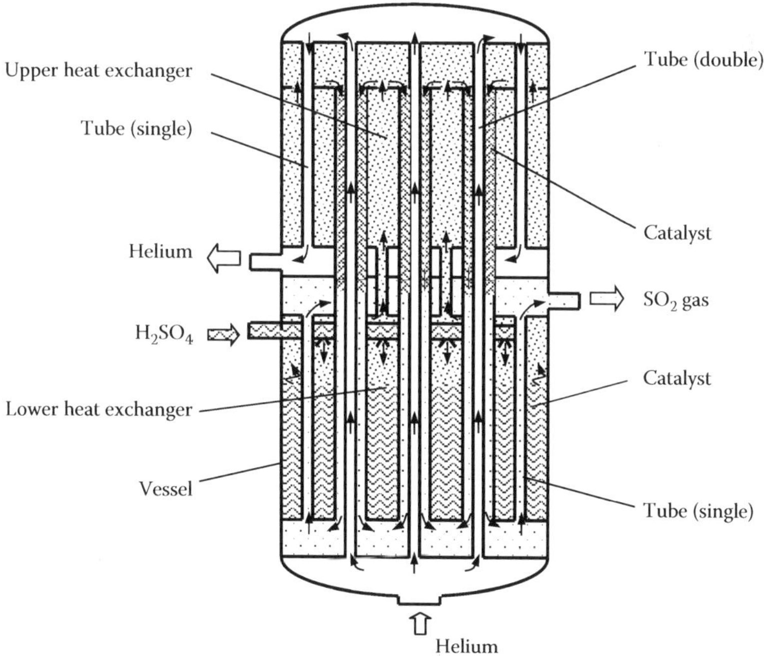

FIGURE 10.56
The integral H_2SO_4 decomposer concept.

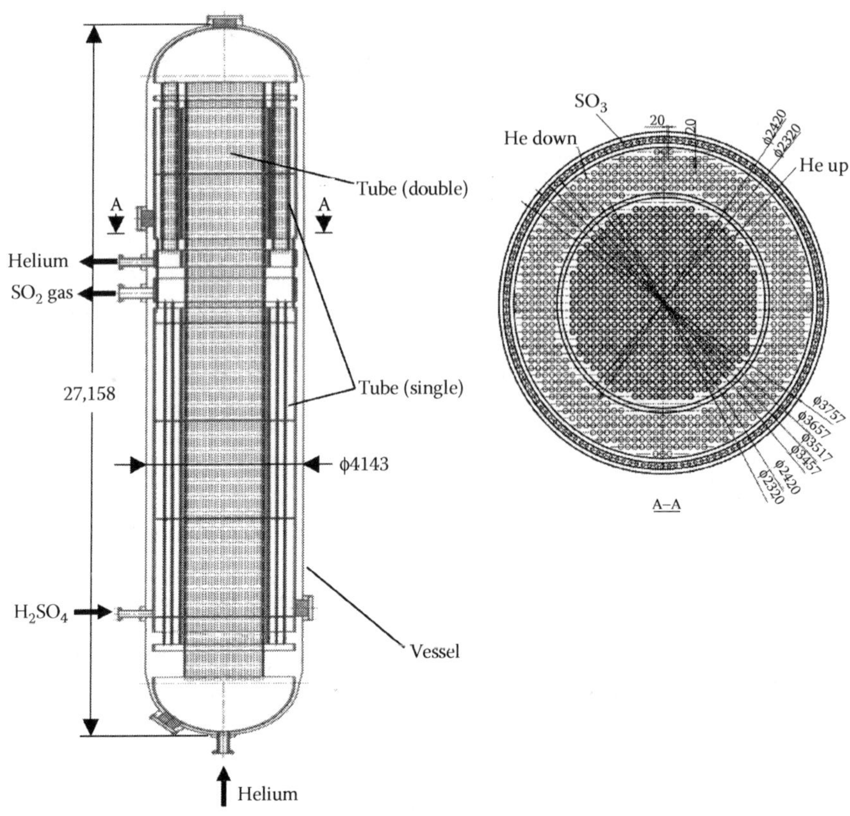

FIGURE 10.57
A 5.7 MWt integrated unit design for H_2SO_4 decomposition.

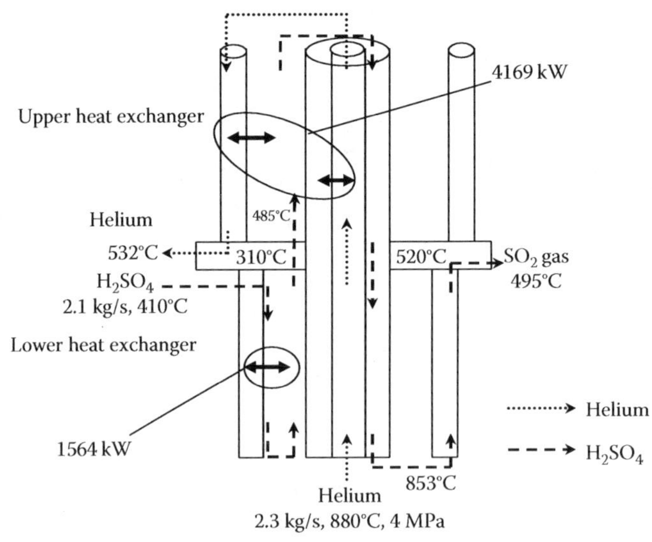

FIGURE 10.58
Heat and mass balance in the 5.7 MWt H_2SO_4 decomposer unit.

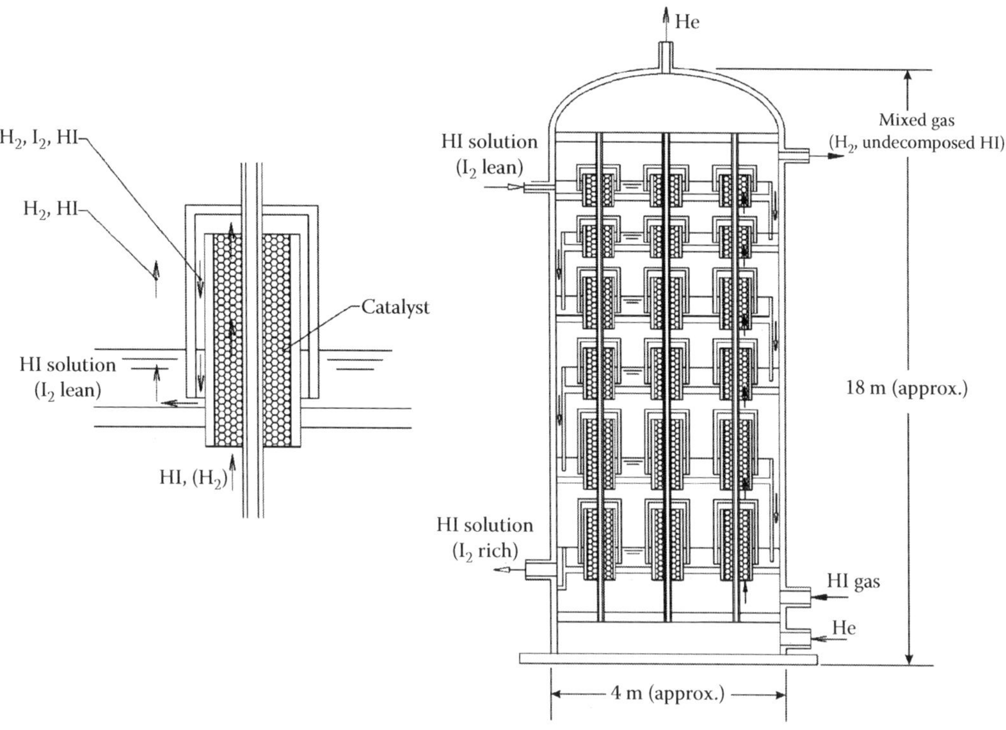

FIGURE 10.59
Schematic of Multistage HI decomposer.

of recurring reaction-removal stages on decomposition rate of HI per pass. The results of the performance analysis indicate that the HI section flow rate can be reduced by more than 30% and the heat exchanging surface area of the section by 50% by the six-stage design unit relative to a single-stage HI section.

The efforts of equipment design optimization discussed in the preceding three paragraphs have produced a much simplified flow sheet in Figure 10.61 from the initial one in Figure 10.53. JAEA plans to evaluate the flow sheet design through the construction and

TABLE 10.16
Major Design Data of the Multistage HI Decomposer

		Pt/γ-Alumina		
Catalyst	Reaction time (kg • s/mol)	60	40	20
	Weight (kg)	3000	2000	1000
	Charging height (m)	1.1	0.7	0.4
	Stage (–)	1–2	3–4	5–10
Reaction tube	Outer diameter (mm)	—	508	—
	Thickness (mm)	—	20.6	—
	Number (each stage) (–)	—	19	—
Reaction vessel	Outer diameter (m)	—	4 (approx.)	—
	Height (m)	—	18 (approx.)	—

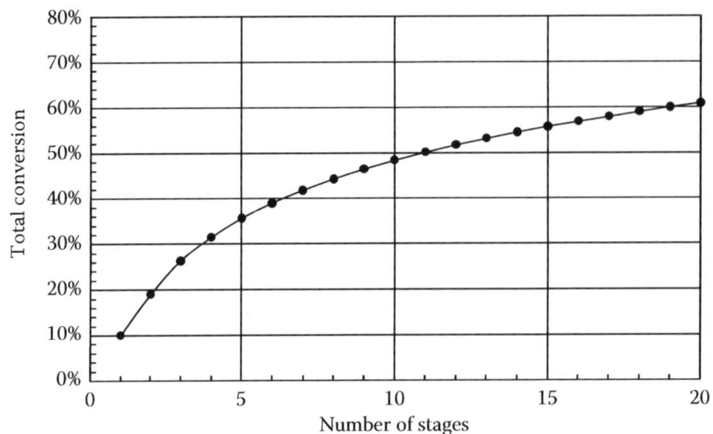

FIGURE 10.60
The estimated conversion (decomposition) rate of the HI multistage decomposer unit design.

operation of an integrated pilot-scale process test, followed by a demonstration of 1000 Nm^3/h-scale IS-process hydrogen production in the HTTR-IS program (Chapter 29).

JAEA with industry collaboration has designed a family of the GTHTR300 commercial plant variants deployable in the near future and capable of producing competitive electricity, hydrogen or variably both, as indicated in Figure 10.62 to adapt to future market demands.

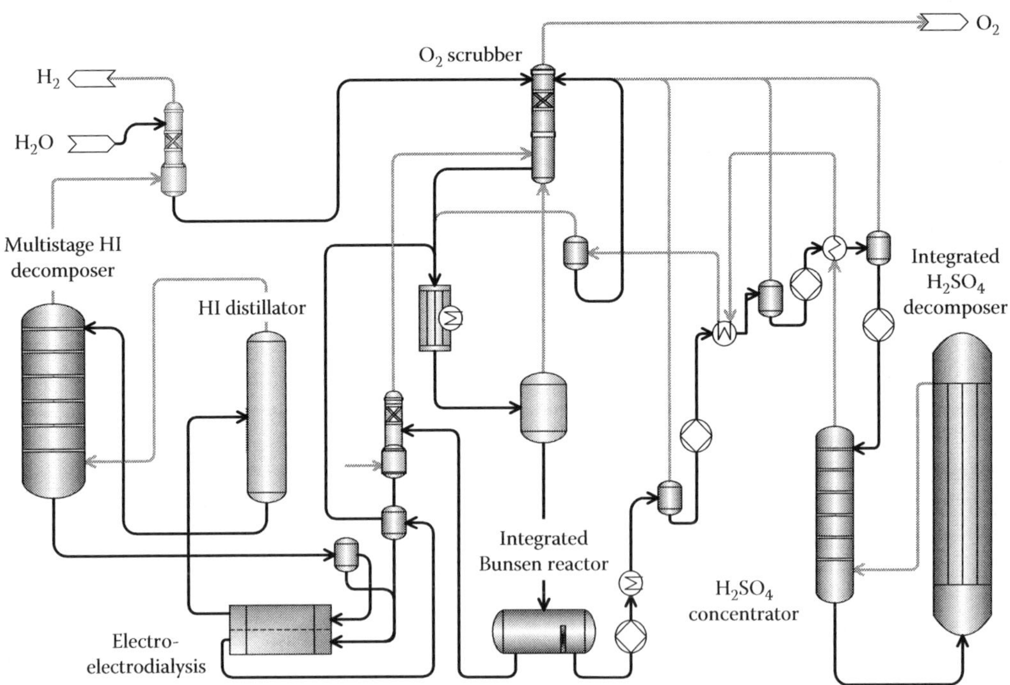

FIGURE 10.61
Schematic of JAEA's optimum IS process flowsheet design for the GTHTR300C.

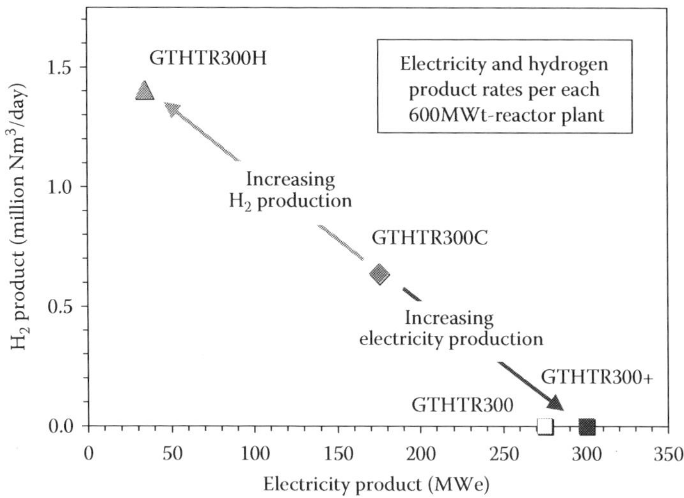

FIGURE 10.62
Variable production ratios of electricity and hydrogen by GTHTR300C.

The baseline system GTHTR300 combines a 600 MWt rated reactor and a DC gas turbine for electricity generation [29]. The reactor coolant temperature of 850°C is selected. The net generating efficiency is 45.6%. The basic design and engineering began in 2001 and completed in 2004. The required component and system technologies up to one-third scale were tested [57]. The plant passive safety performance and cost of electricity target of ¥4/kWh (US¢3.5/KWh) were confirmed. The plant designs that followed include a growth power generation system GTHTR300+ (950°C reactor outlet temperature and 50% generating efficiency) and two hydrogen cogeneration systems, GTHTR300C and GTHTR300H. The "C" design is a substantial cogeneration system, employs the fullest spectrum of common technologies of the GTHTR300 plant family, and is thus suitable for prototypical demonstration. The "H" design yields the highest hydrogen output and is intended mainly for hydrogen production while the limited electricity produced is meant mainly to supply hydrogen plant production and the reactor plant operation. Table 10.17 lists the major design specifications for the GTHTR300 plant design variants.

In the GTHTR300C design point, 170 MWt of the reactor thermal power is extracted via the IHX as 900°C process heat supply to the hydrogen production process and the balance of reactor thermal rate is used for electricity generation. The design development for the hydrogen process flow sheet has been discussed above. A fraction of the 200 MWe electricity generated supplies in-house operations, mainly the hydrogen plant operations to power electrolyzers, circulators, pumps, and other utilities, while exporting a more significant part to grid. The reactor coolant pressure for the cogeneration is 5 MPa, a reduction from the 7 MPa selected for the power reactor systems due to two design considerations. The first is to reduce pressure loads on a cascade of high-temperature heat exchangers including the IHX and chemical reactors to secure economical life time of these cost-intensive components. Although the lower pressure increases the specific size of the gas turbine equipment necessary for cogeneration, the benefits gained in the IHX's life span, cost saving, and for gas turbine technology simplification offer more compelling design advantages. The second consideration is to maintain geometric (thus aerodynamic performance) similarity of the gas turbines to the baseline unit of the GTHTR300, so that there is no need

TABLE 10.17

Major Design Specifications for GTHTR300 Plant Design Variants

	GTHTR300 Power Generation Only	GTHTR300C Power and H_2 Cogeneration	GTHTR300H Mainly H_2 Generation
Reactor thermal power	600 MWt/module	600 MWt/module	600 MWt/module
Reactor lifetime	60 years	60 years	60 years
Plant availability	90% +	90% +	90% +
Reactor fuel cycle	LEU, MOX, others	LEU, MOX, others	LEU, MOX, others
Reactor fuel design	TRISO coated particles	TRISO coated particles	TRISO coated particles
Reactor pressure vessel	SA508/SA533 steel	SA508/SA533 steel	SA508/SA533 steel
Reactor core coolant	Helium gas	Helium gas	Helium gas
Core coolant flow	439/403 kg/s	324 kg/s	324 kg/s
Core inlet temperature	587/663°C	594°C	594°C
Core outlet temperature	850/950°C	950°C	950°C
Core coolant pressure	6.9 MPa	5.1 MPa	5.1 MPa
Core power density	5.4 W/cc	5.4 W/cc	5.4 W/cc
Average fuel burnup	120 GWd/ton	120 GWd/ton	120 GWd/ton
Refueling interval	24/18 months	18 months	18 months
GT conversion cycle	Non-intercooled direct Brayton cycle	Non-intercooled direct Brayton cycle	Non-intercooled direct Brayton cycle
GT cycle pressure ratio	2.0	2.0	1.5
Power generation efficiency	47%/51%	47%	38%
Net electricity output	274/300 MWe	174 MWe	34 MWe
H_2 plant effective heat rate	N/A	220 MWt	505 MWt
H_2 conversion process	N/A	Thermochemical (e.g. S/I) or hybrid (thermal-electro)	Thermochemical (e.g. S/I) or hybrid (thermal-electro)
H_2 conversion efficiency	N/A	43%	41%
H_2 Production	N/A	58 tons/day (0.64 million m^3/day)	126 tons/day (1.41 million m^3/day)
Total plant efficiency (net)	45~50%	45%	40%

to develop separate gas turbine design. The other potential benefit would be to enable the GTHTR300C to generate power in the range of full reactor thermal rate simply by an increase in the primary system coolant inventory. This provides a significant degree of production flexibility, for example, to enable electric load follow and reduce peak electricity demand while producing hydrogen during period of off-peak electricity demand. For example, the results presented below confirm a robust operation capability of the GTHTR300C cogeneration load follow.

Figure 10.63 shows the simulated results of simultaneous follow of 64–100% electric load and 100–0% heat load, achieved at 5% electric load change per minute, the maximum rate that an electric utility would require. Starting from the rated cogeneration point, power generation is increased by increasing primary system coolant inventory, which raises the coolant pressure. This is accompanied by reducing the heat transfer rate from 170 MWt to 0 in the IHX, which is performed by varying the IHX secondary flow circulation with the variable-speed helium circulator (refer to Figure 10.52). This causes the primary exit temperature of the IHX to increase. A bypass control valve is used to maintain turbine inlet temperature constant by directing a colder coolant from the reactor inlet location to mix

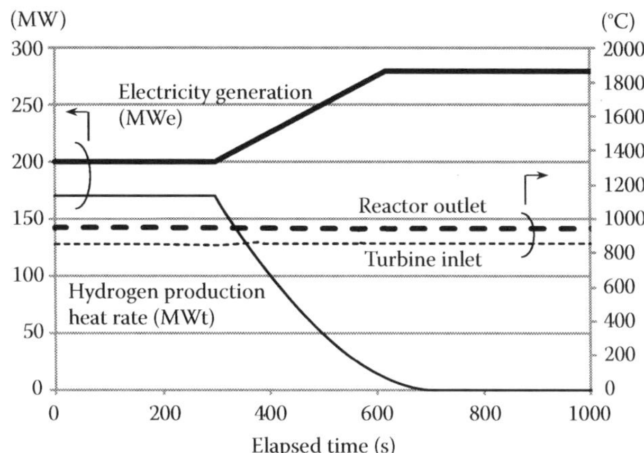

FIGURE 10.63
Simulated results of the GTHTR300C rapid cogeneration load follow.

with the hotter gas exiting the IHX so that the turbine inlet temperature of 850°C is maintained constant as shown. The maximum power sent out to external grid is increased to 276 MWe from 178 MWe, achieved in as little as seven minutes. The reactor coolant pressure increases to 7 MPa from 5 MPa. To return to the rated cogeneration mode, the above process is reversed by reducing coolant inventory in the primary coolant circuit and closing the bypass control valve while increasing the heat load in the process heat exchanger.

The load follow control approach discussed above provides several salient features. First, the joint control of coolant bypass and inventory during load follow can maintain the reactor operation at full power such that the reactor control rod position is not actively changed. This maximizes the reactor operation economics and simplifies the reactor operation control. Second, as shown in Figure 10.63, this control is performed without a change in the reactor outlet coolant temperature, which is the key to technical feasibility of the rapid reactor load follow because fast temperature transients are neither feasible, because of the large thermal inertia of the reactor core, nor desirable, because they would generate excessive thermal stress in the reactor structure and equipment. Third, the control makes no change to the key gas-turbine operating parameters including turbine inlet temperature (shown in Figure 10.63) and pressure ratio such that the aerodynamic performance of both the turbine and compressor remains at their optimum design conditions. This means that the reactor system power generation efficiency remains optimum during load follow.

To develop the multiple GTHTR300 plant variants does not necessarily suggest having the investment and risk multiplied. Rather, the development is minimized thanks to a three-tier SECO (simplicity, economical, competitiveness, and originality) design philosophy. The design simplification builds upon the premise that all plant variants share a common reactor system, an aerodynamically and mechanically similar line of helium gas turbines used for electricity production and the IS process system for hydrogen production in a common plant arrangement shown in Figure 10.64. Since the technologies are shared, the benefit of developing any equipment and system is multiplied. Note that the hydrogen plant and the second heat-transport loop including the IHX, isolation valves, and helium circulator are absent in the plant should electricity be the only product desired. To achieve competitive economics, the original design attributes that are less demanding on the system technologies are adopted, such as, conventional steel RPV, horizontal gas turbine, and

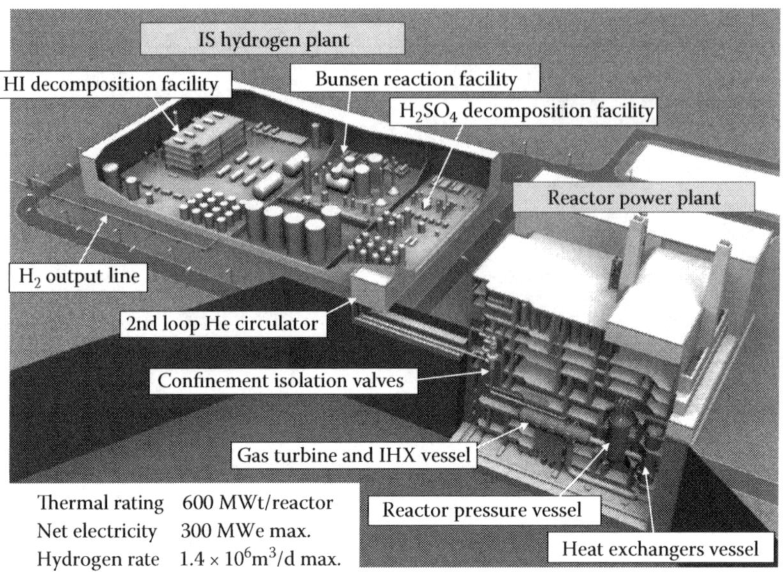

FIGURE 10.64
The GTHTR300 design variants enabling variable cogeneration of electricity by gas turbine in the reactor power plant and hydrogen in the thermochemical IS plant.

distributed modular system arrangement. The pursuit of the technology and design simplification has focused the development activities in JAEA with reduced risk and investment.

10.4.4 Hybrid Process

Hybrid process as introduced in Chapter 5 splits water molecules through a closed series of thermochemical and electrochemical reactions. Water is the only material feed, and heat and electricity are energy input. The hybrid sulfur (HyS) cycle originally invented by Westinghouse is the currently most developed hybrid process. Chapter 18 presents the HyS technology and engineering development in the United States.

The preconceptual reactor plant design NGNP (Figure 10.65) by a Westinghouse-led team selects the HyS [56]. The plant process cogenerating hydrogen is shown in Figure 10.66. The nuclear heat supply system (NHSS) is based on the PBMR (Figure 10.28). Heat is transferred in the two IHXs from the primary heat transport system (PHTS) to the SHTS. The latter transports heat to the HyS HPS and the steam turbine power conversion system (PCS) in serial with the HPS first in the series to be at the highest temperature. A reactor outlet coolant temperature of 950°C and an IHX temperature drop of 50°C are assumed. Gas pressure is 9 MPa in the PHTS and between 8.1 and 8.5 MPa in the SHTS. The rational of selecting a higher pressure in the PHTS is unclear to the authors in this design as the radioactive primary coolant could easily enter the SHTS in case a leak develops in an IHX unit.

IHX-A is designed to operate between 710 and 950°C and to be replaceable because the high-temperature limits lifetime, and IHX-B below 710°C and to last the 60-year plant lifetime. IHX-A is assumed to be made of heat resistant alloy such as Inconel 617 or Haynes 230. Ceramic material is a longer term alternative. IHX-B is only metallic with more material options available. Both IHX sections select the type of PCHE heat exchanger (presented earlier in the chapter) because the tubular type was said by the designer to be too large to

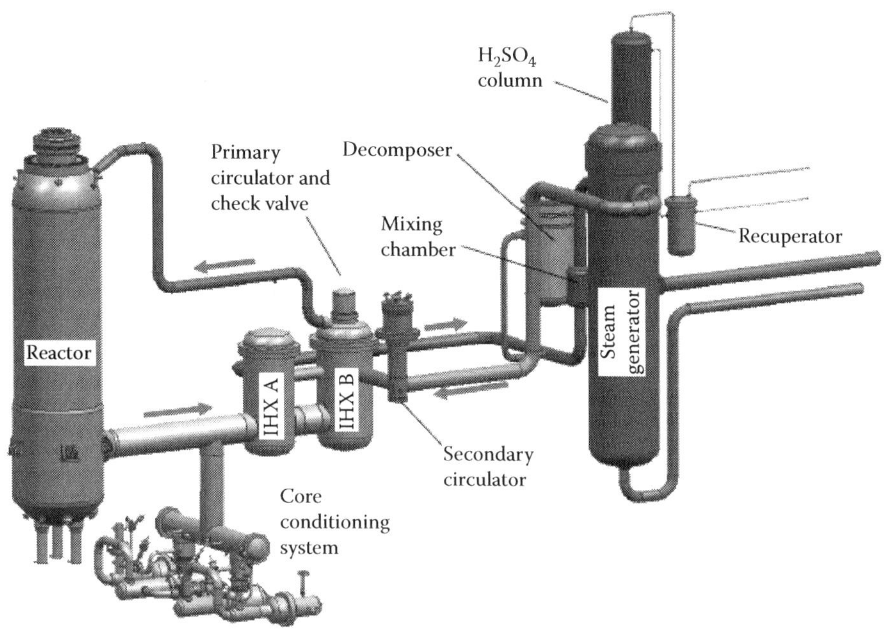

FIGURE 10.65
The plant layout of the pebble bed NGNP reactor for cogeneration of electricity and hydrogen.

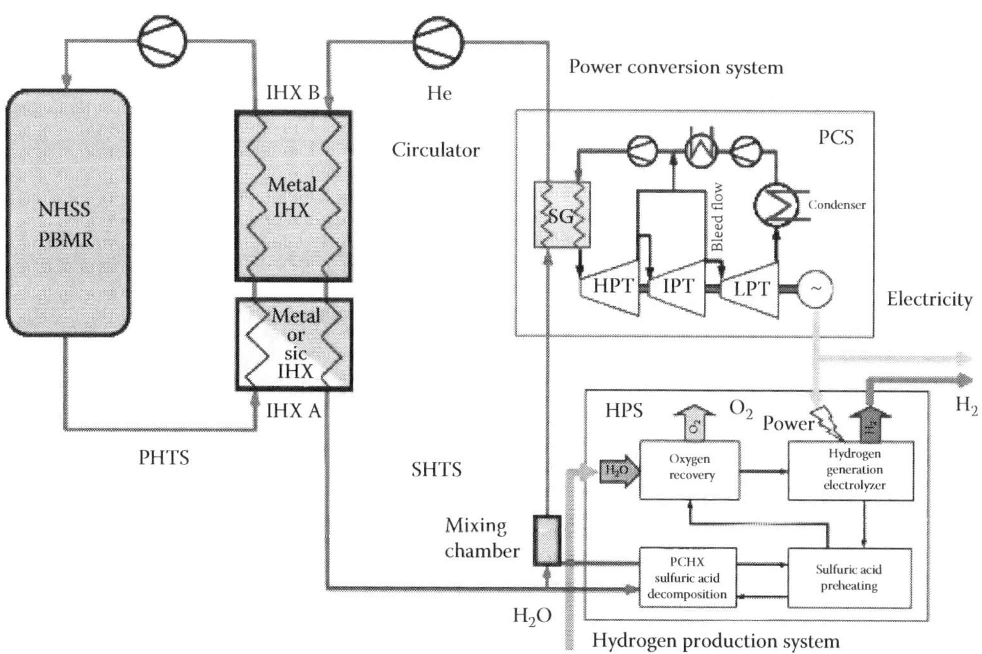

FIGURE 10.66
The pebble bed NGNP hybrid process hydrogen production process.

be economical. On the other hand, the PCHE IHX development is needed to be investigated on the technical issues identified earlier in the chapter.

The Westinghouse team selected the HyS over the HTE because it considered the HyS, although having more technical challenges than the HTE, is likely to have lower capital and hydrogen costs and thus be more commercially attractive. Meeting the HyS technical challenges requires additional component development, studies, and alternative designs. The closed-loop cycle operation of the HyS poses challenges. Selection and development of construction materials are also a major task in the development of this cycle technology.

The HyS cycle consists of two essential reactions. One reaction feeds a solution of sulfur dioxide (SO_2) to depolarize the anode of a water electrolyzer at 80–120°C. The depolarization can significantly reduce the voltage of the electrolysis cell. The voltage accounting for losses is reduced to about 0.6–0.7 V from about 1.7–1.8 V in typical alkaline electrolysis cell. Chapter 18 presents in detail the SO_2 depolarized electrolyzer concept based on a proton exchange membrane and the engineering proposal for a commercial design with graphite and Teflon®-coated steel structures.

The electrolyzer generates the product H_2 gas and a process reagent acid H_2SO_4. The latter is concentrated, as shown by preheating with the recuperated heat, and sent to the other reaction for thermochemical decomposition of the acid at 700–900°C. The thermal energy to this reaction is provided via the helium flow of the SHTS. This reaction also appears in the IS process as detailed earlier in the chapter.

The sulfuric acid decomposition reaction conditions are: the maximum temperature exceeding 850°C, the maximum pressure of about 9 MPa, and the sulfuric acid concentration ranges from 75% to 100% by weight. In these conditions, the sulfuric acid becomes extremely corrosive. In the Westinghouse recuperative decomposer design shown schematically in Figure 10.67, SiC is construction material in all parts that come in contact with acid streams at temperatures greater than 300°C [58]. The sulfuric acid is fed from the top such that the traces of unreacted sulfuric acid in the discharge stream can be captured by the cooled liquid feed. Since the sulfuric acid decomposition rate is 50–75% on a pass, internal capture and recycle lowers recirculation rate of H_2SO_4 stream in overall HyS cycle.

The tubular wall of the ceramic heat exchanger is subjected to pressure difference below 0.3 MPa between the helium and sulfuric acid fluids. Teflon-lined steel is installed in the low-temperature regions of the vessel and supports all the SiC members. All seals are also exposed to low temperatures, allowing for the use of Teflon or similar liner material. The design minimizes cost of construction by using carbon or stainless steel for the pressure boundary of the vessel. For the low-temperature H_2SO_4 wetted streams, Teflon-coated carbon steel may be used. Teflon-coated pipe and corrosion resistant alloys such as zirconium are used in the remainder of the design.

Table 10.18 lists the performance parameters of the Westinghouse H_2SO_4 bayonet tubular decomposer design that is made up of two vessel units for a combined thermal rating of 340 MWt. Table 10.19 provides the construction data of the design. The estimated equipment cost for the two vessel units is US $30 million.

10.4.5 Steam Reforming of Methane

Steam reforming of methane (SRM) is the most economical and widespread process for hydrogen production today. The process requires heat, which is met by combustion of additional methane in a nonnuclear plant. The temperature range of heating the process is 800–850°C, which an HTGR ideally covers. Using nuclear-heat energy source eliminates 37% of methane feed and reduces a similar fraction of CO_2 emission from the conventional

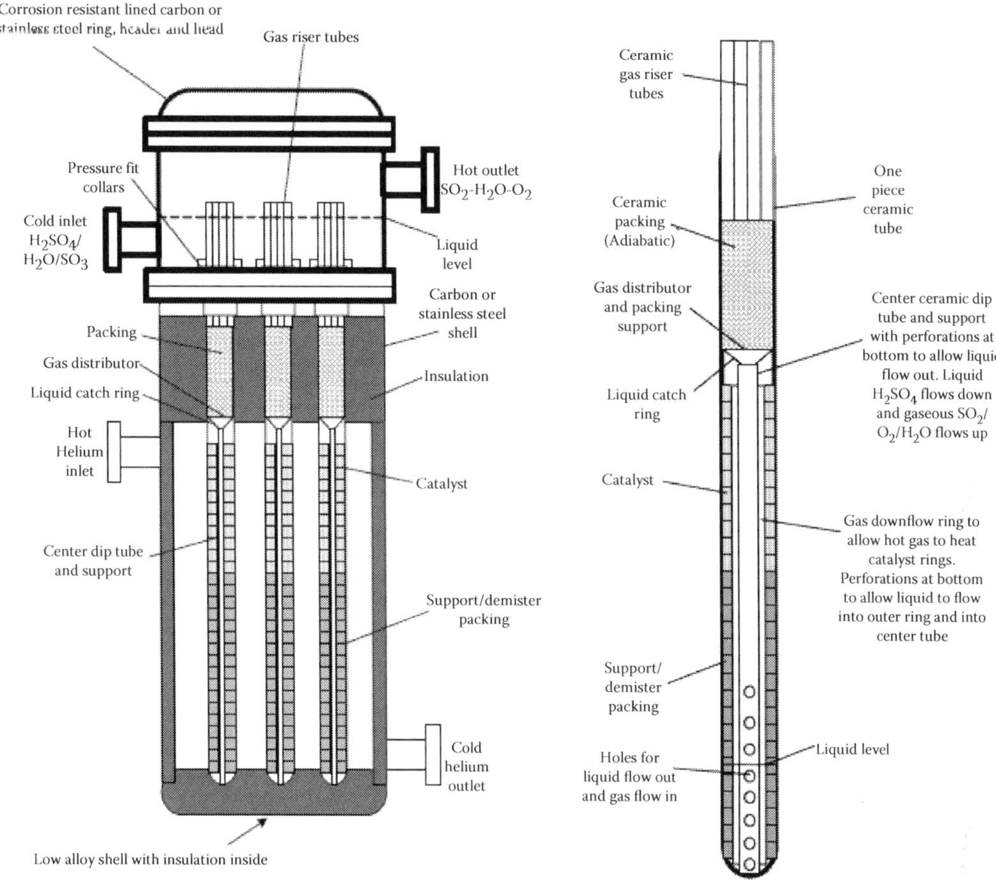

FIGURE 10.67
Integrated recuperative sulfuric acid decomposer design schematic.

TABLE 10.18
H_2SO_4 Decomposer Operating Parameters of HyS Hybrid Process

H_2 generation rate	1.98	kg/s
H_2SO_4 feed solution concentration	75	% by wt.
H_2SO_4 feed solution rate	264	kg/s
Thermal input	388	MWt
H_2SO_4 feed temperature	116	°C
H_2SO_4 product temperature	245	°C
Helium inlet temperature	900	°C
Helium outlet temperature	509	°C
Helium mass flow rate	166	kg/s
H_2SO_4 solution feed rate	0.165	kg/s per bayonet tube
Helium mass flow rate	0.222	kg/s per bayonet tube
Total bayonet tubes	750	Bayonets
Number of reactor vessel	2	Vessels

TABLE 10.19

H_2SO_4 Decomposer Design Specification of HyS Hybrid Process

Bayonet Tube	Material	Length (m)	Inner Diameter (m)	Outer Diameter (m)
Gas riser tube	SiC	0.75	0.0344951	0.042
Concentrator packing	13 mm Berl Saddle	0.23	—	0.103
Bayonet tube	SiC	7.24	0.1032	0.144
Support tube (inner)	SiC	6.7	0.02	0.028
Support tube (outer)	SiC	6.7	0.048	0.056
Gas downflow tube	SiC	6.7	0.061	0.073
Catalyst	Pt	1.20	0.0732	0.103
Support/demister packing	Flexeramic	5.50		
Liquid catch ring	SiC	0.05	0.02	0.1032
Gas distributing plate	SiC	0.01		0.1032
Supporting insulation	ZirCar Ceramics	6.7	0.028	0.048
Teflon coating	Teflon	Surface area	107.03	m^2
Vessel	Material	Height (m)	ID (m)	OD(m)
Gasket	Grafoil		0.1032	0.15
Vessel hemispherical head	Carbon Steel SA 516 70	2.18	4.11	4.36
Vessel plenum	Carbon Steel SA 516 70	1.5	4.11	4.52
Supporting flanged plate	Carbon Steel SA 516 70			4.7
Vessel steel	Carbon Steel SA 516 70	8.74	4.11	4.52
Structured packing	Ceramic	—	—	—
Vessel insulation	BNZ 20 Refractory Brick	8.74	3.65	4.11
Metal supporting spring	Superalloy Inconel 722	—	0.11	0.132

combustion route. The potential cost of hydrogen production estimated for a 400 MWt-scale SRM plant based on an HTGR is estimated to be competitive to the conventional system production [59,60].

Because the HTTR was built with an earlier goal of demonstrating nuclear SRM in mind, JAEA has performed the design evaluation and an out-of-pile pilot test for the system in Figure 10.68. The nuclear heat of 10 MW plant is transferred via the IHX to the secondary helium circuit which is at a slightly higher pressure in order to prevent release of radioactivity from the primary system in case of an IHX tube failure. The reactor provides high-temperature helium gas of 880°C at the inlet of steam reformer. Inside the steam reformer, the helium flows outside the catalyst tubes transferring heat by forced convection flow. The catalyst tubes contain pellets of Ni/A1203 reforming catalysts, through which the process feed gases of methane and steam are routed. The helium enters the steam reformer at the bottom, flows upwards outside the catalyst tubes, squeezed by multiple plates of orifice baffles, and finally exits at a temperature of 585°C. The process feed gas mixture preheated to 450°C at a pressure of 4.5 MPa enters the steam reformer at the top and then flows downwards in an annular flow between the walls of outer and inner tube through the catalyst bed, where the methane and other lighter hydrocarbons together with steam are reformed. The flow rate of methane feed is 1290 kg/h and that of steam is 5160 kg/h at the inlet of the steam reformer. The reformed gas having reached a maximum temperature of 830°C then flows upwards inside the inner tubes at the same time transferring heat to the feed gas, and eventually leaving the steam reformer at a temperature of 580°C and a pressure of 4.1 MPa. This gas is cooled down by the water cooler and separated into steam and dry gas compositions including hydrogen, carbon monoxide, carbon dioxide, and residual methane

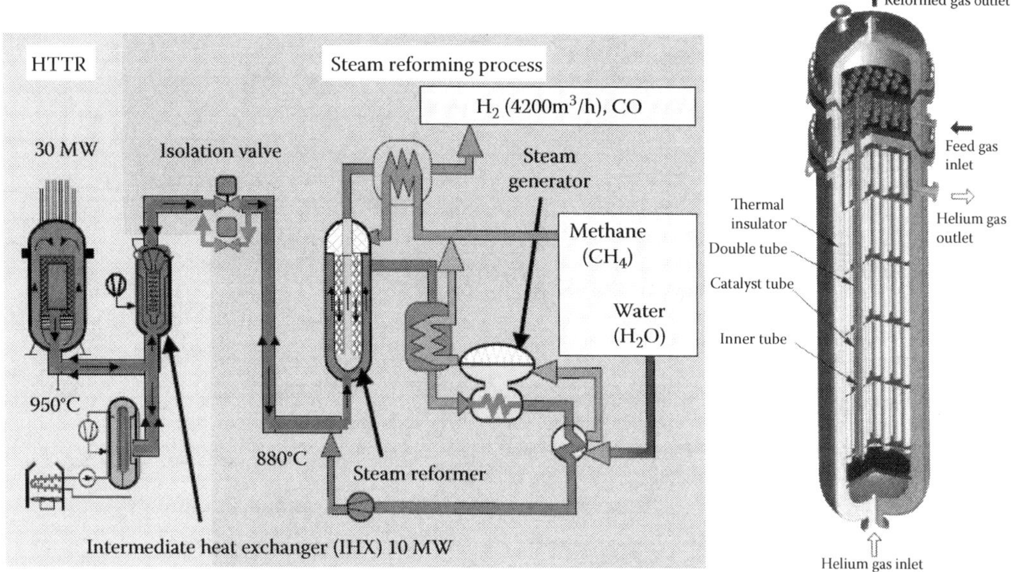

FIGURE 10.68
The HTTR demonstration system of nuclear-steam reforming of methane.

in the separator. The conversion ratio from methane to hydrogen is expected to be 68% in the system and 80% in a larger-scale commercial system.

A reactor safety regulation in Japan stipulates that the ultimate heat sink of an NPP (nuclear power plant) is limited to dedicated cooling system, not to electricity or chemical energy that the NPP produces. Therefore, it is feasible in the nuclear safety concept that the production process should be designed to be conventional to reduce construction and operation cost. In case of a reactor scram, the reactor power falls immediately to a low level, and the abrupt cut in heat input to the SRM is usually followed by an instantaneous disconnection of the feedstock supply and the filling of the process loop with nitrogen to prevent carbon from deposition on the reformer catalyst. The thermal transient in the secondary loop would cease.

On the other hand, a change in the feed rate of either methane or water to the reformer induces a thermal disturbance of the helium outlet temperature from the reformer due to energy imbalance in the reforming reaction. If the temperature of helium returning to the IHX exceeds the allowable limit, the reactor would scram, which is unacceptable because of the high probability of a malfunction or failure to be expected on the process side. The potential disturbances originated from the HPS should be mitigated to avoid reactor scram. A proposed control scheme works as follows: In case of a methane supply malfunction, the helium temperature at the steam reformer outlet would increase. A steam generator installed downstream of the steam reformer in the secondary loop has the thermal capacity large enough to keep the helium gas flow through it from exceeding the saturation temperature of the steam or the rated inlet helium temperature of the IHX secondary side, thus preventing a reactor scram. In case of a loss of feed water to steam reformer, the hot helium gas is cooled by the same steam generator and the generated steam is directed to a natural air cooler for thermal discharge and steam condensation. The condensed water is then recycled to the steam generator as feed water. The steam generator is designed to maintain the water inventory constant during normal operation and have a heat capacity of about 8.8 MW totaling the combined duties of the reformer, super heater, and steam generator.

The control scheme has been tested in an out-of-pile facility. The performance of the reforming HPS has also been tested in the same facility. Chapter 20 provides details of the test. In a particular test case of a loss of methane feed to the steam reformer, the helium temperature measured at the steam generator outlet fluctuates within the range of – 5.5 to +4.0 K, which falls in the specified range of ±10°C for the HTTR nuclear SRM design specification.

Safety aspects such as fire and explosion hazard of flammable gases present near the reactor building have been investigated in the HTTR program, providing the basis for a comprehensive safety analysis of the nuclear/chemical facility (see Chapter 23).

10.5 Summary and Future Outlook

The worldwide development for the HTGR in half a century, including construction and operation of a number of reactors, has advanced the technology that it is possible to develop and deploy the Generation IV HTGR systems in the very near future. In fact, the test reactors are being operated in China and Japan. The HTTR in Japan successfully completed the long-term operation test in high coolant temperature (>930°C) and full 30 MWt power, in which the fuel performed better than the design expectation. Several countries have projects underway to commercialize the Generation IV systems, which include the mission of hydrogen production. These modular and passively—safe reactors are economically sized (100–600 MWt) to provide independent energy for large industry users (fertilizers, steelmaking, oil refining, petrochemical), distributed energy in remote areas, or centralized large plant by combining several reactor modules to serve power grids and in densely populated regions.

The high coolant-outlet temperature of 950°C as demonstrated in the HTTR expands the nuclear reactor technology reach beyond generation of electricity. It is an attractive source for large-scale production of hydrogen as transportation fuel and industrial feedstock. Several HTGR-based hydrogen production processes have been discussed, including electrolysis of water and steam, thermochemical and hybrid processes to decompose water and steam reform of methane.

Table 10.20 summarizes the features of several commercial plant designs that include hydrogen production. As discussed in the chapter, these designs offer a wide range of components and system design options to suit the demonstration and economic objectives of commercial plants. The technical risks in these plants are managed with the selection of the particular process configurations and are being addressed by those developers identified in the table.

Three designs are related to the NGNP project in the United States. The project includes two phases with Phase 1 comprised of research and development, conceptual design, and development of licensing requirements and Phase 2 comprised of detailed design and license review, which will lead to construction in 2017 and startup operation by 2021. The NGNP up-to-date development is reported in Chapter 31. Other national and regional development roadmaps for nuclear hydrogen production exist in the world, some of which are presented in Section V.

Figure 10.69 presents the technical roadmap of nuclear hydrogen production development in Japan. JAEA has formulated its Nuclear Energy Vision 2100, which presents the technical feasibility and nuclear fuel sustainability toward a low-carbon society. The

TABLE 10.20

Proposed Hydrogen Cogeneration Plant Configuration and Operating Conditions

	Westinghouse NGNP Steam Cycle	AREVA NGNP Combined Cycle	General Atomics NGNP Direct Cycle	JAEA GTHTR300C Direct Cycle
Power level	500 WMt	565 MWt	550–600 WMt	600 MWt
Reactor core design	Pebble bed	Prismatic	Prismatic	Prismatic
Reactor outlet temperature	950°C	900°C	Up to 950°C	950°C
Reactor inlet temperature	400°C	500°C	490°C	594°C
Power generation cycle	Indirect cycle steam turbine	Indirect combined cycle	DC gas turbine; direct combined cycle	DC gas turbine
Hydrogen generation	Hybrid cycle	Initial: THE longer term: IS	Initial: THE longer term: IS	IS
Cogeneration cycle	Serial power + hydrogen	Parallel power and hydrogen	Parallel power and hydrogen	Serial power + hydrogen
Power conversion level	100% Reactor power	100% Reactor power	100% Reactor power	100% Reactor power
Hydrogen plant power	10% Reactor power	10% Reactor power	5 MWt—THE; 60 MWt—IS	35% Reactor power
Secondary fluid	He	He-N_2 to PCS; He to H_2 plant	He	He
Fuel	SiC TRISO UO_2	SiC TRISO UCO	SiC TRISO variable	SiC TRISO UO_2
RPV design	Actively cooled by primary coolant	Not cooled; potentially insulated	Not cooled	Passively cooled by primary coolant
RPV material	SA533 (LWR steel)	9Cr-1 Mo	2 ¼Cr – Mo; 9Cr-1Mo	SA533 (LWR steel)
Reactor cavity cooling	Water jacket	Water jacket	Natural convection	Natural convection
IHX power	100% Reactor power	110% Reactor power	10% Reactor power	30% Reactor power
IHX design	PCHE Inconel 617	Power IHX: helical coil tube H_2 IHX: PCHE or plate-fin	PCHE Inconel 617	Helical coil tube Hastelloy-XR

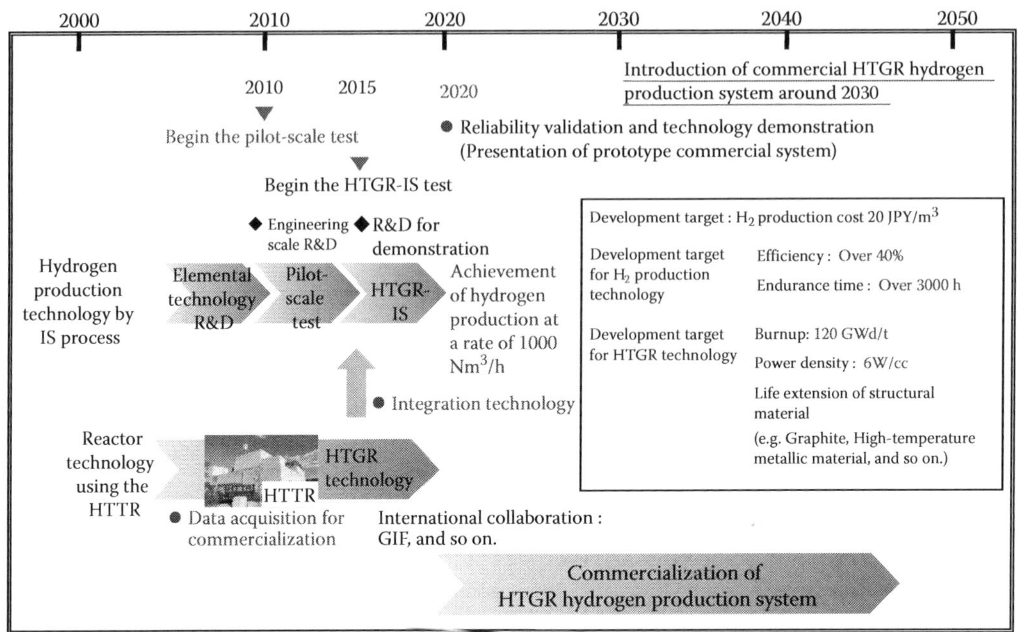

FIGURE 10.69
Technical Development Roadmap of Nuclear Hydrogen Production by the Atomic Energy Commission of Japan (July 2008).

Present (2010) → Future (2020–2030)

JAEA's HTTR technology → HTGR commercial ready technology

	Technology items	Present status	Development pathway	Commercial requirement
Reactor	Fuel	Core power density: 2.5 W/cc * Fuel burnup: 90 GWd/t	HTTR irradiation; Other test reactors	6W/cc 120GWd/t
Reactor	Graphite	Lifetime: 3 years	Nondestructive ISI; HTTR and Joyo irradiation	Extended lifetime of 6 years
Reactor	Metal	Lifetime: 20 years	HTTR irradiation; Extended creep data	Extended lifetime of 30 years
Reactor	Control rod	Rod cladding: Alloy 800H To be replaced after exposure to high temperature in accident	C/C composite rod cladding; HTTR irradiation	Reusable after exposure to high temperature in accident
BOP	Hydrogen production	Technology validation	Engineering validation; HTTR–IS nuclear hydrogen plant demonstration	$0.2/m³ ROM of hydrogen production cost; Non-nuclear hydrogen plant design
BOP	Gas turbine power conversion	30 MWe scale validation	Prototype validation Hot function test P	50–300MWe

FIGURE 10.70
Major development pathway moving from the present technology basis (JAEA's HTTR) to future HTGR commercial-ready technology for nuclear hydrogen production.

technical feasibility is based on the present status of the itemized technologies of reactor and BOP in Figure 10.70, which JAEA has obtained through several long-term research and development programs in the HTTR Project. The commercial technology requirements are fully satisfied by extending the HTTR technologies, that is, without a need of new technology, through the development pathway identified. It is technically feasible that the commercial technology can be developed in 2020s. The nuclear fuel sustainability for the HTGR is based on the flexible fuel cycle options as discussed earlier in Section 10.1.2.3. The Nuclear Energy Vision 2100 calls for installation of 120 commercial HTGRs (a total of 72 GWt) in Japan, beginning in 2030, for use in the production of hydrogen to meet the projected demand in the nation's transportation, residential, and industrial sectors and to contribute to achieve a national CO_2 reduction target of 50% and 90% by the year 2050 and 2100, respectively, from the year 2000 level.

References

1. Evaluation of the gas turbine modular helium reactor, U.S. DOE-GT-MHR-100002, February 1994, prepared by GCRA for the U.S. Department of Energy.
2. Williams, P.M. et al., MHTGR development in the United States, *Progress in Nuclear Energy*, 28(3), 265–346, 1994.
3. Yan, X. et al., GTHTR300 design variants for production of electricity, hydrogen or both, In *Proc. of the OECD/NEA 3rd Information Exchange Meeting on the Nuclear Production of Hydrogen*, OECD/NEA, p. 121, 2005.
4. Matzner, D., PBMR Project Status and the Way Ahead. *Proc. of the 2nd International Topical Meeting on High Temperature Reactor Technology*, Beijing, September 22–24, 2004.
5. A Technology Roadmap for Generation IV Nuclear Energy Systems, GIF-002–00, U.S. DOE Nuclear Energy Research Advisory Committee and the Generation IV International Forum, December 2002.
6. Energy balances of OECD countries (2009 edition)/*Energy Balances of Non-OECD Countries (2009 edition)*, International Energy Agency, ISBN: 978-92-64-06120-0, OECD, 2009.
7. Takei, M. et al., Economical evaluation on gas turbine high temperature reactor 300 (GTHTR300), *Trans. At. Energy Soc. Japan*, 5(2), 109–117, 2006.
8. Greneche, D., HTR fuel cycles: A comprehensive outlook of past experience and an analysis of future options. In: *Proceedings of the ICAPP* 2003.
9. Nickel, H. et al., Long time experience with the development of HTR fuel elements in Germany, *Nucl. Eng. Des.* 217, 141–151, 2002.
10. Richards, M. et al., VHTR deep burn applications, *16th Pacific Basin Nuclear Conference (16PBNC)*, Aomori, Japan, Oct. 13–18, 2008, Paper ID P16P1238.
11. Kuijper, J.C. et. al., HTGR reactor physics and fuel cycle studies, *Nucl. Eng. Des.*, 236, 615–634, 2006.
12. Oak Ridge National Laboratory, Deep burn: Development of transuranic fuel for high-temperature helium-cooled reactors, *ORNL/TM-2008/205*, November 2008.
13. Greneche, D. and Szymczakb, W.J., The AREVA HTR fuel cycle: An analysis of technical issues and potential industrial solutions, *Nucl. Eng. Des.*, 236, 635–642, 2006.
14. Masson, M. et al., Block-type HTGR spent fuel processing: CEA investigation program and initial results, *Nucl. Eng. Des.*, 236, 516–525, 2006.
15. Wu, Z. et al., The design features of the HTR-10, *Nucl. Eng. Des.* 218, 25–32, 2002.
16. Saito, S. et al., Design of high temperature engineering test reactor (HTTR), Technical report JAERI-1332, September 1994, Japan Atomic Energy Research Institute, Tokai-mura, Japan.

17. Ueta, S. et al., Fabrication of uniform ZrC coating layer for the coated fuel particle of the very high temperature, *J. Nucl. Mater.*, 376, 146–151, 2008.
18. Tochio, D. et al., Evaluation of fuel temperature on high temperature test operation at the high temperature gas-cooled reactor HTTR, *Trans. At. Energy Soc. Japan*, 5(1), 57–67, 2006; Takamatsu, K. et al., High temperature continuous operation in the HTTR (HP-11)—Summary of the test results in the high temperature operation mode, JAEA-Research 2010-038, Japan Atomic Energy Agency, November 2010.
19. Sawa, K. et al., Investigation of irradiation behavior of SiC-coated fuel particle at extended burnup, *J. Nucl. Tech.* 142, 250, 2002.
20. Sawa, K. and Ueta, S., Research and development on HTGR fuel in the HTTR project, *Nucl. Eng. Des.*, 233, 163–172, 2004.
21. Katanishi, S. et al., Feasibility study on high burnup fuel for gas turbine high temperature reactor (GTHTR300) (I), *Trans. At. Energy Soc. Japan*, 1(4), 373–383, 2002.
22. Katanishi, S. et al., Feasibility study on high burnup fuel for gas turbine high temperature reactor (GTHTR300) (II), *Trans. At. Energy Soc. Japan*, 3(1), 67–75, 2004.
23. Grover, B. AGR-1 irradiation results, VHTR R&D FY10 Technical Review Meeting, Denver, CO, April 27–29, 2010.
24. Demkowicz, P. AGR-1 post-irradiation examination and safety testing preparations and planning, *VHTR R&D FY10 Technical Review Meeting*, Denver, CO, April 27–29, 2010.
25. JAERI-Data/Code 2004–012, JAERI 2004.
26. Lathrop, K.D. and Brinkley, F.W., TWOTRAN-II: An interfaced exportable version of the TWOTRAN code for two dimensional transport, LA-484-MS, 1973.
27. Fowler, T.B. et al., Nuclear reactor core analysis code CITATION, ORNL-TM-2496, ORNL, 1971.
28. JAERI-Data/code 94–007, JAERI 1994.
29. Yan, X. et al., GTHTR300 design and development, *Nucl. Eng. Des.* 222, 247–262, 2003.
30. Gauthier, J. et al., Antares: The HTR/VHTR project at Framatome ANP. in *Proc. 2nd International Topical Meeting on High Temperature Reactor Technology*. Beijing, September 22–24, 2004.
31. Ohashi, K. et al., Conceptual core design study of the very high temperature gas cooled reactor (VHTR) upgrading the core performance by using multihole-type fuel, *Trans. At. Energy Soc. Japan*, 7(1), 32–43, 2008.
32. Yan, X. et al., A study of air ingress and its prevention in HTGR, *Nucl. Technol.*, 163, 401–415, 2008.
33. Yurko, J. et al., Effect of helium injection on natural circulation onset time in a simulated pebble bed reactor, *Proc. 4th International Topical Meeting on High Temperature Reactor Technology (HTR2008)*, HTR2008–58230, Washington, DC, USA, 2008.
34. Reutler, H. and Lohnert, G.H., The modular high-temperature reactor, *Nucl. Technol.* 62, 22–30, July 1983.
35. Wang, D.Z., Untersuchung eines Hochtemperatarreaktors mit einem Mittleren Graphitkugel-Bereich Jul-1809-RG 1982.10.
36. Zhang, Z. et al., Current status and technical description of Chinese 2×250 MWt HTR-PM demonstration plant, *Nucl. Eng. Des.*, 239, 1212–1219, 2009.
37. Next Generation Nuclear Plant Project—Preliminary Project Management Plan, INL/EXT-05-00952 Rev.1 Idaho National Laboratory, March 2006.
38. Matzner, D., PBMR project status and the way ahead. *Proc. of the 2nd International Topical Meeting on High Temperature Reactor Technology*, Beijing, September 22–24, 2004.
39. Yan, X. et al., Cost and performance design approach for GTHTR300 power conversion system, *Nucl. Eng. Des.*, 226, 351–373, 2003.
40. Weisbrodt, I.A., *Summary Report on Technical Experiences from High Temperature Helium Turbomachinery Testing in Germany*, IAEA, Vienna, Austria, 1994.
41. Zenker, P., 10 Jahre Betriebserfahrung mit der Heliumturbinenanlage Oberhausen, VGB Kraftwerkstechnik, Issue 7/1988.

42. Yan, X. et al., Aerodynamic design, model test and CFD analysis for multistage axial helium compressor, *J. Turbomachinery*, 130/031018, 1–12, 2008.
43. Status of metallic materials development for application in advanced high-temperature gas-cooled reactors, *Nucl. Technol.*, Vol. 66, No. 1: Material Selection (July); No. 2: Properties (August); No. 3: Performance; (September) 1984.
44. Nickel, H. et al., Development and qualification of materials for structural components for the high temperature gas-cooled reactor, *Nucl. Eng. Dec.*, 121, 183–192, 1990.
45. Tachibana, Y. and Iyoku, T., Structural design of high temperature metallic components, *Nucl. Eng. Dec.*, 233, 261–272, 2004.
46. Inagaki, Y. et al., R&D on high temperature components, *Nucl. Eng. Des.*, 233, 211–223, 2004.
47. Kato, R., Design of the intermediate heat exchanger for the high temperature gas cooled reactor hydrogen cogeneration system (I), *Trans. At. Energy Soc. Japan*, 6, 2, 141–148, 2007.
48. INTERATOM, Nukleare Prozesswärmeanlagen mit Hochtemperaturreaktor-Modul zur Kohleveredlung, Technical Report ITB 78.10122.8, Kraftwerk Union AG, INTERATOM/GHT, Bergisch-ladbach, Germany, 1983.
49. Kanzaka, M. et al., Fundamental study on high temperature heat transfer tube for IHX of HTGR, High temperature Heat Exchangers, pp. 433–444, Hemisphere Publishing Corporation, New York, NY, 1986.
50. Shimizu, A. et al. Recent research and development of intermediate heat exchanger for VHTR plant, Mitsubishi Heavy Industries, Ltd. Japan, IAEA Specialists' Meeting on heat exchanging components of gas-cooled reactors, Germany April 16–19, 1984.
51. Tokunaga, K. et al., Technology development of high efficiency and high capacity gas to gas heat exchanger which is necessary to practical application of gas-cooled reactor. In *Proc. GLOBAL 2005*, Tsukuba, Japan, October 9–13, 2005.
52. Mizokami, Y. et al., Development of plate-fin heat exchanger for intermediate heat exchanger of high temperature gas cooled reactor, *AESJ Trans.* 9, 2, 219–232, 2010.
53. Wieting, A.R., Empirical correlation for heat transfer and flow friction characteristics of rectangular offset-fin plate-fin heat exchangers, *J. Heat Transfer*, 97, 488–490, 1975.
54. Anna Swalla, Advanced Alkaline Electrolysis, U.S. DOE Hydrogen Program FY 2008 Annual Progress Report Section II.B.7.
55. Herring, S. *Hydrogen Production Methods, VHTR R&D FY10 Technical Review Meeting*, Denver, CO, April 27, 28, 29, 2010
56. Next Generation Nuclear Plant Pre-Conceptual Design Report, INL/EXT-07-12967, Revision 1, November 2007.
57. Kunitomi, K. et al., Japan's future HTR–the GTHTR300. *Nucl. Eng. Des.*, 233, 309, 2004.
58. Lahoda, E.J. et al. Optimization and costs of the Westinghouse Hybrid Sulfur Process with the PBMR for the production of hydrogen, Paper P16P1241 in *Proc. 16PBNC*, Aomori, Japan, October 13–18, 2008.
59. Shiina, Y., Sakuragi, Y., and Nishihara, T., Cost estimation of hydrogen and DME produced by nuclear heat utilization system, *JAERI-Tech* 2003-076, 2003.
60. Shiina, Y. and Nishihara, T. Cost estimation of hydrogen and DME produced by nuclear heat utilization system II, JAERI-Tech 2004-057, 2004.

11

Sodium Fast Reactor

Takamichi Iwamura and Yoshiyuki Inagaki

CONTENTS

11.1 Overview

Sodium-cooled fast reactor (SFR) features fast neutron spectrum core and enables a closed fuel cycle. In addition to energy production, the additional primary missions for the SFR are management of high-level wastes and management of plutonium and other actinides. The Generation IV Technology Roadmap identifies the SFR as a promising technology to perform in particular the missions of sustainability, actinide management, and electricity production if enhanced economics for the system could be realized [1].

The main characteristics of the Generation IV SFR that make it especially suitable for the missions identified in the roadmap are

1. High potential to operate with a high-conversion fast spectrum core with the resulting benefits of increasing the utilization of fuel resources
2. Capability of efficient and nearly complete consumption of *trans*-uranium as fuel, thus reducing the actinide loadings in the high-level waste with benefits in disposal requirements and potential nonproliferation
3. High level of safety obtained with the use of inherent and passive means that allow the accommodation of transients and bounding events

4. Enhanced economics achieved with the use of high burn-up fuels, reduction in power plant capital costs with the use of advanced materials and innovative design options, and lower operating costs achieved with improved operations and maintenance

The Gen-IV SFR development can rely on technologies already used for the SFRs that have successfully been built and operated in France (Rapsodie in 1967, Phenix in 1973, Super Phenix in 1985), Germany (KNK-II in 1977), the United Kingdom (DFR in 1959, PFR in 1974), Russia (BOR-60 in 1969, BN-600 in 1980), the United States (EBR-II in 1963, Fermi in 1966, FFTF in 1980), Japan (JOYO in 1977, MONJU in 1994), and India (FBTR in 1986). As a result of these previous investments in technology, the majority of the R&D needs that remain for the Gen-IV SFR are related to performance rather than viability of the system.

A range of plant size options are available for the SFR, ranging from modular systems of a few hundred MWe to large monolithic reactors of 1500–1700 MWe. Core outlet temperatures are typically 530–550°C, high enough to heat efficient power conversion cycles (steam and supercritical carbon-dioxide systems) and hydrogen production processes (water splitting and methane reforming). The primary coolant system can either be arranged in a pool layout where all primary system components are housed in a single vessel, or in a compact loop layout. Typical layouts of a pool-type SFR and a loop-type SFR are shown in Figures 11.1 and 11.2, respectively.

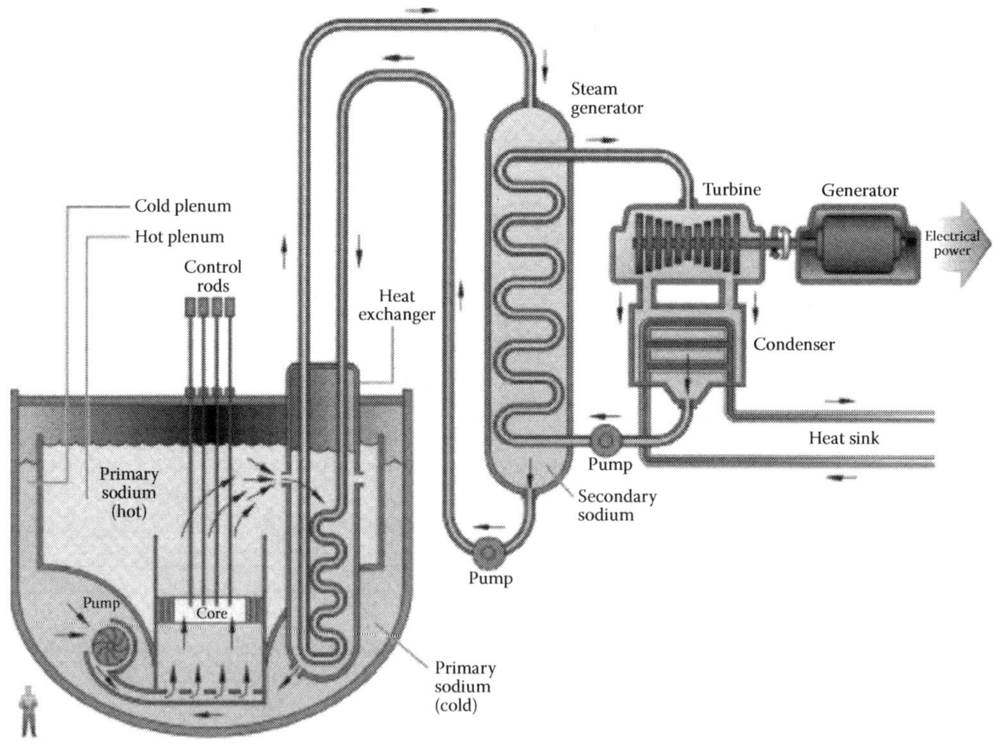

FIGURE 11.1
Layout of pool-type SFR.

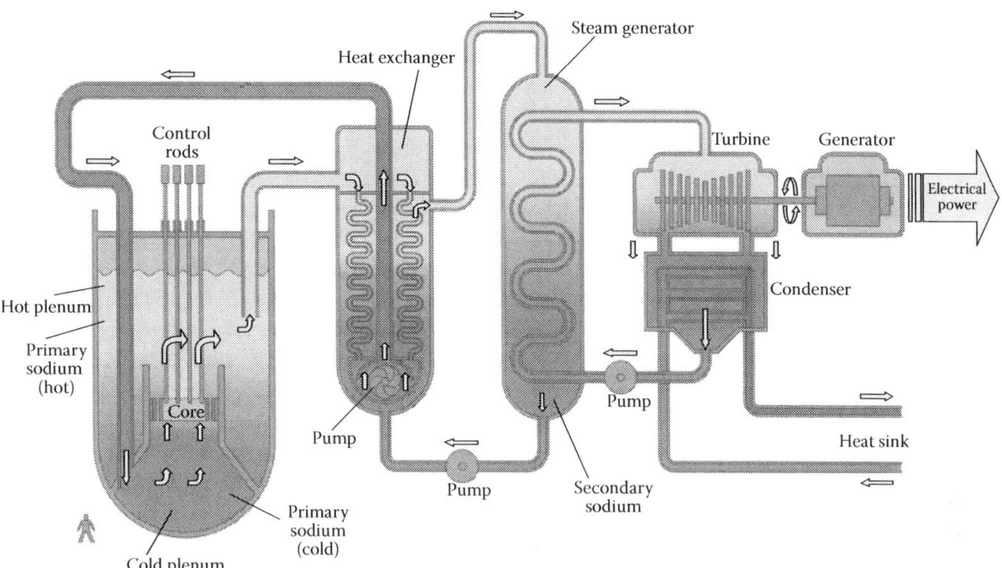

FIGURE 11.2
Layout of loop-type SFR.

For both options, there is a relatively large thermal inertia of the primary coolant. A large margin to coolant boiling is achieved by design, and is an important safety feature of these systems. Another major safety feature is that the primary system operates at essentially atmospheric pressure, pressurized only to the extent needed to move fluid. Sodium reacts chemically with air, and with water, and thus the design must limit the potential for such reactions and their consequences. Two fuel options exist for the SFR: (1) MOX and (2) mixed uranium–plutonium–zirconium metal alloy (metal). The experience with MOX fuel is considerably more extensive than with metal.

As an example of innovative SFR designs, a concept of loop-type SFR with MOX fuel which has been subject of substantial design and development in Japan is described in Section 11.2. Several potential hydrogen production systems that operate in the temperature range of the SFR and which have been investigated for application to the SFR are reported in Section 11.3.

11.2 Reactor and Major System Designs

11.2.1 Japanese Sodium-Cooled Fast Reactor Design

The Japan Atomic Energy Agency (JAEA) and domestic electric utilities initiated a feasibility study (FS) in July 1999, in collaboration with the Central Research Institute of Electric Power Industry (CRIEPI) and manufacturers, in order to effectively utilize the accumulated knowledge from the demonstration fast breeder reactor (DFBR) design, as well as the construction/operation experience from an experimental fast reactor, JOYO and a prototype fast reactor, MONJU. The objective of this study is "to present both an appropriate picture of commercialization of the FR cycle and the research and development (R&D) programs leading up to the commercialization in approximately 2015" [2,3].

An innovative concept, named the Japan sodium-cooled fast reactor (JSFR) was proposed through the FS and achieved full satisfaction of development targets. The concept is recognized as a promising Generation IV SFR system [4,5].

JSFR is a sodium-cooled, MOX-fueled, advanced loop-type design that evolves from Japanese fast reactor technologies and experience. There are two optional reactor sizes, that is, a medium and a large scale generating 750 and 1500 MWe, respectively, and both are of similar plant configuration.

The large-scale sodium-cooled reactor can take the further advantage of "economy of scale" by designing as a twin-reactor commercial plant (total electricity output of 3000 MWe) and reducing the amount of materials by design improvement. The large-scale JSFR design with major specifications is presented in Table 11.1, a bird's-eye view in Figure 11.3, and a schematic of the reactor and cooling system in Figure 11.4.

11.2.2 Reactor Vessel and Internal Structures

Figure 11.5 shows a conceptual configuration of the reactor vessel and internals [6,7]. The diameter and wall thickness of the reactor vessel are minimized and the reactor internal structures are simplified in order to reduce the quantity of construction materials.

TABLE 11.1

Major Specifications of JSFR Design

	Sodium-Cooled Reactor (1500 MWe) MOX Fuel (Metallic Fuel)	
Design Requirement	**Breeding Core**	**Break-Even Core**
Safety	Out-of-pile and in-pile experiments are underway, concerning the passive safety mechanism and measures to avoid recriticality	
Efficient utilization of nuclear fuel resources		
Breeding ratio (1.0~1.2)	1.10(1.11)	1.03(1.03)
Fissile fuel inventory required for the initial loading core	5.7(4.9) t/GWe	5.8(5.1) t/GWe
Time required to replace all nuclear power reactors with FRs	~60 years	—
Reduction of environmental burden		
MA burning	Can accept MA content up to ~5% that is recycled from LWR spent fuel under low decontamination condition (with FP content of 0.2 vol%)	
FP transmutation	Having a possibility of transmuting self-generating LLFP (^{129}I and ^{99}Tc), by installing the FP both inside the core and the radial blanket region	
Economic competitiveness		
In-core average (150 GWd/t or higher)	147(149) GWd/t	150(153) GWd/t
Whole-core average (60 GWd/t or higher)	90(134) GWd/t	115(153) GWd/t
Operation period (18 months or longer)	26(22) months	26(22) months
Availability (calculated value) (90% or more)	~95(94)%	~95(94)%
Reactor outlet temperature	550°C	
Thermal efficiency/onsite load factor	42.5%/4%	
Unit construction cost (¥200,000/kWe or less)	Relative value: ~90%	

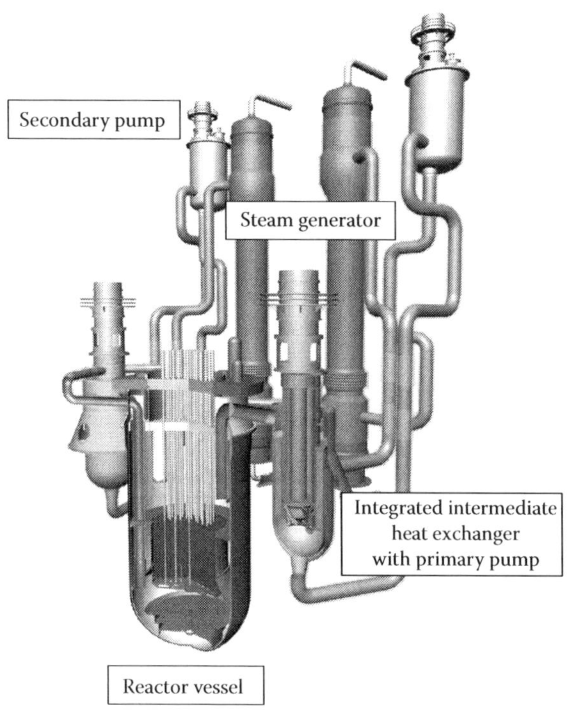

FIGURE 11.3
Japan sodium-cooled fast reactor (JSFR).

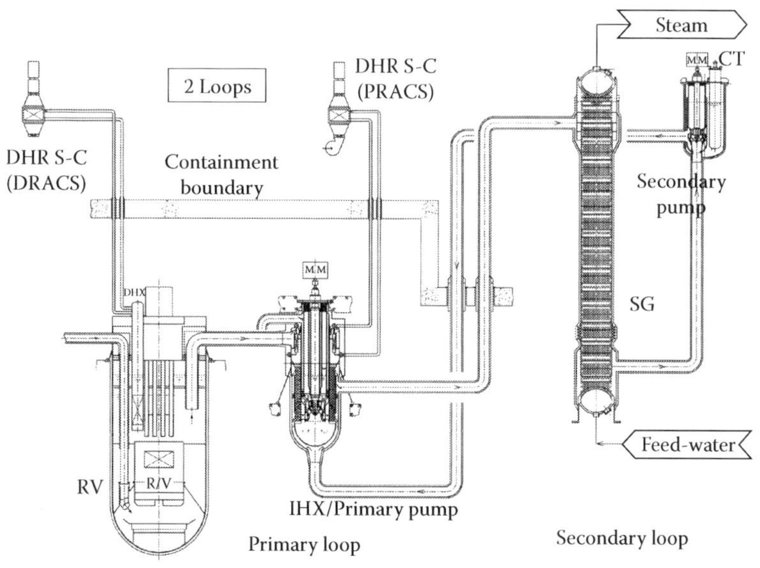

FIGURE 11.4
Conceptual scheme of JSFR and cooling system.

The reactor vessel diameter of 10.7 m and height of 21.2 m are realized by the compact design of the core and closure head structure. In this plant, a shell-less column-type upper inner structure (UIS) is adopted. The fuel handling machine (FHM) can enter into the center of this UIS, which has many control rod-guide tubes. As this type of UIS makes it possible to conduct fuel handling with the UIS remaining just above the core, the diameter of a single rotating plug sets smaller, and this leads to the compact design of the closure head structure.

The 60-mm wall thickness is realized by introducing the recriticality free technology that can mitigate mechanical energy release resulting from hypothetical core disruptive accident (CDA) and by introducing the seismic isolation technology that can mitigate seismic load.

In addition, a simple hot vessel design has been employed. Structural integrity against a thermal stress caused by an axial temperature profile around the liquid surface of the reactor vessel is ensured by the establishment of an advanced elevated temperature structural design standard, which makes it possible to use an inelastic analysis, and by a direct assumption method of the thermal load. Further simplification of the structure is planned to reduce the amount of materials. Such simplification is attained by adopting a simple skirt-type core support structure, eliminating the vessel wall cooling system, eliminating the liquid–surface level control, and so on. A dipped plate is employed to suppress a cover-gas entrainment in the coolant from the liquid surface of the reactor vessel because of an increased sodium velocity due to the compact design of the reactor vessel.

11.2.3 Reactor Core and Fuel

Table 11.2 shows the major core design conditions [8]. The design is focused on the following requirements.

1. Core coolant void reactivity should be low enough to prevent the super-prompt criticality in the initiating phase of CDA. Measures of early discharge of molten fuel should be considered in the core and fuel design to prevent the recriticality in the transition phase of CDA.
2. The target of core average discharge burn up is 150 GWd/t, and the total average discharge burn up (including blankets) is 60 GWd/t or more. The burn up target aims at the reduction of fuel cycle cost due to the economic competitiveness requirement. The high burn up also contributes to reduce the fuel mass capacity requirement of fuel cycle facilities.
3. Core should have flexible breeding capability in a viewpoint of uranium resource utilization. The target of the maximum breeding ratio is from 1.1 to 1.2. The requirement of breeding capability is not only for the fuel breeding itself, but also for other requirements such as economic advantage, environmental burden reduction, and so on. The core with high breeding capability possesses characteristic of high internal conversion ratio, which means the advantages for flexible core design management such as less blanket, enhanced capability for environmental burden reduction, longer reactor operation cycle duration.
4. From economics and proliferation resistance viewpoints, the core should have the capability of loading the low decontamination fuel associated with economical fuel recycle system. It should also have the capability of loading the fuel with a few percents of minor actinides (MA) from uranium resource utilization and environmental burden reduction viewpoints.

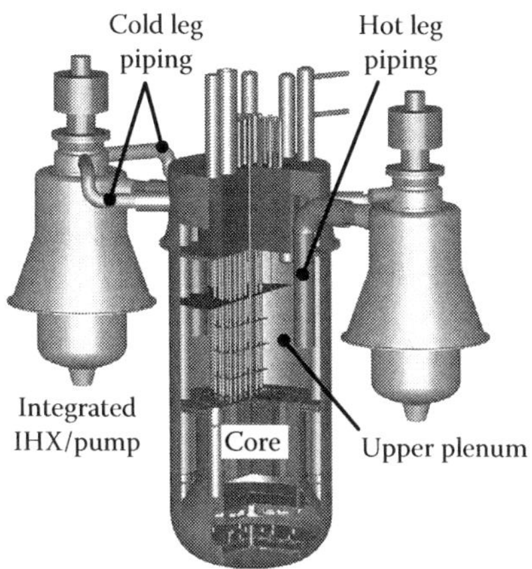

FIGURE 11.5
Reactor vessel and primary piping system.

TABLE 11.2
Major Design Condition

Items		Conditions
Safety Requirement		
Sodium void reactivity ($)		≤6
Core specific power (kW/kg-MOX)		≥40
Core height		≤100
Re-criticality-free		FAIDUS type subassembly
Design Target		
Discharge burn up (GW/d/t)	Core	150
	Core + blanket	≥60
Breeding ratio	Breeding core	1.1
	Break-even core	1.03
Operation cycle length (months)		≥18
Core and Fuel Spec.		
Fuel composition	TRU	FR multi recycle
	FP containment (vol%)	0.2
Core fuel smeared density (%TD)		82
Core material	Cladding	ODS
	Wrapper tube	PNS-FMS
Design Limit		
Maximum linear power (W/cm)		≤430
Maximum neutron dose[a] (n/cm^2)		$\leq 5 \times 10^{23}$
Cladding maximum temperature[b] (°C)		≤700
CDF (steady state)		≤0.5
Others		
Pin bundle pressure drop (MPa)		≤0.2

[a] E >0.1 MeV.
[b] Mid-wall.

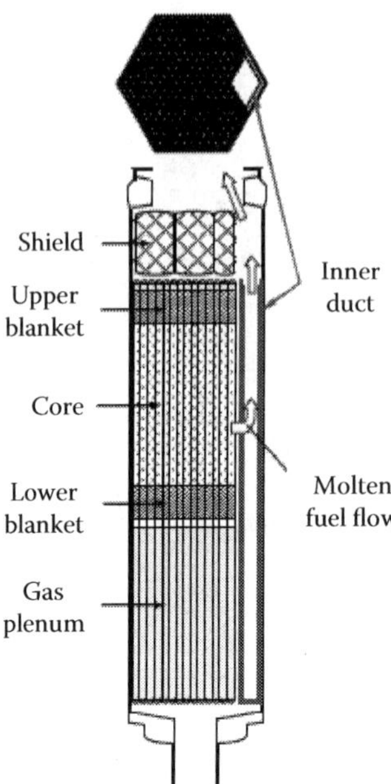

FIGURE 11.6
FAIDUS-type fuel subassembly.

Figure 11.6 shows a conceptual view of fuel assembly with inner duct structure (FAIDUS) type subassembly. In this subassembly, an inner duct is installed at the corner of the subassembly and a part of upper shielding element is removed. The molten fuel enters the inner duct channel and goes out directly into the outside without any interference by the upper shielding.

Table 11.3 shows the major core and fuel specifications of the large-scale breeding core and break-even core. The configurations of each core are shown in Figure 11.7.

An advanced concept of large diameter fuel pin is applied into the oxide fuel core in order to obtain high internal conversion ratio. This leads to the break-even breeding core, in which the breeding ratio is just above 1.0, without radial blanket. Total discharge average burn up (including blanket) achieves as high as about 100 GWd/t in refueling batch average including blanket. The achieved averaged burn up is extremely higher than conventional design with small diameter fuel pin. The cores have flexibility to be modified to breeding core with 1.1 of breeding ratio by adopting radial blanket. This advanced concept has economical advantage by consistently achieving high burn up and breeding with small amount of blanket. The major performance parameters of the large-scale core include;

- The total average discharge burn up (including blanket) of breeding core is 90 GWd/t and break-even core is 115 GWd/t.
- The operation cycle length is able to attain 26 months.
- The breeding ratio of breeding core achieves over 1.1 and break-even core is 1.03, so that the present core design has breeding flexibility corresponding to future fast reactor (FR) deployment scenario.

TABLE 11.3
Core and Fuel Specifications of Oxide Fuel Core

Items	Breeding Core	Break-Even Core
Nominal full power (MWe/MWt)	1500/3570	←
Coolant temperature (outlet/inlet) (°C)	550/395	←
Primary coolant flow (kg/s)	18,200	←
Core height (cm)	100	←
Axial blanket thickness (upper/lower) (cm)	20/20	15/20
Number of fuel assembly (core/radial blanket)	562/96	562/ –
Envelope diameter of radial shielding (m)	6.8	←
Fuel pin diameter (core) (mm)	10.4	←
Fuel pin cladding thickness (core) (mm)	0.71	←
Number of fuel pin per assembly (core)	255	←
Wrapper tube outer flat-flat width (mm)	201.6	←
Wrapper tube thickness (mm)	5.0	←

11.2.4 Heat Transport System

In order to select the most advantageous concept of the cooling system, the loop number, primary piping system, and applicability of the integration of components have been comprehensively examined from the viewpoint of the amount of materials, safety, maintainability, fabricability, and so on. As a result of the examinations, the following concepts were selected: two-loop cooling system, top-entry piping, integrated design of an intermediate heat exchanger (IHX) and a primary pump, and separated design of an steam generator (SG) and a secondary pump [5,7].

The drastically shortened primary and secondary piping layout results in a compact plant configuration through a close arrangement of components, as well as a reduction of the amount of piping materials. In particular, the primary hot-leg piping has been

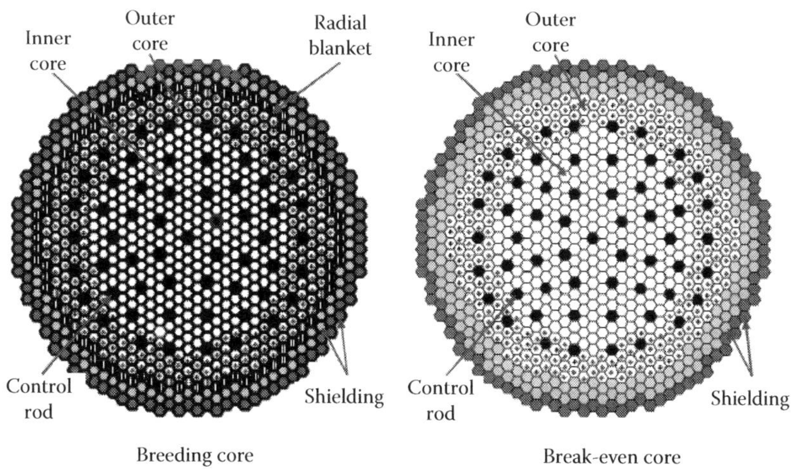

FIGURE 11.7
Core configuration of oxide fuel core.

simplified. It is designed as a simple L-shaped piping and has a structure that absorbs a thermal expansion by only one elbow. Such a shortened piping layout with one elbow is realized with the aid of an adoption of Mod.9Cr-1Mo-steel which embodies high strength and low thermal expansion coefficient.

The sodium-cooled system has some desirable characteristics: that is, the high boiling point and the high heat conductivity of the sodium coolant that can raise the heat transfer capacity per unit volume of heat exchangers under low-pressure condition. Owing to these desirable characteristics, it is possible to design heat exchangers that have a larger capacity by putting compact heat transfer structures in a vessel with a large diameter and a thin wall thickness. The adoption of Mod.9Cr-1Mo-steel with high strength and high heat conductivity makes it possible to enlarge the heat transfer capacity further than that of the conventional design because a thinner heat transfer tube is available and it is easier to ensure the structural integrity against thermal stress by a large tube sheet. In this design, the capacity of the heat exchangers (i.e., the heat transfer capacity per one-cooling loop) has been enlarged as far as the fabricability and the structural integrity of the components are expected to be ensured, and then the number of loops has been reduced to two. According to the estimation on this reduction effect, the two-loop system can reduce ~13% of the nuclear steam supply system (NSSS) weight and ~10% of the reactor building volume compared with the four-loop system.

Although the fuel integrity in a primary pump stick accident and the structural integrity of the large tube sheets of the heat exchangers against a thermal stress become critical issues with employing the two-loop cooling system, prospects for solution have been obtained through analytical examinations.

According to the comparison between the integrated and separated design of an IHX and a primary pump, the integrated design can reduce ~13% of the reactor and primary system weight and ~9% of the containment volume compared with the separated design. Thus, by integrating the IHX and the primary pump, the primary cooling system can be remarkably simplified.

In addition, in the case of the integrated design, improvements of maintainability are expected. That is, middle-leg piping between an IHX and a primary pump, on which it is difficult to do maintenance, can be eliminated, and in-place maintenance of the IHX tubes is easy by accessing the IHX tubes through the center hole after the withdrawal of the primary pump. For the above reasons, the integrated design has been selected. Figure 11.8 shows the conceptual scheme of the integrated components.

The integrated component is formed by installing the primary pump at the center of the IHX tube bundle. The IHX is designed as a no-liquid-surface type in order to allow high velocity of sodium coolant (~10 m/s) in the primary hot-leg piping without raising the cover-gas pressure in the reactor vessel. A straight heat transfer tube has been adopted. The primary sodium flows within the tubes, and the secondary sodium flows zigzag outside of the tube. With the adoption of thinner heat transfer tubes made of Mod.9Cr-1Mo-steel, the heat transfer performance has been improved, and the tube bundle has become compact. The diameter and height of the integrated component are 4.3 and 14.2 m, respectively.

JSFR adopts the integrated once-through-type SG with double-wall straight heat transfer tubes. Straight tube-type SG has an advantage in construction cost due to fewer amounts of material and simpler fabrication (no bending of heat transfer tube) compared with helical type. The major specifications of SG are shown in Table 11.4. The detailed structure of the double-walled tube SG is shown in Figure 11.9.

Compact design of tube bundle is achieved by the improved performance in heat transfer that benefits from thinner heat transfer tubes made of Mod.9Cr-1Mo-steel. As a result,

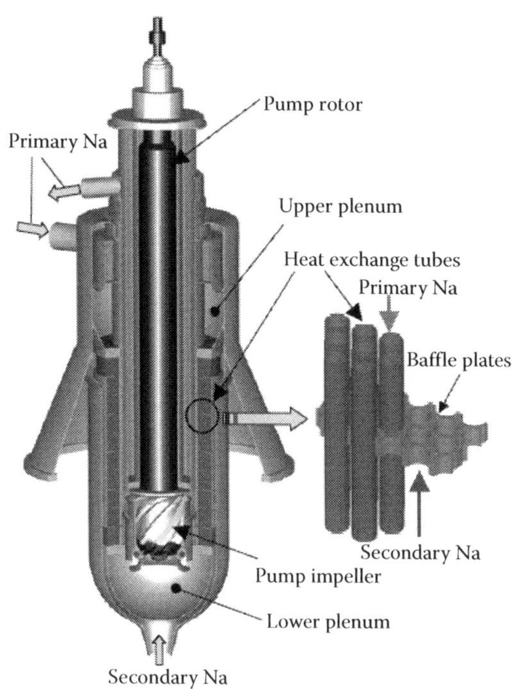

FIGURE 11.8
Concept of the integrated IHX/pump component.

the effective length of heat transfer tube is about 29 m for the SG with capacity of 1765 MWt. The diameter and height of the SG are about 3.5 and 38 m, respectively. As the thermal capacity of the SG is increased in order to pursue the economics of scale, the diameter of the SG shell becomes large. Then, it requires adoption of tube-plate with semispherical shape for the water-side pressure. A single plate-type tube-plate is suitable in order to simplify both structure and plugging process of the tube-plate.

In the thermal hydraulic viewpoints, the provision of baffle-plates and a sodium flow down-up structure in a sodium inlet plenum of the SG can unify the radius sodium flow into the tube bundle, and the subsequently unified sodium flow contributes to flatten a horizontal temperature distribution in the bundle region. This is an indispensable factor for preventing a tube buckling or tube to tube-plate junction failure. The shell bellows can compensate the thermal expansion difference between the SG shell and tube bundle. Sodium fluids flow in the counter direction against the water fluids and in parallel with the heat transfer tubes to reduce the pressure loss and to avoid tube-fretting. No water/steam hydraulics instability occurs by increasing waterside pressure without orifice. This orifice-less method serves to enhance the tube reliability in terms of preventing unpleasant phenomena like erosion or blockage at the orifice.

11.2.5 Safety System

The reactor shutdown system (RSS) consists of two independent active subsystems (i.e., main and backup RSSs) with different design specifications; for example, the principle of

TABLE 11.4
Major Specification of SG

Items		Specifications
SG type		Double-walled straight tube type
Thermal capacity		1765 MWt
Heat transfer area		12,515 m^2
Heat transfer tube	Inner tube (diameter/thickness)	16.0 mm/1.1 mm
	Outer tube (diameter/thickness)	19.0 mm /1.5 mm
	Heat transfer length	29.0 m
	Tube number	7230
	Pitch	38 mm
Tube material		Mod.9Cr-1 Mo steel
Flow rate	Sodium side	2.70×10^4 t/h
	Water/steam side	2.884×10^3 t/h
Temperature	Sodium	520°C/335°C
	Steam	497°C (19.2 MPa)
SG size	Height	38.0 m
	Diameter	3.51 m
	Weight	605 ton

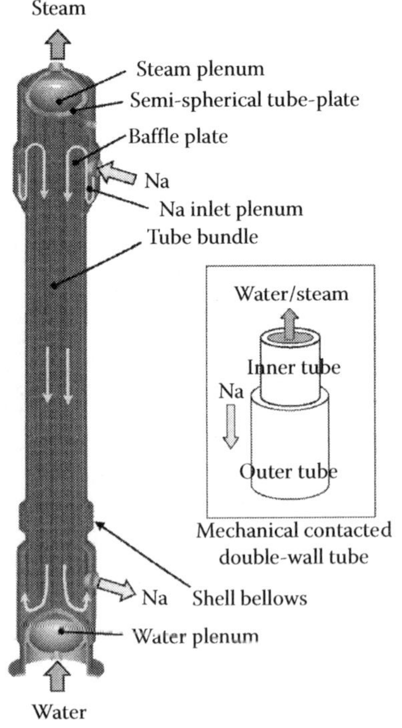

FIGURE 11.9
Detailed structure of double-walled SG.

detectors, driving mechanism of rods, and so on [9–11]. Each of them is designed to prevent fuel failure against design base events (DBEs). In addition, the Curie-point magnet-type self-actuated shutdown system (SASS, shown in Figure 11.10), which is additionally introduced in the backup RSS as a passive shutdown feature ensures the prevention capability against the anticipated transients without scram (ATWS) combined with a robust structural core design. The ATWS is however addressed as design extension conditions (DEC). SASS has been developed by both out- and in-pile experiments so far.

The combination of two primary reactor auxiliary cooling systems (PRACSs) and one direct reactor auxiliary cooling system (DRACS) has been selected from the view point of safety and fluid stability of sodium-coolant at the condition of natural circulation as shown in Figure 11.4.

The analytical results of station blackout are shown in Figure 11.11. It was also confirmed that the decay heat removal capacity with the single failure condition such as one of the air cooler damper system malfunction is ensured.

11.2.6 Plant Layout

Figure 11.12 shows the layout of the reactor building of JSFR [5–7]. The reactor building is designed as a seismic isolated building, and the layout reflects the rationalization and compaction of the reactor and cooling system. In addition, communization of the balance-of-plant facilities has proceeded through the twin-plant design.

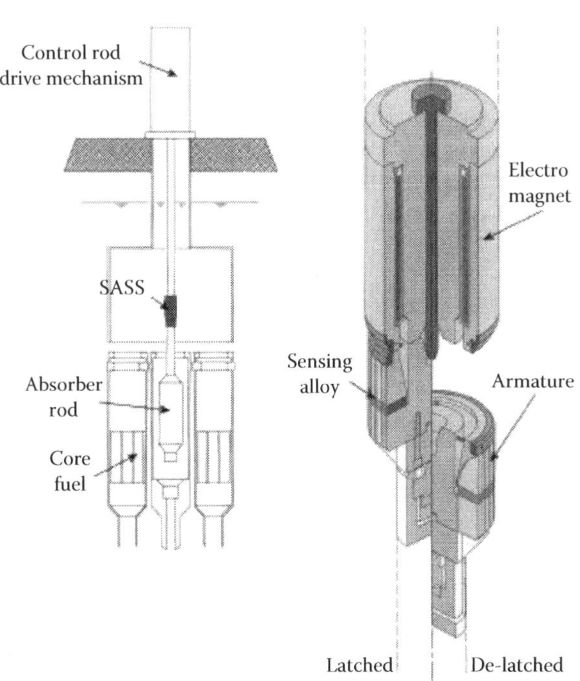

FIGURE 11.10
Self-actuated shutdown system (SASS).

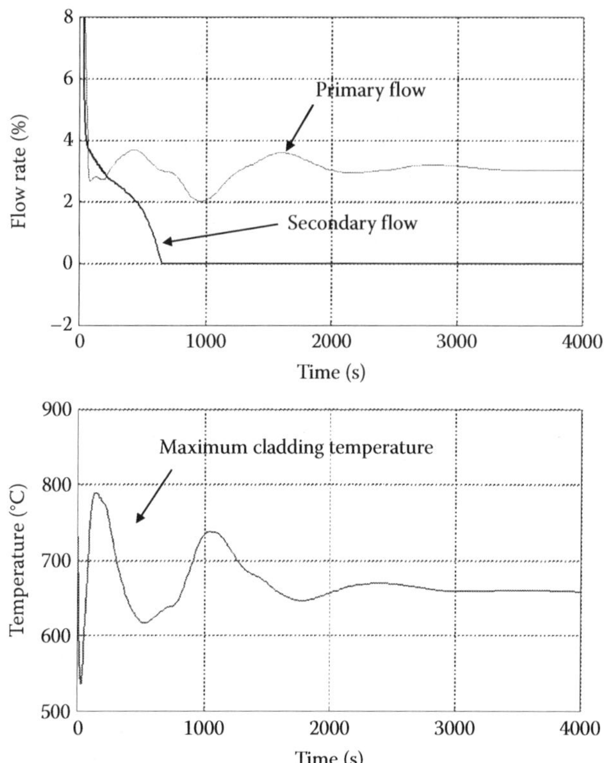

FIGURE 11.11
Analytical Result of Station Blackout (2PRACSs + DRACS).

The influence of sodium leakage coming from defects in materials is localized within the guard vessel or guard piping in this plant. Another cause of sodium leakage is overpressure to the wall. But, adoption of recriticality free measures eliminates such sodium leakage from the reactor vessel caused by a significant mechanical energy discharge in the CDA. Therefore, a simplified design of the reactor containment facility (reinforced concrete structure with an inner lining) is adopted from the point of view of radionuclide confinement, which is more important than pressure resistance.

11.3 Hydrogen Production Systems

SFR enables hydrogen production via the following routes:

1. The conventional electrolysis of water with the electric power that the SFR generates.
2. Alternatively, it can be applied to hydrogen production from water by coupling to some of the thermochemical and hybrid processes which are introduced in Chapter 5, provided the process temperature range is within the maximum 550°C coolant outlet temperature of the SFR.

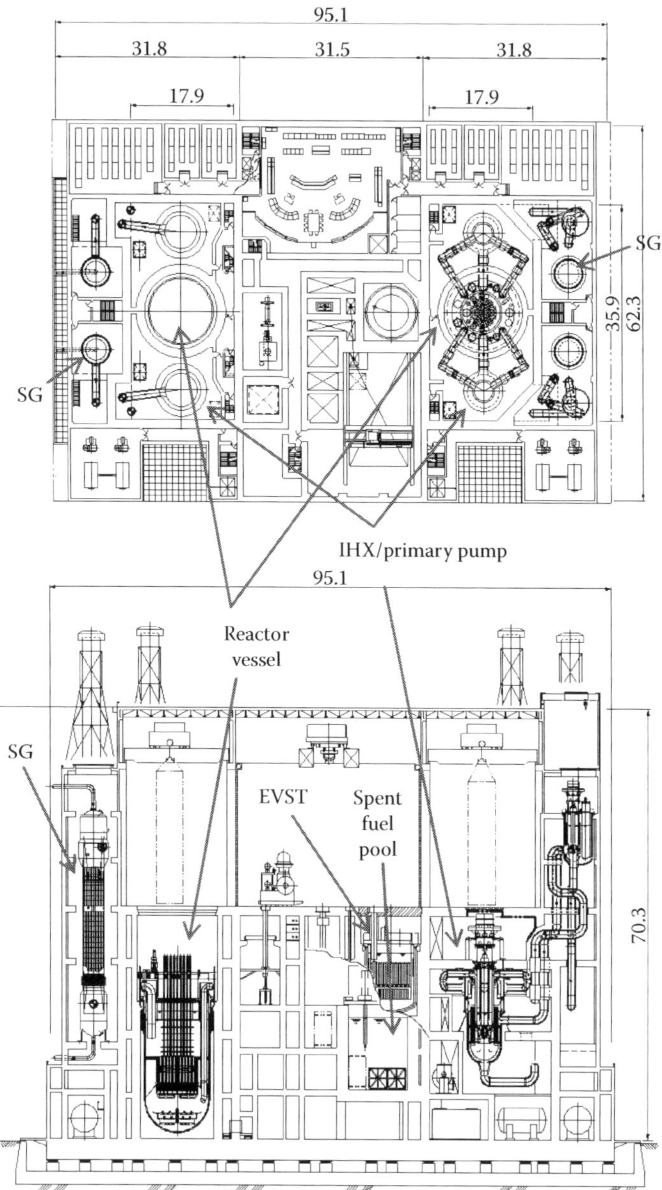

FIGURE 11.12
Layout of the JSFR Reactor Building (twin-reactor plant).

3. Finally, the SFR has been considered in industry studies for application to the steam reforming of methane with a concurrent reforming-shift process in the temperature of interest to the SFR.

The application of the conventional electrolysis to the SFR via the first route above is similar to that for the water reactor, about which Chapter 9 discusses in detail. The other two routes have been investigated specifically for the SFR-based hydrogen production and are described with the examples of the selected processes in this section.

11.3.1 Hybrid Process

A sulfur process and a copper chloride process are described below as they have been considered promising and studied most extensively among the reported hybrid processes (Chapter 5).

11.3.1.1 Sulfur Process

A particular hybrid hydrogen production process by thermochemical and electrolytic decomposition of water for application of the FBR-generated electricity and medium-range temperature (500–550°C) heat has been researched in JAEA. The process is modified from the Westinghouse HyS process (Chapter 18) and is referred to the HHLT (hybrid hydrogen production in lower temperature range) process [12].

Figure 11.13 shows the chemical reaction scheme of the HHLT process. The basic chemical reactions of the HHLT process are those of the Westinghouse process, except for the third key reaction condition, as follows:

$$SO_2\,(g) + 2H_2O(l) \rightarrow H_2SO_4\,(l) + H_2(g) \text{ (electrolytic reaction, <100°C)}, \qquad (11.1)$$

$$H_2SO_4(l) \rightarrow H_2O(g) + SO_3(g) \text{ (thermal reaction, ~400°C)}, \qquad (11.2)$$

$$SO_3(g) \rightarrow SO_2(g) + \frac{1}{2}O_2(g) \text{ (thermal and electrolytic reaction, ~500°C)}, \qquad (11.3)$$

where l and g within parentheses indicate liquid and gas phases of the process reactants and products.

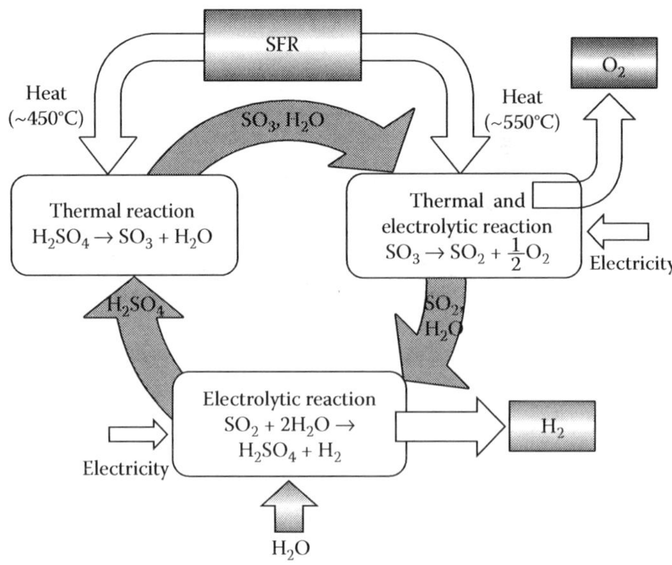

FIGURE 11.13
Chemical scheme of the HHLT process.

In the Westinghouse process, Reaction 11.3 is performed by the thermal decomposition in the high-temperature range of around 800°C, which requires high-temperature nuclear heat source such as a gas faster reactor. If the same reaction of thermal decomposition takes place in the medium-temperature heat range of the FBR, the decomposition rate of sulfur trioxide is only 7.8% in equilibrium condition at 500°C. Therefore, new technology is necessary to promote decomposition of sulfur trioxide. The HHLT process circumvents this problem by using electrolysis with ionic oxygen conductive solid electrolyte.

The theoretical voltage of Reaction 11.1 is 0.17–0.29 V, and that of Reaction 11.3 is 0.13 V. Therefore, the total theoretical voltage of the HHLT process is less than 0.5 V, which is less than a half of the voltage of 1.23 V of water electrolysis. Unlike in the thermochemical IS process, iodine is not used in the HHLT process. This results in less corrosion of structural materials and simplifies the hydrogen production system in comparison to the IS process. On the other hand, the thermal efficiency of the HHLT process becomes lower than that of the IS process, because larger input of electricity is needed into the HHLT process.

At present, the basic performance tests on the SO_2 solution electrolysis cell for Reaction 11.1 and the SO_3 electrolysis cell for Reaction 11.3 is being carried out. Figure 11.14 shows the flow sheet of the experimental apparatus producing hydrogen at the rate of about 1 L/min [13]. The raw material feed of water is supplied to the cathode side and sulfurous acid to the anode side of SO_2 solution electrolysis cell to synthesize sulfuric acid. As a result of the transportation of split proton to the cathode side through the membrane, hydrogen is produced by electrolysis at the room temperature. To increase cell current and its current density, flow-type cell with MEA (Membrane Electrode Assembly) is employed. Sulfuric acid passing through the anode electrolyte solution tank is decomposed to sulfur trioxide and water by the sulfuric acid vaporizer at the temperature of around 400°C. Sulfur trioxide and water vapor are transported to the SO_3 electrolysis cell. Sulfur trioxide is decomposed

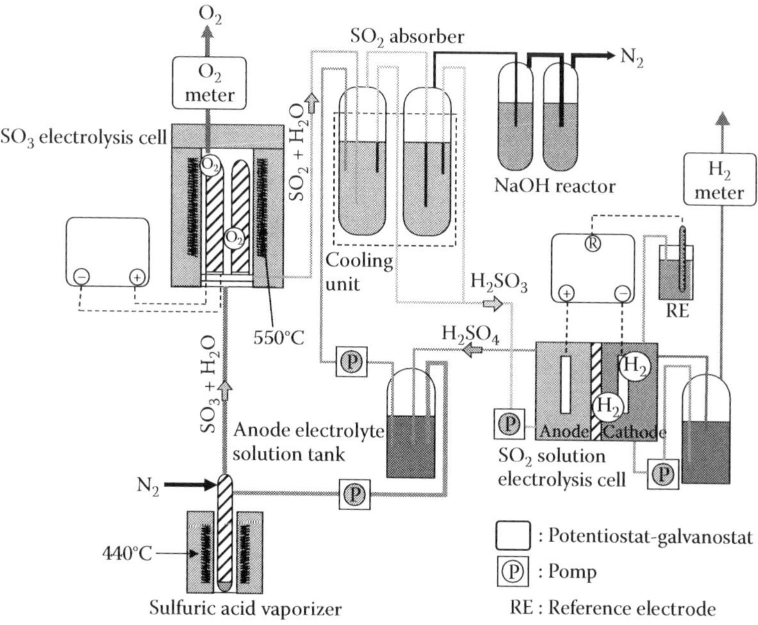

FIGURE 11.14
Flow sheet of experimental apparatus of the HHLT process.

into sulfur dioxide and oxygen by electrolysis at the temperature of around 500°C. For the decomposition of sulfur trioxide, the YSZ (yttria-stabilized zirconia) tube plated with platinum in thickness of about 1 μm is employed. Sulfur dioxide and water vapor are transported to the SO_2 absorber to produce sulfurous acid.

After the performance tests of the electrolysis cells, the research of the system control technology will be carried out by the continuous hydrogen production operation in the future plan.

A hydrogen production plant system based on a small-scale sodium-cooled reactor was designed by JAEA [14]. Figure 11.15 shows the simplified flow sheet of the plant, and Table 11.5 list the major design and performance parameters. The thermal output of the reactor is 395 MWt with the outlet temperature at 550°C. The hydrogen production rate of the plant is 49,000 Nm^3/h, and the efficiency of the hydrogen production is estimated to be 44%. The small system may be up-scaled to the medium- and large-scale SFRs including the JSFR described in Section 11.2 to increase the hydrogen production output per plant.

The primary sodium heat is transferred to the secondary (intermediate) sodium circuit by the IHX. The secondary sodium heat is input in process heat exchangers from 540 to 455°C to chemical reactions and from 455 to 350°C to steam turbine power generation.

The process heat exchangers for hydrogen production are arranged in the secondary sodium circuit. The hot-leg temperature of the secondary sodium circuit is 540°C since the solid electrolyte for SO_3 electrolysis works at a temperature above 500°C.

The steam generator in the secondary sodium circuit is rated at 224 MWt capacity. The superheated steam conditions are 400°C and 6 MPa. The power generation efficiency at these steam conditions is estimated to be 38%, and the total power generation in the plant is estimated to be 85.2 MWe. All of the electricity generated is supplied in-house to the sulfur trioxide electrolysis at 18.5 MWe, the sulfur dioxide solution electrolysis at 58.5 MWe and the plant load at 8.25 MWe. The schematic arrangement of a commercial plant based on these HHLT process parameters for hydrogen production is shown in Figure 11.16.

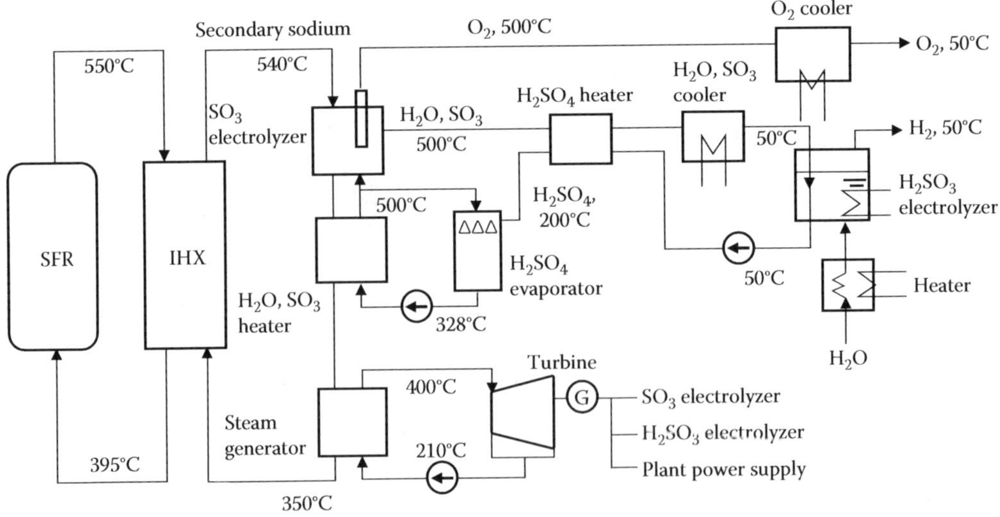

FIGURE 11.15
The SFR sulfur hybrid hydrogen production process parameters.

TABLE 11.5

Process Parameters of the SFR Hydrogen Production Plant Design

Items	Specifications
SFR thermal output	395 MW (thermal)
Primary sodium temperature (hot/cold)	550°C/395°C
Primary sodium mass flow	2012 kg/s
Secondary sodium temperature (hot/cold)	540°C/350°C
Secondary sodium mass flow	1631 kg/s
Sulfuric acid temperature (hot gas phase/cold)	500°C/50°C
Sulfuric acid mass flow	62 kg/s
Pressure of sulfuric acid system	0.1 MPa
Sulfuric acid concentration	95%
Feeedwater mass flow	10 kg/s
Hydrogen production	49000 m^3/h
Electric output (used for electrolysis)	85.2 MW
Electric load at SO_3 electrolysis	18.5 MW
Electric load at SO_2 solution electrolysis	58.5 MW
Electric load for plant power	8.2 MW
Efficiency of electricity generation	38%
Efficiency of SO_3 electrolysis	85%
Efficiency of SO_2 solution electrolysis	90%
Efficiency of hydrogen production	44%

11.3.1.2 Copper Chloride Process

The process is composed of mainly three thermochemical and electrolytic reactions using copper and chloride. Figure 11.17 shows the reaction scheme of the process.

$$2CuCl_2(s) + H_2O(g) \rightarrow Cu_2OCl_2(s) + 2HCl(g) \text{ (thermal reaction, 300–390°C)}, \quad (11.4)$$

$$Cu_2OCl_2(s) \rightarrow 2CuCl(l) + \frac{1}{2}O_2(g) \text{ (thermal reaction, 450–530°C)}, \quad (11.5)$$

$$2CuCl(l) + 2HCl(l) \rightarrow 2CuCl_2(l) + H_2(g) \text{ (electrolytic reaction, <100°C)}, \quad (11.6)$$

where s, l, and g indicate the solid, liquid, and gas phases of the process materials, respectively.

Reaction 11.4 is a solid gas process involving surface area reaction, production, and handling of the solids. It would demand large process equipment and result in difficult economy of scale. For it to react efficiently it requires a large amount of excess water which will have to be separated from HCl(g) product. It produces Cu_2Cl_2O whose thermodynamic properties are not well known. The reaction needs a precise temperature control in order to avoid dissociation of this product. On the other hand, Reaction 11.5 is a high-temperature reaction with liquid and gaseous products, which improves the thermodynamic process design and performance. The potential presence of molecular chlorine formed by $CuCl_2$ dissociation (whose reaction is incomplete in Reaction 11.4), or formed according Deacon, reaction starting from HCl traces must be checked [15]. The basic experimental

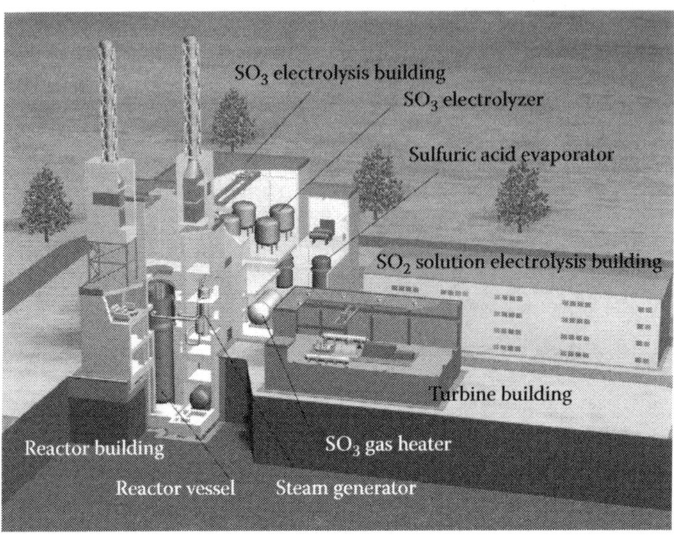

FIGURE 11.16
SFR-based sulfur hybrid process plant arrangement.

study on each of the reactions has been carried out in a joint research of United States, France, and Canada.

11.3.2 Steam Reforming of Methane

As introduced in Chapter 6, the conventional steam reforming of methane is carried out in the process temperature range of 600–900°C, which is above the SFR reactor coolant temperature limit. A membrane reforming concept that lowers process temperature has recently been proposed for the SFR application. The membrane reformer consists of membrane modules of palladium-based alloy and nickel-based catalyst, as shown in

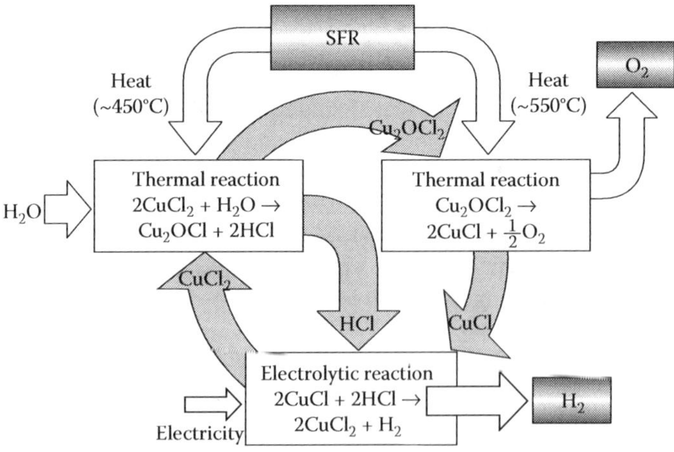

FIGURE 11.17
Chemical scheme of the copper chloride hybrid process.

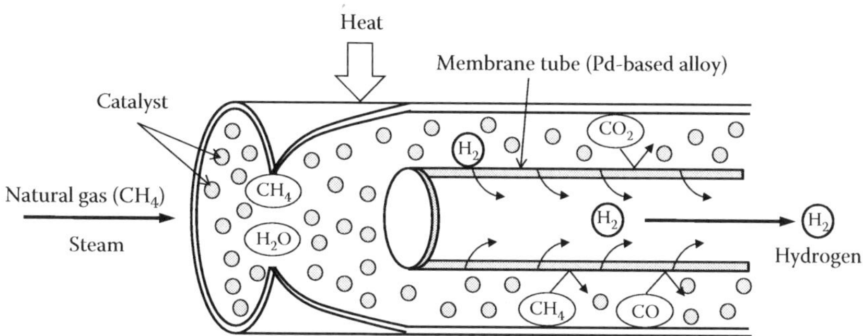

FIGURE 11.18
Conceptual membrane steam reformer of methane.

Figure 11.18. It simultaneously performs the reforming reaction, shift reaction and hydrogen separation, which are otherwise performed sequentially in the high-temperature conventional process. By the simultaneous generation and separation of hydrogen, the membrane reformer system can be more compact and thermally efficient than the conventional process. The continuous extraction of product hydrogen breaks down chemical equilibrium of the reaction and results in a lower heating temperature of 500–550°C, which can be met by the SFR. The lower temperature places less demand on construction materials [16].

The technical feasibility of the membrane reforming system is demonstrated in a facility producing pure hydrogen at 15 Nm^3/h for about 1500 h with 40 start-up and shut-down cycles [17,18]. A typical set of operation results from this reformer demonstration include the reaction temperature and pressure of 550°C and 0.83 MPa, natural gas conversion of 76.2%, hydrogen production rate of 21.8 Nm^3/h, hydrogen purity of 99.999%, hydrogen production efficiency of 73.6% including 3.83 kW auxiliary power. Follow-on demonstration in another 40 Nm^3/h hydrogen production facility was done to verify system efficiency, long-term durability, and reliability of the membrane reformer system.

A schematic of a nuclear-heated recirculation-type membrane reformer is shown in Figure 11.19 [18]. Merits of the nuclear-heated recirculation-type membrane reformer,

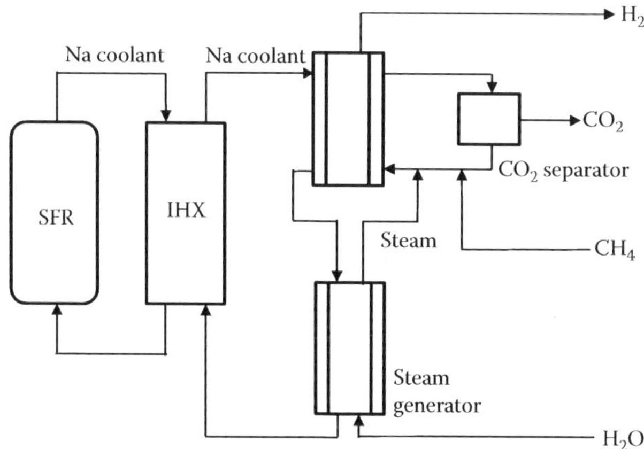

FIGURE 11.19
Schematic of the SFR-heated steam reforming of methane with a membrane reformer.

relative to the other steam reforming processes, include: (1) supply of nuclear heat at the medium temperature range of several nuclear reactor types including the SFR; (2) about 30% reduction of overall carbon dioxide emission relative to the conventional reforming process with methane or natural gas combustion; (3) separated carbon dioxide product from within the process to simplify carbon sequestration requirement; and (4) compact surface area of membrane modules through the recirculation of reaction product gases including residual hydrogen in a closed loop.

11.4 Conclusions

SFR is a Generation IV technology that utilizes fast neutron spectrum and enables a closed fuel cycle. The feasibility studies performed by JAEA and electric utilities of Japan select the SFR with MOX fuel to be the most promising concept of an FR system.

There are several SFR system development programs in the world. France has decided to launch the design work of a prototype FR, which will be commissioned in 2020. Russia, China, and India have ambitious plans for commercial deployment of FR systems by the middle of this century. JAEA launched a new fast reactor cycle technology development (FaCT) project in cooperation with the Japanese electric utilities in 2006 [19]. In line with the directions given by the government of Japan, design and experimental studies regarding the SFR and fuel cycle will be implemented in order to present the conceptual designs of demonstration and commercial FR cycle facilities by 2015. The development plans that have already been drawn call for the demonstration FR to become operational around 2025 and the commercialized FR cycle system to be deployed before around 2050.

SFR allows hydrogen production through (1) conventional electrolysis with electricity it generates, (2) any of the thermochemical and hybrid processes (Chapter 5) from water with a heat input requirement compatible with the SFR outlet temperature of 550°C. The hybrid process being specifically investigated by JAEA for the SFR application is discussed in this chapter, and (3) the membrane steam reforming of methane proposed for the SFR and introduced in the chapter.

References

1. The U.S. DOE Nuclear Energy Research Advisory Committee and the Generation IV International Forum, a technology roadmap for Generation IV nuclear energy system, December 2002.
2. Japan Atomic Energy Agency and The Japan Atomic Power Company, Phase II final report of feasibility study on commercialized fast reactor cycle systems-executive summary, March, 2006.
3. Sagayama, Y., Feasibility study on commercialized fast reactor cycle systems (1) current status of the phase-II study, *GLOBAL 2005*, paper no. 380, Tsukuba, Japan, October 9–13, 2005.
4. Kotake, S. et al., Feasibility study on commercialized fast reactor cycle systems: Current status of the FR system design, *GLOBAL 2005*, paper no. 435, Tsukuba, Japan, October 9–13, 2005.
5. Ichimiya, M. et al., A promising sodium-cooled fast reactor concept and its R&D plan, *GLOBAL 2003*, paper 0434, Louisiana, USA, November 16–20, 2003.

6. Hishida, M. et al., Progress on the plant design concept of sodium-cooled fast reactor, *GLOBAL 2005*, paper no. 068, Tsukuba, Japan, October 9–13, 2005.
7. Shimakawa, Y. et al., An innovative concept of a sodium-cooled reactor to pursue high economic competitiveness, *Nucl. Technol.*, 140, 1–16, 2002.
8. Mizuno, T. et al., Advanced oxide fuel core design study for SFR in the "Feasibility Study" in Japan, *GLOBAL 2005*, paper no. 434, Tsukuba, Japan, October 9–13, 2005.
9. Kubo, S. et al., Status of conceptual safety design study of Japanese sodium-cooled fast reactor, *GLOBAL 2005*, paper no. 221, Tsukuba, Japan, October 9–13, 2005.
10. Niwa, H. et al., LMFBR design and its evolution: (3) Safety system design of LMFBR, *GENES4/ANP2003*, no. 1154, Kyoto, Japan, September 15–19, 2003.
11. Morihata, M. et al., Development of self actuated shutdown system for FBR in Japan, *Proc. 5th Int. Conf. Nuclear Engineering*, Nice, France, May 26–30, 1997.
12. Nakagiri, T. et al., Development of a new thermochemical and electrolytic hybrid hydrogen production system for sodium cooled FBR, *JSME Int. J. Ser. B*, 49, 302–308, 2006.
13. Takai, T. et al., Development of a new thermo-chemical and electrolytic hybrid hydrogen production process utilizing the heat from medium temperature heat source—development of the 1 NL/h hydrogen production experimental apparatus, *Proc. Hydrogen and Fuel Cells 2007*, April 29–May 2, Vancouver, Canada, pp. 233–242.
14. Chikazawa, Y. et al., A system design study of a fast breeder reactor hydrogen production plant using thermochemical and electrolytic hybrid process, *Nucl. Technol.*, 155(3), 340–349, 2006.
15. Doizi, D. et al., Study of the hydrolysis reaction of the CuCl thermochemical cycle using optical spectrometries, *OECD/NEA Fourth Information Exchange Meeting on the Nuclear Production of Hydrogen*, Chicago, 2009.
16. Hori, M. et al., Synergistic hydrogen production by nuclear-heated steam reforming of fossil fuels, *Progr. Nucl. Energy*, 47(1–4), 519–526, 2005.
17. Tashimo, M., Advanced design of fast reactor-membrane reformer (FR-MR), *OECD/NEA Second Information Exchange Meeting on Nuclear Production of Hydrogen*, Argonne, 2003.
18. Yasuda, I. and Shirasaki, Y., Development of membrane reformer for highly-efficient hydrogen production from natural gas, *15th World Hydrogen Energy Conference*, Yokohama, 2004.
19. Sagayama, Y., Launch of fast reactor cycle technology development project in Japan, *GLOBAL 2007*, No. 175710, Boise, Idaho, September 9–13, 2007.

12

Gas Fast Reactor

Yoshiyuki Inagaki and Takamichi Iwamura

CONTENTS

12.1 Overview

The gas-cooled fast reactor (GFR) is one of the six promising Generation-IV nuclear energy system concepts under international development. In accordance with the Generation-IV Technology Roadmap [1], the GFR system features a fast-neutron spectrum and closed fuel cycle for efficient conversion of fertile uranium and management of actinides. A full actinide-recycled fuel cycle with on-site fuel-cycle facilities is envisioned. The fuel-cycle facilities can minimize transportation of nuclear materials and will be based on either advanced aqueous, pyrometallurgical, or other dry processing options. Several fuel forms are being considered for their potential to operate at very high temperatures and to ensure an excellent retention of fission products: composite ceramic fuel, advanced fuel particles, or ceramic clad elements of actinide compounds. Core configurations are being considered based on pin- or plate-based fuel assemblies or prismatic blocks.

The reference reactor design is a 600 MWt/288 MWe, helium-cooled system operating with an outlet temperature of 850°C using a direct Brayton cycle gas turbine. The high outlet temperature of the helium-cooled GFR system makes it possible to produce multiple energy products at high thermal efficiency. Electricity is generated at greater than 45% net thermal efficiency. Hydrogen can be produced by a wide range of processes as will be discussed in this chapter.

The reactor system designs have been scaled up to thermal power of 2400 MWt in Japan [2] and in France [3,4]. The GFR is coupled to a direct-cycle helium turbine for electricity generation in the case of Japanese design and an indirect combined cycle through a compact intermediate heat exchanger in the case of French design. It can utilize its high-temperature process heat for cogeneration or dedicated production of hydrogen via various process routes including high-temperature electrolysis of steam, thermochemical, and hybrid processes. The GFR through the combined fast spectrum and full actinide-recycled fuel cycle minimizes the production of long-lived radioactive waste. The GFR's fast spectrum also makes it possible to use available fissile and fertile materials including depleted uranium considerably more efficiently than thermal spectrum gas reactors with once-through fuel cycles.

During 1999 through 2005, Japan Atomic Energy Agency (JAEA) and domestic electric utilities had been conducting Feasibility Studies on commercialized fast reactor (FR) Systems [2]. In these studies, the preliminary concepts of various types of fast breeder reactors such as sodium-cooled, heavy metal-cooled, and gas-cooled reactors, and so on had been designed and evaluated. For the GFRs, the preliminary design concepts of all the possible combination of carbon dioxide and helium coolants, steam turbine and gas turbine power generation cycles, sealed pin and coated particle fuels (CPFs) had been examined incorporating many innovative technologies. The feasibility studies were focused on three GFR concepts based on (1) carbon dioxide-cooled FR using pin-type fuel, (2) helium-cooled FR using pin-type fuel, (3) helium-cooled FR using CPF for the final selection of the most promising concept. The studies concluded by selecting the helium-cooled FR using CPF as the promising GFR reference concept due to its important advantages of economic competitiveness and reactor safety. Figure 12.1 shows the plant concept of the selected helium-cooled GFR.

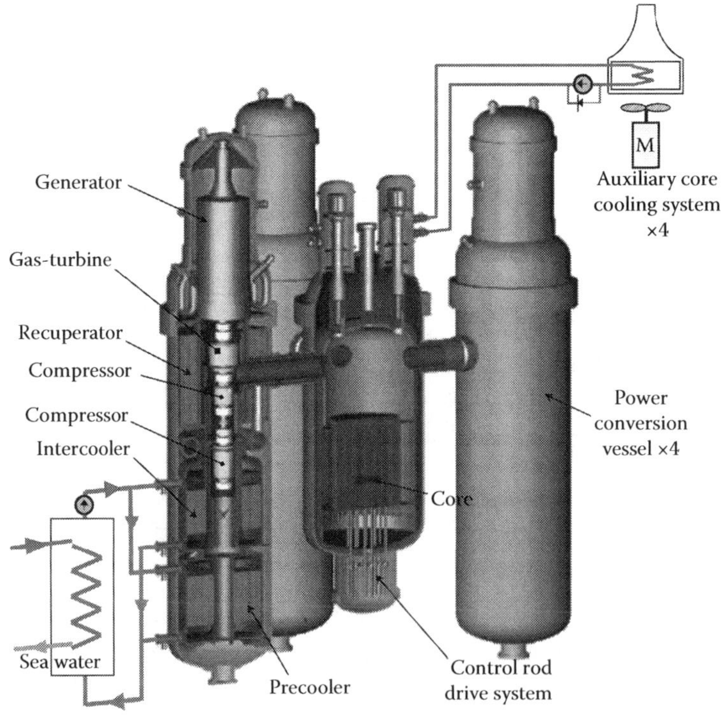

FIGURE 12.1
Plant layout of helium-cooled GFR.

12.2 Reactor and Major System Designs

12.2.1 Plant System

The main characteristics and system layout of the GFR plant concept in the Japanese proposal are shown in Table 12.1 and Figure 12.2, respectively [5].

The thermal power is 2400 MWt and the electric power is 1120 MWe. A reactor pressure vessel (RPV) and four power conversion vessels (PCVs) are installed in a container. A gas turbine power generation system is housed in a PCV. CPF is employed as a fuel in this design. The CPFs are cooled directly by the helium coolant. The direct cooling of the CPFs enables a good coolability performance and a large fuel to core volume ratio. The reactor inlet/outlet temperature is 460°C/850°C. Compared to sodium-cooled reactors, intermediate heat transfer systems and cover gas systems and so on are not necessary and these systems are reduced.

Helium (He) gas is chemically inert and has good compatibility with fuels and structures even in the high-temperature range and is not activated by fast neutron radiations. By using He gas, the sodium–water reaction which poses a problem at sodium-cooled FR can be eliminated, and He gas can make inspections and surveillances easily because of its transparency.

12.2.2 Reactor Vessel and Internal Structures

The reactor structural design concept is shown in Figure 12.3. The reactor consists of two vessels, a RPV and an inner vessel, which form a coolant inlet flow path. Four PCVs are connected to the RPV by concentric cross ducts. The nozzles protruded from the RPV at each 90° angle support the RPV. At the upper head of the RPV, auxiliary core cooling system heat exchangers (ACCS HXs) and standpipes for fuel exchange are installed. At the lower head of the RPV, control rod drive assemblies are installed. Heat-treated Mn–Mo–Ni steel, which is generally used for PWR, is employed for the RPV considering a cost reduction. The inner vessel is made of modified 9Cr steel.

Helium coolant flows from turbine through the annulus of cross ducts, flows down between the RPV and the inner vessel, goes up through the core and returns to gas turbine

TABLE 12.1

Plant Main Characteristics of He-Cooled GFR using CPF

Electric power	1120 MWe
Thermal power	2400 MWt
Coolant temperature (in/out)	460°C/850°C
Coolant flow rate	4262 ton/h
Coolant pressure	6 MPa
Primary circuit loops	4 loops
Fuel type	Coated particle fuel (nitride fuel)
Average fuel burn up	100 GWd/t
Breeding ratio	1.2
Core power density	52 MW/m^3 (Av.)
Reactor pressure vessel	Steel
Core auxiliary cooling system	Direct core cooling 50% (FC) × 4
Reactor containment	Steel

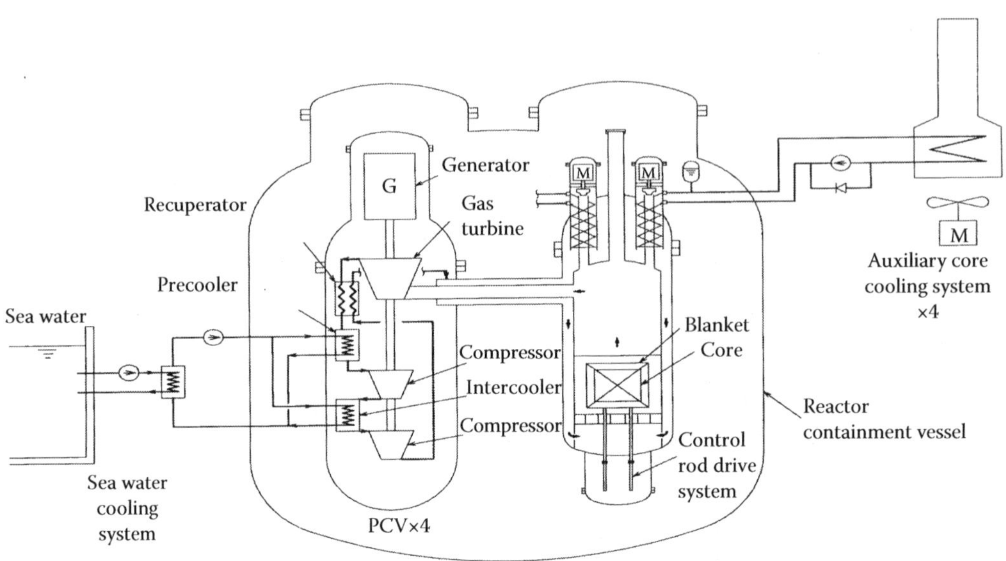

FIGURE 12.2
Plant system concept.

power-generating systems through the inner pipe of cross ducts. In order to thermally insulate inlet coolant from outlet one, a thermal insulator is installed at the inside of the inner vessel. As a candidate thermal insulator, KAOWOOL which is made of ceramic fiber is considered. A B4C layer is also installed under the thermal insulator to shield fast neutron streaming from the core.

12.2.3 Reactor Core and Fuel

The core structure and core main characteristics are shown in Figure 12.4. The height and the equivalent diameter of the active core are 1.8 and 5.6 m, respectively. The core consists of 543 core fuel assemblies, 34 main and back up control rods and six passive control shutdown equipments. Mixed nitride fuel is adopted as fuel material. The core burn up is 100 GWd/t and the breeding ratio is 1.2. The fuel inventory of the CPFs is restricted by the CPF diameter and the coated layer thickness. Therefore, the plutonium enrichment that expands the burn up and the quantity of uranium-238 that raises the breeding ratio cannot be increased simultaneously. The reactor power is determined based on the diameter of the steel reactor pressure vessel and the core power density.

The CPF and fuel assembly concepts and coolant flow paths are shown in Figure 12.5. TiN is selected as a reference coating material. Mixed nitride fuel is employed as the fuel kernel material considering the increase of the fuel inventory and the core fuel performance such as the reactivity and the breeding ratio. In this design, the CPFs put in a fuel compartment without any compound and these are directly cooled in order to increase the fuel to core volume ratio and heat transfer areas.

The fuel assembly consists of a cylinder rod element called fuel compartment, axial blankets and shields at the upper and lower part of the rod. The fuel compartment consists of an inner and an outer cylinder and the CPFs are packed between these two cylinders. The two cylinders are made of SiC/SiC composite material, which is the fabric weaved by SiC fibers and coated by a thin layer of SiC. Helium coolant flows in from the bottom of the fuel

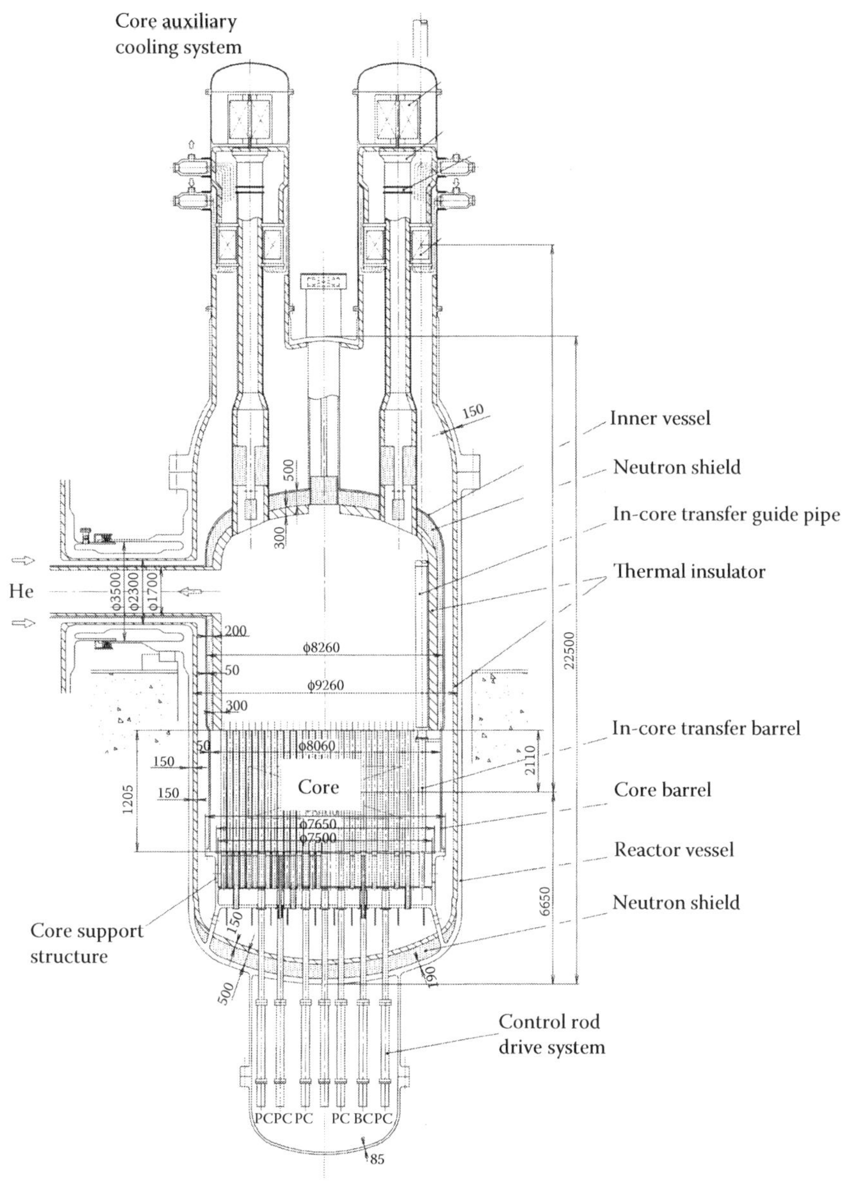

FIGURE 12.3
Reactor internal structure.

compartment and flows up axially in the inner side of the inner cylinder, and flows radially through the porous inner cylinder, CPF-packed beds, the outer porous cylinder, and flows up axially to an upper plenum. The radial flow in the CPF-packed beds with a small axial flow rate under normal and accident natural circulation conditions is realized by the larger pressure loss coefficient of the inner cylinders compared to the CPF-packed beds and the outer cylinders. The fuel assembly structure is designed so as to maintain its integrity under high-temperature accident conditions by the ceramics structures and to obtain the smallest total pressure loss (about 0.4 MPa) by short flow paths.

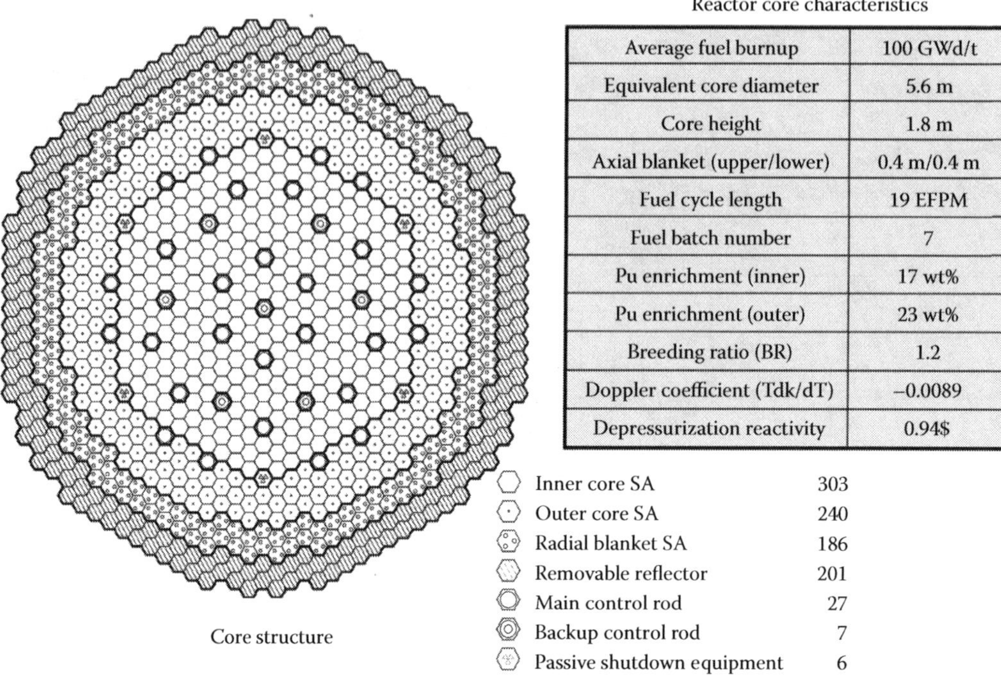

Average fuel burnup	100 GWd/t
Equivalent core diameter	5.6 m
Core height	1.8 m
Axial blanket (upper/lower)	0.4 m/0.4 m
Fuel cycle length	19 EFPM
Fuel batch number	7
Pu enrichment (inner)	17 wt%
Pu enrichment (outer)	23 wt%
Breeding ratio (BR)	1.2
Doppler coefficient (Tdk/dT)	−0.0089
Depressurization reactivity	0.94$

FIGURE 12.4
Core structure and core characteristics.

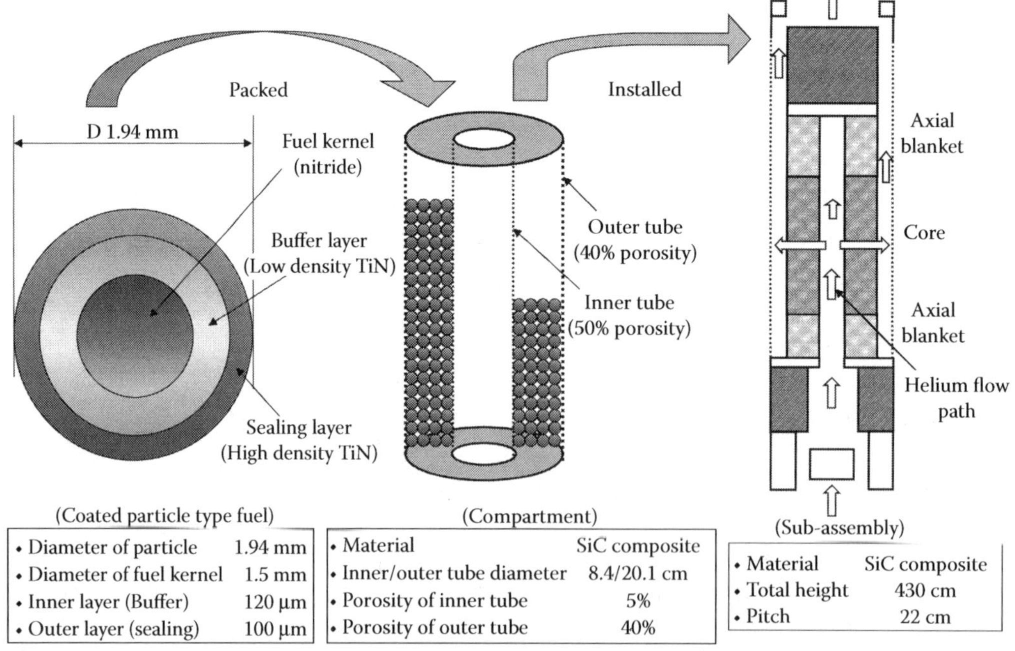

(Coated particle type fuel)	
• Diameter of particle	1.94 mm
• Diameter of fuel kernel	1.5 mm
• Inner layer (Buffer)	120 μm
• Outer layer (sealing)	100 μm

(Compartment)	
• Material	SiC composite
• Inner/outer tube diameter	8.4/20.1 cm
• Porosity of inner tube	5%
• Porosity of outer tube	40%

(Sub-assembly)	
• Material	SiC composite
• Total height	430 cm
• Pitch	22 cm

FIGURE 12.5
Coated particle fuel, fuel assembly, and coolant flow path.

In the horizontal flow cooling concept, since a CPF touches a coolant directly, radioactive contamination problem of a primary cooling system may be posed according to the defect of a coating layer. In order to prevent the release of fuel materials, an alternative fuel assembly concept (block-type vertical flow cooling concept) with second boundary is considered. The block-type vertical flow cooling concept is shown in Figure 12.6. It is a concept which is filled up with CPF in a hexagonal prism vessel in which cooling tubes are arranged. The coolant passes through a cooling tube, and removes the heat of CPF. Improvement in the core performance is a target of the concept which fills up gap between CPFs with SiC.

12.2.4 Main Cooling System (Gas Turbine)

The gas turbine power generation system is composed of a generator, a low-pressure compressor, a high-pressure compressor, a recuperator, a precooler, and an intercooler. Turbomachines are coupled to a single-shaft. The height and the inner diameter of the PCV are 43 and 8 m, respectively. The heat cycle is closed, recuperated, and intercooled cycle, of which the plant efficiency reaches as high as 47%.

12.2.5 Auxiliary Core Cooling System

An ACCS is shown in Figure 12.7. The ACCS removes heat from the core at normal operation, and at accident conditions. In the case of loss-of-forced circulation events due to multiple failures of emergency diesel power generation systems, the core heat is removed by the natural circulation in the reactor through HXs then transferred to secondary loops and thrown away to the air.

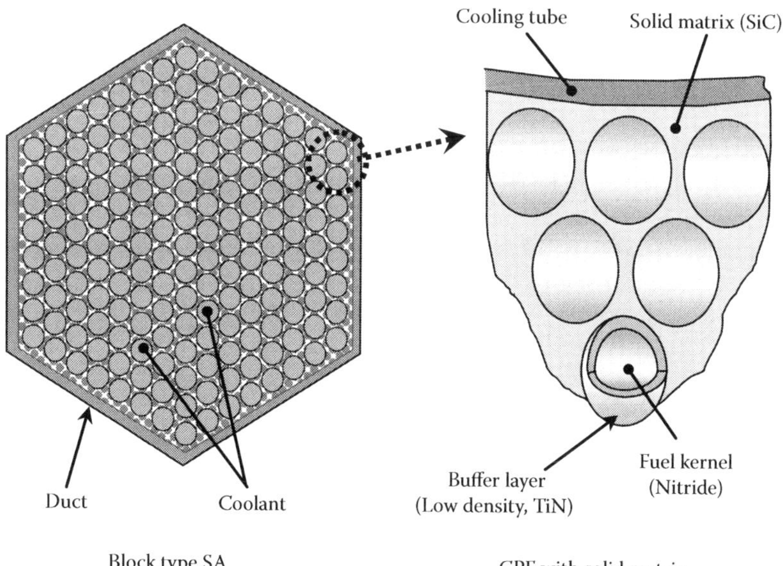

FIGURE 12.6
Block-type vertical flow cooling.

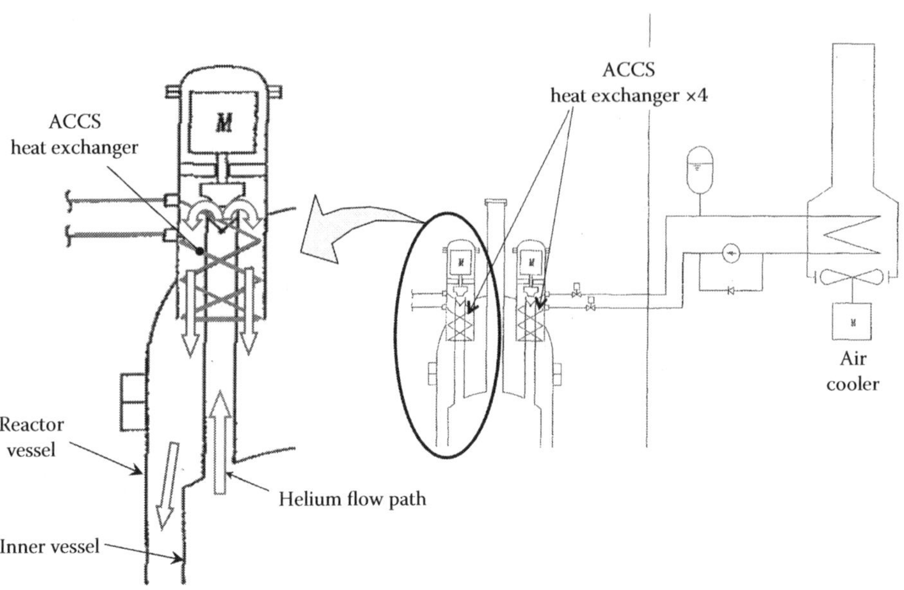

FIGURE 12.7
Auxiliary core cooling system concept.

When ACCS is operating, directional control check valves at the inlet of HXs open and coolant flows from reactor upper plenum into the HXs, down the path between the RPV and the inner vessel and flows into the core from the lower plenum. The inlet valves open automatically due to the pressure difference by a gas turbine generator trip between the inner and the outer side of the inner vessel. The coolant of the secondary side of the ACCS is pressurized water. The entire ACCS comprises four systems and each system has 50% heat rejection capacity by forced circulation.

12.2.6 Safety Analysis

The safety goal in this design is set to the accomplishment of the core meltproof characteristics. The core melt and re-criticality should be avoided without any active component actuations even under severe depressurized accident conditions, namely coolable geometries of the core are kept and the core is cooled even with a depressurized accident, without scram and without active component actuation. However, this is not a requirement of safety rules. FP release from the core is allowed in core destructive accident (CDA) condition and it is retained in the containment.

An allowable limit temperature of the core is assumed to be 2200°C based on literature as the maximum temperature at that SiC/SiC fuel compartments can maintain the coolable geometry. Safety analysis of the depressurized accident, without scram and natural circulation is performed.

Temperature of the calculated result is shown in Figure 12.8. As a result of safety analysis, the core-melt would be avoided without any active component actuations even under severe depressurized accident conditions, namely, coolable geometries of the core should be kept and the core should be cooled only with natural circulations. The temperature of CPF and fuel compartment reaches the maximum temperature 2184°C after 4500 s. The maximum temperature is below the assumed allowable temperature 2200°C.

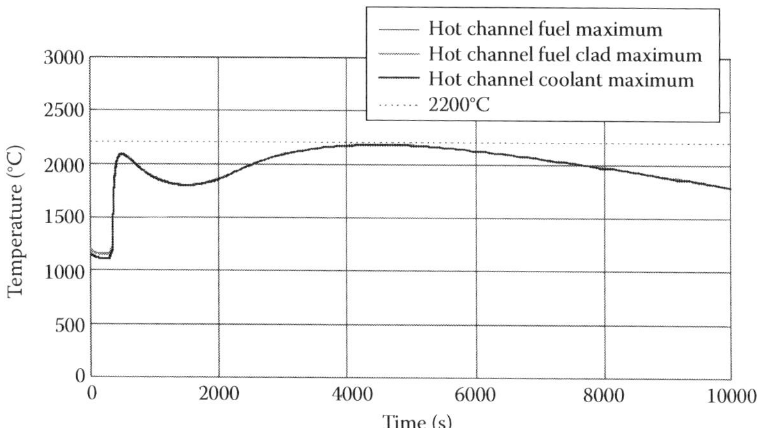

FIGURE 12.8
Severe accident analysis results (loss of coolant without scram and natural circulation).

12.2.7 Technology Requirements

Many innovative technologies must be developed in future in order to realize this plant concept. Therefore, future R&D is essential, in particular, the following are the most important items influencing the feasibility of the design concept.

1. *Coated particle fuel productivity, strength, and irradiation characteristics:* It is necessary to establish the technical development of fabrication of large sphere-coated particle fuel and mass-production technology of TiN coating layer. The examination of high-temperature capability and irradiation performance should be performed.
2. *Fuel assembly strength and irradiation characteristics:* The examination of high-temperature capability and irradiation performance of ceramics-composite, which is fuel assembly structure, should be performed. The fabrication of fuel assembly (uniform porosity, pressure drop characteristic, etc.) which is the formation conditions of a horizontal flow cooling concept should be checked.
3. *Thermal insulator and liner material:* The high-temperature-proof material should be developed, and the examination of structure is performed under high-temperature conditions.
4. *Gas turbine:* Since it is a large-sized gas turbine with a single-shaft in closed cycle, it should be necessary to perform element development of each apparatus (turbine, magnetic bearing, regenerative heat exchanger, etc.). The performance and characteristic examination of the gas turbine power generation system should be performed using helium loop equipment.
5. *Passive shutdown equipment:* About self-actuated shutdown system (SASS) of an electromagnetic method utilizing the Curie point characteristic, a heat sensor alloy should be developed, and the SASS operation characteristic should be corroborated.
6. *ACCS and directional-control check valve:* It is necessary to develop the high-temperature-proof material SiC. The examination of structure and reliability (high-temperature durability test, etc.) of the check valve should be performed.

7. *Natural circulation core cooling characteristics in the reactor:* The neutron flux distribution in a core and the cooling characteristic of a primary system, should be checked when an auxiliary reactor core cooling system is operated. The whole characteristic of natural circulation should be grasped by a scale model.

12.3 Hydrogen Production Systems

Because the GFR reactor design conditions including the coolant, primary pressure, and reactor outlet coolant temperature are similar to those of the high temperature gas-cooled reactor (HTGR), the hydrogen production processes such as the high-temperature steam electrolysis and the thermochemical process that are generally considered for use with the HTGR (see details in Chapter 10) can also be applied to the GFR. The major difference between the two reactors is that the GFR outlet coolant temperature is limited to 850°C while the HTGR to 950°C. As a result, the efficiency of hydrogen production with the GFR is slightly lower than that with the HTGR, given the same hydrogen production process.

Furthermore, the hydrogen production technologies available to the medium temperature (550–600°C) SFR such as the hybrid thermochemical process and the methane steam reforming (see descriptions on these medium-temperature processes in Chapter 11) are also available to the high-temperature GFR at improved thermal efficiency.

12.4 Conclusions

GFR is a Generation IV technology that utilizes fast neutron spectrum and enables closed fuel cycle and on-site full actinide recycle. The feasibility studies performed by JAEA and domestic electric utilities selected the GFR with MOX fuel to be the promising concept of fast reactor system [2]. The French Commissariat à l'Energie Atomique (CEA) considers it to be a promising concept as it offers the benefits of fast spectrum for fuel resources saving, very limited use of fertile blankets for proliferation resistance, and high-temperature reactor using helium as coolant [3].

GFR is an innovative system and its development requirements are given in Section 12.2.7. No demonstration GFR has been built. The ALLEGRO project aims at building and operating the first GFR, an 80 MWt experimental reactor with the main objective to validate GFR specific technologies including fuel element and subassembly, safety systems. The start of the operation of ALLEGRO is planned by 2020 [4].

GFR allows hydrogen production via various potential routes that could be supported by the reactor's high temperature (850°C) and wide power range (600–2400 MWt). These routes include (1) conventional water electrolysis (Chapter 3) and steam electrolysis (Chapter 4) with electricity it generates, (2) medium-temperature thermochemical and hybrid processes (among the processes in Chapter 5) from water with the process heat it generates, and (3) GFR-heated fossil fuel conversion (Chapter 7) at reduced carbon dioxide emission.

References

1. The U.S. DOE Nuclear Energy Research Advisory Committee and the Generation IV International Forum, a technology roadmap for Generation IV nuclear energy system, December 2002.
2. Japan Atomic Energy Agency and The Japan Atomic Power Company, Phase II final report of feasibility study on commercialized fast reactor cycle systems—executive summary, March 2006.
3. Malo, J. et al., Gas cooled fast reactor 2400 MWth, status on the conceptual design studies and preliminary safety analysis, *Proc. of 2009 International Congress on Advances in Nuclear Power Plants (ICAPP '09)*, Tokyo, Japan, May 10–14, 2009.
4. Poette, C. et al., GFR Demonstrator ALLEGRO Design Status, *Proc. of 2009 International Congress on Advances in Nuclear Power Plants (ICAPP '09)*, Tokyo, Japan, May 10–14, 2009.
5. Konomura, M. et al., A promising gas-cooled fast reactor concept and its R&D plan, GLOBAL 2003, New Orleans, USA, November 2003.

13

Fluoride Salt Advanced High-Temperature Reactor

Per F. Peterson and Edward D. Blandford

CONTENTS

13.1 Overview of the Advanced High-Temperature Reactor Technology

13.1.1 Introduction

The thermophysical and chemical properties of liquid fluoride salts [1,2] create unique capabilities for use as primary coolants for high-temperature reactors (HTRs). One of the most important properties is their uniquely high volumetric heat capacity, as shown in Table 13.1, which enables significantly more compact liquid salt primary loop components than equivalent equipment in pressurized water reactors (PWRs).

While fluoride salts were developed extensively for the fluid-fueled molten salt reactor (MSR) in the 1960s, only recently have these salts been studied extensively for application as coolants for solid-fueled HTRs, termed Advanced High-Temperature Reactors (AHTRs) [3–5]. This chapter provides an overview of the fundamental aspects of the proposed reactor systems where high-temperature hydrogen production can be performed.

TABLE 13.1

Comparison of PB-AHTR Primary (Flibe) and Intermediate (Flinak) Coolant Thermophysical Properties with Alternative Coolants and Materials (Values at 700°C, Except for 290°C for Water)

Material	T_{melt} (°C)	T_{boil} (°C)	ρ (kg/m³)	ρcp (kJ/m³ C)	k (W/m² C)
$^7Li_2BeF_4$ (flibe)	459	1430	1940	4540	1.0
LiF–NaF–KF (flinak)	454	1570	2019	4060	0.60
Sodium	97.8	883	790	1000	62.0
Lead	328	1750	10,540	1700	16.0
Helium (7.5 MPa)	—	—	3.8	20	0.29
Water (7.5 MPa)	0	100	732	4040	0.56
Graphite	—	—	1700	3230	200.0

Source: Adapted from D. F. Williams, L. M. Toth, and K. T. Clarno, Assessment of candidate molten salt coolants for the advanced high-temperature reactor (AHTR), Oak Ridge National Laboratory, ORNL/TM-2006/12, March 2006; D. F. Williams, Assessment of candidate molten salt coolants for the NGNP/NHI heat-transfer loop, Oak Ridge National Laboratory, ORNL/TM-2006/69, June 2006.

13.1.2 History of MSR Technology

The history of molten salts as working fluids for nuclear reactors begins with Ed Bettis and Ray Briant of Oak Ridge National Laboratory (ORNL) post-World War II. They were in charge of designing a nuclear-powered aircraft using molten salts due primarily to the high-temperature performance and overall chemical stability of salts. It was in 1954 that the first small MSR, the aircraft reactor experiment (ARE), was built and achieved a power of 2.5 MWt. The circulating fuel was comprised of a $NaF–ZrF_4–UF_4$ mixture. The maximum operating temperature of the fuel was 882°C. The military application demand for nuclear powered aircraft decreased sharply toward the latter half of the 1950s as attention shifted toward ballistic missile technology [6,7].

Following the closing of the ARE in 1956, Alvin Weinberg wanted to see whether this technology could be adapted for civilian power reactors and so began the MSR Program. Shortly after, the MSR experiment (MSRE) was approved and design started in the summer of 1960 at ORNL. The MSRE reactor was a cylinder measuring 1.37-m-diameter × 1.62-m-high in order to minimize neutron leakage. The core operated at a power of 8 MWt, which allowed it to still be classified as an experimental reactor. It was intended to simulate only the fuel stream of a two-fluid breeder reactor. Ultimately, the MSRE was built for just over $8 Million (1961$) [7] and took approximately 3 years to construct. The initial fuel for the MSRE was $^7LiF–BeF_2–ZrF_4–UF_4$ [8], while the intermediate coolant was clean $^7LiF–BeF_2$. In 1968, the original fuel was replaced with ^{233}U where it was the first reactor to have been run on with this fissile fuel. The moderator used was graphite where the structural materials selected was primarily Hastelloy N. The MSRE ran from 1965 to 1969 at a typical operating temperature of 600°C [9]. During operation the concentrations of CrF_2 in the fuel salt were observed to rise by a level indicating an average corrosion rate of 4 mills per year, and after shutdown it was found that fission products had caused intergranular attack. In contrast, the coolant salt loop experienced no detectable corrosion after over 26,000 h of operation [10], which has important implications for the use of this salt as a primary coolant in the AHTR.

For a variety of reasons, the MSR program in the United States was ultimately shut down in the middle of the 1970s. The objectives of the MSR program were shifting toward an efficient thorium breeder program known as the MSR breeder (MSBR) program that competed with the uranium liquid metal fast breeder reactor (LMFBR) program being developed at

ANL [7]. It was not until the early 2000s with the introduction of the liquid salt very high temperature reactor (LS-VHTR), that research in molten salts as reactor primary fluids was renewed in the United States. The LS-VHTR was essentially a modified helium-cooled VHTR using liquid salt as the working fluid, which operates at near atmospheric pressure and substantially greater power density. This reactor configuration with the fuel being separated from the coolant represented a significant departure from the liquid fuel MSR technology being developed in the 1960s. Most recently in the Generation IV roadmap [11], the term AHTR has been used to describe fluoride salt-cooled HTR technology that uses solid fuel.

13.2 AHTR System Concept

The PB-AHTR is the latest design based on the original AHTR concept to use liquid fluoride salt to cool coated-particle HTR fuel [12]. Coated-particle fuel has recently undergone rapid design evolution and is currently undergoing extensive irradiation testing [13]. The PB-AHTR utilizes key technological features developed for gas-cooled high-temperature thermal and fast reactors, sodium fast reactors, and MSRs. The modular 900-MWt PB-AHTR is the reference design for this chapter [14], but it is expected that the core and system configuration will evolve with time (Figure 13.1).

13.2.1 System Design

In Figure 13.1, the primary loop of the most recent design of PB-AHTR connects the core and the intermediate heat exchanger (IHX) modules. During a loss of forced circulation (LOFC) transient (i.e., after a primary pump trip), a natural circulation flow loop is formed between the core and a set of direct reactor auxiliary cooling system (DRACS)

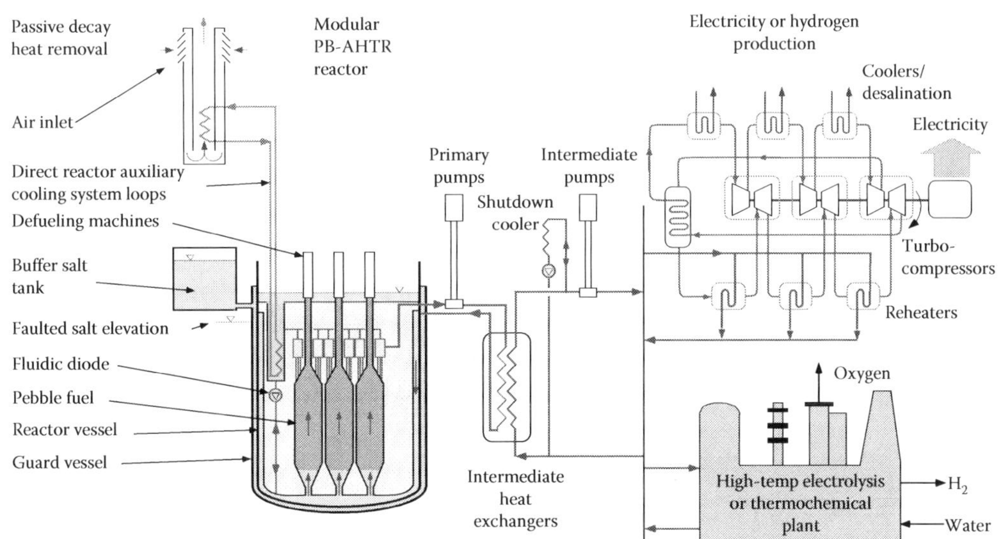

FIGURE 13.1
Simplified schematic of modular PB-AHTR system and possible applications.

heat exchangers (DHX modules). The DRACS heat exchangers transfer heat by natural circulation flow of a DRACS salt from the DHX modules to air cooled natural draft heat exchangers (NDHXs) cooled by outside ambient air. Under forced circulation the reverse bypass flow through the DHX is minimized by a fluidic diode. It should be noted that the PB-AHTR, like all liquid salt technologies, is susceptible to overcooling transients where after a substantial time period the salt can freeze in the primary loop. As shown in Figure 13.1, the IHX's intermediate loop which can be used to deliver thermal power to a variety of applications such as process heat for hydrogen generation or electricity generation. The various hydrogen production options for the AHTR are discussed further in Section 13.4 where HTE and IS are the preferred technology options (Figures 13.2 and 13.3).

The PB-AHTR also implements a novel buoyant shutdown rod design for passive reactivity control. The insertion of its shutdown rod elements provide negative temperature feedback in order to augment the negative feedback already provided by the negative coolant and fuel temperature reactivity coefficients [15]. Insertion of the shutdown rods occurs due to buoyancy forces generated by the difference between the density of the control element and the reactor coolant during an unexpected reactor transient. A heavy metallic driver element is suspended by a magnetic latch system above each shutdown rod but is not physically connected to the shutdown rod. In the event of a reactor scram signal, the electromagnetic coupling holding the drive elements are de-energized thus causing the elements to drive the shutdown rods into the active core region via gravity. If this active insertion mechanism does fails to operate, buoyancy forces cause the shutdown rods to insert anyhow.

13.2.2 Fuel and Pebble Channel Assemblies

Due to the fact that liquid salts have very high volumetric heat capacity, the necessary recirculating power is two to three orders of magnitude smaller than for equivalent modular helium reactors (MHRs). In the current modular PB-AHTR design, the pebbles are selected to have 3-cm diameter, with a 2.5-mm thick graphite shell and a low-density inert graphite kernel at the center of the pebble, creating an annular fuel region in the pebble. Reducing the pebble diameter by a factor of two compared to helium-cooled reactor pebbles doubles the pebble surface area per unit of core volume, and halves the thermal conduction length scale in the pebble. The annular fuel configuration reduces the temperature difference by another factor of approximately two, allowing the power density to be increased by a factor of 8 while maintaining the same temperature difference from the surface to the center of the pebble.

The modular PB-AHTR also adopts a considerably different core design than alternative pebble bed reactors. In the modular PB-AHTR, the pebble fuel moves through a large number of channels inside seven hexagonal Pebble Channel Assemblies (PCAs), shown in Figure 13.4. The PCAs are designed to be removable and replaceable, and the modularity of the plant design allows the pilot plant to use a single PCA. The advantages of the PCAs include

1. The moderation provided by the PCA's graphite allows for an increase in the heavy metal loading in the pebbles, reducing the number of required pebbles and the spent fuel volume potentially by a factor of around 2.

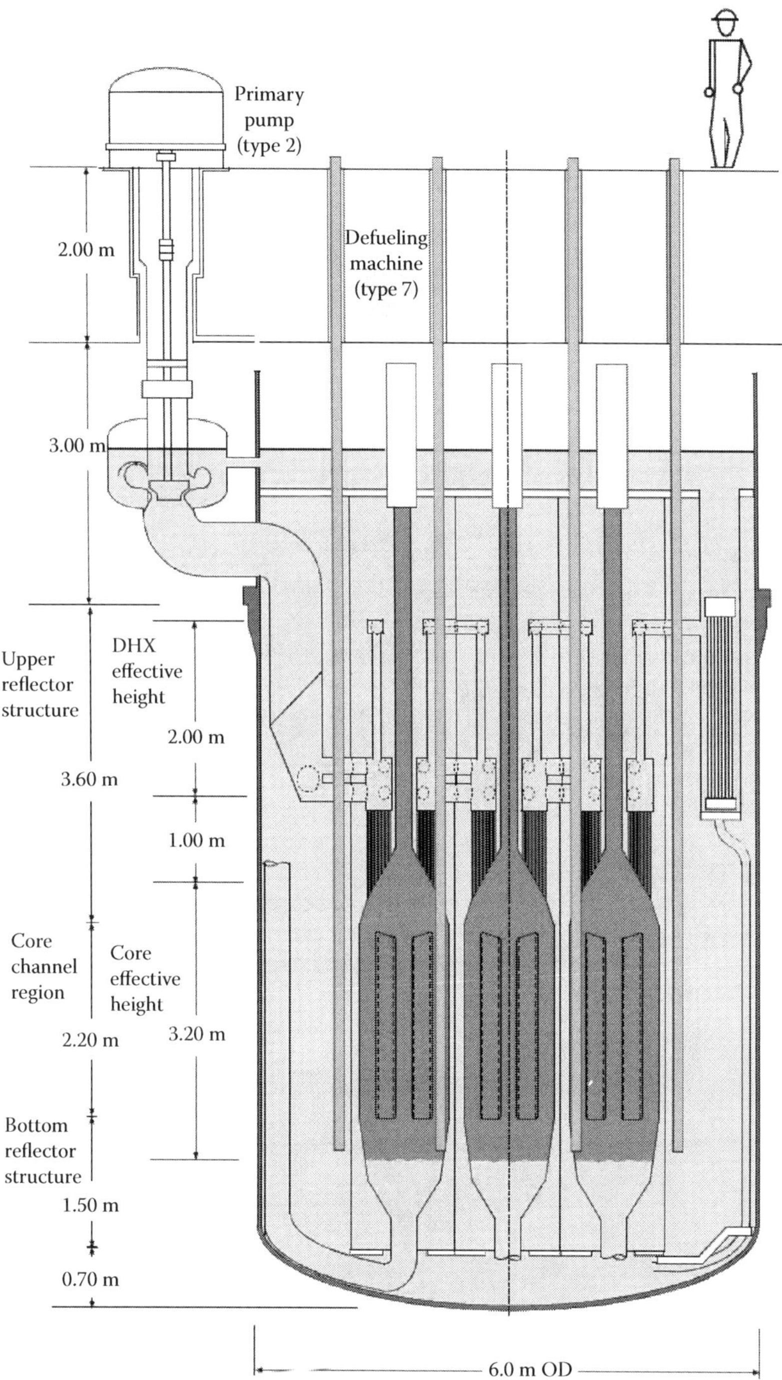

FIGURE 13.2
Elevation view of the modular PB-AHTR. The orange region indicates where the pebble fuel is recirculated through a series of channels consisting of graphite pebble channel assemblies.

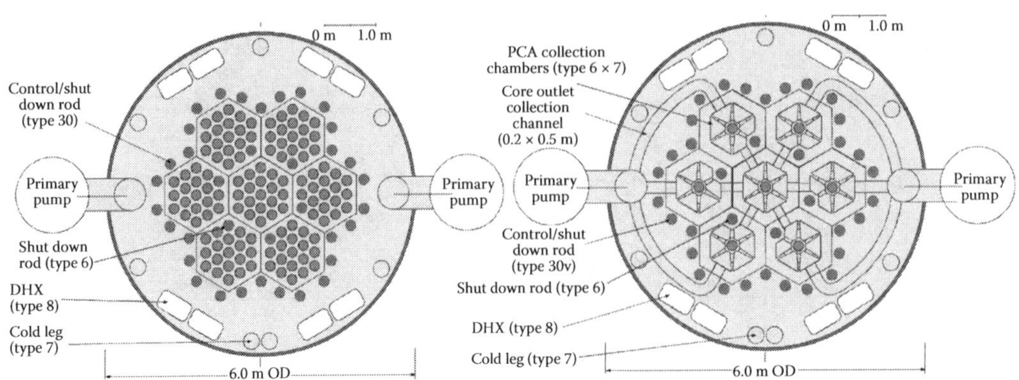

FIGURE 13.3
Plan view of the modular PB-AHTR at the pebble channels elevation (left) and at the core outlet plenum elevation (right).

2. The coolant void fraction in the core is reduced by approximately a factor of 2, reducing parasitic neutron absorption in the coolant and increasing the discharge burn up.
3. The multiple channel configurations allow a simple approach to a 2-zone core, where pebbles discharged from the outer zone are then circulated in the inner zone to drive them to higher burn up, flattening the power distribution in the core.
4. The solid reflectors provide locations for insertion of control elements.

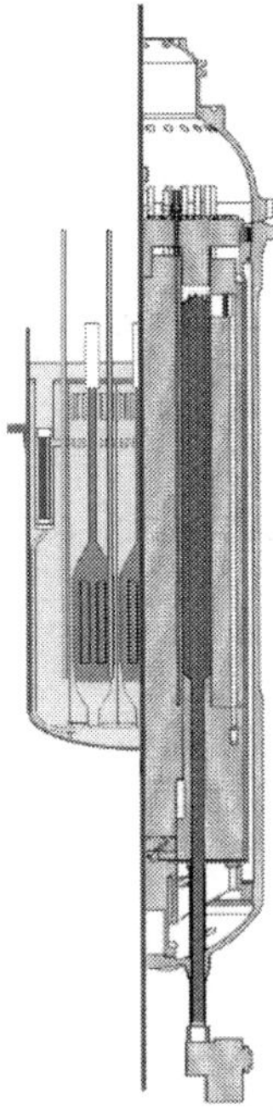

FIGURE 13.4
Scaled elevation views of the 900 MWt PB-AHTR (left) and 400 MWt PBMR (right).

5. The pebble channel configuration addresses the question of pebble bed motion under seismic loading (although the PB-AHTR is a seismically base isolated plant). It is simpler to design and qualify for seismic loading than the tall solid central reflector of the PBMR.

In the conceptual design shown in Figures 13.1 through 13.3, the core-average power density is 30 MW/m^3. This results in a reference plant design with a 6.0-m diameter, 11-m high reactor vessel with a mass of approximately 150 MT that can be readily transported and that operates at near atmospheric pressure. This can be compared favorably to the 9-m diameter, 31-m high reactor vessel for the 600 MWt GT-MHR that operates at nearly 7 MPa. More recent studies have considered other potential core configurations for the PB-AHTR, including a radially zoned, annular configuration that enables the use of thorium blankets.

Initial studies of the major AHTR reactor and balance of plant subsystems concluded that AHTR construction costs may be 55–60% of the costs for MHRs (Figure 13.5) and sodium fast reactors [12]. Likewise, material take off studies have indicated that, per MWe of reactor capacity, the construction of the AHTR would require 43% of the steel and 56% of the concrete used in the construction of 1970s vintage PWRs [16].

Due to its high discharge burn up and power conversion efficiency, a LEU-fueled PB-AHTR requires 64% of the natural uranium and 86% of the enrichment separative work needed for a typical PWR [5]. The reduction in size of primary loop components results in an AHTR reactor building with 35-m height, approximately half that of corresponding PWRs and MHRs, and so the construction time is also projected to be lowered significantly.

13.3 Major Materials and Equipment Designs

13.3.1 Primary Reactor Coolant

The baseline PB-AHTR design uses the beryllium-based salt flibe ($^7Li_2BeF_4$) [1] as its primary coolant, and flinak (LiF–NaF–KF) [2] as its intermediate coolant. The use of a beryllium-based coolant involves a trade-off between fuel utilization, neutronics, activation, salt cost, corrosion, and chemical safety. Flibe is the only fluoride salt that has sufficiently low parasitic neutron capture, and sufficiently high moderating capability, to allow the design of reactor cores with negative void reactivity [1,17,18]. This characteristic is particularly important for the modular PB-AHTR, where the average coolant volume fraction in the core is between 20% and 40%. The resulting negative coolant-temperature reactivity coefficient plays a critical role in limiting the peak temperatures reached during anticipated transient without scram (ATWS) [19]. Flibe has exceptional material compatibility and extremely low corrosion rates with high nickel alloys, when proper chemistry control is used. Its activation products also have very short half lives, so that radiation levels associated with the coolant are very low.

However, due to the cost of beryllium and enriched lithium, flibe is more expensive than alternative working fluids. The PB-AHTR is therefore designed to minimize the primary salt inventory, and uses a separate buffer salt, discussed later, to assure inventory control if the reactor vessel or primary loops are compromised. Detailed estimates of the cost of flibe are not possible, because its constituents are not readily available commodity chemicals, however, the cost can be estimated and bound. Studies conducted for the historic MSBR program at ORNL estimated the salt cost to be \$26/kg (1971\$) ([1], p. 44). The

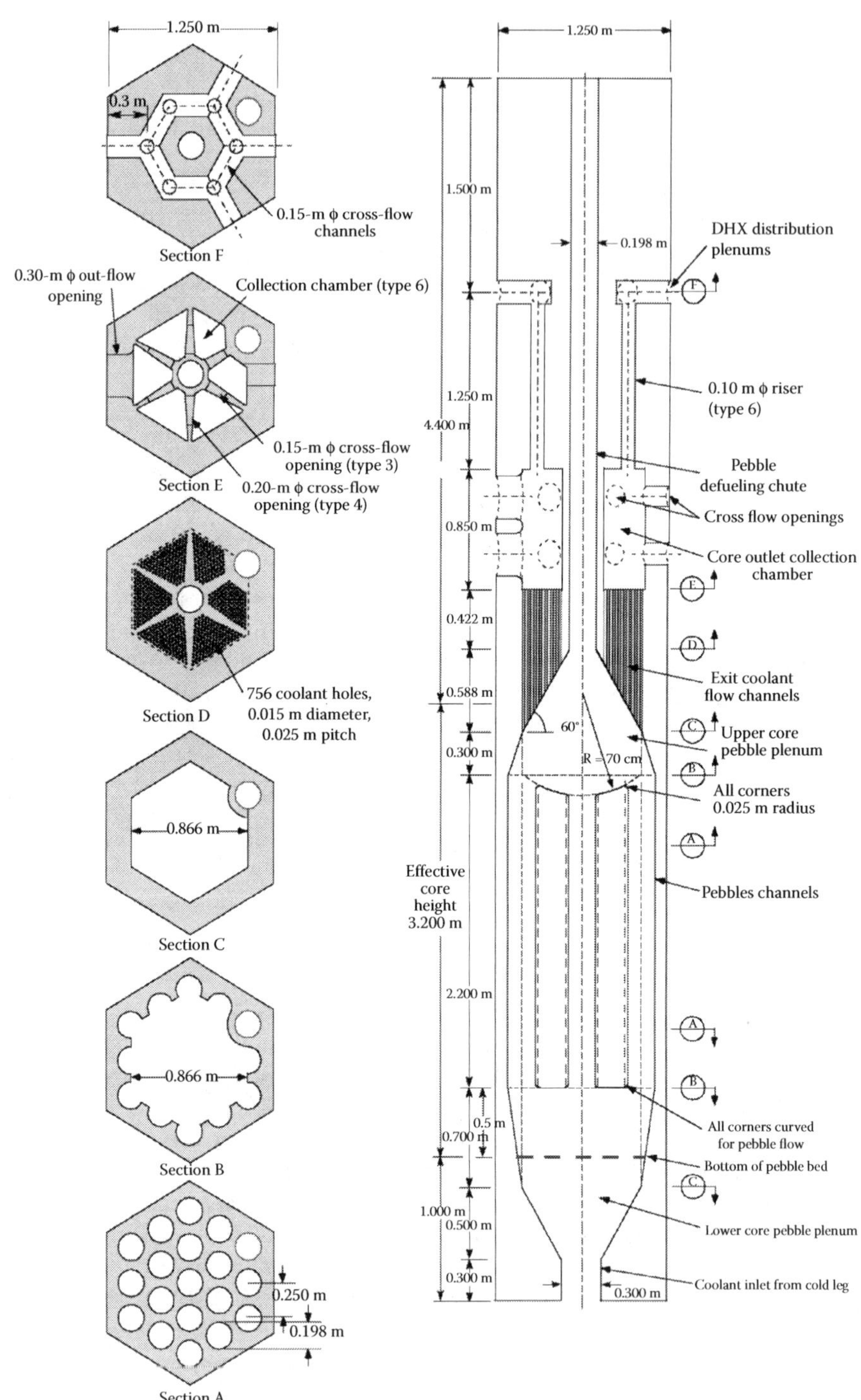

FIGURE 13.5
Detailed design of a PB-AHTR Pebble Channel Assembly.

total salt volume in the 900 MWt modular PB-AHTR is estimated to be between 30 and 40 m^3, or 58–78 metric tons, of which 5.2–7.0 MT is beryllium, and 4.1–5.5 MT is lithium-7. Commodity costs for beryllium and lithium metals are $770 and $63/kg (2002$), respectively [1], and so the total cost of the beryllium and lithium inputs in the primary salt would be $4.2 to $5.7 million, or $10 to $14 per kWe installed capacity. Lithium enrichment costs are however uncertain. Lithium enrichment to provide Li-7 for the MSRE used a chemical process contacting lithium chloride in ethylene diamine with a lithium–mercury amalgam, but modern lithium-enrichment processes would more likely use ion exchange. The potential costs for Li-7 enrichment, conversion of Li-7 and Be to fluorides and purification may increase these costs by a factor of a few. In any case, these costs would be reasonable as a fraction of total capital costs, and are likely to be fully offset by the cost savings from improved fuel utilization due to reduced parasitic neutron capture compared to non-beryllium based salts. Because U.S. beryllium consumption averaged 226 t per year from 2000 to 2004 [20], compared with the 5.2–7.0 t required for a PB-AHTR reactor, the additional beryllium production and consumption required to deploy significant numbers of modular PB-AHTRs would be reasonable.

The most important disadvantage of beryllium-based salt arises from the toxicity of beryllium, which poses the potential for inflammatory lung disease. Therefore, all activities associated with the production, use, recycling and disposal of the PB-AHTR primary coolant must meet rigorous industrial safety standards. Because considerable quantities of beryllium are used in industry, and beryllium is also used in some research reactors, these industrial hygiene practices are understood and have been demonstrated. In the PB-AHTR design, the beryllium control program will be integrated with the radiation control program where appropriate, so that these efforts are synergistic and costs are minimized. It is possible that one can use tracers in the primary salt in order to generate an activation product that would permit the radiation detection instrumentation to identify beryllium contamination as well. Rubidium fluoride, for example, generates an activation product that produces 1 MeV gamma radiation with an 18-day half-life.

13.3.2 Structural Materials

The baseline PB-AHTR uses Alloy 800 H as the structural material for the reactor vessel and for components that may experience high temperature during licensing basis events LBEs (hot legs, primary pumps, IHXs, DHXs, core support structures), with a cladding of Hastelloy N on surfaces exposed to liquid salts. Hastelloy N has well characterized and excellent corrosion resistance for fluoride salts, and when used as a cladding material does not require ASME code qualification, even though an old ASME Section III code case does exist for its use up to 704°C. The DOE VHTR program is currently sponsoring a review of Alloy 800 H stress allowables in the ASME Section III-NH code and extension of Alloy 800 H's time-dependent allowable stresses to 900°C, and to develop approaches to account for neutron and environmental effects structural design criteria for high-temperature components. Because Alloy 800 H is already used extensively in the chemical and fossil industries and multiple vendors exist, its selection is logical as the primary structural material for the PB-AHTR reactor vessel and primary loop components.

The major components in the reactor core are fabricated from graphite, which is essentially inert to fluoride salts. Extensive experience exists with the design and fabrication of graphite components for HTGRs, but detailed structural design work remains to be done for the PB-AHTR reflector and PCA structures. Likewise, while friction coefficients and abrasion processes have been studied for high-temperature conditions in helium-cooled pebble bed reactors [21–23], liquid salts are now known to be effective lubricants for high-temperature graphite [24].

Graphite components will experience varying levels of neutron irradiation. Neutronic simulations, discussed in the next section, provide the basis to predict these doses and assess the necessary replacement frequency of the PCA graphite and the lifetime of the permanent radial reflector graphite, as well as to assess neutron dose to the reactor vessel and core support structures. The peak core power density in the PB-AHTR will be comparable to the MSBR peak value of 65 MW/m^3, which permitted a projected graphite moderator lifetime of approximately 4 years based on an integrated dose limit of 3×10^{22} neutrons/cm^2 (for $E > 50$ keV) ([25], p. 50).

While the extensive existing base of corrosion data suggests that additional corrosion testing will not be required early in the design phase, a comprehensive experimental program will provide extensive test data for component mechanical performance and corrosion under prototypical thermal and chemical conditions for all of the PB-AHTR salt-contacting components, as well as testing of in service inspection and maintenance techniques (more on the technology development program can be found in Section 13.5).

The liquid-salt-cooled PB-AHTR has design characteristics similar to the MSBR, which had an extensive test and development program including successful operation of the MSRE. Therefore, the PB-AHTR adopts, with appropriate modifications, several components and systems that were originally developed for the MSBR. These include the MSBR primary pump, based on a cantilevered centrifugal pump design developed and tested at multiple scales in the ORNL MSR program ([4], as shown in Figure 13.2).

To facilitate in service inspection and reliability, the PB-AHTR adopts the MSBR shell and tube IHX design. Each IHX had 5900 tubes, 0.95 cm (3/8 in.) OD and 6.6 m (21.5 ft) long. To minimize the salt inventory, the MSBR design proposed knurled tubing to enhance heat transfer, a relatively high salt velocity was maintained in the tubes (3.0 m/s) and shell (2.3 m/s) sides, and a large log mean temperature difference (95°C) was used, giving an impressively high power density of 120 MW/m^3. The tube bundle was designed for replacement [25]. For the PB-AHTR a lower log mean temperature difference (LMTD) is desired to increase the power conversion efficiency, and thus a larger surface area will be used.

Other systems adaptable from the MSBR include the reactor cavity thermal insulation system. In the MSBR, all of the primary loop components were uninsulated and were maintained in a uniform thermal environment by a cell liner thermal insulation system with a thin stainless-steel inner liner, an insulation layer, and a steel thermal shield with air cooling to prevent overheating of the concrete biological shielding ([25], pp. 26, 46). For the PB-AHTR a trade study will be performed to determine whether air or water cooling is preferred for the steel thermal shield, since this cooling system and the surrounding concrete and ground also provide the ultimate heat sink for beyond design basis accidents.

13.4 Hydrogen Production Systems

13.4.1 Overview of Technology Integration

The baseline core inlet and outlet temperatures are 600 and 704°C, which is acceptable for HTE however, typically low for IS applications (both processes are discussed further in Chapters 18 and 19, respectively). One of the major challenges for using nuclear power for

HTE is the ability to produce electricity cheaply. Due to the compact nature of the AHTR, the capital costs for a facility are expected to be quite a bit lower than current LWR technology for the same power output. Therefore, the AHTR could be an ideal candidate for hydrogen produced from HTE. In this section, the role of coupling a hydrogen production facility with an AHTR is examined. The AHTR has several advantages with respect to hydrogen production when compared with alternative HTR technologies however some significant technological gaps must be resolved first before commercial deployment. From a thermodynamic standpoint, the average temperature (650°C) is quite close to that of typical modular HTR designs and thus power conversion efficiency is similar (46%). It is conceivable that the development of superior alloys in the future will allow for higher operating temperatures however, this is a long-term capability.

For high-temperature intermediate heat transport, liquid salts are a desirable heat transfer fluid because of their high volumetric heat capacity and therefore, longer heat transport distances are acceptable for potential hydrogen production options. This ability to increase the distance between the reactor system and the hydrogen plant can help mitigate potential safety hazards for the AHTR and co-located facility. For utilizing high-temperature process heat from a nuclear reactor for hydrogen production there will be a need for a long pipe of 100 m or more for the transport of this liquid-salt coolant from the nuclear power plant to the IHX or thermochemical plant. For this purpose efficient heat transfer fluids are required. In addition to helium, molten fluoride and chloride salt are being considered as heat transport working fluids. The material screening focuses on salt compositions with high chemical stability for $T > 800°C$, melting points $T < 525°C$, low vapor pressure, and compatibility with alloys, graphite, and ceramics as needed for the heat transfer loop. Impurities, temperature gradients, and activity gradients might increase liquid salt corrosion problems. Additional candidate salts include the LiF/NaF/KF salt (FLiNaK) as well as LiCl–KCl–$MgCl_2$ salts which have demonstrated the potential to meet these basic requirements and are relatively inexpensive but require further corrosion studies [26].

13.4.2 Role of Tritium in Hydrogen Production

Tritium production in the AHTR will play a very important role when considering the various hydrogen-production technologies. Tritium production is dominated by the $^{6}Li(n,\alpha)^{3}H$ reaction and the $^{9}Be(n,\alpha)^{6}He$ with subsequent decay to ^{6}Li. At equilibrium, these reactions balance each other and at 900 MWt the tritium production rate is approximately 190 Ci/day ([6], p. 46). For comparison, a typical LWR produces ~50 Ci/day, and a typical heavy water reactor (HWR) produces 3500–6000 Ci/day [27]. Tritium behaves chemically like hydrogen and can at high temperatures readily diffuse through a variety of metallic alloys considered for the AHTR. Tritium has low solubility in flibe and flinak. Fortuitously, for electricity production with the multiple reheat helium Brayton cycle used for the PB-AHTR [28,29], the only interface between tritium-containing high-temperature fluid and water is in the precooler and intercoolers of the power conversion system. Nevertheless, tritium clean up and control is required, and may be accomplished using the MSBR gas separator system ([25], p. 20) or a similarly designed system.

For process heat applications, tritium control requires consideration in design. It is preferred to use process heat exchangers fabricated from SiC or Si/SiC materials, with pyrolytic CVD carbon coating on the salt side, for heat transfer to process fluids due to the extremely low permeability to tritium. SiC is an excellent material for many process fluids of interest (particularly IS process fluids) [26,30].

13.5 Experiment and Analysis

The technology development program proposed for the PB-AHTR is structured to follow the phased approach recommended by the Generation IV roadmap [11], consisting of Viability, Performance, and Demonstration phases. This approach targets research investments to address key viability questions early and to support subsequent decisions to proceed with subsequent phases involving detailed design, licensing, and construction of a 125-MWt Pilot Plant.

Key R&D activities during the Viability phase include the conceptual design and physical arrangement for major systems, structures and components (SSCs); identification of a reference set of LBEs for initial analysis and development of best estimate simulations for the consequences of these reference LBEs, and construction and operation of separate effects test (SET) and integral effects test (IET) experiments using simulant fluids to validate these models. The main Performance phase activities include interactions with the NRC during review of the preapplication submittal, detailed design of a <20-MWt Test Reactor, the construction and operation of a component test facility (CTF), and additional tasks discussed in the following sections. Finally, key Development phase activities then include construction and operation of a test reactor, and detailed design and regulatory design certification review and approval of a commercial-scale pilot plant. The Development phase is followed by construction and initial start-up and operational testing in a pilot plant. The general approach of initiating early interactions with the regulatory authority, which has been adopted for the PB-AHTR development plan, also has more general applicability to other advanced reactor and fuel cycle facility designs.

13.5.1 Experimental and Test Program Overview

The experimental test program for the PB-AHTR provides empirical data to validate models that predict overall plant reliability and safety. Reliability involves quasi-steady phenomena that evolve over substantial periods of time. Reliability-related phenomena, along with human error, sabotage, and external events, can initiate safety-related transients that evolve over relatively short timescales. The PB-AHTR experimental test program is designed to validate models that predict slowly evolving reliability phenomena and rapidly evolving safety phenomena.

The phenomena identification and ranking process provides the basis to identify the dominant phenomena that control the system response to a specific reliability or safety-related transient. Three major types of experiments are used to study these phenomena. SET experiments are designed to study specific phenomena under closely controlled (and potentially idealized) boundary and initial conditions. These experiments are typically highly instrumented, and provide the basis to validate models for local phenomena. IET experiments are scaled experiments designed to reproduce the integral response of a system to multiple phenomena, with acceptably low and quantifiable distortions. IET experiments address the fact that for coupled phenomena the initial and boundary conditions might be significantly more complex than can be generated in SET experiments. IET experiments provide a basis to validate system-modeling codes to confirm that their SET-based phenomenological models adequately reproduce the global system response. Finally, component test (CT) experiments verify the capability of equipment to operate under near-prototypical conditions, and provide one basis for assessing component reliability for probabilistic risk analysis (PRA) models.

Liquid salts are unique among candidate reactor coolants (water, helium, liquid metals) due to the existence of simulant fluids that can replicate salt fluid mechanics and heat transfer phenomena at reduced length scales and temperature, and greatly reduced heater and pumping power. In the late 1990s, UC Berkeley pioneered the use of water as a simulant for liquid salt fluid mechanics, noting that at approximately 50% geometric scale room-temperature water experiments can match the Reynolds and Froude numbers associated with the liquid salt flibe for inertial fusion energy applications [31]. Subsequently, UC Berkeley identified a class of heat transfer oils that have the same Prandtl number as the major liquid salts and that can thus match Reynolds, Froude, Prandtl, and Grashof numbers simultaneously at approximately 40% geometric scale and heater power under 2% of prototypical [32]. While scaled oil experiments do not reproduce thermal radiation heat transfer from the salt to heat transfer surfaces, under most conditions this is a second-order effect that can be corrected in the system model [32].

The availability of these simulant fluids allows a large reduction in the cost and difficulty of performing the IETs required for system modeling code validation for reactor licensing, compared to working at prototypical temperatures and power levels with the actual coolant. These simulant fluids also can be used in SET experiments to develop heat transfer and pressure loss correlations for use in system modeling codes. The implications for IET and SET experiments are discussed in the following two sections, followed by a section discussing CT experiments.

13.5.2 Separate Effects Test

The PB-AHTR SET experiment program covers those phenomena identified in the phenomena identification and ranking table (PIRT) development process for which high-quality experimentally validated models are not yet already available. While detailed Viability phase PIRTs have not yet been performed, several dominant phenomena have already been identified in PB-AHTR modeling efforts where existing experimental data is insufficient.

For the Viability phase, SET experiments include studies of mixed convection heat transfer in pebble beds and in vertical channels using simulant fluids, because relevant experimental data do not exist in the range of Prandtl and Grashof numbers existing in the PB-AHTR. Additionally, friction coefficients and abrasion wear rates for graphite, lubricated by flibe, are not currently known and are of importance in interpreting experimental data for pebble motion with simulant fluids, where friction coefficients are likely different from prototypical, and in verifying models for the required pebble graphite shell thickness and the abrasion wear rate for pebble channels. During the Performance phase, the CTF provides opportunities to collect SET data for prototypical components with the prototypical heat transfer fluid. Thus, the CTF will be instrumented to collect heat transfer, pressure drop, and other SET data of interest (more on the CTF can be found in Section 13.5.3). In particular, this data will address questions about the potential impact of thermal radiation on heat transfer to liquid salts. Very limited data for infrared absorption is available is available for flibe and flinak. Thus, the Performance phase is also expected to have SET experiments to measure IR absorption in the primary, secondary, and DRACS salts.

PB-AHTR fuel operates at much higher power density than PBMR fuel, but at lower temperatures. Performance phase SET experiments also include fuel fabrication and irradiation testing. A benefit of very high power density is that such testing can be performed more rapidly, and still involve the prototypical neutron dose rate.

The PB-AHTR neutronics codes will be validated against a variety of existing SET experimental data, such as from the ASTRA critical facility of the RRC Kurchatov Institute [33]. The Test Reactor has in-core neutron flux mapping and will provide extensive zero power critical testing under fully prototypical conditions. The potential need for an additional zero power critical testing during the Performance phase remains to be determined, recognizing that fully prototypical criticality data will become available in the Demonstration phase.

13.5.3 Integral Effects Test

The liquid-salt-cooled PB-AHTR is distinguished from reactors with that using coolants because the PB-AHTR IETs can be performed with simulant fluids at greatly reduced power levels. Thus the key IET experimental facilities needed to validate the PB-AHTR transient analysis codes are university-scale facilities that are built and operated during the Viability phase. These facilities continue to operate, as needed, during the Performance phase to provide new IET data as the detailed design and safety analysis are finalized.

The two IET facilities required for the PB-AHTR program are the Compact Integral Effects Test (CIET), and the Pebble Recirculation Experiment (PREX). CIET is a scaled integral heat transfer loop, using oil as a simulant fluid. PREX is a scaled experiment to replicate pebble recirculation phenomena, using water as a simulant fluid. During the Performance phase the CTF provides IET data for pebble recirculation under prototypical conditions. During the 1980s and early 1990s, very large experimental programs studied safety-related phenomena that occur in light water reactors (LWRs). These experiments included large, electrically heated integral test facilities to study the transient response of LWRs to large and small break loss of coolant accidents, and other transients [28]. Examples of such facilities include Semiscale in the United States, BETSY in France, and ROSA-IV in Japan. The major design parameters for these facilities are summarized below in Table 13.2. All of the these IET facilities were full, 1:1 height, and used reduced flow area to reduce the electrical heating power and component sizes to economically viable values. Because the reduced flow area generates potentially important distortions in three-dimensional effects and in heat loss from the primary loop boundary, a major trade-off in such facilities involves providing heaters capable of reaching 100% of prototypical power (semiscale) versus only matching decay heat and thus achieving larger flow area scaling (BETHSY and ROSA-IV).

The baseline design for CIET uses a 100-kW, 70-V DC power supply. As shown in Table 13.2, due to the scaling for heat transfer oils as simulant fluids, CIET is effectively equivalent to a 4.7 MW IET facility using the prototypical liquid salt under prototypical

TABLE 13.2

Comparison of CIET, Semiscale, BETSY, and ROSA-IV IET Facilities

Facility	CIET	Semiscale	BETHSY	ROSA-IV
Plant type	AHTR	PWR	PWR	PWR
Effective power (MW)	4.7	2.0	3.0	10
Actual power (MW)	0.1	2.0	3.0	10
Fraction of scaled full power	100%	100%	10%	14%
Effective flow area scaling	1:190	1:1705	1:100	1:48
Actual height scaling	1:2	1:1	1:1	1:1
Time scaling	1:1.4	1:1	1:1	1:1

conditions. The CIET facility has 50% height scaling compared to the prototypical PB-AHTR. Because the PB-AHTR primary system height is half that of a conventional LWR or MHR, and because CIET is half that height, the CIET facility with its DRACS loops can fit inside a 10-m high laboratory.

There exists a major difference in the difficulty of performing IET experiments with a near room-temperature fluid, at greatly reduced power and length scales, compared to equivalent experiments using the prototypical reactor coolant at prototypical temperatures. The very large simplification of the experiment design, and reduction in cost, are the unique attributes of liquid salt coolants that allow major IET facilities to be built and operated during the Viability phase of R&D.

Pebble recirculation is the second key process that requires experimental validation. The PB-AHTR PCAs have a unique design feature, where the pebbles flow through a large number of small channels. The diameter of each channel is over 6 pebble diameters, so bridging and plugging are not expected to occur. But the availability of water as a simulant fluid that can match Reynolds and Froude numbers allows experimental validation of pebble recirculation model predictions to be made during the Viability phase, in a PREX-2 experiment that is geometrically scaled to match the multiple channel configuration of the modular PB-AHTR PCA. The CTF, constructed during the Performance phase, includes pebble injection, PCA, defueling machine, and pebble handling components and provides integral test data for pebble recirculation under prototypical coolant conditions.

13.5.4 Component Tests

A substantial base of experience exists with many of the major components required for the PB-AHTR, including pumps and heat exchangers, from the MSRE and MSBR projects. ORNL design and testing reports for these components provides a major input for the design of the PB-AHTR components. More unique to the PB-AHTR are the pebble injection and core unloading devices, which differ in key design constraints from those for helium-cooled PBMRs. During the viability phase, these components will be fabricated using stereolithography rapid prototyping and tested for functional performance in the PREX facility.

Due to the fact that component reliability has major importance for both plant reliability and safety, a major activity during the Performance phase is the construction and operation of a CTF. The CTF will support steady state and transient tests of component and system functionality in the prototypical salt environments, life cycle tests, and reliability tests. The proposed facility has similar purposes to the PBMR Helium Test Facility (HTF) that became operational in September 2006 in Pelindaba, South Africa. The PBMR HTF is a 40 m high, full-height facility designed for full-scale testing of the PBMR fuel handling system, helium blower, valves, reactivity control, and reserve shutdown systems, coolers, recuperator, valves, helium inventory control system, and other components. The HTF supports component testing at helium pressures up to 9.5 MPa and temperatures up to 1100°C. While the PB-AHTR CTF will perform the same functions as the PBMR HTF, it will physically be a substantially smaller facility—approximately half the height of the HTF—due to the much higher power density of a liquid-salt-cooled reactor. The CTF will test the following components and systems:

- Primary pump (full scale Test Reactor pump)
- Pebble injection and core unloading devices, pebble conveyance system, spent/fresh pebble storage canister system
- Reactor vessel (isothermal)

- Demonstrate procedures for initial heat up and salt filling, pebble fueling and defueling, reflector graphite replacement
- Control/safety rod drive assemblies, maintenance methods (heated)
- Reduced area IHX (heated)
- DRAC heat exchangers and heat removal system (heated)
- Seismic snubbers
- Reactor cavity insulation and heating/cooling system
- Salt chemistry control system
- Cover gas chemistry and thermal control system
- In service inspection equipment and methods
- Temperature, pressure, and flow and control instrumentation

The CTF will also serve as a center for training of PB-AHTR plant operations and maintenance staff.

13.6 Conclusion

The liquid fluoride-salt-cooled high-temperature nuclear reactor concept using pebble fuel is an attractive option for producing hydrogen utilizing the HTE and IS process. The very high volumetric heat capacity of liquid fluoride salts enables the design of high power density HTRs with passive safety. From a thermodynamic standpoint, the AHTR average temperature (650°C) is quite close to that of typical MHR designs and thus power conversion efficiency is similar (46%). With materials advances higher temperatures are possible, but remain a long-term capability. Due to the fact that the intermediate loop uses a fluoride salt, longer heat transport distances are acceptable, which mitigates potential safety impacts from co-location with hydrogen or chemical plants. Tritium control, however, remains an important issue for process heat applications for hydrogen production and will require further research in suitable heat exchanger materials.

Preliminary top-down economics estimates indicate the potential for quite attractive construction and fuel costs, compared to current alternative HTR and LWR technologies. The licensing approach for the PB-AHTR can be adapted from that for MHRs, with the main change being that PB-AHTR fuel is immersed in a chemically inert coolant, and thus the design criteria for controlling chemical attack by air and steam is replaced by an design criterion to control the coolant inventory.

The experimental test program for the PB-AHTR considers two major categories of phenomena, quasi-steady phenomena that evolve over long periods of time and affect component reliability, and more rapid transients that could follow initiating events such as the failure of a component or components. The PB-AHTR baseline design adopts a conservative approach to quasi-steady, reliability-related phenomena, by selecting existing materials and fuels that are already well characterized, and using them at a conservatively low operating temperature. For the more rapid transient phenomena that follow a LBE, the PB-AHTR development plan takes advantage of the existence of simulant fluids that allow PB-AHTR fluid mechanics and heat transfer phenomena to be studied at greatly reduced power, temperature, and cost.

References

1. D. F. Williams, L. M. Toth, and K. T. Clarno, Assessment of candidate molten salt coolants for the advanced high-temperature reactor (AHTR), Oak Ridge National Laboratory, ORNL/TM-2006/12, March 2006.
2. D. F. Williams, Assessment of candidate molten salt coolants for the NGNP/NHI heat-transfer loop, Oak Ridge National Laboratory, ORNL/TM-2006/69, June 2006.
3. C. W. Forsberg, P. Pickard, and P. F. Peterson, Molten-salt-cooled advanced high-temperature reactor for production of hydrogen and electricity, *Nucl. Technol.*, 144, 289–302, 2003.
4. P. F. Peterson and H. Zhao, A flexible baseline design for the advanced high temperature reactor utilizing metallic reactor internals (AHTR-MI), *Proceedings of the 2006 International Congress on Advances in Nuclear Power Plants, ICAPP'06*, pp. 650–661, 2006.
5. C. W. Forsberg, P. F. Peterson, and R. A. Kochendarfer, Design options for the advanced high-temperature reactor, *International Congress on Advanced Nuclear Power Plants*, Anaheim, CA, June 8–15, 2008.
6. J. Uhlir, Chemistry and technology of molten salt reactors—History and perspectives, *J. Nucl. Mater.*, 360, 6–11, 2007.
7. H. G. MacPherson, The molten salt adventure, *Nucl. Sci. Eng.*, 90, 374–380, 1985.
8. J. H. Shaffer, Preparation and handling of salt mixtures for the molten salt reactor experiment, ORNL-4616, January 1971.
9. MSRE Systems and Components Performance, Oak Ridge National Laboratory, ORNL-TM-3039, June 1973.
10. The Development Status of Molten-Salt Breeder Reactors, ORNL-4812, pp. 200–201, 207–211, August 1972.
11. U.S. DOE Nuclear energy research advisory committee and the Generation IV international forum, "A Technology Roadmap for Generation IV Nuclear Energy Systems," Report No. GIF002–00, December 1, 2002.
12. D. T. Ingersoll, C. W. Forsberg, L. J. Ott, D. F. Williams, J. P. Renier, D. F. Wilson, S. J. Ball, et al., Status of Preconceptual Design of the Advanced High-Temperature Reactor (AHTR), Oak Ridge National Laboratory, ORNL/TM-2004/104, May 2004.
13. D. A. Petti, R. R. Hobbins, J. M. Kendall, and J. J. Saurwein (Eds.), Technical program plan for the advanced gas reactor fuel development and qualification program, Idaho National Laboratory, Idaho Falls, ID, Tech. Rep. INL/EXT-05–00465 Revision 1, August. 2005.
14. P. Bardet, E. D. Blandford, M. Fratoni, A. Niquille, E. Greenspan, and P. F. Peterson, Design, Analysis and Development of the Modular PB-AHTR, 2008 International Congress on Advances in Nuclear Power Plants (ICAPP '08), Anaheim, CA, June 8–12, 2008.
15. E. D. Blandford and P. F. Peterson, A novel buoyantly-driven shutdown rod design for passive reactivity control of the PB-AHTR, *4th International Topical Meeting on High Temperature Reactor Technology (HTR-2008)*, September 28–October 1, 2008, Washington, DC.
16. P. F. Peterson, H. Zhao, and R. Petroski, Metal and concrete inputs for several nuclear power plants, Report UCBTH-05-001, UC Berkeley, February 4, 2005.
17. S. J. de Zwaan, B. Boer, D. Lathouwers, and J. L. Kloosterman, Static design of a liquid-salt-cooled pebble bed reactor (LSPBR), *Annals Nucl. Energy*, 34, 83–92, 2007.
18. M. Fratoni, F. Koenig, E. Greenspan, and P. F. Peterson, Neutronic and depletion analysis of the PB-AHTR, *GLOBAL 2007*, Boise, ID, September 9–13, 2007.
19. A. Griveau, F. Fardin, H. Zhao, and P. F. Peterson, Transient thermal response of the PB-AHTR to loss of forced cooling, *GLOBAL 2007*, Boise, ID, September 9–13, 2007.
20. U.S. Geological Survey, Mineral Commodity Summaries, p. 32, January 2005.
21. A. P. Semenov, Tribology at high temperatures, *Tribol. Int.*, 28(1), 45–50, 1995.
22. H. Zajdi, D. Paulmier, and J. Lepage, The influence of the environment on the friction and wear of graphitic carbons, *Appl. Surf. Sci.*, 44, 221–233, 1990.

23. L. Xiawei, Y. Suyuan, S. Xuanyu, and H. Shuyan, The influence of roughness on tribological properties of nuclear grade graphite, *J. Nucl. Mater.*, 350, 74–82, 2006.
24. R. Hong, S. Huber, K. Lee, P. Purcell, S. Margossian, and J. D. Seelig, Reactor safety and mechanical design for the annular pebble-bed advanced high temperature reactor, 2009 NE 170 Senior Design Project, U.C. Berkeley, Report UCBTH09-001, May 19, 2009.
25. J. R. McWherter, Molten salt breeder experiment design bases, ORNL-TM-3177, p. 50, November 1970.
26. P. F. Peterson, C. Forsberg, and P. Pickard, Advanced CSiC composites for high-temperature nuclear heat transport with helium, molten salts, and sulfur–iodine thermochemical hydrogen process fluids, *Second Information Exchange Meeting on Nuclear Production of Hydrogen*, Argonne National Laboratory, IL, USA, October 2–3, 2003.
27. R. B. Briggs, Molten salt reactor program semiannual progress report for the period ending July 31, 1963, ORNL-3529, p. 125, 1963.
28. P. F. Peterson, Multiple-reheat Brayton cycles for nuclear power conversion with molten coolants, *Nucl. Technol.*, 144, 279–288, 2003.
29. H. Zhao and P. F. Peterson, Low-temperature multiple-reheat closed gas power cycles for the AHTR and LSFR, *Proc. 2006 International Congress on Advances in Nuclear Power Plants (ICAPP '06), Embedded International Topical Meeting 2006 American Nuclear Society Annual Meeting*, Reno, NV, USA, June 4–6, 2006.
30. J. Schmidt, M. Scheiffele, M. Crippa, P. F. Peterson, K. Sridharan, Y. Chen, L. C. Olson, M. H. Anderson, and T. R. Allen, Design, fabrication, and testing of silicon infiltrated ceramic plate-type heat exchangers, *33rd International Conference on Advanced Ceramics and Composites (ICACC)*, Daytona Beach, FL, January 18–23, 2009.
31. C. J. Cavanaugh and P. F. Peterson, Scale modeling of oscillating sheet jets for the HYLIFE-II inertial confinement fusion reactor, *Fusion Technol.*, 26, 917–921, 1994.
32. P. Bardet and P. F. Peterson, Options for scaled experiments for high temperature liquid salt and helium fluid mechanics and convective heat transfer, *Nucl. Technol.*, 163, 344–357, 2008.
33. N. E. Kukharin, E. S. Glushkov, G. V. Kompaniets, V. A. Lobyntsev, D. N. Polyakov, and O. N. Smirnov, Investigation of criticality parameters of high temperature reactors at Kurchtov Institute's ASTRA critical facility, HTR-2002 Conference on High-Temperature Reactors, Petten, NL, April 22–24, 2002.

14

STAR-H2: A Pb-Cooled, Long Refueling Interval Reactor for Hydrogen Production

David C. Wade

CONTENTS

14.1 Technology Overview

The secure transportable autonomous reactor for hydrogen production (STAR-H2), is an element of a proposed, sustainable global, mid-twenty-first century hierarchical hub–spoke nuclear energy supply architecture. This energy architecture will use uranium as the energy resource; will use nuclear fuel, hydrogen, and electricity as the energy carriers; and will contribute to *all* primary energy requirements—not electricity alone. STAR-H2 power plants will operate in a fissile self-sufficient (core conversion ratio equal one) mode on very long (15–20 years) refueling interval of entire core refueling cassettes, and will produce hydrogen, oxygen, and potable water to service cities and their surrounding regions under an assumed distributed electrical-generation network based on fuel cells and micro turbines and an assumed transportation sector using hydrogen-fueled vehicles.

Front and back end (including waste management) fuel-cycle services for the refueling cassettes will be provided by regional fuel-cycle centers owned by consortia from client nations and will operate under international nonproliferation oversight.

The STAR-H2 reactor is a Pb-cooled, mixed U–TRU–Nitride-fueled, fast spectrum reactor delivering 400 MWt of heat at 800°C core outlet temperature. The primary coolant circulates by natural circulation; the 400 MWt heat rating is set by dual requirements for natural circulation and for rail shippability of the vessel. STAR-H2 heat-source reactors will be factory fabricated and delivered to the site of a preprepared balance of plant (BOP) as a turnkey (battery) heat supply reactor for production of hydrogen by thermochemical water cracking. A low-pressure molten salt intermediate heat-transport loop carries the heat to a Ca–Br thermochemical water-cracking cycle for the manufacture of H_2 (and O_2). The water-cracking cycle rejects heat at 550°C and that heat is used in a supercritical CO_2 Brayton cycle turbogenerator to provide for on-site electricity needs. A thermal desalination plant receives discharge heat at 125°C from the Brayton cycle and the brine provides for ultimate heat rejection from the cascaded thermodynamic cycles.

The *modified* UT-3 cycle used in STAR-H2, called the Ca–Br cycle, operates at atmospheric pressure and 750–725°C, uses solid/gas separation, and achieves about 44% efficiency. It employs a single-stage HBr-dissociation step based on a plasma chemistry technique operating near ambient conditions.

The hierarchical hub-spoke infrastructure relies on recycle of energy conversion products and wastes at each link of the energy supply chain and will have the favorable energy security, and ecological and nonproliferation features needed to meet the requirements of sustainability.*

14.2 Mid-Century Energy Needs

Global energy demand forecasts for the twenty-first century project massive growth in demand for energy services, and they show that the dominant capacity additions by 2030

* The author expresses his appreciation to the International Atomic Energy Agency for permission to extensively use material from IAEA-TECDOC-1536, "Status of Small Reactor, Designs without Onsite Refueling" ANNEX-24 (Jan 2007).

and beyond will occur in the currently developing countries. They predict that demographic migrations will lead to a majority of global population living and working in urban centers by mid century. Thus, the global reach of the nuclear client base must be expanded to include cities in developing countries. The range of demanded energy products will also expand; an emerging need for process heat conversion of water or hydrocarbon feedstocks to hydrogen is foreseen. Manufacture of potable water may also be needed as cities increasingly outsource municipal water supply contracts to profit-making entities.

Developing economies enjoy the opportunity to "leapfrog" to new sustainable energy infrastructures which meet their special needs. Population and economic activity which is focused primarily in cities will require an energy supply architecture having high-energy density. Rapid economic growth rates will require emplacement of energy infrastructures having a short energy-payback period. These two requirements preclude a major role for renewables for the clients targeted here, but are well suited to the innate features of nuclear energy.

14.3 Reconfiguring the World's Energy Architecture to Exploit Nuclear Energy's Innate Features

Nuclear energy has much to offer to fuel a sustainable development [1] revolution on the scale of the Industrial Revolution, but to do so it must be reconfigured to meet the twenty-first century's market situation. The fact that much of the future growth will be in cities of developing nations, means that market conditions facing future nuclear deployment will be different from historical conditions where deployment occurred primarily in industrialized countries under regulated electricity-market conditions. The proposed STAR-H2 energy supply architecture has been optimized to exploit all of nuclear energy's innate features for the new market situation. STAR-H2 power plants are 400 MWt turnkey plants which manufacture hydrogen and electricity as energy carriers and potable water. They are targeted for worldwide deployment and especially for urban centers in developing countries. To break the energy security/nonproliferation dilemma, they are designed with 20-year refueling interval and they fit within a proposed hierarchical hub–spoke energy supply architecture using regional fuel-cycle centers; using nuclear fuel and hydrogen as long distance energy carriers—and supporting distributed electricity generation as the local energy carrier. In this way, the new architecture will *mesh seamlessly with existing and imminent urban energy-distribution infrastructures using grid delivery of electricity, hydrogen, potable water, and communications (and sewage return) through a common grid of easements.* This will facilitate incremental market penetration. The small sizing and outsourced fuel cycle and waste management configuration allows for plant deployment at modest initial capital outlay for the client. Turnkey plants are transported to the client's site and rapidly connected to a preconstructed nonnuclear safety grade BOP to achieve a rapid start of the revenue stream. Figure 14.1 illustrates the STAR-H2 energy-supply infrastructure.

STAR-H2 is intended to meet the needs of two categories of customers: (1) utilities seeking to provide all primary energy and potable water needs of cities in developing countries including those in the early stages of economic development and having limited infrastructure; and (2) independent power producer (IPP) customers in developed

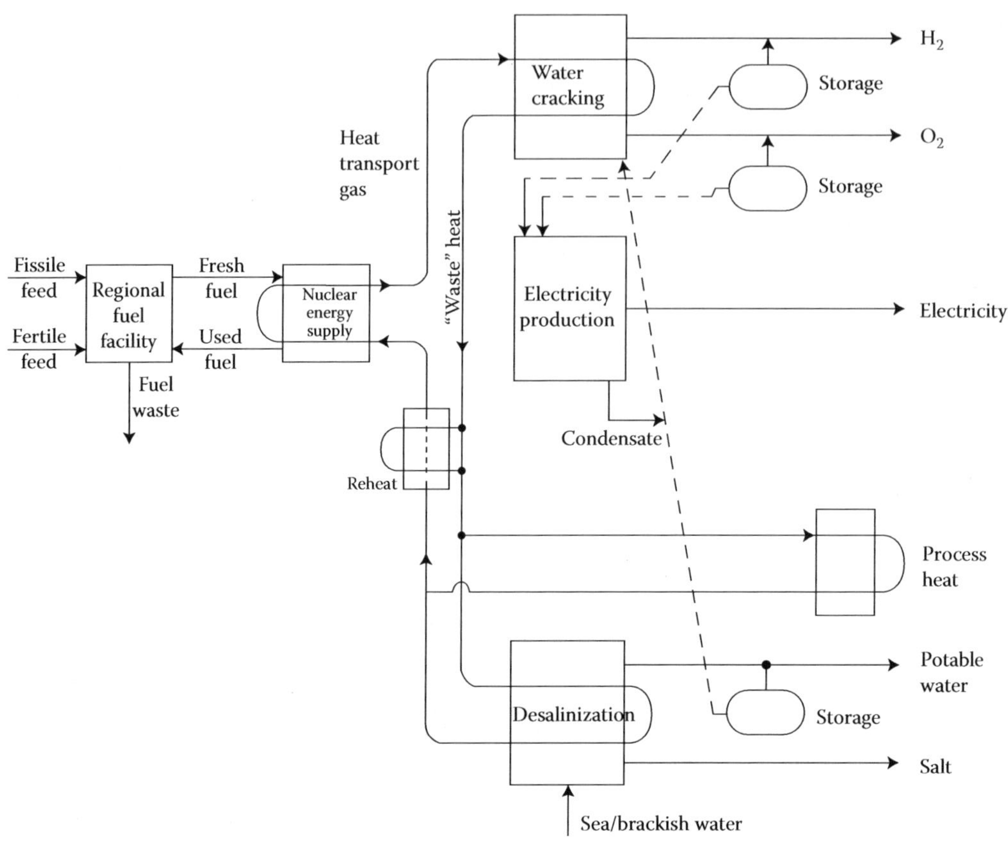

FIGURE 14.1
STAR-H2-based energy supply infrastructure.

countries who wish to enter emerging markets for hydrogen and/or potable-water production.*

Both categories of customers desire limited capital outlay, rapid site assembly and early initiation of a revenue stream, outsourcing of the front end and back end fuel cycle and waste management services, reduced operational staffing requirements, near-urban siting based on unprecedented levels of safety and robustness with respect to equipment malfunction and/or human error, and a nonnuclear safety grade BOP.

Developing-country customers may additionally seek the energy security afforded by very long refueling interval and of legally binding assurances of access to fuel cycle and waste services from the regional fuel-cycle centre. Finally, developing-country customers would welcome the job creation and economic growth opportunities, which could derive from the STAR-H2 design approach for a nonsafety grade BOP, which can be constructed and operated by local companies to local standards and using local labor.

The 400 MWt STAR-H2 is being designed to achieve 44% conversion of heat to lower heating value (LHV) of H_2 in a thermochemical water cracking cycle [2]—making 160 MWt

* Independent power producers (IPPs) are merchant generation companies who operate outside the regulatory framework of regulated utilities and sell their product on a competitive market (i.e., they receive no guarantee of profitability in exchange for a guarantee of providing service to consumers).

days/day of H_2 (LHV). It uses a supercritical CO_2 Brayton cycle to generate electricity for on-site needs [3], and a feed-forward multieffect distillation (MED) bottoming cycle to manufacture 8000 m^3/day of water—enough to support primary energy and potable water needs of a city of 25,000 using primary energy at 4 toe/capita/year—the level of use in Western Europe. Overall, 85% of the reactor's 400 MWt is to be converted to energy products; 15% would be rejected in the form of heated brine.

14.4 STAR-H2 Design Description

14.4.1 Reactor Design Overview

Table 14.1 summarizes the reactor operating parameters for STAR-H2 at the conceptual stage of development.

14.4.1.1 Reactor Core and Fuel

The STAR-H2 reactor uses a radially heterogeneous core layout of ductless assemblies. The pin lattice is open with a large coolant volume fraction. Table 14.2 shows the neutronics design parameters and performance results while Figure 14.2a and b shows radial power-density profiles at beginning and end of cycle.

14.4.1.2 Thermal-Hydraulics

The STAR-H2 reactor is cooled by natural circulation of the Pb primary coolant; Table 14.3 gives the relevant design data and Table 14.4 gives the full power results for thermal hydraulics.

14.4.2 BOP Design

14.4.2.1 Intermediate Heat Transport Circuit

A forced circulation, ambient pressure fused salt (flibe) intermediate heat-transport loop carries the heat from the in-vessel intermediate heat exchanger (IHX) to the BOP. Figure 14.3 shows the overall heat flow for the reactor and BOP.

14.4.2.2 Water Cracking Cycle

The STAR-H2 BOP is comprised of three cascaded cycles (water cracking, Brayton cycle, and desalination) operating at successively lower temperatures—and with the heat rejected from each cycle used to drive the succeeding cycle, see Figure 14.3. The reactor supplies 400 MWt of heat between 800 and ~650°C. The strategy for BOP plant design is to use that heat as follows:

1. Use as much of the heat as possible to maximize hydrogen production—consistent with (2).
2. Use only as much of the heat to make electricity in the Brayton cycle as is required to run the BOP (i.e., neither off-site electricity sales nor reliance on off-site power is required).

TABLE 14.1
Summary Table of Major Design and Operating Characteristics

Characteristic	Value
Installed capacity	400 MWt
Mode of operation	Autonomous load follow based on passive feedbacks
Load factor/availability	The targeted capacity factor (CF) is 90% for the operation in (base load → storable product) mode, with refueling once every 20 years. Possible need to regenerate the coolant every N years.
Type of fuel	Uranium/transuranic (TRU) nitride in clad cylindrical fuel rods. Enriched in nitrogen isotope ^{15}N.
Fuel enrichment (TRU/HM) (Fast neutron spectrum, internal conversion ratio = 1.0)	Reference core design: radially heterogeneous core layout; ductless assemblies. 3 enrichment zones: 13.14% inner and mid zones; 1.4 × 13.14% outer zone.
Coolant	Lead (T_{Inlet} = 663.7°C; T_{Outlet} = 793.4°C)
Moderator	None
Core structural materials	Unknown; calculations assume SiC/SiC composite
In-vessel structural materials	Unknown; calculations assume SiC/SiC composite
Core	9 rows of fuel (or blanket) assemblies; 2 rows of reflector assemblies
Assembly flat to flat size	16.24 cm
Active core height	2 m
Above core fission gas plenum height	2.0 m
Below core axial reflector height	0.25 m
Open-lattice of cylindrical fuel rods on a triangular pitch (optional square pitch) lattice	
Cladding outer diameter	1.905 cm (driver and internal blanket)
Cladding thickness	1 mm
Coolant volume fraction	0.667
Fuel volume fraction	0.248
Fuel pellet-cladding bond	Pb
Reflector	50 volume % ferritic–martensitic stainless steel and 50 volume % Pb
Reactor vessel	Cylinder with hemispherical lower head
Outer diameter	5.5 m
Height	16.9 m
Thickness	Unknown
Design lifetime	60 years
Cycle type	Indirect cycle: intermediate loop of forced circulation molten salt
	Ca–Br thermochemical water cracking cycle/supercritical CO_2 Brayton cycle with a feed-forward MED bottoming cycle
Number of circuits	2
	Primary circuit: natural circulation, ambient pressure Pb; intermediate circuit: forced circulation ambient pressure flibe molten salt
Neutron physical characteristics	
Refueling cycle length	20 full power years
Number of batches	1; whole core cassette refueling
Burn-up reactivity swing	K_{eff} = 1.00 (BOL); 1.013 EOL
Peaking factor	~1.5 except ~1.8 near BOL and EOL
Power flattening	Three-zone radial enrichment zoning and internal blankets
Reactivity control mechanism	Shutdown rod for start-up and shutdown

TABLE 14.1 (continued)

Summary Table of Major Design and Operating Characteristics

Characteristic	Value
	During operation, reactor power autonomously adjusts to load by means of inherent physical processes without the need for any motion of control rods or any operator actions
	System temperatures change corresponding to reactivity feedbacks from fuel Doppler, fuel and cladding axial expansion, core radial expansion, and coolant density effects
	Control rods for possible fine reactivity compensation during cycle (tentative)
	(Control rods would also provide for diverse and independent shut down)
Energy conversion cycle type	The reactor heat drives a Ca–Br thermochemical
	Water cracking cycle and a supercritical CO_2
	Brayton cycle (sized for on site needs)
Conversion efficiency	
(LHV of H_2)/(reactor heat)	~44%
Bottoming cycle	Feed forward MED desalination
Thermal-hydraulic characteristics	Primary coolant based on natural circulation of lead. No primary coolant pumps
Core inlet temperature	663.7°C
Core outlet temperature	793.4°C
Primary coolant flow rate	21,770 kg/s
Primary coolant cover gas pressure	Slightly below 1 atm
Temperature limit for cladding	~950°C (tentative)
Average fuel temperature	970°C
Average cladding inner surface temperature	803°C
Maximum fuel temperature	995°C (hot channel)
Maximum cladding inner surface temperature during normal operation	878°C (hot channel)
Maximum/average discharge burn-up of fuel	Average = 82 MWd/kg (drivers); 28 (internal blankets)
	Peak = 126 MWd/kg (driver)
Fuel lifetime/period between refuelings	20 full power years
Mass balances/flows of fuel:	
Initial loading	29,600 kg of heavy metal
Initial TRU loading	1700 kg TRU
Internal conversion ratio	~1.0
	Best estimate calculation using DIF3D and
	REBUS-3 computer codes
Design basis lifetime:	
Core refueling cassette	20 years
Reactor vessel	60 years (tentative)
In-vessel structures	60 years (tentative)
Design and operating characteristics of systems for nonelectric applications	STAR-H2 is dedicated to H_2 production with desalinated water production using reject heat
Economics	To be determined

TABLE 14.2

Neutronic Design Parameters and Calculated Performance Results for the Reference Heterogeneous Core Layout (Internal Blanket; Cladding Material Sic; Fuel Residence Time 15 Years; Capacity Factor 90%)

Characteristic	Value	
Design Parameters		
Enrichment Pu/HM, %	13.14	
Inner core	×1.0	
Middle core	×1.0	
Outer core	×1.4	
Driver fuel pins		
Fuel pin diameter (clad), cm	1.905	
Fuel volume fraction	0.247657	
Cladding volume fraction	0.078858	
Coolant volume fraction	0.666785	
Blanket fuel pins		
Fuel pin diameter (clad), cm	1.905	
Fuel volume fraction	0.247657	
Cladding volume fraction	0.078858	
Coolant volume fraction	0.666785	
Number of fuel driver pins in the core	4638	
Number of blanket pins in the core	2301	
Number of inner core driver assemblies	1	
Number of medium core driver assemblies	48	
Number of outer core driver assemblies	84	
Number of blanket assemblies	66	
Number of control rods locations	12	
Number of reflector locations	54	
Number of core barrel locations	60	
Calculation Results		
K_{eff}; Beginning of cycle (BEOC)	1.000	
K_{eff}; End of cycle (EOEC)	1.013	
Peaking factor, BOEC	1.77	
Peaking factor, EOEC	1.84	
Power split BOEC	94.21/4.59	
Power split EOEC	75.14/23.96	
Average discharge burn-up, MW day/kg	82.17/28	
Peak discharge burn-up, MW day/kg; <150	126.0	
Peak fast fluence, 10^{23}n/cm^2; <4.0	2.70	
Breeding ratio	1.0078	
Reactivity swing, %Dk	−1.27	
Maximum temperature at the center of driver fuel pin, (BOEC/EOEC), °C	1362.9	1259.0
Maximum temperature of the driver pin cladding, (BOEC/EOEC), °C	951.4	914.6

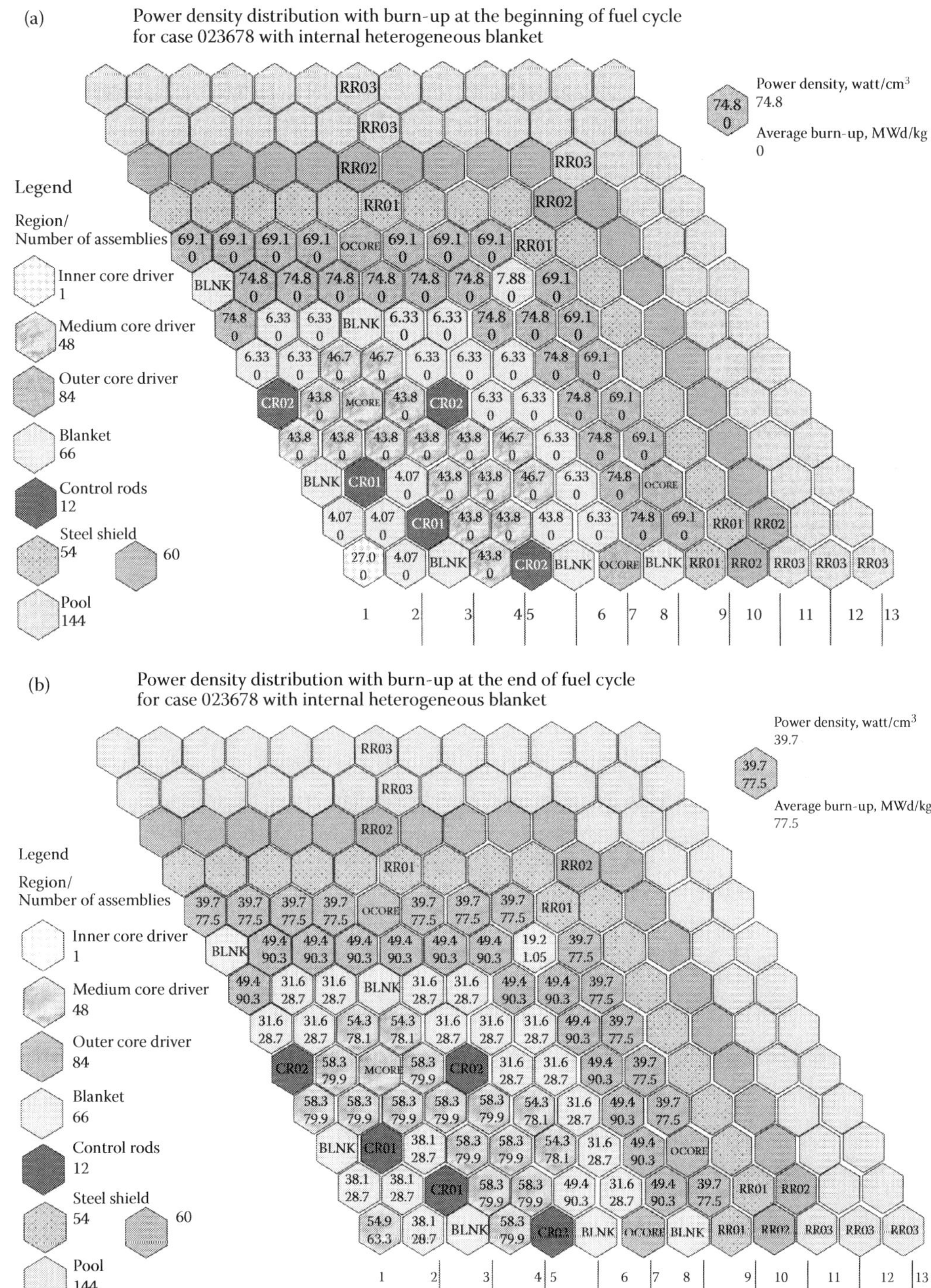

FIGURE 14.2
Power density distribution with burn-up in the core for reference core configuration (a) at the beginning of the cycle and (b) at the end of the cycle.

TABLE 14.3

STAR-H2 (400 MWt) Design Conditions for Thermal-Hydraulic Analyses[a]

Core thermal power, MW	400
Coolant	Pb
Core diameter, m	2.5
Core active (heated) zone height, m	2.00
Fission gas plenum height, m	2.00
Total core (frictional) height, m	4.00
Fuel rod/cladding outer diameter, cm	1.905
Fuel rod triangular pitch-to-diameter ratio	1.50
Cladding thickness, cm	0.10
Fuel material	(92%U–8%Pu)N
Fuel smeared density	0.78
Fuel porosity	0
Fuel pellet diameter, cm	1.51
Cladding-fuel pellet gap thickness, cm	0.0996
Gap bond material	Lead
Core hydraulic diameter, cm	2.82
Number of spacer grids in core	3
Core-wide fuel volume fraction	0.252
Core-wide cladding volume fraction	0.0802
Core-wide bond volume fraction	0.0710
Core-wide coolant volume fraction	0.597
Core fuel mass, kg	29,600
Core uranium mass, kg	27,900
Core flow area, m^2	2.93
Number of fuel rods	6940
Number of support and flow distributor plates below core	2
Plate open area fraction	0.6
Core coolant-to-fuel rod volume ratio	1.48
Core specific power of uranium, kW/kg	12.0
Core power per volume, MW/L	0.044
Core mean heat flux, MW/m^2	0.520

[a] Owing to the iterative process of design, the volume fractions used for neutronic and for thermal-hydraulic calculations differ by a few percent at the conceptual design stage.

TABLE 14.4

Thermal-Hydraulic Results Calculated with the Natural Circulation Model

Mean temperature rise across core, °C	129
Core outlet temperature, °C	793
Core inlet temperature, °C	664
Total core coolant flow rate, kg/s	21,770
Coolant velocity in IHX tubes, m/s	0.369
Coolant Reynolds number in IHX tubes	58,000

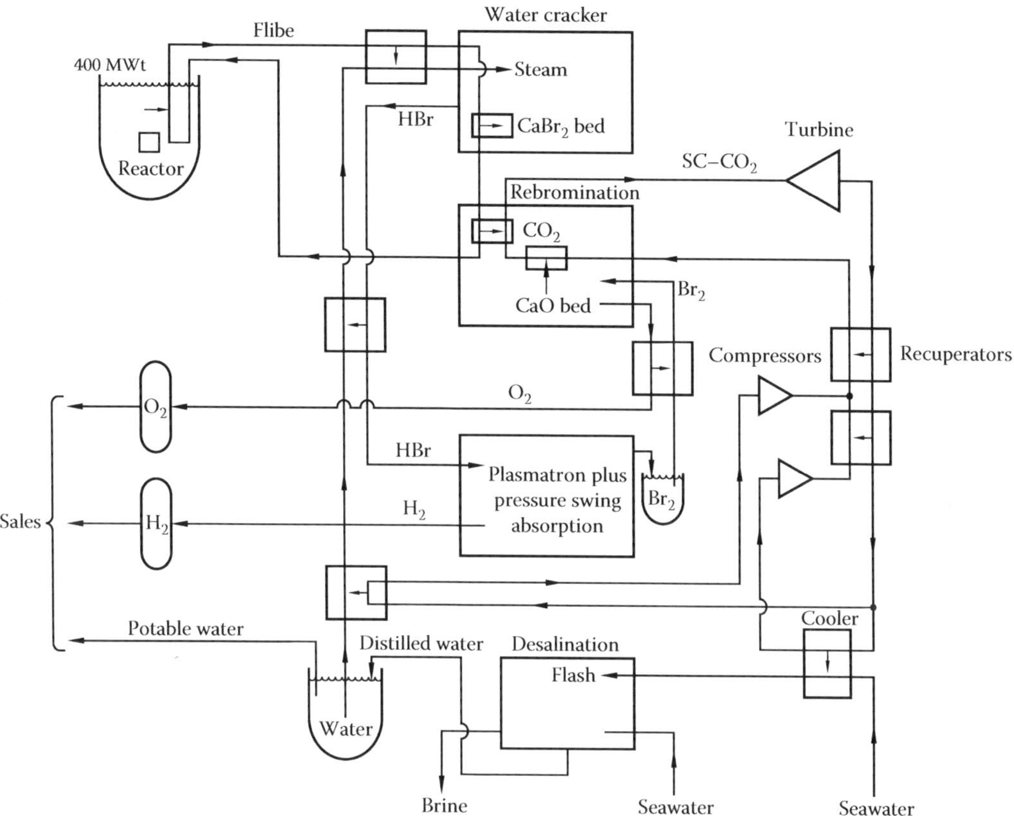

FIGURE 14.3
Reactor and BOP process heat cascade.

3. Use whatever heat is finally left over to desalinate water—first to supply distilled water feedstock to the water cracking plant and second, to use the excess for off-site sales of potable water.

Finally, heat at a temperature near ambient is rejected from the plant in the desalination brine tailings. The plant is designed to minimize this lost heat—both for efficiency and to minimize ecological impacts of heat rejection.

The Ca–Br water-cracking cycle has three main segments: an endothermic "water cracking" segment where $CaBr_2$ and steam react at 700–750°C to produce HBr and CaO; an exothermic Ca rebrominization segment at 600°C where CaO and bromine react to regenerate $CaBr_2$ for recycle and release heat and oxygen; and a plasma chemistry HBr cracking segment at room temperature where electrical driven (radio frequency) energy cracks HBr to regenerate bromine for recycle and to release hydrogen. The plasmatron is followed by a pressure swing absorption cascade, which cleans and pressurizes the hydrogen to meet pipeline-delivery specifications.

The high-temperature endothermic water-cracking segment receives heat from the reactor through the flibe (molten salt) intermediate heat-transport loop. Heat is delivered to the water-cracking plant over the range ~750–700°C via two heat exchangers—one to complete the superheating of steam to 750°C; the other to maintain the $CaBr_2$ beds at 725°C. It must remain at those temperatures—even at partial load—in order to drive the chemical reaction. The final stage of steam superheat includes allowance for heat to overcome thermal inertia of the calcium titanate support for the calcium reagent as it cycles from one reaction segment at 600°C to the other at 725°C.

In the segment for regeneration of $CaBr_2$ from CaO, heat at 600°C is rejected. It is used to help drive the SC–CO_2 Brayton cycle. However, because that heat supply is not sufficient, the Brayton cycle also receives heat from the flibe loop. The flibe then returns to the reactor IHX at ~650°C.

Figure 14.4 indicates mass flows, temperatures, and pressures.

14.4.2.3 Brayton Cycle

After expansion in the Brayton cycle turbine, the SC–CO_2 passes through a high and a low-temperature recuperator. It exits the low-temperature recuperator at 125°C and the heat liberated from cooling it further to 100°C is used to provide vaporization and slight superheat to the distilled-water feedstock destined for the water-cracking cycle. Further, heat is rejected from the SC–CO_2 in the Brayton cycle cooler, which cools the SC–CO_2 to 31°C in preparation for its compression. Seawater provides the cooling fluid for the cooler, and the resulting 100°C seawater, which exits the Brayton cycle cooler, then delivers heat and seawater feedstock to the desalination plant. Finally, as little as is feasible heat at temperature above ambient exits the plant in the form of heated brine tailings from the desalination process.

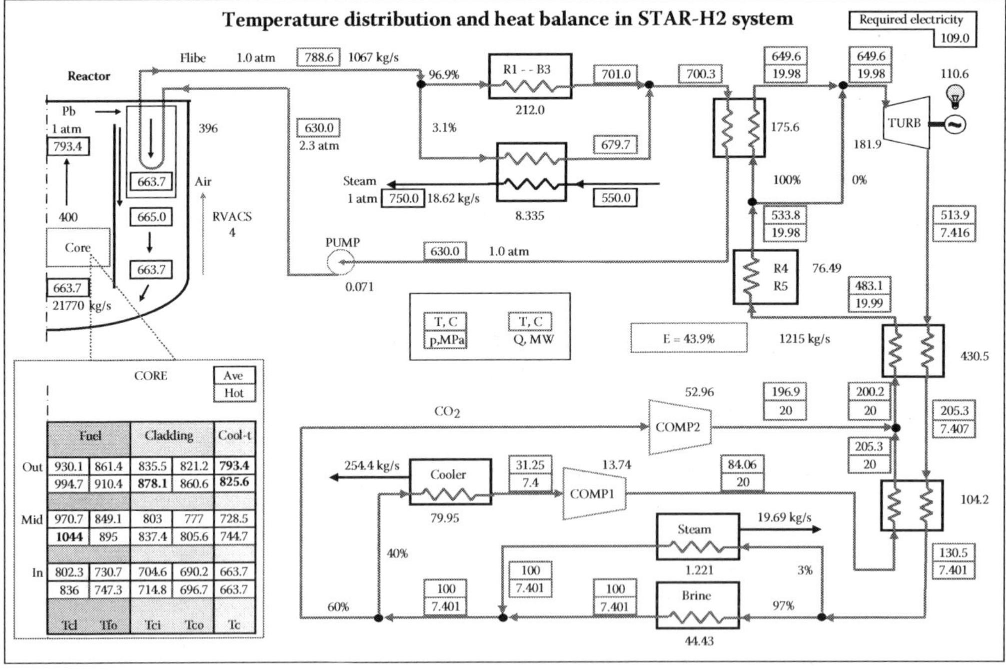

	Fuel		Cladding		Cool-t
Out	930.1	861.4	835.5	821.2	793.4
	994.7	910.4	878.1	860.6	825.6
Mid	970.7	849.1	803	777	728.5
	1044	895	837.4	805.6	744.7
In	802.3	730.7	704.6	690.2	663.7
	836	747.3	714.8	696.7	663.7
	Tcl	Tfo	Tci	Tco	Tc

FIGURE 14.4
Simplified schematic diagram of the STAR-H2 plant, illustrating the cascade of processes: water cracking → Brayton cycle → desalination.

Regenerative heating is used in the water-cracking plant to increase efficiency. The O_2 coming off the $CaBr_2$ regeneration step at 600°C is cooled to room temperature by heating up bromine, which had been recovered in the plasmatron. The bromine is reheated up to 600°C for driving the Ca rebrominization reaction. The HBr from the water-cracking reaction at 700°C is cooled to room temperature for introduction into the plasmatron, and its rejected heat is used to superheat the distilled-water feedstock to near reaction temperature.

The Brayton cycle turbogenerator is sized to meet on-site demands; it is not intended for electricity sales. The electricity generated by the SC–CO_2 Brayton cycle drives the flibe pump, the plasmatron, reagent pumps, pressure swing absorption compressors, and the desalination plant's brine pumps. The Brayton cycle is run at constant temperature and pressure at the turbine inlet and at constant temperature and pressure at the cooler outlet. For partial load, its power output is adjusted via SC–CO_2 mass flow rate.

14.4.2.4 Desalination Plant

The desalination plant to produce potable water from seawater feedstock (assumed at 25°C) is a feed forward MED design, which is driven by the heat recovered in cooling the SC–CO_2 from 100°C down to the SC–CO_2 critical temperature of 31°C. The brine from the desalination plant is rejected at 35°C.

Two reagent buffers are used so that mass flows throughout the BOP need not always be in perfect quasi-equilibrium. After regeneration in the plasmatron, the bromine is stored as liquid in an ambient temperature and pressure buffer tank from which it can be withdrawn as needed to feed the $CaBr_2$ regeneration step. Similarly, the distilled water produced by the desalination plant—which is substantially in excess of requirements for water-cracking feedstock needs—goes to an atmospheric pressure, 35°C holding tank for off-site sales of potable water. The distilled-water feedstock for the water-cracking segment of the Ca–Br cycle is drawn from this buffer tank as needed.

Many alternative options can be considered for productive use of reject heat at any of the three temperatures. Moreover, the passive safety/passive load-follow design of the reactor facilitates siting it in industrial parks near urban areas, and/or close to cities—facilitating cogeneration opportunities.

Several options for yet further extraction of marketable product from the brine tailings have been identified for study in the future. Similarly, several alternative bottoming cycles for use at landlocked sites have been identified for consideration in the future.

14.4.3 Plant Safety Design

14.4.3.1 Overall Safety Considerations

The plant is designed at low power density and large operating margins with natural circulation cooling, innate load following, and passive safety for high reliability and forgiving robustness with respect to BOP failures or operator/maintenance personnel mistakes.

The reactor is a low-pressure vessel filled with a low chemical potential coolant—explosions and fire hazards are small for the reactor itself. The chemical plant, where explosive chemicals are handled, and industrial hazards exist is decoupled from the reactor by distance and by the molten salt intermediate heat-transport circuit operating at ambient pressure; since the reactor can remain within a safe operating regime while innately adjusting its power production to any heat demand communicated through the intermediate circuit—intended or spurious—events in the industrial chemical plant would not influence reactor safety performance.

The long refueling interval reduces time at risk for refueling accidents.

14.4.3.2 Structure of the Defense in Depth

The reactor is designed for a near-zero reactivity burn-up swing such that the safety rod system is vested with minimal positive reactivity at beginning of life (BOL) full power. A safety rod scram system provides a first line of defense for reactivity initiators. Moreover, passive-reactivity feedbacks and passive self-adjustment of natural circulation flow will maintain reactor power to flow ratio in a safe-operating range even with failure to scram; this safe passive response applies for all out-of-reactor vessel-initiated events, that is, for any and all events communicated to the reactor through the flibe intermediate loop. Periodic *in situ* measurements would be made to confirm the operability of these passive feedbacks.

A decay-heat removal path is provided through the heat-transport system through the BOP and ultimately to a seawater heat sink. Additionally, a passive decay-heat removal channel operates continuously carrying ~1% of full power from the pin lattice to the ambient air, using passive natural circulation, conduction, and radiation-heat transport links (see Figure 14.5). This passive path may be periodically tested *in situ* to assure its operability. The thermal inertia of the primary circuit coolant is sufficient to safely absorb the initial decay-heat transient which exceeds the 1% capacity of the passive-heat removal channel.

The fuel, coolant, and internal structural materials are chemically compatible such that clad/coolant chemical interaction is avoided with control of coolant chemistry and such that run beyond clad breach due to manufacturing flaws would not lead to autocatalytic degradation—even for the very long duration of refueling operations. *In situ* monitoring of coolant and cover gas conditions would be used to confirm normality of conditions.

The first line of containment defense is the fuel cladding; the second line of defense is the reactor vessel wall and head cover, and the IHX tube walls. The third line of containment defense is the guard vessel and its cover, and perhaps quick acting valves on the flibe intermediate-loop piping. Since there is no credible high-pressure hazard within the reactor vessel or the intermediate-heat transport loop, the guard vessel and its cover is a low-volume (high surface/volume ratio) containment made of thin-walled steel. The reactor building is a gravel and dirt bunker placed over the silo emplaced reactor to protect the containment from hazards; it has no containment function (see Figure 14.6).

14.4.3.3 Passive Load Follow Capability

The reactor is connected to the BOP through the flibe-heat transport loop and only through the flibe-heat transport loop. The heat demand from the BOP is made known to the reactor through the flibe flow rate and the flibe return temperature. The flibe loop delivers heat from the reactor to three heat exchangers in the BOP: to the $CaBr_2$ bed in the water-cracking vessel; to the last-stage steam superheater, and to the last-stage SC–CO_2 heater. The flibe is then recirculated back to the nuclear reactor heat exchanger, and the rest of the BOP runs on heat cascaded down from these processes, see Figure 14.4.

The goal for passive load follow design is to use the intermediate loop flow and temperature information and only this information to cause the reactor to self-adjust its power

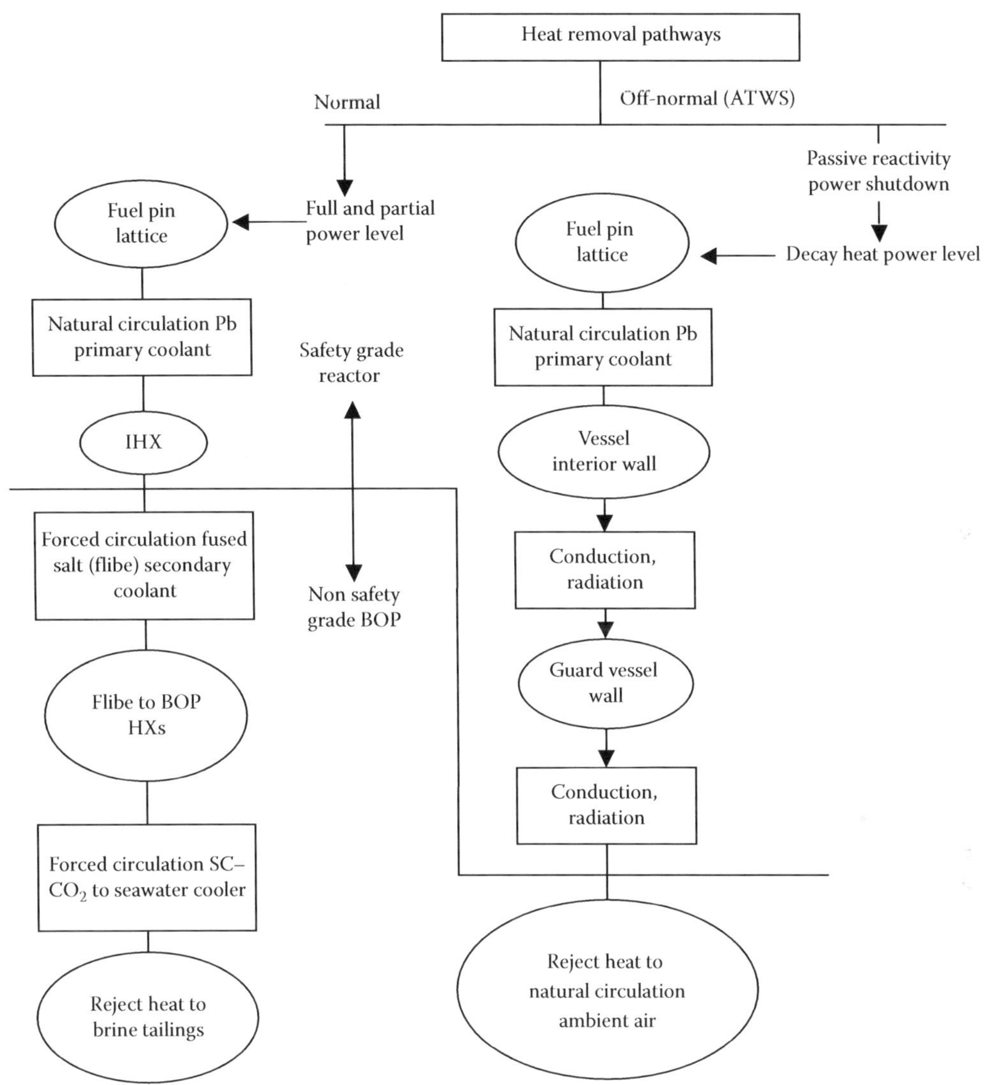

FIGURE 14.5
Heat removal pathways.

level such as to exactly match that heat demand communicated through the flibe loop—and to do so within a safe operating envelope.

Since the Ca–Br water splitting chemical reaction requires a heat source at ~725°C to cause the $CaBr_2 + H_2O \leftrightarrow CaO + 2HBr$ reaction to take place, that temperature must be maintained even if the water-cracking plant is operating at only partial load. Therefore, for STAR-H2, another design constraint was imposed—the core outlet temperature of 800°C at full load must also be maintained at all levels of partial load.

The basic character of the STAR-H2 passive load follow design strategy is as follows: the Pb coolant outlet temperature should remain constant versus fraction of full load, whereas coolant temperature rise across the core should increase versus increasing fraction of full

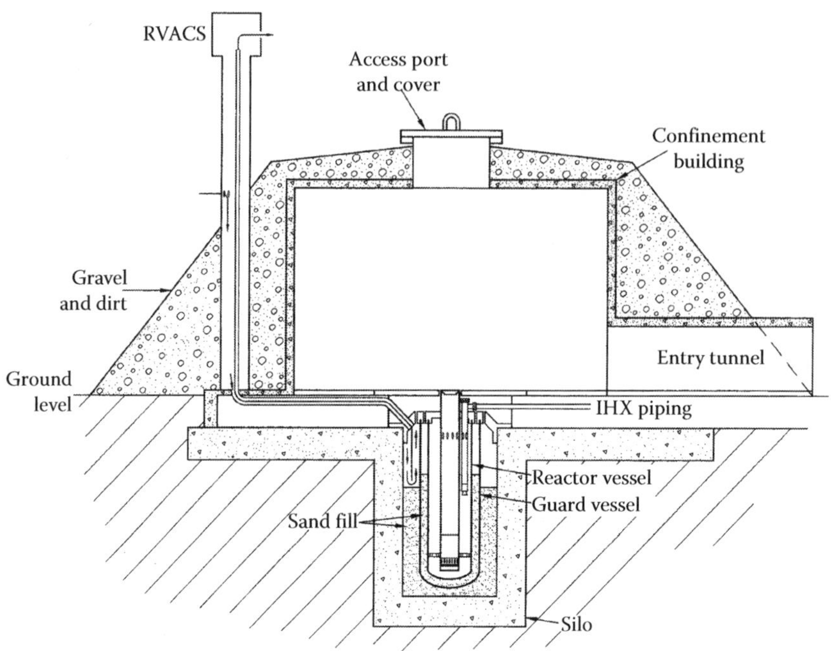

FIGURE 14.6
Protection of the STAR-H2 reactor from external hazards.

power—implying that core inlet temperature T_{Inlet} decreases with increasing fraction of full load. This, in turn, increases the buoyancy driving head for natural circulation, which depends on differences in the temperature of coolant exiting the core and the heat exchanger.

The core average coolant temperature and fuel temperature—the two temperatures which dominate contribution to reactivity feedbacks—should behave as follows: as fraction of full power is increased, the fuel temperature rise above the coolant will increase, which adds negative reactivity due to Doppler and fuel-axial expansion. On the contrary, so as to offset the negative reactivity of fuel temperature rise, the coolant's average temperature should decrease with increasing fraction of full power, adding positive reactivity. This can happen—given that core outlet temperature remains fixed—by causing coolant inlet temperature to decrease as heat demand increases. The design challenge is to create a core design and a reactor natural circulation cooling circuit such that the two reactivities cancel at every value of partial load and the flow readjusts to produce the required core temperature rise ΔT versus power level.

The coupled neutronics/thermo-hydraulic/thermo-structural reactivity feedback design approach for the STAR-H2 reactor has achieved the proper ratio between that reactivity which is vested in the coolant temperature rise relative to inlet temperature *vis-à-vis* that reactivity which is vested in the fuel temperature rise above the coolant, and at the same time in having designed an overall coolant flow-circuit pressure drop tailored to cause coolant flow rate to adjust properly to changes in pressure driving head caused by source/sink temperature difference. A nonconventional open-pitch ductless fuel assembly structural design coupled with a nonconventional core support approach (the assemblies tend to neutral buoyancy in the dense Pb coolant) has been proposed to simultaneously provide

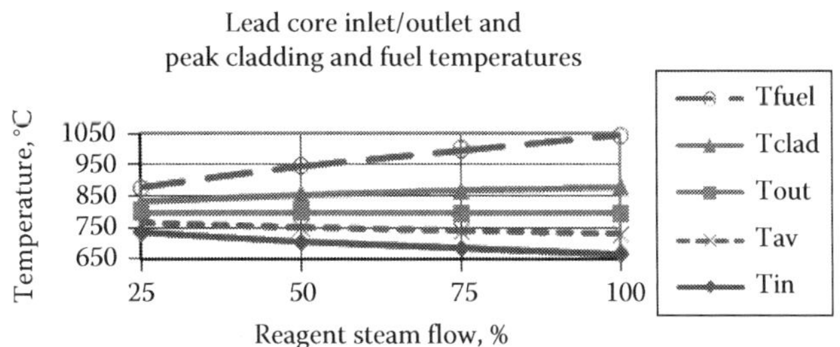

FIGURE 14.7
Reactor partial load schedule.

low-pressure drop, structural reliability of grid spacers, and an appropriate value for coolant power/flow reactivity temperature coefficient.

The resulting STAR-H2 partial load schedule is shown in Figure 14.7. The interplay of coolant ΔT and natural circulation coolant flow rate is shown in Figure 14.8.

Coupled neutronics/thermal-hydraulics stability analyses of the STAR reactor at these plant equilibrium states at full and partial load will be required. Such analyses have been conducted already for the secure transportable autonomous reactor—liquid metal (STAR-LM) which shares the neutronics and thermal-hydraulics properties of STAR-H2 reactor—and stability has been demonstrated.

14.4.3.4 Passive Safety Response

Given that the STAR-H2 reactor will passively self-adjust its power level to meet the heat request from the BOP communicated via the flibe flow rate and return temperature (when the BOP is under purposeful control), it is still necessary to show that the reactor would not self-adjust itself to damaging power or power/flow levels when adverse conditions exist in the BOP.

The STAR-H2 reactor has a central safety rod and an active scram circuit. It also will have a decay-heat removal path through the flibe loop. However, in order to achieve the levels

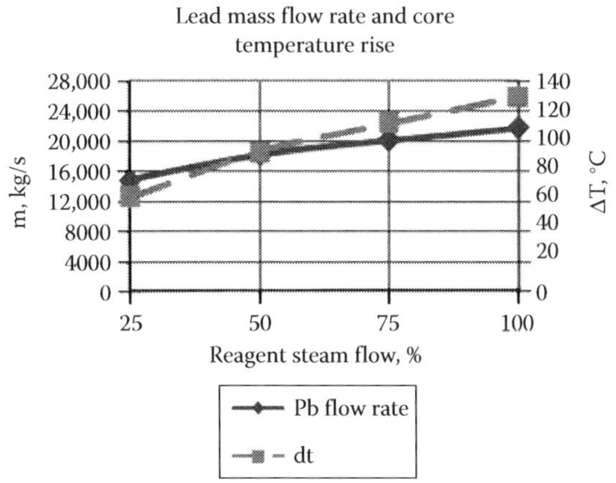

FIGURE 14.8
Reactor coolant flow rate and temperature rise versus percent (%) of full load.

of safety required for deployments of tens of thousands of STAR-H2 plants worldwide—sited near cities, it will be essential to avoid core damage even in the absence of a scram. The technology to achieve passive safety response to anticipated transients without scram (ATWS) events was well developed for the integral fast reactor (IFR) concept [4] and was famously demonstrated in tests conducted at the EBR-II reactor [5].

The quasi-static reactivity-balance theory of designing for passive safety response to ATWS initiators has been applied in the design of STAR-H2. The efficacy of this design approach is confirmed by the evaluation results to be presented next.

14.4.3.4.1 Bounding BOP and Control Room Events

Since the only communication channel from the BOP to the reactor is through the flibe loop, it might be possible to bound all possible BOP conditions—whether intentional or spurious—via limiting flibe conditions at the reactor IHS. These off-normal conditions could arise from BOP equipment failure or from maintenance errors or from operator error. The point is that the reactor "sees" the external world through only one window—the flibe loop. It could thereby be possible to deterministically span the space of all possible externally initiated accident events that the reactor will face and to determine whether the reactor's passive response (without scram) will hold the reactor in a safe condition. The flibe flow rate can change; but it cannot decrease to less than zero nor can it increase to more than that which cavitates the pump—taken here to be 115% of full flow. The flibe return temperature can change, but it cannot increase to above the Pb-delivery temperature at the reactor IHX nor can it decrease to below the lowest temperature it encounters in the heat exchangers it passes through in the BOP.

Finally, the flibe pressure can change, but it cannot go below ambient pressure (which is the normal condition) nor can it go above the pressure of fluids, which it encounters in heat exchangers (should a tube rupture occur). Since all processes except for the Brayton cycle operate at ambient pressure, this upper pressure bound is 20 MPa—should a tube in the flibe-to-SC–CO_2 heat exchanger rupture.

All physically feasible conditions of the flibe communication channel from the external world to the reactor are bounded by the physically limited extremes listed in Table 14.5. These are innate physical bounds—they span the space of all possible conditions communicated to the reactor from outside the vessel.

TABLE 14.5

Physical Bounds on Flibe Loop Parameters

	Physical Limits	
Category of Disruption	**Lower Limit**	**Upper Limit**
Flibe flow rate disruptions	Zero	Pump cavitation
Flibe return temperature disruptions	Lowest temperature encountered in the three heat exchangers in BOP: Reagent steam super-heater; Water cracking heat exchanger; CO_2 heat exchanger	Pb temperature in reactor heat exchanger (HX)
Flibe pressure disruptions	Ambient (normal conditions)	CO_2 pressure (Flibe/CO_2 HX tube rupture)

TABLE 14.6

ATWS Events, Which Span the Space of BOP-Initiated Accidents

		Flibe Loop Condition			
Category	**Name**	**Flibe Flow**	**Flibe Return Temperature**	**Flibe Pressure**	**Description**
Base case	Nominal 100% power	Nominal	~630°C Nominal	Ambient	Normal 100% power condition
Pump disruptions	LOHS	0		Ambient	LOHS; flibe flow stops
	POS	115% of normal	Nominal	Ambient	Pump over-speed; flibe pump over-speeds to cavitation (assumed 115%)
Return Temperature Disruptions					
Under-cooling	LOCP	Nominal	>630°C	Ambient	LOCP heat sink; Brayton cycle continues to run
	LOBC	Nominal	>630°C	Ambient	LOBC heat sink; chemical plant heat sink continues to run (assumes off-site or emergency electricity source)
Overcooling	COS	Nominal	<630°C	Ambient	CO_2 compressor over-speed (assumes CO_2 compressors at 155% of normal flow rate)
	SBD	Nominal	<630°C	Ambient	Reagent steam line blow down (taken to be 300% normal flow through flibe-to-steam heat exchanger)
Pressure disruption		Nominal	Nominal	Overpressure	Flibe-to-CO_2 heat exchanger tube rupture

14.4.3.4.2 Reactor Passive Safety Response to Bounding BOP Events

The resulting reactor power level and the coolant, cladding, and fuel temperatures which result from the ensemble of the bounding conditions in the flibe loop have been calculated using the quasi-static methodology used previously for passive load follow analysis. The calculations were made in response to the specific BOP conditions enumerated in Table 14.6. These BOP conditions represent plausible off-normal events in the categories of off-normal flibe flow disruptions (overcooling and undercooling) and off-normal flibe-return temperature disruptions—undercooling and overcooling events. Table 14.6 defines a mnemonic name for each off-normal anticipated transient without scram (ATWS) event analyzed, see Figure 14.9 showing the results.

a. *Loss of heat sink* Considering the flibe pump-speed disruptions to the limits of their physical bound, the first case (see Table 14.6) is the reactor loss of heat sink (LOHS) case where the flibe flow rate stops. As shown in Figure 14.9a, the reactor power level is driven to decay-heat level by the action of the negative temperature

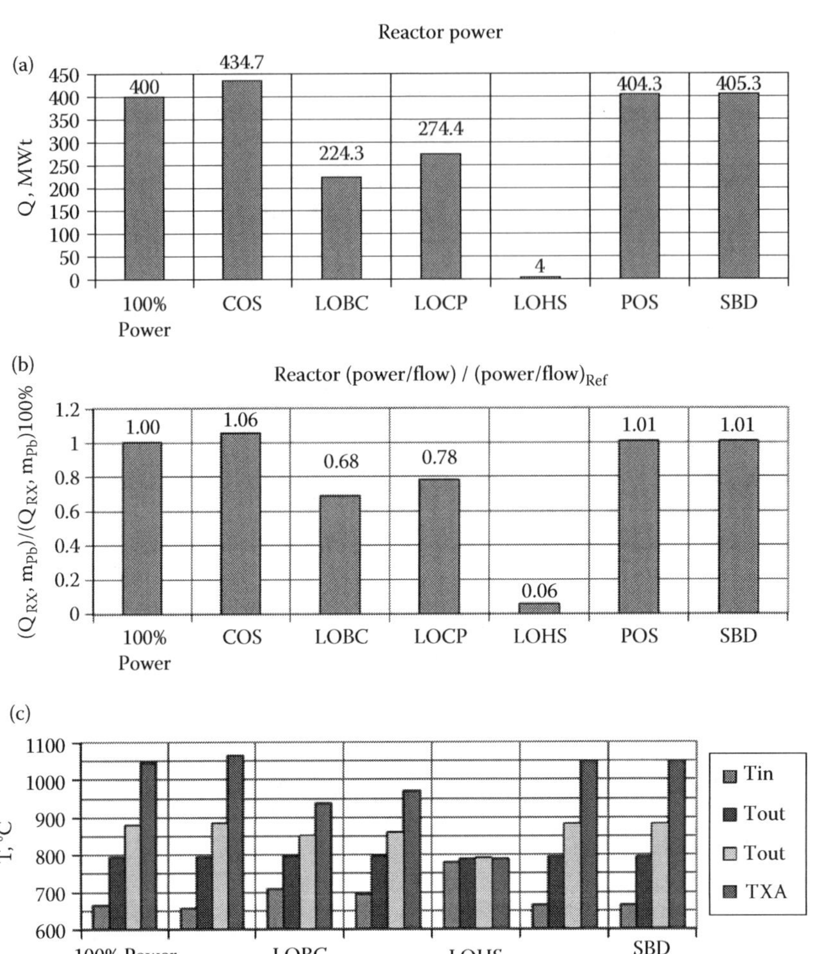

FIGURE 14.9
(a) ATWS event asymptotic power levels. (b) ATWS event asymptotic power/flow ratios. (c) ATWS event asymptotic temperature levels.

coefficient operating on a reactor coolant inlet-temperature increase. Although fuel temperature decrease adds positive reactivity, negative reactivity from coolant temperature increase dominates and leads to net reactivity decrease. The Pb outlet temperature decreases slightly and, as the power is zero, the coolant temperature rise collapses to essentially zero. Clad and fuel temperatures each drop (see Figure 14.9c), and the reactor becomes essentially isothermal at about 790°C. There is no core damage and the reactor's asymptotic equilibrium state is delayed critical at decay-heat level with the reactor vessel air cooling system (RVACS) passively removing decay heat.

b. *Flibe pump over-speed* Second, the overcooling accident (POS) caused by flibe pump speed increase to cavitation assumed at 115% of full flow leads to an asymptotic state with power and power-to-flow ratio at about 101% of nominal (Figure 14.9a and b). Since the power/flow remains essentially unchanged, the coolant average temperature raises only slightly; clad and fuel temperatures remain very near nominal.

In summary, the two flibe flow disruption accidents—which bound all other flibe flow disruptions—cause the reactor to self-adjust asymptotically to a safe operating state—even without scram.

c. *Off-normal flibe return temperature—Too high* Looking next at the flibe return temperature disruptions, it is clear that the flibe return temperature can never exceed the reactor's Pb-outlet temperature (which would consist of a LOHS case already discussed) nor can it lie below the lowest temperature it encounters in the BOP heat exchangers it passes through. It passes through three heat exchangers: the reagent steam superheater, the heat exchanger to the $CaBr_2 + H_2O$ water cracking bed, and the CO_2 heat exchanger. An off-normal condition in any heat exchanger could arise from too little heat removal from the flibe (partial loss of load) or from too much heat removal (overcooling accident).

The first category (flibe return temperature increase) results from partial LOHS; two plausible "partial LOHS" cases have been examined—loss of chemical plant (LOCP) heat sink and loss of Brayton cycle (LOBC) heat sink. The resulting clad and fuel temperatures are shown in Figure 14.9c. When the flibe return temperature increases due to partial loss of load, the reactor power self-adjusts downward to match the decreased load under action of the coolant average reactivity coefficient. Clad and fuel temperatures decrease and the reactor passively adjusts to a safe operating state.

d. *Off-normal flibe return temperature—Too low* The second category (flibe return temperature decrease) results when too much heat is removed in one of the three heat exchangers—reagent steam, water cracking, or CO_2 heat exchangers. In the case of the water cracking $CaBr_2$ bed-heat exchangers, the $CaBr_2$ bed is a fixed solid bed at atmospheric pressure, and the only plausible overcooling would come from excess reagent steam flow. Thus, this case is bounded by the analysis of overcooling in the reagent steam heat exchanger. Overcooling by reagent steam flow could occur from feed water pump over-speed, from loss of function of the regenerative HBr to reagent steam heat exchanger or from steam line blow down following a pipe rupture. However, as the reagent steam is at atmospheric pressure and lacking a driving force, the "blow down" would be benign. Detailed studies of such a blow down or of loss of the HBr to reagent steam regenerative heating have not yet been completed and so a conservative case (SBD) of 300% nominal reagent steam flow has been used.* The results of this case are shown in Figure 14.9c. Due to a decrease in flibe return temperature, the positive reactivity change from lowering reactor inlet temperature causes power to rise from 400 to 405.3 MWt. Power-to-flow ratio increases by only 1%; thus, negligible increase in clad and fuel temperatures occur. The reactor passively accommodates this accident sequence without damage, even without scram.

The other possibility for flibe overcooling is in the flibe-to-CO_2 heat exchanger. That case has been examined assuming the CO_2 compressors both over-speed to 115% of nominal capacity (COS). The resulting reactor power and temperatures are shown in Figure 14.9a and c. Here, power increase is substantial—from 400 to 434.7 MWt. However, natural circulation again passively increases to match

* This will also bound a feedwater pump over-speed to cavitation—estimated to occur at 115% of normal flow.

increased power; power-to-flow ratio increases by only 6% and the clad and fuel temperature increases are small. Again the reactor passively adjusts to a safe operating state—even without scram.

e. *Flibe loop over-pressurization—SC–CO_2 HX tube rupture* The last category of a bounding off-normal BOP condition affecting the reactor is flibe loop over-pressurization. A tube rupture in the flibe-to-SC–CO_2 heat exchanger would subject the flibe, which is normally at atmospheric pressure, to a ~3000 psi pressure source. Absent some intervention, a compression wave would travel at the speed of sound in flibe through the flibe loop piping to the in-vessel flibe-to-Pb IHX where the thin-walled tubes designed for ambient-to-ambient pressure heat transfer would likely rupture and expose the reactor vessel to an abrupt pressure increase. This highly undesirable scenario has been faced and handled for Na reactors in the case of an intermediate Na loop/steam generator tube rupture. The design solution is to put a large diameter rupture disk in the intermediate loop. A small overpressure will rupture the disk—allowing the pressure to release to the atmosphere. This terminates the over-pressurization transient and leads to the LOHS case, which has been shown above to lead to a benign passive shutdown. Alternately, a more advanced design for the flibe-to-SC–CO_2 heat exchanger itself could be considered, with flibe and SC–CO_2 tubes both immersed in a common pool of high-conductivity fluid, such as flibe or Pb. A SC–CO_2 tube rupture would cause CO_2 venting to ambient through the fluid in the tank, and the flibe tubes would remain intact.

14.4.3.4.3 *Summary of ATWS Event Passive Response*

To summarize, the asymptotic (bounding event) analyses performed have shown all bounding cases of ATWS initiators originating from a BOP disruption to be passively accommodated within safe asymptotic temperature conditions. Decay-heat removal was assumed to rely on a passive RVACS. Both, the innate thermostructural reactivity feedbacks and the innate decay-heat removal pathway to ambient could be nonintrusively monitored to assure their continued capability to provide safe response. Thus, no matter what happens in the BOP, the reactor will self-adjust itself to a safe asymptotic condition even without scram. This could make it possible to construct and operate the BOP to ordinary industrial standards.

Given that the asymptotic states in response to ATWS initiators are safe, it remains to show that the dynamic transition to the asymptotic state will not engender damaging conditions on the in-core structures. A plant dynamic code, which can model the STAR-H2 BOP, was not available at the time when this report was prepared. Such a code is being developed first for the STAR-LM, which has a simpler (SC–CO_2 Brayton cycle) BOP. In the future, after further refinement of the Ca–Br water cracking cycle, that dynamics code will be modified for applicability to the STAR-H2.

Moreover, coupled neutronics/thermal-hydraulics stability analyses would be required for the ending equilibrium states from the passive accommodation of ATWS initiators. Work for the STAR-LM suggests that these states are indeed stable ones.

14.4.3.5 Beyond Design Base Events and Elimination of Need for Off-Site Emergency Response

As shown in the previous section, the designed in and always operating passive reactivity feedback response and passive decay-heat removal pathways are capable to close off all conceivable pathways to core disruption using innate, *in situ* testable processes. These innate responses apply to any event originating outside the reactor vessel.

Accident initiators, which might originate inside the vessel, could only have come from initial manufacturing or assembly flaws during construction or from long-term neglect of coolant-chemistry control. Such initiators would (in future) be addressed by probabilistic risk assessment (PRA) methods and are expected (based on similar PRAs performed for sodium-cooled systems) to represent triple-fault events of such low probability as to lie in the beyond-design basis range. Moreover, as there are no credible mechanisms for high ramp rate, many such noncredible events would be gradual and could be annunciated early by coolant and cover gas monitoring before gross fuel pin disruption occurs.

Even in the event of fuel pin disruption, since the specific gravity of the nitride fuel and the Pb coolant are nearly identical, one could expect fuel particle dispersal and dilution in the vast Pb inventory, thus precluding recriticality concerns.

14.4.3.6 Probability of Unacceptable Radioactivity Release beyond Plant Boundaries

The STAR-H2 safety strategy is adapted from that used for the IFR, which was demonstrated in full-scale tests at the EBR-II 62 MWt power plant in 1986 tests [5]; loss of heat sink without scram (LOHSWS) and loss of flow without scram (LOFWS), both from full power as well as run beyond cladding breach, were all demonstrated to yield benign results. The Level 1 PRA conducted for the EBR-II showed that probability of technical specification violation with marginal loss of fuel pin lifetime came in at a slightly lower frequency (~10^{-6}/year) than the probability for core disruption and overall loss of the reactor for the PWR PRAs reported in NUREG-1150, see Figure 14.10.

14.4.4 Fuel Cycle and Sustainability

14.4.4.1 Fuel Cycle Technology

STAR-H2 employs a closed fuel cycle. The reactor is fueled with uranium/transuranic (TRU) nitride fuel enriched in ^{15}N and it operates on a 20-year whole core cassette-refueling interval; it is fissile, self-sufficient with an internal core conversion ratio of one.

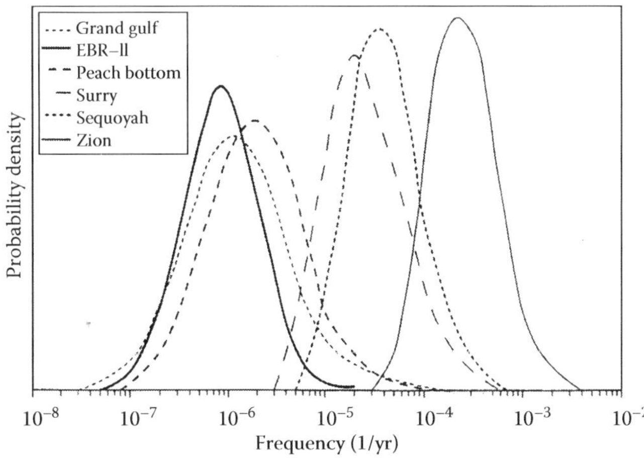

FIGURE 14.10
Comparison of EBR-II damage frequency with core damage frequency at commercial LWRs (LWR data from NUREG-1150) [6].

All fuel cycle services and waste management is assumed to be outsourced to a regional fuel-cycle center, which is owned and operated under the control of consortia of its customers while under international nonproliferation oversight.

Current thinking is that pyro recycle and vibropack remote refabrication technology would be developed and used for the TRU/U-based nitride fuel.

The pyro recycle technology is assumed to produce a commixed stream of all TRUs and achieve incomplete fission product removal such that the fissionable materials during processing and during fresh and used cassette shipping are always at least as unattractive for weapon use as is LWR spent fuel. All fuel cassette shipments and used cassette returns would be conducted by regional center personnel who bring the refueling equipment with them and take it away with the spent cassette. No refueling equipment would remain at the site. No spent fuel would be stored at the site for cooling. Because of derating the specific power (kW/kg) to achieve long refueling intervals only a short cooling time is required before fuel handling can commence.

The fuel cycle feedstock will be natural or depleted uranium, and multirecycle through sequential cassette reload cycles could achieve near total fission consumption of the ^{238}U feedstock; only fission product waste forms (and trace losses of TRU) would go to a geologic repository operated by the regional center.

Once deployed, each STAR will be fissile self-sufficient. Initially, fuel for STAR-H2 new deployments could come from TRU recovered from spent LWR fuel or from enriched uranium sources. Later, when that source is exhausted, fast breeder reactors could be sited at the regional fuel cycle centers. Their function will be to manufacture excess fissile material so as to fuel the initial working inventories of new STAR deployments in a growing economy. The heat from their operation could be converted to hydrogen for shipment to regional consumers.

14.4.4.2 Sustainability, Waste Management, and Minimum Adverse Environmental Impacts

The hierarchical energy architecture utilizing the STAR-H2 concept is being devised to meet both the energy resource sustainability and the environmental compatibility tenants of sustainable development [1].

As to resource sustainability, the known plus speculative economically recoverable ore of ~15 million tons of U, when fully fissioned, could supply the world's entire primary energy needs for a millennium.

The application of nuclear heat to produce hydrogen as a replacement for fossil fuels might achieve an essentially greenhouse gas free energy supply chain extending from resource to end use and it allows nuclear energy to move beyond electricity to service all sectors of primary energy usage.

Processes convert one energy carrier into another at the hubs of the hub-spoke architecture optimized for nuclear energy (e.g., nuclear fuel to hydrogen; hydrogen to electricity) or to energy services (nuclear heat to potable water, electricity to motive force, etc.). These conversion processes generate wastes. The proposed architecture provides for an ecologically neutral closure of the entire energy supply enterprise through recycle of these wastes as illustrated in Figure 14.11. Referring to Figure 14.11, closure is obtained on electricity production and use by electron return through ground. Closure is obtained on thermochemical water cracking hydrogen production and its use in fuel cells or micro-turbines by nature's oxygen and water cycles.

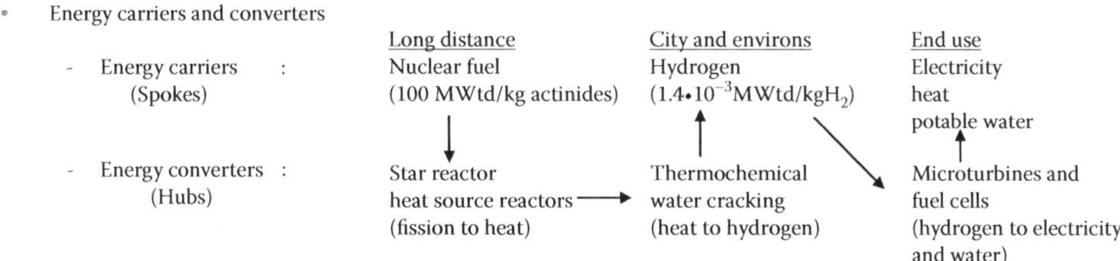

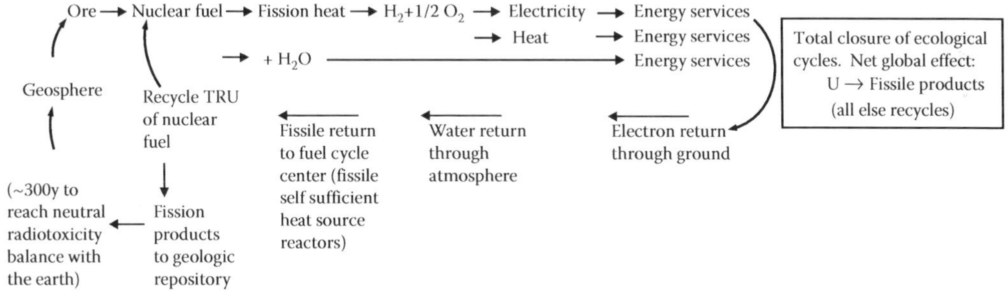

FIGURE 14.11
Ecologically neutral energy supply chain.

14.5 STAR-H2 Development and Funding Status

14.5.1 Development Strategy

STAR* concept development is being conducted for a portfolio of reactor and BOP designs to enable an incremental market penetration that is time-phased according to the degree of R&D required. The STAR portfolio of designs was initiated in the 1997 time frame and is comprised of SSTAR, STAR-LM, and STAR-H2. They employ technological innovations of increasing lead-time to deployment (see Figure 14.12).

STAR-LM, is a Pb-cooled, 400 MWt, natural circulation reactor of 565°C core outlet temperature driving a supercritical CO_2 Brayton cycle. It draws on many proven technologies and will be ready for market in 15–20 years. The SSTAR takes the STAR-LM design features down to ~25–50 MWt to provide for secure energy supply to remote small villages. It is targeted for early prototyping of the technology and institutional features of the STAR concept. STAR-H2, raises the Pb outlet temperature to 800°C to drive a thermochemical

* STAR = Secure, Transportable, Autonomous Reactor. The STAR reactors are referred to as "Batteries" because they store 20 years worth of heat and they load followed by passive means—delivering heat when it is requested by the balance of the plant and passively shutting off when the request stops.

Portfolio member	Power	Coolant	Tout	Converter	Products	Client*	Deployment target
SSTAR S=small	20–50 MW_{th}	Pb Nat'l Circ.	550°C	Rankine steam – or $SC\text{-}CO_2$ Brayton ↓ Desalination	Electricity +potable water or potable water	Electricity for remote town of ~ 6500	~2015 Potential 1st prototype
Star-LM LM=Liquid Metal	400 MW_{th}	Pb Nat'l Circ.	550°C–580°C	$SC\text{-}CO_2$ Brayton ↓ Desalination	Electricity +potable water	Electricity for city of ~115,000	~2020
STAR-H2 H2 = Hydrogen	400 MW_{th}	Pb Nat'l Circ.	800°C	Ca-Br Thermo-chemical cycle ↓ $SC\text{-}CO_2$ Brayton ↓ Desalinatation	H_2 +potable water	All primary energy and potable water for city of 25,000	~2030

* Assume 4 toe/capita year primary energy ≡ 12 kw th year/person year
Assumes 1/3 of primary energy converted to electricity

FIGURE 14.12
The STAR Portfolio.

water cracking cycle and will require additional R&D. It is targeted for deployment by 2030.

The STAR-H2 concept is at the conceptual design stage of development. However, its safety strategy and its fuel cycle and waste management strategies have been adapted from the 10 years of development for the IFR [4]. Furthermore, the reactor structural, refueling, neutronics, and thermal-hydraulics design approaches have been adapted from the STAR-LM project [7], which has a several year head start in design effort relative to STAR-H2.

The salient features of the new reactor design of STAR-H2 relative to STAR-LM are the related material—choice and qualification of cladding and structural materials for 800°C service conditions in Pb. Also, the fabrication technologies for low-cost serial factory fabrication of reactor modules, and of refueling cassettes, using the new structural materials. Material screening tests have been conducted and among the materials tested or to be tested in the corrosion/mass transport convection harps are composites including SiC and ZrC and refractory metal alloys. If the ceramic composites prove out, it may be possible to bring aerospace fabrication technologies to bear on STAR manufacture.

STAR-H2 relies on passive safety accommodation of ATWS initiators and passive decay-heat removal—technologies, which are already well developed in the IFR program. Several safety-related issues require more work, however. Potential for degradation of cooling capacity by sludge build-up in the event of loss of control of coolant chemistry and/or by coolant solidification in the event of local system cool down (327°C Pb freezing temperature) need to be addressed. The phenomenology and consequences of nitride fuel dissociation under high-temperature accident conditions must be understood; on the one hand it may provide a fuel dispersal/HCDA* quenching mechanism; on the other it might produce significant reactor tank overpressurization.

The nonaqueous recycle technology for nitride fuel, while under development in Japan and in the Russian Federation, is not as well advanced as it is for metallic alloy, neither is the Russian-developed vibropack remote fabrication technology as available outside the Russian Federation as is the Argonne-developed remote casting fabrication used for

* HCDA is for hypothetical core disruptive accident.

metallic alloy. Development and prototype testing of these technologies and resulting waste forms is needed. A major nitride fuel irradiation test program is required, and a fast spectrum fuel irradiation test facility is needed to conduct it.

The development of the Ca–Br thermochemical water-cracking process must be taken beyond the bench scale, which has been achieved in Japan. Reliable thermodynamic data (Gibbs free energies) are available for all reactions. However, reaction kinetics data on prototypic reaction bed configurations are lacking and are being researched. Significant and cost-effective proposed modifications of the flow sheet based on plasma chemistry must be researched starting at the bench scale.

The supercritical (SC) CO_2 Brayton cycle has been optimized on paper and control strategies are under current study. Testing is being initiated on printed circuit heat exchanger components, which hold potential for reduction of recuperator size and cost. An entire SC–CO_2 Brayton cycle prototype will have to be built and tested to bring the technology to a state of commercial availability.

The desalination bottoming cycle uses commercial technology and off-the-shelf components.

A capital cost containment strategy based on simplification, component elimination, serial factory fabrication, and rapid site assembly and a nonnuclear safety grade BOP has been devised. And an operating cost containment strategy based on ultrahigh capacity factor, high energy conversion efficiency, product diversification and operating staff reductions based on simplification, passive load following, and passive safety and elimination of components has been devised. Whether these strategies can overcome the economic penalty of derating power density to achieve 20 year refueling interval and reducing plant rating to lower initial capital outlay is not yet known. Capital cost estimates will not be determinable until further engineering refinement of the concept is completed.

14.5.2 Status of R&D Funding

Development and design of the STAR-H2 concept was funded by a U.S. Department of Energy (DOE) NERI grant from years 2000 through 2003. SSTAR development at Argonne and Livermore National Laboratories is ongoing—supported by the U.S. Department of Energy R&D funds.

Currently, funding is not available at a level sufficient to make feasible the design, construction, and initial operation of a demonstration test reactor within a 2015 timeframe; and uncertainty remains as to what funding priorities the U.S. DOE would place on the STAR concept.

14.6 Other Battery-Type Systems for Hydrogen Production

The IAEA has recently conducted an in-depth survey of battery-type reactor plants under development throughout the world (IAEA-TECDOC-1536, "Status of Small Reactor Designs without Onsite Refueling," Jan 2007). Thirty such concepts were identified. Of these, only five of them have targeted core outlet temperatures in excess of 700°C; two used Pb or Pb–Bi coolant; two used molten salt; and one used He gas. They range in power from 100 kWth to 450 MWt, and only two explicitly targeted H_2 productions—STAR-H2 and the

Pb-Bi-cooled thermal spectrum reactor CHTR under development at the Bhabha Institute in India.

The CHTR is a very small 5 MWe Pb-Bi-cooled thermal neutron spectrum reactor being developed to service the electricity needs of the many off-grid villages and towns in the mountainous regions of India and on its numerous islands. The current configuration employs metallic U^{235}/ Th^{232}/Zr alloy fuel and delivers heat through heat pipes at 600°C to a thermoelectric generator. It is cooled by natural circulation and has a refueling interval of 10 years. Although, it is initially intended to drive a passive thermoelectric generator, some consideration has been given to hydrogen production as a later mission.

Acknowledgment

The author expresses his appreciation to the International Atomic Energy Agency for permission to extensively use material from IAEA-TECDOC-1536, "Status of Small Reactor, Designs without Onsite Refueling" ANNEX-24 (Jan 2007).

Disclaimer

References

1. Bruntland, G., *World Commission on Environment and Development, Our Common Future (The Brundtland report)*, Oxford University Press, Oxford, 1987.
2. Doctor, R., Matonis, D., and Wade, D., Hydrogen generation using a calcium–bromine thermochemical water-splitting cycle, *OECD-NEA Second Information Exchange Meeting on Nuclear Production of Hydrogen*, Argonne National Laboratory, October 2–3, 2003 (Proceedings) OECD NEA No. 5308 2004.
3. Moisseytsev, A., Sienicki, J., and Wade, D., Cycle analysis of supercritical carbon dioxide gas turbine Brayton cycle power conversion system for liquid metal cooled fast reactors, *ICONE11 Proc. of Int. Conf. on Nuclear Energy*, Tokyo April 20–23, 2003, paper ICONE 11-36023, ASME.
4. Hannum, W., guest editor, The technology of the integral fast reactor and its associated fuel cycle, Special issue, 31 (1–2), *Progress in Nuclear Energy*, ISSNO149-1970 1997.
5. Fistedis, S. H. *The Experimental Breeder Reactor-II Inherent Safety Demonstration*, Elsevier Science Publishers B. V., North-Holland Publishers, Amsterdam 1987.

6. Hill, D., Ragland, W., and Roglans, J., *The EBR-II Probabilistic Risk Assessment: Lessons Learned Regarding Passive Safety, Reliability Engineering and System Safety*, Vol. 62, pp. 43–50, Elsevier Science Limited 1998.
7. Sienicki, J. J., Moisseytsev, A. V., Wade, D. C., Farmer, M. T., Tzanos, C. P., Stillman, J. A., Holland, J. W., Petkov, P. K., et. al., The STAR-LM lead-cooled closed fuel cycle fast reactor coupled to a supercritical carbon dioxide Brayton cycle advanced power converter, *GLOBAL 2003* (Paper presented at *Int. Winter Meeting*, New Orleans, November 16–20, 2003) ANS/ENS.

15

Fusion Reactor Hydrogen Production

Yican Wu and Hongli Chen

CONTENTS

15.1 Overview

15.1.1 Principles of Fusion

Fusion power is produced by nuclear fusion, which causes two lighter atomic nuclei to form a heavier nucleus. The latter weighs slightly less than the total of the two nuclei. According to Albert Einstein's famed formula $E = mc^2$, the small difference in mass m is transformed into energy E, where c is the speed of light approximating 3×10^8 m/s. Hence fusion has great potential for energy production, which means practical production of net usable power from a fusion system. Most design studies for fusion power plants thus involve in enabling fusion reactions to produce heat, which is used to rotate electric generator to produce electric power. Fusion power plant with the exception for the use of a thermonuclear heat source works in general similar to nuclear fission power station or fossil-fired power station [1,2].

The basic concept behind any fusion reaction is to bring two or more atoms close enough together so that the residual strong force (nuclear force) in their nuclei will pull them together into one larger atom. If two light nuclei fuse, they will generally form a single nucleus with a slightly smaller mass than the sum of their original masses, and a mass is released as energy as said above. However, if the reactant atoms are sufficiently heavy or massive, the fusion product will be heavier than the reactants, in which case the reaction requires an external source of energy to convert to that mass, that is, unable to produce energy. The dividing line between "light" and "heavy" is iron-56. Above this atomic mass, energy will generally be released by nuclear fission reactions; below it, by fusion [1].

Fusion between the atoms is opposed by their shared electrical charge, specifically the net positive charge of the nuclei. In order to overcome this electrostatic force, or "Coulomb barrier," some external source of energy must be supplied. The easiest way to do this is to heat the atoms, which has the side effect of stripping the electrons from the atoms and leaving them as bare nuclei. In most experiments the nuclei and electrons are left in a fluid state known as plasma. The temperatures required to provide the nuclei with enough energy to overcome their repulsion is a function of the total charge, so hydrogen, which has the smallest nuclear charge, therefore, reacts at the lowest temperature. Helium has an extremely low mass per nucleon and therefore, is energetically favored as a fusion product. As a consequence, most fusion reactions combine isotopes of hydrogen (protium, deuterium, or tritium) to form isotopes of helium (^{3}He or ^{4}He). This is depicted in Figure 15.1. If a nucleus of deuterium fuses with a nucleus of tritium, an α-particle is produced and a neutron released. The nuclear rearrangement results in a reduction in total mass and a consequent release of energy in the form of the kinetic energy of the reaction products. The energy released in 17.6 MeV per reaction. In macroscopic terms, just 1 kg of this fuel would release 10^8 kWh of energy and would provide the requirements of a 1 GW (electrical) power station for a day [3]. D-T reaction has the highest cross-section for fusion at a lower temperatures, it is quite possible that D-T fusion will have been fully developed.

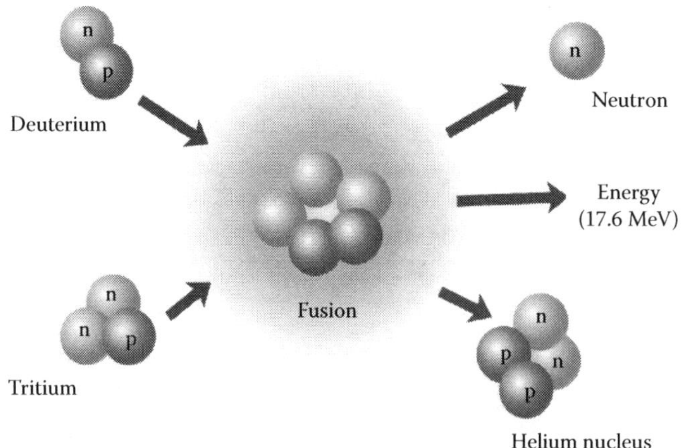

FIGURE 15.1
The fusion reaction of deuterium and tritium.

15.1.2 Benefits of Fusion as Energy Source and for Hydrogen Production

Fusion energy has important benefits, and some are in principle more attractive than nuclear fission energy. The most important benefits are summarized below.

1. Fusion fuel is abundantly available. Two isotopes of hydrogen, deuterium, and tritium are well suited for fusion. Deuterium is available from seawater (and can be extracted by electrolysis) and tritium is expected to be produced in-house in fusion power station from small quantities of lithium. Lithium has a range of commercial uses, including, importantly, in modern batteries. Despite increasing demand, lithium supplies remain abundant. The long-term fuel security of fusion exceeds that of fission power and by far that of fossil-fuel energy. A fusion station would use about 100 kg of deuterium and 3 tons of lithium to produce the same amount of energy as a coal-fired power plant using three million tons of fuel [1,2].
2. Fusion has a low environmental impact. The reaction products and most of the radioactivity induced in a fusion reactor vessel would be short-lived, making disposal easy. The radioactivity of tritium is short-lived, with a half-life of around 12 years. If the structural materials are chosen appropriately, the radiotoxicity of a fusion power station's waste materials decay very rapidly; after less than 100 years it is equal to the radiotoxicity of the waste from a coal-fired power station. Thus, fusion wastes present no accumulating or long-term burden on future generations. They would not need guaranteed isolation from the environment for very long time spans.
3. Fusion is inherently safer than fission in that it does not rely on a critical mass of fuel. This means that there are only small amounts of fuel in the reaction zone, making nuclear meltdown impossible.
4. Fusion power plants would present no opportunity for terrorists to cause widespread harm (no greater than a typical fossil-fuelled plant) owing to the intrinsic safety of the technology. Fusion in a tokamak relies on a continuous supply of fuel, without which the process soon dies away. Furthermore, the process is only sustained via careful use of the controlling magnetic fields. While the magnets contain some limited stored energy, the fusion reactor does not. This is in contrast with other low-carbon electricity sources, fission and conventional hydropower, which require the safe control of large amounts of stored energy, even when not operating.
5. Fusion involves low proliferation risks. Fusion power plants would not produce fissile materials and make no use of uranium and plutonium, the elements associated with nuclear weapons. This reduces proliferation concerns associated with these elements, although fusion is not completely free from proliferation risks (such as tritium).
6. Finally, fusion source being inherently of high temperature enables variable and economical energy products including hydrogen from water. Energy is the basic requirement for national economy, it is very important to the society, and the development of the economy changes the lives of the people. Combustion of fossil fuels provides 86% of the world's energy [4], the fossil fuel is the main source of energy, it is limited and is getting reduced every year. Fossil fuels generate air pollution and emit greenhouse gases. So a new clean energy source is required instead, which is nonfossil, and nongreenhouse gas emitting.

As with fission, fusion power plants would provide energy at a constant rate, making them suitable for base-load electricity supply. The cost structure of electricity based on fusion will be similar to that based on fission; in comparison a fusion power plant will require complex and expensive engineering, while fuel costs will be negligible. Staffing levels will be roughly constant whether the plant is generating or not. As such, the majority of costs will be capital costs and almost all will be fixed. The cost of electricity generation will be very small.

Hydrogen energy is the energy we should consider. Hydrogen has captured the imagination of the technical community recently, with visions of improved energy security, reduced global warming, improved energy efficiency, and reduced air pollution. Hydrogen is an environmentally friendly fuel that has the potential to displace fossil fuels, but hydrogen is an energy carrier, not an energy source [5]. In spite of the fact that hydrogen element is richly available on the earth, it exists as hydrocarbons or carbon hydrates, so must be procured from the hydrocarbons, carbon hydrates or water. Furthermore, hydrogen is nonpolluting, compared to the fossil fuels, and combustion of H_2 produces water, hence it is a clean energy.

Production of hydrogen is a potential opportunity for nuclear energy. Water electrolysis needs much more electrical energy, with low efficiency and high cost, but nuclear reactors can provide the high-temperature heat source required for hydrogen production. Hydrogen could potentially be produced from water using fusion energy by direct interaction of fusion products, and by electrolytic or thermochemical means [6]. With the pressures of environment protection and the energy problems, using nuclear fusion energy for hydrogen production is a hot topic in the world today.

15.1.3 History of Fusion Energy Research

The first fusion experiment was conducted at the University of Cambridge, UK, during the 1930s, but it was not until the following decade that fusion's potential as an energy source was recognized. Fusion research for energy generation has had a turbulent and complex history.

In early 1950s, four magnetic fusion concepts were pursued internationally: Tokamak, Stellarator, Pinch, and Mirror. The 1950s saw misplaced optimism with the operation in the United Kingdom of the Harwell Laboratory's Zero-Energy Thermonuclear Assembly (ZETA) [1]—a stabilized toroidal pinch machine. It had a toroidal shape but the region of plasma physics' interest was restricted to a particular toroidal segment—"the pinch," where the plasma was magnetically squeezed to increase its "magnetic density." After the 1958, second International UN Conference on Peaceful Uses of Atomic Energy, magnetic fusion researchers (in the United States, Russia, and the United Kingdom) shared results and discussed the challenges. In 1960s, numerous theoretical analyses and experiments were conducted to understand and advance fusion physics and engineering.

Throughout the early years of fusion research, plasma stability in the magnetic confinement fusion (MCF) system presented an ongoing difficulty. While some in Britain were calling for an end to fusion-energy research, a breakthrough came in 1968 from the Kurchatov Institute in the Soviet Union. A new approach known as the tokamak was discovered. Researchers there were able to achieve temperature levels and plasma confinement times—two of the main criteria to achieving fusion in tokamak device—that had never been attained before. From then on, the tokamak was to become the dominant concept in fusion research, and tokamak devices multiplied across the globe. Tokamaks require several sets of magnetic coils shown in Figure 15.2. Physically, the largest coils are

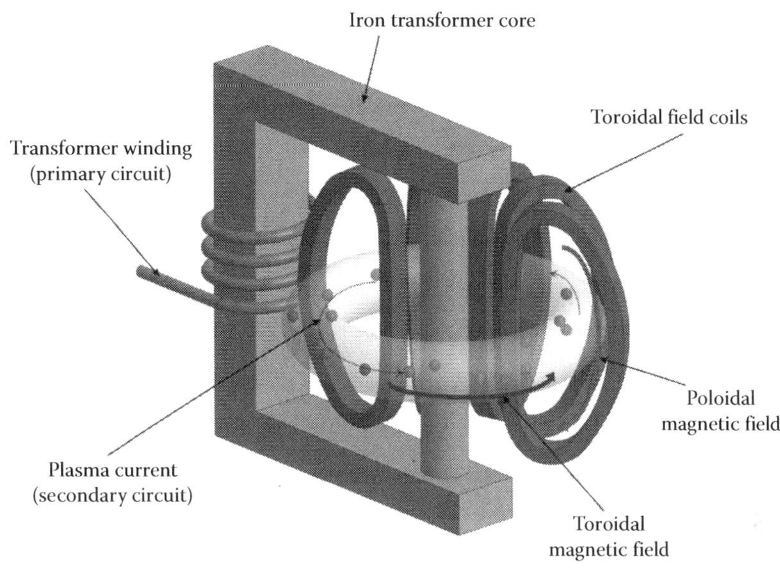

FIGURE 15.2
Schematic of a tokamak.

the toroidal field (TF) coils—superconducting (SC) magnets. The magnetic fields required for a tokamak are very large, yet space around and within the machine is at a premium. These constraints lead to the need for SC magnets in all future large-scale tokamaks.

Since 1970, more than 50 conceptual power plant (PP) design studies have been conducted worldwide, covering a wide range of new and old design approaches, which include tokamaks, and other alternate concepts such as Stellarators, spherical tori (ST), field-reversed configurations (FRC), reversed-field pinches (RFP), Spheromaks and Tandem mirrors (TM) [7]. But most of the large scale experimental fusion devices are based on tokamak. In the 1970s, the construction of large-scale fusion-research machines was approved, including a European collaboration to build the biggest machine to date—the Joint European Torus (JET) [8] shown in Figure 15.3.

Steady progress has been made since then in fusion devices around the world. The Tore Supra [9] tokamak that is part of the Cadarache nuclear research centre holds the record for the longest plasma duration time of any tokamak: 6 min 30 s. The Japanese JT-60 [10] achieved the highest value of fusion triple product—density, temperature, confinement time—of any device to date. U.S. fusion installations have reached temperatures of several hundred million degrees Celsius. Achievements like these have led fusion science to an exciting threshold: the long sought-after plasma energy breakeven point. Breakeven describes the moment when plasmas in a fusion device release at least as much energy as is required to produce them. Plasma energy breakeven has never been achieved: the current record for energy release is held by JET. In 1997, JET produced a peak of 16.1 MW of fusion power (which succeeded in generating 70% of input power), as shown in Figure 15.4, with fusion power of over 10 MW sustained for over 0.5 s [1,11].

Since 2000, the full-superconducting fusion-research tokamak, such as China's experimental advanced SC tokamak (EAST) [12] and Korea Superconducting Tokamak Advanced Research (KSTAR) [13] were constructed. The EAST experimental system has been set up at the Institute of Plasma Physics, of the Chinese Academy of Sciences (ASIPP) in January

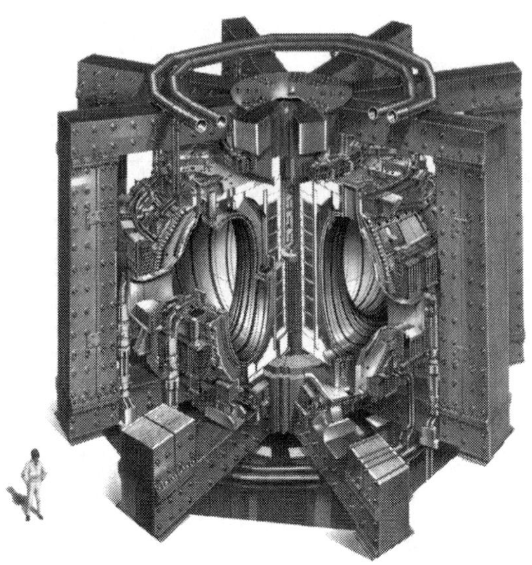

FIGURE 15.3
The JET tokamak. (From http://www.jet.efda.org/wp-content/gallery/graphics/j82–348c.jpg. With permission.)

2006 and its engineering commissioning was finished successfully in March 2006. The scientific and engineering missions of the EAST project have been defined to study physics issues of the advanced steady-state tokamak operations and to establish technology basis of full SC tokamaks. The EAST SC tokamak is the core of the project. It is a full SC tokamak, which has not only SC TF magnet system but also SC poloidal field (PF) magnet system, with a noncircle cross-section of the vacuum vessel (VV) and active cooling plasma facing

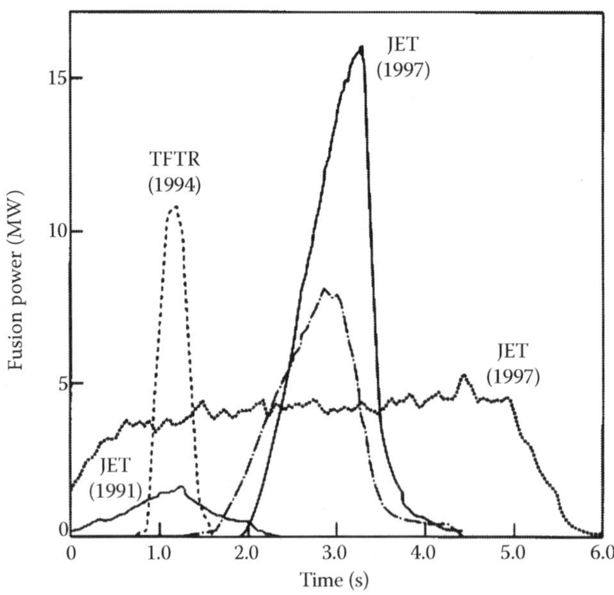

FIGURE 15.4
Record breaking deuterium–tritium fusion energy production at the JET facility.

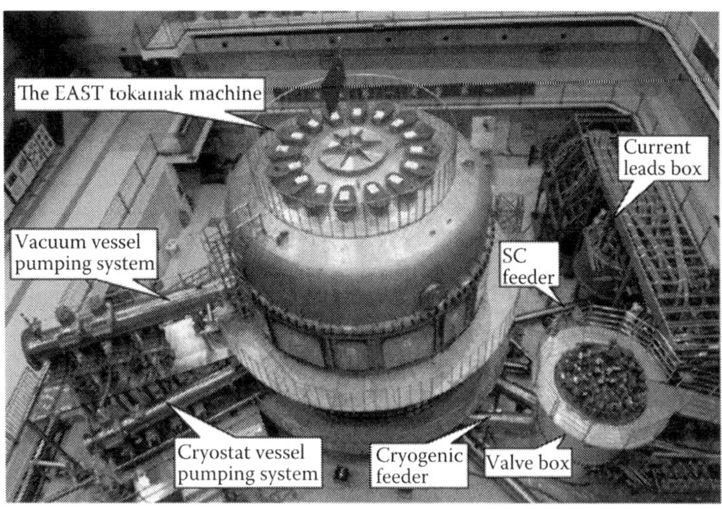

FIGURE 15.5
The EAST superconducting tokamak with vacuum pumping systems, feeders, and valve box.

components (PFCs). The EAST SC tokamak machine with the vacuum pumping systems, the feeders, the current lead boxes and the valve box is shown in Figure 15.5. The main parameters of the EAST are listed in Table 15.1. The mission of the KSTAR project is to develop a steady-state-capable advanced SC tokamak to establish the scientific and technological bases for an attractive fusion reactor as a future energy source. The KSTAR's machine construction was completed in July 2007, and got the first plasma in June 2008.

15.1.4 The International Thermonuclear Experimental Reactor

In the 1980s, Soviet general secretary Mikhael Gorbachev proposed to the then U.S. president Ronald Reagan that the superpowers might collaborate to build an international thermonuclear experimental reactor (ITER, "the way" in Latin). In the 1990s, however,

TABLE 15.1
The Main Parameters of the EAST

Toroidal field, B_0	3.5 T
Plasma current, I_P	1 MA
Major radius, R_0	1.7 m
Minor radius, a	0.4 m
Aspect ratio, R/a	4.25
Elongation, K_x	1.6–2
Triangularity, δx	0.6–0.8
Heating and driving (in the first phase)	
ICRH	3 MW
LHCD	3.5 MW
ECRH	0.5 MW
Pulse length	1–1000 s
Configuration	Double-null divertor; single-null divertor; pump limiter

policy-makers' enthusiasm for grand energy research projects wavered against a background of sustained low oil prices. A key step was taken in November 2006 when a much-revised ITER plan was finally agreed as a seven-party international collaboration among China, EU, India, Japan, Russia, South Korea, and the United States. Following intense international competition, the global partners agreed to locate the machine at Cadarache in southern France.

ITER is to be the next step on the main trajectory towards a fusion power station, combining fusion science and technology. It aims to demonstrate that it is possible to produce commercial energy from fusion, which will produce more power than it consumes: for 50 MW of input power, 500 MW of output power will be produced (Figure 15.6) [14]. ITER will cost at least €10 bn over its 30-year lifetime. Roughly half of this will be used to build the machine and half to operate and decommission it.

The goal for ITER is to produce roughly 500 MW of fusion energy in long pulses of at least 400 seconds. The main parameters of ITER are shown in the Table 15.2. The reactor is experimental; there is no intention of using ITER as a power station. The ITER construction works have started and the first plasma operation is expected in 2018.

ITER aims to gain the data necessary to the design and operation of the first electricity-producing fusion power plant. Scientists will study plasmas under conditions similar to those expected in a future power plant. ITER will be the first fusion experiment to produce net power. It will also test key technologies, including: heating, control, diagnostics, and remote maintenance, and so on.

ITER is not an end in itself: it is the bridge toward a first plant that will demonstrate the large-scale production of electrical power and tritium fuel self-sufficiency. This is the next step after ITER: the demonstration power plant, or DEMO for short. If all goes well, DEMO will lead fusion into its industrial era, beginning operations in the early 2030s, and putting fusion power into the grid as early as 2040. While ITER is being constructed and DEMO is in its conceptual phase, several fusion installations, with different characteristics and objectives, will be operating around the world to conduct complementary research and

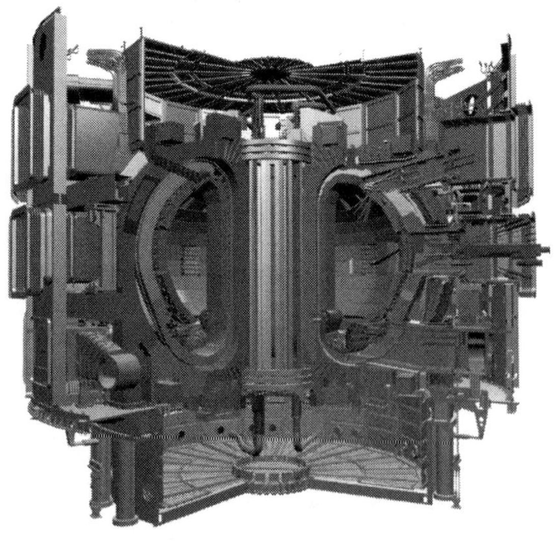

FIGURE 15.6
ITER: The World's Largest Tokamak under construction. (From http://www.iter.org/mach/Pages/Tokamak.aspx. With permission.)

TABLE 15.2
The Main Parameters of ITER

Parameters	ITER
Fusion power (MW)	500
Major radius (m)	6.2
Minor radius (m)	2
Aspect ratio	3.1
Plasma elongation	1.85
Triangularity	0.33
Toroidal magnetic field on axis (T)	5.3
Safety factor/q-95	3
Plasma current (MA)	15
Average neutron wall load (MW/m^2)	0.57
Average surface heat load (MW/m^2)	0.27
Fusion gain β_N	>10
Normalized beta (%)	2.5

development in support of ITER. By the last quarter of this century, if ITER and DEMO are successful, our world will enter the "Age of Fusion"—an age when mankind covers a significant part of its energy needs with an inexhaustible, environmentally benign, and universally available resource.

15.2 International High-Temperature Fusion Reactor Concepts

Fusion power has a potential to produce the high-temperature heat and it can be applied to the hydrogen production. So the fusion blanket design, especially the high-temperature fusion blanket design has been extensively investigated in the world. Some high-temperature fusion blanket concepts based on different materials and technologies have been developed.

15.2.1 Advanced Ceramic Breeder Blanket Concepts

The helium cooled pebble bed (HCPB) blanket (shown in Figure 15.7) is one of the two European DEMO blanket concepts, which is made from low activation martensitic ferritic steel (Eurofer) as structure material. Beryllium or beryllide in pebbles with a diameter in the range of 1 mm is used as neutron multiplier material and a lithium ceramic such as Li_4SiO_4 in pebbles with a typical diameter of 0.4–0.6 mm is used as breeding material. Helium gas is used as coolant to remove the heat deposited in the structure and breeding zones [15,16]. But the coolant outlet temperature of HCPB blanket is about 500°C. So the advanced HCPB blanket concept (A-HCPB) is designed to get high-temperature heat.

For A-HCPB design, several additional modifications will be necessary. Already for the advanced HCPB blanket based on a SiC_f/SiC-composite as structural material [17] a rupture disk was proposed to handle the accidental situation of high-pressure coolant leakage, which may cause pressurization of blanket modules. In this concept two SiC_f/SiC components were suggested, cooling plates formed by long meanders separating the

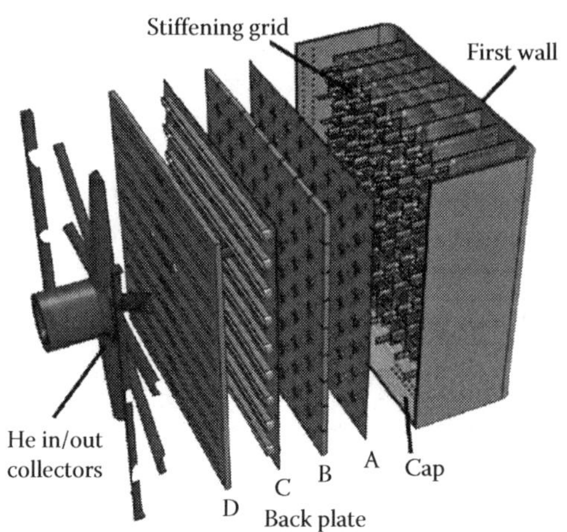

FIGURE 15.7
The exploded scheme of HCPB blanket design.

breeder ceramic pebbles from the beryllium pebbles and a first wall (FW) made from a series of parallel tubes from SiC_f/SiC. This design allowed outlet temperatures in the range of 700°C. Figure 15.8 shows the scheme of A-HCPB blanket design.

For the heat transfer from the ceramic and multiplier pebbles in hightemperature HCPB concepts there are two possibilities to be considered. In the first case, the high pressure coolant itself is directed through the pebbles and the bred tritium will be collected within the main helium coolant flow. This principle was suggested for the advanced steady-state tokamak reactor (A-SSTR-2) blanket design, shown in Figure 15.9, where it was combined with a structure made from SiC_f/SiC [18]. In general this arrangement combined with a SiC_f/SiC structure could allow helium outlet temperatures as high as 900°C, when combined with a FW made from advanced ferritic steel the inlet temperature has to be reduced and the outlet temperature achievable will be reduced as well, but 800°C could be still possible.

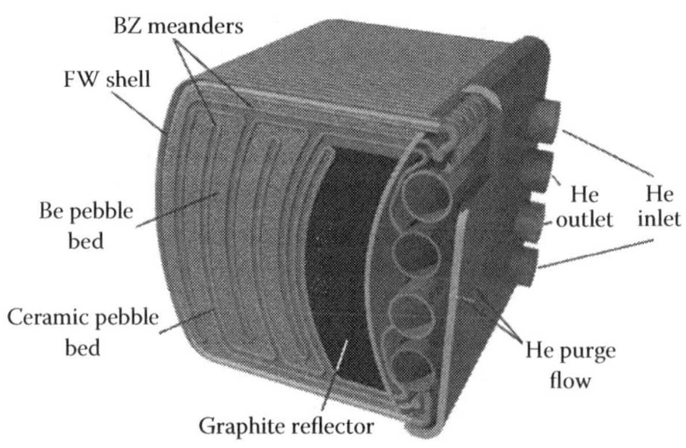

FIGURE 15.8
The schematic A-HCPB blanket design.

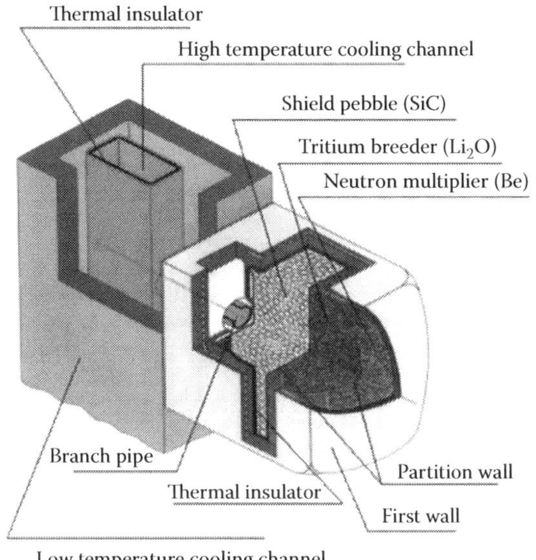

FIGURE 15.9
The schematic A-SSTR-2 blanket design.

In the second case the breeder material is placed in tubes. This principle was investigated, for example, in Europe known as breeder inside tubes (BIT) concept [19]. It was proposed to place the breeder pebbles in small packed tube arrays (Figure 15.10) which on their outside are cooled by the main helium coolant flow. Inside the small tubes containing the ceramic pebbles tritium is collected by a separate helium purge flow as in the case of the EU HCPB DEMO reference design. The small circular breeder-pebble containing tubes are made from an advanced high-temperature steel allowing operational temperatures up to

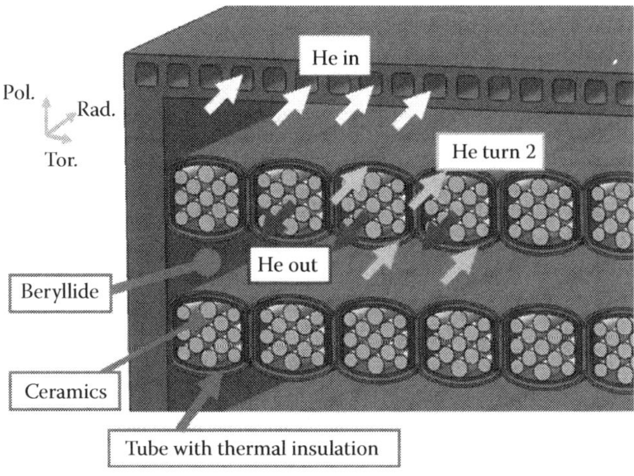

FIGURE 15.10
HCPB blanket with concentric flow channels and small packed breeder tubes.

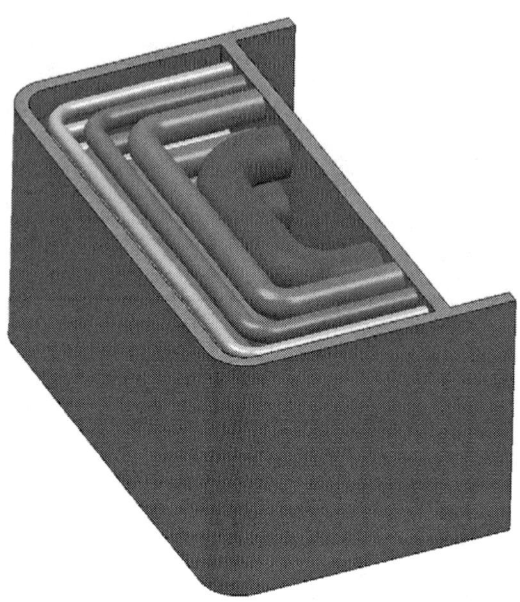

FIGURE 15.11
HCPB concentric BIT version blanket box.

800°C, such as ferritic ODS (oxide disperse strengthened) steels. Therefore, for this concept the helium outlet temperature could be very close to 800°C. While the achievable helium outlet temperature could be very attractive, the large number of pebble-containing tubes is the disadvantage of this configuration.

Another helium coolant pebble bed blanket option for outlet temperature around 700°C is illustrated in Figures 15.11 and 15.12. In this concept concentric pipes are arranged in

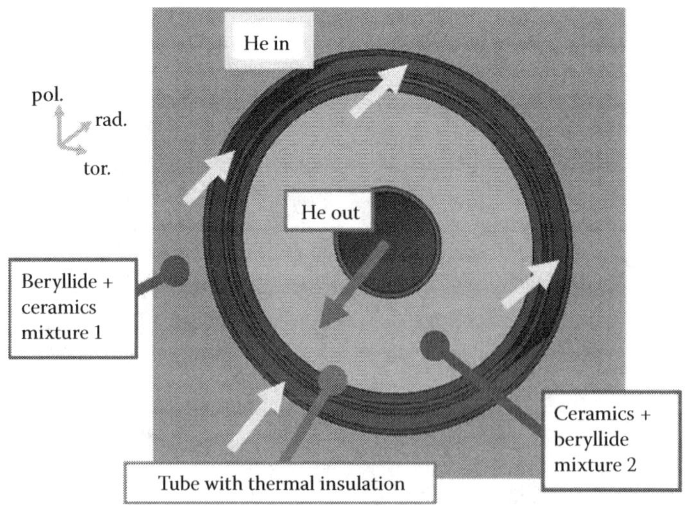

FIGURE 15.12
Concentric BIT arrangement.

U-shape inside the blanket boxes and fed from the manifolds at the back of the boxes. The diameter of the pipes is varied to accommodate for the reduction of the volumetric-heat generation in radial direction. The concentric pipes consist of three thin steel tubes and a thermal insulation inside the second tube. Helium enters the pipes in the annular gap between the outmost tubes. The volume around the concentric pipes is filled with a mixed bed containing mostly beryllide and a small fraction of ceramic pebbles. The volume between the insulation layer and the innermost tube is filled with a mixture of ceramic and beryllium pebbles operated at a temperature up to 900°C, while the coolant reaches an outlet temperature of 700°C. As the thermal conductivity in mixed beds is better than that in pure ceramic beds, the number of single tubes can be kept relatively low. The layout of the system is quite flexible and could be optimized for specific reactor conditions.

15.2.2 Dual Coolant Blanket Concepts

A large variety of lithium-lead (LiPb) breeding concepts have been proposed up to now. A completely helium-cooled lithium–lead blanket concept (HCLL, Figure 15.13), has been assessed in the PPCS/DEMO studies as one of the blanket options for DEMO and is proposed by EU for the ITER test blanket module (TBM) program [20]. The HCLL concept is the short-term variant of the LiPb concepts and is considered suitable for DEMO. In contrast, dual coolant lithium-lead (DCLL) concepts utilize helium only to cool a ferritic steel structure (including the FW) and slowly flowing LiPb acts as self-cooled breeder in the inner channels (Figure 15.14) [21]. The type of dual coolant concept considered allows for operating the breeder at a higher temperature than the structural blanket walls by means of flow channel inserts (FCI from SiC, Figure 15.15) to maximize the power cycle efficiency and to expand the nonelectric applications. Use of a liquid breeder also provides the possibility of active tritium breeding control during operation by adjusting the ^{6}Li enrichment. Use of He coolant as a separate coolant for the FW/structure also facilitates the preheating of the

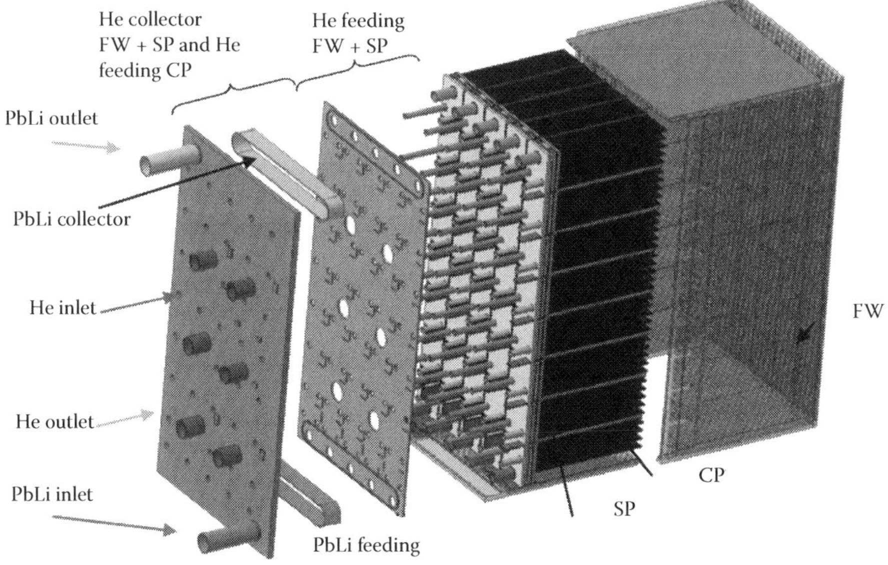

FIGURE 15.13
Exploded view of HCLL blanket design.

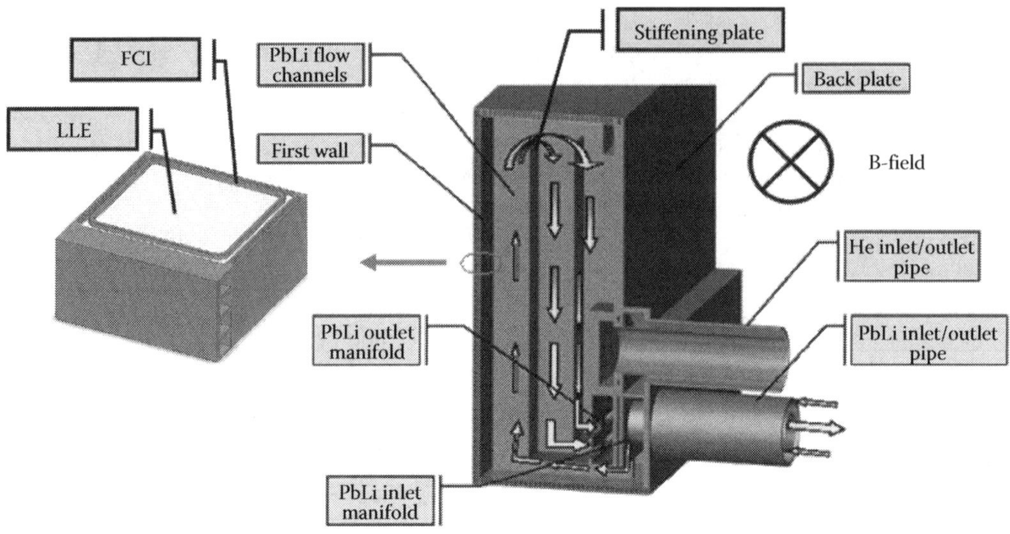

FIGURE 15.14
The view of DCLL blanket design.

blankets before the liquid breeder is filled in, serves as guard heating in case the liquid breeder cannot be circulated, and provides independent and redundant decay-heat removal in case the liquid metal (LM) loop is not operational. In addition, cooling the FW region of the blanket (where the heat load is highest) with helium (instead of a LM) avoids the need for an electrically insulating coating in this high velocity region to prevent the large magnetohydrodynamics (MHD) pressure drop associated with LM flow.

Such a dual coolant concept with FCI was originally developed as part of the ARIES-ST study [22,23] and then at FZK in Germany [24]. It has been further developed as part of the ARIES-CS [25] and FDS series blankets study [26].

The DCLL blanket module concept is illustrated in Figures 15.16 and 15.17, with both helium and LiPb fed to the back of the blanket module through concentric pipes. The

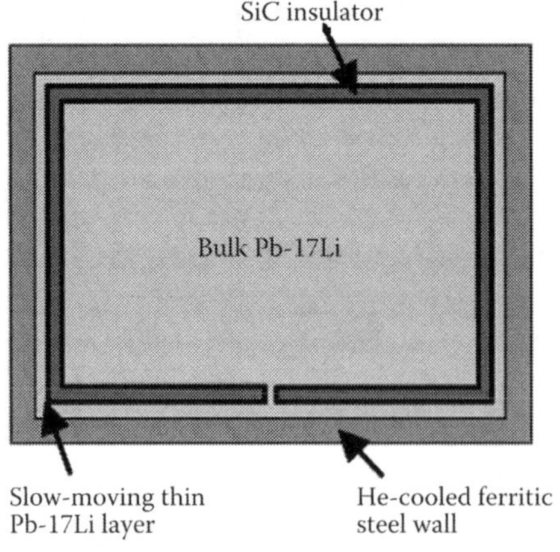

FIGURE 15.15
FCIs in DCLL for enhanced breeder temperatures and low MHD pressure drops.

FIGURE 15.16
ARIES-CS DCLL blanket module.

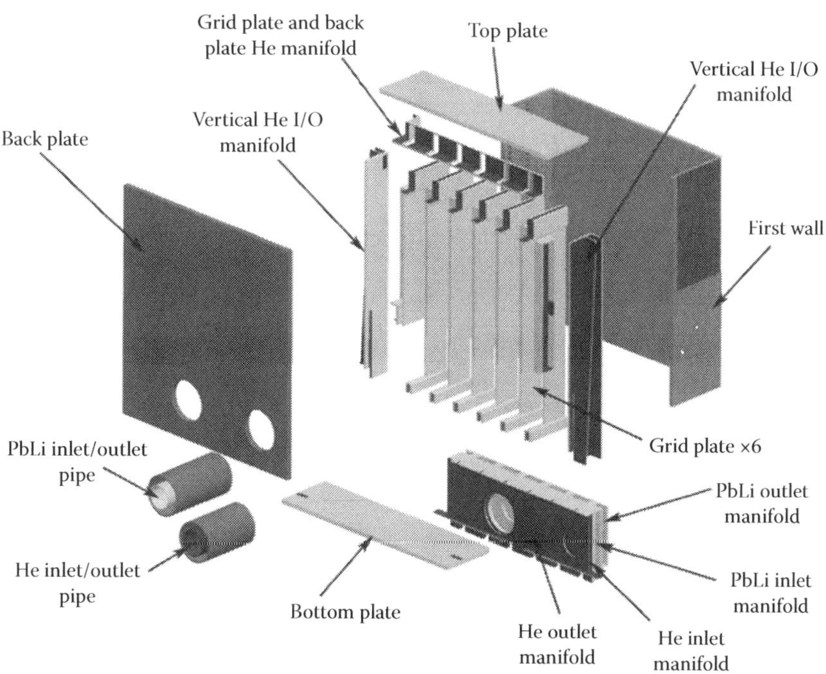

FIGURE 15.17
Exploded view of ARIES-CS DCLL blanket module.

helium coolant is routed to first cool the FW in a single pass with an alternating toroidal flow configuration to create a more uniform temperature (and reduce thermal stresses); it is then routed in a combination of series and parallel flow to cool the other structural walls. The LiPb flows slowly (10 cm/s or less) in the large inner channels in a two-pass poloidal configuration. The LiPb channels are lined with a SiC FCI which has no structural function as the thin LiPb layer and the bulk LiPb are pressure-balanced through a thin slot or holes in the SiC, as illustrated in Figure 15.15. This FCI plays a key thermal insulation function to allow high-temperature (700°C) LiPb in the channel while the LiPb/steel interface temperature is maintained below its compatibility limit. It also provides the electrical insulation needed to minimize the MHD effects on the LiPb pressure drop and flow distribution. To help increase the coolant temperature at the FW (and thermal efficiency) while maintaining the reduced activation ferritic steel (RAFS) temperature within its 550°C limit, the FW is made up of a layer of ODS ferritic steel diffusion bonded to the thin RAFS module wall.

Several variations of the lithium-lead-based blankets with dual-coolants have been designed for a series of fusion power plants (named FDS series) in China, that is the fusion-driven subcritical reactor (FDS-I), the fusion power reactor (FDS-II), the fusion-based hydrogen production reactor (FDS-III) and the spherical tokamak-based compact reactor (FDS-ST) [26–32].

FDS-I, a fusion-driven subcritical system, in which a fusion core is used as a neutron source to drive the subcritical blanket, has very attractive advantages, which ease the requirements for the plasma and enable adequate excess neutrons available for breeding fissile fuels, transmuting long-lived fission products (FPs) and actinides. In addition, there are no risk of critical accident and less danger to nuclear proliferation, as compared to the critical fission systems. If an optimized blanket design was adopted, the requirement for neutron source intensity and subsequently fusion plasma technologies could be lowered. FDS-I is designated to transmute the long half-life nuclear wastes from fission power plants and to produce fissile nuclear fuels for feeding fission power plants as an intermediate step and early application.

An overview of the FDS-I reference model is shown in Figure 15.18. And a set of plasma-related parameters of FDS-I are given in Table 15.3.

The FDS-I blanket design focuses on the technology feasibility and concept attractiveness to meet the requirement for fuel sustainability, safety margin, and operation economy. A design and its analysis on the dual-cooled waste transmutation (DWT) blanket with Carbide heavy nuclide Particle fuel in circulating Liquid LiPb coolant (named DWT-CPL) has been studied for years. Other concepts such as the DWT blanket with Oxide heavy nuclide Pepper pebble bed fuel in circulating helium-Gas (named DWT-OPG) and with Nitride heavy nuclide Particle fuel in circulating helium-Gas (named DWT-NPG) are also being investigated. For the DWT-CPL blanket concept, helium gas was adopted to cool the structural walls and long half-life FP transmutation zones (FP-zones), LM LiPb eutectic with tiny particles of long half-life fuel to self-cool actinide (AC) zones including minor actinides (MA) transmutation zones (MA-zones) and uranium-loaded fissile breeding zones (U-zones). U-zones may be replaced with AC-zones if fertile free concept is considered. The reference module, basic material compositions, and radial sizes of DWT-CPL blanket at the tokamak mid-plane of FDS-I are given as in Figure 15.19 and Table 15.4. The details on these blanket designs can be found in Refs. [27–29].

FDS-II, a fusion power reactor, is designated to exploit and evaluate potential attractiveness of pure fusion-energy application, that is, obtaining a high-grade heat for generation of electricity on the basis of conservatively advanced-plasma parameters, which can be

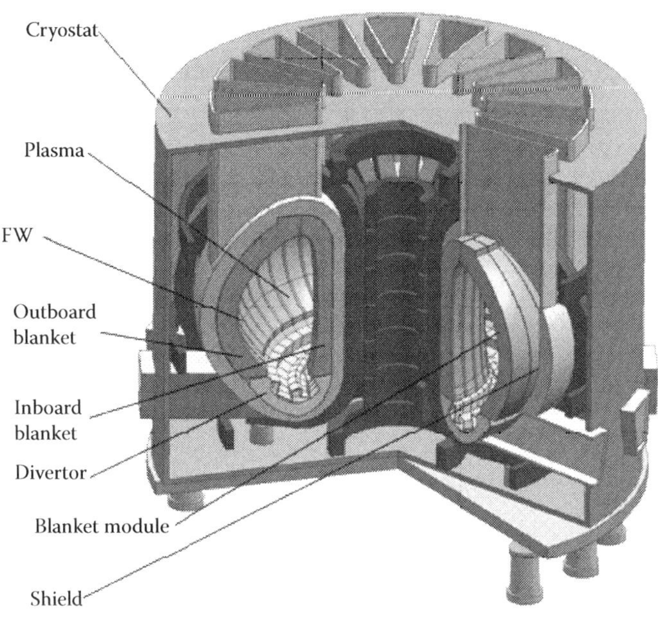

FIGURE 15.18
Overview of FDS-I reference model.

extrapolated in a limited way from the successful operation of ITER. The configuration of the FDS-II reactor is shown in Figure 15.20.

Plasma physics and engineering parameters of FDS-II are selected on the basis of the progress considered in recent experiments and associated theoretical studies of magnetic confinement fusion plasma (as in Table 15.5). Both the feasibility and attractiveness of technology are of concern to the FDS-II blanket design, which must meet the requirement

TABLE 15.3
Main Core Parameters of FDS-I

	Design	
Parameters	**FDS-I**	**ITER**
Fusion power (MW)	150	500
Major radius (m)	4	6.2
Minor radius (m)	1	2
Aspect ratio	4	3.1
Plasma elongation	1.78	1.70
Triangularity	0.4	0.33
Plasma current (MA)	6.3	15
Toroidal field on axis (T)	6.1	5.3
Safety factor /$q_{\_95}$	3.5	3
Auxiliary power /P_{add}(MW)	50	73
Energy multiplication /Q	3	≥ 10
Average neutron wall load (MW m^{-2})	0.5	0.57
Average surface heat load (MW m^{-2})	0.1	0.2

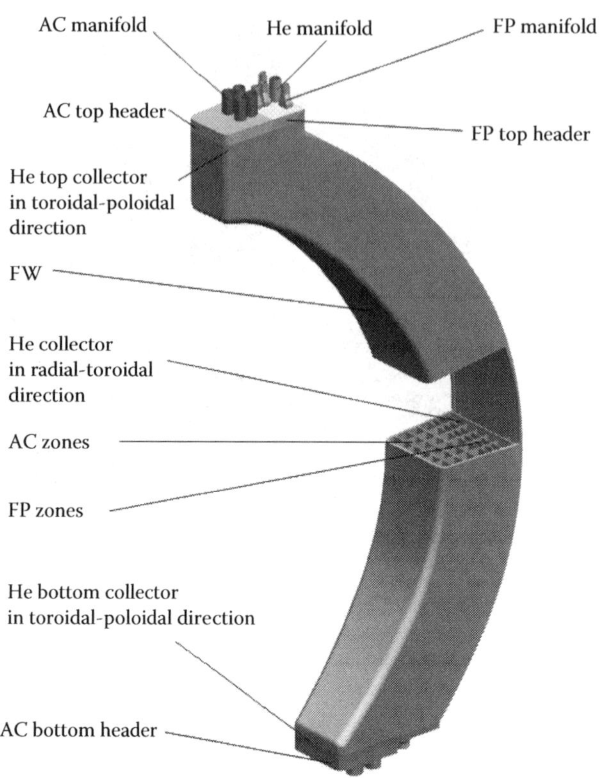

FIGURE 15.19
Material compositions and radial sizes of DWT-CPL blanket.

TABLE 15.4
Reference Module Design Parameters of DWT-CPL Blanket

Zones	Material Component	Thickness(cm)
Inboard Blanket		
FW	RAFM steel (45%) + He (55%)	3
Tritium breeding zone	LiPb (100%)	11 × 2
Structural walls	RAFM steel (60%) + He (40%)	1 × 2
Neutron reflector	Graphite (100%)	13
Helium mainfold	RAFM steel (40%) + He (60%)	10
Shield layer	RAFM steel (75%) + H_2O (25%)	30
Outboard Blanket		
FW	RAFM steel (45%) + He (55%)	3
AC1/AC2/AC3/AC4	MAC (0.6%) + PuC (4.3%) + LiPb (95.1%)	11/11/11
Structural walls	RAFM steel (60%) + He (40%)	1 × 6
FP(CsCl/NaI/Tc)	FP (17.6%/1.7%/1.5%) + graphite (60%) + He (22.4%/38.3%/38.5%)	7/7/7
Helium mainfold	RAFM steel (45%) + He (55%)	11
Shield layer	RAFM steel (75%) + H_2O (25%)	60

FIGURE 15.20
Configuration of the fusion power reactor FDS-II.

TABLE 15.5
The Core Physics Parameters of FDS-II

Parameters	
Fusion power (MW)	2500
Major radius (m)	6
Minor radius (m)	2
Aspect ratio	3
Elongation	1.9
Triangularity	0.6
Plasma current (MA)	15
Toroidal field (B_t)	5.93
Safe factor (q_{-95})	5.0
Normalized (β_N)	5
Average density (10^{20} m^{-3})	2.2
Average temperature (keV)	11.1
Assistant heat power (MW)	80
Bootstrap current fraction	0.69
Power gain (Q)	31
FW neutron wall load (MW m^{-2})	2.63
FW surface heat flux (MW m^{-2})	0.54

for tritium self-sufficiency, safety margin, operation economy, and environment protection, and so on. Two optional concepts of liquid LiPb blankets including the reduced activation ferritic/martensitic (RAFM) steel-structured He-cooled LiPb tritium breeder (SLL) blanket and the RAFM steel-structured He-gas/liquid LiPb dual-cooled (DLL) blanket (Figure 15.21) are adopted for FDS-II. The structural material is made of China low activation martensitic (CLAM) steel [33]. For the DLL design, He gas is used to cool the FW and blanket structure and liquid LiPb is to be the self-cooled tritium breeder with a high outlet temperature up to 700°C in order to achieve high thermal efficiency. The FCIs, for example, SiC_f/SiC composite or other refractory materials, are designed and used inside the LiPb coolant channel and manifold, which act both as thermal and electrical insulators to keep the temperature of RAFM structure below the maximum allowable temperature. Coating (e.g., Al_2O_3) on RAFM structure is also considered in the design to reduce tritium permeation and prevent corrosion of LiPb.

The SLL concept is another option of FDS-II blanket considering the SLL blanket could be developed relatively easily with lower LiPb outlet temperature and slower LiPb flow velocity and that it allows the utilization of relatively mature material technology. Coating is probably needed to protect the structure and to reduce tritium permeation and also MHD effects. The details on this design can be found in Ref. [30]. The thermal-hydraulic parameters of DLL/SLL blankets for FDS-II are listed in Table 15.6.

FDS-III, a high-temperature fusion reactor for hydrogen production, aims to obtain the high-temperature heat in the blanket of fusion reactor for efficient production of hydrogen using thermo-chemical sulfur–iodine cycles technology based on the current status or promising extrapolation of material technology. This innovative blanket design with "multilayer flow channel inserts (MFCI)" is considered to obtain high-temperature heat while using the relatively mature and most promising RAFM steel (such as CLAM) as structural material, refractory material with low thermal conductivity, such as SiC_f/SiC

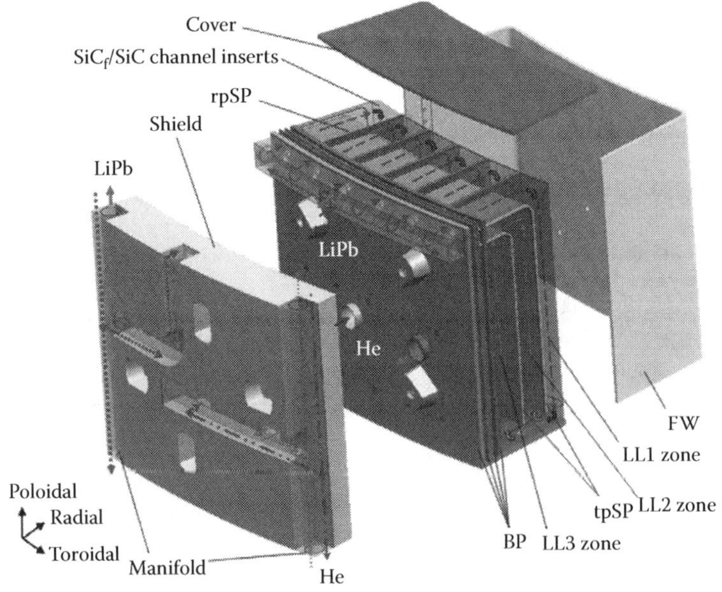

FIGURE 15.21
The exploded 3D view of DLL blanket.

TABLE 15.6
The Reference Design Parameters of DLL/SLL Blankets

Blanket		DLL	SLL
Structural material		RAFM (CLAM) ODS	RAFM (CLAM) ODS
Functional material(s)		FCI : e.g., SiC_f/SiC Coating: e.g., Al_2O_3	Coating: e.g., Al_2O_3
Plasma core	Fusion power /MW	2500	2500
	Major radius/m	6	6
	Minor radius/m	2	2
Heat source	Average/Maximum heat flux / $MW \cdot m^{-2}$	0.54/0.7	0.54/0.7
	Average/Maximum neutron wall load/$MW \cdot m^{-2}$	2.63/3.54	2.63/3.54
	Nuclear heat deposition (FW/breeder zone)[a] /MW	5.7/15.3	~ 5.7/15
He system	In/Out temperature /°C	300/450	300/450
	FW/SP velocity /m s^{-1}	115/40	
	Pressure/MPa	8	8
LiPb system	In/Out temperature /°C	480/700	/450
	Average velocity of LiPb in breeder zone/mm s^{-1}	99/78/12	~ 1

[a] Only for one equatorial blanket module.

composite material, or other components as the functional material inserted in the flow channel. Low temperature LiPb flows into the channel, and then meanders through the MFCI. The temperature of the coolant LiPb is improved gradually, and at last it is exported from the blanket to high-temperature heat of above 900°C. The details on the blanket design can be found in Refs. [31,34,35]. The configuration of FDS-III is shown in Figure 15.22. And a set of main plasma parameters of FDS-III are listed in Table 15.7.

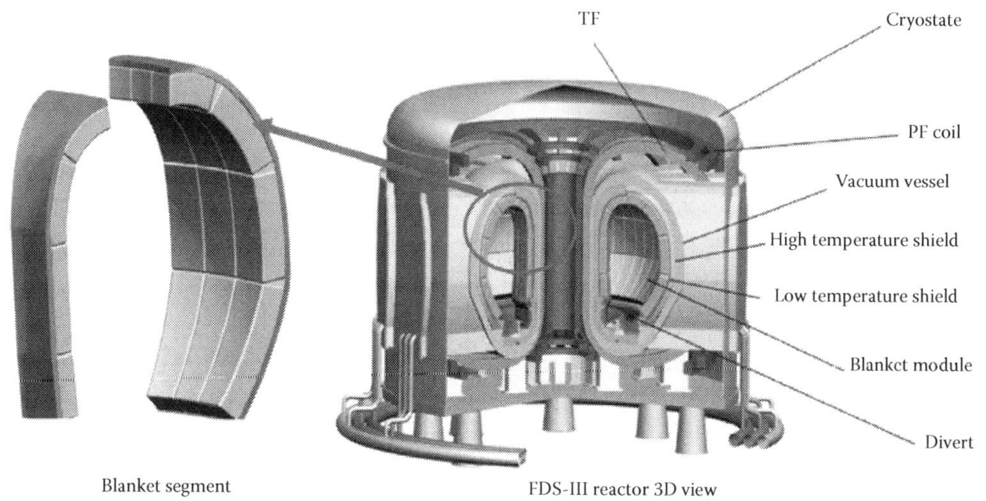

FIGURE 15.22
Configuration of FDS-III.

TABLE 15.7

Main Core Parameters of FDS-III Reactor

Major radius	R (m)	5.10
Minor radius	A (m)	1.70
Aspect radio	A	3.00
Plasma current	Ip (MA)	16.00
Toroidal field	B_0 (T)	8.00
Elongation	κ	1.90
Triangularity	δ	0.53
Safe factor	Q	8.03
Edge safe factor	Q_{9s}	3.30
Toroidal β	β	5.65
Poloidal β	βp	1.88
Normalized β	βN	4.80
Average density	$\langle n_e \rangle$ (10^{20}m^{-3})	1.00
Average temperature	<Te> (KeV)	10.00
Bootstrap current fraction	f_b	0.65
Fusion power	P_{fu} (MW)	2600
Drive power	P_d (MW)	20

An advanced dual coolant blanket concept named HTL (high-temperature liquid) blanket for FDS-III has been developed to obtain a high LiPb outlet temperature of about 1000°C with the RAFM steel as main structural material and using SiC_f/SiC MFCI in the LiPb flow channels. Figure 15.23 shows the HTL blanket structure and LiPb flow scheme. The breeder zone is configured of three concentric LiPb channels and MFCI. The directions

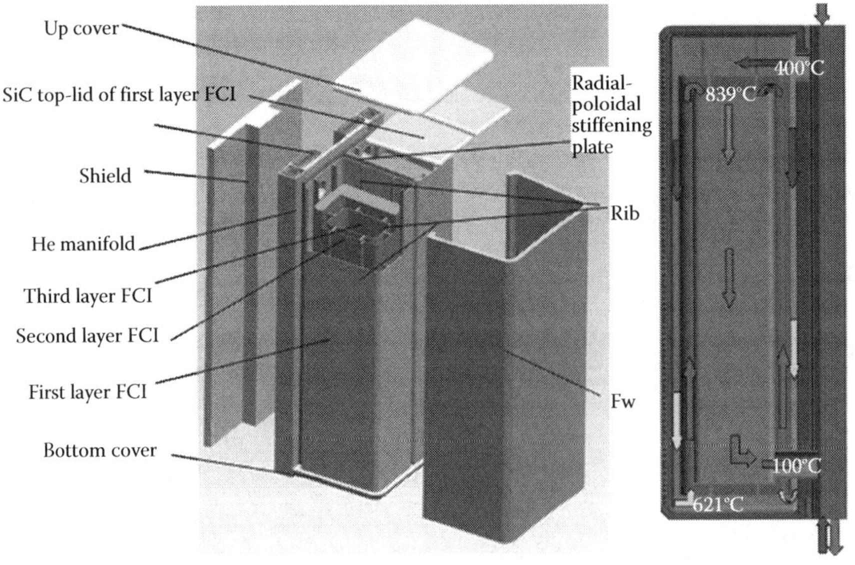

FIGURE 15.23
The HTL blanket module structure (left) and LiPb flow scheme inside module (right).

provide toroidal and radial restrictions between layers of FCI formed LiPb channels. Helium cools the structural components to keep them below the RAFM steel temperature limit of 550°C and LiPb is heated to high-outlet temperature when meandering through the MFCI, as no additional cooling is foreseen for the breeding zones. The 400°C LiPb is fed into the module from the pipe outside the manifold, and then flows in series on the outside, middle, and inside of the FCI channel, and finally, at about 1000°C LiPb flows out of the module into a pipe inside the manifold. The design parameters of blanket module are listed in Table 15.8.

15.2.3 Self-Cooled LiPb Blanket Concepts

The ARIES-AT power plant was evolved to assess and highlight the benefits of advanced technologies and of new physics understanding and modeling capabilities on the performance of advanced tokamak power plants [36,37]. Figure 15.24 shows the ARIES-AT power core and Table 15.9 summarizes the typical geometry and power parameters of the reactor, emerging from the parametric system studies [38].

The self-cooled LiPb blanket design utilizes LiPb as breeder and coolant, and SiC_f/SiC composite as structural material. SiC_f/SiC is attractive based on its high-temperature compatibility and its low decay heat. However, there are some key issues influencing its attractiveness, including thermal conductivity, swelling under irradiation, compatibility with the LM, fabrication, and joining procedures [39]. A typical self-cooled LiPb blanket concept is developed by ARIES-AT which uses SiC_f/SiC as structure material. Figure 15.25 shows ARIES–AT FW and blanket design. The SiC_f/SiC parameters and properties used in the ARIES-AT analysis are summarized in Table 15.10 [38].

Another self-cooled blankets using SiC_f/SiC as structural material has been proposed for TAURO [40,41]. The TAURO design (Figure 15.26) is based on the principle of coaxial flow which allows having a maximum LiPb outlet temperature of 1100°C without exceeding

TABLE 15.8

Design Parameters of HTL Blanket

Neutron wall load (MW m^{-2})		~ 4
FW surface heat load (MW m^{-2})		~ 1.04
FW channel	mm^2	18 × 20
	T_{in}/T_{out}(°)	350/366.5
	V_{He}(m/s)	100
Cover channel	mm^2	12 × 18
	T_{in}/T_{out}(°)	350/368
	V_{He}(m/s)	88
Radial-poloidal	mm^2	7 × 14
Stiffening plate	T_{in}/T_{out}(°)	350/363.5
Channel	V_{He}(m/s)	81
Helium pressure (MPa)	8	
	$T_{LiPb\ in}/T_{LiPb\ out}$(°)	400/1000
	LL1 $V_{LiPb\ 1}$(m/s)	0.041
	LL2 $V_{LiPb\ 2}$(m/s)	0.028
	LL3 $V_{LiPb\ 3}$(m/s)	0.030

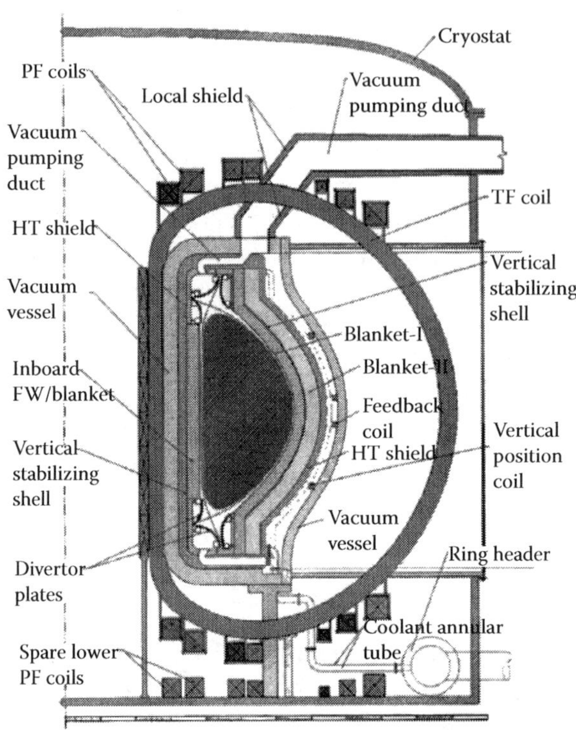

FIGURE 15.24
ARIES-AT power core (radial dimension in m).

the limit of 1000°C for SiC_f/SiC [39]. The 6-mm-thick FW is protected by a 2-mm-thick layer of tungsten. Each outboard segment is formed by five modules attached to a thick back plate that gives sufficient strength for segment transport during replacement and allows its attachment to the back components. The LiPb enters from the bottom in the thin external layer, turns down at the top and flows down at low velocity in the central region.

TABLE 15.9
Typical ARIES-AT machine and Power Parameters

Machine geometry	
Major radius (m)	5.2
Minor radius (m)	1.3
On-axis magnetic field (T)	5.9
Power parameters	
Fusion power (MW)	1719
Neutron power (MW)	1375
Alpha power (MW)	344
Blanket multiplication factor	1.1
Maximum thermal power (MW)	1897
Average neutron wall load (MW/m^2)	3.2
Outboard maximum wall load (MW/m^2)	4.8
Inboard maximum wall load (MW/m^2)	3.1

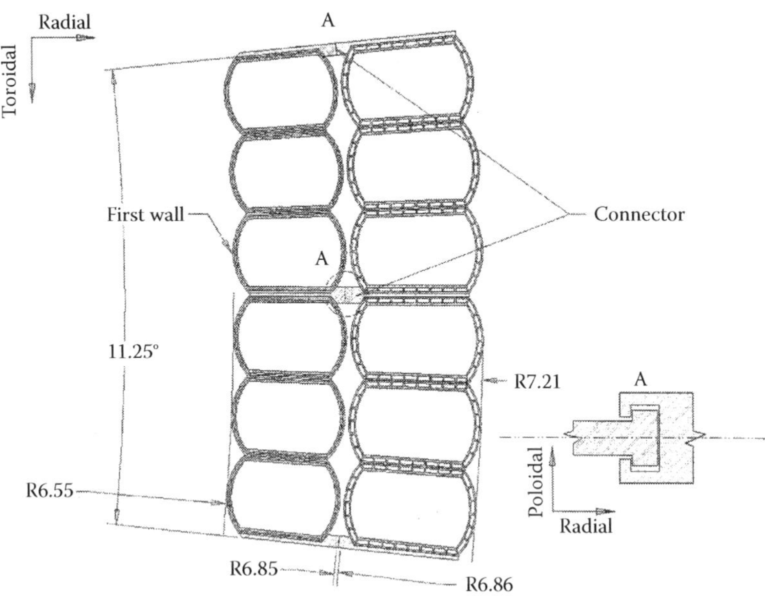

FIGURE 15.25
The ARIES-AT FW and blanket.

The self-cooled lithium–lead blanket mainly addressed here also has been studied as model D in the framework of the European Power Plant Conceptual Study (PPCS) [42]. The blanket is formed of only two materials: the SiC_f/SiC structure and the LiPb which acts as breeder/multiplier and coolant. Associated with the use of LiPb (a material that is easily recyclable) as breeder, coolant, neutron multiplier, and tritium carrier, this system achieves high plant efficiency and has the potential for hydrogen production.

15.2.4 Self-Cooled Flibe Blanket Concepts

The molten salt Flibe, especially the composition $(LiF)_2(BeF_2)$, offers good tritium breeding characteristics, low MHD pressure drop due to its low electrical conductivity, and in

TABLE 15.10
SiC_f/SiC Properties and Parameters Assumed in This Study

Density (kg/m^3)	3200
Density factor	0.95
Young's modulus (GPa)	200–300
Poisson's ratio	0.16–0.18
Thermal expansion coefficient (ppm/°C)	4
Thermal conductivity through thickness (W/m K)	20
Maximum allowable combined stress (MPa)	190
Maximum allowable operating temperature (°C)	1000
Maximum allowable SiC/LiPb interface temperature (°C)	1000
Maximum allowable SiC burn-up (%)	3

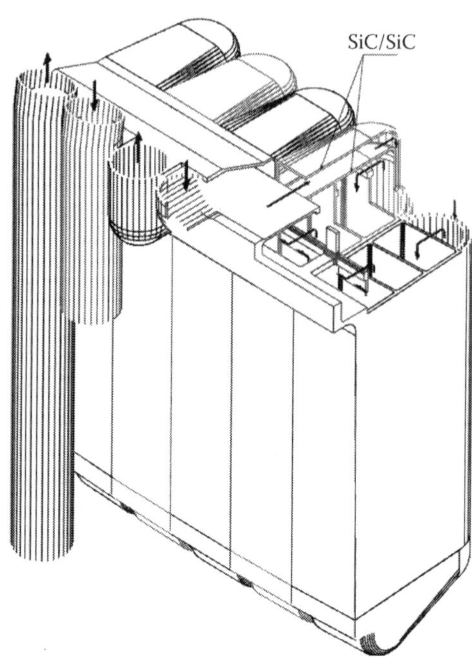

FIGURE 15.26
Self-cooled Pb-17Li TAURO blanket.

general better safety features than LM breeders due to the low chemical reactivity with air and water, and the exceptional low tritium inventory. For these reasons, it became the reference blanket in the force free helical reactor (FFHR) [43].

Major drawbacks for the use of Flibe in the breeding blankets of power plants are the extremely low thermal conductivity (about 1 W/m K) at 500°C, compared to 15 W/m K for LiPb), an extremely high viscosity (100 times higher than of water in a pressurized water reactor), and a melting point of 459°C, requiring advanced ODS steels with an operating temperature of >650°C as structural material. To mitigate the corrosion of the structural material, FT has to be reduced to T_2 through contact of the Flibe with Be which at the same time serves as neutron multiplier.

A novel Flibe blanket concept has been proposed in the US APEX study [44], named recirculating flow concept [45] as shown in Figure 15.27. In this concept, only about 25% of the entire FW flow is sent to the power conversion system, and the larger part (75%) of the flow is recirculated directly through the blanket in order to combine efficient cooling of the FW with a high exit temperature for an efficient Brayton cycle power conversion system. By this measure, it was possible to design a self-cooled Flibe blanket with a maximum neutron wall load of 5.4 MW/m^2 and a maximum FW surface heat flux of 1 MW/m^2, leading to an efficiency of 45% in the power-conversion system.

Concerning tritium permeation from the point of view of safety, double-walled tube concepts have been evaluated and the He gas or pure Flibe with low flow rates are shown to be sufficient and feasible as a permeation barrier [15,46], where the tritium leakage at the heat exchanger remains as a serious issue. Regarding the operation temperature window limited by Flibe and structural materials, a Flibe/V alloy blanket concept has been proposed [47], where, in spite of Be redox, a W coating using WF_6 mixed in Flibe is a candidate and disengagement of HF from Flibe becomes a new key issue.

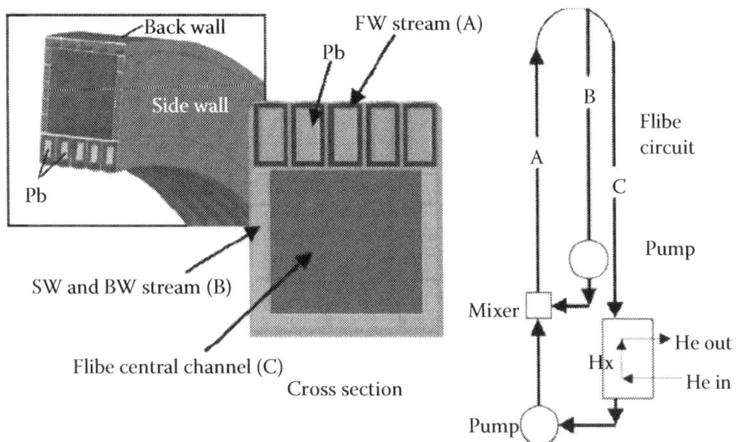

FIGURE 15.27
Schematics of the Flibe FW/blanket design.

15.2.5 Evaporation of Lithium and Vapor Extraction Blanket Concept

In the framework of the US APEX study [44] a very ambiguous goal of a maximum neutron wall load of 10 MW/m^2 together with a FW surface heat flux of 2 MW/m^2 had been set as the criteria for the selection of blanket concepts named evaporation of lithium and vapor extraction (EVOLVE). It desired to achieve both high power density and high power conversion efficiency leading to several required features of a FW and blanket concept. Achieving high power density means that the coolant heat removal capability must be high and the FW material should have attractive thermophysical properties (high thermal conductivity, low thermal expansion, etc.). Achieving high-power conversion efficiency means that the FW and blanket should operate at very high temperatures. Materials operating at very high temperatures generally have limited strength and, therefore, such a concept should operate at low primary stresses. This means that the coolant pressure should be as low as possible, and the temperatures throughout the blanket should be as uniform as possible to reduce thermal stresses. In EVOLVE, a significant fraction of the fusion power is released in the liquid by nuclear-volumetric heating and is removed from the blanket by evaporation of lithium. The large latent heat of vaporization allows removal of high-heat fluxes while the velocities in the liquid phase are very small. In an early design (Figure 15.28) the volumetrically deposited heat is supposed to be removed by vapor bubbles rising through layers of liquid lithium [48].

This was the starting point for a novel concept, based on the use of the exceptional high heat of vaporization of lithium (about 10 times higher than water) for an effective heat extraction from the FW as well as the breeding zone. A reasonable range of boiling temperature of this alkali metal is 1200–1400°C, corresponding with a saturation pressure of 0.035–0.2 MPa. For this temperature range, tungsten alloys are suitable structural materials due to the high strength, the high thermal conductivity, and the excellent compatibility with boiling lithium. Calculations indicate that such an evaporating system with Li at 1200°C as breeder/coolant can remove a neutron wall load of >10 MW/m^2 with an accompanying FW surface heat flux of >2 MW/m^2.

Thermal-hydraulics and neutronics calculations have shown, that the system provides adequate tritium breeding and shielding, very high efficiency of the Brayton cycle power

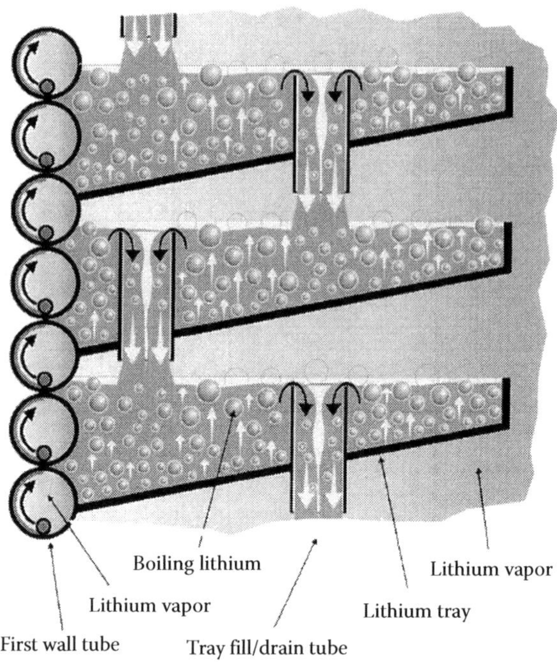

FIGURE 15.28
Sketch of early EVOLVE design.

conversion system (>55%), and low stresses in the blanket structure due to the low coolant pressure.

The main issues are the fabric ability of the tungsten alloy structure, and the performance under high fluence neutron irradiation. The minimum operational temperature of >1200°C is well above the ductile–brittle transition temperature, and the temperature variations inside the entire structure are <100 K due to the high thermal conductivity and the excellent heat transfer under boiling conditions. Because of the demanding operating conditions and the level of material R&D required, this concept is considered more as a longer term, later generation concept.

15.3 Hydrogen Production Systems

Hydrogen can be produced from nuclear energy by several means, such as water electrolysis, steam reforming, and thermochemical water-splitting process, and so on [5,49]. Water electrolysis can be served by electrical energy, the steam reforming reaction, and the thermochemical water-splitting process can be served by high-temperature thermal energy. The work about steam reforming with biomass waste, radiation chemical process and plasma chemical processes for hydrogen production also have been reported recently [50,51]. Three potential hydrogen production processes under development for the industrial production of hydrogen using nuclear energy are introduced below.

15.3.1 Steam Methane Reforming

Steam methane reforming (SMR) is a thermochemical process that is commonly used in industry for the production of hydrogen, now almost 48% of the world's hydrogen is produced from SMR. The process consists of reacting methane and steam at high temperature. The conventional process occurs in a chemical reactor at temperatures of about 800–900°C. Typical thermal efficiencies for steam reforming processes are about 70% [52] because this process requires oil or natural gas for the source of hydrocarbon, it cannot be expected to reduce carbon dioxide emission. Recent Japanese results showed the possibility of hydrogen-production process by endothermic reaction above 900°C with biomass [50]. A process that yields hydrogen from biomass is formulated as $(C_6H_{10}O_5)n + mH_2O\text{–}6nH_2 + 6nCO - 814$ kJ. Namely, the process is such that it decomposes cellulose, a principal component of biomass, mostly to H_2 and CO by supplying steam and heat generated by nuclear energy. An additional shift reaction contributes to yield $12H_2$. This reaction is superior in hydrogen-production efficiency by factor of 2–5 because of the chemical energy of biomass itself. Increase in reaction temperature results in the decomposition of residual char and tar, improving gasification efficiency. Biomass is regarded as carbon-neutral material and thus this process does not emit any net carbon dioxide.

The SMR process consists of the following two steps [5,52]:

1. Reforming reaction. The first step of the SMR process is highly endothermic and to produce a synthesis gas (syngas), a mixture is primarily made up of hydrogen and carbon monoxide.
2. Shift reaction. The second step, is exothermic, known as a water gas shift (WGS) reaction, the carbon monoxide produced in the first reaction is reacted with steam over a catalyst to form hydrogen and carbon dioxide. This process occurs in two stages, consisting of a high-temperature shift at 350°C and a low-temperature shift at 190–210°C.

Hydrogen produced from the SMR process includes small quantities of carbon monoxide, carbon dioxide and hydrogen sulfide as impurities and may require purification.

15.3.2 Water Electrolysis

Hydrogen can be produced by electrolysis of water. This electrolysis technology can get pure hydrogen, but the electrolysis needs much electricity energy. Water electrolysis is the splitting of water molecules by electricity, and is the most well known of the methods for producing hydrogen. The cell reaction is as follows [5]:

$$H_2O \rightarrow H_2 + \frac{1}{2}O_2 - 242\,\text{kJ}$$

It is considered a candidate for hydrogen production with nuclear energy because it may be combined with either existing nuclear electrical generating plants or with new, high efficiency nuclear generating plants. Although it is a conservative and established technology, electrolysis is not generally considered viable for larger plants because of the low efficiency. High-temperature steam electrolysis (HTSE) process can offer the high efficiency about 50%, but it needs high temperature above 900°C. According to the research, hydrogen-production efficiency is 3 kW h m^{-3} at the 900°C. And at the 1000°C, the electricity energy is reduced by 30% [53]. So if the fusion reactor can generate high-temperature heat, the HTSE technology would likely be used in the future hydrogen economy. Because this advanced technology for electrolysis is not only superior in efficiency due to lower resistance loss, it also has an endothermic nature to utilize part of the heat itself for splitting water. Also this technology is a reverse reaction of solid oxide fuel cell (SOFC) that converts hydrogen into electricity again, and thus is advantageous.

15.3.3 Thermochemical Cycle

Because of water splitting directly at temperature exceeding 2500°C, such process is not feasible for industrial practice. Thermochemical water-splitting is decomposition of water molecule into hydrogen and oxygen by one or more cyclic thermally driven chemical reactions, whose reaction temperatures are greatly reduced from that of direct water splitting. The temperature requirement for heat source is usually below 900°C to yield high cycle performance. The gain of thermochemical cycle efficiency follows the principle of the Carnot cycle, that is, the higher the temperature, the higher is the efficiency [5,6].

The sulfur–iodine (S–I) cycle is an example of the thermochemical cycle, and it is also a better thermochemical cycle, and has been evaluated by many authorities and institutes [49,54,55]. The S–I cycle consists of three chemical reactions.

$$H_2SO_4 \rightarrow SO_2 + \frac{1}{2}O_2 + H_2O\ (850°C)$$

$$SO_2 + I_2 + 2H_2O \rightarrow H_2SO_4 + 2HI\ (120°C)$$

$$2HI \rightarrow H_2 + I_2\ (450°C) \quad 1$$

$$H_2O \rightarrow H_2 + \frac{1}{2}O_2\ \text{(net effect of the three reactions)}$$

The sulfuric acid decomposition reaction proceeds above 700°C and more efficiently at higher than 850°C. The reaction is endothermic and the high-temperature heat can be provided by high-temperature fusion as well as fission reactor. The S–I cycle has been validated in closed-cycle continuous operation with the water as the only material feed and with the recycling of all other process chemicals.

Energy, as heat, is input to a thermochemical cycle via one or more endothermic high-temperature chemical reactions. Heat is rejected via one or more exothermic low-temperature reactions. In the S–I cycle most of the input heat goes into the dissociation of sulfuric acid. Sulfuric acid and hydrogen iodide are formed in the exothermic Bunsen reaction of H_2O, SO_2, and I_2 and hydrogen is generated in the decomposition of hydrogen iodide. The net reaction is the decomposition of water into hydrogen and oxygen. The whole process takes in only water and high-temperature heat and releases only hydrogen, oxygen and low-temperature heat. All the reactants, other than water, are regenerated and recycled. There are literally no effluents. Each of the major chemical reactions of this process was demonstrated in the laboratory at GA (General Atomics Inc.) and JAEA (Japan Atomic Energy Research Institute).

The S–I cycle does require high temperatures, but offers the prospects for high efficiency conversion of heat energy to hydrogen energy. Its efficiency is expected to be ~50%, because it is free from heat to electricity conversion efficiency limit, it can be regarded as another kind of heat cycle that discards low-temperature heat at the end.

There are some key problems in the S–I cycle to be solved, such as: (1) the decomposition of hydrogen iodine acid. At the 400°C, the reaction of the HI decomposition is reversible, it needs to get H_2 and I_2 steam from the HI gas for the higher efficiency. (2) The evaporation of the H_2SO_4 is a strong corrosion process, it demands special material.

15.4 Potential Couple of Fusion Reactor to Hydrogen Process System

HTSE process and thermochemical water-splitting S–I cycle can offer the prospect for high heat-to-hydrogen efficiency of about 50%. However, these processes need high-temperature heat more than 850°C for the higher efficiency. So the high-temperature fusion reactor which can offer high-temperature heat is a good way for hydrogen production. Till now, the designs which can have potential ability to obtain high-temperature heat of more than 850°C are only several blanket concepts. And most of them are based on the advanced structural materials, such as SiC_f/SiC composition and W alloy. Only the high-temperature HTL LiPb blanket based on the MFCI developed by China adopts the RAFM steel as its structural material. RAFM steel remains presently the most promising structural material for breeding blanket with great technological maturity [56]. It is considered to be the primary structural material candidate for future fusion power reactors. The recommended blanket concepts for hydrogen production should have mature technology development. And the recommended process for the hydrogen production should have high efficiency and good environment effect.

One potential arrangement of fusion reactor and hydrogen process system is considered, that is the HTL blanket concept combine with the thermochemical water-splitting S–I cycle. About 1000°C temperature heat is obtained from this HTL blanket, which can offer high enough temperature for thermochemical S–I cycle. However, some important coupling technical issues are specified for this arrangement, including material selections for hydrogen production system and intermediate heat exchanger, inter-contamination issues between hydrogen production system and fusion reactor, and R&D on the subsystems like power conversion system, tritium control system, and so on.

15.4.1 Material Selections for Heat Exchanger and Thermochemical Hydrogen Process

The high-temperature heat exchanger (HTHE) between HTL blanket coolant circulation system and hydrogen production system will operate in high-temperature windows up to ~1000°C. The material for this heat exchanger will encounter tough challenge to sustain high temperature and harsh, aggressive and corrosive environments. The refractory metals (e.g., W, Nb, Ta, etc.), high-temperature alloy (e.g., ODS–FeCrAl), ceramic (e.g., SiC), and ceramic matrix composite (CMC) materials (e.g., SiC_f/SiC composites), might be the most potential candidate structural materials for the heat exchanger. However, the refractory metals have not enough oxidation properties at very high temperature [57]. The high-temperature alloy has critical issues, for example, low melting point, low oxidation resistance, and creep under high temperature, and so on. Ceramic is a kind of good high-temperature material with high strength and stability at high temperature, but it is difficult to fabricate because of its inherent brittleness. Toughness improved ceramic based composite is attractive as potential structural material in the future. And also its fabrication technology is not mature at present [58].

CMC has been used for gas–gas HTHE concept design in 500 ~ 1000°C [59]. A ceramic plate–fin heat exchanger based on the offset strip fin (OSF) design was presented in Ref. [60]. It can be used for the HTHE in the externally fired combined cycle (EFCC) or other applications that need operational material temperatures up to 1250°C.

The I–S thermochemical cycle is recommended for the hydrogen production process. It involves aggressive corrosive chemical environment of H_2SO_4, HI, and I_2 at high temperature of 950–1000°C. Hence, the material for this process should be carefully chosen to resist corrosion. In Japan, corrosion test of materials has been carried out. Among these candidate materials, heat-resisting alloy showed good corrosion resistance in the gas–phase environments, such as Hastelloy. On the other hand, in the liquid–phase environments, only special material, for example, ceramic (SiC, etc.), and rare metals (Zr, Ta, etc.) showed resistance to corrosion [61]. Therefore, SiC ceramic is the preferred recommended material in contact with the process fluid at high temperature in the hydrogen process plant, other parts use the carbon steel as much as possible to reduce cost.

15.4.2 Integrated Fusion Reactor Hydrogen Production System

The major concern with integrating fusion reactor with hydrogen production system is chemical reactivity of LiPb with water. The reaction can release large amounts of unintended energy and hydrogen. H_2O has to be removed in the LiPb exchanger in the power conversion system. To reduce the risk of the LiPb–H_2O reaction, an intermediate helium loop may be used to transport the heat from the LiPb blanket coolant loop to the hydrogen production system. The intermediate loop is designed to prevent any leakage from the reactor coolant loop from contaminating the hydrogen production system and any corrosive process chemical from entering the fusion reactor. Furthermore, the inter-contamination about tritium and hydrogen between fusion reactor and hydrogen process can be reduced with such intermediate helium loop separating them.

The heat exchanger interface sets the boundary conditions for the high-temperature blanket system. The principal requirement is the temperature requirement, which must account for the temperature drop between the coolant outlet and the point of application in the hydrogen production system. We assumed the HTL blanket can provide outlet temperature of 1000°C. This should give a peak-process temperature of 950°C and a process efficiency of 51% for S–I cycle.

Figure 15.29 shows the scheme of an integrated fusion reactor hydrogen production system that employs the intermediate helium loop heat exchanger. HE1 is a He–He heat exchanger constructed of Incoloy800 alloy to heat the intermediate helium from 300 to 350°C. HE2 is a LiPb–He heat exchanger constructed of SiC_f/SiC composite to continue to heat the intermediate helium from 350 to 975°C. And HE3 is a He–H_2O heat exchanger constructed of SiC_f/SiC composite to provide as high as 950°C heat to the water-splitting hydrogen production loop from the intermediate helium loop. All the heat exchangers have been designed to adopt antipermeation barrier coating, such as oxide in HE1 and SiC coating in the HE2 and HE3, to minimize the inter-contamination between the reactor coolant loop and the hydrogen production system.

15.4.3 Tritium Control System

Tritium is radioactive and violent active, and very easy to permeate out of the container. The reduction of tritium permeation is one of key issue for the high-temperature LiPb blanket. Tritium permeation barrier should be considered in the He and LiPb coolant channels

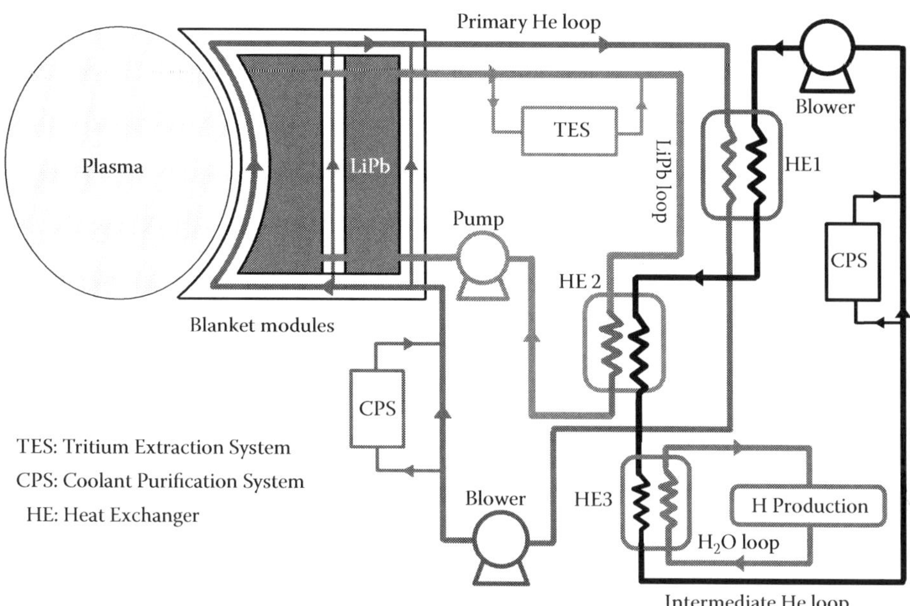

FIGURE 15.29
The scheme for the power conversion system with intermediate helium loop heat exchanger.

besides in the heat exchanger to reduce tritium permeation into hydrogen production system, for example, by using coating. The high tritium permeation reduction factor (TPRF) is very critical for these coating both on high-temperature blanket and heat exchanger.

Coating materials and fabrication technologies need to be investigated and developed in the future. The materials are needed to have good compatibility with fundamental material and high-temperature LiPb, to achieve low T-permeation rate and to provide self-healing function on defects that might occur in normal operation or accidental conditions. The development of fabrication technology is more required for intermediate heat exchanger due to large area and thin wall for efficient heat transfer from the primary coolant LiPb system to the secondary system.

A tritium extraction system is proposed based on tritium permeation through tube walls made of a refractory metal from the flowing liquid breeder into a vacuum chamber. This method seems to have the potential to control the maximum tritium partial pressure in the LiPb less than 1 Pa. However, using refractory metals for large components could not be realized without strong R&D programs, thus the concept needs experimental verification. In addition, reliable permeation barriers should be fabricated on the outside of the tritium extraction system to reduce tritium permeation.

15.5 Summary

Hydrogen promises to be an important energy carrier for the future. Nuclear thermal energy is not only feasible but attractive heat source to produce hydrogen in quantity and thermal efficiency that can meet future demand. Hydrogen can be produced by three conventional processes, steam reforming, water electrolysis, and thermochemical water-splitting process. However, the high efficiency hydrogen production is not always available. Thermochemical water-splitting S–I cycle can offer the prospect for high heat-to-hydrogen efficiency of ~50% when the temperature reaches 850°C. HTSE can also achieve similar efficiency when combined with high efficiency generation with advanced gas turbine.

Some high temperature DEMO blanket concepts suitable for providing high-temperature heat for hydrogen production have been reviewed. The recommended blanket concepts for hydrogen production should have mature technology development to meet high temperature requirement.

The HTL blanket has high-temperature capability and relatively mature technology base to satisfy hydrogen thermal processes. It has been developed to obtain 1000°C high-temperature heat with RAFM steel as structural material by using MFCI technology in LiPb flow channels. The potential combination of HTL blanket concept coupling to the thermochemical water-splitting S–I cycle is recommended in China.

For integrated fusion reactor hydrogen production system, some important technical issues, such as material characterization and selections for hydrogen production system and intermediate heat exchanger, inter-contamination issues between hydrogen production system and fusion reactor and the area of subsystems like power conversion system, tritium control system, and so on have been specified and should be developed in next step.

References

1. W.J. Nuttall, 2008. Fusion is an energy source: Challenges and opportunities. Institute of Physics Report.
2. A.A. Harm, K.F. Schoepf, G.H. Miley, D.R. Kingdon, 2002. *Principles of Fusion Energy: An Introduction to Fusion Energy for Students of Science and Engineering*. Singapore: World Scientific.
3. J.A. Wesson. 1997. *Tokamaks*. Oxford University Press, New York.
4. Energy Information Administration. 1999. Annual Energy Outlook 2000: With projections to 2020. DOE/EIA-0383(2000).
5. K.R. Schultz, 2003. Production of hydrogen by fusion energy: A review and perspective. *Fusion Science and Technology* 44(2): 393–399.
6. L.M. Crosbie, D. Chapin. 2003. Hydrogen production by nuclear heat. GENES4/ANP 2003, Kyoto, Japan.
7. C.M. Braams, P.E. Stott. 2002. *Nuclear Fusion: Half a Century of Magnetic Confinement Fusion Research*. London: Institute of Physics Publishing.
8. J. Jacquinot (for the Jet Team). 1995. JET relevance to ITER, new trends and initial results. *Fusion Engineering and Design* 30: 67–84.
9. B. Saoutic, M. Chatelier, C. De Michelis, 2009. Tore Supra: Toward steady state in a superconducting tokamak. *Fusion Science and Technology* 56(3): 1079–1091.
10. Y. Miyo, J. Yagyu, T. Nishiyama, M. Honda, H. Ichige, A. Kaminaga, T. Sasajima, T. Arai, A. Sakasai, 2008. Tokamak machine monitoring and control system for JT-60. *Fusion Engineering and Design* 83(2–3): 337–340.
11. M. Keilhacker and Jet Team. 1999. Fusion physics progress on JET. *Fusion Engineering and Design* 46(2–4): 273–290.
12. Songtao Wu, the EAST Team. 2007. An overview of the EAST project. *Fusion Engineering and Design* 82: 463–471.

13. G.S. Lee, J. Kim, S.M. Hwang, C.S. Chang, H.Y. Chang, M.H. Cho, B.H. Choi, K. Kim, 2000. The KSTAR project: An advanced steady state superconducting tokamak experiment. *Nuclear Fusion* 40(3 Y): 575–582.
14. http://www.iter.org/proj/Pages/ITERMission.aspx.
15. T. Ilhli, T.K. Basu, L.M. Giancarli, S. Konishi, S. Malang, F. Najmabadi, S. Nishio, et al., 2008. Review of blanket designs for advanced fusion reactors. *Fusion Engineering and Design* 83(7–9): 912–919.
16. L.V. Boccaccini, L. Giancarli, G. Janeschitz, S.Hermsmeyer, Y. Poitevin, A. Cardella, E. Diegele, 2004. Materials and design of the European DEMO blankets. *Journal of Nuclear Materials* 329–333: 148–155.
17. L.V. Boccaccini, U. Fischer, S. Gordeev, S. Malang, 2000. Advanced helium cooled pebble bed blanket with SiC_f/SiC as structural material. *Fusion Engineering and Design* 49–50: 491–497.
18. S. Nishio. 1998. Prototype fusion reactor based on SiCf/SiC composite material focusing on easy maintenance. *Proceedings of the IAEA-TCM on Fusion Power Plant Design*, Culham.
19. M.D. Donne, L. Anzidei, H. Kwast, F. Moons, E. Proust. 1995. Status of EC solid breeder-blanket designs and R&D for DEMO fusion reactors. *Fusion Engineering and Design* 27: 319–336.
20. A. Li-Puma, J. Bonnemason, L. Cachon, J.L. Duchateau, F. Gabriel, 2009. Consistent integration in preparing the helium cooled lithium lead DEMO-2007 reactor. *Fusion Engineering and Design* 84: 1197–1205.
21. N.B. Morley, Y. Katoh, S. Malang, B.A. Pint, A.R. Raffray, S. Sharafat, S. Smolentsev, G.E. Youngblood, 2008. Research and development for the dual-coolant blanket concept in the U.S. *Fusion Engineering and Design* 83: 920–927.
22. M.S. Tillack, S. Malang. 1997. High performance LiPb blanket. *Proceedings of the 17th Symposium on Fusion Engineering*, San Diego, CA.
23. M.S. Tillack, X.R. Wang, J. Pulsifer, S. Malang, D.K. Sze, M. Billone, I. Sviatoslavsky, ARIES Team, 2003. Fusion power core engineering for the ARIES-ST power plant. *Fusion Engineering and Design* 65(2): 215–261.
24. P. Norajitra, L. Bühler, A. Buenaventura, E. Diegele, U. Fischer, S. Gordeev, E. Hutter, R. Kruessmann, 2003. Conceptual design of the dual-coolant blanket within the framework of the EU power plant conceptual study (TW2-TRP-PPCS), Final Report, FZKA 6780, Forschungszentrum Karlsruhe GmbH, Germany.
25. A.R. Raffray, L. EL-Guebaly, S. Malang, X. R. Wang, L. Bromberg, T. Ihli, B. Merrill, L. Waganer, and ARIES-CS team, 2008. Engineering design and analysis of the ARIES-CS power plant. *Fusion Science and Technology* 54: 725–746.
26. Y. Wu, FDS team. 2006. Conceptual design activities of FDS series fusion power plants in China. *Fusion Engineering and Design* 81(23–24): 2713–2718.
27. Y. Wu, J. Qian, J. Yu. 2002. The fusion-driven hybrid system and its material selection. *Journal of Nuclear Materials* 307–311(2): 1629–1636.
28. Y. Wu, S. Zheng, X. Zhu, W. Wang, H. Wang, S. Liu, Y. Bai, et al., 2006. Conceptual design of the fusion-driven subcritical system FDS-I. *Fusion Engineering and Design* 81: 1305–1311.
29. Y. Wu, 2002. Progress in fusion-driven hybrid system studies in China. *Fusion Engineering and Design* 63–64: 73–80.
30. Y. Wu, FDS team. 2008. Conceptual design of the China fusion power plant FDS-II. *Fusion Engineering and Design* 83(10–12): 1683–1689.
31. Y. Wu, FDS team. 2009. Fusion-based hydrogen production reactor and its material selection. *Journal of Nuclear Materials* 386–388: 122–126.
32. Y. Wu, L. Qiu, Y. Chen. 2000. Conceptual study on liquid metal center conductor post in spherical tokamak reactors. *Fusion Engineering and Design* 51–52: 395–399.
33. Q. Huang, C. Li, Y. Li, M. Chen, M. Zhang, L. Peng, Z. Zhu, Y. Song, S. Gao, 2007. Progress in development of China low activation martensitic steel for fusion application. *Journal of Nuclear Materials* 367–370: 142–146.
34. H. Chen, Y. Wu, S. Konishi, J. Hayward, 2008. A high temperature blanket concept for hydrogen production. *Fusion Engineering and Design* 83: 903–911.

35. S. Liu, Y. Wu, H. Chen, S. Zheng, Y. Bai, 2007. Progress in conceptual study of China fusion-based hydrogen production reactor. *The 2nd IAEA Technical Meeting on First Generation of Fusion Power Plants*, Vienna, Austria.
36. http://aries.ucsd.edu/
37. F. Najmabadi, 2000. Impact of advanced technologies on fusion power plant characteristics—The ARIES-AT study. *14th ANS TOFE*, Park City, USA.
38. A.R. Raffray, L. El-Guebaly, S. Gordeev, S. Malang, E. Mogahed, F.Najmabadi, I. Sviatoslavsky, et al., 2001. High performance blanket for ARIES-AT power plant. *Fusion Engineering and Design*, 58–59: 549–553.
39. L. Giancarli, H. Golfier, S. Nishio, R. Raffray, C. Wong, R. Yamada, 2002. Progress in blanket designs using SiCf/SiC composites. *Fusion Engineering and Design* 61–62: 307–318.
40. L. Giancarli, J.P. Bonal, A. Caso, G. Le Marois, N.B. Morley, J.F. Salavy, 1998. Design requirements for SiC/SiC composites structural material in fusion power reactor blankets. *Fusion Engineering and Design* 41: 165–171.
41. A.R. Raffray, R. Jones, G. Aiello, M. Billone, L. Giancarli, H. Golfier, A. Hasegawa, et al., 2001. Design and material issues for SiCf/ SiC-based fusion power cores. *Fusion Engineering and Design* 55: 55–95.
42. L. Giancarli, L. Bühler, U. Fischer, R. Enderle, D. Maisonnier, C. Pascal, P. Pereslavtsev, et al., 2003. In-vessel component design for a self-cooled lithium–lead fusion reactor. *Fusion Engineering and Design* 69: 763–768.
43. A. Sagara, O. Motojima, K. Watanabe, S. Imagawa, H. Yamanishi, O. Mitarai, T. Satow, H. Tikaraishi, FFHR Group. 1995. Blanket and divertor design for force free helical reactor (FFHR). *Fusion Engineering and Design* 29: 51–56.
44. M.A. Abdou, The APEX TEAM, A. Ying, N. Morley, K. Gulec, S. Smolentsev, M. Kotschenreuther, S. Malang, 2001. On the exploration of innovative concepts for fusion chamber technology. *Fusion Engineering and Design* 54: 181–247.
45. C.P.C. Wong, S. Malang, M. Sawan, I. Sviatoslavsky, E. Mogahed, S. Smolentsev, S. Majumdar, et al., 2004. Molten salt self-cooled solid firstwall and blanket design based on advanced ferritic steel. *Fusion Engineering and Design* 72(1–3): 245–275.
46. A. Sagara, T. Tanaka, T. Muroga, H. Hashizume, T. Kunugi, S. Fukada, A. Shimizu, 2005. Innovative liquid breeder blanket design activities in Japan. *Fusion Sci. Technol.* 47: 524–529.
47. T. Muroga, T. Tanaka, A. Sagara, 2006. Blanket neutronics of Li/vanadium-alloy and Flibe/vanadium-alloy systems for FFHR. *Fusion Engineering and Design* 81: 1203–1209.
48. R.F. Mattas, S. Malang, H. Khater, S. Majumdar, E. Mogahed, B. Nelson, M. Sawan, D.K. Sze, 2000. EVOLVE—An advanced firstwall/blanket system. *Fusion Engineering and Design* 49–50: 613–620.
49. G.E. Besenbruch, 1982. General atomic sulfur–iodine thermochemical water-splitting process. *Am. Chem. Soc. Div. Pet. Chem.* 271: 48–53.
50. S. Konishi, 2005. Potential fusion market for hydrogen production under environmental constraints. *Fusion Sci. Technol.* 47: 1205–1209.
51. R. Schiller, G. Nagy, J. Hayward, D. Maisonnier, 2006. Radiation chemical and plasma chemical processes for hydrogen production from water. *24th Symposium on Fusion Technology*, Warsaw, Poland.
52. C.E.G. Padro, V. Putsche, 1999. Survey of the Economies of Hydrogen Technologies. National Renewable Energy Laboratory, NREL/TP-570-27079.
53. R. Hino, K. Haga, H. Aita, K. Sekita, 2004. R&D on hydrogen production by high-temperature electrolysis of steam. *Nuclear Engineering and Design* 233(1–3): 363–375.
54. Seiji Kasahara, Shinji Kubo, Ryutaro Hino, Kaoru Onuki, Mikihiro Nomura, Shin-ichi Nakao, 2007. Flowsheet study of the thermochemicalwater-splitting iodine–sulfur process for effective hydrogen production. *International Journal of Hydrogen Energy* 32: 489– 496.
55. Junhu Zhou, Yanwei Zhang, Zhihua Wang, Weijuan Yang, Zhijun Zhou, Jianzhong Liu, Kefa Cen, 2007. Thermal efficiency evaluation of open-loop SI thermochemical cycle for the production of hydrogen, sulfuric acid and electric power. *International Journal of Hydrogen Energy* 32(5): 567–575.

56. Y. Kohno, A. Kohyama, T. Hirose, M.L. Hamilton, M. Narui, 1999. Mechanical property changes of low activation ferritic/martensitic steels after neutron irradiation. *Journal of Nuclear Materials* 271–272: 145–150.
57. Alain Lasalmonie, 2006. Intermetallics: Why is it so difficult to introduce them in gas turbine engines? *Intermetallics* 14: 1123–1129.
58. J.C. Zhao, J.H. Westbrook, Guest Editors. 2003. Ultrahigh-temperature materials for jet engines. *MRS Bulletin* 9: 622–627.
59. C. Luzzatto, A. Morgana, S. Chaudourne, T. O'Doherty, G. Sorbie, 1997. A new concept composite heat exchanger to be applied in high temperature industrial processes. *Applied Thermal Engineering* 17: 789–797.
60. J. Schulte-Fischedick, V. Dreißigacker, R. Tamme, 2007. An innovative ceramic high temperature plate-fin heat exchanger for EFCC processes. *Applied Thermal Engineering* 27: 1285–1294.
61. B. Yildiz, M.S. Kazimi, 2006. Efficiency of hydrogen production systems using alternative nuclear energy technologies. *International Journal of Hydrogen Energy* 31: 77–92.
62. http://www.jet.efda.org/wp-content/gallery/graphics/j82–348c.jpg.
63. http://www.iter.org/mach/Pages/Tokamak.aspx.

Section IV

Applied Science and Technology

16

High-Temperature Electrolysis of Steam

James E. O'Brien, Carl M. Stoots, and J. Stephen Herring

CONTENTS

16.1 Fundamentals

16.1.1 Thermodynamics of Thermal Water Splitting Processes

A basic thermodynamic analysis can be applied to a general thermal water-splitting process in order to determine the overall process efficiency limits as a function of temperature. Consider the process diagram for thermal water splitting shown in Figure 16.1. Water enters the control volume from the left. Since the ultimate feedstock for any large-scale water-splitting operation will be liquid water, it is reasonable to consider the case in which water enters the control volume in the liquid phase at a specified temperature T and pressure P, typically near ambient conditions. Pure hydrogen and oxygen streams exit the control volume on the right, also at T and P. Two heat reservoirs are available, a high-temperature reservoir at temperature T_H and a low-temperature reservoir at temperature T_L. Heat transfer between these reservoirs and the control volume is indicated in Figure 16.1

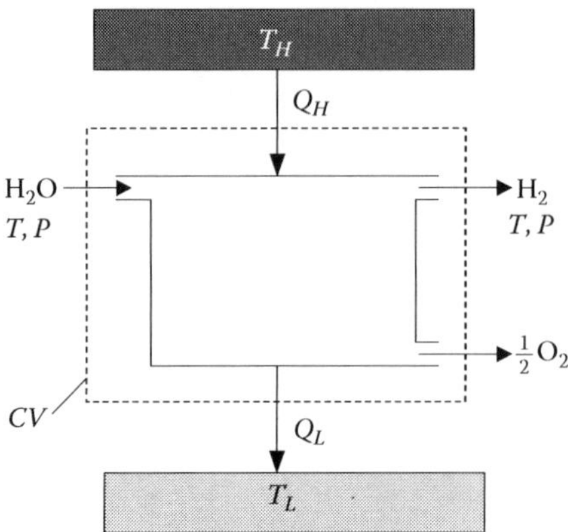

FIGURE 16.1
Schematic of a generic thermal water-splitting process operating between temperatures T_H and T_L.

as Q_H and Q_L. Note that there is no work crossing the control volume boundary. Therefore, if the process under consideration is high-temperature electrolysis (HTE), both the power cycle (based on a heat engine for the purposes of this discussion) and the electrolyzer are located inside the control volume.

From an overall chemical reaction standpoint, the water-splitting process corresponds to the dissociation or reduction of water:

$$H_2O \rightarrow H_2 + \frac{1}{2}O_2 \tag{16.1}$$

The first and second laws of thermodynamics can be applied to this process as follows:

$$\text{First law: } Q_H - Q_L = \Delta H_R, \tag{16.2}$$

$$\text{Second law: } \Delta S_R \geq \frac{Q_H}{T_H} - \frac{Q_L}{T_L}, \tag{16.3}$$

where ΔH_R is the enthalpy of reaction and ΔS_R is the entropy change of the reaction. The overall thermal-to-hydrogen efficiency of thermal water-splitting processes can be defined in terms of the net enthalpy increase of the reaction products over the reactants (can also be thought of as the energy content or heating value of the produced hydrogen), divided by the high-temperature heat added to the system:

$$\eta_H = \frac{\Delta H_R}{Q_H} \tag{16.4}$$

Combining the first and second law equations for the reversible case and substituting into the efficiency definition yields:

$$\eta_{H,\max} = \frac{1 - T_L/T_H}{1 - T_L\Delta S_R/\Delta H_R}. \tag{16.5}$$

Note that the water-splitting process defined in Figure 16.1 is simply the reverse of the combustion reaction of hydrogen with oxygen. Therefore, the enthalpy of reaction for the water-splitting process is the opposite of the enthalpy of combustion, which by definition is equal to the "heating value" of the hydrogen. Since for our process, we have assumed that the water enters the control volume in the liquid phase,

$$\Delta H_R = \text{HHV}, \tag{16.6}$$

where HHV is the "high heating value" of hydrogen. If we further assume that T and P represent standard conditions, and that $T_L = T_o$,

$$\Delta H_R - T_L \Delta S_R = -\Delta G^o_{f,H_2O} \tag{16.7}$$

such that the efficiency expression can be rewritten as

$$\eta_{H,\max} = \left(1 - \frac{T_L}{T_H}\right)\left(\frac{\text{HHV}}{-\Delta G^o_{f,H_2O}}\right) = \left(1 - \frac{T_L}{T_H}\right)\left(\frac{1}{0.83}\right). \tag{16.8}$$

The HHV of the hydrogen and the standard-state Gibbs energy of formation for water are fixed quantities such that the second factor on the right-hand side is a constant. This efficiency limit has also been derived for the sulfur-iodine thermochemical process based on an exergy analysis [1].

Comparing Equation 16.8 to 16.4, the high-temperature heat requirement for the process can be stated as

$$Q_H \geq \frac{T_H}{T_H - T_L}\left(-\Delta G^o_{f,H_2O}\right). \tag{16.9}$$

This result was derived for thermochemical cycles by Abraham and Schreiner [2], and applied to solar thermal dissociation of water by Fletcher and Moen [3], who noted that the maximum efficiencies of all thermochemical processes can be related to the efficiencies of Carnot engines operating between the same upper and lower temperatures. It is necessary only to add, conceptually, a reversible fuel cell which converts the hydrogen and oxygen to liquid water at the lower temperature, performing an amount of electrical work given by the Gibbs free energy of the reaction.

A plot of thermal water-splitting efficiencies is presented in Figure 16.2 for $T_L = 20°C$. The top curve represents the maximum possible water-splitting efficiency result given by Equation 16.8. The exergetic efficiency of the thermal water-splitting process is given by the ratio of the actual efficiency to the maximum possible efficiency. A reasonable value to assume for exergetic efficiency is 65%, which is represented by the bottom curve in Figure 16.2. The 65% value is based on a typical percentage of Carnot efficiency that can be achieved with a well-engineered modern power cycle. The first conclusion to be drawn is that high temperature is needed for efficient hydrogen production based on thermal water splitting, regardless of the specific method used. If we assume that 65% of the maximum possible efficiency might also be achievable with a well-engineered thermal water-splitting process, then efficiencies of the magnitude given in the lower curve of Figure 16.2 should be expected.

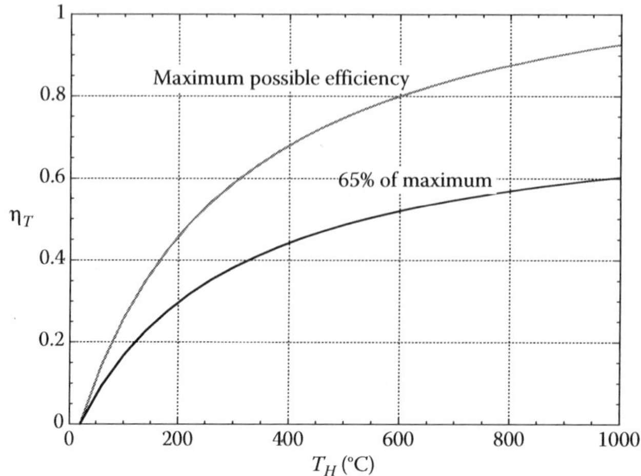

FIGURE 16.2
Theoretical thermal water-splitting efficiencies.

Detailed process analyses have been performed [4] to analyze HTE-based hydrogen production systems coupled to advanced nuclear reactors. Results from this study are presented in Figure 16.3. This figure shows overall hydrogen production efficiencies, based on HHV, plotted as a function of reactor outlet temperature. The figure includes the curve that represents 65% of the thermodynamic maximum efficiency, again assuming $T_L = 20°C$.

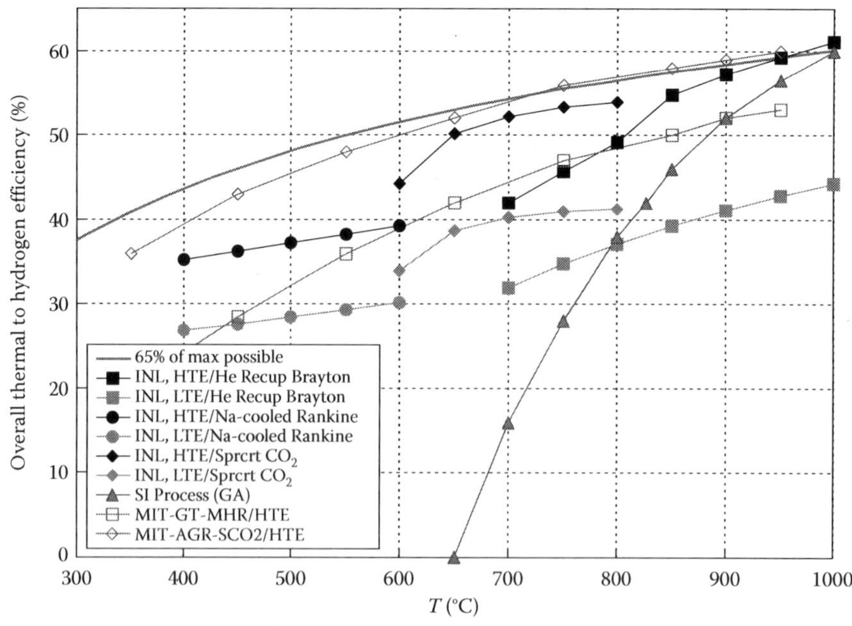

FIGURE 16.3
Overall thermal-to-hydrogen production efficiencies based on HHV for several reactor/process concepts, as a function of reactor outlet temperature.

Three different advanced-reactor/power-conversion combinations were considered: a helium-cooled reactor coupled to a direct recuperative Brayton cycle, a supercritical CO_2-cooled reactor coupled to a direct recompression cycle, and a sodium-cooled fast reactor coupled to a Rankine cycle. The system analyses were performed using UniSim [5] software. Each reactor/power-conversion combination was analyzed over an appropriate reactor outlet temperature range. The figure shows results for both HTE and low-temperature electrolysis (LTE). Results of system analyses performed at MIT [6] are also shown. The lower MIT curve, labeled MIT-GT-MHR/HTE represents overall efficiency predictions for a helium-cooled reactor with a direct Brayton cycle power conversion unit. The upper MIT curve, labeled MIT-AGR-SCO2/HTE represents overall efficiency predictions for a CO_2-cooled advanced gas reactor with a supercritical CO_2 power conversion unit. For reactor outlet temperatures of 600–800°C, the supercritical CO_2/recompression power cycle is superior to the He-cooled/Brayton cycle concept. This conclusion is consistent with results presented by Yildiz et al. [6]. Finally, an efficiency curve for the SI thermochemical process [7] is also shown. The results presented in Figure 16.3 indicate that, even when detailed process models are considered, with realistic component efficiencies, heat exchanger performance, and operating conditions, overall hydrogen production efficiencies in excess of 50% can be achieved for HTE with reactor outlet temperatures above 850°C. The efficiency curve for the SI process also includes values above 50% for reactor outlet temperatures above 900°C, but it drops off quickly with decreasing temperature, and falls below values for LTE coupled to high-temperature reactors for outlet temperatures below 800°C. Note that even LTE benefits from higher reactor outlet temperatures because of the improved power conversion thermal efficiencies associated with higher reactor outlet temperatures. Current planning for the next generation nuclear plant (NGNP) [8] indicates that reactor outlet temperatures will be at or below 850°C, which favors HTE.

16.1.2 Thermodynamics of HTE

Focusing now on electrolysis, consider a control volume surrounding an isothermal electrolysis process, as shown in Figure 16.4. In this case, both heat and work interactions cross the control volume boundary. The first law for this process is given by

$$Q - W = \Delta H_R. \tag{16.10}$$

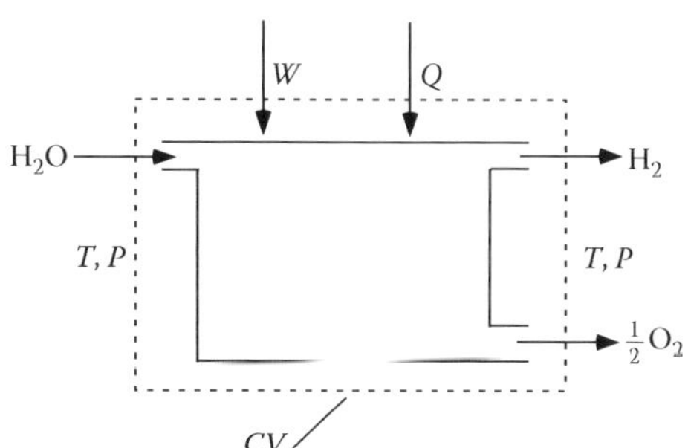

FIGURE 16.4
Schematic of a water electrolysis process operating at temperature T.

For reversible operation,

$$Q_{rev} = T\Delta S_R. \tag{16.11}$$

Such that

$$W_{rev} = \Delta H_R - T\Delta S_R = \Delta G_R. \tag{16.12}$$

The thermodynamic properties appearing in Equation 16.12 are plotted in Figure 16.5 as a function of temperature for the H_2–H_2O system from 0 to 1000°C at standard pressure. This figure is often cited as a motivation for HTE versus LTE. It shows that the Gibbs free energy change, ΔG_R, for the reacting system decreases with increasing temperature, while the product of temperature and the entropy change, $T\Delta S_R$, increases. Therefore, for reversible operation, the electrical work requirement decreases with temperature, and a larger fraction of the total energy required for electrolysis, ΔH_R, can be supplied in the form of heat, represented by $T\Delta S_R$. Since heat-engine-based electrical work is limited to a production thermal efficiency of 50% or less, decreasing the work requirement results in higher overall thermal-to-hydrogen production efficiencies. Note that the total energy requirement, ΔH_R, increases only slightly with temperature, and is very close in magnitude to the lower heating value (LHV) of hydrogen. The ratio of ΔG_R to ΔH_R is about 93% at 100°C, decreasing to only about 70% at 1000°C. Operation of the electrolyzer at high temperature is also desirable from the standpoint of reaction kinetics and electrolyte conductivity, both of which improve dramatically at higher operating temperatures. Potential disadvantages of high-temperature operation include the limited availability of very high-temperature process heat and materials issues such as corrosion and degradation.

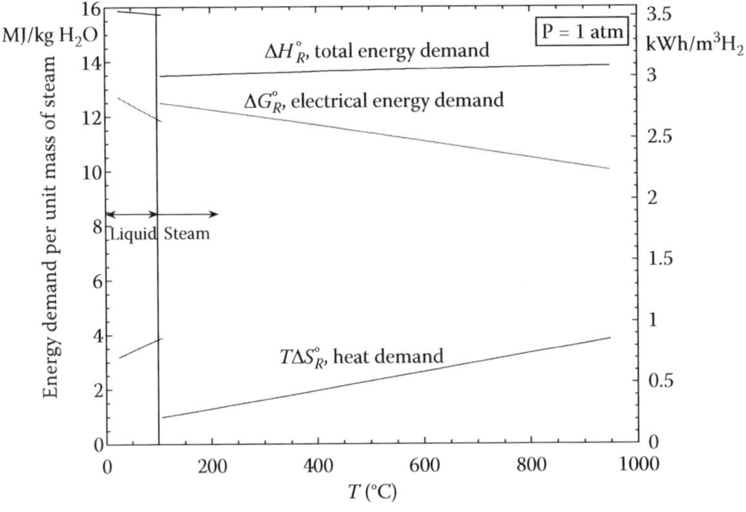

FIGURE 16.5
Standard-state ideal energy requirements for electrolysis as a function of temperature.

In order to accomplish electrolysis, a voltage must be applied across the cell that is greater in magnitude than the open-cell potential. The standard-state open-cell potential is given by

$$V^o = \frac{\Delta G_R^o}{jF}, \tag{16.13}$$

where j is the number of electrons transferred per molecule of hydrogen produced. For the steam–hydrogen system, in which the $O^=$ ions are transported through the solid-oxide electrolyte, $j = 2$. The standard-state open-cell potential applies to the case in which pure reactants and products are separated and at one standard atmosphere pressure. In most practical HTE systems, the incoming steam is mixed with some hydrogen and possibly some inert gas. Some inlet hydrogen is normally required in order to maintain reducing conditions on the steam-side electrode, typically a nickel cermet. Also, it is not desirable to run the electrolyzer to 100% steam utilization, because localized steam starvation will occur, severely degrading performance. Therefore, the outlet stream will include both steam and hydrogen. Residual steam can be removed from the product by condensation. On the oxygen-evolution side of the cells, air is often used as a sweep gas; so the oxygen partial pressure is only about 21% of the operating pressure. In addition, the electrolysis system can operate at elevated pressure. In order to account for the range of gas compositions and pressures that occur in a real system, the open-cell (or Nernst) potential can be obtained from the Nernst equation, which can be written as

$$V_N = V^o - \frac{R_u T}{jF} \ln\left[\left(\frac{y_{H_2O}}{y_{H_2} y_{O_2}^{1/2}}\right)\left(\frac{P}{P_{std}}\right)^{-1/2}\right]. \tag{16.14}$$

Operation of a solid-oxide stack in the electrolysis mode is fundamentally different than operation in the fuel-cell mode for several reasons, aside from the obvious change in the direction of the electrochemical reaction. From the standpoint of heat transfer, operation in the fuel-cell mode typically necessitates the use of significant excess air flow in order to prevent overheating of the stack. The potential for overheating arises from two sources: (1) the exothermic nature of the hydrogen oxidation reaction and (2) ohmic heating associated with the electrolyte ionic resistance and other loss mechanisms.

Conversely, in the electrolysis mode, the steam reduction reaction is endothermic. Therefore, depending on the operating voltage, the net heat generation in the stack may be negative, zero, or positive. This phenomenon is illustrated in Figure 16.6. The figure shows the respective internal heat sink/source fluxes in a planar solid-oxide stack associated with the electrochemical reaction and the ohmic heating. The ohmic heat flux (W/cm^2) is given by

$$q''_{Ohm} = i^2 \text{ASR} = i(V_{op} - V_N), \tag{16.15}$$

where i is the current density (A/cm^2) and V_N is the mean Nernst potential for the operating cell. The reaction heat flux is given by

$$q''_R = \frac{i}{2F}(T\Delta S_e) = \frac{i}{2F}(\Delta G_e - \Delta H_R), \tag{16.16}$$

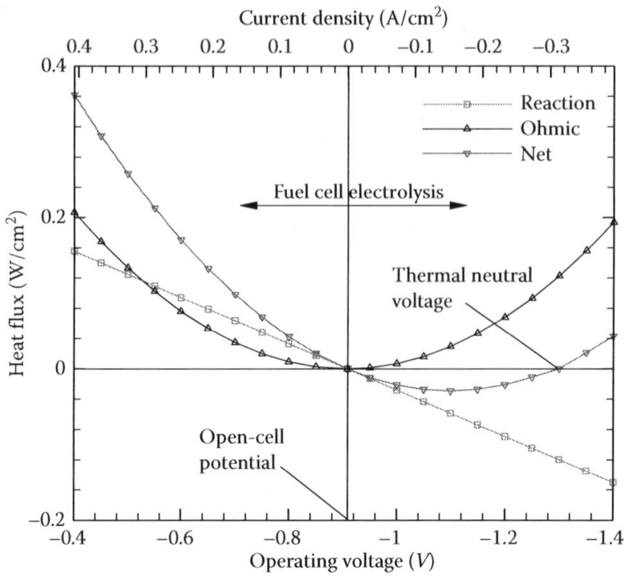

FIGURE 16.6
Thermal contributions in electrolysis and fuel cell modes of operation.

where ΔS_e is the entropy change for the actual electrolysis process, accounting for the reactant and product partial pressures.

The net heat flux is also shown in Figure 16.6. An area-specific resistance (ASR) of 1.25, an operating temperature of 1200 K, and hydrogen mole fractions of 0.1 and 0.95 at the inlet and outlet, respectively, were assumed for these calculations. In the fuel-cell mode, the net heat flux is always positive and increases rapidly with operating voltage and current density. In the electrolysis mode, the net heat flux is negative for low operating voltages, increases to zero at the "thermal-neutral" voltage, and is positive at higher voltages and current densities. The thermal-neutral voltage can be predicted from direct application of the rate-based First Law to the isothermal system shown in Figure 16.4:

$$\dot{Q} - \dot{W} = \Delta\dot{N}_{H_2}\Delta H_R, \quad (16.17)$$

where, from Faraday's law,

$$\Delta\dot{N}_{H_2} = \frac{I}{2}F. \quad (16.18)$$

Letting $\dot{Q} = 0$ (no external heat transfer), $\dot{W} = IV_{tn}$, yielding

$$V_{tn} = \frac{\Delta H_R}{2F}. \quad (16.19)$$

Note that the reaction heat flux of Equation 16.16 can also be written in terms of the thermal-neutral voltage as

$$q_R'' = i(V_N - V_{\text{tn}}). \tag{16.20}$$

Since the enthalpy of reaction, ΔH_R, is strictly a function of temperature (ideal gas approximation), the thermal-neutral voltage is also strictly a function of temperature, independent of cell ASR and gas compositions. The particular values of net cell heat flux at other operating voltages do however depend on cell ASR and gas compositions. The thermal-neutral voltage increases only slightly in magnitude over the typical operating temperature range for solid-oxide cells, from 1.287 V at 800°C to 1.292 V at 1000°C. At typical solid-oxide electrolysis cell (SOEC) stack temperatures and ASR values, operation at the thermal-neutral voltage yields current densities in the 0.2–0.6 A/cm^2 range, which is very close to the current density range that has yielded successful long-term operation in solid-oxide fuel cell (SOFC) stacks.

Operation at or near the thermal-neutral voltage simplifies thermal management of the stack since no significant excess gas flow is required and component thermal stresses are minimized. In fact, in the electrolysis mode, since oxygen is being produced, there is also no theoretical need for air flow to support the reaction at all. In a large-scale electrolysis plant, the pure oxygen produced by the process could be saved as a valuable commodity. However, there are several good reasons to consider the use of a sweep gas on the oxygen side. First, the use of a sweep gas will minimize the performance degradation associated with any small leakage of hydrogen from the steam/hydrogen side to the oxygen side of the cell. Second, there are serious material issues associated with the handling of pure oxygen at elevated temperatures. Finally, the use of a sweep gas (especially one that does not contain oxygen) on the oxygen side of the electrolysis cell reduces the average mole fraction and partial pressure of oxygen, thereby reducing the open-cell and operating potentials, resulting in higher electrolysis efficiencies, as we shall see shortly.

There are some additional thermodynamic implications related to the thermal-neutral voltage. In particular, electrolyzer operation at or above the thermal-neutral voltage negates the argument that is often stated as a motivation for HTE that a fraction of the total energy requirement can be supplied in the form of heat. In fact, for isothermal operation at voltages greater than thermal neutral, heat rejection is required.

Electrolysis efficiency, η_e, can be defined for HTE, analogous to the definition of fuel cell efficiency [9]. The electrolysis efficiency quantifies the heating value of the hydrogen produced by electrolysis per unit of electrical energy consumed in the stack. Based on this definition,

$$\eta_e = \frac{\dot{N}_{H_2} \Delta H_R}{VI} \tag{16.21}$$

and since the stack electrical current is directly related to the molar production rate of hydrogen via Faraday's law, the electrolysis efficiency can be expressed strictly in terms of cell operating potentials as

$$\eta_e = \frac{\Delta H_R / 2F}{V_{\text{op}}} = \frac{V_{\text{tn}}}{V_{\text{op}}}. \tag{16.22}$$

The efficiency for the fuel-cell mode of operation is the inverse of Equation 16.22. A fuel utilization factor is often included in the fuel-cell efficiency definition, but it is not needed in the electrolysis definition since no fuel (only steam) is wasted at low utilization.

It should be noted that the value of the efficiency defined in this manner for electrolysis is greater than 1.0 for operating voltages lower than thermal neutral. As an example, for the reversible standard-state reference case, from Equation 16.12, on a rate basis,

$$\dot{W}_{\text{rev}} = \dot{N}_{\text{H}_2}\Delta G_R^o = IV^o. \tag{16.23}$$

Invoking Faraday's law, the operating cell potential for this case approaches the reference open-cell value, $V^o = \Delta G_R/2F$, yielding

$$\eta_{e,o} = \frac{\Delta H_R^o}{\Delta G_R^o}, \tag{16.24}$$

which for steam electrolysis at 850°C is equal to 1.34. For cases with variable gas composition or partial pressure, the open-cell potential is given by the Nernst Equation 16.14 and the corresponding efficiency limit varies accordingly. It is not desirable to operate an electrolysis stack near the efficiency limit, however, because the only way to approach this limit is to operate with very low current density. There is a trade-off between efficiency and hydrogen production rate in selecting an electrolysis stack operating voltage. This trade-off is illustrated in Figure 16.7. The upper curve in the figure shows the decrease in electrolysis efficiency that occurs as the per-cell operating voltage is increased above the open-cell voltage, V_N, according to Equation 16.22. Operation at the thermal-neutral voltage yields an electrolysis efficiency of 1.0. ASR represents the net effect of all the loss

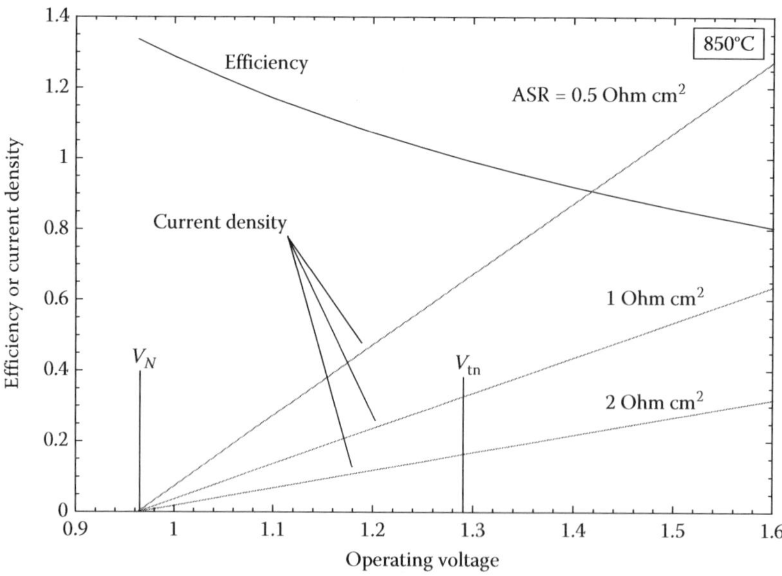

FIGURE 16.7
Effect of operating voltage and ASR on electrolysis efficiency.

mechanisms in the electrolysis stack including, ohmic losses, activation and concentration over potentials, and so on. The bottom curves show the effect of operating voltage and ASR on the current density. Noting that

$$V_{op} = \bar{V}_N + i \times \text{ASR}, \tag{16.25}$$

if a target current density (and corresponding hydrogen production rate) is selected, lower ASR values allow for stack operation at lower voltages and correspondingly higher efficiencies. Similarly, in the fuel-cell mode, there is a trade-off between efficiency and maximum power production. Maximum power production for SOFCs occurs for operation at around 0.5 V, whereas maximum efficiency occurs at the open-cell potential, around 1.1 V for hydrogen-dominated SOFC fuel cell inlet gas compositions. Depending on cell performance and optimization parameters, a good operating point usually occurs at around 0.7 V in the fuel-cell mode of operation. In the electrolysis mode, a good trade-off between efficiency and hydrogen production rate will occur at operating voltages below $\Delta H_R/2F$, around 1.1 V. The challenge is to develop electrolysis stacks with low ASR such that a reasonable current density will be achievable at lower operating voltages. Low operating voltages can also be maintained at a specified current density if the mean Nernst potential, $\bar{V}_{\text{Nernst}}$, is low. The mean Nernst potential can be reduced by increasing the cell operating temperature, increasing the steam content and flow rate in the feed stream, or by decreasing the oxygen content on the sweep-gas side (anode) of the electrolysis cell. Of course, as the cell current density and hydrogen production rate is increased, the average steam content on the cathode side decreases and the average oxygen content on the anode side increases. These considerations indicate that, for maximum cell efficiency at a specified current density, steam utilization should be kept low and a high flow rate of a non-oxygen-containing sweep gas should be used. Unfortunately, results of large-scale system analyses [29] show that operating with low steam utilization results in low overall hydrogen production system efficiencies. For the system, the thermodynamic benefit of excess steam (lower average Nernst potential) is outweighed by the penalties associated with handling of the excess steam and incomplete heat recuperation. Similar conclusions were drawn when considering the use of a non-oxygen-containing sweep gas (e.g., steam) on the oxygen side. Again, the thermodynamic benefits were outweighed by system considerations. In fact, the highest overall efficiencies for pressurized electrolyzers were achieved with no sweep gas, where the oxygen is allowed to evolve from the cells undiluted.

16.1.3 Thermal Requirements for Isothermal and Nonisothermal SOEC Operation

The analyses presented so far have all assumed isothermal electrolysis operation such that the outlet temperature of the products is the same as the inlet temperature of the reactants. For operating voltages between the open-cell potential and thermal neutral, isothermal operation requires heat addition during the electrolysis process. For operating voltages above thermal neutral, heat rejection is required to maintain isothermal operation. The enthalpy change for the electrolysis process under isothermal conditions is, by definition, the "enthalpy of reaction," ΔH_R. The enthalpy of reaction for steam reduction is a weak function of temperature, with a numerical value very close to the low heating value of hydrogen over a wide range of temperatures, as shown in Figure 16.5. The magnitude of the heat transfer required to achieve isothermal operation, $\dot{Q}_T(T)$, can be calculated directly from the following form of the first law:

$$\dot{Q}_T(T) = \Delta\dot{N}_{H_2}\Delta H_R(T) - IV_{op} \tag{16.26}$$

and since the hydrogen production rate, ΔN_{H_2}, is equal to $I/2F$, and the thermal-neutral voltage, $V_{tn} = \Delta H_R(T)/2F$,

$$Q_T(T) = I(V_{tn} - V_{op}). \tag{16.27}$$

Note that this result predicts positive heat transfer to the electrolyzer for operating voltages less than thermal neutral and negative heat transfer (i.e., heat rejection from the electrolyzer) for operating voltages greater than thermal neutral. Since there is no sensible enthalpy change, this result is valid for all isothermal cases, even if excess reactants and/or inert gases are present. A graphical interpretation of the isothermal heat requirement on $V - i$ coordinates is shown in Figure 16.8. The figure shows the heat fluxes required to maintain isothermal operation for a target current density of 0.3 A/cm² for two values of ASR: 0.5 and 1.5 Ω cm² represented by the area enclosed between the vertical line at $V = V_{tn}$, the vertical line $V = V_{op}$ ($V_{op} = 1.113$ V for ASR = 0.5 Ω cm²; $V_{op} = 1.413$ V for ASR = 1.5 Ω cm²), and the horizontal lines at $i = 0$ and at $i = 0.3$ A/cm². Note that the higher ASR case requires an operating voltage that is above V_{tn} in order to achieve the target current density of 0.3 A/cm². Consequently, the associated isothermal heat transfer requirement is negative, indicating that heat rejection is needed to maintain isothermal operation at that condition.

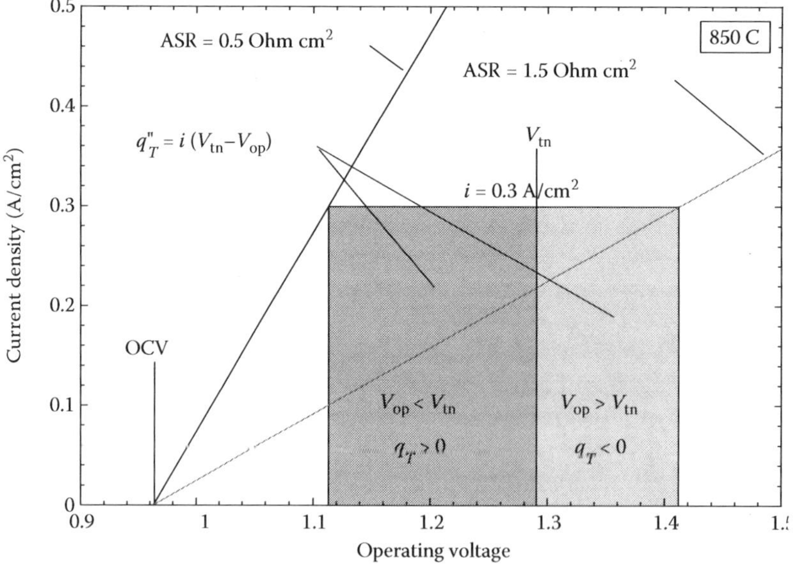

FIGURE 16.8
Graphical interpretation of isothermal heat requirements for two values of ASR.

Equation 16.27 can also be used to show that the maximum isothermal heat addition operating point corresponds to an operating voltage equal to the average of the open-cell potential and the thermal-neutral voltage. Accordingly, the maximum isothermal heat addition is given by

$$\dot{Q}_{\max}(T) = I\left(\frac{V_{\text{tn}} - V_N}{2}\right), \tag{16.28}$$

where V_N is the open-cell potential. The total stack current, I, at any operating voltage is dependent on the stack ASR value, which is typically temperature dependent.

Actual HTE processes will generally not operate isothermally unless the operating voltage is very close to the thermal-neutral voltage. For nonisothermal cases, the first law for electrolysis process must be written as

$$\dot{Q} - \dot{W} = \sum_P \dot{N}_i[\Delta H^o_{fi} + H_i(T_P) - H^o_i] - \sum_R \dot{N}_i[\Delta H^o_{fi} + H_i(T_R) - H^o_i]. \tag{16.29}$$

In this form, all reacting and nonreacting species included in the inlet and outlet streams can be accounted for, including inert gases, inlet hydrogen (introduced to maintain reducing conditions on the steam/hydrogen electrode), and any excess unreacted steam. In general, determination of the outlet temperature from Equation 16.29 is an iterative process [10]. The heat transferred during the process must first be specified (e.g., zero for the adiabatic case). The temperature-dependent enthalpy values of all species must be taken into account. The solution procedure begins with specification of the cathode-side inlet flow rates of steam, hydrogen, and any inert carrier gas such as nitrogen (if applicable). The inlet flow rate of the sweep gas (e.g., air or steam) on the anode side must also be specified. Specification of these flow rates allows for the determination of the inlet mole fractions of steam, hydrogen, and oxygen that appear in the Nernst equation. The steam mole fraction is expressed in terms of the hydrogen mole fraction as $\Delta\dot{N}_{H_2}$. The desired current density and active cell area are then specified, yielding the total operating current. The corresponding hydrogen production rate is obtained from Faraday's law.

Once the per-cell hydrogen production rate is known, the outlet flow rates of hydrogen and steam on the cathode side and oxygen on the anode side can be determined. The flow rates of any inert gases, the anode-side sweep gas, and any excess steam or hydrogen are the same at the inlet and the outlet. Once all these flow rates are known, the summations in Equation 16.29 can be evaluated. The product summation must be evaluated initially at a guessed value of the product temperature, T_P.

The operating voltage corresponding to the specified current density is obtained from Equation 16.25, where the stack ASR must be estimated and specified as a function of temperature. To account for the variation in temperature and composition across an operating cell, the mean Nernst potential can be obtained from an integrated version of the Nernst equation:

$$\bar{V}_N = \frac{1}{2F(T_P - T_R)(y_{o,O_2,A} - y_{i,O_2,A})(y_{o,H_2,C} - y_{i,H_2,C})} \times \int_{T_R}^{T_P} \int_{y_{i,O2,A}}^{y_{o,O2,A}} \int_{y_{i,H2,C}}^{y_{o,H2,C}} \Delta G_R(T) + R_u T \ln\left(\frac{1 - y_{H_2} - y_{N_2}}{y_{H_2} y_{O_2}^{1/2}}\right) dy_{H_2} dy_{O_2} dT, \tag{16.30}$$

where $y_{i,O2,A}$ is the anode-side inlet mole fraction of oxygen, and so on. Note that the upper limit of integration on the temperature integral is initially unknown. Once the ASR and the mean Nernst potential are known, the operating voltage is obtained from Equation 16.25 and the electrical work term in Equation 16.29 is obtained from $\dot{W} = -V_{op}I$. An algorithm then must be developed to iteratively solve for the product temperature, T_P, in order to satisfy Equation 16.29.

The procedure described above was formulated as an integral electrolyzer model [10]. The results of sample parametric calculations based on this procedure are presented in Figure 16.9. The inlet mass flow rates of steam–hydrogen and sweep air per cm^2 of active cell area are indicated in the caption. The calculations were performed for an inlet hydrogen mole fraction of 0.1 and an inlet temperature of 800°C (1073 K). Figure 16.9a shows the heat flux required to maintain isothermal operation as a function of per-cell operating voltage for three different ASR values. This heat flux is positive (heat addition required) for voltages between open-cell and thermal neutral and negative for higher operating voltages. The peak heat flux requirement occurs halfway between the open-cell potential and the thermal-neutral voltage. The magnitude of the peak heat flux is highest for the lowest ASR value since the current density (and hydrogen production rate) corresponding to each voltage value is highest for the lowest ASR value. Figure 16.9b shows the mean outlet gas temperature as a function of per-cell operating voltage for adiabatic operation for three different ASR values. For adiabatic conditions, outlet temperatures are lower than inlet temperatures for voltages between open-cell and thermal neutral. For higher voltages, outlet temperatures increase rapidly with voltage. Again, the low-ASR case exhibits the largest effect due to its higher current density at each operating voltage.

Actual electrolyzers will generally operate at conditions that are neither isothermal nor adiabatic. These two cases represent limits. For optimal performance, isothermal operation at an operating voltage below thermal neutral is desirable. In this case, some of the electrolysis energy is indeed supplied in the form of heat. One way to supply the required heat directly to the stack is through the use of a heated sweep gas. This strategy is just the opposite of the situation encountered in the fuel cell mode in which excess air is used for cell cooling.

16.2 SOEC Materials

The SOEC is a solid-state electrochemical device consisting of an oxygen-ion-conducting electrolyte (e.g., yttria- or scandia-stabilized zirconia) with porous electrically conducting electrodes deposited on either side of the electrolyte. The standard electrolyte material is formed by doping zirconia (ZrO_2) with 8 Molar percent of yttria (yttrium oxide, Y_2O_3). A dopant composition of 8% or higher yields a "fully stabilized" electrolyte. The dopant serves two purposes. It "stabilizes" the cubic (or fluorite) crystal structure over a wide temperature range. Undoped zirconia exhibits a monoclinic crystal structure at room temperature and a tetragonal phase above 1170°C. Zirconia doped with yttria is called yttria-stabilized zirconia (YSZ). In addition to stabilizing the crystal structure, when trivalent Y is substituted for tetravalent Zr, holes (unfilled positions) in the oxygen sublattice are introduced at the same time. This makes it possible for oxygen ions to move through the solid by hopping from hole to hole in the lattice. YSZ is therefore, a good oxygen ion conductor. Other compounds such as Scandia (Sc_2O_3) can also be used as the dopant. Scandia-stabilized

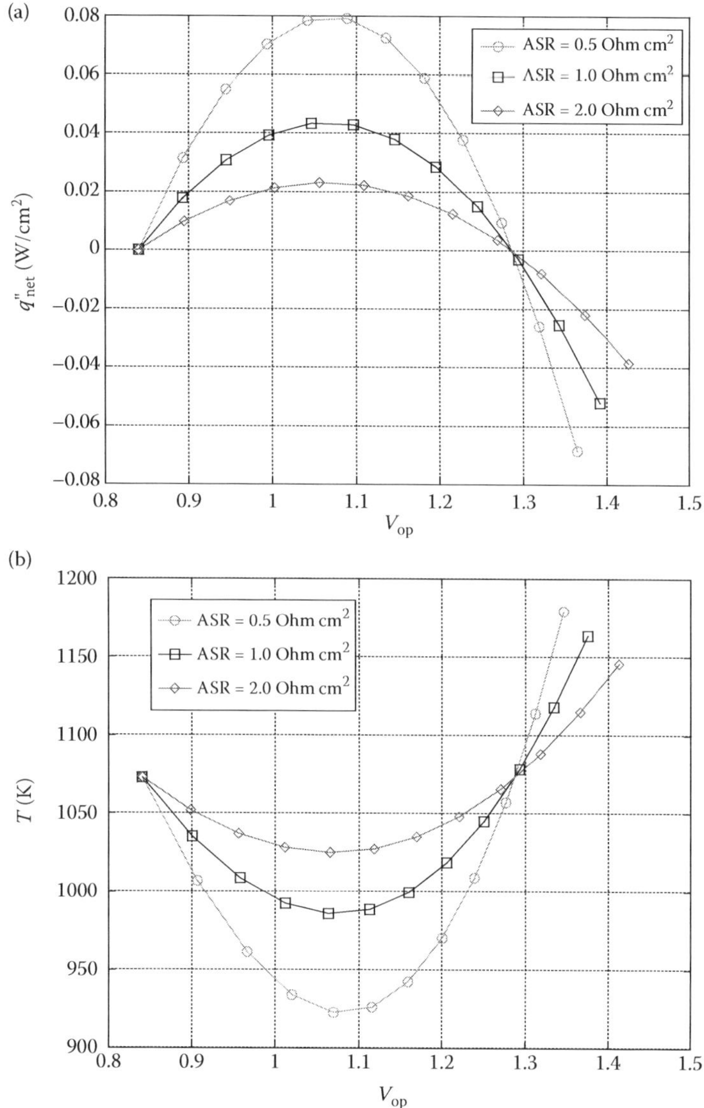

FIGURE 16.9
(a) Heat flux required for isothermal operation; (b) Outlet temperature for adiabatic operation; steam–hydrogen inlet flow rate: 0.0085 g/min/cm², $y_{H_2i} = 0.1$, sweep air inlet flow rate: 0.00561 g/min/cm², and $T_{in} = 1073$ K.

zirconia (ScSZ) has a significantly higher ionic conductivity than YSZ, but is more expensive. Other potential electrolyte materials include samaria-doped ceria and calcia-doped lanthanum gallate.

The most common steam–hydrogen electrode material is porous nickel-zirconia (YSZ) cermet. A nickel-ceria cermet can also be used. In the electrolysis mode, this electrode serves as the cathode. Because the cathode contains nickel metal, reducing conditions must be maintained on this electrode during cell operation. This is typically accomplished by including ~10% or higher mole fraction hydrogen in the inlet flow. The nickel in the cathode acts as an electronic conductor and as a catalyst for steam reduction. The zirconia in the cermet provides ionic conductivity. Porosity allows the steam to migrate to the active electrochemical reaction sites and hydrogen to migrate away from the sites. The active reaction sites correspond to what is typically termed the triple-phase boundary (TPB) where the electronic, ionic, and gas phases coexist. These sites occur at locations where a pore structure intersects with nickel and zirconia particles.

Several materials have been studied for the air-oxygen electrode. This electrode must operate in a highly oxidizing environment. The most common material used is strontium-doped lanthanum manganite, $La_{0.8}Sr_{0.2}MnO_3$, or LSM. This material provides good electronic conductivity and good catalytic activity and tolerance to the oxidizing environment. LSM is an example of a class of materials called perovskites, which have the general chemical formula ABO_3, where "A" and "B" are two cations of very different sizes (A much larger than B), and O is the anion that bonds with both. These perovskites exhibit p-type electrical conductivity that is enhanced by the introduction of lower-valence dopant cations such as Sr^{2+} to replace La^{3+} cations. Strontium and cobalt-doped lanthanum ferrites (LSF, lanthanum strontium cobalt ferrite (LSCF)) have also received lots of attention recently. These materials are more catalytically active than LSM and therefore, yield generally better performance, especially at temperatures below 800°C.

16.3 Electrolysis Cell Designs and Stack Configurations

Several basic cell designs have been developed for SOFC applications including electrolyte-supported, electrode-supported, and porous ceramic or metal substrate-supported cells. A full discussion of these various cell designs and the various fabrication techniques is beyond the scope of this chapter. Common cell characteristics include a dense gas-tight electrolyte layer, with porous electrodes on either side. In an electrolyte-supported cell, the electrolyte layer is thicker than either of the electrodes and must have sufficient mechanical strength to withstand any stresses. However, as a result of the relatively thick electrolyte, ionic resistance across the electrolyte is large for this design. The best performing SOFC cells at present are the anode-supported cells in which the mechanical strength is provided by a thick (~1.5 mm) layer of anode (usually nickel-YSZ cermet) material (e.g., [11]). Thin electrolyte and cathode layers are deposited on the anode material by screen printing or other techniques. This design has exhibited very high performance in SOFC tests. Some researchers have suggested that the best performance for the electrolysis mode of operation could be obtained using air-side (e.g., LSM) electrode-supported cells [12]. In the SOEC mode, air-electrode-supported cells should exhibit a lower concentration polarization than steam–hydrogen-electrode supported cells, due to the direction of steam and oxygen diffusion in the two modes. A wealth of information on materials, configurations, and designs of solid-oxide electrochemical systems is available in [13].

The highest energy density configuration for a HTE stack is the planar geometry. A cross-section of a planar stack design is shown in Figure 16.10. The design depicted in the figure shows an electrolyte-supported cell with a nickel cermet cathode and a perovskite anode such as strontium-doped LSM. The flow fields conduct electrical current through the stack and provide flow passages for the process gas streams. The separator plate or bipolar plate separates the process gas streams. It must also be electrically conducting and is usually metallic, such as a ferritic stainless steel. The electrochemical half-cell reactions

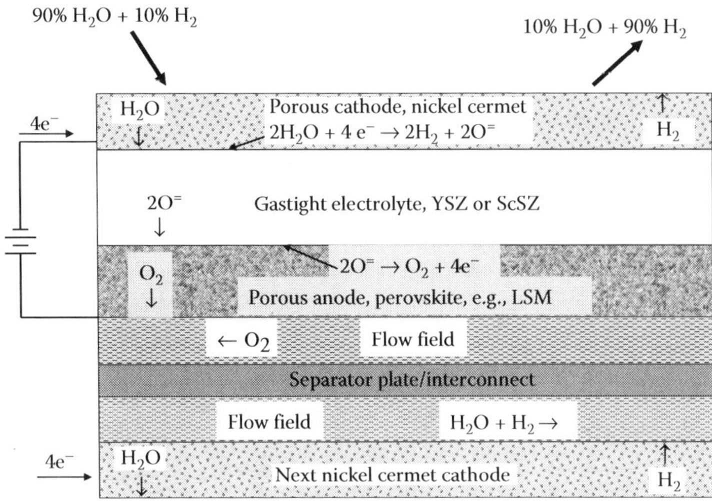

FIGURE 16.10
Cross-section of a planar HTE stack.

are shown in the figure. At the steam-hydrogen electrode–electrolyte interface (cathode in the electrolysis mode), the steam is electrochemically reduced according to

$$2H_2O + 4e^- \rightarrow 2H_2 + 2O^{2-} \tag{16.31}$$

This net reaction can be further subdivided as

$$2H_2O \rightarrow 4H^+ + 2O^{2-} \tag{16.32}$$

$$4H^+ + 4e^- \rightarrow 2H_2 \tag{16.33}$$

At the oxygen electrode–electrolyte interface (anode in the electrolysis mode), oxygen is produced according to

$$2O^{2-} \rightarrow O_2 + 4e^- \tag{16.34}$$

The electrons are driven from the anode to the cathode by means of an external power source. The electrolyte conducts the oxygen ions from the cathode to the anode where they liberate their extra electrons to the external circuit. These electrochemical reactions occur at the TPB.

As shown in the figure, a mixture of steam and hydrogen at 750–950°C is supplied to the cathode side of the electrolyte (note that cathode and anode sides are opposite to their fuel-cell-mode roles). The half-cell electrochemical reactions occur at the TPB near the electrode–electrolyte interface, as shown in the figure. Oxygen ions are drawn through the electrolyte by an applied electrochemical potential. The ions liberate their electrons and recombine to form molecular O_2 on the anode side. The inlet steam–hydrogen mixture composition may be as much as 90% steam, with the remainder hydrogen. Hydrogen is included in the inlet stream in order to maintain reducing conditions at the cathode. This is necessary in order to avoid oxidation of the elemental nickel in the electrode. The exiting mixture may be as much as 90% H_2. It is not desirable to attempt to operate the stack with a higher outlet composition of hydrogen because there is a risk of local steam starvation which can lead to significantly reduced cell performance. Furthermore, based on detailed analyses of large-scale HTE systems, the overall hydrogen production efficiency is almost flat above about 50% steam utilization [14]. Product hydrogen and residual steam is passed through a condenser or membrane separator to purify the hydrogen. The condensate can then be recycled through the system again.

A research program on HTE at the Idaho National Laboratory (INL) has demonstrated the feasibility of HTE for efficient hydrogen production from steam. The majority of the electrolysis stack testing that has been performed at INL to date has been with planar stacks fabricated by Ceramatec, Inc. of Salt Lake City, UT. A magnified view of the internal components of one of these stacks is shown in Figure 16.11. The cells have an active area of 64 cm^2. The stacks are designed to operate in cross flow, with the steam–hydrogen gas mixture flowing from front to back in the figure and air flowing from right to left. Air flow enters at the rear though an air inlet manifold and exits at the front directly into the furnace. The power lead attachment tabs, integral with the upper and lower interconnect plates are also visible in Figure 16.11. Stack operating voltages were measured using wires that were directly spot-welded onto these tabs. The interconnect plates are fabricated from ferritic stainless steel. Each interconnect includes an impermeable separator plate (~0.46 mm thick) with edge rails and two corrugated "flow fields," one on the air side and one on the steam/hydrogen side. The height of the flow channel formed by the edge rails and flow fields is 1.0 mm. Each flow field includes 32 perforated flow channels across its width to provide uniform gas-flow distribution. The steam/ hydrogen flow fields are

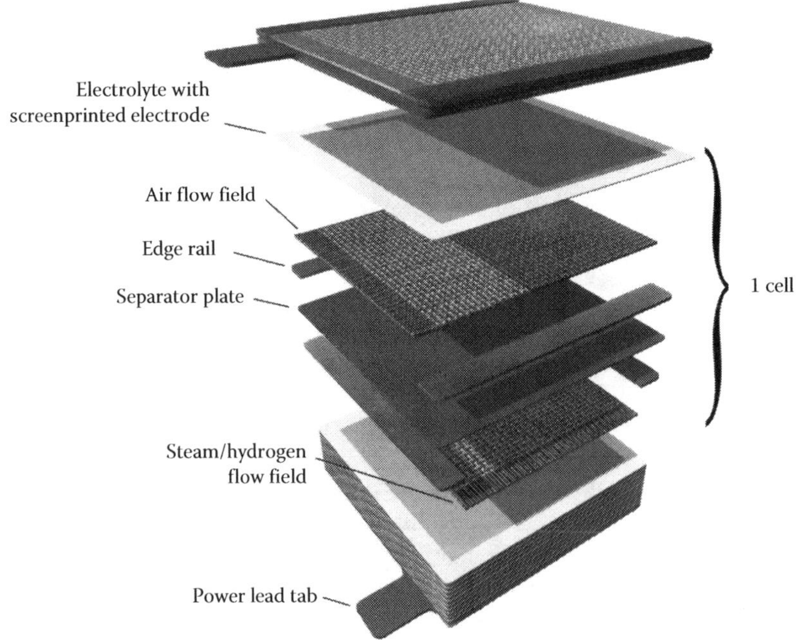

FIGURE 16.11
Magnified view of electrolysis stack.

fabricated from nickel foil. The air-side flow fields are ferritic stainless steel. The interconnect plates and flow fields also serve as electrical conductors and current distributors. To improve performance, the air-side separator plates and flow fields are pre-surface-treated to form a rare-earth stable conductive oxide scale. A perovskite rare-earth coating is also applied as a bond layer to the separator-plate oxide scale by either screen printing or plasma spraying. On the steam/hydrogen side of the separator plate, a thin (~10 μm) nickel metal coating is applied as a bond layer.

The stack electrolytes are ScSZ, about 140 μm thick. The air-side electrodes (anode in the electrolysis mode) are strontium-doped manganite. The electrodes are graded, with an inner layer of manganite/zirconia (~13 μm) immediately adjacent to the electrolyte, a middle layer of pure manganite (~18 μm), and an outer bond layer of cobaltite. The steam/hydrogen electrodes (cathode in the electrolysis mode) are also graded, with a nickel-zirconia cermet layer (~13 μm) immediately adjacent to the electrolyte and a pure nickel outer layer (~10 μm).

Planar stacks can also be assembled using electrode-supported cells. Advanced technology SOFC stacks based on anode-supported cell technology have been developed by several manufacturers under the Solid-State Energy Conversion Alliance (SECA) [15] program. For example, Versa Power Systems has developed anode-supported planar cells with dimensions as large as 33 × 33 cm [16]. Their stacks are internally manifolded, as shown in Figure 16.12.

Additional cell and stack configurations are under development for SOFC applications, including tubular and integrated planar designs. INL participated in a research agreement with Rolls Royce fuel cell systems (RRFCS) to evaluate the performance of their cells operating in the electrolysis mode. The RRFCS cells utilize a segmented-in-series integrated planar (IP) SOFC design [17] in which thin electrode and electrolyte layers are screen printed on the surface of a flattened ceramic tube, as shown in Figure 16.13. Figure 16.13a

FIGURE 16.12
Planar stack with anode-supported cells, developed by Versa Power.

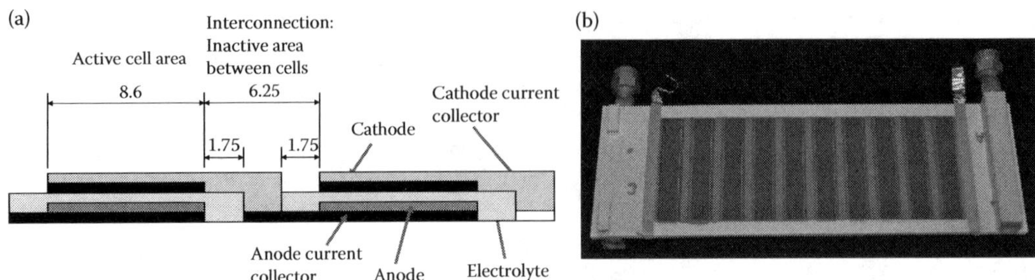

FIGURE 16.13
Rolls Royce integrated planar cells: (a) the schematic of cells layers and (b) the photograph of IP-SOFC tube.

is a schematic representation of the electrochemical layers. Figure 16.13b is a photograph of one of an IP-SOFC tube that was tested at INL in the SOEC mode.

16.4 HTE Experiments

16.4.1 Small-Scale Tests

The results of initial single (button) cell HTE tests completed at the INL were documented in detail in Ref. [18]. Button cell tests are useful for basic performance characterization of electrode and electrolyte materials and of different cell designs (e.g., electrode-supported, integrated planar, tubular). Polarization curves for several representative DC potential sweeps are presented in Figure 16.14a. Both the applied cell potentials and the corresponding power densities are plotted in the figure as a function of cell current density. Positive current densities indicate fuel cell mode of operation and negative current densities indicate electrolysis mode. Cell potential values at zero current density correspond to open-circuit potentials, which depend on the furnace temperature and the gas composition. The three sweeps acquired at 800°C (sweeps 1, 3, and 5) have a steeper $E - i$ slope, due to the lower zirconia ionic conductivity at the lower temperature. The continuous nature of the $E - i$ curves across the zero-current-density (open-circuit) point provides no indication of significant activation overpotential for these electrolyte-supported cells. In the electrolysis mode, the voltage data vary linearly with current density up to a value that depends on the inlet steam flow rate, which for a fixed dry-gas flow rate depends on the inlet dew point temperature. For low inlet dew point values (sweeps 1 and 2), the voltage begins to increase rapidly at relatively low values of current density (~ −0.15 A/cm^2), due to steam starvation. For higher inlet dew points, the steam starvation effect is forestalled to higher current densities. The single-cell results demonstrated the feasibility of HTE for hydrogen production linear operation from the fuel cell to the electrolysis mode.

Results of initial short-stack HTE tests performed at INL are provided in Refs. [19,20]. A good summary of our experience is provided by the results plotted in Figure 16.14b, from [20]. Results of several representative sweeps are shown in the form of polarization curves, representing per-cell operating voltage versus current density. Test conditions for each of the seven sweeps are tabulated in the figure. Five of the sweeps were obtained from a 10-cell stack (sweeps 10-1 through 10-5) and two were obtained from a 25-cell stack (25-1

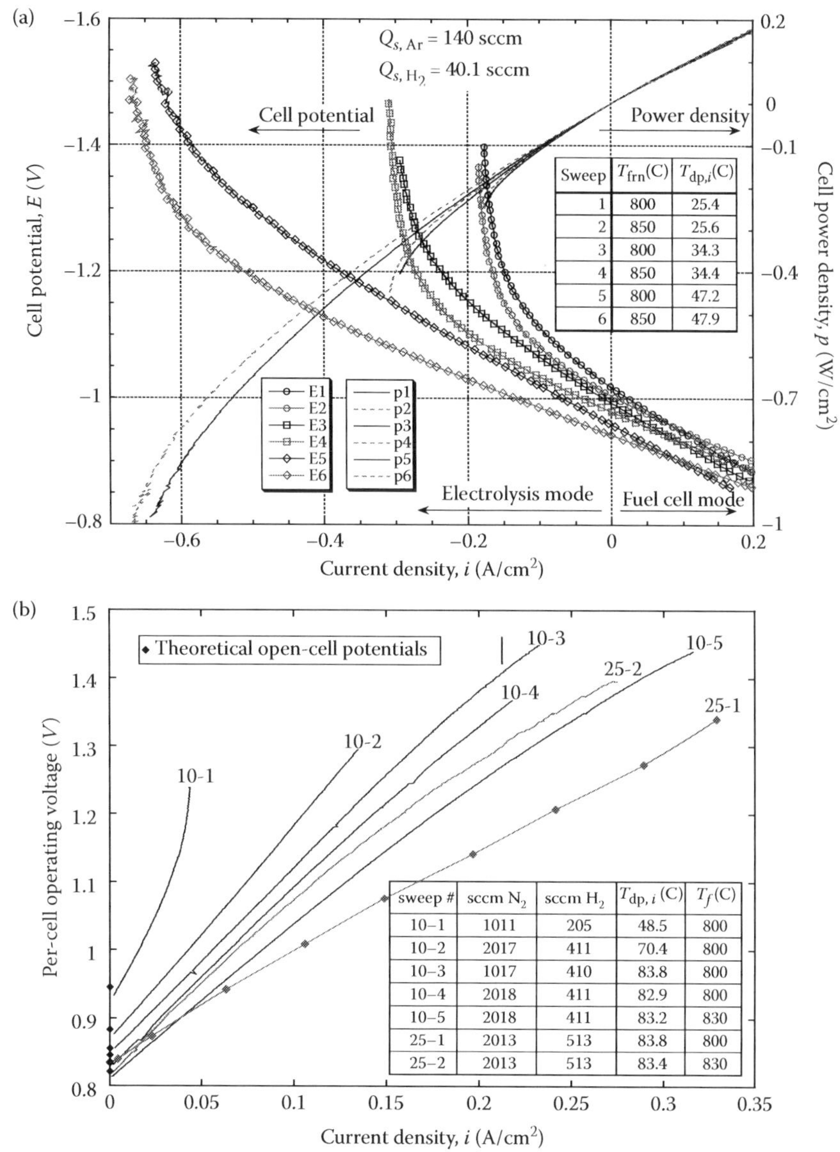

FIGURE 16.14
Polarization curves: (a) button cell and (b) planar stack.

and 25-2). Theoretical open-cell potential values are shown in the figure for each sweep using a single data point at zero current density. Note that the measured open-cell potentials are in excellent agreement with the theoretical values for each sweep. Sweep 10-1 was performed with a relatively low inlet steam flow rate, corresponding to the low inlet dew point value of 48.5°C and relatively low nitrogen and hydrogen flow rates. This sweep has a relatively high slope on $i-V$ coordinates, indicating a relatively high ASR. This sweep also clearly shows the effects of steam starvation; the slope of the $i-V$ curve increases dramatically as the current density is increased. The outlet dew point temperature corresponding to the highest current density shown in this figure was only 4°C for this sweep.

Sweep 10-2 was performed at an intermediate steam concentration, with an inlet dew-point temperature of 70°C. This sweep exhibits nearly linear behavior over the range of current densities shown, with a much smaller slope than sweep 10-1. Sweeps 10-3 and 10-4 are nearly linear at low current densities, then slightly concave-down at higher current densities. Sweep 10-5 has a shallower slope than the others, consistent with the higher operating temperature of 830°C. Sweep 25-1 was performed in a stepwise fashion, rather than as a continuous sweep. This was done in order to ensure sufficient time for the internal stack temperatures to achieve steady-state values at each operating voltage. Note that the slope of this sweep is small, indicating low ASR (~1.5 Ω cm^2). This sweep was performed at the beginning of a 1000-h long-duration 25-cell stack test. Sweep 25-2 was acquired at the end of the long-duration test. The stack operating temperature was increased from 800 to 830°C part way through the test. Note that the slope of sweep 25-2 is higher than that of sweep 25-1, despite the higher temperature, due to performance degradation over 1000 h of operation.

16.4.2 Large-Scale Demonstrations

One of the objectives of the INL HTE program is technology scale-up and demonstration. To this end, the INL has developed a 15 kW HTE test facility, termed the Integrated Laboratory Scale (ILS) HTE test facility. Details of the design and initial operation of this facility are documented in Refs. [21–23]. A condensed description of the facility will be provided here. The ILS includes three electrolysis modules, each consisting of four stacks of 60 cells, yielding 240 cells per module and 720 cells total. The cells are similar to those discussed earlier. Each electrolysis module utilizes an independent support system supplying electrical power for electrolysis, a feedstock gas mixture of hydrogen and steam (and sometimes nitrogen), a sweep gas, and appropriate exhaust handling. Each module includes a controlled inlet flow of deionized water, a steam generator, a controlled inlet flow of hydrogen, a superheater, inlet and outlet dew-point measurement stations, a condenser for residual steam, and a hydrogen vent. All three modules were located within a single hot zone. Heat recuperation and hydrogen product recycle were also incorporated into the facility. An overview photograph of the ILS is provided in Figure 16.15.

FIGURE 16.15
INL 15 kW ILS HTE test facility.

A magnified view of one of the ILS module assemblies including the recuperative heat exchanger, base manifold unit, and four-stack electrolysis unit is presented in Figure 16.16. For each four-stack electrolysis module, there were two heat exchangers and one base manifold unit. Each base manifold unit has nine flow tubes entering or exiting at its top and only four flow tubes entering or exiting at the bottom of the unit and at the bottom of the heat exchangers, thereby reducing the number of tube penetrations passing through the hot zone baseplate from nine to just four. This feature also reduces the thermal load on the hot zone baseplate. An internally manifolded plate-fin design concept was selected for this heat recuperator application. This design provides excellent configuration flexibility in terms of selecting the number of flow elements per pass and the total number of passes in order to satisfy the heat transfer and pressure drop requirements. Theoretical counterflow heat exchanger performance can be approached with this design. This design can also accommodate multiple fluids in a single unit. More details of the design of the recuperative heat exchangers are provided in [24].

Figure 16.17 shows a cut-away design rendering of the three ILS electrolysis modules with their base manifolds and heat exchangers beneath. This illustration also shows the instrumentation wires for intermediate voltage and temperature readings. Each module is instrumented with 12 1/16 in. sheathed thermocouples for monitoring gas temperatures in the electrolysis module manifolds and in the base manifold. These thermocouples are attached to the manifolds using compression fittings. There are also 12 miniature 0.020 in. diameter inconel-sheathed type-K thermocouples per module that are used for monitoring internal stack temperatures. Access to the internal region of the stacks is provided via the

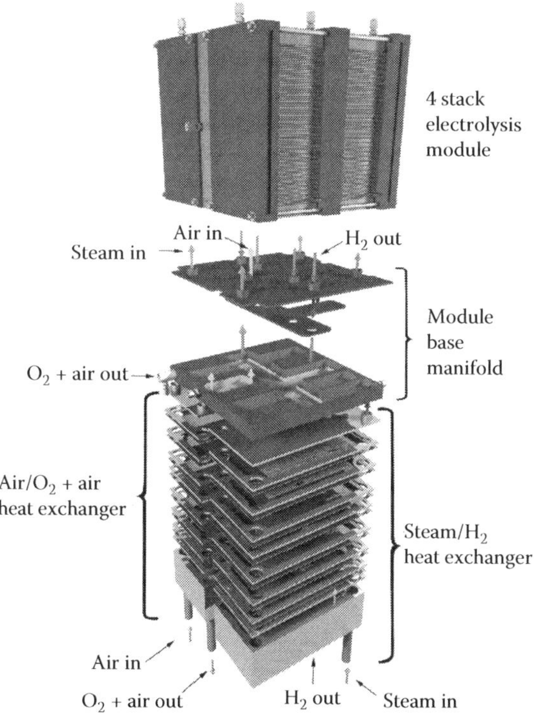

FIGURE 16.16
Magnified view of a heat exchanger, a base manifold unit, and a four-stack electrolysis unit.

FIGURE 16.17
ILS modules, mounted in the hot zone.

air outlet face. The internal thermocouples are inserted into the small exit air flow channels. Similarly, seven intermediate voltage tap wires per module are inserted into the air flow channels of the four stacks.

Two compression bars are shown across the top of each module in Figure 16.17. These bars are used to maintain compression on all of the stacks during operation in order to minimize electrical contact resistance between the cells, flow fields, and interconnects. The bars are held in compression via spring-loaded tie-downs located outside of the hot zone under the baseplate.

Note that the heat exchangers are partially imbedded in the insulation thickness. The top portion of each heat exchanger is exposed to the hot zone radiant environment, which helps to insure that the inlet gas streams achieve the desired electrolyzer operating temperature prior to entering the stacks. The temperature at the bottom of each heat exchanger will be close to the inlet stream temperature, minimizing the thermal load on the hot zone baseplate in the vicinity of the tubing penetrations. A photograph of the three ILS electrolysis modules installed in the hot zone is shown in Figure 16.18.

Performance degradation with the ILS system is documented in Figure 16.19. Over a period of 700 h of test time, module-average ASR values increased by about a factor of 5, from an initial value near 1.5 Ohm cm^2. Some of the observed degradation was related to balance-of-plant issues. For example, prior to about 480 h of operation, unanticipated condensation occurred in the hydrogen recycle system which led to erratic control of the hydrogen recycle flow rate due to the intermittent presence of liquid water in the mass flow controllers. This problem resulted in time periods during which there may have been no hydrogen flow to the ILS stacks, leading to accelerated performance degradation associated with oxidation of the nickel cermet electrodes. Despite the problems with the ILS, the test successfully demonstrated large-scale hydrogen production with heat recuperation and hydrogen recycle, as would be required in a large-scale plant. A plot of the time history of ILS hydrogen production is given in Figure 16.19. Peak electrolysis power consumption

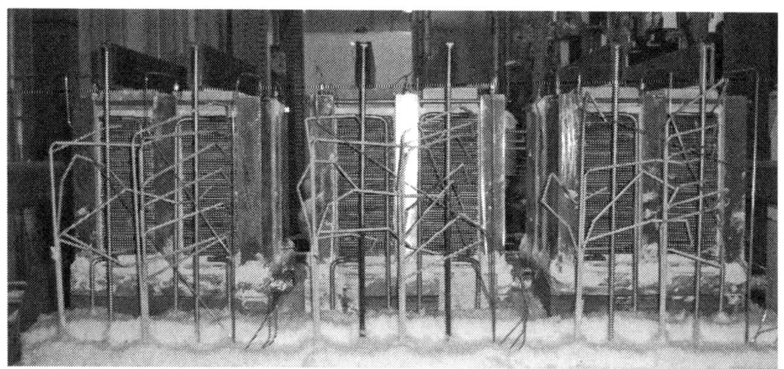

FIGURE 16.18
HTE ILS hot zone with three modules installed.

and hydrogen production rate were 18 kW and 5.7 Nm^3/h, respectively, achieved at about 17 h of elapsed test time.

16.5 Computational Analysis of SOECs

Significant effort has been spent in recent years in the development of numerical modeling methods for detailed thermal and electrochemical analysis of SOFCs. Relatively, little work has been completed for SOEC modeling. Fluent Inc. was funded by the U.S. Department of Energy National Energy Technology Laboratory (DOE-NETL) to develop an SOFC module

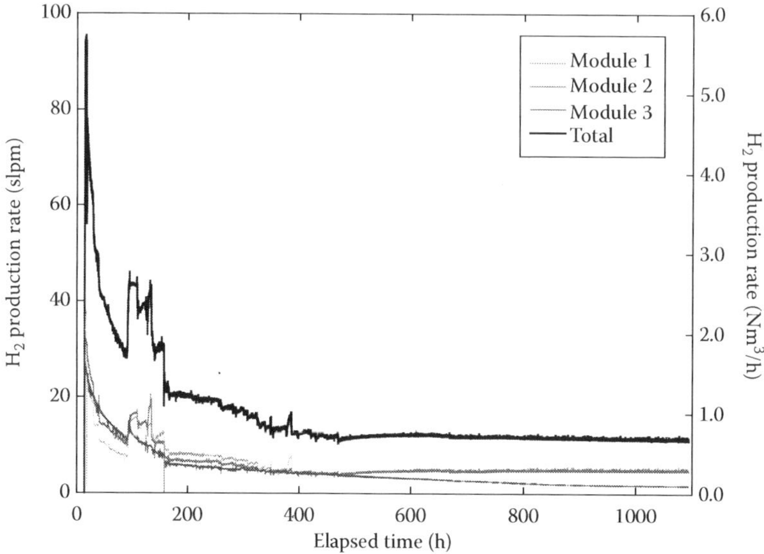

FIGURE 16.19
Time history of ILS hydrogen production rate.

for coupling to the core mass, momentum, energy, and species conservation and transport features of the FLUENT computational fluid dynamics code [25]. The SOFC module adds the electrochemical reactions, loss mechanisms, and computation of the electric field throughout the cell. Under a cooperative effort between INL and Fluent Inc., the FLUENT SOFC user-defined subroutine was modified to allow for operation in the SOEC mode [26]. Model results provide detailed profiles of temperature, Nernst potential, operating potential, anode-side gas composition, cathode-side gas composition, current density, and hydrogen production over a range of stack operating conditions. Results of the numerical model have been compared with experimental results obtained from a ten-cell planar stack.

16.5.1 Numerical Model of a Planar SOEC Stack

Complete details of the FLUENT electrolysis stack model developed at INL are provided in [26]. A condensed description is presented here. The numerical model developed for this study was based on the geometry of a single SOEC taken from a planar stack similar to the stack described in detail in [19]. The numerical domain extends from the center plane of one separator plate to the center plane of the next separator plate. Symmetry boundaries are applied at the top and bottom of the model. Three representations of the numerical model are presented in Figure 16.20a and b. In the top left portion of this figure, the full model is shown to scale. Since the model includes only one cell, the model geometry is quite thin in the vertical (z) direction. To show more details, the model is shown in the bottom left portion of Figure 16.20 with a vertical exaggeration of 10× in the z-direction. A magnified view with the 10× vertical exaggeration is shown in Figure 16.20b.

In the magnified view, the element at the bottom is the bottom separator plate. Since we are trying to represent a unit cell extracted from a larger stack, the bottom and top separator

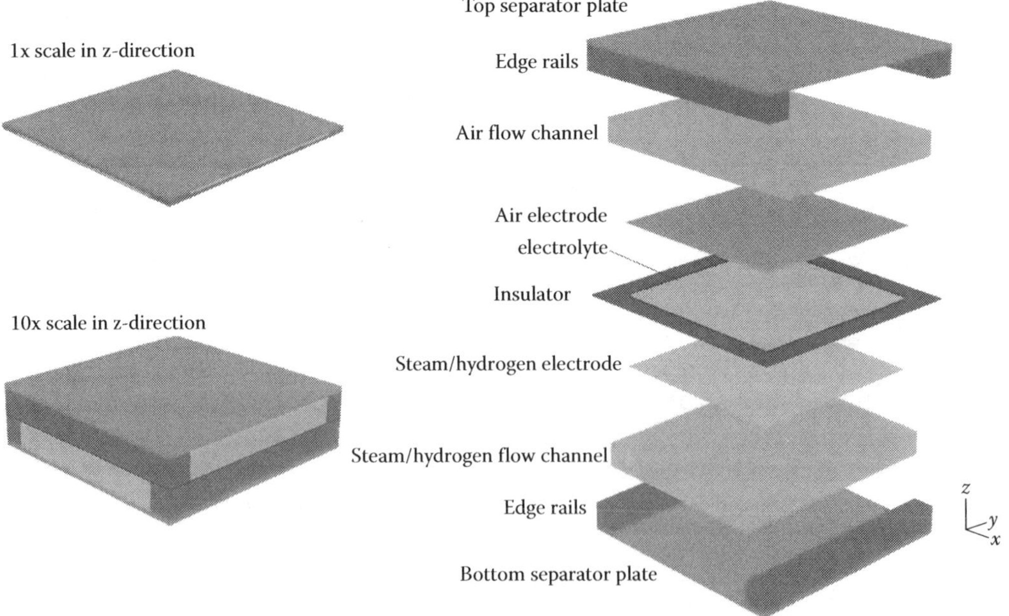

FIGURE 16.20
Fluent single-cell SOEC model.

plates in the numerical model are only half as thick (i.e., 0.19 mm) as the hardware separator plates. Therefore, the top and bottom boundaries of the numerical model represent symmetry planes and the boundary conditions on those faces are set accordingly. The edge rails are shown attached to the bottom separator plate. In the stack hardware, the edge rails are fabricated from the same material as the separator plates, but they are separate pieces.

The next element in the numerical model is the steam/hydrogen flow channel. The flow channels are the regions in the stack between the separator plate, the edge rails, and the electrodes in which the corrugated/perforated "flow fields" are located. In the FLUENT model, the steam/hydrogen flow channel has been specified as a high-porosity porous-media region with metallic nickel as the solid material and with anisotropic permeability, much higher in the primary flow direction than in the cross flow directions. The height of the flow channel is set by the thickness of the edge rails, 1.019 mm.

The next three layers in the numerical model are associated with the electrolyte/electrode assembly, as shown in Figure 16.20b. The FLUENT SOFC module treats the electrolyte as a 2D planar element with the properties of YSZ. Therefore, the electrolyte in the model has geometrical thickness of zero. On the either side of the electrolyte are the electrodes which are created with 3D elements. Therefore, the electrolyte/electrode assembly in the model is only as thick as the two electrodes. Around the outer periphery of the electrolyte/electrode assembly, we have included an "insulator" with the properties of YSZ. The insulator prevents an electrical short-circuit between the top and bottom edge rails. No ionic transport occurs through this insulator.

The next element in the numerical model is the air/oxygen flow channel. It has also been specified as a high-porosity media region with ferritic stainless steel as the solid material and with the same anisotropic permeabilities and flow channel height used in the steam/hydrogen flow channel. The top separator plate and edge rails are identical to those on the bottom, but the edge rails are oriented perpendicular to the bottom edge rails to allow for the cross-flow arrangement. The bottom separator plate in the FLUENT model serves as the electrical ground and the top separator plate serves as the current source.

Additional parameters specified in the numerical model include the electrode exchange current densities and several gap electrical contact resistances. These quantities were determined empirically by comparing FLUENT predictions with stack performance data. The FLUENT model uses the electrode exchange current densities to quantify the magnitude of the activation overpotentials via a Butler–Volmer equation [25].

The gas flow inlets are specified in the FLUENT model as mass-flow inlets, with the gas inlet temperatures are set at 1103 K and the inlet gas composition determined by specification of the mass fraction of each component. The gas flow rates used in the model were the same as those used for the experimental base case, on a per-cell basis. For example, the base case for the steam/hydrogen inlet used a total inlet mass flow rate of 8.053×10^{-6} kg/s, with nitrogen, hydrogen, and steam mass fractions of 0.51, 0.0074, and 0.483, respectively. The base case air flow rate was 4.33×10^{-6} kg/s.

Details of the core mass, momentum, energy, and species conservation and transport features of FLUENT are documented in detail in the FLUENT theory manual [27]. An SOFC model adds the electrochemical reactions, loss mechanisms, electric field computation, and electrode porous media constitutive relations [25]. This reference also documents the treatment of species and energy sources and sinks arising from the electrochemistry at the electrode–electrolyte interfaces. The FLUENT SOFC user-defined subroutine was modified for our HTE work to allow for operation in the SOEC mode. Model results provide detailed profiles of temperature, Nernst potential, operating potential, anode-side gas composition, cathode-side gas composition, current density, and hydrogen production over a range of stack operating conditions.

16.5.2 Representative Computational Results

Representative results obtained from the integral electrolyzer model for an adiabatic case are presented in Figure 16.21, along with results obtained from FLUENT. Figure 16.21 shows predicted voltage–current characteristics and predicted gas outlet temperatures. Predictions obtained from the integral model described earlier in this report (Section 16.1.3) have also been compared with results obtained from a full 3D FLUENT simulation. The integral model predicts somewhat higher operating voltages compared with the FLUENT results. This makes the model conservative since higher operating voltages correspond to lower electrolysis efficiencies. The disparity can be explained by noting that the computational fluid dynamics (CFD) model can more accurately account for the variation in local Nernst potential and local current density associated with the cross-flow geometry of the planar stack. Note that, for an operating voltage near the thermal minimum (~1.06 V), both models predict outlet temperatures for this particular adiabatic case that are about 30°C lower than the inlet temperatures. This temperature depression is due to the fact that the endothermic heat requirement of the steam dissociation reaction is larger than the ohmic heating in the operating voltage range between open-cell potential and the thermal-neutral voltage. Per-cell gas flow rates for this case were based on the flow rates used in planar HTE stack tests [19,28]. The integral model also predicts the correct value of the thermal-neutral voltage for 800°C, 1.287 V. At this operating voltage, the outlet temperatures are equal to the inlet temperatures under adiabatic conditions. The integral model is also useful for assessing the effect of using a steam sweep rather than an air sweep on the oxygen side. Use of a sweep gas that does not contain oxygen is advantageous because it reduces the Nernst potential, thereby increasing the electrolysis efficiency for a specified current density. We are considering the use of steam for the sweep gas since it would be relatively easy to separate the steam from the oxygen produced by condensation. The produced oxygen then could be sold as a commodity. Incorporation of the integral model into our UniSim system simulations enabled a broad range of parametric studies.

Results obtained from FLUENT were also compared with experimental results. One set of representative results is shown in Figure 16.22. The results shown correspond to sweep

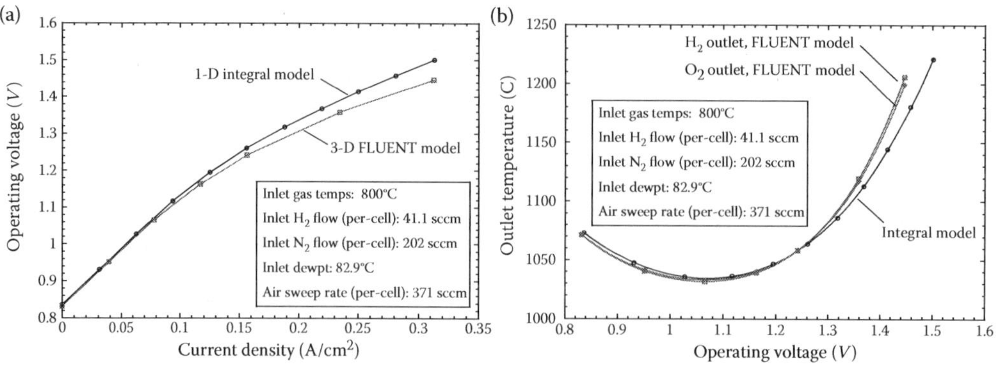

FIGURE 16.21
Predicted operating voltage and gas outlet temperatures for adiabatic electrolyzer operation; comparison of integral model with full 3D FLUENT simulation.

25-1 of Figure 16.14b. This sweep was performed in a stepwise fashion in order to allow sufficient time at each operating condition for steady-state thermal conditions to be achieved in the stack. The figure shows experimentally measured voltage–current characteristics (Figure 16.22a) and internal stack temperatures obtained during a DC potential sweep (Figure 16.22b), along with FLUENT predictions. The FLUENT model included empirical values for internal stack contact resistances, scaled to match the measured voltage–current values of sweep 25-1 of Figure 16.14b. This scaling is necessary because it is not possible to predict these contact resistance values from first principles. The resulting voltage–current CFD prediction is shown in Figure 16.22a, along with the measured data. Corresponding predicted and measured internal stack temperatures are shown in Figure 16.22b. The experimental internal stack temperatures were obtained from miniature (inconel-sheathed, 0.020-inch (500 μm) OD, mineral-insulated, ungrounded, type-K) thermocouples that were inserted into selected air flow channels. The comparison between the experimentally obtained stack internal temperatures and the FLUENT mean electrolyte temperature is quite good, and serve to validate the numerical methods and models used.

Detailed CFD analyses also provide a means of visualizing temperature and current density distributions with operating cells and stacks. A series of contour plots representing local FLUENT results for temperature, current density, Nernst potential, and hydrogen mole fraction is presented in Figures 16.23 through 16.26. In these figures, the steam/hydrogen flow is from top to bottom and the air flow is from left to right. Figure 16.23 shows electrolyte temperature contour plots for amperages of 10, 15, and 30 amps. These current values correspond to operating voltage regions near the minimum electrolyte temperature (10 amps), near thermal-neutral voltage (15 amps), and in the region dominated by ohmic heating (30 amps). The radiant boundary condition at 1103 K tends to hold the outside of the model at a higher temperature for the 10-amp case (Figure 16.23a) while the endothermic heat requirement maintains the center of the electrolyte at a lower temperature. Minimum and maximum temperatures for this case are 1091 and 1100 K, respectively. Figure 16.23b shows a temperature difference across the electrolyte of only 1K, with values very near 1103 K; this current density is very near the thermal-neutral voltage. Figure 16.23c shows that ohmic heating in the electrolyte is dominating and the thermal boundary condition is keeping the edges cooler than the inside. Minimum and maximum temperatures are 1139 and 1197 K, respectively, for this case.

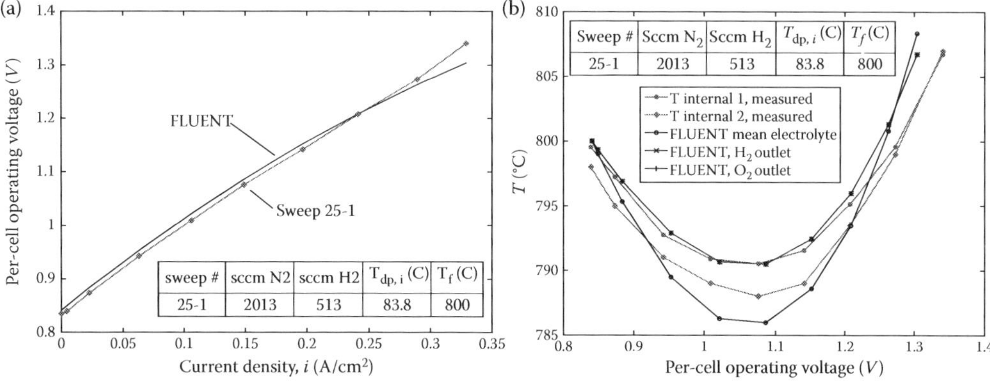

FIGURE 16.22
(a,b) Comparison of internal stack temperature predictions with experimentally measured values.

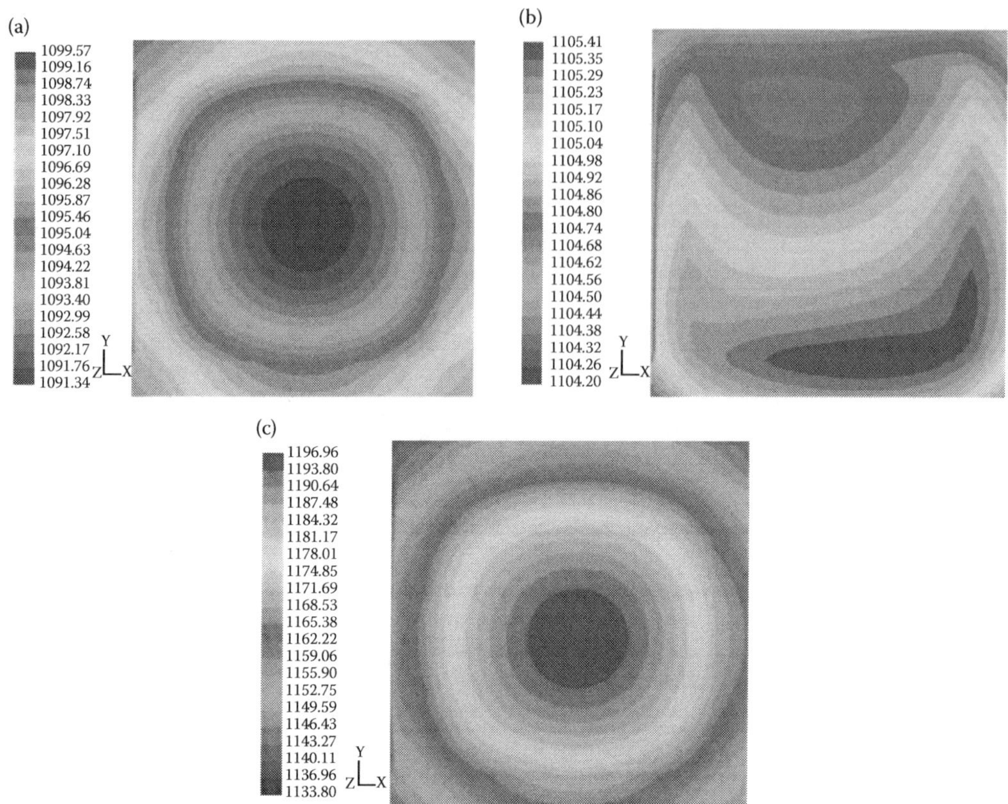

FIGURE 16.23
(a–c) Temperature (K) contours on the electrolyte and insulator for currents of 10, 15, and 30 amps.

Contour plots of local current density on the electrolyte are shown in Figure 16.24 for 10, 15, and 30 amps. Mean current densities for these three cases are 0.156, 0.234, and 0.469 A/cm^2, respectively. These plots correlate directly with local hydrogen production rates. Since FLUENT is being run in electrolysis mode, the current density values are all negative and hence the blue values have the largest magnitudes. Highest current density magnitudes occur near the steam–hydrogen inlet (the top of the figures). This corresponds to the location of the greatest steam concentration. The orange areas show where the current density is lowest because the available steam concentration is lower.

Figure 16.25 shows the local variation in Nernst potential for currents of 10, 15, and 30 amps. The minimum Nernst voltage occurs at the top left of the plots where the steam and oxygen concentrations are the highest and hydrogen concentration the lowest. The minimum value for the Nernst voltage in these three plots is 0.84 V, while the maximum increases from 0.91 to 0.93 to 0.99 V as the current increases from 10 to 30 amps, respectively. Maximum Nernst voltage occurs in the bottom right where the steam concentration is the lowest. The highest Nernst potential regions correspond to the lowest current density regions. Note that the variation in Nernst potential indicated in these plots is dominated by gas concentration effects, rather than thermal effects.

Molar hydrogen fraction contours are shown in Figure 16.26 for currents of 10, 15, and 30 amps. These contours show the entire steam/hydrogen flow channel, including the top and bottom regions adjacent to the edge rails where no hydrogen production is occurring.

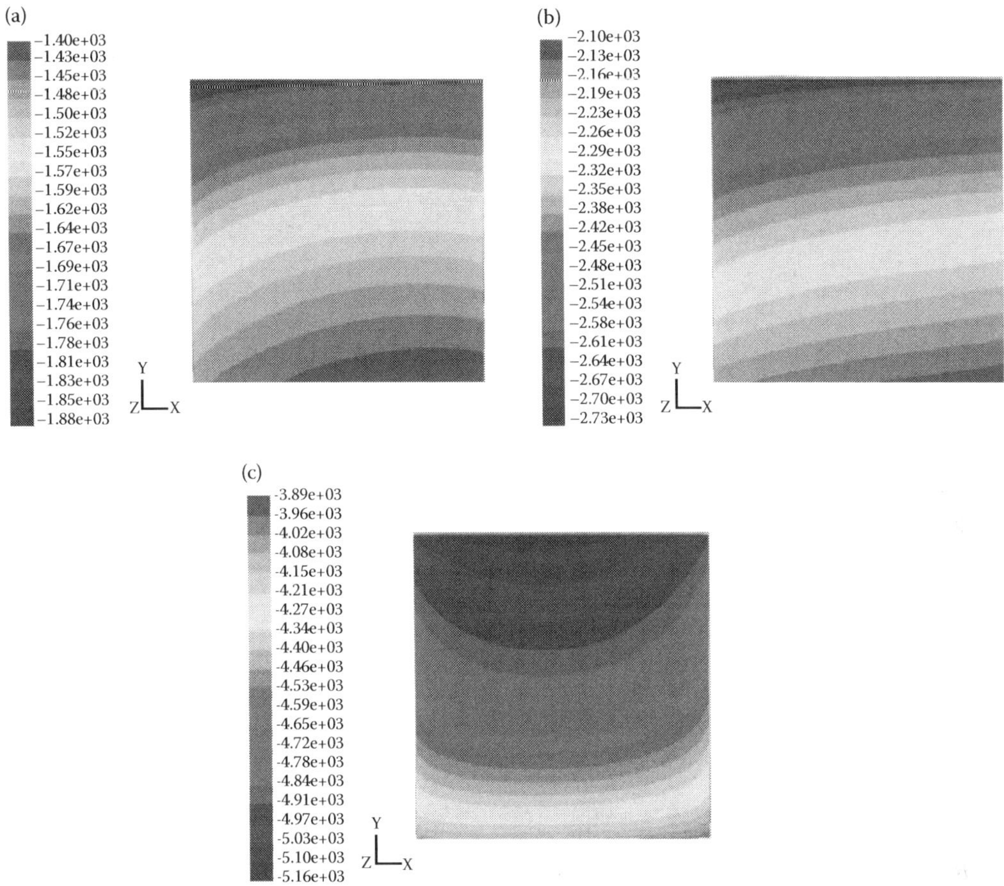

FIGURE 16.24
(a–c) Current density (A/m^2) contours on the electrolyte for currents of 10, 15, and 30 amps.

Hydrogen concentration increases as the flow progresses through the channel from top to bottom. There is a slight bump of higher concentration at the left side of the flow channel for the first two plots and in the center for the third plot. This corresponds to the local variation in current densities. The hydrogen concentrations at the outlet are 0.21, 0.28, and 0.48 for the three cases.

16.6 System Analysis of Large-Scale Hydrogen Production Plants

16.6.1 System Models

When considering the development of any new technology for possible large-scale application, it is important to study the technology from a variety of perspectives including fundamental experimental and computational studies as well as large-scale system simulation. Accordingly, a number of detailed process models have been developed at INL for

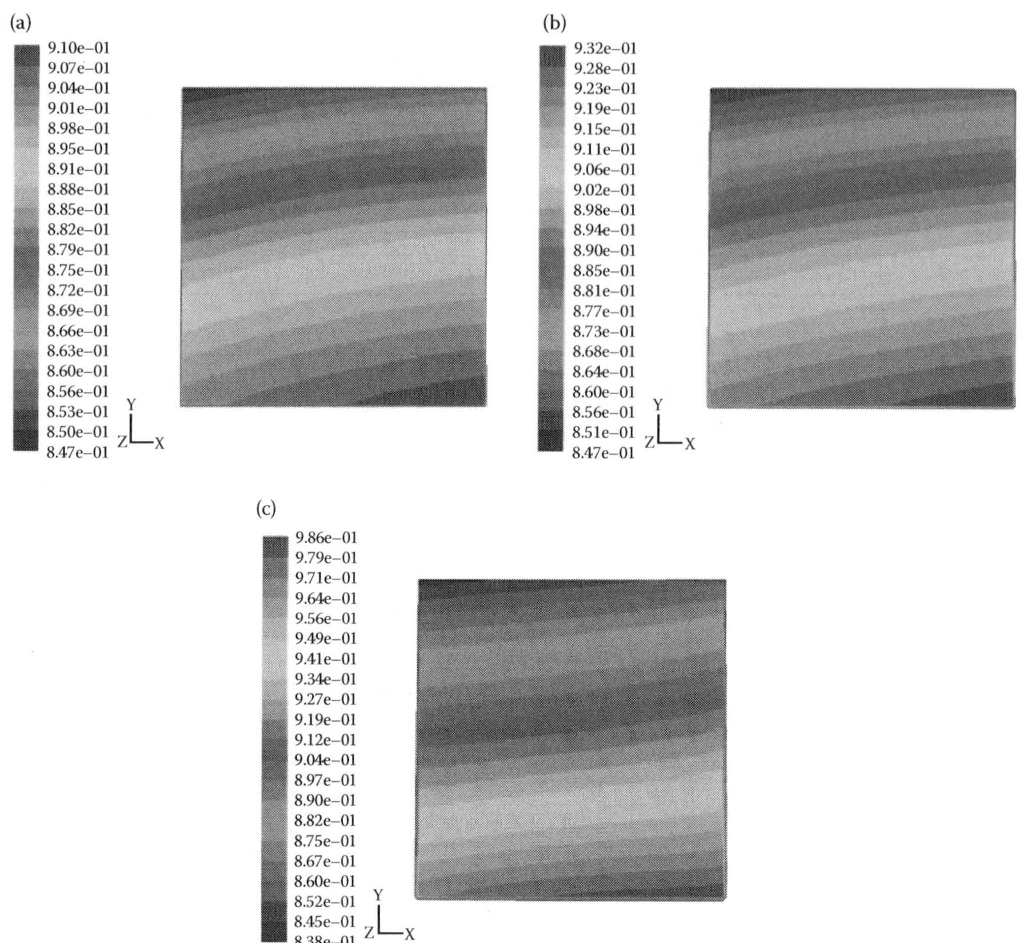

FIGURE 16.25
(a–c) Nernst potential (V) contours on the electrolyte for currents of 10, 15, and 30 amps.

large-scale system analysis of HTE plants. These analyses have been performed using UniSim process analysis software [5]. UniSim is a derivative of HYSYS. The software inherently ensures mass and energy balances across all components and includes thermodynamic data for all chemical species. The overall process flow diagram for a helium-cooled very high-temperature reactor (VHTR) coupled to the direct helium recuperated Brayton power cycle and the HTE plant with air sweep is presented in Figure 16.27 [4]. The reactor thermal power assumed for the high-temperature helium-cooled reactor is 600 MWt. The primary helium coolant exits the reactor at 900°C. This helium flow is split at T1, with more than 90% of the flow directed toward the power cycle and the remainder directed to the intermediate heat exchanger (IHX) to provide process heat to the HTE loop. Within the power-cycle loop, helium flows through the power turbine where the gas is expanded to produce electric power. The helium, at a reduced pressure and temperature, then passes through a recuperator and precooler where it is further cooled before entering the low-pressure compressor. To improve compression efficiencies, the helium is again cooled in an intercooler heat exchanger before entering the high-pressure compressor. The helium

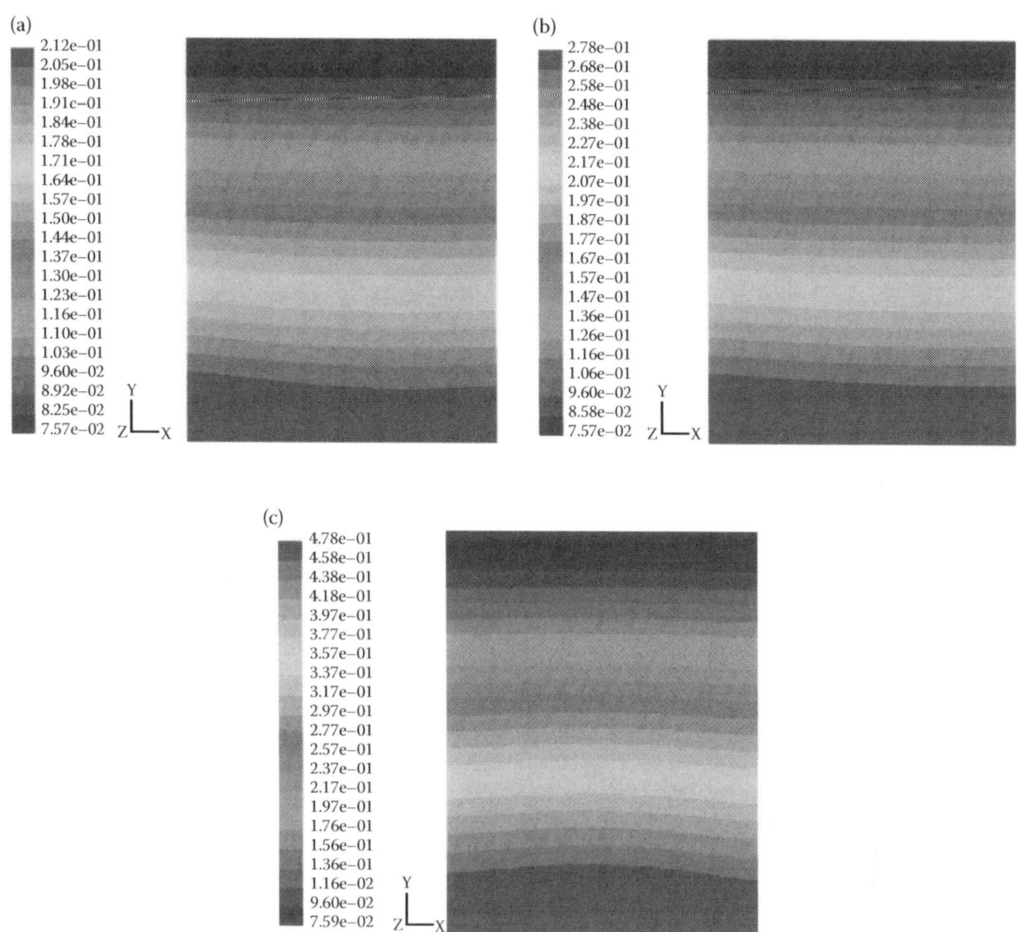

FIGURE 16.26
(a–c) Contours of hydrogen mole fraction in hydrogen flow channel for currents of 10, 15, and 30 amps.

exits the high-pressure compressor at a pressure that is slightly higher than the reactor operating pressure of 7 MPa. The coolant then circulates back through the recuperator where the recovered heat raises its temperature to the reactor inlet temperature of 647°C, completing the cycle.

Liquid water feedstock to the HTE process enters at the left in the diagram (Figure 16.27). The water is compressed to the HTE process pressure of 3.5 MPa in the liquid phase using a pump. The HTE process is operated at elevated pressure for two reasons. Elevated pressure supports higher mass flow rates for the same size components. Furthermore, the gaseous hydrogen product will ultimately be delivered at elevated pressure either for storage or pipeline. Therefore, from the standpoint of overall process efficiency, it is logical to compress the liquid-water feedstock at the process inlet since liquid-phase compression work is very small compared to compression of the gaseous product.

Downstream of the pump, condensate from the water knockout tank is recycled back into the inlet stream at M3. The water stream is then vaporized and preheated in the electrolysis recuperator, which recovers heat from the postelectrolyzer process and sweep-gas outlet

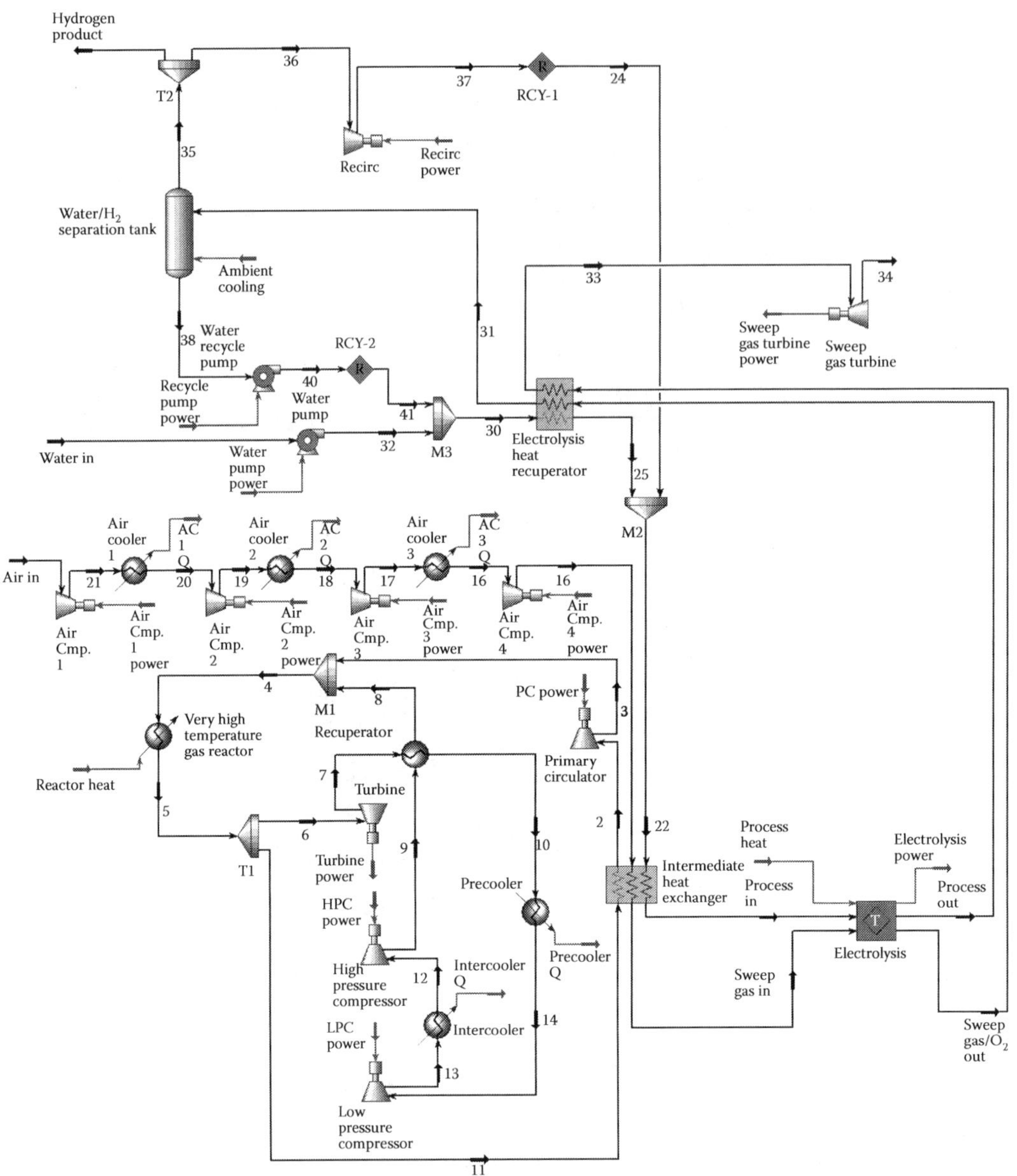

FIGURE 16.27
Process flow diagram for a helium-cooled reactor/direct Brayton/HTE system with air sweep.

streams. Downstream of the recuperator, at M2, the steam is mixed with recycled hydrogen product gas. A fraction of the product gas is recycled in this way in order to assure that reducing conditions are maintained on the steam/hydrogen electrode. Downstream of the mixer, the process gas mixture enters the IHX, where final heating to the electrolysis operating temperature occurs, using high-temperature process heat from the nuclear reactor. The process stream then enters the electrolyzer, where the steam is electrolytically reduced yielding hydrogen on the cathode side of each cell and oxygen on the anode side. Most of the components included in the process flow diagram are standard UniSim components. However, a custom electrolyzer module was developed at INL for direct incorporation into the UniSim system analysis code, as described in detail in Ref. [10].

Downstream of the electrolyzer, the hydrogen-rich product stream flows through the electrolysis recuperator where it is cooled and the inlet process stream is preheated. The cooled product stream is split at T2 and a fraction of the product gas is recycled into the inlet process stream, as discussed previously. A recirculating blower is required to repressurize the recycle stream to the upstream pressure at M2. The remainder of the product stream is cooled further at the water knockout tank, where the majority of any residual steam is condensed and separated yielding dry hydrogen product.

The process flow diagram shows air in use as a sweep gas, to remove the excess oxygen that is evolved on the anode side of the electrolyzer. For the air-sweep cases, inlet air is compressed to the system operating pressure of 3.5 MPa in a four-stage compressor with intercooling. The final compression stage is not followed by a cooler; so the air enters the IHX at about 120°C. The sweep gas is heated to the electrolyzer operating temperature of 800°C via the IHX which supplies high-temperature nuclear process heat directly to the system. The sweep gas then enters the electrolyzer, where it is combined with product oxygen. Finally, it passes through the electrolysis recuperator to help preheat the incoming process gas. Some of the sweep-gas compression work is recovered using a sweep-gas turbine located at the sweep-gas exit.

In order to avoid the work requirement associated with compression of the sweep gas, it is possible to operate with no sweep gas, and to allow the system to produce pure oxygen, which could potentially be supplied to another collocated process such as an oxygen-blown gasifier. For this mode of operation, the four-stage air compressor would not be included in the process flow diagram and there would be no air flow through the IHX. Air preheat at the IHX is no longer needed. Oxygen would simply be evolved from the anode side of the electrolyzer at the electrolysis operating pressure and temperature. It would flow through the electrolysis heat recuperator and the outlet turbine. The results of our system analyses have shown that this concept is desirable from the standpoint of overall process efficiency, but there are significant technical issues associated with handling high-temperature pure oxygen that would have to be addressed.

For these simulations, the per-cell active area for electrolysis was assumed to be 225 cm^2. This cell size is well within the limits of current technology for planar cells. ASR was used to characterize the performance of the electrolysis cells. This parameter incorporates the loss mechanisms in the cells. The ASR value used in the electrolyzer module is temperature dependent using an empirically developed Arrhenius equation [29]. In order to show the trends that can be expected with higher or lower ASR, two values of $ASR_{1100\,K}$ have been included in this study. The $ASR_{1100\,K}$ value of 1.25 represents a stack-average ASR value at 1100 K that is achievable in the short term with existing technology. The $ASR_{1100\,K}$ value of 0.25 is an optimistic value that has been observed in button cells, but will be difficult to achieve in a stack in the short term. The temperature dependence of the ASR is important for the adiabatic cases (since the outlet temperature in these cases is generally different than the inlet temperature) and for evaluating the effect of electrolyzer inlet temperature on overall process efficiency.

The total number of cells used in the process simulations was determined by specifying a maximum current density for each ASR value considered that was large enough to ensure that the operating voltage would just exceed the thermal-neutral voltage. For the higher nominal ASR value of 1.25 Ω cm^2, the maximum current density was set at 0.25 A/cm^2 and an adiabatic thermal boundary condition was assumed. The total number of cells for this base case was adjusted until the total remaining power was zero. In other words, the full power-cycle output at this operating point is dedicated to electrolysis. This procedure resulted in 1.615×10^6 cells required. At lower current densities, the power-cycle output exceeds the value required for electrolysis and this excess power would be supplied to the grid. For the case of ASR = 0.25 Ω cm², the maximum current density was set at 1.0 A/cm². A much higher maximum current density was required for the lower ASR case, again in order to assure that the thermal-neutral voltage was just exceeded.

Two thermal boundary condition limits were considered for the electrolyzer: isothermal and adiabatic. Actual electrolyzer operation will generally lie between these limits. For the isothermal cases, heat from the reactor was directly supplied to the electrolyzer to maintain isothermal conditions for operation below the thermal-neutral voltage. Heat rejection from the electrolyzer is required to maintain isothermal operation at operating voltages above thermal neutral. For the adiabatic cases, the direct electrolyzer heater shown in Figure 16.27 was not used.

To allow for comparisons between the performance of the HTE processes and alternate hydrogen production techniques, we have adopted a general efficiency definition that can be applied to any thermal water-splitting process, including HTE, LTE, and thermochemical water splitting. Since the primary energy input to the thermochemical processes is in the form of heat, the appropriate general efficiency definition to be applied to all of the techniques is the overall thermal-to-hydrogen efficiency, η_H. This efficiency is defined as the heating value of the product hydrogen divided by the total thermal input required to produce it. In this report, the LHV of the products has been used:

$$\eta_H = \frac{\sum_i \dot{N}_i \mathrm{LHV}_{H_2}}{\sum_i \dot{Q}_i}. \tag{16.35}$$

The denominator in this efficiency definition quantifies all of the net thermal energy that is consumed in the process. For a thermochemical process, this summation includes the direct nuclear process heat as well as the thermal equivalent of any electrically driven components such as pumps, compressors, and so on. The thermal equivalent of any electrical power consumed in the process is the power divided by the thermal efficiency of the power cycle. The power-cycle thermal efficiency for the helium-cooled direct Brayton cycle concept described in this chapter was 52.6%. For an electrolysis process, the summation in the denominator of Equation 16.35 includes the thermal equivalent of the primary electrical energy input to the electrolyzer and the secondary contributions from smaller components such as pumps and compressors. In addition, any direct thermal inputs are also included. Direct thermal inputs include any net (not recuperated) heat required to heat the process streams up to the electrolyzer operating temperature and any direct heating of the electrolyzer itself required for isothermal operation.

16.6.2 Representative System Analysis Results

A summary of results obtained from the hydrogen production system analyses is presented in Figures 16.28 and 16.29. The results presented in these figures were obtained for a fixed steam utilization of 89% (i.e., 89% of the inlet steam was converted to hydrogen). In order to maintain fixed steam utilization, the flow rates of the process streams were adjusted with lower flow rates for lower current densities and higher flow rates for higher current densities. Results of eight cases are presented in Figure 16.28: low and high ASR, adiabatic and isothermal electrolyzer operation, and air-sweep and no-sweep. The figure provides overall hydrogen production efficiencies (Equation 16.35) as a function of per-cell operating voltage. Recall that electrolyzer efficiency is inversely proportional to operating voltage [30]. Higher operating voltages yield higher current densities and higher hydrogen production rates, but lower overall efficiencies; so the selection of electrolyzer operating condition is a trade-off between production rate and efficiency. For a specified target production rate, higher production efficiency requires a higher capital cost, since more cells would be required to achieve the target production rate. In general, a good trade-off between production rate and efficiency occurs for cell operating voltages near or slightly below the thermal neutral value, around 1.29 V. This operating voltage is also desirable from the standpoint that the electrolysis stack operates nearly isothermally at this voltage. Predicted overall thermal-to-hydrogen efficiency values shown in Figure 16.28 are generally within 8 percentage points of the power-cycle efficiency of 52.6%, decreasing with operating voltage. It is interesting to note that the overall process efficiencies for these fixed-utilization cases collapse onto individual lines, one for the air-sweep cases and another for the no-sweep cases, when plotted as a function of cell operating voltage, regardless of the electrolyzer mode of operation (adiabatic or isothermal) and ASR value. Note that the highest operating voltages shown are just above the thermal-neutral voltage of 1.29 V. Note also that the highest overall efficiency plotted in Figure 16.28 (for no-sweep, ASR = 0.25, isothermal, V = 1.06 A/cm²) exceeds 51%.

An additional line, based on a simple thermodynamic analysis [30], is also shown in Figure 16.28. This analysis considers a control volume drawn around the electrolysis process, with the process consuming the electrical work from the power cycle, and heat from a high-temperature source. If the inlet and outlet streams are assumed to be liquid water,

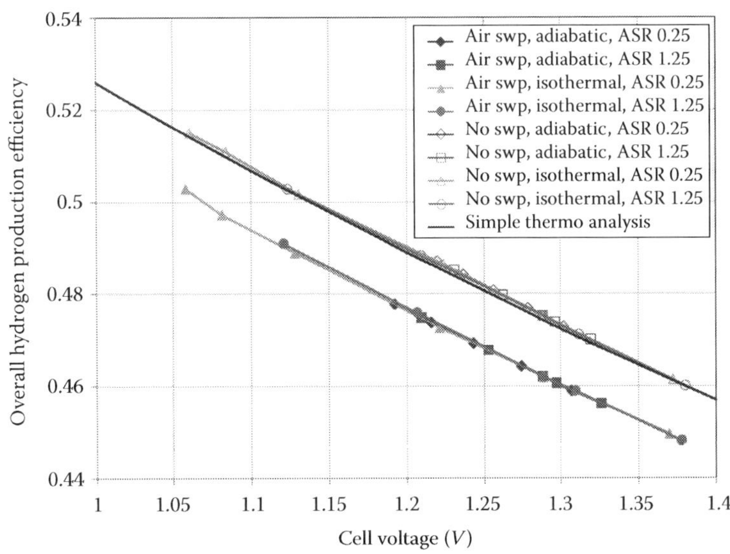

FIGURE 16.28
Overall HTE hydrogen production efficiencies for the VHTR/recuperated direct Brayton cycle, as a function of per-cell operating voltage.

and gaseous hydrogen and oxygen, respectively, at $T = T°$, $P = P°$, direct application of the first law, Faraday's law, and the definition of the overall thermal-to-hydrogen efficiency yields:

$$\eta_H = \frac{\text{LHV}}{2FV_{\text{op}}(1/\eta_{\text{th}} - 1) + \text{HHV}}. \tag{16.36}$$

The curve labeled "simple thermo analysis" in Figure 16.28 represents Equation 16.36. This equation provides a useful reference against which detailed system analyses can be measured. The simple thermodynamic analysis agrees quite closely with the detailed system analysis results for the no-sweep cases, which correspond directly with the conditions of simple analysis since it does not include consideration of a sweep gas. Overall hydrogen efficiency results of the air-sweep cases are about 1% lower than the no-sweep cases.

Hydrogen production efficiencies can also be plotted as a function of hydrogen production rate, as shown in Figure 16.29. As expected, efficiencies decrease with production rate since higher production rates require higher current densities and higher per-cell operating voltages, for a fixed number of cells. For this plot, the full 600 MWt output of the reactor is assumed to be dedicated to hydrogen production. Under this assumption about four times as many electrolysis cells are required for the high-ASR cases than for the low-ASR cases, with a correspondingly higher associated capital cost. Figure 16.29 shows that hydrogen production rates in excess of 2.3 kg/s (92,000 standard cubic meters per hour (SCMH), 78×10^6 standard cubic feet (SCF)/day). could be achieved with a dedicated 600 MWt hydrogen production plant. This rate is the same order of magnitude as a large hydrogen production plant based on steam-methane reforming. Figure 16.29 indicates similar

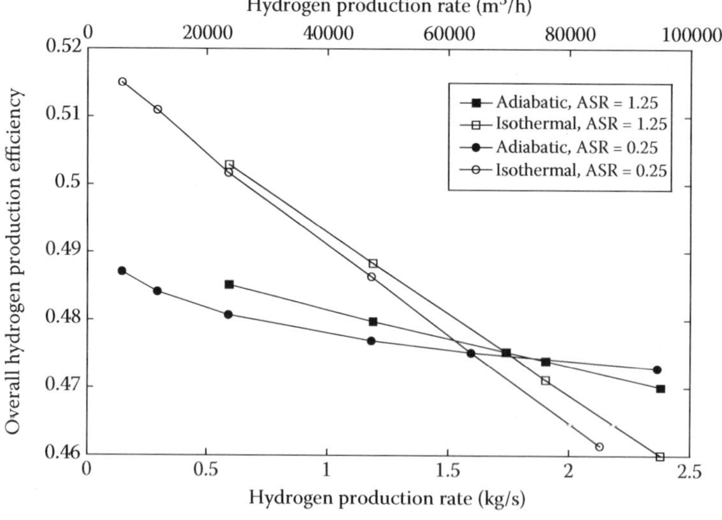

FIGURE 16.29
Overall hydrogen production efficiency as a function of hydrogen production rate, with air sweep.
Note: The high-ASR cases shown here require ~ four times as many cells.

overall efficiencies for the low and high-ASR cases at a specified electrolyzer thermal operating condition (adiabatic or isothermal) and hydrogen production rate.

The effect of steam utilization was examined by fixing the electrolyzer inlet process gas flow rates at the values corresponding to the highest current density achievable with each ASR value, then varying the current density over the full range of values considered for the fixed-utilization cases. Low current densities for this case yield low values of steam utilization since the inlet steam flow rate is fixed at a value that yields 89% utilization at the highest current density. Results of this exercise are presented in Figure 16.30. The overall efficiency results for the variable-utilization cases nearly collapse onto a single curve when plotted versus utilization. The plot indicates a strong dependence on utilization, with overall hydrogen production efficiencies less than 25% at the lowest utilization values shown (~5.5%), increasing to a maximum value of ~47% at the highest utilization value considered (89%). So, from the overall system perspective, low steam utilization is undesirable. This is an interesting result because, from the perspective of the electrolyzer alone, low utilization yields high electrolyzer (not overall) efficiency values. Excess steam in the electrolyzer keeps the average Nernst potential low for each cell, which assures a low operating voltage for a specified current density (or hydrogen production rate). However, from the overall system perspective, low steam utilization means that the system is processing lots of excess material, resulting in relatively high irreversibilities associated with incomplete heat recuperation, pumping and compression of excess process streams, and so on. Above ~50% utilization, however, the efficiency curves are relatively flat, even decreasing slightly for the isothermal cases. Regarding very high utilization values, achievement of steam utilization values much above 90% is not practical from an operational standpoint because localized steam starvation can occur on the cells, with associated severe performance penalties and possible accelerated cell lifetime degradation.

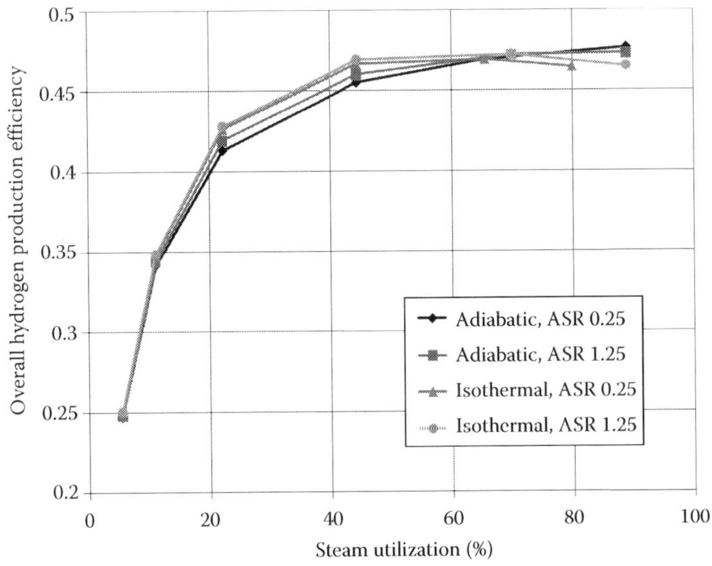

FIGURE 16.30
Effect of steam utilization on overall hydrogen production efficiency.

16.7 Future Key Technical Challenges for HTE

16.7.1 Overall Challenges

Hydrogen production by means of HTE using solid oxide cells is not yet a commercial-ready technology. The state-of-the-art of SOFCs have advanced significantly in recent years, thanks in part to financial support from the DOE SECA program. The development of advanced anode-supported cells represents a major milestone for SOFCs. SOEC technology has also benefited from the advancement of SOFCs, but relatively little direct funding has been available for SOEC research. There are some fundamental differences in the two modes of operation, such that different cell materials, cell designs, and stack configurations will be required for optimal long-term performance in the electrolysis mode. Some of these issues were discussed in Section 16.2 on cell materials.

The ultimate cost of hydrogen production by any technology is dependent on both capital and operating costs. In order to achieve competitive capital costs, HTE cells and stacks must exhibit both high performance and low degradation rates. High performance can be quantified in terms of the ASR. Per-cell ASR values of 0.40 Ω cm^2 or lower in a stack configuration, with very low degradation rates of less than 1% per 1000 h will be required for viable large-scale hydrogen production. These performance values have already been demonstrated by several SOFC manufacturers in the fuel-cell mode of operation. Cell ASR values lower than 0.4 Ω cm^2 have been observed at INL in single-cell tests, but not yet in a stack configuration. SOEC stack testing performed at INL to date has exhibited high degradation rates. Several mechanisms have been proposed as contributors to the accelerated degradation observed in the electrolysis mode. INL sponsored a workshop on SOEC performance degradation in October 2008. INL also participated in an invitation-only international workshop in June 2009 on this topic. The INL research focus is in fact on minimizing SOEC degradation. We are working with Ceramatec Inc., and MSRI, both in Salt Lake City, toward development of improved electrolysis cells and stacks. We are also supporting research activities at MIT and the University of Utah on understanding degradation mechanisms. A comprehensive summary of the state of knowledge of SOEC degradation and a summary of the topics presented at the INL SOEC degradation workshop is available in Ref. [31]. A summary of the major observations and conclusions from this report is provided below.

16.7.2 Degradation in SOECs

At present, a complete understanding and general agreement on the causes of degradation in SOECs and the electrochemical and physical mechanisms behind them does not exist. Experimental data on degradation can be classified into three main categories: (a) baseline progressive constant-rate degradation, (b) degradation corresponding to transients caused by thermal or redox (**red**uction and **ox**idation) cycling phenomena occurring in a cell, and (c) degradation resulting from a sudden incident or a failure/malfunction of a component or a control in a stack system.

Degradation data have been obtained both in single cells as well as in stacks. Degradation mechanisms in a stack are not identical to those in a single cell [32]. Also, degradation in an SOEC is not identical to that in an SOFC. Long-term single-cell tests show that SOEC operation generally exhibits greater degradation rates than SOFC operation. Therefore, SOFC degradation can be used for background information and guidance. But for specific

SOEC stack development, more studies have to be done on SOEC stacks. Some researchers have observed that higher operating temperatures increase degradation rates in SOECs, but higher current density does not increase degradation. However, researchers at Argonne National Laboratory (ANL) observed higher degradation in higher current flow regions of O_2-electrodes. Such unconfirmed and conflicting opinions need to be resolved via future research and collaboration.

There is general agreement that degradation of the O_2-electrode is more severe than that of the H_2-electrode. Therefore, it was proposed to focus initially on the degradation of the O_2-electrodes in a stack. Posttest examination at ANL of an SOEC stack operated by INL for ~1500 h showed that the O_2-electrode delaminated from the bond layer/electrolyte. However, the causes of the delamination can be termed as speculative, because confirmative tests proving the fundamental cause(s) have not been performed. In the SOEC mode, oxygen evolution in over sintered layers can result in a buildup of high pressure in regions of closed porosity, increasing the chances of delamination. Therefore, high porosity is very important in the oxygen electrode. This opinion has been expressed by many workshop participants and hence deserves further examination. Per-ANL observations, delamination is more likely to occur in cell areas with high current flows.

One potential degradation mechanism that is common between the SOFC and SOEC modes of operation is chromium poisoning originating from the interconnects or from balance-of-plant piping. Volatile chrome oxide may deposit at the electrode–electrolyte interface or TPB. This deposition can result in deactivation of electrochemical reaction sites and/or separation of the bond layer from the oxygen electrode.

An oxygen electrode side bond layer is often used. Because it is next to the oxygen electrode, it encounters similar electrochemical phenomena that lead to cell degradation. ANL found an average of 1–8% (~30% maximum) chromium contamination in the bond layer, probably originating from interconnects. Chromium contaminants were found in association with LSC. In the oxygen bond layer, a secondary phase may form. However, there are conflicting opinions about the severity of bond-layer chromium contamination. ANL observed delamination and a weak interface between the oxygen electrode and the LSC bond layer, which can prevent solid-state Cr from diffusing into the O_2-electrode. For this reason, the oxygen electrode can remain stable. However, such a weak interface is not desirable from electrical conductivity point of view.

In electrolytes, the main cause of degradation is loss of ionic conductivity. Müller et al. [33] showed that during first 1000 h of testing, "fully-stabilized" yttria- and scandia-doped zirconia (8 mol% Y_2O_3 or Sc_2O_3) electrolytes showed ~23% performance degradation. For the next 1700 h of testing, the decrease in conductivity was as high as 38%.

Both Steinberger–Wilckens [34] and Hauch [35] reported formation of impurities at the TPBs as a degradation mechanism. A substantial amount of SiO_2 was detected at the Ni/YSZ H_2-electrode–electrolyte interface during electrolysis, while no Si was detected in other reference cells. The Si-containing impurities probably originated from the albite glass seal. ANL also observed Si as a capping layer on steam/H_2-electrode. It probably was carried by steam from SI-containing seals. SiO_x also emanates from interconnect plates. Mn also diffuses from interconnects, but the significance of Mn diffusion is not known. Hauch [35] observed contaminants containing Si to segregate to the innermost few microns of the H_2-electrode near the electrolyte. The impurities that diffused to and accumulated at the TPBs of the H_2-electrode are believed to be a significant cause of performance degradation in SOECs [35].

Interconnects can also be a source of serious degradation. Sr, Ti, and Si segregate and build-up at interfaces. Sr segregates to the interconnect–bond layer interface. Mn segregates to the interconnect surface. Si and Ti segregate to the interconnect–passivation layer interface. Cr contamination can originate from interconnects and it can interact with the O_2-electrode surface or even diffuse into the O_2-electrode. Chromium reduction (Cr^{6+} to Cr^{3+}) can take place at the electrode–electrolyte interface [36]. Under the sponsorship of the US-DOE SECA program, coatings for the interconnects have been developed. Coated stainless steel interconnects have shown reduced degradation rates.

References

1. Nomura, M., Kasahara, S., and Onuki, K., Estimation of thermal efficiency to produce hydrogen from water through IS process, *Proceedings, 2nd Topical Conference on Fuel Cell Technology*, AIChE Spring National Meeting, New Orleans, 2003.
2. Abraham, B. M. and Schreiner, F., General principles underlying chemical cycles which thermally decompose water into the elements, *Industrial Engineering and Chemistry Fundamentals*, 13(4), 1974.
3. Fletcher, E. A. and Moen, R. L., Hydrogen and oxygen from water, *Science*, 197, 1050–1056, 1977.
4. O'Brien, J. E., McKellar, M. G., and Herring, J. S., Performance predictions for commercial-scale high-temperature electrolysis plants coupled to three advanced reactor types, *ANS International Congress on Advances in Nuclear Power Plants (ICAPP08)*, Anaheim, CA, June 8–12, 2008.
5. UniSim Design, R360 Build 12073, Copyright ©2005–2006 Honeywell International Inc.
6. Yildiz, B. and Kazimi, M. S., Efficiency of hydrogen production systems using alternative nuclear energy technologies, *International Journal of Hydrogen Energy*, 31, 77–92, 2006.
7. Brown, L. C., Lentsch, R. D., Besenbruch, G. E., Schultz, K. R., and Funk, J. E., Alternative Flowsheets for the Sulfur-Iodine Thermochemical Hydrogen Cycle, (General Atomics, San Diego, CA.) Sponsor: Department of Energy, Washington, DC. Report: GA-A24266, 24p, Feb 2003.
8. Southworth, F., Macdonald, P. E., Harrell, D. J., Park, C. V., Shaber, E. L., Holbrook, M. R., and Petti, D. A., The next generation nuclear plant (NGNP) project, *Proceedings, GlOBAL 2003*, 276–287, New Orleans, 2003.
9. Larminie, J. and Dicks, A., *Fuel Cell Systems Explained*, New York, NY: John Wiley & Sons, 2003.
10. O'Brien, J. E., Stoots, C. M., and Hawkes, G. L., Comparison of a one-dimensional model of a high-temperature solid-oxide electrolysis stack with CFD and experimental results, *Proceedings, 2005 ASME International Mechanical Engineering Congress and Exposition*, Orlando, November 5–11, 2005.
11. Steinberger-Wilkens, R. Tietz, F., Smith, M. J., Mougin, J., Rietveld, B., Bucheli, O., Van Herle, J., Mohsine, Z., and Holtappels, P., Real-SOFC—A joint European effort in understanding SOFC degradation, *ECS Transactions, 7(1) PART 1, ECS Transactions—10th International Symposium on Solid Oxide Fuel Cells, SOFC-X*, Lucerne, Switzerland, pp. 67–76, 2007.
12. Ishihara, T., Hiroyuki, E, Zhong, H., and Matsumoto, H., Intermediate temperature solid oxide fuel cells using $LaGaO_3$-based perovskite oxide for electrolyte, *Electrochemistry*, 77(2), 155–157, February 2009.
13. Singhal, S. C. and Kendall, K. *Solid Oxide Fuel Cells*, Oxford, UK: Elsevier Advanced Technology, 2003.
14. O'Brien, J. E., McKellar, M. G., Harvego, E. A., and Stoots, C. M., High-temperature electrolysis for large-scale hydrogen and Syngas production from nuclear energy—System simulation and economics, *International Conference on Hydrogen Production, ICH2P-09*, Oshawa, Canada, May 3–6, 2009.

15. Williams, M. C., Strakey, J. P., Surdoval, W. A., and Wilson, L. C., Solid oxide fuel cell technology development in the US, *Solid-State Ionics*, 177(19–25), 2039–2044, October 2006.
16. Borglum, B., Development of solid oxide fuel cells at Versa Power Systems, *2008 Fuel Cell Seminar*, Phoenix, October 28, 2008.
17. Nichols, D. K., Agnew, G. D., and Strickland, D., Outlook and application status of the Rolls-Royce fuel cell systems SOFC, *IEEE Power and Energy Society 2008 General Meeting*, Pittsburgh, July 20–24, 2008.
18. O'Brien, J. E., Stoots, C. M., Herring, J. S., Lessing, P. A., Hartvigsen, J. J., and Elangovan, S., Performance measurements of solid-oxide electrolysis cells for hydrogen production, *Journal of Fuel Cell Science and Technology*, 2, 156–163, August 2005.
19. O'Brien, J. E., Stoots, C. M., Herring, J. S., and Hartvigsen, J. J., Hydrogen Production performance of a 10-cell planar solid-oxide electrolysis stack, *Journal of Fuel Cell Science and Technology*, 3, 213–219, May 2006.
20. O'Brien, J. E., Stoots, C. M., Herring, J. S., and Hartvigsen, J. J., Performance of planar high-temperature electrolysis stacks for hydrogen production from nuclear energy, *Nuclear Technology*, 158, 118–131, May 2007.
21. Housley, G., Condie, K., O'Brien, J. E., and Stoots, C. M., Design of an integrated laboratory scale experiment for hydrogen production via high temperature electrolysis, paper no. 172431, *ANS Embedded Topical: International Topical Meeting on the Safety and Technology of Nuclear Hydrogen Production, Control, and Management*, Boston, MA, USA, June 24–28, 2007.
22. Stoots, C. M. and O'Brien, J. E., Initial operation of the high-temperature electrolysis integrated laboratory scale experiment at INL, *2008 International Congress on Advances in Nuclear Power Plants*, Anaheim, CA, June 8–12, 2008.
23. Stoots, C. M., O'Brien, J. E., Condie, K., Moore-McAteer, L., Housley, G. K., Hartvigsen, J. J., and Herring, J. S., The high-temperature electrolysis Integrated Laboratory Experiment, *Nuclear Technology*, 166, 32–42, April 2009.
24. Housley, G. K., O'Brien, J. E., and Hawkes, G. L., Design of a compact heat exchanger for heat recuperation from a high temperature electrolysis system, *2008 ASME International Congress and Exposition*, paper no. IMECE2008-68917, Boston, MA, November 2008.
25. Prinkey, M., Shahnam, M., and Rogers, W. A., *SOFC FLUENT Model Theory Guide and User Manual*, Release Version 1.0, FLUENT, Inc., 2004.
26. Hawkes, G. L., O'Brien, J. E., Stoots, C. M., Herring, and J. S., Shahnam, M., CFD model of a planar solid oxide electrolysis cell for hydrogen production from nuclear energy, *11th International Topical Meeting on Nuclear Reactor Thermal-Hydraulics NURETH-11*, Popes Palace Conference Center, Avignon, France, October 2–6, 2005.
27. *FLUENT Theory Manual, Version 6.1.22*, Fluent Inc., Lebanon, New Hampshire, 2004.
28. O'Brien, J. E., Herring, J. S., Stoots, C. M., and Lessing, P. A., High-temperature electrolysis for hydrogen production from nuclear energy, *11th International Topical Meeting on Nuclear Reactor Thermal-Hydraulics NURETH-11*, Popes Palace Conference Center, Avignon, France, October 2–6, 2005.
29. Stoots, C. M., O'Brien, J. E., McKellar, M. G., Hawkes, G. L., and Herring, J. S., Engineering process model for high-temperature steam electrolysis system performance evaluation, *AIChE 2005 Annual Meeting*, Cincinnati, OH, October 30–November 4, 2005.
30. O'Brien, J. E., Thermodynamic considerations for thermal water splitting processes and high-temperature electrolysis, *2008 ASME International Congress and Exposition*, paper no. IMECE2008-68880, Boston, MA, November 2008.
31. Sohal, M. S., Degradation in solid oxide cells during high temperature electrolysis, INL/EXT-09-15617, May 21, 2009.
32. Virkar, A. V., A model for solid oxide fuel cell (SOFC) stack degradation, *Journal of Power Sources*, 172, 713–724, 2007.
33. Müller, A. C., Weber, A., Herbstritt, D., and Ivers-Tiffée, E., Long-term stability of yttria and scandia doped zirconia electrolytes, *Proceedings, 8th International Symposium on SOFC*, S. C. Singhal and M. Dokiya (Eds), PV 2003–07, The Electrochemical Society, Daytona Beach, FL, 196–199, 2003.

34. Steinberger-Wilckens, R., Degradation issues in SOFCs, Presented at the workshop on degradation in solid oxide electrolysis cells and Strategies for its Mitigation, *Fuel Cell Seminar and Exposition*, Phoenix, AZ, October 27, 2008.
35. Hauch, A., Solid oxide electrolysis cells—Performance and durability, PhD Thesis, Technical University of Denmark, Risø National Laboratory, Roskilde, Denmark, 2007.
36. Singh, P., Pederson, L. R., Stevenson, J. W., King, D. L., and McVay, G. L., Understanding degradation processes in solid oxide fuel cell systems, Presented at the Workshop on Degradation in Solid Oxide Electrolysis Cells and Strategies for its Mitigation, *Fuel Cell Seminar and Exposition*, Phoenix, AZ, October 27, 2008.

17

Thermochemical Iodine–Sulfur Process

Kaoru Onuki, Shinji Kubo, Nobuyuki Tanaka, and Seiji Kasahara

CONTENTS

17.1 Fundamentals of the Process

Thermochemical water-splitting cycles transform nuclear or solar heat energy into hydrogen energy that works as an energy carrier. Reserves of fossil resources in the world are limited and will be depleted in future. In addition, it is concerned that carbon dioxide produced by use of fossil fuel is a major causative factor of global warming [1]. Therefore, carbon-free energy system is attracting lots of interest. Hydrogen energy system is one of such future energy systems if hydrogen is produced carbon dioxide free. Among technical challenges for realizing the hydrogen energy system, technologies for the production and the storage of hydrogen represent the most important ones. As for the former, concept of thermochemical water-splitting cycle offers an attractive candidate for the large-scale and economical hydrogen production. The direct thermal decomposition of water requires high temperature heat of exceeding a few thousand Kelvin. On the contrary, thermochemical water-splitting cycles make it possible to decompose water at lower temperatures by combining high-temperature endothermic chemical reactions and low-temperature exothermic ones, where the net chemical change is water splitting [2].

Research and development of thermochemical water-splitting cycles was started by Funk and Reinstrom [3] in 1960s. Since then, active studies have been carried out in various countries and a number of thermochemical water-splitting cycles have been proposed to date [2,4]. It is desired that the cycles possess such characteristics as high thermal efficiency of hydrogen production, good matching in T–Q relationship with high-temperature heat sources, easy plant operation, easy scale up, and so on. The iodine–sulfur process (or sulfur–iodine process, hereafter termed as the IS process), is the most famous and well-studied version which was proposed by General Atomics [5] called the GA process. The IS process is one of the most promising processes regarding the utilization of nuclear heat sources, high-temperature gas-cooled reactors (HTGRs), which can supply heat at temperatures close to 1273 K. The IS process requires only fluid handling, which provides the process with significant advantages in operability and plant scaling up.

The IS process consists of the following three chemical reactions (Equations 17.1 through 17.3) (Figure 17.1):

$$SO_2(g) + I_2(l) + 2H_2O(l) \rightarrow H_2SO_4(aq) + 2HI(aq) \quad \Delta H = -98\ kJ \tag{17.1}$$

$$H_2SO_4(aq) \rightarrow SO_2(g) + H_2O(g) + \frac{1}{2}O_2(g) \quad \Delta H = 329\,kJ \tag{17.2}$$

$$2HI(aq) \rightarrow H_2(g) + I_2(g) \quad \Delta H = 119\ kJ \tag{17.3}$$

$$H_2O(l) \rightarrow H_2(g) + \frac{1}{2}O_2(g) \quad \Delta H = 286\,kJ \tag{17.4}$$

In the calculation of reaction enthalpy, $_{\Delta}H$, $H_2SO_4(aq)$ means the H_2SO_4 in the aqueous solution which has molar ratio of $H_2SO_4{:}H_2O = 1{:}4$. HI(aq) is the HI in the solution where the molar ratio is $HI{:}I_2{:}H_2O = 1{:}4{:}5$. Temperature is 298.15 K.

Equation 17.1 known as the Bunsen reaction, is a low-temperature exothermic reaction. Raw materials of the reaction, that is, water, iodine, and sulfur dioxide, react spontaneously to produce an aqueous solution of sulfuric acid and hydriodic acid at about 373 K. The produced solution exhibits liquid–liquid phase separation (LL separation) in the presence of excess iodine, by which the sulfuric acid and hydriodic acid can be separated

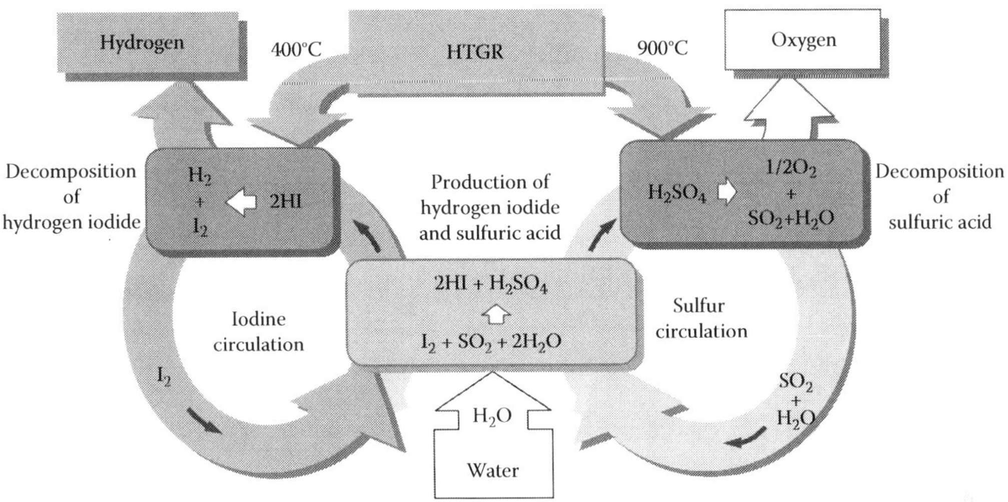

FIGURE 17.1
Scheme of the iodine–sulfur (IS) process.

easily. One liquid phase contains mainly sulfuric acid and water and the other contains mainly hydrogen iodide, iodine, and water.

The thermal decomposition of sulfuric acid, Equation 17.2, proceeds in the following two steps.

$$H_2SO_4(aq) \rightarrow SO_3(g) + H_2O(g) \quad \Delta H = 363 \text{ kJ } (673.15 \text{ K}) \tag{17.5}$$

$$SO_3(g) \rightarrow SO_2(g) + \frac{1}{2}O_2(g) \quad \Delta H = 97 \text{ kJ } (1123.15 \text{ K}) \tag{17.6}$$

H_2SO_4(aq) is in the same condition as in Equation 17.2. Both steps are highly endothermic and proceed smoothly without side reactions and with high equilibrium conversion ratio. Reaction (Equation 17.5) proceeds at 673–773 K in the course of the concentration and vaporization of sulfuric acid. Reaction (Equation 17.6) proceeds at 1073–1123 K with the help of catalyst. The temperature range matches well with the temperature distribution of the heat supplied by HTGR, which is transferred by the sensible heat of helium gas whose temperature range is ca. 673–1223 K. Because of these attractive characteristics, sulfuric acid decomposition reaction has been utilized as a high–temperature endothermic reaction in various thermochemical water-splitting cycles [2,4].

The thermal decomposition of hydrogen iodide, Equation 17.3, can be performed either in gas phase or in liquid phase by using suitable catalysts. The reaction proceeds endothermically with a small heat of reaction. The equilibrium conversion ratio of the gas phase decomposition reaction is 20–30 mol% in the temperature range up to 1273 K. Prior to the decomposition, hydrogen iodide must be separated from the mixture of hydrogen iodide, iodine, and water that is supplied from the Bunsen reaction section, since iodine is one of the products of the decomposition reaction and, therefore, suppresses the reaction.

Sulfur dioxide and iodine produced by these decomposition processes are reused in the Bunsen reaction. The net chemical change (Equation 17.4) resulting from the chemical reaction cycle consisting of reactions (Equations 17.1 through 17.3) is water splitting. Reaction enthalpy of the reaction (Equation 17.4) does not equal the sum of the reaction enthalpy of the reactions (Equations 17.1 through 17.3) because phase shift of water and iodine is not counted in these equations. The cyclic operation creates the free energy change that is required to perform the water splitting using the heat with temperature of lower than 1273 K.

17.2 Analytical and Experimental Studies

17.2.1 Bunsen Reaction

The Bunsen reaction (Equation 17.1) proceeds as the chemical absorption of sulfur dioxide gas by iodine and water. In the mixture of sulfur dioxide, iodine, and water, various reactions occur in addition to the Bunsen reaction and also the reaction products exhibit complicated phase behavior. Figure 17.2 shows the phase states of the reaction products at 333 K [6].

In the reaction mixture whose composition locates in the hatched area shown in the figure, sulfur formation takes place, in addition to the Bunsen reaction, which may be expressed as follows [7,8].

$$SO_2 + 4HI \rightarrow S + 2I_2 + 2H_2O \tag{17.7}$$

$$H_2SO_4 + 6HI \rightarrow S + 3I_2 + 4H_2O \tag{17.8}$$

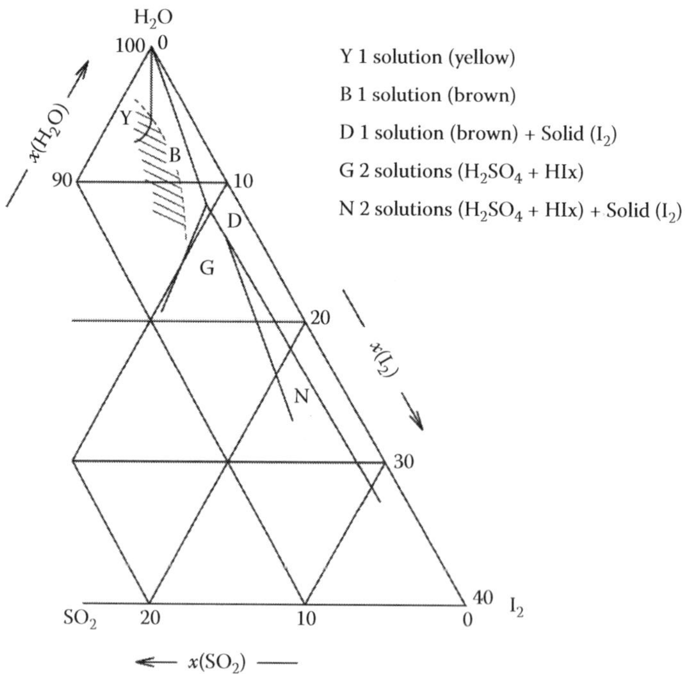

FIGURE 17.2
Phase states of products in ternary reaction system of $SO_2 + I_2 + H_2O$ at 333 K.

Also, hydrogen sulfide formation reactions may occur as follows [7,8]:

$$SO_2 + 6HI \rightarrow H_2S + 3I_2 + 2H_2O \tag{17.9}$$

$$H_2SO_4 + 8HI \rightarrow H_2S + 4I_2 + 4H_2O \tag{17.10}$$

In the region denoted by Y, the iodide ion, I^-, produced by the Bunsen reaction forms a complex with sulfur dioxide, whereas it does with iodine in the region denoted by B. In the region denoted by D, equilibrated solid phase mainly composed of iodine appears. In the region denoted by G, the product solution exhibits two liquid phases. The products from the reaction mixture in region N exhibit three phase equilibrium of two liquids and one solid.

The LL separation was found by researchers at GA in their efforts of searching for the separation methods of hydrogen iodide and sulfuric acid produced by the Bunsen reaction in the thermochemical water-splitting cycle [5]. The LL separation occurs spontaneously in the presence of excess iodine. Figure 17.3 shows an example of the composition of phase-separated solutions [9]. The phase-separated solutions differ very much in density. The heavier solution mainly consists of hydrogen iodide, iodine and water, and is called HIx solution. The main components of the lighter phase are sulfuric acid and water.

HIx solution obtained by the LL separation contains small amounts of sulfur compounds such as sulfuric acid and sulfur dioxide. These sulfur compounds cause undesired side reactions such as the reactions (Equations 17.7 through 17.10) at iodine-lean environments in the following distillation operation of the hydrogen production process. In some cases, elemental sulfur precipitates in the condenser of the distillation column. Therefore,

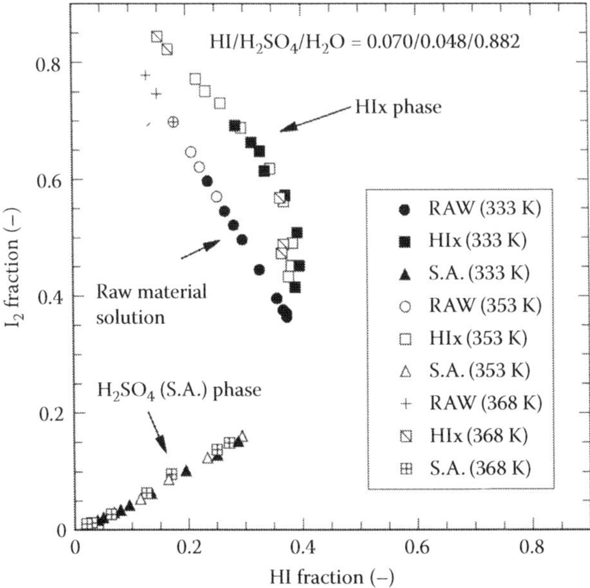

FIGURE 17.3
Phase diagram of LL separation in four component system of $HI + I_2 + H_2SO_4 + H_2O$ with $HI/H_2SO_4/H_2O$ molar ratio of 0.070/0.048/0.882 at 333, 353, and 368 K. The ordinate and abscissa denote molar ratio of I_2 and HI in $I_2 + HI + H_2SO_4$, respectively.

complete separation of the sulfur compounds is required. This "purification" of HIx solution can be carried out by converting sulfuric acid to sulfur dioxide gas utilizing the reverse reaction of the Bunsen reaction,

$$H_2SO_4 + 2HI \rightarrow SO_2 + I_2 + 2H_2O \tag{17.11}$$

and then separating it from the HIx solution by flushing or stripping operation. Figure 17.4 shows an experimental result of sulfur dioxide gas evolution from HIx solution containing small amount of sulfuric acid in the course of heating up the solution while bubbling it with an inert gas (nitrogen) [10]. As seen in the figure, the sulfur dioxide gas evolution occurred at and above 353 K (80°C). The material balance of the experiment suggested that reaction (Equation 17.11) proceeded quantitatively. It was noted that the solution temperature should be lower than 373 K to avoid sulfur formation.

As for the purification of the lighter phase, the sulfuric acid phase, it can easily be performed in the course of concentration operation.

In optimizing the operating condition of the Bunsen reaction section featuring the LL separation, it should be taken into consideration to avoid the "side" reactions, and to attain high concentrations of HI and H_2SO_4 in the heavier phase and in the lighter phase, respectively, as much as possible so as to reduce the excess inventory of water and iodine within the process. From these points of view, parametric studies on the reaction and separation have been carried out to deepen the understanding of the complicated behavior. As for the side reactions, Kumagai et al. [11] found that the proton concentration governed the occurrence of the sulfur and/or hydrogen sulfide formation reaction in the mixture of sulfuric acid and hydrogen iodide by the experiments at 343 K. Sakurai et al. [12] examined the side reactions in the mixture of sulfuric acid and hydriodic acid dissolving iodine at 295–368 K,

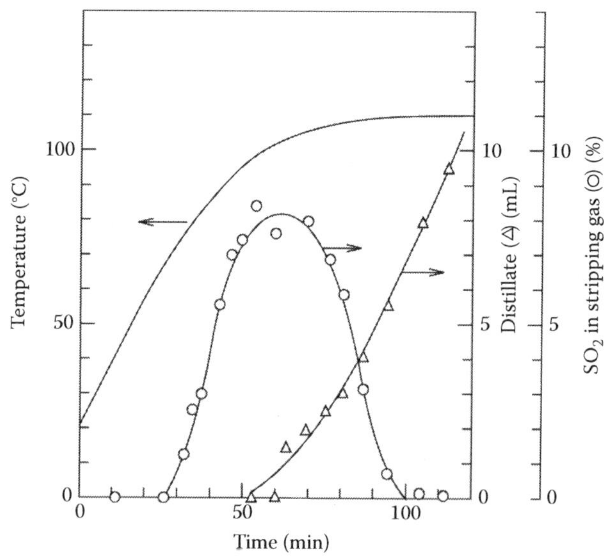

FIGURE 17.4
Time profile of desulfurization of HIx solution (200 mL), using nitrogen (100 SCCM) as the stripping gas. Triangles (△) denote the volume of condensed liquid obtained by cooling the tail gas from the reactor. Circles (○) denote the SO_2 gas concentration in the uncondensed tail gas consisting of SO_2 and N_2. SO_2 in stripping gas with the ordinate means the SO_2 concentration in the tail gas.

and found that the major side reaction was the sulfur formation and that these undesired reactions occurred under the conditions of high temperature and low iodine concentration. Within the conditions where the Bunsen reaction proceeds quantitatively, increasing the iodine concentration of the mixture is effective for enriching the acid concentration, and enhancing the LL separation as well [9,13,14]. At constant temperature, iodine saturation condition gives the highest acid concentration and the best acid separation of the reaction products. As for the effect of temperature, although lower temperature favors both of the exothermic Bunsen reaction and the LL separation, increase of iodine solubility at elevated temperatures results in the more concentrated acid solution and the better separation. As for the other reactant, sulfur dioxide, the effect of its partial pressure on hydrogen iodide concentration of HIx solution produced by the Bunsen reaction was examined under iodine saturated condition [15]. Figure 17.5 shows the molar ratio of $HI/(HI + H_2O)$ in an "ideally-desulfurized" HIx solution as a function of sulfur dioxide partial pressure. Here, the molar ratio was calculated by assuming a hypothetical operation, where all the sulfuric acid was reduced to sulfur dioxide by the reverse Bunsen reaction and the total sulfur dioxide gas was separated from the HIx solution by stripping or flushing. As seen in the figure, increasing the sulfur dioxide pressure enhances the Bunsen reaction even though the effect is not so large.

An interesting method was proposed by Norman et al. [16] to enrich the sulfuric acid in the lighter phase. The idea is to expose the lighter phase to liquid iodine that is saturated with sulfur dioxide, so that the Bunsen reaction proceeds consuming the water in the lighter phase. Increase of sulfuric acid concentration was demonstrated from 50 to 57 wt% by this method named "sulfuric acid boost reaction."

It is worth mentioning here that, aside from the LL separation, several methods have been pursued for the separation of the Bunsen reaction products. Dokiya et al. [17] proposed to carry out the Bunsen reaction using an electrochemical cell equipped with an ion-exchange membrane. With this method, the product separation can easily be attained because hydriodic acid is produced at cathode side while sulfuric acid is at anode side.

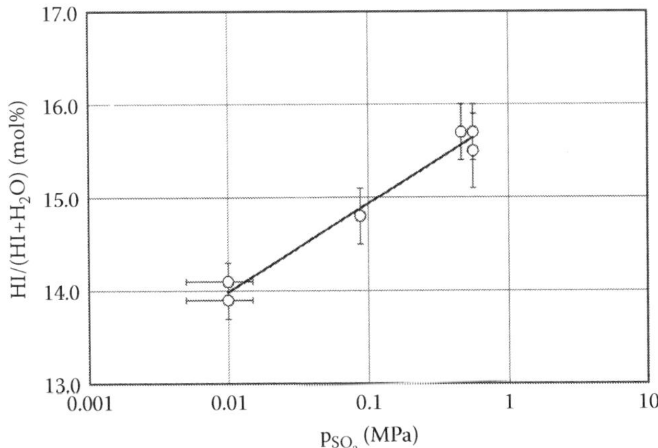

FIGURE 17.5
Effect of sulfur dioxide pressure on the Bunsen reaction under iodine saturation at 323 K. The ordinate indicates molar ratio of $x_{HI}/(x_{HI} + x_{H_2O})$ in an "ideally-desulfurized" HIx solution and the abscissa denotes sulfur dioxide partial pressure.

Additional merit of this method is that it needs no excess iodine for separation. Further, higher concentration than the quasi-azeotropic one may be obtained with the help of electric energy. Nomura et al. [18] examined this method to obtain the concentrated acidic solutions using Nafion® membrane. However, the membrane allowed such electrically neutral species as sulfur dioxide to permeate. The sulfur dioxide caused the formation of elemental sulfur. Introduction of additional elements such as Ni [19], Mg [20] was attempted to replace the acid separation with the separation of corresponding salts, which could be carried out utilizing the difference in their solubility. This method allows easier separation of the Bunsen reaction products than the LL separation. The main problem associated with this method is that it requires handling of solids. Solvent extraction was examined to separate the Bunsen reaction products, as well [21]. Recently, Taylor et al. [22] examined the possibility of using ionic liquids as extracting agents of hydrogen iodide from the heavier phase. Applicability of ionic liquids as reaction field of the Bunsen reaction is also an interesting research topic, since some ionic liquids show high solubility of iodine and sulfur dioxide.

Optimization of the performance of this section as a total considering heat/mass balance is described in Section 17.3.1.

17.2.2 Sulfuric Acid Decomposition

Although the SO_3 decomposition reaction can be carried out as a homogeneous gas phase reaction, the reaction rate is very small at temperatures lower than 900°C which can be achieved by using HTGRs. Therefore, catalysts are required to achieve technically feasible reaction rate, and have been studied since 1980s. Table 17.1 lists the main catalysts studied so far [16,23–29].

Norman et al. [16] found that platinum on TiO_2, ZrO_2, and SiO_2 showed good catalytic performance in wide temperature range. In addition, Fe_2O_3 and CuO showed good catalytic activity in the condition where the corresponding metal sulfates were thermodynamically unstable, that is, high-temperature and low-pressure condition. Yannopoulos and Pierre [25] noted that hematite (α-Fe_2O_3) was an efficient, chemically stable and economically acceptable catalyst for high temperature service. Norman et al. [16] reported that Al_2O_3 was a poor support for the platinum metal group discussing that aluminum sulfate poisoned platinum catalyst. The durability of catalyst is important for the economics of hydrogen production. However, few reports refer to the long-term durability [23,28]. Further efforts must be devoted to the study of catalyst stability.

Optimization of the performance of this section as a total considering heat/mass balance is described in Section 17.3.2.

17.2.3 Hydrogen Iodide Decomposition

17.2.3.1 Separation of Hydrogen Iodide from HIx Solution

In the hydrogen iodide (HI) decomposition section, prior to the decomposition reaction, HI must be separated from the HIx solution ($HI + I_2 + H_2O$ mixture, heavier phase), which is obtained as one of the LL-separated Bunsen reaction products. The separation is necessary to facilitate the decomposition of HI, since iodine is one of the reaction products and its presence hinders the decomposition reaction. This separation can be performed by distillation. However, such thermodynamic data of hydriodic acid and HIx solution as vapor–liquid equilibrium (VLE) and enthalpy are insufficient for optimizing the process design, in contrast to the case of sulfuric acid, which has been widely used in the chemical industry.

TABLE 17.1

List of Main Catalysts Studied for SO_3 Decomposition Reaction

Pt[a]				Pd–Ag	Al_2O_3	Fe_2O_3		$Fe_{2(1-x)}Cr_2O_3$	Cr_2O_3		CuO		$NiFe_2O_4$	$ZnFe_2O_4$	
Al_2O_3[a]	TiO_2	ZrO_2	SiO_2	–	–	Al_2O_3	–	–	Al_2O_3	–	Al_2O_3	–	Al_2O_3	Al_2O_3	Ref.
+					+	+					+				23
+						+									24
+	+	+	+		+	+			+		+				16
						+							+	+	25
+					+		+			+		+			26
				+			+								27
+	+	+													28
							+	+		+					29

Note: "+" mark denotes the catalyst is described in the references.

[a] The upper column indicates the catalyst components, whereas the lower column does the support.

In particular, in the VLE of HI + H_2O binary system and of HI + I_2 + H_2O ternary system, it is known that there exists azeotrope and quasi-azeotrope (or pseudo-azeotrope), respectively, where the molar ratio of HI and H_2O in equilibrium vapor is the same as that in liquid. These phenomena affect a lot the heat consumption of the distillation and so the thermal efficiency of hydrogen production. When HIx solution of lower HI concentration than that of quasi-azeotrope is fed to a distillation column, azeotropic mixture of HI + H_2O is distilled. This distillation consumes large heat energy for the vaporization of excess water and impairs the thermal efficiency of hydrogen production. However, precise experimental data are scarce, especially at elevated temperature and pressure range that is interested for the process.

Efforts have been devoted to acquire such data, some of which are summarized as follows. Total vapor pressure of HI + I_2 + H_2O ternary system was measured by Engels et al. [30] in the HI mole fraction range of up to 0.193 and temperatures of up to 553 K. Neumann [31] correlated the data using an NRTL equation modified with the "solvation model." Recently, Doizi et al. [32] built measurement apparatuses and their group reported vapor composition at high temperature and pressure conditions [33]. Hodotsuka et al. [34] and Liberatore et al. [35] reported isobaric VLE data at elevated pressure. Figure 17.6 shows the T–x–y curves of HI + H_2O binary system [34]. As seen in the figure, in the lower concentration range than the azeotrope, the main component of equilibrium vapor is H_2O, whereas it is HI in the higher concentration range. Here, it is noted that the binary azeotropic concentration shifts to the left, or to the more dilute side, as pressure increases. Concerning the HI + I_2 + H_2O ternary system, the quasi-azeotropic H_2O/HI molar ratio is about five at any I_2 composition. The drastic change of vapor composition of HI and H_2O between the solutions of lower concentration than the azeotrope and of the higher one described above for the binary system occurs also at the quasi-azeotrope of the ternary system.

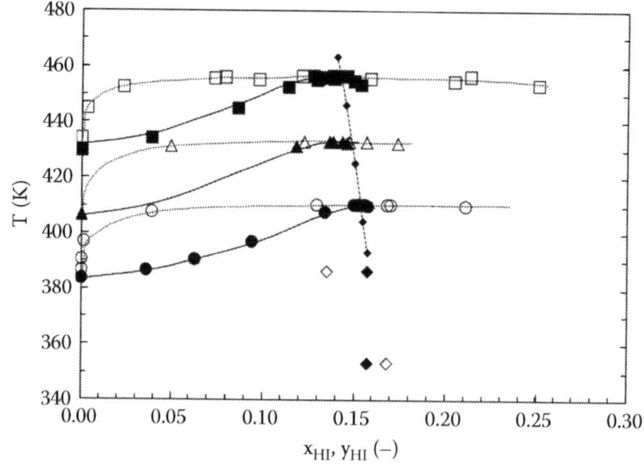

FIGURE 17.6
Isobaric VLE of HI + H_2O binary system. The abscissa denotes molar composition of HI in liquid (x) and gas (y) phases. (□, ■): 0.58 MPa, (△, ▲): 0.30 MPa, (○, ●): 0.15 MPa. Open symbols denote dew point and filled symbols denote bubbling point. The data shown with (◇, ◆) are those at 0.016 MPa and 0.059 MPa, which were reported by Doizi et al. [32]. The data shown with small diamonds that are connected with broken line indicate the azeotropic compositions evaluated by Engels et al. [30] based on the total vapor pressure measurements. (Adapted from M. Hodotsuka et al., *J. Chem. Eng. Data*, 53, 2008, 1683–1687. With permission.)

The HI concentration of HIx solution produced by the Bunsen reaction hardly exceeds that of the quasi-azeotrope according to the studies up to now. Distillation of such solution consumes large heat as described above and decreases the thermal efficiency of hydrogen production. Therefore, to overcome this difficulty, following methods have been examined so far.

1. *Extractive distillation using phosphoric acid* [13]. The HIx solution is first to be exposed to concentrated phosphoric acid so as to separate iodine by precipitation. Then, the remaining solution is processed by distillation. The solution, $HI + H_2O + H_3PO_4$ mixture also exhibits pseudo-azeotropy. However, as seen in Figure 17.7, the pseudo-azeotropy diminishes in the HI-free phosphoric acid concentration range of higher than 85 wt%. Therefore, dehydrated HI can be separated from the $HI + H_2O + H_3PO_4$ solution by distillation. The residue, diluted phosphoric acid solution, is concentrated and used again. In this extractive distillation concept, the heat duty for the distillation is greatly reduced. Instead, the heat of reconcentration of phosphoric acid is required. However, the heat duty requirement for the latter (reconcentration) is, in principle, lower than the heat duty reduction for the former (distillation), because the reconcentration needs no reflux. Additionally, in the reconcentration process, conventional energy saving techniques can be used such as the auto-vapor compression method to reduce the net input of thermal energy.
2. *Preconcentration of HIx solution by membrane techniques*. Membrane techniques may be applied, as a pretreatment of distillation, to increase the HI concentration of the HIx solution to over quasi-azeotropic so as to facilitate the separation of pure HI in the following distillation and reduce the reboiler duty.

   Electro-electrodialysis (EED) has been studied for this purpose [36–38]. In the EED, the redox reactions of iodine–iodide ion at electrodes,

$$\text{Cathode: } I_2 + 2e^- \rightarrow 2I^- \quad (17.12)$$

$$\text{Anode: } 2I^- \rightarrow I_2 + 2e^- \quad (17.13)$$

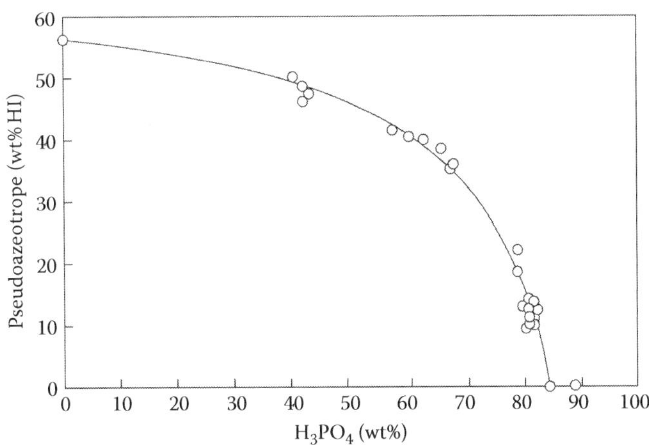

FIGURE 17.7
Pseudo-azeotrope line in $HI + H_2O + H_3PO_4$ system. Abscissa denotes the HI-free H_3PO_4 concentration. Ordinate denotes HI concentration in the pseudoazeotropic HI-H_2O solution of the $HI + H_2O + H_3PO_4$ solution. (From J.H. Norman, G.E. Besenbruch, and D.R. O'Keefe, GRI-80/0105, 1981. With permission.)

coupled with selective proton permeation through an ion-exchange membrane enable to increase the HI concentration of catholyte. The concentrated catholyte solution is used as the feed for the following distillation, by which pure HI can be obtained as distillate. The distillation requires only the similar reboiler duty as in the extractive distillation. EED requires electric energy to drive the redox reaction and the proton permeation. The principle energy requirements may just be the EMF of the relevant concentration cell. Concentrating the HIx solution to over quasi-azeotropic composition has been demonstrated in beaker-level experiments [38].

Application of pervaporation has also been studied to concentrate HIx solution [39–41]. The feasibility of aimed concentration was demonstrated using Nafion membrane. In this method, in principle, the same amount of heat is required as in the extractive distillation to vaporize the water. However, the required temperature level of heat may well be lower than that of the extractive distillation. Furthermore, membrane distillation was examined and the concentration over quasi-azeotrope was demonstrated by using polytetrafluoroethylene membrane [42].

17.2.3.2 Decomposition of Hydrogen Iodide

The distilled HI is decomposed into hydrogen and iodine. Catalysts for the gas phase thermal decomposition were investigated at GA [43] and at the National Chemical Laboratory for Industry, Japan, in the 1970s and 1980s [44–46]. Oosawa et al. [45] reported that platinum and active carbon showed high catalytic activity, and platinum on active carbon support (Pt/C) showed the highest one.

A rate equation of the catalytic reaction was derived which featured the Langmuir-Hinshelwood type adsorption isotherm. It was recommended the Pt/C catalyst should be used at the temperature range of 550–700 K from the view points of lowering the effects of reaction inhibition by iodine and of the water-gas reaction that consumed active carbon.

Equilibrium conversion of HI decomposition is limited to a low value of ca. 20 mol% that depends little on temperature. The low conversion induces large circulation of HI and its associated chemicals, such as iodine in the hydrogen production process. In order to enhance the one-pass conversion and to reduce the excess circulation, three ideas have been examined so far. The first idea is to carry out the gas phase HI decomposition while removing the product iodine from the reaction zone using some solid adsorbents. The one-pass conversion as high as 70 mol% was demonstrated by performing the reaction in batch mode at 450 K using Pt/C catalyst. The active carbon support acted as the iodine adsorbent [47]. The problems of this interesting idea were reported to be the low reaction rate and the lack of idea of energy efficient ways to recover the adsorbed iodine.

The second idea developed at GA is to carry out the decomposition in liquid HI [43,48]. When vapor and liquid coexist in $HI + H_2 + I_2$ system, equilibrium vapor pressure of I_2 is far lower than those of HI and H_2. The researchers of GA discussed that, in this condition, higher HI conversion was expected than that in the reaction system of homogeneous gas phase, since the liquid phase acted as an iodine absorber. Using noble metal catalysts such as platinum on TiO_2, ruthenium on TiO_2, the HI conversion of ca. 50 mol% was demonstrated with this concept by beaker-level experiments at 303 K. The extrapolation of experimental data suggested a technically feasible reaction rate might be attained at 423 K, the critical temperature of HI. The concept was further developed to include liquid phase homogeneous catalysts, for which palladium iodide was reported to be the candidate chemical. This idea is very promising, but it needs experimental verification in the expected practical operation conditions.

Application of ceramic membranes for hydrogen separation is the third idea, with which the reaction may be enhanced by removing the product hydrogen from the reaction field. In 1980s, use of porous membranes was discussed for this purpose [49]. Recently, in accordance with the progress of the study on ceramic membranes, application of amorphous silica membranes has been investigated and the one-pass conversion as high as 61 mol% was demonstrated in beaker-scale experiments at 723 K [50]. Concerning this idea, further development of the membrane preparation techniques is required aiming at achieving higher permeation rate and stability in the service environments.

17.2.3.3 Direct Dissociation of Hydrogen Iodide

In the total vapor pressure measurements of HIx solution using a quartz ampoule at temperatures higher than about 473 K, Engels and Knoche [30] observed evolution of hydrogen whose pressure was one order of magnitude higher than that expected from the homogeneous gas phase HI decomposition reaction equilibrium (Figure 17.8). The phenomenon was observed when the HI content in liquid was higher than pseudo-azeotrope and that of iodine was lean. Therefore, the phenomenon was considered to occur due to the iodine absorption by the liquid. This is the same as in the case of liquid HI decomposition described above in terms of the reaction enhancement by iodine separation from the reaction field. Based on this finding, they proposed a reactive distillation of HIx solution which they called "direct dissociation of HI" [51]. The idea is to carry out the distillation under such high pressure condition, where the enhanced hydrogen evolution realizes by the iodine absorption at the top of the distillation column. The reactive distillation performs two functions, that is, the separation of HI from HIx solution and the decomposition of HI into hydrogen and iodine, with one column. Therefore, the concept is very attractive due to its simplicity. It is noted for further study that, concerning the observed hydrogen evolution, it is not clear at present the reason why the decomposition reaction proceeded so fast. Berndhäuser and Knoche [52] measured the hydrogen evolution rate using an

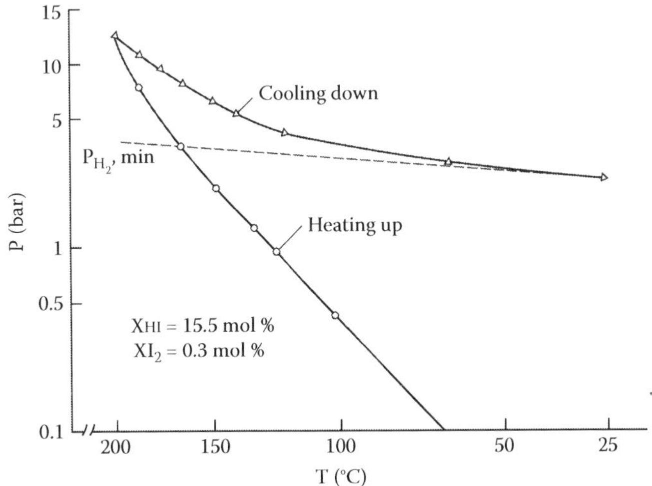

FIGURE 17.8
Total pressure measurement of a $HI + I_2 + H_2O$ solution. (From H. Engels and K.F. Knoche, *Int. J. Hydrogen Energy*, 11, 1986, 703–707. With permission.)

apparatus made of enamel and found that it was some orders of magnitude higher than that of homogeneous gas phase reaction.

Optimization of the performance of HI decomposition section as a total considering heat/mass balance is described in Section 17.3.3.

17.3 Process Performance Optimization

The heat/mass balance of the IS process has been studied on various flow sheets using thermal efficiency as a performance index. The thermal efficiency was defined as the ratio of the combustion heat of produced hydrogen to the overall heat input to the hydrogen production process. Higher heating value (HHV) of hydrogen, 286 kJ/mol H_2, has usually been used as the heat of combustion. The standard enthalpy changes of endothermic chemical reactions in the IS process featuring LL separation in the Bunsen reaction section are roughly estimated [53]. Here, the molalities of reactants, H_2SO_4 and HI, are assumed to be 12 and 10 mol/kg H_2O, respectively.

$$H_2SO_4(aq) \rightarrow SO_2(g) + H_2O(g) + 1/2O_2(g) \quad \Delta H = 325.4 \text{ kJ/mol } H_2 \tag{17.14}$$

$$2HI(aq) \rightarrow H_2(g) + I_2(g) \quad \Delta H = 157.8 \text{ kJ/mol } H_2 \tag{17.15}$$

The enthalpy changes may represent the target figures in the flow sheet optimization in terms of thermal efficiency.

17.3.1 Bunsen Reaction

The Bunsen reaction (Section 17.2.1) is an exothermic reaction and needs no heat input. However, the composition of the reaction products is an important parameter for the heat/mass balance of the hydrogen production process, since it governs the conditions of the main energy consuming operations such as the concentration of sulfuric acid and the distillation of HIx solution. When high concentration of HI in heavier phase and high concentration of H_2SO_4 in lighter phase are obtained, circulation amount within the HI and H_2SO_4 decomposition sections is small and high thermal efficiency is expected. From this viewpoint, high iodine concentration in the reaction mixture is favorable since it enhances the Bunsen reaction. However, too high iodine concentration induces too much iodine inventory and impairs the economics of hydrogen production. Recently, Giaconia et al. [54] studied the LL separation behavior in concentrated acidic solution with modest iodine concentration. Lee et al. [55] discussed the optimal reaction condition from the viewpoints of water distribution between the phases, side reactions, iodine solubility, operating costs, and so on. Figure 17.9 shows the proposed conditions.

17.3.2 Sulfuric Acid Decomposition

In the sulfuric acid decomposition section, sulfuric acid supplied from the Bunsen reaction section is first concentrated from 50–57 wt% to ca. 90 wt%. The concentrated sulfuric acid is then evaporated at temperatures of 573–773 K. In the course of evaporation, decomposition of H_2SO_4 occurs spontaneously to produce water (H_2O) and sulfur trioxide

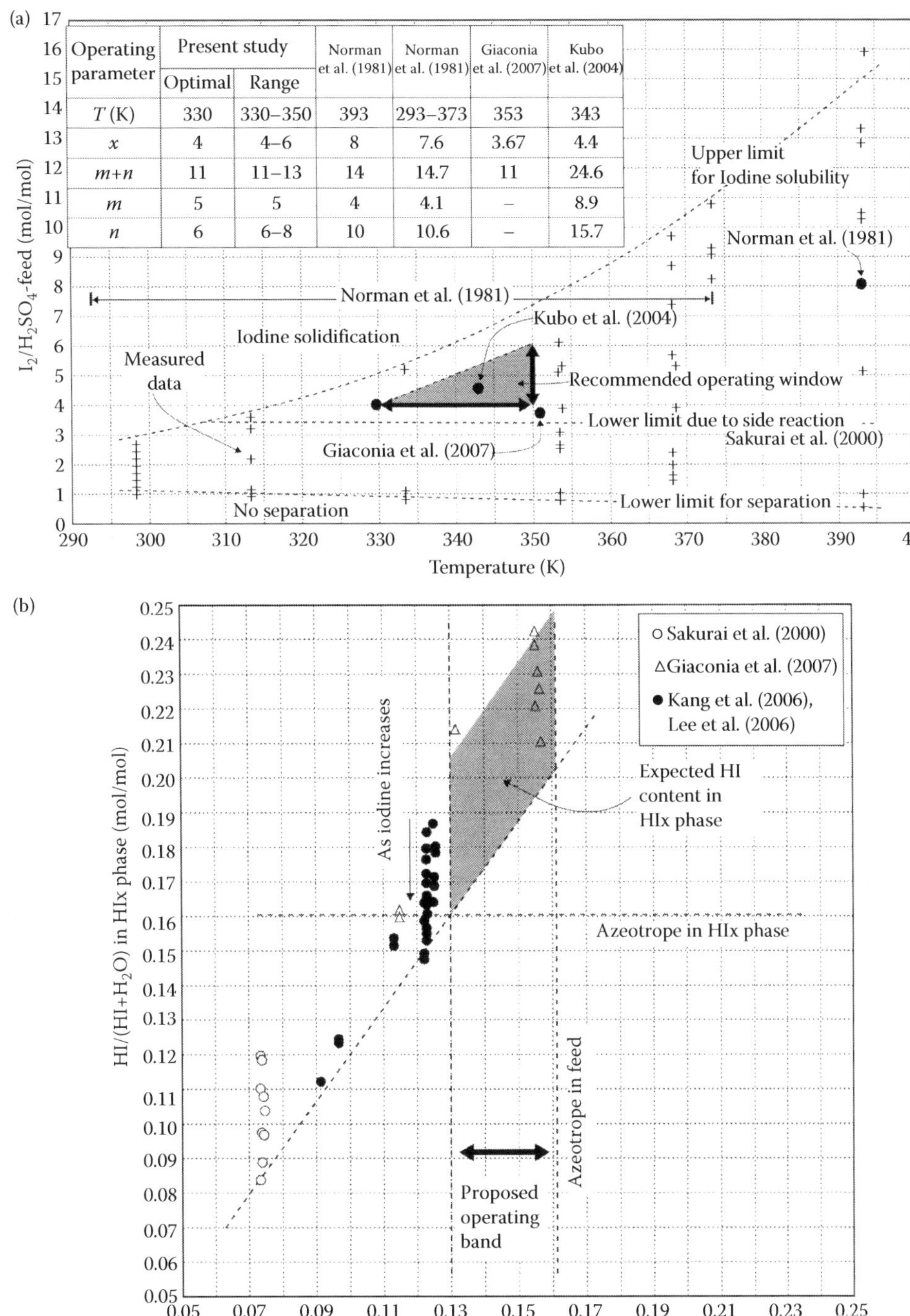

Operating parameter	Present study		Norman et al. (1981)	Norman et al. (1981)	Giaconia et al. (2007)	Kubo et al. (2004)
	Optimal	Range				
T (K)	330	330–350	393	293–373	353	343
x	4	4–6	8	7.6	3.67	4.4
m+*n*	11	11–13	14	14.7	11	24.6
m	5	5	4	4.1	–	8.9
n	6	6–8	10	10.6	–	15.7

FIGURE 17.9
Optimal operating window for the Bunsen reaction: (a) recommended reaction condition; (b) Expected HI content in the HIx phase produced. The abscissa and ordinate of the plot indicate the molar ratio of $x_{HI}/(x_{H_2O} + x_{HI})$ in the feed and HIx phase solution, respectively. (From B.J. Lee et al., *Int. J. Hydrogen Energy*, 33, 2008, 2200–2210. With permission.)

(SO_3) as shown in reaction (Equation 17.5). Finally, SO_3 is decomposed to sulfur dioxide (SO_2) and oxygen (O_2) at 1073–1123 K as described in reaction (Equation 17.6). The produced gaseous mixture (SO_2 and O_2) are sent back to the Bunsen reaction section, where the gases are separated by the selective adsorption of SO_2 into the Bunsen reaction solution. Investigation of each component in this section is described in Section 17.2.2.

Development of an energy-efficient concentration scheme is a main subject in this section. The optimization of heat-mass balance has been studied as described below. Here, it should be noted that these studies assumed the initial sulfuric acid concentration of 57 wt% that was obtained by the "boost reaction" mentioned in Section 17.2.1. In order to reduce the thermal burden for the concentration operation, application of conventional energy-saving methods has been studied such as auto-vapor compression [56] and multi-effect evaporation [57]. Both techniques use two or more vaporizers. In auto-vapor compression, vapor produced in one vaporizer is compressed to make its condensation temperature higher so that the heat of condensation is recovered in other evaporator(s). In multieffect evaporation, the vaporizers are operated under different pressures so that the heat of condensation of the high pressure vapor is recovered in the low pressure vaporizer. Figure 17.10 shows the flow sheet proposed by Knoche et al. [57]. Later, Öztürk et al. [58] proposed a flow sheet that incorporated two kinds of heat recovery techniques, three-stage evaporation and direct contact heat exchange. In the three-stage evaporation, the operation pressure was varied in the range of 1.2 MPa to 0.008 MPa in order to recover the heat of vaporization of water. In the direct contact heat exchange, hot product gaseous mixture of sulfur trioxide decomposition reaction was allowed to contact directly with the sulfuric acid for the recovery of unreacted SO_3 and preheating of the solution to be

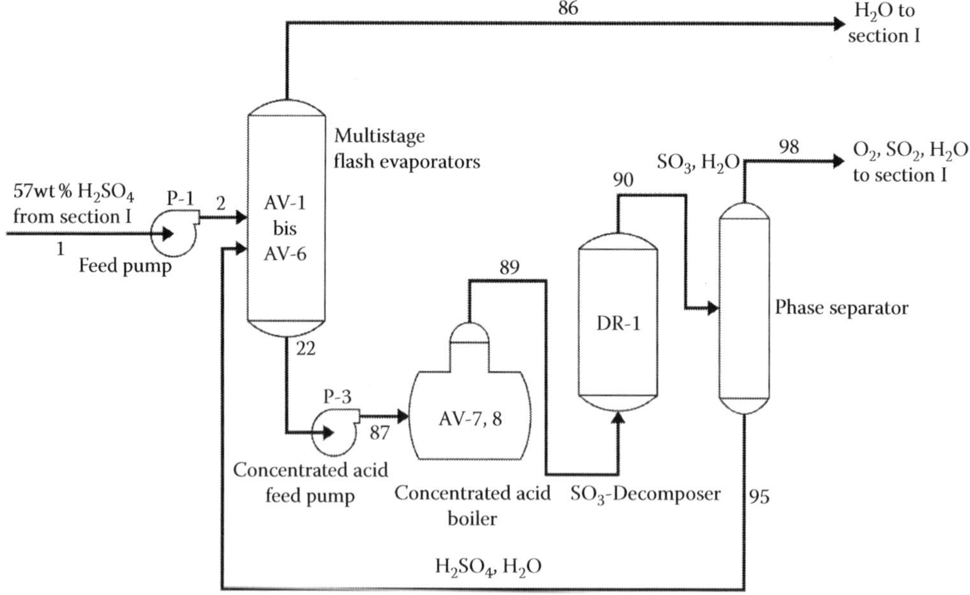

FIGURE 17.10
Simplified flow sheet of sulfuric acid decomposition section featuring multieffect evaporation considered by Knoche et al. AV-1 bis AV-6, acid vaporizers Nos. 1 through 6; AV-7, 8, acid vaporizers Nos. 7 and 8; DR-1, SO_3 decomposition reactor; P-1, pump No. 1; P-3, pump No. 3. Section I denotes Bunsen reaction section. (From K.F. Knoche, H. Schepers, and K. Hesselmann, *Proc. 5th World Hydrogen Energy Conf.*, Toronto, July 1984, **2**, 487–502. With permission.)

vaporized. It was reported that total heat demand for the sulfuric acid decomposition section featuring this concept was estimated to be 366.7 kJ/mol-H_2.

17.3.3 Hydrogen Iodide Decomposition

In this section, the separation of HI from HIx solution and the decomposition of HI are performed. As described in Section 17.2.3, many options have been proposed for these operations, and various process flow sheeting studies have been carried out to optimize the heat/mass balance. The following are the outline of the representative studies, where the feed HIx solution was assumed to have $HI/I_2/H_2O$ molar ratio of 0.1/0.39/0.51 which was selected by Norman et al. [56] in their initial flow sheeting study.

Norman et al. [56] investigated a flow sheet that adopted extractive distillation and liquid phase catalytic HI decomposition. One simplified flow sheet resulted from this study was reported as shown in Figure 17.11. Here, the HIx solution (line No. 301) is fed to the column C-302 and mixed with phosphoric acid to separate most of the iodine (ca. 95%). The overhead solution (line No. 312) is fed to the extractive distillation (C-303), by which most of HI and small amount of iodine are separated from the phosphoric acid. The distillate is further processed by another distillation column (C-304) to obtain pure HI. The HI thus obtained (line No. 333 in Figure 17.11a which connects to the line No. 401 in Figure 17.11b) is catalytically decomposed at ca. 393 K and 5 MPa at R-401 with the conversion ratio of about 30 mol%. The diluted phosphoric acid is concentrated and recycled to C-302. Multi-effect evaporation (E-318, -319, -320, and -321) is adopted for the reconcentration. By adopting intensive heat recovery network including power recovery system, the net power input was 104 kJ/mol H_2 and the net heat input was 151 kJ/mol H_2.

Goldstein et al. [59] discussed a flow sheet featuring reactive distillation. Figure 17.12 shows the schematic flow sheet. Here, the HIx solution is fed to the distillation column by the pump, P, which raises the system pressure to 5 MPa. H_2 is produced by reactive distillation in the column. The residue of distillation with temperature of 584 K is cooled down to 548 K, the same temperature with the distillate, by the heat exchanger, E2, and mixes with the distillate at the separator, S. Some part of HI and H_2O in the distillate is condensed into the liquid. The product gas containing H_2 is taken from Exit 1, whereas the separated liquid returns to the Bunsen reaction section from Exit 2 after reducing its pressure to 0.2 MPa. It was assumed to recover the sensible heat released at the heat exchangers, E2, E3, and E4, for the preheating of HIx solution at the heat exchanger, E1. Also, latent heat recovery was assumed using a water heat pump system by which part of the heats released at the condenser, Ec, and the separator, S, were used at the reboiler, Eb. The net power input was 132 kJ/mol H_2 and the net heat input was 111 kJ/mol H_2, assuming minimum temperature difference of 10 K for the heat exchangers.

Kasahara et al. [60] studied a flow sheet that adopted an EED cell for concentrating the HIx solution and a hydrogen separation membrane for enhancing the gas-phase HI decomposition. Figure 17.13 shows the conceptual flow sheet. Here, the HIx solution (line No. 6) supplied from the Bunsen reaction section is mixed with the residue solution (line No. 13) of the HI distillation column. The mixture is separated into two streams and fed to the EED cell. The HI-rich stream (line No. 10) from the cell is fed to the distillation column. HI gas (line No. 16) is assumed to be taken as the distillate and sent to the HI decomposer featuring hydrogen perm-selective ceramic membranes. The net power input was 102 kJ/mol H_2 and the net heat input was 167 kJ/mol H_2 based on the analysis using the experimental values of EED operation parameters and assuming minimum temperature difference of 10 K for heat exchangers.

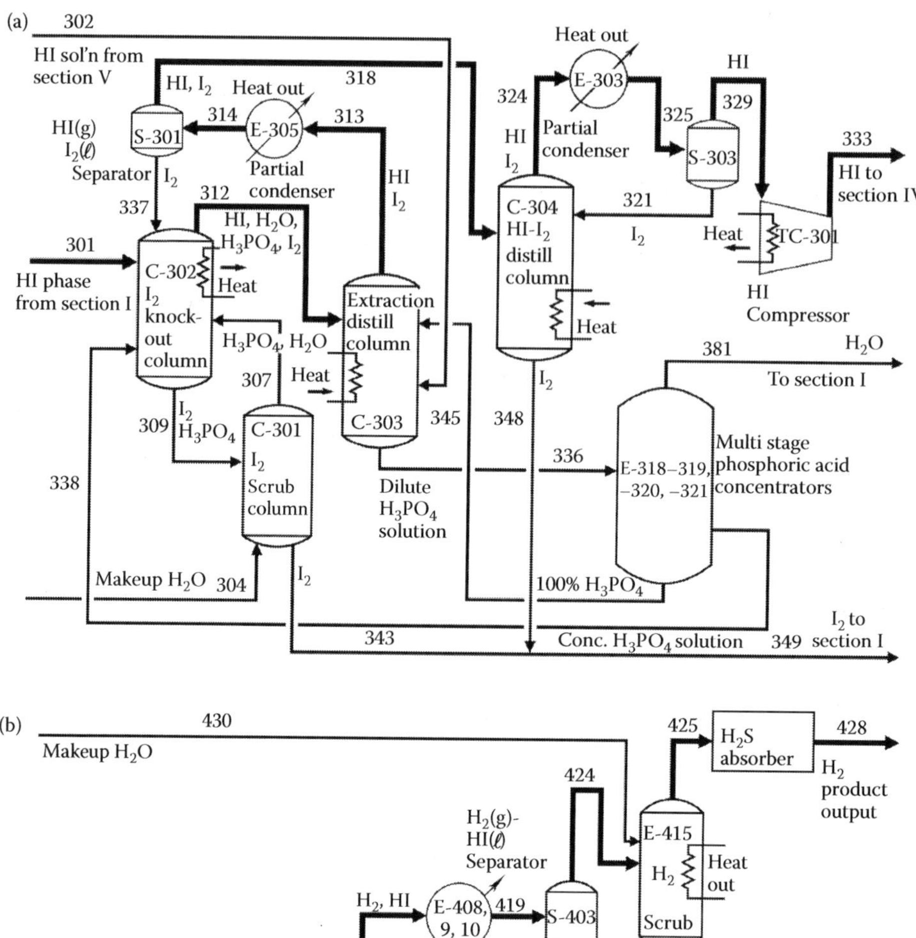

FIGURE 17.11
Flow sheet of hydrogen iodide decomposition section featuring extractive distillation and liquid phase catalytic HI decomposition designed by Norman et al. The upper figure (a) shows the HI separation using phosphoric acid as the extractive agent, whereas the lower figure (b) does the HI decomposition part. Sections I, III, and IV in these figures denote the Bunsen reaction section, the HI separation subsection (the upper figure), and the HI decomposition subsection (the lower figure). HI sol'n from section V in the upper figure is an error of the original figure. This flow is actually from line No. 431 in the lower figure. S-303 is the HI(g) I_2(l) separator. (From J.H. Norman et al., GA-A16713, 1982. With permission.)

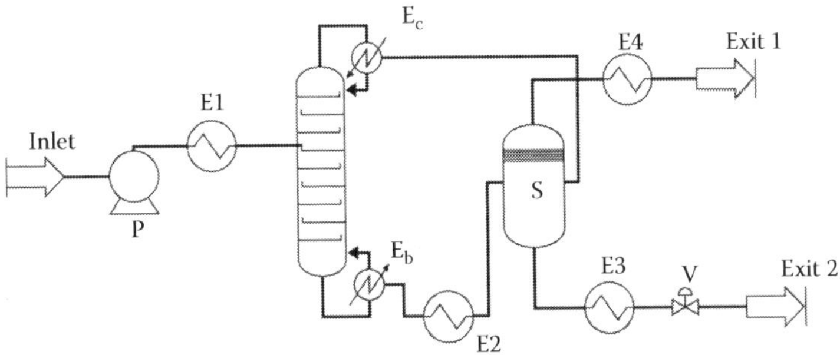

FIGURE 17.12
Conceptual flow sheet of hydrogen iodide decomposition section featuring reactive distillation considered by Goldstein et al. E1, E2, E3, and E4, heat exchanger; E_b, reboiler; E_c, condenser; P, pump; S, separator; V, valve. Inlet flow is from the Bunsen reaction section. (From S. Goldstein, J. M. Borgard, and X. Vitart, *Int. J. Hydrogen Energy*, 30, 2005, 619–626. With permission.)

17.3.4 Overall Process Efficiency

Heat/mass balance of the hydrogen production process integrated with a HTGR was examined by Norman et al. [56]. They estimated the thermal efficiency of 47% on hydrogen production process featuring LL separation in the Bunsen reaction section and extractive

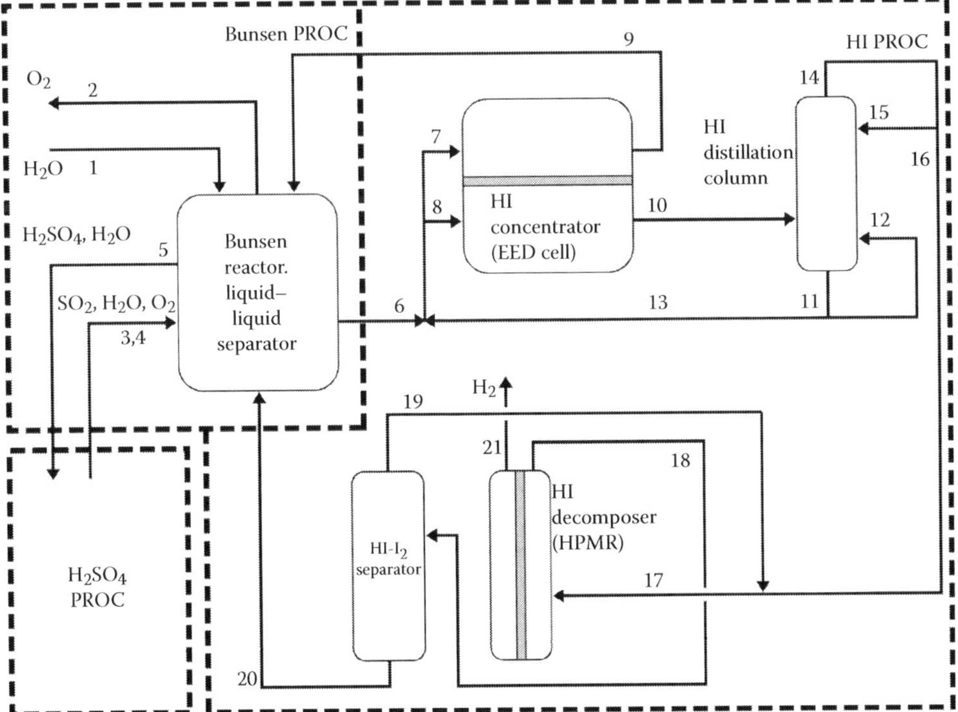

FIGURE 17.13
Conceptual flow sheet of hydrogen iodide decomposition section (denoted as HI PROC) featuring electro-electrodialysis and membrane hydrogen separation considered by Kasahara et al. "PROC" is the abbreviation of "procedure," denoting the same thing as "section" in the text.

distillation and liquid phase HI decomposition in the HI decomposition section. Considering the intensive heat recovery using complicated heat exchanger network adopted in their study, the reported figure may represent a kind of upper bound of thermal efficiency expected in the combination of the assumed unit operations. In the following studies, no optimization was carried out for the overall process. However, the researchers discussed the overall thermal efficiency by combining the optimization results of their studied section with those of other sections reported in literature. Goldstein et al. [59] reported an upper bound of 51% and a best estimate efficiency of 33–36% for a system featuring reactive distillation. Kasahara et al. [60] estimated an efficiency of 57% for the system that featured EED assuming the ideal membrane separation performance and no over potential, and 34% based on the experimental parameters. It was noted that over 40% could be expected by improving the EED cell performance. In summary, the realistic thermal efficiency of hydrogen production by the IS process featuring LL separation may be around 35% based on the experimental data till now, and increasing the value to 40% is possible by improving key unit operations such as membrane processes. It should be reminded here that there exists a space of further study concerning the Bunsen reaction and the HIx distillation as discussed above. The experimental and analytical studies on these key reaction and separation may lead to further improvement as suggested by the recent study on the Bunsen reaction by Lee et al. [61]. They summarized the efficiency estimated in literature as Table 17.2.

17.4 Hydrogen Production Tests by Combinations of Unit Operations

17.4.1 GA Closed-Loop Cycle Demonstration

In 1979, GA demonstrated the IS process in a closed-loop using recycled materials to provide feasibility under recycle conditions [56]. The subsection summarizes the demonstration operation. A closed-loop cycle demonstrator (CLCD) was designed and constructed by using glass and quartz, which was not in duplicate conditions of a process engineering flow sheet to meet the purpose of the feasibility demonstration. The CLCD was furnished with a set of equipments in order to complete the loop of the IS cycle. A Bunsen reactor at temperature of 45°C produced the two separate liquid phases, which were fed to the phase separator intermittently. The lighter phase was then purified and H_2SO_4 was concentrated from 30 wt% to about 95 wt% in a H_2SO_4 purification boiler; the liquid H_2SO_4 was vaporized and decomposed to produce O_2 in a reactor filled with pellets of Fe_2O_3 catalyst at 850°C. The undecomposed H_2SO_4 recycled to the boiler and the resulting SO_2–O_2 mixture recycled directly to the Bunsen reactor. The heavier phase was degassed (SO_2 removal) under a vacuum condition. The solution ($HI + I_2 + H_2O$) was then fed directly to a decomposition reactor without catalyst at 900°C to produce hydrogen. The resulting H_2O, I_2, and unreacted HI were condensed and recycled to the Bunsen reactor. The rate of H_2 production was about 1.2 L/h. Operation of the CLCD was accomplished subsequently in a complete recycle mode: the Bunsen reaction products were formed by reacting the recycled H_2O, I_2, and SO_2. No operational problems happened: all valves, joints, pumps, and temperature and flow controls worked as designed. On the other hand, problems were recognized that a small quantity of sulfur was observed in the recycled liquids in the condenser placed after the HI decomposition reactor and a small amount of H_2S was collected in the H_2 purification trap placed after the

TABLE 17.2
Thermal Efficiency Evaluation Results Reported the on IS Process

		Norman et al. (1981)	Roth and Knoche (1989)	Öztürk et al. (1995)	Buckingham et al. (2004)	Goldstein et al. (2004)		Kasahara et al. (2007)		Sakaba et al. (2007)	Present Work[c]
Energy Demand (kJ/mol)		**BE[a]**	**BE[a]**	**BE[a]**	**BE[a]**	**Max[a]**	**BE[a]**	**Max[a]**	**BE[a]**	**BE[a]**	**BE[a]**
Sulfuric acid section	Heat	460	411.4	366.7		352	420	411.4	411.4	514	411–420
HI section	Heat	148	237			187	375–454	119.5	166.7		167
Electricity	Required thermal energy[b]							–27.4	255.6	142	
Pumping						6 + 11	6 + 11				17
Efficiency	% (HHV)	47		56	39	51	33–36	57	34	44	47–48

Source: Adapted from B.J. Lee et al., *Int. J. Hydrogen Energy*, 34, 2009, 2133–2143.

[a] BE denotes the best-estimated efficiency, whereas Max denotes the maximum efficiency.

[b] Required thermal energy means to heat demand to generate the power for electricity and pumping.

[c] The work of Lee et al., the authors of the source.

condenser. These problems were caused by omitting the purification step which removes the sulfur-containing species (SO_2 and H_2SO_4) from the heavier phase.

17.4.2 GA Bench-Scale Investigations

After the CLCD operation, GA studied a bench-scale system [56,62] under continuous operation conditions as a part of an engineering evaluation. The system consisting of three subunits (HI–H_2SO_4 production and separation, H_2SO_4 concentration and decomposition, and HI concentration and decomposition) was designed to produce H_2 at 60 L/h, constructed, and operated. The operations of each of the subunits were all tested except the redesigned HI distillation column from H_3PO_4 solution; the system was basically considered operational for an integrated test which was not programmed. Following sebsections summarize the bench-scale system operation.

17.4.2.1 Subunit for HI–H_2SO_4 Production and Separation

The subunit for HI–H_2SO_4 production and separation consisted of a Bunsen reactor, a gas-liquid–liquid phase separator, a SO_2 delivering system and an I_2 delivering system, which were made of the Pyrex glass, the Teflon fitting and valves, and some tubes (Hastelloy B) and valves (Hastelloy C) for the I_2 delivering system. The Bunsen reactor and the separator were placed in a thermostated forced-convection oven. The Bunsen rector was a co-current reactor with a finned tube (10 mm I.D., 340 mm long) in a jacket to adjust temperature. The residence time of the SO_2 gas in the reactor was below 1 s. The reactor operated at about 200 kPa and mixed the reactants of SO_2, I_2 and H_2O to produce HI and H_2SO_4. The desired iodine flow rate was established by a positive displacement ceramic metering pump and SO_2 flow was delivered from a pressurized gas cylinder through a metering valve and a flow meter. The separator which accepted the fluids from the Bunsen reactor was an invented gravity-overflow-separator system. The lighter H_2SO_4 phase overflowed to a H_2SO_4 reservoir and also the heavier HIx phase overflowed to a HIx reservoir.

The subunit was operated by delivering known flow rates of I_2, H_2O, and SO_2 using design weight ratio of I_2 to H_2O of 7.0. A series of experiments was carried out in this condition (I_2 flow rate of 112 g/min, H_2O flow rate of 16 g/min, and SO_2 flow rate of over 3.2 g/min). Analyses of the lighter phase after degassing indicated H_2SO_4 concentration of between 46 and 53 wt% which were near or over the designed concentration. The heavier phase was not analyzed. These experiments verified the concept of the Bunsen reactor and the operability of this subunit.

17.4.2.2 Subunit for H_2SO_4 Concentration and Decomposition

The subunit for H_2SO_4 concentration and decomposition received the lighter phase liquid (at an absolute pressure from ambient to 3.1×10^2 kPa and at a temperature from ambient to 115°C) from the subunit for HI–H_2SO_4 production and separation. This unit mainly consisted of an iodine stripping column, a sulfuric acid concentration fractionation-still and a sulfur trioxide decomposer. Materials of construction were glass, quartz, alumina ceramic, Teflon, a few kinds of plastics, and 316 stainless steel. Quartz was used for the decomposer, the ceramic was used for a metering plunger pump to supply precise-flow feed for the still and the decomposer.

Iodine stripping experiments were carried out for the iodine removal from the lighter phase solution, which will reduce potential corrosion problems in concentration of the H_2SO_4.

Prior to the stripping experiments, compositions of synthetic feed solutions were examined to prevent formations of iodine-crystals and solid sulfur, which caused partial flow blockage and feed composition changes. To determine the highest levels of iodine concentration, four different solutions were prepared, and kept for observations; results were as follows:

1. 0.2 wt% I_2, 1.2 wt% HI, 50 wt% H_2SO_4, no precipitation in 18 h.
2. 0.4 wt% I_2, 1.2 wt% HI, 50 wt% H_2SO_4, no precipitation in 18 h.
3. 0.6 wt% I_2, 1.2 wt% HI, 50 wt% H_2SO_4, I_2 precipitation occurred in the night of the prepared day.
4. 0.8 wt% I_2, 1.2 wt% HI, 50 wt% H_2SO_4, immediate precipitation.

Accordingly, all feed solutions for the I_2 stripping experiment were prepared in composition of 0.6 wt% I_2, 1.2 wt% HI, 50 wt% H_2SO_4. The iodine stripping column was worked to make the feed solutions contacted against the counter current of heated water vapor generated from the solutions itself. The contaminated I_2 was vaporized and the contaminated HI returned into I_2, H_2O, and SO_2 by reverse Bunsen reaction with H_2SO_4 as reaction (Equation 17.11). This column consisted of a quartz tube, 60 cm in length and 3 cm in diameter, which was filled with glass beads of 3 mm in diameter. This tube was heated externally by heating tapes, and insulated with fiber frays. Results of the stripping experiments indicated that feed flow rates of 10 and 17 cm^3/min showed very good stripping of the I_2; I_2 concentration of below 2 ppm was obtained with a operation temperature over about 150°C for 10 cm^3/min feeding, and with a operation temperature over about 170°C for 17 cm^3/min feeding. Moreover, these results suggested the I_2 concentration in the liquid outflow was probably controlled by the rate of the back-reaction between HI and H_2SO_4 to generate I_2; also the residence time of the fluid flow in the column and the rate of steam flow were probably important parameters with respect to this reaction.

H_2SO_4 concentration experiments were carried out just to demonstrate the operation of the sulfuric acid concentration fractionation still with respect to accepting the upstream output and meeting the downstream input concentration because extensive VLE data had already existed for the H_2SO_4-H_2O system and theoretical possibility was already clear. A temperature of about 330°C at a still reboiler was maintained, which is necessary to produce H_2SO_4 concentration levels of greater than 96 wt%, on feeding of 5–10 cm^3/min of 60 wt% H_2SO_4. This experiment proved attainment of satisfactory operation at initial design levels.

Sulfur trioxide decomposition experiments were carried out with a catalyst of a platinum on commercial-zirconia-support, that is, right cylinder pellets in 3.2 mm diameter loaded with 0.2 wt% platinum. A reactor, 60 mm in outer diameter and 610 mm in length, was used at a temperature between 800°C and 900°C; the gas plug flow residence time in the reactor was 0.3–1 s. The catalyst was filled in a half region of the reactor at downstream side. Conversion of the decomposition was calculated by measuring the amount of unreacted H_2SO_4 which condensed in a decomposer effluent and the amount of SO_2 absorbed in alkali solution. At feed flow rates of 3.0 and 5.3 cm^3/min, the difference of conversions from experimental results were within 2% of the theoretical prediction using maximum catalyst temperature and the chemical equilibrium composition calculated from thermodynamic data tables. A data showed a low conversion compared to the theoretical one, which was caused by the significant back reaction. This effect occurred in the case that the catalyst was filled in low temperature region of the reactor. It should be noted that the zirconia bed at the boiling interface of the concentrated H_2SO_4 feed was chemically attacked.

Overall simultaneous operation of all this subunit components was demonstrated at a flow rate equivalent to 2 standard L/min of H_2 production. All process steps functioned as designed until after 12–14 h operation.

17.4.2.3 Subunit for HI Concentration and Decomposition

The subunit for HI concentration and decomposition processed the heavier phase to produce hydrogen. Unit operations in this subunit were: (1) a removal of dissolved SO_2 from the heavier phase obtained from the subunit for HI–H_2SO_4 production and separation, (2) a I_2 knockout with a concentrated H_3PO_4 to separate HI and H_2O, (3) a washing of I_2 for recycle to the Bunsen reaction, (4) a distillation of HI from H_3PO_4 solution, (5) a reconcentration of H_3PO_4, (6) a catalytic HI decomposition, and (7) a recovery for recycle or distribution reactant HI and products I_2 and H_2.

The removal of SO_2 from the heavier phase was performed by a packed column which was a countercurrent inert gas purging system. This unit was tested with the heavier phase made in the subunit for HI–H_2SO_4 production and separation. The success of this gas purge was evidenced by the apparent lack of appearance of sulfur downstream. The I_2 knockout was performed by counter current contact of the degassed solution with concentrated H_3PO_4 in a Raschig ring packed column. The exiting impure I_2 was pumped to a second column for the washing of I_2. This second column was operated at 2 atm to keep the liquid water from vaporization at the operating temperature of 115°C. The system was run in conjunction with the subunit for HI-H_2SO_4 production and separation and performed as intended, although this test was shortened by a HIx pump malfunction.

The distillation of HI from the H_3PO_4 solution was performed by a packed column with small Raschig rings; this column worked at low flow rate. However, as flow rate was increased, the flooding occurred and separation was lost. Large packing materials improved the flow rate for this relatively viscous working fluid without flooding; nevertheless, column efficiency was decreased. An attempt at mass balancing indicated that higher flow rate was possible and that behavior was roughly as desired.

The H_3PO_4 concentration column concentrated the 85 wt% H_3PO_4 outflow from the HI distillation by heating up to 222°C to remove water. Highly dehydrated high-temperature phosphoric acid was formed in the vapor space from contact of H_3PO_4 with heating element. The dehydrated phosphoric acid attacked both the heater and the glass vessel in the vapor region. A deep vessel to accept full immersion heater below the liquid line solved this problem. Moreover, the column without any packing ran for many hours and produced 96 wt% H_3PO_4 at design level rates because H_3PO_4, one of the products is basically nonvolatile.

The catalytic gas phase HI decomposition was performed using cooled outflow from the HI distillation. The HI vapor containing I_2 from the HI distillation column was first cooled to the temperature slightly exceeding the melting point of iodine (114°C). This condensable I_2 in the distillate was separated from the vapor before the decomposition. A thermal decomposition reactor as a vertical tubular heated chamber employing an activated charcoal catalyst bed was operated at temperatures in 177–440°C and ambient atmospheric pressure. The temperature controlled volume for HI decomposition was 4×10^{-3} m^3, including catalyst volume. The HI decomposition gaseous products were cooled to a temperature below the melting point of iodine to separate condensates from the vapor. The residual gas was chilled by a stainless-steel HI trap with liquid nitrogen or dry ice to remove all condensable vapors from the gas. The H_2 purification step was countercurrent scrubbing with water to remove low concentration of HI from the residual gas of the HI trap. This unit was operated with no major engineering problem; the conversion of HI decomposition reaction was near equilibrium conversion at measured reactor outlet temperature.

17.4.3 JAEA Continuous and Closed-Cycle Operations

The IS process features a unique characteristic, namely, the operation under continuous and closed-cycle conditions. This feature is important for running the IS process with no waste generation. However, actual closed-loop operation is technically difficult.

To ensure the IS process be a practical operational chemical plant, Japan Atomic Energy Agency (JAEA) implemented studies on the process operation [63] by laboratory-scale and bench-scale test facilities. Figure 17.14 shows photographs of the facilities. Techniques for recycling chemicals were investigated through laboratory scale tests to complete the closed-cycle in the first stage. The test facility was mainly made of glass, with the required heat supplied by electricity. By developing chemical recycling techniques, Nakajima et al. succeeded in producing 1 L/h of hydrogen over 48 h [64]. For stable and durable hydrogen production, control methods for the closed-cycle operation were essential. Hence JAEA constructed a larger test facility [65] to develop such methods in the second stage. Figures 17.15 and 17.16 show the view of equipments and the simplified flow sheet of the facility, respectively. The facility had sufficient size to allow instruments for the measurements and controls to be integrated with. By further developing the process control method, stable hydrogen production of 31 L/h was successfully accomplished for 175 h [66].

17.4.3.1 Techniques for Recycling Chemicals during the Lab-Scale Stage

All chemicals except water input and hydrogen and oxygen output circulate through the IS process, changing their chemical forms under the closed-cycle condition, and are thus

Lab–scale

Bench–scale

FIGURE 17.14
Continuous and closed-cycle operation test facilities of the iodine–sulfur process.

FIGURE 17.15
The hydrogen production bench-scale test facility of the iodine–sulfur process.

completely recycled. However, conducting such an operation involves various practical difficulties. During the lab-scale stage, key techniques for recycling chemicals were investigated.

Purification, to remove and recycle impurities, was one of the key techniques making up part of the closed-cycle, which targeted the complete separation of the residual acid to prevent any side reactions forming sulfur or hydrogen sulfide. Although the solution produced from the Bunsen reaction can be separated into lighter and heavier phases in the presence of an excess amount of iodine, they are contaminated by each other's components. The lighter phase (sulfuric acid phase) is rich in sulfuric acid, which contains impurities of HI and I_2 as minor components. The heavier phase (HIx phase) is rich in HI and I_2, which contains impurities of sulfur compounds as its minor component. To purify the phase separated acids, wetted-wall columns were adopted for the purifiers using nitrogen as a stripping gas for the evolved SO_2. Based on examinations of the purifier [9], complete separation of the sulfuric acid involved in the HIx phase was attained above a wall temperature of 170°C. By adapting the purifier, the sulfur, separated from the HIx phase in the form of sulfur dioxide, could be recycled back to the Bunsen reaction step. In addition, the purification of the sulfuric acid phase was possible in the same manner.

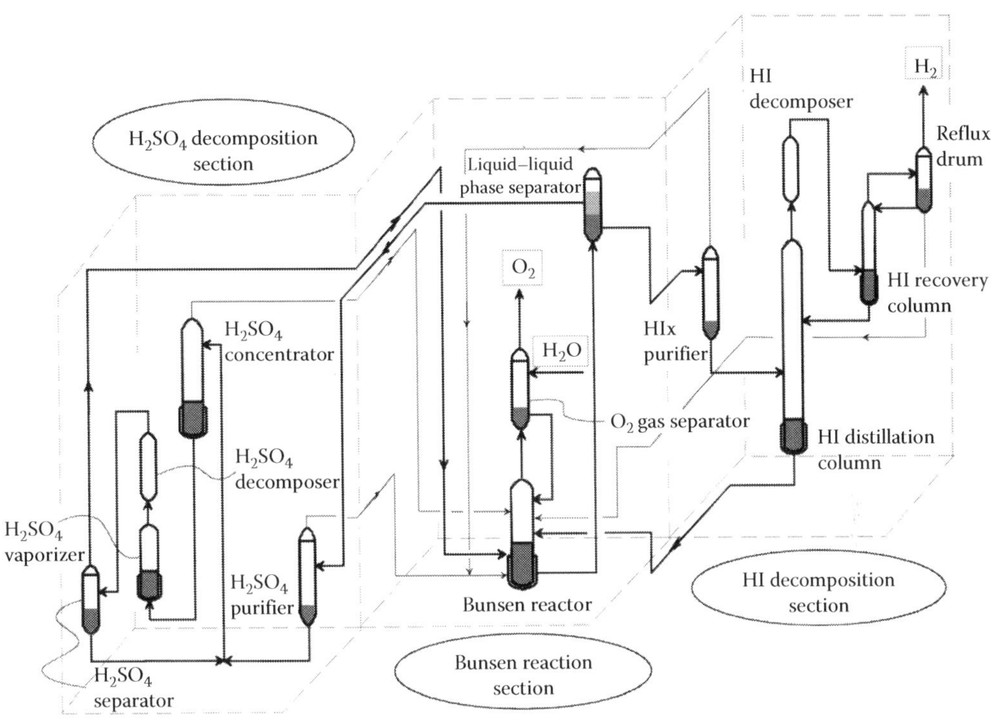

FIGURE 17.16
Simplified flow sheet of the hydrogen production bench-scale test apparatus.

Compositions of the Bunsen reaction and the liquid–liquid separation (LL separation) solutions represent basic design data, which was used to determine the precise material balances for the closed-cycle operation. The compositions, which depended upon the Bunsen reaction, were investigated experimentally [14,67] for the $HI–H_2SO_4–I_2–H_2O$ mixture produced by absorbing SO_2 until saturation. In addition, the compositions of both phases in the LL separation were investigated. The compositions of solutions were quantified by a chemical titration method, devised for the multicomponent mixed solutions [83]. For the Bunsen reaction, the acid concentration in the product solution under iodine saturation increased with increasing the temperature, due to increase of the iodine solubility. For the phase-separated solutions, high concentrations of acids and iodine were vital to attain the high separation factor. In contrast, the high acid concentrations caused side reactions, forming sulfur and/or hydrogen sulfide, and high iodine concentration resulted in the precipitation of solid iodine. Based on the investigations, the acquired compositions of the phase-separated solutions were adopted to evaluate the flow rates from the separation section into the decomposition sections. Besides, the conditions of the solutions for the laboratory-scale hydrogen production test were determined so as to prevent side reactions and solid iodine precipitation and thus, facilitate handling.

Transportation of the melting iodine involved technical difficulties, linked to the need to recycle iodine from the HI decomposition section into the Bunsen reaction section. The iodine during transportation had to be maintained in liquid phase, namely between melting and boiling points. In the case of pure iodine, the temperatures were 113.6°C and 184.4°C, respectively. The materials used for pumps to transport this chemical also had to be capable of resisting such a corrosive environment; hence, nonseal structures were desirable for the pumps. In order to grow the low volume flow rate for the laboratory-scale hydrogen production test, a pump system was devised [68], employing a glass capillary tube with an injection nozzle for the carrier gas. The melting iodine was forced by the gas pressure to rise in the tube until it gained a level sufficient to facilitate flow down into the destination vessels.

17.4.3.2 Process Control Methods during the Bench-Scale Stage

The IS process had to have the desirable and unique feature of being operable on a closed cycle condition. All chemicals involved circulated through the process as the chemical forms changed by several reactions, that is to say, they had to be recycled. Because of influences on the recycling chemicals in the closed-loop, establishing stable hydrogen production was quite difficult during practical operations. Therefore, to develop process control methods for stable hydrogen production, which maintained the process in a stable state, was vital. To achieve this, the process had to be controlled to produce hydrogen and oxygen at the molar ratio of 1:0.5. To realize this object, a fundamental concept of a process control method [66] was developed to maintain a balance between the reacting constituents of chemical reactions and the raw material supply. The method involved both the manipulated and controlled parameters, with the latter controlled by the former, being created in the process.

To construct the bench-scale test facility, techniques for recycling chemicals and technical knowledge collected at the lab-stage, such as the means of preventing side reactions, were incorporated into the facility. Moreover, the experimental examinations of the process fluid at atmospheric pressure and at the temperature and compositions of the solutions for the Bunsen reaction were carried out [65] using a small glass apparatus. Throughout the examinations, the molar flow rate for the detailed design of the facility was fixed. Equipment and operational designs were established based on the molar flow rate from which the throughput of process fluids was estimated.

The operating temperature of the Bunsen reaction was investigated during the examination. In processing concentrations, it is advantageous to obtain rich acids, high concentration HI and H_2SO_4 as the products of the Bunsen reaction. Reducing impurities, namely HI and I_2 in the lighter phase and H_2SO_4 in the heavier phase, facilitated the processing of purifications. Examinations showed the I_2 fraction of the heavier phase increased in correlation to the reaction temperature increase, and the concentrations of acids also increase with the I_2 fraction. The concentrations of impurities decreased as the I_2 fractions gained. With these results and optimized handing of the solutions in mind, a temperature of 70°C was chosen to operate the Bunsen reaction for the bench-scale hydrogen production test.

Maintaining the compositions of the Bunsen reaction solutions was investigated throughout the examinations. The followings were essential to work the Bunsen reaction satisfactorily. First, the SO_2 had to be entirely reacted to ensure the sulfur compounds was involved in the process at atmospheric pressure without emission to environment. Next, I_2 had to be completely dissolved in the solution to prevent any clogging of pipes due to I_2 forming a solid cake. Additionally, the solution had to be separated into two phases to obtain two rich acids of HI and H_2SO_4. Based on the examinations, the target compositions of the Bunsen reaction to be maintained during the bench-scale hydrogen production test were determined which allowed the demands for the Bunsen reaction to be satisfied.

17.4.3.3 Continuous and Closed-Cycle Hydrogen Production Tests

Both the laboratory- and bench-scale facilities were made of fluorine resin, glass, and quartz; comprising fundamental reactors and separators and operated at atmospheric pressure. The heat required for the operations was supplied by electricity. The laboratory-scale facility adopted techniques for recycling chemicals, while the bench-scale facility employed automated control methods for longer term hydrogen production.

Figure 17.17 shows the results of production by the laboratory- and bench-scale test facilities. The two facilities carried out continuous and closed-cycle hydrogen production tests to demonstrate the techniques and control methods, respectively. These tests were successfully completed for 48 and 175 h, respectively. The respective levels of hydrogen production were maintained at virtually constant rates of 1 and 31 L/h, while the production ratio of oxygen to hydrogen correlated almost exactly to the volume ratio 0.5:1, the evidence of stoichiometric water-splitting.

Regarding the evaluation of operating stability, the number of times for consumption of circulatory chemicals was determined; the number, *N*, was given by

$$N_i = \frac{r_i}{n_i}, \tag{17.16}$$

where r_i is the total amount of the reacted chemical during the experiment, n_i is the total contents of the chemical in the test facility, and *i* represents the components of HI, H_2SO_4, I_2, and H_2O. r_i was estimated based on the volumes of hydrogen and oxygen production. N_i was counted in Table 17.3 for both the hydrogen production tests. In the lab-scale test, N_{HI} and $N_{H_2SO_4}$ were 1. The bench-scale test achieved a level of 1 for N_{H_2O}, the largest amount of constituents in the facility. Therefore, the test span was longer than that required for the single replacement of all the water contained in the facility by chemical reactions. Based on this evaluation, the stability of this control method was confirmed.

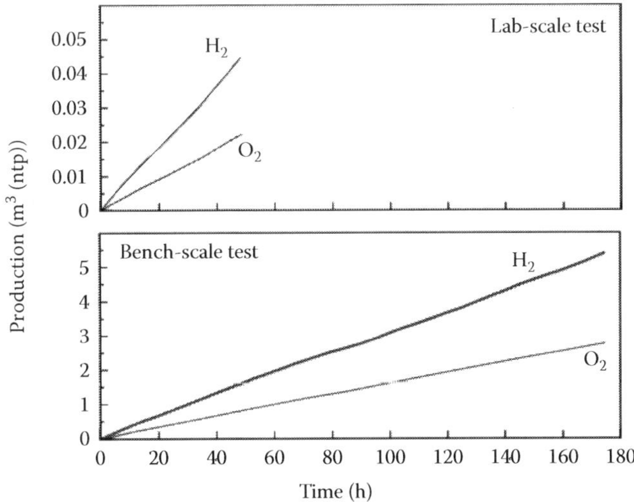

FIGURE 17.17
Hydrogen and oxygen production over time in the lab- and bench-scale tests.

TABLE 17.3

Number of Times for Consumption of Circulatory Chemicals

N_i	Laboratory	Bench
HI	1	15
H_2SO_4	1	15
I_2	0.2	2
H_2O	0.1	1

Source: Adapted from S. Kubo et al., *Proc. Nuclear production of hydrogen 3rd Information exchange meeting*, Oarai, October 2005, 197–204.

17.4.4 International Nuclear Energy Research Initiative Integrated Laboratory-Scale Experiment

Integrated laboratory scale (ILS) experiments [69,70] for the IS process were projected to provide the technical basis to evaluate the potential of the IS process for large-scale hydrogen production using nuclear energy. The ILS experiments were performed to demonstrate the closed-loop operation, and to provide the technical basis to support selection decisions for nuclear hydrogen production [71] through 100–200 L/h hydrogen production operations under process conditions expected at the pilot plant-level and beyond, that is, pressure up to 20 bar using engineering materials of construction. The ILS experiments for the IS process were conducted as an International Nuclear Energy Research Initiative (INERI) project and were collaborative efforts involving the Sandia National Laboratories (SNL) for the section of sulfuric acid decomposition, GA for the section of HI decomposition, and the French Commissariat à l'Énergie Atomique (CEA) for the section of the Bunsen reaction. Each participant was responsible for the design, construction, standalone testing, and integrated operation. The process sections were constructed on skids and were interconnected through a chemical storage skid. All chemicals used in the experiment, with the exception of sulfur dioxide, were transferred to and from tanks located in the chemical storage skid. Because of the concern over its high toxicity, sulfur dioxide was not stored and was transferred directly from the sulfuric acid decomposition section to the Bunsen reaction section. The Bunsen reaction section employing a counter-current Bunsen reactor was constructed by CEA, which was made of glass-lined steels or tantalum-tungsten alloys; the operation was performed at below 130°C and 4–6 bar. SNL covered the sulfuric acid decomposer. The component was constructed from commercially available silicon carbide bayonet tubing which was able to withstand the corrosive nature of the acid at 850°C. The component was operated at a pressure of 2 bar absolute with 41 mol% acid feed. The HI decomposition section constructed by GA, which was operated at pressure ranging 0.31–1.4 MPa gauge, employed the liquid–liquid extraction process utilizing phosphoric acid.

The results [71,72] from the project included several partially integrated runs conducted to produce H_2 at up to 75 L/h. The Bunsen reaction section demonstrated successful control of both the lighter and heavier phase product flows, reactor pressure and temperature control and the reactant flow control (water, sulfur-dioxide and iodine). Analysis for samples of the process fluids from the experiment showed the lighter phase was dilute, at

12 wt% H_2SO_4, while the heavier phase was close to the target at 40 wt% HI. The operation of the HI decomposition section showed the extractive distillation worked well, however it seems this distillation is complex and too costly for large-scale applications. The SiC bayonet reactor for the sulfuric acid decomposition operated routinely with no material problems with sulfur dioxide production at rate of 100–200 L/h.

17.5 Process Engineering

17.5.1 Materials of Construction

Sulfuric acid and HIx solution are very corrosive. Therefore, the selection of materials of construction from the viewpoint of corrosion resistance is one of the most important problems for the development of large-scale plants, and various corrosion tests have been conducted on commercially available materials. Most of the tests were performed as the constant temperature immersion tests and/or exposure tests in the simulated process environments. The test duration was about 1000 h. The test results were published elsewhere, which are summarized as follows [14,73–78].

17.5.1.1 Bunsen Reaction

In the Bunsen reaction environment, the mixture of sulfuric acid and HIx solution is to be processed at temperatures of around 373 K. In these environments, Ta, Zr, SiO_2, SiC, Si_3N_4, fluorocarbon resins, and so on showed excellent corrosion resistance. These materials may be used as corrosion-resistant lining/coating materials.

17.5.1.2 Sulfuric Acid Decomposition

The environments of the sulfuric acid decomposition section consisted of the liquid phase with H_2SO_4 concentration of 50–98 wt% and temperature of up to around 673 K and the gas phase of mainly SO_3, SO_2, O_2, and H_2O up to around 1123 K. In the former liquid environments, silicon containing materials such as SiC and Fe–Si showed excellent corrosion resistance. Soda glass also showed corrosion resistance in the environments and is a promising lining material. In the latter gas environments, refractory alloys that have been used in conventional industrial plants such as Incoloy 800, Hastelloy C-276, and Inconel 625 showed corrosion resistance.

17.5.1.3 Hydrogen Iodide Decomposition

In the HI decomposition section, mixture of HI, I_2, and H_2O (HIx solution) is to be processed in liquid phase with temperature range of up to around 573 K. The corrosion tests clarified that Ta, Zr, and SiC showed corrosion resistance in the environments. In the gaseous mixture of HI, H_2O, I_2, and H_2 at temperature range of 473–673 K, good corrosion resistance was observed in Ti, Ta, Mo, and Ni–19Cr–19Mo–1.8Ta alloy. Conventional refractory alloys such as Hastelloy C-276 and Incoloy 625 also showed corrosion resistance. However, Ta suffered hydrogen embrittlement.

17.5.2 Development of Components with Thermal and Corrosion Resistance

Liquid sulfuric acid makes the severest environment in the IS process. Researchers at JAEA and Toshiba Corp. proposed the concept of a sulfuric acid decomposer, in which the sulfuric acid of concentration of 90 wt% was evaporated and, simultaneously, part of H_2SO_4 decomposed spontaneously into gaseous sulfur trioxide and water using the heat of high-pressure helium gas [79,80]. The decomposer was equipped with a block-type heat exchanger made of SiC which showed excellent corrosion resistance and mechanical strength in the assumed operation condition. Figure 17.18 shows the cut-away view of the decomposer designed for the test plant with a hydrogen production capacity of 30 Nm^3/h. Its integrity was confirmed by preliminary thermal stress analysis and the fabricability was confirmed by test fabrication of SiC block components, as well. Also, helium leak tests using a mock up model of the decomposer confirmed its good seal performance against horizontal loading simulating earthquake motion.

Researchers at SNL developed a bayonet-type heat exchanger made of SiC for the sulfuric acid processing in the ILS experiments [81]. The exchanger was designed to integrate a sulfuric acid vaporizer, a super-heater, a catalytic SO_3 decomposer, and a recuperator into a single bayonet unit. Figure 17.19 shows the concept. This design eliminated all high temperature connections, making the use of glass- or Teflon-lined steel only in lower temperature areas was enough. The components were fabricated and its performance was confirmed by test operations in the range of 200–300 L/h of SO_2 production.

It should be worth noting that researchers at Westinghouse Electric [82] have developed a similar concept of bayonet-type reactor made of SiC for the sulfuric acid processing in the hybrid sulfur cycle.

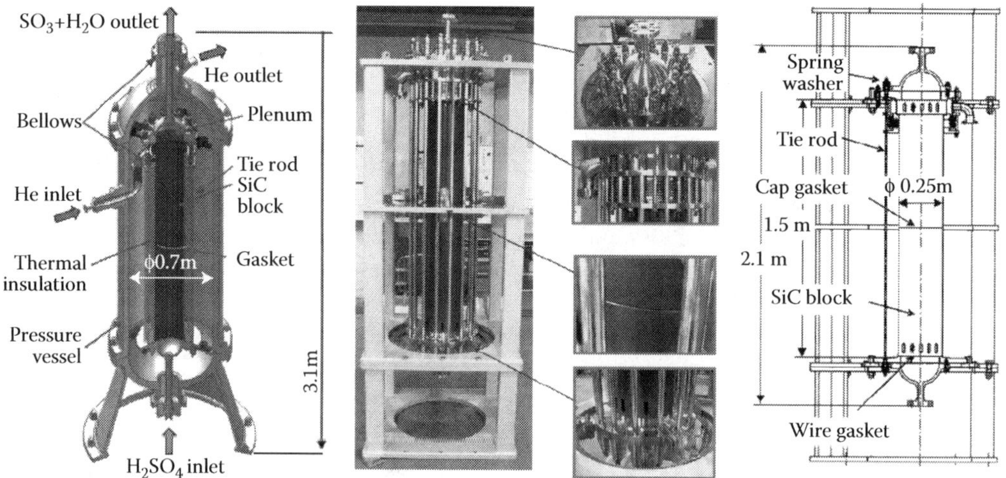

FIGURE 17.18
A heat exchanger made of SiC ceramic (leftmost one, conceptual design; the other ones, test-fabricated heat exchanger blocks).

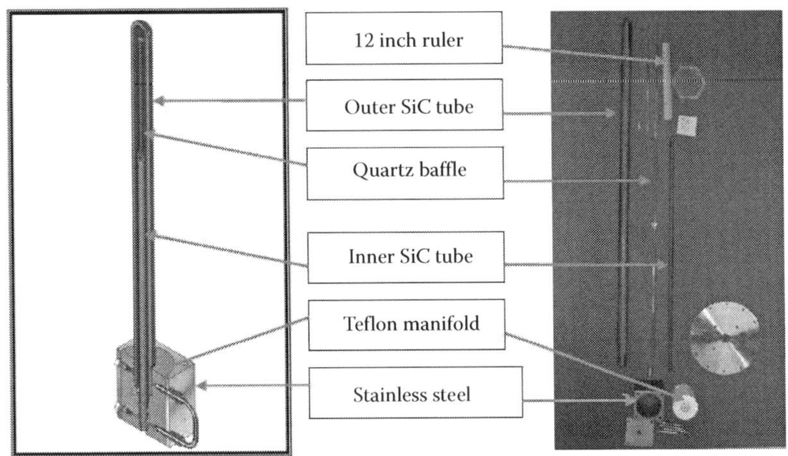

FIGURE 17.19
Schematic of the half-scale (686 mm) silicon carbide bayonet decomposer for the ILS experiments. (From F. Gelbard, R. Moore, and E. Parma, SAND2007-7841, 2007. With permission.)

17.6 Future Technical Challenges

The IS process offers an attracting concept of carbon-free large-scale hydrogen production. Many efforts to date have resulted in the steady accumulation of chemical and technical knowledge on the IS process. Based on these achievements, it is desired to carry out the hydrogen production experiments using an apparatus made of industrial materials and driven by helium gas heating. Figure 17.20 shows the schematic of a test plant considered [79]. The high pressure helium gas of about 4 MPa is heated by an electric heater, which simulates the intermediate heat exchanger of a HTGR to about 1173 K and flows into the IS process test plant, where such main endothermic chemical reactors of IS process as the SO_3 decomposer, the sulfuric acid decomposer, and the HI decomposer are heated by the helium gas in series. The lowest temperature helium-gas-heated component is the reboiler of the HI distillation column. The helium gas is cooled down to about 673 K by these four IS components and returns to the electric heater. The plant is constructed of process equipment made of industrially available materials. Optimum process conditions and performance can be evaluated and validated. The test operations include startup, shutdown, and automated control of steady-state hydrogen production over a long period of time. The experiments can provide useful information for the process development such as

1. Technical basis for constructing and operating IS process plants connected to HTGR by verifying the integrity of process components and the process controllability.
2. Technical data for verifying the estimated process performance such as thermal efficiency.

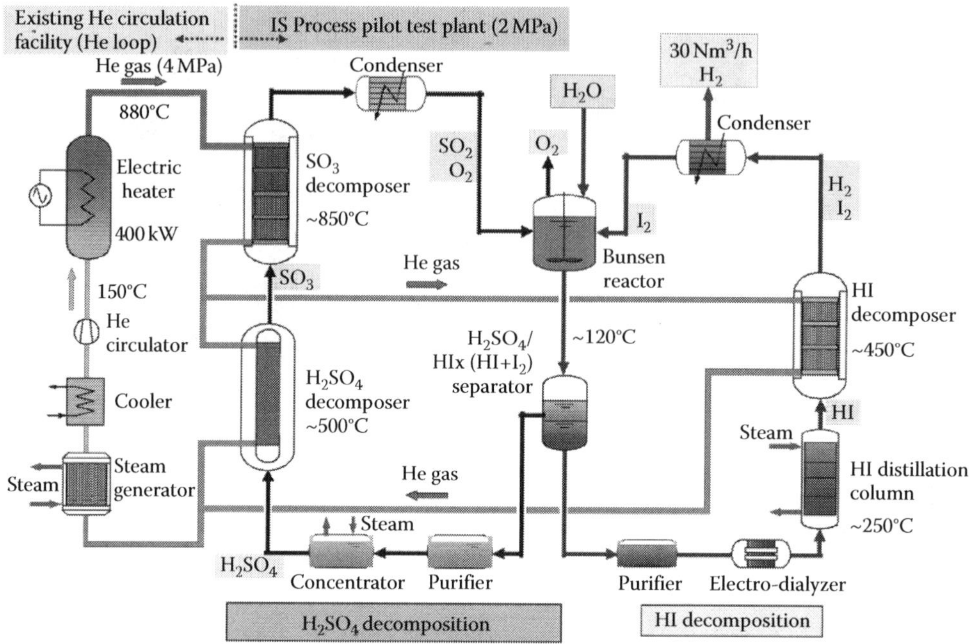

FIGURE 17.20
Schematic flow sheet of a helium-heated IS plant.

At the same time, efforts should be devoted for the development of economical hydrogen production process by further improving the process heat/mass balance and also by reducing the plant cost.

References

1. S. Solomon, D. Qin, M. Manning, Z. Chen, M. Marquis, K.B. Averyt, M. Tignor, and H.L. Miller (ed.), Contribution of Working Group I to the Fourth Assessment Report of the Intergovernmental Panel on Climate Change, In Intergovernmental Panel on Climate Change (eds.), *IPCC Fourth Assessment Report: Climate Change 2007*, Cambridge and New York, Cambridge University Press, 2007.
2. S. Sato, Thermochemical hydrogen production. In T. Ohta (ed.), *Solar-Hydrogen Energy Systems*, Pergamon Press, Oxford, 1979, 81–114.
3. J.E. Funk and R.M. Reinstrom, Energy requirements in the production of hydrogen from water, *I&EC Process Design and Development*, 5, 1966, 336–342.
4. J.E. Funk, Thermochemical hydrogen production: Past and present, *Int. J. Hydrogen Energy*, 26, 2001, 185–190.
5. J.L. Russell, Jr., K.H. McCorkle, J.H. Norman, J.T. Porter II, T.S. Roemer, J.R. Schuster, and R.S. Sharp, Water splitting—A progress report, *Proc. 1st World Hydrogen Energy Conf*, Miami Beach, March 1976, 1, 1A105–1A124.
6. S. Shimizu, K. Onuki, H. Nakajima, and S. Sato, The nickel–iodine–sulfur process for thermochemical hydrogen production IV. Bunsen Reaction, *Denki Kagaku*, 53, 1984, 114–118 (in Japanese).

7. L. Gmelin, *Gmelins Handbuch der Anorganischen Chemie (8th edition), J*, Verlag Chemie, Berlin, 1933 (in German).
8. M. Atterer, M. Becke-Goehring, K.-C. Buschbeck, L. Gmelin, R.J. Meyer, F. Peters, and E.H.E. Pietsch, *Gmelins Handbuch der Anorganischen Chemie (8th edition), S*, Verlag Chemie, Berlin, 1960 (in German).
9. M. Sakurai, H. Nakajima, K. Onuki, and S. Shimizu, Investigation of 2 liquid phase separation characteristics on the iodine–sulfur thermochemical hydrogen production process, *Int. J. Hydrogen Energy*, 25, 2000, 605–611.
10. K. Onuki, S. Shimizu, H. Nakajima, S. Fujita, Y. Ikezoe, S. Sato, and S. Machi, Studies on an iodine–sulfur process for thermochemical hydrogen production, *Proc. 8th World Hydrogen Energy Conf.*, Honolulu and Waikoloa, July 1990, 2, 547–556.
11. T. Kumagai, C. Okamoto, and S. Mizuta, Thermal decomposition of magnesium sulfate and separation of the product gas mixture, *Denki Kagaku*, 52, 1984, 812–819 (in Japanese).
12. M. Sakurai, H. Nakajima, R. Amir, K. Onuki, and S. Shimizu, Experimental study on side-reaction occurrence condition in the iodine–sulfur thermochemical hydrogen production process, *Int. J. Hydrogen Energy*, 25, 2000, 613–619.
13. J.H. Norman, G.E. Besenbruch, and D.R. O'Keefe, Thermochemical water-splitting for hydrogen production, GRI-80/0105, 1981.
14. K. Onuki, H. Nakajima, I. Ioka, M. Futakawa, and S. Shimizu, IS process for thermochemical hydrogen production, *JAERI-Review* 94–006, 1994.
15. H. Nakajima, Y. Imai, S. Kasahara, S. Kubo, and K. Onuki, Effect of sulfur dioxide partial pressure on the reaction of iodine, sulfur dioxide and water, *Kagaku Kogaku Ronbunshu*, 33, 2007, 257–260 (in Japanese).
16. J.H. Norman, K.J. Mysels, R. Sharp, and D. Williamson, Studies of the sulfur–iodine thermochemical water-splitting cycle, *Int. J. Hydrogen Energy*, 7, 1982, 545–556.
17. M. Dokiya, T. Kameyama, and K. Fukuda, Thermochemical hydrogen preparation—Part V. A feasibility study of the sulfur iodine cycle, *Int. J. Hydrogen Energy*, 4, 1979, 267–277.
18. M. Nomura, S. Nakao, H. Okuda, S. Fujiwara, S. Kasahara, K. Ikenoya, S. Kubo, and K. Onuki, Development of an electrochemical cell for efficient hydrogen production through the IS process, *AIChE J.*, 50, 2004, 1991–1998.
19. S. Shimizu, S. Sato, H. Nakajima, and Y. Ikezoe, The nickel–iodine–sulfur process for thermochemical hydrogen production. I. Proposal, *Denki Kagaku*, 49, 1981, 699–704 (in Japanese).
20. S. Mizuta and T. Kumagai, Thermochemical hydrogen production by the magnesium–sulfur–iodine cycle, *Bull. Chem. Soc. Jpn.*, 55, 1982, 1939–1942.
21. G. De Beni, G. Pierini, and B. Spelta, The reaction of sulfur dioxide with water and a halogen. The case of iodine: Reaction in presence of organic solvent, *Int. J. Hydrogen Energy*, 5, 1980, 141–149.
22. M.L. Taylor, P. Styring, and K. Allen, Novel organic solvents for the Bunsen reaction, *Proc. AIChE 2008 Annual Meeting*, Philadelphia, November 2008, 201a.
23. H. Ishikawa, E. Ishii, I. Uehara, and M. Nakane, Catalyzed thermal decomposition of H_2SO_4 and production of HBr by the reaction of SO_2 with Br_2 and H_2O, *Proc. 3rd World Hydrogen Energy Conf.*, Tokyo, June 1980, 1, 297–309.
24. S. Sato, Y. Ikezoe, T. Suwa, S. Shimizu, H. Nakajima, and K. Onuki, Studies on closed-cycle processes for hydrogen production, V (Progress Report for the F. Y. 1980), JAERI-M 9724, 1981, 53–67 (in Japanese).
25. L.N. Yannopoulos and J.F. Pierre, Hydrogen production process: High temperature-stable catalysts for the conversion of SO_3 to SO_2, *Int. J. Hydrogen Energy*, 9, 1984, 383–390.
26. H. Tagawa and T. Endo, Catalytic decomposition of sulfuric acid using metal oxides as the oxygen generation reaction in thermochemical water splitting process, *Int. J. Hydrogen Energy*, 14, 1989, 11–17.
27. V. Barbarossa, S. Brutti, M. Diamanti, S. Sau, and G. De Maria, Catalytic thermal decomposition of sulphuric acid in sulphur–iodine cycle for hydrogen production, *Int. J. Hydrogen Energy*, 31, 2006, 883–890.

28. D.M. Ginosar, L.M. Petkovic, A.W. Glenn, and K.C. Burch, Stability of supported platinum sulfuric acid decomposition catalysts for use in thermochemical water splitting cycles, *Int. J. Hydrogen Energy*, 32, 2007, 482–488.
29. A.M. Banerjee, M.R. Pai, K. Bhattacharya, A.K. Tripathi, V.S. Kamble, S.R. Bharadwaj, and S.K. Kulshreshtha, Catalytic decomposition of sulfuric acid on mixed Cr/Fe oxide samples and its application in sulfur–iodine cycle for hydrogen production, *Int. J. Hydrogen Energy*, 33, 2008, 319–326.
30. H. Engels and K.F. Knoche, Vapor pressures of the system $HI/H_2O/I_2$ and H_2, *Int. J. Hydrogen Energy*, 11, 1986, 703–707.
31. D. Neumann, Phasengleichgewichte von $HJ/H_2O/J_2$-Loesungen, Diplomarbeit, RWTH Aachen, Mai 1987.
32. D. Doizi, V. Dauvois, J.L. Roujou, V. Delanne, P. Fauvet, B. Larousse, O. Hercher, P. Carles, C. Moulin, and J.M. Hartmann, Total and partial pressure measurements for the sulphur-iodine thermochemical cycle, *Int. J. Hydrogen Energy*, 32, 2007, 1183–1191.
33. B. Larousse, P. Lovera, J.M. Borgard, G. Roehrich, N. Mokrani, C. Maillault, D. Doizi, et al. Experimental study of the vapour–liquid equilibria of $HI–I_2–H_2O$ ternary mixtures, Part 2: Experimental results at high temperature and pressure, *Int. J. Hydrogen Energy*, 34, 2009, 3258–3266.
34. M. Hodotsuka, X. Yang, H. Okuda, and K. Onuki, Vapor-liquid equilibria for the $HI + H_2O$ system and the $HI + H_2O + I_2$ system, *J. Chem. Eng. Data*, 53, 2008, 1683–1687.
35. R. Liberatore, A. Ceroli, M. Lanchi, A. Spadoni, and P. Tarquini, Experimental vapor-liquid equilibrium data of $HI–H_2O–I_2$ mixtures for hydrogen production by sulphur-iodine thermochemical cycle, *Int. J. Hydrogen Energy*, 33, 2008, 4283–4290.
36. K. Onuki, G. J. Hwang, and S. Shimizu, Electrodialysis of hydriodic acid in the presence of iodine, *J. Membr. Sci.*, 175, 2000, 171–179.
37. S.D. Hong, J.K. Kim, K.K. Bae, S.H. Lee, H.S. Choi, and G.J. Hwang, Evaluation of the membrane properties with changing iodine molar ratio in HIx ($HI–I_2–H_2O$ mixture) solution to concentrate HI by electro-electrodialysis, *J. Membr. Sci.*, 291, 2007, 106–110.
38. M. Yoshida, N. Tanaka, H. Okuda, and K. Onuki, Concentration of HIx solution by electro-electrodialysis using Nafion 117 for thermochemical water-splitting IS process, *Int. J. Hydrogen Energy*, 33, 2008, 6913–6920.
39. C.J. Orme, M.G. Jones, and F.F. Stewart, Pervaporation of water from aqueous HI using Nafion®-117 membranes for the sulfur–iodine thermochemical water splitting process, *J. Membr. Sci.*, 252, 2005, 245–252.
40. F.F. Stewart, C.J. Orme, and M.G. Jones, Membrane processes for the sulfur–iodine thermochemical cycle, *Int. J. Hydrogen Energy*, 32, 2007, 457–462.
41. C.J. Orme, F.F. Stewart, Pervaporation of water from aqueous hydriodic acid and iodine mixtures using Nafion® membranes, *J. Membr. Sci.*, 304, 2007, 156–162.
42. G. Caputo, C. Felici, P. Tarquini, A. Giaconia, and S. Sau, Membrane distillation of HI/H_2O and H_2SO_4/H_2O mixtures for the sulfur-iodine thermochemical process, *Int. J. Hydrogen Energy*, 32, 2007, 4736–4743.
43. D.R. O'keefe, J.H. Norman, and D.G. Williamson, Catalyst research in thermochemical water-splitting processes, *Catal. Rev.*, 22, 1980, 325–369.
44. Y. Oosawa, Y. Takemori, and K Fujii, Catalytic decomposition of hydrogen iodide in the magnesium-iodine thermochemical cycle–Catalyst search, *Nippon Kagaku Kaishi*, 7, 1980, 1081–1087 (in Japanese).
45. Y. Oosawa, T. Kumagai, S. Mizuta, W. Kondo, Y. Takemori, and K. Fujii, Kinetics of the catalytic decomposition of hydrogen iodide in the magnesium-iodine thermochemical cycle, *Bull. Chem. Soc. Jpn.*, 54, 1981, 742–748.
46. Y. Shindo, N. Ito, K. Haraya, T. Hakuta, and H. Yoshitome, Kinetics of the catalytic decomposition of hydrogen iodide in the thermochemical hydrogen production, *Int. J. Hydrogen Energy*, 9, 1984, 695–700.

47. Y. Oosawa, The decomposition of hydrogen iodide and separation of the products by the combination of an adsorbent with catalytic activity and a temperature-swing method, *Bull. Chem. Soc. Jpn.*, 54, 1981, 2908–2912.
48. D.R. O'Keefe, J.H. Norman, L.C. Brown, and G.E. Besenbruch, The application of a homogeneous catalysis concept to the liquid hydrogen-iodide decomposition step of the General Atomic sulfur–iodine water-splitting cycle, *Proc. 4th World Hydrogen Energy Conf.*, California, June 1982, 2, 687–701.
49. N. Ito, Y. Shindo, T. Hakuta, and H. Yoshitome, Enhanced catalytic decomposition of HI by using a microporous membrane, *Int. J. Hydrogen Energy*, 9, 1984, 835–839.
50. M. Nomura, S. Kasahara, H. Okuda, and S. Nakao, Evaluation of the IS process featuring membrane techniques by total thermal efficiency, *Int. J. Hydrogen Energy*, 30, 2005, 1465–1473.
51. H. Engels, K.F. Knoche, and M. Roth, Direct dissociation of hydrogen iodide—an alternative to the General Atomic proposal, *Int. J. Hydrogen Energy*, 12, 1987, 675–678.
52. C. Berndhäuser and K.F. Knoche, Experimental investigations of thermal HI decomposition from H_2O–HI–I_2 solutions, *Int. J. Hydrogen Energy*, 19, 1994, 239–244.
53. M. Nomura, S. Kasahara, and Y. Okuda, Evaluation of thermal efficiency to produce hydrogen through the IS process by thermodynamics, *JAERI-Research*, 2002-039, 2003.
54. A. Giaconia, G. Caputo, A. Ceroli, M. Diamanti, V. Barbarossa, P. Tarquini, and S. Sau, Experimental study of two phase separation in the Bunsen section of the sulfur–iodine thermochemical cycle, *Int. J. Hydrogen Energy*, 32, 2007, 531–536.
55. B.J. Lee, H.C. NO, H.J. Yoon, S.J. Kim, and E.S. Kim, An optimal operating window for the Bunsen process in the IS thermochemical cycle, *Int. J. Hydrogen Energy*, 33, 2008, 2200–2210.
56. J.H. Norman, G.E. Besenbruch, L.C. Brown, D.R. O'Keefe, and C.L. Allen, Thermochemical water-splitting cycle, bench-scale investigations, and process engineering, GA-A16713, 1982.
57. K.F. Knoche, H. Schepers, and K. Hesselmann, Second law and cost analysis of the oxygen generation step of the General Atomic sulfur–iodine cycle, *Proc. 5th World Hydrogen Energy Conf.*, Toronto, July 1984, 2, 487–502.
58. I.T. Öztürk, A. Hammache, and E. Bilgen, An improved process for H_2SO_4 decomposition step of the sulfur–iodine cycle, *Energy Convers. Manage.*, 36, 1995, 11–21.
59. S. Goldstein, J.M. Borgard, and X. Vitart, Upper bound and best estimate of the efficiency of the iodine–sulphur cycle, *Int. J. Hydrogen Energy*, 30, 2005, 619–626.
60. S. Kasahara, S. Kubo, R. Hino, K. Onuki, M. Nomura, and S. Nakao, Flowsheet study of the thermochemical water-splitting iodine–sulfur process for effective hydrogen production, *Int. J. Hydrogen Energy*, 32, 2007, 489–496.
61. B.J. Lee, H.C. NO, H.J. Yoon, H.G. Jin, Y.S. Kim, and J.I. Lee, Development of a flow sheet for iodine–sulfur thermo-chemical cycle based on optimized Bunsen reaction, *Int. J. Hydrogen Energy*, 34, 2009, 2133–2143.
62. D. O'Keefe, C. Allen, G. Besenbruch, L. Brown, J. Norman, and R. Sharp, Preliminary results from bench-scale testing of a sulfur–iodine thermochemical water-splitting cycle, *Int. J. Hydrogen Energy*, 7, 1982, 381–392.
63. S. Kubo, S. Shimizu, H. Nakajima, and K. Onuki, Studies on continuous and closed-cycle hydrogen production by a thermochemical water-splitting iodine-sulfur process, *Proc. Nuclear production of hydrogen 3rd Information exchange meeting*, Oarai, October 2005, 197–204.
64. H. Nakajima, M. Sakurai, K. Ikenoya, G. Hwang, K. Onuki, and S. Shimizu, A study on a closed-cycle hydrogen production by thermochemical water-splitting IS process, *Proc. 7th Int. Conf. Nuclear Engineering*, Tokyo, April 1999, ICONE-7104.
65. S. Kubo, H. Nakajima, S. Kasahara, S. Higashi, T. Masaki, H. Abe, and K. Onuki, A demonstration study on a closed-cycle hydrogen production by the thermochemical water-splitting iodine-sulfur process, *Nucl. Eng. Des.*, 233, 2004, 347–354.
66. S. Kubo, H. Nakajima, S. Shimizu, K. Onuki, and R. Hino, A bench scale hydrogen production test by the thermochemical water-splitting iodine-sulfur process, *Proc. GLOBAL 2005*, Tsukuba, October 2005, No. 474.

67. K. Onuki, H. Nakajima, M. Futakawa, I. Ioka, and S. Shimizu, A study on the iodine-sulfur thermochemical hydrogen production process. In K. Yoshida (eds.), *Principle of Exergy Reproduction, Reports of Project Research (No. 256)*, Ministry of Education, Japan, Tokyo, 1998, 253–260.
68. H. Nakajima, S. Shimizu, and K. Onuki, *Japan Patent*, 3605252, 2004 (in Japanese).
69. R. Buckingham, B. Russ, R. Moore, P. Pickard, M. Helie, and P. Carles, Status of the INERI sulfur-iodine integrated-loop experiment, *Proc. 17th World Hydrogen Energy Conf.*, Brisbane, June 2008, No. 310.
70. R. Moore, E. Parma, B. Russ, W. Sweet, M. Helie, N. Pons, and P. Pickard, An integrated laboratory-scale experiment on the sulfur–iodine thermochemical cycle for hydrogen production, *Proc. 4th Int. Topical Meeting High Temperature Reactor Technology*, Washington, DC, September–October 2008, HTR2008–58225.
71. B. Russ and P. Pickard, 2009 DOE hydrogen program review sulfur–iodine thermochemical cycle, U.S. Department of Energy Hydrogen Program 2009 Annual Merit Review Proceedings, Arlington, May 2009 (http://hydrogendoedev.nrel.gov/pdfs/review09/pd_12_pickard.pdf).
72. R. Moore, G. Naranjo, B.E. Russ, W. Sweet, M. Hele, and N. Pons, Nuclear hydrogen initiative, results of the phase II testing of sulfur–iodine integrated lab scale experiments, Final Report, GA-C26575, 2009.
73. P.W. Trester and H.G. Staley, Assessment and investigation of containment materials for the sulfur–iodine thermochemical water-splitting process for hydrogen production, GRI-80/0081, 1981.
74. Y. Imai, S. Mizuta, and H. Nakauchi, Material problems associated with hydrogen production from water—with special interest in thermochemical method, *Boshoku Gijutsu*, 35, 1986, 230–240 (in Japanese).
75. Y. Kurata, K. Tachibana, and T. Suzuki, High temperature tensile properties of metallic materials exposed to a sulfuric acid decomposition gas environment, *J. Japan Inst. Metals*, 65, 2001, 262–265.
76. S. Kubo, M. Futakawa, I. Ioka, K. Onuki, S. Shimizu, K. Ohsaka, A. Yamaguchi, R. Tsukada, and T. Goto, Corrosion test on structural materials for iodine-sulfur thermochemical water splitting cycle, *Proc. AIChE 2003 Spring National Meeting*, New Orleans, March 2003, 153b.
77. B. Wong, R.T. Buckingham, L.C. Brown, B.E. Russ, G.E. Besenbruch, A. Kaiparambil, R. Santhanakrishnan, and A. Roy, Construction materials development in sulfur–iodine thermochemical water-splitting process for hydrogen production, *Int. J. Hydrogen Energy*, 32, 2007, 497–504.
78. N. Tanaka, J. Iwatsuki, S. Kubo, A. Terada, and K. Onuki, Research and development on IS process components for hydrogen production (II) Corrosion resistance of glass lining in high temperature sulfuric acid, *Proc. ICAPP'09*, Tokyo, May 2009, No. 9291.
79. A. Terada, H. Ota, H. Noguchi, K. Onuki, and R. Hino, Development of sulfuric acid decomposer for thermochemical hydrogen production IS process, *Trans. Atom. Energy Soc. Jpn.*, 5, 2006, 68–75 (in Japanese).
80. H. Noguchi, H. Ota, A. Terada, S. Kubo, K. Onuki, and R. Hino, Development of sulfuric acid decomposer for thermo-chemical IS process, *Proc. ICAPP'06*, Reno, June 2006, No. 6166.
81. F. Gelbard, R. Moore, and E. Parma, Status of initial testing of the H_2SO_4 section of the ILS experiment, SAND2007-7841, 2007.
82. S.M. Connolly, E.J. Lahoda, and E.F. Mclaughlin, Design of recuperative sulfuric acid decomposition reactor for hydrogen generation processes, *Proc. AIChE 2008 Annual Meeting*, Philadelphia, November 2008, 328e.
83. H. Nakajima, *Japan Patent*, 4521527, 2010 (in Japanese).

18

The Hybrid Sulfur Cycle

Maximilian B. Gorensek and William A. Summers

CONTENTS

18.1 Introduction

The Hybrid Sulfur (HyS) cycle (Figure 18.1) is one of the simplest thermochemical water-splitting processes, comprising only two reactions with all reactants and products in the fluid state. Known also as the Westinghouse Sulfur or Ispra Mark 11 cycle, it entails coupled reactions that cycle sulfur between its +4 (sulfite) and +6 (sulfate) oxidation states. One of the two reactions is thermochemical (high-temperature, endothermic) while the other is electrochemical, making it a hybrid thermochemical cycle.

The HyS cycle was first proposed by Brecher and Wu at Westinghouse Electric Corp. [1]. Westinghouse engineers and scientists invested considerable effort developing the cycle from its inception in the mid-1970s into the mid-1980s, publishing numerous reports [2–5], conference proceedings [6–27], papers [28–35], and patents [1,36–37]. They referred to it as the "Westinghouse Sulfur Cycle," which explains the name by which it is still called by many. Simultaneous development also took place in Europe at the Joint Research Center in Ispra, Italy, where it was known as the Mark 11 cycle [38], and at the Nuclear Research Center (KFA) in Jülich, Germany [39–47], as well as elsewhere [48–58]. As a result of waning

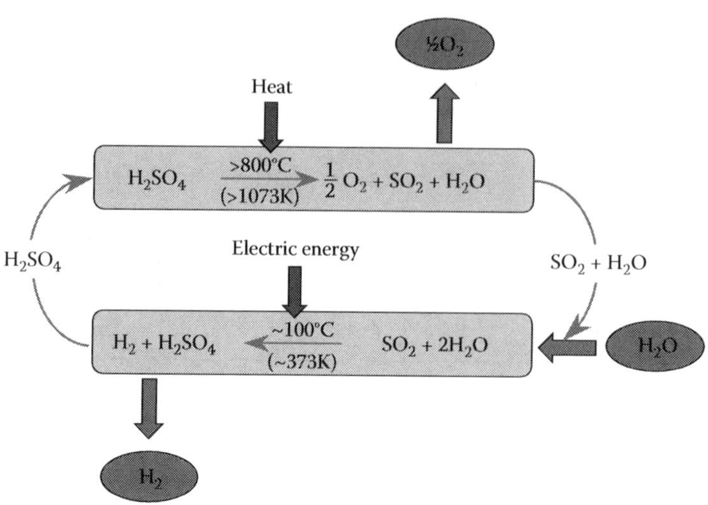

FIGURE 18.1
The HyS Cycle.

interest in advanced nuclear technologies beginning in the early 1980s, development activities essentially ceased. After a nearly twenty-year hiatus, interest in this process was revived following a study of thermochemical cycles that ranked HyS first among the 115 cycles considered [59]. Westinghouse resumed their investigations [60–64] and the Savannah River National Laboratory (SRNL) assumed the lead development role in the United States under the Department of Energy, Office of Nuclear Energy's (DOE-NE's) Nuclear Hydrogen Initiative (NHI) [65–66]. Since then, researchers in the European Union (EU) [67], France [68–71], the Republic of Korea (ROK) [72–73], Japan [74–82], and South Africa have also begun taking a serious look at HyS. U.S. work on HyS has been centered at SRNL [83–102] and the University of South Carolina (USC) [103–109].

As already noted, two reaction steps are comprised in the cycle. The thermochemical step, which is common to all sulfur cycles, is the high-temperature decomposition of sulfuric acid into water, sulfur dioxide, and oxygen,

$$H_2SO_4(aq) \xrightarrow{\text{Heat, } T>800°C} H_2O(g) + SO_2(g) + \frac{1}{2}O_2(g). \tag{18.1}$$

This is an equilibrium-limited reaction that requires a catalyst, as well as heat input at relatively high temperatures, such as those that an advanced, high-temperature gas-cooled reactor (HTGR) heat source could supply. The electrochemical step is the SO_2-depolarized electrolysis of water,

$$SO_2(aq) + 2H_2O(l) \xrightarrow{\text{Power, } T\approx 80-120°C} H_2SO_4(aq) + H_2(g), \tag{18.2}$$

which takes place at much lower temperatures and requires electric power. Adding together the two reactions, the net effect is the splitting of one mole of water into one mole of hydrogen and one-half mole of oxygen. The same HTGR that serves as the heat source for Reaction 18.1 could supply the power for Reaction 18.2 with cogeneration. Power could also be used from the grid or from an independent source. The choice should be dictated by economic considerations.

The fact that the electrochemical step is an electrolysis process begs the question, why not split water directly using electrolysis,

$$H_2O(l) \xrightarrow{\text{Power}} H_2(g) + \frac{1}{2}O_2(g), \tag{18.3}$$

and avoid the need to handle sulfuric acid at high temperatures? The answer lies in the much lower power requirement of the SO_2-depolarized electrolysis step, which was first proposed for large-scale electrolytic hydrogen production by Juda and Moulton in 1967 [110]. As is well-known, the standard potential at 25°C for water electrolysis (Reaction 18.3) is –1.229 V. The standard potential at 25°C for SO_2-depolarized electrolysis (Reaction 18.2), on the other hand, is only –0.158 V [99]. Operating potentials are somewhat higher in both cases because of the nonstandard conditions present in real-world cells and the inevitable overpotentials resulting from kinetic and mass transfer limitations, as well as Ohmic losses. Alkaline electrolyzers used to conduct Reaction 18.3 typically operate at –1.8 to –2.0 V, while SO_2-depolarized electrolyzers (SDEs) are expected to be capable of achieving –0.6 V at practical current densities (500 mA/cm^2). Therefore, the HyS cycle requires only one-third as much electricity for the electrochemical step as conventional water electrolysis. Of course, the feasibility of the HyS cycle will depend on the sum of the primary energy required for both steps, Reactions 18.1 and 18.2, being less than that required for water electrolysis, Reaction 18.3.

18.2 Fundamentals

The two steps that constitute the HyS cycle are considered in turn, beginning with the electrochemical sulfite oxidation reaction that makes sulfuric acid and hydrogen from SO_2 and water. The high-temperature sulfate reduction reaction, which regenerates SO_2 from sulfuric acid while giving off oxygen and water, is discussed subsequently.

18.2.1 SO_2-Depolarized Electrolysis

The SO_2-depolarized electrolysis step distinguishes HyS from the other sulfur cycles. Early work at Westinghouse was based on a parallel plate electrolyzer concept that used a microporous diaphragm separator. With the advent of proton exchange membranes (PEMs) such as Nafion®, the focus shifted toward PEM designs. Nearly all of the recent SDE development has been based on PEM technology, leveraging the tremendous R&D investment for fuel-cell applications.

The original patent [1] described an electrolyzer built up from parallel banks of individual cells connected in series and contained in a pressurized container. Each cell (Figure 18.2) consisted of an enclosure, 13, separated into two compartments by a membrane or diffusion barrier, 19, capable of passing protons or electrons but not sulfate or sulfite ions. An aqueous sulfurous acid (H_2SO_3) solution, 21, made by continually feeding water and SO_2 filled the cells. Both compartments held a hollow electrode equipped with a sparger, 15 and 17 (presumably for feeding SO_2), and a direct current (DC) potential of a few tenths of a V was applied between the electrode pair in each cell. Sulfuric acid was produced at the anode,

$$SO_2(aq) + 2H_2O(l) \rightarrow H_2SO_4(aq) + 2H^+ + 2e^-, \tag{18.4}$$

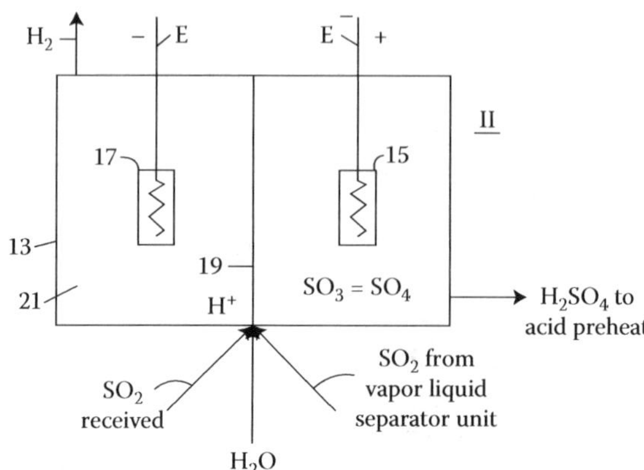

FIGURE 18.2
SDE cell in the original Westinghouse patent. (From Brecher, L. E. and Wu, C. K. Electrolytic decomposition of water. U.S. Patent 3,888,750: Westinghouse Electric Corp., June 10, 1975. With permission.)

while hydrogen gas was evolved at the cathode.

$$2H^+ + 2e^- \rightarrow H_2(g) \tag{18.5}$$

The actual design used by Westinghouse was quite different. Sulfurous acid at the cathode is not desirable because it can be reduced to elemental sulfur or hydrogen sulfide, taking away from current efficiency, potentially poisoning the electrocatalyst, and causing other problems. High sulfuric acid concentrations in the product are desirable because they result in more energy-efficient operations downstream because of smaller quantities of water having to be vaporized by the acid concentration and decomposition processes. This led to the design shown in Figure 18.3.

Both the catholyte and anolyte contain sulfuric acid at similar concentrations (up to 75 wt% H_2SO_4) and are separately recirculated in the early Westinghouse design [2]. The anolyte feed is also saturated with dissolved SO_2. A pressure difference is imposed that forces a steady flow of acid from the cathode side to the anode side, through a microporous diaphragm separator. This is intended to counteract diffusional flux of SO_2 across the separator to the cathode. The cell is pressurized (up to 25 bar) to maximize the solubility of SO_2 in the anolyte. Hydrogen product is separated from the catholyte effluent. A portion of the anolyte effluent is removed and the unconverted SO_2 stripped off. Some of this stripped acid is fed to the cathode, enough to make up for losses across the diaphragm, while the remainder is passed on to the high-temperature decomposition step.

The development of PEM fuel cells and electrolyzers in the 1960s and 1970s led to the availability of PEM electrolyte separators, which quickly replaced the diaphragm in SDE designs. Use of membrane-electrode assemblies (MEAs) incorporating PEM electrolyte separators offered the promise of the shortest possible current path between anode and cathode, minimizing IR losses and reducing the SDE footprint. Much of the early HyS work was focused on identifying PEM formulations that had good conductivity (low resistance) at high H_2SO_4 concentrations together with low SO_2 permeability [4,44].

Figure 18.4 illustrates the SDE configuration used by SRNL. The MEA (anode catalyst layer, PEM, and cathode catalyst layer) is at the "heart" of the SDE. A carbon cloth or paper diffusion layer is pressed against each side by a graphite flow field which is designed to

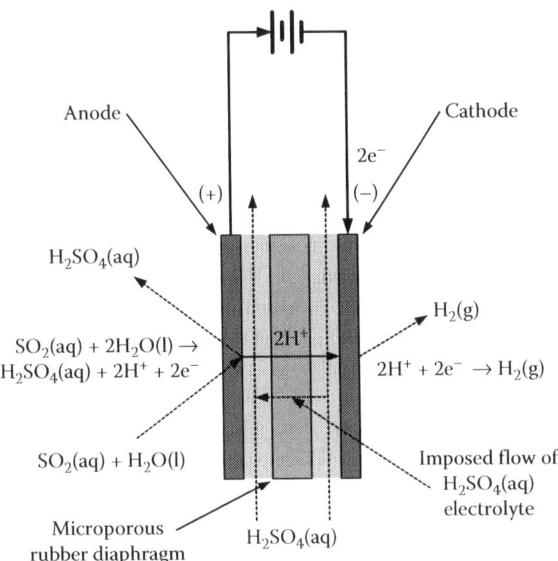

FIGURE 18.3
Westinghouse SDE cell.

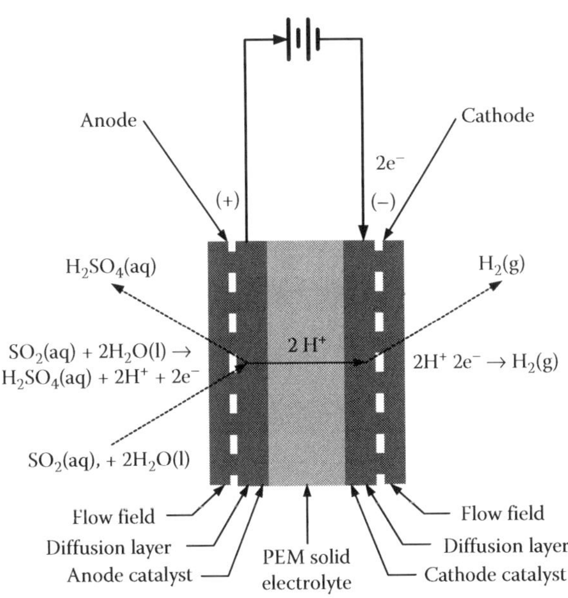

FIGURE 18.4
SRNL PEM SDE cell; SO_2 dissolved in sulfuric acid under pressure is fed to the anode, while water is fed to the cathode.

force fluid flow through the diffusion layer. The anolyte is sulfuric acid saturated with dissolved SO_2. Unlike the Westinghouse design, the catholyte is water. This helps the membrane remain hydrated and also serves to sweep out any undesirable sulfur species that may form at the cathode because of SO_2 crossover. SO_2 reacts with water at the anode to form sulfuric acid, protons, and electrons according to Reaction 18.4. The protons pass through the PEM under an applied potential and recombine with the electrons at the cathode to make hydrogen via Reaction 18.5. Essentially, no undissociated H_2SO_4 remains in solution at typical anolyte concentrations, 30–75 wt% H_2SO_4. (Experimental measurements and thermodynamic analyses have shown that nearly all of the sulfuric acid is present as sulfate and bisulfate ion under these conditions [111–116].) Consequently, both the applied potential and the anionic character of the membrane provide an effective barrier to sulfate and bisulfate ion diffusion, keeping sulfuric acid out of the catholyte. More details concerning the SRNL SDE are given below.

The use of a PEM imposes some limitations on the SDE. The conductivity of the most common PEM material, Nafion, depends on the water content of the membrane; water content, in turn, depends on the activity of water at the membrane surfaces. When in contact with sulfuric acid, it has been shown that the resistivity of Nafion increases with the H_2SO_4 content of the acid. The practical upper limit for the acid concentration in a Nafion-equipped SDE is about 50 wt% H_2SO_4 [95]. To achieve higher acid–product concentrations will require using a PEM other than Nafion—preferably one that has good conductivity and low SO_2 permeability at high acid strength [109]. Furthermore, Nafion is limited to an operating temperature less than 100°C. Use of a PEM that can operate at higher temperatures is one way to improve the reaction kinetics and increase the electrolyzer efficiency.

Recirculation of the anolyte and catholyte in the SRNL SDE design serves several purposes. On the anolyte side, flow through the porous flow field helps increase mass transfer, including transport of dissolved SO_2 from the bulk solution to the anode surface as well as removal of H_2SO_4 product from the anode surface to the bulk phase. Furthermore, anolyte recirculation is inevitable because of SO_2 solubility and conversion limitations. On the cathode side, the flow of water helps sweep hydrogen bubbles from the cathode surface and keeps the membrane hydrated. It also removes any sulfur fines that can build up because of the reduction of any SO_2 that diffuses across the PEM from the anode. In the full-scale plant, recirculation of both flows through heat exchangers will be used to help control temperature and recover waste energy from the electrode overpotentials and Ohmic losses in the form of heat to use elsewhere in the cycle.

An alternative SDE design was concurrently developed and tested at USC [103]. As illustrated in Figure 18.5, gaseous SO_2 is fed to the anode while liquid water is fed to the cathode, which is maintained at a higher pressure than the anode. This pressure differential drives a net water flux across the PEM and keeps it hydrated. Water transported across the membrane reacts with gaseous SO_2 at the anode to make sulfuric acid, which accumulates and is removed as the anode product. When compared to the all-liquid SRNL design, a significantly higher catholyte (water) recirculation rate will be needed in order to control the SDE temperature. More importantly, a HyS flow sheet incorporating this gaseous anode design will differ in the way the products of the high-temperature sulfuric acid decomposition step are handled in order to produce a gaseous SO_2 feed to the anode instead of a sulfuric acid feed containing dissolved SO_2. Since SO_2 and O_2 are quite easily separated based on their solubility difference, the gaseous SO_2 anode flow sheet will be somewhat more complex because it cannot use simple preferential dissolution to do this separation. Otherwise, the performance of this design is expected to be comparable if the same PEM and electrode materials are used.

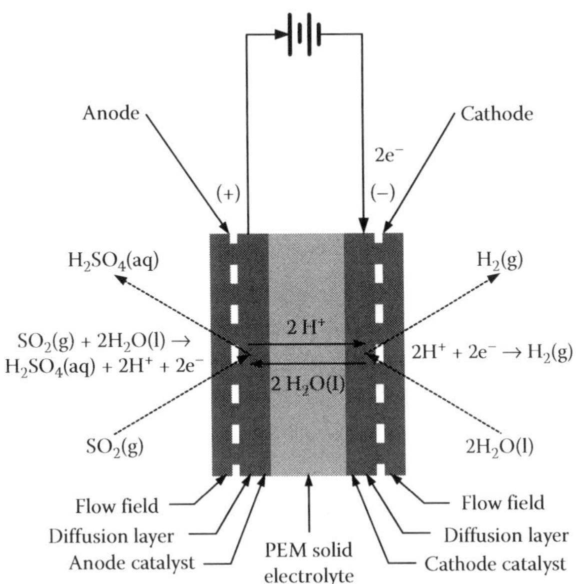

FIGURE 18.5
USC PEM SDE cell; gaseous SO_2 is fed to the anode, while water is fed to the cathode.

18.2.2 High-Temperature Sulfuric Acid Decomposition

As noted earlier, the high-temperature decomposition of sulfuric acid into water, sulfur dioxide, and oxygen, Reaction 18.1, is common to all sulfur cycles. Consequently, this topic has already been addressed at length as part of the sulfur–iodine (SI) cycle. However, several aspects of relevance to HyS are worth mentioning here.

In their second-law analysis of the HyS cycle, Knoche and Funk [58] showed that the net thermal efficiency is limited by the temperature range over which heat is exchanged between the primary energy source (the HTGR) and the water-splitting process, among other things. This is true for any heat-driven water-splitting process. The higher the HTGR coolant outlet temperature and the smaller the difference between the outlet and inlet temperatures, the higher the reversible or limiting efficiency will be. Thus, it is clear that the design of the high-temperature step of the HyS cycle will have a significant impact on the performance of the overall process.

A generic schematic of the way in which high-temperature heat would be transferred from the HTGR to the HyS process is illustrated in Figure 18.6. The primary coolant is assumed to be helium under pressure (40–90 bar). Indirect heating by means of a secondary helium loop will be inevitable for isolation purposes. This is accomplished by means of an intermediate heat exchanger (IHX). The process heat exchanger in the HyS cycle is the acid decomposition reactor, which will achieve a peak process fluid temperature about 75°C lower than the reactor outlet temperature. As the coolant temperature decreases, additional heat may be transferred to an acid vapor superheater or acid vaporizer (not shown). Depending on the HTGR design, the coolant return temperature may be too low for all of the heat to be used for sulfuric acid decomposition. In that case, steam may also be made, as shown in Figure 18.6, for a bottoming power cycle or heating needs elsewhere in the process.

Early flow sheets proposed by SRNL [83,86] featured a high-temperature sulfuric acid decomposition process design adapted from one proposed by Öztürk et al. [117] for the SI

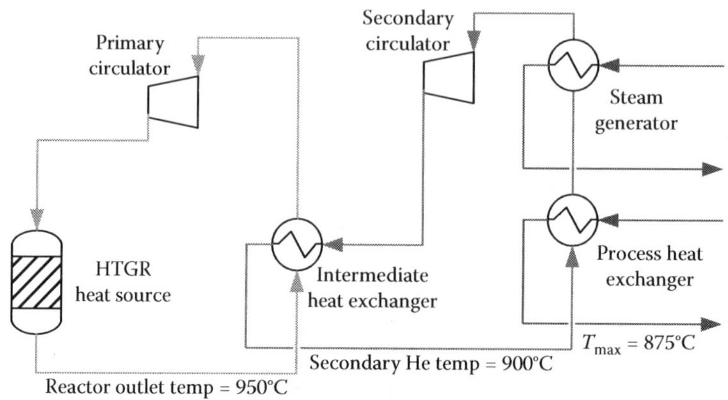

FIGURE 18.6
Schematic diagram for an HTGR-heated water-splitting process showing the relationship between the heat source and the process; in this example, the reactor outlet temperature is 950°C, with a 50°C temperature difference between the primary and secondary loops, and a minimum temperature difference for helium-to-process fluid heat transfer of 25°C, resulting in a peak process temperature of 875°C.

cycle. The key feature was a direct contact heat exchanger/sulfuric acid scrubber as shown in Figure 18.7. This scrubber served as a sulfuric acid trap since the total vapor overhead product could only contain very small amounts of H_2SO_4 because of the highly nonideal behavior of the H_2O–H_2SO_4–SO_3 system. Consequently, it offered the promise of high thermal efficiency since essentially all of the liquid H_2SO_4 fed would be converted to SO_2 vapor. A simple energy balance then shows that the high-temperature heat requirement would be equal to the enthalpy difference between the SO_2–O_2–water vapor product stream and the liquid–acid feed stream. For fully converting an equimolar H_2O–H_2SO_4 liquid feed (84.5 wt% H_2SO_4) to an H_2O–SO_2–O_2–vapor product stream at 300°C and 12 bar, the enthalpy difference would only be 317 kJ/mol SO_2 (a low value representing high efficiency).

However, Moore et al. [118] cited serious corrosion problems and difficulties in making and maintaining the high-temperature connections required for this design. As a result,

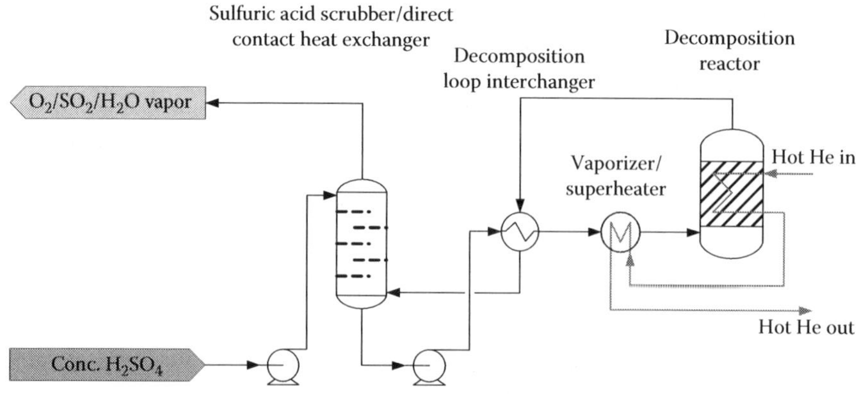

FIGURE 18.7
High-temperature sulfuric acid decomposition loop.

they abandoned the direct contact heat exchanger/sulfuric acid scrubber approach in favor of a recuperative bayonet reactor for the NHI Integrated Laboratory Scale (ILS) SI cycle demonstration. [118] A simplified schematic of this design is shown in Figure 18.8. (While the ILS unit actually used electrical resistance heaters on the outer surface, the full-scale reactor would be heated by secondary helium coolant flowing past the exterior.) The bayonet reactor embodies all of the operations shown in Figure 18.7 in a single unit, albeit without direct contact heat exchange and requiring only two fluid conduit connections (inlet and outlet) at much lower temperatures. A detailed analysis by Gorensek and Edwards [98] indicates that the energy requirements could be low enough to be incorporated in a practical HyS process that would compare favorably with water electrolysis. Consequently, the bayonet reactor was chosen as the basis for subsequent SRNL HyS flow sheets [95,97].

18.2.3 Energy Considerations

The HyS process will be practical as a means for producing hydrogen from nuclear energy only if it will cost less to do so than by alternate means. The obvious benchmark is nuclear water-splitting by alkaline electrolysis using electricity from a nuclear power plant. Unless HyS has the potential to make hydrogen at a lower total cost, there is no reason to pursue its development any further. After all, the technology to make electricity with a pressurized water reactor (PWR) or boiling water reactor (BWR) and use that power to drive an array of alkaline electrolysis cells to make hydrogen is already well-established.

Arguably the most expensive component of a hydrogen production process driven by nuclear energy would be the nuclear reactor heat source. This implies that the net thermal efficiency of a nuclear water-splitting process is a key discriminator. If one process has significantly higher efficiency than another, it is likely to cost less. For example, if a HyS

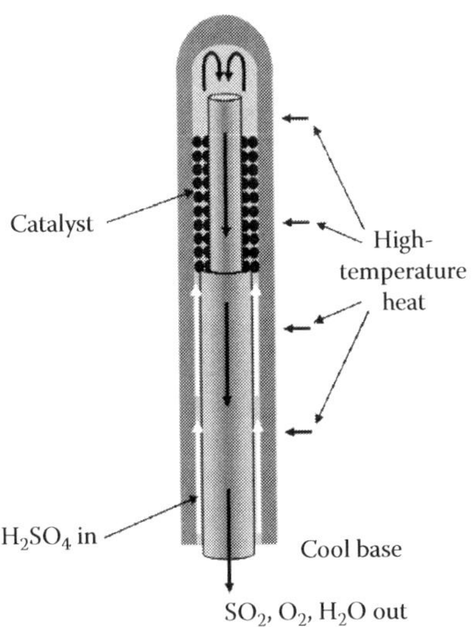

FIGURE 18.8
High-temperature sulfuric acid decomposition Bayonet reactor. The exterior is heated by secondary helium heat transfer fluid flowing from the top, with the lower end insulated to facilitate recuperation and allow low-temperature fluid connections. (From Gorensek, M. B. and Edwards, T. B. *Ind. Eng. Chem. Res.* **48**(15), 7232–7245, 2009. With permission.)

plant had a net thermal efficiency of 45% compared to an alternative plant with 30% efficiency, this would mean that only two reactors would be needed to produce the same quantity of hydrogen using the HyS process that would require three reactors with the alternative process.

The electrical generating efficiency for a typical BWR/PWR power plant is approximately 33%, while alkaline electrolysis can make hydrogen from electricity with a lower heating value (LHV) efficiency of about 68%. This gives a net thermal efficiency of 22.4% (LHV basis) for nuclear-powered electrolysis [95]. Of course, if HTGRs are available for hydrogen production, they could be used to generate electricity too. The conversion efficiency for an HTGR-heated closed loop helium Brayton power cycle has been projected to range between 41% [119] and 52% [120], depending on the design and underlying assumptions. Taking 45% as a conservative estimate, using HTGR power would give alkaline electrolysis a net thermal efficiency of 30.6% (LHV basis) [95]. For a HyS process to be practical, it must exceed these efficiency benchmarks by a considerable margin. As discussed in Section 18.4, SRNL has devised process flow sheets that demonstrate the potential of a HyS process to achieve acceptable efficiencies.

18.3 Analytical and Experimental Studies

Of the thermochemical and hybrid cycles, only the SI has been more extensively studied than the HyS cycle. SDE component development efforts will be addressed first, followed by a review of electrolyzer testing results. This section will conclude with a discussion of high-temperature sulfuric acid decomposition reactor developments, which are, for all practical purposes, shared with the SI cycle.

18.3.1 SDE Component Development

The SDE is composed of several major components, each of which requires careful selection, analysis, characterization, and experimental study. The use of a PEM as the cell electrolyte dictates the general geometry and design approach for the SDE. A schematic showing the major components of a PEM-type SDE is given in Figure 18.9. Experimental study at the component level has primarily focused on the PEM electrolyte and the electrocatalyst for both the anode and cathode.

Selection of a suitable catalyst is important in order to (1) increase the electrical efficiency by minimizing kinetic overpotential losses and allowing the SDE to operate at conditions closer to the reversible cell potential, and (2) maintain long-term stable cell performance. Prior work by Westinghouse Electric in the 1980s investigated the use of precious metal blacks (Pd and Pt) as electrocatalysts for the SDE reactions [4]. The results indicated that palladium was a more effective electrocatalyst than platinum for SO_2-depolarized electrolysis. More recently, SRNL conducted experimental work to compare high-surface-area catalysts using palladium on carbon (Pd/C) and platinum on carbon (Pt/C) [91].

The catalyst activity and stability were evaluated using a three-electrode cell consisting of a glass vial with a Teflon™ cap and a water jacket. Electrochemical characterization of catalysts consisted of cyclic voltammograms (CVs) in the solution purged with nitrogen and linear sweep voltammograms (LSVs) in SO_2-saturated sulfuric acid solutions. The catalytic activity and stability of Pt/C and Pd/C were studied in 3.5–10.4 M (30–70 wt%)

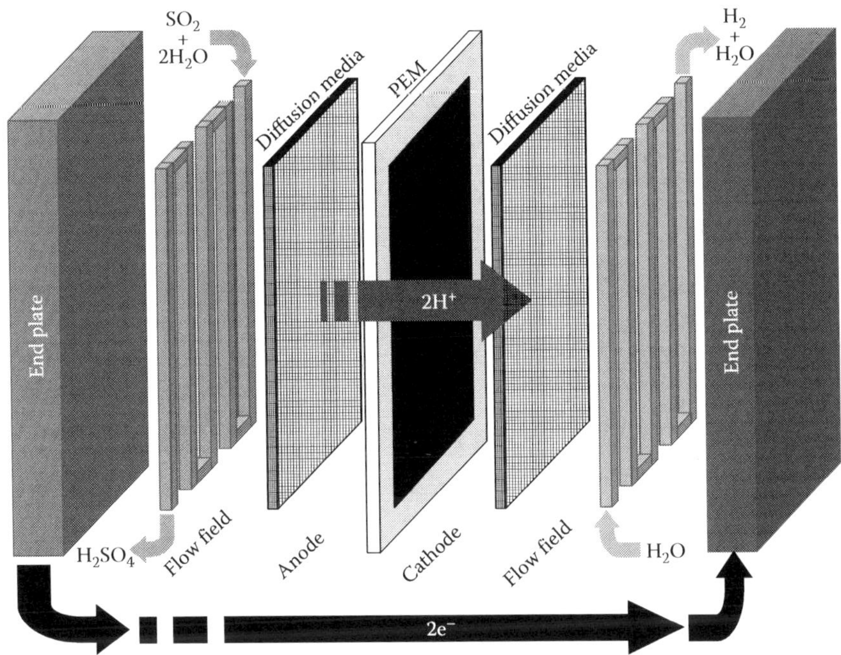

FIGURE 18.9
Major components of a PEM-type SDE.

sulfuric acid solutions and at temperatures ranging from 30 to 70°C. The results are shown in Figure 18.10 [91]. Pd/C showed somewhat higher initial activity than Pt/C, but was much less stable. After a few cycles, the activity of the Pd/C catalyst decreased significantly, whereas the Pt/C exhibited very good stability and activity for the oxidation of SO_2. Pt/C did exhibit instability in very high H_2SO_4 concentrations (10.4 M) at temperatures of 50°C and above. Tafel plots showed lower potentials (ca. 100 mV) and much higher exchange currents (ca. 1000 times greater) for the oxidation of SO_2 on Pt/C compared to Pd/C. Furthermore, the activation energy for the oxidation of SO_2 on Pt/C is at least half of that on a Pd/C surface. As a result, SRNL focused its SDE work on the use of Pt electrocatalysts.

Experiments were also carried out by SRNL to measure the performance of Pt-alloy catalysts created by alloying Pt with various transition metals [101]. The potential versus current relationship is shown in Figure 18.11 [101] in the form of Tafel plots. As shown there, when the potential approaches the open circuit potential and the exchange current density, it varies, depending on the transition metal that is alloyed with the Pt. It is interesting to note that the best performance is observed when Pt is alloyed with nonnoble metals such as cobalt and chromium. Note, however, that when Pt is alloyed with noble metals such as iridium or ruthenium, the performance is decreased. The best performance was measured with a catalyst featuring Pt alloyed with Co and Cr. This catalyst produced a 20-mV decrease in the potential compared to the baseline Pt/C catalyst. However, a further decrease in the potential is needed to achieve the commercial performance target. Thus, there is a need to further increase the anode catalyst performance for the oxidation of sulfur dioxide. This can be accomplished through a combination of more active catalysts and increased operating temperature, which improves the overall kinetics.

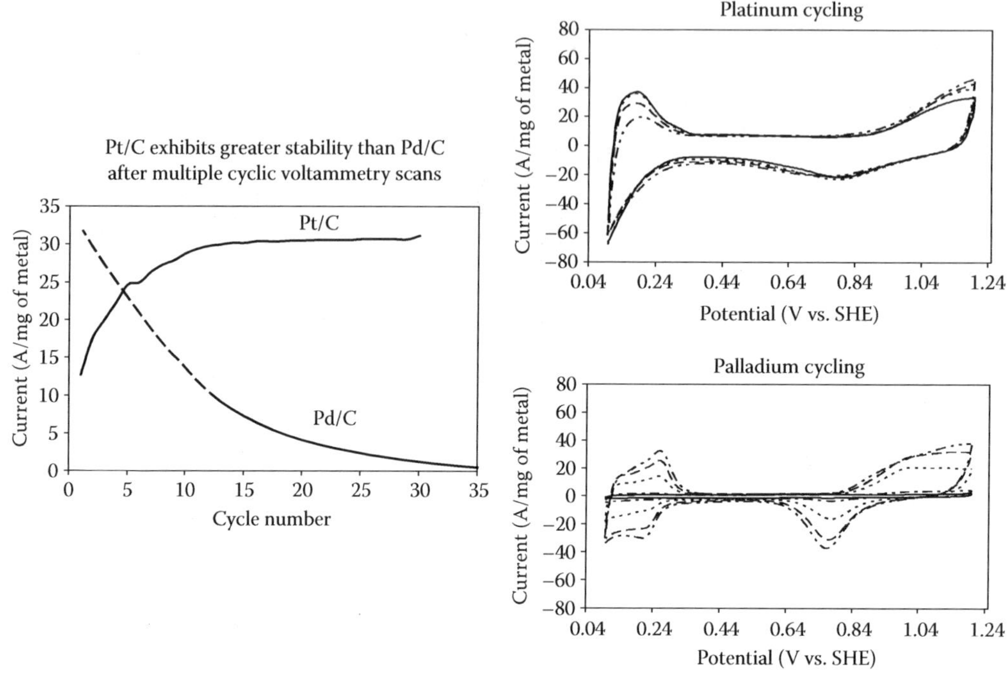

FIGURE 18.10
Catalytic activity and stability of Pt/C and Pd/C for SO_2 oxidation. (From Colon-Mercado, H. R. and Hobbs, D. T. *Electrochem. Commun.* **9**(11), 2649–2653, 2007. With permission.)

In addition to the electrocatalyst, the other key component of an SDE is the PEM electrolyte. There are many requirements of a PEM for the successful functioning of the electrolyzer. The PEM must be stable in highly corrosive solution (>30 wt% H_2SO_4 saturated with SO_2) and at high-operating temperature (80–140°C), allow minimal transport of SO_2, and must maintain high ionic conductivity under these conditions. These requirements allow the electrolyzer to perform at high current density and low cell potential, thus maximizing the energy efficiency for hydrogen production. Lastly the PEM serves to separate the anolyte from the hydrogen output in order to prevent the production of undesired sulfur-based reaction products and poisoning of the cathode catalyst.

The SRNL has investigated numerous candidates for use as membranes in the SDE [93]. Both commercially-available and experimental membranes were selected and evaluated for chemical stability, sulfur dioxide transport and ionic conductivity characteristics. An array of thicknesses, equivalent weights, chemistry, and reinforcements were considered. Commercial membranes included perfluorinated sulfonic acid (PFSA) membranes, such as DuPont Nafion series, which is widely used in PEM fuel cells. Nafion series membranes exhibited excellent chemical stability and ionic conductivity in sulfur dioxide-saturated sulfuric acid solutions. Sulfur dioxide transport in these membranes varied proportionally with the thickness and equivalent weight of the membrane. Although the SO_2 transport was higher than desired, the excellent chemical stability and conductivity indicated this membrane type as the best commercially available membrane at this time. The Nafion series membranes were therefore chosen as the benchmark for comparing membrane performance.

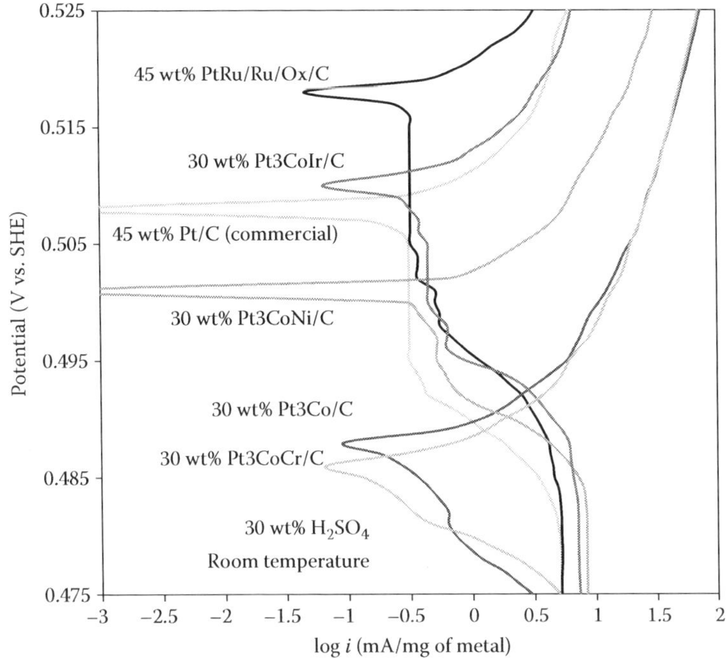

FIGURE 18.11
Catalytic activity of Pt-based catalysts for the oxidation of SO_2 in 30-wt% sulfuric acid and at room temperature. (From Colón-Mercado, H. R., Elvington, M. C., and Hobbs, D. T. Component development needs for the hybrid sulfur electrolyzer. Savannah River National Laboratory, Report No. WSRC-STI-2008-00291, May 2008 (doi: 10.2172/935436). With permission.)

Many of the membranes considered are also being utilized or are under development for use in PEM fuel cells, PEM electrolyzers and/or direct methanol fuel cells (DMFCs). This permits the developers of an SDE to leverage much of the extensive work being done by others for these related PEM applications. Membranes being developed for DMFC applications are particularly relevant, since a major challenge for this technology is the reduction of methanol crossover. Methanol, like sulfur dioxide, is a neutral species that is soluble in water. The methanol crossover issue is similar to the SDE problem with the crossover of dissolved SO_2. A list of some of the membranes selected and characterized by SRNL is given in Table 18.1. All of these membranes exhibited excellent short term chemical stability in concentrated sulfuric acid solutions at temperatures up to 80°C.

The conductivity and SO_2 transport for each membrane was evaluated using a custom-made permeation cell consisting of two glass chambers joined by a Teflon bridge where the membrane is secured. During measurements both chambers were filled with the concentrated acid of interest (typically 30 wt % H_2SO_4) and purged of oxygen by flowing nitrogen. A constant potential was applied and the SO_2 transport through the membrane was determined by measurement of the current as a function of time. Each membrane was also tested for through-plane ionic conductivity using electrochemical impedance spectroscopy. Representative results are shown in Figure 18.12 [93].

The lowest SO_2 transport coefficient was measured for membranes of the DuPont bilayer and high equivalent weight PFSA type. However, these membranes also showed unacceptably low ionic conductivity, negating their use in the SDE. Several of the advanced membranes had promising combined performance (low SO_2 transport and high conductivity).

TABLE 18.1

Candidate PEM Materials for the SDE Application

Membrane Description	Source
Perfluorinated sulfonic acid (PFSA)	DuPont
Polybenzimidizole (PBI)	BASF (Germany)
Sulfonated Diels-Alder polyphenylene (SDAPP)	Sandia National Laboratories
Stretched recast PFSA	Case Western Reserve University
Nafion/fluorinated ethylene propylene (FEP) blends	Case Western Reserve University
Treated PFSA	Giner Electrochemical Systems, LLC
Perfluorocyclobutane-biphenyl vinyl ether (BPVE)	Clemson University
Perfluorocyclobutane-biphenyl vinyl ether-hexafluoroisopropylidene (BPVE-6F)	Clemson University

Both the PFSA/FEP blend and the BPVE-6F membranes exhibited lower SO_2 transport than Nafion, but with somewhat lower ionic conductivity. The SDAPP membrane had high conductivity, but also a somewhat higher SO_2 transport coefficient. Results for the PBI membrane (not shown in Figure 18.12) were also encouraging, with low SO_2 transport and reasonable conductivity.

The reported test results were determined at 67°C and atmospheric pressure. However, several of the more promising membranes, such as the PBI, Clemson BPVE and Sandia SDAPP, are expected to perform significantly better at higher temperature (120–140°C), which is not possible with the PFSA membrane. The higher temperature operation is also expected to improve reaction kinetics and lower the cell potential. Preliminary testing of these membranes at higher temperature using anhydrous SO_2 as the anolyte has shown promising results [105]. Operation with liquid anolyte (SO_2 dissolved in sulfuric acid) at

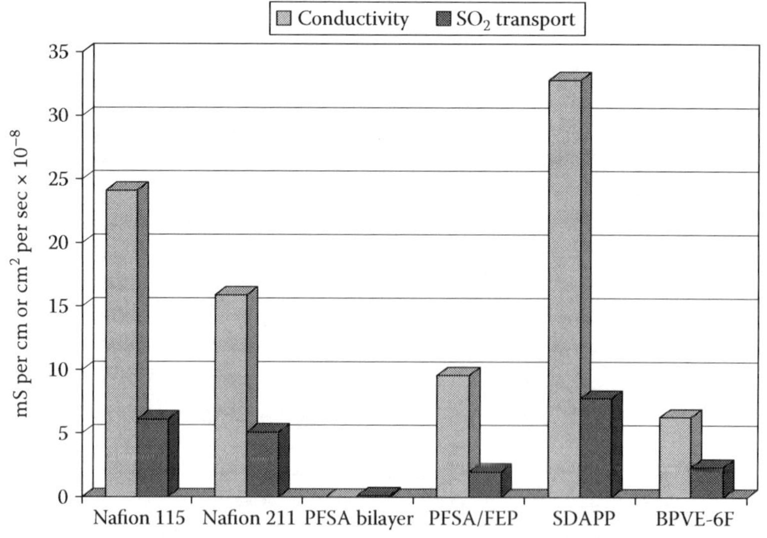

FIGURE 18.12
PEM results showing ionic conductivity and SO_2 diffusion. (From Hobbs, D. *Pittcon Conference & Expo 2009*. Chicago, IL, USA, 2009. With permission.)

the higher temperature requires a higher pressure test facility in order to maintain adequate SO_2 concentrations in the feed. No results for higher-temperature/higher-pressure testing with a liquid-fed SDE have been reported yet, although this is an obvious future research direction.

18.3.2 Single-Cell SDE Testing

Previous electrolyzer tests using SO_2 depolarization were reported by Westinghouse in the 1970s [4,20,28]. More recent tests have been reported by USC [104–109] and SRNL [85,89, 100]. As discussed previously, Westinghouse employed a two-compartment filter-press type cell rather than the more compact fuel-cell-type PEM arrangement used by current investigators. The USC tests employed a 10-cm² cell based on conventional PEM architecture. Liquid water was fed to the cathode and gaseous SO_2 was fed to the anode. A Nafion membrane electrode assembly (MEA) using Pt black catalyst on both the anode and cathode was employed. Polarization curves for Nafion membranes N212, N115, and N117 are shown in Figure 18.13 [108]. These membranes have thicknesses of 51, 127, and 178 μm, respectively. A differential pressure of 600 kPa across the membrane (higher on the cathode) and an operating temperature of 80°C were maintained. The thinner membranes have lower cell potentials and permit higher current densities. At a current density of 500 mA/cm², the N212 cell achieved a cell potential of 720 mV. The measured sulfuric acid concentration leaving the cell was 4.2 M. However, the thinner membranes are also expected to permit more SO_2 crossover from the anode to the cathode. Furthermore, the authors concluded that Nafion membranes were not a suitable choice for an SDE because of the significant reduction in conductivity in the presence of concentrated sulfuric acid. Alternatively, they recommended the development of a novel PEM in which the conductivity is not adversely affected by sulfuric acid.

Extensive SDE testing has been reported by SRNL using both a button-cell electrolyzer and a larger single-cell electrolyzer. Photographs of the single cell electrolyzer are shown

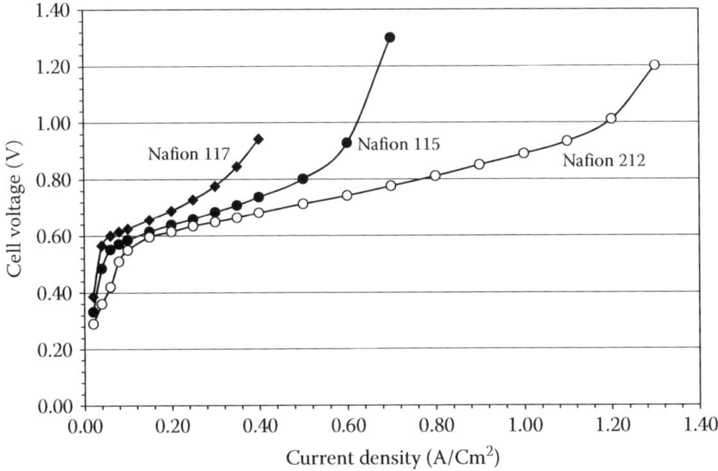

FIGURE 18.13
Polarization curves for Nafion 212, 115, and 117 membranes. (Reproduced from Staser, J. A. and Weidner, J. W. *J. Electrochem. Soc.* **156**(1), B16–B21, 2009. With permission.)

in Figure 18.14. It is constructed of steel support plates, graphite plates, graphite flow fields, graphite diffusions layers, PVDF gaskets and PFA-lined steel tubing. Copper current collectors are sealed from contacting the process fluids. The cell is designed to permit easy replacement of the MEA, consisting of the PEM bonded between platinized carbon anode and cathode layers. The cell has a nominal active cell area of 60 cm^2. A test facility capable of automated unattended testing of SDE units at pressures up to 600 kPa and temperatures up to 80°C was designed and constructed. Most of the single-cell tests were operated to generate approximately 10–20 L/h of hydrogen. The SDE was operated in the liquid-fed mode, with the anolyte consisting of sulfuric acid saturated with dissolved sulfur dioxide. Water was also added on the cathode side of the cell in order to maintain membrane hydration and to remove any sulfur species that could result from SO_2 crossover.

Thirty-seven different MEA configurations were tested, with representative polarization curve results shown in Figure 18.15 [94]. The test results show a cell potential of approximately 760 mV at a current density of 500 mA/cm^2 during operation at 80°C with 30 wt% sulfuric acid. Operation at higher temperature (120–140°C) and use of advanced membranes are expected to lower the cell potential to a value nearer the commercial goal of 600 mV at 500 mA/cm^2. The advanced membranes will also be required to maintain high conductivity in the presence of high sulfuric acid concentrations.

A major concern of SDE operation is the diffusion of sulfur dioxide through the PEM to the cathode where it can be reduced to hydrogen sulfide and eventually elemental sulfur. Earlier tests performed by SRNL revealed the build-up of a sulfur layer inside of the cell's MEA [96]. More recent testing indicated that such a sulfur build-up could be avoided through the proper selection of membranes and the choice of operating conditions. In this case, the SO_2 crossover may be reduced to the point where a manageable amount of H_2S (part-per-million level) is generated and leaves the cell with the hydrogen gas, but no sulfur build-up occurs inside the SDE. SRNL conducted two tests of 212 h each using a Nafion membrane verifying this mode of operation [102]. Longer term testing is planned

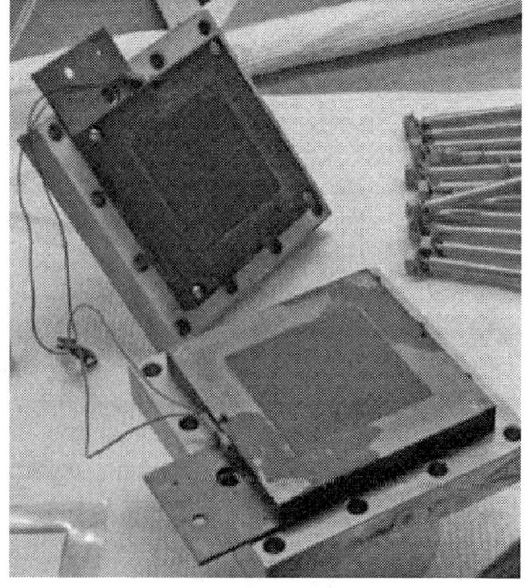

FIGURE 18.14
The SRNL single-cell SDE.

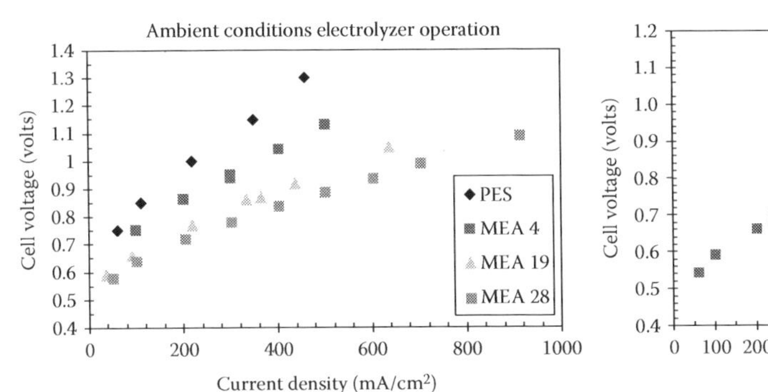

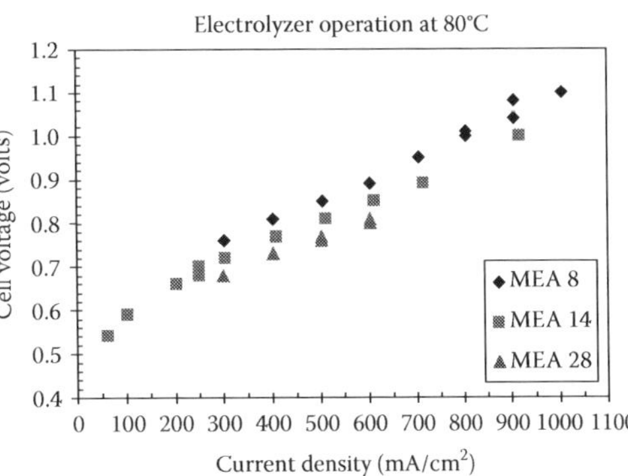

FIGURE 18.15
Representative polarization curve for the SRNL single-cell electrolyzer. (From Steimke, J. L., Steeper, T., and Herman, D. *Hybrid Sulfur Electrolyzer Workshop and Information Exchange*, Aiken, SC, United States, April 20–21, 2009, http://srnl.doe.gov/hse_workshop/Steimke%20Electrolyzer%20Performance.pdf (accessed October 21, 2009). With permission.)

in the future, including operation at higher temperature and pressure using advanced membranes.

SRNL has also demonstrated multicell SDE stack operation [92]. A photograph of a three-cell SDE installed in the SRNL test facility is shown in Figure 18.16 [92]. This unit utilized round plates with 160 cm² active area per cell stacked in a bipolar arrangement. It was successfully operated at 80°C and 700 kPa for 72 h with a constant hydrogen output of 86 L/h. Development of higher capacity SDE units, consisting of larger area plates in multicell stacks, will be necessary in order to construct commercial-scale electrolyzers.

18.3.3 High-Temperature Sulfuric Acid Decomposition

The decomposition of sulfuric acid to release oxygen and regenerate sulfur dioxide is the thermochemical step in the HyS cycle. It is also the high-temperature step of the SI

FIGURE 18.16
The SRNL Three-cell SDE. (From Summers, W. A. Hybrid sulfur thermochemical process development, *DOE 2008 Hydrogen Program Annual Merit Review*, Arlington, VA, United States, June 9–13, 2008. http://www.hydrogen.energy.gov/pdfs/review08/pd_26_summers.pdf (accessed October 21, 2009). With permission.)

thermochemical cycle. Much of the research and testing for this step has been performed as part of the work on the SI cycle, and it can mostly be directly applied to HyS. Therefore, in order to minimize redundancy, the reader is referred to the section of this book regarding the SI cycle for a review of the analytical and experimental studies of high temperature sulfuric acid decomposition. However, a recent analysis of the bayonet design concept for sulfuric acid decomposition [98] will be briefly reviewed here.

The recuperative bayonet reactor design was evaluated using pinch analysis in conjunction with statistical methods. The primary objective was to establish the minimum energy requirements without having to perform a detailed heat transfer design. This meant calculating the minimum heating target, or the lowest possible heat input if all thermodynamically feasible recuperation were to be fully utilized. Another objective was to determine the heating utility pinch temperature for the helium heat transfer fluid, that is, the lowest temperature to which the secondary helium loop could be cooled by heat exchange with the bayonet without establishing inefficient, cross-pinch heat transfer. Aspen Plus [121] and Aspen Energy Analyzer [122] were used for the bayonet reactor simulations and pinch analyses, respectively, while JMP [123] was used to suggest simulation "experiments" and perform statistical analyses of the results.

The lowest value of the minimum heating target, 320.9 kJ/mol SO_2, was found at the highest pressure (90 bar) and peak process temperature (900°C) considered, and at a feed concentration of 42.5 mol% H_2SO_4. This assumed a 10°C minimum temperature difference for recuperation, a catalyst bed inlet temperature of 675°C, and local attainment of the decomposition equilibrium in the catalyst bed. Feed temperature was found not to affect the heating target.

For the overall HyS process to be competitive with PWR/BWR-powered alkaline electrolysis, it has been shown that the acid decomposition step should consume no more than about 450 kJ/mol SO_2 (which would provide a 33% net thermal efficiency advantage) [95]. This means that the bayonet design is capable of achieving a heating requirement low enough for a HyS water-splitting process that should be able to compete with electrolysis, even taking into account the additional energy needed to concentrate the acid feed.

Above 800°C peak process temperature, lower pressure gave higher heating targets, while the effect was reversed below 800°C. Lower temperatures consistently gave higher minimum heating targets. This behavior is illustrated in Figure 18.17 [98]. The lowest peak process temperature that could meet the 450-kJ/mol SO_2 benchmark turned out to be 750°C (allowing up to 75 kJ/mol SO_2 for preconcentration of the acid feed), although operation at a pressure significantly lower than that of the helium heat transfer fluid would be required. If the decomposition reactor were to be heated indirectly by an advanced gas-cooled reactor heat source with a 50°C temperature difference between the primary and secondary coolant loops, and a 25°C minimum temperature difference between the secondary coolant and the process, then a HyS cycle using the bayonet concept could be competitive with alkaline electrolysis provided the reactor outlet temperature is at least 825°C, preferably higher. The bayonet design will likely not be practical for use with reactor outlet temperatures below 825°C. In this case, a design based on the direct contact heat exchanger/sulfuric acid scrubber concept, as discussed in Section 18.2.2, may be required.

The temperature of the HTGR heat source can also have a significant impact on the helium heating pinch temperature, which is the minimum temperature at which heat can be extracted from the secondary helium heat source. From Figure 18.17, the lowest heating target (best efficiency) at the highest temperature (900°C) was obtained at the highest pressure (90 bar), while the lowest heating target at the lowest temperature (700°C) was obtained at the lowest pressure (15 bar). Comparing the helium heating pinch temperatures at these

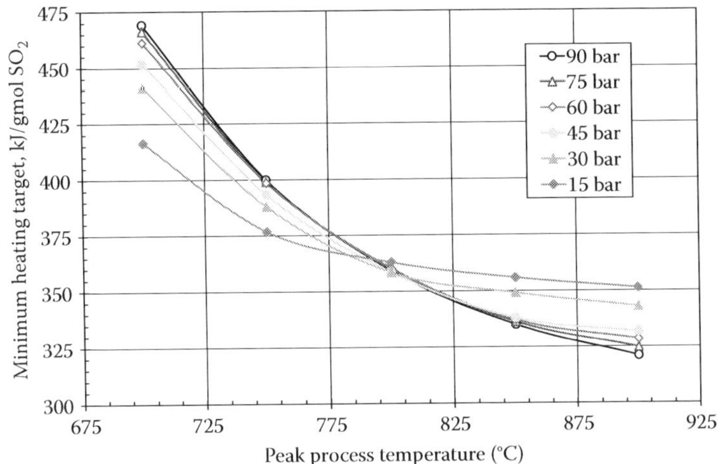

FIGURE 18.17
Effect of pressure and peak process temperature on bayonet reactor minimum heating target [98]; 80.1 wt% H_2SO_4 feed concentration, 675°C catalyst bed inlet temperature, thermodynamic equilibrium in catalyst bed, 10°C minimum temperature difference for recuperation.

two extremes as a function of feed concentration in Figure 18.18 [98], it can be seen that the lower heat source temperature gives the higher helium heating pinch temperature, greatly narrowing the temperature range over which heat can be efficiently transferred from the HTGR to the bayonet reactor. This is a consequence of the assumption that catalyst activity is insufficient below a certain threshold temperature (here assumed to be 675°C). As a result, the same heat of reaction must be supplied over a narrower temperature range, causing the utility pinch temperature to be determined by the catalyst bed inlet conditions instead of the recuperation pinch point. Therefore, lowering the HTGR operating temperature not

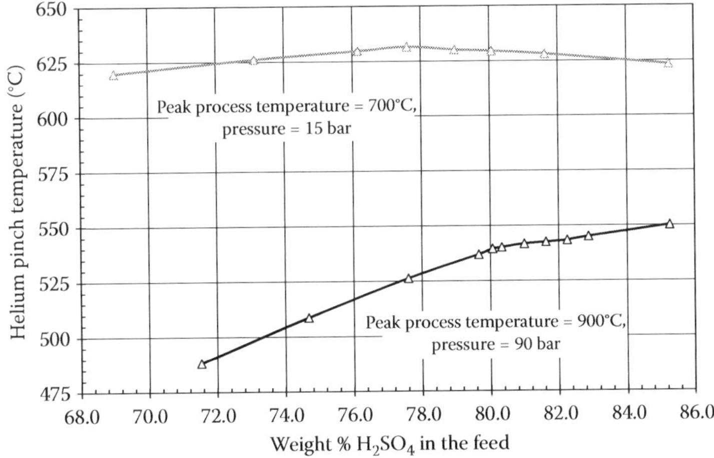

FIGURE 18.18
Bayonet reactor helium heating pinch temperature dependence on feed concentration at stated conditions [98]; 675°C catalyst bed inlet temperature, thermodynamic equilibrium in catalyst bed, 10°C minimum temperature difference for recuperation.

only lowers the upper end of the temperature range over which heat can be transferred from the secondary helium loop to the bayonet reactor, but it may also raise the lower end, greatly reducing the fraction of HTGR heat output that can be applied to sulfuric acid decomposition.

More details are available elsewhere [98].

18.4 Process Flow Sheet

One of the first complete HyS flow sheets was presented in a conceptual design report prepared by Westinghouse for NASA [2]. The best illustration of this flow sheet appears in Knoche and Funk's second-law analysis [58], and is reproduced in Figure 18.19. The salient features include an HTGR heat source operating at 1010°C, a microporous diaphragm SDE operating at 90°C and 26 bar to produce 75 wt % sulfuric acid, and a sulfuric acid decomposer operating at 871°C and 3.8 bar. SO_2 is separated from O_2 by first compressing the decomposition product to 52 bar (in stages, with intercoolers and condensers to remove liquid water) and then cooling while expanding to 25 bar to achieve −97°C temperature and a relatively clean split between liquid SO_2 and gaseous O_2. The cell potential was assumed to be 450 mV.

The net thermal efficiency of this flow sheet was estimated to be 45% on a higher heating value (HHV) basis, which is equivalent to 38% (LHV basis) [2]. This value was confirmed in Knoche and Funk's subsequent analysis [58]. With the over 30 years' perspective from which these results are now viewed, it is clear that some of the underlying assumptions are unrealistically optimistic and should be changed. However, the diminishing effect of these changes has been offset by advances in process integration, particularly pinch technology.

A more recent HyS flow sheet has been presented by SRNL and others [97]. It is based on the use of a PEM-type SDE and a bayonet decomposition reactor. Many earlier efforts to prepare flow sheets for HyS and other thermochemical processes used somewhat simplified and idealized designs. The SRNL-led team, which included experienced industrial partners, attempted to develop a flow sheet reflecting normal industrial practices, including realistic operating conditions, such as heat exchanger approach temperatures, conservative ambient design values, and the need for blow-down and make-up streams. A summary of this flow sheet is presented in the following.

18.4.1 Integrating a PEM SDE with a Bayonet Decomposition Reactor

A PEM-type electrolyzer using a Nafion membrane was assumed as the basis for the SDE design. Use of Nafion membranes limits the SDE anolyte product to 50 wt% H_2SO_4 at 100°C and 20 bar. Higher temperatures give better kinetics, but cannot be tolerated by Nafion PEMs. SDEs using advanced, higher temperature membranes are being investigated, and successful implementation of these components may permit an improved flow sheet design using higher temperature SDE operation and more concentrated sulfuric acid anolyte. The pressure was selected to provide product hydrogen under pressure, which minimizes downstream compression requirements. The operating pressure was limited, however, by phase equilibrium considerations. Higher pressures favor SO_2 solubility, but are limited by SO_2 vapor pressure—two liquid phases could otherwise form in the SDE in some circumstances.

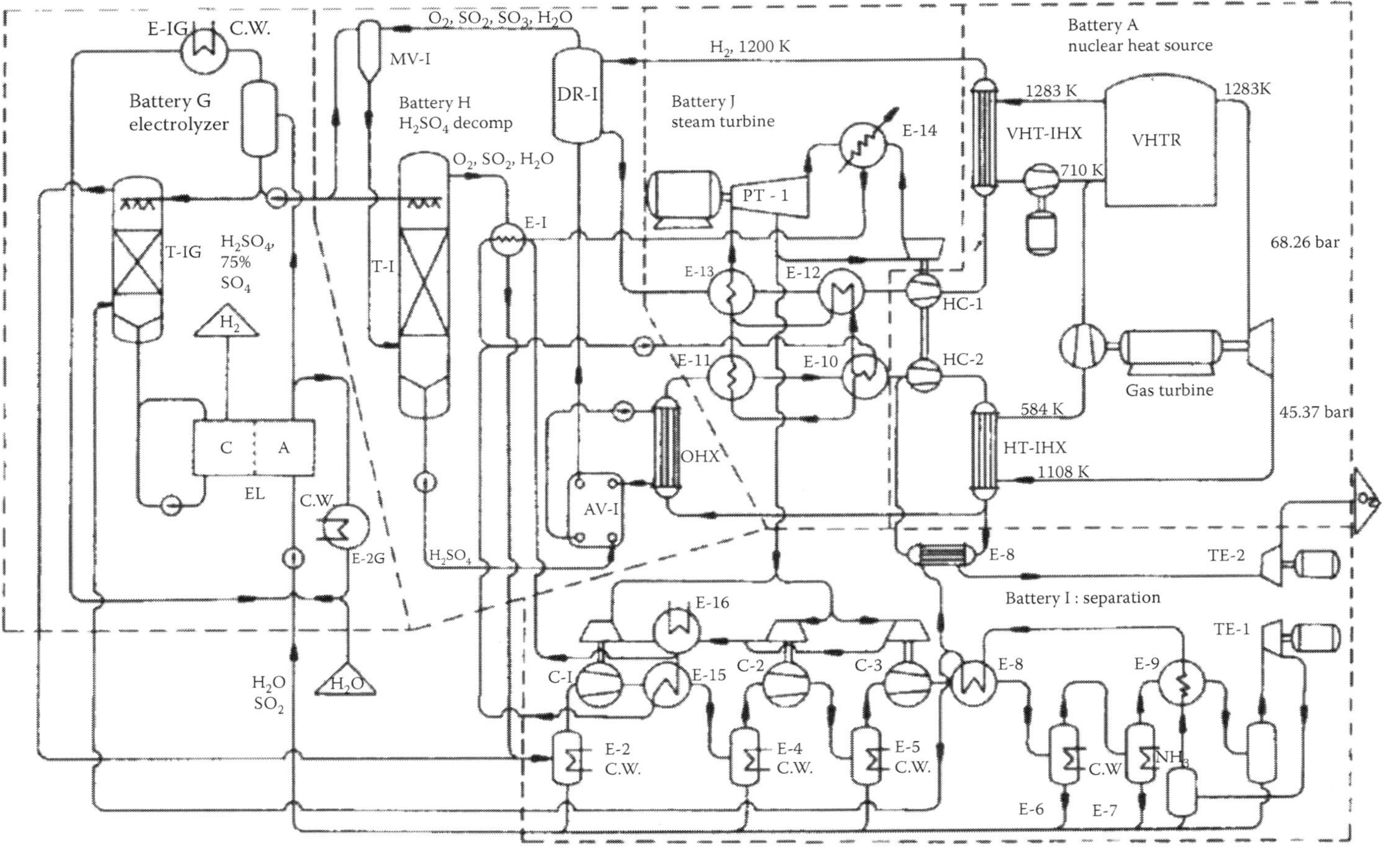

FIGURE 18.19
Early Westinghouse HyS flow sheet with microporous diaphragm SDE and low-temperature (−100°C) SO_2–O_2 separation. (Adapted from Knoche, K. F. and Funk, J. E. *Int. J. Hydrogen Energy* **2**(4), 377–385, 1977.)

Anolyte effluent contains unreacted SO_2 as well as traces of O_2. To integrate the SDE with a high-temperature decomposition reactor, most of the anolyte needs to be recycled. The remaining product stream, containing one mole of H_2SO_4 for each mole of H_2 produced in the SDE, is passed on. Unreacted SO_2 and trace O_2 first need to be removed from this stream, along with about one-third (to yield 60 wt% H_2SO_4) to three-quarters (to yield 80 wt% H_2SO_4) of the water content. A detailed analysis of the bayonet reactor has shown that a 65–80 wt % H_2SO_4 feed concentration gives the best energy utilization [98]. Similarly, the bayonet product has to be cooled, unreacted H_2SO_4 recycled, O_2 removed, and the remaining SO_2 and water condensed and combined with recycled, spent anolyte to form fresh anolyte feed.

The simplest way to remove unreacted SO_2 and trace O_2 from the SDE product is by dropping the pressure. This requires recompressing the predominantly SO_2 and trace O_2 vapor stream that is out-gassed so that it can be recycled to the SDE feed system. The shaft work required is not excessive and the separation can be made without any additional heat input.

A variety of methods are available to concentrate degassed SDE product. All consume energy, some more than others. For example, dropping the pressure of a 50% H_2SO_4 solution at 100°C isothermally, from 20 bar to just under 0.05 bar, can concentrate the liquid residue to 75% H_2SO_4 while consuming 169 kJ/mol H_2SO_4 heat duty. Water boils off in the vapor phase. Heating 50% H_2SO_4 at 20 bar from 100 to 317°C could just as easily boil off the water instead and would need 246 kJ/mol H_2SO_4 heat duty. In either case, excessive heating would be required, causing the sum of the acid concentration and decomposition duties to exceed the 450-kJ/mol H_2 practical upper limit. A recuperative method is clearly needed.

SRNL has chosen vacuum distillation with recuperative preheating/partial vaporization of the feed streams to concentrate the SDE product. Vacuum is maintained using a two-stage steam ejector, keeping temperatures high enough to allow use of cooling water in the condenser, yet low enough to permit metallic materials of construction. Water removed in the concentration process is condensed and recycled to the SDE feed system. The vacuum column bottoms, containing concentrated sulfuric acid at 75 wt% H_2SO_4, can be pumped to the necessary pressure and fed directly to the bayonet reactor.

Effluent from the bayonet reactor is readily separated into unreacted H_2SO_4 feed and SO_2/O_2 product by means of a vapor/liquid split. The acid is simply recycled to the vacuum still. The vapor is cooled further and let down to the pressure of the SDE feed system, resulting in a three-phase mixture: wet liquid SO_2, a saturated solution of SO_2 in H_2O, and wet O_2 gas contaminated with SO_2. The O_2 gas can be scrubbed with the water collected in the concentration process to remove most of the SO_2. The two liquid phases can then be combined with that water and with recycled, spent anolyte to form fresh anolyte feed.

Integration of the PEM SDE and the bayonet reactor as described above results in a flow sheet that uses only proven technology, with the sole exception of the SDE and the decomposition reactor. Thus, the flow sheet itself does not introduce any additional technical hurdles. This should help give the performance projections greater credibility.

18.4.2 The PEM SDE/Bayonet Reactor HyS Flow Sheet

An Aspen Plus flow sheet was prepared according to the integration approach described in the preceding section. The flow sheet is shown in Figure 18.20. Stream data for this flow sheet are tabulated in Table 18.2. The basis is a 1-kmol/sec H_2 production rate.

The flow sheet simulations used OLI Systems, Inc.'s mixed solvent electrolyte (MSE) properties model [124], which has been shown to represent the H_2SO_4–H_2O system very

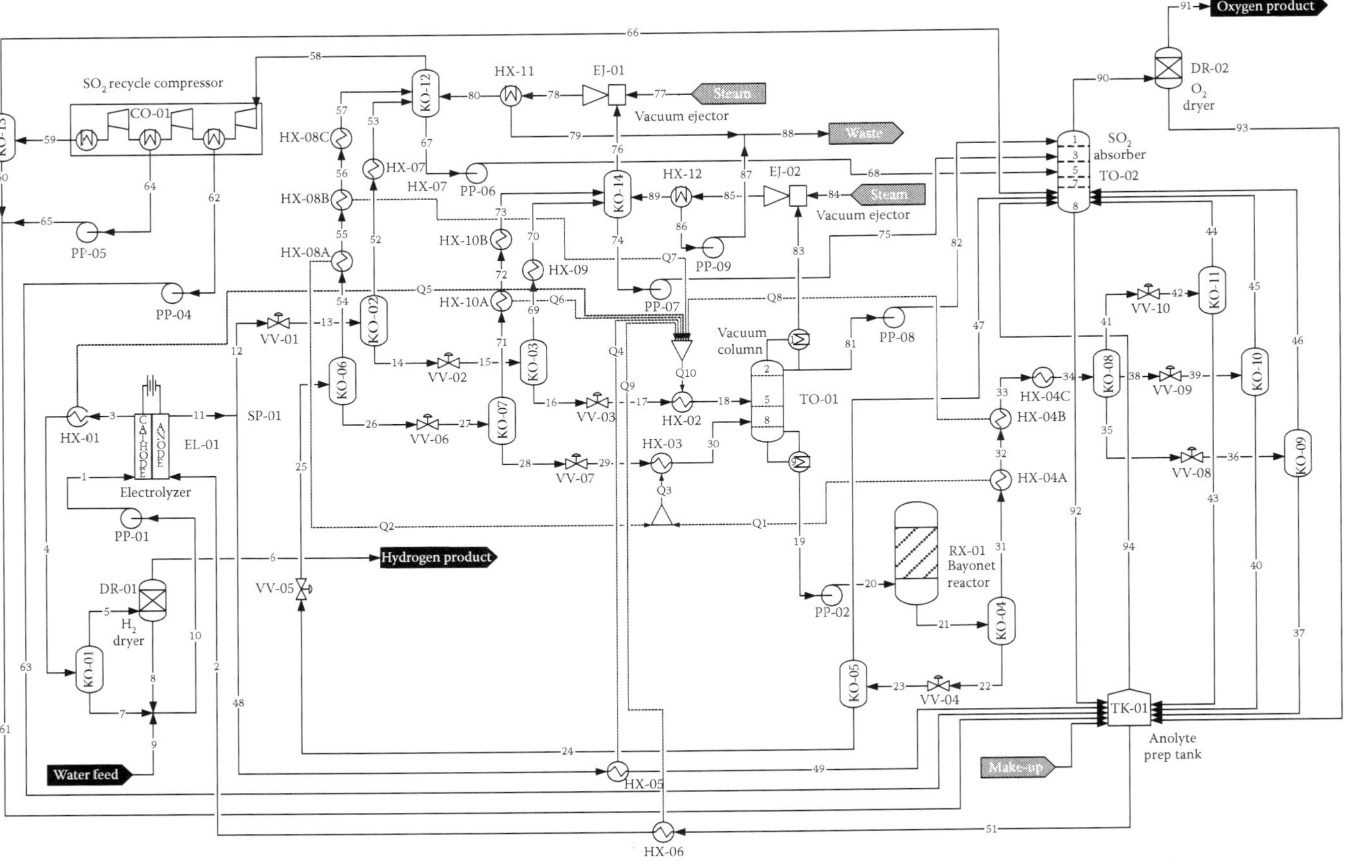

FIGURE 18.20
HyS Flow sheet using a PEM SDE and a Bayonet decomposition reactor. (From Gorensek, M. B. et al. Hybrid sulfur process reference design and cost analysis. Savannah River National Laboratory, Report No. SRNL-L1200-2008-00002, June 12, 2009 (doi: 10.2172/956960). With permission.)

TABLE 18.2
HyS Flow Sheet Stream Table

Stream ID	Molar Flow Rates, kmol/s						Temperature,		Pressure, bar	Phase
	H_2O	H_2SO_4	SO_2	O_2	H_2	Total	°C	°K		
1	10	0	0	0	0.00235	10.0024	73.20	346.35	21	L
2	27.4794	3.88405	2.5	0.00527	0	37.7528	78.77	351.92	21	L
3	9	0	0	0	1.00235	10.0024	100.00	373.15	20	L + V
4	9	0	0	0	1.00235	10.0024	76.96	350.11	20	L + V
5	0.02185	0	0	0	1.00004	1.02189	76.96	350.11	20	V
6	0	0	0	0	1.00004	1.00004	40.00	313.15	20	V
7	8.97815	0	0	0	0.00231	8.98046	76.96	350.11	20	L
8	0.02185	0	0	0	0	0.02185	40.00	313.15	20	L
9	1	0	0	0	0	1	40.00	313.15	20	L
10	10	0	0	0	0.00231	10.0023	73.19	346.34	20	L
11	26.4794	4.88405	1.5	0.00527	0	37.7528	100.00	373.15	20	L + V
12	5.42394	1.00043	0.30725	0.00108	0	7.73314	100.00	373.15	20	L + V
13	5.42394	1.00043	0.30725	0.00108	0	7.73314	84.08	357.23	1.01325	L + V
14	5.33174	1.00043	0.02004	3.9E − 07	0	7.35265	84.08	357.23	1.01325	L
15	5.33174	1.00043	0.02004	3.9E − 07	0	7.35265	80.38	353.53	0.30198	L + V
16	5.29406	1.00043	0.00263	2.9E − 10	0	7.29756	80.38	353.53	0.30198	L
17	5.29406	1.00043	0.00263	2.9E − 10	0	7.29756	66.96	340.11	0.11	L + V
18	5.29406	1.00043	0.00263	2.9E − 10	0	7.29756	94.41	367.56	0.11	L + V
19	3.7702	2.07782	6.7E − 12	2E − 21	0	7.92584	122.87	396.02	0.13	L
20	3.7702	2.07782	6.7E − 12	0	0	7.92584	123.63	396.78	86	L
21	4.77063	1.07739	1.00043	0.50021	0	8.42605	254.70	527.85	86	L + V
22	4.30695	1.07739	0.23587	0.00844	0	6.70604	254.70	527.85	86	L
23	4.30695	1.07739	0.23587	0.00844	0	6.70604	233.42	506.57	21	L + V
24	4.03002	1.07739	0.04733	8.9E − 05	0	6.23222	233.42	506.57	21	L
25	4.03002	1.07739	0.04733	8.9E − 05	0	6.23222	147.33	420.48	1.01325	L + V
26	3.13019	1.07739	0.00035	1.4E − 08	0	5.28532	147.33	420.48	1.01325	L
27	3.13019	1.07739	0.00035	1.4E − 08	0	5.28532	118.57	391.72	0.30198	L + V
28	2.87261	1.07739	3.7E − 06	2.4E − 12	0	5.02739	118.57	391.72	0.30198	L

29	2.87261	1.07739	3.7E – 06	2.4E – 12	0	5.02739	100.50	373.65	0.13	L + V
30	2.87261	1.07739	3.7E – 06	2.4E – 12	0	5.02739	114.62	387.77	0.13	L + V
31	0.46368	5.5E – 06	0.76455	0.49177	0	1.72001	254.70	527.85	86	V
32	0.46368	5.5E – 06	0.76455	0.49177	0	1.72002	110.50	383.65	86	L + V
33	0.46368	5.5E – 06	0.76455	0.49177	0	1.72002	76.96	350.11	86	L + V
34	0.46368	5.5E – 06	0.76455	0.49177	0	1.72002	40.00	313.15	86	L + V
35	0.08967	1.8E – 07	0.65507	0.00123	0	0.74597	40.00	313.15	86	L
36	0.08967	1.8E – 07	0.65507	0.00123	0	0.74597	41.78	314.93	21	L + V
37	0.08966	1.8E – 07	0.65459	0.00025	0	0.74451	41.78	314.93	21	L
38	0.37282	5.3E – 06	0.03946	0.00066	0	0.41296	40.00	313.15	86	L
39	0.37282	5.3E – 06	0.03946	0.00066	0	0.41296	40.99	314.14	21	L + V
40	0.37282	5.3E – 06	0.03922	0.00014	0	0.41218	40.99	314.14	21	L
41	0.00119	0	0.07002	0.48988	0	0.56109	40.00	313.15	86	V
42	0.00119	0	0.07002	0.48988	0	0.56109	15.50	288.65	21	L + V
43	0.0006	0	4.7E – 05	3.8E – 07	0	0.00065	15.50	288.65	21	L
44	0.00059	0	0.06997	0.48988	0	0.56044	15.50	288.65	21	V
45	3.5E – 06	4.3E – 26	0.00025	0.00052	0	0.00077	40.99	314.14	21	V
46	6.9E – 06	1E – 25	0.00048	0.00098	0	0.00146	41.78	314.93	21	V
47	0.27692	1.3E – 06	0.18854	0.00835	0	0.47382	233.42	506.57	21	V
48	21.0554	3.88362	1.19275	0.00419	0	30.0196	100.00	373.15	20	L + V
49	21.0554	3.88362	1.19275	0.00419	0	30.0196	80.00	353.15	20	L
50	21.0554	3.88362	1.19275	0.00419	0	30.0196	80.01	353.16	21	L
51	27.4794	3.88405	2.5	0.00524	0	37.7527	84.71	357.86	21	L
52	0.0922	2E – 11	0.28721	0.00108	0	0.38049	84.08	357.23	1.01325	V
53	0.0922	2E – 11	0.28721	0.00108	0	0.38049	40.00	313.15	1.01325	L + V
54	0.89983	6.6E – 07	0.04698	8.9E – 05	0	0.9469	147.33	420.48	1.01325	V
55	0.89983	6.6E – 07	0.04698	8.9E – 05	0	0.9469	110.50	383.65	1.01325	L + V
56	0.89983	6.6E – 07	0.04698	8.9E – 05	0	0.9469	76.96	350.11	1.01325	L + V
57	0.89983	6.6E – 07	0.04698	8.9E – 05	0	0.9469	40.00	313.15	1.01325	L + V
58	0.02674	2.3E – 24	0.33759	0.00117	0	0.3655	40.00	313.15	1.01325	V
59	1.9E – 05	0	0.00541	0.00116	0	0.00659	40.00	313.15	21	L + V

continued

TABLE 18.2 (continued)
HyS Flow Sheet Stream Table

Stream ID	Molar Flow Rates, kmol/s						Temperature,		Pressure, bar	Phase
	H_2O	H_2SO_4	SO_2	O_2	H_2	Total	°C	°K		
60	1.9E − 05	0	0.0048	1.2E − 06	0	0.00482	40.00	313.15	21	L
61	0.00936	0	0.33626	1E − 05	0	0.34562	41.20	314.35	21	L
62	0.01738	0	0.00073	3.5E − 09	0	0.01811	40.00	313.15	2.78324	L
63	0.01738	0	0.00073	3.5E − 09	0	0.01811	41.11	314.26	21	L
64	0.00934	0	0.33146	9.2E − 06	0	0.3408	40.00	313.15	7.64513	L
65	0.00934	0	0.33146	9.2E − 06	0	0.3408	41.22	314.37	21	L
66	3.2E − 07	0	0.00061	0.00116	0	0.00177	40.00	313.15	21	V
67	0.96665	6.6E − 07	0.01365	6.1E − 08	0	0.9803	40.00	313.15	1.01325	L
68	0.96665	6.6E − 07	0.01365	6.1E − 08	0	0.9803	40.31	313.46	21	L
69	0.03768	6.4E − 12	0.01741	3.9E − 07	0	0.0551	80.38	353.53	0.30198	V
70	0.03768	6.4E − 12	0.01741	3.9E − 07	0	0.0551	40.00	313.15	0.30198	L + V
71	0.25758	1E − 07	0.00035	1.4E − 08	0	0.25793	118.57	391.72	0.30198	V
72	0.25758	1E − 07	0.00035	1.4E − 08	0	0.25793	76.96	350.11	0.30198	L + V
73	0.25758	1E − 07	0.00035	1.4E − 08	0	0.25793	40.00	313.15	0.30198	L
74	0.28961	1E − 07	0.00106	2.8E − 11	0	0.29067	40.61	313.76	0.30198	L
75	0.28961	1E − 07	0.00106	2.8E − 11	0	0.29067	41.14	314.29	21	L
76	0.0059	1.2E − 25	0.01745	4.1E − 07	0	0.02335	40.61	313.76	0.30198	V

77	0.02335	0	0	0	0	0.02335	169.99	443.14	7.91	L + V
78	0.02925	0	0.01745	4.1E − 07	0	0.0467	93.77	366.92	1.01325	V
79	0.0279	0	0.0004	1.2E − 11	0	0.0283	40.00	313.15	1.01325	L
80	0.00135	0	0.01706	4.1E − 07	0	0.0184	40.00	313.15	1.01325	V
81	4.39285	6.9E − 35	0.00185	4.9E − 13	0	4.3947	40.02	313.17	0.09	L
82	4.39285	0	0.00185	4.9E − 13	0	4.3947	40.21	313.36	21	L
83	0.00361	5.2E − 35	0.00079	3E − 10	0	0.0044	40.02	313.17	0.09	V
84	0.0044	0	0	0	0	0.0044	169.99	443.14	7.91	L + V
85	0.00801	0	0.00079	3E − 10	0	0.0088	95.05	368.20	0.30198	V
86	0.00777	0	2.9E − 05	1.3E − 14	0	0.0078	40.00	313.15	0.30198	L
87	0.00777	0	2.9E − 05	1.3E − 14	0	0.0078	40.04	313.19	1.01325	L
88	0.03567	0	0.00042	1.2E − 11	0	0.03609	40.03	313.18	1.01325	L
89	0.00025	0	0.00076	3E − 10	0	0.001	40.00	313.15	0.30198	V
90	0.00228	1.4E − 34	0.00413	0.50024	0	0.50665	41.12	314.27	21	V
91	0	0	0	0.50024	0	0.50024	40.00	313.15	21	V
92	5.92438	2.1E − 06	0.27327	0.00116	0	6.19882	82.33	355.48	21	L
93	0.00228	0	0.00413	0	0	0.00641	40.00	313.15	21	L
94	2.9E − 05	7.6E − 16	0.00098	0.00051	0	0.00152	84.71	357.86	21	V
MAKEUP	0.00749	0.00042	0	0	0	0.00834	40.00	313.15	21	L

Source: Data from Gorensek, M. B. et al. Hybrid sulfur process reference design and cost analysis. Savannah River National Laboratory, Report No. SRNL-L1200-2008-00002, June 12, 2009 (doi: 10.2172/956960).

well over its entire composition range and at temperatures as high as 500°C [116]. (The Aspen-OLI interface allows OLI Engine use from within Aspen Plus.) Spot checks of the OLI MSE model's representations of SO_2–H_2O and SO_2–H_2SO_4–H_2O vapor–liquid and liquid–liquid equilibria against the available data showed generally good agreement (e.g., see Figures 18.21 [125–127] and 18.22 [128–130]).

Block EL-01, on the left-hand side of Figure 18.20, is the SDE. Stream 1 is the H_2O feed to the cathode, while stream 2 is the anode feed, containing 15.5 wt% SO_2 dissolved in 43.5 wt% H_2SO_4. Both streams are at 21 bar, and the SDE imposes a frictional pressure drop of 1 bar. The quantity of SO_2 dissolved in stream 2 is just below the saturation point (beyond which a separate liquid SO_2 phase would form) and the target concentration of H_2SO_4 in the anolyte product following the reaction is 50% by weight (excluding SO_2).

A detailed unit operation model of the SDE has yet to be developed, so a "black box" approach was used instead as described below:

- A net flux of H_2O is imposed from cathode to anode (arbitrarily set at 1 kmol H_2O/kmol SO_2 reacted) because of the difference in water activity across the PEM.
- Reaction 18.2 occurs with an assumed SO_2 conversion of 40%.
- All H_2 exits the cathode with the remaining water from stream 1 via stream 3.
- Everything else exits the anode via stream 11.
- The temperatures of streams 3 and 11 are identical and calculated to make the enthalpy change between feed and product streams equal to the electrical work performed with an assumed cell potential of 0.6 V.
- Streams 3 and 11 exit the SDE at a pressure 1 bar lower than streams 1 and 2, respectively.

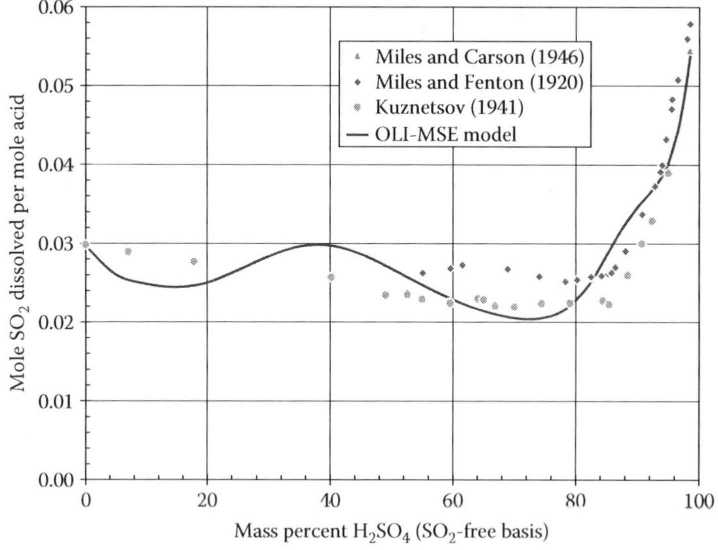

FIGURE 18.21
Solubility of SO_2 in sulfuric acid at 1.013 bar partial pressure—comparison of OLI MSE model with data of Miles and Carson [127], Kuznetsov [126], and Miles and Fenton [125]. (From Gorensek, M. B. and Summers, W. A. *Int. J. Hydrogen Energy* **34**(9), 4097–4114, 2009. With permission.)

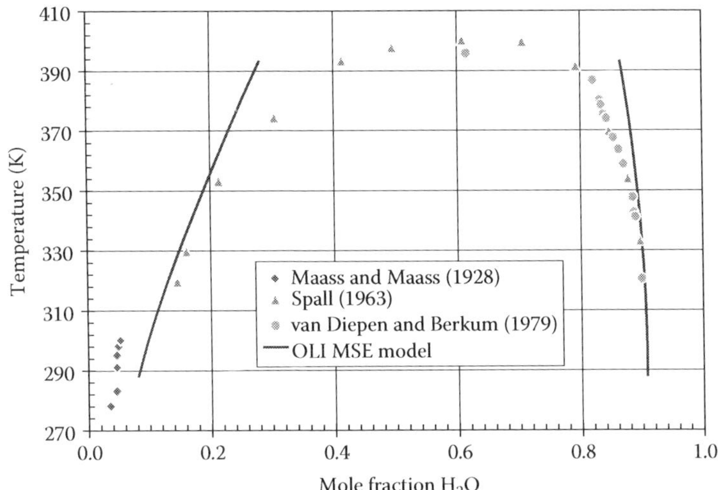

FIGURE 18.22
Liquid-liquid phase equilibrium in the SO_2–H_2O system—comparison of OLI MSE Model with Data of Maass and Maass [128], Spall [129], and van Diepen and Berkum [130]. (From Gorensek, M. B. and Summers, W. A. *Int. J. Hydrogen Energy* **34**(9), 4097–4114, 2009. With permission.)

The temperature of streams 3 and 11 is controlled to 100°C by adjusting the temperature of anolyte feed stream 2 using heat exchanger HX-06. Experimental evidence for the diffusion of water from the cathode to the anode in a PEM SDE has been provided by Staser and co-workers [104] at the University of South Carolina (USC). The assumed flux of 1 kmol H_2O/kmol SO_2 reacted is an estimate consistent with USC data that needs to be refined with further experiments.

The assumed SO_2 conversion of 40% is a design goal based on experimental data from SRNL and USC. Higher conversion means less unconverted SO_2 to recycle, but then cell potential will likely increase because of lower average SO_2 concentration at the anode. Conversions in excess of 20% are routinely achieved in SDE development experiments at SRNL, and over 50% at USC using gaseous SO_2 feed. The trade-off between cell potential and recycle effects will ultimately determine the optimum value of conversion once the relationship between conversion and cell potential has been established.

Stream 3 is cooled by interchange with stream 17 in heat exchanger HX-01, maintaining a 10°C minimum temperature difference. The remaining vapor is removed as stream 5 in knock-out KO-01 and fed to the H_2 dryer (DR-01), which removes all of the H_2O as stream 8. Pure, dry H_2 product exits as stream 6 at 40°C and 20 bar. H_2O "feedstock" (stream 9) is combined with recycled H_2O (stream 7) and that removed by the dryer (stream 8) and pumped back to the cathode using pump PP-01. Since the flux of water across the PEM from the cathode to the anode because of diffusion was deliberately set at exactly 1 kmol H_2O/kmol SO_2, the water-splitting material balance closes.

The H_2 dryer was not designed in detail. Regeneration requirements were not taken into account. Simple cooling to 40°C should condense most of the water; the remainder could be absorbed by a molecular sieve. This should not significantly affect energy efficiency.

Anolyte product (stream 11) is split at SP-01, sending enough H_2SO_4 to decomposition via stream 12 to make all of the SO_2 needed for the SDE. The remainder (about 80% of stream 11) is recycled to the anolyte prep tank (TK-01) via stream 48, which is cooled by interchange with stream 17 in heat exchanger HX-05 with a 10°C minimum temperature difference.

The anolyte product pressure is dropped in three stages before entering the vacuum column (TO-01). Throttling valve VV-01 drops the pressure to atmospheric, vaporizing over 90% of the SO_2 and all but a trace of the dissolved O_2 for removal in knock-out KO-02 via stream 52, along with some H_2O. The pressure is then dropped to slightly over 0.3 bar via throttling valve VV-02. A steam ejector (EJ-01) provides the necessary suction. Nearly 90% of the remaining SO_2 and a little more H_2O are vaporized and removed in knock-out KO-03 via stream 69. Pressure is dropped once more to match the feed stage pressure in the vacuum column (0.11 bar) by means of throttling valve VV-03. The SO_2–free acid (stream 17) is first heated and partially vaporized by interchange with six other streams in heat exchanger HX-02, maintaining a 10°C minimum temperature difference with all of them, before being fed to the 5th equilibrium stage of TO-01.

The second-stage steam ejector, EJ-02, maintains the vacuum column overhead pressure at 0.09 bar. Pressure drop from bottom to top is set at 0.04 bar, placing the bottoms at 0.13 bar. This is feasible with ceramic or metallic structured packing, as can be demonstrated with Aspen Plus RadFrac Pack Sizing calculations. TO-01 has 9 equilibrium stages, including a partial condenser (stage 1) and a kettle reboiler (stage 9). The reflux/distillate ratio is 0.1, with a distillate vapor fraction of 0.001. A bottoms concentration of 75% H_2SO_4 can be achieved by adjusting the distillate rate to just under 4.4 kmol/s. At these conditions, the bottoms temperature is 122.9°C and the reboiler requires 75.5 kJ/mol H_2 duty, which can be supplied with a low-pressure steam utility. The overhead is at 40°C and requires 210.5 kJ/mol H_2 cooling duty, which can be provided using a conventional cooling water system. (This may not be possible during summertime in some climates, because it would require the cooling water supply temperature not to exceed 30°C. In that case, the column pressure could be increased if necessary.)

Bottom product from the vacuum column (stream 19), containing 75 wt% H_2SO_4 at 122.9°C and 0.13 bar is pumped directly to the decomposition reactor (bayonet reactor, RX-01) by means of pump PP-02, which raises the pressure to 86 bar. A separate simulation (detailed in References [95] and [98]) was used to establish the minimum high-temperature heating target for RX-01, 340.3 kJ/mol H_2, which was set equal to the actual high-temperature heat requirement (with the implicit assumption of perfect recuperation). An HTGR outlet temperature of 950°C was assumed, leading to a peak process temperature of 875°C, as shown in Figure 18.6. At these conditions, fractional conversion of H_2SO_4 was 48.1%, and the effluent temperature was 254.7°C with a vapor fraction of 20.4% (assuming 10°C minimum temperature difference for recuperation).

Bayonet reactor product (stream 21) is separated into its liquid (stream 22) and vapor (stream 31) components by knock-out KO-04. The pressure of the liquid product is then dropped to 21 bar using throttling valve VV-04. This allows about 80% of the dissolved SO_2 and 99% of the dissolved O_2 to be removed in knock-out KO-05. The remaining liquid (stream 24) is a 59.3 wt% H_2SO_4 solution at 233.4°C that can be recycled to the vacuum column. The vapor that was removed (stream 47) contains some of the SO_2 and O_2 products of the decomposition reaction. It is sent to the SO_2 absorber (TO-02) for separation.

Recycled unconverted H_2SO_4 (stream 24) is depressurized in three stages before being fed to the vacuum column in a manner identical to the SDE anolyte product. Throttling valves VV-05, VV-06, and VV-07 serve this purpose, and act with knock-outs KO-06 and KO-07 to strip out all but a trace of the dissolved SO_2 and O_2 along with some H_2O via streams 54 and 71. The resulting 67.1 wt% H_2SO_4 stream (number 29) is heated and partially vaporized by interchange with two other streams in heat exchanger HX-03, maintaining a 10°C minimum temperature difference with both. It is then fed to equilibrium stage 8 of the vacuum column.

Hot vapor from the bayonet reactor product (stream 31) is cooled to 40°C and partially condensed in three steps (by heat exchangers HX-04A, HX-04B, and HX-04C). It is then separated into one vapor and two liquid phases in knock-out decanter KO-08. The lighter liquid phase consists of H_2O with 27.3 wt% dissolved SO_2 and a small amount of O_2, while the heavier phase is liquid SO_2 containing 3.7 wt % H_2O and a small amount of O_2. HX-04A is cooled by interchange with the recycled acid feed to the vacuum column (HX-03), while HX-04B interchanges with the anolyte product feed to the vacuum column (HX-02). HX-04C rejects heat to cooling water. In practice, it would not be necessary to separate the two liquid phases in KO-08, since both are ultimately sent to the anolyte prep tank. However, the split is shown to demonstrate that separate aqueous and SO_2 phases can and do exist at these conditions.

The vapor and liquid effluents from KO-08 are let down in pressure from 86 bar to 21 bar via throttling valves VV-08, VV-09, and VV-10, which leads to further phase changes. Vapor is separated from residual liquids in knock-outs KO-09, KO-10, and KO-11, and fed to the bottom of the SO_2 absorber, while the liquid phases are collected by the anolyte prep tank.

Second-stage ejector (EJ-02) effluent (stream 85) is cooled to 40°C with cooling water in heat exchanger HX-12. The condensate (stream 86) is pumped to waste by pump PP-09 to help ensure that any contaminants that may have entered with the steam (stream 84) are purged from the system. Remaining vapors (stream 89) are sent to knock-out KO-14, which is maintained at 0.3 bar by means of the first-stage ejector, EJ-01. KO-14 also receives partially condensed vapors from knock-outs KO-03 (stream 70) and KO-07 (stream 73); these streams are cooled with cooling water to 40°C by heat exchangers HX-09 and HX-10B. Some of the heat released by condensing vapor (heat exchanger HX-10A) is transferred by interchange to HX-02. The aqueous phase removed by KO-14 (stream 74) is pumped to the 3rd equilibrium stage of the SO_2 absorber by pump PP-07.

First-stage ejector (EJ-01) effluent (stream 78) is cooled to 40°C with cooling water in heat exchanger HX-11. This ejector's condensate (stream 79) is also sent to waste. The remaining vapors (stream 80) are sent to knock-out KO-12, which receives partially condensed vapors from knock-outs KO-02 and KO-06 as well. Vapors removed in knock-out KO-02 (stream 52) are first cooled to 40°C with cooling water by heat exchanger HX-07, while those removed in knock-out KO-06 (stream 54) are cooled to 40°C in a series of three heat exchangers: HX-08A, by interchange with HX-03; HX-008B, by interchange with HX-02; and HX-08C, using cooling water.

A dilute solution of SO_2 in H_2O is recovered as liquid in knock-out KO-12 (stream 67). It is pumped to the fifth equilibrium stage of the SO_2 absorber by pump PP-06. The remaining vapor (stream 58) is a moist SO_2 stream containing about 0.3 mol% O_2. It must be compressed to 21 bar before it can be fed to the bottom of the SO_2 absorber. The three-stage SO_2 recycle compressor, CO-01, is used for this purpose. It has intercoolers and knock-outs that cool the outlet of each stage to 40°C and remove any condensate.

The first stage intercooler and knock-out condense and remove about 5% (by volume) of the feed as a 12.9% (by weight) solution of SO_2 in H_2O (stream 62) at 2.78 bar. Pump PP-04 sends this stream to the anolyte prep tank. The second stage intercooler condenses most of the remaining vapors, which are removed by the knock-out (stream 64) as nearly pure liquid SO_2 containing 0.8% H_2O by weight and a trace of O_2 at 7.65 bar. Pump PP-05 sends this stream to the anolyte prep tank. Partially condensed effluent from the last stage (stream 59) is fed to knock-out KO-13. Less than 1% (by volume) of the original feed to

CO-01 (stream 58) remains in the vapor phase (stream 66) after compression. It is fed to the bottom of the SO_2 absorber. Slightly more than twice that quantity is removed as condensate (stream 60) and sent to the anolyte prep tank. Stream 66 is roughly 65% O_2/35% SO_2 (by volume), while stream 60 is 99.9% SO_2 (by weight), with minor amounts of O_2 and H_2O.

The SO_2 absorber has eight equilibrium stages and operates at 21 bar pressure with an assumed negligible pressure drop. Liquid distillate from the vacuum column (stream 81, 99.9% H_2O by weight, with 0.1% SO_2 at 40°C) is fed by pump PP-08 to the top stage. Effluent from the second-stage ejector knock-out (stream 75), containing 98.7% H_2O by weight, with 1.3% SO_2 at 41°C is fed to stage 3. The first-stage ejector knock-out effluent (stream 68) is routed to stage 5. It contains 95.2% H_2O by weight, with 4.8% SO_2 at 40°C. Vapor effluents from five knock-outs and the anolyte prep tank vent are fed to the bottom. SO_2 is scrubbed from the vapor phase as it rises up the absorber, encountering water with progressively less dissolved SO_2. The overhead product (stream 90) is 98.7% O_2 (by volume) at 41°C and 21 bar, with 0.5% H_2O and 0.8% SO_2. It is fed to the O_2 dryer (DR-02), where the H_2O and SO_2 are removed (stream 93) and routed to the anolyte prep tank, leaving a pure O_2 product (stream 91) at 40°C and 21 bar. The SO_2 absorber bottoms stream (92) contains 14.1% (by weight) SO_2 in H_2O at 82°C and 21 bar, with traces of O_2 and H_2SO_4. It is sent to the anolyte prep tank.

As was the case with the H_2 dryer, a detailed design for the O_2 dryer was not made, and regeneration requirements were not taken into account. Selective absorption of SO_2 and H_2O by a molecular sieve is one possible method. This simplification is likewise not expected to alter the outcome significantly.

All of the SO_2 produced by the decomposition of H_2SO_4, as well as that recovered from the SDE anolyte effluent and any liquids end up in the anolyte prep tank. The resulting liquid (stream 51) is a 43.5% H_2SO_4 solution (by weight), containing 15.5% SO_2 at a temperature of 84.7°C and a pressure of 21 bar. This is fed to the SDE anode after adjusting the temperature in heat exchanger HX-06 (by interchange with stream 17 via HX-02) to achieve a 100°C SDE outlet temperature. The anolyte feed (stream 2) temperature is 78.8°C.

Make-up sulfuric acid is added to the anolyte prep tank as needed to compensate for SO_2 and H_2O losses because of the ejector blow-downs. (More water is wasted via stream 88, which also contains some SO_2, than enters with streams 77 and 84.) The quantity of make-up needed is very small (about 0.008 kmol total/kmol H_2 product) and the sulfur content, at 23.6% H_2SO_4 (by weight), corresponds to only about 0.0004 kmol S/kmol H_2 product.

18.4.3 Flow Sheet Energy Requirements

Heat exchanger specifications are presented in Table 18.3. External heat input is required in only two places: the bayonet reactor (RX-01) and the vacuum column (TO-01) reboiler. The bayonet reactor receives 340.3 kJ/mol H_2 high-temperature heat from the HTGR. The reboiler, however, does not need such high temperatures since it operates in the 100–125°C range. The amount required, only 75.5 kJ/mol H_2, can be supplied with low-pressure steam. Heat rejection to cooling water adds up to 252.9 kJ/mol H_2, with over 80% attributable to the vacuum column condenser. Recuperation takes place in exchangers HX-02 and HX-03, accounting for 130.5 and 31.8 kJ/mol H_2 heat exchange, respectively.

The vacuum ejectors require a small quantity of steam for their operation (streams 77 and 84, 1:1 molar entrainment ratio basis). This corresponds to an equivalent 1.31-kJ/mol H_2 heat duty for the estimated 0.0277-kmol/sec flow of low-pressure steam. At two orders

TABLE 18.3
HyS Flow Sheet Heat Exchangers

Block ID	Duty (MWt)	Temperature, °C (K)		Heat Exchanged With
		Inlet	Outlet	
CO-01/Stage 1 Cooler	– 2.290	138.02 (411.2)	40.0 (313)	Cooling water
CO-01/Stage 2 Cooler	– 9.109	137.79 (410.9)	40.0 (313)	Cooling water
CO-01/Stage 3 Cooler	– 0.132	143.76 (416.9)	40.0 (313)	Cooling water
DR-01	– 2.045	76.96 (350.1)	40.0 (313)	Cooling water
DR-02	– 0.197	41.12 (314.3)	40.0 (313)	Cooling water
HX-01	– 17.688	100.00 (373.0)	76.96 (350.1)	HX-02
HX-02	130.486	66.96 (340.1)	94.41 (367.6)	HX-01, HX-04B, HX-05, HX-06, HX08B, HX-10A
HX-03	31.777	100.50 (373.7)	114.62 (387.8)	HX-04A, HX-08A
HX-04A	– 30.543	254.70 (527.9)	110.50 (383.7)	HX-03
HX-04B	– 8.400	110.50 (383.7)	76.96 (350.1)	HX-02
HX-04C	– 6.243	76.96 (350.1)	40.0 (313)	Cooling water
HX-05	– 47.903	100.00 (373.0)	80.00 (353.0)	HX-02
HX-06	– 18.686	84.71 (357.9)	78.77 (351.9)	HX-02
HX-07	– 3.702	84.08 (357.2)	40.0 (313)	Cooling water
HX-08A	– 1.234	147.33 (420.5)	110.50 (383.7)	HX-03
HX-08B	– 37.440	110.50 (383.7)	76.96 (350.1)	HX-02
HX-08C	– 4.021	76.96 (350.1)	40.0 (313)	Cooling water
HX-09	– 1.476	80.38 (353.5)	40.0 (313)	Cooling water
HX-10A	– 0.368	118.57 (391.7)	76.96 (350.1)	HX-02
HX-10B	– 11.503	76.96 (350.1)	40.0 (313)	Cooling water
HX-11	– 1.310	93.77 (366.9)	40.0 (313)	Cooling water
HX-12	-0.354	95.05 (368.2)	40.0 (313)	Cooling water
RX-01	340.280	123.63 (396.8)	254.70 (527.9)	High-temp. source
TO-01 Reboiler	75.482	102.74 (375.9)	122.87 (396.0)	Low-temp. source
TO-01 Condenser	– 210.542	44.80 (318.0)	40.0 (313)	Cooling water

Source: Data from Gorensek, M. B. et al. Hybrid sulfur process reference design and cost analysis. Savannah River National Laboratory, Report No. SRNL-L1200-2008-00002, June 12, 2009 (doi: 10.2172/956960).

of magnitude smaller than the bayonet reactor duty, it has little effect on the efficiency calculation. Nevertheless, ejector design should be more closely examined to confirm the validity of this simplification.

Electric power requirements are detailed in Table 18.4. The SDE accounts for the majority (115.8 kJ/mol H_2) of the 120.9 kJ/mol H_2 electric energy consumed by the flow sheet. Most of the rest, about 2.8 kJ/mol H_2, is attributable to SO_2 recycle compressor. The actual pumping requirement will be somewhat higher because frictional losses because of flow through equipment have been ignored. Assuming a thermal-to-electric conversion efficiency of 33% (BWR/PWR power), the total electric power requirement is equivalent to a

TABLE 18.4

HyS Flow Sheet Electrolyzers, Pumps, and Compressors

Block ID	Work, MW_e
EL-01	115.782
CO-01/Stage 1	1.464
CO-01/Stage 2	1.357
CO-01/Stage 3	0.025
PP-01	0.022
PP-02	1.836
PP-03	0.071
PP-04	0.002
PP-05	0.028
PP-06	0.055
PP-07	0.021
PP-08	0.212
PP-09	0.00003

Source: Data from Gorensek, M. B. et al. Hybrid sulfur process reference design and cost analysis. Savannah River National Laboratory, Report No. SRNL-L1200-2008-00002, June 12, 2009 (doi: 10.2172/956960).

heat input of 366.4 kJ/mol H_2. If HTGR power were to be used instead (45% conversion efficiency), the equivalent heat input would be 268.7 kJ/mol H_2.

The total heat consumed by the process is 340.3 + 75.5 + 1.3 = 417.1 kJ/mol H_2, which meets the ≤450 kJ/mol H_2 goal discussed in Section 18.3.3. Adding the thermal equivalent of the power consumption (assuming a BWR/PWR power source), this flow sheet requires a total of 783.5 kJ/mol H_2 of primary energy, comparing favorably with BWR/PWR-powered water electrolysis at 1080 kJ/mol H_2 [95]. A significant portion of the heat could be supplied by another source. The 76.8-kJ/mol H_2 heat duty of the vacuum column reboiler and steam ejectors could be provided by a low-pressure steam utility that recovers low-grade, possibly waste haste from another primary source. In that case, the primary (nuclear) energy (HTGR heat + BWR/PWR power) requirement for this flow sheet would be only 706.7 kJ/mol H_2, making the comparison with direct electrolysis even more favorable.

Using an HTGR power source instead would reduce the primary energy requirement from 783.5 to 685.8 kJ/mol H_2, which still compares favorably with HTGR-powered water electrolysis at 791 kJ/mol H_2 [95]. The differential would be even greater if the vacuum column reboiler and steam ejectors could make use of low-grade heat from an alternative source.

18.4.4 Flow Sheet Discussion

The HyS flow sheet presented above combines a PEM SDE with a bayonet decomposition reactor and otherwise uses only proven chemical process technology. If the SDE and high-temperature decomposition reactor perform as projected (600-mV cell potential, 40% conversion, 50-wt % H_2SO_4 product; 950°C HTGR operating temperature, 50°C drop between the primary and secondary helium coolants, 25°C minimum temperature difference between helium coolant and process stream, 10°C minimum temperature difference for recuperation, adequate heat transfer characteristics) the flow sheet process will split water into H_2 and O_2 while consuming 340.3 kJ/mol H_2 high-temperature heat, 76.8 kJ/mol H_2 low-temperature heat, and 120.9 kJ/mol H_2 electric power. Should the ultimate source of the electric power be a BWR/PWR (electricity from heat at 33% conversion efficiency), the net thermal efficiency would be 30.9%, LHV basis, excluding the power needed for helium circulators. This is significantly more efficient than alkaline electrolysis coupled with BWR/PWR power (22.4%, LHV basis). A higher net thermal efficiency, 35.3%, LHV basis would be attainable with HTGR power, which is still better than HTGR-powered electrolysis (30.6%, LHV basis).

Recent developments indicate that there may be considerable upside for HyS net thermal efficiency. The results of preliminary experiments with PBI membranes at USC show that significantly higher acid concentrations, in excess of 65% H_2SO_4 by weight may be feasible with PBI PEM SDEs [131]. If this work is successful, the 75.5-kJ/mol H_2 heat duty for the vacuum column would be eliminated (recuperative heating would suffice) and the bayonet heat requirement could possibly be lowered to 321 kJ/mol H_2 (80 wt% acid feed). Using HTGR power with an assumed conversion efficiency of 45%, the primary energy requirement would then be only $321 + 1.3 + 120.9/0.45 = 591$ kJ/mol H_2. This would give a net thermal efficiency of 40.9%, LHV basis, which is more efficient than the original Westinghouse design. It also compares very favorably with water electrolysis at 30.6%. If the more optimistic estimate of 52% HTGR power conversion efficiency used in some hydrogen production scenarios [120] were to be adopted instead, the net thermal efficiency would increase to 43.6%, LHV basis (51.6%, HHV basis).

The unit cost of H_2 production (\$/kg H_2) will ultimately determine whether the HyS process will be commercialized. HyS plants will be the preferred choice for H_2 production using nuclear power only if they can split water into H_2 and O_2 more economically than simple electrolysis or other competing process.

It should be noted that moving hot, pressurized helium through ductwork and heat exchangers will consume a significant amount of energy, which is not accounted for in Table 18.4. The primary and secondary loop circulators for the HTGR heat source (see Figure 18.6) will require on the order of 20 kJ/mol H_2 electric power; the exact figure depends on the HTGR design details. While the work performed by these circulators is eventually recovered as additional heat, it comes with a conversion loss penalty that should be included in the efficiency calculation.

Finally, some additional work will be needed to reject 252.9 kJ/mol H_2 waste heat to cooling water (i.e., to pump water to and from the cooling tower). However, the amount should be much smaller than that needed for the circulators and should not materially affect the results.

18.5 Process Engineering

The process engineering challenges for the HyS cycle are somewhat less than the challenges for either the other thermochemical water-splitting processes or even a high-temperature steam electrolysis (HTSE) plant. HyS has two major chemical reactions, which involve only compounds of hydrogen, oxygen and sulfur. This minimizes material challenges and simplifies separation steps, unit operations and other process engineering requirements. All process streams contain only fluids, and except for the two reaction operations, the chemical processing steps consist mainly of conventional vapor–liquid separations, such as flash evaporation, distillation, and gas/liquid absorption. Since only sulfuric acid and its derivative compounds are recycled between the two main reaction steps, there are no requirements for separating different chemical species (e.g., sulfur compounds from iodine or bromine compounds, etc.) and the issues of cross-contamination between steps are eliminated. The single gas/gas separation involves the removal of sulfur dioxide from the by-product oxygen stream. However, since the majority of the SO_2 can be removed by condensation under the proper temperature and pressure conditions, the final purification processing of the oxygen is minimized.

The relative simplicity of the HyS cycle, however, does not mean that the process engineering design is not challenging. Processing of sulfuric acid at moderate temperatures is a well-known industrial practice, but processing it under the extreme temperatures and pressures required for the HyS process presents unique challenges with regard to materials of construction and equipment design. Furthermore, portions of the flow sheet involve sulfuric acid containing dissolved sulfur dioxide, which results in additional material corrosion concerns. The process engineering challenges were addressed in a detailed study performed by SNRL and a team of industrial partners [97]. A comprehensive commercial-type flow sheet resulting from this work was presented in the previous section. All major equipment was sized, materials of construction selected, and capital costs estimated.

The two most important pieces of equipment are the SDE and the acid decomposer. The SDE is a modular component by its nature, and a nuclear hydrogen plant requires a large number of SDE units operating in parallel. Each SDE in turn is composed of a stack of electrochemical cells. Many of the SDE equipment design and process engineering issues can be addressed by recognizing the similarities between the requirements for a large SDE module and the electrochemical cells used in commercial chlor-alkali plants. Like the SDE, modern chlor-alkali cells use a proton-exchange membrane cell design, and they can be built with membranes containing up to 4 m^2 of active cell area. A proposed conceptual design of a HyS SDE is shown in Figure 18.23. It consists of 200 cells connected in series and constrained between steel end plates using tie-rods. The cells consist of circular bipolar plates and an MEA using a PEM electrolyte with 1 m^2 of active area. Gaskets between cells provide sealing for operating pressures up to 20 bar. An alternative approach would be use a stack design with a small nominal differential pressure across the seals by enclosing the entire stack inside a pressure vessel. Such an approach had been suggested by Westinghouse [2].

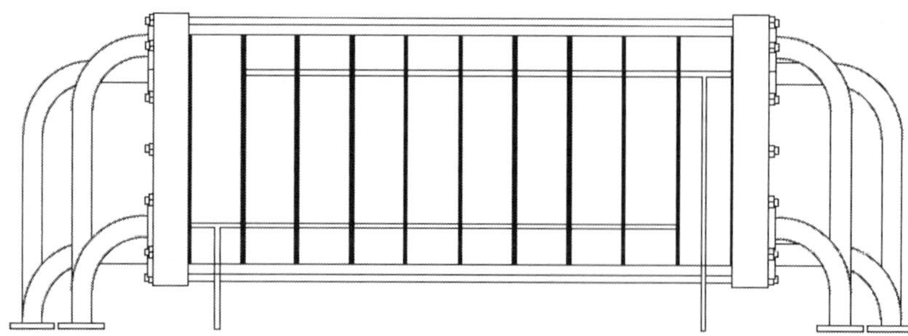

FIGURE 18.23
Commercial SDE module employing 200 Cells of 1 m^2 active area each rated hydrogen production rate of 1.5 MW (HHV).

Based on operation at 500 mA/cm^2 current density and an average cell voltage of 600 mV, the module would produce 38 kg/h of hydrogen with an energy content of 1.5 MW, HHV basis. The module is approximately 1.22 m (48 in.) in diameter and has an overall length, including end plates and other structural components, of 3.15 m (124 in.). A nuclear hydrogen plant will require multiple modules, including extra modules for capacity margin and to provide for maintenance. A 500-MWt HTGR plant would require approximately 182 SDE modules, containing a total of 38,400 individual cells. This can be contrasted with a similarly sized HTSE plant, which will need over one million high-temperature ceramic cells. The SDE modules might be arranged in four separate cell rooms or buildings, with 48 modules per cell room. A plot plan for a typical cell room is shown in Figure 18.24.

Anolyte and catholyte tanks, pumps and electrical equipment are located near the cell rooms. Sulfuric acid piping will be needed to connect the cell rooms to the acid concentration and feed preparation portions of the HyS plant. The piping system, however, will only convey liquids at temperatures of approximately 100°C. This is a much simpler process engineering challenge than that faced by HTSE plants, which are required to convey a mixture of hot steam and hydrogen at 800–900°C from a central point to hundreds and perhaps thousands of individual HTSE modules distributed over a large physical area.

Various designs have been proposed for the acid decomposer portion of the HyS plant. The Öztürk and bayonet decomposer designs were discussed previously. An artist's drawing of a potential large-scale bayonet type decomposer is shown in Figure 18.25 [98]. Westinghouse Electric has also proposed a somewhat different recuperative-type decomposition reactor that combines both acid concentration and acid decomposition in a single piece of equipment [62–63]. The Westinghouse design would require two large decomposers for a single 500-MWt HTGR nuclear hydrogen plant.

As mentioned previously, the balance of the process design, with the exception of the SDE and acid decomposer, utilizes mostly conventional process equipment, although the material issues can be unique. It should be recognized that system integration resulting in an efficient process design is critical to obtaining a high overall HyS process efficiency. For example, the sulfuric acid product leaving the SDE section must be concentrated prior to its introduction into the acid decomposition section. The efficient use of heat recuperation

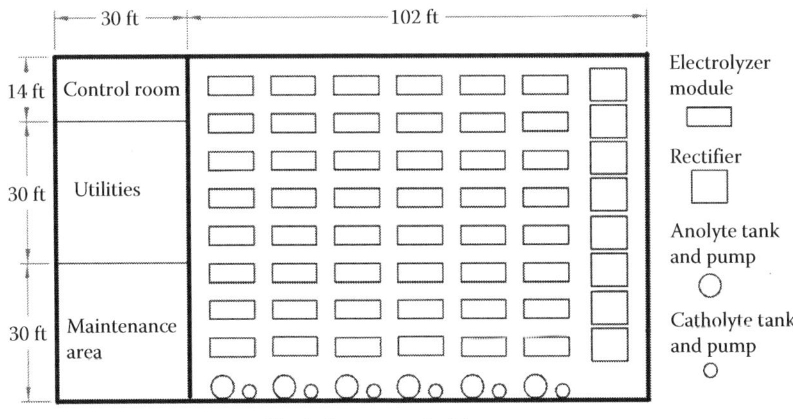

FIGURE 18.24
Plot plan of SDE cell room comprising 48 SDE modules rated at 1.5 MW of hydrogen output each.

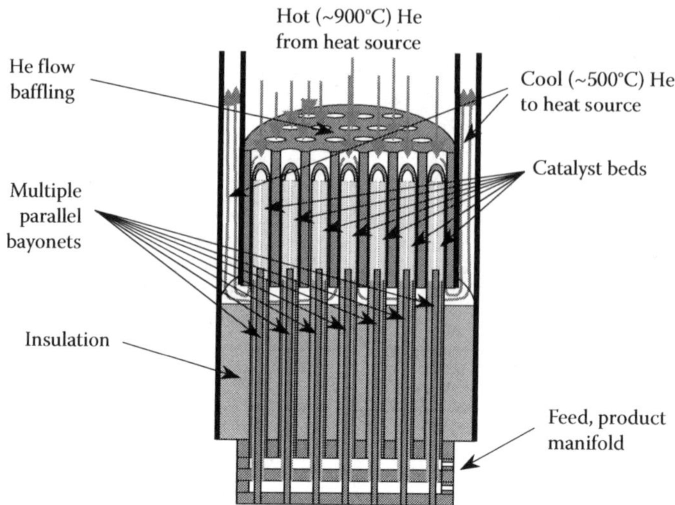

FIGURE 18.25
Commercial-scale acid decomposer of Bayonet design (From Gorensek, M. B. and Edwards, T. B. *Ind. Eng. Chem. Res.* **48**(15), 7232–7245, 2009. With permission.)

and an optimized design based on pinch analysis can help to minimize the additional energy needed for this step. Furthermore, any increase in acid concentration exiting the SDE will minimize the amount of water that needs to be evaporated in the acid concentration step, thus reducing energy demand. The use of advanced, high-temperature membranes in the SDE is one means of permitting electrolysis operation with higher acid concentration.

The conceptual design reported in Reference [97] used traditional engineering procedures to size all the major process equipment of the HyS plant, aiming at the best possible design from an economic point of view. Major equipment lists were prepared showing equipment type, physical size, mechanical design conditions and materials of construction. Shell-and-tube heat exchangers were selected in most cases as heat transfer devices. They were sized in detail, following standard TEMA classification, rules and suggestions and varying the most important parameters and degrees of freedom (e.g., tube diameters, tube length, baffle typology and arrangement, etc.) to get to the best solution. Knock-out pots, flash drums, and so on, were sized adopting traditional methodologies choosing vertical or horizontal equipment depending on the specific working conditions. Pumping equipment (centrifugal pumps, rotary pumps, and diaphragm pumps) was selected and sized to provide the most suitable pump models accounting for flow rates and pressure head to be supplied. A three-stage centrifugal compressor was adopted to recycle SO_2 in the acid decomposition section, internally recovering (in the H_2SO_4 section) the heat available from the inter-cooling process. The vacuum H_2SO_4 concentration column and the SO_2 absorber were sized as packed towers considering traditional sizing methods.

The selection of materials of construction for the equipment represents a fundamental aspect of the plant design, which can lead to wide variations of the total capital cost of the HyS plant. All the components (with the exception of the SDE and the acid decomposition reactor) were sized considering only well-known, traditional metal materials capable of resisting the actual working conditions. In the case of the product purification section of the plant, almost all of the equipment is made of carbon steel (CS) material. Likewise, components of the feed and utility supply section, which are in contact with water, are assumed to be made of CS material.

As for the other parts of the plant, where sulfur compounds and hydrogen (and water) circulate, various corrosion-resistant materials were selected. In some cases commercial experience with conventional sulfuric acid plant operation may lead to somewhat different selections than those used in the SRNL study, and this should be investigated. However, as cautioned previously, the presence of dissolved SO_2 in the sulfuric acid for some streams and the need to operate at higher temperatures and pressures than conventional sulfuric acid plants, introduce additional materials of construction concerns. Therefore, 316 L stainless steel (SS) was adopted as the baseline material, because of its capability to resist aggressive conditions, such as contact with H_2SO_4 over wide concentration ranges. To resist more difficult environments (in terms of concentrations and operating temperatures) which would attack 316 L SS type materials, Carpenter 20-Cb3 and, where appropriate, Hastelloy B3 were adopted. Carpenter 20-Cb3 is an austenitic SS with good resistance to concentrated H_2SO_4 at high temperatures. Hastelloy B3 is a nickel-based alloy which shows good resistance to H_2SO_4 mixtures over a broad range of temperatures and concentrations, with high thermal stability and resistance to pitting corrosion and stress corrosion. 20-Cb3 and Alloy B3 materials were adopted inside the H_2SO_4 section for the most aggressive environments.

The sulfuric acid concentration process equipment is made of 20-Cb3 and Alloy B3 materials. In particular, the shell of the H_2SO_4 vacuum column (with ceramic packing internals) has been assumed to be made of CS material, with Hastelloy B3 internal cladding. The two concentration flashes are made of 20-Cb3 and Alloy B3 (for higher temperatures and concentrations) and the H_2SO_4 decomposer feed pump has been assumed to be made of Alloy B3. Likewise, the two flashes, operating at temperatures of 255°C (528 K) and 234°C (507 K) respectively, which separate unreacted H_2SO_4 from the other compounds (SO_2, H_2O, and O_2) exiting the bayonet reactor, are made of Hastelloy B3. The remainder of the equipment (i.e., SO_2 separation, SO_2 recirculation, etc.) has been assumed to be made of 316 L SS. Particularly aggressive conditions exist in the balance of the electrolysis section equipment, which led to the adoption of 20-Cb3 and Hastelloy B3 as the materials of construction for both heat exchangers and recirculation pumps.

18.6 Future Technical Challenges

The current status of the research and development for the HyS thermochemical process has been reviewed, and a process flow sheet and conceptual design for a full-size commercial HyS plant has been discussed. The future technical challenges include the following:

- SDE development: (1) reduction in cell potential through the use of higher temperature operation and improved electrocatalyst; (2) high-temperature membranes with low SO_2 crossover; (3) operation with higher H_2SO_4 product; (4) demonstration of long-time stable operation with minimal operating potential degradation; (5) scale-up to larger cell sizes and multicell modules.

- Acid Decomposer: (1) demonstration of a long-life catalyst; (2) operation with convective-heating using a helium heat source; (3) decomposer design and material selections; (4) SiC joining; (5) catalyst replacements methods.
- HyS Process and System Development: (1) integrated system demonstrations at increasingly larger scale; (2) materials of construction validation; (3) capital cost reduction and evaluations.

The majority of the work presented here was the result of activities by the SRNL and its partners, which was part of the US DOE-NE's NHI program. Increasingly, other entities have begun to perform research on the HyS cycle, and international collaboration has increased. A HyS work package has been added to the research subjects under the framework of very-high temperature reactor (VHTR) system in the Generation IV International Forum. These activities are expected to address the remaining technical challenges for the HyS cycle, and to lead to pilot-plant facilities and demonstration of a commercial-scale plant.

References

1. Brecher, L. E. and Wu, C. K. Electrolytic decomposition of water. US Patent 3,888,750: Westinghouse Electric Corp., June 10, 1975.
2. Farbman, G. H. Conceptual design of an integrated nuclear hydrogen production plant using the sulfur cycle water decomposition system. Westinghouse Electric Corp., Astronuclear Lab., NASA Contractor Report, NASA-CR-134976, April 1, 1976.
3. Farbman, G. H., Krasicki, B. R., Hardman, C. C., Lin, S. S. and Parker, G. H. Economic comparison of hydrogen production using sulfuric acid electrolysis and sulfur cycle water decomposition. Final report. Westinghouse Electric Corp., Advanced Energy Systems Division., Electric Power Research Institute Report No. EPRI-EM-789, June 1978.
4. Lu, W. T. P., Ammon, R. L. and Parker, G. H. A study on the electrolysis of sulfur dioxide and water for the sulfur cycle hydrogen production process. Westinghouse Electric Corp., Advanced Energy Systems Division, NASA Contractor Report, NASA-CR-163517, July 1980.
5. Parker, G. H. Solar thermal hydrogen production process: final report, January 1978-December 1982. Westinghouse Electric Corp., Advanced Energy Systems Division., U.S. Department of Energy Report DOE/ET/20608-1, January 21, 1983.
6. Brecher, L. E., Spewock, S. and Warde, C. J. The Westinghouse sulfur cycle for the thermochemical decomposition of water, *Proceedings, 1st World Hydrogen Energy Conference*, Miami Beach, FL, USA, March 1–3, 1976 Vol. 1, pp. 9A, 1–16.
7. Carty, R., Cox, K., Funk, J., Soliman, M., Conger, W., Brecher, L. and Spewock, S. Process sensitivity studies of the Westinghouse sulfur cycle for hydrogen generation, *Proceedings, 1st World Hydrogen Energy Conference*, Miami Beach, FL, USA, March 1–3, 1976, Vol. 1, pp. 9A, 17–28.
8. Farbman, G. H. and Brecher, L. E. Hydrogen production by water decomposition using a combined electrolytic-thermochemical cycle, *Proceedings, 1st World Hydrogen Energy Conference, Miami Beach, FL, USA, March 1–3,* 1976 Vol. 1, pp. 9A, 29–50.
9. Spewock, S., Brecher, L. E. and Talko, F. The thermal catalytic decomposition of sulfur trioxide to sulfur dioxide and oxygen, *Proceedings, 1st World Hydrogen Energy Conference*, Miami Beach, FL, USA, March 1–3, 1976 Vol. 1, pp. 9A, 53–68.
10. Parker, G. H., Brecher, L. E. and Farbman, G. H. Nuclear driven water decomposition plant for hydrogen production, *Proceedings, 11th Conference on Intersociety Energy Conversion Engineering, State Line*, NV, USA, September 12–17, 1976 Vol. 2, pp. 1095–1101.

11. Farbman, G. H. Sulfur cycle water decomposition system, Proceedings of the ERDA contractors' review meeting on chemical energy storage and hydrogen energy systems, Airlie House, VA, USA, November 8–9, 1976, pp. 123–128.
12. Farbman, G. H., Ammon, R. L., Hardman, C. C. and Spewock, S. Development progress on the sulfur cycle water decomposition system, *Proceedings of the 12th Intersociety Energy Conversion Engineering Conference*, Washington, DC, USA, August 28-September 2, 1977 Vol. 1, pp. 928–932.
13. Farbman, G. H. The sulfur cycle water decomposition system, Proceedings of the DOE Chemical/Hydrogen Energy Systems Contractor Review, Hunt Valley, MD, USA, November 16–17, 1977, pp. 63–70.
14. Farbman, G. H. Hydrogen production with the sulfur cycle water decomposition system, Proceedings of Condensed Papers, Miami International Conference on Alternative Energy Sources, Miami Beach, FL, USA, December 5–7, 1977, pp. 529–531.
15. Farbman, G. H. The Westinghouse sulfur cycle hydrogen production process, *Symposium Papers: Hydrogen for Energy Distribution*, Chicago, IL, USA, July 24–28, 1978, pp. 317–337.
16. Farbman, G. H. The Westinghouse sulfur cycle hydrogen production process: program status, Hydrogen Energy System: Proceedings of the Second World Hydrogen Energy Conference, Zurich, Switzerland, August 21–24, 1978 Vol. 5, pp. 2485–2504.
17. Ammon, R. L., Parker, G. H. and Farbman, G. H. Materials considerations for the Westinghouse sulfur cycle hydrogen production process, *Proceedings: Reliability of Materials for Solar Energy Workshop*, Denver, CO, USA, December 18, 1978, *Vol. 2, Part 2*, pp. 371–394.
18. Farbman, G. H. and Parker, G. H. The sulfur-cycle hydrogen production process, *Hydrogen: Production and marketing; Proceedings of the Symposium, Honolulu, Hawaii, USA, April 2–6, 1979.* American Chemical Society, ACS Symposium Series, No. 116, 1980, pp. 359–389.
19. Parker, G. H. and Lu, P. W. T. Laboratory model and electrolyzer development for the sulfur cycle hydrogen production process, *Proceedings of the 14th Intersociety Energy Conversion Engineering Conference*, Boston, MA., USA, August 5–10, 1979 Vol. 1, pp. 752–755.
20. Lu, P. W. T. and Ammon, R. L. Development status of electrolysis technology for the sulfur cycle hydrogen production process, *Hydrogen Energy Progress III: Proceedings of the 3rd World Hydrogen Energy Conference*, Tokyo, Japan, June 23–26, 1980 Vol. 1, pp. 439–461.
21. Irwin, H. A. and Ammon, R. L. Status of materials evaluation for sulfuric acid vaporization and decomposition applications, *Hydrogen Energy Progress III: Proceedings of the 3rd World Hydrogen Energy Conference*, Tokyo, Japan, June 23–26, 1980 Vol. 4, pp. 1977–1999.
22. Summers, W. A., Ammon, R. L. and Parker, G. H. Recent progress on the sulfur cycle hybrid hydrogen production process, *Proceedings of the 15th Intersociety Energy Conversion Engineering Conference*, Seattle, WA., USA, August 18–22, 1980 Vol. 3, pp. 2008–2014.
23. Lu, P. W. T. Technological aspects of sulfur dioxide depolarized electrolysis for hydrogen production, *Hydrogen Energy Progress IV: Proceedings of the 4th World Hydrogen Energy Conference*, Pasadena, CA, USA, June 13–17, 1982 Vol. 1, pp. 203–214.
24. Lin, S. S. and Flaherty, R. Design studies of the sulfur trioxide decomposition reactor for the sulfur cycle hydrogen production process, *Hydrogen Energy Progress IV: Proceedings of the 4th World Hydrogen Energy Conference*, Pasadena, CA, USA, June 13–17, 1982 Vol. 2, pp. 599–610.
25. Ammon, R. L. Status of materials evaluation for sulfuric acid vaporization and decomposition applications, *Hydrogen Energy Progress IV: Proceedings of the 4th World Hydrogen Energy Conference*, Pasadena, CA, USA, June 13–17, 1982 Vol. 2, pp. 623–644.
26. Pierre, J. F. and Ammon, R. L. The Westinghouse sulfur cycle: sulfur trioxide reduction catalyst screening test program, *Hydrogen Energy Progress IV: Proceedings of the 4th World Hydrogen Energy Conference*, Pasadena, CA, USA, June 13–17, 1982 Vol. 2, pp. 703–712.
27. Lin, S. S. Parker, G. H. and Stella, M. E. Solar hydrogen system design considerations, *Proceedings of the 16th Intersociety Energy Conversion Engineering Conference*, Atlanta, GA, USA, August 9–14, 1981 Vol. 2, pp. 1430–1435.
28. Brecher, L. E., Spewock, S. and Warde, C. J. The Westinghouse sulfur cycle for the thermochemical decomposition of water, *Int. J. Hydrogen Energy* **2**(1), 7–15, 1977.

29. Carty, R., Cox, K., Funk, J., Soliman, M., Conger, W., Brecher, L. and Spewock, S. Process sensitivity studies of the Westinghouse sulfur cycle for hydrogen generation, *Int. J. Hydrogen Energy* **2**(1), 17–22, 1977.
30. Farbman, G. H. Hydrogen production by the Westinghouse sulfur cycle process: program status, *Int. J. Hydrogen Energy* **4**(2), 111–122, 1979.
31. Lu, P. W. T., Garcia, E. R. and Ammon, R. L. Recent developments in the technology of sulfur dioxide depolarized electrolysis, *J. Appl. Electrochem.* **11**(3), 347–355, 1981.
32. Lu, P. W. T. and Ammon, R. L. Sulfur dioxide depolarized electrolysis for hydrogen production: development status, *Int. J. Hydrogen Energy* **7**(7), 563–575, 1982.
33. Lin, S. S. and Flaherty, R. Design studies of the sulfur trioxide decomposition reactor for the sulfur cycle hydrogen production process, *Int. J. Hydrogen Energy* **8**(8), 589–596, 1983.
34. Lu, P. W. T. Technological aspects of sulfur dioxide depolarized electrolysis for hydrogen production, *Int. J. Hydrogen Energy* **8**(10), 773–781, 1983.
35. Yannopoulos, L. N. and Pierre, J. F. Hydrogen production process: High temperature-stable catalysts for the conversion of SO_3 to SO_2, *Int. J. Hydrogen Energy* **9**(5), 383–390, 1984.
36. Lu, W. T. P. and Ammon, R. L. Carbon cloth-supported electrode. U.S. Patent 4,349,428: U.S. Department of Energy, September 14, 1982.
37. Lu, W. T. P. System using sulfur dioxide as an anode depolarizer in a solid oxide electrolyte electrolysis cell for hydrogen production from steam. U.S. Patent 4,412,895: Westinghouse Electric Corp., November 1, 1983.
38. Beghi, G. E. A decade of research on thermochemical hydrogen at the Joint Research Centre, Ispra, *Int. J. Hydrogen Energy* **11**(12), 761–771, 1986.
39. Struck, B. D., Junginger, R., Boltersdorf, D. and Gehrmann, J. Problems concerning the electrochemical step of the sulfuric acid hybrid cycle, *Proceedings, Seminar on Hydrogen as an Energy Vector: Its Production, Use and Transportation*, Brussels, Belgium, October 3–4, 1978, pp. 109–123.
40. Struck, B. D., Junginger, R., Boltersdorf, D., Dujka, B., Neumeister, H. and Triefenbach, D. Development of an electrolytic cell for the anodic oxidation of sulfur dioxide and the cathodic production of hydrogen within the sulfuric acid hybrid thermochemical cycle, *Hydrogen as an Energy Vector: Proceedings of the International Seminar*, Held in Brussels, Belgium, February 12–14, 1980. pp. 80–100.
41. Struck, B. D., Junginger, R., Boltersdorf, D., Dujka, B., Neumeister, H. and Triefenbach, D. Development of an electrolytic cell for the anodic oxidation of sulfur dioxide and the cathodic production of hydrogen within the sulfuric acid hybrid cycle. KFA Juelich, Juelich, Federal Republic of Germany, Report No. EUR-6961, 1980.
42. Struck, B. D., Junginger, R., Boltersdorf, D. and Gehrmann, J. The anodic oxidation of sulfur dioxide in the sulfuric acid hybrid cycle, *Int. J. Hydrogen Energy* **5**(5), 487–497, 1980.
43. Struck, B. D., Neumeister, H., Dujka, B. and Siebert, U. Electrolytic cell for the sulfuric acid hybrid cycle, *Hydrogen Energy Progress IV: Proceedings of the 4th World Hydrogen Energy Conference*, Pasadena, CA, USA, June 13–17, 1982 Vol. 2, pp. 483–492.
44. Junginger, R. and Struck, B. D. Separators for electrolytic cells of the sulfuric acid hybrid cycle, *Int. J. Hydrogen Energy* **7**(4), 331–340, 1982.
45. Struck, B. D., Neumeister, H., Dujka, B., Siebert, U. and Triefenbach, D. Development and scaling up of an electrolytic cell for the anodic oxidation of sulfur dioxide and cathodic production of hydrogen for the sulfuric acid hybrid cycle, in *Hydrogen As an Energy Carrier*. Imarisio, G. and Strub, A. S., eds. Hingham, MA, USA: Kluwer Academic Publishers, 1983, pp. 35–45.
46. Struck, B. D., Neumeister, H., Dujka, B., Siebert, U. and Triefenbach, D. Development of an electrolytic cell for the anodic oxidation of sulfur dioxide and cathodic production of hydrogen for the sulfuric acid hybrid cycle. Commission of the European Communities, Luxembourg, Report No. EUR-8300EN, 1983.
47. Struck, B. D., Neumeister, H. and Naoumidis, A. Electrolytic cell for the thermochemical-electrochemical sulfur hybrid cycle for water splitting, *Hydrogen Energy Progress VI: Proceedings of the 6th World Hydrogen Energy Conference*, Vienna, Austria, July 20–24, 1986 Vol. 2, pp. 739–743.

48. Takehara, Z., Nogami, M. and Shimizu, Y. New electrolytic process in hybrid sulfur cycle for hydrogen production from water, *Int. J. Hydrogen Energy* **14**(4), 233–239, 1989.
49. Merkulova, N. D., Boikova, G. V., Zhutaeva, G. V., Tarasevich, M. R. and Shumilova, N. A. Selectivity of tungsten carbide and platinum in cathodic reactions of the sulfuric acid cycle, *Zhurnal Prikludnoi Khimii (Leningrad)* **57**(10), 2264–2268, 1984.
50. Gorbachev, A. K. Corrosion resistance of electrode graphites during anodic polarization in sulfuric acid solutions, *Zashchita Metallov* **20**(3), 436–438, 1984.
51. Schepers, H. Thermodynamic analysis of the sulfur cycle water splitting process, dissertation, Fakultät für Maschinenwesen, Technische Hochschule, Aachen, Germany, 1983.
52. Onstott, E. I. Measurement of reversible and irreversible energy requirements for hydrogen production in hybrid cycles, *Hydrogen Energy Progress IV: Proceedings of the 4th World Hydrogen Energy Conference*, Pasadena, CA, USA, June 13–17, 1982, Vol. 1, pp. 159–165.
53. Appleby, A. J. and Pinchon, B. Electrochemical aspects of the H_2SO_4-SO_2 thermoelectrochemical cycle for hydrogen production, *Int. J. Hydrogen Energy* **5**(3), 253–267, 1980.
54. Sedlak, J. M., Russell, J. H., LaConti, A. B., Gupta, D. K., Austin, J. F. and Nugent, J. S. "Hydrogen production using solid-polymer-electrolyte technology for water electrolysis and hybrid sulfur cycle." General Electric Co., Electric Power Research Institute Report No. EPRI-EM-1185, September 1, 1979.
55. Feuillade, G. and Caradec, J. Y. Study of sulfur dioxide electrocatalytic oxidation during the sulfur cycle. Compagnie générale d'électricité, Marcoussis, France, Commission of the European Communities, Luxembourg, Report No. EUR 6603, 1979.
56. Appleby, A. J. and Pichon, B. The mechanism of the electrochemical oxidation of sulfur dioxide in sulfuric acid solutions, *J. Electroanal. Chem.* **95**(1), 59–71, 1979.
57. Appleby, A. J. and Pichon, B. Electrochemical aspects of the sulfuric acid-sulfur dioxide thermoelectrochemical cycle for hydrogen production, *Hydrogen Energy System: Proceedings of the Second World Hydrogen Energy Conference*, Zurich, Switzerland, August 21-24, 1978 Vol. 2, pp. 687–707.
58. Knoche, K. F. and Funk, J. E. Entropy production, efficiency and economics in the thermochemical generation of synthetic fuels: I. The hybrid sulfuric acid process, *Int. J. Hydrogen Energy* **2**(4), 377–385, 1977.
59. Brown, L. C., Funk, J. F. and Showalter, S. K. Initial screening of thermochemical water-splitting cycles for high efficiency generation of hydrogen fuels using nuclear power. General Atomics, Report No. GA–A23373, April 2000.
60. Matzie, R., Lahoda, E. J. and Botma, A. Interfacing the Westinghouse sulfur cycle with the PBMR for the production of hydrogen, *Conference Proceedings, AIChE Spring National Meeting*, New Orleans, LA, USA, April 25–29, 2004, pp. 1034–1043.
61. McLaughlin, D. F., Paletta, S. A., Lahoda, E. J. and Kriel, W. Revised capital and operating HyS hydrogen production costs, *Proceedings of the 2006 International Congress on Advances in Nuclear Power Plants, Embedded Topical Meeting*, Reno, NV, USA, June 4–8, 2006, pp. 2263–2269.
62. Connolly, S. M., Lahoda, E. J. and McLaughlin, D. F. Design of recuperative sulfuric acid decomposition reactor for hydrogen generation processes, *Conference Proceedings, AIChE Annual Meeting*, Philadelphia, PA, USA, November 16–21, 2008, pp. 88/1–88/16.
63. Connolly, S. M., Zabolotny, E., McLaughlin, D. F. and Lahoda, E. J. Design of a composite sulfuric acid decomposition reactor, concentrator and preheater for hydrogen generation processes, *Int. J. Hydrogen Energy* **34**(9), 4074–4087, 2009.
64. Lahoda, E. J. and McLaughlin, D. F. Hydrogen generation process with dual pressure multi stage electrolysis. Patent Application, Europe, No. 2008–12190: Westinghouse Electric Company LLC, USA, July 5, 2008.
65. Sink, C. J. An overview of the U.S. Department of Energy's research and development program on hydrogen production using nuclear energy, *Presentation, AIChE Spring National Meeting*, Orlando, FL, USA, April 23–27, 2006, http://www.aiche-ned.org/conferences/aiche2006spring/session_51/AICHE2006spring-51b-Sink.pdf (accessed October 14, 2009).

66. Hydrogen posture plan: An integrated research, development and demonstration plan. United States Department of Energy, United States Department of Transportation, December 2006, http://www.hydrogen.energy.gov/pdfs/hydrogen_posture_plan_dec06.pdf (accessed October 14, 2009).
67. Le Duigou, A., Borgard, J.-M., Larousse, B., Doizi, D., Allen, R., Ewan, B. C., Priestman, G. H., et al. HYTHEC: An EC funded search for a long term massive hydrogen production route using solar and nuclear technologies, *Int. J. Hydrogen Energy* **32**(10–11), 1516–1529, 2007.
68. Feraud, J. P., Jomard, F., Ode, D., Duhamet, J., Dehaudt, P., Morandini, J., Couvat, Y. D. and Caire, J. P. Modeling a filter press electrolyzer by using two coupled codes within nuclear Gen. IV hydrogen production, *Proceedings, Global 2007: Advanced Nuclear Fuel Cycles and Systems*, Boise, ID, USA, September 9–13, 2007, pp. 837–844.
69. Jomard, F., Feraud, J. P., Morandini, J., Du Terrail Couvat, Y. and Caire, J. P. Hydrogen filter press electrolyser modelled by coupling Fluent and Flux Expert codes, *J. Appl. Electrochem.* **38**(3), 297–308, 2008.
70. Jomard, F., Feraud, J. P. and Caire, J. P. Numerical modeling for preliminary design of the hydrogen production electrolyzer in the Westinghouse hybrid cycle, *Int. J. Hydrogen Energy* **33**(4), 1142–1152, 2008.
71. Charton, S., Rivalier, P., Ode, D., Morandini, J. and Caire, J. P. Hydrogen production by the Westinghouse cycle: modelling and optimization of the two-phase electrolysis cell, *WIT Trans. Eng. Sci.* **65**(Electrochemical Process Simulation III), 11–22, 2009.
72. Lee, S.-K., Kim, C.-H., Cho, W.-C., Kang, K.-S., Park, C.-S. and Bae, K.-K. The effect of Pt loading amount on SO_2 oxidation reaction in an SO_2-depolarized electrolyzer used in the hybrid sulfur (HyS) process, *Int. J. Hydrogen Energy* **34**(11), 4701–4707, 2009.
73. Kim, N. and Kim, D. SO_2 permeability and proton conductivity of sPEEK membranes for SO_2-depolarized electrolyzer, *Int. J. Hydrogen Energy* **34**(19), 7919–7926, 2009.
74. Nakagiri, T., Hoshiya, T. and Aoto, K. Investigation on a new hydrogen production process for FBR, *Proceedings, 2nd Information Exchange Meeting, Nuclear Production of Hydrogen, Argonne, IL, USA*, October 2–3, 2003, pp. 131–143.
75. Nakagiri, T., Ohtaki, A., Hoshiya, T. and Aoto, K. A new thermochemical and electrolytic hybrid hydrogen production process for FBR, *Nihon Genshiryoku Gakkai Wabun Ronbunshu* **3**(1), 88–94, 2004.
76. Kawamura, H., Mori, M. and Uotani, M. Basic study of application of electronic conductive ceramic anode to sulfur-based hybrid hydrogen production, *Suiso Enerugi Shisutemu* **30**(1), 42–48, 2005.
77. Kawamura, H., Mori, M. and Uotani, M. Electron-conductive ceramic powders, cubic titanium-oxide pyrochlore sintered bodies, and electrolysis vessels for sulphur-cycle hybrid hydrogen manufacture. Japan Patent 2005330133: Central Research Institute of Electric Power Industry, Japan, December 2, 2005.
78. Kawamura, H., Chu, S.-Z., Mori, M. and Uotani, M. Electrochemical properties of ceramic electrodes for sulfur-based hybrid cycle -development of Pd-coated electronic conductive ceramics, *Suiso Enerugi Shisutemu* **31**(2), 32–41, 2006.
79. Nakagiri, T., Kase, T., Kato, S. and Aoto, K. Development of a new thermochemical and electrolytic hybrid hydrogen production system for sodium cooled FBR, *JSME Int. J., Ser. B: Fluids Thermal Eng.* **49**(2), 302–308, 2006.
80. Kawamura, H., Mori, M. and Uotani, M. Hydrogen preparation using sulfur cycle hybrid method, *Electrochemistry (Tokyo, Japan)* **75**(3), 289–293, 2007.
81. Nakagiri, T., Yamaki, T., Asano, M. and Tsutsumi, Y. Development of water electrolysis cell for hydrogen production utilizing sulfur dioxide, *Nihon Genshiryoku Gakkai Wabun Ronbunshu* **7**(1), 58–65, 2008.
82. Shimomiya, O. Apparatus for production of hydrogen by sulfuric acid hybrid process. Japan Patent 2008223098: Ishikawajima-Harima Heavy Industries Co., Ltd., Japan, September 25, 2008.

83. Gorensek, M. B., Buckner, M. R., Summers, W. and Qureshi, Z. H. Conceptual design for a hybrid sulfur thermochemical process plant, *Conference Proceedings, AIChE Spring National Meeting*, Atlanta, GA, USA, April 10–14, 2005, p. 76F/1.
84. Summers, W. A. and Buckner, M. R. Hybrid sulfur thermochemical process development, *DOE Hydrogen Program FY 2005 Progress Report*. May 23, 2005. http://www.hydrogen.energy.gov/pdfs/progress05/iv_g_7_summers.pdf (accessed October 21, 2009).
85. Steimke, J. and Steeper, T. Generation of hydrogen using electrolyzers with sulfur dioxide depolarized anodes, *Conference Proceedings, AIChE Annual Meeting*, Cincinnati, OH, USA, October 30-November 4, 2005, pp. 581d/1–581d/19.
86. Eargle, J. A., Gorensek, M. B. and Knight, T. W. Optimization of the thermal efficiency of a hybrid sulfur thermochemical hydrogen process, *Trans. Am. Nucl. Soc.* **93**(1), 917–918, 2005.
87. Colon-Mercado, H., Ekechukwu, A. and Hobbs, D. Electrolyzer component testing for hybrid sulfur process, *Abstracts, 58th Southeast Regional Meeting of the American Chemical Society*, Augusta, GA, USA, November 1–4, 2006, p. SRM06–235.
88. Steimke, J. and Steeper, T. J. Generation of hydrogen using hybrid sulfur process electrolyzer, *Abstracts, 58th Southeast Regional Meeting of the American Chemical Society*, Augusta, GA, USA, November 1–4, 2006, p. SRM06–231.
89. Steimke, J. L. and Steeper, T. J. Generation of hydrogen using electrolyzer with sulfur dioxide depolarized anode, *Conference Proceedings, AIChE Annual Meeting*, San Francisco, CA, USA, November 12–17, 2006, pp. 126c/1–126c/10.
90. Summers, W. A. and Gorensek, M. B. Nuclear hydrogen production based on the hybrid sulfur thermochemical process, Proceedings of the 2006 International Congress on Advances in Nuclear Power Plants, Embedded Topical Meeting, Reno, NV, USA, June 4–8, 2006, pp. 2254–2256.
91. Colon-Mercado, H. R. and Hobbs, D. T. Catalyst evaluation for a sulfur dioxide-depolarized electrolyzer, *Electrochem. Commun.* **9**(11), 2649–2653, 2007.
92. Summers, W. A. Hybrid sulfur thermochemical process development," *DOE 2008 Hydrogen Program Annual Merit Review*, Arlington, VA, United States, June 9–13, 2008. http://www.hydrogen.energy.gov/pdfs/review08/pd_26_summers.pdf (accessed October 21, 2009).
93. Hobbs, D. Challenges and progress in the development of a sulfur dioxide-depolarized electrolyzer for efficient hydrogen production, *Pittcon Conference & Expo 2009*. Chicago, IL, USA, 2009.
94. Steimke, J. L., Steeper, T. and Herman, D. Results from testing hybrid sulfur single-cell and three-cell stack, *Hybrid Sulfur Electrolyzer Workshop and Information Exchange*, Aiken, SC, United States, April 20–21, 2009. http://srnl.doe.gov/hse_workshop/Steimke%20Electrolyzer%20Performance.pdf (accessed October 21, 2009).
95. Gorensek, M. B. and Summers, W. A. Hybrid sulfur flowsheets using PEM electrolysis and a bayonet decomposition reactor, *Int. J. Hydrogen Energy* **34**(9), 4097–4114, 2009.
96. Summers, W. A. Hybrid sulfur thermochemical cycle, 2009 DOE Hydrogen Program and Vehicle Technologies Program Annual Merit Review and Peer Evaluation Meeting, Arlington, VA, United States, May 18–22, 2009. http://www.hydrogen.energy.gov/pdfs/review09/pd_13_summers.pdf (accessed October 21, 2009).
97. Gorensek, M. B., Summers, W. A., Bolthrunis, C. O., Lahoda, E. J., Allen, D. T. and Greyvenstein, R. Hybrid sulfur process reference design and cost analysis. Savannah River National Laboratory, Report No. SRNL-L1200-2008-00002, June 12, 2009 (doi: 10.2172/956960).
98. Gorensek, M. B. and Edwards, T. B. Energy efficiency limits for a recuperative bayonet sulfuric acid decomposition reactor for sulfur cycle thermochemical hydrogen production, *Ind. Eng. Chem. Res.* **48**(15), 7232–7245, 2009.
99. Gorensek, M. B., Staser, J. A., Stanford, T. G. and Weidner, J. W. A thermodynamic analysis of the SO_2/H_2SO_4 system in SO_2-depolarized electrolysis, *Int. J. Hydrogen Energy* **34**(15), 6089–6095, 2009.
100. Hobbs, D. T., Summers, W. A., Colon-Mercado, H. R., Elvington, M. C., Steimke, J. L., Steeper, T. J., Herman, D. T. and Gorensek, M. B. Recent advances in the development of the hybrid

sulfur process for hydrogen production, *Abstracts of Papers, 238th ACS National Meeting, Washington, DC, USA, August 16–20, 2009*, p. NUCL-160.

101. Colón-Mercado, H. R., Elvington, M. C. and Hobbs, D. T. Component development needs for the hybrid sulfur electrolyzer. Savannah River National Laboratory, Report No. WSRC-STI-2008-00291, May 2008 (doi: 10.2172/935436).
102. Steimke, J. L., Steeper, T. J. and Herman, D. T. Prevention of sulfur formation in hybrid sulfur electrolyzer, paper presented at the *2009 AIChE Annual Meeting*, Nashville, TN, USA, November 8–13, 2009.
103. Sivasubramanian, P., Ramasamy, R. P., Freire, F. J., Holland, C. E. and Weidner, J. W. Electrochemical hydrogen production from thermochemical cycles using a proton exchange membrane electrolyzer, *Int. J. Hydrogen Energy* 32(4), 463–468, 2007.
104. Staser, J., Ramasamy, R. P., Sivasubramanian, P. and Weidner, J. W. Effect of water on the electrochemical oxidation of gas-phase SO_2 in a PEM electrolyzer for H_2 production, *Electrochem Solid-State Lett.* **10**(11), E17–E19, 2007.
105. Staser, J. A., Norman, K., Fujimoto, C. H., Hickner, M. A. and Weidner, J. W. Transport properties and performance of polymer electrolyte membranes for the hybrid sulfur electrolyzer, *J Electrochem. Soc.* **156**(7), B842–B847, 2009.
106. Staser, J. A. and Weidner, J. W. Effect of SO_2 crossover on hydrogen and sulfuric acid production in a PEM electrolyzer, *ECS Transactions* **19**(10), 67–75, 2009.
107. Staser, J. A. and Weidner, J. W. Sulfur dioxide crossover during the production of hydrogen and sulfuric acid in a PEM electrolyzer, *J. Electrochem. Soc.* **156**(7), B836–B841, 2009.
108. Staser, J. A. and Weidner, J. W. Effect of water transport on the production of hydrogen and sulfuric acid in a PEM electrolyzer, *J. Electrochem. Soc.* **156**(1), B16–B21, 2009.
109. Staser, J. A., Gorensek, M. B. and Weidner, J. W. Quantifying individual potential contributions of the hybrid sulfur electrolyzer, *J. Electrochem. Soc.* **157**(6), B952–B958, 2010.
110. Juda, W. and Moulton, D. M. Effect of low-cost electrolytic hydrogen on the generation of basic chemicals, *Chem. Eng. Progr., Symp. Ser.* **63**(71), 59–62, 1967.
111. Hood, G. C. and Reilly, C. A. Ionization of strong electrolytes. V. Proton magnetic resonance in sulfuric acid, *J. Chem. Phy.* **27**(5), 1126–1128, 1957.
112. Young, T. F., Maranville, L. F. and Smith, H. M. Raman spectral investigations of ionic equilibria in solutions of strong electrolytes, in *The structure of electrolytic solutions*. Hamer, W. J. ed., Wiley, New York, 1959, pp. 35–63.
113. Clegg, S. L., Rard, J. A. and Pitzer, K. S. Thermodynamic properties of 0–6 mol kg^{-1} aqueous sulfuric acid from 273.15 to 328.15 K, *J. Chem. Soc., Faraday Transactions* **90**(13), 1875–1894, 1994.
114. Clegg, S. L. and Brimblecombe, P. Application of a multicomponent thermodynamic model to activities and thermal properties of 0–40 mol kg^{-1} aqueous sulfuric acid from <200 to 328 K, *J Chem. Eng. Data* **40**(1), 43–64, 1995.
115. Walrafen, G. E., Yang, W. H., Chu, Y. C. and Hokmabadi, M. S. Structures of concentrated sulfuric acid determined from density, conductivity, viscosity, and Raman spectroscopic data, *J. Sol. Chem.* **29**(10), 905–936, 2000.
116. Wang, P., Anderko, A., Springer, R. D. and Young, R. D. Modeling phase equilibria and speciation in mixed-solvent electrolyte systems: II. Liquid-liquid equilibria and properties of associating electrolyte solutions, *Journal of Molecular Liquids* **125**(1), 37–44, 2006.
117. Öztürk, I. T., Hammache, A. and Bilgen, E. An improved process for H_2SO_4 decomposition step of the sulfur-iodine cycle, *Energy Conversion and Management* **36**(1), 11–21, 1995.
118. Moore, R. C., Gelbard, F., Parma, E. J., Vernon, M. E., Lenard, R. X. and Pickard, P. S. A laboratory-scale sulfuric acid decomposition apparatus for use in hydrogen production cycles, *Proceedings: International Topical Meeting on Safety and Technology of Nuclear Hydrogen Production, Control, and Management*, Boston, MA, USA, June 24–28, 2007, pp. 161–166.
119. Koster, A., Matzner, H. D. and Nicholsi, D. R. PBMR design for the future, *Nucl. Eng. Des.* **222**(2–3), 231–245, 2003.

120. Richards, M., Shenoy, A., Schultz, K., Brown, L., Harvego, E., McKellar, M., Coupey, J.-P., Reza, S. M. M., Okamoto, F. and Handa, N. H2-MHR conceptual designs based on the sulphur–iodine process and high-temperature electrolysis, *Int. J. Nucl. Hydrogen Prod. Appl.* **1**(1), 36–50, 2006.
121. *Aspen Plus. Version 7.0* (22.0.4191), Aspen Technology, Inc., Burlington, MA, USA, 1981–2008.
122. *Aspen Energy Analyzer. Version 7.0* (22.0.0.5705), Aspen Technology, Inc., Burlington, MA, USA, 1995–2008.
123. *JMP. Version 7.0.2*, SAS Institute, Inc., Cary, NC, USA, 1989–2007.
124. *OLI Engine in Aspen Plus. Version 8.0*, Revision 0, April 2008, ESP Build 58, OLI Systems, Inc., Morris Plains, NJ, USA, 1997–2008.
125. Miles, F. D. and Fenton, J. The solubility of sulfur dioxide in sulfuric acid, *J. Chem. Soc., Trans.* **117**, 59–61, 1920.
126. Kuznetsov, D. A. Study of the equilibrium pressures of sulfur dioxide over water and aqueous solutions of sulfuric acid (transl.), *Zhurnal Khimicheskoi Promyshlennosti* **18**(22), 3–7, 1941.
127. Miles, F. D. and Carson, T. Solubility of sulfur dioxide in fuming sulfuric acid, *J. Chem. Soc.*, **1946**, 786–790.
128. Maass, C. E. and Maass, O. Sulfur dioxide and its aqueous solutions. I. Analytical methods, vapor density and vapor pressure of sulfur dioxide. Vapor pressure and concentrations of the solutions, *J. Am. Chem. Soc.* **50**(5), 1352–1368, 1928.
129. Spall, B. C. Phase equilibriums in the system sulfur dioxide-water from 25–300°, *Canad. J. Chem. Eng.* **41**, 79–83, 1963.
130. Van Berkum, J. G. and Diepen, G. A. M. Phase equilibria in $SO_2 + H_2O$: The sulfur dioxide gas hydrate, two liquid phases, and the gas phase in the temperature range 273 to 400 K and at pressures up to 400 MPa, *J. Chem. Therm.* **11**(4), 317–334, 1979.
131. Weidner, J. W. Personal Communication. Email dated June 17, 2009.

19

Nuclear Coal Gasification

Karl Verfondern

CONTENTS

19.1 Introduction

Germany has played in the past a leading role in the development of the essential coal-refining processes. Extensive studies included coal gasification, liquefaction, and advanced combustion systems aiming at the generation of substitute natural gas (SNG), liquid hydrocarbons, and other raw materials for the chemical industries. Numerous coal gasification projects were launched to investigate on pilot plant scale various methods and reactor types, and optimize operational conditions. With the first oil crisis at the beginning of the 1970s, comprehensive research and development (R&D) activities were conducted in Germany as a contribution to an away-from-oil policy to investigate also the introduction of nuclear energy in the heat-intensive processes of coal gasification. The utilization of nuclear primary energy in these processes represents a significant contribution to the saving of resources and thus the lowering of specific carbon emissions to the environment.

The Prototype Nuclear Process (PNP) Heat project in Germany was a cooperation between the high temperature gas-cooled reactor (HTGR) industries (Hochtemperatur-Reaktorbau GmbH, Mannheim, and Gesellschaft für Hochtemperaturreaktortechnik mbH, Bensberg), the coal industries (Bergbauforschung GmbH, Essen, and Rheinische Braunkohlenwerke AG, Cologne), and the nuclear research center Kernforschungsanlage Jülich (today: Forschungszentrum Jülich (FZJ)). The project was funded by the Federal Government, the State Goverment of Northrhine Westphalia, and the participating industries. Main objective was the development, design, and construction of an energy system based on a combination of German coal and nuclear power, including the developing and prototype testing of a nuclear heat-generating system to be operated at up to 950°C gas outlet temperature with intermediate circuit, heat extraction, coal gasification processes, and nuclear energy transport.

19.2 Nuclear Process Heat Reactor

The concept for a nuclear process heat plant with a pebble bed high-temperature reactor was originally based on thermal power sizes of 500 MW (PR-500) and 3000 MW (PNP-3000), respectively. The PR-500 pebble bed reactor was designed to produce 523 t/h of steam at a temperature of 265°C and a pressure of 2 MPa plus an electric power of 55 MW. The coolant helium was heated up from 265 to 865°C. The reactor was placed in a prestressed concrete pressure vessel surrounded by three units each containing a heat exchanger and a blower [1]. The large-size reactor concept of the PNP-3000 was foreseen to be connected to steam reforming with 1071 MW heat input (to eight units in four loops) and electricity cogeneration with 540°C/19.5 MPa turbine steam.

As most chemical processes are performed at lower pressures, some adaptation of the reactor design and the chemical process has been necessary. Therefore, the reactor pressure had been fixed in the PNP project to 4 MPa being well below the pressure for electricity-generating plants (~7 MPa). The choice of the pressure is also important to reduce the loads on the high-temperature barriers in case of depressurization accidents either in the primary or in the secondary circuit. Other important aspects of reactor design are the amount of cogenerated electricity, high availability, and an optimization toward significant simplification of the nuclear island. Heat transfer under varying operational load conditions, hot gas mixing in the core bottom, or the lifetime of hot gas thermal insulation have been comprehensively investigated in experiments.

Figure 19.1 shows a flow sheet of the PNP nuclear steam coal gasification process. For this gasification method, the heat from the reactor coolant was foreseen to be transferred to an additional intermediate circuit via a helium–helium intermediate heat exchanger (He–He IHX). The main reason was to avoid the handling of coal and ash in the primary system of the reactor, and a much more complex way for repair and maintenance work. Primary helium of 950°C flowing on the outside of the IHX tubes passes its heat to the secondary

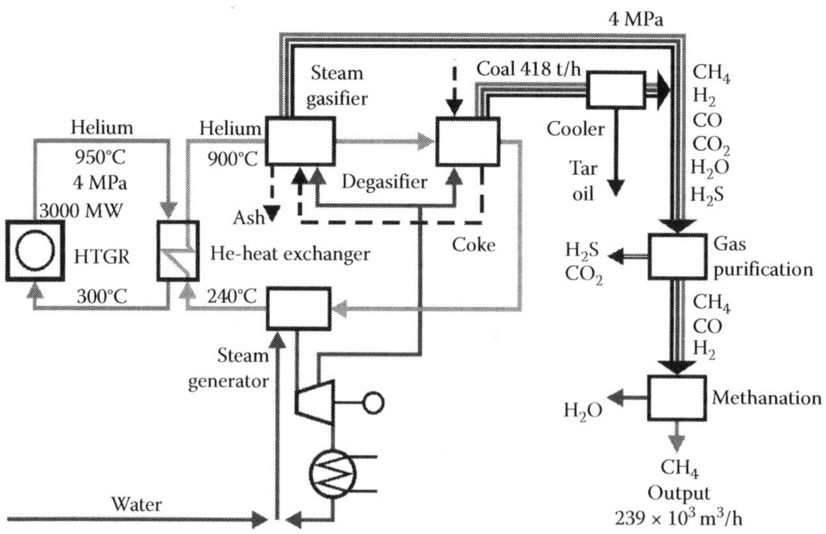

FIGURE 19.1
Schematic of steam gasification of hard coal with nuclear process heat plant PR-3000.

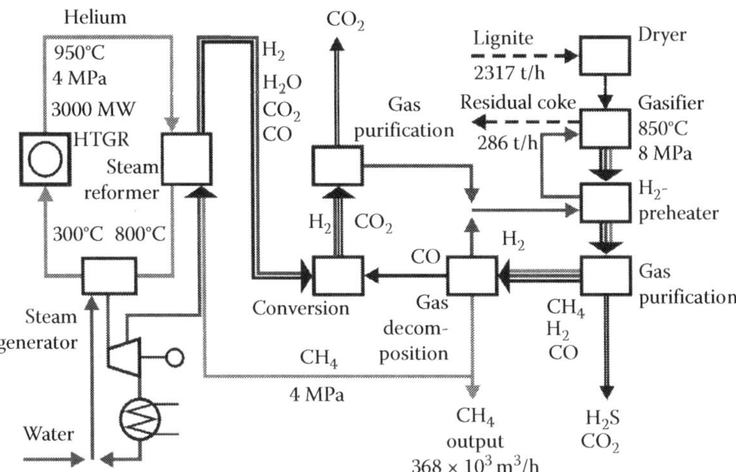

FIGURE 19.2
Schematic of hydrogasification of lignite with nuclear process heat plant PR-3000.

helium entering the steam gasifier at 900°C. Also pressure is slightly higher than on the primary side for the purpose of preventing radioactivity to enter the secondary circuit in case of a leak. The hot steam produced is routed into the coal bed to be gasified.

Unlike the steam-coal gasification, in the hydrogasification process, the nuclear heat is not coupled directly into the gasification reactor. Since for this gasification method, two variants of H_2 production exist, there are two alternatives of nuclear involvement [2]. The one variant is steam methane reforming, as is shown in the flow sheet in Figure 19.2, using a part of the product gas SNG as feedstock. The hydrogen from the reforming step is routed at a relatively low temperature (~400°C) to the reactor where the gasification of coal takes place in an exothermic reaction at 800–900°C and 8 MPa. In the second variant, the nuclear heat is taken to preheat the hydrogen produced during the gasification process itself, that is, by the water gas shift reaction which occurs at 100%. Due to the endothermic character of the shift reaction resulting in a much lower heat production in the reactor, the gasification agent hydrogen needs to be preheated to 800–950°C. Nuclear heat is also used for steam production and the residual power for electricity generation. Compared to the first variant, it has a simpler process scheme, but it requires high gasification pressures (8 MPa), whereas the PR-3000 reactor would be operated at 4 MPa.

19.3 Nuclear Process Heat for Coal Gasification

19.3.1 Nuclear-Assisted Steam-Gasification of Hard Coal

In the process of nuclear steam coal gasification, a new component to be developed was the gas generator with allothermal heating. In 1973, a first device was tested on a small technical scale (~1 kg/h) to investigate kinetics and heat transfer characteristics, gas composition, and other parameters. The follow-on plant on semitechnical scale operated from 1976 to 1984 at the Bergbau-Forschung, Essen, was a first-of-its-kind gas generator with a fluidized

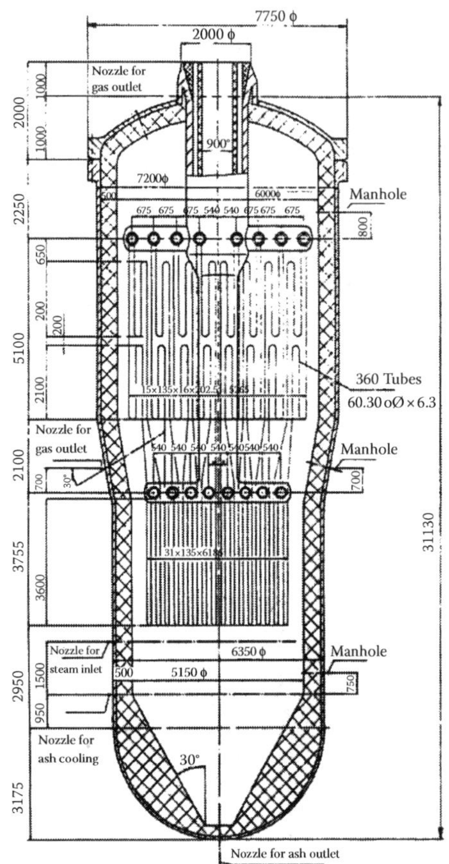

FIGURE 19.3
Schematic of allothermal gas generator (left) and tube-type immersion heat exchanger bundle.

bed of about 1 m^2 base area and a height of up to 4 m, laid out for a coal throughput of ~200 kg/h. This unit was constructed as a vertically arranged cylindrical vessel (Figure 19.3, left) [3] with the outer dimensions of 7.75 m (max.) diameter and 21.13 m height, designed for pressures up to 4 MPa. Its concept differed from the conventional one in that the coal was gasified indirectly by means of a tube-type immersion heater (Figure 19.3, right) which was placed into the fluidized bed to transfer heat from a separate helium circuit. The helium was electrically heated up to 950°C with the heat transferred at a power of 1.2 MW. Characteristic data of the semitechnical plant are listed in Table 19.1 [4].

The semitechnical plant was used for testing components, feeding devices, insulation, investigating broad ranges of operating conditions, and applying different types of coal. Reaction rates were observed to decrease with height of the fluidized bed which can be explained with the inhibiting effect of the product gases whose concentration increases with height. Compared to the conventional case, the temperature provided by the helium is limited. Consequently, reaction rates are slower which, however, could be enhanced by adding a catalyst. The catalytic coal gasification was also tested in the plant. The addition of 4 wt% of the catalyst potassium carbonate enhanced coal throughput by 44% [4]. Furthermore, residence times could be reduced from 7 to 9 h (anthracite) down to about

TABLE 19.1

Characteristic Data of Semitechnical Gas Generator for Steam Coal Gasification

Parameter	Value
Thermal power (MW)	1.2
Helium inlet temperature (°C)	<1000
Helium flow (kg/s)	1.1
Heat-exchanging surface (m^2)	33
Height (m)	<4
Cross section (m^2)	0.8 × 0.9
Fluidized-bed density (kg/m^3)	344
Coal input (kg/h)	233
Coal particle size (mm)	<1
Steam velocity (m/s)	1.13
Gasification temperature (°C)	700–850
Pressure (MPa)	4
Raw gas production rate (Nm^3/h)	816
Conversion rate (%)	83

1.5 h. The catalyst, however, was found to be not effective until a certain threshold value of ~2 wt% due to bonding on the coal. Also, corrosion effects were enhanced observing a strong inner oxidation at temperatures >800°C.

The semitechnical plant was in hot operation for ~26,600 h with more than 13,600 h under gasification conditions (750–850°C, 2–4 MPa). Maximum capacity was 0.5 t/h of coal, the total quantity of coal gasified was 2400 t [4,5]. The overall result of the semitechnical scale operation was that an industrial scale gas generator in connection with a nuclear heat source was considered feasible.

The commercial size gas generator was foreseen to have a thermal power of 340 MW requiring three units for a 1000 MW nuclear process heat plant [6]. It was designed (Figure 19.4), unlike the semitechnical plant, as a horizontal pressure vessel to contain a fluidized

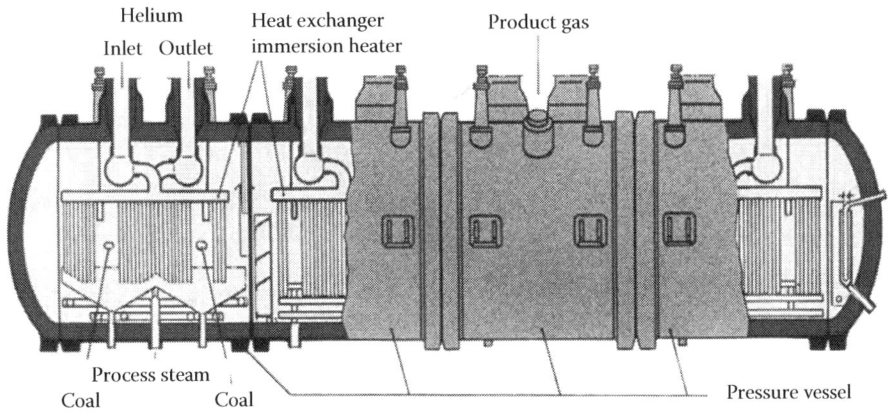

FIGURE 19.4
Schematic of commercial size gas generator with 340 MWt to be connected to nuclear process heat plant.

bed with the shape of a long-stretched channel to allow for long residence times. It consisted of four sections (modules) plus the two ending pieces. The coal is introduced through several inlets in the first module where mainly the pyrolysis process takes place. It passes through the reactor as a plug flow. The gasification zone spreads over the other three modules. In the fourth module, the remaining ash is cooled and removed. Each module contains steam inlets in the bottom section and an immersion heat exchanger bundle, through which heat is transferred from the hot helium to the fluidized bed. The dust formed in the pyrolysis zone is collected in a sieve before passing on to the gasification zone and recirculated.

19.3.2 Nuclear-Assisted Hydrogasification of Lignite

Between 1976 and 1982, the Rheinische Braunkohlenwerke, Wesseling, investigated the hydrogasification process in a 1.5 MW semitechnical test facility with both lignite and hard coal [2,7,8]. The reactor of 8 m height contained a fluidized bed with 0.2 m diameter where the gasification agent hydrogen was injected. The hydrogen was electrically preheated to 750°C and could, if necessary, be further heated to 1000°C by partial combustion. System parameters could be varied in a broader range. A part of the hydrogen was used as a carrier medium for the coal input. Due to the exothermic character of the reactor, a direct heat input is not required. With residence times varied between 20 and 40 min, gasification degrees (for lignite) were up to 75%. The test facility was operated for about 27,000 h with more than 12,000 h under gasification conditions. The throughput was 320 kg/h of lignite or 160 kg/h of hard coal, the total quantity gasified was 1800 t.

From 1983 to 1985, a follow-up pilot plant was operated over 8300 h, with half of the time under coal gasification conditions with high availability. It included, unlike the semitechnical plant, all postprocessing components of gas treatment up to the stage of SNG production. The plant had a throughput of 9.6 t/h corresponding to a total power of 50 MW. Gasification of more than 17,000 t of brown coal was made to yield a total of 11 million Nm^3 of SNG, whose fraction in the raw gas was between 22% and 36%. The SNG production was at a rate of up to 6400 Nm^3/h.

Both the semitechnical and the pilot plant were, apart from lignite, also used for testing whether or not hydrogasification is feasible with the mostly caking hard coal using a gasification agent of either pure hydrogen or mixtures of H_2, CO, and steam. The results have shown that the reaction capability of hard coal is significantly smaller than for lignite. Due to the specific kinetics of hydrogasification, residence times for complete gasification were comparatively long. Therefore a degree of gasification of ~65% was considered reasonable requiring residence times of about 30 min at 9 MPa and 900°C.

19.4 Summary and Conclusions

Within the frame of the PNP project in the 1970s and 1980s, comprehensive R&D activities on nuclear coal gasification were conducted in Germany in a joint cooperation between the Research Center Jülich and partners from the coal and nuclear industries. The results can be summarized as follows:

1. Nuclear coal gasification is a process using nuclear primary energy to convert coal into SNG for direct use, synthesis gas for the production of methanol or other

hydrocarbons, heating (town) gas for residential heating purposes, reduction gas, for example, for the direct reduction of iron ore, or hydrogen as a raw material or universal energy carrier. It is considered a diverse component in the energy supply which may help to significantly reduce the dependency on oil and natural gas imports.

2. The technology of coal gasification to produce SNG or liquid fuels can be used economically under certain conditions depending mainly on the location, the cost of coal, and the product and its price compared to the competitors natural gas and crude oil. The use of nuclear primary energy promises a saving of ~35% of the coal resources, increases the specific SNG output, and reduces specific emissions of carbon and other pollutants to the atmosphere, respectively.
3. Allothermal steam coal gasification with both main types of coal in Germany, hard coal and lignite, was examined in numerous experiments to investigate heat transfer from the helium circuit to the fluidized bed and to optimize operating conditions and components. A wide variety of coal gasification processes has been developed and demonstrated, which are differently appropriate depending on user requirements, plant size, and coal characteristics.
4. For the nuclear coal gasification process, the only new major component that needed to be newly developed was the gas generator. The operation of the semi-technical plants for both coal gasification methods considered confirmed the technical feasibility of allothermal, continuous coal gasification under the nuclear conditions of a process heat high-temperature reactor.
5. Steam methane reforming for hydrogen production is a conventional process widely practiced in industry at a large scale. Its application to the HTGR to be used in hydrogasification of coal is judged technically simple and feasible for near-term commercialization. Studies have also concluded that the cost performance of hydrogen production with an HTGR can be economically competitive to the industrial fossil fuel combustion process [9,10]. Technology development for nuclear-assisted steam reforming has been undertaken in Germany and Japan.
6. A key problem remained is the selection of appropriate high-temperature materials for the heat-exchanging components such as the nuclear steam reformer and the He–He intermediate heat exchanger, but also for the gas generator in the case of catalytic coal gasification.

The plan of construction of a PNP demonstration plant was eventually abandoned at the beginning of the 1990s, when economic analyses had shown that competitiveness of nuclear SNG from expensive German coal with the cheap oil and gas available on the world markets was not given. Still, as was pointed out in a study by the Lurgi company in 1988 [11], alternatives were seen in various industrial sectors such as oil refineries and petrochemical industries (recovery of heavy oil and oil shale/sand, naphtha cracking), iron production, or aluminum oxide production, where nuclear energy could be used to meet their needs of process heat/steam, electricity and/or hydrogen.

Future activities could take benefit from a reevaluation of the studies conducted in the past on HTGR process heat applications by comparing against current nuclear (and conventional) technologies that have current strong interests and are covered in other chapters of Section IV. The comparison shall consider fossil fuel market conditions and environmental effects. The goal should be to select promising applications under the current industrial practice within existing and evolving markets. Superior safety features and high reliability are considered prerequisites for the introduction of nuclear process heat and nuclear combined heat and power.

References

1. Schulten, R., Krieb, K.H., Kugeler, K., Schroeder, G., Barnert, H., Dannenbaum, J., Drescher, H.P., et al., Industriekraftwerk mit Hochtemperaturreaktor PR 500—OTTO-Prinzip zur Erzeugung von Prozeßdampf, Report Jül-941-RG, Jülich, Germany, 1973.
2. Schrader, L., Strauss, W., and Teggers, H., The application of nuclear process heat for hydrogasification of coal, *Nuclear Engineering and Design* **34**:51–57, 1975.
3. Barnert, H. and Singh, Y., Design evaluation of a small high-temperature reactor for process heat applications, *Nuclear Engineering and Design* **109**:245–251, 1988.
4. Kirchhoff, R., Van Heek, K.H., Jüntgen, H., and Peters, W., Operation of a semi-technical pilot plant for nuclear aided steam gasification of coal, *Nuclear Engineering and Design* **78**:233–239, 1984.
5. Kubiak, H., Van Heek, K.H., and Ziegler, A., Nukleare Kohlevergasung—Erreichter Stand, Einschätzung und Nutzung der Ergebnisse, Fortschritte in der Energietechnik, Monographien des Forschungszentrums Jülich, Vol. 8, Jülich, Germany, 1993.
6. Heek, K.H. van, Jüntgen, H., and Peters, W., Wasserdampfvergasung von Kohle mit Hilfe von Prozeßwärme aus Hochtemperatur-Kernreaktoren, *Atomkernenergie/Kerntechnik* **40**:225–246, 1982.
7. Fladerer, R. and Schrader, L., Hydrierende Vergasung von Kohle—Neuere Betriebsergebnisse, *Chemie Ingenieur Techik* **54**:884–892, 1982.
8. Scharf, H.-J., Schrader, L., and Teggers, H., Results from the operation of a semitechnical test plant for brown coal hydrogasification, *Nuclear Engineering and Design* **78**:223–231, 1984.
9. Shiina, Y., Sakuragi, Y., and Nishihara, T., Cost estimation of hydrogen and DME produced by nuclear heat utilization system, Report JAERI-TECH 2003-076, Japan Atomic Energy Research Institute, Oarai, Japan, 2003.
10. Shiina, Y. and Nishihara, T., Cost estimation of hydrogen and DME produced by nuclear heat utilization system—II, Report JAERI-TECH 2004-057, Japan Atomic Energy Research Institute, Oarai, Japan, 2004.
11. Schad, M., Didas, U., Ebeling, F., Kreuzkamp, G., and Renner, H., Project study on utilization of process heat from the HTGR in the chemical and related industries, Report Lurgi GmbH, Frankfurt, Germany, 1988.

20

Nuclear Steam Reforming of Methane

Yoshiyuki Inagaki and Karl Verfondern

CONTENTS

20.1 Introduction

Steam-methane reforming (SMR) takes place in an endothermic reaction (Equation 6.5 in Chapter 6) and performs efficiently at high temperature. While the heat requirement for the reaction is met by methane combustion in conventional systems, it can be met by the heat generated and supplied from high-temperature capable nuclear fission and fusion reactors. Several suitable types of nuclear reactors are presented in the chapters of Section III.

This chapter discusses the technology and engineering development of the SMR process and equipment specifically oriented for nuclear reactor application. Figure 20.1 shows schematically an application example of the SMR process coupled to a high temperature gas-cooled reactor (HTGR) (Chapter 10). The heat generated in the reactor core is exchanged from primary to secondary helium gas in the intermediate heat exchanger (IHX), and then transported to the hydrogen production system by the secondary loop hot gas duct. The transported heat is used in the hydrogen production system for the endothermic reaction of hydrogen production and steam generation. Alternatively, the steam reforming system has been considered for placement in the secondary circuit, thereby eliminating the IHX and the intermediate heat transport loop.

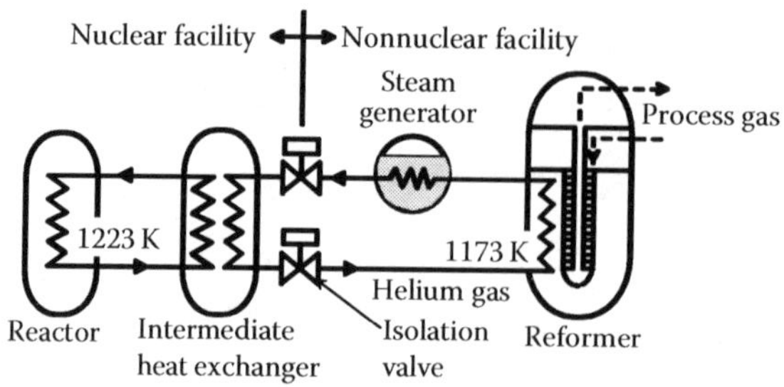

FIGURE 20.1
Schematic flow diagram of HTGR steam reforming system.

20.2 Technology and Engineering Development

The combination of SMR with the HTGR is considered a feasible technology for near-term commercialization. Several studies have been performed to analyze the cost performance of this nuclear production technology [1–3]. These studies concluded that the SMR hydrogen production with an HTGR is economically competitive to the industrial fossil fuel combustion process.

A development step taken to commercialize the nuclear SMR system is out-of-pile (simulated nuclear reactor heating) testing, in which the engineering design and the system operation can be simulated and confirmed. The results from the significant efforts made in this kind of test in Germany and Japan are described below.

20.2.1 Development in Germany

Nuclear SMR was subject of extensive R&D activities in Germany in the 1970–1980s. In the German "Prototype Nuclear Process Heat" project, PNP, the steam reformer was an essential component in the nuclear-assisted hydro-gasification of coal. In this concept, the nuclear steam reformer was included directly within the HTGR primary circuit. The employment of an IHX was, in those days, deemed unnecessary. Such an arrangement poses much more stringent requirements to this component and its reliability, than is, in the case of an indirect cycle via an IHX as was pursued in the Japanese high temperature engineering test reactor (HTTR) project. HTTR is a HTGR with thermal output of 30 MW and outlet temperature of 950°C currently installed in JAEA. The direct coupling to the steam reformer resulting in a simplified design of a process heat HTGR may, from today's safety perspective, be regarded as a long-term option.

Within the German Nuclear Long-Distance Energy Transportation (Nukleare Fernenergie, NFE) project, test facilities in the MW scale were constructed and successfully operated at Forschungszentrum Jülich (FSJ) (research center Jülich) with the general aim to demonstrate the operation of a helium-heated system for steam reforming under HTGR typical conditions and to study the technical performance of the thermochemical pipeline system and its efficiency. With regard to the operational conditions, the ranges of interest were temperatures of 400–900°C, pressures of 2–4 MPa, and H:C:O ratios of 8:1:2–12:1:4 [4].

20.2.1.1 Tubular Reformer Concept

A schematic of a nuclear-heated splitting tube and temperature profiles along the tube length are shown in Figure 20.2. Different from conventional steam reforming, the splitting

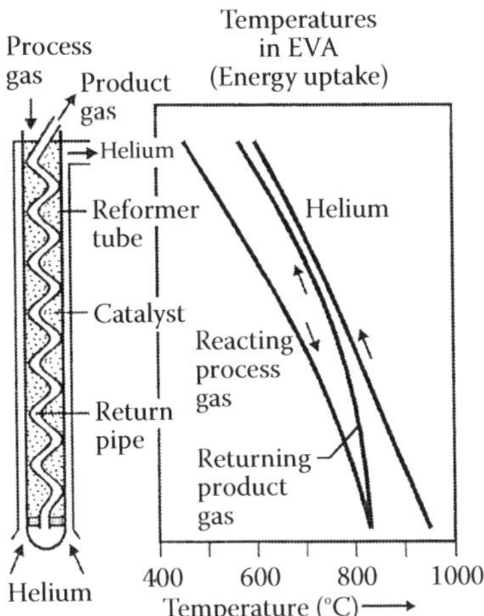

FIGURE 20.2
Schematic of a nuclear-heated splitting tube and temperature profiles along the tube.

tube is heated from outside (allothermal) by hot helium with an inlet temperature at the bottom of ~950°C. The process feed gas (methane) enters the splitting tube from top at a temperature of typically ~450°C, passes through the catalyst bed, takes up heat to a final value of ~830°C, and converts into a mixture of H_2, CO, CO_2, and still unreformed methane. A special feature is the recirculation tube which is a thin tube located inside the splitting tube to guide the process gas back to the top. By this it means, not only does the counter-current flow allow cooling of the process gas and heating of the feed gas, it also leaves the splitting tube in the area with highest thermal impact completely closed without connecting pipes and assures free thermal expansion movement in axial direction.

20.2.1.2 EVA-I Reformer Single Tube Test

The first test facility to demonstrate steam reforming of natural gas under simulated nuclear conditions was EVA-I, German acronym for single splitting tube, starting operation in 1972 as shown in Figure 20.3 [4]. Helium was heated up to 950°C by a 1 MW electric heater and introduced at 4 MPa into the annular gap around the splitting tube. Splitting tubes of various geometries (tube diameter: 80–160 mm; heated tube length: 10–15 m) and tube materials were subjected to operational conditions covering broad ranges as indicated in Table 20.1. The heat transferring inner surface of the splitting tube was between 3 and 8 m^2. After initial tests without inner return pipe for the product gas, the comparison has clearly proven the advantage of the concept of internal recirculation and was applied since. Differently shaped inner recirculation tubes were investigated (straight, helical), also the arrangement of more than one per splitting tube which, however, was at the expense of ease of catalyst replacement. As a catalyst, various types of Ni Raschig rings, discs, and other novel systems were examined especially with regard to easy and rapid exchange.

The test with a double-walled splitting tube revealed the expected result of practically no transfer of hydrogen to the helium system and allowed a direct quantification of H_2 permeation rates through the Incoloy 800 tube wall to be initially 0.26×10^{-3}, later reduced to 0.013×10^{-3} Nm^3/(m^2 h) at 830°C and 1.3 MPa of H_2 partial pressure on the process side.

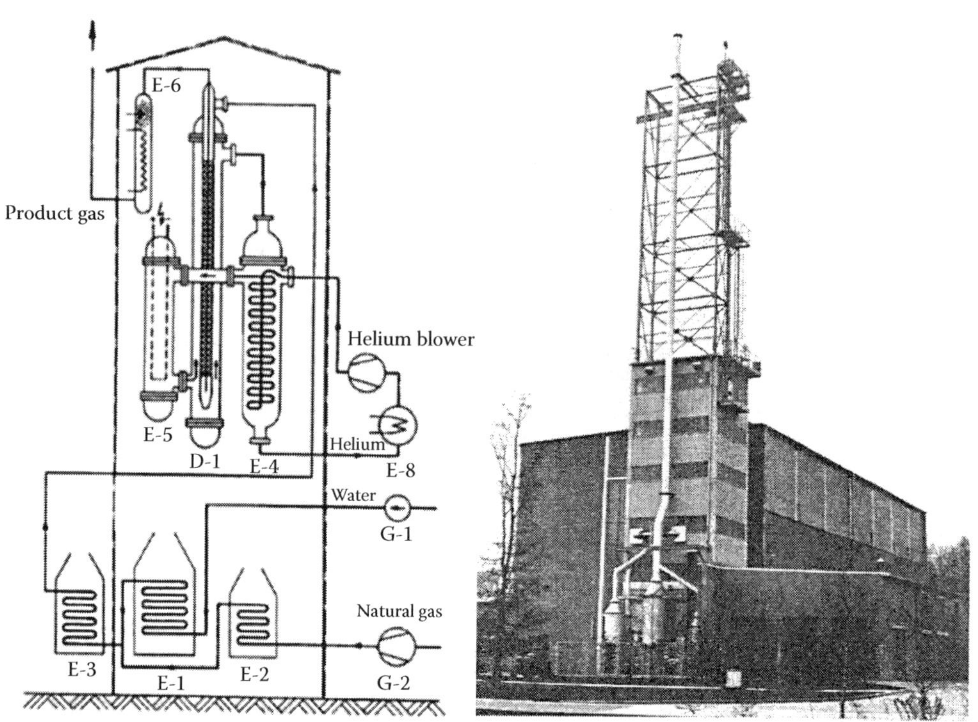

FIGURE 20.3
Single splitting tube test facility EVA-I at the Research Center Jülich.

Also methanation, the reverse process of steam reforming, where the product gas was reconverted to methane and steam, was investigated to study in a closed cycle the complete system of a long distance energy transportation system based on the energy carrier hydrogen. From 1979, the pilot plant ADAM-I was operated where high-temperature methanation takes place in a three-stage process with adiabatic fixed bed reactors. In conjunction with EVA-I, the ADAM-I plant was tested over 2344 h and in a broad range of operational parameters [5].

20.2.1.3 EVA-II Reformer Tube Bundle Test

The follow-on test facility, EVA-II, represented a complete helium circuit containing a bundle of reformer tubes [6]. The nuclear heat source was simulated by an electrical heater with a power of 10 MW to heat up helium gas to a temperature of 950°C at 4.0 MPa. As can be seen from the schematic in Figure 20.4, the principal components of heater, reformer bundle and steam generator were housed in separate steel pressure vessels in a side-by-side arrangement. The vessels of helium heater and steam reformer were connected by a coaxial helium duct of 5 m length.

Two reformer bundles both with convective helium heating have been investigated in the EVA-II facility. They differed in the way of channeling the helium flow. The first bundle tested was a baffle design tube bundle for a power of 6 MW consisting of 30 tubes. The bundle was characterized by baffle structures (discs and doughnuts) on the helium side as shown in Figure 20.5, and each splitting tube connected at the top with a separate feed and product gas line. Inside the splitting tube, the heat between 950°C and 650°C was used to

TABLE 20.1

Comparison of Data of Steam Reformer in EVA Test Facilities and in a Commercial-Size Nuclear Application

Parameter	Test Facilities			Nuclear Steam Reformer for 170 MWt HTR Module
		EVA-II		
	EVA-I	Annulus Design	Baffle Design	
Nuclear heat input (MWt)	**~0.3**	5	~ 6	60.2
Catalyst tube				
Outer tube diameter (mm)	80–160	120	130	120
Wall thickness (mm)	10–21	10	15	10
Length (m)	10.0–15.7	13	11.5	14
Number of tubes	1	18	30	199
Tube material	Incoloy 800 H and others	Incoloy 800 H Incoloy 802 Manaurite 36 X IN 519	Incoloy 800 H	Inconel 617
Primary helium				
Inlet temperature (°C)	800–950	950	950	950
Outlet temperature (°C)	~600–750	700	650	720
Inlet pressure (MPa)	4	4	4	4.987
Flow rate (kg/s)	0.15–0.45	~3.8	~3.8	50.3
Process feed gas				
Temperature inlet recuperator (°C)	400–600	330	330	347
Temperature outlet catalyst (°C)	700–850	810	800	810
Pressure inlet recuperator (MPa)	2–4	4.0	4.0	5.6
Raw gas flow rate (kg/s)	~0.01–0.04	0.62	0.62	34.8
Steam-methane ratio	2–4	4	4	4
Product gas				
Temperature inlet inner tube (°C)	~830	810	800	810
Temperature outlet recuperator (°C)	~520	457	450	480

run the steam reforming process. The internal helical tube for recirculation of the product gas had an outer diameter of 20 mm and a wall thickness of 2.1 mm, the helix itself had a diameter of 70 mm.

The second steam reformer bundle tested, was of annulus design for a power of 5 MW. It consisted of 18 tubes with each splitting tube placed inside a guiding tube channel, where the hot helium was flowing upwards through the annular gap. For both designs, tubes were hanging on a supporting plate which allowed an easy exchange of single tubes. The specific data of both tubes and catalytic systems were very similar compared to components planned for nuclear applications (see Table 20.1). Also, the loads imposed on the supporting structures were characteristic to the nuclear case.

In the connected steam generator, an upward boiler with forced convection and direct superheating and composed of helical tubes for a total power of 4 MW, the remaining helium heat was used up to 350°C. Via a helium circulator, the cold helium was routed back to the electrical heater. The circuit was operated under nuclear conditions at a lower power level, but with a full-scale steam reformer component. Also the process gas handling system was the same as it would have been in a commercial-scale nuclear plant.

The catalytic steam reforming reaction allows reaction rates of 10^3 Nm^3 CH_4/(m^3 catalyst h) (related to the catalyst volume in the reformer tubes) at 800°C, 4 MPa and $H_2O/CH_4 = 4$.

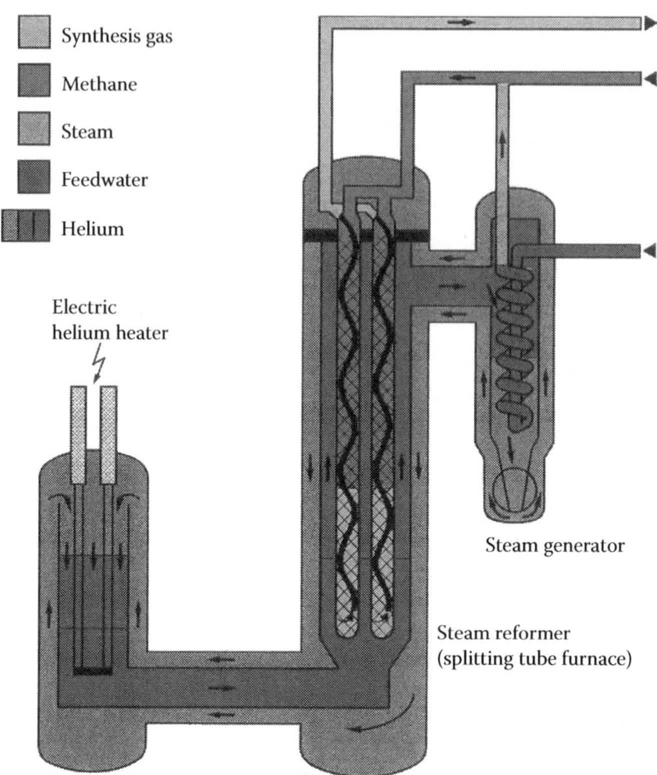

FIGURE 20.4
Flow sheet of steam reforming test facility EVA-II.

The overall heat transport coefficient in connection with the chosen temperature distribution on the helium side and on the process side allows an average heat flux of 60 kW/m^2 for the reformer tubes. The heat transfer coefficient on the process side was observed to be rather high. More than 1000 $W/(m^2\ K)$ were realized. The heat transfer coefficient on the helium side was around 500 $W/(m^2\ K)$, which is also relatively high. It is, however, limited by the allowable pressure drop of around 40 kPa.

In the steam reforming system of EVA-II, no damage was observed on reformer tubes or guiding tubes, internal recuperators or inner return pipes. The components behaved very well in terms of thermal expansion, bending of tubes, friction, fretting, and vibrations caused by flow effects. The gas ducting structures, insulations, and supporting structures were operated without difficulties, too. Measurement and control of process parameters during operation was easy.

The efficiency of the catalyst was practically not changed during some 1000 h of operation. There was no carbon deposition on the catalyst because of the chosen H_2O/CH_4 ratio of 4 in the process. The handling of different catalysts was tested with procedures which can be applied to a larger component.

Components were also tested at transient conditions with changing rates for the temperature on the helium side of >10°C/min and for the pressure of >4 MPa/min. The rates of changes of parameters on the process side were even larger. No difficulties were encountered.

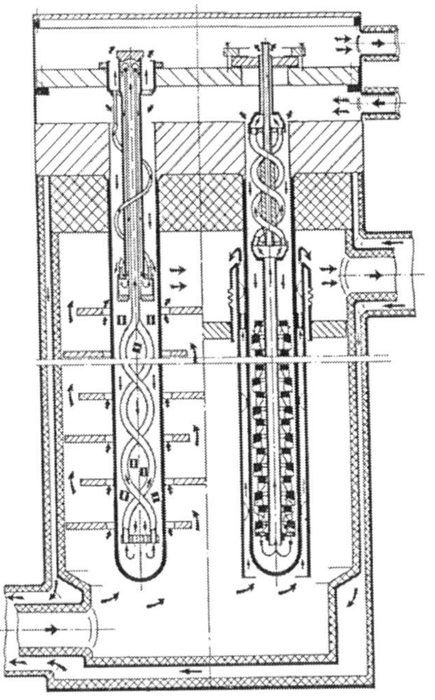

FIGURE 20.5
EVA-II steam reformer bundles of baffle design (left) and annulus design (right).

The switch-off of single splitting tubes within the bundle was observed to result in enhanced heat transfer for the process gas which, however, did not compensate for the reduced heat area. The consequence was a slight reduction in the splitting temperature connected with an increased portion of unreformed methane in the product gas.

Instability in the operational behavior was observed in the case of low helium flow rates, where an inhomogeneous temperature distribution was developing with differences of up to 150°C causing several centimeters buckling of the vessel wall. Another observation was the back-circulation of the helium flow in the entrance area, which resulted in inhomogeneities in the inlet-temperature profile. In the reformer tube bundle of the baffle design, these differences were quickly equalized, whereas in the annulus design, differences in the inlet temperature would result in different power rates of the single tubes.

The EVA-II helium system was operated over a total time of 13,000 h with 7750 h at a temperature of 900°C. Both types of steam reformer were operated for more than 6000 h each without any difficulties. The operation included both steady-state, partial load, transient conditions, and also special tests with tube blockage at full power operation. Postoperation inspection confirmed integrity of all tubes.

A respective methanation plant, ADAM-II completing a closed cycle without any CO_2 emissions, was operated over almost 6000 h. The conversion of methane at the reaction conditions mentioned above is in the order of 65%. It corresponds nearly to the thermodynamic equilibrium. From the power input of 10.78 MWe into the helium, the transported energy in the pipeline was 5.72 MWt. The heat eventually released in the ADAM plant was 5.42 MWt in form of 650°C heat. With an additionally utilized heat of 2.34 MWt, the total energetic efficiency achieved was 72%.

Within the frame of a comprehensive research program of materials for splitting tubes, the experimental work was concentrating on the four iron-based alloys Incoloy 800 H, Incoloy 802, Alloy IN 519, and Manaurite 36 X. Also considered were the nickel-based wrought alloys Inconel 617, NIMONIC 86, Hastelloy X, and Hastelloy S.

For the purpose of measuring creep resistance and the 1% time-elongation limit of all materials, long-term tests were done over 10,000 h in air and HTGR helium, and over 5000 h in product gas of the methane-reforming process. Longest test duration was 34,000 h for Incoloy 800 H and Inconel 617. The creep resistance for Incoloy 800 H after 30,000 h at 850°C was measured to be 16.5 N/mm^2 with an expected value of 11 N/mm^2 after 100,000 h. The respective data for Inconel 617 after 30,000 h at 950°C were 13 and 7 N/mm^2, respectively.

20.2.1.4 Nuclear Steam Reformer Design

A variant of the HTR-Modul pebble-bed reactor for process heat production has been designed by SIEMENS-INTERATOM for a thermal power of 170 MW to deliver helium temperatures of 950°C. The thermal power was reduced compared to the electricity generating variant due to the requirement of self-acting decay-heat removal from the core, that is, for the maximum fuel temperature to stay below 1600°C in case of a loss-of-forced-convection accident. Without IHX, the helium coolant is directly fed to the steam reformer which consumes 71 MW and to the steam generator which is operated with 99 MW.

Details of the concept of the steam reformer as a novel component in nuclear process heat plants are shown in Figure 20.6. It is a bundle consisting of straight tubes connected to an upper supporting plate. The tubes filled with a catalyst contain an internal recuperator and an inner return duct for the hot process gas. The upper part of the component includes the collector structures for the feed gas composed of steam and methane, and for the product gas containing H_2, CO, CH_4, CO_2, and steam. From the total heat transferred into the steam reformer, 85% are used for the reforming process, while 15% are taken to heat up the feed gas. The main characteristic data of the steam reformer component as was designed by INTERATOM on the basis of a simple cylindrical pebble bed core with a power of 170 MWt are listed in Table 20.1.

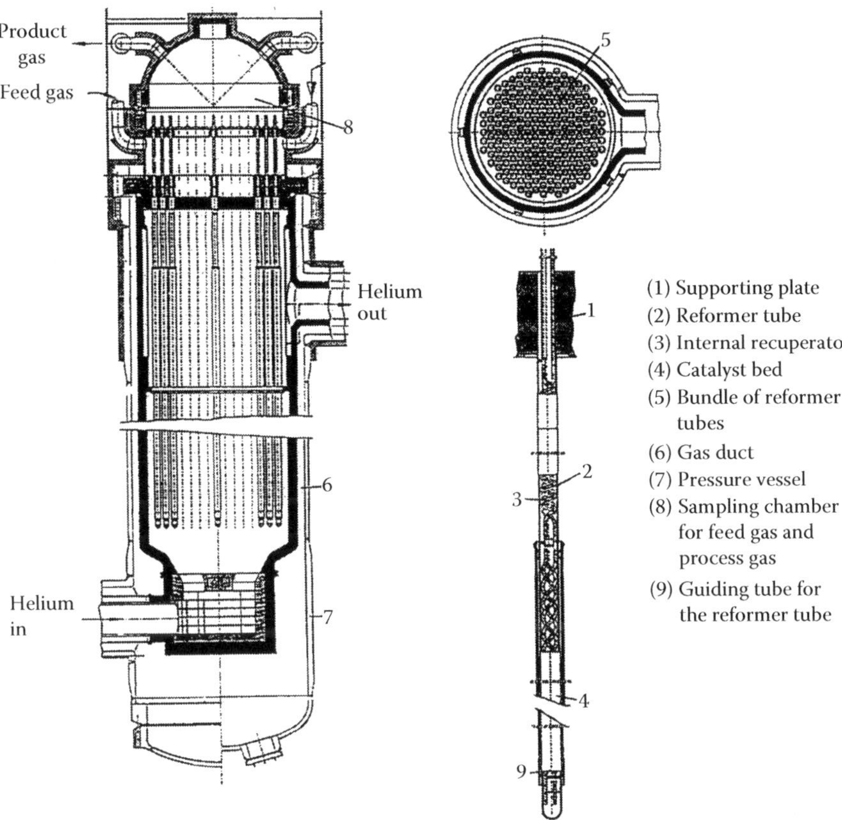

FIGURE 20.6
Technical concept of a helium heated steam reformer connected to a modular process heat HTGR.

20.2.1.5 Summary

On the basis of a broad experimental program for nuclear steam reforming (EVA-I and EVA-II facilities, additional testing of kinetics, heat transfer, materials) it can be stated that the allothermal steam-reforming process with hot helium to be supplied in future by a process heat HTGR was successfully tested on a large scale and is well understood. With EVA-II/ADAM-II, it was worldwide the first time that a plant was designed, constructed, and successfully operated which, in all components, process steps, and auxiliary systems, represented a nuclear long-distance energy transportation system and proved its operability and reliability over 7800 h with 5660 h under nuclear heat transport conditions. Results achieved should allow, in terms of power, the extrapolation by a factor of around 10 from the EVA-II plant to a helium heated reformer connected to a modular HTGR of 170–200 MWt. For larger powers, more reformer loops should be coupled to the reactor.

20.2.2 Development in Japan

JAEA plans to demonstrate a hydrogen production system coupled to the nuclear test reactor HTTR. To enable successful coupling, an out-of-pile test by SMR was carried to investigate system integration requirements and technologies including high-temperature gas

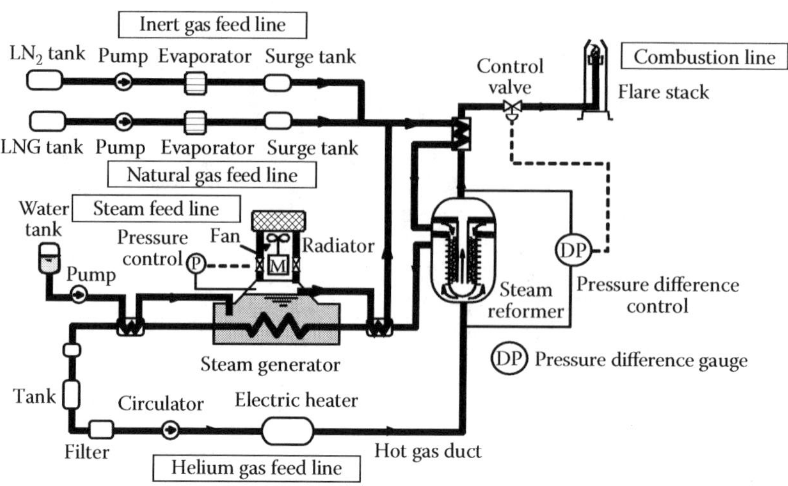

FIGURE 20.7
Schematic flow diagram of mock-up test facility.

isolation valve and system control technology [7]. Of course, this result can be applied to the thermal fluid coupling of not only steam reforming system but also other hydrogen production processes (Chapters 16 through 19) to nuclear reactor.

20.2.2.1 Nuclear Steam Reforming Mockup Test Facility

Figure 20.7 shows a schematic flow diagram of the mock-up test facility which had an approximate hydrogen production capacity of 120 Nm^3/h and simulated key components downstream from the IHX of the HTGR [7]. A 400 kW electric heater was used as a heat source instead of the nuclear heat in order to heat helium gas up to 880°C at a steam-reformer inlet. The hot helium gas was used as a heat source of the reforming reaction, the steam generation and preheating of the steam and methane. Table 20.2 lists the main specifications and Figure 20.8 shows the overview of the test facility.

TABLE 20.2
Design Specifications of the Mock-up Test Facility

Items	Test Facility
Pressure	
Process gas/helium gas	4.1 MPa
Temperature inlet at steam reformer	
Process gas/helium gas	450/880°C
Temperature outlet at steam reformer	
Process-gas/helium-gas	600/650°C
Natural gas feed	43.2 kg/h (27 kmol/h)
Helium gas feed	327.6 kg/h
Steam-carbon ratio (S/C)	3–4
Hydrogen product	120 Nm^3/h
Heat source	Electric heater (400 kW)

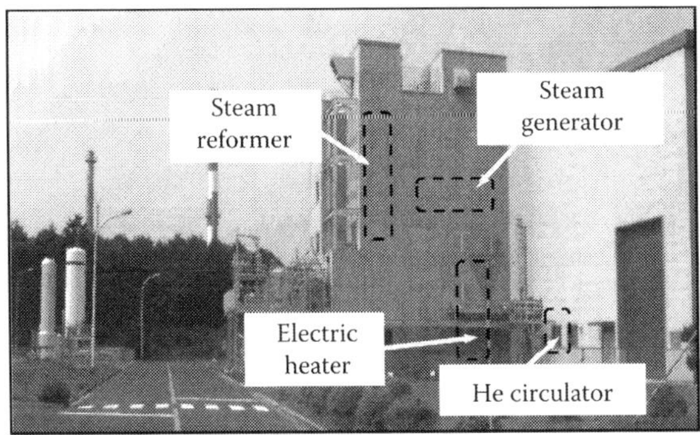

FIGURE 20.8
Site of constructed mock-up test facility.

20.2.2.2 Bayonet Tube Steam Reformer Design

Figure 20.9 shows the schematic view of the steam reformer which has one bayonet-type catalyst tube made of Alloy 800 H (146 mm O.D., 10 mm thick, and 7000 mm length). The process gas, a mixture of steam and methane, flows in the catalyst tube at inlet temperature of 450°C and helium gas flows in a channel between catalyst and guide tubes at inlet temperature of 880°C. In the fossil-fuel-fired system, the process gas receives heat from combustion air of about 1200°C by radiation, and the heat flux at the outer surface of the catalyst tube reaches 70,000–87,000 W/m^2. In order to achieve the same heat flux as that of

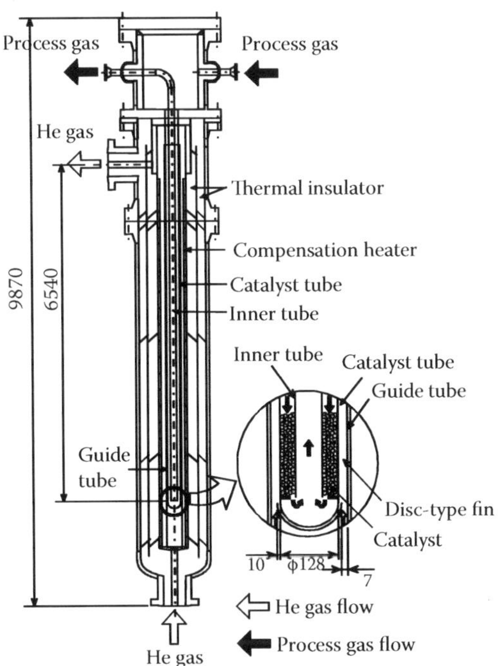

FIGURE 20.9
Schematic view of steam reformer with one bayonet-type catalyst tube.

the fossil-fuel-fired system, it is very important to promote heat transfer of helium gas by forced convection because the temperature of helium gas, that is, the temperature of heat source is lower compared with the fossil-fuel-fired system. So, disc-type fins, 2 mm in height, 1 mm in width and 3 mm in pitch, are arranged around outer surface of the catalyst tube in the test facility in order to increase the heat transfer coefficient of helium gas by a factor of 2.7, 2150 W/(m^2 K) with the fins, and the heat transfer area by a factor of 2.3 compared with those of a smooth surface, respectively. As the result, the heat transfer performance of the catalyst tube in the test facility becomes competitive to that of the fossil-fuel-fired system. Figure 20.10 shows the overview of the catalyst tube. The catalyst tube is a pressure boundary between the gas and the process gas. Its wall thickness is decided considering the pressure difference between helium and process gases. In order to assure the structural integrity of the catalyst tube in all conditions such as not only a normal start-up and shut-down but also a loss of the chemical reaction, a control system is required to keep the pressure difference within an allowable range. In this system, the pressure of the process gas is to be controlled according to the pressure change of the helium gas.

FIGURE 20.10
Fabricated bayonet-type catalyst tube (full production scale tube).

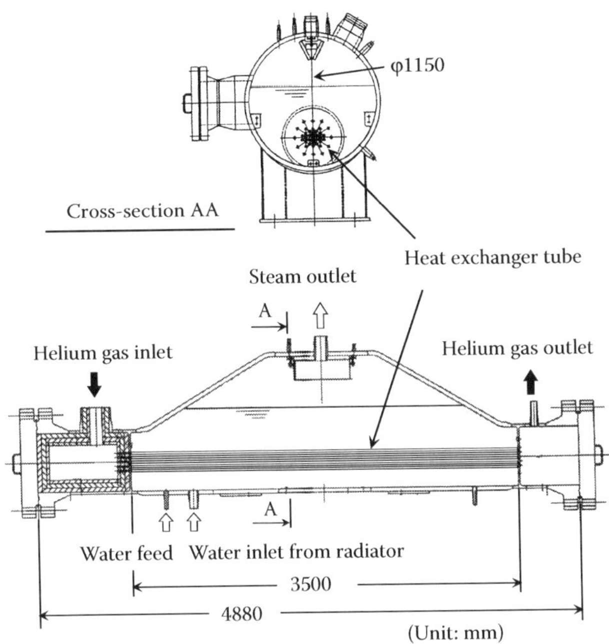

FIGURE 20.11
Schematic view of steam generator.

20.2.2.3 Steam Generator as Thermal Shock Absorber

Figure 20.11 shows a schematic view of the steam generator, which is a kettle type, with 27 heat exchanger tubes (25.4 mm O.D., 4.0 mm thick, and 3500 mm length). Helium gas flows inside the heat exchanger tubes and the heat of the helium gas is transferred to water, whose holding quantity is 1.7 m^3 in the rated design condition, in which the heat exchange rate between the helium gas and the water is 135 kW. The radiator was installed above the steam generator to cool the large quantity of steam produced in the steam generator in case of loss of chemical reaction and circulation of steam and condensed water circulate between the steam generator and the radiator. The steam generator installed downstream, the steam reformer works as a thermal absorber to mitigate the temperature fluctuation of helium gas caused by the steam reformer in the normal start-up and shut-down operations and accidents such as loss of the chemical reaction. Chapter 25 presents details of the performance of the steam generator as an effective thermal shock absorber at a loss of chemical reaction accident.

20.2.2.4 Test Operation of the Mockup Facility

Figure 20.12 shows the flow rates of each gas, the pressure difference between the helium and the process gases in the steam reformer (DP), helium gas pressure at the inlet of the steam reformer and helium gas temperature change at the inlet and outlet of the steam reformer and the steam generator in the start-up operation at the inlet helium gas temperature of the steam reformer from 700 to 880°C [8]. At time 0 h, steam supply to the steam reformer was started and supply flow rate was increased gradually to the rated value of

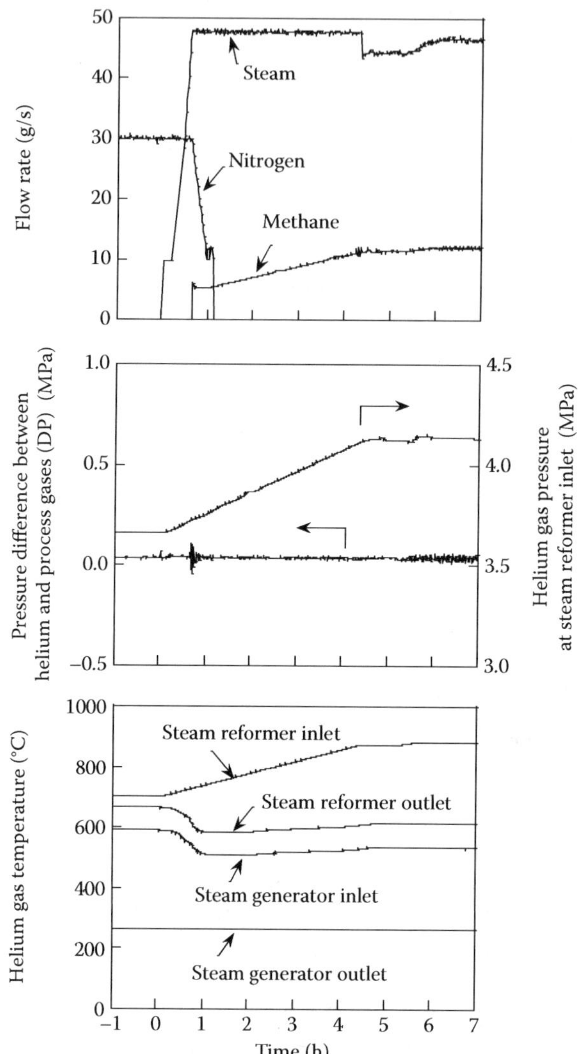

FIGURE 20.12
Flow rate of each gas, the pressure difference between the helium and the process gases in the steam reformer, helium gas pressure at the inlet of the steam reformer and helium gas temperatures change at the inlet and outlet of the steam reformer and the steam generator in the start-up operation.

47 g/s at 0.75 h. After that, methane was supplied up to 12.0 g/s at 880°C and nitrogen supply was stopped. At 0 h, when water feed was started, DP increased from around 0.04 to 0.065 MPa due to the increase of total flow rate in the steam reformer and then it decreased immediately to around 0.04 MPa. At elapsed time of 0.75 h when the steam reforming reaction was started by feed of methane, DP fluctuated in the range of −0.04 and 0.10 MPa due to the increase of total flow rate by produced hydrogen. However, its fluctuation range was extremely small compared to the design pressure range of −0.5 to 1.0 MPa. The outlet helium gas temperature of the steam reformer decreased drastically at 0.75 h because the steam reforming reaction occurred. The helium gas temperature at the inlet of the steam generator showed almost the same temperature profile as at the outlet of the steam reformer. On the

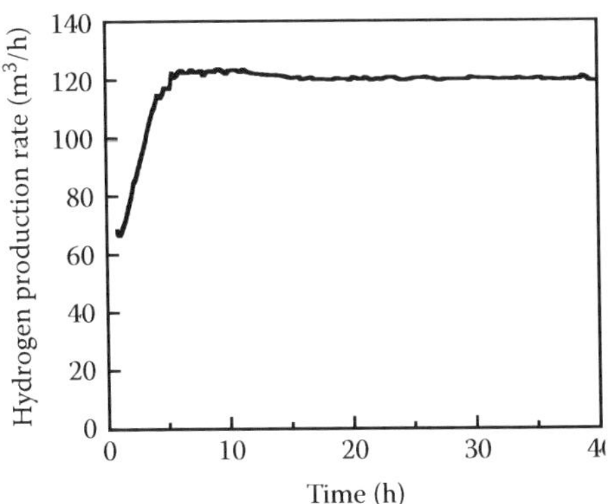

FIGURE 20.13
Hydrogen production rate during start-up and steady operations.

other hand, the helium gas temperature at the outlet of the steam generator decreased only 1.5°C during the start-up operation although it changed at the inlet from 591°C at 0 h to 508°C at 1.5 h. This result shows that fluctuations of the helium gas temperature caused by the steam reformer were mitigated by the steam generator.

Figure 20.13 shows hydrogen production rate during the start-up and the steady operation. In the steady operation, average flow rates of methane, steam, and helium gas were 12.0, 46.6, and 91.0 g/s, respectively [8]. The hydrogen production rate was stable and an average value between 14 and 40 h was 120.2 m^3/h. At that time, the maximum reaction temperature and the pressure of the process gas were around 756°C and 4.0 MPa, and the methane conversion rate, which is the ratio of produced CO quantity to supplied methane quantity, was 55%.

The operation of the mock-up test facility was started in 2002 and finished with fulfilling the objective of the test in 2004.

20.3 Future Technical Issues

The target of the process optimization is increase of the hydrogen-production efficiency. For this purpose, the improvement of methane conversion is very important. The lower the conversion rate, the larger becomes the loss of the steam generated for the reforming reaction. As a result, the hydrogen production efficiency becomes lower. The methane conversion rate in the mock-up test facility of JAEA is low with about 60% and the conversion rate is ruled by the reaction pressure. The steam reforming reaction is promoted in the low pressure condition. For example, the methane conversion rate calculated by chemical equilibrium in conditions at 800°C of a reaction temperature and 4 of a steam carbon ratio is about 64% at 4.5 MPa of reaction pressure, about 81% at 2 MPa, and about 92% at 1 MPa.

As mentioned above, the steam reforming reaction at low pressures of less than 2 MPa is one of the methods to increase the hydrogen-production efficiency. However, the reaction in a low pressure is technically difficult at present. The helium pressure will be more than 4 MPa to cool effectively the reactor core, component and so on. In this case, the pressure difference between helium and process gases is too high, more than 2 MPa, to manufacture the catalyst tube which is the pressure boundary between helium and process gases. This is because the current metal material is not strong enough to sustain such high pressure differences at around 900°C. In the case of the mock-up test facility, the pressure resistant design value of the catalyst tube was 0.5 MPa against the outer pressure and 1.0 MPa against the inner one. The development of the catalyst tube made of ceramics which has enough strength in a high temperature is one of solutions for low pressure reaction.

Another promising technology to improve the hydrogen-production efficiency is a membrane reactor which stores the hydrogen permeation membrane. The methane conversion can be greatly improved as a result of promotion of the steam reforming reaction by removing hydrogen in the reaction process. The membrane reactor is one of the key technologies for the future challenge. It requires lower process temperatures than the conventional reforming process. While the conventional process is applicable to high-temperature fission and fusion reactors, the membrane process can be applied to nuclear reactors of medium temperature ranges such as Generation IV sodium-cooled fast reactor and supercritical water reactors. Chapter 11 reports on SFR-based membrane hydrogen production.

References

1. M. Correia, R. Greyvenstein, F. Silady, and S. Penfield, PBMR as an ideal heat source for high-temperature process heat applications, *Proc. ICONE-14*, July 17–20, 2006, Miami, FL, USA, No. 89473.
2. W. Reiner, R.W. Kuhr, C.O. Bolthrunis, and M. Corbett, Economics of nuclear process heat application, *Proc. ICAPP' 06*, June 4–8, 2006, Reno, NV, USA, No. 6302.
3. C. O. Bolthrunis, R. W. Kuhr, and A. E. Finan, Using a PBMR to heat a steam-methane reformer: Technology and economics, *Proc. HTR2006*, October 1–4, 2006, Johannesburg, South Africa, No. I00000118.
4. NFE, Nukleare Fernenergie, Zusammenfassender Bericht zum Projekt Nukleare Fernenergie (NFE), Report Jül-Spez-303, Research Center Jülich, Germany, 1985.
5. B. Höhlein, H. Niessen, J. Range, H. J. R. Schiebahn, and M. Vorwerk, Methane from synthesis gas and operation of high-temperature methanation, *Nucl. Eng. Des.*, 78, 241–250, 1984.
6. H. F. Niessen, A. T. Bhattacharyya, M. Busch, K. Hesse, and A. Zentis, Erprobung und Versuchsergebnisse des PNP-Teströhrenspaltofens in der EVA II-Anlage, Report Jül-2231, Research Center Jülich, Germany, 1988.
7. Y. Miyamoto, S. Shiozawa, M. Ogawa, Y. Inagaki, T. Nishihara, and S. Shimizu, Research and development program of hydrogen production system with high temperature gas-cooled reactor, *Proc. HYFORUM 2000*, Munich, Germany, September 11–15, Vol. 2, 271, 2000.
8. H. Ohashi, Y. Inaba, T. Nishihara, Y. Inagaki, T. Takeda, K. Hayashi, S. Katanishi, S. Takada, M. Ogawa, and S. Shiozawa, Performance test results of mock-up test facility of HTTR hydrogen production system, *J. Nucl. Sci. Technol.*, 41, 385–392, 2004.

21

Hydrogen Plant Construction and Process Materials

Shinji Kubo and Hiroyuki Sato

CONTENTS

21.1 Thermochemical Corrosion Environments

One of the important issues for the iodine–sulfur process is the corrosion of the construction materials since highly corrosive chemicals such as halogens and sulfur compounds would be used in the iodine–sulfur process as working fluids in the various phases and over a wide temperature range.

Figure 21.1 shows a brief scheme and the corrosive environment of the iodine–sulfur process. The Bunsen reaction, which provides a springboard for the thermochemical cycle, is an exothermic reaction with absorbing sulfur dioxide in an aqueous phase. A poly-hydriodic

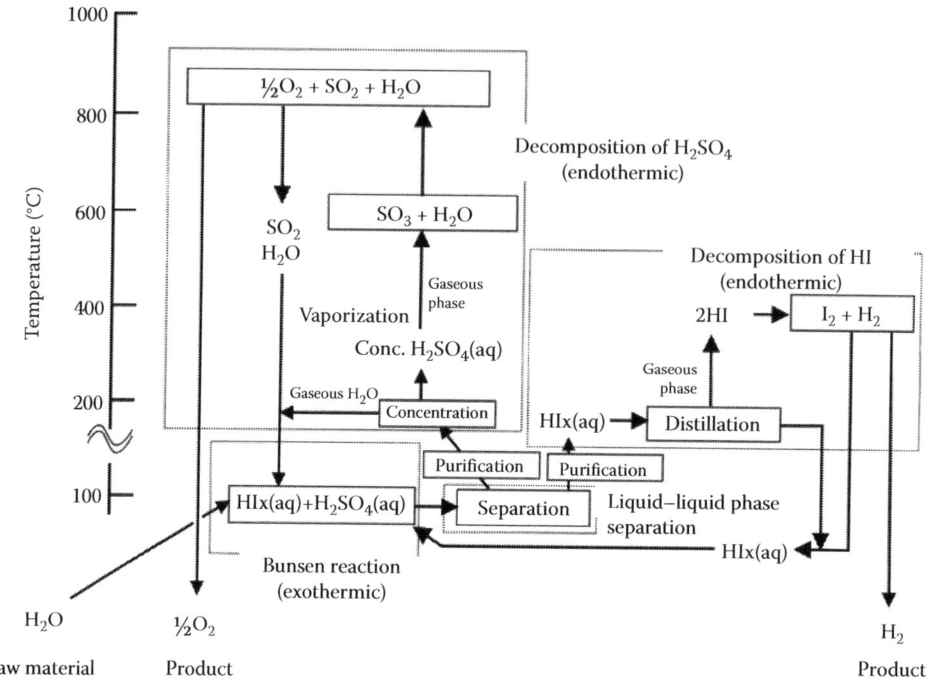

FIGURE 21.1
Brief process scheme of iodine–sulfur process; rough temperature standards are illustrated on corresponding each unit operations. Temperature of vaporization and distillation can vary with each operating pressure. HIx (aq) and H_2SO_4 (aq) represent poly-hydriodic acid and sulfuric acid produced by Bunsen reaction, in which molar fractions are about $HI:I_2:H_2O = 1:3.8:5.3$ and $H_2SO_4:H_2O = 1:4.1$. Concentrated H_2SO_4 is $H_2SO_4:H_2O = 1:0.58$. (From Kubo, S. et al. 2004. *Nucl. Eng. Des.* 233: 347–354. With permission.)

acid, hydriodic acid with dissolved iodine, and sulfuric acid, which are formed by the Bunsen reaction, are separated by a liquid–liquid phase separation phenomenon that occurs in the presence of excess iodine, and divided into two solutions with a clear boundary. After purification, hydrogen iodide is separated from the poly-hydriodic acid by distillation, and catalytically decomposed to produce hydrogen. Similarly, the separated sulfuric acid is purified, concentrated by removing water, vaporized, and decomposed to produce oxygen. The decomposition reaction proceeds endothermically in two stages. First, the sulfuric acid is decomposed into sulfur trioxide and gaseous water at about 600°C. Second, the sulfur trioxide is decomposed into sulfur dioxide and oxygen in the presence of a solid catalyst at higher temperatures. As a consequence, these chemical processes form a chemical cycle which performs as an energy converter from the heat to hydrogen. In this way, the iodine–sulfur process for hydrogen production takes place in severe corrosive environments.

21.2 Construction Material Requirements

A typical corrosive environment is illustrated in Figure 21.2. The materials should satisfy the requirements of corrosion resistances, thermal resistances, material strengths, and

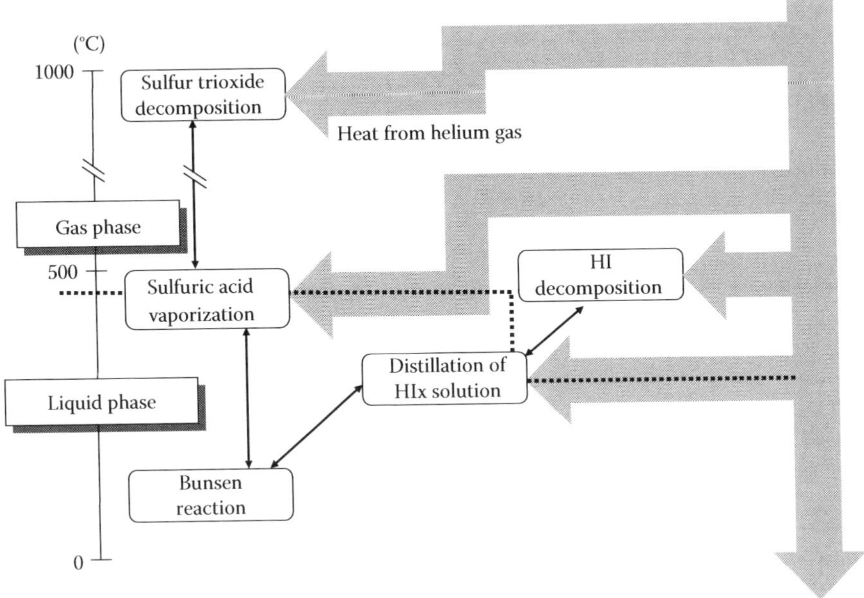

FIGURE 21.2
Typical corrosive environments in which equipment materials are required to be resistant. Environments cover from liquid phase to gaseous phase in wide temperature range. Corrosive fluids consist of sulfur compounds or iodine compounds, or both. Required heat to drive process probably supplied from high-temperature and high-pressure helium gas.

practicable manufacturing. The corrosive environment is roughly classified as liquid and gaseous phase. A summary of the corrosive environments conditions are as follows: (1) gas phase, sulfuric trioxide, sulfuric dioxide, oxygen, water, bubble point to 850°C; (2) gaseous phase, hydrogen iodide, iodine, hydrogen, water: bubble point to 500°C; (3) liquid phase, sulfuric acid, sulfuric acid with contaminated iodine and hydrogen iodide, room temperature to bubble point; (4) liquid phase, poly-hydriodic acid, poly-hydriodic acid with contaminated sulfuric acid, room temperature to bubble point. The environments of liquid phase are more severe than those of the gaseous phase. On the other hand, the materials for the gaseous phase require high thermal resistance. The endothermic heat for sulfur trioxide decomposition is supplied by the high temperature pressurized helium gas, and therefore the materials for the decomposition step should endure high-temperature creep behaviors. A design consideration for the hydrogen embrittlement of the materials is needed in both the gaseous and liquid phases. Structural design contrivances are required to apply brittle materials to equipments in industrial plants.

21.3 Material Screening for Heat and Corrosion Resistance

Candidates of the materials of large-scale chemical plants should be selected based on previous researches from the viewpoint of the material requirement and commercial availability. Till now, many investigations and screening tests of structural materials have been conducted [3–14] in representative process environments in order to fabricate corrosion-resistant equipment. The following section mainly discusses on the efforts made by JAEA.

It is a common practice to evaluate the corrosion-resistance performances of construction materials from corrosion weight-losses and corrosion penetration-depths of the material surfaces. One of the general criteria [15,16] for the corrosion resistances appears in Table 21.1. In the case of general corrosion behavior, the materials with the corrosion rate less than 0.05 mm year^{-1} are rated as good durability, while the materials with the corrosion rate more than 1.0 mm year^{-1} are rated as poor, and not acceptable for the construction material used for industrial plants.

Figure 21.3 shows an outline of the candidate construction materials. For the gas phase environment, refractory alloys that have been used in conventional chemical plants showed good corrosion resistance; there are few concerns on structural materials. On the other hand, the liquid-phase environments are more severe than the gas-phase environments. Figure 21.4 shows a typical example of a specimen which is examined in a corrosion test; these harsh environments seriously harm the construction materials. For the liquid phase, exclusive metals, glasses, and ceramics have corrosion resistance. The following are brief details of the construction materials for each corrosive environment.

21.3.1 Environments of Gaseous Sulfuric Acid Decomposition

The gaseous sulfuric acid decomposition produces oxygen in catalytic layers at around 800–900°C. Tables 21.2 and 21.3 show corrosion test results [3] of materials in the front and the rear of the catalytic layers. The test pieces were placed in a heated quartz tube where the sulfuric acid (98 wt%) runs through in order to set the corrosive environment. From the result of 50 h exposure test at both 800 and 900°C, thin and closeness scales were observed on surface of the all test pieces, however, most of the surface conditions at the materials were sound after descaling. The Hastelloy® B2, the JIS SUS304, and the zirconium (Zr) were remarkably corroded, and these materials showed the intergranular corrosion. Materials with better results in 50 h test were examined for 1000 h at 850°C; higher penetration of the front were observed in comparison with that of the rear for of the catalyst differences between the front and rear of the catalyst on each material. The Incoloy® 800, the Hastelloy

TABLE 21.1

Acceptability Criteria of Corrosion Rates for Plant Construction Materials, which are Used for Representing Corrosion Resistance

Rating of Corrosion Resistance	Criteria of Rates		
	Weight Loss[a,c] [15] (g m^{-2} day^{-1})	Corrosion[a,c] [16] (mm year^{-1})	Penetration[b] [15] (mm year^{-1})
Good	0–10.0	0–0.05	0.05–0.1
Fair	1.0–10	0.05–1.0	0.1–1.0
Poor	> 10	> 1.0	1.0

Source: Adapted from Hatano, S., 1980. *Kagaku-sochi zairyo taisyoku-hyo, Kagaku-kougyo sya*, p. 194 and *Kagaku-sochi benran, Kagaku-kogaku kyokai*, p. 499, 1970.

[a] General corrosion.

[b] Maximum value.

[c] Conversion factor of weight loss (g m^{-2} day^{-1}) into corrosion depth (mm year^{-1}) is 0.365 ρ^{-1}, ρ is a density (g cm^{-3}) of materials.

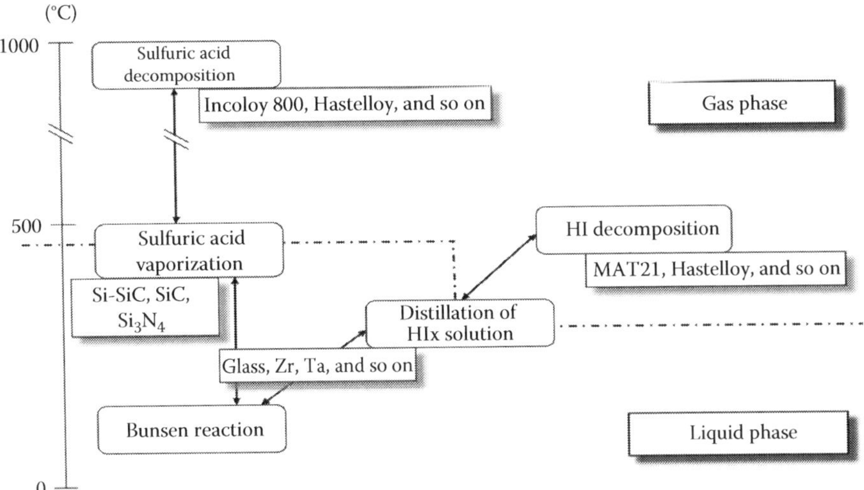

FIGURE 21.3
Candidate materials in prototypical environments of iodine-sulfur process (MAT21 is Ni–Cr–Mo–Ta alloy made by Mitsubishi Materials Corp). Refractory alloys, exclusive metals, glasses, and ceramics showed good corrosion resistance. (Adapted from Kubo, S. et al., *Proceedings of International Conference on Advances in Nuclear Power Plants* (ICAPP'06), Paper 6170, 2006.)

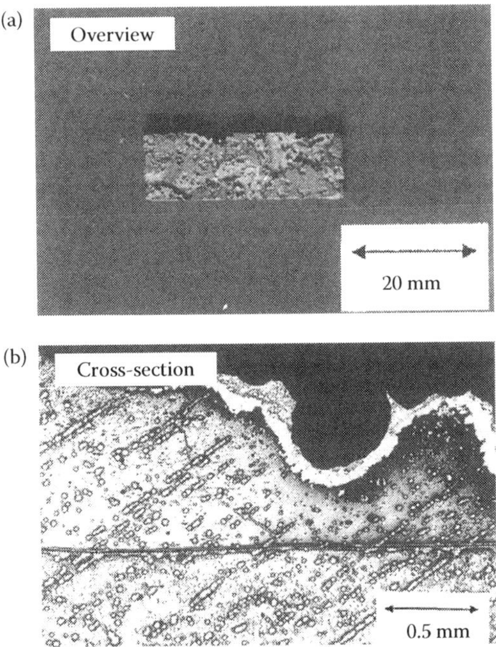

FIGURE 21.4
Appearance of Fe–Si alloy specimen which contained 9 wt% Si in boiling 95 wt% sulfuric acid for 25 h at atmospheric pressure. Specimen suffered from severe corrosion rate of 65 g m^{-2} h^{-1}. (From Ioka, I. et al., 1997. *Journal of Society Materials Science, Japan*, 46: 1041–1045 (in Japanese). With permission.)

TABLE 21.2

Corrosion Rates of Materials in Gaseous Sulfuric Acid Decomposition Reaction for 50 h at Temperatures of 800 and 900°C

Material[a]	Maximum Penetration[b] ($mm\ year^{-1}$), Exposure Duration 50 h			
	800°C		900°C	
	Gaseous $H_2SO_4 + H_2O + SO_3$[c] (Before Catalyst)	Gaseous $H_2SO_4 + H_2O + SO_3 + SO_2 + O_2$[d] (After Catalyst)	Gaseous $H_2SO_4 + H_2O + SO_3$[c] (Before Catalyst)	Gaseous $H_2SO_4 + H_2O + SO_3 + SO_2 + O_2$[d] (After Catalyst)
Incoloy 800	3.5	0	6.1	5.3
Hastelloy XR	4.4	0	14	0
Hastelloy C276	1.8	0	6.1	0.7
JIS SUS329J1	3.5	0	8.8	1.8
Inconel 625	0.9	0	7.0	0
JIS SUS304	7.0	0	8.8	1.4
JIS SUS310S	3.5	0	7.9	1.4
Incoloy 825	2.6	0	3.5	2.6
SZ				
(Ni–27Cr–5Mo–5 W)	4.4	0	17.5	0
NSC1				
(Ni–20Cr–13 W)	0	0	8.8	0
Inconel 600	5.3	0	11	2.6
IN 657				
(50Cr–50Ni–Nb)	8.8	0.5	7.0	3.5

SSS113MA (Ni–23Cr–18 W)	2.6	0	17.5	2.6
JIS SUS316L	5.3	0	8.8	7.0
Hastelloy G3	1.8	0	11	0
JIS SUS405	3.5	0	28.9	29.8
Zr	12	17.5	43.8	47.3
Hastelloy B2	140	87.6	40.3	29.8

Source: From Onuki, K. et al. 1993. *Journal of the Hydrogen Energy Systems Society of Japan*, 18: 49–56 (in Japanese). With permission.

Note: Materials were exposed to both front and rear of the catalytic layer.

[a] Specimen configuration, $10 \times 40 \times 3–5$ mm^3.

[b] Values were measured by microscopic cross-section observations; thicknesses of surface scales are included.

[c] Vaporized sulfuric acid (98 wt%) were supplied to test field in which specimens were placed; atmospheric pressure.

[d] Decomposition rates of sulfuric acid decomposition reaction were about 70% though they vary depending on temperature; atmospheric pressure.

TABLE 21.3

Corrosion Rates of Materials in Gaseous Sulfuric Acid Decomposition Reaction for 1000 h at Temperature of 850°C

	Maximum Penetration[b] (mm year^{-1}), Exposure Duration 1000 h, 850°C	
Material[a]	**Gaseous $H_2SO_4 + H_2O + SO_3$[c] (Before Catalyst)**	**Gaseous $H_2SO_4 + H_2O + SO_3 + SO_2 + O_2$[d] (After Catalyst)**
Incoloy 800	0.79	0.53
Hastelloy XR	0.79	0.61
Hastelloy C276	0.83	0.35
JIS SUS329J1	0.88	0.53
Inconel 625	0.88	0.70
JIS SUS304	1.01	0.74
JIS SUS310S	1.05	0.61
Incoloy 825	1.10	0.61
SZ		
(Ni–27Cr–5Mo-5 W)	1.14	0.61
NSC1		
(Ni–20Cr–13 W)	1.31	0.11
Inconel 600	1.49	0.88
IN 657		
(50Cr–50Ni–Nb)	2.10	0.53
SSS113MA		
(Ni–23Cr–18 W)	1.75	0.11

Source: From Onuki, K. et al. *Journal of the Hydrogen Energy Systems Society of Japan*, 18: 49–56 (in Japanese). With permission.

Note: Materials were exposed to both front and rear of the catalytic layer.

[a] Specimen configuration, 10 × 40 × 3–5 (mm^3).

[b] Values were measured by microscopic cross-section observations; thicknesses of surface scales are included.

[c] Vaporized sulfuric acid (98 wt%) were supplied to test field in which specimens were placed; atmospheric pressure.

[d] Decomposition rates of sulfuric acid decomposition reaction were about 70% though they vary depending on temperature; atmospheric pressure.

XR, and the Hastelloy C276 showed relatively good corrosion resistance with the penetration of 1.0 mm year^{-1} or less. In addition, high-temperature tensile test was conducted for the candidate materials under the sulfur trioxide decomposition environment. The results showed that the tensile strengths did not decrease after the 1000 h exposure [17].

21.3.2 Environments of Gaseous Hydrogen Iodide Decomposition

Gaseous hydrogen iodide decomposition produces hydrogen in catalytic layers at around 400°C. Tables 21.4 and 21.5 show corrosion test results [5] of materials in a gaseous $HI–I_2–H_2O$ mixture which was intended to represent the environment in front of the catalytic layer. The test pieces were placed in a heated vessel where vaporized poly-hydriodic acid runs through in order to set the corrosive environment.

The test of 96 h exposure at 200, 300, and 400°C resulted: for the ceramics of the Si_3N_4 and the SiC, no evidences of corrosion were observed; for zirconium, tantalum, and titanium, the scaling, discolorations, and weight losses were hardly observed; for the Hastelloy C276, scales were observed at each temperature, while surface conditions were sound after

TABLE 21.4

Corrosion rates of Materials in Gaseous Hydrogen Iodide with Iodine for 96 h at Temperatures of 200, 300, and 400°C

	Corrosion Rate[b] (mm $year^{-1}$), Exposure Duration 96 h Gaseous $HI + I_2 + H_2O$[c]		
Material[a]	**200°C**	**300°C**	**400°C**
Si_3N_4	0.1	0	0
SiC	0	0	0
Zr	0	0.1	0
Ta	0	0	0
Ti	0	0	0
Hastelloy C276	0	0	0.1
Inconel 600	0.1	0.1	0.3
Incoloy 800	0.7	0.3	0.4
JIS SB42 (carbon steel)	0.7	0.5	0.6
JIS SCMV3 (Fe–1 Cr–0.5 Mo)	0.4	0.4	0.6
JIS SUSXM27 (Fe–26Cr–1 Mo)	0.6	0.5	0.6
JIS SUS444	0.4	0.6	0.6
JIS SUS405	0.6	0.5	0.6
Incoloy 825	0.2	0.2	0.7
JIS SUS329J1	0.9	0.4	0.7
JIS SUS310S	1.2	0.3	0.7
JIS SUS316L	0.4	0.3	0.7
JIS SUS304	0.8	0.4	0.8

Source: Onuki, K. et al. 1997. *Zairyo-to-Kankyo* 46: 113–117. With permission.

Note: Test was intended to represent the surroundings in front of the catalytic layer for hydrogen production reaction.

[a] Specimen configuration, 10 × 30 × 3–10 (mm^3).

[b] Values were calculated from weight losses.

[c] Molar ratio, $HI:I_2:H_2O = 1:1:6$; atmospheric pressure.

descaling; for other materials, scales and roughened surfaces were observed. Considering the 96 h exposure test, candidate materials were examined for 1000 h at 200, 300, and 400°C. This test results showed: for the titanium in 400°C, there were no weight losses, while thin scale blackened surface was observed after descaling; the Hastelloy C276 and the Inconel® 600 were in second place; corrosion rates of all materials were less than 0.2 mm $year^{-1}$, and showed good corrosion resistance. Since the corrosion rates of 1000 h test were less than those of the 96 h test, protective effects of the scales are suggested in spite of high-vapor pressures of common metal iodides.

A corrosion test considering hydrogen embrittlement was conducted for SUS316, MAT21 (Ni-alloy, 19Cr–19Mo–1.8 Ta), Ti, Ta [18]. The specimens were placed in a heated quartz tube where the mixture gas consisted of HI, I_2, H_2O, H_2 runs through in order to set the corrosive environment. The results of corrosion rate are shown in Figure 21.5. A hydride in the titanium was observed by a cross-section observation with a laser microscope. For tantalum, ductility deterioration in a surface thickness of 200 μm was observed. The Ni-alloy showed good corrosion resistance, and hydrogen embrittlement was hardly observed.

TABLE 21.5

Corrosion Rates of Materials in Gaseous Hydrogen Iodide with Iodine for 1000 h at Temperatures of 200, 300, and 400°C

	Corrosion Rate[b] (mm year^{-1}), Exposure Duration 1000 h Gaseous HI + I_2 + H_2O[c]		
Material[a]	**200°C**	**300°C**	**400°C**
Ti	0.0 NS, N	0.0 NS, N	0.0 GR, N
Hastelloy C276	0.0 BK, N	0.0 DB, N	0.05 BK, R
Inconel 600	0.0 BK, N	0.05 DB, N	0.1 BK, R
JIS SUS316L	0.1 BK, N	0.2 DB, R	0.2 BR, R
JIS SUS444	0.1 BK, P	0.2 DB, R	0.2 GB, R
JIS SCMV3 (Fe–1 Cr–0.5 Mo)	0.1 BK, R	0.1 DB, R	0.2 BK, R

Source: Onuki, K. et al. 1997. *Zairyo-to-Kankyo* 46: 113–117. With permission.

Note: Test was intended to represent the surroundings in front of the catalytic layer for hydrogen production reaction.

Scale appearance—NS, no scale; BK, black; DB, dark brown; GB, greenish brown; BR, brown; GR, gray.

Surface appearance after descaling—N, no change; P, pitted; R, roughened.

[a] Specimen configuration, 10 × 30 × 3–10 (mm^3).

[b] Values were calculated from weight losses.

[c] Molar ratio, HI:I_2:H_2O = 1:1:6; atmospheric pressure.

21.3.3 Environments of Bubbling Sulfuric Acid

Bubbling sulfuric acid environments, in which a gaseous and a liquid phase coexist, of various concentrations appear in the purification, concentration, and vaporization processes. Tables 21.6 and 21.7 show corrosion test results [3] of materials in such bubbling conditions. Corrosive environments were set with a heated glass flask, which contained

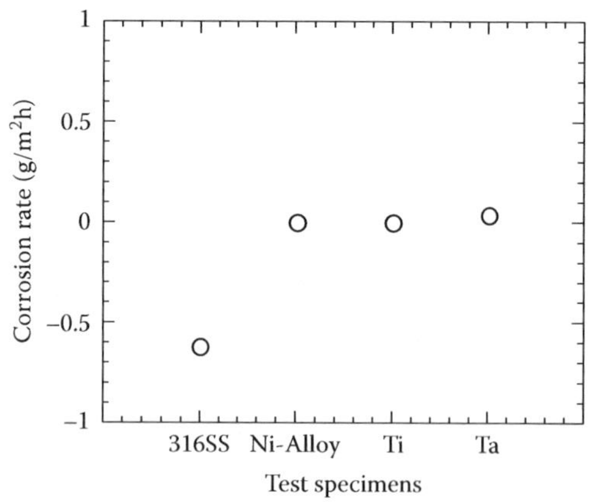

FIGURE 21.5
Corrosion rates of materials in simulated HI decomposition environment considering hydrogen embrittlement, gaseous mixture of HI, I_2, H_2O, and H_2 at 450°C for 100 h. Vaporized HI–I_2–H_2O solution (molar ratio, 1:1:6) was blended with gaseous hydrogen in 20% to make corrosive gas. Corrosion rates were calculated by weight-loss measurements. (From Iwatsuki, J., 2008. *Spring Meeting of Atomic Energy Society of Japan*, Osaka, Japan, March 2008. With permission.)

TABLE 21.6

Corrosion Rates of Materials in Bubbling Sulfuric Acid Environments, in Which Gaseous and Liquid Phase Coexist, for 24 h in 40 and 98 wt% Sulfuric Acid

	Corrosion Rate[b] (mm year^{-1}), Exposure Duration 24 h							
	H_2SO_4, 40 wt%[c]				H_2SO_4, 98 wt%[dc]			
	HI, 0 wt%; I_2, 0 wt%		HI, 1 wt%; I_2, 0.5 wt%		HI, 0 wt%; I_2, 0 wt%		HI, 0.5 wt%; I_2, 0.2 wt%	
Material[a]	**Gas Phase[e]**	**Liquid Phase[c]**	**Gas Phase[e]**	**Liquid Phase[c]**	**Gas Phase[e]**	**Liquid Phase[d]**	**Gas Phase[e]**	**Liquid Phase[d]**
High Si cast iron (15 wt% Si)	0.5 S	0.5 S	12 R	3.2 S	0.1 S	0.1 S	0 S	0.1 S
Ta	0.0 S	0.0 S	0.0 S	0.0 S	2.3 R	27 R	1.3 R	15 R
Hastelloy B2	0.2 S	0.2 S	7.7 R	18 R	8.1 R	63 R	8.2 C	56 C
Zr	0.0 S	0.0 S	0.1 S	0.0 S	38 C	126 C	16 C	83 C
Pb	0.2 S	9.4 S	3.7 S	1.5 S	D C	D C	D C	D C
20Cr–25Ni–6 Mo	2.8 R	13 R	20 R	0.2 S	—	—	—	—
Hastelloy G3	0.2 R	1.9 R	14 R	0.9 S	—	—	—	—
Hastelloy C276	2.4 S	0.3 S	17 R	19 R	—	—	—	—
Monel 400	0.1 R	0.1 S	91 C	23 C	—	—	—	—
Inconel 625	0.7 S	7.6 R	16 R	24 R	—	—	—	—
Inconel 825	0.5 R	1.6 R	112 C	190 C	—	—	—	—
Carpenter 20Cb	0.5 S	0.1 S	112 C	218 C	—	—	—	—
JIS SUS316L	10 C	D C	105 C	277 C	—	—	—	—
Inconel 600	19 R	191 C	132 C	323 C	—	—	—	—
Inconel 800	356 C	62 C	116 C	D C	—	—	—	—
JIS SUS304	4.5 C	D C	57 C	D C	—	—	—	—
JIS SUS310S	15 C	D C	65 C	D C	—	—	—	—

Source: Onuki, K. et al. 1993. *Journal of the Hydrogen Energy Systems Society of Japan*, 18: 49–56 (in Japanese). With permission.

Note: Sulfuric acids with and without HI and I_2 were used for corrosive solutions. Symbol "D" denotes that specimens were dissolved and symbol "—" denotes that tests were not carried out. Surface appearances—S, smooth; R, rough; C, severely corroded.

[a] Specimen configuration, 10 × 40 × 3–5 (mm^3).

[b] Values were calculated from weight losses.

[c] Bubbling sulfuric acid solutions, roughly 120°C; atmospheric pressure.

[d] Bubbling sulfuric acid solutions, roughly 320°C; atmospheric pressure.

[e] Atmospheres on bubbling solutions.

sulfuric acids, with a condenser. Test pieces were placed at both the liquid phase and the gaseous phase in the flask. For the 20 h exposure test using 40 wt% sulfuric acids, tantalum and zirconium showed good corrosion resistance, regardless of the presence or absence of HI and I_2. In case of the absence of HI and I_2, Monel®, Carpenter 20 Cb, Hastelloy B2, high-silicon cast-iron, and lead showed good corrosion resistance, however, the corrosion resistances of these materials except lead remarkably declined in presence of HI and I_2. Materials with better results in the test using 40 wt% sulfuric acids were examined for 24 h in 98 wt% sulfuric acids. Tantalum and zirconium were severely corroded. High-silicon cast-iron showed good corrosion resistances. Considering the 24 hours exposure tests, candidate materials were examined for 800 h in 40, 70, and 98 wt% sulfuric acid. Zirconium (and probably tantalum) in the bubbling conditions in 40 and 70 wt% sulfuric

TABLE 21.7

Corrosion Rates of Materials in Bubbling Sulfuric Acid Environments for 800 h in 40, 70, and 98 wt% Sulfuric Acid

	Corrosion Rate[b] (mm year^{-1}), Exposure Duration 800 h											
	H_2SO_4, 40 wt%[c]				H_2SO_4, 70 wt%[d]				H_2SO_4, 98 wt%[e]			
	HI, 0 wt%; I_2, 0 wt%		HI, 1 wt%; I_2, 0.5 wt%		HI, 0 wt%; I_2, 0 wt%		HI, 0.5 wt%; I_2, 0.2 wt%		HI, 0 wt%; I_2, 0 wt%		HI, 0.5 wt%; I_2, 0.2 wt%	
Material[a]	Gas Phase[f]	Liquid Phase[c]	Gas Phase[f]	Liquid Phase[c]	Gas Phase[f]	Liquid Phase[d]	Gas Phase[f]	Liquid Phase[d]	Gas Phase[f]	Liquid Phase[e]	Gas Phase[f]	Liquid Phase[e]
High Si cast iron (15 w% Si)	—	—	—	—	—	—	—	—	0	0	0	0
Ta	—	—	—	—	—	—	—	—	1.6	P	1.2	P
Hastelloy B2	0.1	0.1	1.5	D	2.3	1.2	2.5	D	—	—	—	—
Zr	0	0	0	0	0.1	0.2	0	0.2	—	—	—	—
Pb	0.3	0.5	0.9	0.1	6.1	D	4.2	6.1	—	—	—	—

Source: From Onuki, K. et al. 1993. *Journal of the Hydrogen Energy Systems Society of Japan*, 18: 49–56 (in Japanese). With permission.

Note: Sulfuric acids with and without HI and I_2 were used for corrosive solutions. Symbol "D" denotes that specimens were dissolved and symbol "P" denotes that specimens went to pieces. Symbol "–" denotes that tests were not carried out.

[a] Specimen configuration. 10 × 40 × 3–5 (mm^3).

[b] Values were calculated from weight losses.

[c] Bubbling sulfuric acid solutions, roughly 120°C; atmospheric pressure.

[d] Bubbling sulfuric acid solutions, roughly 170°C; atmospheric pressure.

[e] Bubbling sulfuric acid solutions, roughly 320°C; atmospheric pressure.

[f] Atmospheres on bubbling solutions.

acid showed good corrosion resistances, regardless of the presence or absence of HI and I_2. In the bubbling condition in 98 wt% sulfuric acid the test resulted: high-silicon cast-iron showed good corrosion resistances, regardless of the presence or absence of HI and I_2; hydrogen embrittlement occurred in tantalum, since hydrogen was observed in the specimens that broke into pieces.

Regarding higher pressure conditions at near a bubbling temperature of sulfuric acid, results of material immersion test at 2 MPaG [14] are shown in Table 21.8. Test pieces were encapsulated in quartz ampoules with 5 mL sulfuric acid. High-pressure test conditions were achieved by using a gas-pressurized and thermostated autoclave in which the quartz ampoules were placed. In the cases of the Si–SiC and Si_3N_4, formation of silicon oxide layers, whose thickness was several micrometers, were observed on the surface of test pieces by electron probe microanalyser (EPMA) line analysis. In the case of the SiC, the silicon oxide layer was also detected, though it did not contain sulfur. As for the Fe–Si, the prepared sample suffered from corrosion and exhibited weight loss. Also, microscopic observation revealed a formation of a number of cracks. Annealing was found to be effective to improve the corrosion resistance. The fact suggests that the stress imposed on the sample preparation was the main reason for the corrosion. In the previous experiments performed at bubbling conditions under atmospheric pressure, the Fe–Si containing silicon of about 20% showed good corrosion resistance, however, the present results suggested that Fe–Si might not be suitable for use as the structural materials in this high pressure and temperature environments. As for the Sandvik SX®, the only ductile material among the tested materials, it suffered from uniform corrosions. The surface layer formed after the immersion test contained Si, O, Fe, and S. The corrosion rate could be decreased by preoxidation

TABLE 21.8

Corrosion Rates of Materials in Pressurized Sulfuric Acid (2 MPaG) at Near Bubbling Temperature

	Corrosion Rate[b] ($g\ m^{-2}\ day^{-1}$) Aqueous H_2SO_4 95 wt%[c]	
Material[a]	**Exposure Duration 100 h**	**Exposure Duration 1000 h**
SiC	– 2	– 0.05
Si–SiC		
(SiC, 80 wt%; Si, 20 wt%)	0	– 0.1
Si_3N_4	0	– 0.2
Fe–Si		
(Si, 19.7 wt%, as-prepared)	26	3.1
Fe–Si		
(Si, 19.7 wt%; annealed[d])	– 2.9	1.6
Sandvik SX		
(Fe–19Ni–17Cr–5Si–2Cu, preoxidized[e])	– 6.7	23[f]

Source: From Kubo, S. et al. 2003. *Proc. of 2nd Topical Conf. on Fuel Cell Technology, AIChE 2003 Spring National Meeting*, New Orleans, AIChE, NY. With permission.

[a] Specimen configuration, 4 × 4 × 40 (mm^3).
[b] Values were calculated from weight losses.
[c] Test conditions: 460°C, 2 MPa.
[d] Annealed at 1100°C under vacuum for 100 h.
[e] Oxidized in air at 800°C for 90 h.
[f] Quartz ampoule containing specimen broke after 800 h.

treatment described in the table caption, although the corrosion rate was not small enough to use as the structural material in contact with the process solution.

21.3.4 Environments of Poly-Hydriodic Acid Distillation

Distillations of the poly-hydriodic acid separate gaseous hydrogen iodide from the poly-hydriodic acid, which is to be decomposed to produce hydrogen. Table 21.9 shows corrosion test results [4] of materials in such liquid phase at pressurized conditions. Test pieces were encapsulated in glass ampoules with 20 mL of a poly-hydriodic acid or an iodine solution. High-pressure test conditions of 20 atm were achieved by using a gas-pressurized and thermostated autoclave containing silicon-oil in which the quartz ampoules were placed. From the test results of 20 h exposure, tantalum, niobium, zirconium, and silicon carbide showed good corrosion resistances, while the other materials were severely corroded. The materials with better results in 20 h test were subjected to 100 h exposure test. In the poly-hydriodic acid, the weight of tantalum and zirconium specimens remained unchanged; weight of the niobium specimen was slightly reduced and surface was slightly roughened. Hydrogen contents of these materials before and after the exposure test were unchanged. For SiC, evidences of surface attacks at the micro level

TABLE 21.9

Corrosion Rates of Materials in Poly-Hydriodic Acid and Iodine Solution at 20 bar for 20 and 100 h

	Corrosion Rate[b] (mm year^{-1})			
	Exposure Duration 20 h		Exposure Duration 100 h	
Material[a]	Poly-Hydriodic Acid[c]	Iodine Solution[d]	Poly-Hydriodic Acid[c]	Iodine Solution[d]
Ta	0	0	0 S BR	0 S
Nb	1.1	0	0.5 R	0.0 S BK
Zr	0	3.5	0 S PL	0.6 P
SiC	0	0	I N[e]	0.1 N[e]
Ti	18	D	—	—
ZrO_2	269[f]	0[g]	—	—
Hastelloy C276	334	67	—	—
Incoloy 825	D	43	—	—
JIS SUS316	D	140	—	—
SN-3 (Fe–11Cr –17Ni–6Si)	D	D	—	—

Source: Onuki, K. et al. 1994. *Journal of Hydrogen Energy Systems Society of Japan*, 19: 10–16 (in Japanese). With permission.

Note: Tests were intended to represent environments in distillation of poly-hydriodic acid. "D" denotes that specimens were dissolved and "I" denotes that specimens gained weight. "—" denotes that tests were not carried out. Surface appearances—N, no change; S, smooth; R, rough; P, pitting; BR, brown; PL, purple; BK, black.

[a] Specimen configuration, column shape, $d = 5, l - 5$ (mm).

[b] Values were calculated from weight losses.

[c] Molar ratio, $HI:I_2:H_2O = 1:0.38:6.3$, 240°C, 20 atm.

[d] Molar ratio, $HI:I_2:H_2O = 1:94:5$, 300°C, 20 atm.

[e] Micro-level corrosions were observed by microscopic cross-section observations.

[f] Specimens changed into porous form, by microscopic cross-section observations.

[g] Surfaces changed into porous form, by microscopic cross-section observations.

were observed by a cross-sectional observation. In the iodine solution, weights of tantalum and niobium specimens were unchanged; zirconium specimen suffered from slight pitting corrosion. For SiC, evidences of surface attacks at the micro level were observed.

Table 21.10 shows corrosion test results [19] of materials in boiling hydriodic acid. Tests were intended to represent environment in a condenser in which iodine is already removed from the poly-hydriodic acid by the distillation process. Corrosive environments were set within a heated glass vessel, which contained the hydriodic acid, with a condenser; and the test pieces were placed in the solution. Tantalum, zirconium, niobium, SiC, and ZrO_2 showed good corrosion resistances, and Hastelloy C276 showed relatively good corrosion resistance.

21.3.5 Environments in Bunsen Reaction Solution

Bunsen reaction produces hydriodic acid and sulfuric acid which separate spontaneously into two types of solutions by a liquid–liquid phase separation. An upper phase is rich in sulfuric acid; a lower phase is rich in hydrogen iodide with iodine. Corrosive environments were set within a heated glass-vessel, which contained corrosive solutions, with a condenser; and the test pieces were placed in liquid phase. For the lower phase, preliminary examinations of materials were conducted using a hydriodic acid and poly-hydriodic acid without the sulfuric acid at 95°C for 100 h. Table 21.11 shows results of corrosion tests [4]. Tantalum, zirconium, SiO_2, SiC, and Si_3N_4 were unchanged in weights and appearances. Titanium showed relatively good result. Organic materials, like poly-perfluoroalkoxy copolymer resin (PFA), showed good corrosion resistance; and poly-phenylene sulfide (PPS) was in second place. Overall, the environment in the poly-hydriodic acid was more severe than that of the hydriodic acid. Table 21.12 shows the results of corrosion

TABLE 21.10

Corrosion Rates of Materials in Boiling Hydriodic Acid for 20 and 100 h Exposure

	Corrosion Rate[b] (mm year^{-1}), Hydriodic-Acid[c]	
Material[a]	**Exposure Duration 20 h**	**Exposure Duration 100 h**
Ta	0	0
Nb	0	0
Zr	0	0
SiC	0	0.1
Ti	39.9 P	—
ZrO_2	0	—
Hastelloy C276	0.5	0.2
Incoloy 825	4.7	8.5
JIS SUS316	14.8	—
SN-3 (Fe–11Cr–17Ni–6Si)	129	

Source: From Tanaka N., Iwatsuki, J., Kubo, S. et al. 2009. *Proc. of 2009 International Congress on Advances in Nuclear Power Plants* (ICAPP2009), Tokyo, Japan, May 10–14, Paper 9291. With permission.

Note: Tests were intended to represent environments in condenser where iodine is removed from poly-hydriodic acid by distillation. "—" denotes that tests were not carried out. Surface appearances—P, pitting.

[a] Specimen configuration, column shape, $d = 5$, $l = 10$ (mm).

[b] Values were calculated from weight losses.

[c] Molar ratio, $HI:H_2O = 1:5.4$; bubbling state, 136°C; atmospheric pressure.

TABLE 21.11
Corrosion Rates of Materials in Hydriodic Acid and Poly-Hydriodic Acid at 95°C for 100 h Exposure

	Corrosion Rate[b] (mm year^{-1}), Exposure Duration 100 h	
Material[a]	**Hydriodic Acid[c]**	**Poly-Hydriodic Acid[d]**
Ta	0 S	0 S
Zr	0 S	0 S
Ti	0.2 S	0.1 S
Hastelloy C276	0.1 S	2.8 R
High Si cast iron (15 wt% Si)	7.9 RS	4.5 R
Inconel 625	0.2 S	13.5 R
Incoloy 825	1.4 RM	D
Carpenter 20Cb	4.1 RM	D
Monel 400	D	D
JIS SUS304	WS	D
Pb	D	D
SiO_2	0 S	0 S
SiC	0 S	0 S
Si_3N_4	0 S	0 S
PFA	0 LBR	0 BR
PPS	0.8 S	1.5 S
Vinylester CFRP[e]	3.2 WP	4.0 DL
Polyimido CFRP[e]	4.8 WP	6.3 WP
Polyimido	6.7 WP	16 R

Source: Onuki, K. et al. 1994. *Journal of the Hydrogen Energy Systems Society of Japan*, 19: 10–16 (in Japanese). With permission.

Note: D: denotes that specimens were dissolved. Surface appearances—S, smooth; R, rough; RS, rough (severely); RM, rough (mildly); WS, wastage (severely); WP, wastage (partially); BR, brown; LBR, light brown; DL, delamination.

[a] Specimen configuration, 10 × 40 × 3–5 (mm^3).
[b] Values were calculated from weight losses.
[c] Molar ratio, $HI:H_2O = 1:5.4$ (57 wt%); 95°C; atmospheric pressure.
[d] Molar ratio, $HI:I_2:H_2O = 1:1:6$; 95°C; atmospheric pressure.
[e] Carbon fiber-reinforced plastics.

tests using simulated Bunsen reaction solutions at 120°C for 100 and 1000 h [4,5]. In the poly-hydriodic acid with the sulfuric acid, tantalum and zirconium were unchanged in weights and appearances; titanium showed relatively good result. Despite good results, careful applications of zirconium are required since severe corrosion at 125°C in a poly-hydriodic acid have been reported [20]. For ceramics, although appearances were unchanged, the specimens of SiC and Si_3N_4 gained weight significantly, moreover iodine segregations on the surface by EPMA analyses and slight roughened surface by scanning electron microscope (SEM) observations were observed. These results suggest that binder materials of SiC and Si_3N_4 may flow out and the test solution may penetrate the ceramics. For PFA specimens changed color to pink internally, since iodine easily penetrates through the PFA. In sulfuric acid with hydrogen iodide, tantalum and zirconium were unchanged in weights and appearances; lead showed relatively good result. Titanium was not examined since a severe corrosion was found during a trial corrosion test. PPS slightly gained weight of 2% after 100 h immersion. The specimens of SiC and Si_3N_4 lost weight slightly

TABLE 21.12

Corrosion Rates of Materials in Poly-Hydriodic Acid with Sulfuric Acid and Sulfuric Acid with Hydrogen Iodide at 120°C for 100 and 1000 h Exposure

	Corrosion Rate[b] (mm year^{-1})			
	Exposure Duration 100 h		Exposure Duration 1000 h	
Material[a]	Poly-Hydriodic Acid + H_2SO_4[c]	Sulfuric Acid + HI[d]	Poly-Hydriodic Acid + H_2SO_4[c]	Sulfuric Acid + HI[d]
Ta	0 NC	0.005 NC	0 NC	0 NC
Zr	0.02 NC	0.025 NC	0 NC	0 NC
Ti	0.28 NC	—	0.17 NC	—
Pb	—	0.045 NC	—	0.05 R
Hastelloy C276	13 W	21.6 W	—	—
High Si cast iron (15% Si)	35 WS	1.42 R	—	—
SiO_2	0 NC	0.005 NC	IS NC	0 NC
SiC	0 NC	0.075 NC	I NC[e]	0.05 NC
Si_3N_4	0.11 NC	0.085 NC	I NC[e]	0.01 NC
PFA (perfluoroalkoxy copolymer resin)	0 RBR	0 RBR	I RBR[f]	0 NC
PPS (polyphenylene sulfide)	—	I NC	—	I NC

Source: From Onuki, K., et al., 1994. *Journal of the Hydrogen Energy Systems Society of Japan*, 19: 10–16 (in Japanese). With permission and Adapted from Onuki, K., et al., 1997. *Zairyo-to-Kankyo*, 46: 113–117.

Note: Tests were intended to represent environments of Bunsen reaction solutions. "—" denotes that tests were not carried out. "I" denotes that specimens gained weight and "IS" denotes that specimens gained weight slightly. Surface appearances—NC, no change; W, wastage; WS, wastage (severely); RBR, reddish brown.

[a] Specimen configuration, 10 × 40 × 3–5 (mm^3).
[b] Values were calculated from weight losses.
[c] Poly-Hydriodic acid (molar ratio, HI:I_2:H_2O = 1:1:6) + H_2SO_4 (0.1 wt%); 120°C; atmospheric pressure.
[d] Sulfuric acid solution (molar ratio, H_2SO_4:H_2O = 1:5.4, 50 wt%) + HI (0.1 wt%); 120°C; atmospheric pressure.
[e] Decreases of binder ingredients (Al, Mg) and segregation of iodine were observed nearby surfaces by EPMA.
[f] Coloration ranged to central part.

and sulfur segregations and slightly roughened surface were observed by EPMA analyses and SEM observations, respectively. These results suggest that binder materials of SiC and Si_3N_4 may flow out. Tantalum and glass which showed good results can be applied using lining techniques for plant equipment.

21.3.6 Anticorrosion Lining Techniques

One of the promising applications is a soda glass lining on base metallic materials. Two types of soda glass were examined which were used in commercial glass-lining materials supplied by Asahi Techno Glass Corporation and Hakko Sangyo Co. Ltd. Immersion tests were performed in 47, 75 and 90 wt% sulfuric acid for 5–100 h at temperatures ranging from 200 to 400°C, under inert gas atmosphere with pressure of 2 MPaG. As a result, no localized corrosions such as defect formations or pitting corrosions were detected on the surface of the test pieces in all test conditions. Corrosion rates decreased with the elapse of immersion time and was low enough (0.1 mm year^{-1}) after immersion for 60–90 h [21]. Figure 21.6 shows a commercially available glass-lined pipe, which is widely used in corrosive-chemical industries. The possibility of using glass-lined pipes in sulfuric acid

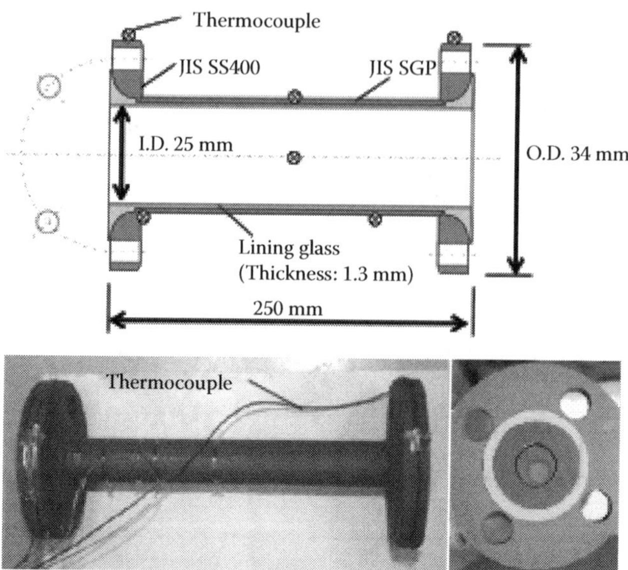

FIGURE 21.6
Commercially available glass-lined pipe (Asahi Techno Glass Corporation). Carbon steel pipe is coated with soda-lime glass. These pipes were examined for thermal cyclic test. (From Onuki, K. et al. 1994. *JAERI-Review*, 94-006. With permission.)

processing was examined [22]. Thermal cyclic tests using a carbon-steel pipe coated with soda-lime glass was conducted with a temperature range from ambient temperature to 325°C and temperature change rate of 100°C year^{-1}. From the results of 20 thermal cycle tests of two pipes, breakages of the pipes were hardly observed on visual inspection. As the maximum temperature increased in steps of 25°C, a dye penetrant test showed no cracks at a maximum temperature of 400°C.

The other example of the application is a gold lining on base materials [23]. Two types of gold plating materials were examined which differed in the base metal, JIS SUS304 and Alloy B2. Figure 21.7 shows a test specimen processed by the gold plating technique. Figure 21.8 shows all the measured corrosion rates as a function of the temperature and the concentration of sulfuric acid. The observed weight changes were within the detection limit of the present experiments. The results indicate that gold lining shows good corrosion resistance of the materials irrespective of the base materials, JIS SUS304 and Alloy B2.

21.4 R&D on Catalysts and Performance

Catalysts are principally used in order to increase chemical reaction rates. The iodine–sulfur process crucially requires heterogeneous catalysts for the gaseous SO_3 and HI decomposition reaction, and therefore, studies have been carried out to identify highly active, low-cost, and long-term stable catalyst. Notice that the Bunsen reaction does not require a catalyst.

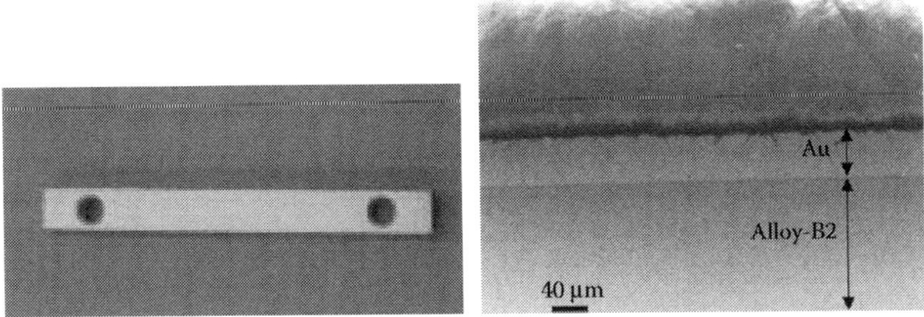

FIGURE 21.7
Test specimen with gold plating with a thickness of 30 μm, Base material of Hastelloy B2, and SEM image of the cross section of the gold plating of Alloy-B2 after 5 h immersion test at 400°C, 90 wt% sulfuric acid. (Adapted from Tanaka, N., et al., 2007. *Proceedings of 15th International Conference on Nuclear Engineering (ICONE-15)*, Nagoya, Japan, April 22–26, ICONE15-10331.)

21.4.1 HI Decomposition Reaction

HI decomposition reaction produces hydrogen which is a main product of the iodine–sulfur process. A plausible process for HI decomposition reaction is a homogeneous gas-phase chemical reaction described as follows:

$$2HI\ (g) = I_2\ (g) + H_2\ (g) + 13\ kJ.$$

The gaseous HI splits into I_2 and H_2 at around 700 K; an enthalpy change is rather low, and the molar amounts are equal throughout this reaction. Since a reaction rate of this reaction

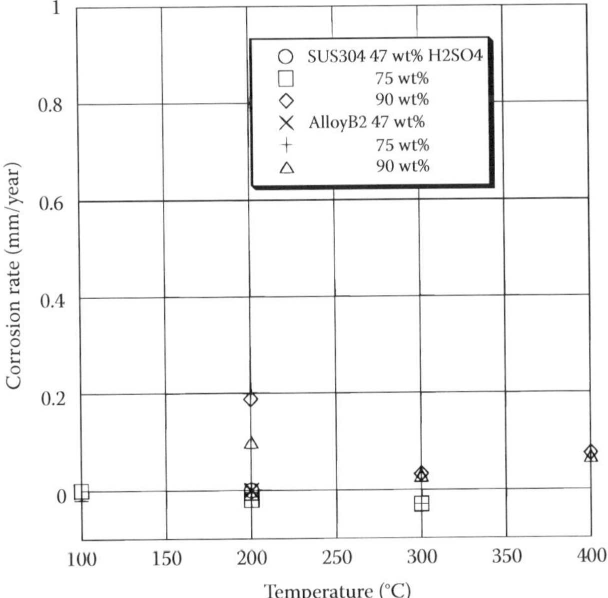

FIGURE 21.8
Corrosion rates of gold-plating material at 47, 75, and 90 wt% sulfuric acid. (From Trester, P. W. and Staley, H. G., 1981. Assessment and investigation of containment materials for the sulfur–iodine thermochemical water-splitting process for hydrogen production, GRI-80/0081. With permission.)

at temperature around 700 K is very slow, applications of catalyst are indispensable to design acceptable sized catalytic reactors.

The catalytic activity was surveyed [24] for metal elements, Pd, Pt, Ni, Mo and so on, and compounds using a variety of support materials such as alumina, silica, zeolites, active carbon, and so on. The platinum (Pt) group metal showed high-catalytic activity, Ni and Mo followed that. Figures 21.9 and 21.10 illustrate performances of Pt catalysts. A combination of Pt with active carbon supports (Pt/C) showed the highest activity, which made the equilibrium conversion within the contact time of 0.1 s at 650–700 K [25]. It is conceivable that the solo-application of the active carbon will have an economic advantage over the use of expensive metal, on the other hand a contact time using the active carbon to reach the chemical equilibrium demanded 0.6 s at 650–700 K [25], which requires larger reactors than that using the combination catalyst. Recently, CeO_2 and Ni catalysts were explored [26] to remove the precious metals; various active carbons were examined [27] to understand their properties, catalytic activities, and stabilities.

A study of the decomposition reaction in the liquid phase was also conducted. Platinum supported on TiO_2, ZrO_2 and so on, catalysts for the liquid-phase decomposition reaction, showed catalytic activity with a conversion ratio of over 50% at 303 K. In this operation, the use of liquid-phase homogeneous catalyst was also considered [28].

21.4.2 H_2SO_4 Decomposition Reaction

H_2SO_4 decomposition reaction produces oxygen which is half of the product as a consequence of water splitting. H_2SO_4 decomposition, which is common reaction in the sulfur family thermochemical cycle [29], for example, the sulfur–iodine cycle, the hybrid sulfur cycle, and so on, proceeds at highest temperature in the iodine–sulfur processes with large

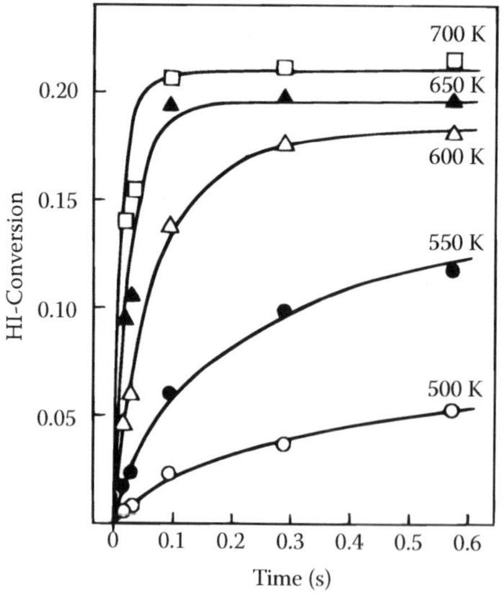

FIGURE 21.9
Contact time–conversion relationship of hydrogen-iodide over Pt/C catalyst. (From Oosawa, Y., Yakemori, Y., and Fujii, K., 1980. *Nippon Kagaku Kaishi*, 7: 1081–1087. With permission.)

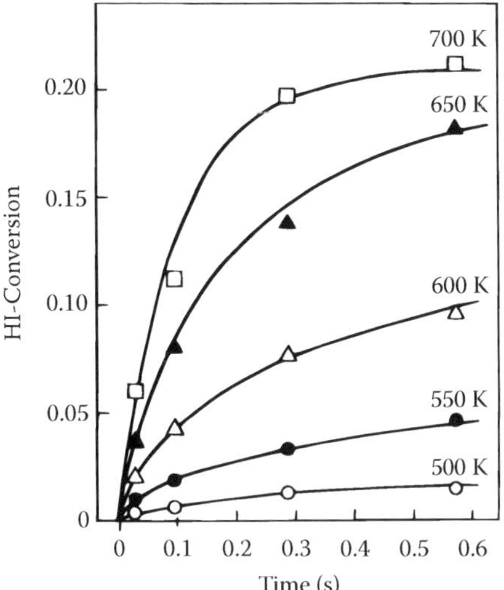

FIGURE 21.10
Contact time–conversion relationship of hydrogen-iodide over active carbon catalyst. (From Oosawa, Y., Yakemori, Y., and Fujii, K., 1980. *Nippon Kagaku Kaishi*, 7: 1081–1087. With permission.)

endothermic heat. This thermal decomposition reaction of sulfuric acid proceeds in the following two stages:

$$H_2SO_4\ (g) = H_2O\ (g) + SO_3\ (g) + 92\ kJ,$$

$$SO_3\ (g) = SO_2\ (g) + 0.5O_2\ (g) + 98\ kJ.$$

The former reaction is a thermal decomposition reaction without catalysts proceeding at 670–770 K, while the later reaction, which proceeds at 1020–1170 K, requires catalysts to achieve high-level of conversions with a limited reactor volume. High values of the chemical equilibrium generally favors high-temperature and low-pressure environments.

Precious metals and metal oxides catalyze the SO_3 decomposition reaction. Figure 21.11 shows catalyst performances of Pt and metal oxides. An order of catalytic activities [30] of precious metals and of transition metal oxides was presented:

$$Pt \sim Cr_2O_3 > Fe_2O_3 > (CuO) > CeO_2 > NiO > Al_2O_3.$$

Platinum is a promising candidate since it performs in the high level of activity. With Pt supported Al_2O_3, the decomposition conversion reached chemical equilibrium at space velocity of 20,000 h^{-1} at 1153 K [31]; a reaction rate was apparently first order against the SO_3 concentration. The catalytic activity of metal oxides was closely related with their thermodynamic properties [30]; in other words, the catalytic activity of the oxide increased with decreasing thermodynamic stability of the sulfate.

Durability of the catalysts is important since the catalysts are exposed to high temperatures and corrosive chemicals. Regarding the stability of 0.1–0.2 wt% Pt supported on several metal oxides [32] such as the Pt/Al_2O_3 and Pt/ZrO_2, which have large surface areas performed in highest activity, time scales for deactivation were short as 6 h, and $Pt/A12O_3$

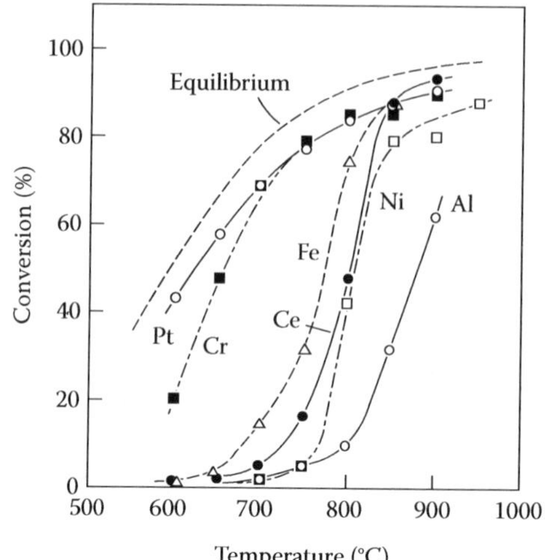

FIGURE 21.11
Relationship between conversion and temperature for metal oxides and Pt catalysts in a N_2 flow containing 4 mol% SO_3 at a space velocity of 430 h^{-1}. (From Beghi, G.E. 1981. *International Journal of Hydrogen Energy*, 6(6): 555–566. With permission.)

lost the activity likely by solid-state transformation of the gamma–alumina support; a lower surface area catalyst of Pt/TiO_2 showed lower initial activity, which was relatively stable instead. A 200 h stability test results of Pt/TiO_2 catalyst at 1123 K is given in Figure 21.12. Liquid sulfuric acid (96 wt%) was fed to the catalytic reactor with a weight hour space velocity (WHSV) of 52 h^{-1}. The catalyst exhibited significant levels of activity but a constant loss in activity for more than 200 h of the continuous operation, and lost about 2.6% per day of its initial activity.

Other catalysts are the metal oxides which have the potential of high economy and long-term stability. Regarding the choice of metal oxides catalysts, deactivation by sulfate formation poisoning [33] should be considered. For catalytic activities of the oxides [30],

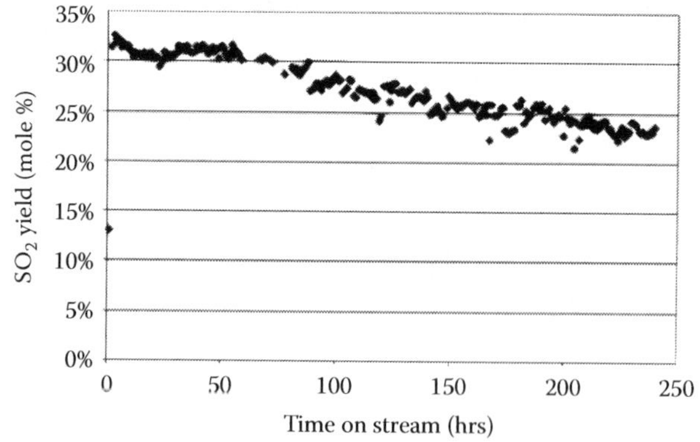

FIGURE 21.12
Transient activity of 0.1 wt% Pt/TiO_2 catalyst for SO_3 decomposition at 1123 K. (From Sato, S. et al. 1980. Studies on closed-cycle, process for hydrogen production, V, Progress Report for the F. Y. *JAERI-M 9724*, 1980. With permission.)

Ce(IV), Cr(III), Fe(III), AI, Ni, Cu, and Pt in the range 600–950°C at space velocities of 4300–20,000 h^{-1}, the catalytic activity of metal oxides could be determined from thermodynamic stabilities of metal oxide in the atmosphere of sulfur oxide—the catalysts could work in states of the metal oxide. Sulfate formation was more significant in lower temperature regions and in high-pressure regions. Regarding the stabilities of the complex metal oxides, catalysts in examination [34] showed shortcomings including material sintering, phase changes, low activities at moderated temperatures due to sulfate formation, and decompositions to their individual oxides. A $CuFe_2O_4$ catalyst appeared relatively promising due to the high activity and lack of any leaching issues; however it deactivated in week-long stability experiments [34].

21.5 R&D on Membranes and Performance

21.5.1 Separation Process Membrane

Separation processes accounts for most unit operations in the sulfur–iodine process. Membrane separation processes are able to be applied to the separations of an objective substance from mixtures. Until now, the development of separation techniques are the major challenges in the iodine–sulfur process, since the process principally consists of three homogeneous chemical reactions, gaseous phase for HI and H_2SO_4 decomposition reactions and liquid phase for Bunsen reaction, not that these are not heterogeneous chemical reactions such as gas–solid reactions [35,36]. The mixture separations are processes of decrease in entropy, and therefore energy inputs (or separating agents) are required.

A base phenomenon in separation techniques are broadly divided into differences of the composition equilibrium relationships and the transfer rates. Generally, separation performances are limited depending on physical properties—the equilibrium relationships—of the object substances. On the other hand, the membrane separation processes, which employ permselective membranes taking advantage of permeability rate differences in the membranes, can be designed by choices or manufacturers of membrane materials in spite of the fact that the object substances are concreted. Membrane separation processes for the iodine–sulfur process have been studied as advanced separation techniques which are alternatives to the equilibrium separation such as the gas–liquid phase separation.

Diverse membrane separation processes are available, which are classified by phases of supply sides and permeated sides—liquid–liquid, liquid–gas, and gas–gas—plus driving forces arising from differences—concentrations, pressures, electric potentials, and temperatures; some of these have possibilities of applications for the iodine–sulfur process. Development of membrane materials, which express, permeability rate, selectivity, durability, and cost, are the key factors to the membrane separation process.

21.5.1.1 Hydrogen-Permselective Membrane Reactor for Gaseous Hydrogen Iodide Decomposition

A chemical equilibrium of HI decomposition reaction at 723 K is only 22%. A membrane reactor can improve [37] one-pass conversion of the HI decomposition reaction, so that the thermal efficiency of hydrogen production may increase with an increase of the conversion since a quantity of unreacted HI reused for the hydrogen production reaction can be reduced.

Figure 21.13 shows schematic of the membrane reactor. The reactor employs a hydrogen-permselective membrane with the catalysts. The gaseous HI catalytically decomposes into I_2 and H_2, and the produced H_2 is removed from the reaction field by permeation through the membrane in order to enhance the conversion.

There are no metallic materials found to be stable under HI, H_2, and H_2O vapor. The silica membrane prepared by a chemical vapor deposition (CVD) of tetraethyl orthosilicate [38] was effective for this application. This silica membrane was stable under H_2, HI, and H_2O vapors at 723 K for 48 h [39]. Performances of a membrane reactor were examined [40], which is fabricated from the silica membranes prepared by the CVD method and a membrane substrate of gamma–alumina tube having 4 nm pores. The HI conversion was improved to 76.4% from equilibrium value (25.0%) at 873 K, so that the hydrogen-permselective silica membranes were successfully applied to the HI decomposition reaction.

21.5.1.2 Pervaporation Dewatering of Poly-Hydriodic Acid and Sulfuric Acid

One of the most unique characteristics of a pervaporation method is an accompanying liquid–gas phase change, which requires a latent heat as well as ordinary distillations; moreover the permeated side maintains low pressure conditions, which makes the recovery of latent heat difficult. Accordingly, this process is inferior in an efficiency of energy saving than the other membrane separation processes. However, an application of high-selectivity membrane probably surpasses the distilled separation, since selective permeability of the membrane has a great impact on the separating performance which is normally restricted by the vapor–liquid equilibrium. In addition, if HIx solution can be concentrated over a pseudo-azeotrope composition [41], this subsequent distillation can produce gaseous HI without water. This may lead an energy saving since a quantity of the water, which is accompanied with unreacted HI reused for the hydrogen production reaction, can be eliminated.

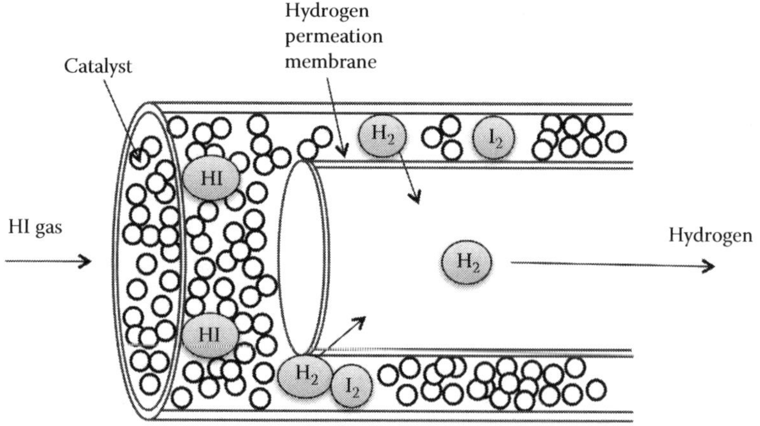

FIGURE 21.13
Schematic diagram of membrane reactor for HI decomposition. Products of H_2 are removed from reaction field by permselective membrane in order to enhance conversion.

Nafion®, a perfluoroethylene sulfonate polymer, is able to be applied to the pervaporation dewatering devices. This membrane is widely used commercially and has a high degree of ionic conductivity, thermal resistance, and hydrophilic pores. Its shortcoming of high water permeability for the fuel cell turns into an advantage for dewatering applications. Figure 21.14 shows schematic of a dewatering application. Nafion and sulfonated poly ether ether ketone (SPEEK) membranes were applied [42,43] to the pervaporation dewatering for concentrations of poly-hydriodic acid and sulfuric acid. Both membranes effectively concentrated HI at temperatures as high as 407 K without any significant degradations of transport behavior. Nafion membranes concentrated sulfuric acid at 373 K. Measured fluxes of water and separation factors were commercially competitive. Besides, pervaporation examinations [44] showed almost pure water permeation, dewatering of HIx solutions, and breaking of the azeotrope.

21.5.2 Electrochemical Process Solid Electrolyte Membrane

Electrochemical processes as typified in the water-electrolysis produces valuable materials with electrochemical devices. In return for consumption of the electric energy, the electrochemical processes readily develop chemical reactions which are hard to progress by only the thermal energy. Various electrochemical processes with solid electrolyte membranes have possibilities of applications for the iodine–sulfur process. A selection of membrane materials which express resistance, selectivity, durability, and cost, is important in the electrochemical processes.

21.5.2.1 Electro-Electrodialysis of Poly-Hydriodic Acid

Regarding the thermal efficient separations of HI from HI–I_2–H_2O mixture (poly-hydriodic acid, HIx solution), an HI enrichment technique by a electro-electrodialysis of HIx solution [45] was applied, since an usual distillation under atmospheric pressure, which results in a distillate of hydriodic azeotropic composition (about HI:H_2O = 1:5), requires a large thermal burden. This issue is caused by the solution in the process that has a HI molality of about 10 mol/kg, which is very close to but less concentrated than the quasi-azeotropic composition.

Figure 21.15 shows the concept [46] of the electro-electrodialysis employing the cation-exchange membrane. Electrode reactions, the redox reaction of iodine–iodide ions in HIx

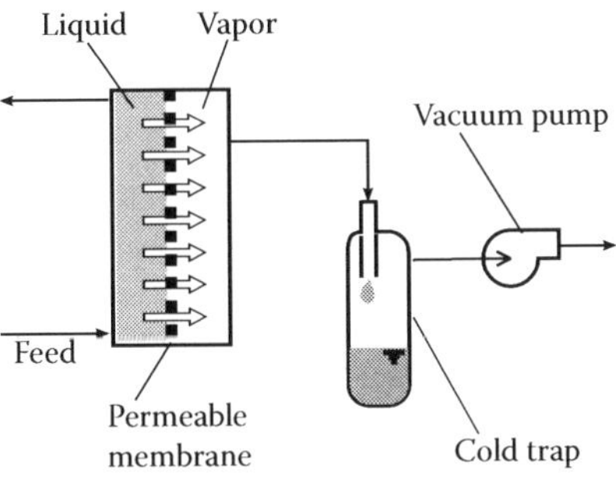

FIGURE 21.14
Diagram of pervaporation with vacuum collection system. Constituent part of feed solution permeates through permeation membrane and evaporates with aids of vacuum pump to perform separations.

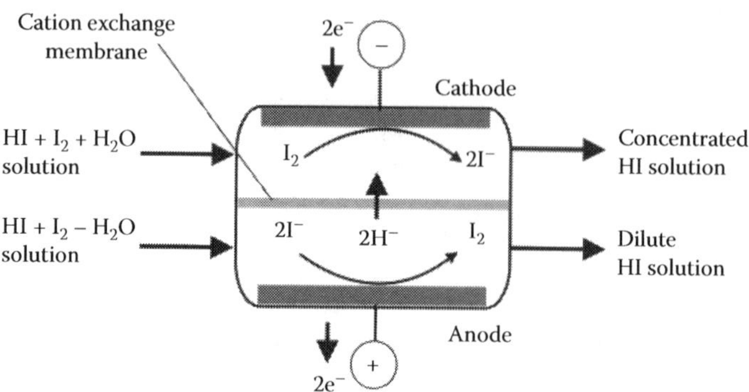

FIGURE 21.15
Concept of an electro-electrodialysis for concentrating HIx solution. HI enrichment in catholyte is achieved by redox reactions.

solution, progressed on the glassy carbon electrodes. Consequently, the HI enrichment in the catholyte was achieved with the help of selective proton permeation through the membrane, while HI molality of the anolyte decreased.

Nafion, a perfluoroethylene sulfonate polymer, is able to be applied to the electro-electrodialysis device. This membrane is widely used commercially and has a high degree of ionic conductivity and thermal resistance. Cell voltages to operate the electro-electrodialysis device represent a primary performance, since it is directly connected with the electric power consumption. The cell voltage is a sum of a theoretical decomposition voltage, an over-voltage of electrode reaction, an ohmic loss in solution, and a membrane voltage drop. Experimental measurement using Nafion 117 [47] illustrated that the cell voltages were 0.26–0.58 V with current density of 100 mA cm^{-2} and a main contributor to the cell voltage was the membrane voltage drop, which suggested the improvements of membrane performance, especially the membrane resistivity, are effective to reduce the electric-power consumption. Alternative polymer-electrolyte membranes were applied [48,49] to the electro-electrodialysis device, which have been developed for fuel cells by radiation-induced graft polymerization and cross-linking methods [50], in order to explore more efficient electro-electrodialysis process.

21.5.2.2 Electrochemical Membrane Reactor for Bunsen Reaction

An electrochemical membrane reactor for Bunsen reaction [51] is of the type where a Bunsen reactor is housed in the same device as a two liquid–liquid phase separator and HI concentrator such as the electro-electrodialysis cell. These reactors can be operated on low level of I_2 concentration, so that a heat duty for the separation of I_2 from water may be reduced. In addition, the heat traces of pipes, which are installed to maintain the temperature of HIx solution over the iodine solidification point in order to prevent the pipe clogging, may be removed.

Figure 21.16 shows the schematic diagram of the reactor. HI and I_2 solution is introduced into a cathode side, and SO_2 solution is supplied to an anode side. The reactions occur through the following equations:

$$\text{Anode side, } SO_2 + 2H_2O = H_2SO_4 + 2H^+ + 2e^-;$$

$$\text{Cathode side, } I_2 + 2H^+ + 2e^- = 2HI.$$

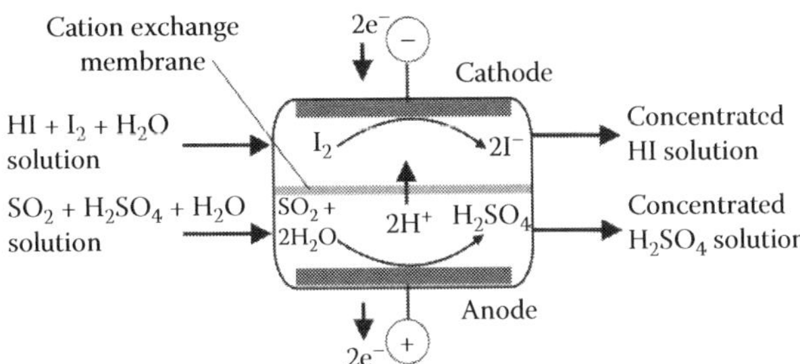

FIGURE 21.16
Schematic diagram of electrochemical membrane reactor for Bunsen reaction. This reactor is housed in same device as two liquid–liquid phase separator and HI-concentrator such as electro-electrodialysis cell.

H_2SO_4 solution and HI solution is separately obtained at the both sides of the cation exchange membrane.

Nafion, a perfluoroethylene sulfonate polymer, is able to be applied [52] to the electrochemical membrane reactor. This membrane is widely used commercially and has a high degree of ionic conductivity and thermal resistance. Experimental examinations of the electrochemical membrane reactor for Bunsen reaction employing Nafion 117 as a cation exchange membrane and glassy carbon as electrodes were demonstrated. The results showed that the Bunsen reaction was successfully carried out, I_2 concentration can be reduced by 93%, and the lower I_2 concentration is efficient from the viewpoint of electric energy consumption.

21.5.2.3 Electrolysis of Gaseous Sulfur Trioxide

Electrolysis of gaseous sulfur trioxide enables to decrease operating temperature of sulfur trioxide decomposition reaction down to 500°C, which regularly progresses around 900°C with thermal energy. The electrolysis may enable to adapt the reaction to be conducted with the sodium-cooled fast reactor which is possible to supply heat at about 500°C. In addition, high-temperature components are able to be eliminated from such harsh environments.

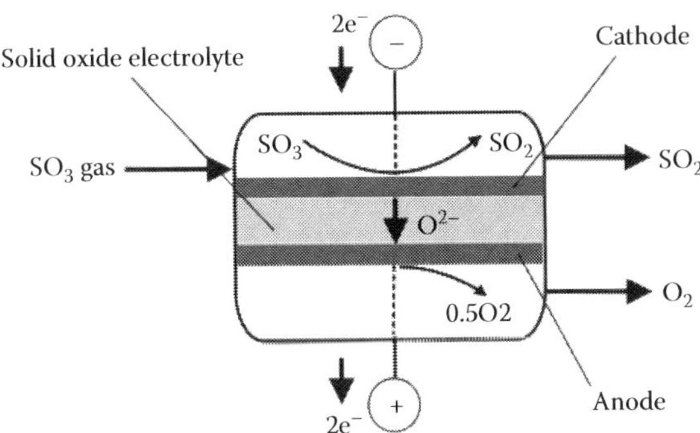

FIGURE 21.17
Schematic of sulfur trioxide electrolyzer to produce oxygen. Electrochemical reaction reduces operation temperature of sulfur trioxide decomposition than pure thermal reaction.

Figure 21.17 shows a schematic of the electrolyzer. Ionic oxygen conductive solid electrolytes made of the yttria-stabilized zirconia were applied [53] to a laboratory-scaled cylindrical electrolyzer with platinum electrodes on inner and outer surfaces. The gaseous sulfuric acid was supplied to the inside of the cylindrical electrolyzer and was decomposed into sulfur dioxide and oxygen by the electrolysis process at the temperature of 450–600°C. A product of oxygen permeated through the electrolyte from inside to outside. Through an experiment, an observed amount of oxygen agreed well with the value calculated from current input, so that the possibility of the sulfur trioxide electrolysis was confirmed.

21.6 Future Material Challenges

More efforts are needed in the research and developments of materials in order to achieve the economical hydrogen production by iodine–sulfur process. Metal construction materials will be examined in actual plant conditions, which are crevice corrosion, stress corrosion cracking, welding area corrosion, hydrogen embitterment, galvanic corrosion, and contaminated solution by both the hydriodic-acids and the sulfuric acids for a long-term confirmation. The lining and coating techniques should be reviewed, which employ the precious metal such as tantrum, resins, for example, fluoroplastic, and so on and ceramics. Sealing materials should be explored for high-temperature gaseous and liquid phases. An up-sizing and developments for strength evaluation methods are required for the components made of ceramic materials. More stable, economical catalysts should be identified in further tests. Improvements of the membrane performance are desired especially for the permselectivity, electric resistance, thermal resistance, permeability rate, and energy consumptions. Further developments in the field of materials would open the way to the commercialization of the hydrogen production by sulfur–iodine process.

References

1. Norman, J. H., Besenbruch, G. E., and O'Keefe, D. R., 1981. Thermochemical water-splitting for hydrogen production, Final Report (January 1975–December 1980), GRI-80/0105, March 1981.
2. Kubo, S., Nakajima, H., Kasahara, S., Higashi, S., Masaki, T., Abe, H., and Onuki, K., 2004. A pilot test plan of the thermochemical water-splitting iodine–sulfur process, *Nuclear Engineering and Design*, 233: 347–354.
3. Onuki, K., Nakajima, H., Shimizu, S., Sato, S., and Tayama, I., 1993. Materials of construction for the thermochemical IS process, (I), *Journal of the Hydrogen Energy Systems Society of Japan*, 18: 49–56 (in Japanese).
4. Onuki, K., Ioka, I., Futakawa, M., Nakajima, H., Shimizu, S., Sato, S., and Tayama, I., 1994. Materials of construction for the thermochemical IS process, (II), *Journal of the Hydrogen Energy Systems Society of Japan*, 19: 10–16 (in Japanese).
5. Onuki, K., Ioka, I., Futakawa, M., Nakajima, H., Shimizu, S., and Tayama, I., 1997. Screening tests on materials of construction for the thermochemical IS process, *Zairyo-to-Kankyo*, 46: 113–117.
6. Ioka, I., Onuki, K., Futakawa, M., Kuriki, K., Nagoshi, M., Nakajima, H., and Shimizu, S., 1997. Corrosion resistance of Fe–Si alloys in boiling sulfuric acid, Zairyo, *Journal of Society Materials Science, Japan*, 46: 1041–1045. (in Japanese)

7. Futakawa, M., Onuki, K., Ioka, I., Nakajima, H., Shimizu, S., Kuriki, Y., and Nagoshi, M., 1997. Corrosion test of compositionally graded Fe–Si alloy in boiling sulfuric acid, *Corrosion Engineering*, 46: 811–819.
8. Nishiyama, N., Futakawa, M., Ioka, I., Onuki, K., Shimizu, S., Eto, M., Oku, T., and Kurabe, M., 1999, Corrosion resistance evaluation of brittle materials in boiling sulfuric acid, Zairyo, *Journal of Society Materials Science, Japan*, 48: 746–752 (in Japanese).
9. Ioka, I., Mori, J., Kato, C., Futakawa, M., and Onuki, K., 1999. The characterization of passive films on Fe–Si alloy in boiling sulfuric acid, *Journal of Materials Science Letter*, 18: 1497–1499.
10. Kurata, Y., Tachibana, K., and Suzuki, T., 2001. High temperature tensile properties of metallic materials exposed to a sulfuric acid decomposition gas environment, *Journal of the Japan Institute of Metals*, 65: 262–265 (in Japanese).
11. Ammon, R. L., 1982. Status of materials evaluation for sulfuric acid vaporization and decomposition application. *Proc. 4th World Hydrogen Energy Conf., California, USA*, June 1982, vol. 2, pp. 623–644.
12. Porisini, F. C., 1989. Selection and evaluation of materials for the construction of a pre-pilot plant for thermal decomposition of sulfuric acid, *International Journal of Hydrogen Energy*, 14: 267–274.
13. Imai, Y., Kanda, Y., Sasaki, H., and Togano, H., 1982. Corrosion resistance of materials in high temperature gases composed of iodine, hydrogen iodide and water, *Boshoku Gizyutsu*, 31: 714–721 (in Japanese).
14. Kubo, S., Futakawa, M., Ioka, I., Onuki, K., Shimizu, S., Ohsaka, K., Yamaguchi, A., Tsukada, A., and Goto, T., 2003. Corrosion test on structural materials for iodine–sulfur thermochemical water splitting cycle, *AIChE 2003 Spring National Meeting*, New Orleans.
15. Hatano, S., 1980. *Kagaku-sochi zairyo taisyoku-hyo, Kagaku-kougyo sya*, p. 194 (in Japanese).
16. *Kagaku-sochi benran, Kagaku-kogaku kyokai*, p. 499, 1970 (in Japanese).
17. Kurata, Y., Tachibana, K., and Suzuki, T., 2000. High temperature tensile properties of metallic materials exposed to a sulfuric acid decomposition gas environment, *Journal of Japan Institute of Metals*, 65: 262–265.
18. Futakawa, M., Kubo, S., Wakui, T., Onuki, K., Shimizu, S., and Yamaguchi, A., 2003. Mechanical property evaluation of surface layer corroded in thermochemical-hydrogen-production process condition. *Jikenn rikigaku*, 3: 109–114 (in Japanese).
19. Onuki, K., Nakajima, H., Ioka, I., Futakawa, M., and Shimizu, S., 1994. IS process for thermochemical hydrogen production, *JAERI-Review*, 94–006.
20. Trester P. W., and Staley, H. G., 1981. Assessment and investigation of containment materials for the sulfur–iodine thermochemical water-splitting process for hydrogen production, GRI-80/0081.
21. Tanaka N., Iwatsuki, J., Kubo, S., Terada, A., and Onuki, K., 2009. Research and development on IS process components for hydrogen production (II). Corrosion resistance of glass lining in high temperature sulfuric acid, *Proc. of 2009 International Congress on Advances in Nuclear Power Plants (ICAPP2009)*, Tokyo, Japan, May 10–14, Paper 9291.
22. Iwatsuki, J, 2008. Hydrogen production with high-temperature gas-cooled reactors (1) Application of glass lining material for thermochemical IS process, *Spring Meeting of Atomic Energy Society of Japan*, Osaka, Japan, March 2008 (in Japanese).
23. Tanaka, N., Suwa, H., and Terada, A., 2007. Study on corrosion resistance of glass lining and gold plating in high temperature sulfuric acid for thermochemical hydrogen production IS process, *Proc. of 15th International Conference on Nuclear Engineering (ICONE-15)*, Nagoya, Japan, April 22–26, ICONE15–10331.
24. Oosawa, Y., Yakemori, Y., and Fujii, K., 1980. Catalytic decomposition of hydrogen iodide in the magnesium–iodine thermochemical cycle, *Nippon Kagaku Kaishi*, 7: 1081–1087 (in Japanese).
25. Oosawa, Y., Kumagai, T., Mizuta, S., Kondo, W., Yakamori, Y., and Fujii, K., 1981. Kinetics of the catalytic decomposition of hydrogen iodide in the magnesium–iodine thermochemical cycle, *Bull. Chem. Soc. Jpn.*, 54: 742–748.

26. Zhang, Y. W., Zhou, J. H., Chen, Y., Wang, Z. H., Liu J. Z., and Cen, K. F., 2008. Hydrogen iodide decomposition over nickel-ceria catalysts for hydrogen production in the sulfur–iodine cycle, *International Journal of Hydrogen Energy*, 33: 5477–5483.
27. Petkovic, L. M., Ginosar, D. M., Rollins, H. W., Burch, K. C., Deiana, C., Silva, H. S., Sardella, M. F., and Granados, D., 2009. Activated carbon catalysts for the production of hydrogen via the sulfur–iodine thermochemical water splitting cycle, *International Journal of Hydrogen Energy*, 34: 4057–4064.
28. O'Keefe, D. R., Norman, J. H., Brown, L. C., and Besenbruch, G. E., 1982. The application of a homogeneous catalysis concept to the liquid hydrogen iodine decomposition step of the General atomic sulfur–iodine water-splitting cycle, *Proc. of 4th World Hydrogen Energy Conf.*, CA, June 1982, 2: 687–701.
29. Beghi, G. E. 1981. Review of thermochemical hydrogen production, *International Journal of Hydrogen Energy*, 6(6): 555–566.
30. Tagawa, H. and Endo, T., 1989. Catalytic decomposition of sulfuric acid using metal oxides as the oxygen generating reaction in thermochemical water splitting process, *International Journal of Hydrogen Energy*, 14(1): 11–17.
31. Sato, S., Ikezoe, Y., Suwa, T., Shimizu, S., Nakajima, H., and Onuki, K., 1981. Studies on closed-cycle, process for hydrogen production, V, Progress Report for the F. Y. *JAERI-M 9724*, 1980.
32. Ginosar, D. M., Petkovic, L. M., Glenn, A. M., and Burch, K. C., 2007. Stability of supported platinum sulfuric acid decomposition catalysts for use in thermochemical water splitting cycles, *International Journal of Hydrogen Energy*, 32: 482–488.
33. Norman, J. H., 1980. Studies of the sulfur–iodine thermochemical water-splitting cycle, *Proc. of 3rd World Hydrogen Energy Conference*, Tokyo, June, 1: 257–275.
34. Ginosar, D. M., Rollins, H. W., Petkovic, L. M., Burch, K. C., and Rush, M. J., 2009. High-temperature sulfuric acid decomposition over complex metal oxide catalysts, *International Journal of Hydrogen Energy*, 34: 4065–4073.
35. Ota, K., and Conger, W. L., 1977. Thermochemical hydrogen production via a cycle using barium and sulfur: Reaction between barium sulfide and water, *International Journal of Hydrogen Energy*, 2: 101–106.
36. Nakayama, T., Yoshioka, H., Furutani, H., Kameyama, H., and Yoshida, K., 1984. MASCOT—A bench-scale plant for producing hydrogen by the UT-3 thermochemical decomposition cycle, *International Journal of Hydrogen Energy*, 9(3): 187–190.
37. Hwang, G. and Onuki, K., 2001. Simulation study on the catalytic decomposition of hydrogen iodide in a membrane reactor with a silica membrane for the thermochemical water-splitting IS process, *Journal of Membrane Science*, 194: 207–215.
38. Hwang, G., Onuki, K., Shimizu, S., and Ohya, H., 1999. Hydrogen separation in H_2–H_2O–HI gaseous mixture using the silica membrane prepared by chemical vapor deposition, *Journal of Membrane Science*, 162: 83–90.
39. Hwang, G., Onuki, K., and Shimizu, S., 2000. Separation of hydrogen from a H_2–H_2O–HI gaseous mixture using a silica membrane, *AIChE Journal*, 46: 92.
40. Nomura, M., Kasahara, S., and Nakao, S., 2004. Silica membrane reactor for the thermochemical iodine–sulfur process to produce hydrogen, *Industrial Engineering and Chemical Research*, 43: 5874–5879.
41. Engels, H. and Knoche, K. H., 1986. Vapor pressures of the system $HI/H_2O/I_2$ and H_2, *International Journal of Hydrogen Energy*, 11 (11): 703–707.
42. Orme, C. J., Jones, M. G., and Stewart, F. F., 2005. Pervaporation of water from aqueous HI using Nafion-117 membranes for the sulfur–iodine thermochemical water splitting process, *Journal of Membrane Science*, 252: 245–252.
43. Orme, C. J., Klaehn, J. R., and Stewart, F. F., 2009. Membrane separation processes for the benefit of the sulfur–iodine and hybrid sulfur thermochemical cycles, *International Journal of Hydrogen Energy*, 34: 4088–4096.
44. Elder, R. H., Priestman, G. H., and Allen, R. W. K., 2009. Dewatering of HIx solutions by pervaporation through Nafion(R) membranes, *International Journal of Hydrogen Energy*, 34: 6129–6136.

45. Onuki, K., Hwang, G., and Shimizu, S., 2000. Electrodialysis of hydriodic acid in the presence of iodine, *Journal of Membrane Science*, 175: 171–179.
46. Onuki, K., Hwang, G., and Arifal, Shimizu, S., 2001. Electro-electrodialysis of hydriodic acid in the presence of iodine at elevated temperature, *Journal of Membrane Science*, 192: 193–199.
47. Yoshida, M., Tanaka, N., Okuda, H., and Onuki, K., 2008. Concentration of HIx solution by electro-electrodialysis using Nafion 117 for thermochemical water-splitting IS process, *International Journal of Hydrogen Energy*, 33: 6913–6920.
48. Okuda, H., Yamaki, T., and Kubo, S., 2006. Improvement of the thermal efficiency of hydrogen iodine concentration in I–S process by using radiation-grafted membrane in electrolysis system, AIChE 2006 Spring National Meeting, Orlando, USA.
49. Tanaka, N., Yamaki, T., Asano, M., Maekawa, Y., and Onuki, K., 2010. Electro-electrodialysis of $HI–I_2 H_2O$ mixture using radiation-grafted polymer electrolyte membranes, *Journal of Membrane Science*, 346: 136–142.
50. Yamaki, T., Kobayashi, K., Asano, M., Kubota, H., and Yoshida, M., 2004. Preparation of proton exchange membranes based on crosslinked polytetrafluoroethylene for fuel cell applications, *Polymer*, 45: 6569–6573.
51. Dokiya, M., Fukuda, K., Kameyama, T., Kotera, Y., and Asakura, S., 1977. The study of thermochemical hydrogen preparation (II) electrochemical hybrid cycle using sulfur–iodine system, *Denki Kagaku*, 45: 139–143.
52. Nomura, M., Fujiwara, S., Ikenoya, K., Kasahara, S., Nakajima, H., Kubo, S., Hwang, G., Choi, H., and Onuki, K., 2004. Application of an electrochemical membrane reactor to the thermochemical water splitting IS process for hydrogen production, *Journal of Membrane Science*, 240: 221–226.
53. Nakagiri, T., Kase, T., Kato, S., and Aoto, K., 2006. Development of a new thermochemical and electrolytic hybrid hydrogen production system for sodium cooled FBR, *JSME International Journal Series B*, 49(2): 302–308.

22

Nuclear Hydrogen Production Process Reactors

Atsuhiko Terada and Hiroaki Takegami

CONTENTS

22.1 Process Equipment Requirements

Nuclear hydrogen production processes such as steam reforming of methane [1–4], thermochemical water decomposition [5–8], and high-temperature steam electrolysis [9–11] employ a wide range of thermal and chemical reactors and other equipment, for example a steam reformer (SR) or chemical decomposer, which are designed to meet safe and economical performance requirements. Typical requirements are structural integrity at service temperature and pressure that can approach to 900°C and 4 MPa, heat and resistance to corrosive process fluids, affordable material and fabrication cost, and sufficient life time. Since chemical reactors to utilize nuclear heat and electricity for hydrogen production are often heat-exchanger type, a usual requirement for economical design is efficient performance of heat transfer between heating medium such as helium gas and process fluids.

Steam reforming of methane and thermochemical IS decomposition of water present some of the most significant equipment design requirements and challenges as described in the following.

22.1.1 Steam Reforming Process

Although steam reforming of methane is established and widely practiced by conventional process industry, it requires new safety-related design and operational consideration for its application in nuclear plant.

In an high-temperature gas-cooled reactor (HTGR)-based steam reforming hydrogen production system, controllability of pressure difference between the secondary helium and process gases is an important requirement. The reformer reaction tube constitutes pressure boundary between the heat transport secondary helium gas from HTGR and the process gas of a mixture of methane and steam. The reformer tube wall thickness is decided by considering the controllability of minimum pressure difference between helium and process gases in not only normal operations including start-up and shutdown but also off-normal conditions such as a loss of chemical reaction.

JAEA investigated the design of a steam reforming hydrogen production system with the objective of connecting it to the high-temperature test reactor (HTTR) nuclear test reactor (Chapter 20). Figure 22.1 shows the flow scheme of the HTTR-SR hydrogen production system [1]. The system is designed to utilize the nuclear heat effectively and achieve economical hydrogen production with operability, controllability, and safety. The major design specifications of the system are listed in Table 22.1 [2].

Relative to conventional process, the HTTR steam-reforming conditions are at higher pressure but lower temperature as shown in Table 22.2, resulting in reduced process thermal

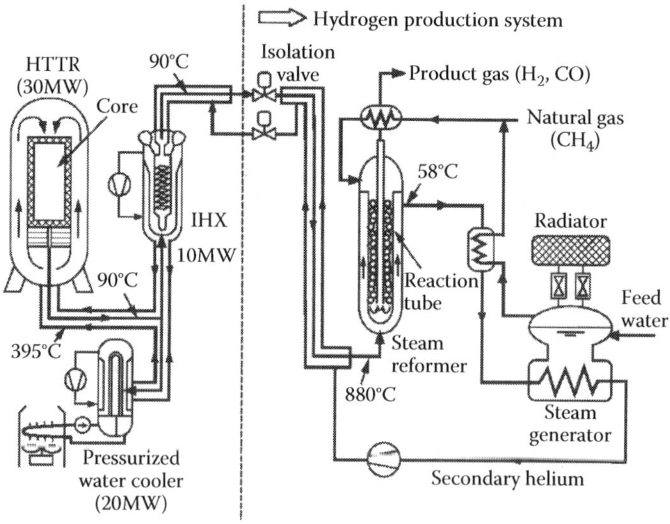

FIGURE 22.1
Flow scheme of the HTTR-SR hydrogen production system. (From H. Ohashi et al., *J. Nucl. Sci. tech.* 41, 385–392, 2004. With permission.)

TABLE 22.1

Design Specifications of HTTR-SR Hydrogen Production system and Mock-Up Test Facility

Items	HTTR Hydrogen Production System	Mock-Up Lest Facility
Pressure Process gas/helium gas	4.5/4.1 MPa	
Temperature at Inlet of Steam Reformer Process-gas/helium-gas	450/880°C	
Temperature at Outlet of Steam Reformer Process gas/helium gas	580/585°C	600/650°C
Methane (natural gas) feed	1,400 kg/h	43.2 kg/h
Helium gas circulation	9,070 kg/h	327.6 kg/h
Steam–carbon ratio (5/C)	35	
Hydrogen production ability	4,240 m^3/h	110 m^3/h
Heat source	Reactor (10 MW)	Electric heater (0.42 MW)

Source: Adapted from H. Ohashi et al., *J. Nucl. Sci. Tech.*, 41, 385–392, 2004.

efficiency [3,4]. The reformer design for nuclear hydrogen system requires increased rate of heat input to process gas for given capacity of hydrogen production.

22.1.2 Thermochemical IS Process

The iodine–sulfur (IS) process is the leading thermochemical process of hydrogen production for application to high-temperature nuclear fission and fusion reactors. JAEA has designed a pilot-scale of 30 Nm^3/h hydrogen production IS process plant based on industrial materials such as glass coated steel, SiC ceramics for process equipment specification [7,12]. Figure 22.2 shows a candidate flow scheme of the test facility consisting of the IS process plant and a helium gas (He) circulation loop that heats the process [7]. The He loop designed to simulate the high temperature and pressure conditions of the HTTR heat supply consists of a 400 kW-electric heater, a He circulator and a steam generator working as a helium gas cooler.

TABLE 22.2

Comparison of Operational Conditions and Performance of Steam Reformers in Fossile-Fired Plant and in HTTR System

Reformer Type	Fossil-Fired Plant	HTTR System
Process gas pressure	1~3 Mpa depending upon final products	4.5 Mpa at the inlet of steam reformer (>Helium pressure P_{Ho} of 4.1 Mpa)
Maximum process gas temperature	850~900°C	800°C
Maximum heat flux to catalyst zone	50~80 kW/m^2	40 kW/m^2
Thermal energy utilization of steam reformer	80–85%	78%
CO_2 emission from heat source for heat source power	3t-CO_2/h/10MW	0

Source: Adapted from Y. Miyamoto et al., *JAERI, IAEATECDOC-988*, 1997.

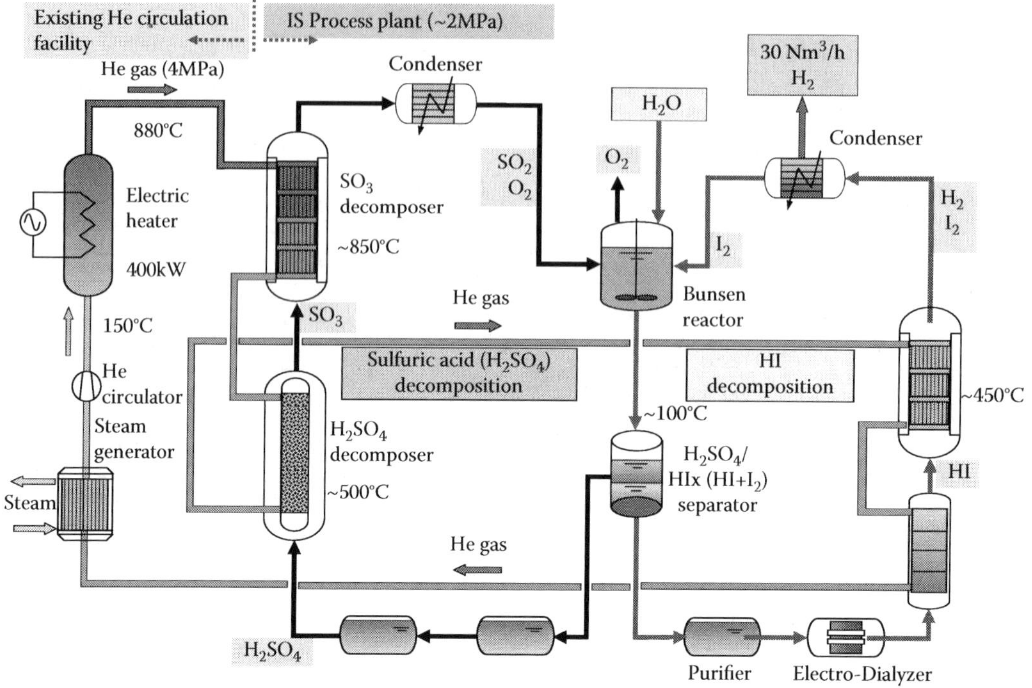

FIGURE 22.2
Schematic flow diagram of 30 Nm³/h scale IS process plant. (From J. Iwatsuki et al., *Proc. ICONE14-89267, 14th International Conference on Nuclear Engineering*, FL, USA, 2006. With permission.)

The IS process plant consists of three sections, that is, Bunsen reaction section, hydrogen iodide (HI) decomposition section and sulfuric acid (H_2SO_4) decomposition section. The equipment requirements in each section are described separately.

22.1.2.1 Bunsen Reaction Section

The flow sheet will reflect the main features of the original flow sheet proposed by General Atomics [13]. The functions of the Bunsen reaction section are to reproduce two types of acids and to feed the acids into the decomposition sections. The candidate materials of construction that show corrosion resistance in the concerned process environments are glass, tantalum, and so on. These materials may be considered as the lining materials of equipments. The main equipments to be considered are as follows:

1. Bunsen reactor
2. H_2SO_4/HIx separator
3. HIx purifier
4. Sulfuric acid purifier

Bunsen reactor produces mixed acid, which contains HI and H_2SO_4, through the chemical absorption of SO_2 gas in aqueous solution of HI, I_2, and H_2SO_4. In order to prevent the side reactions forming sulfur and H_2S, Bunsen reactor should have mixer of the solution which are to suppress increasing of temperature and to homogenize temperature and concentration of the solution. The mixed acids are separated with H_2SO_4/HIx separator, which uses a liquid–liquid separation phenomenon due to the difference of density. The separated hydriodic acid dissolves the iodine and is denoted as the HIx phase. After purification, HIx phase is fed into the HI decomposition section. Similarly, the separated sulfuric acid denoted as the sulfuric acid phase is fed into the H_2SO_4 decomposition section after purification.

Figure 22.3 shows the liquid–liquid phase separation phenomenon in a glass vessel. The two acids divide into upper (sulfuric acid phase) and lower (HIx phase) solutions with a clear boundary [12].

22.1.2.2 HI Decomposition Section

Flow sheet of the HI decomposition section will include essential equipments that are required for the selected energy efficient treatment of HIx solution. In the case of electro-electrodialysis (EED), as shown in Figure 22.2, the cell will be placed in between the purifier of HIx solution and the HI distillation column. The candidate materials of construction are glass, tantalum, and so on, for the liquid phase service, and Ni alloy for the gas phase service. The main equipment to be considered are as follows:

1. HI distillation column
2. HIx solution concentrator
3. HI decomposer
4. Condenser

A main reactor of this section is the HI decomposer, which is a heat exchange type reactor using He gas. Since the gas-phase thermal decomposition of HI is slightly endothermic and the reaction temperatures higher than about 400°C are desired from the viewpoint of the reaction rate, HI decomposer may have to be heated directly by the helium gas. Also, from the viewpoint of corrosion resistance, it may be required that the condenser is made of ceramics.

Moreover, JAEA has been studying the application of membrane technology to achieve efficient separation of HI from the HIx solution and enhancement of HI conversion, which

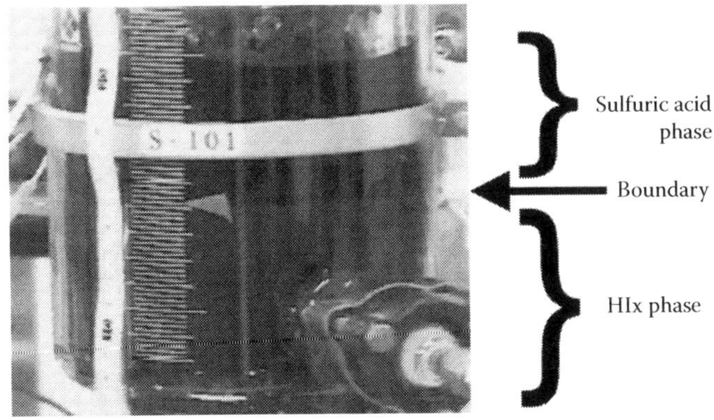

FIGURE 22.3
Liquid–liquid phase separation phenomenon in a glass vessel. (From S. Kubo et al., *Nucl. Eng. Des.* 233, 355–362, 2004. With permission.)

improve thermal efficiency of the section [14,15]. The application of EED has also been studied to concentrate the HIx solution over the pseudo-azeotropic composition, which enables attainment of highly concentrated HI in the subsequent distillation [5].

22.1.2.3 H_2SO_4 Decomposition Section

The flow sheet of this section will be determined considering the conditions proposed by RWTH Aachen [16], Ecole Polytechnique de Montreal [17], and GA [18]. The candidate materials of construction, which show corrosion resistance in the concerned process environments, are silicon-containing ceramics, such as silicon carbide (SiC), for the liquid-phase service, and refractory alloys for the gas-phase service. Main equipment to be considered are as follows:

1. Concentrator
2. H_2SO_4 decomposer
3. SO_3 decomposer
4. Condenser

The commercial IS process plant will be driven by sensible heat of high-pressure helium gas, whose highest temperature may reach 900°C. The process fluids of IS process are very corrosive in nature. Therefore, the chemical reactors have to be of heat-exchanging types and be made of refractory materials that exhibit corrosion resistance in the process environments.

H_2SO_4 decomposer and SO_3 decomposer are heated directly by the helium gas of 4 MPa and up to 880°C. Based on the corrosion test results [19–23], these decomposers have to be made of silicon contained ceramics. In developing a ceramic component, it is required specific design studies on connection with metal components and seal. The seal performance will deteriorate due to the difference of thermal expansion between SiC and metal at the high-temperature condition.

22.2 Engineering and Industrial Designs

22.2.1 Steam Reformer

Prior to the construction of the HTTR hydrogen production system, the mock-up test facility was fabricated to investigate transient behavior of the hydrogen production system and establish system controllability as well as design verification of performance of high-temperature components such as SR and SG (Steam Generator). Figure 22.4 shows a schematic flow diagram of the test facility [4]. And design specifications of the test facility are also shown in Table 22.1.

Figure 22.5 shows the schematic view of a SR which has one bayonet-type catalyst tube made of Allow 800H [4]. Design specifications of the test facility dimensions such as diameter and length are approximately the same as those of the catalyst tube of the HTTR hydrogen-production system. By the way, the SR of the HTTR hydrogen-production system has 30 bayonet-type catalyst tubes made of Hastelloy XR. Process gas flows in the catalyst and guide tubes at inlet temperature of 880°C. In the fossil-fuel fired plant, the process gas

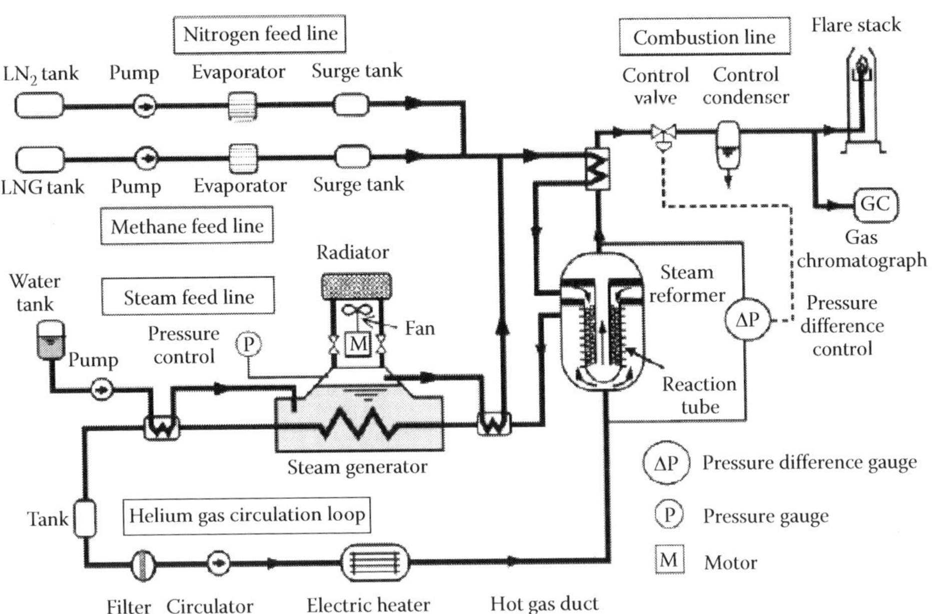

FIGURE 22.4
Schematic flow diagram of the mock-up test facility. [From Inagaki et al., *Proc. Technical Committee Meeting*, Beijing, People's Republic of China, November 2–4, 1998. *International Atomic Energy Agency, International Working Group on Gas-Cooled Reactors*, Vienna (Austria) IAEATECDOC 1210, pp. 213–221. With permission.]

receives heat from combustion air of about 1200°C by radiation, and the heat flux at the outer surface of the catalyst tube reaches 70,000–87,000 W/m^2. In order to achieve the same heat flux as that of the fossil-fuel-fired plant, it is very important to promote heat transfer of helium gas by forced convection because the temperature of helium gas, that is, the temperature of heat source is too low compared with that of the fossil-fuel-fired plant. So, disc-type fins, 2 mm in height, 1 mm in width, and 3 mm in pitch, are arranged around outer surface of the catalyst tube in the test facility in order to increase a heat transfer coefficient of helium gas by 2.7 times, 2150 W/m^2 K with the fins, and a heat-transfer area by 2.3 times larger than those of smooth surface, respectively. As a result, the heat-transfer performance of the catalyst tube in the test facility becomes competitive to that of the fossil-fuel fire plant.

An electric heater is used as a heat source instead of the nuclear heat in order to heat helium gas up to 880°C at inlet of the SR. The mock-up test facility, using an electric heater as a reactor substitute, simulates key components downstream an intermediate heat exchanger of the HTTR hydrogen-production system on a scale of 1–30 with a hydrogen–production rate of 110 Nm^3/h. A bayonet-type catalyst tube was applied to the SR of the mock-up test facility in order to enhance the heat utilization rate. In order to promote heat transfer, the thickness of the catalyst tube should be decreased to 10 mm while augmenting heat transfer by fins formed on the outer surface of the catalyst tube. Therefore, the catalyst tube was designed on the basis of pressure difference between helium and process gases instead of total pressure of them. This design method was authorized for the first time in Japan. Furthermore, an explosion proof function was applied to the SR because it contains inflammable gas and electric heater. This report describes the structure of the SR as well as the authorization both of the design method of the catalyst tube and the explosion proof function of the SR.

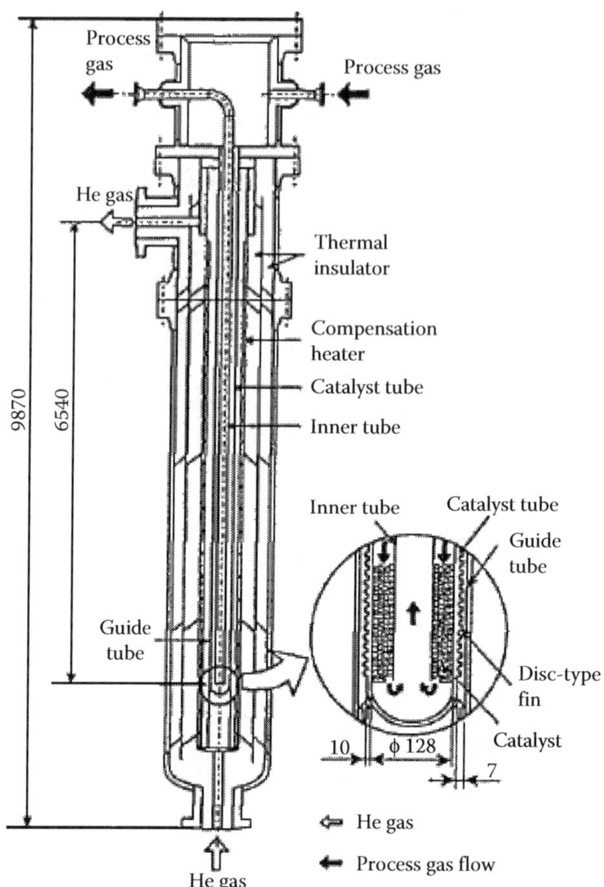

FIGURE 22.5
Schematic view of a SR. (From Y. Inagaki et al., Proc. Technical Committee Meeting, Beijing, People's Republic of China, November 2–4, 1998. International Atomic Energy Agency, International Working Group on Gas-Cooled Reactors, Vienna (Austria) IAEATECDOC-1210, pp. 213–221. 2001. With permission.)

On the other hand, in the German "Prototype Nuclear Process Heat" project, PNP, the SR was an essential component in the nuclear-assisted hydro-gasification of coal. Unlike Japan, in this concept, the nuclear SR was included directly within the HTGR primary circuit. The direct coupling to the SR resulting in a simplified design of a process heat HTGR may, from today's safety perspective, be regarded as a long-term option.

The follow-on test facility, EVA-II, represented a complete helium circuit containing a bundle of reformer tubes [24]. The nuclear heat source was simulated by an electrical heater with a power of 10 MW to heat up helium gas to a temperature of 950°C at 4.0 MPa. The principal components of heater, reformer bundle, and steam generator were housed in separate steel pressure vessels in a side-by-side arrangement as shown in Figure 22.6 [24]. The vessels of helium heater and SR were connected by a coaxial helium duct of 5 m length.

Two reformer bundles both with convective helium heating have been investigated in the EVA-II facility. They differed in the way of channeling the helium flow. The first bundle tested was a baffle-design tube bundle for a power of 6 MW consisting of 30 tubes. The bundle was characterized by baffle structures (discs and doughnuts) on the helium side as

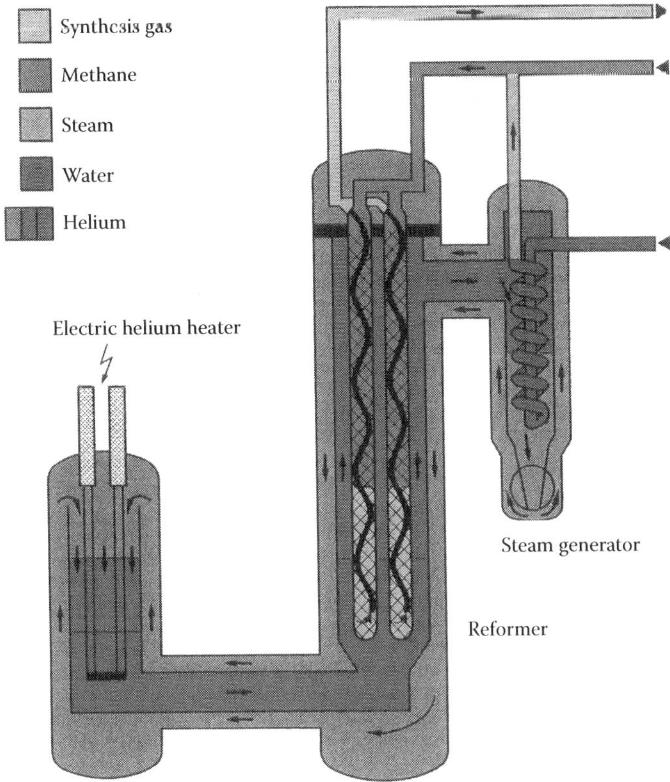

FIGURE 22.6
Flow sheet of steam reforming test facility EVA-II. (From H. F. Niessen et.al., Erprobung und Versuchsergebnisse des PNP-Teströhrenspaltofens in der EVA II-Anlage, Report Jül-2231, Research Center Jülich, Germany, 1988. With permission.)

shown in Figures 22.7 and 22.8 [24], and each splitting tube connected at the top with a separate feed and product gas line. Inside the splitting tube, the heat between 950 and 650°C was used to run the steam reforming process. The internal helical tube for recirculation of the product gas had an outer diameter of 20 mm and a wall thickness of 2.1 mm, the helix itself had a diameter of 70 mm.

The second SR bundle tested was of annulus design for a power of 5 MW. It consisted of 18 tubes with each splitting tube placed inside a guiding tube channel, where the hot helium was flowing upwards through the annular gap. For both designs, tubes were hanging on a supporting plate which allowed an easy exchange of single tubes. The specific data of both tubes and catalytic systems were very similar compared to components planned for nuclear applications (Table 22.3) [24,25]. Also the loads imposed on the supporting structures were characteristic to the nuclear case.

In the steam reforming system EVA-II, no damage was observed on reformer tubes or guiding tubes, internal recuperators or inner return pipes. The components behaved very well in terms of thermal expansion, bending of tubes, friction, fretting, and vibrations caused by flow effects. The efficiency of the catalyst was practically not changed during some 1000 h of operation.

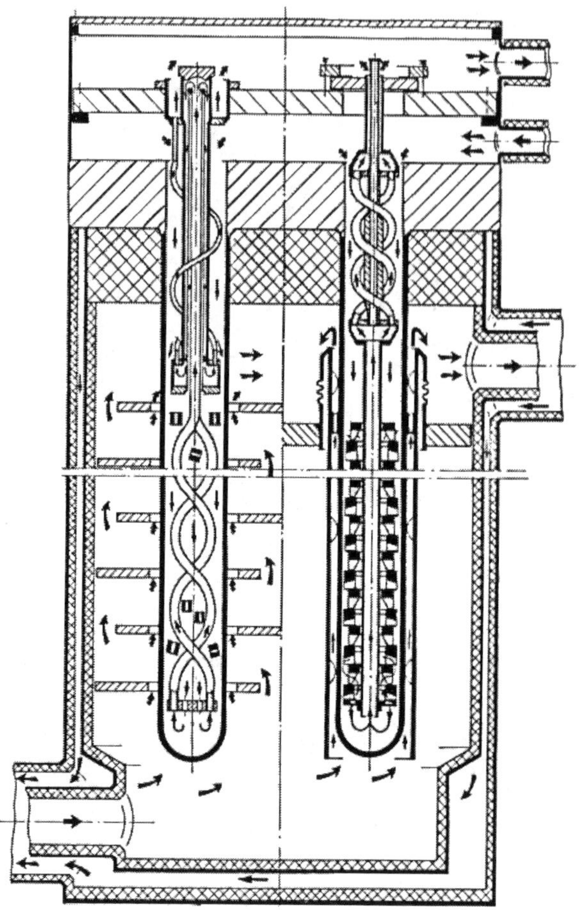

FIGURE 22.7
EVA-II steam reformer bundles of baffle design (left) and annulus design (right). (From H. F. Niessen et. al., Erprobung und Versuchsergebnisse des PNP-Teströhrenspaltofens in der EVA II-Anlage, Report Jül-2231, Research Center Jülich, Germany, 1988. With permission.)

22.2.2 Bunsen Reactor in IS Process

A Bunsen reactor is the key component of IS process hydrogen-production system. Equipment designs are carried out based on the molar flow rate from which throughput of the process fluids are estimated. Several conceptual designs of the reactor have been proposed to satisfy the condition of a Bunsen reaction. Fundamental design of the Bunsen reactor for the 30 Nm^3/h plant is turbulent gas–liquid stirred system, which consists of glass-lined pipe, tank, mixer, and temperature controller. Mixing devices are used—in-line static mixer and mechanically agitated equipment such as impellers—which makes temperature and concentration of the solution homogenous. The temperature controller is used to suppress the increasing of temperature caused by the Bunsen reaction which is an exothermic reaction. Meanwhile, Sakaba et al. proposed a combined design with Bunsen reactor and liquid–liquid separator [26]. The detail of the design will be described in Section 22.3. The French Alternative Energies and Atomic Energy Commission (CEA) designed the counter-current reactor as part of the demonstration loop of IS process in the frame work of the INERI project [27].

FIGURE 22.8
Appearances of steam reformer annulus design (left) and bundles of baffle design (right). (From H. F. Niessen et. al., Erprobung und Versuchsergebnisse des PNP-Teströhrenspaltofens in der EVA II-Anlage, Report Jül-2231, Research Center Jülich, Germany, 1988. With permission.)

For stable continuous operation of hydrogen production by the IS process, regulating the composition of Bunsen reaction solution is one of the key techniques. Kubo et al. developed a simple and easy method for the composition control on plant operation [28,29]. The control method was confirmed by a bench-scaled closed-cycle hydrogen production test, in which the compositions were retained at around regular molar fraction.

22.2.3 HI Section Components in IS Process

From the viewpoint of thermal efficiency of IS process, an improvement of heat balance in the HI separation from HIx solution is the most important issue and many studies have been carried out. Thermal efficiency of IS process is defined as the ratio of higher heating value (HHV) of hydrogen to net external heat input to the process. Efficiency has been expected to be above 40% with IS process. The extractive distillation using phosphoric acid [30], the reactive distillation under pressurized condition [31], and EED was proposed and its feasibility has been studied [32,33]. In addition to these techniques, new ideas have been proposed such as to utilize pervaporation [34]. The 30 Nm^3/h scale IS process plant was designed so that selected options can be examined. A study was carried out to develop the element technologies for the commercial IS process that exhibits competitive performance with other hydrogen-production process. A configuration of the EED was planned to adopt

TABLE 22.3

Comparison of Data of Steam Reformer in EVA Test Facilities and in a Commercial-Size Nuclear Application

	Test Facilities			Nuclear Steam Reformer for 170 MWt HTR Module
		EVA-II		
Parameter	**EVA-I**	**Annulus Design**	**Baffle Design**	
Nuclear heat input [MWt]	~ 0.3	5	~ 6	60.2
Catalyst Tube				
Outer tube diameter [mm]	80–160	120	130	120
Wall thickness [mm]		10	15	10
Length [m]	10–15	13	11.5	14
Number of tubes	1	18	30	199
Tube material	Incoloy 800 H and others	Incoloy 800 H Incoloy 802 Manaurite 36 X IN 519	Incoloy 800 H	Inconel 617
Primary Helium				
Inlet temperature [°C]	800–950	950	950	950
Outlet temperature [°C]	~ 600–750	700	650	720
Inlet pressure [MPa]	4	4	4	4.987
Flow rate [kg/s]	0.15–0.45	~ 3.8	~ 3.8	50.3
Process Feed Gas				
Temperature. inlet recuperator. [°C]	950	330	330	347
Temp. outlet catalyst [°C]	700	810	800	810
Pressure inlet recuperator. [MPa]	2–4	4.0	4.0	5.6
Raw gas flow rate [kg/s]	~ 0.01–0.04	0.62	0.62	34.8
Steam-methane ratio	2–4	4	4	4
Product Gas				
Temperature. inlet inner tube [°C]	~ 830	810	800	810
Temperature. outlet recuperator. [°C]	~ 520	457	450	480

Source: Adapted from H. F. Niessen et al., Erprobung und Versuchsergebnisse des PNP-Teströhrenspaltofens in der EVA II-Anlage, Report Jül-2231, Research Center Jülich, Germany, 1988; Nukleare Fernenergie, Zusammenfassender Bericht zum Projekt Nukleare Fernenergie (NFE), Report Jül-Spez-303, Research Center Jülich, Germany, 1985.

a glass-coated electrolyzer in the plant. The electrolyzer system has been proved industrially in ion-exchange membrane-process electrolysis system used for manufacturing caustic soda and chlorine [35].

Improvements in the processing method of the HIx solution and evaluation of thermal efficiency have been conducted [32,36,37]. In beaker-scale experiments, it was revealed that the concentration of the HIx solution using an electrodialysis device was of an over-quasi-azeotropic composition that separated pure HI from the HIx solution. There are prospects for improvement of the thermal efficiency to adapt these technologies. Theoretical studies on the evaluation and optimization of thermal efficiency of the process system are also ongoing.

Hydrogen permselective membrane reactors (HPMRs) have been investigated to improve the HI one-pass conversion ratio by extracting hydrogen from the reaction field. So far, enhancement of HI conversion ratio to 61.3% at 723 K has been demonstrated using a silica membrane prepared by chemical vapor deposition of tetraethyl orthosilicate (TEOS) on γ-alumina [14].

As for the design of HI decomposer, Ohashi et al. proposed a multistage type HI decomposer [38]. The total heat transfer area of heat exchangers in the proposed HI processing section could be reduced to less than about half of that in the conventional HI processing section. The detail of the decomposer will be described in Section 22.3.2.

22.2.4 Sulfuric Acid Decomposer in IS Process

22.2.4.1 Sulfuric Acid Decomposer Design

The design conditions of the sulfuric-acid decomposer using in the IS process test plant are shown in Table 22.4 [7], which are defined based on the heat-mass balance of the sulfuric-acid decomposition section performed by Knoche et al. [16]. The evaporation temperature of H_2SO_4 is assumed to be 455°C. Heat-exchanger type reactor made of pressureless sintered SiC was adopted for the 30 Nm^3/h test plant. Figure 22.9 shows a concept design of the sulfuric-acid decomposer featuring the SiC ceramic block as the heat exchanger [39]. Considering the fabricability and manufacturability, the shape of heat exchanging component was determined to provide the flow channels of He and sulfuric acid. He gas flows through the inner channels of the SiC ceramic blocks, and exchanges the heat with counter-current H_2SO_4 flow.

Figure 22.10 shows the appearance of the sulfuric acid decomposer featuring multihole heat exchanger blocks made of SiC ceramic [39,40]. The SiC ceramic blocks of 0.25 m outer diameters and 0.75 m height were piled vertically. From the viewpoint of keeping high yield rate of a 0.25 m-diameter SiC ceramic block, the ceramic block was divided into 2 parts in the axial direction. Pure gold, which has excellent corrosion resistance to high-temperature H_2SO_4, was chosen as the seal material between the SiC ceramic blocks. The flow channels of He and sulfuric acid were arranged side-by-side in the block. The number of the flow channels for sulfuric acid and He were 32 and 38, respectively; diameter of these flow channels was 14.8 mm. The mock-up model of sulfuric-acid decomposer was test-fabricated to confirm fabricability and structural integrity as shown in Figure 22.11 [40]. Other new conceptual designs have been proposed in the world, which will be described in Section 22.3.

TABLE 22.4

Design Conditions of Sulfuric Acid Decomposer in the IS Process Test Plant

Items	Conditions
Thermal rating	82.7 kW
He inlet/outlet temperature.	710/535°C
Process inlet/outlet temperature.	435/460°C
He pressure	4 MPaG
Process pressure	2 MPaG
He flow rate	0.091 kg/s
Process flow rate	0.066 kg/s

Source: Adapted from J. Iwatsuki et al., *Proc. of ICONE14-89267, 14th International Conference on Nuclear Engineering*, FL, USA, 2006.

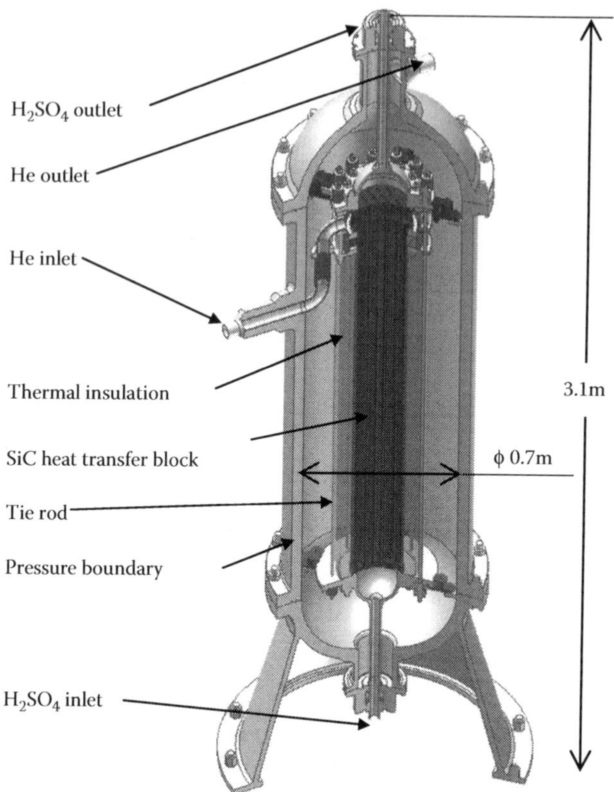

FIGURE 22.9
Concept design of H_2SO_4 decomposer. (From H. Ota et al., *Proc. of ICONE13-50494, 13th International Conference on Nuclear Engineering*, Beijing, China, 2005. With permission.)

22.2.4.2 Mechanical Strength Analyses of SiC Blocks

Temperature and pressure distribution in the ceramic block was analyzed by FEM using ABAQUS Ver.6.4. As for the heat transfer analysis, axial temperature distribution were specified for the surfaces of both He and sulfuric flow cannels, considering heat-transfer coefficients, and outer surface of block is modeled as adiabatic condition.
Properties of SiC were as follows:

- Density: 3.1×10^3 kg/m^3
- Young's modulus: 3.98×10^5 N/mm^2
- Thermal conductivity: 0.12 W/mm K(RT)
  0.08 W/mm K (800°C)
- Thermal expansion coefficient: 3.47×10^{-6}/K(RT)
  3.90×10^{-6} /K (800°C)

Figure 22.12 shows the axial temperature distributions of fluids under the normal operating condition. Figure 22.13 shows the temperature and the stress distributions in the block, respectively [8]. The temperature difference between He and sulfuric acid flow channels was about 250°C at maximum.

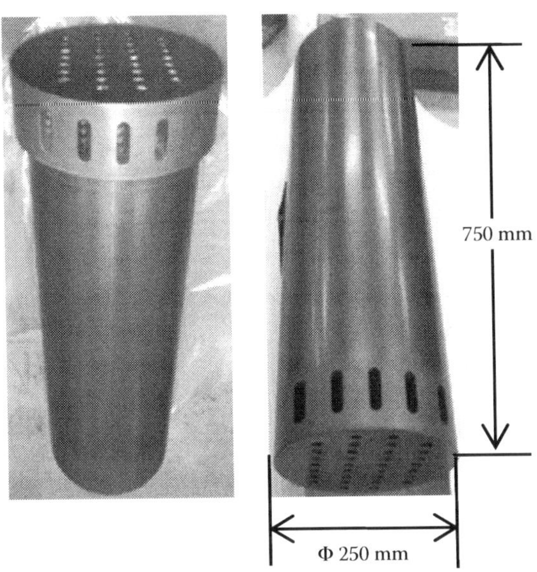

FIGURE 22.10
Heat exchanger blocks made of SiC. (From H. Ota et al., *Proc. of ICONE13-50494, 13th International Conference on Nuclear Engineering*, Beijing, China, 2005; and H. Noguchi et al., *JAEA-Technology*, 2007-041, 2007 (in Japanese). With permission.)

A high-temperature region was located above the dryout region, as shown in Figure 22.13a. The stress shown in Figure 22.13b is a coupled stress of thermal and static stress caused by the operating-pressure difference between He and sulfuric acid. The maximum stress, about 124 MPa, was generated around the dryout region of the sulfuric-acid flow channels. This value is much lower than the average bending strength of high-strength pressureless sintered SiC, 450 MPa [8].

22.2.4.3 Seal Performance of Sulfuric Acid Decomposer

The SiC blocks of the sulfuric acid decomposer were connected with a special seal. This seal technology, between the ceramic blocks, is useful for decreasing fabrication costs of large-scale ceramic structures. However, it is well known that the seal performance will deteriorate due to the difference of thermal expansion between SiC and metal under the high-temperature condition.

Seal performances at the interfaces of metal/metal, metal/SiC, and SiC/SiC, which are the boundary of He–He and He–H_2SO_4, were investigated with scaled models under hydrostatic and He gas conditions [40]. Figures 22.14 and 22.15a show the schematic and an outer view of the test apparatus, respectively. A SiC block model (120 mm in diameter and 50 mm in thickness) shown in Figure 22.15b was installed in the test apparatus. Wire gaskets and the cap gaskets made of pure gold were examined for the test plant. The wire gasket was used at the connections of metal/metal and metal/SiC, and the cap gasket was used to the SiC block-parts connection. Figure 22.15c shows the cap gasket mounted in a SiC block model.

Helium leak tests were carried out under load pressure up to 4 MPa and surface temperature of the models up to 500°C. The SiC block model was heated with a vacuum

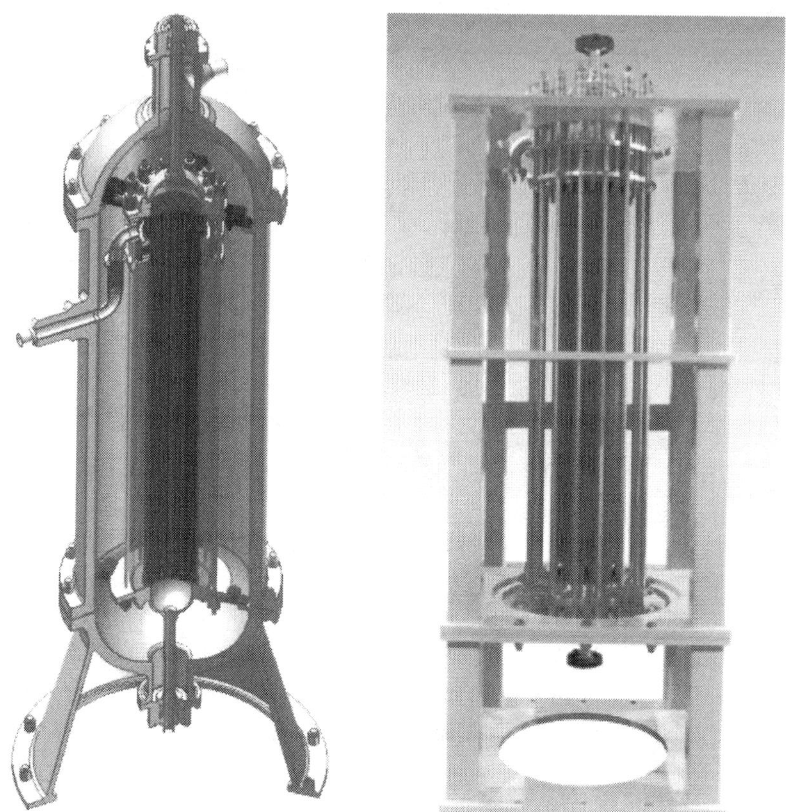

FIGURE 22.11
Test fabrication of H_2SO_4 decomposer. [From H. Noguchi et al., *JAEA-Technology*, 2007-041, 2007 (in Japanese). With permission.]

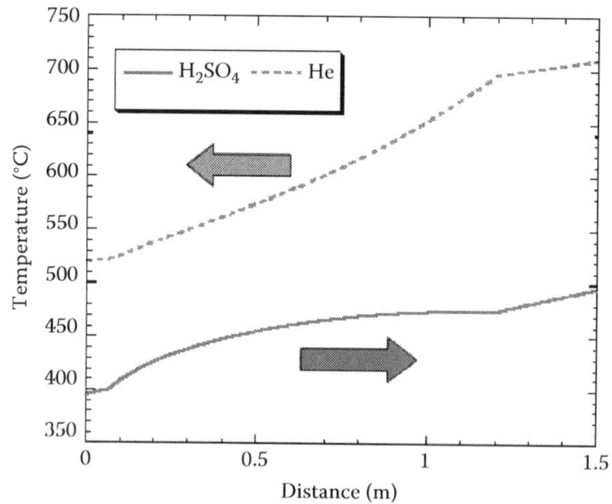

FIGURE 22.12
Axial temperature distribution. [From A. Terada, J. Iwatsuki, and S. Ishikura, *J. Nucl. Sci. Tech.*, 44(3), 477–482, 2007. With permission.]

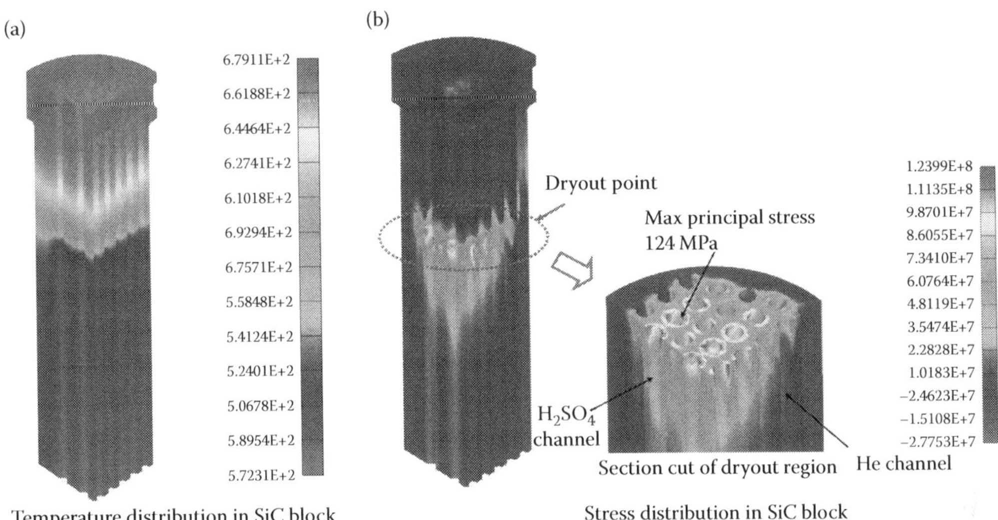

FIGURE 22.13
Temperature and stress distribution in SiC block. [From A. Terada, J. Iwatsuki, and S. Ishikura, *J. Nucl. Sci. Tech.*, 44(3), 477–482, 2007. With permission.]

electric furnace, and leak rates were measured with a helium leak detector. Figure 22.15d shows the appearance of the wire gasket after the seal performance test. Table 22.5 shows the leak rates obtained at 500°C in the three times of heat cycle up to 500°C [40]. Table 22.6 shows the results obtained at 20°C after the leak measurement at 500°C [40]. In the tables, line loads and seating stresses (tightening forces) are also indicated. Although the tightening forces under 500°C decreased down to less than half of the forces under 20°C, the leak rates under 500°C were decreased in metal/SiC and SiC/SiC connections. This might be due to softening and adhesion effect of gold.

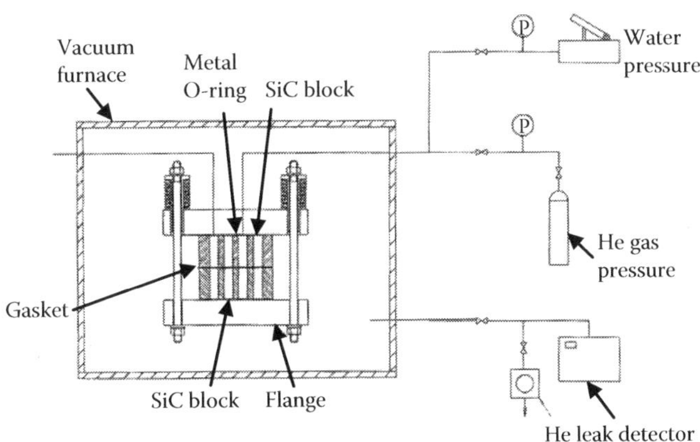

FIGURE 22.14
Schematic of leak test apparatus. (From H. Noguchi et al., *Proc. of ICAPP2006-6166*, 2006. With permission.)

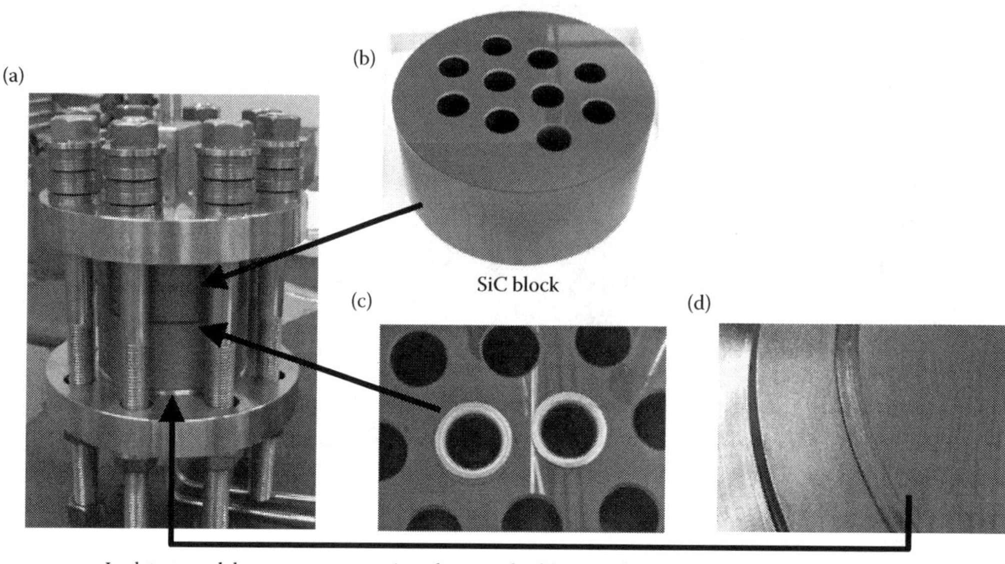

FIGURE 22.15
Leak test model of H_2SO_4 decomposer. (From H. Noguchi et al., *Proc. of ICAPP2006-6166,* 2006. With permission.)

Furthermore, in order to confirm the seal performance under earthquake conditions, two kinds of horizontal loading tests, 130 cyclic loading of 612 N (0.31 G) and one cycle loading up to 1224 N (0.61 G), were carried out by using the mock-up model [41]. The above horizontal loads are mentioned in Japan's high-pressure gas regulations. A hydraulic lifter, strain gauges, displacement gauges, and a digital manometer were installed on the mock-up model. Helium gas was filled up to 0.2 MPaG in the model, then, horizontal load was added to the near centroid of the SiC ceramic blocks by the lifter. In the tests, the following were measured: strain of the SiC ceramic blocks and the tie rods to be within 3×10^{-5}%, horizontal displacement of the SiC ceramic blocks to be within 30 μm, gap variation between the upper and lower SiC ceramic block not to be detected, and He gas pressure. Their values were negligible for the function and strength of the decomposer.

The maximum value in the cyclic horizontal loading test results was 1.37×10^{-3} and 2.86×10^{-3} Pa m^3/s at 1224 N in one cycle loading test results. Based on these test results,

TABLE 22.5
Helium Leak Rate at 500°C in the Seal Performance Test of the Sulfuric Acid Decomposer

Connection	He Leak Rate at 3 Heat Cycle Pa m³/s	Maximum Pressure	Line Load or Seating Stress
Metal/Metal	3.6×10^{-4}	3.6 MPa	1.0×10^5 N/m (3.2×10^4 N)
Metal/SiC	2.2×10^{-6}	2.6 MPa	1.2×10^5 N/m (4.0×10^4 N)
SiC/SiC	7.5×10^{-8}	4.0 MPa	24 MPa (5.9×10^3 N)

Source: Adapted from H. Noguchi et al., *Proc. of ICAPP2006-6166,* 2006.

TABLE 22.6

Helium Leak Rate at 20°C in the Seal Performance Test of the Sulfuric Acid Decomposer

Connection	He Leak Rate After 3 Heat Cycle Pa m³/s	Maximum Pressure	Line Load or Seating Stress
Metal/Metal	6.4×10^{-8}	4.0 MPa	2.8×10^5 N/m (8.9×10^4 N)
Metal/SiC	2.7×10^{-4}	3.4 MPa	3.0×10^5 N/m (9.3×10^4 N)
SiC/SiC	2.8×10^{-7}	4.0 MPa	58 MPa (1.4×10^4 N)

Source: Adapted from H. Noguchi et al., *Proc. of ICAPP2006-6166*, 2006.

the mock-up model involves four wire gaskets and 70 hat gaskets. The leak rare per one gasket will be estimated to be about 3.9×10^{-5} Pa m³/s.

22.2.5 SO_3 Decomposer in IS Process

22.2.5.1 Conceptual Design of SO_3 Decomposer

As for the SO_3 decomposer, a concept of SiC plate-type SO_3 decomposer has been proposed [42]. The plate-type decomposer can realize relatively large heat transfer surface relatively and consequently can make the decomposer more compact. It also presents a possibility to add gas-separation function as a membrane reactor to enhance conversion rates of SO_3 decomposition under high operation pressure of 2 MPa (equilibrium conversion rates is less than 50% at 850°C).

Table 22.7 shows the design conditions for the SO_3 decomposer in the test plant [42], based on the literature [43]. SiC is presently the most promising material for the SO_3 decomposer to meet the following requirements:

- High thermal conductivity necessary to transfer heat effectively by high temperature He gas (max 880°C).
- Anticorrosiveness against the process gases such as the SO_3, SO_2, H_2O, and O_2.
- High mechanical strength to withstand the static pressure difference between He gas and process gas.

TABLE 22.7

Design Conditions of SO_3 Decomposer in the IS Process Test Plant

Item	Conditions for Hydrogen Production 30 Nm³/h
Process gas pressure	2.0 (MPaG)
Process gas temperature (inlet/outlet)	527/850 (°C)
He pressure (inlet)	4.0 (MPaG)
He flow rate	100 (g/s)
He temperature (inlet)	880 (°C)
Heat exchange	100(kW)

Source: Adapted from A. Kanagawa et al., *Proc. 13th Int. Conf. Nucl. Eng., ICONE13-50451*, Beijing, China, May 16–20, 2005.

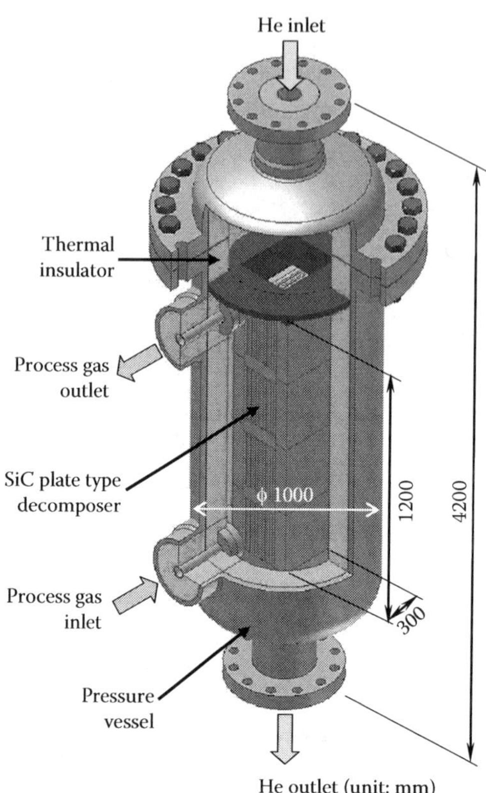

FIGURE 22.16
Concept design of the SiC plate type SO_3 decomposer. (From A. Kanagawa et al., *Proc. 13th Int. Conf. Nucl. Eng., ICONE13-50451*, Beijing, China, May 16–20, 2005. With permission.)

Figure 22.16 shows concept design of the SiC plate-type SO_3 decomposer for the test plant [42]. The SO_3 decomposer was installed in a pressure vessel lined internally with thermal insulator. One unit of SO_3 decomposer was a stack of SiC plate-type heat exchanger segment. He gas (heating source) came from the top of pressure vessel, and was introduced to the upper unit of the SO_3 decomposer by way of partition plate. Process gases flowing up in the unit were heated by He gas flowing down to the unit's bottom, so that the SO_3 decomposer worked as a counter-flow-type heat exchanger. The space between the heat insulator and the heat exchanger unit was also filled with He.

Required heat transfer area of the SO_3 decomposer was more than 3.6 m^2, which was decided on the basis of conservative heat-transfer evaluation. This value was verified with the numerical analytical works.

The dimension of a SO_3 decomposer unit consists of 4-staged stack of the heat exchangers, which ensured about 1.2 m^2 of the heat transfer area. Each stage of the stack was around 250 mm wide, 300 mm long and 200 mm high. For 30 Nm^3/h scale component, it was necessary to integrate 3 units of the stack. Figure 22.17 shows the structure of the decomposer unit [42]. He gas flow channels had many arranged cylindrical ribs to keep mechanical strength of the channel, which worked as heat-transfer promoters.

The rated space velocity (SV) in a SO_3 catalyst layer was necessary to be kept less than 10000 hr^{-1} in order to achieve high conversion ratio of SO_3 decomposition. In this unit, the

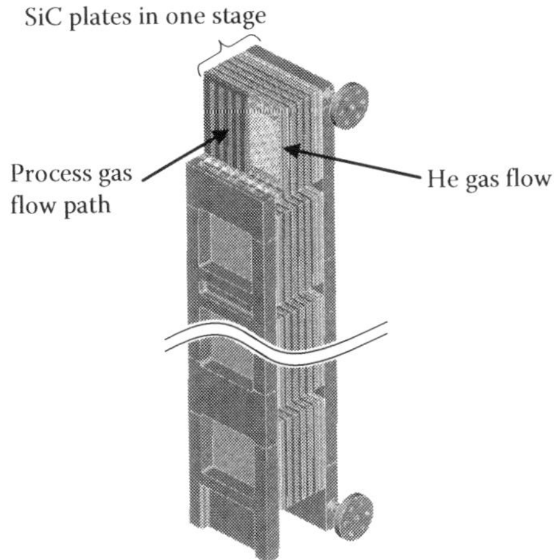

FIGURE 22.17
Structure of the SO_3 decomposer unit. (From A. Kanagawa et al., *Proc. 13th Int. Conf. Nucl. Eng., ICONE13-50451,* Beijing, China, May 16–20, 2005. With permission.)

internal volume charged with catalyst was around 17.64 L. Then, the SV became 1.71 (mol/s) × 22.4 (NL/mol) × 3600/17.64 (L) = 7817 h^{-1}.

A mock-up model of plate-type SO_3 decomposer was test fabricated to verify manufacturing feasibility such as a stacked structure and pipe connections with flanges. Figure 22.18 shows an appearance of the prototype SO_3 decomposer. It was confirmed that the prototype of the heat exchanger could be produced as designed by an appearance inspection (crack, omission) and a dimensional inspection. Preliminary stress analysis and test fabrication were also carried out on SO_3 decomposer, the results of which indicated the structural feasibility of the SiC plate-type heat exchanger [42].

FIGURE 22.18
Appearance of prototype SO_3 decomposer. (From A. Kanagawa et al., *Proc. 13th Int. Conf. Nucl. Eng., ICONE13-50451,* Beijing, China, May 16–20, 2005. With permission.)

22.2.5.2 *Catalyst Test for SO_3 Decomposition*

JAEA had carried out a preliminary test of SO_3 decomposition by using of a high-pressure catalytic reaction test apparatus. 98 wt.% H_2SO_4 is vaporized by an electric heater, and sent to the catalytic reactor with Argon (Ar) gas as carrier, and then sulfuric-acid vapor was decomposed. At normal operation, sulfuric acid is supplied from the H_2SO_4 tank by the H_2SO_4 pump, and vaporized to the SO_3 gas at the evaporator, and decomposed to SO_2 and O_2 gases. The decomposed gases are cooled to around room temperature (40°C) and condensed at a condenser. SO_2, O_2, unreacted H_2SO_4, and carrier Ar-gas are separated to gas and liquid at the condenser, and the separated gases are sent to the SO_2 absorber. The trapped SO_2 gas is neutralized with NaOH and removed.

The catalytic reactor is composed of double-pipe structure. The inner pipe (ID: 30 mm) is a catalytic reactor made of SiC filled with SiO_2 loading Pt catalyst (size is about φ1 mm). The outer pipe (ID: 48 mm) is a pressure retaining pipe made of Alloy 800H. In the gap between inner pipe and outer pipe (width 5.5 mm), Ar gas is vented in order to prevent process gases from mixing in the gap. The performance of the catalyst is evaluated by SO_3 gas decomposing rate to the SV in a catalyst bed, and the SO_3 gas decomposing rate is evaluated by SO_2 concentration and O_2 flow rate. The SO_2 concentration is measured by nondispersive SO_2 concentration meter, and O_2 flow rate is measured by wet gas meter.

The main testing condition is as follows. The SO_3 decomposing test of the parameters SV, pressure, and temperature, are performed and the performance of the catalyst evaluated [7].

- Maximum test temperature: 850°C
- Maximum test pressure: 0.75 MPa
- SV in a catalyst layer: 1000–10,000 h^{-1}

It was difficult to evaluate the conversion ratio form of the concentration of SO_2, because of the SO_2 gas solubility into H_2O which is decomposed from sulfuric acid. Therefore, the conversion ratio was evaluated from the flux of O_2. In the preliminary test for about 11 h, almost equivalent value to equilibrium conversion ratio was attained under the SV of 1000 h^{-1}.

22.3 Future Technical Issues

22.3.1 Steam Reformer

Development of high efficiency and high capacity equipment is one of the most important technical issues for economical improvement and practical application of HTGR. The compactness and the performance upgrading of equipments in a nuclear hydrogen-production plant, such as an IHX and a decomposer, determine the feasibility of a whole plant.

A survey by JAEA of heat-exchanger technologies with potential to augment hydrogen-production rate resulted in the SR design-improvement measures illustrated in Figure 22.19 [3]. These measures were reported to have contributed to increasing overall steam reforming thermal efficiency to 78%, which approaches to the 80–85% range of the conventional fissile-fuel-fired plant process. The technologies developed to improve the SR design are applicable to other HTGR-hydrogen production systems that share the heat-exchanging technology of endothermic chemical reactor.

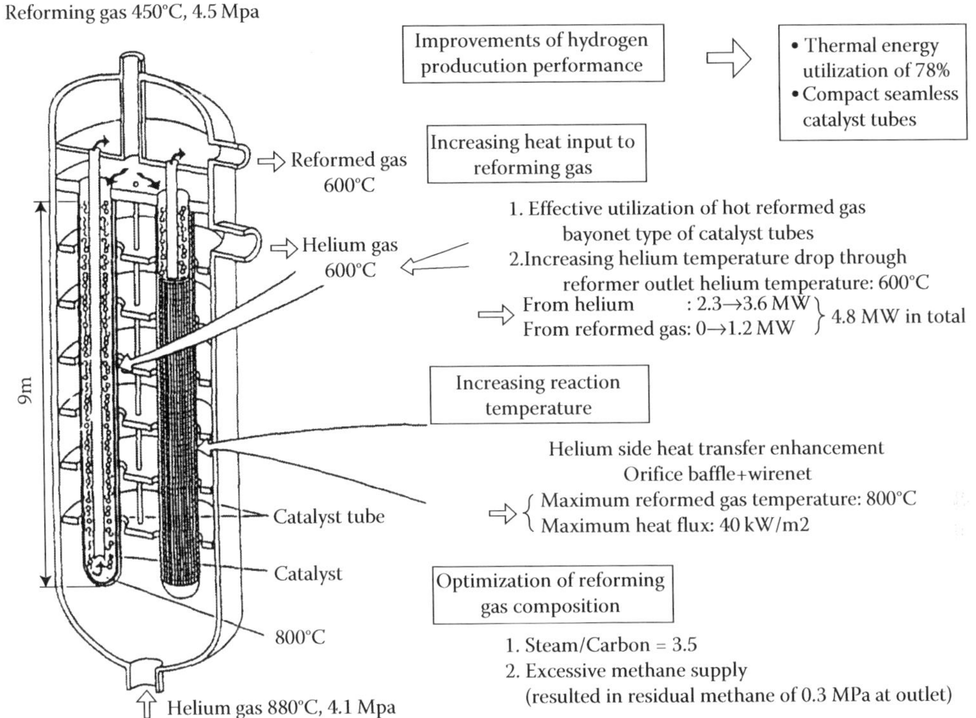

FIGURE 22.19
Improvements of helium-heated SR performance. (From Y. Miyamoto et al., *JAERI, IAEA-TECDOC-988*, 1997. With permission.)

In Germany, details of the concept of the SR have been proposed as a novel component in nuclear process heat plants as shown in Figure 22.20 (private communication). It is a bundle consisting of straight tubes connected to an upper supporting plate. The tubes filled with a catalyst contain an internal recuperator and an inner return duct for the hot process gas. The upper part of the component includes the collector structures for the feed gas composed of steam and methane, and for the product gas containing H_2, CO, CH_4, CO_2, and steam. From the total heat transferred into the SR, 85% are used for the reforming process, while 15% are taken to heat up the feed gas. The main characteristic data of the SR component as was designed by INTERATOM on the basis of a simple cylindrical pebble-bed core with a power of 170 MWt are listed in Table 22.3.

22.3.2 IS Process Components

22.3.2.1 *Design Concept of Chemical Reactors*

Various promising conceptual designs of the key IS process components have been proposed. For the Bunsen reactor, a design of mixer–settler reactor integrating the Bunsen reaction function with the liquid–liquid separator function is proposed [44]. Figure 22.21 illustrates the design concept based on a mechanically agitated reaction vessel section and a long settler vessel section. In the mixer, sulfur dioxide, iodine, and water are mixed and react to form acids by using multiple impellers on a shaft. The chambers and baffles are

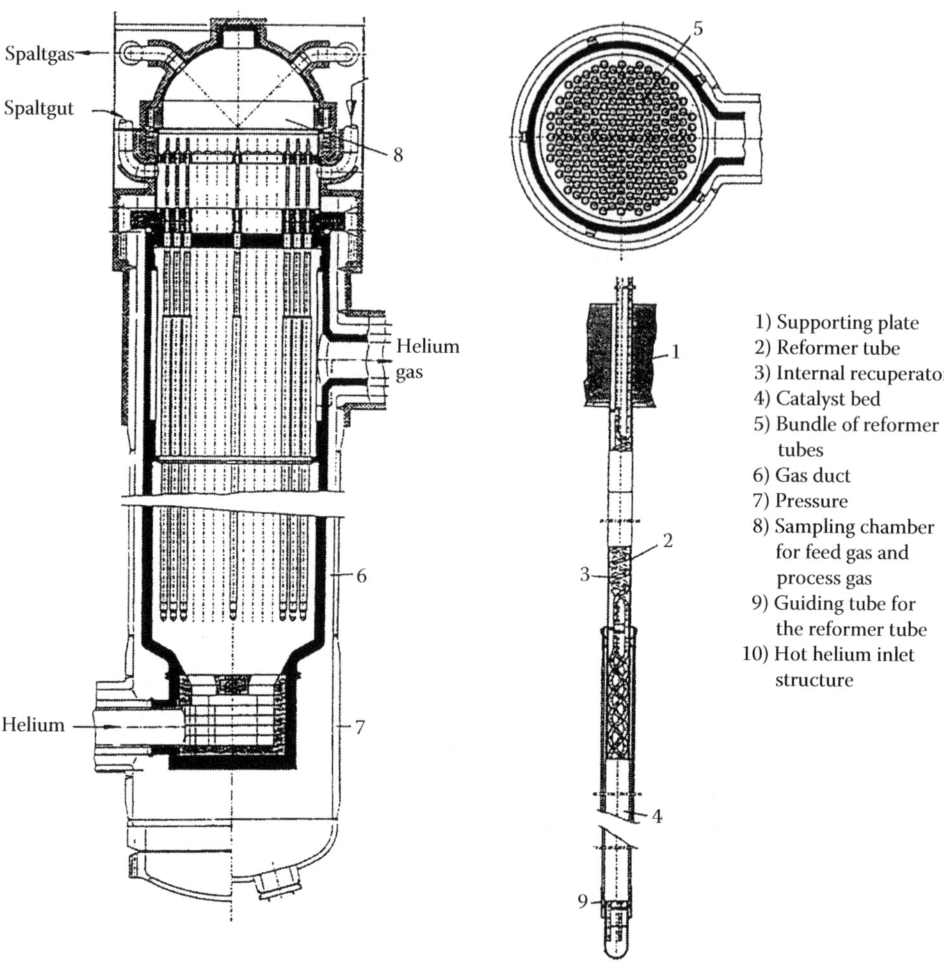

FIGURE 22.20
Technical concept of a helium heated SR connected to a modular process heat HTGR. *Private communication.*

designed to retain unreacted solutions while facilitating the flow of the acids into the settler chamber and subsequent phase separation of the acids.

A proposed design of the HI decomposer allows internal multistage of HI decomposition reaction and removal of iodine from HIx solution [38]. The design aims at improving equipment cost performance by reducing total heat-transfer area of heat exchangers and by reducing the recycle duty of undecomposed HIx and eliminating separation components. The new concept design was numerically evaluated, especially in terms of the flow rate of undecomposed HIx and product iodine at the outlet of the decomposer.

A suitable configuration of the multistage HI decomposer was countercurrent rather than cocurrent, and the HIx solution from an EED at a low temperature was a favorable feed condition for the multistage HI decomposer. Figures 22.22 and 22.23 show the conceptual design and system of the multistage HI decomposer, which is a countercurrent type decomposer for 1000 Nm^3/h scale IS process plant [38,44]. The flow rate of undecomposed HI and product iodine at the outlet of the multistage HI decomposer was

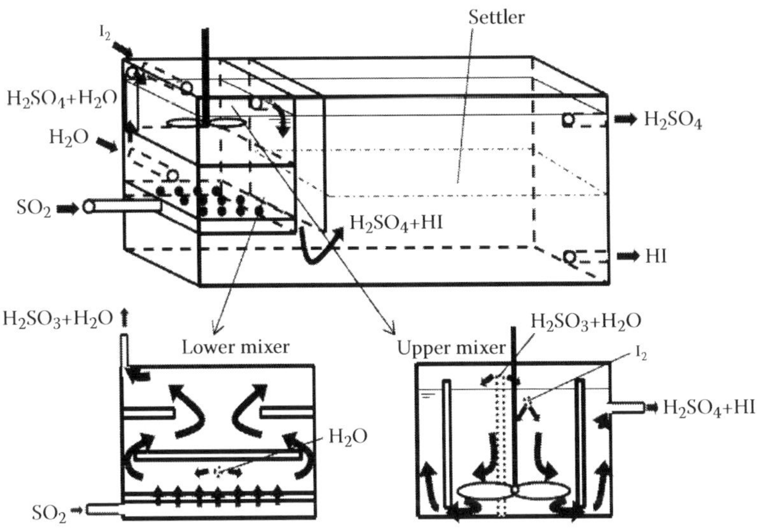

FIGURE 22.21
Mixer-settler type Bunsen reactor. [From N. Sakaba et al., *Trans. Atom. Energy. Soc. Japan.*, 7(3), 242–256, 2008 (in Japanese). With permission.]

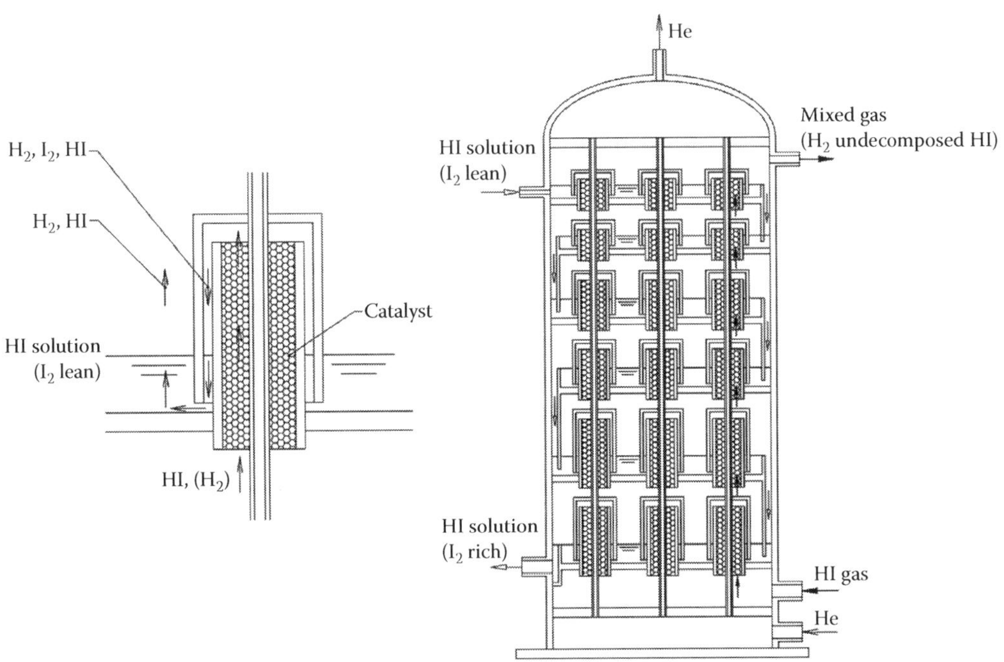

FIGURE 22.22
Conceptual design of a multistage HI decomposer. [From H. Ohashi et al., *Trans. Atom. Energy. Soc. Japan.*, 29, 68–82, 2009 (in Japanese). With permission.]

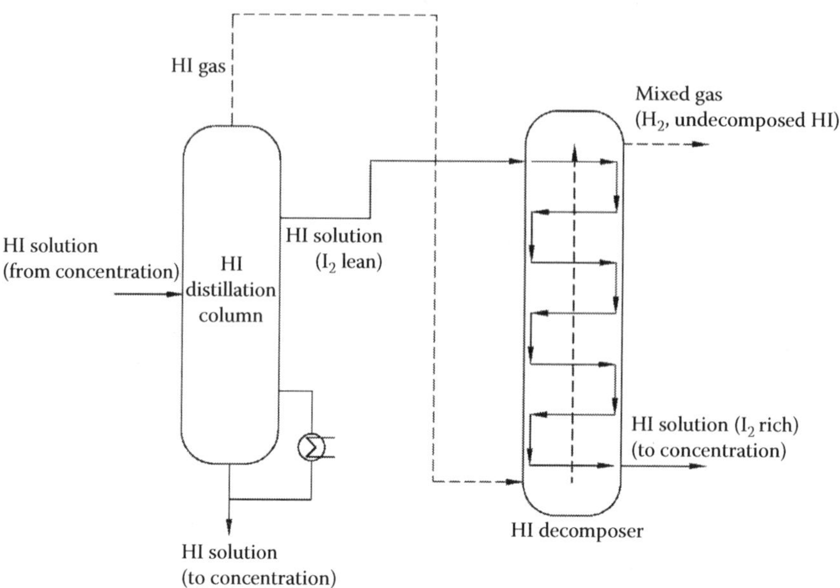

FIGURE 22.23
Conceptual system of a multistage HI decomposer. [From N. Sakaba et al., *Trans. Atom. Energy Soc. Japan*, 7(3), 242–256, 2008 (in Japanese). With permission.]

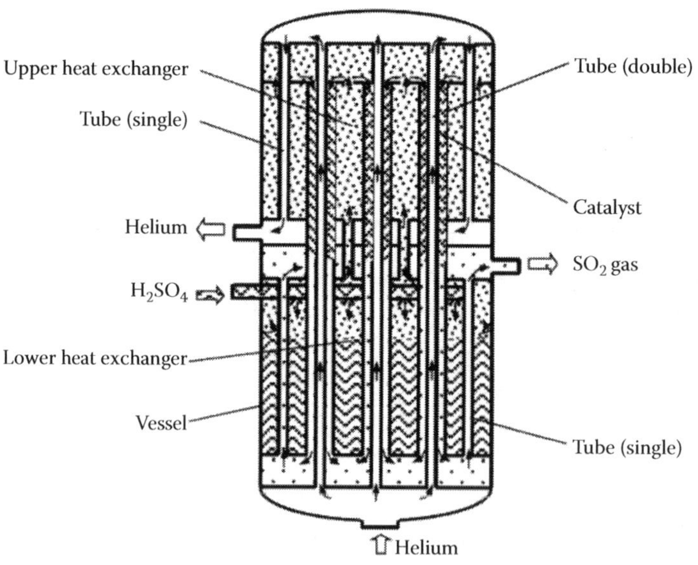

FIGURE 22.24
Combined sulfuric acid decomposer. [From N. Sakaba et al., *Trans. Atom. Energy. Soc. Japan.*, 7(3), 242–256, 2008 (in Japanese). With permission.]

significantly lower than that of the conventional HI decomposer, because the conversion was increased, and HI and iodine were removed by the HIx solution. Based on this result, an alternative HI processing section using the multistage HI decomposer and eliminating some recuperators, coolers, and components for the separation was proposed and evaluated. The total heat transfer area of heat exchangers in the proposed HI processing section could be reduced to less than about 1/2 that in the conventional HI processing section.

As for the H_2SO_4 and SO_3 decomposers, Sakaba et al. proposed combined H_2SO_4 decomposer [44], as shown in Figure 22.24, which was designed based on shell and tube-type heat exchanger integrating SO_3 decomposer, H_2SO_4 vaporizer, and process heat exchanger. The vessel consists of two heat exchanger parts which evaporate sulfuric acid and decompose sulfur trioxide by the heat of sulfur dioxide and helium, respectively. Heat exchanger parts are connected with double pipes in which helium flows through inner pipe and sulfur trioxide flows through outer pipe. Regarding the material, SiC film coating is used against corrosive circumstances. Integration of components can reduce number of equipment such as reactor vessels, connecting piping, and

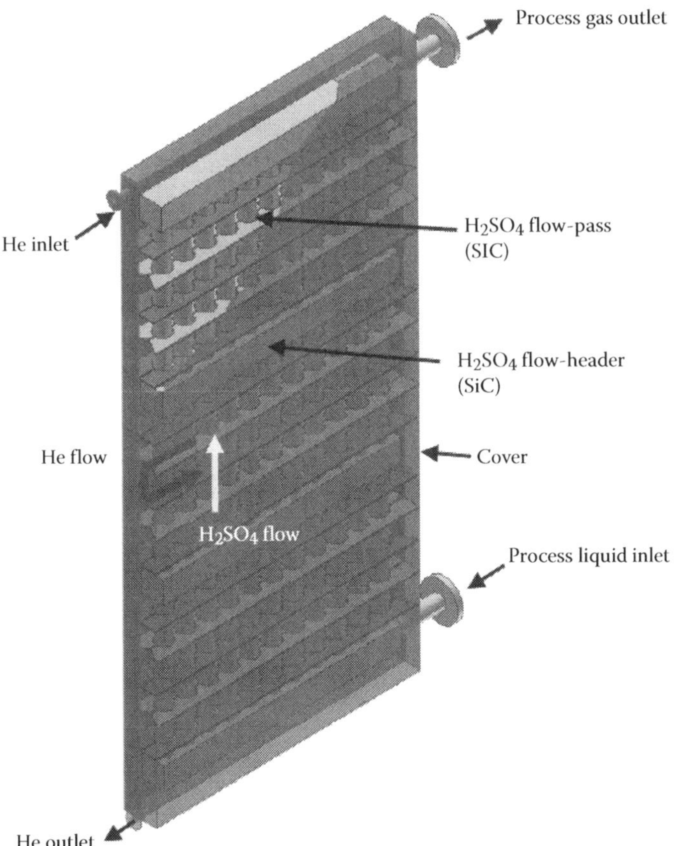

FIGURE 22.25
Concept of evaporation part of H_2SO_4 decomposer. [From I. Minatsuki et al., *J. Power and Energy Systems*, 181, 36–48, 2007. With permission.]

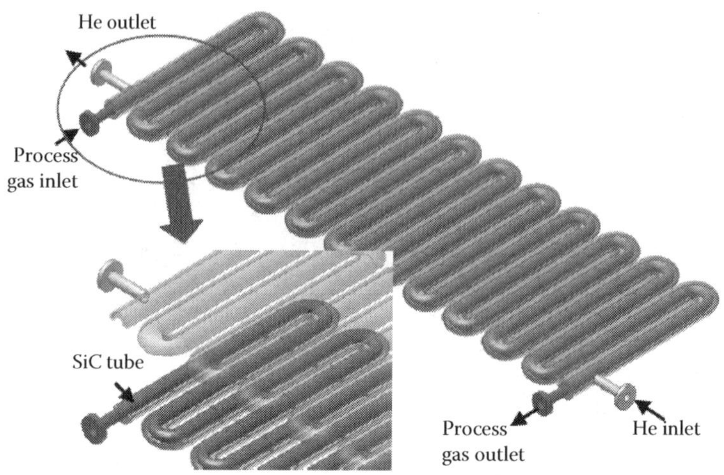

FIGURE 22.26
Concept of gas phase super-heating part of H_2SO_4 decomposer. [From I. Minatsuki et al., *J. Power and Energy Systems*, 1(1), 36–48, 2007. With permission.]

transfer pumps, so that the amount of material is expected to be reduced substantially. Also, the number of connections was reduced and it results in the reduction of risks of process-fluid leakage.

Mitsubishi Heavy Industries Ltd. (MHI) developed a cylinder type sulfuric-acid decomposer made of SiC [45]. The constitution of the decomposer dividing into two regions

FIGURE 22.27
Outer view of prototype model. [From I. Minatsuki et al., *J. Power and Energy Systems*, 1(1), 36–48, 2007. With permission.]

due to the difference of heat-transfer forms in the evaporation part and gas phase super-heating part. Figures 22.25 and 22.26 show the concept design of the evaporation part and gas phase super-heating part, respectively, in a sulfuric acid decomposer. Structural feasibility of the heat exchanger was analytically verified by the thermal analyses and thermal stress analysis. Figure 22.27 shows the test-fabricated prototype model.

Sandia National Laboratories (SNL) has developed an integrated decomposer unit that utilizes a SiC-based bayonet-type heat-exchanger tube for acid decomposition and effective heat recuperation [46–48]. Two series of the SiC integrated decomposer (SID) have been tested. The first series utilizes a 686 mm long bayonet for the Integrated Lab-Scale (ILS) experiments. The second series utilizes a 1372 mm long SiC bayonet which is a full scale ILS. Figure 22.28 shows the half-scale SiC bayonet decomposer. The upper third region of the unit is a catalyst region [47]. The upper two thirds of the unit are externally heated. Heat from the output stream is redirected to the inlet stream. Figure 22.29 shows

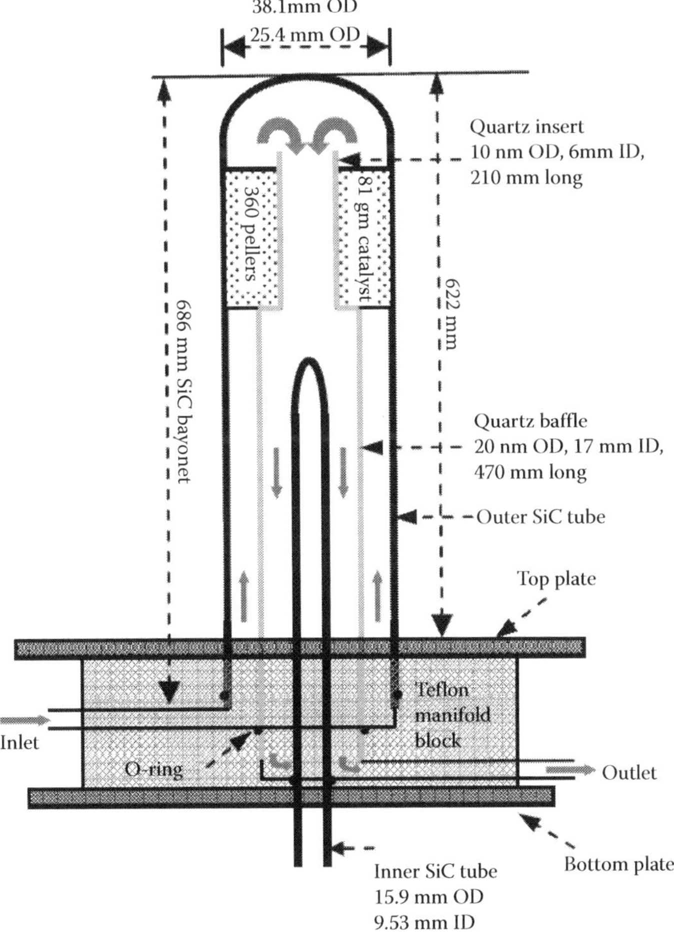

FIGURE 22.28
Schematic of half-scale bayonet-type sulfuric acid decomposer. [From F. Gelbard et al., *ANS Embedded Topical Meeting on the Safety and Technology of Nuclear Hydrogen Production, Control and Management (ST-NH2)*, Boston, MA, June 24–28, 2007. With permission.]

FIGURE 22.29
The sulfuric acid decomposition apparatus. [From R. Moore et al., *ANS Embedded Topical Meeting on the Safety and Technology of Nuclear Hydrogen Production, Control and Management (ST-NH2)*, Boston, MA, June 24–28, 2007. With permission.]

the completed sulfuric acid apparatus [48]. The SID is shown on the left side of the photograph with the heaters removed. The performance of the bayonet decomposers have been confirmed by experiments. The half-scale (686 mm long) unit readily produced 100 L/h of SO_2, even for dilute (20 mol%) acid. The full-scale (1372 mm long) unit achieved over 250 L/h SO_2 production.

Westinghouse Electric Company has proposed several designs of sulfuric-acid decomposer including a two-vessel concentrator/decomposition reactor configuration [49]. The decomposer utilizes bayonet tubes with internal feed tubes for the hot $H_2SO_4/H_2O/SO_3$ coming down from the top in the center of the finger as shown in Figure 22.30. Westinghouse has also developed an advanced version of the decomposition reactor that utilizes SiC for all surfaces that are in contact with sulfuric acid or decomposition products above temperatures of about 300°C. For temperatures below 300°C, carbon steel with suitable coatings are used. This concept is shown in Figure 22.31 [50].

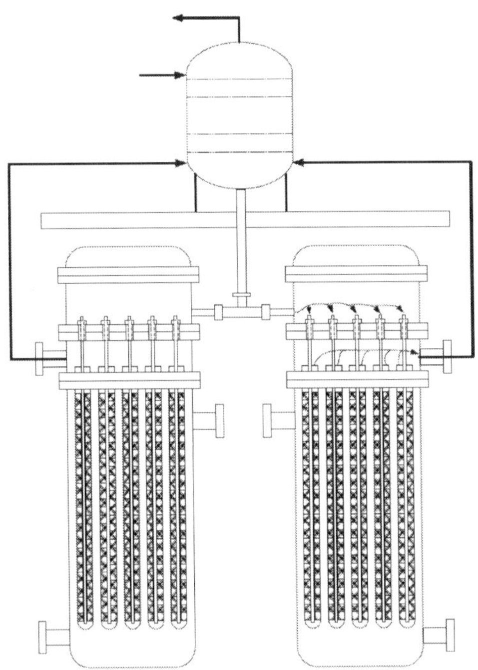

FIGURE 22.30
Top down inside feed flow finger tube decomposition reactor and concentrator. [From D. McLaughlin et al., *ANS Embedded Topical Meeting on the Safety and Technology of Nuclear Hydrogen Production, Control and Management (ST-NH2)*, Boston, MA, June 24–28, 2007. With permission.]

CEA has focused on two kinds, printed circuit heat exchanger (PCHE) and plate fin heat exchanger for a decomposer. Its materials were Incoloy 800 with aluminide coating and SiC [51].

Kim et al. developed a hybrid heat exchanger for the SO_3 decomposer [52]. The heat exchanger has a hot gas channel and a SO_3 mixture channel. Hot gas channel is a semicircular shape like a PCHE type, which allows for a large differential pressure between the hot gas and SO_3 mixture. The surface of SO_3 mixture channel is coated with SiC by ion-beam technology to improve its corrosion resistance [53].

22.3.2.2 Analytical System for Developmental Components

Following analyses and estimations are necessary for designing the IS process plant:

- Process simulation to optimize state functions in the process and effective functions necessary for improving process efficiency.
- Dynamic process simulation to construct of control system.
- Examination of probabilistic damage evaluation technique to confirm structural integrity for SiC ceramics.
- Thermal–hydraulic analyses for interaction chemical reaction and multiphase flow.
- Mechanical strength of plant in case of earthquakes.

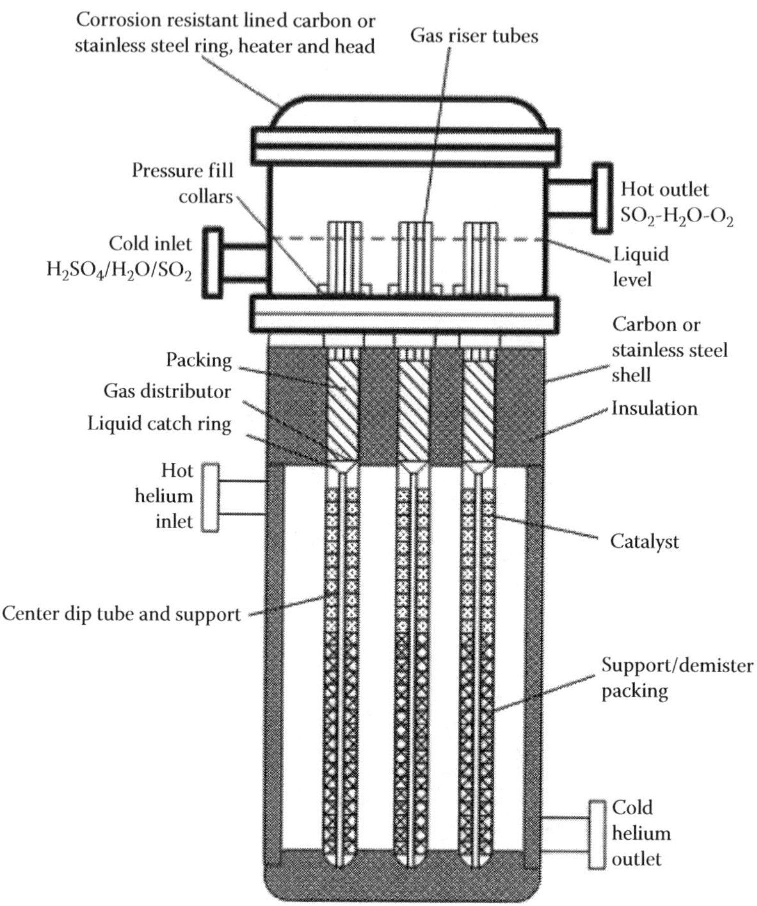

FIGURE 22.31
Advanced H_2SO_4 decomposition reactor design. (From S. M. Connollya et al., *Int. J. Hydrogen Energy*, 34, 4074–4087, 2009. With permission.)

To corroborate above analyses and estimations efficiently, a design and analysis system based on almost all codes using the CAE, are commercial codes such as PRO/II [54], Object-DPS, OLI [55], ABAQUS, CARES-Life, and so on.

Analyses models for designing apparatus used in these codes needs be improved in order to increase their estimation accuracy by using fundamental experimental results. In order to verify analytical models necessary for designing a complicated internal flow structure with chemical reactions such as in a Bunsen reactor, preliminary experiments are being conducted under a simulated Bunsen reaction condition with CO_2 gas–water two-phase bubbly flow.

Flowing CO_2 gas bubbles and velocity distribution in CO_2 gas–water two-phase bubbly flow was measured with a (particle imaging velocimetry) PIV system as shown in Figure 22.32 [8]. Reflecting results of the preliminary experiment, the thermal–hydraulics code developed based on α-Flow/MISTRAL [56] was improved to its highest efficiency.

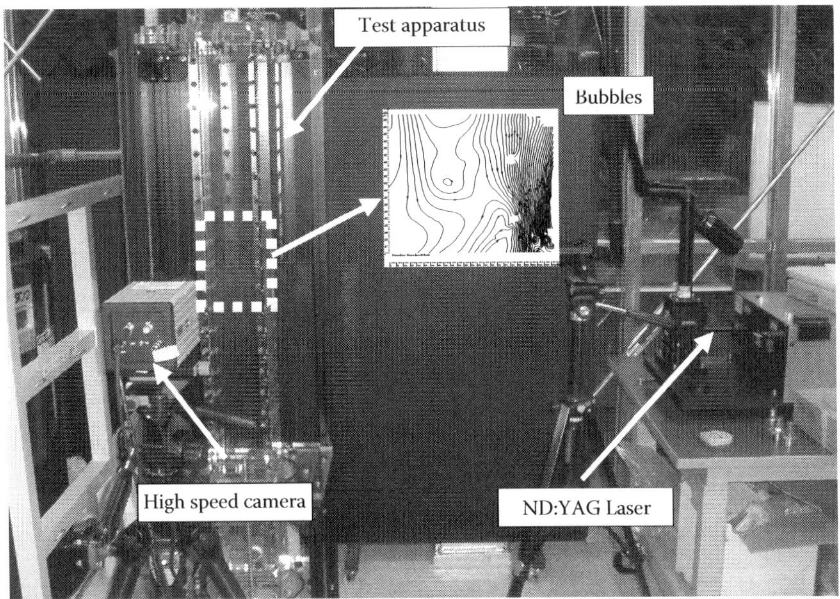

FIGURE 22.32
PIV apparatus for flow visualization. (From A. Terada, J. Iwatsuki, and S. Ishikura, *J. Nucl. Sci. Tech.*, 44(3), 477–482, 2007. With permission.)

References

1. K. Hada et al., Design of steam reforming system to be connected to HTTR, *Proc. 3rd JAERI Symp. on HTGR Technologies*, Oarai, Japan, February 15–16, 229, 1996.
2. H. Ohashi et al., Performance test results of Mock-up test facility of HTTR hydrogen production system, *J. Nucl. Sci. Tech.*, 41, 385–392, 2004.
3. Y. Miyamoto et al., Overview of HTGR utilization system developments, *JAERI, IAEA-TECDOC-988*, 1997.
4. Y. Inagaki et al., Development programme on hydrogen production in HTTR, *Proc. Technical Committee Meeting*, Beijing, People's Republic of China, November 2–4, 1998. *International Atomic Energy Agency, International Working Group on Gas-Cooled Reactors*, Vienna (Austria) IAEA-TECDOC-1210, pp. 213–221, 2001.
5. S. Kasahara et al., Flowsheet study of the thermochemical water-splitting iodine–sulfur process for effective hydrogen production, *Int. J. Hydrogen Energy*, 32, 489–496, 2007.
6. S. Kubo et al., A demonstration study on a closed-cycle hydrogen production by the thermochemical water-splitting iodine-sulfur process, *Nucl. Eng. Desi.*, 233, 347–354, 2004.
7. J. Iwatsuki et al., Development program of IS process pilot test plant for hydrogen production with high-temperature gas-cooled reactor, *Proc. ICONE14-89267, 14th International Conference on Nuclear Engineering*, FL, USA, 2006.
8. A. Terada, J. Iwatsuki, and S. Ishikura, Development of hydrogen production technology by thermochemical water splitting IS process pilot test plan, *J. Nucl. Sci. Tech.*, 44(3), 477–482, 2007.
9. R. Hino et al., R&D on hydrogen production by high-temperature electrolysis of steam, *Nucl. Eng. Desi.*, 233, 363–375, 2004.

10. R. Hino et al., Present status of R&D on hydrogen production by high temperature electrolysis of steam, *JAERI-Res.*, 95–057, 1995.
11. W. Doenits et al., Electrochemical high temperature technology for hydrogen production or direct electricity generation, *Int. J. Hydrogen Energy*, 13, 283–287, 1988.
12. S. Kubo et al., A pilot test plan of the thermochemical water-splitting iodine–sulfur process, *Nucl. Eng. Des.*, 233, 355–362, 2004.
13. J. H. Norman et al., Thermochemical water-splitting cycle, bench-scale investigations and process engineering, *GA-A16713, General Atomics*, 1982.
14. M. Nomura, S. Kasahara, and S. Nakao, Silica membrane reactor for the thermochemical iodine–sulfur process to produce hydrogen, *Ind. Eng. Chem. Res.* 43(18), 5874–5879, 2004.
15. M. Nomura et al., Evaluation of the IS process featuring membrane techniques by total thermal efficiency, *Int. J. Hydrogen Energy* 30, 1465–1473, 2005.
16. K. F. Knoche, H. Schepers, and K. Hesselmann, Second law and cost analysis of the oxygen generation step of the General Atomic Sulfur-Iodine cycle, *Proc. 5th World Hydrogen Energy Conf.*, Toronto, Canada, July 15–20, 487–502, 1984.
17. I. T. Oeztuerk, A. Hammache, and E. Bilgen, A new process for oxygen generation step for the hydrogen producing sulfur–iodine thermochemical cycle,' *Trans. IChemE.*, 72 (Part A), 241–250, 1989.
18. L. C. Brown, G. E. Besenbruch, R. D. Lentsch, K. R. Schultz, J. F. Funk, P. S. Pickard, A. C. Marshall, and S. K. Showalter, High efficiency generation of hydrogen fuels using nuclear power, *GA-A24285, General Atomics*, 2003.
19. J. H. Norman, G. E. Besenbruch, and D. R. O'Keefe, Thermochemical water-splitting for hydrogen production, *GRI-80/0105, Gas Research Institute*, 1981.
20. S. Kubo et al., Corrosion test on structural materials for iodine–sulfur thermochemical water splitting cycle. *Proc. AIChE 2003 Spring National Meeting*, New Orleans, LA, USA; March 30–April 3, 153b, 2003.
21. K. Onuki et al., Screening tests on materials of construction for the thermochemical IS process, *Corrosion Eng.*, 46, 113–117, 1997.
22. N. Nishiyama et al., Corrosion resistance evaluation of brittle materials in boiling sulfuric acid, *J. Soc. Mat. Sci., Japan*, 48, 746–752, 1999.
23. N. Tanaka, H. Suwa, and A. Terada, Study on corrosion resistance of glass lining and gold plating in high temperature sulfuric acid for thermochemical hydrogen production IS process, *Proc. ICONE15–10331, 15th International Conference on Nuclear Engineering*, Aichi, Japan, 2007.
24. H. F. Niessen et al., Erprobung und Versuchsergebnisse des PNP-Teströhrenspaltofens in der EVA II-Anlage, Report Jül-2231, Research Center Jülich, Germany, 1988.
25. Nukleare Fernenergie, Zusammenfassender Bericht zum Projekt Nukleare Fernenergie (NFE), Report Jül-Spez-303, Research Center Jülich, Germany, 1985.
26. N. Sakaba et al., Nuclear demonstration program of hydrogen production using the HTTR (HTTR-IS Program), *16th Pacific Basin Nuclear Conference (16PBNC)*, Aomori, Japan, October 13–18, 2008.
27. X. Vitart, P. Carles, and P. Anzieu, A general survey of the potential and the main issues associated with the sulfur–iodine thermochemical cycle for hydrogen production using nuclear heat, *Progress in Nuclear Energy*, 50, 402–410, 2008.
28. S. Kubo et al., Closed cycle and continuous operations by a thermo-chemical water-splitting IS process, *Int. Conf. on Non-electric Applications of Nuclear Power*, Oarai, Japan, April 16–18, 2007.
29. S. Kubo et al., Control techniques for Bunsen reaction solution to regulate process condition, *AIChE 2007 Annual Meeting*, Philadelphia, USA, 2007.
30. J. H. Norman, G. E. Besenbruch, and D. R. O'Keefe, Thermochemical water splitting for hydro gen production, *GRI-80/0105, Gas Research Institute*, 1981.
31. B. Belaissaoui, R. Thery, X. M. Meyer, V. Gerbaud, and X. Joulia, Vapour reactive distillation process for hydrogen production by HI decomposition from $HI\text{-}I_2\text{-}H_2O$ solutions, *Chem. Eng. Process.*, 47, 396–407, 2008.
32. K. Onuki et al., Electro-electrodialysis of hydriodic acid in the presence of iodine at elevated temperature, *J. Membr. Sci.*, 192, 193–199, 2001.

33. M. Yoshida, N. Tanaka, H. Okuda, and K. Onuki, Concentration of HIx solution by electro-electodialysis using Nafion 117 for thermochemical water-splitting IS process, *Int. J. Hydrogen Energy*, 33, 6913–6920, 2008.
34. Rachael H. Elder, Geoffrey H. Priestman, and Ray W. K. Allen, Dewatering of HIx solutions by pervaporation through Nafion(r) membranes, *Int. J. Hydrogen Energy*, 34, 15, 6129–6136, 2009.
35. http://www.uhde.eu/competence/index.en.html.
36. G. J. Hwang et al., Improvement of the thermochemical water-splitting (iodine–sulfur) IS process by electro-electrodialysis. *J. Membr. Sci.*, 220, 129–136, 2003.
37. S. Kasahara et al., Effects of process parameters of the IS process on total thermal efficiency to produce hydrogen from water. *J. Chem. Eng. Jpn.*, 36(7), 887–899, 2003.
38. H. Ohashi et al., Hydrogen iodide processing section in a thermochemical water-splitting iodine–sulfur process using a multistage hydrogen iodide decomposer. *Trans. Atom. Energy Soc. Japan*, 29, 68–82, 2009 (in Japanese).
39. H. Ota et al., Conceptual design study on sulfuric acid decomposer for thermochemical iodine sulfur process. *Proc. of ICONE13–50494, 13th International Conference on Nuclear Engineering*, Beijing, China, 2005.
40. H. Noguchi et al., Test fabrication of sulfuric acid decomposer applied for thermo-chemical hydrogen production IS process. *JAEA-Technology*, 2007-041, 2007 (in Japanese).
41. H. Noguchi et al., Development of sulfuric acid decomposer for thermo-chemical IS process, *ICAPP2006-6166*, 2006.
42. A. Kanagawa et al., Conceptual design of SO_3 decomposer for thermo-chemical iodine–sulfur process pilot plant, *Proc. 13th Int. Conf. Nucl. Eng., ICONE13-50451*, Beijing, China, May 16–20, 2005.
43. L. C. Brown et al., High efficiency generation of hydrogen fuels using nuclear power, *GA-A24285*, 2003.
44. N. Sakaba et al., Hydrogen production by high-temperature gas-cooled reactor; conceptual design of advanced process heat exchangers of the HTTR-IS hydrogen production system, *Trans. Atom. Energy Soc. Japan*, 7(3), 242–256, 2008 (in Japanese).
45. I. Minatsuki, H. Fukui, and K. Ishino, A development of ceramics cylinder type sulfuric acid decomposer for thermo-chemical iodine-sulfur process pilot plant, *J. Power Energy Systems*, 1(1), 36–48, 2007.
46. E. Parma et al., Modeling the sulfuric acid decomposition section for hydrogen production, *ANS Embedded Topical Meeting on the Safety and Technology of Nuclear Hydrogen Production, Control and Management (ST-NH2)*, Boston, MA, June 24–28, 2007.
47. F. Gelbard et al., Sulfuric acid decomposition experiments for thermochemical hydrogen production from nuclear power, *ANS Embedded Topical Meeting on the Safety and Technology of Nuclear Hydrogen Production, Control and Management (ST-NH2)*, Boston, MA, June 24–28, 2007.
48. R. Moore et al., A laboratory-scale sulfuric acid decomposition apparatus for use in hydrogen production cycles, *ANS Embedded Topical Meeting on the Safety and Technology of Nuclear Hydrogen Production, Control and Management (ST-NH2)*, Boston, MA, June 24–28, 2007.
49. D. McLaughlin et al., Modeling of the sulfuric acid decomposition reactor, *ANS Embedded Topical Meeting on the Safety and Technology of Nuclear Hydrogen Production, Control and Management (ST-NH2)*, Boston, MA, June 24–28, 2007.
50. S. M. Connollya et al., Design of a composite sulfuric acid decomposition reactor, concentrator, and preheater for hydrogen generation processes, *Int. J. Hydrogen Energy*, 34, 4074–4087, 2009.
51. G. Rodriguez et al., Development program of a key component on the iodine sulfur thermochemical cycle: The SO_3 decomposer, *16th World Hydrogen Energy Conference*, Lyon, France, June 13–16, 2006.
52. Y. Kim et al., High temperature and high pressure corrosion resistant process heat exchanger for a nuclear hydrogen production system, R.O.K. Patent submitted, 10-2006-0124716, assigned to Korea Atomic Energy Research Institute, 2006.

53. C. Kim et al., Thermal sizing of a lab-scale SO_3 decomposer for nuclear hydrogen production, *ANS Embedded Topical Meeting on the Safety and Technology of Nuclear Hydrogen Production, Control and Management (ST-NH2)*, Boston, MA, June 24–28, 2007.
54. http://www.simsci.jp/home.shtml.
55. http://www.olisystems.com/.
56. http://www.mizuho-ir.co.jp/.

23

Nuclear Hydrogen Production Plant Safety

Tetsuo Nishihara, Yujiro Tazawa, and Yoshiyuki Inagaki

CONTENTS

23.1 Safety Requirements and Approaches

Safety concept is common for all types of nuclear reactors that are thermodynamically coupled with a hydrogen production plant. The economical design requirements for commercial nuclear hydrogen plant include the design of a hydrogen production plant as a conventional chemical plant and the proximity of such a hydrogen production plant to the nuclear reactor. To meet these requirements, the significant design issues including prevention and control of anticipated operational occurrences and accidents in both the nuclear reactor and the hydrogen production plant must be addressed to meet the regulatory requirements for safety.

Section III of this handbook includes a number of nuclear reactor systems suitable for hydrogen production. The light water reactor (LWR) is conventional type of nuclear reactor used to generate electricity by a steam turbine system. It operates at about 300°C of coolant temperature such that hydrogen production is limited to water electrolysis. The sodium faster reactor (SFR) is being demonstrated for generation of electricity efficiently by a steam turbine system. Since the SFR operates around 600°C, some thermal hydrogen production processes whose temperatures of chemical reaction are within this range can be supplied by the heat of the SFR for hydrogen production. The gas faster reactor (GFR) and the fusion reactor are under development and can produce higher-temperature process heat for hydrogen production.

The high temperature gas reactor (HTGR) can supply process heat of about 900°C to a wide range of hydrogen production processes. The design of the HTGR-based hydrogen production plant has been extensively investigated by JAEA. The following subsections describe the safety requirements and the corresponding design approaches taken for nuclear reactor plants, hydrogen production plants, and interaction between the nuclear reactor plants and the hydrogen production plants. The technical concepts described for the HTGR are generally applicable to the safety design considerations of other nuclear reactors for thermal process hydrogen production.

23.1.1 Nuclear Reactor Plant

The safety objective to use nuclear power peacefully is to protect people and the environment from harmful effects of ionizing radiation. The safety design concept of the nuclear plant is based on the defense in depth. A nuclear plant is designed to operate the plant safely and to prevent anticipated operational occurrences and accidents and mitigate their consequences so that public and operators are protected from radiation exposure. Safety items with high level of confidence to prevent accidents and mitigate their consequences are provided in the nuclear plant. Validity of the safety items to prevent accidents and to mitigate their consequences is assessed in the safety design review. International Atomic Energy Agency (IAEA) safety series are very useful documents to consider how to accomplish nuclear plant safety.

Sitting of a nuclear plant is evaluated to reduce radiation exposure on the public in accident as low as reasonably achievable in all operational states. Then, most nuclear plants are placed in a low population region. Inherent safety feature of the reactor and safety items shall be provided to reduce the risk of radiation exposure on the operators of the nuclear reactor and the hydrogen production plant to place both plants adjacently.

23.1.2 Hydrogen Production Plant

A hydrogen production plant is basically designed to meet the conventional design codes and rules to operate safely in normal operational state and to prevent accidents. Significant safety issues of the hydrogen production plant are fire, explosion, and toxic gas release. Fire and explosion in the hydrogen production plant originate strong heat and blast overpressure. They may cause severe damage to the hydrogen production plant and affect the nuclear plant's safe operation.

Hydrogen explosion in a closed space becomes detonation which generates shock waves. Strong blast overpressure originated by detonation can propagate to several hundreds of meters such that a hydrogen production plant may be damaged. Hydrogen gas cloud may move toward the nuclear plant if hydrogen gas does not ignite. Adequate safety items or passive safe operation shall be provided to reduce the effect of blast overpressure and to avoid internal hydrogen explosion in the nuclear plant so that the nuclear plant can maintain safe operation.

Thermochemical water splitting hydrogen production process uses some toxic materials as reactants in their process. Low-airborne-density toxic materials can harm the health of operators and public. Toxic gases may flow into the building of the nuclear plant through the air intake of air conditioning systems so that operators in the control room of the nuclear plant are damaged. The hydrogen production plant shall be designed to prevent accidental release of hydrogen and toxic materials, to detect leakage of them, and to shut and isolate the failed system to reduce an amount of leakage. Operators and public must be protected against toxic gas attack.

23.1.3 Coupling a Hydrogen Plant to a Nuclear Plant

In a nuclear hydrogen production plant, an intermediate loop is needed to transfer heat from the nuclear plant to the hydrogen production plant and to separate the hydrogen plant physically from the nuclear plant. The nuclear plant has radioactive materials in the coolant of the core. Tritium, one of the radioactive materials in the coolant, can permeate the heat exchanger tubes of the intermediate loop in high-temperature operational state. Contamination of the coolant of the intermediate loop shall be kept as low as acceptable. To reduce the tritium concentration of the coolant in the intermediate loop and to meet an acceptable limit in the product hydrogen, safety items are provided in the primary helium loop or in the intermediate loop if needed.

The hydrogen production plant contains flammable, toxic, and corrosive materials. These materials affect significant damage on the component in the nuclear plant. The intermediate loop has a function to prevent these materials from flowing into the nuclear plant through the loop.

The hydrogen production plant is one of the heat sinks of the nuclear plant in normal operation state. Thermal load variation of the hydrogen plant affects heat balance of the nuclear plant. Increase or decrease of core inlet coolant temperature affects reactivity control of the core. As a result, reactor power and reactor outlet coolant temperature change. Large thermal load variation caused an emergency reactor shutdown. The reactor coolant system or the intermediate loop should provide items or functions to mitigate thermal variation of the reactor coolant.

23.2 Nuclear Reactor Safety

Safety objectives of a nuclear plant are to prevent accidents and to mitigate their effects. Basic safety functions of the nuclear power plant are

a. To control reactivity in the core
b. To remove heat from the core
c. To confine radioactive materials released from the primary coolant and the core

All nuclear plants are designed to meet safety functions. The HTGR hydrogen production plant provides safety items to fulfill these safety functions. The safety items shall maintain their safety functions against internal and external events. External events include fire, explosion, asphyxiant and toxic gases, and corrosive gases and liquids which originate in the hydrogen plant.

There in no difference between the HTGR power generation system, the HTGR hydrogen cogeneration system, and the HTGR hydrogen production system, except the existence of an intermediate heat transfer loop for the hydrogen production plant. These safety items are applied to all types of HTGR systems.

23.2.1 Reactivity Control System

The reactor core withstands static and dynamic loading in normal operation and accident to ensure the safe shutdown of the reactor, to keep the reactor subcritical and to cool the core. Reactivity control system consists of two different systems to provide redundancy, diversity, and independency.

JAEA operates the test reactor of HTGR, named HTTR [1]. The HTTR provides the control rod system and the reserve shutdown system to meet above safety requirement. Figures 23.1 and 23.2 show the control rod system and the reserved shutdown system in the HTTR. The drive mechanisms of both systems are installed in the stand pipe mounted on the top dome of the reactor vessel. Control rods are inserted into the channels of the control rod columns. Core arrangement of the HTTR is shown in Figure 23.3. The reserved shutdown system drops B_4C pellets into the core to shut down the reactor when the control rod system fails. Commercial scaled hydrogen cogeneration HTGR designed by JAEA, named GTHTR300C, provides the same reactivity control system [2].

The control rod system and the reserved shutdown system ensure the function to shut down the reactor in all operational states. All operational states in the hydrogen plant do not affect the reactivity control system.

23.2.2 Reactor Coolant System

A reactor coolant system of GTHTR300C provides an intermediate heat exchanger to connect a secondary helium loop to transfer high-temperature heat from the nuclear reactor as shown in Figure 23.4. The reactor coolant system including the intermediate heat exchanger shall withstand the static and dynamic loading in all operational states including accident states to prevent release of radioactive materials with the coolant. Residual heat of the core is removed by the reactor coolant system in normal operation state.

When reactor scram occurs, the reactor coolant system shuts down to prevent overcooling of the core. Residual heat of the core should be removed by an auxiliary cooling system. In the loss of the coolant, cooling by forced helium circulation is not available. GTHTR300C provides the reactor cavity cooling system (RCCS) as shown in Figure 23.5 instead of the emergency core cooling system (ECCS) which is provided in the water-cooled reactor to remove the residual heat from the core in the loss of the coolant. The RCCS removes passively the residual heat after the reactor scram. The residual heat of the core can be transferred to the reactor vessel by thermal conduction and convection in the core graphite components, and then the reactor vessel is cooled by the cooling panel of RCCS by thermal radiation. The cooling panel is cooled by the air which flows passively in the flow channels in the cooling panel by natural circulation. This function shall be maintained in all operational states. Air intake and stacks of the RCCS should be protected against explosion of hydrogen.

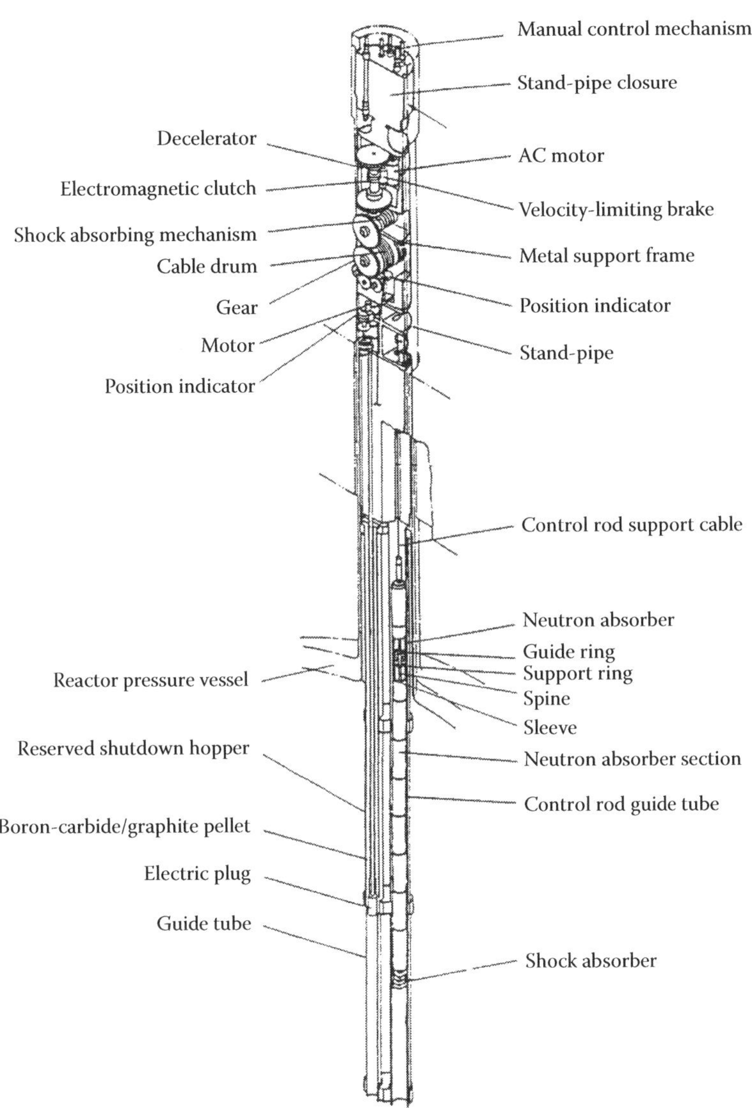

FIGURE 23.1
Control rod system in the HTTR.

23.2.3 Reactor Containment System

A reactor containment system is provided to mitigate the amount of radioactive material release into the environment in accident. The reactor containment is a leak-tight structure to reduce the release rate of radioactive materials and withstands internal pressure rising in the loss of coolant to ensure the structural integrity. GTHTR300C has evaluated that the amount of radioactive material released in the loss of coolant is not large so that it is designed that the reactor confinement is enough to meet the safety requirement of radiation exposure. System configuration of the reactor containment is shown in Figure 23.6. Reactor building and stacks perform the reactor confinement function.

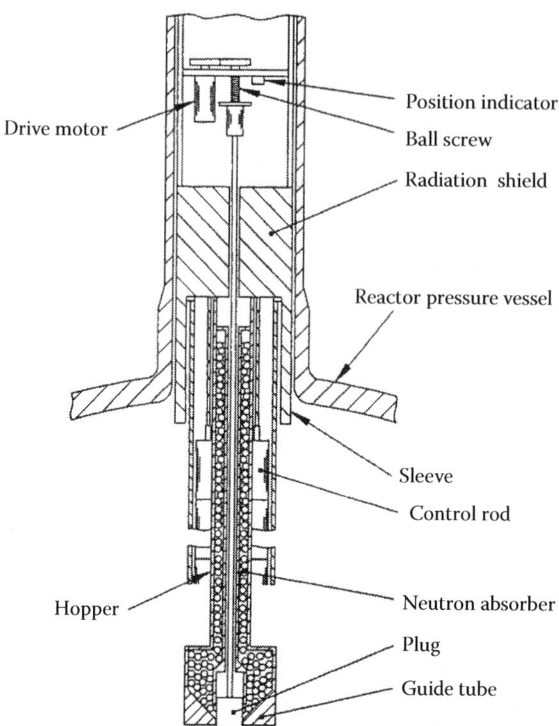

FIGURE 23.2
Reserved shutdown system in the HTTR.

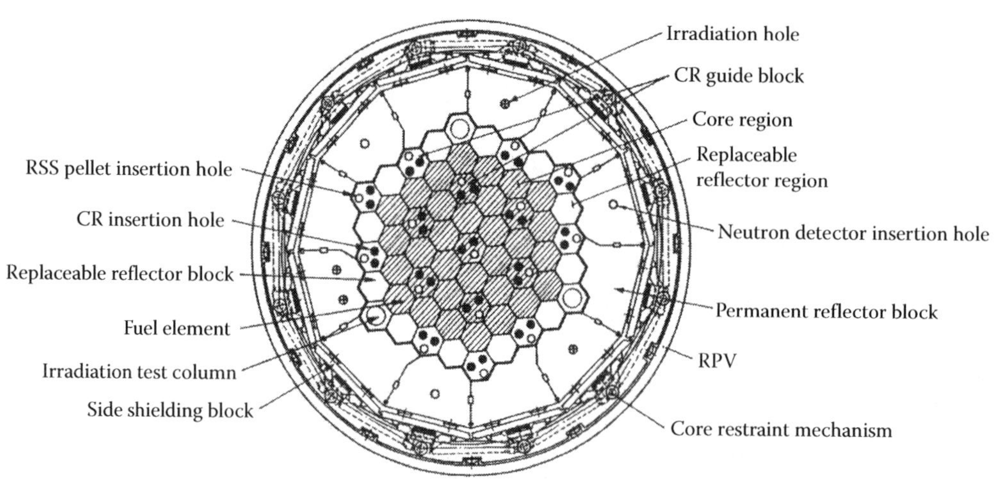

FIGURE 23.3
Core arrangement of the HTTR.

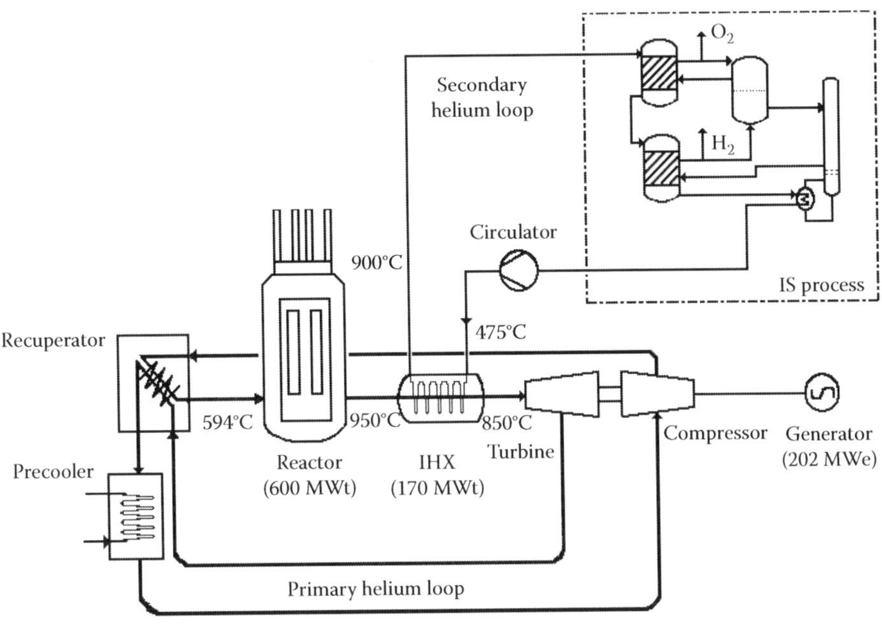

FIGURE 23.4
Coolant system of the GTHTR300C.

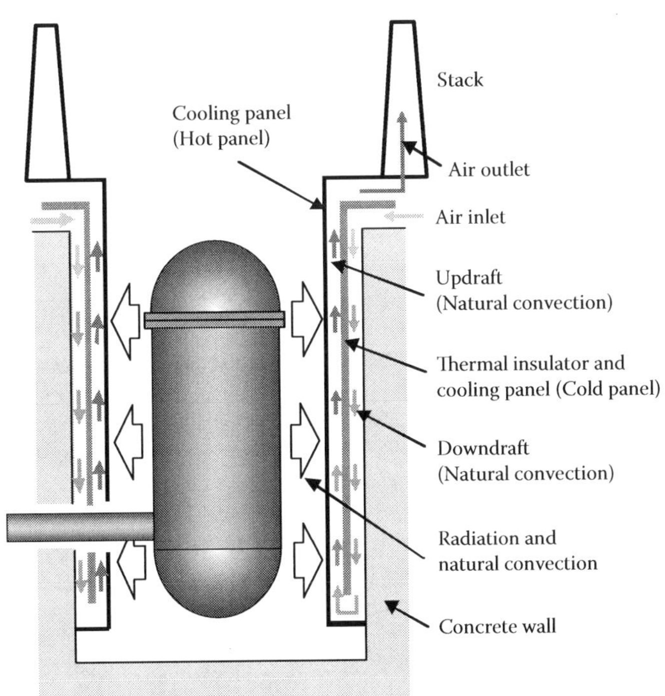

FIGURE 23.5
Reactor cavity cooling system.

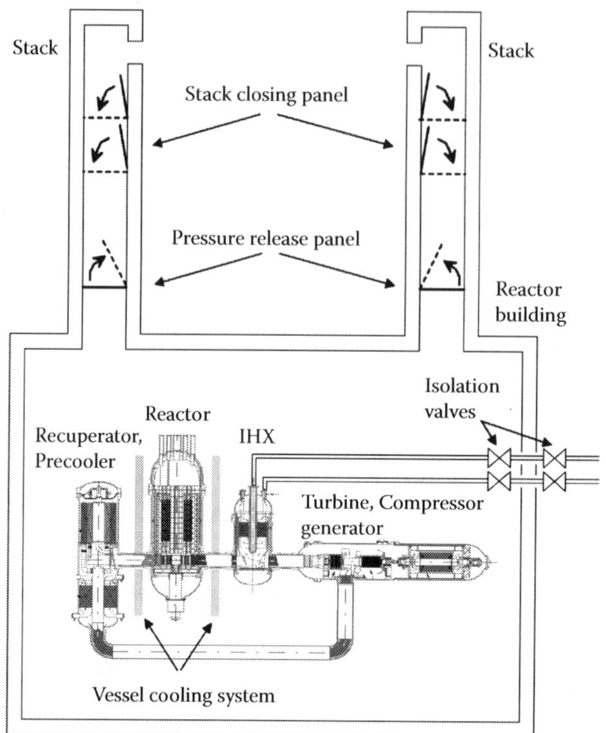

FIGURE 23.6
Reactor confinement system.

The amount of coolant released into the environment from the stacks in the design basis event can be acceptable. But air ingress into the core is strictly protected to prevent core oxidation and consequent fuel failure or core damage. The closing mechanism of the stacks is provided to close the stacks after the initial release of the coolant. Inventory of the confinement and thickness of the concrete wall shall be designed to ensure the structural integrity so that pressure load and temperature of the concrete structure is lower than the acceptable limit in accident states.

The hot gas duct of the heat transfer loop penetrates the concrete wall of the reactor building to connect the nuclear plant with the hydrogen production plant to transfer hot helium gas for hydrogen production. Release of radioactive materials and inflow of air through the heat transfer loop must be protected so that isolation valves shall be provided close to the penetration in the tube failure of the intermediate heat exchanger and hot gas duct rupture outside reactor confinement.

23.2.4 Reactor Safety Protection from the Hydrogen Plant

Postulated initiating event (PIE) in the hydrogen plant should be considered in the safety design of the HTGR hydrogen production plant as an external event because the hydrogen plant is a ternary cooling loop of the HTGR and is categorized as a non-nuclear graded system.

Failure of equipment and piping of the hydrogen production plant leads to flammable or toxic gas release accident. In the hydrogen production plan, the gas detecting system to detect gas release as early as practicable and the emergency isolation system to shut off the piping and reduce the amount of gases released are provided. The safety items in the hydrogen production plant, designed on the basis of conventional industrial design codes and rules, cannot be expected to mitigate the consequences of events in the nuclear safety assessment because safety items in the nuclear plant have redundancy, diversity, or independence to increase their reliability but safety items in the hydrogen production plant are not required to provide those functions.

Passive safety items such as physical barriers and safe distance are applicable for protection of the nuclear plant. Physical barriers accelerate dilution of released gas and mitigate the direct effect of thermal radiation and blast from fire and explosion. Barriers must withstand design-based earthquake and anticipated maximum blast so that the required safe distance can be reduced. The safe distance is evaluated in consideration of gas release condition enough to reduce the effect of gas dispersion and gas explosion.

23.3 Hydrogen Plant Safety

A hydrogen plant connected to the HTGR will handle a large amount of combustible gas and toxic gas. The risk from fire, explosion, and acute toxic exposure caused by chemical material release in a hydrogen production system should be assessed from the viewpoint of plant safety and human safety. Preliminary safety analysis of the GTHTR300C with thermochemical iodine–sulfur process described in this section provides us useful information on the separation distance between a nuclear plant and a hydrogen production system, and a prospect that the accidents in a hydrogen production system does not significantly increase the risks of the public [3].

23.3.1 Fire and Explosion

An accidental release of hydrogen gas from the hydrogen production plant is one of the most important external events in the HTGR hydrogen production plant. Pressurized hydrogen gas can blow out as a jet when a hydrogen transfer pipeline is ruptured and the generated hydrogen gas cloud can move a long distance. General safety means are to take an adequate separation distance between the hydrogen production plant and the nuclear safety items or to install a physical barrier between them. Accelerating the diffusion of hydrogen gas can reduce an amount of explosive hydrogen mass and minimize the separation distance. A concept of the safety distance is shown in Figure 23.7.

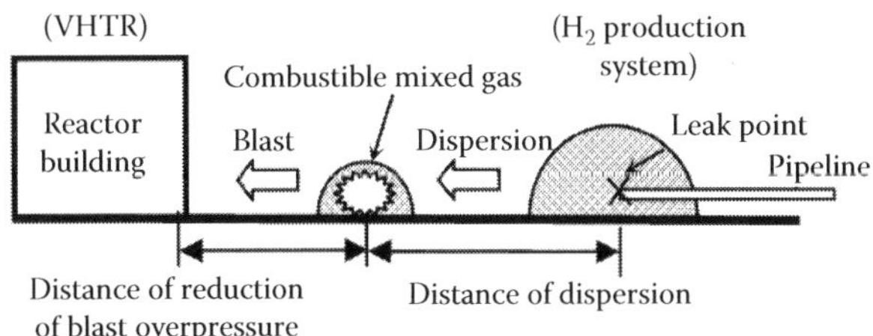

FIGURE 23.7
Concept of the safe distance against the combustible gas leakage and explosion.

The maximum horizontal distance between the release point and the explosive hydrogen–air mixture is defined as "moving distance" in this safe distance evaluation. The moving distance is analyzed by numerical calculation. The mass of explosive hydrogen gas in the cloud is also evaluated. In the next step to determine the effect of blast overpressure, the trinitrotoluene (TNT) equivalent method or the multienergy method which assumes the explosion of semispherical stoichiometric gas cloud is applicable [4,5]. In this section, the multienergy method is selected to calculate the distance for blast. The available energy of hydrogen explosion is calculated by the analyzed explosive hydrogen mass. The separation distance between the hydrogen production plant and the safety items in the HTGR can be decided by combining the moving distance and the safe distance against blast overpressure to be below the acceptable limit of safety items. For example, the acceptable limit of overpressure on the reactor building is tentatively defined as 30 kPa.

Variation of the mass fraction of combustible hydrogen corresponding to the elapsed time from the initiation of gas release is shown in Figure 23.8. Concentration of combustible hydrogen is in the range 4–75 vol%. Calculation conditions include hydrogen release of 97 kg, a pipe of diameter 100 mm, height of release point 2.5 m, and wind speed 1 m/s. The release duration is 5 s. The mass of explosive hydrogen cloud increases for first 5 min, after which it decreases gradually.

Figure 23.9 shows the evaluation result of the moving distance of combustible hydrogen gas and the overall separation distance. The separation distance depends on the amount of hydrogen release as shown in Figure 23.10. Hydrogen inventory of 97 kg is required in a separation distance of 150 m. The effect of the protection wall is investigated. A reasonable protection wall effectively reduces the separation distance as shown in Figure 23.10.

The separation distance should be shorter from the viewpoint of hydrogen economy because a hot gas connecting the nuclear plant and the hydrogen production plant is an expensive component. Measurements to reduce the separation distance are considered and proposed as follows.

a. To provide a protection wall near the hydrogen production plant to prevent the hydrogen dispersion by diffusion.
b. To avoid the installation of packed obstacles between the hydrogen production plant and the HTGR to prevent strong blast.

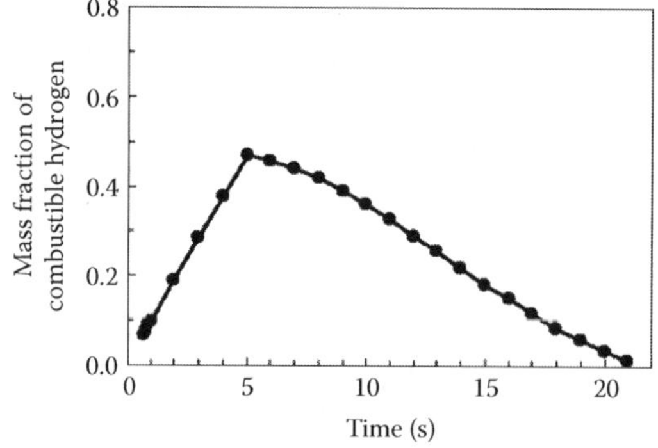

FIGURE 23.8
Trend of combustible hydrogen mass (case of 97 kg).

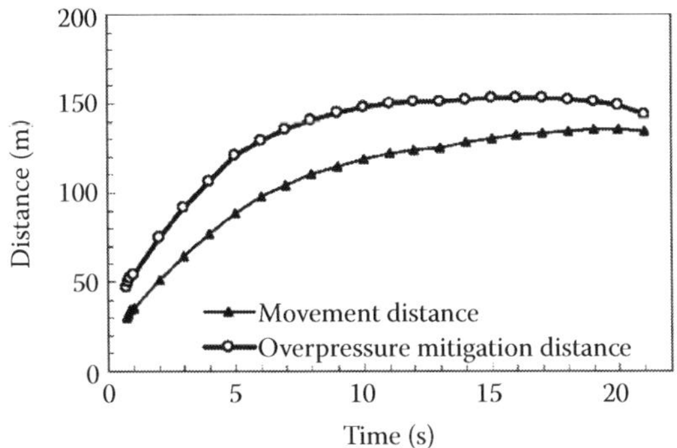

FIGURE 23.9
Movement distance of the combustible hydrogen gas and the overpressure mitigation distance (H_2: 97 kg).

In general, the wall is not required to be wide and tall. The analysis shows that a wall of 2 m higher than the elevation of hydrogen pipeline can effectively reduce the moving distance of combustion by half, making economical separation feasible between the hydrogen plant and the nuclear reactor.

23.3.2 Toxic Gas Release

Thermochemical processes use some toxic materials as a reactant for hydrogen production. Iodine–sulfur process uses sulfur acid and hydrogen iodide. Some of them are gaseous so that they easily disperse. Toxic gas has potential hazard even in low airborne concentration. An accidental release of toxic gas from the hydrogen production plant harms operators and public in wide area around the plant. It is important to maintain the control room of the nuclear plant in safe condition after the accidental release of toxic gas to continue safe operation of the nuclear reactor.

HTGR shall provide safety measures like an isolation system against intake of toxic gas in the control room, a self-contained air breathing apparatus, and a purification system of toxic gas to reduce hazard against toxic gas. The functions of ventilation in the control room

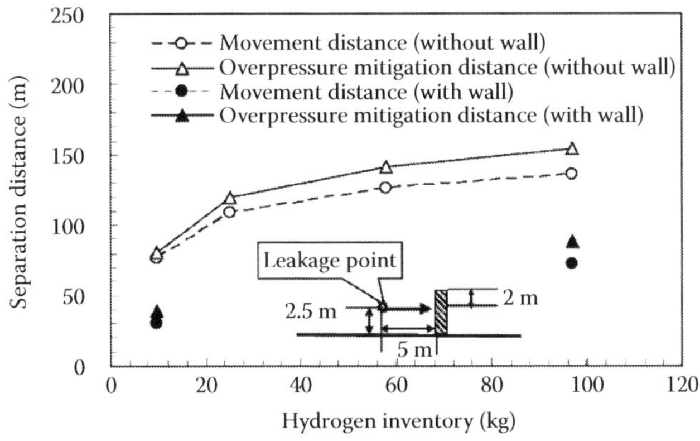

FIGURE 23.10
Effect of a wall to reduce separation distance.

under the accident condition requires the following items referred in NRC Regulatory Guide 1.78 [6].

1. Monitoring the concentration of toxic materials involved in the air supplied to the reactor building
2. Isolating against inleakage into the control room and automatically changing operation of ventilation to the closed circuit system when the concentration is above the acceptable limit
3. Restart the ventilation system when the concentration of toxic materials outside the reactor building is decreased below both the acceptable limit and the concentration in the control room

The acceptable limit of the gas concentration in the control room should be determined adequately in reference with the industrial regulation for the toxic materials.

The separation distance against toxic materials are needed to consider the safety design of an HTGR hydrogen production plant.

23.3.3 Public Risk

Potential risk of public shall be low enough when a hydrogen production plant be coupled with HTGR. It is considered that an HTGR hydrogen production plant can maintain the safety function of a nuclear plant even though the hydrogen production plant is in abnormal state such as thermal disturbance, combustible gas release, and so on. Abnormal events initiated in the hydrogen production plant shall not increase the risk of radiation exposure. And the risk of acute toxic exposure of public should be as low as reasonably achievable.

Preliminary risk evaluation of GTHTR300C is performed in [3]. The public risk level for the hydrogen production plant is tentatively determined below: 10^{-6}/year and 10^{-7}/year for each chemical material release in this evaluation. The evaluated limit amount of chemical material release compared to the inventory in the iodine–sulfur process in which hydrogen production rate is 24,000 N m^3/h by using 170 MW nuclear heat from HTGR is shown in Figure 23.11. This evaluation result provides us with a prospect that an accidental release

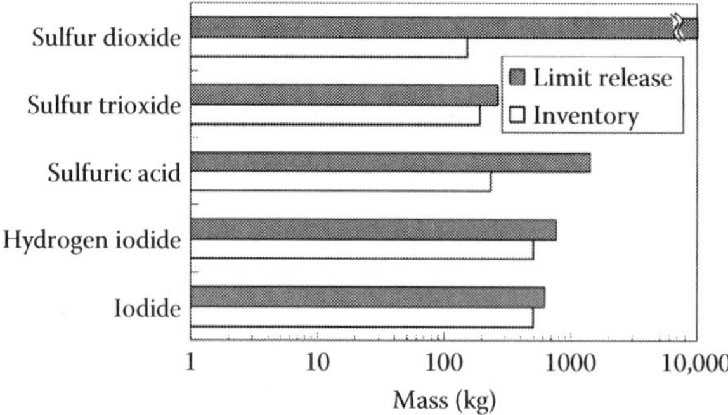

FIGURE 23.11
Evaluated limit amount of release and inventory in the IS process in which hydrogen production rate is 24,000 N m^3/h .

of chemical materials from a hydrogen production system does not significantly increase the risk for the public.

23.3.4 Safety for Steam Reforming System

JAEA had once developed a demonstration test plan of a hydrogen production system by steam reforming of methane coupling with the HTTR. A large amount of liquefied natural gas (LNG) will be handled in the system. The LNG tank shall be set several hundred meters away from the HTTR reactor building. JAEA developed an analysis code system to simulate the fire and explosion accidents assumed in the HTGR, and simulated preliminarily for the HTTR coupled with a steam reforming system [7]. In this result, it was shown that the reactor building, the structures in the containment vessel, and the air flow around the reactor building were hardly affected by the explosion or the fire in those analytical conditions.

23.4 Reactor and Hydrogen Plant Coupling

Figure 23.12 shows the flow diagram of the HTGR hydrogen production system and research and development (R&D) items for coupling the hydrogen production system to the nuclear reactor:

1. Control technology to keep reactor operation against thermal disturbance caused by the hydrogen production system.
2. Development of high-temperature isolation valve (HTIV) to separate nuclear reactor from hydrogen production plant in accidents [8].

23.4.1 Control Technology

The nuclear reactor and the hydrogen production plant are connected by the helium gas loop. The temperature fluctuation of the secondary helium gas is caused by the fluctuation of the chemical reaction which can occur at normal start-up and shutdown operation as well as during malfunction or accident of the hydrogen production plant.

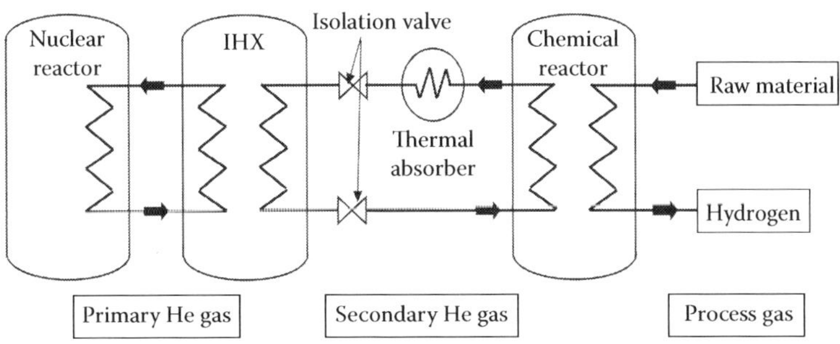

FIGURE 23.12
Flow diagram of the HTGR hydrogen production system and R&D items for coupling.

Large temperature fluctuation of the nuclear reactor results in a reactor scram. Therefore, the control technology should be developed to mitigate the temperature fluctuation in the nuclear reactor.

23.4.1.1 Thermal Absorber Concept

JAEA proposed to use a steam generator (SG) as the thermal absorber which is installed downstream the chemical reactor in the secondary helium gas loop [9]. A simulation test with the mock-up test facility was carried out to investigate performance of the SG for mitigation of the temperature fluctuation and transient behavior of the hydrogen production plant and to obtain experimental data for verification of a dynamic analysis code [10,11]. In the test, the steam reforming process of methane, $CH_4 + H_2O = 3H_2 + CO$, instead of the IS process is used for hydrogen production.

Figure 23.13 shows one of the test conditions of the feed flow rate of each gas to the chemical reactor, the steam reformer (SR), and hydrogen production rate, simulating loss of chemical reaction which is the most severe accident for the temperature fluctuation of the secondary helium gas [12]. Figure 23.14 shows the test results of the helium gas temperatures of the SR and the SG, the steam flow rate of the radiator, and water pressures in the SG and the radiator. The fluctuation of helium gas temperature could be mitigated successfully in the range from −0.2 to +2.0°C at the SG outlet, whereas SG inlet helium gas temperature increased to 248°C at loss of chemical reaction. It was within the target fluctuation range from −10 to +10°C at SG outlet. As a result, it was confirmed that the SG can be used as a thermal absorber to mitigate the temperature fluctuation of the secondary helium gas caused by the hydrogen production plant. This technology can keep reactor operation at normal start-up and shutdown operation as well as at malfunction or accident of the hydrogen production plant.

Impact of the loss of heat load on the reactor operation in the HTTR-IS system was assessed by the numerical calculation. Figure 23.15 shows the calculation results of the outlet temperature at the thermal load mitigation system and reactor power. Calculation

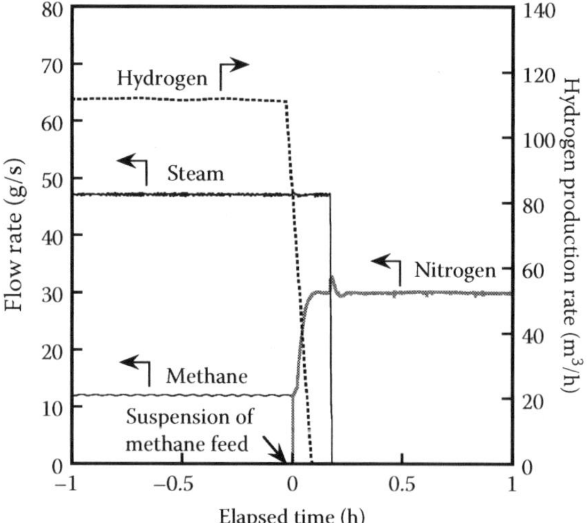

FIGURE 23.13
Test condition of feed flow rate in simulating loss of chemical reaction.

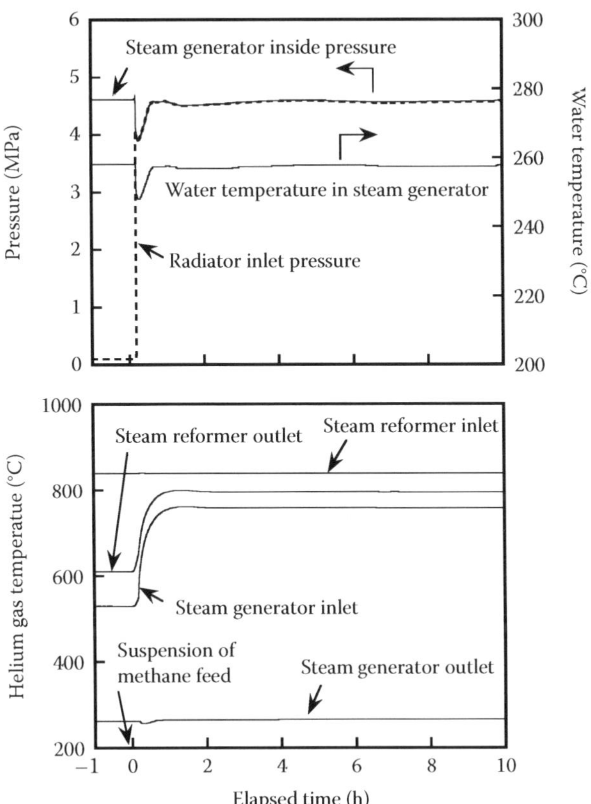

FIGURE 23.14
Steam pressures at the SG inside and the radiator inlet, water temperature in the SG and helium gas temperatures at the loss of chemical reaction.

result shows that the proposed system mitigates the impact on the primary cooling system so that the reactor can operate normally during the event [13].

23.4.1.2 Cogeneration System Temperature Control

Future commercial HTGR hydrogen cogeneration plants like GTHTR300C used a helium gas turbine system for electricity generation. In this system, the turbine system will work as the thermal absorber instead of SG. In the GTHTR300C design, the control valves installed in the primary cooling system mitigates the loss of load in the hydrogen production plant [14]. Figure 23.16 shows the configuration of primary cooling system of the GTHTR300C.

Loss of load in hydrogen production plant results in temperature increase at the gas turbine inlet. The turbine inlet temperature control valve intends to keep the temperature at 850°C by mixing cold gas from the compressor outlet. This flow rate control action induces flow rate decrease at the turbine and the compressor. Then, turbine by-pass flow control valve acts to increase the flow rate in order to recover the rated rotational speed. These control valve actuations lead to decrease of core flow rate. Reactor power is adjusted by the reactor outlet temperature control system so that the reactor and the power

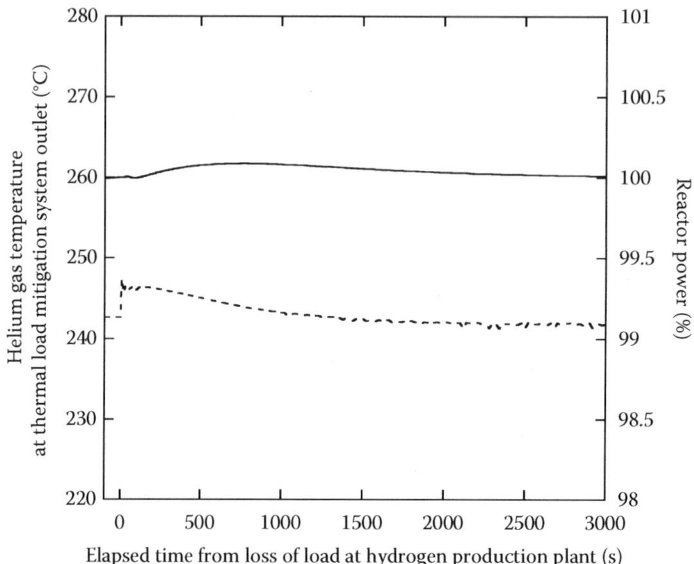

FIGURE 23.15
Calculation results of helium gas temperature at the thermal load mitigation system outlet and reactor power in the HTTR-IS system during loss of heat load at the hydrogen production plant.

generation system enable to maintain safe and steady operation. Detailed technical discussion including simulation on operation control strategies for a commercial cogeneration system is continued in Chapter 24.

23.4.2 High-Temperature Isolation Valve

The isolation valve is a key component to protect radioactive material release from the reactor to the hydrogen production system and combustible gas ingress to the reactor at

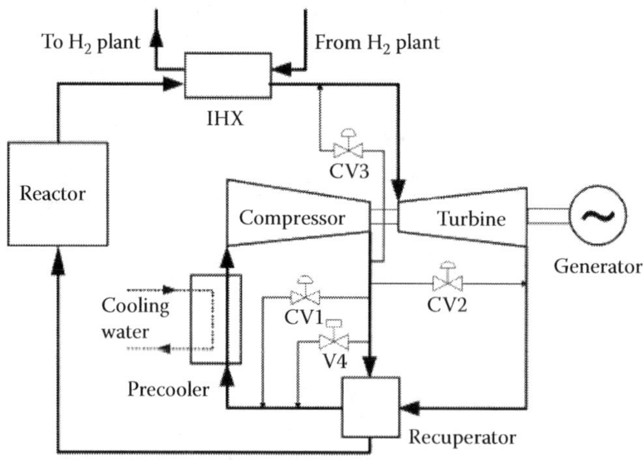

FIGURE 23.16
Simplified plant process diagram of the GTHTR300C.

the accident of fracture of the IHX and the chemical reactor. A HTIV must be developed to apply in the helium condition over 900°C. The research and development of a valve for high-temperature and high-pressure helium gas were carried out in the PNP project in Germany [15] and in Engineering Research Association of Nuclear Steel-making (ERANS [16]) in Japan. A practical HTIV has not been made yet. Main technical issues of the high-temperature valve are to mitigate thermal deformation and to protect the self-welding of the seat and the disk so that seal performance and structural integrity are maintained.

23.4.2.1 Design Concept

JAEA has been conducting design and component test of an HTIV based on the results of ERANS for the demonstration test with HTTR. An angle valve was selected from the viewpoint of workability of the inner thermal insulator and the detailed structure was decided by the thermal stress analysis to prevent the valve seat deformation. Figure 23.17 shows the structure of HTIV. A coating material of the valve seat, which can keep hardness and wear resistance at high temperature over 900°C, is necessary to maintain the leak-tightness for helium gas and to prevent the self-welding of the seat and the disk. A new coating material was developed by adding 30 wt%-Cr_3C_2 to the coating material which is used for the valve at around 500°C [17].

23.4.2.2 Design Confirmation Test

A component test was carried out with a 1/2 scale model of the HTIV as shown in Figure 23.18 to confirm the structural integrity and the seal performance of the valve seat. An experimental apparatus is composed of the 1/2 scale model, electric heaters, gas supply systems, an actuator, and a concentration measurement system as shown in Figure 23.19. Helium gas at 4 MPa was supplied to the 1/2 scale model and the leaked helium gas from the closed valve seat was transported to a helium gas detector with argon carrier gas

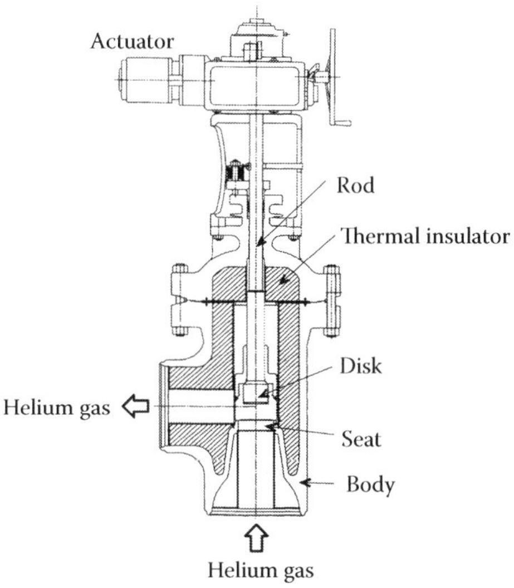

FIGURE 23.17
Schematic view of an HTIV.

FIGURE 23.18
Overview of 1/2 scale model.

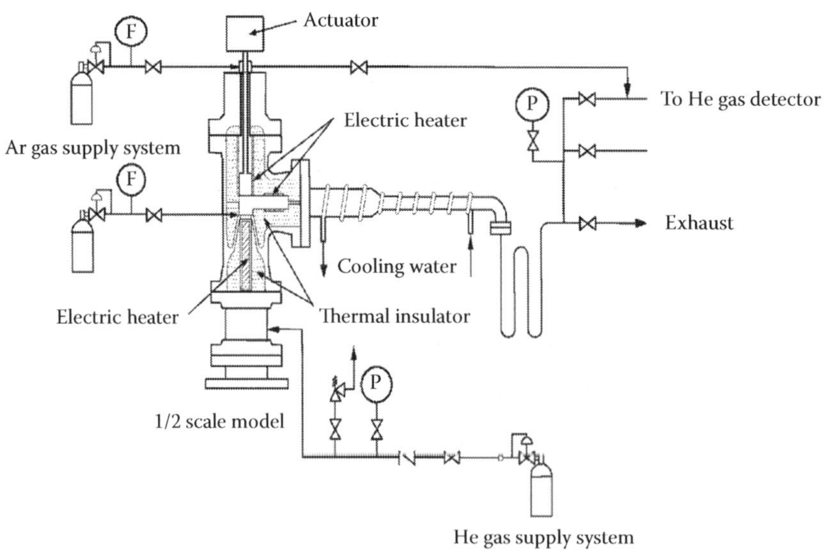

FIGURE 23.19
Experimental apparatus for an HTIV.

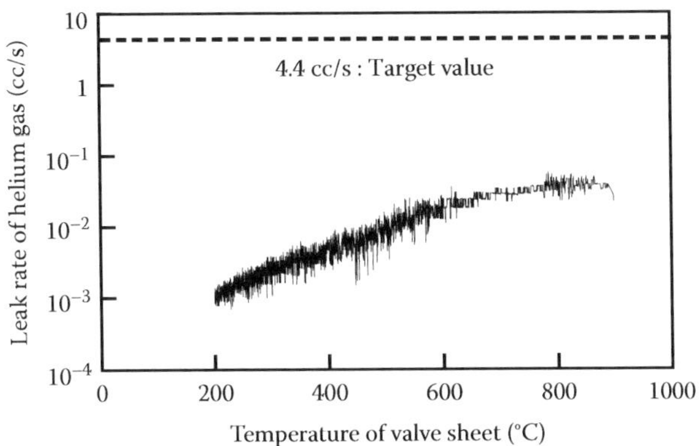

FIGURE 23.20
Leak test result of the valve seat with 1/2 scale model of an HTIV.

to measure the leak rate of helium gas. The carrier gas pressure downstream of model was atmospheric pressure.

Helium gas leak rate from the valve seat at room temperature was less than 1 cc/s before conducting the leak test at 900°C. This satisfies the design target of 4.4 cc/s. Leak rate decreased less than 10^{-1} cc/s at 900°C as shown in Figure 23.20 and then it increased up to around the design target at room temperature after once opening the valve [8]. Leak rate recovered less than 10^{-1} cc/s when the valve reclosed at 900°C. The following mechanism is considered from the above results:

1. Plastic deformation slightly of the coating material of the valve seat occurred by decline of the hardness at 900°C.
2. Face roughness of the valve seat increased by opening the valve at a room temperature.

By fitting the valve seat, the leak rate at room temperature became less than 1 cc/s again. The current technology can be applied to HTTR. Valve seat fitting is effective to recover the seal performance of HTIV after closing at high temperature. Improvement of durability of the valve seat by refinement of the coating metal is needed for future R&D.

23.5 Conclusions and Future Technical Challenges

Prior experience to connect process heat applications, particularly high-temperature chemical processes for hydrogen production, to a nuclear reactor is limited. JAEA study whose major results are presented in this chapter is intended to establish basic safety design rules for reactor protection and mitigation against explosion of hydrogen, toxic gas exposure and others.

Safety design of HTGR will be carried out according to the codes and rules for a nuclear plant. And a hydrogen production plant will be designed based on the conventional industrial design codes and rules. Then safety for the public can be achieved in all operation states. Coupling of these plants is the most significant safety issue for an HTGR hydrogen production plant. Safe distance must be provided between the HTGR and a hydrogen production plant to maintain safe operation of a nuclear reactor to mitigate the effect of accident in each other.

A high-temperature gas isolation valve is an essential safety device located in the secondary gas loop (see Figure 23.12). It provides physical separation between the HTGR and the conventional hydrogen production plant. Its safety functions are to be a confinement barrier of a nuclear reactor which prevents radioactive coolant release in case of accidental rupture of the IHX tubing and to be a protection of the IHX heat transfer surfaces from chemicals accidently flowing into the intermediate loop in case of helium-chemical boundary failure. Design and testing of the scaled-down isolation valve (see Figure 23.18) in 900°C and 4 MPa helium condition confirmed the safety performance of such devices in high-temperature structural integrity, leak-tightness, and maintainability.

A commercial HTGR hydrogen production plant requires a large-sized isolation valve with economic installation. Figure 23.21 shows two design approaches. They are based on the angle-type isolation valve in consideration of seal performance at high temperature and the test result of R&D conducted by JAEA. The single unit is of very large size so that the multiple unit will be applied to the commercial plant. Size reduction of the multiple unit is a technical challenge.

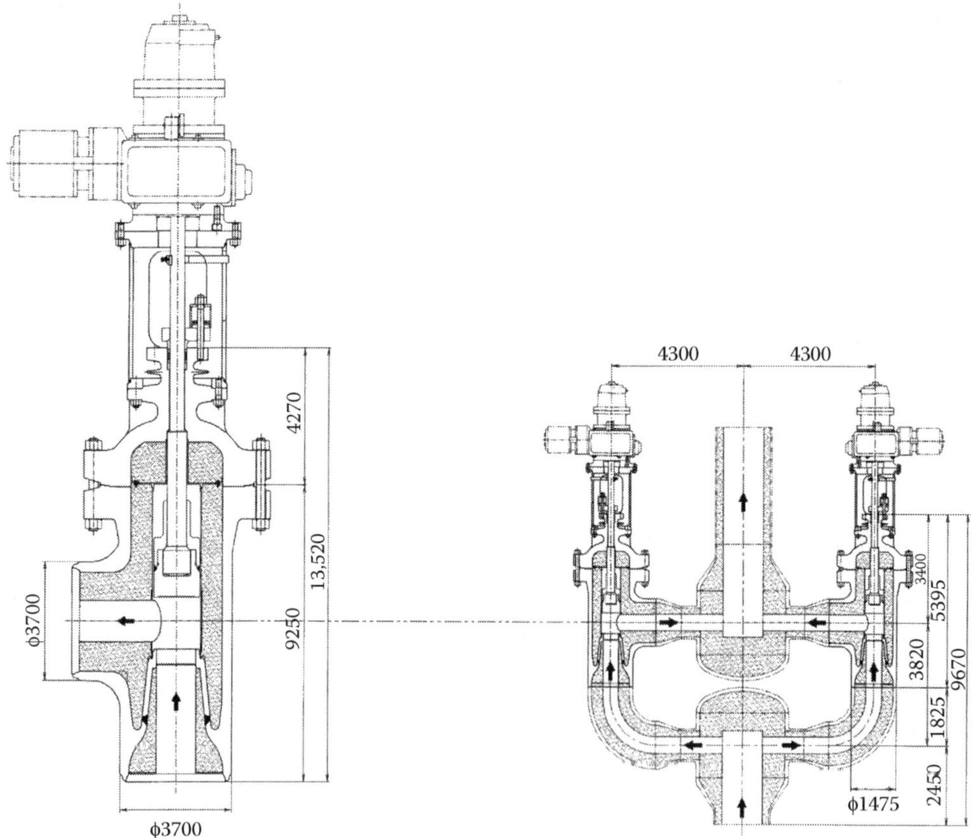

FIGURE 23.21
Alternative designs of a high-temperature gas isolation valve for commercial plants.

Size of the valve seat is determined at 600 mm by the manufacturability in present technical level. R&D on the fabrication of a large seat for a high-temperature valve is required. Helium flow rate in the valve is determined at 70 m/s by the experience of the hot gas duct of the HTTR. Then six valves are necessary for a 170 MW thermal load intermediate loop. Isolation valves are arranged around the hot gas duct so that the connecting pipe can be shortened and configuration space can be minimized. Size of the valves is listed in Figure 23.21. Valve size of the multiple unit is smaller by 40% in inner diameter, 40% in height, and 7% in volume.

Another important safety issue is tritium permeated from the reactor through the IHX and chemical reactors to end up in the final product of hydrogen. JAEA has tested underlying tritium permeation characteristics and is investigating design countermeasures and their performance to reduce tritium contamination in product hydrogen to acceptable limits that satisfy the requirements of consumer safety and environment protection. This investigation is being carried out by using HTTR. Chapter 24 describes details of this current research interest.

References

1. Design of high temperature engineering test reactor (HTTR), JAERI-1332, 1994.
2. K. Kunitomi et al., JAEA'S VHTR for hydrogen and electricity cogeneration: GTHTR300C, *Nucl. Eng. Technol.*, 39(1), 9–20, 2007.
3. T. Murakami et al., Safety assessment of VHTR hydrogen production system against fire, explosion and acute toxicity, *Trans. At. Energy Soc. Japan*, 7(3), 231, 2008 (in Japanese).
4. Committee for the prevention of disasters, Netherlands, *Yellow Book* 3rd Edition, Method for the calculation of physical effects, CPR 14E, 1997.
5. T. Murakami et al., Analysis on characteristic of hydrogen gas dispersion and evaluation method of blast overpressure in VHTR hydrogen production system, *Trans. At. Energy Soc. Japan*, 5(4), 316, 2006 (in Japanese).
6. U.S. Nuclear Regulatory Commission Office of Nuclear Regulatory Research, USA, Evaluating the habitability of a nuclear power plant control room during a postulated hazardous chemical release, *Regulatory Guide 1.78 Revision 1*, 2001.
7. Y. Inaba, T. Nishihara, and Y. Nitta, Analytical study on fire and explosion accidents assumed in HTGR hydrogen production system, *Nucl. Technol.*, 146(1), 49–57, 2004.
8. Y. Inagaki et al., Research and development on system integration technology for connection of hydrogen production system to an HTGR, *Nucl. Tech.* 157, 111, 2007.
9. S. Uchida et al., A conceptional design study on the hydrogen production plant coupled with the HTTR, *Proc. 11th Int. Conf. Nucl. Eng.*, Tokyo, Japan, April 20–23, 2003, ICONE11–36319, 2003.
10. H. Ohashi et al., Performance test results of mock-up test facility of HTTR hydrogen production system, *J. Nucl. Sci. Tech.*, 41(3), 385, 2004.
11. Y. Inaba et al., Study on control characteristics for HTTR hydrogen production system with mock-up test facility system controllability test for fluctuation of chemical reaction, *Nucl. Eng. Des.*, 235, 111, 2004.
12. H. Ohashi et al., Experimental and analytical study on chemical reaction loss accident with a mock-up model of HTTR hydrogen production system, *Proc. 12th Int. Conf. Nucl. Eng.*, Virginia, USA, April 25–29, 2004, ICONEC12-49419, 2004.
13. H. Sato et al., Control methods for the HTTR-IS nuclear hydrogen production system *Trans. At. Energy Soc. Japan*, 7(4), 328–337, 2008 (in Japanese).

14. T. Nishihara et al., Safety design philosophy of hydrogen cogeneration high temperature gas cooled reactor (GTHTR300C), *Trans. At. Energy Soc. Japan*, 5(4), 325–333, 2006 (in Japanese).
15. K. Tsuruoka et al., Design status of nuclear steel making system, *Trans. ISIJ*, 23, 1091, 1983.
16. P. Bröcherhoff et al., Status of design and testing of hot gas duct, *Nucl. Eng. Design*, 78, 215, 1984.
17. T. Nishihara et al., Development of high temperature isolation valve for the HTTR hydrogen production system, *Trans. At. Energy Soc. Japan*, 3(4), 69, 2004 (in Japanese).

24

Nuclear Hydrogen Plant Operations and Products

Hiroyuki Sato and Hirofumi Ohashi

CONTENTS

24.1 Introduction

Nuclear hydrogen plant to be considered in this chapter is defined as a system which generates hydrogen utilizing the heat supplied by nuclear reactor. The nuclear reactor such as a very high temperature reactor (VHTR), sodium-cooled fast reactor (SFR), and gas-cooled fast reactor (GFR) is coupled with a hydrogen production plant using thermochemical or hybrid (thermochemical + electrolytic) method. Hence, the commercial plant would be designed for cogeneration of both electricity and hydrogen. Nuclear reactors and hydrogen production plants of a commercial nuclear hydrogen plant would be built as a nuclear service area and a conventional service area, respectively, and operated according to different safety and operation requirements and by separate plant operators. Thus, intrinsic issues arise in the stability of operation in the nuclear hydrogen plant. Development of control strategy and its evaluation are one of the key items for the nuclear hydrogen demonstration.

Another potential problem in using a thermal coupling to transmit high-temperature heat from a nuclear reactor to a hydrogen production plant is the transmission of tritium from the nuclear reactor primary coolant to the hydrogen plant products as a consequence of the permeation of tritium through the intermediate heat exchanger (IHX) surfaces and then the bulk transport of tritium within the heat transport loop that connects the nuclear reactor to the hydrogen production plant. On the other hand, no other fission products would in general contaminate the process chemicals in the hydrogen plant and in the hydrogen plant products, because only tritium among fission products can penetrate the metallic or ceramic heat transfer surface wall of a primary system IHX.

The following two subsections discuss some details about the currently developing issues on hydrogen production plant control operations and the limitation to the process and product contamination from a nuclear heat source. The nuclear heat source of interest is a nuclear reactor such as the high-temperature gas-cooled reactor (HTGR) or fast reactors that connects and supplies nuclear generated heat to a hydrogen production facility.

24.2 Control of Plant Operations

Plant operation of nuclear hydrogen plants is different from that of traditional nuclear power plants. This section describes the particular operation issues and control strategies of the nuclear hydrogen plant. Sections 24.2.1 through 24.2.3 describe the common issues related to the operation of general nuclear hydrogen production plant. Section 24.2.4 describes the control strategy which enables to overcome the issues raised. Also, results of control method evaluation using simulation tools are described.

24.2.1 Normal Operation

Daily hydrogen production is assumed to vary in response to market demands and therefore assignment of the load to the hydrogen production plant should be changed. The immediate load change would induce the instability of nuclear hydrogen plant depending on the dynamic characteristics of nuclear reactor, hydrogen production plant, and the interface system. Consider a typical configuration shown in Figure 24.1. In case of load variation in hydrogen production plants, the variation transfers to reactor inlet. The reactivity would be increased or decreased by reactor core temperature variations and therefore, the reactor power varies. Assuming the reactor, IHX, and hydrogen production plant

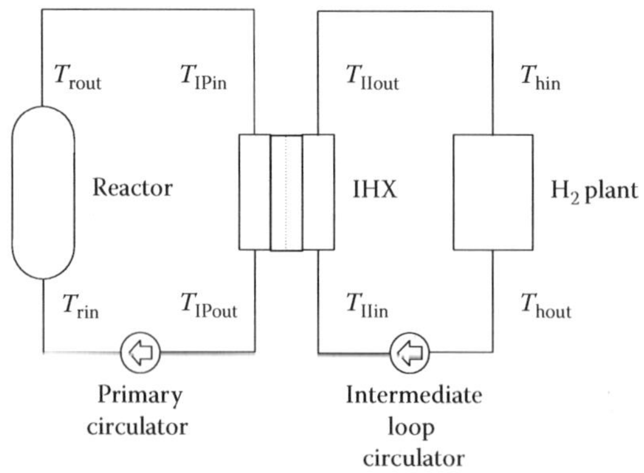

FIGURE 24.1
A typical configuration of a nuclear hydrogen plant.

characteristics as simple expressions, outlet temperature responses at reactor and hydrogen production plant are shown as follows:

$$\Delta T_{\text{rout}} = \alpha \Delta T_{\text{rin}} \tag{24.1}$$

$$\Delta T_{\text{IPout}} = \beta \Delta T_{\text{IPin}} \tag{24.2}$$

$$\Delta T_{\text{IIout}} = \gamma \Delta T_{\text{IIin}} \tag{24.3}$$

$$\Delta T_{\text{hout}} = \delta \Delta T_{\text{hin}} \tag{24.4}$$

where ΔT_{rout}, ΔT_{rin}, ΔT_{IPout}, ΔT_{IPin}, ΔT_{IIout}, ΔT_{IIin}, ΔT_{hout}, and ΔT_{hin} are variations of outlet and inlet temperature at a reactor, a primary side of IHX, an intermediate loop side of IHX, and a hydrogen production plant, respectively. The outlet responses depend on the signs and magnitudes of coefficients. If all the coefficients are positive or negative, and their absolute values are larger than one, the temperature variation would be amplified and induces harmonics between the reactor and hydrogen production plant.

Meanwhile, potential concern for oscillation arises if the coefficients have opposite signs. Cyclic loading due to the oscillation and harmonics in the nuclear hydrogen plant would greatly impact on integrity of components used in high temperature circumstances such as reactor pressure vessels and IHXs by the repeated stress. To be exact, these coefficients should be considered in the dynamics of configuration factors, for example, integral of relativities in the core, chemical reaction kinetics, control system characteristics, attenuation effect of components, and so on. In addition, interface system with an appropriate control strategy would perform as a damping system. Thus, detail assessment is required with reliable evaluation methods.

24.2.2 Start-Up and Shut-Down

Start-up and shut-down of the nuclear hydrogen plant is complicated due to the multi-operational stages of each component for example, a nuclear reactor, a power conversion unit and hydrogen production plants, and so on during the procedures. Nuclear reactors start with subcritical state, attain criticality, and increase power with the addition of control rod reactivity. Steam turbine requires superheated steam generated by steam generators which switch from the stand-by operation to rated operation during the start-up. Helium gas turbine initially performs as a compressor in order to circulate coolant followed with connection to the generator. Regarding the hydrogen production plants, an inappropriate control sequence would damage the main components for example, reactors, cells, and so on. Start-up and shut-down procedures should be considered at all the stages involved in the components with appropriate steps. In general, detailed assessment would be conducted by system analysis software.

In addition, restart-up operation of hydrogen production plant should be considered since the electricity generation would be maintained during the short-term maintenance or suspension of hydrogen production in the commercial cogeneration plant of electricity and hydrogen. In case of restart-up operation of the hydrogen production plant, the primary cooling system maintains high temperature in contradiction to low temperature in hydrogen production plant, since the components such as cells and process heat exchangers in the hydrogen production plant are greatly affected by immediate temperature increase. The increase would shorten the lifetime of the components. Thus, prevention of the thermal shock should be considered in the restart-up operation of the hydrogen production plant.

24.2.3 Loss of Hydrogen Production Plant Heat Load

Loss of heat load in hydrogen production plants occurs when endothermic reactions occupied in chemical reactors stop because of the accident, such as a process feed stoppage. In nuclear hydrogen production plants, chemical reactors act as heat sinks of a nuclear reactor during normal operation. Hence, loss of heat load in the hydrogen production plant directly impacts on the cooling ability of the nuclear reactor and results in a reactor scram. It is undesirable that shut down of hydrogen production plants affect the operation of the nuclear reactor from the economical and safety point of view. In addition, immediate temperature increase would damage the components installed, at downstream of chemical reactors. Therefore, a mitigation method of abrupt load changes should be taken into consideration for the control strategy.

24.2.4 Establishment of System Control Scheme

This section is focused on the control scheme and its assessment for HTGR nuclear hydrogen plants. Basic control strategies are discussed. Also, control schemes which could meet the requirements are reported in line with the proposed strategy. Furthermore, the control scheme is assessed by using a system analysis tool.

24.2.4.1 HTTR-IS System

Chapter 29 presents the design of the high temperature engineering test reactor-iodine-sulfur (HTTR-IS) system. One of the objectives for the HTTR-IS system is to show the economical feasibility of the commercial VHTR hydrogen production plant, for example, gas turbine high temperature reactor 300-cogeneration (GTHTR300C), where the reactor and hydrogen production plants would be managed by different companies. This requires high operating rates, and therefore the HTTR-IS system should demonstrate a control method applicable to commercial plants which enables to (1) operate the reactor normally against abnormal events of the IS process, and (2) restart the IS process during the reactor normal operation.

Japan Atomic Energy Agency (JAEA) reported a control strategy for the HTTR-IS system which could meet these requirements [1]. In order to mitigate the heat load disturbance due to the abnormal events of the hydrogen production plant, JAEA proposed a thermal load mitigation system which consists of a steam generator and an air cooler [2]. Figure 24.2 shows the schematic diagram of the thermal load mitigation system at the rated operation in the HTTR-IS system. Figure 24.3 described the control scheme of the thermal load mitigation system [2]. When abnormal events occur in the hydrogen production plant, diverter "Valve A" opens and "Valve B" closes to isolate the hydrogen production plant. This flow-path change causes immediate increase of inlet helium gas temperature at the steam generator, so that the peak value of outlet helium gas temperature at steam generator exceeds the target value transiently. Therefore, the following scheme was proposed as a countermeasure:

1. Interface the actuation of "Valve A," "Valve B," and "Valve D"
2. Actuates "Valve F" as the pressure difference between the steam generator bottom and the down comer from the air cooler to the steam generator approaches zero

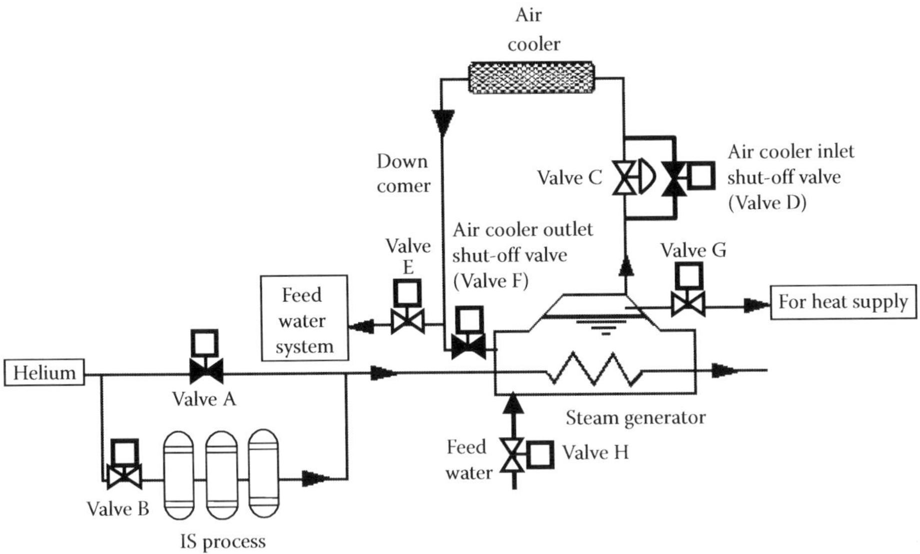

FIGURE 24.2
A schematic diagram of the thermal load mitigation at rated operation in the HTTR-IS system. (Adapted from H. Sato et al., *Trans. Atom. Energ. Soc. Jpn*, 7(4), 328–337, 2008.)

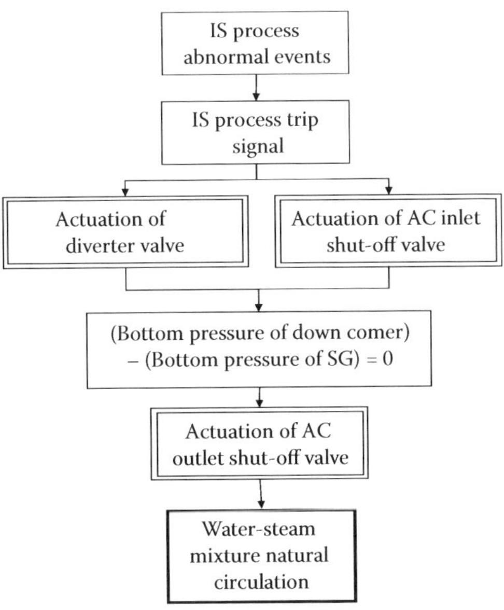

FIGURE 24.3
A control scheme of the thermal load mitigation system in the HTTR-IS system. (Adapted from H. Sato et al., *Trans. Atom. Energ. Soc. Jpn*, 7(4), 328–337, 2008.)

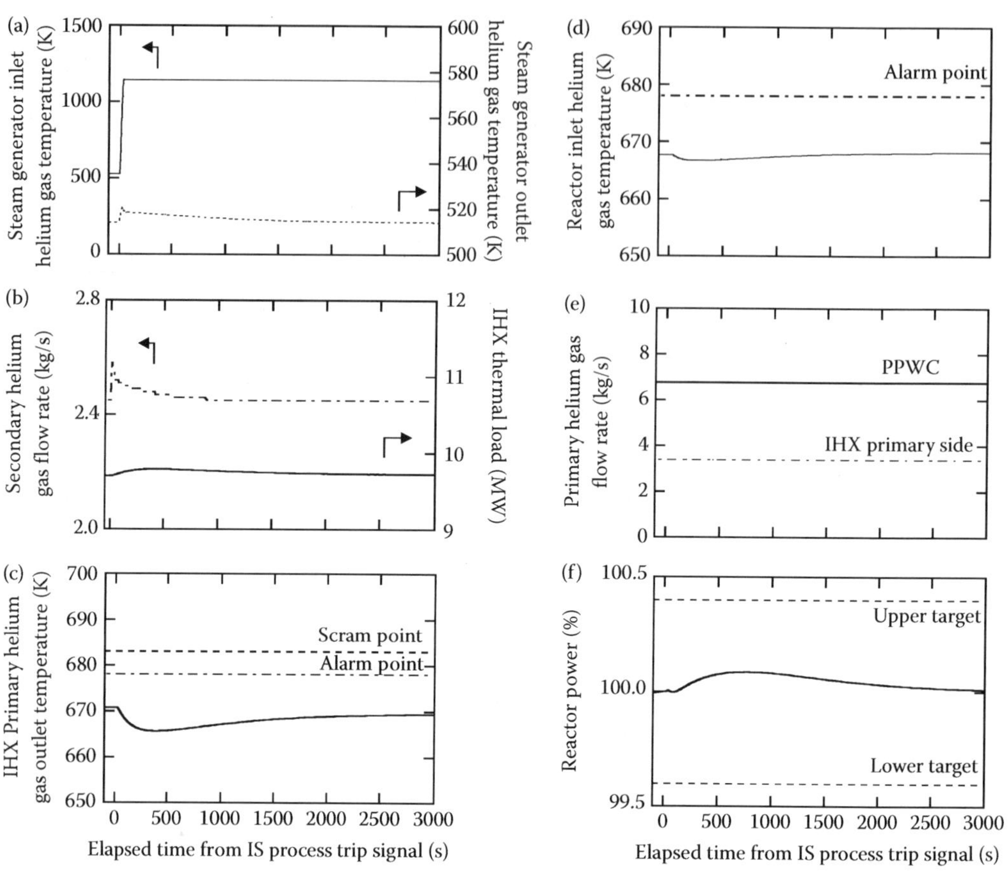

FIGURE 24.4
Calculation results of the HTTR-IS system behavior during abnormal events in the hydrogen production plant. (a) Steam generator inlet and outlet helium gas temperature, (b) secondary helium gas flow rate and IHX thermal load, (c) IHX outlet primary helium gas temperature, (d) reactor inlet helium gas temperature, (e) primary helium gas flow rate, and (f) reactor power. (Adapted from H. Sato et al., *Trans. Atom. Energ. Soc. Jpn*, 7(4), 328–337, 2008.)

Figure 24.4 shows the evaluation results of system behavior of the HTTR-IS system with employed control scheme. The simulation is conducted by using extension version of the RELAP5 code [2] developed by JAEA [3]. Inlet temperature at steam generator increases at about +12°C/s due to the actuation of diverter valves. Air cooler inlet shut-off valve and outlet shut-off valve actuates according to the proposed control scheme, so that helium gas temperature increase at steam generator outlet can be mitigated within target values. Also, calculation results show that outlet helium gas temperature at primary and secondary side of IHX, inlet helium gas temperature at reactor, and reactor power can be mitigated within evaluation criteria. The results clearly showed availability of the proposing scheme which can mitigate the impact of abnormal events in the hydrogen production plant on reactor operation.

Also, control scheme for the restart procedure of a hydrogen production plant was investigated by JAEA [1]. JAEA proposed the following scheme as a countermeasure:

1. Decrease the reactor power and control the primary helium flow-rate to decrease the thermal load of IHX

2. Increase the secondary helium flow-rate to decrease the IS process inlet helium temperature and decrease the flow-rate of the IS process using the control valve installed at downstream of IS process
3. Decrease the secondary helium flow-rate to increase the IS process inlet helium temperature as well as increase the flow-rate of the IS process to supply heat
4. Increase the flow-rate of IHX to supply heat and increase the reactor power to full power

Figure 24.5 shows the evaluation results of system behavior of the HTTR-IS system with employed control scheme. The simulation is conducted by using extension version of the RELAP5. The simulation starts from reactor power of 71%, helium gas flow-rate of primary pressurized water cooler (PPWC) primary side of 8.6 kg/s, IHX primary side and secondary side flow-rate of 1.5 kg/s, and 4.9 kg/s, respectively. Firstly, the helium gas flow-rate in the hydrogen production plant increases in small steps by using control valve installed at

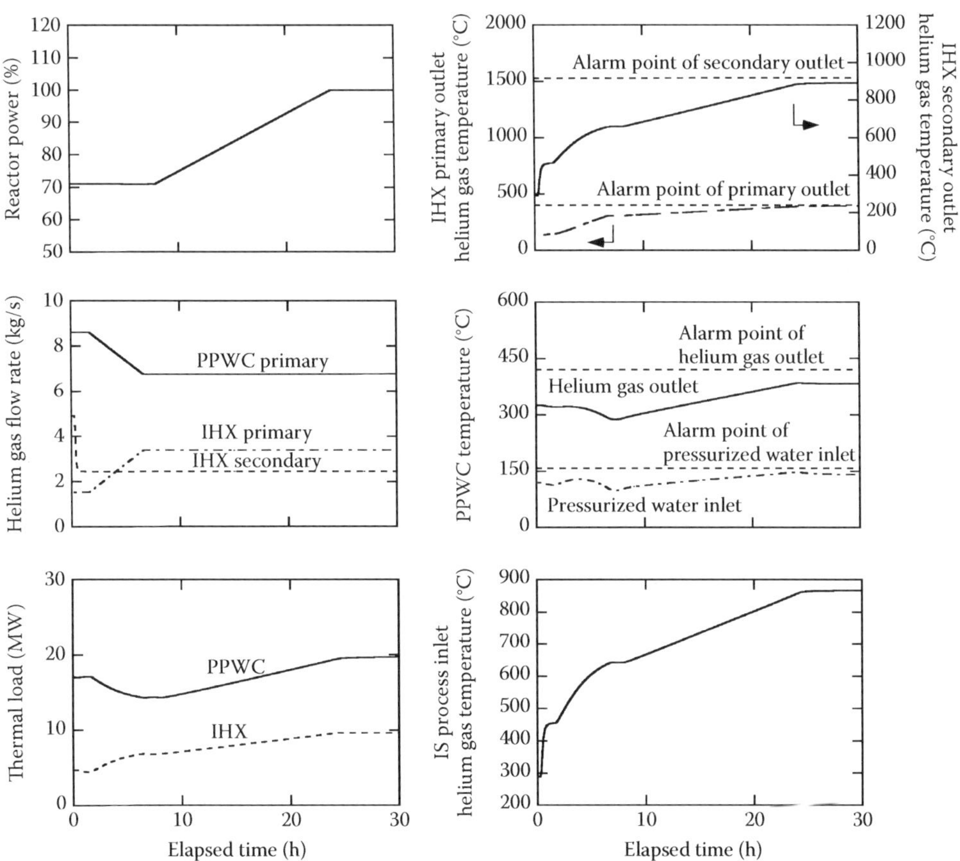

FIGURE 24.5
Calculation results of the HTTR-IS system behavior during the restart-up operation of hydrogen production plant. (Adapted from H. Sato et al., Proceedings of the 4th International Topical Meeting on High Temperature Reactor Technolgy HTR2008, HTR2008-58044, 2008.)

downstream of the hydrogen production plant, so that the structures of process heat exchangers are gradually warmed up. Secondly, helium gas flow-rate of IHX secondary side decreases to 2.4 kg/s in order to increase inlet helium gas temperature at IS process from 290 to 455°C. Allocation of the thermal load is controlled by helium gas flow-rate at the primary side of PPWC and IHX from 8.6 to 6.8 kg/s and from 1.5 to 3.4 kg/s, respectively, so that the heat supplied to the IHX increases from 4.7 to 6.8 MW. Finally, reactor power increases from 71 to 100% and rated operation can be conducted. The power rising rate is restricted to 15°C/h in terms of safety operation regulated by the safety regulation. Calculation results show that all process values are within the allowable criteria, so that the reactor enables the normal operation.

24.2.4.2 GTHTR300C Plant

JAEA reported the control strategy of the GTHTR300C design [4]. Chapter 10 includes a description of the plant design. Basically, the same concept of the control scheme is employed, though GTHTR300C is a cogeneration system producing both electricity and hydrogen configuration that is different from that of the HTTR-IS system. Figure 24.6 shows the schematic diagram of the GTHTR300C and its control scheme is shown in Figure 24.7. The hydrogen production output from the plant would be adjusted in order to follow change in market demand. The adjustment would require flow-rate change in the hydrogen process heat exchangers so that the heat load input to the hydrogen production plant would be adjusted to match the production output. The heat load change would be detected by subsequent change in the IHX outlet helium gas temperature. The control valves are actuated to maintain constant gas turbine inlet temperature and rotational speed so that the reactor and gas turbine electricity generation

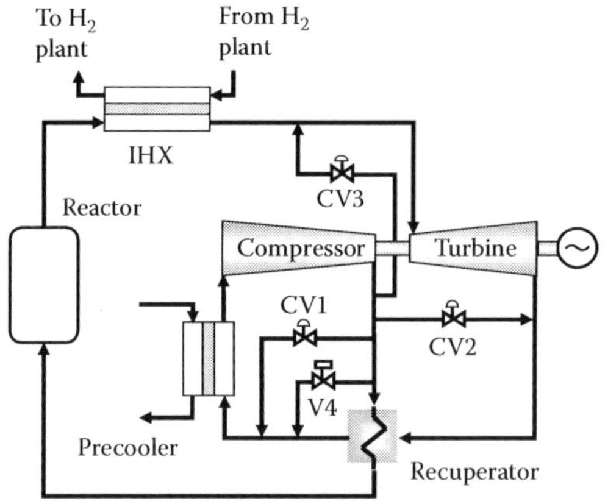

CV1 : Turbine bypass flow control valve

CV2 : Recuperator inlet temperature control valve

CV3 : Turbine inlet temperature control valve

V4 : Turbine bypass valve

FIGURE 24.6
Schematic diagram of the GTHTR300C.

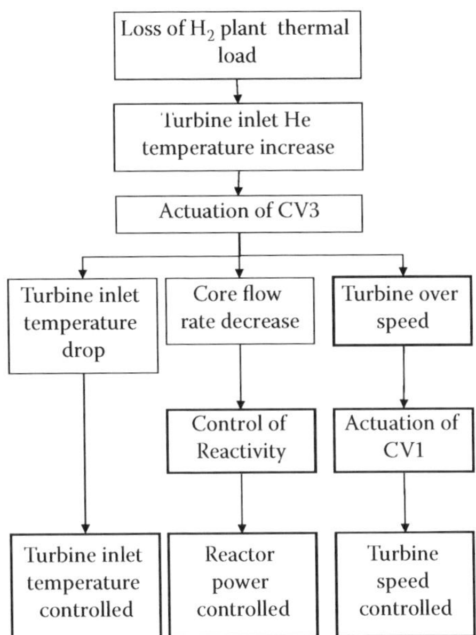

FIGURE 24.7
Control scheme of the GTHTR300C.

continue their normal operations. Any surplus heat load during the temporary thermal transient as a result of the hydrogen plant load change would be discharged in the precooler heat sink.

Figure 24.8 shows the calculation results of loss of heat load of the hydrogen production plant in the GTHTR300C. The simulation is conducted by using extension version of the RELAP5. In this calculation, stepwise increase of IHX secondary helium gas temperature is set as a boundary condition. At the beginning of the event, turbine inlet temperature increases, and therefore, turbine inlet temperature control valve "CV3" actuates to mitigate the temperature increase. As can be seen, the increase is limited within 20°C. The actuation of CV3 causes the over speed of turbine bypass flow control valve because of the volume flow increase, turbine rotational speed control valve "CV1," opens. As a result, turbine over speed is limited within 5 rpm. Due to the decrease of inlet flow rate of the core, the reactivity control system reduces reactor power to keep the reactor outlet temperature constant. Reactor core flow rate decreases to 88% due to the actuation of CV1 and CV3 and reactor power decreases to 84% of the rated thermal power. The simulation results revealed that the load assignment of hydrogen production plant could be smoothly conducted with a proposed control scheme.

24.2.4.3 VHTR-HTE Plant

Argonne National Laboratory reported the control strategy for the typical VHTR-HTE (high temperature electrolysis) plant configuration [5]. The strategy applies inventory control in the VHTR plant and flow control in the hydrogen production plant and simulation results showed that the hot-side temperatures in VHTR plant can likely be maintained near constant. In addition, start-up procedure was proposed. The procedure initiates with attaining

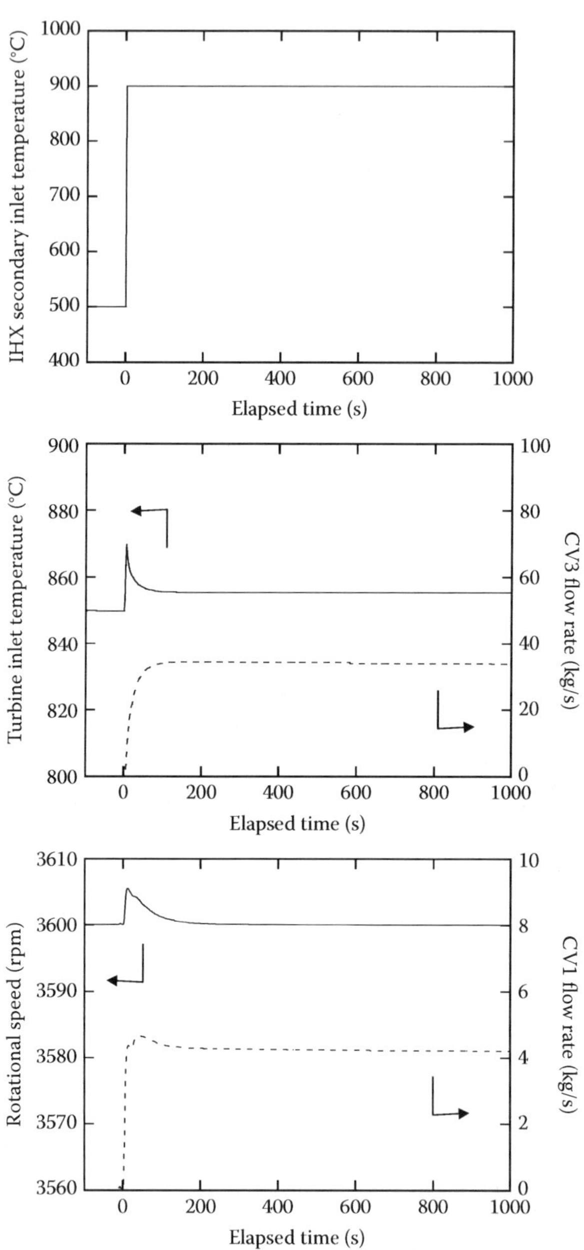

FIGURE 24.8
Calculation results of the GTHTR300C system behavior during abnormal events in the hydrogen production plant.

criticality of reactor, and increase reactor power by adding rod reactivity followed with ascendance of electricity load and thermal load. Full-power operation could be achieved for about four hours without considering thermal stress on structure components. The proposed scheme would be evaluated in the future by using GAS-PASS/H code [6] which models major components of VHTR-HTE plant.

24.3 Contamination of Process and Hydrogen

24.3.1 Tritium Source and Pathway to Hydrogen Product

The tritium behavior in HTGRs was evaluated in several countries in the 1970s (e.g., the Dragon Reactor in the United Kingdom [7], the Peach Bottom HTGR in the United States [8], the AVR in Germany [9]). Data from the operation of HTGRs and laboratory experiments revealed the mechanisms of tritium production, transport, and release to the environment. The primary tritium birth mechanism is ternary fission of fuel (e.g., 233U, 235U, 239Pu, and 241Pu) because of thermal neutrons. Tritium is also generated in HTGR from 6Li, 7Li, 3He, and 10B by neutron capture reactions. 6Li and 7Li are impurities in the core graphite material (e.g., sleeve, spine, reflector, and fuel matrix). 3He is an impurity in the reactor coolant helium. Because the helium coolant leaks from the primary loop to the containment vessel, 3He is supplied along with helium to the primary coolant. 10B exists in the control rods, burnable poison, and reflector.

Tritium generated in the fuel particles by ternary fission mechanisms can escape into the primary coolant after permeating the barrier layers of the fuel particles. In addition, tritium born from 10B and 6Li can pass into the primary coolant. Figure 24.9 shows a schematic diagram of tritium behavior in the nuclear hydrogen plant. The principal chemical form for tritium in the reactor coolant was reported as HT due to the isotope exchange reaction between T_2 and H_2 [8].

Some of the tritium in the primary coolant is removed by a purification system installed in the primary loop. Some of the tritium can escape to the outside environment by permeation through the components and piping and by leakage of the bulk primary helium coolant. The remainder of the tritium in the primary coolant permeates through the heat transfer tubes heat exchanger and is mixed into the secondary coolant. According to a review by Gainey [10], calculations with some of the HTGR data showed that tritium releases should be well within the U.S. Federal guidelines contemporary with the earlier HTGR operations, and so further laboratory-scale work related to tritium transport in HTGRs was discontinued. However, HTGRs at that time only produced electricity and were not targeted toward high-temperature heat applications, and the potential for tritium migration into downstream processes was not evaluated.

Finally, tritium accumulation in the process chemicals of the hydrogen plant may become a critical issue in seeking to operate the hydrogen plant associated with the HTGR

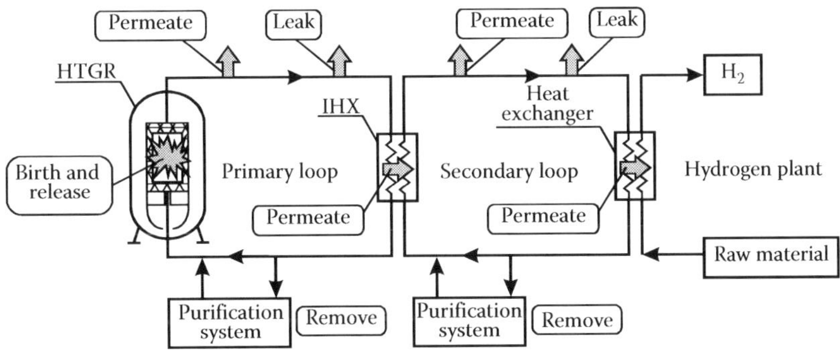

FIGURE 24.9
Schematic diagram of tritium behavior in the nuclear hydrogen plant.

as a nonnuclear facility and to provide hydrogen to customers that meets specifications in regard to tritium. Note that the issue of the tritium contamination in the hydrogen product arises not only for the HTGR but also the fast reactors, because tritium production mechanisms by ternary fission of fuel and by neutron capture reaction of 10B in the control rods in the fast reactors are same as in the HTGR. Tritium released to the primary coolant (e.g., He for the GFR, Na for the SFR) migrates to the hydrogen product by the permeation through the heat transfer tube of the IHX and the chemical reactors.

24.3.2 Regulatory Constraints on Tritium

The U.S. Nuclear Regulatory Commission (NRC) is continuously evaluating the latest radiation protection recommendations from international and national scientific bodies to ensure the adequacy of the standards the agency uses. Among those standards, the NRC and the U.S. Environmental Protection Agency (EPA) have established radiation protection limits to protect the public against potential health risks from exposure to radioactive liquid discharges (effluents) from nuclear power plant operations. Table 24.1 summarizes the regulatory constraints on tritium. The NRC's final layer of protection of public health and safety is a dose limit of 100 mrem per year to individual members of the public. For gas and liquid effluents, including water contaminated with tritium, any NRC licensee can demonstrate compliance with the 100 mrem per year dose standard by not exceeding the concentration values specified in Table 2 of Appendix B to 10 CFR Part 20. For tritium, the effluent concentration in air and water are 3.7×10^{-3} and 37 Bq/mL, respectively. In 1976, the EPA also established a dosed-based drinking water standard of 4 mrem per year to avoid the undesirable future contamination of public water supplies as a result of controllable human activities related to nuclear industries. To achieve this result, the EPA set a maximum contamination level of 0.74 Bq/mL (20 n Ci/L). In Japan, same level of the limit for the effluent tritium concentration in air and water are defined as in the United States. (e.g., 3.0×10^{-3} and 20 Bq/cm^3 for organic bonded tritium, respectively). Hydrogen produced

TABLE 24.1

Regulatory Constraints on Tritium in the United States

		Annual Radiation Dose		Effluent Concentration			
				Air		Water	
	Regulation	(mrem)	(mSv)	(μCi/mL)	(Bq/mL)	(μCi/mL)	(Bq/mL)
Limit	10 CFR 20.1301(a)1	100	1	—	—	—	—
	Table 2 of Appendix B to 10 CFR 20	50	0.5	1×10^{-7}	3.7×10^{-3}	1×10^{-3}	37
Standard	10 CFR 20.1301(e)	25	0.25	(5×10^{-8})[a]	(1.85×10^{-3})[a]	(5×10^{-4})[a]	(18.5)[a]
ALARA	Appendix I to 10 CFR 50	15	0.15	(3×10^{-8})[a]	(1.11×10^{-3})[a]	—	—
		3	0.03	—	—	(6×10^{-5})[a]	(2.22)[a]
Drinking Water	EPA standard	4	0.04	—	—	2×10^{-5}	0.74

[a] Calculated by assuming the linear relationship between the annual dose of 50 mrem and the values in Table 2 of Appendix B of 10 CFR 20.

ALARA = as low as reasonably achievable.

CFR = Code of Federal Regulations.

EPA = U.S. Environmental Protection Agency.

by HTGR may be used in fuel cell applications, a chemical production system, or in a petroleum refiner, and legal limits for tritium in the hydrogen for these applications do not exist. In the absence of industry-established tritium limits, the dose to people must be the governing limit. Therefore, tritium levels and limits will be evaluated and defined by considering hydrogen usage and a tritium pathway to the human body.

24.3.3 Current Status of R&D Activities

JAEA has researched tritium behavior in the HTGR hydrogen production system. The transportation of tritium to the product hydrogen was numerically evaluated for the HTTR hydrogen production system by steam reforming of natural gas [11]. It was concluded that the tritium activity concentration in the product hydrogen is strongly dependent on the permeability of heat transfer tubes in IHX and the chemical reactors. Therefore, the permeability of IHX material, Hastelloy XR, was investigated experimentally [12] and a model for the permeation mechanism was established [13]. It was determined that the permeation rate of hydrogen isotope was affected by the hydrogen existing on the surface of the heat transfer tubes. Also, the actual hydrogen permeability through IHX was evaluated using the HTTR data [14].

In order to evaluate the tritium behavior in the HTGR hydrogen production system, JAEA developed a numerical analysis code, tritium and hydrogen transportation analysis code (THYTAN), which can calculate the mass balance of tritium taking into account the following phenomena: tritium generation in the core, tritium release to the coolant, tritium permeation at the heat transfer tubes of the heat exchangers, tritium removal by purification systems, tritium permeation to the atmosphere, and tritium leakage to the atmosphere or another loop with helium leakage, and isotope exchange reaction between tritium containing chemicals and hydrogen containing chemicals [15]. Mass balances and a node-link computational scheme were used to represent the various HTGR process units and flow streams. A node is used to represent a process unit or a section of a process unit, while links are used to indicate the flow of materials into and out of a node.

Using the above computational scheme and the THYTAN algorithms, a representation of the Peach Bottom HTGR, a power station with a steam turbine, was analyzed, and the results were compared to the data available from Peach Bottom operations in regard to tritium concentration in the primary coolant. The comparison of THYTAN results with Peach Bottom data showed that the THYTAN code provided results that were relatively good; agreement with the observed data in most cases is shown in Figure 24.10 [16].

For the HTGR hydrogen production system, once in the hydrogen plant fluids, tritium can react with hydrogen-containing process chemicals through isotope exchange reactions. The IS process contains H_2O, H_2SO_4, and HI as circulating chemicals, and these can undergo isotope exchange mechanisms to produce HTO, $HTSO_4$, and TI. Since, tritium losses from the SI process occur only from the occasional leaks of process chemicals and contamination of the hydrogen and oxygen products, these tritium-containing chemicals accumulate in the process until the tritium input rate by permeation equals the tritium loss rate through these other mechanisms.

Tritium migration behavior in the IS process was numerically evaluated using the THYTAN code [15]. Figure 24.11 shows the schematic flow diagram of the IS process including the flow of the process and tritium-containing chemicals. Tritium permeated the H_2SO_4 vaporizer and the SO_3 decomposer, migrated to the product hydrogen, and changed its form from HT to HTO in the components of the H_2SO_4 section, from HTO to TI in the Bunsen reactor and from TI to HT in the HI decomposer by isotope exchange reactions. It

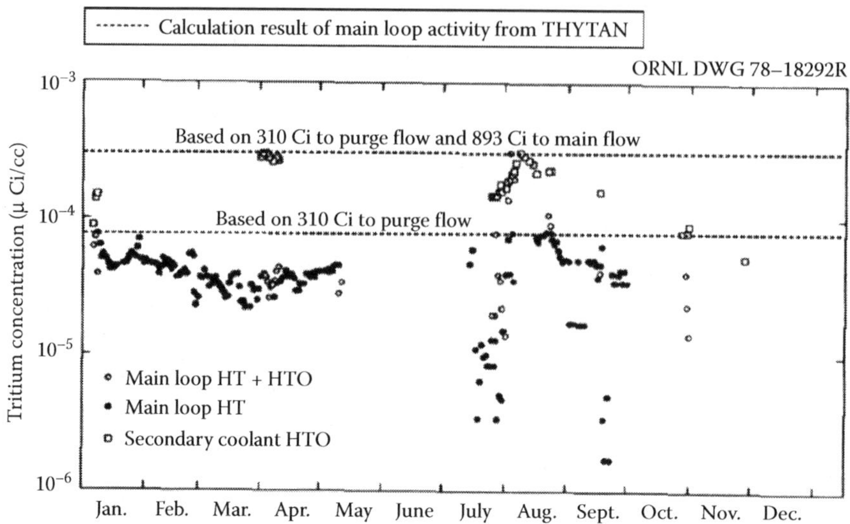

FIGURE 24.10
Comparison of tritium concentration observed in the Peach Bottom HTGR (From R.P. Wichner and F.F. Dyer, Distribution and transport of tritium in the peach bottom HTGR, ORNL-5497, Oak Ridge National Laboratory, August 1979. With permission) and computed results of THYTAN (From H. Ohashi and S.R. Sherman, Tritium movement and accumulation in the NGNP system interface and hydrogen plant, INL/EXT-07-12746, June 2007. With permission).

was confirmed that the numerically evaluated tritium activity concentration in the product hydrogen increased by considering the isotope exchange reactions.

Tritium concentrations in the hydrogen product and in process chemicals in the hydrogen plant associated with the GTHTR300C designed by JAEA and the Next Generation Nuclear Plant (NGNP) designed in the United States were estimated using the THYTAN code [16–18]. Estimated tritium concentrations in the hydrogen product and in process chemicals in the hydrogen plant of the NGNP employing the high-temperature electrolysis are slightly higher than the drinking water limit defined by the EPA and the limit in the effluent at the boundary of an unrestricted area of a nuclear plant as defined by the NRC. The tritium concentrations in the hydrogen product and in process chemicals in GTHTR300C employing the IS process are also slightly higher than the effluent limit in Japanese regulation. However, these concentrations can be reduced to within the limits by the design (i.e., increasing the capacity of a helium purification system as shown in Figure 24.12). The increase of the capacity of a helium purification system results in an increase of the reactor facility cost. Therefore, JAEA proposed alternative countermeasure to decrease the tritium permeation rate through the heat transfer tubes by water injection in the intermediate loop, which produces HTO from HT by an isotope exchange reaction [18]. Analytical results showed that tritium concentration in the hydrogen and in process chemicals in the hydrogen plant decreases by the water injection as shown in Figure 24.13.

24.4 Conclusion and Future Works

This section introduces the fundamentals and application examples of nuclear hydrogen plant operation and product contaminations. Operational issues are classified depending

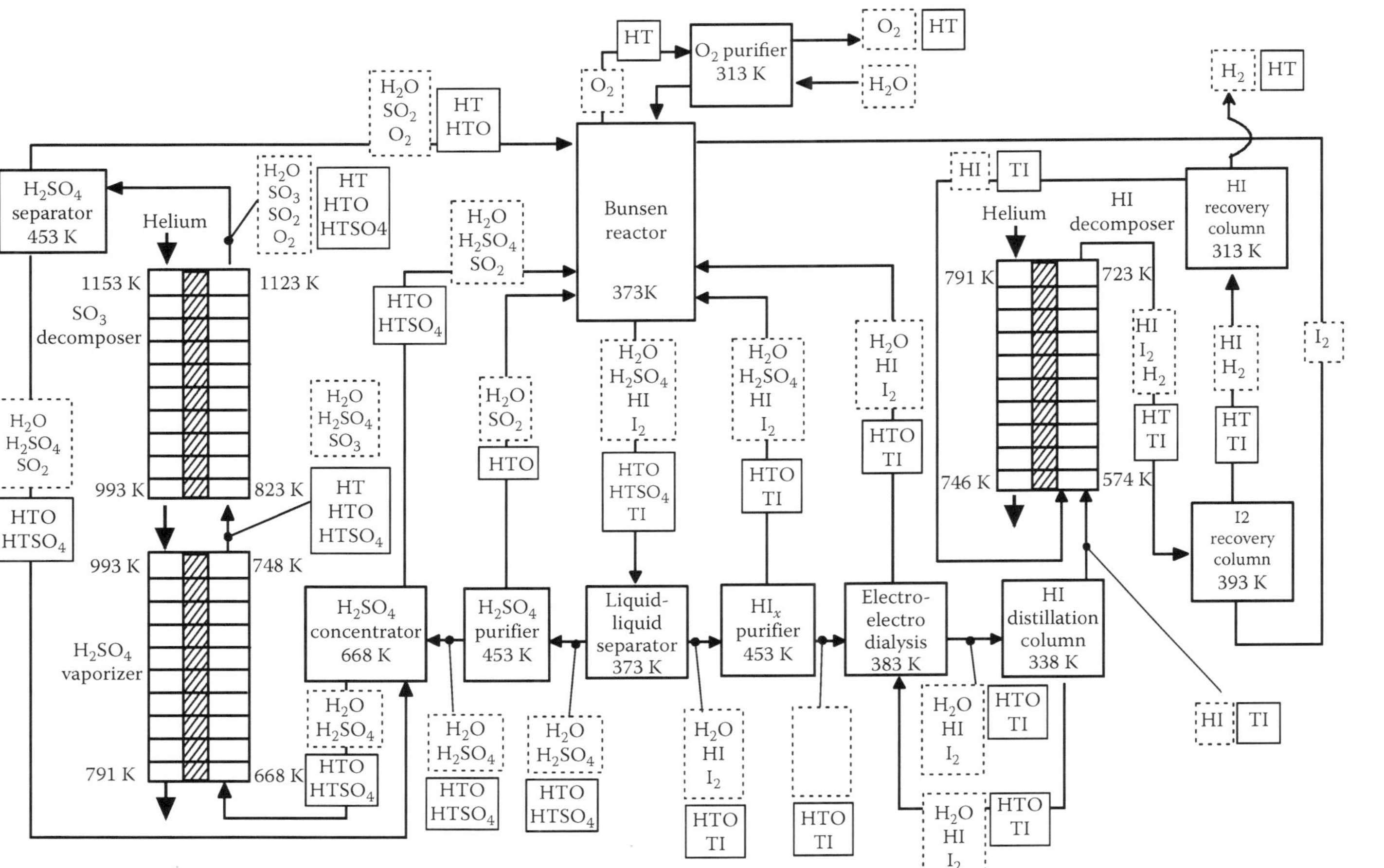

FIGURE 24.11
Schematic flow diagram of the IS process including the flow of the process and tritium-containing chemicals. (From H. Ohashi, et al., *J. Nucl. Sci. Technol.*, 44(11), 1407–1420, 2007. With permission.)

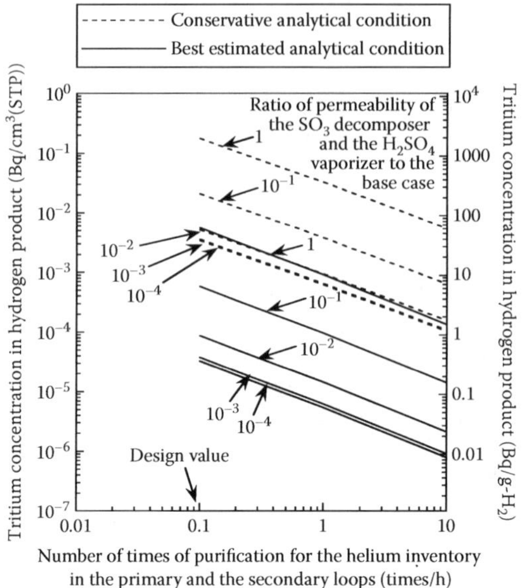

FIGURE 24.12
Effects of varying helium flow rates in the purification systems for the primary and secondary coolants on the tritium concentration in the hydrogen product of GTHTR300C. (From H. Ohashi et al., *J. Nucl. Sci. Technol.*, 45(11), 1215–1227, 2008. With permission.)

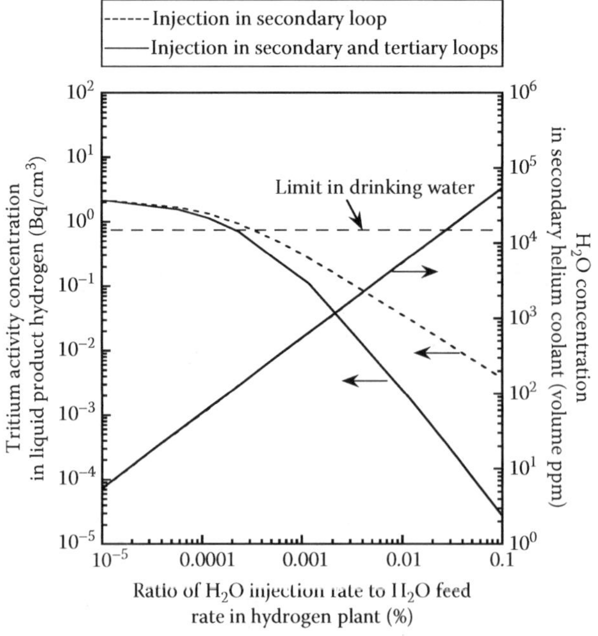

FIGURE 24.13
Effect of varying water injection rate in helium coolant on the tritium concentration in the hydrogen product for the NGNP using the HTE process. (From H. Ohashi and S.R. Sherman, *Trans. Atom. Energ. Soc. Jpn.*, 7(4), 439–451, 2008 (in Japanese). With permission.)

on the operational stages of the plant. Immediate load change in nuclear hydrogen production plants would induce instability or shut down of nuclear hydrogen plants. Several approaches to mitigate the issues are explained by examples from HTGR nuclear hydrogen plants. In addition, restart up operation issues and control scheme of the hydrogen production plant, which is an inherent requirement to nuclear hydrogen plant operation, are discussed. The control schemes are evaluated by using system analysis codes and the results are shown to illustrate the procedure of the control scheme evaluation.

These topics are unique characteristics comparing to the typical nuclear power plant and are still in the research stages. Further progression is required for the nuclear hydrogen demonstration.

As for the plant operation of nuclear hydrogen production system, further developments are required especially for the evaluation methods which can consider the characteristics of the control system in order to evaluate detailed system behavior. Developed method would be validated by the data obtained from a pilot plant scale experiment. In addition, demonstration is necessary to ensure the reliability of the control strategy employed by using actual nuclear reactor, for example, HTTR.

As for the tritium contamination, additional data is needed to determine more accurately the tritium concentrations in the hydrogen product and in the process chemicals in the hydrogen plant in order to design the commercial plant by using proper countermeasures or develop alternative countermeasures if the current countermeasures are not sufficient enough. These data include information about the tritium permeability of nonmetallic heat exchanger materials, a better understanding of tritium release rates from HTGR core graphite and other materials, and measurements of the tritium isotope exchange rates and equilibrium constants for the interactions between tritium and key process chemicals (e.g., H_2SO_4, HI). The THYTAN code will be validated using the tritium data in the HTTR.

References

1. H. Sato, N. Sakaba, N. Sano et al., Demonstration of nuclear hydrogen production utilizing the Japan's HTTR; Control scheme evaluation of the HTTR-IS nuclear hydrogen production system, *Proceedings of the 4th International Topical Meeting on High Temperature Reactor Technology HTR2008*, September 28–October 1, Washington, DC, USA, HTR2008-58044, 2008.
2. H. Sato, H. Ohashi, N. Sakaba, T. Nishihara, and K. Kunitomi, Thermal load control methods for the HTTR-IS nuclear hydrogen production system, *Trans. Atom. Energ. Soc. Jpn*, 7(4), 328–337, 2008 (in Japanese).
3. U.S. NRC, RELAP5/MOD3 Code Manuals, NUREG/CR-5535, 1995.
4. K. Takamatsu, S. Katanishi, S. Nakagawa et al., Development of plant dynamics analytical code named Conan-GTHTR for the gas turbine high temperature gas-cooled reactor, (I), *Trans. Atom. Energ. Soc. Jpn.* 3(1), 76–87, 2004 (in Japanese).
5. T. Nishihara, K. Ohashi, T. Murakami et al., Safety design philosophy of hydrogen cogeneration high temperature gas cooled reactor (GTHTR300C), *Trans. Atom. Energ. Soc. Jpn.*, 5(4), 325–333, 2006 (in Japanese).
6. R.B. Vilim, Initial Assessment of the Operability of the VHTR-HTSE Nuclear-Hydrogen Plant, ANL-08/01, 2007.
7. R.B. Vilim, J. Cahalan, and U. Mertyurek, GAS-PASS/H: A simulation code for gas reactor plant systems, *Proc. of ICAPP 2004*, Pittsburgh, PA, June 2004.

8. N. Forsyth, Tritium production and distribution in high temperature gas-cooled reactors Part 1: Tritium production, migration and removal in the dragon reactor between Core 5 Charge III and Core 1 Charge IV, DP-REPORT-799, O.E.C.D. High Temperature Reactor Project DRAGON, June 1972.
9. R.P. Wichner and F.F. Dyer, Distribution and transport of tritium in the peach bottom HTGR, ORNL-5497, Oak Ridge National Laboratory, August 1979.
10. W. Steinwarz, H.D. Röhrig, and R. Nieder, Tritium behavior in an HTR-system based on AVR-experience, International Atomic Energy Agency, Vienna (Austria), IWGGCR-2, 153, 1980.
11. B.W. Gainey, A review of tritium behavior in HTGR systems, GA-A13461, General Atomic Company, April 1976.
12. T. Nishihara and K. Hada, Evaluation of tritium transportation to the product hydrogen in the HTGR hydrogen production system, *Nihon-Genshiryoku-Gakkai Shi (J. At. Energy Soc. Jpn.)*, 41(5), 571, 1999 (in Japanese).
13. T. Takeda, J. Iwatsuki, and Y. Inagaki, Permeability of hydrogen and deuterium of Hastelloy XR, *J. Nucl. Mater.*, 326, 47, 2004.
14. T. Takeda and J. Iwatsuki, Counter-permeation of deuterium and hydrogen through inconel 600, *Nucl. Technol.*, 146, 83, 2004.
15. N. Sakaba, H. Ohashi, and T. Takeda, Hydrogen permeation through heat transfer pipes made of Hastelloy XR during the initial 950°C operation of the HTTR, *J. Nucl. Mater.*, 353, 42, 2006.
16. H. Ohashi, N. Sakaba, T. Nishihara et al., Numerical study on tritium behavior by using isotope exchange reactions in thermochemical water-splitting iodine–sulfur process, *J. Nucl. Sci. Technol.*, 44(11), 1407–1420, 2007.
17. H. Ohashi and S.R. Sherman, Tritium movement and accumulation in the NGNP system interface and hydrogen plant, INL/EXT-07–12746, June 2007.
18. H. Ohashi, N. Sakaba, T. Nishihara et al., Analysis of tritium behavior in very high-temperature gas-cooled reactor coupled with thermochemical iodine–sulfur process for hydrogen production, *J Nucl. Sci. Technol.*, 45(11), 1215–1227, 2008.
19. H. Ohashi and S.R. Sherman, Analytical study on tritium migration in NGNP and countermeasures to reduce tritium contamination, *Trans. Atom. Energ. Soc. Jpn.*, 7(4), 439–451, 2008 (in Japanese).

25

Licensing Framework for Nuclear Hydrogen Production Plant

Yujiro Tazawa

CONTENTS

25.1 Introduction

Before a civilian nuclear reactor can be built and operated, approval is usually obtained as in Japan and the United States from such authorities as designated by the government of a state as being responsible by law for licensing and regulating nuclear reactors and materials for their intended safe and peaceful purposes while protecting the safety of people and the environment. A nuclear hydrogen production plant will require such approvals. Several types of nuclear reactors as described in Section III enable nuclear hydrogen production. This chapter introduces the current licensing practices, while available, for several of the reactor types. A number of the designs in Section III involve thermodynamic coupling of nuclear reactor to hydrogen plant. Since such nuclear system has not been licensed to date, a licensing framework is discussed based on the safety design and licensing approaches proposed at the moment in the nuclear heated hydrogen plant development.

25.2 Light Water Reactors Licensing Practice

About 390 light water reactors (LWRs) of 330 GWe total capacity are presently operational or under construction in the world for commercial power generation.

In Japan, 53 LWRs generating 47 GWe in total are in operation and 13 new reactors are planned [1]. The Nuclear and Industrial Safety Agency (NISA) of Ministry of Economy, Trade and Industry (METI) is responsible for licensing and regulating the construction and operations of commercial power reactors including the LWRs and developmental reactors while the Ministry of Education, Culture, Sports, Science and Technology (MEXT) licenses and regulates research and test reactors in Japan. The Nuclear Safety Commission (NSC) of Japan directs and audits the safety review activities of the NISA and MEXT.

The top level regulatory criteria for safety design of LWRs in Japan are summarized in Table 25.1. This table shows dose limits for each event category, which is determined in Japanese licensing of nuclear power plant. To meet the top level criteria, the design review is carried out by the Japanese government under design guidelines including the "Regulatory Guide for Reviewing Safety Design of Light Water Nuclear Power Reactor Facilities," [2] "Regulatory Guide for Reviewing Safety Assessment of Light Water Nuclear Power Reactor Facilities," [3] and so on. These guidelines involve the design criteria for the fuel design, the reactor pressure boundary, and containment boundary, in addition to the above dose limits to public.

United States has 104 operational LWRs generating a total of 100.3 GWe [4]. The U.S. Nuclear Regulatory Commission (NRC) has been licensing commercial LWRs under a two-step process governed by Title 10 of the Code of Federal Regulations under Part 50, simply known as 10CFR50. This process requires separate construction permit and operating license.

TABLE 25.1

Summary of Top Level Regulatory Criteria for LWR in Japan

Category	Top Level Regulatory Criteria (Dose Limits)
Normal operation	50 μSv of annual radiation exposure outside the site boundary (Based on "Regulatory Guide for the Annual Dose Target for the Public in the Vicinity of Light Water Nuclear Power Reactor Facilities")
Accident	No significant risk of radiation exposure to the public. Effective dose shall not exceed 5 mSv (Based on "Regulatory Guide for Reviewing Safety Assessment of Light Water Nuclear Power Reactor Facilities")
Major accident	Effective dose to the whole body shall not exceed 0.25 Sv outside the site boundary Effective dose to the thyroid shall not exceed 1.5 Sv for a child outside the site boundary (Based on "Regulatory Guide for Reviewing Nuclear Reactor Site Evaluation and Application Criteria")
Hypothetical accident	Effective dose to the whole body shall not exceed 0.25 Sv outside the site boundary Effective dose to the thyroid shall not exceed 3.0 Sv for an adult outside the site boundary The whole-population dose shall not exceed 2×10^4 man Sv (Based on "Regulatory Guide for Reviewing Nuclear Reactor Site Evaluation and Application Criteria")

Since 2002, the U.S. DOE has carried out a major initiative of Nuclear Power 2010 to work with the nuclear industry to develop projects that would lead to construction of new nuclear plants by 2010. As of May 2009, combined construction and operating license (COL) applications under a recently established process stipulated in the 10CFR52 regulations [5] have been filed with the NRC for 18 projects totaling 28 LWRs involving the designs of AP1000, ESBWR, USEPR, US-APWR, and ABWR [6]. Seven of the projects selected AP1000, the most preferred by number. The AP1000 is a two-loop 1000 MWe pressurized water reactor (PWR). The AP1000 design contains many features that are not found in current operating reactors. The most significant improvement to the design is the use of safety systems that employ passive means, such as gravity, natural circulation, condensation and evaporation, and stored energy, for accident mitigation. These passive safety systems perform safety injection, residual heat removal, and containment cooling functions [7]. Westinghouse Electric Company submitted the application for the AP1000 design in March 2002. The technical review of the AP1000 standard nuclear reactor design by the U.S. NRC was performed [8], and the NRC certified the AP1000 design in January 2006 [9].

25.3 Fast Breeder Reactors Licensing Practice

Experimental and demonstration fast breeder reactors (FBRs) have been built and operated around the world. They include JOYO and Monju in Japan and Phenix and Super Phenix in France. USA, Russia, and India have also had operational FBRs.

Monju is a prototype FBR of sodium-cooled, MOX-fueled, loop-type reactor rated at 714 MWt and generating 280 MWe. Its construction began in 1985 and first criticality was reached in 1994. The construction and operation licenses for Monju were granted by the Ministry of International Trade and Industry (the former organization of METI) in consideration of the conventional LWR safety guidelines and the particular design features of FBR. The design features that were considered include use of liquid metal sodium for the primary coolant, breeding core design with fast neutron, high power density and high burn-up of the MOX fuel, intermediate cooling loop, cover gas system closed to sodium, and so on. For dose limit in addition to the top level regulatory criteria for safety design of LWRs, the reference dose for plutonium in site evaluation was prepared as 2.4 Sv to bone surfaces, 3 Sv to lung and 5 Sv to liver [10].

Recent developments for Fast Reactors were carried out through international collaborations, such as the Generation IV International Forum (GIF) [11]. In Japan, JAEA has been leading the work to design for a larger size commercial fast reactor in the Fast Reactor Cycle Technology (FaCT) Development project [12].

25.4 High-Temperature Gas-Cooled Reactors Licensing Practice

Several test and demonstration high-temperature gas-cooled reactors (HTGRs) have been built and operated in the United Kingdom, Germany, the United States, Japan, and China, of which two test reactors are in operation today and their licensing practices that are closely relevant to future HTGRs are presented in this section. In addition, the safety design and licensing development for three planned demonstration reactors will be reviewed.

25.4.1 Test Reactors

25.4.1.1 High Temperature Engineering Test Reactor

25.4.1.1.1 Licensing Strategy

The high temperature engineering test reactor (HTTR) is an operating test reactor for the HTGR. The HTTR has inherent safety features such as strong negative feedback coefficient, large heat capacity of the core, inert helium gas as coolant, and so on. It ensures safety with a sufficient margin without depending much on highly sophisticated active safety systems. Thus, innovative HTGRs being designed in the world have the strong economical advantage because they can achieve the same level of safety required for current reactors with simple passive safety systems. In licensing approach for the HTGRs, safety design philosophies of the HTGRs should be established by properly considering these physical design features.

In Japan, Japan Atomic Energy Research Institute (JAERI) (the former name of Japan Atomic Energy Agency (JAEA)) established the safety philosophy of the HTTR in 1990s [13]. Although the safety design philosophy of the HTTR is based on that of the LWR, the LWR safety design philosophy is not directly applicable. Therefore, the safety design philosophy considering several unique design features such as coated fuel particles, indirect heat removal system of the core, graphite core, and so on, was established. Due to the large heat capacity of the graphite core and the strong negative temperature coefficient, the reactor safety can be maintained without using direct forced cooling of the reactor core even in the depressurization accident. On the other hand, oxidation of the fuel due to air ingress in the depressurization accident is typically an issue of the HTGR safety design. The safety design philosophy of the HTTR was determined considering these merits and drawbacks to ensure safety in all operational states.

The safety evaluation procedure of the HTTR is also different from that of current reactors. No criteria for the fuel, core structures and high-temperature components had been established and no design base events for the safety evaluation were determined. Acceptance criteria for these structures, components and systems were established and method to select evaluation events was developed.

The safety design of the HTTR [14] is described in Section 25.4.1.1.2 and safety evaluation [15] in Section 25.4.1.1.3.

25.4.1.1.2 Safety Design

a. *Basic safety design philosophy:* Figure 25.1 shows a logical flow to establish the basic safety design philosophy of the HTTR. The top level regulatory criteria for the HTTR is identical to those of the LWR (Table 25.1). In addition, according to the International Commission on Radiological Protection (ICRP) recommendation, the principle of ALARA (as low as reasonably achievable) was applied to reduce the radiation dose to plant personnel and members of the public around the HTTR. To meet the top level criteria and apply the principle of ALARA, the basic safety design philosophy of the HTTR was determined based on that of the LWR stipulated in "Regulatory Guide for Reviewing Safety Design of Light Water Nuclear Power Reactor Facilities" considering inherent safety characteristics of the HTGR. In order to ensure safety, the well-known fundamental safety functions such as

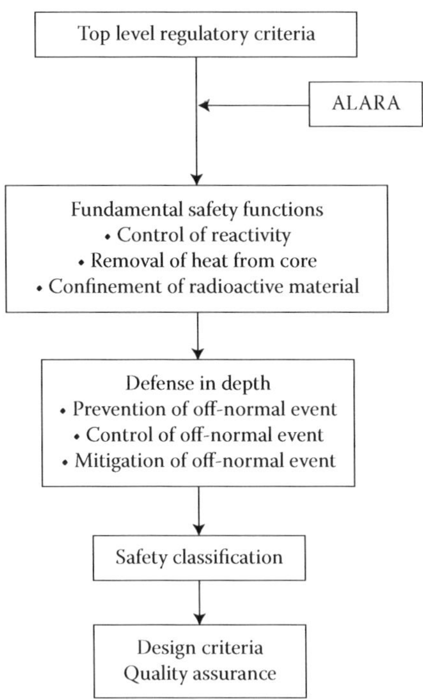

FIGURE 25.1
Logical flow to establish a safety design philosophy of the HTTR.

control of reactivity, removal of heat from the core, and confinement of radioactive materials shall be performed in normal and off-normal states. The strategy of the defense-in-depth [16] that provides a series of level of defense is implemented to ensure that the fundamental safety functions shall be reliably achieved in normal and off-normal states. The level of the defense in depth consists of prevention of off-normal events, control of off-normal events and mitigation of off-normal events. The newly considered premise to establish the basic safety design philosophy of the HTTR is that it considers an air ingress accident and following oxidation of the core. The HTTR safety design shall prevent the excessive oxidation of the core and fission products release to the environment.

b. *Safety design criteria:* In the development of the HTTR, several basic design criteria were established, such as fuel design criteria for the fuel, graphite design criteria [17], high-temperature metallic component design criteria [18].

c. *Safety classification:* In compliance with the strategy of the defense in depth, all systems having safety functions in the HTTR were classified as prevention system (PS)-1, PS-2, PS-3, mitigation system (MS)-1, MS-2, or MS-3 depending on their roles and significance.

d. *Quality assurance:* Quality assurance for metals in the HTTR components is basically the same as that of LWR. Various nondestructive inspections are conducted on them. Inspection items are determined dependent on their component class. However, no rule had been stipulated for high-temperature resistant material Hastelloy XR. JAERI established quality assurance for the high-temperature metals. Quality assurance for graphite is determined in the graphite design criteria. Regarding the fuel, no experience had been obtained. Therefore, JAERI determined the inspection items for the fuel.

25.4.1.1.3 *Safety Evaluation*

a. *Evaluation procedure:* Figure 25.2 shows the safety evaluation procedure of the HTTR. The safety evaluation of the HTTR was performed basically in a deterministic methodology. Design basis event (DBE) was selected, and its plant behavior and effect on the environment around the reactor site were evaluated.

   The safety evaluation of the HTTR was conducted taking into account its specific design features of the HTTR. The categorization of the events to be evaluated is based on "regulatory guide for reviewing safety assessment of light water nuclear power reactor facilities." That is, in confirming the adequacy of basic design principles of reactor facilities, it is necessary to evaluate anticipated operational occurrences and then to evaluate the events beyond the scope of anticipated operational occurrences, that is, accidents. On the other hand, it is necessary to evaluate "major accidents" and "hypothetical accidents" to judge the appropriateness of reactor siting condition based on the "regulatory guide for reviewing nuclear reactor site evaluation and application criteria" [19].

   The anticipated operational occurrences shall be evaluated for events which include conditions beyond normal reactor operation resulting from a single failure or malfunction, or a single operator error anticipated to occur during the life time of the reactor facility. The accidents shall be postulated from the view point of the possibility of release of radioactive materials from the facility, though the frequencies of those occurrences are smaller than the anticipated operational occurrences.

b. *Acceptance criteria:* Acceptance criteria for the HTTR are established fundamentally reflecting the safety requirements for LWR power plants and taking into account major features of HTGRs and the HTTR. Acceptance criteria for the anticipated operational occurrences and the accidents for the HTTR and LWRs are shown in Tables 25.2 and 25.3. The maximum fuel temperature is restricted to 1600°C to avoid fuel failure during the anticipated operational occurrences. Criteria for the temperature and the pressure of the primary pressure boundary and the containment vessel are determined considering safety margins. In the

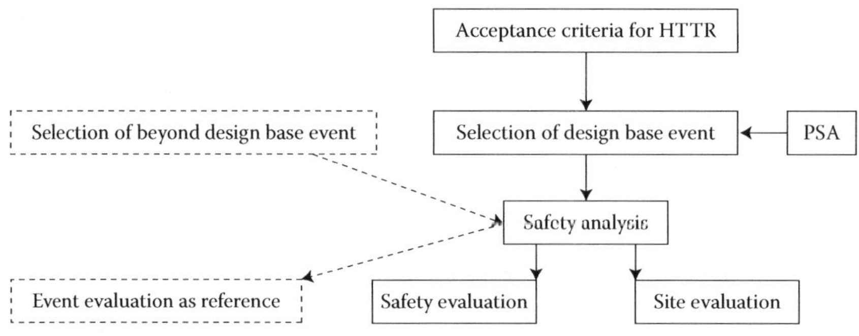

FIGURE 25.2
Logical flow of safety evaluation procedure of the HTTR.

TABLE 25.2

Comparison of Acceptance Criteria for Anticipated Operational Occurrence for the HTTR and LWR in Japan

HTTR	LWR
The peak fuel temperature shall be less than 1600°C	Minimum critical heat flux (MCHF) or MCHF ratio shall not exceed the limited value Fuel cladding shall not fail mechanically Fuel enthalpy shall not exceed the limited value
Pressure on reactor pressure boundary is less than 1.1 times of maximum pressure in service	Pressure on reactor pressure boundary is less than 1.1 times of maximum pressure in service
Maximum temperature of reactor pressure boundary 9/4Cr–1 Mo steel <500°C Austenite stainless steel <600°C Hastelloy XR <980°C	—

case of accidents, the core shall not be seriously damaged and shall maintain its geometry for sufficient coolability. The radiation exposure is limited to 5 mSv as effective dose equivalent outside the site boundary of the HTTR.

Acceptance criteria for the major accident and the hypothetical accident are the same as that for the LWRs. These acceptance criteria are established in "Regulatory Guide for Reviewing Nuclear Reactor Site Evaluation and Application Criteria" as follows: (1) effective dose to whole body shall not exceed 0.25 Sv in a major accident or a hypothetical accident; (2) effective dose to thyroid shall not exceed 1.5 Sv for a child in a major accident and 3.0 Sv for an adult in a hypothetical accident; (3) whole-population dose shall not exceed 2×10^4 man Sv in a hypothetical accident.

c. *Selection of events:* Abnormal events to be postulated as anticipated operational occurrences and accidents have been selected considering their frequencies of

TABLE 25.3

Comparison of Acceptance Criteria for Accident for the HTTR and LWR in Japan

HTTR	LWR
The reactor core shall not be seriously damaged and can be cooled sufficiently	Minimum critical heat flux (MCHF) or MCHF ratio shall not exceed the limited value Fuel enthalpy shall not exceed the limited value
Pressure on reactor pressure boundary is less than 1.2 times of maximum pressure in service	Pressure on reactor pressure boundary is less than 1.2 times of maximum pressure in service
Maximum temperature of reactor pressure boundary 9/4Cr–1 Mo steel <550°C Austenite stainless steel <650°C Hastelloy XR <1000°C	—
Maximum pressure on containment boundary is less than maximum pressure in service	Maximum pressure on containment boundary is less than maximum pressure in service
No significant risk of radiation exposure to public	No significant risk of radiation exposure to public

occurrence and based on the investigation of main causes which affect each item of the acceptance criteria identified for the HTTR; that is, (1) fuel temperature, (2) core damage, (3) temperature of reactor coolant pressure boundary, (4) pressure at reactor coolant pressure boundary, (5) pressure at containment vessel boundary, and (6) risk of radiation exposure for the public. The initiating abnormal events have been classified into similar event groups according to "Regulatory Guide for Reviewing Safety Assessment of Light Water Nuclear Power Reactor Facilities." Then, the most severe events with respect to the acceptance criteria within each similar event group are selected as the representative postulated events.

A major accident and a hypothetical accident are evaluated to ensure the safety of the public in the case of serious accidents. "Major accidents" are postulated assuming the occurrence of the worst-case accident from a technical standpoint considering the reactor characteristics and engineered safety features. "Hypothetical accidents" are postulated assuming the occurrence of an accident more serious than a "major accident," which is unlikely to occur from a technical standpoint, and shall be based on the assumption that one or more engineered safety features fail to function. A double-ended rupture of coaxial double pipes of the primary cooling system (depressurization accident) is postulated for the HTTR as the major accident and the hypothetical accident with respect to the risk of radiation exposure for the public.

25.4.1.2 HTR-10

The 10 MW (HTR-10) is an high temperature gas-cooled test reactor (HTGR) test reactor developed by the Institute of Nuclear Energy Technology (INET) of Tsinghua University in China. The HTR-10 uses TRISO SiC-coated UO_2 fuel particles dispensed in the graphite matrix of a 6-cm spherical fuel element. The core comprises a cylindrical pebble bed of thousands of the fuel elements, which are recycled through the core multiple times with online refueling, and thick graphite wall all around the bed. The graphite built into the core plays the roles of neutron moderator and reflector. Reactor control neutron absorber systems are in the side reflector wall. However, these systems are not required for reactor shutdown in the HTR-10 safety design. In addition, the decay heat is passive removable requiring no active emergency cooling system. The reactor and a steam generator are arranged side-by-side in two steel pressure vessels.

The HTR-10 safety philosophy [20] is based on defense in depth with multiple barriers against radioactivity release, similar to the HTTR approach. The HTR-10 multiple barriers, which are important measures of defense in depth, include (1) the spherical fuel elements with coated particles, (2) primary cooling system pressure boundary, and (3) a vented reactor safety concrete building confinement. The last barrier is an important difference from the HTTR safety design, which employs the conventional type of leak-tight steel pressure vessel containment.

Because the HTR-10 is the first HTGR type of reactor developed in the country, the above said safety design features presents a licensing challenge to both the developer and the nuclear reactor regulator [21,22]. The site-permit application was preceded by compiling and submitting, in mid-1992, an Environmental Impact Report (EIR) of the reactor for review and approval by the National Environmental Protection Administration (NEPA). The Siting and Seismic Report of the reactor was then submitted for examination by the National Nuclear Safety Administration (NNSA), which granted the site permit in

December 1992. Concurrently, two sets of technical documentation prepared by the INET for the NNSA, including the Design Criteria of HTR-10 and the Standard Content and Format of the Safety Analysis Report. These documents meant to establish the basis of licensing for this first kind of the reactor in the country were examined and accepted by the NNSA in August 1992 and March 1993, respectively. The HTR-10 detailed design then began and was supported by domestic institutes and industries.

The administrative procedure required for licensing the HTR-10 follows the two-step process of a construction permit followed by an operation commissioning permit generally used for other reactors in China and most other countries. The Preliminary Safety Analysis Report (PSAR) was completed and submitted in December 1993 to the NNSA to apply for HTR-10 construction permit. The activities of the licensing procedure ensued and more than 700 technical questions were raised and answered in writing between the licensing examiners and the designers, who met frequently to discuss and resolve the questions and issues and consulted for special issues with the Nuclear Safety Expert Committee. The main safety licensing issues were related to the specific fuel design, the mechanistic determination for radioactive source terms, component and system classification, selection of the design basis accidents, and the unique safety confinement design. In December 1994 the licensing activities were completed with the issuing of the construction permit by the NNSA.

First concrete was poured in June 1995, the pressure vessel went up in November 1998, and the HTR-10 construction was completed in May 2000.

In October 1999 the Final Safety Analysis Report (FSAR) and a second EIR—this time for the application of the operation permit—were submitted to the NSAA and the NEPA, respectively [23]. The NNSA approved the FSAR and issued the operation commissioning permit in November 2000. The HTR-10 attained official criticality in the following month and began full power operation in 2003.

25.4.2 Demonstration Reactors

There are a lot of new projects for the HTGRs, such as the Next Generation Nuclear Plant (NGNP) in the United States, PBMR and process heat plant (PHP) in South Africa, HTR-PM in China, and GTHTR300C in Japan. Licensing strategy and states for NGNP, PBMR, and HTR-PM are shown in the next section.

25.4.2.1 Next Generation Nuclear Plant

The U.S. DOE and U.S. NRC introduced to Congress the NGNP Licensing Strategy Report in August 2008 [24]. This report describes the license approach, the analytical tools, the R&D activities and estimated resources. The report provides the key descriptions below.

a. *Recommended strategy:* The best alternative for licensing the NGNP prototype will be for the applicant to submit a combined construction and operating license (COL) application under 10 CFR Part 52. And the best approach to establish the licensing and safety basis for the NGNP will be to develop a risk-informed and performance-based technical approach that adapts existing NRC LWR technical licensing requirements in establishing NGNP design-specific technical licensing requirements.

b. *Technical approach:* The technical approach to establishing the NGNP licensing basis and requirements is expected to include the following:
   - Establishment of licensing-basis event categories and selection of basis events within each category.
   - Selection of the safety-significant systems, structures, and components (SSCs).
   - Establishment of conservative design and acceptance criteria for core and safety-significant SSCs.
   - Verification of adequate safety margins to the integrity and performance of core and safety-significant SSCs.
   - Establishment of special treatment requirements to ensure the required performance capability and reliability of the safety-significant SSCs.
   - Use of consequence acceptance limits for on-site or off-site releases for licensing-basis events, and assessment of radiological consequences for licensing-basis events on the basis of event-specific mechanistic source terms.
   - Consideration of containment-functional-performance requirements as a radionuclide barrier.
   - Establishment of defense-in-depth requirements.

c. *Development of analytical tools:* It is expected that certain analytical tools will need to be developed or modified in a number of technical areas to enable the review of NGNP license application, evaluate the safety case, and assess the safety margin. The major technical areas for NGNP include accident analysis, fuel performance and fission product transport, high temperature materials and graphite performance, and process heat applications. Other areas that may require limited development of tools include structural analysis, human factors and human–machine interface, and probabilistic risk assessment.

25.4.2.2 Pebble Bed Modular Reactor

The electrical utility in South Africa (Eskom) was planning to construct a first of a kind pebble bed modular reactor (PBMR). In July 2000, Eskom submitted a nuclear installation license application for the PBMR to the national nuclear regulator (NNR). In order to determine whether such a design is licensable, the NNR developed requirements for licensing of a HTGR and elaborated the processes that must be undertaken to demonstrate compliance with the requirements [25].

The scope of regulatory assessment for licensing of the PBMR is based on the licensing requirements and safety criteria defined by the NNR in the regulatory documents. In addition, guidance is provided on selected issues in appropriate NNR license guides. The principal licensing requirements comprise, besides the general requirements to respect good engineering practice, ALARA and defense-in-depth principle, and specific radiation dose limits. These are categorized for normal operation and operational occurrences as well as for design basis events for workers and the public. The safety criteria also stipulate occupational risk limits for the workers as well as risk limits for the public for all possible events that could lead to radioactive exposure. The dual nature of the NNR safety criteria implies that the safety analyses for demonstration of compliance of the Safety Case with the licensing criteria for the PMBR have to comprise both deterministic and probabilistic analyses.

The PBMR design is also in preapplication discussions with the U.S. NRC that will culminate in the submittal of a formal application for design certification of the PBMR reactor design.

25.4.2.3 High Temperature Gas-Cooled Reactor-Pebble Bed Module

The high temperature gas-cooled reactor-pebble bed module (HTR-PM), based on the technology and experiences on the HTR-10, was developed in China. Detailed design and site preparation for the HTR-PM are underway. Licensing review by the NNSA is under preparation.

It is described that the challenge for the HTR-PM is to improve the economy features while maintaining the inherent safety [26]. The HTR-PM has a 450 MWt pebble bed annular core, which adopts a movable graphite ball zone in the core center, and use standard steam turbine proven in coal plant as the conventional island. There are still some technical and engineering problems required to be solved, such as how to maintain the boundary between fuel ball zone and graphite ball zone, how to ensure the reflector graphite to withstand the whole reactor life time, how to mix the hot helium out from the fuel zone and the cold helium out from the graphite zone, and so on.

25.5 Nuclear Hydrogen Plant Licensing Approach

The designs of the NGNP, the PHP, and the GTHTR300C are considered coupling with a hydrogen production plant as commercial plant scale. The key issues of the licensing approach to these plants will include (1) definition of design criteria of fuel, graphite, and high-temperature components, (2) technical basis for the use of a confinement, (3) definition of source term, and (4) combined plant safety. These issues are described in the next subsection.

25.5.1 Fuel Development

The performance of coated fuel particles is central to the economic viability in the helium-cooled nuclear system concepts and directly establishes the operational and accident mitigation design features for the HTGR. The TRISO fuel is used for the operating demonstration reactors (the HTTR in Japan, the HTR-10 in China), and for the plants currently under development.

For the future commercial HTGR plants, it is necessary to develop fuel to meet the high power density, high burn up, and high outlet gas temperature. Therefore, qualification of coated fuel particle technology is an important development task for the licensing. Simultaneously the standards of quality assurance and inspection should be developed.

25.5.2 Graphite Development

Graphite will be an essential material for HTGR core structural and fuel components. There are a number of potential sources of graphite that either have been or are being developed. These include IG-110, which is being used for both the Japanese HTTR and the Chinese HTR-10. Other grades of graphite may be available for use in structural applications.

The qualification of these sources of graphite will require an extensive R&D or a new qualification program. While there are a number of sources of graphite available, the degree of qualification is variable though several programs are in progress to develop qualified databases. ASME Code of graphite design basis under development [27].

25.5.3 High Temperature Components

There are several undeveloped components for the HTGR, including the intermediate heat exchanger (IHX), the hot gas isolation valves, the reactor pressure vessel, the reactor internals, the reactor inlet/outlet pipes, the helium circulator, and instrumentation for high-temperature application. The significant components especially are the IHX and associated isolation valves. Qualification of materials for the high-temperature and extended lifetime conditions is critical and alternatives must be carefully evaluated.

It is required for alternative materials to establish the relevant thermomechanical performance data to support the development of IHX and other high-temperature components. Depending on the outlet temperature selected, additional high-temperature data may be needed to support relevant ASME code cases for the material.

25.5.4 Reactor Confinement

Most of the planned HTGRs do not employ a leak-tight pressure containment vessel as in the LWR. These designs consider use of confinement, including ventilation and filter, to limit the fission product releases of normal operation and post-accident.

The coated fuel particles are used for the core designs. The ceramics layers surrounding the fuel kernel act as the primary barrier for the fission product release. It is necessary to ensure the design of confinement that the fuel performance model is developed, the fission product released to the environment and the public dose are calculated, and identified to satisfy the criteria. And also, the safety evaluation of the air ingress accident is necessary to show the ability to limit air ingress.

25.5.5 Source Terms

Source terms are used for the assessment of dose to workers and the public and comparison against regulatory dose criteria, and for the examination of maintenance methods and procedures for the primary components. The applicant will be needed to establish such source terms for the safety design and their use should be justified in licensing. The assessment of radiological consequences for licensing-basis events on the basis of event-specific mechanistic source terms will be considered on the each safety design.

25.5.6 Combined Plant Safety

Upset conditions in either the reactor system or hydrogen production plant must not affect the safe operation of either plant. Reactor safety and hydrogen plant safety should be confirmed by the methods described in Chapter 23. Generally, for reactor safety analysis, the hydrogen plant aspects may be considered in the same category as external events with a coupling to the reactor plant.

References

1. Federation of Electric Power Companies of Japan, Present Status of Nuclear Power Generation in Japan, January 2009.

2. Nuclear Safety Commission of Japan, Regulatory Guide for Reviewing Safety Design of Light Water Nuclear Power Reactor Facilities (latest issue, 2001).
3. Nuclear Safety Commission of Japan, Regulatory Guide for Reviewing Safety Assessment of Light Water Nuclear Power Reactor Facilities (latest issue, 2001).
4. U.S. Energy Information Administration, Nuclear Energy Overview, Monthly Energy Review September, 2009.
5. U.S. Regulations, Title 10 of the Code of Federal Regulations (10CFR) Part 52, *Licenses, Certification, and Approvals for Nuclear Power Plants*.
6. http://www.nrc.gov/reactors/new-reactors/col.html.
7. Schulz, T.L., Westinghouse AP1000 advanced passive plant, *Nucl. Eng. Des.* 236, 1547–1557, 2002.
8. NRC, Final Safety Evaluation Report Related to Certification of the AP1000 Standard Design, NUREG-1793, 2004.
9. http://www.nrc.gov/reactors/new-reactors/design-cert/ap1000.html.
10. Nuclear Safety Commission of Japan, Reference Dose for Plutonium in Site Evaluation of Nuclear Reactors Using Plutonium Fuels (latest issue, 2001).
11. Generation IV International Forum (GIF), GEN IV International Forum 2008 Annual Report, 2009.
12. JAEA, Assessment Report of Research and Development Activities in FY2008 Activity: Fast Reactor Cycle Technology Development Project (Interim Report), JAEA-Evaluation 2009-004, 2009 (in Japanese).
13. Saito, S. et al., Design of high temperature engineering test reactor (HTTR), JAERI-1332, 1994.
14. Kunitomi, K. and Shiozawa, S. Safety design, *Nucl. Eng. Des.* 233, 45–58, 2004.
15. Kunitomi, K., Nakagawa, S. and Shiozawa, S. Safety evaluation of the HTTR, *Nucl. Eng. Des.* 233, 235–249, 2004.
16. IAEA, Basic Safety principles for Nuclear Power Plants 75-INSAG-3 Rev. 1, INSAG-12, 1999.
17. Iyoku, T., Ueta, S., Sumita, J., Umeda, M. and Ishihara, M., Design of core components, *Nucl. Eng. Des.* 233, 71–79, 2004.
18. Tachibana, Y. and Iyoku, T., Structural design of high temperature metallic components, *Nucl. Eng. Des.* 233, 261–272, 2004.
19. Nuclear Safety Commission of Japan, Regulatory Guide for Reviewing Nuclear Reactor Site Evaluation and Application Criteria (latest issue, 1989).
20. Wu, Z. and Xi, S., Safety functions and component classification for the HTR-10. *Nucl. Eng. Des.* 218, 103–110, 2002.
21. Sun, Y. and Xu, Y., Licensing experience of the HTR-10 test reactor, Technical committee meeting on design and development of gas cooled reactors with closed cycle gas turbines. Beijing, China. 30 October–2 November 1995, International Atomic Energy Agency, Vienna (Austria), IAEA-TECDOC–899, pp. 157–162.
22. Zhong, D. and Xu, Y., Progress of the HTR-10 Project, Technical committee meeting on design and development of gas cooled reactors with closed cycle gas turbines, Beijing, China, 30 October–2 November 1995, International Atomic Energy Agency, Vienna (Austria), IAEA-TECDOC–899, pp. 25–29.
23. Xu, Y., The HTR-10 project and its further development, *Proc. of the 1st International Topic Meeting on High Temperature Reactor Technology*, Petten, NL, HTR-2002, pp. 1–8, April 22–24, 2002.
24. Next Generation Nuclear Plant Licensing Strategy, A Report to Congress, August 2008.
25. Bester, P. and Hill, T. South African licensing framework for the pebble bed modular reactor, *Proceedings of HTR2008 Conference,* Washington, DC, USA, September 28–October 1, 2008.
26. Zuoyi Zhang et al., Design of Chinese high-temperature gas-cooled reactor HTR-PM, HTR-2004, Beijing, September 22–24, 2004.
27. Bratton, R.L. Status of ASME Section III Task Group on Graphite Core Support Structures, INL/EXT-05–00552, 2005.

Section V

Worldwide Research and Development

26

Hydrogen Production and Applications Program in Argentina

Ana E. Bohé and Horacio E.P. Nassini

CONTENTS

26.1 Introduction

The Comisión Nacional de Energía Atómica (CNEA) of Argentina is a governmental research and development (R&D) institution that investigates all aspects of peaceful uses of nuclear energy in the country, and also acts as adviser of the national government in nuclear policy. Furthermore, according to present regulations, CNEA is responsible for the decommissioning of domestic nuclear power plants, as well as for the management of spent fuels and radioactive wastes generated by the nuclear activities.

R&D activities carried out in CNEA are categorized in four main areas: (1) nuclear energy; (2) nuclear applications; (3) nuclear safety and environmental protection; and (4) nonnuclear research and applications. In relation with production, purification, storage and applications of hydrogen in Argentina, CNEA's scientists are working in the framework of a Hydrogen Energy Program designed on the basis of an Argentine Law that promotes the use of hydrogen in the country.

The so-called Hydrogen Law was dictated by the Argentine Congress in 2006 and it declares of national interest the development of technologies needed for the progressive introduction of hydrogen as a clean fuel and energy carrier that can be used to meet increasing residential, commercial, and industrial demands.

The law promotes and regulates the use of hydrogen in industry, transport, and electric power stations, with the objective of diversifying the current energy matrix of the country which is largely based on fossil fuels, that is, petroleum oil and natural gas. For achieving this objective, the national government is promoting all scientific activities related with the production, purification, safe storage and applications of hydrogen for which the formation of qualified human resources and the scientific cooperation with other countries are greatly encouraged.

Based on own capabilities developed for more than 50 years in CNEA, related with both nuclear and hydrogen technologies, and taking into account the new legal framework that promotes the development of a hydrogen economy in Argentina, the production of hydrogen by using nuclear energy is being seriously considered as an alternative for the country, with the conviction that nuclear production of hydrogen has the potential to contribute significantly to the national energy supply in a sustainable, competitive and environmentally friendly manner.

In the following paragraphs, the main R&D activities that are being performed in Argentina in relation with the production, purification, storage and applications of hydrogen are summarized.

26.2 Hydrogen Production from Thermochemical Cycles

26.2.1 Overview

Four different approaches have been identified for the efficient production of hydrogen by using nuclear energy. The first one, nuclear-assisted steam reforming of natural gas uses nuclear heat to reduce the amount of natural gas needed to produce a given quantity of hydrogen. The second approach, hot electrolysis, involves the electrolysis of water at high temperatures. The third approach consists of coal gasification with steam to produce hydrogen by two reactions, the direct gasification of coal and, then, the water-shift reaction, both of them giving hydrogen as by product. Finally, thermochemical cycles use a series of high-temperature chemical reactions for the water splitting, and they are expected to get an overall efficiency of about 50%, receiving then the most attention.

Even though many types of thermochemical processes exist, the Argentine Hydrogen Program is addressed to be one of the leading long-term methods: the metallic chlorides cycles. A lot of studies have been performed in the past on these methods, but the kinetics and mechanisms of reactions are not completely understood yet, and this project is expected to contribute to a better understanding of the critical problems identified for each cycle.

The economics of hydrogen production strongly depends on the efficiency of the method used. Production efficiency can be defined as the energy content of the resulting hydrogen divided by the energy expended to produce it. Nuclear-assisted hydrogen production by electrolysis is relatively efficient (about 80%) but when this factor is combined with the electrical conversion efficiency, which ranges from approximately 34% in current light-water reactors to 50% for advanced reactor designs, the overall efficiency would be approximately 25–40%. Moreover, a significant capital investment in electrolytic cells is required. On the opposite, thermochemical processes for producing hydrogen with the input of heat and water appear to be much more efficient, since for the most developed iodine–sulfur cycle, an overall efficiency higher than 50% has been projected. However, for achieving low production costs in thermochemical water splitting processes, high operating temperatures are required to ensure fast chemical kinetics, that is, small plant size with low capital costs and high conversion efficiencies. If this technology is able to be developed at commercial scale, hydrogen production from nuclear power plants would be expected to become competitive with that from natural gas. In addition, refineries and chemical plants have nearly a constant demand for hydrogen that matches the base-load capabilities of nuclear power plants. Finally, a special attraction of thermochemical processes for hydrogen production is that there is no associated carbon dioxide emission to the atmosphere.

The Argentine research activities on thermochemical cycles for hydrogen production are expected to contribute in this mainstream through a well-designed theoretical and experimental program addressed to elucidate the kinetics and mechanisms of thermochemical reactions at laboratory scale, in order to find the optimum conditions for increasing the efficiency of these cycles with the objective of a future scaling up of the experimental facilities.

The theoretical and experimental program is principally associated with the understanding of the basic mechanisms of thermochemical water splitting reactions as well as the determination of the kinetic parameters of reactions which are essential for the design of the chemical reactors. Furthermore, the program is expected to provide a significant knowledge about the behavior of corrosive compounds at high temperatures in relation with the life assessment of the chemical reactor materials and components.

The first stage of the program was addressed to find suitable thermochemical reaction cycles by means of thermodynamic calculations and verification of their theoretical feasibility. For this purpose, a complete thermodynamic analysis was done in order to determine whether the iron chloride cycle may be replaced by another one that enhances the hydrogen production. Using the software HSC for thermodynamic calculations, the equilibrium amount of the different species was evaluated for several experimental conditions.

26.2.2 Iron Chloride Cycle

Firstly, the well-known iron chloride cycle was analyzed considering that the most important reactions to be taken into account are [1–4]:

$$3FeCl_2(s,l) + 4H_2O(g) \rightarrow Fe_3O_4(s) + 6HCl(g) + H_2(g) \tag{26.1}$$

$$Fe_3O_4(s) + 8HCl(g) \rightarrow FeCl_2(s,l) + 2FeCl_3(g) + 4H_2O(g) \tag{26.2}$$

$$2FeCl_3(g) \rightarrow FeCl_2(s,l) + Cl_2(g) \tag{26.3}$$

$$Cl_2(g) + H_2O(g) \rightarrow 2HCl(g) + \frac{1}{2}O_2(g) \tag{26.4}$$

$$2HCl(g) \rightarrow H_2(g) + Cl_2(g) \tag{26.5}$$

The analysis began with stoichiometric amounts of $FeCl_2$ and H_2O (1:1 in moles), and the equilibrium composition in the gaseous phase showed that the amount of H_2 increases with temperature from 2.32% at 500°C to 7.24% at 1000°C. The total pressure considered was 1 atm, so using the ideal gas equation, $p_{H2} = X_{H2} \times P_T$, where p_{H2} is the partial pressure of hydrogen, X_{H2} is the hydrogen mole fraction, and P_T the total pressure of the system, these percentages mean that the partial pressure of H_2 is 0.02 and 0.7 atm, respectively. The other species in the gaseous phase are mostly unreacted water and undissociated HCl; while less than 1% the $FeCl_2$, $FeCl_3$, Cl_2, and O_2 are also present as impurities. In the condensed phase, only $FeCl_2$ has a minimum reaction while a small amount of Fe_3O_4 is also formed, but no Fe_2O_3 was detected in equilibrium.

If more water is added to the system, increasing the ratio between H_2O and $FeCl_2$ to 5:1 in moles, there are changes in the relative amounts of the different species. The changes with temperature of H_2O, H_2, HCl, $FeCl_2$, $FeCl_3$, and Cl_2 are less pronounced than in the previous case; for example, they are from 83.72 to 64.33% (instead of 83.72 to 21.21%) for H_2O, and from 2.33 to 4.75% (instead of 2.32 to 7.24%) for H_2 at 500 and 1000°C, respectively. The relative change of O_2 increases from 3×10^{-23} to 2.5×10^{-11} compared with 3×10^{-23} to 5×10^{-12} for the previous case. It is expected that the percentage of water increases and the percentage of the other species decreases. Nevertheless, the final equilibrium amount of O_2 at 1000°C is greater than in the previous case. In the condensed phase, it is observed a less amount of $FeCl_2$ and a greater amount of magnetite, at all temperatures.

These results show that the presence of water in excess creates an oxidative atmosphere that is not convenient for the formation of hydrogen. Therefore, it is indicating that the dissociative reaction of HCl is not shifted by the presence of more quantity of water. Meanwhile, the reactions given by Equations 26.1 through 26.4 are applied in the Le Chatellier principle, which establishes that, in a chemical reaction when a reactive is augmented, the equilibrium is shifted to the formation of the products. On the other hand, if the amount of $FeCl_2$ is increased, there is no change in the fraction of the gaseous phase. This result means that the condensed chloride does not affect the equilibrium amount without beneficiating the hydrogen reaction.

If the partial pressure of O_2 is diminished by its capture through an oxidation reaction, it is possible to produce more HCl that enhances the generation of hydrogen by dissociation. The scavenger should react with O_2 but it should be almost nonreactive with HCl. As it is well known that refractory metals have this characteristic, it was considered to incorporate metallic titanium or zirconium to the system and additional thermodynamic calculations were carried out for determining the possible species that may be formed with either O_2 or Cl_2. Consequently, a considerable amount of O_2 is generated but if a refractory metal is added to the system for capturing the gas produced, the partial pressure of O_2 diminishes markedly and it enhances the production of hydrogen.

Considering a starting system with 5 moles of $FeCl_2$, 1 mole of H_2O and 10 moles of Ti, the thermodynamic calculations showed that the equilibrium partial pressure of H_2 in the gaseous phase is 100 to 32% in the temperature range between 500 and 1000°C, while the other gaseous species are: 10^{-3} to 68% for $FeCl_2$, 10^{-11} to 10^{-10} for HCl, 10^{-12} to 10^{-7} for H_2O, 10^{-18} to 10^{-7} for $FeCl_3$, 10^{-27} to 0 for $TiCl_4$, 0 to 10^{-30} for Cl_2, and 0 to 10^{-29} for O_2 at 500 and 1000°C, respectively. Calculations showed that all H_2O reacts and HCl is dissociated, indicating that O_2 is completely consumed and the reactions given by Equations 26.1 through 26.4 shift toward the right-hand side.

In the condensed phase, these hypotheses are confirmed because the amount of $FeCl_2$ is practically the same as at the beginning at low temperature as expected by reactions given by Equations 26.2 and 26.3. Otherwise, the decrease of Ti and the increase in the amount of TiO_2 confirm the evidence that the O_2 is captured by the metallic titanium. The null amount of Fe_3O_4 evidences that reaction (Equation 26.2) is very effective in the formation of $FeCl_2$.

If Fe is used as reactive specie instead of Ti and the formation of Fe_3O_4 is considered, the performance of the cycle is less efficient for the hydrogen generation and a greater amount of O_2 is present in the closed system. But this O_2 is accompanied by smaller amounts of $FeCl_2$ and greater amounts of H_2O, HCl, and $FeCl_3$. In the condensed phase, a small quantity of Fe_3O_4 is formed.

FIGURE 26.1
Experimental setup for studies on metallic chlorides thermochemical cycles.

In order to verify the predictions of thermodynamic calculations, an experimental setup was designed and built for producing hydrogen through metallic chlorides' thermochemical reactions in a batch operating mode. A general view of the experimental setup is given in Figure 26.1 while Figure 26.2 shows a detailed view of the quartz capsule where thermochemical reactions are produced in well-controlled atmosphere and temperature

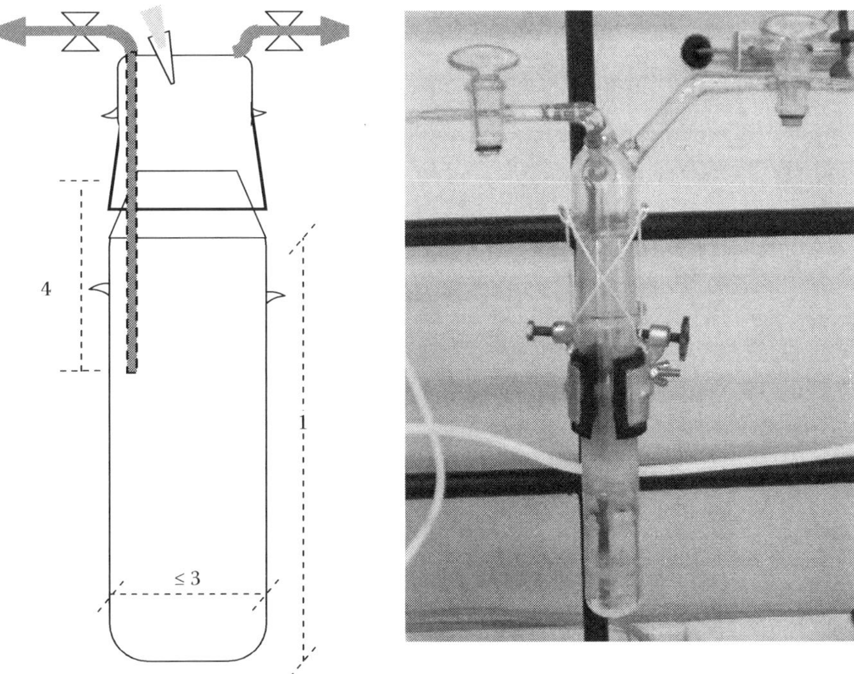

FIGURE 26.2
Capsule design for thermochemical reaction studies in the Fe–Cl cycle.

conditions. In each experimental run, the reactants are introduced inside the capsule, air is evacuated and the capsule is sealed under vacuum. The sealed capsule is then heated in an electric furnace at a given temperature and for a certain time, and the thermochemical reactions are then produced in a closed system. After cooling to room temperature, the capsule containing the chemical reaction products is introduced in a bath of air liquid for condensing all gaseous species (mainly O_2, HCl, and Cl_2) excepting the hydrogen which remains in gaseous state. Finally, the gaseous hydrogen is removed from the capsule using argon as carrier gas and it is conducted to a gas chromatograph for analysis, while the condensed species are analyzed by atomic absorption, UV/visible infrared and Mössbauer spectroscopy, and x-ray diffraction (XRD).

Experiments on the well-studied four-step Fe–Cl thermochemical cycle given by Equations 26.1 through 26.4, which is known in literature as Mark-15 cycle, were performed in this experimental setup and main results are summarized in Figures 26.3 through 26.5 [5].

Experiments on the original Fe–Cl cycle were carried out by introducing hydrated ferrous chloride ($FeCl_2 \cdot 4H_2O$) in the capsule as chemical reactants. Figure 26.3a shows the aspect of the capsule after heating at 650°C for 48 h, and the characteristic red color observed in the condensed phase is indicative of the presence of hematite (Fe_2O_3) instead

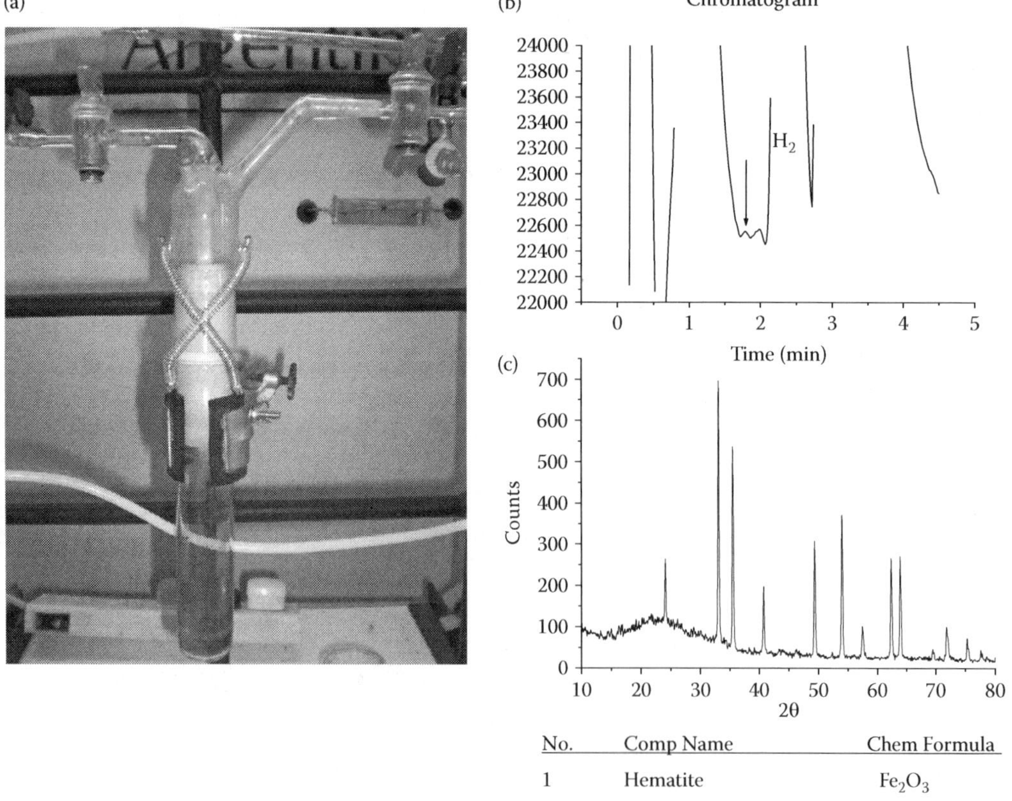

FIGURE 26.3
Experiment on the conventional Fe–Cl thermochemical cycle: (a) capsule aspect after heating at 650°C for 48 h; (b) chromatogram of the gaseous phase; (c) XRD pattern of the condensed phase.

of magnetite (Fe_3O_4), as it was predicted by thermodynamic analysis [4]. The presence of hematite was also confirmed by XRD analysis performed on the condensed species, as shown in Figure 26.3c. As can be observed in Figure 26.3b, a small amount of hydrogen was produced with this cycle despite the long exposure of reactants at high temperatures, and it can be explained by the presence of an oxidative atmosphere in the reaction site.

For enhancing the hydrogen production through the removal of the O_2 from the reaction bed, experiments were performed in which a metallic Ti sheet as O_2 scavenger was introduced inside the capsule along with the chemical reactants. Results of experimental after heating the capsule at 650°C for 5 h and 24 h are presented in Figures 26.4 and 26.5, respectively. Figures 26.4a and 26.5a show the aspect of capsules after experiments, while in chromatograms given in Figures 26.4b and 26.5b, it can be clearly appreciated that the production of hydrogen is increased when O_2 is removed from the reaction site as soon as formed, according to thermodynamic predictions. The sharp increment of hydrogen production when the reaction time is increased from 5 to 24 h also reveals that the Fe–Cl thermochemical reactions are quite slow and the kinetics had to be drastically accelerated,

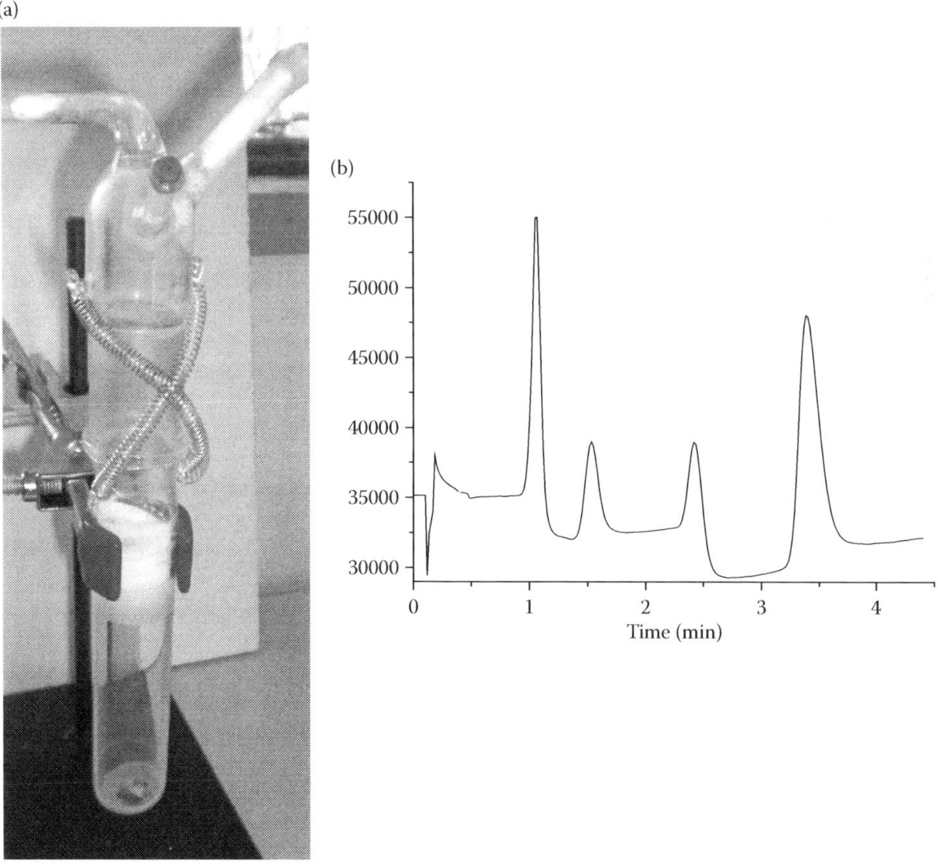

FIGURE 26.4
Experiment on the Fe–Cl thermochemical cycle with an O_2 scavenger (metallic Ti sheet): (a) capsule aspect after heating at 650°C for 5 h; (b) chromatogram of the gaseous phase.

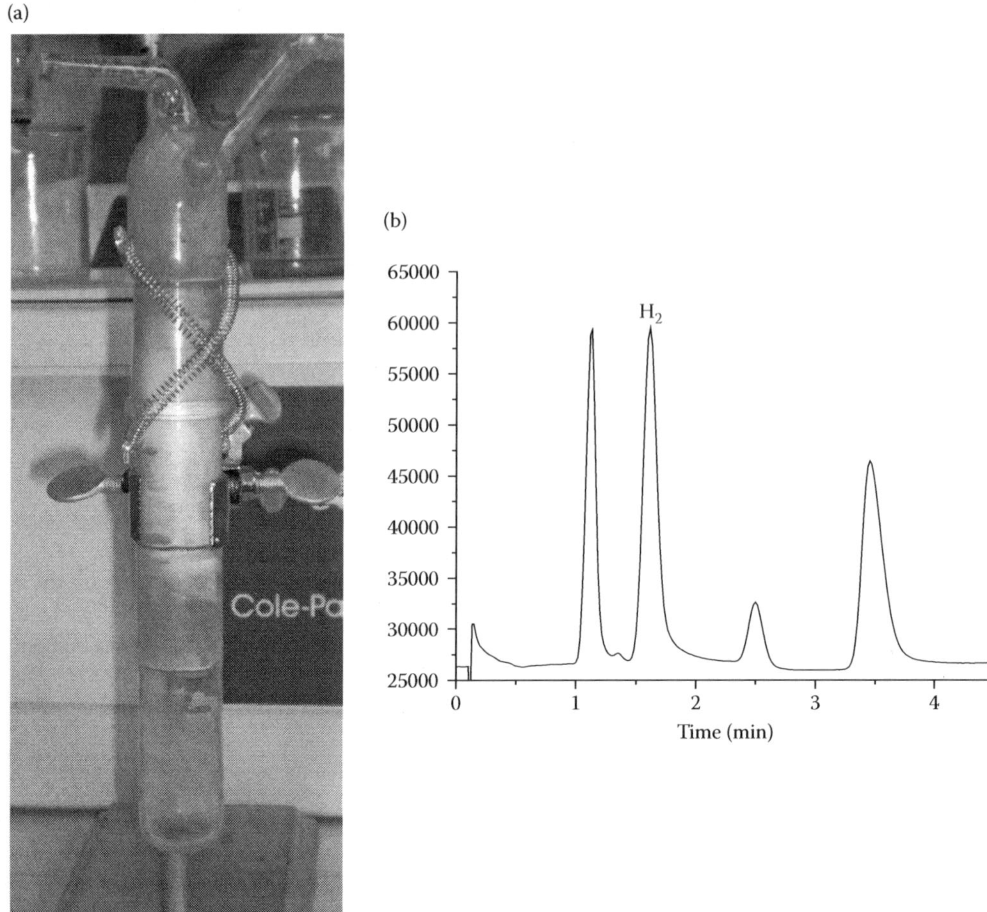

FIGURE 26.5
Experiment on the Fe–Cl thermochemical cycle with an O_2 scavenger (metallic Ti sheet): (a) capsule aspect after heating at 650°C for 24 h; (b) chromatogram of the gaseous phase.

that is, by using catalysts, if this hydrogen production process has to be implemented to a commercial scale.

26.2.3 Alternative Chloride Cycles

Other chloride cycles were also theoretically analyzed like the copper chloride cycle (CuCl), the zinc chloride cycle ($ZnCl_2$), and the manganese chloride cycle ($MnCl_2$) which are based on the following reactions [6–11]:

- Copper chloride cycle:

$$2CuCl(s,l) + 2H_2O(g) \rightarrow Cu_2O(s) + 2HCl(g) + H_2(g) + \frac{1}{2}O_2(g) \tag{26.6}$$

$$Cu_2O(s) + 2HCl(g) \rightarrow 2CuCl(s,l) + H_2O(g) \tag{26.7}$$

$$2HCl(g) \rightarrow H_2(g) + Cl_2(g) \tag{26.8}$$

- Zinc chloride cycle:

$$ZnCl_2(s,l) + H_2O(g) \rightarrow ZnO(s) + 2HCl(g) \quad (26.9)$$

$$ZnO(s) + 8HCl(g) \rightarrow ZnCl_2(s,l) + H_2O(g) \quad (26.10)$$

$$2HCl(g) \rightarrow H_2(g) + Cl_2(g) \quad (26.11)$$

- Manganese chloride cycle:

$$3MnCl_2(s,l) + 4H_2O(g) \rightarrow Mn_3O_4(s) + 6HCl(g) + H_2(g) \quad (26.12)$$

$$Mn_3O_4(s) + 6HCl(g) \rightarrow 3MnCl_2(s,l) + 3H_2O(g) + \frac{1}{2}O_2(g) \quad (26.13)$$

$$2HCl(g) \rightarrow H_2(g) + Cl_2(g) \quad (26.14)$$

In the case of the copper chloride cycle, the amount of hydrogen obtained is various orders of magnitude less than in the iron chloride cycle. It was also observed that the zinc chloride cycle is not promising because the partial pressure of hydrogen is various orders of magnitude lower than the corresponding to the iron chloride cycle. Finally, when $MnCl_2$ is taken into account the situation is improved as a less oxidative environment is reached, but in any case, the production of hydrogen is not so satisfactory when compared with the iron chloride cycle.

With these ideas in mind, new thermodynamic calculations were performed, including the starting reacting species and metallic titanium like O_2 scavenger. The presence of titanium enhances the production of hydrogen more than in the iron chloride cycle, as can be seen from the following results: 100–99.9% for Cu, 98.9–33.3% for Zn, 100–85.8% for Mn, at 500°C and 1000°C, respectively. Therefore, it was concluded that the copper chloride cycle is the best process for the hydrogen production since there is the smallest amount of its chloride in the gaseous phase and it condenses at room temperature with minimum tendency to its hydration.

In the three cycles analyzed, the amount of the other gaseous species such as HCl, O_2, $TiCl_4$ and H_2O are insignificant, but several others reactions may take place due to the different interactions between gaseous and condensed reaction product phases. Then, a detailed study of one of the most important interactions: the reaction between $TiCl_4$ and Fe_2O_3, was done and results showed that different Fe–Ti compounds like rutile, ilmenite and pseudorutile can be obtained under different temperature, Cl_2 pressure, and atmosphere conditions.

26.2.4 Current Research Objectives and Activities

Following the first stage of the research program and based on the promising preliminary experimental results, additional steps were planned for obtaining a deeper understanding of the mechanisms of thermochemical cycles. These research activities are currently in progress and involve:

1. Design and construction of a laboratory-scale reactor for continuous operation. Measurement of the reaction rates at different temperatures and partial pressures of gaseous species.

2. Development of catalytic surfaces with rare earth and refractory metals oxides [12], in order to make the reactions faster and to improve the efficiency of the process by using O_2 capture in the form of blocks or as deposits on inert substrates [13].
3. Determination of the best way for the *in situ* production and regeneration of dry metallic chlorides.
4. Monitoring of evolution of chemical composition of reaction products by atomic absorption and UV/visible spectroscopy, Mössbauer analysis, XRD and conventional analytical procedures, in order to detect the intermediate products of the different chemical species and to find the reaction rate dependence with partial pressure of gases, temperature, and microstructural characteristics of solids. This task will allow to determine the rate equation at all experimental conditions and to determine the most efficient way to produce hydrogen by these reactions.
5. Economic evaluations, mainly on a comparative basis, to provide guidance in cycle selection and process-flow sheet development.

It is expected that results from these research activities would give impulse to the growth of hydrogen economy in Argentina, which is associated with a cleaner technology and a sustainable energy source, especially applicable for the remote desert regions of the country like the vast Patagonia in the south and the Puna in the north.

26.3 Hydrogen Production from Coal Gasification

26.3.1 Overview

Gasification technologies offer the potential of clean and efficient energy. Even they have been commercially applied for more than a century for the production of both fuels and chemicals. A renewed interest on gasification technologies has emerged recently as a consequence of more stringent environmental standards and the need of improving the overall efficiency of energy production systems [14]. Gasification has an additional advantage over conventional combustion processes since it is able to accommodate a wide range of feed stocks, including coal and low-cost fuels as petroleum coke, biomass, and municipal wastes [15].

In this new favorable framework, the gasification of coal to produce hydrogen for use either in power generation or/and for synthesis applications and transport is attracting considerable interest worldwide [16]. In Argentina, CNEA is investigating the application of gasification technologies for hydrogen production using a domestic low-rank coal extracted from the Río Turbio minefield, which is located in the southwest region of Argentina and represents, by far, the main coal reserve of the country.

Coal gasification is defined as a set of exothermic and endothermic reactions of coal particles with air, oxygen, steam, carbon dioxide, or a mixture of these gases at a temperature exceeding 700°C, to yield a gaseous product suitable for use either as a source of energy or as a raw material for the synthesis of chemicals, liquid fuels, or other gaseous fuels such as hydrogen. The combustible or synthetic gas produced from coal gasification is a mixture of CO and H_2, with minor amounts of CO_2, H_2O, and CH_4 [17]. In gasification processes, coal particles are fed in either dry or in a slurry form into the gasification reactor where they are subjected to heat, pressure and reactant gases (air or oxygen and steam) for converting the carbonaceous solid to a gaseous state. In contrast to combustion processes that work with air in excess, the gasification processes operate in the absence of oxygen or with a limited amount of oxygen (generally 20–70% of the amount of O_2 theoretically required for complete combustion) such that both heat and a new gaseous fuel are produced as the feed coal particles are consumed. Some gasification processes also use indirect heating to replace the partial combustion of the feed material in the gasification reactor, and one alternative that is being evaluated for indirect heating is the use of high-temperature nuclear reactors [18].

The chemical composition, the heating value and, then, the future use of the gas produced by coal gasification is variable with the gasification technology employed, depending on a lot of factors such as coal composition and rank; coal preparation and feeding methods; gasification agents; operational conditions in the gasifier (temperature, pressure, heating rate, residence time); and plant configuration characteristics (flow geometry, ash removal, gas cleaning system). There are a large number of gasification processes implemented at commercial level and the choice of a given gasification technology is difficult because it depends on diverse factors such as coal availability, type and cost; size constraints; and production rate of energy. Even, in principle, all types of coals can be gasified, the properties of the coal to be processed are the least flexible factor to be considered in the analysis and, then, the gasification technology should be primarily matched to the properties of the coal [19].

26.3.2 Current Research Objectives and Activities

The comprehensive research program implemented at CNEA is addressed to characterize the behavior of Río Turbio coal under typical gasification conditions with the objective of identifying the most suitable gasification technology to be implemented for hydrogen production. The research program comprises both theoretical and experimental studies on laboratory scale which are designed to simulate the operational conditions of a large-scale gasification plants and to provide the necessary data about the mechanisms and kinetics of gasification reactions.

It is known that coal gasification is a two-step process. In the first step, pyrolysis, volatile components of coal are rapidly released at temperatures between 300 and 500°C, leaving residual char and mineral matter as byproducts. The second step, the char conversion, involves the gasification of residual char that is much slower than the devolatilization step. Earlier studies demonstrated that the reactivity of chars to gasifying agents is very dependent on their formation conditions, particularly temperature, pressure, heating rate, time at peak temperature, and the gaseous environment [20]. For this reason, to get meaningful data about kinetics of gasification reactions, it is very important to produce chars in laboratory that replicate, as close as possible, the real conditions of char formation in large-scale gasifiers.

When introduced into a high-temperature atmosphere in a gasifier, coal particles are heated at high heating rates (over 10^3°C s^{-1}). Then, nascent char particles react with steam or steam–oxygen mixtures under more or less intensive gas convection around the particles. Hence, in order to experimentally simulate the gasification conditions and achieve the kinetics/mechanism of the gasification, it is essential to reproduce, at least, (1) the rapid heating of coal particles and (2) gas convection around individual char particles with definable intensity. According to that, the most popular experimental approach followed to achieve both objectives consists of using the so-called "two-stages experiments" in which the gasification reactivities are determined using char samples prepared in a

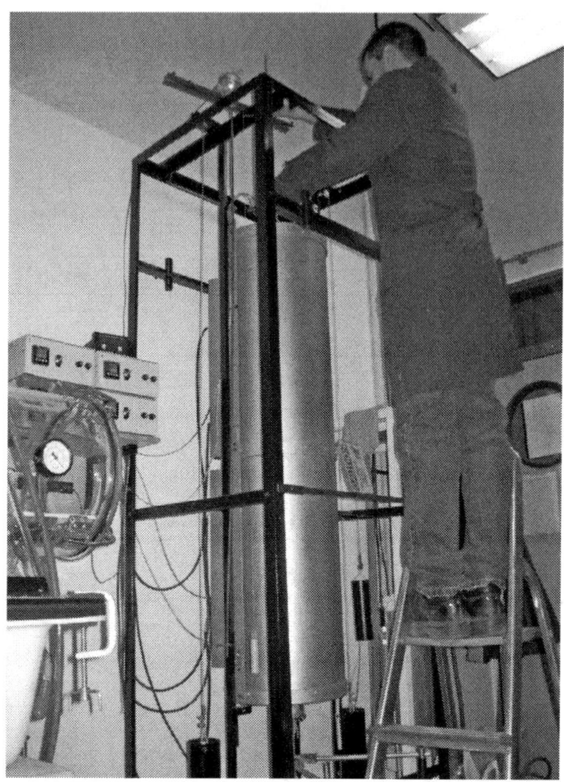

FIGURE 26.6
Drop tube furnace for high heating rate pyrolysis studies on the Río Turbio coal.

previous pyrolysis step, where coal particles are heated at high-heating rates and short residence times at high temperatures [21].

For the preparation of chars at different conditions of temperature, heating rate and high-temperature residence time, a drop tube furnace (DTF) was constructed as shown in Figure 26.6. The reactor has an electric furnace able to operate at 1100°C, which surrounds two concentric quartz tubes of 41 and 26 mm inner diameter, 1.30 and 1.20 m long, respectively. Inert gas (N_2) is injected at the bottom of the outer cylinder and is preheated while flowing upwards. When at the top of the outer cylinder, the gas is forced into the inner tube through a flow rectifier and the gas flows downwards and leaves the reactor through a water-cooled collection probe. The coal particles are entrained by a jet of nonpreheated inert gas to a water-cooled injection probe placed on top of the inner tube. The heating rate is expected to be higher than 10^{3}°C s^{-1} and the residence time of particles in the reactor less than 0.3 s. The chars leave the reactor through the collection probe, and an extra nitrogen flow is added to the exhaust gases in order to quench the reaction and improve the collection efficiency in the cyclone [22].

Kinetic measurements on chars prepared with the DTF are performed with a thermogravimetric analyzer (TGA), which has been extensively described elsewhere [23]. The experimental setup, shown in Figure 26.7, consists of an electrobalance, a gas line, and an acquisition system, having a sensitivity of ±5 μg while operating at 950°C under a flow of 8 L h^{-1}. In a typical TGA run, the weight of a char sample is measured as a function of time

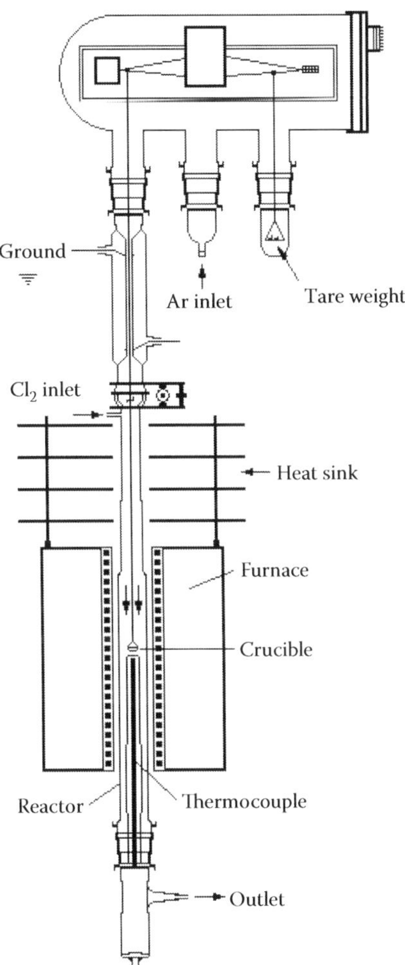

FIGURE 26.7
Scheme of the TGA system for isothermal and nonisothermal kinetic measurements on char samples from the Río Turbio coal.

and temperature as it is subjected to a controlled temperature program. TGA tests are usually carried out in two ways: (1) isothermal, where the sample is heated at a constant temperature, and (2) nonisothermal with linear heating, where the sample is heated at a constant temperature rate.

Following a complete microstructure, chemical and thermal characterization of the raw coal material, preliminary tests on CO_2 gasification of char samples from Río Turbio coal were carried out with TGA experimental setup, using isothermal and nonisothermal measurements to study the influence on the reaction rate of several parameters like gaseous flow rate, sample mass, temperature, and carbon dioxide partial pressure [24].

Tests were made with char prepared in a fixed bed tubular reactor heated by an electric furnace under an argon stream. Coal samples of around 300 mg were placed inside the tubular reactor on a flat quartz crucible, adopting a loose packed bed. They were heated at 950°C for 4 h to release the volatile matter. The structure of the char obtained was

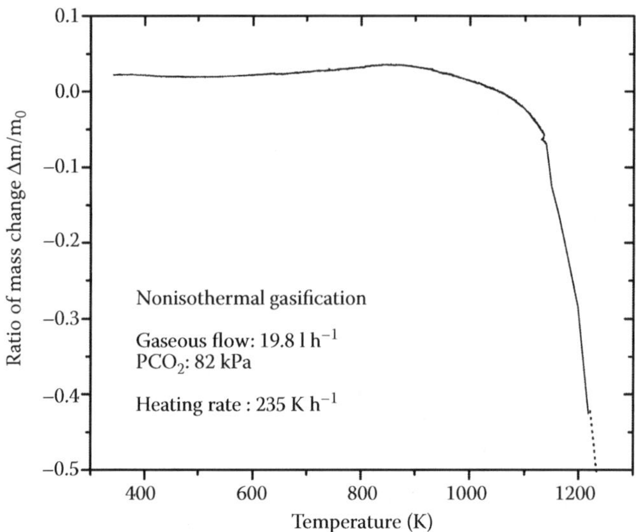

FIGURE 26.8
Nonisothermal thermogravimetric curve of CO_2 gasification of char produced from the Río Turbio coal.

well-characterized by scanning electron microscopy (SEM), while the pore size distribution and BET area were measured by N_2 adsorption/desorption.

CO_2 gasification behavior of char samples was analyzed for temperatures between 800 and 950°C, and CO_2 partial pressures ranging from 28 to 82 kPa. Figure 26.8 shows a typical nonisothermal thermogravimetric curve where it is observed that the reaction starts slowly at approximately 600°C and the reaction rate increases significantly at about 850°C. The experimental conditions under which the reaction rate is controlled by chemical reaction were established for the setup used, and a theoretical model to describe the evolution of the reaction degree as a function of time was developed. An intrinsic activation energy of 45.6 ± 1.7 kcal mol^{-1} (190 ± 7 kJ mol^{-1}) was calculated with an isoconversional method. Finally, for low partial pressures of CO_2 (up to 61 kPa), a first order reaction was obtained for CO_2. A Langmuir–Hinshelwood-type expression was also applied to fit the whole range experimental data [24].

Steam gasification of char from Río Turbio coal was also preliminary investigated, using a specially designed experimental setup that is shown in Figure 26.9. The experimental setup is able to produce different partial pressures of steam. Gasification reactions are produced using a tubular flow reactor and an electrical furnace that can reach temperatures up to 950°C. The overall reaction rates under several experimental conditions of gaseous flow rate and sample mass are measured by following the temporal evolution of the gaseous product concentration, that is, CO, H_2, CO_2, by gas chromatography.

26.4 Hydrogen Purification

Currently, over 90% of hydrogen is produced from the steam–methane reforming process which requires invariably an additional step of purification for satisfying the purity

FIGURE 26.9
Experimental setup for steam gasification of char produced from the Río Turbio coal.

requirements of main hydrogen applications. Since the hydrogen purification is a costly step in the whole process, a cost-effective method for impurities removal (mainly CO_2 and CO) from hydrogen is of paramount importance. Compared to traditional energy-intensive processes of pressure swing adsorption (PSA) and cryogenic distillation, membrane technology is a promising alternative to achieve high hydrogen separation and purification [25].

Membranes are layered materials with special transport properties of interest for gas and liquid separation and purification. With an ideally permeable (nonresistive) membrane there are no irreversible energy losses if a molecule is transferred from the feed side to the permeate side at the same chemical potential. Since real membranes are resistive and often not 100% selective for one species, the selectivity of membranes is addressed by selecting a suitable nanostructure and chemical composition tailored for the target application.

The viability of membranes depends critically on their stability at operating conditions, and their operational costs that must be less than the energy savings they offer. When membranes are used to separate and purify hydrogen from hydrogen-rich mixtures, a variety of materials have been investigated, including ceramic ion-transport membranes, carbon microporous membranes, crystalline alloy membranes, and amorphous alloy membranes [26].

In ceramic ion-transport membranes, hydrogen is transported across the membrane in the ionic form and catalysts are usually used to facilitate the hydrogen ionization [27]. Carbon microporous membranes are prepared by pyrolysing a variety of organic precursors and they can be used in a reducing environment with temperatures in the range of 500–900°C even though they are brittle and still expensive [28]. Crystalline alloy membranes can be of many types such as pure metals, binary alloys, and complex alloys. The most widely studied palladium (Pd) and Pd-alloy membranes pertain to the crystalline alloy membranes. Even high-purity hydrogen can be obtained through Pd and Pd-alloy membranes, these membranes are not suitable for operation when the temperature is below 300°C, as a consequence of hydrogen embrittlement [29]. Membranes produced from some low-cost elements, such as Zr, Ni, Cu, and Al, are inherently resistant to embrittlement due to the fact that the amorphous structure resists crystalline hydride formation. For this reason, amorphous alloy membranes are presently being developed to tackle the problems of hydrogen embrittlement, sintering, and cost that occur with crystalline alloy membranes [30].

Pd and Pd-alloy membranes have been studied for a long time for hydrogen purification due to their high permeability and selectivity properties. However, conventional technology is limited by the high cost of the precious metal combined with manufacturing methods which result in thick Pd films, having a membrane thickness of 50 μm or more that reduces the hydrogen flux and inhibits their application for large-scale separation units due to the high investment costs. For this reason, a substantial research effort has been carried out in the last years on achieving higher hydrogen fluxes by depositing Pd or Pd/Ag membranes on porous support ceramic materials. However, the selectivity of these composite membranes is often poor as a consequence of bad step coverage of Pd(Ag) on porous support, probably due to the fact that the Pd(Ag) has a tendency to deposit on top of the support material or on the top of previously deposited layer, and not to enter the pores, resulting in the formation of pin-hole type defects. On the other hand, the different elongation of the metal layer with respect to the ceramic support triggers shear stresses that can cause the rupture of the metallic film during thermal cycling and hydrogen loading [31].

Based on development of inorganic tubular membranes for separating ^{235}U from ^{238}U started in Argentina more than 25 years ago, a research project is being carried out presently for hydrogen purification using membrane technology. The project consists of the development of a supported membrane with three different layers: a thin separative film of Pd and Pd-alloy (Pd/Ag, Pd/Ni, and Pd/Cu are being tested), a nonseparative support macroporous alumina layer, and an intermediate nickel film between separative and supporting layers, deposited by electroless plating.

Planar and tubular membrane samples are being extensively characterized in their microstructure and physical properties including: (1) measurements of thickness and uniformity of film layers; (2) detection of possible inter-diffusion effects among different layers during high temperature processing; (3) determination of adherence and integrity of films, and (4) verification of mechanical strength and structural stability of the whole composite membrane. After completing the characterization of membranes, tubular samples are expected to be tested in a gas permeation system especially designed to determine the hydrogen permeability and selectivity. Permeation tests will be performed by feeding high purity hydrogen gas and hydrogen-rich gas mixtures inside the membrane tube, at temperatures and inlet gas pressures within ranges from 150°C to 400°C and 10–200 kPa, respectively. The permeating gas stream will be collected in a capsule surrounding the membrane tube and it will be further analyzed by gas chromatography.

26.5 Hydrogen Storage

In relation with safe storage of hydrogen in the form of metallic hydrides, CNEA's scientists are investigating on the design of a variety of Mm–Ni–Al alloys (Mm: mischmetal) for compression. The incorporation of Al to the Mm–Ni alloy reduces the reaction pressure, diminishes the capacity of storage, increases the hysteresis and the dependence of pressure with the reaction degree, and promotes the degradation during cycling. Nevertheless, the equilibrium pressure from 1 to 100 bar widely affected by the concentration of Al in the alloy make these materials as candidates for application in the compression of hydrogen at low pressure, in the temperature range between 10 and 90°C. In the case of purification, the reversibility of the hydride formation is relevant, following cycles of selective absorption and desorption of hydrogen and then allowing its separation from the other gases [32].

The preparation of alloys is made by two different procedures: by fusion (in arc or induction furnace) and by mechanical milling (in a low-energy ball mill). Following, an exhaustive microstructure characterization is done in order to define the elemental composition of alloys, the presence of metastable and stable phases, degree of defects, morphology, and so on. The analysis of equilibrium properties of hydride reaction is obtained by pressure–temperature behavior, enthalpy of formation of hydrides, and hysteresis. The dependence of the reaction rate with pressure, temperature, microstructure and the rate determining step of these reactions are important for the application of these types of alloys. The reaction of hydrogen with elements or alloys yields to the formation of the respective metallic hydrides, which let the concentration of a high amount of hydrogen per unit of volume and final mass. These compounds may be very important for an efficient and safe storage of hydrogen. For this application the most remarkable properties to be attained are: maximum capacity of absorption of hydrogen, lowest range of pressure and temperature during operation, fast kinetics behavior during incorporation and release of hydrogen, low degradation rate, and competitive cost.

Actually, there is a great interest in those systems based on magnesium, the family of $LaNi_5$, and inter-metallic compounds called Friau–Laves phases which is AB_2 (with A: Zr, Ti, and so on and B: Cr, Mn, Fe, and so on). The first compounds based on Mg were widely developed and characterized in CNEA. It was observed that the kinetics of hydrogen absorption and desorption is enhanced by the addition of metals like Ge, Cu, Zn, Fe, and Ni, which act as catalysts. Furthermore, the incorporation of Fe and Ge diminishes the reaction temperature in 100 and 150°C, respectively, while the incorporation of Cu increases the capacity of absorption compared with pure MgH_2.

Moreover, some metastable phases were synthesized by ball milling at high pressures and temperatures. For example, by milling Mg in hydrogen atmosphere two metastable phases like β- and γ-MgH_2 were obtained while in the Mg–Sn system the Mg_2Sn hexagonal metastable phase was produced. The presence of metastable phases and microstructure modification attained by milling in hydrogen atmosphere, have a cooperative effect on the kinetics and the temperature reduction of desorption [33]. An alternative route for synthesis of materials addressed to hydrogen storage was the preparation of $MgFeH_6$ compound with a capacity of 5 wt% by a one step procedure consisting of milling the powder mixture of pure metals in hydrogen. Another way to improve the properties of hydrogen absorption and desorption was observed in the MgNi alloy by the partial substitution of nickel or magnesium by Ge, obtaining a cubic phase which is an isostructural compound with Mg_3Ni_2M (with M = Al, Ti), very stable in argon up to 410°C [34].

Another focus of interest in research on hydrogen applications in CNEA is the development of nickel metallic hydride batteries. The importance of this kind of storage device is the replacement of cadmium in the Ni–Cd rechargeable batteries. Two remarkable aspects were studied in this area: (a) preparation of new nanophase materials for the absorption and desorption of hydrogen; and (b) proposal of a physicochemical model that describes the kinetics of the process during the incorporation and release of hydrogen [35].

The preparation of new alloys were based on the AB_2 type (with a structure type $MgCu_2$), AB_3 (structure type $PuNi_3$), AB (structure type CsCl) and AB_5 (structure type $MgCu_2$). The AB_2 alloy type and the new AB_3 alloy type like LaCaMgNi present a charge capacity of 400 mAh/g. In this kind of battery, it improved the way of intensifying the capacity of absorption of hydrogen during the charge step, the increment of the surface catalytic effects in order to obtain faster discharge velocities diminishing the self discharging, and the development of new cell designs for reducing the global cost of these devices [36].

The determination of kinetic parameters of the electrochemical steps involves the charge and discharge cycles, and the catalytic activity of the electrodes. They are determined by relaxation techniques using cyclic voltametry, impedance spectroscopy, potentiostatic pulses, potentiodynamic and galvanostatic curves, and conventional stationary methods [37].

26.6 Use of Hydrogen in Proton Exchange Membrane Fuel Cells

Two projects related to proton exchange membrane (PEM) fuel cells are currently being developed in CNEA, both of them with co-financial support of the Argentine National Agency for the Promotion of the Science and Technology. The first project began in 2007 and it is a start up project related to the development of a PEM mini-fuel cell fed with methanol and oxygen (air) with power between 500 and 1000 mW for use as power source for secondary batteries of cellular phones. The second project is part of the program for production, purification, and applications of hydrogen as a fuel and energy vector, considered by the Argentine Ministry of Science and Productive Innovation as a strategic program in the area of energy. The subproject Fuel Cells of this program is being developed with the following main objectives: (1) design and characterization of materials (electrolytes, cathodes, anodes) for PEM fuel cells; (2) development of membrane electrode assemblies (MEAs) for PEM fuel cells; (3) development of a stack PEM of 1–5 kW; and (4) integration and operation of prototypes of PEM fuel cells with the hydrogen produced by reforming of ethanol or other technologies.

The main R&D activities carried out in the framework of the PEM fuel cells and the corresponding advances and achievements are summarized below:

Related to the development of membranes for PEM fuel cells, the low-temperature thermal behavior of Nafion 117 membranes with different water contents and equilibrated with water–methanol mixtures of different concentrations were studied by differential scanning calorimetry (DSC) [38]. The DSC curves revealed a weak transition at temperatures below –100°C related to the γ peak reported in mechanical dynamics studies, and associated to the true-glass transition temperature of the polymer. The amount of freezable water was determined from the area of the ice-fusion peak and it increased from 1 to 23% as the relative humidity increases from 84 to 100%. The presence of methanol in the membrane increased the amount of freezable water up to 65%, which takes place at temperatures in the range of –10 to –20°C (see Figure 26.10), being relevant to the operation of direct methanol PEM fuel cells in low-temperature environments. The amount of freezable water could be the criteria for choosing membranes for this type of fuel cells, beyond the methanol permeation characteristics.

Moreover, modified polybenzoimidazole (ABPBI) was prepared by condensation of 3,4-diaminobenzoic acid (DABA) monomer in polyphosphoric acid (PPA). This polymer is not charged, but it becomes proton conductor when doped with phosphoric acid. Thus, phosphoric acid-doped membranes based in poly [2,5-benzimidazole] (ABPBI) were obtained by high-temperature casting from methanesulphonic acid and by a new low-temperature casting procedure [39]. These membranes, which can be suitable for application in direct

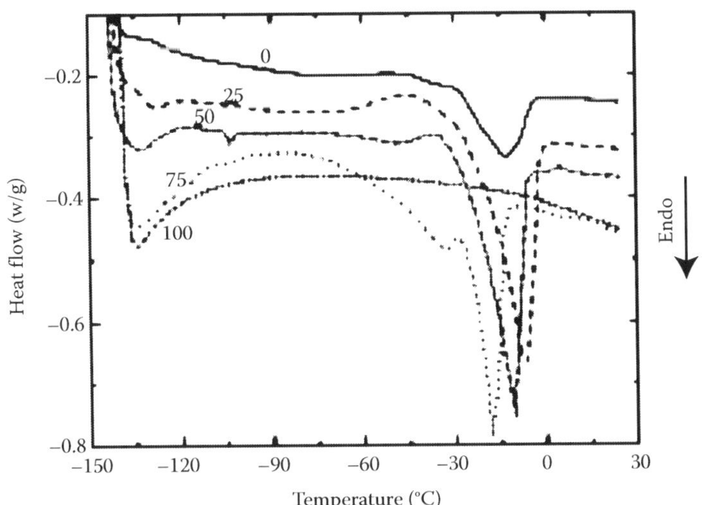

FIGURE 26.10
DSC curves of Nafion 117 membranes equilibrated in methanol–water solutions. Scan rate, 10 K min^{-1}. The numbers indicate the percentage (w/w) of methanol in the mixture.

methanol PEM fuel cells, were studied in relation to their phosphoric acid doping level by measuring the free and bonded acid. The water isotherms were also determined for the low and high temperature cast ABPBI membranes. Both properties were compared with those determined in poly [2–2′-(m-fenylene)-5-5′ bibenzimidazole] (PBI). It was observed by atomic force microscopy (AFM) imaging that the morphology of the ABPBI membranes strongly depends on the casting procedure (Figure 26.11). The elastic modulus of the

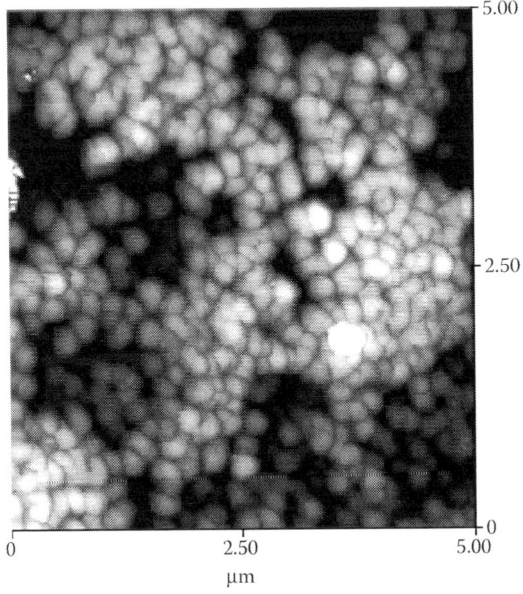

FIGURE 26.11
AFM images of undoped ABPBI obtained by high-temperature casting.

undoped and phosphoric-acid doped membranes were determined using the AFM force-spectroscopy technique and the differences observed with doped and undoped PBI and Nafion membranes, were discussed in terms of the electrostatic and swelling forces between polymer chains [40]. The analysis of the force curves indicates differences in the mechanical behavior of doped PBI and ABPBI membranes compared to Nafion, which could have practical consequences on the stability of the membrane electrode assemblies.

In parallel, ammonium quaternized polymers such as poly (arylene ether sulfones) are being developed and studied as candidates of ionomeric materials for application in alkaline fuel cells, due to their low cost and promissory electrochemical properties. A quaternary ammonium polymer synthesized by chloromethylation of a commercial polysulfone followed by amination process was analyzed in relation to the water and water–methanol uptake, electrical conductivity and Young's modulus, and compared with those properties obtained for Nafion 117. The anionic polysulfone membrane sorbs more water than Nafion all over the whole range of water activities, but it uptakes much less methanol as compared with Nafion. The specific conductivity of the fully hydrated polysulfone membrane equilibrated with KOH solutions at ambient temperature increases with the KOH concentration, reaching a maximum for 2M KOH, slightly less conductive than Nafion 117. The elastic modulus of the polysulfone membranes immersed in water is similar to that reported for Nafion membranes under the same conditions. It is concluded that quaternized polysulfone membranes are good candidates as electrolytes in alkaline direct methanol fuel cells.

In relation with catalyst development, nanoparticulated platinum catalysts have been prepared on carbon Vulcan XC-72 using three methods with chloroplatinic acid as a precursor: (1) formic acid as a reducing agent; (2) impregnation method followed by reduction in hydrogen atmosphere at moderated temperature; (3) microwave-assisted reduction in ethylene glycol [41]. The catalytic and size studies were also performed on a commercial platinum (Pt) catalyst (E-Tek, De Nora). The characterization of the particle size and distribution were performed by means of transmission electron microscopy (TEM) and XRD (Figure 26.12). The characterizations of the catalytic and electro-catalytic properties of the catalysts were determined by studying the cyclohexane dehydrogenation reaction (CHD) and the behavior under cyclic voltammetry (CV) in sulfuric acid solutions. The catalysts prepared by reduction with formic acid and ethylene glycol (microwave assisted) show electrochemical activities very close to those of the commercial catalyst, and are almost insensitive to the Pt dispersion or Pt particle size. The chemical activity in CHD correlates

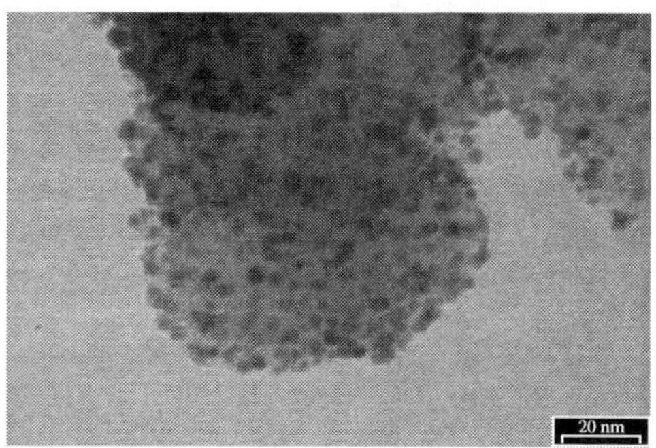

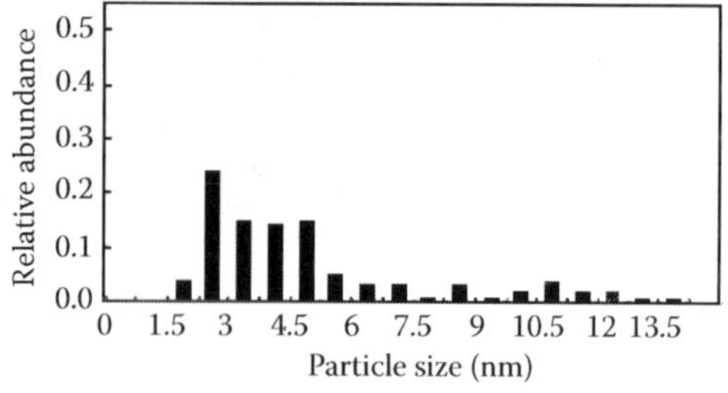

FIGURE 26.12
TEM micrographs of a Pt catalyst on Vulcan prepared by the microwave-assisted method (left) and histogram of particle size distribution (right).

well with the metallic dispersion determined by hydrogen chemisorption, indicating similar accessibility of H_2 and cyclohexane to the catalyst surface.

Mesoporous films of Pt catalysts were electrodeposited over carbon with hierarchical porous structure [42]. The liquid crystal used as a template allowed the electrodeposition of the catalyst on the outer region of the carbon with low penetration in the porous structure. The Pt hexagonal mesostructured deposits exhibit an excellent stability enhanced by the roughness of the carbon support. The mass activity for the electro-oxidation of methanol of the mesoporous Pt catalyst supported on the hierarchical carbon is similar to that observed on gold and to that reported for commercial Pt nanoparticulated catalysts, even when this catalyst has a smaller Pt load than the commercial one. Also, the poisoning rate of the mesoporous catalyst is lower than that observed for the commercial catalyst. Mesoporous Pt and Pt/Ru catalysts with cubic structures were synthesized using a F127 template, on a gold support. Large electrochemical surface area was observed for the catalysts prepared at high over potentials. Compared to the Pt catalyst, the Pt/Ru alloy containing 3 wt% of Ru exhibits a lower onset potential and more than twice the limit current density for methanol oxidation. The large pore size of the Pt/Ru catalyst seems to improve the methanol accessibility as compared to other mesoporous Pt/Ru catalysts of similar composition and having smaller pores.

Related with catalysts support and gas diffusion layer, the application of mesoporous carbon as catalysts support in gas diffusion electrodes to be used in fuel cells, requires a careful consideration of the gas and/or liquid micro fluidic through the material. A way to improve the flow of fluids while maintaining a high surface area involves the preparation of highly porous carbon structures but having a hierarchical distribution of pore size, as shown in Figure 26.13. The performance of these systems could be improved taking into account different aspects such as mass transport, volume reservoir, and selective control of pores' surface hydrophobicity/hydrophilicity. In the case of carbon formation from a polymer precursor, a wide variety of methods supply different alternatives for the structural design of the material in each production step, including synthesis, drying (polymer stage), and carbonization.

The structure of the carbon material can be controlled to nanometric scale starting from the polymerization step. The use of additives such as templates or structuring agents (soft

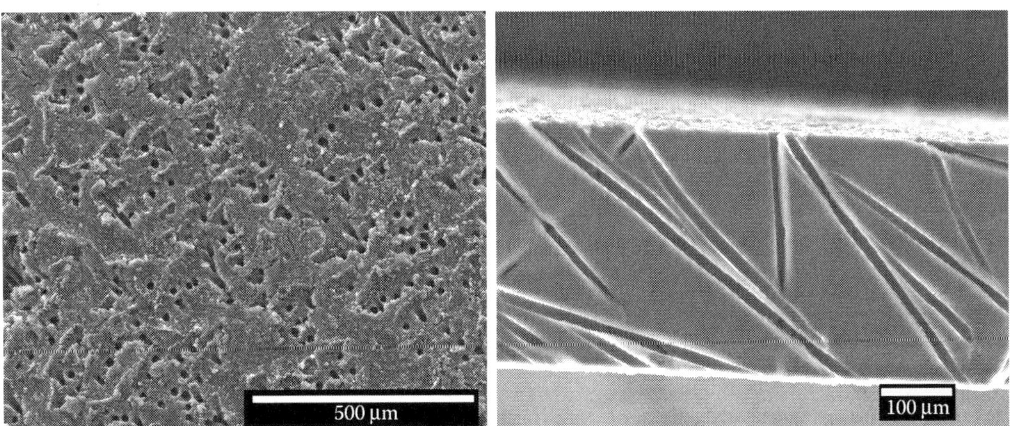

FIGURE 26.13
Mesoporous carbon with hierarchical porous structure. Mesoporous structure with capillar channels (left) and details of the capillar distribution (right).

and hard templates) in the polymerization media generates controllable structures of the polymer precursor with pore formation. Moreover, the macroscopic shape of the piece can be molded by a suitable choice of the polymerization conditions. The second step, drying to get the polymer stage, allows obtaining a material with suitable mechanical properties to ensure an easier machining than the final carbon material. Finally, during the carbonization step, an optional activation process can be used to increase the surface area [43].

26.7 Use of Hydrogen in Solid Oxide Fuel Cells

The basic construction of fuel cell electrochemical device, that is, cathode–electrolyte–anode, is the most efficient technology for the chemical to electrical energy conversion. There are different types of fuel cells under development and the main differences among them are the electrolyte type, that is, H^+, OH^-, or O^{2-} conductors, and the operating temperature. Solid oxide fuel cells (SOFCs) are characterized by the highest operation temperatures (800–1000°C) and all their basic components are ceramic oxides. These cells are widely studied due to their potential for applications as high efficient stationary compact systems. Commercial SOFC usually use $La_{1-x}Sr_xMnO_{3-\delta}$ (LSM) cathodes, zirconia stabilized yttria (YSZ) electrolytes, and Ni–YSZ cermet anodes, that are very efficient at operation temperatures over 900°C. The employment of new electrolyte materials like gadolinium-doped ceria and lanthanum gallate allowed the development of intermediate-temperature solid oxide fuel cells (IT-SOFCs), which represent an alternative to conventional SOFC that uses YSZ due to its lower operation temperature (500–700°C) and cost [44].

Several projects related to the new materials to improve the performance of SOFCs and IT-SOFCs are being developed at CNEA. These studies include thermodynamic properties, crystal structures, material defects, and transport properties with special emphasis on the effect of oxygen defects in physical properties of nonstoichiometric oxides. Some of these projects are developed in collaboration with several other research groups of different countries like France and the United States.

The main R&D activities carried out by in the framework of SOFCs and the corresponding advances and achievements are summarized below:

One of the problems to overcome in IT-SOFCs is that cathode over-potential becomes important at lower temperatures, decreasing the cell performance. Most of research work focused to enhance cathode performance only deal with chemical composition, although microstructure is also relevant. For instance, nanostructured materials with a high surface area would further improve the cathode performance [45]. Nevertheless, the high temperatures involved for phase formation of IT-SOFC cathode materials hinder the possible achievements of nanostructures. Many researchers have been searching new cathode materials, like the cobaltite $La_{1-x}Sr_xCo_{1-y}Fe_yO_{3-\delta}$, but in order to further increase SOFC performance, it is possible to modify the microstructure of the electrodes that also affects the performance of the system. In CNEA, powders and films prepared by different methods but with the same composition, that is, $La_{0.4}Sr_{0.6}Co_{0.8}Fe_{0.2}O_{3-\delta}$ (LSCFO), were studied (see Figure 26.14):

1. Acetic acid-based gel and hexamethylenetetramine (HTMA) routes to obtain powders with grain sizes ranging from 10 to 250 nm, depending on sintering

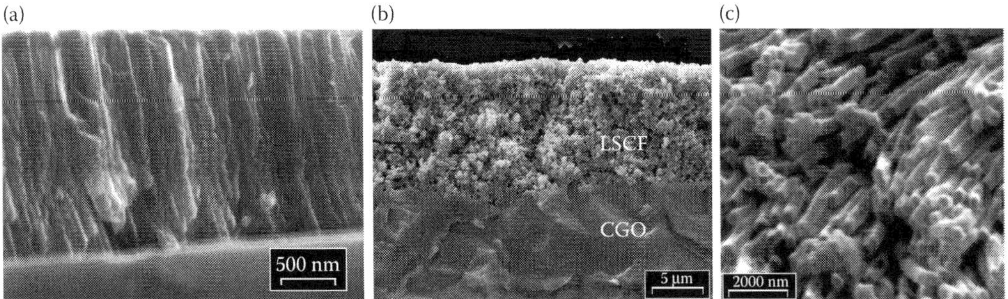

FIGURE 26.14
SEM images of LSCF films prepared by (a) PLD, (b) HMTA deposited by spin coating, and (c) nanostructured tubes.

conditions. Cathodes films from powders were deposited onto CGO (pressed $Ce_{0.9}Gd_{0.1}O_{1.95}$ disks) ceramic substrates by serigraphy, spin coating, and dip-coating [46,47]

2. Nanostructured cathode thin films with vertically aligned nanopores (VANP) processed using a pulsed laser deposition (PLD) technique deposited on various substrates (YSZ, Si, and CGO) [48]
3. Nanostructured tubes prepared by a porous polycarbonate membrane approach with tube diameters ranging from 100 to 500 nm and ~20 nm crystallite size [49]

The structure, morphology and composition of the powders and films were characterized by XRD, transmission and scanning electron microscopies, and energy-dispersive spectroscopy, respectively. The influence of the microstructure on the transport properties (ionic and electronic conductivities) was evaluated by means of impedance spectroscopy. It was found that electrochemical properties of LSFC cathodes films strongly depend on the micro/nanostructure that is mainly determined by synthesis parameters and technique used for the film deposition. Results indicate that perovskite oxide cobaltites (prepared by an acetic acid-based gel route or PLD) have excellent performance to be used as cathodes and their characteristics mainly depend on the microstructure but the composition [47].

Another focus of research is devoted to the study of new attractive materials to replace traditional compounds used in SOFCs and the study of the cathode reaction processes. Two main lines can be highlighted: mixed conductor oxides (MIEC: ionic and electronic conductors) as cathode material, and LAMOX family as electrolyte material.

MIEC such as the already mentioned $La_{1-x}Sr_xCo_{1-y}Fe_yO_{3-\delta}$ (LSCF), layered perovskites as the Ruddlesden–Popper families $Ln_2NiO_{4+\delta}$ (Ln = La, Nd, Pr), and $Sr_3FeMO_{6+\delta}$ (M = Fe, Co, Ni, Mn) or the double perovskites $LnBaCo_2O_{5+\delta}$ are good candidates for IT-SOFC cathodes. These materials present high ionic and electronic conductivity which should reduce the polarization resistance and so improve the cell performance at intermediate temperature. However, phase transitions, high values of thermal expansion coefficients, and chemical reaction could affect the chemical and thermomechanical compatibility with electrolyte and interconnection materials. Therefore, the study of high-temperature properties and the understanding of the oxygen reduction reaction are fundamental issues to develop new cathode materials.

Thus, for example, despite the high-quality of ionic and electrical conductivity of $SrCo_{0.8}Fe_{0.2}O_{3-\delta}$ compound, a cubic to orthorhombic (C/O) brownmillerite phase transition

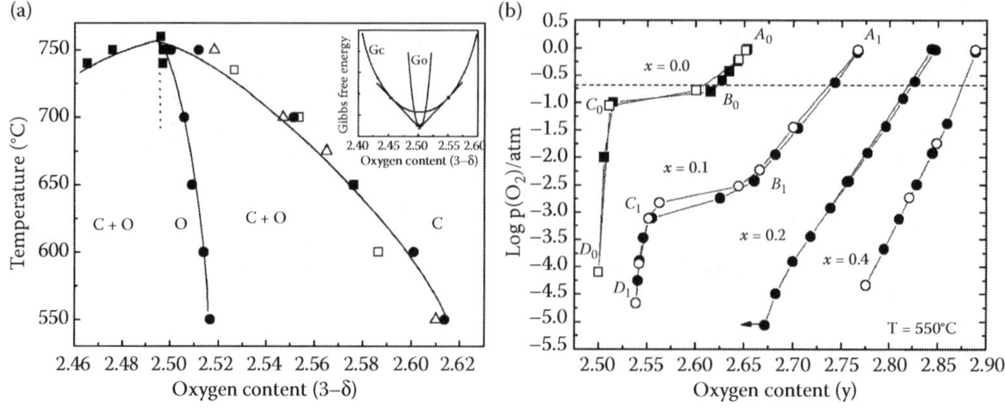

FIGURE 26.15
(a) Phase diagram of $SrCo_{0.8}Fe_{0.2}O_{3-\delta}$. The inset shows the schematic free energy-composition curves for the cubic and orthorhombic phases at a given T lower than 750°C; (b) Variations of the oxygen content with the equilibrium $p(O_2)$ at 550°C for the $Sr_{1-x}La_xCo_{0.8}Fe_{0.2}O_{3-\delta}$ ($0.0 \leq x \leq 0.4$) samples.

has been found (see Figure 26.15a) [50]. This phase transition affects the good performance of this material, however, the partial substitution of Sr by La and Fe by Co in $La_{1-x}Sr_xCo_{1-y}Fe_yO_{3-\delta}$ prevents the C/O phase transition (see Figure 26.15b) and improves the long-term stability of this material [51]. Then, the electrode reaction of porous $La_{1-x}Sr_xCo_{1-y}Fe_yO_{3-\delta}$ films deposited on CGO was investigated by complex impedance spectroscopy and experimental data provide evidence of which are the limiting steps of the oxygen-reduction reaction, which depend on the temperature and oxygen partial pressure (for details see [52]).

In the same way, the characterization of thermodynamic, oxygen permeation, electrical, and structural properties at high temperature of $Sr_3FeMO_{6+\delta}$ (M = Fe, Co, Ni, Mn) family could be related to their oxygen-reduction mechanism through the evaluation of its resistance of polarization [53,54].

Currently, an exploratory study is being performed on high-temperature properties of new mixed conductors $Ln_2NiO_{4+\delta}$ (Ln = La, Nd, Pr) and $LnBaCo_2O_{5+\delta}$ (Ln = Pr, Nd, Sm, Gd). The $Ln_2NiO_{4+\delta}$ has tetragonal crystal structure (space group I4/mmm or F4/mmm) consisting of alternating LnO rock salt and $LnNiO_3$ perovskite layers. Interstitial oxygen can be accommodating in their crystalline structure promoting the ionic conductivity, and the p-type electronic conductivity. The main advantage of these compounds are their thermal expansion coefficients (TEC ~ 13.0 10^{-6} K^{-1}), which are close to those of most widely used SOFC electrolytes, guarantying the thermomechanical compatibility between cell components. Despite this, we are finding that a chemical reaction between $La_2NiO_{4+\delta}$ and several electrolyte materials (YSZ, CGO) takes place affecting the long-term stability at high operation temperatures (>700°C). Thus, for example, the La rich electrode materials react with YSZ electrolyte leading to the formation of an insulating $La_2Zr_2O_7$ pyrochlore phase. However, this reaction may be precluded when La is substituted by Nd.

The discovery, a few years ago, of a new oxide-ion conductor, the lanthanum molybdate $La_2Mo_2O_9$ (LAMOX), has promoted great interest due to its high oxide-ion conductivity. LAMOX undergoes a structural phase transition at around 580°C, from an α-monoclinic form to a β-cubic one which exhibits higher conductivity than YSZ. The applicability of LAMOX materials as electrolytes for SOFCs is still quite limited since hexavalent

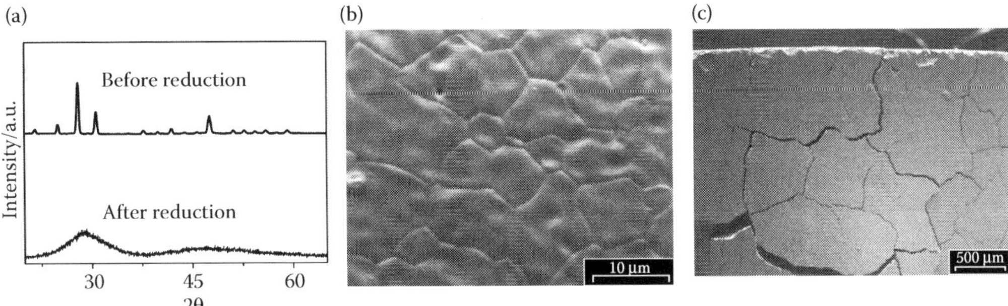

FIGURE 26.16
(a) XRD patterns at room temperature of a $La_2Mo_2O_9$ pellet before and after reduction; (b) and (c) SEM images of $La_2Mo_2O_9$ pellet before and after reduction, respectively.

molybdenum is easily reduced under low oxygen partial pressure pO_2. Regarding this fact, partial substitution of Mo^{6+} by W^{6+} increases the redox stability of LAMOX materials but, in contrast, tends to slightly decrease their ionic conductivity [55]. Research activities on the $La_2Mo_{2-x}W_xO_9$ ($x = 0$, 0.5, 1.0 *y* 1.3) compounds are mainly focused to determine the stability range of the β phase and reduced phases, $La_2Mo_{2-x}W_xO_9$, $La_7(Mo_{1-x}W_x)_7O_{30}$ and amorphous $La_7(Mo_{1-x}W_x)_7O_y$, as a function of temperature and oxygen partial pressure pO_2. This information is crucial in order to study the electrical transport properties of β and reduced phases (ionic and electronic conductivities) as a function of the oxygen content. For example, LAMOX phase was found to be unstable at 608°C under Ar–H_2 flowing atmosphere, and reduction leads to an amorphous phase with composition close to $La_2Mo_2O_{6.88}$ (see Figure 26.16).

26.8 Summary and Future Perspective

Based on domestic capabilities developed for more than 50 years in Argentina, related to both nuclear energy and hydrogen production and applications technologies, the production of hydrogen using high-temperature nuclear reactors is being seriously considered as a sustainable and environmentally friendly alternative for the long term. In accordance with this hydrogen energy roadmap, the national government is promoting all scientific activities related to the production, purification, safe storage, transportation, distribution, and applications of hydrogen for which the formation of qualified human resources and the scientific cooperation with other countries are greatly encouraged. These scientific activities are being promoted through a National Hydrogen Law dictated by the Argentine Congress in 2006, which declares of national interest the development of technologies needed for the progressive introduction of hydrogen as a clean fuel and energy carrier that can be used to meet increasing residential, commercial, and industrial demands.

In the framework of a Hydrogen Energy Program designed on the basis of the National Hydrogen Law, about 100 scientists and engineers are now working in CNEA on different issues of hydrogen technology. On the other hand, cooperation research agreements were established by CNEA with R&D groups of national universities like the Buenos Aires

University (UBA), the National Technological University (UTN), the National University of Comahue (UNCo), and the National University of Cuyo (UNC).

Related with nuclear energy supported hydrogen production, research activities currently underway in Argentina are focused on two processes: (1) coal gasification using carbon dioxide and steam; and (2) metallic chlorides water-splitting thermochemical cycles. Theoretical and experimental investigations are addressed to elucidate the kinetics and fundamental mechanisms of gasification and thermochemical reactions at laboratory scale, in order to find the optimum conditions for increasing the efficiency of these processes with the objective of a future scaling up of the experimental facilities.

Experimental results at laboratory scale have shown that the sub-bituminous Río Turbio coal appears to be very suitable for hydrogen production through gasification, since it is very reactive to gasifying agents, that is, the gasification occurs at quite low temperatures and with a fast kinetics, and the hydrogen can be easily separated from the synthesis gas with a double-layer ceramic separating membrane.

Water-splitting thermochemical cycles studied at 650°C have shown a promising response with iron chlorides and coupled iron–copper chlorides, both in the presence of an oxygen scavenger. Development of a continuous separating system for the separation of oxygen and hydrogen from the gaseous stream, constitute the next stage of the project with the objective of enhancing the hydrogen production and then improving the overall efficiency of these thermochemical cycles.

Safe storage and transportation of hydrogen are some of the determining steps in the feasibility of hydrogen uses in Argentina, taking into account that the hydrogen has to be transported over long distances which is cost intensive and would pose safety-related problems. In this sense, research activities in CNEA are focused on the use of metallic hydrides with a high efficiency in absorption/desorption processes which are being manufactured by novel manufacturing processes.

Even present hydrogen applications are mainly related with the chemical industry, the studies developed in CNEA are strictly associated with the generation of energy by means of two kinds of fuel cells: (a) PEM; and (b) SOFC, for operation at low and high temperatures, respectively. The diminishing of over potential in the cathode is one of the most important issue to be solved for the better performance of SOFC fuel cells and, in this way, the development of new materials for the cathode produced with innovative methods that allow modifying their microstructure will continue in the future. Main research activities planned to be executed in the framework of PEM fuel cell are focused on development of better membranes and the manufacturing of carbon supported nanophase catalysts using three methods of deposition. The main goal of these studies is to apply them in the construction and operation of a prototype PEM fuel cell with *in situ* hydrogen production by reforming of ethanol or other technologies.

References

1. Gaviría, J.P., Bohé, A.E., and Pasquevich, D.M., Hematite to magnetite reduction followed by Mössbauer spectroscopy and X-ray diffraction, *Physica B: Condensed Matter*, 389, 198–201, 2007.
2. Alvarez, F.J., Bohé, A.E., and Pasquevich, D.M., Comparative analysis of the chlorination of mixtures iron–aluminum and the binary alloy $FeAl_3$, *J. Alloys Compounds*, 424, 78–87, 2006.

3. Pasquevich, D.M., Gaviría, J.P., Esquivel, M.R., and Bohé, A.E., Intrinsic kinetics of the chlorination of hematite, *Metallurg. Trans. B*, 37, 589–597, 2006.
4. Bohé, A.E., Studies developed in Argentina for hydrogen production, storage and uses, *IAEA's Technical Meeting on Status of Hydrogen Production using Nuclear Energy*, Vienna, 1–3 September, 2008.
5. Bohé, A.E., Bosco, M., and Nassini, H.E.P., Studies on metallic chlorides thermochemical cycles using nuclear energy for hydrogen production in Argentina, *IAEA's Technical Meeting on Status of Hydrogen Production using Nuclear Energy*, Mumbai, India, October 12–14, 2009.
6. De Micco, G., Bohé, A.E., and Pasquevich, D.M., Kinetic study of the $\beta \rightarrow \alpha + \gamma$ transformation reaction in a CuZnAl alloy, *Int. J. Mater. Res.*, 97, 1093–1097, 2006.
7. De Micco, G., Bohé, A.E., and Pasquevich, D.M., A thermogravimetric study of copper chlorination, *J. Alloys Compd.*, 437, 351–359, 2007.
8. De Micco, G., Pasquevich, D.M., and Bohé, A.E., Chlorination of aluminum-copper alloys, *Thermochim. Acta*, 457, 83–91, 2007.
9. De Micco, G., Fouga, G.G., and Bohé, A.E., The chlorination of zinc oxide between 723 and 973 K. A kinetic model, *Metall. Mater. Trans. B*, 38, 6853–6862, 2007.
10. De Micco, G., Pasquevich, D.M., and Bohé, A.E., Interactions during the chlorination of a copper–zinc alloy, *Thermochim. Acta*, 470, 83–90, 2008.
11. Fouga, G.G., De Micco, G., and Bohé, A.E., The chlorination of manganese oxide, *Themochim. Acta*, 494, 141–146, 2009.
12. Gaviría, J.P. and Bohé, A.E., The kinetics of the chlorination of yttrium oxide, *Metall. Mater. Trans.* B, 40, 45–53, 2009.
13. Alvarez, F.J. and Bohé, A.E., Direct chlorination of nickel containing materials. Recovery of the metal from different sources, *Industrial Engineering and Chemistry Research*, 47, 8184–8191, 2008.
14. Rezaiyan, J. and Cheremisinoff, N., *Gasification Technologies: A Primer for Engineers and Scientists*, CRC Press, Taylor & Francis, Boca Raton, FL, 2005.
15. Higman, C. and van der Burgt, M., *Gasification*, Gulf Professional Publishing, Elsevier, New York, NY, 2003.
16. Yamashita, K. and Barreto, L., Energyplexes for the 21st century: Coal gasification for co-producing hydrogen, electricity and liquid fuels, *Energy*, 30, 2453–2473, 2005.
17. Minchener, A., Coal gasification for advanced power generation, *Fuel*, 84, 2222–2235, 2005.
18. Belghit, A. and El Issami, S., Hydrogen production by steam gasification of coal in gas–solid moving bed using nuclear heat, *Energy Conversion and Management*, 42, 81–99, 2001.
19. Collot, A., Matching gasification technologies to coal properties, *International Journal of Coal Geology*, 65, 191–212, 2006.
20. Yu, J., Lucas, J., and Wall, T., Formation of the structure of chars during devolatilization of pulverized coal and its thermoproperties: A review, *Progress in Energy and Combustion Science*, 33, 135–170, 2007.
21. Megaritis, A., Messenbock, R., Collot, A., Zhuo, Y., Dugwell, D., and Kandiyoti, R., Internal consistency of coal gasification reactivities determined in bench-scale reactors: Effect of pyrolysis conditions on char reactivity under high-pressure CO_2, *Fuel*, 77(13), 1411–1420, 1998.
22. Nassini, H.E.P., Gaviría, J.P., Fouga, G.G., De Micco, G., Venaruzzo, J., and Bohé, A.E., Desarrollo de tecnologías avanzadas para el aprovechamiento integral del carbón de Río Turbio en la generación de hidrógeno, Primer Congreso Nacional: Hidrogeno y Fuentes sustentables de Energía, 8–10 June, 2005, Bariloche, Rio Negro, Argentina (in Spanish).
23. Pasquevich, D., and Caneiro, A., A thermogravimetric analyzer for corrosive atmospheres and its application to the chlorination of ZrO_2-C mixtures, *Thermochim. Acta*, 156(1), 275–283, 1989.
24. De Micco, G., Gorena, C., Fouga, G., and Bohé, A.E., Coal gasification studies applied to hydrogen production, *Int. J. Hydrogen Energy*, 35, 6012–6018, 2010.
25. Mottern, M., Shqau, K., Shi, J., Yu, D., and Verweij, H., Thin supported inorganic membranes for energy-related gas and water purification, *Int. J. Hydrogen Energy*, 32, 3713–3723, 2007.
26. Dolan, M., Dave, N., Ilyushechkin, A., Morpeth, L., and McLennan, K., Composition and operation of hydrogen-selective amorphous alloy membranes, *J. Membr. Sci.*, 285, 30–55, 2006.

27. Lin, Y.S., Microporous and dense inorganic membranes: current status and prospective, *Separation Purif. Technol.*, 25, 39–55, 2001.
28. Kusakabe, K., Yamamoto, M., and Morooka, S., Gas permeation and micropore structure of carbon molecular sieving membranes modified by oxidation, *J. Membr. Sci.*, 149, 59–67, 1998.
29. Paglieri, S. and Way, J., Innovations in palladium membrane research, *Separation Purif. Methods*, 31, 1–169, 2002.
30. Yamaura, S., Shimpo, Y., Okouchi, H., Nishida, M., Kajita, O., and Inoue, A., Hydrogen permeation characteristics of melt-spun Ni–Nb–Zr amorphous alloy membranes, *Mater. Trans.*, 44, 1885–1890, 2003.
31. Tosti, S., Supported and laminated Pd-based metallic membranes, *Int. J. Hydrogen Energy*, 28, 1445–1454, 2003.
32. Rodríguez, D., and Meyer, G., Improvement of the activation stage of MmNi4.7Al0.3 hydride forming alloys by surface fluorination, *J. Alloys Compounds*, 293–295, 374–378, 1999.
33. Urretavizcaya, G. and Meyer, G., Metastable hexagonal Mg 2Sn obtained by mechanical alloying, *J. Alloys Compounds*, 334, 211–215, 2002.
34. Gennari, F., Urretavizcaya, G., Andrade Gamboa, J., and Meyer, G., New Mg-based alloy obtained by mechanical alloying in the Mg–Ge–Ni system, *J. Alloys Compounds*, 354, 187–192, 2003.
35. Cuscueta, D., Ghilarducci, A., Salva, H., and Peretti, H., Low cobalt content alloy for Ni-MH battery electrodes, *J. New Mater. Electrochem. Systems*, 10, 213–216, 2007.
36. Ruiz, F., Castro, E., Real, S., Peretti, H., Visintin, A., and Triaca, W., Electrochemical characterization of AB_2 alloys used for negative electrodes in Ni/MH batteries, *Int. J. Hydrogen Energy*, 33, 3576–3580, 2008.
37. Visintin, A., Peretti, H., Ruiz, F., Corso, H., and Triaca, W., Effect of additional catalytic phases imposed by sintering on the hydrogen absorption behavior of AB_2 type Zr-based alloys, *J. Alloys Compounds*, 428(1–2), 244–251, 2007.
38. Corti, H., Nores-Pondal, F., and Buera, M., Thermal properties of Nafion 117 membranes in water and methanol–water mixtures, *J. Power Sources*, 161, 799–805, 2006.
39. Díaz, L., Abuin, G., and Corti, H., Water and phosphoric acid uptake of poly [2,5-benzimidazole] (ABPBI) membranes prepared by low and high temperature casting, *J. Power Sources*, 188, 45–50, 2009.
40. Franceschini, E., and Corti, H., Elastic properties of Nafion, PBI and poly [2,5-benzimidazole] (ABPBI) membranes determined by AFM tip nano-indentation, *J. Power Sources*, 188, 379–386, 2009.
41. Nores-Pondal, F.J., Vilella, I.M.J., Troiani, H., Granada, M., de Miguel, S.R., Scelza, O.A., and Corti, H.R., Catalytic activity vs. size correlation in platinum catalysts of PEM fuel cells prepared on carbon black by different methods, *Int. J. Hydro. Energy*, 34, 8193–8203, 2009.
42. Franceschini, E., Bruno, M., Planes, G., and Corti, H., Electrodeposited platinum catalysts over hierarchical carbon monolithic support, *J. Appl. Electrochem.*, 40, 257–263, 2010.
43. Bruno, M., Corti, H., and Barbero, C., Hierarchical porous materials: Capillaries in mesoporous carbon, *Funct. Mater. Lett.*, 2, 135–138, 2009.
44. Teraoka, Y., Zhang, H.M., Okamoto, K., and Yamazoe, N., Mixed ionic-electronic conductivity of $La_{1-x}Sr_xCo_{1-y}Fe_yO_{3-\delta}$ perovskite-type oxides, *Mater. Res. Bull.*, 23, 51–58, 1988.
45. Baqué, L., Serquis, A., Grunbaum, N., Prado, F., and Caneiro, A., Preparation and characterization of solid oxide fuel cells cathode films, *Mater. Res. Soc. Symp. Proc.*, 928, GG16–03, 2006.
46. Baqué, L. and Serquis, A., Microstructural characterization of $La_{0.4}Sr_{0.6}Co_{0.8}Fe_{0.2}O_{3-\delta}$ thin films deposited by dip coating, *Appl. Surf. Sci.*, 254, 213–218, 2007.
47. Baqué, L., Caneiro, A., Moreno, M.S., and Serquis, A., High performance nanostructured IT-SOFC cathodes prepared by novel chemical method, *Electrochem. Commun.*, 10, 1905–1908, 2008.
48. Yoon, J.S., Araujo, R., Grunbaum, N., Baqué, L., Serquis, A., Caneiro, A., Zhang, X., and Wang, Y., Nanostructured cathode thin films with vertically-aligned nanopores for thin film SOFC and their characteristics, *Appl. Surf. Sci.*, 254, 266–269, 2007.

49. Napolitano, F., Baqué, L., Troiani, H., Granada, M., and Serquis, A., Synthesis and characterization of cobaltite nanotubes for solid-oxide fuel cell cathodes, *J. Phys. Conf. Ser.*, 012042 (5pp), 2009.
50. Grunbaum, N., Mogni, L., Prado, F., and Caneiro, A., Phase equilibrium and electrical conductivity of $SrCo_{0.8}Fe_{0.2}O_{3-\delta}$, *J. Solid State Chem.*, 177(7), 2350–2357, 2004.
51. Prado, F., Grunbaum, N., Caneiro A., and Manthiram, A., Effect of La3+ doping on the perovskite-to-brownmillerite transformation in $Sr_{1-x}La_xCo_{0.8}Fe_{0.2}O_{3-\delta}$ $(0 < x < 0.4)$, *Solid State Ionics*, 167, 147–154, 2004.
52. Grunbaum, N., Dessemond, L., Fouletier, J., Prado, F., and Caneiro A., Electrode reaction of $Sr_{1-x}La_xCo_{0.8}Fe_{0.2}O_{3-\delta}$ with $x = 0.1$ and 0.6 on $Ce_{0.9}Gd_{0.1}O_{1.95}$ at $600 \leq T \leq 800°C$, *Solid State Ionics*, 177, 907–913, 2006.
53. Mogni, L., Prado, F., Cuello, G., and Caneiro, A., Study of the crystal chemistry of the $n = 2$ ruddlesden-popper phases $Sr_3FeMO_{6+\delta}$ (M = Fe, Co, and Ni) using *in situ* high temperature neutron powder diffraction, *Chem. Mater.*, 21, 2614–2623, 2009.
54. Mogni, L., Prado, F., and Caneiro, A., Electrochemical characterization of the $n = 2$ ruddlesden-popper $Sr_3FeMO_{6+\delta}$ (M = Fe, Co, Ni) phases by electrochemical impedance spectroscopy, *ECS Trans.*, 6, 233–243, 2008.
55. Georges, S., Bohnk´e, O., Goutenoire, F., Laligant, Y., Fouletier, J., and Lacorre, P., Effects of tungsten substitution on the transport properties and mechanism of fast oxide-ion conduction in $La_2Mo_2O_9$, *Solid State Ionics*, 177, 1715–1720, 2006.

27

Nuclear Hydrogen Production Development in China

Jingming Xu, Ping Zhang, and Bo Yu

CONTENTS

27.1 Introduction

China has enjoyed a rapid growth of economy since 1980s. This trend of growth can be expected to continue for some decades. In the mean time, a greater energy demand and environmental problem caused by burning larger quantity of fossil fuels will pose significant challenges to a sustainable development and economical expansion in China. During the period 2000–2007, China's average annual growth rate of energy consumption was 8.9% and that of electricity consumption was as high as 13.0%. In 2007, China's total energy consumption was 2.66 billion tce (ton-coal equivalent). The primary energy production was 2.23 billion tce. The sources of primary energy production (as coal equivalent calculation) was 76.63% raw coal, 11.31% crude oil, 3.91% natural gas, 7.24% hydro power, and 0.91% nuclear power [1]. Such primary energy portfolios resulted in large amounts of SO_2 and CO_2 atmospheric emissions. In 2006, the emission of CO_2 was 5.61 billion tons [2] while that of SO_2 from industry sector was 22.35 million tons [3].

To meet the formidable energy and environmental challenges, China is formulating a clean energy strategy that includes nuclear energy and renewable energy of wind, sun, and so on. In the recent years, China's nuclear electricity production has been increasing with a present total of 9.11 GWe installed capacity [4]; new PWRs with a total capacity of over 10 GWe are being constructed. Future construction of Gen-III+ PWR (AP1000 and EPR) are planned. According to the "State Medium–Long Term (2005–2020) Development

Program of Nuclear Power" published in October 2007, the total capacity of operating nuclear power plants will be 40 GWe with additional 18 GWe under construction by 2020. Considerable increase in the application of nuclear energy will greatly improve China's primary energy mix and effectively improve air quality.

Hydrogen is a clean energy carrier. China is taking a very active approach to developing hydrogen energy technology related to production, storage, and application of hydrogen. In the "Tenth Five-Year Plan (2001–2005)," funding for the electric vehicle (EV) & hydrogen and fuel cell (H/FC)-related programs accounted for up to 40% of total energy research budget. It focuses mainly on basic and technical aspects of hydrogen energy and related demonstration projects [5] such as fuel cell city bus, refueling station and Hydrogen Park. An example is the successful service provided by hydrogen powered automobiles during the 2008 Beijing Olympic Games.

Since 1970s, the high-temperature gas-cooled reactor (HTGR) technology has been under development in China. The 10 MWt test reactor (HTR-10) with spherical fuel elements was constructed in 2000 and is now operational. A number of safety-related experiments have been conducted on the HTR-10. The research and development (R&D) on direct cycle helium turbine technology is being carried out. Coupling a helium turbine system to the reactor is foreseen [6]. The construction of commercial demonstration plant of HTR-PM is one of the National Major Science and Technology Special Projects. The construction of 200 MWe HTR-PM will be completed around 2013. Its design, construction and R&D work has been started.

Nuclear hydrogen (NH) production is a promising way for industrial scale production comparing with other developing production methods. Among all nuclear reactors, the HTGR is most suitable for NH process due to its potential of high-efficiency electricity generation and providing high-temperature process heat. Therefore, R&D on NH, as a part of the HTR-PM project, has started in the Institute of Nuclear and New Energy Technology (INET), Tsinghua University. The I–S thermo-chemical cycle for splitting water and the high-temperature steam electrolysis (HTSE) are selected as potential processes of NH production. Since 2005, INET has conducted preliminary study on the iodine–sulfur (IS) process and HTSE process. The laboratory of NH with facilities for process studies has been established. The HTR-10 constructed in INET will provide a suitable nuclear reactor facility for future research and development of NH production technology.

Development of NH in China will proceed in three phases:

Phase I (2008–2014):	Process verification of NH process and bench-scale test
Phase II (2015–2020):	R&D on coupling technology with reactor, NH production safety, and pilot-scale test
Phase III (2020~):	Commercialization demonstration of NH production

The current R&D activities on NH in INET are focused on achieving the target of the first phase above.

27.2 Research and Development Program on IS Process

27.2.1 Program Plan for IS process

The IS process is generally considered as one of the best thermo-chemical water-splitting processes for hydrogen production, and it has been widely investigated in many institutes across the world, including INET. According to the general schedule, the final objective of

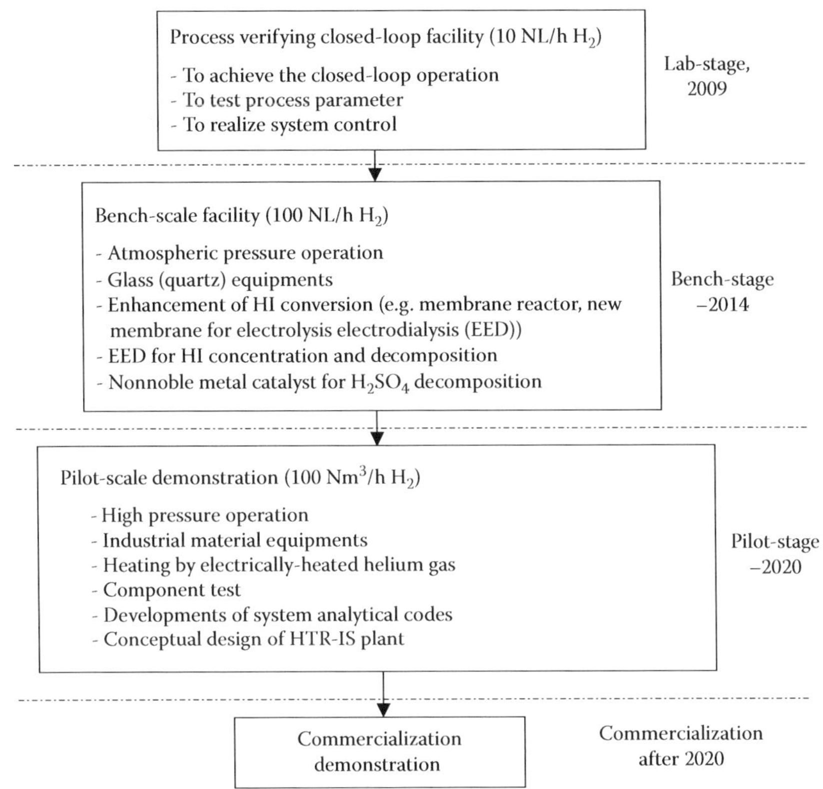

FIGURE 27.1
R&D plan of NH production through IS process in INET.

the Chinese NH program is to achieve the commercialization of NH production after 2020. With this consideration, a four-stage plan has been formulated as shown in Figure 27.1, in which the features and main research topics in the different stages are shown.

Currently, the R&D activities involve in (1) fundamental research on the three chemical reactions, that is, Bunsen reaction, hydriodic (HI) acid decomposition, and sulfuric acid decomposition, and related technologies and (2) design, construction, and operation of a closed-loop facility for verifying the process.

27.2.2 Fundamental Research

The fundamental research has focused on the topics and the produced major results as described in the following [7–10]:

1. *Process study on Bunsen reaction and separation characteristics of sulfuric acid phase and HIx phase.* For Bunsen reaction, the involved reactions are assumed as follows:

$$SO_2 + H_2O = H_2SO_3 \leftrightarrow H^+ + HSO_3^- \leftrightarrow 2H^+ + SO_3^{2-} \tag{27.1}$$

$$I_2 + SO_3^{2-} + H_2O \leftrightarrow 2H^+ + I^- + SO_4^{2-} \tag{27.2}$$

$$\left(\frac{x-1}{2}\right)I_2 + HI \leftrightarrow HI_x \tag{27.3}$$

The products of the Bunsen reaction, that is, H^+, I^-, and SO_4^{2-} in the solution are formed stoichiometrically, that is, $[H^+] = 2[SO_4^{2-}]+[I^-]$; the polyiodide ions are negligible.

The separation characteristics of the two phases, sulfuric acid and polyhydriodic acid (HI_x), formed were investigated. The effects of temperature of the solution and molar fraction of iodine in the mixture of H_2SO_4 and HI_x on the separation characteristics were studied. The influence of the molar ratio of iodine to water in the Bunsen reaction on the phase separation and saturation of I_2 was studied. The results show that addition of excess iodine causes formation of two insoluble liquid phase, the amount of excess iodine increases with the increasing of temperature and reaches the highest at 60°C. The separation characteristics are notably affected by molar ratio of iodine to water ($n(I_2)/n(H_2O)$). If $n(I_2)/n(H_2O) < 1/8$, the two acids cannot be separated; if the $n(I_2)/n(H_2O) > 1/5$, iodine in the products may solidify at room temperature.

2. *Purification of the two phases by reverse Bunsen reaction.* Because the separated H_2SO_4 and HI_x phases are cross-contaminated, side reactions shown as formulae 27.4 and 27.5 will occur, causing negative after-effects. Therefore, it is necessary to purify the two acids.

$$H_2SO_4 + 6HI = S + 3I_2 + 4H_2O \tag{27.4}$$

$$H_2SO_4 + 8HI = H_2S + 4I_2 + 4H_2O \tag{27.5}$$

The purification process is carried out by reverse Bunsen reaction shown in Equation 27.6.

$$H_2SO_4 + 2HI = SO_2 + I_2 + 2H_2O \tag{27.6}$$

The effects of temperature, flow rate of carrier gas, and feed rate on the purification (denoted with removal efficiency of impurities) were investigated. As Bunsen reaction is an exothermic one, raising temperature will be beneficial to the reverse reaction. In addition, a gaseous product, SO_2 forms in the reverse Bunsen reaction, therefore, removal of the gas will enhance the reverse reaction. These assumptions were confirmed by experiments. The results show that the two acid phases could be purified with higher removal efficiency of impurities under suitable conditions, such as higher temperature and suitable carrier gas flow rate.

3. *Preconcentration of HI acid by electro-electrodialysis (EED) to exceed azeotropic composition, and concentration of HI acid by conventional distillation.* The concentration of HI acid in HIx phase produced in Bunsen reaction is about 10 mol/kg H_2O, closing to the azeotropic composition of HI acid solution, which is $HI:H_2O = 1:5$ (molar ratio), that is, $[HI] = \sim 11.1$ mol/kg. Therefore, it is difficult to concentrate HI solution by conventional distillation. The EED technique was employed to concentrate HI solution, the influences of different materials, such as electrode and membrane, and operational parameters, such as temperature and initial composition of HI acid, on the concentration effects were investigated. Results show that HI acid could be effectively concentrated through EED, and the concentration of HI could exceed the azeotropic composition. Higher operation temperature leads to lower cell volt, which is beneficial to lower the energy consumption. However, the concentration efficiency will be lower as temperature increases.

4. *Catalytic decomposition of HI acid; development of efficient catalyst/supporter and novel preparation method, such as electro-plating of Pt on various supporters.* HI decomposition reaction is usually catalyzed by Pt catalysts, which were prepared by an impregnation–calcination method or impregnation–H_2 reduction method. At INET, Pt/supporters catalysts were prepared by various methods, including impregnation–H_2 reduction at high temperature, impregnation–calcination at high temperature, impregnation–hydrazine reduction at room temperature, and Pd-inducing electroless plating, and so on. Much attention was paid on the Pd-inducing electroless plating method. The prepared catalysts were characterized by x-ray diffraction (XRD), transmission electron microscopy (TEM), Brunauer–Emmett–Teller adsorption (BET), and so on, and their catalytic performances were evaluated in a fixed bed reactor. The catalytic activities of the catalyst prepared by the Pt-inducing electroless plating method presents best activity on the HI decomposition reaction in the temperature range tested.
5. *Catalytic decomposition of sulfuric acid on Pt and non-Noble metal catalysts, including Fe/Cu oxides and Cr/Cu oxides.* In this section, our work focused on the development of non-Pt catalysts. Two composite metal oxides, $CuFe_2O_4$ and $CuCr_2O_4$, were prepared by sol–gel, vacuum freeze drying (VFD), and following calcination. These oxides were characterized by XRD, TEM, and BET analyses; and their catalytic performances to the decomposition reaction of SO_3 were evaluated in a fixed bed reactor, the results show that both copper ferrite and copper chromate show catalytic activities close to that of Pt/Al_2O_3. However, the stability and lifetime of these oxides need to be further explored.
6. *Correlation between density and composition of $HI/I_2/H_2O$ and $HI/H_2O/I_2/H_2SO_4$ system.* The relationship between the solution density and the concentrations of HI, I_2, and H_2O was investigated, expecting to develop a convenient method for the analysis of HIx phase. For $HI–I_2–H_2O$ solution, degree of freedom is 2 at fixed pressure and temperature (1 atm and 20°C in our work), according to the Gibbs phase law. Therefore, once solution density and the concentration of proton (the concentration of HI) are measured by density meter and titration, respectively, the concentration of iodine and water could be calculated directly. A linear regression is conducted based on the following equation,

$$\rho = C_{HI}M_{HI} + C_{I2}M_{I2} + C_{H_2O}M_{H_2O},$$

where ρ represents density (g/L), C_i represents concentration of component *i* (mol/L), and Mi is the molar mass of component *i*. Applying the obtained equation,

$$\rho = 0.9710 + 0.0967{*}C_{HI} + 0.2038{*}C_{I2}$$

C_{I2} can be easily determined, and the relative error between the experimental value and the calculated value achieved is within 1.0%.

27.2.3 Closed-Loop Process Verification Operations

To verify the data obtained from the fundamental studies and to acquire the operating experience of the closed-loop facility, a process verifying facility (IS-10) has been designed and established at INET. The main specifications of the facility are shown in Table 27.1.

TABLE 27.1

Main Specifications of IS-10

Specifications	Contents
Capacity	Hydrogen production rate 10 NL/h
Main parts	Bunsen reactor, H_2SO_4 decomposition reactor, HI decomposition reactor, EED, control system, pumps, and so on.
Heating	Electricity
Materials	Quartz glass, B–Si glass, Teflon
Size	3000 mm (*L*) × 2000 mm (*W*) × 2000 mm (*H*)
Features	• Automated temperature control and record • Liquid/gas flow rate control/record (diaphragmatic metric pump, mass flow controller (MFC)) • Multirunning mode (section continuous, open cycle, closed cycle) • Protect I_2 from solidifying

Figure 27.2a and b shows the simplified flow sheet and the photo of IS-10, which consists of the three main sections and the control device.

Many experiments have been done on IS-10, including continuous unit (refers to each step of a section) operation, continuous section (Bunsen section, HI section, and sulfuric acid section) operation, open-loop continuous operation (Bunsen section plus HI section, Bunsen section plus sulfuric acid experiment), and finally, the closed-loop experiment has been carried out; some of the preliminary results are summarized as follows:

1. The closed-loop experiment lasted for 8 h, HI, and H_2SO_4 sections run in continuous mode, while Bunsen section run in batch mode. All temperatures and rates were kept constant.
2. The hydrogen production rate was almost stable at 10 NL/h, and the hydrogen and oxygen production ratio was approximately 2:1. The compositions of Bunsen reaction products, which were taken as the main standard for the evaluation of the operating stability, were almost constant after three cycles.
3. Most of the reactors and processes were successfully operated. For purification of the two acids, 95% above impurities were removed. For the decomposition of sulfuric acid and HI acid, the conversion yield reached to 75 and 20% above, respectively.
4. However, under some unoptimized conditions, problems such as clogging caused by iodine and formation of sulfur arose, leading to mass unbalance or failure of long-time operation. To achieve longer and stable operation, we need further investigation.

27.3 Research and Development Program on HTSE Process

27.3.1 Program Schedule

The R&D on HTSE development are divided into four stages: (1) from 2008 to 2009, construction of HTSE test facilities and process verification, (2) from 2010 to 2012, bench-scaled

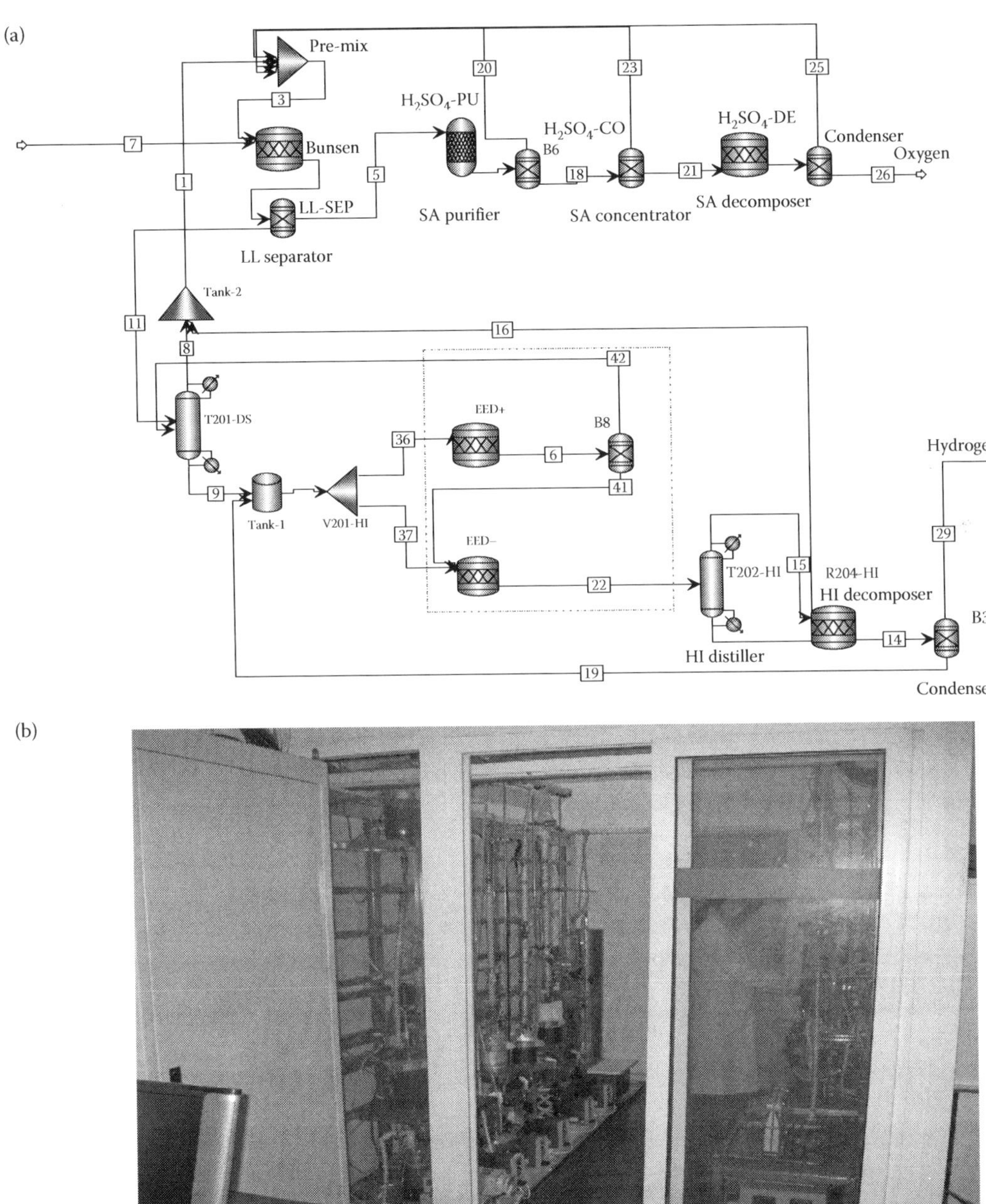

FIGURE 27.2
(a) Simplified flow sheet and (b) lab-scale installation of IS-10 for the IS closed-loop process verification operation at INET.

experimental study with hydrogen production yield of 60 L/h, (3) from 2013 to 2020, the design of pilot-scaled equipments and the pilot scale test with a hydrogen production yield of 5 Nm^3/h as well as R&D on the coupling technology with HTGR, and (4) commercial demonstration after 2020 [11].

Currently, the research activities are mainly focused on (1) demonstration of the feasibility of using planar solid oxide electrolyser cell (SOEC) technology for high-temperature electrolysis; (2) development of new materials with corrosion-resistant and high-performance HTSE; (3) analysis of the degradation mechanisms of SOEC cells used in HTSE mode; (4) HTSE cell and stack optimization; and (5) system design studies to support cycle life assessment and cost analysis for the HTSE plant.

27.3.2 Research and Development on HTSE

The research and development of HTSE technology was initiated at INET in 2005. In the past four years, researchers mainly conducted on preliminary investigation, feasibility study, equipment development, and fundamental researches. Currently, two testing systems, one is for HTSE cell online testing and another is for high-temperature electrochemical performance evaluation of SOEC components have been designed and constructed as shown in Figure 27.3 [12]. In addition, the research on novel anode materials has obtained excellent results. The theoretical analysis of hydrogen production efficiency of HTSE coupled with HTGR has been carried out [13].

27.3.2.1 Study on Conventional Planar LSM-SOEC System

The lab-scale hydrogen production on conventional LSM (lanthanum strontium manganite)-SOEC system was investigated. The electrolyte layer made of yttria-stabilized zirconia (YSZ) (containing 8% mol% of Y_2O_3) was sandwiched between the porous cathode (Ni/YSZ) and the anode layer (LSM). Under the same current density, the overpotential voltage decreases and the electrolysis performance improves with increasing temperature. When the input voltage is 1.0 V and the temperature is 850°C, the hydrogen production density is 0.315 mL/min cm^2. When the voltage increases to 1.3 V, the hydrogen production density increases to 0.98 mL/min cm^2 correspondingly [14].

Area-specific resistance (ASR) is one of the most important characteristic parameters in measuring the electrolysis performance of SOEC for hydrogen production. Testing results of LSM electrodes under SOEC and SOFC modes show that ASR value of a Ni-YSZ/YSZ/

(a) (b)

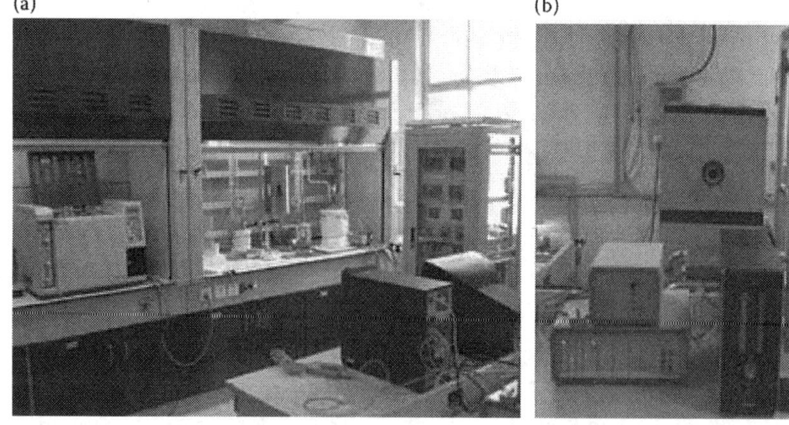

FIGURE 27.3
(a) HTSE online testing system and (b) material electrochemical performance evaluation system.

LSM cell was only 0.76 Ω cm^2 while it increased about five times, that is, 3.7 Ω cm^2 when operating in SOEC mode. Therefore, we can see that although HTSE is essentially a reverse process of SOFC in principle, the conventional materials of SOFC are not suitable for operation in SOEC mode.

27.3.2.2 Development of Novel Anode Materials with Low ASR

The feasibility of novel conductive membrane $Ba_xSr_{1-x}Co_{0.8}Fe_{0.2}O_{3-\delta}$ used as the oxygen electrode of SOEC was studied from several aspects of the Goldschmidt tolerance factor, critical radius, lattice free volume, average bond energy, and variable valence capability of B elements, as shown in Figure 27.4a and b. A strategy for the systematic selection of oxygen electrode material was explored and the optimum combination of the A site ($x = 0.5$), that is, $Ba_{0.5}Sr_{0.5}Co_{0.8}Fe_{0.2}O_{3-\delta}$(BSCF) was selected out.

ASR can directly represent the level of electrochemical performance of BSCF oxygen electrodes. The lower the ASR value, the higher the performance of the anode electrode, which means the stronger the oxygen permeability and conductivity of BSCF. Compared with other oxygen electrode materials, ASR data of the electrode BSCF/YSZ are 0.66 Ω cm^2 at 750°C, 0.27 Ω cm^2 at 800°C, and only 0.077 Ω cm^2 at 850°C, remarkably lower than the commonly used oxygen electrode materials LSM as well as the current focused materials strontium doped lanthanum cobaltite (LSC) and lanthanum strontium cobalt iron oxide (LSCF), as shown in Figure 27.5.

Figure 27.6 shows the hydrogen production rate of SOEC prepared by BSCF and LSM anodes at various electrolysis voltages (half-cell with the cathode and the electrolyte of YSZ/Ni-YSZ is kindly supplied by Shanghai Institute of Ceramics, Chinese Academy of Sciences) under the same current density of 300 mA, respectively. From Figure 27.6, it can be seen that the hydrogen production rate of both the BSCF/YSZ/Ni-YSZ cell and the LSM/YSZ/Ni-YSZ cell increases with the increasing electrolysis voltage. When the voltage is up to 1.4 V, the hydrogen production rate of the BSCF-cell is 147.2 mL $cm^{-2}h^{-1}$, about three times as that of the LSM-cell (about 49.8 mL $cm^{-2}h^{-1}$), which indicates that BSCF could be a potential candidate for the application of SOEC anode [15].

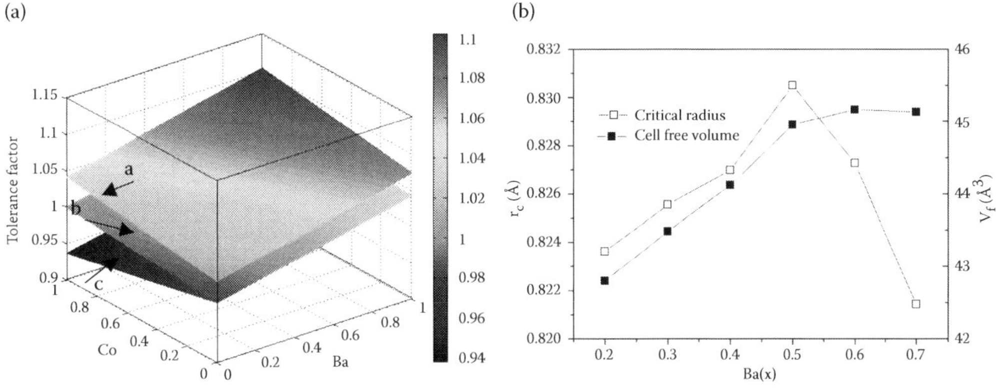

FIGURE 27.4
(a) Calculated results of the tolerance factor of $Ba_{1-x}Sr_xCo_{1-y}Fe_yO_{3-\delta}$ and (b) critical radius and cell free volume of $Ba_xSr1-_xCo_{0.8}F_{e0.2}O_{3-\delta}$ with various Ba contents (b).

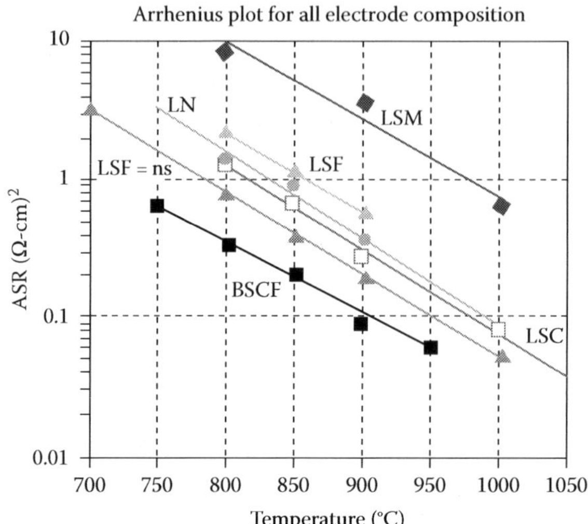

FIGURE 27.5
ASR of BSCF in comparison with other oxygen electrodes.

27.3.2.3 Microstructure Control of Cathode

Previous analysis indicated that the coarsening and the oxidation of nickel particles as well as the diffusion of steam were the limited step in the whole electrolysis reaction. As gas permeability and electrical conductivity of SOEC cathodes are strongly dependent on the cathode microstructure, the reasonable control of the microstructure is crucial for the optimization of the electrochemical performance of the cathode.

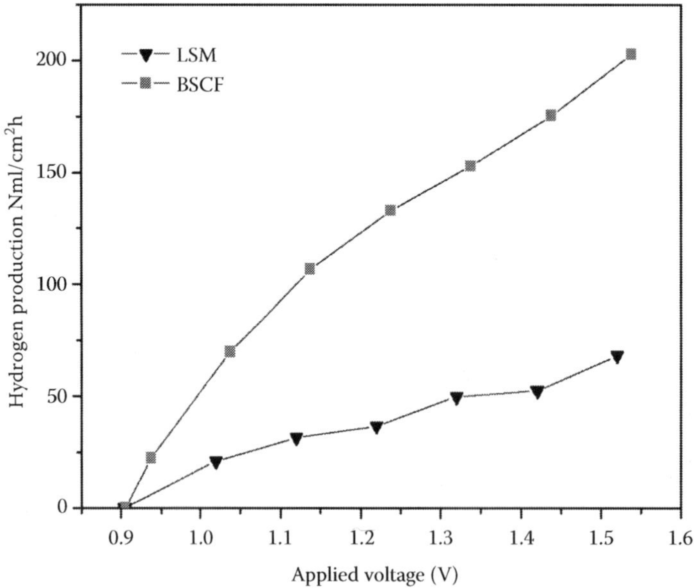

FIGURE 27.6
Hydrogen production rate of SOEC prepared by BSCF and LSM anodes at various electrolysis voltages, respectively.

In order to enhance performance, the study of hydrogen electrodes focuses on two aspects: (1) prepare the materials of the functional layer with the microstructure as fine as possible via a novel *in situ* coating combustion method to extend the length of three-phase-boundary (TPB) and (2) screen novel pore formers to get suitable porosity, pore shape, and distribution of supporting cathodes for SOEC application.

Nanosized NiO powder on submicron-sized YSZ particles of the functional layer was prepared via the *in situ* coating combustion method. XRD and field emission scanning electron microscope (FESEM) analyses showed that the products were well crystallized with NiO coating on YSZ particles. The optimized addition ratio of $CO(NH_2)_2$ to $Ni(NO_3)_2$ was 2:1. An SOEC single cell made from NiO-YSZ with the molar ratio of 2:1 composite powder exhibited better performance than the other samples with the electrolytic voltage of 0.98 V and showed excellent durability (zero degradation) under an electrolytic current density of 0.33 A/cm^2, an input stream composition of 80% H_2O + 20% H_2 and a temperature of 900°C for 50 h (see Figure 27.7) [16].

Four different pore formers, including polymethyl methacrylate (PMMA), potato starch, ammonium oxalate, and ammonium carbonate, were considered for the optimization. Their influences on the amount of porosity and on the pore shape and distribution as well as the effect on the electronic conductivity were analyzed. The results showed that PMMA was the most promising pore former, which had high porosity and uniform pore size distribution. The optimum weight percent concentration was 10%; correspondingly, porosity was 45% and electronic conductivity was 6726 S cm^{-1}, which was suitable for supporting cathodes for SOEC application.

27.3.2.4 Design of the Stack and the HTSE Stack Online Testing System

Figure 27.8 shows the design of the HTSE stack online testing system. The whole system mainly consists of three parts: measure and control part, gas loop part, and hydrogen monitoring part. Figure 27.9 shows the design of the planar stack. The short stack is designed to be composed of three planar individual cells with a hydrogen production rate of 1 L/h. Table 27.2 indicates some of the cell configuration details that have been adopted for this conceptual design. Figure 27.10 are the photographs of the running testing system and the assembled stack.

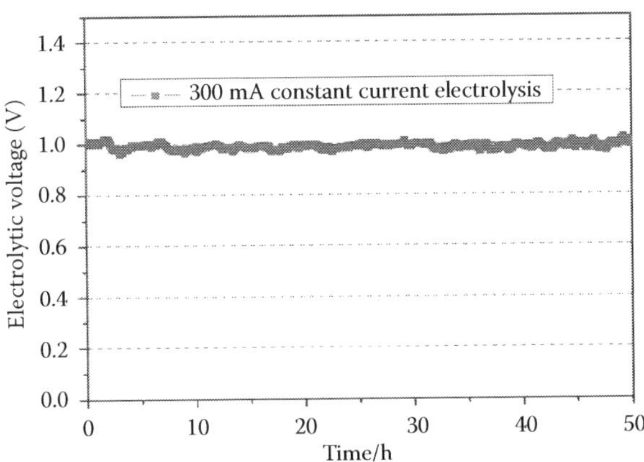

FIGURE 27.7
Stability of the single button cell.

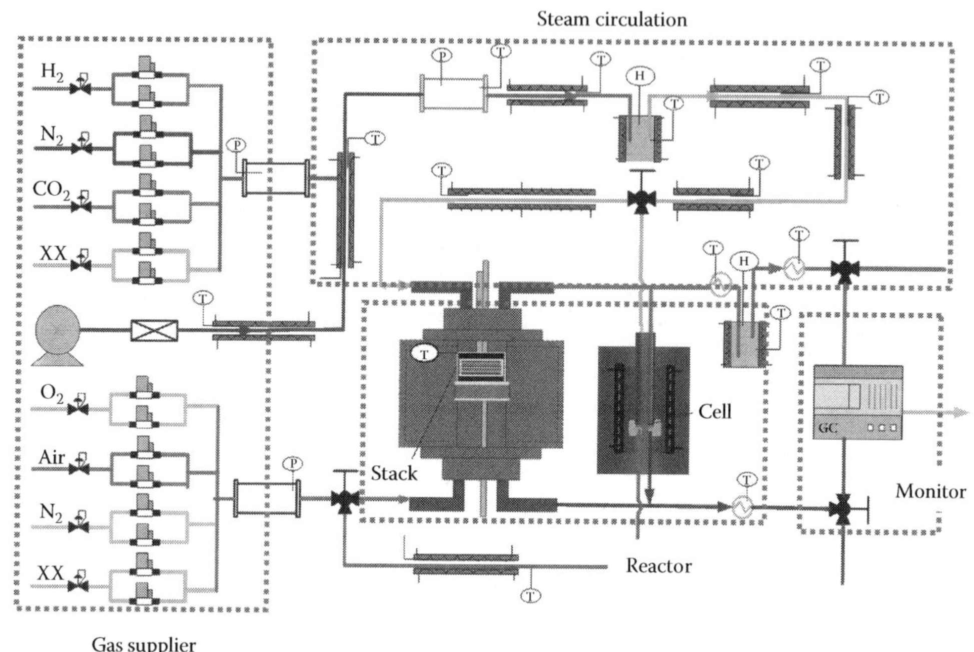

FIGURE 27.8
Design of the modular HTSE testing loop.

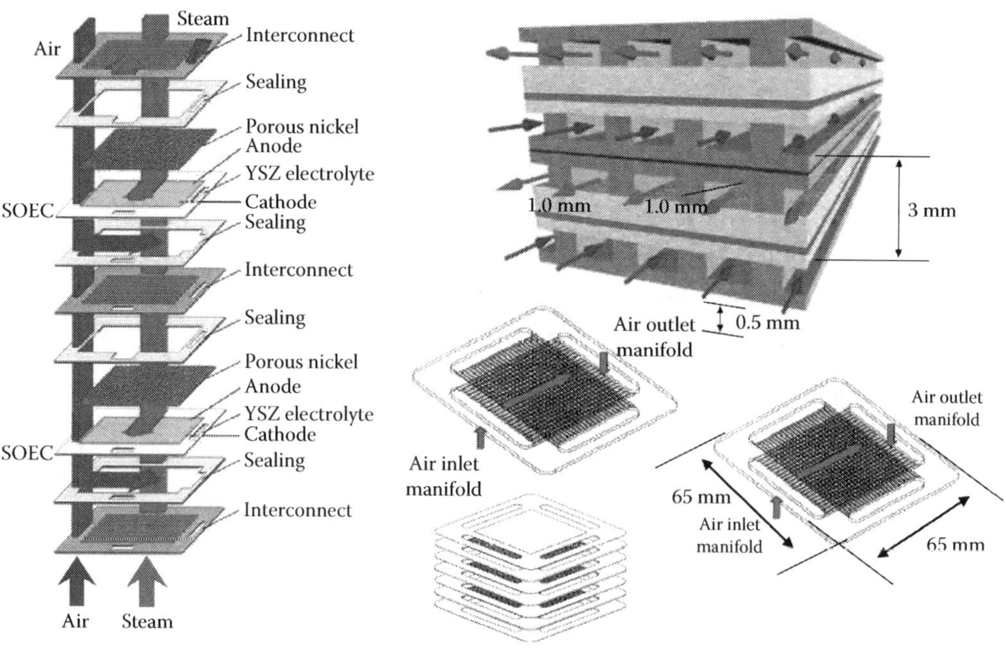

FIGURE 27.9
Design of the stack.

TABLE 27.2
Cell Configuration and Technological Specifications

	Cell Configuration				Specifications	
	Composition	Thickness	Size	Porosity/ Density	Temp.	850°C
Electrolyte	YSZ	10–20 μm	6.5 × 60.5	$D > 95\%$	Input steam	> 70%
Anode	LSM	30–50 μm	5 × 5	$P > 20\%$		
Cathode	Ni-YSZ	1000 μm	6.5 × 60.5	$P > 35\%$	Efficiency	> 90%
Seal	Glass-ceramic	3000 μm		$D > 95\%$	ASR-C	< 1 Ω cm^2
Bipolar plate	Ferrite		6.5 × 6.5		ASR-U	< 1.5 Ω cm^2
Channel width	1.0 mm	Bipolar thickness		3.0 mm	Degradation rate	< 0.05%/5 h
Channel width	0.5 mm	Ridge width		1.0 mm		

FIGURE 27.10
Photographs of the running testing system and the assembled stack.

27.4 Summary

China has launched development of NH production technology. The R&D on the technology was initiated as a component of China's HTR-PM Demonstration Nuclear Power Plant Project. The IS process and HTSE were selected as potential production processes of hydrogen. The R&D on both processes is being conducted at INET. It is expected to commercialize nuclear production of hydrogen after 2020, and therefore the coming decade is a critical period to realize the target. Many challenges exist and comprehensive international cooperation is desired.

Acronyms

BET	Brunauer–Emmett–Teller adsorption
EED	electro-electrodylysis

EV	electric vehicle
FESEM	field emission scanning electron microscope
H/FC	hydrogen and fuel cell
LSC	strontium doped lanthanum cobaltite
LSCF	lanthanum strontium cobalt iron oxide
LSM	lanthanum strontium manganite
MFC	mass flow controller
SOEC	solid oxide electrolyser cell
TEM	transmission electron microscopy
TPB	three-phase-boundary
XRD	x-ray diffraction
YSZ	yttria-stabilized zirconia

References

1. *China Energy Statistical Yearbook 2008*, China Statistics Press, Beijing, 2009.
2. IEA report: Key world energy statistics, 2006. www.iea.org/textbase/nppdf/free/2006/key2006.pdf
3. *China Statistical Yearbook 2008*, China Statistics Press, Beijing, 2009.
4. Y. Ouyang. Development strategy and process of world nuclear power states and nuclear power development in China (in Chinese), *China Nuclear Power*, 1(3):194–201, 2008.
5. S. Dinghuan. *5 IPHE Steering Committee Meeting*, 28–29 March, Canada, 2006.
6. S. Yuliang et al. *Int. Nucl. Hydrogen Production Appl.*, 1:104–111, 2006.
7. Y. Bai, P. Zhang, and Y.S. Qu. Bunsen reaction in the thermochemical iodine sulfur cycle. *Appl. Chem. (in Chinese)*, 26(3):292–296, 2009.
8. S.Z. Chen, P. Zhang, T.Y. Yao, L.J. Wang, and J.M. Xu. HI Concentration of HIx (HI–H_2O–I2) solution in iodine–sulfur water-splitting cycle by electro-electrodialysis. Xi'an Jiaotong University Xuebao, 42:252–255, 2008.
9. M.M. Jin, P. Zhang, and J.C. Wang. Catalysts for decomposition of sulfuric acid in the iodine–sulfur process. *Ind. Catal. (in Chinese)*, 15(8):15–19, 2007.
10. Y. Bai, P. Zhang, S.Z. Chen, L.J. Wang, and J.M. Xu. Purification of H_2SO_4 and HI phases in IS process. *Chinese J. Chem. Eng*. 17(1):160–166, 2009.
11. B. Yu, W.Q. Zhang, J. Chen, and J.M. Xu. Research advance on highly efficient hydrogen production by high temperature steam electrolysis, *Sci. China Ser. B: Chemistry*, 51:289–304, 2008.
12. B. Yu, W.Q. Zhang, J. Chen, and J.M. Xu. Status and research of highly efficient hydrogen production through high temperature steam electrolysis, *INET—The Fourth International Hydrogen Forum*, Changsha, China 2008.
13. M.Y. Liu, B. Yu, J. Chen, and J.M. Xu. Two-dimensional simulation and critical efficiency analysis of high-temperature steam electrolysis system for hydrogen production. *J. Power Source*, 183:708–712, 2008.
14. M.Y. Liu, B. Yu, J. Chen, and J.M. Xu. Thermodynamic analysis of the efficiency of high temperature steam electrolysis (HTSE) system for hydrogen production, *J. Power Source*, 177:493–499, 2008.
15. B. Yu, W.Q. Zhang, J. Chen, and J.M. Xu. Microstructural characterization and electrochemical properties of $Ba_{0.5}Sr_{0.5}Co_{0.8}Fe_{0.2}O_{3-\delta}$ and its application for anode of SOEC, *Int. J. Hydrogen Energy*, 33:6873–6877, 2008.
16. M.D. Liang, B. Yu, and J.M. Xu. Preparation of LSM–YSZ composite powder for anode of solid oxide electrolysis cell and its activation mechanism. *J. Power Source*, 190(2):341–345, 2009.

28

European Union Activities on Using Nuclear Power for Hydrogen Production

Karl Verfondern

CONTENTS

28.1 Energy Situation in the European Union

The European Union (EU) comprises highly industrialized countries with extended urban agglomerations, and therefore needs to rely on a secure and economic supply with energy. As of 2007, the EU holding 7.5% (or 496 million) of the world population consumed 15% of the total energy or 1.757 billion TOE (tons of oil equivalent), and 18% (3325 TWh) of the total electricity [1], and is responsible for 14% (4100 Mt) of the total CO_2 emissions. The situation in the European Union as predicted for the next 30 years is characterized by a growing demand for energy by 2% annually and, at the same time (after 2010), a decreasing domestic energy production. In 2030, if no additional measures are taken, 70% of the energy demand will have to be covered by imports. In addition, this development will push CO_2 emissions to a plus of 14% compared to the 1990 level, far off the Kyoto commitment of an 8% reduction. For these reasons, all energy options should be left open for the next generations [2].

In 2007, principal energy and climate policy targets for the European Union have been redefined by the European Council (the decision-making organ of the EU) to be attained by the year 2020, which are characterized by the "Three Twenties": 20% reduction of greenhouse gases compared to the 1990 level; 20% share of renewable energies of end-use (compared to the present 8.5%); and 20% efficiency of energy use.

Nuclear power is one of the major energy sources in Europe covering 32% of the total electricity demand and providing a significant source of reliable and secure base load power. As of October 2010, a total of 195 nuclear reactors (including the 32 plants in the Russian Federation, of which 5 units are located in the Asian part of Russia) were operated in Europe with a net capacity of 170 GWe, and 19 more units with 16.9 GWe under construction. Within the European Union, 143 nuclear plants with 131 GWe capacity were operated, located in 14 of the 27 Member States are being operated representing 35% of the installed nuclear capacity in the world. The nuclear share, however, varies significantly between countries: while some countries completely refrain from nuclear power, in other countries the nuclear shares range between ~3% for the Netherlands to ~78% for France. In many EU countries, the possibility of a "nuclear renaissance" is discussed and first new construction projects (Generation III+ European Pressurized Reactors (EPR)) are underway in Finland and France. There exists already a long-term intensive cooperation among the nuclear vendors, utilities and research organizations, not only aiming at an evolutionary development of existing nuclear technology, but also searching for innovative concepts of power plants and components with improved safety characteristics and different applications.

With the recent worldwide increased interest in hydrogen as a clean fuel of the future, Europe has also embarked on comprehensive research, development, and demonstration activities with the main objective of the transition from a carbon-based economy toward a CO_2 emission free energy structure as the ultimate goal. The near and medium term, however, due to the growing demand for hydrogen in the petrochemical, fertilizer, and refining industries, will be characterized by a coexistence between the energy carriers hydrogen and hydrocarbons.

28.2 EU Research Policy

EU energy policy is characterized by a diversity in national energy policies and the tendency among the EU member states to consider their energy strategies as a matter of national security. Therefore a closer collaboration among the EU countries is required supported by the preference of many decision makers to search for European solutions. One of the main goals is to provide a balanced choice of energy supply technologies, which will meet the energy needs of the future while achieving the principal objectives of energy supply security, sustainable development, and Europe's competitiveness.

Of particular importance are Europe's research and development (R&D) efforts in the field of nuclear fission, which is deemed an essential part of the solution in a balanced energy mix, at the same level as other developments, such as clean fossil and renewable sources as well as rational use of energy. Since a worldwide increase in the use of nuclear energy is expected, a potential for economic benefits in maintaining and developing the technological lead of the European Union is seen in this field.

The four areas selected by the Generation IV International Forum as technology goals (namely, sustainability, economics, safety, and reliability, proliferation resistance, and physical protection) correspond exactly to the priorities set above by the energy policy for Europe [3]. An important issue for any energy policy is the correct evaluation of the new industrial needs. Besides electricity production, other industrial requirements are addressing high-temperature heat production for the (petrochemical) industry, synthetic fuel production for transportation, and hydrogen production from water. In this way, nuclear energy will penetrate the global energy market through cogeneration of heat and power, and thus reducing the dependence on fossil fuel supply.

Also in the light of the 2006 Ukrainian gas crisis, the EU has set energy security on top of the agenda with the objectives to strengthen energy markets, secure imports, improve efficiencies, and increase the degree of self-sufficiency.

At the European Commission (EC), nuclear-related research is managed either indirectly via multipartner projects usually on a shared-cost basis, or directly in the laboratories of the Joint Research Centers (JRCs) with the latter covering the safety of existing plants as well as innovative reactor systems and fuel cycles. Community research established in Framework Programmes (FPs) is mainly aimed at acquiring scientific "knowledge" in support of EU policy decisions, in particular, in the area of energy.

Between the 1960s and 1980s, Europe has played a leading role in the development of HTGRs with the successful construction and operation of the test reactors DRAGON (UK) and AVR (Germany), the extensive exploration of the possibilities for process heat applications, and later the introduction of a modular concept. Starting in 2000 with the foundation of the European High Temperature Reactor Technology Network (HTR-TN), by industrial and research actors, major activities were resumed dedicated to the development of base generic HTR technologies, to the exploration of advanced solutions for improving the performances for future VHTR.

Under FP-6 (2003–2006), two new instruments have been created and further developed in FP-7 (2007–2013): Network of Excellence (NOE) with long-term joint planning, and Integrated Projects (IPs) to develop new knowledge, new technologies, and demonstration activities with a planned total budget of 48 billion Euro for EU research and 3 billion for EURATOM, a 40% increase compared to the predecessor FP-6. The so-called "European Technological Platforms" are another new element, of which numerous have been founded in the meantime. These TPs are forums bringing together stakeholders of a strategic area and defining strategic research agendas, set up industrial deployment strategies, analyze potential markets, and preparing legal and political framework. The Platforms propose so-called joint technology initiatives (JTIs), large projects in public–private partnership with a great autonomy. The first JTIs became active in 2008, among them is the JTI on fuel cells and hydrogen.

28.3 Nuclear and Hydrogen Research Efforts in the European Union

Various countries have initiated ambitious programs with the goal to bring nuclear hydrogen production to the energy market. In the European Union, there are no explicit research activities dedicated to nuclear hydrogen production. Respective research programs are either concentrating on the nuclear aspects or on the hydrogen aspects. There is, however, a little overlap of both areas in such a way that research on innovative nuclear reactor designs also takes into consideration one of their most pronounced features, which is the possibility of penetration of the nonelectricity market with hydrogen production as a major issue. On the other hand, research projects which deal with large-scale production methods of the future may also include the option of nuclear power to provide the required primary energy.

France announced in 2006 to pursue the aim of constructing a nuclear reactor of the fourth generation which may be a sodium- or helium-cooled fast reactor. France and other countries put a high priority on the long-term management of their radioactive waste and sustainable nuclear fuel supply.

In the European Union, it is recognized that for the introduction of hydrogen energy into the future energy market, political support, for example, by public funding of research projects, is an essential key to long-term success. Hydrogen production technologies are strongly focusing on CO_2-neutral or CO_2-free methods as represented by, for example, biomass conversion or thermochemical water-splitting processes or reforming of fossil fuels plus CO_2 sequestration. Primary energy sources include nuclear and renewable energies.

28.3.1 EURATOM Activities on Nuclear Process Heat

28.3.1.1 Nuclear Technology Networks and Platforms

The main incentive for the foundation of the MICHELANGELO Network (2001–2005) within FP-5 was to move away from the fragmentation and isolation of national research efforts and elaborate a common European position on the priorities of future R&D for a sustainable use of nuclear energy within the worldwide activities in this area. Principal results were the establishment of a consistent strategy of the European nuclear R&D, a dynamic, long-term R&D partnership between the main European organizations of the nuclear industry, and research in the form of a stable network in an international frame. Essential requirements are the following:

- To pursue innovative approaches of nuclear designs which may help to obtain—apart from technical and economical issues—also political and social acceptability.
- To establish a long-term stable partnership not only among the European projects dealing with innovative nuclear fission energy systems, but also between Europe and the nuclear industries worldwide.

In particular, a strong European partnership to the U.S. initiative "Generation IV" was deemed crucial in order to obtain benefit for Europe. But this holds also for other international initiatives like International Project on Innovative Nuclear Reactors and Fuel Cycles (INPRO) (IAEA) or the Three Agency Study International Atomic Energy Agency (IAEA), International Energy Agency (IEA), Organisation for Economic Cooperation and Development (OECD)/Nuclear Energy Agency (NEA) or the US-DOE Nuclear Energy Research Initiative (NERI), and in addition, respective national programs like the HTTR project in Japan or the HTR-10 in China.

Due to the fact that fossil fuels will still be used to a significant amount over the next decades, it is highly recommended to follow also the more pragmatic way of a CO_2-reduced economy by providing nuclear energy for cogeneration and crude oil processing including evolution in present industrial practice like steam reforming of natural gas. Medium and long-term strategies for introducing nuclear hydrogen production into the market will start from the present use of hydrogen and the present methods. Technologies with reduced CO_2 emissions and those substituting natural oil and gas resources will have to be included in addition to the puristic CO_2 emission free approaches.

Therefore the network has made proposals of orientation for future EURATOM R&D FPs including new aspects of nuclear energy like combined heat and power (CHP), desalination, and hydrogen or other fuel production as a complement to other CO_2-free energy sources. Nuclear driven steam reforming or coal gasification may well serve as a bridge to the water-splitting processes such as thermochemical cycles or high-temperature electrolysis. In both cases, the complexity of the processes implies that capital and maintenance costs would be higher than for alkaline electrolysis and even further increased, because they are nonproven processes. The efficiency of the production of electricity and hydrogen is generally increased by raising the process temperatures. But they are limited by the existing high-temperature alloys unless ceramic heat exchangers become available.

The activities of the MICHELANGELO Network are now continued by the "Sustainable Nuclear Energy-Technology Platform" (SNE-TP) which puts priorities on the deployment of Generation III reactors and on the development of Generation IV systems, both fast neutron reactor systems with fuel multirecycling for sustainable electricity-generating capability and (very) high-temperature reactors for nonelectric applications of nuclear energy, such as production of hydrogen or biofuels.

28.3.1.2 The RAPHAEL IP

Among the nuclear projects of FP-6 with a certain relationship to hydrogen, the most important was the IP RAPHAEL, acronym for "Reactor for Process Heat, Hydrogen, and Electricity Generation." Starting in 2005 and terminated in 2010, the IP consisted of 33 partners from 10 countries, with the main objectives are, on the one hand, a study of advanced gas-cooled reactor technologies, which are needed for industrial reference designs, but also taking benefit from the existing demonstrator projects in Japan and China. On the other hand, it aimed at exploring options for the new nuclear generation with "very high temperature" applications, that is, at coolant exit temperatures of 800–1000°C, and also addressed the coupling of a very high-temperature reactor (VHTR) with hydrogen production systems. The IP comprised efforts in all VHTR sections including reactor physics and thermodynamics, fuel, back-end of the fuel cycle, materials and components development, safety, and system integration, building upon the successful HTGR projects initiated within FP-5.

While RAPHAEL was fully concentrating on the development of the VHTR, five more activities (Table 28.1) were launched in the form of "Specific Targeted Research Projects" to deal with the other Generation IV reactor systems.

28.3.1.3 The EUROPAIRS Project Proposal

After several years of joint European research dedicated to the development of base HTGR technologies, HTR-TN has proposed to launch the development of a demonstrator plant coupling an HTGR with industrial process heat applications. Such a development requires a close partnership with the end-user industries, since their needs will certainly be different from those of utilities in terms of power and temperature. Typically, nuclear designers are not familiar with industrial process heat needs. But also the end-users themselves, accustomed to integrate heat supply needs in the global optimization of their processes through heat recovery from these processes, burning of exhaust gases, etc., are not used to consider the availability of an external massive cheap (nuclear) heat source [4]. For the

TABLE 28.1
Current European Nuclear Technology Projects

RAPHAEL	Very high-temperature reactor (VHTR)	2005–2010
GCFR	Gas-cooled fast reactor (GFR)	2005–2009
HPLWR	High-performance light water reactor (SCWR)	2006–2010
ELSY	European lead-cooled system (LFR)	2006–2010
EISOFAR	Road map for a European innovative sodium-cooled fast reactor (SFR)	2007–2008
ALISIA	Assessment of liquid salts for innovative applications (MSR)	2007

demonstrator plant, the coupling should involve a full-scale proven industrial process, with high reliability of the nuclear heat supply to the process.

The project EUROPAIRS standing for "End-User Requirements for Industrial Process Heat Applications with Innovative Nuclear Reactors for Sustainable Energy Supply" is a so-called Coordination and Support Action (CSA) within FP-7 which started in 2009. This project intends to take into account the advice of RAPHAEL's Industrial User Advisory Group (IUAG) and could eventually transfer valuable input to the so-called "Confirmation of Key Technologies" phase, as defined by the SNE-TP, considering VHTR systems. EUROPAIRS has the following main objectives:

1. To identify the main applications for nuclear process heat
2. To determine the viability of combining a nuclear heat source with conventional industrial processes and CHP applications
3. To elaborate a program for the development of a coupled demonstrator between a (V)HTR and industrial processes that require heat supply
4. To form a strategic alliance between nuclear industry and process industries

Essential prerequisite for the success of this CSA is a significant involvement of private companies in the form of industrial participation to develop and deploy innovative energy supply systems is thereby established.

28.3.2 EU Activities on Hydrogen

The Hydrogen Network (HYNET) is the analog to MICANET on the hydrogen side. HYNET became active from 2001 to 2004 as a "Thematic Network" within FP-5 with 12 contractors and more than 70 interested partners. It was working on the development of strategies for the introduction of a European hydrogen fuel infrastructure road map (Figure 28.1).

In 2002, a so-called "High Level Group on Hydrogen and Fuel Cells (HLG)" has been established by the EC. Its principal task is to initiate strategic discussions for the development of a European consensus on the introduction of hydrogen energy. The group strongly recommended the development of an integrated European strategy on hydrogen energy by the creation of a political framework consisting of a partnership of major private and public hydrogen stakeholders. The International Partnership for a Hydrogen Economy (IPHE) was launched in 2003 representing 15 countries and the European Union. Work is done here on governmental level to foster international collaboration on policy and research programs and thus to accelerate the transition to a hydrogen economy

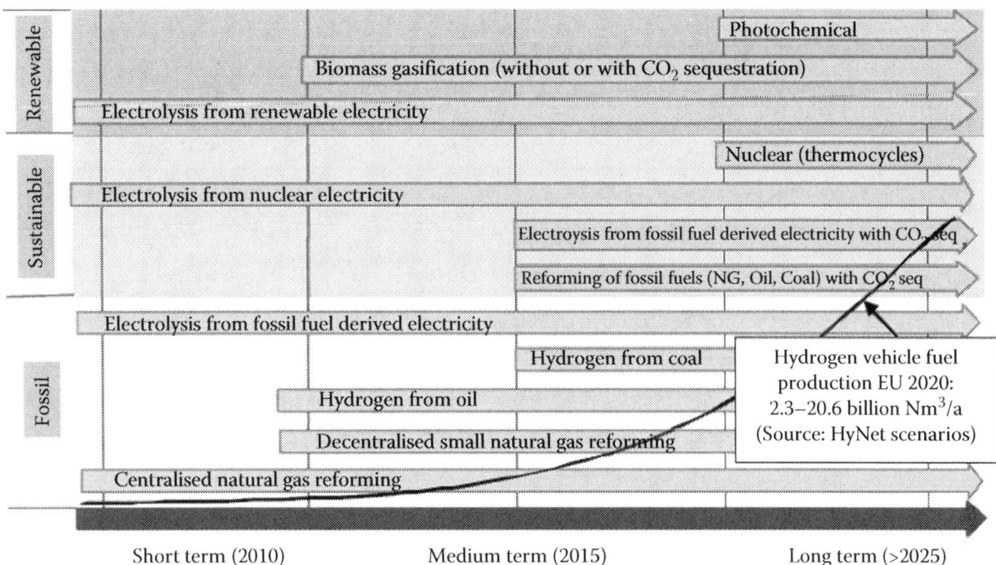

FIGURE 28.1
HYNET timeline for hydrogen production technologies. (From Hy-Net, toward a European Hydrogen Energy Roadmap, preface to HyWays—The European Hydrogen Energy Roadmap Integrated Project, Executive Report, May 12, 2004. With permission.)

with the H_2 to come from multiple sources: renewables, nuclear energy, fossils plus sequestration.

In 2004, the EC started another policy group, the "European Hydrogen and Fuel Cell Technology Platform" (HFP). Key elements of the integrated European strategy to be deployed include a strategic research agenda with performance targets, timelines, lighthouse demonstration projects, and a deployment strategy or road map for Europe. General EU targets by 2020 are a 10–20% supply of the hydrogen energy demand by CO_2-free or lean sources and a 5% hydrogen fuel market share.

HyWays is an IP proposed by HYNET which has elaborated a fully validated European Hydrogen Energy Roadmap as a synthesis of national road maps from the participating member states. It comprises a comparative analysis of regional hydrogen supply options and energy scenarios including renewable energies. The study includes the investigation of the technical, socioeconomic, and emission challenges and impacts of realistic hydrogen supply paths as well as of the technological and economical needs, and details the steps of an action plan necessary to move toward greater use of hydrogen. According to the HyWays road map, an estimated 16 million hydrogen cars will be existing in 2030. The study has also found that introducing hydrogen into the energy system would reduce the total oil consumption by the road transport sector by 40% between today and 2050. Regarding the large-scale hydrogen production, in the early phase up to 2020, hydrogen production will rely on steam reforming of natural gas, electrolysis, and by-product contribution. On the longer term, by 2050, production will be based on centralized electrolysis and thermochemistry from renewable feedstock and CO_2-free or lean sources (coal and natural gas with carbon capture and sequestration, and nuclear energy) [5].

28.4 Hydrogen Production

Hydrogen production is deemed a crucial element for the introduction of hydrogen into the energy sector. Hydrogen production from fossil feedstock, mainly natural gas, is a mature technology for the chemical industry. Research efforts were to be concentrating on the further improvement of basically known reforming and gasification methods, also with regard to high-temperature primary energy systems such as Generation IV nuclear reactors and solar-thermal concentrating systems, on the development of CO_2 sequestration systems, on gas separation technologies, and on the efficiency improvement of hydrogen liquefaction technologies and system integration with hydrogen production facilities [6]. With CHRISGAS, SOLREF, HYTHEC, and Hi2H2, projects have started in 2004 dedicated to the hydrogen production by biomass gasification, steam reforming, thermochemical cycles, and high-temperature electrolysis, respectively. Another IP, HYVOLUTION, started in 2006 to deal with biological processes of hydrogen production.

28.4.1 The CHRISGAS IP

The IP CHRISGAS with a duration of five years had the goal to develop and optimize an energy-efficient and cost-efficient method to produce hydrogen-rich gases from biomass. This gas can then be upgraded to commercial quality hydrogen or to synthesis gas for liquid fuels production. The core of the project is a (solid) biomass-fueled integrated gasification combined cycle (IGCC) pilot plant facility in Värnamo, Sweden. New process equipment has been developed and tested, and implemented in a pilot facility to produce hydrogen-enriched gas. Studies included the conditioning of the synthesis gas to the quality required for the production of transportation fuels.

28.4.2 The Hi2H2 STREP

In the Specific Targeted Research Project (STREP) Hi2H2 with a duration of three years, it was proposed to develop a compact high-temperature water electrolyzer of low cost and with very high electrical efficiencies of more than 90%. The project makes use of technological developments that have been made in the field of high-temperature fuel cells and evaluate a solid oxide water electrolyzer with target cost 400 Euro/kWe that is considered feasible. Experimental work was concentrating on the performance of two types of planar SOFC designs. Steam electrolysis was successfully demonstrated at high current densities and high efficiency on single cells. Record high values of ~2 A/cm^2 at a cell voltage below thermal-neutral voltage and ~3.6 A/cm^2 at 1.48 V in a reversely operated YSZ electrolyte based SOFC cell have been achieved so far [7]. In durability tests up to 2500 h operation of single cells, a maximum degradation of about 2% per 1000 h at 0.5 A/cm^2 and temperatures between 800 and 950°C has been observed. Performance tests were also conducted with electrolyzer stacks of 250 and 600 cm^2 active area, respectively.

28.4.3 HYTHEC-HycycleS Project

HYTHEC was a STREP with six partners (lead: CEA, France) starting in 2004 and running over almost four years. Its main objective was to evaluate the potential of thermochemical processes, focusing on the sulfur–iodine (S–I) cycle to be compared with the Westinghouse hybrid (HyS) cycle. Nuclear and solar energy were considered as the primary energy sources with a maximum temperature of the process limited to 950°C. The HYTHEC activities comprised, apart from the coupling to high-temperature heat sources, the modeling, chemical analysis, process flow sheeting, and experimental studies of the H_2 and O_2 production steps. The project included also industrial scale-up studies for the investigation of the feasibility of the main components, the safety aspects, and cost [8]. During the course of the project, intensive interaction was given with the other EU projects RAPHAEL on the VHTR concept, HYSAFE on safety issues, and EXTREMAT on high-temperature material questions.

A preliminary reference sheet of the S–I cycle has been reviewed and optimized to a "reference" flow sheet with coupling to an indirect cycle VHTR (Figure 28.2) delivering both heat and electricity ("self-sustaining concept") to run the H_2 production process at a rate of 110 t/d and an overall plant efficiency of ~35%. The heat of the 890°C hot primary helium is decoupled in an IHX to secondary helium which partially serves a Brayton power cycle at 5 MPa and the endothermic sections of the S–I cycle. A specific study was dedicated to the performance of candidate membranes for separation processes in the HI_x section to further improve the S–I cycle efficiency.

A large-scale solar furnace was used as an experimental tool in HYTHEC to study the chemical reactions common to both cycles at VHTR temperatures (~900°C) and also at a higher level (1100–1200°C). The receiver reactor was qualified for the solar decomposition of H_2SO_4. Maximum conversion rates were observed with a Pt catalyst, but significant catalytic activity was also found for uncoated SiSiC absorbers. Both nuclear and solar energy can be adjusted to the typical thermal power sizes of present modular power plant designs of up to 600 MW. It will thus support the design of novel components at an industrial scale, which are required in allothermal reforming processes as well as in high-temperature electrolysis, or in thermochemical processes. A preliminary evaluation of the hydrogen production costs based on solar, nuclear, and hybrid operation led to following results: small plants are powered most favorably by solar energy, while nuclear plants are most economic at high power levels >300 MW(th); hybrid systems may have their niche in the mid-range of 100–300 MW(th).

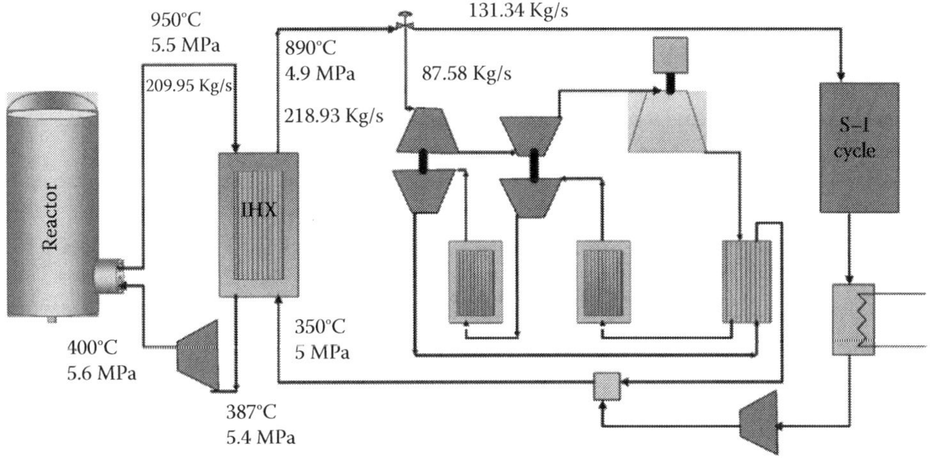

FIGURE 28.2
Coupling of a VHTR nuclear reactor to the S–I cycle. (From DLR, http://www.hycycles.eu, 2008. With permission.)

The 2015 targets defined for high-temperature thermoelectrical–chemical processes with solar/nuclear heat sources are a reduction of CO_2 emissions for fossil reforming by >25% and hydrogen production cost of <2 EUR/kg [9].

Starting in 2008, HycycleS is a new European project within FP-7 with nine European and four international associated partners involved [10,11]. Following the footsteps of the HYTHEC project, HycycleS aims at the qualification of materials and reliability of components for the essential reactions in thermochemical cycles. The focus is on the decomposition of sulfuric acid as the central step of the hybrid-sulfur (HyS) cycle and the S–I cycle. The final goal is to bring thermochemical water splitting closer to realization by improving the efficiency, stability, practicability, and economy. Works include the design, modeling, construction, and testing of a large compact plate SiC heat exchanger for thermal SO_3 decomposition. A criterion of thermal efficiency of more than 85% has been set, and the mock-up facility will be tested under realistic thermal conditions (850°C). The results will lead to recommendations for a scale-up in the several kW range.

28.4.4 INNOHYP Project

INNOHYP "Innovative High Temperature Routes for Hydrogen Production" was a "Coordination Action" within FP6 starting in 2004 to run over four years. Its main objectives were the collection and coordination of information describing the options for hydrogen production at high temperatures and the identification of further research efforts. From all routes to hydrogen considered, 17 were identified as important, for which conversion efficiencies and their influence on various operating parameters have been studied. Some results are shown in Chapter 10. Also a categorization of primary energy sources was made. While nuclear is a "high density" energy source with maximum process temperatures limited to ~900°C, solar energy appears to be a key primary energy source in the "low density" category for high temperatures up to ~1600°C [12].

28.4.5 HYSAFE Program

The HYSAFE project (2004–2009) was conceived as a NOE program comprising 24 partners from research industries and universities. The main objective was to strengthen, integrate, and concentrate existing capacities and fragmented research efforts intending to remove safety-related barriers to the large-scale introduction of hydrogen as an energy carrier. By harmonizing methodologies for safety assessment, the focus was on studies of fire and explosion safety, mitigating techniques, and detection devices. In this way, the network contributed to promoting public awareness and trust in hydrogen technology. Its specific objectives include

- The improvement of common understanding and approaches for addressing hydrogen safety issues
- The integration of experience and knowledge on hydrogen safety in Europe
- The contribution to EU safety requirements, standards, and codes of practice
- The promotion of public acceptance of hydrogen technologies

After termination of HYSAFE, the activities on hydrogen safety of the NOE have been transferred to a newly founded association, the "International Association of Hydrogen Safety."

28.5 Outlook

In the European Union, the large-scale hydrogen production by means of fossil, renewable, or nuclear energy as a significant component in a future European energy economy has been or is treated in a wide variety of projects. It is a consequence of the obvious existence of separate communities—nuclear energy on the one side and hydrogen production by conventional and renewable energy on the other side. In the light of growing environmental concerns and dependencies on energy imports, however, a tendency toward an approach of both sides can be perceived already. A "Strategic Energy Technology Plan" (SET Plan) propounded recently by the EC [13] is aiming at low-carbon energy technologies for the future explicitly including fission energy. In a collective effort of the EU Member States in the form of public–private partnerships, industrial initiatives, and joint intensified international programs, the SET plan is intended to improve coherence and effectiveness of joint actions to prepare for a Trans-European energy network.

Next-generation nuclear reactors (Generation IV) targeting the criteria of sustainability in terms of safety, economics, and waste issues may help to surmount the separation of the communities and take advantage of the synergies resulting from a closer collaboration of the different technologies in the fossil, renewable, and nuclear communities. Paving the way for deployment of nuclear power also in the heat, hydrogen, and cogeneration sectors represents a global challenge and a technological milestone of significant importance. Thus, further international collaboration in order to support the demonstration of nuclear process heat applications will be necessary.

References

1. IEA, *World Energy Outlook 2009*. International Energy Agency, Paris, France, 2009.
2. EC, Green Paper: A European Strategy for Sustainable, Competitive and Secure Energy, Commission of the European Communities, COM(2006) 105 final, {SEC(2006) 317}, Brussels, March 8, 2006.
3. Van Goethem, G., Hugon, M., Bhatnagar, V., Manolatos, P., and Deffrennes, M., Euratom innovation in nuclear fission: community research in reactor systems and fuel cycles, *Nuclear Engineering and Design* **237**:1486–1501, 2007.
4. Basini, V., Bogusch, E., Breuil, E., Buckthorpe, D., chauvet, V., Fütterer, M., van Heck, A., et al., Why HTR/VHTR? a European point of view. *Proc. International Congress on Advances in Nuclear Power Plants ICAPP'08*, Paper 8433, Anaheim, USA, 2008.
5. EC, HyWays The European Hydrogen Energy Roadmap, Final Report, February 22, 2008.
6. HFP, Strategic Research Agenda, European Hydrogen and Fuel Cell Technology Platform, July 2005.
7. Jensen, S.H., Larsen, P.H., and Mogensen, M., Hydrogen and synthetic fuel production from renewable energy sources, *International Journal of Hydrogen Energy* **32**:3253–3257, 2007.
8. Le Duigou, A., Borgard, J.-M., Larousse, B., Doizi, D., Allen, R., Ewan, B.C., Priestman, G.H., et al., HYTHEC: An EC funded search for a long term massive hydrogen production route using solar and nuclear technologies, *International Journal of Hydrogen Energy* **32**:1516–1529, 2007.
9. HFP, Draft Implementation Plan, Status 2006, Implementation Panel, October 2006.
10. DLR, http://www.hycycles.eu, 2008.

11. Poitou, S., Rodriguez, G., Haquet, N., Cachon, L., Bucci, P., Tochon, P., Chaumat, V., et al., Development program of the SO_3 decomposer as a key component of the sulphur-based thermochemical cycles: New steps towards feasibility demonstration, *17th World Hydrogen Energy Conference*, Brisbane, Australia, June 15–19, 2008.
12. Le Naour, F., INNOHYP CA Final Report, INNOHYP CA–FR–CEA/07-05, 2007.
13. EC, Communication from the Commission to the European Parliament, the Council, the European Economic and Social Committee and the Committee of the Regions, Second Strategic Energy Review, An EU Energy Security and Solidarity Action Plan, Commission of the European Communities, COM(2008) 781 final, {SEC(2008) 2870} {SEC(2008) 2871} {SEC(2008) 2872}, Brussels, November 13, 2008.
14. Hy-Net, Towards a European Hydrogen Energy Roadmap, Preface to HyWays—The European Hydrogen Energy Roadmap Integrated Project, Executive Report, May 12, 2004.

29

HTTR-IS Nuclear Hydrogen Demonstration Program in Japan

Nariaki Sakaba, Hirofumi Ohashi, and Hiroyuki Sato

CONTENTS

29.1 Program Overview

Recently, climate changes and effect to the ecosystem are concerning because of the global warming induced by the emissions of greenhouse gases such as carbon dioxide. Nuclear hydrogen production is gathering worldwide attention since it can produce hydrogen, a promising energy carrier, without an environmental burden. In 2008, a massive hydrogen production technology utilizing nuclear heat source was selected as one of the innovative technologies which could achieve a favorable balance between a sustainable development and an environmental protection was undertaken by the Council for Science and Technology Policy [1]. Also, the Japan Atomic Energy Commission addressed a roadmap of nuclear energy research and development (R&D) [2] positioning the nuclear hydrogen production as a candidate of the technologies that can achieve the vision for applying a nuclear reactor as a heat source. In the roadmap, it is declared that the prototype demonstration of commercial nuclear hydrogen plants should start by 2020.

Japan Atomic Energy Agency (JAEA) has been conducting R&D on enabling technologies for nuclear hydrogen demonstration. High-Temperature Gas Reactor (HTGR) was selected as nuclear heat source due to the reactor's inherent safety, economic viability, high efficiency, very high burn up, and wide industrial applicability from electricity generation to hydrogen production. As for the reactor technology, R&D related to fuel, material, thermal-fluid, neutronics, and high-temperature components technologies have been conducted since 1970s. Based on the technology, the HTTR (high-temperature engineering test reactor, which is the first HTGR in Japan, has been constructed and is in operation at the site of JAEA Oarai R&D Center) [3]. The HTTR has successfully delivered 950°C helium coolant to the outside of the reactor vessel [4]. The reactor outlet coolant temperature of 950°C makes it possible to extend the HTGR use beyond the field of electric power such as hydrogen production from water splitting. As for the hydrogen production technology, R&D on the thermochemical iodine–sulfur water-splitting process (IS process) (Chapter 17) has been performed. During the past several years, an autonomous control scheme was developed and validated in continuous closed-loop hydrogen production with a production rate of 31 NL/h for 175 h [5]. Also, successful development of corrosion-resistant process components and high-efficiency process technology has advanced the basis for nuclear hydrogen production demonstration.

JAEA plans to connect an IS process hydrogen production plant to the HTTR in the near future. This will establish the nuclear hydrogen production technology including the first kind of the system integration technology that would be required for safe connection of a hydrogen production system to a Generation IV VHTR. The program aims to be the world's first to demonstrate hydrogen production using heat directly supplied by a nuclear reactor. The demonstration of the HTTR-IS system is a nuclear pilot-plant step in the commercial development of nuclear hydrogen production based on the IS process. The step will demonstrate the licensing of industrial construction and operation of a nuclear heated hydrogen plant and accumulate data to be used to evaluate the economics of hydrogen product produced by a VHTR.

29.2 HTTR-IS System Design

29.2.1 Present Status of the HTTR Reactor

The HTTR is a helium-cooled and graphite-moderated thermal neutron test reactor with thermal power of 30 MW. As the HTTR is the first HTR in Japan and a test reactor, it has following purposes:

- Establishment of basic HTGR technologies
- Demonstration of HTGR safety operations and inherent safety characteristics
- Demonstration of nuclear process heat utilization
- Irradiation of HTGR fuels and materials in an HTGR core condition
- Provision of testing equipment for basic advanced studies

In order to demonstrate the nuclear process heat utilization, the intermediate heat exchanger (IHX) is equipped in the cooling system to supply high-temperature helium gas to a process heat application system. Figure 29.1 shows the flow diagram of the HTTR. Main cooling system is operated at normal operations and an auxiliary cooling system and a vessel cooling system, the engineered safety features, are operated after a reactor scram to remove residual heat from the core. The main cooling system, which consists of a primary, a secondary, and a pressurized water cooling system, removes heat generated in the core and dissipates it to the atmosphere by an air cooler. The primary cooling system consists of an IHX, a primary pressurized water cooler (PPWC), a primary concentric hot-gas-duct, and so on. Primary coolant of helium gas from the reactor at 950°C maximum flows inside the inner-pipe of the primary concentric hot-gas-duct to the IHX and PPWC. The primary helium is cooled to about 400°C by the IHX and PPWC and returns to the reactor

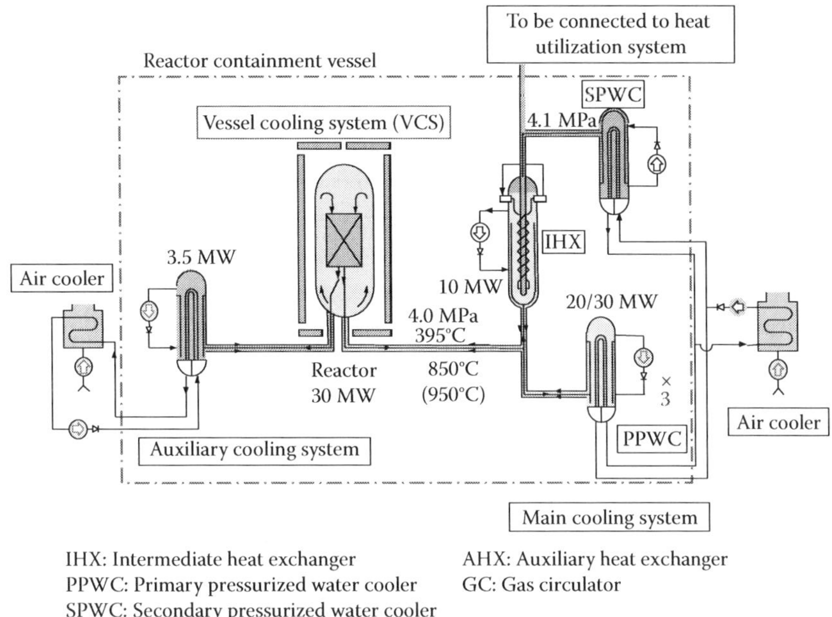

FIGURE 29.1
Flow diagram of the HTTR.

flowing through the annulus between the inner- and outer-pipes of the primary concentric hot-gas-duct.

The HTTR has two operation modes. One is the single-loaded operation mode using only the PPWC for the primary heat exchange. Almost all the basic performance of the HTTR system has been confirmed by the single-loaded operation mode. The other is the parallel-loaded operation mode using the PPWC and IHX. In a single-loaded operation mode the PPWC removes 30 MW of heat and in a parallel-loaded operation mode the PPWC and IHX remove 20 and 10 MW, respectively.

The HTTR experienced several operational tests such as rated operation test and safety demonstration tests. At the rated operation, valuable operational data such as core physics characteristics, thermal hydraulic characteristics, fission product behavior, chemical impurities, and so on, are obtained. As for the safety demonstration test, the control rod withdrawal test and the partial loss of coolant flow test have been carried out [6]. Through these tests, the inherent safety features, which show the slow temperature response during abnormal events due to large heat capacity of the reactor core and have the negative reactivity feedback effect, were demonstrated. Table 29.1 shows the test items and obtained results of major operational tests. In addition, further items of the operational test such as vessel cooling system stop test, to be conducted in the HTTR are shown. The operational data obtained in the future test is expected to contribute the validation of simulation tools which would be used for the design and licensing of commercial VHTRs.

29.2.2 Conceptual Design of the HTTR-IS System

A hydrogen production system based on the IS process is planned to be connected to the actual HTTR in the near future. This will establish the hydrogen production technology

TABLE 29.1

Operational Tests of the HTTR

Rated Power Operation Test	
Items	**Obtained Results**
High-temperature operation test	• Achieved reactor outlet temperature of 950°C • Obtained data related to core physics, thermal hydraulics, and fission product release characteristics • Operational and maintenance experiences
High-temperature continuous operation test (Rated power 50 days)	• Achieved 50 days continuous full power operation • Obtained data related to core physics, thermal hydraulics, and fission product release characteristics • Operational and maintenance experiences
Safety Demonstration Test	
Reactivity insertion test • Control rod withdrawal test	• Reactor power increase due to the control rods withdrawal can be restrained by the negative feedback of reactivity without operating the reactor power control system • Transition of core component temperature was slow due to the large heat capacity
Coolant flow reduction test • Gas circulators trip test (one and two out of three gas circulators)	• Rapid decrease of the coolant flow rate brings the reactor power to a stable without a reactor shutdown • The transition of fuel temperatures was slow
• Gas circulators trip test (three gas circulators)	Planned
Black out test • Vessel cooling system stop test	Planned

with an HTR including the system integration technology for connection of hydrogen production system to HTRs.

There are two main specific objectives of the demonstration program. The first main objective is demonstration of economical hydrogen production plant construction. An effective approach to economical design is simplification in construction and operation of hydrogen plant process and equipment. In the economic evaluation for the JAEA's commercial hydrogen cogeneration plant GTHTR300C (Chapter 10) [7], the IS process cost is assumed as 14 billion ¥ [8] including construction and operating costs. Figure 29.2 shows the breakdown of the commercial IS process hydrogen production plant construction cost at the preconceptual design stage. The study indicates that the cost for the component fabrication absorbs more than 45% of the overall construction cost [9]. Therefore, reduction of the component fabrication cost would significantly contribute to the reduction of the cost of the IS process. In general, reduction of quantities of construction materials is the most effective way to reduce the fabrication cost.

The other main objective of the HTTR-IS program is demonstration of a safety philosophy that requires a nonnuclear grade IS process. The term "nonnuclear grade" means that the IS plant is designed and constructed as a conventional chemical plant even if it is coupled to a nuclear reactor. The cost target of nuclear-produced hydrogen by the GTHTR300C is 20.5 ¥/N m^3 [8], to stay economically competitive to hydrogen produced by other methods in a future hydrogen society when massive quantity of hydrogen is needed. The system applying nuclear grade regulations requires ensuring the reliability in terms of nuclear safety such as multiple components, seismic protection, and so on, such that construction and operation costs are much greater than the cost of conventional chemical plants. In addition, nonnuclear

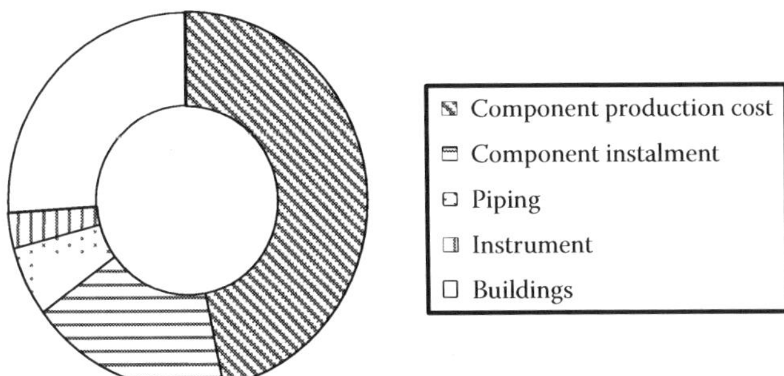

FIGURE 29.2
Breakdown of the commercial IS process hydrogen production plant construction cost at the preconceptual design stage.

grade hydrogen production system can open the door for nonnuclear industries to enter as a constructor and operator. It is expected that the reactor facility will be operated by electric power companies and the hydrogen production system will be managed by gas and oil companies in commercial VHTR-IS hydrogen production plant.

29.2.2.1 Safety Philosophy for Nonnuclear Grade Hydrogen Plant

Safety philosophy establishment for nonnuclear grade hydrogen production plant is one of the important objectives for the nuclear hydrogen demonstration. JAEA proposed development strategy, as well as R&D items, to contribute to the nonnuclear-grade IS process [10]. Figure 29.3 shows the R&D maps for the nonnuclear grade IS process hydrogen production plant to be coupled with the HTTR [10].

Safety philosophy strategy for nonnuclear grade hydrogen production plant is consolidated into the following items [11]:

- Exempt the IS process from Prevention System 3 (PS-3)
- Identify abnormal events initiated in the IS process as external events

Key R&D items for meeting these requirements are:

- Evaluation and mitigation methods for load variation initiated by the hydrogen production plant
- An evaluation method and a countermeasure for combustible gas explosion
- An evaluation method and a countermeasure for toxic gas inflow to reactor control room
- An evaluation method and a countermeasure for tritium migration into product hydrogen
- Separation technology of hydrogen production plant from nuclear facility

Further details of the R&D can be found in Chapters 23 and 24. Established evaluation methods to achieve nonnuclear grade hydrogen production plant would be examined during the safety review from the government. A further direction of the study for the HTTR-IS

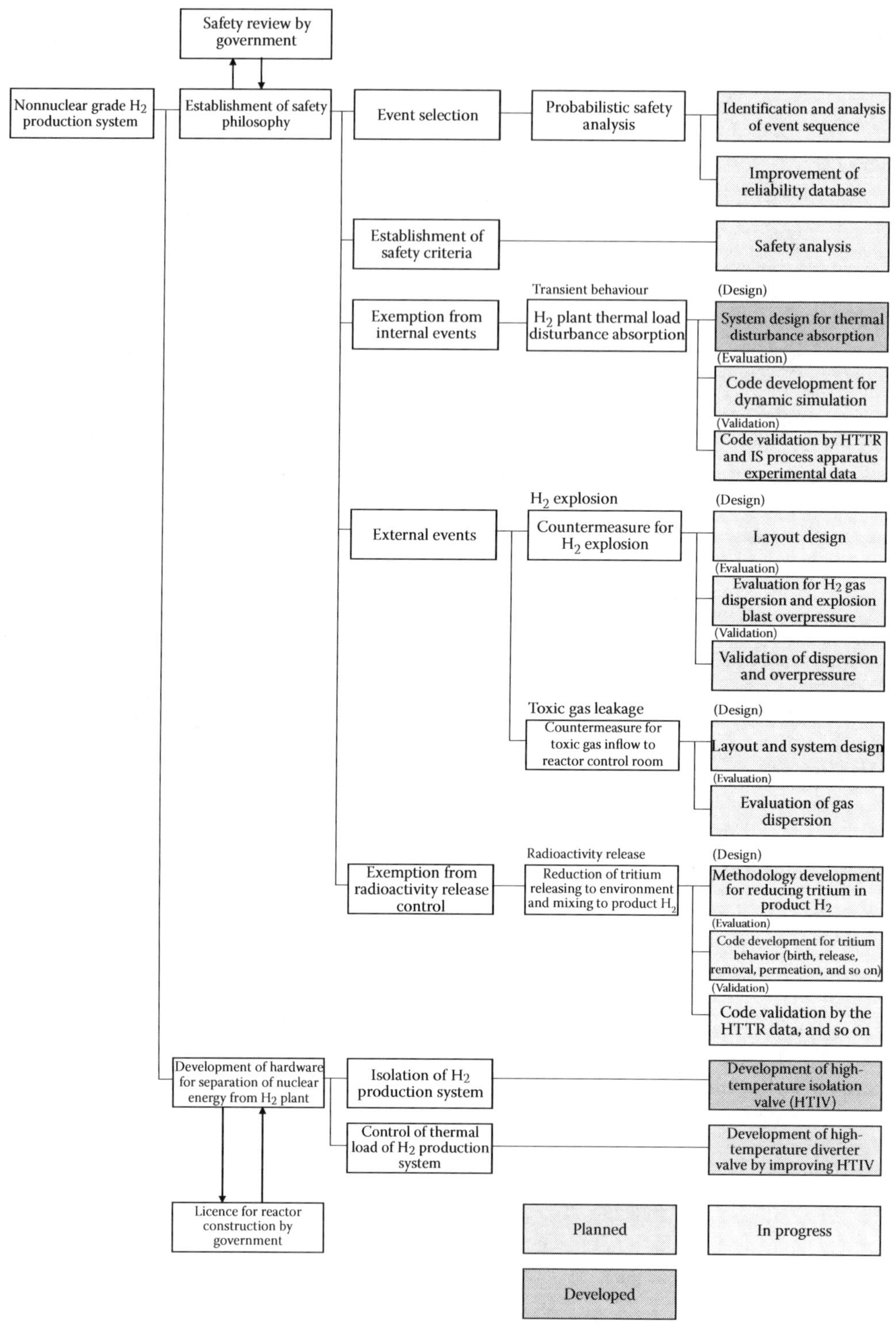

FIGURE 29.3
R&D requirements for nonnuclear grade IS process hydrogen production plant to couple with the HTTR. (Adapted from N. Sakaba, et al., *J. Nucl. Sci. Technol.*, 45(9), 962–969, 2008.)

demonstration would be the validations of the evaluation methods to ensure their reliability.

Once the nonnuclear grade hydrogen production plant is approved in the safety review, the safety evaluation for the internal events would be conducted based on the experience of the HTTR construction. Regarding the external events such as combustible gas explosion, and so on, design parameters such as an inventory of combustible gases in hydrogen production plant, an offset distance from reactor building to hydrogen production plant, and so on, would be defined for the evaluation in order to facilitate the design modification of hydrogen production plant.

The HTTR-IS demonstration would contribute not only for hydrogen production but also for various heat utilizations of nuclear reactors.

29.2.2.2 Conceptual Design

Conceptual design of the HTTR-IS system was performed in order to show a feasibility of nuclear hydrogen demonstration. Following studies have been conducted during the design stage:

- Heat supply system design, for example, secondary cooling system, thermal load mitigation system and so on, considering heat and mass balance
- Flowsheet evaluation which could achieve hydrogen production rate of approximately 1000 Nm^3/h
- Component design which can simplify the system configuration and reduce a fabrication cost of components
- Layout design considering the hydrogen explosion which will be an assumed accident event in the hydrogen production system
- Evaluation methods development which can contribute to safety case studies, product quality assessment, and so on
- Event identification study for the safety evaluation

29.2.2.2.1 Heat Supply System Design

A heat supply system which delivers heat to the hydrogen production plant is designed considering the heat and mass balance of the system. Figure 29.4 shows the candidate flow diagram [12]. The heat generated in the HTTR core is transferred to the IHX, and then the heat transfers from IHX to steam generator and helium cooler which would be replaced from the secondary pressurized water cooler in existing HTTR. The secondary helium flows through the inner-pipe of the concentric hot-gas-duct and a high-temperature isolation valve (HTIV), and supplies heat to the process heat exchangers such as H_2SO_4 decomposer and HI decomposer, and the reboiler of the HI distillation column. Finally, after cooled by the steam generator and helium cooler, secondary helium is pressurized by the helium circulator and returns to the IHX through the outer-pipe of the concentric hot-gas-duct. To isolate the IS process against its abnormal events, secondary cooling system has bypass flow and diverter valves are installed upstream both of the IS process flow and its bypass flow. The pressure of the secondary cooling system and IS process are set at 4.1 and 2.1 MPa, respectively, so that the process fluid will not intrude into the secondary helium loop in case of the boundary failure.

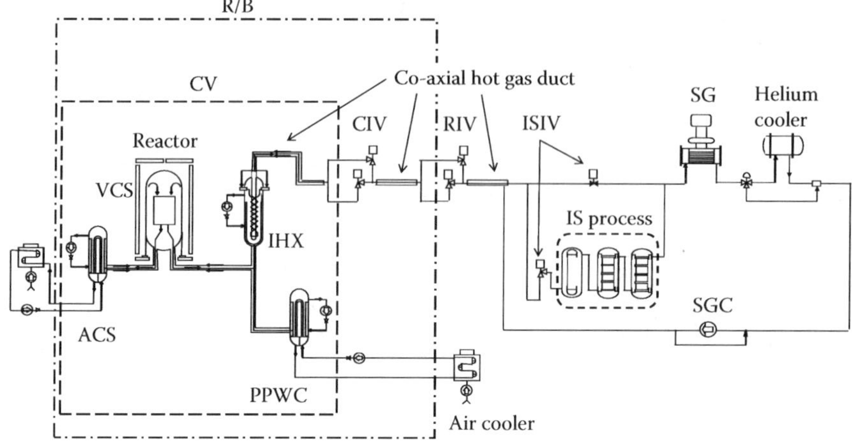

FIGURE 29.4
Tentative flow diagram of the heat supply system in the HTTR-IS system.

29.2.2.2.2 Flow-Sheet Evaluation

Flow-sheet study was conducted in order to show the high potentiality of the VHTR coupled IS process hydrogen plant. Figure 29.5 illustrates a flow-sheet example for the HTTR–IS system [12], which is different from bench-scale test apparatus in which several major differences exist (Chapter 19). A three-stage multieffect vaporizer is used as H_2SO_4 concentrator. A direct contact heat exchanger at upper stream of the H_2SO_4 vaporizer is adapted in order to recover unreacted H_2SO_4 in the SO_3 decomposer and exchange heat efficiently. An electro–eléctrodialysis (EED) cell and a reverse osmosis (RO) membrane are added in order to increase the HI concentration of the HIx solution to over pseudo-azeotropic one. When the concentrated HIx solution is fed to the HI distillation column, HI-rich vapor is obtained from the top. I_2 is removed from the reaction field by reaction with cobalt (Co) in HI decomposition reactor. Single pass conversion ratio is expected to be improved by the shift of reaction equilibrium by the removal of I_2. Fifteen heat exchangers are attached for its heat recovery in this process. The hydrogen production rate of the process of Figure 29.4 is calculated to be about 850 Nm^3/h and its efficiency is about 43%. Figure 29.6 shows the flow-sheet where the hydrogen permselective membrane reactors (HPMRs) are used [12]. When HPMRs are installed in the system, temperature is reduced to about 400–500°C. A preliminary calculation of thermal efficiency and hydrogen production rate is carried out and the hydrogen production rate is calculated as 1100 Nm^3/h at a process efficiency of about 44%.

29.2.2.2.3 Component Design

For the simplification of the system, components integration in the IS process was conducted in terms of reducing the construction cost of IS process. The study is focused on the component installed in H_2SO_4 decomposition and Bunsen reaction procedures. Combined H_2SO_4 decomposer [13] was designed based on shell and tube type heat

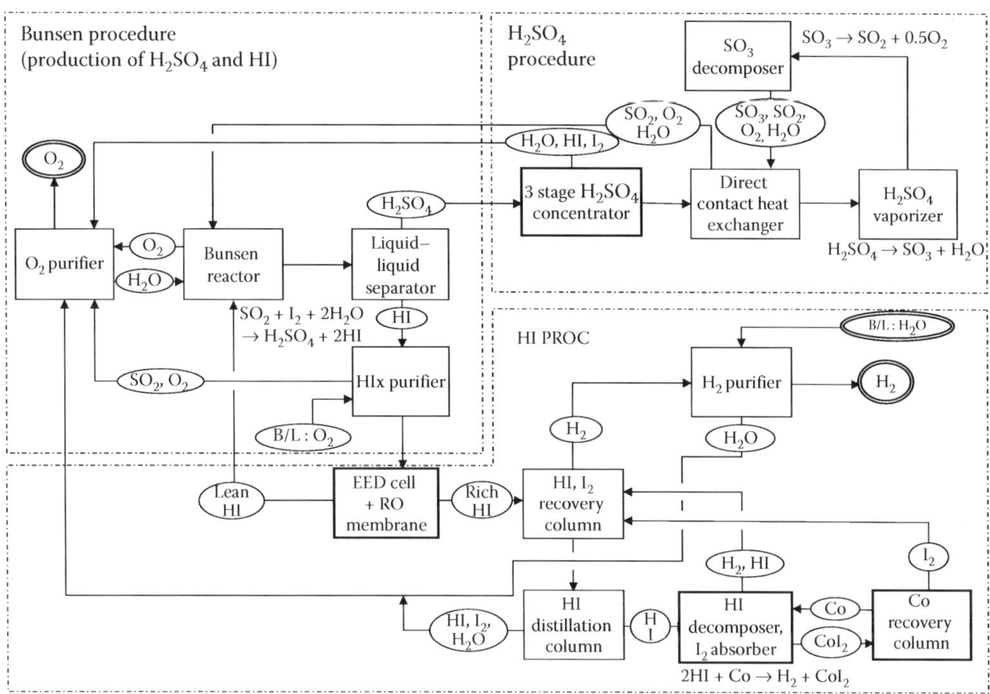

FIGURE 29.5
Tentative simplified flowsheet of the HTTR–IS plant (absorption of I_2 by Co in the HI decomposer).

exchanger integrating SO_3 decomposer, H_2SO_4 vaporizer, and process heat exchanger. The vessel consists of two heat exchanger parts which evaporate sulfuric acid and decompose sulfur trioxide by the sensible heat of sulfur dioxide and helium, respectively. Heat exchanger parts are connected with double pipes of which helium flows through inner pipe and sulfur trioxide flows through outer pipe. Regarding the material, SiC film coating is estimated against corrosive circumstances. Combined Bunsen reactor [13] was designed based on mixer-settler type reactor integrating the Bunsen reaction part and liquid–liquid separator part. The reactor consists of a line mixer, a cylindrical mix vessel, and a rectangular settler. In the mixer part, sulfuric dioxide, iodine, and water are mixed and react to form acids by using static mixer and stirring machine. Mixed solution is distributed to settler with weirs. These weirs carry out the function of retaining separately mixed solution and enhancing the phase separation rate. H_2SO_4 phase and HI phase can be obtained from the top and the bottom of the reactor, respectively. Integration of components can reduce the number of equipment such as reactor vessels, connecting piping, and transfer pumps, so that the amount of material is expected to be reduced considerably. Also, the number of connections was reduced and it results the reduction of risks of process-fluid leakage. Figure 29.7 shows the tentative flow diagram which involves the integrated components [9].

29.2.2.2.4 Layout Design

Figure 29.8 shows the site arrangement of the HTTR-IS system [14]. In order to keep the integrity of reactor building during the abnormal hydrogen explosion assumed in the IS process, the HI section which involves hydrogen is installed with appropriate distance from the reactor building. The heat from the HTTR is delivered to IS process via secondary

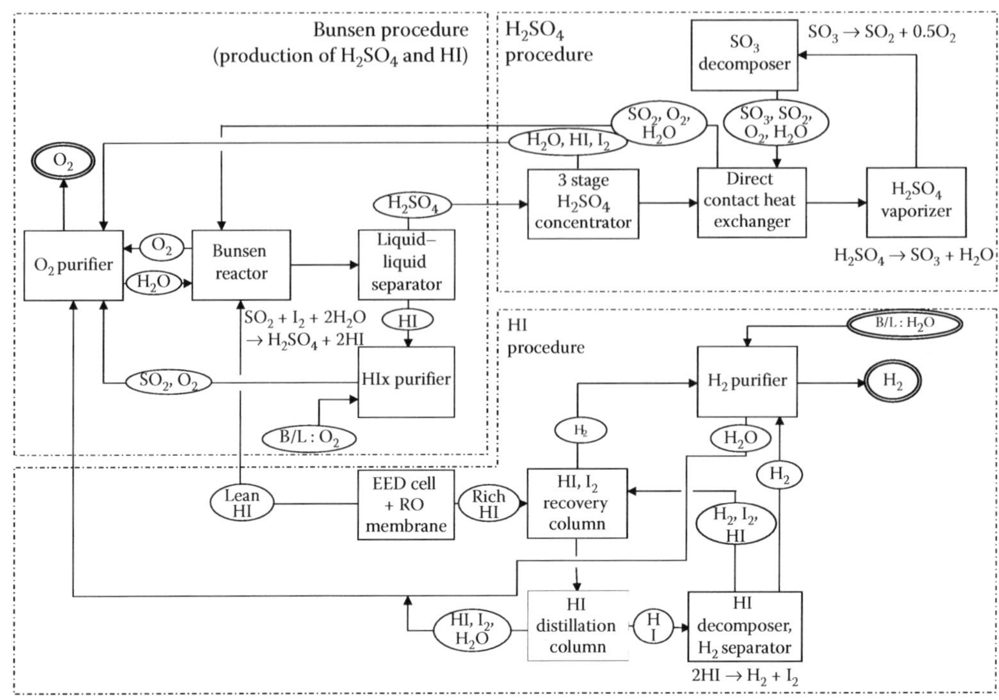

FIGURE 29.6
Tentative simplified flowsheet of the HTTR–IS plant (utilizing an HPMR for separation of H_2 in the HI decomposer).

helium loop which utilizes concentric hot-gas-duct installed underground. Components are not installed between the reactor and HI decomposer considering the effect of a missile phenomenon. From the viewpoint of the prevention of toxic gases inflow to the reactor controlling room, the equipment utilized in some of the H_2SO_4 and HI sections is stored inside the buildings. The conceptual layout design shows that the IS process can be arranged besides the existing HTTR compactly.

29.2.2.2.5 Evaluation Method Development

Evaluation method developments were conducted about areas of dynamic system behavior, product contamination, combustible gas explosion, and toxic-gas inflow to the reactor control room.

Regarding the system analysis method, plant dynamic simulation code has been developed, which can evaluate following characteristics in the HTTR-IS system:

1. Reactor power behavior in the reactor core
2. Thermal hydraulics of noncondensable gases in primary and secondary system
3. Thermal hydraulics of two-phase steam-water mixture in steam generator and air cooler
4. Chemical reactions in the IS process
5. Control systems

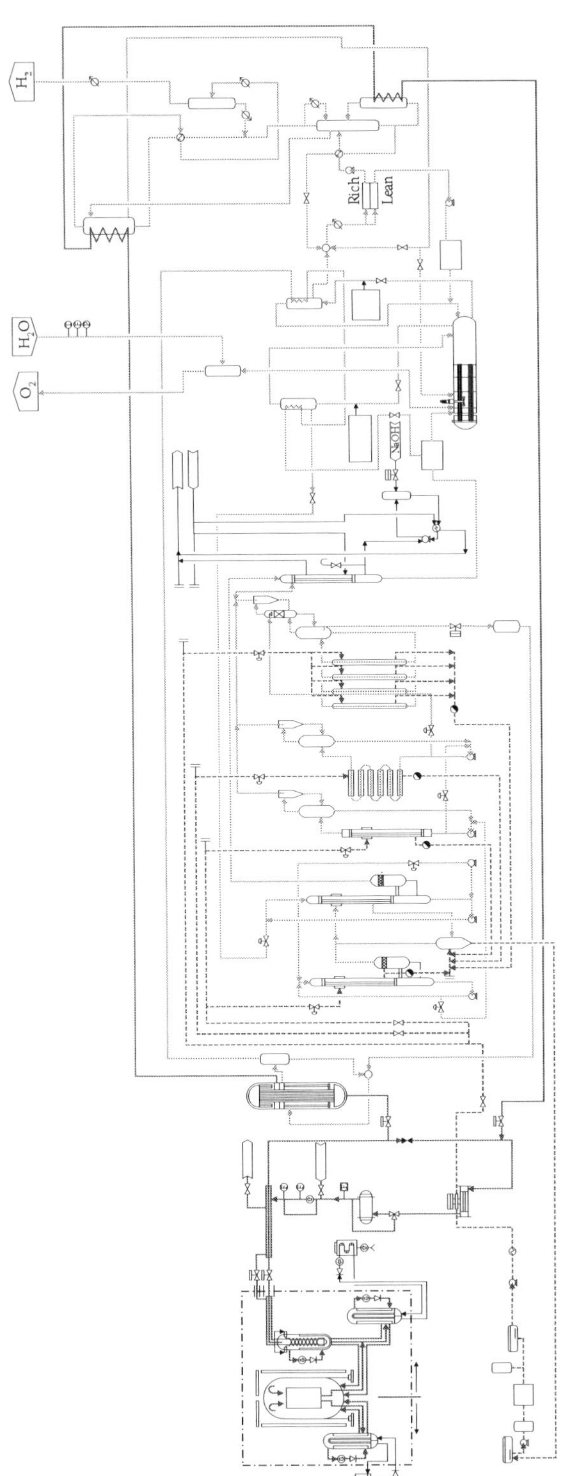

FIGURE 29.7
Tentative flow diagram of the HTTR-IS system with simplified components. (Adapted from N. Sakaba, et al., *J. Nucl. Sci. Technol.*, 45(9), 962–969, 2008.)

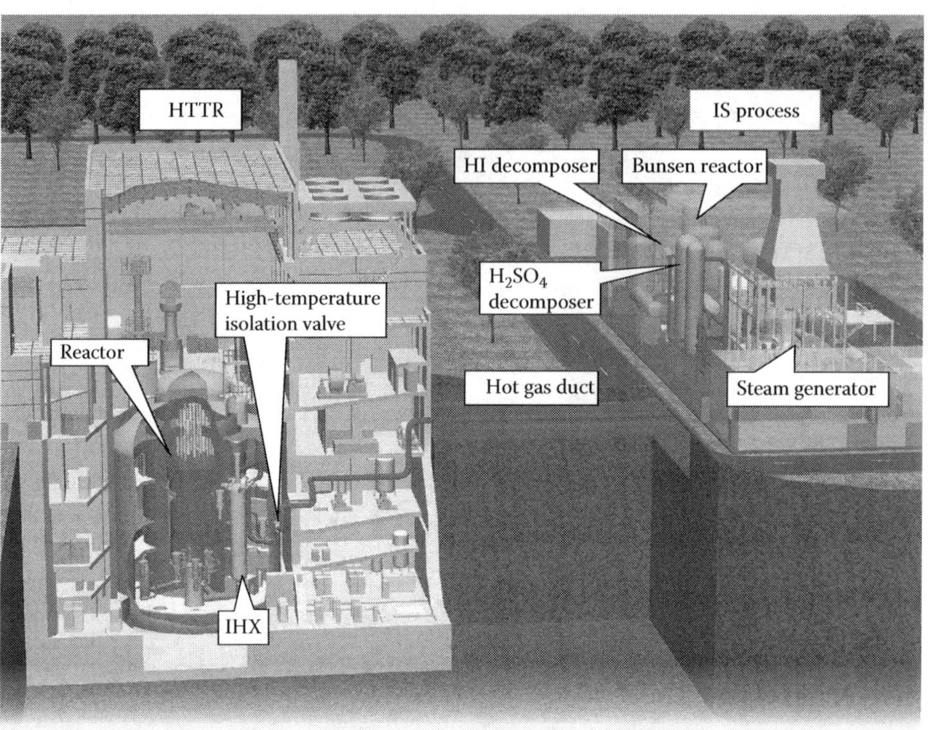

FIGURE 29.8
Site arrangement of the HTTR-IS system.

The code consists of two modules: (1) Reactor module and (2) process heat exchanger module.

Reactor module was developed based on RELAP5 MOD3 [15] and was able to evaluate reactor power behavior in the reactor core, thermal hydraulics of helium gas in primary and secondary system, thermal hydraulics of pressurized water in primary pressurized water cooling system and auxiliary water cooling system, thermal hydraulics of the two-phase water mixture in steam generator and air cooler, and characteristics of the control systems in the HTTR-IS system. Though the development of the RELAP5 aims to evaluate the thermal hydraulic transient behavior of light water reactors, some modifications [16] were performed so that the code enables to use additional properties such as air, graphite, and experimental heat transfer correlations. Field equations consist of mass continuity, momentum conservation and energy conservation, and reactor power is calculated by point reactor kinetics equations. This module has a flexibility to enhance a function using control components or adding subroutines so that it has great advantages to incorporate experimental knowledge into the numerical models. Process heat exchanger module [17] has the following features:

- Eight chemicals (H_2, O_2, SO_2, SO_3, H_2O, HI, H_2SO_4, and I_2) are taken into consideration.
- A modular architecture classified into SO_3 decomposition module, H_2SO_4 decomposition module, and HI decomposition module according to the dominant phenomenon assumed at each process heat exchanger is applied. Thus, configuration

of the process heat exchangers are user definable assigning the interface to helium flow by the input data.
- A lumped parameter system is employed on account of the simplification of the model.

The code will be validated by IS process pilot plant tests, and it will be utilized for the safety assessment.

Regarding the product contamination evaluation, JAEA has developed a numerical analysis code, tritium and hydrogen transportation analysis code (THYTAN) [18] in order to estimate the tritium movement behavior in the HTTR-IS system. Consequently, the THYTAN code can calculate the mass balance of tritium and hydrogen in the HTTR-IS system taking the following phenomena into account:

- Birth of tritium production by the ternary fission reaction in the fuel particle and by the neutron absorption reaction of ^{6}Li, ^{10}B, and ^{3}He in the core, and tritium release into the helium coolant
- Tritium and hydrogen permeation through the heat transfer tube of the heat exchanger, for example, IHX, chemical reactor, recuperator, and so on
- Tritium and hydrogen permeation at concentric hot-gas ducts
- Tritium and hydrogen permeation from the helium systems to atmosphere through the outer wall of the component and piping
- Tritium and hydrogen removal by the purification system installed in the primary and secondary cooling systems
- Tritium and hydrogen leakage to atmosphere and another cooling system
- Isotope exchange reactions between tritium- and hydrogen containing process chemicals, for example, H_2O, H_2SO_4, and HI, in the IS process

The code will be validated to ensure its accuracy, utilizing actual tritium concentration data obtained during the HTTR high-temperature 50-day operation planned in the future.

Regarding the combustible gas explosion evaluation, JAEA proposed a new evaluation scenario [19] which takes into consideration the decrease in hydrogen concentration and the increase in distance during dispersion procedure, and combined with a multienergy method [20]. Computational fluid dynamics simulation tool STAR-CD [21] was utilized for evaluating the behavior of leakage hydrogen dispersion. The calculation results depend on the hydrogen inventory, piping rupture diameters, the height of the leakage point, angles of blowouts, and the arrangement of the partition walls. For further details, see Chapter 25.

Regarding the toxic gas inflow to the reactor building, JAEA has been proposing an evaluation method [22] to determine appropriate offset distance between the reactor facilities and IS process against toxic gas leakage accidents. Density levels of toxic gases in the air are evaluated by using SLAB model [23]. Density levels and exposure time of toxic gases in the reactor control room is evaluated by mass balance differential equation considering density levels of toxic gases in the air and air exchange rate. Toxic gas concentration of incoming air flowing into reactor control room is monitored constantly and the ventilation system of reactor control room isolates the incoming air from outside of the reactor building when the measureable value exceeds the acceptance value. Offset distance will be determined at the point which evaluation results retain below the acceptance value. For further details, see Chapter 25.

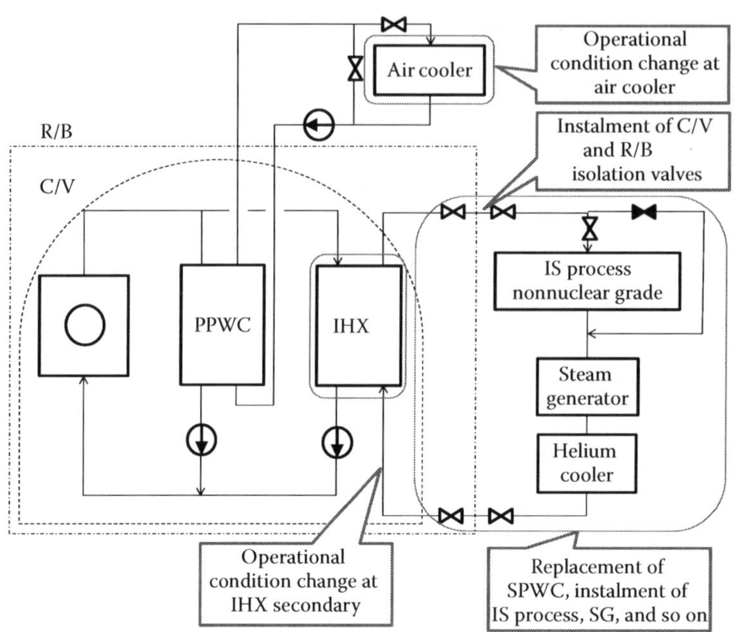

FIGURE 29.9
Major modification of the HTTR-IS system.

29.2.2.2.6 Event Identification

An event identification approach proposed by JAEA is to define abnormal events initiated by the hydrogen production plant as external events, and evaluate the event whose scenarios are changed from those of the HTTR because of the hydrogen production plant coupling. Abnormal events to be evaluated are identified based on failure mode and effect analysis focusing on modifications of the secondary cooling system design in the HTTR such as hydrogen production plant coupling, replacement of the secondary pressurized water cooler, installation of reactor containment isolation valves, and installation of steam generator and helium cooler [24]. In addition, impact of operational condition changes on evaluation items is taken into consideration for the event identification. Figure 29.9 shows the major modifications of the HTTR-IS system. Following events were identified:

1. Increase of IHX primary gas circulator revolution
2. Opening of air cooler bypass flow control valve
3. Closing of air cooler bypass flow control valve
4. Opening of exhaust valve in secondary helium storage and supply system
5. Closing of isolation valve in secondary cooling system
6. Rupture of inner tube in coaxial hot gas duct
7. Rupture of piping in secondary cooling system
8. Rupture of IHX heat transfer tube
9. Rupture of piping in hydrogen production plant

The identified events will be evaluated in the next stage using system analysis code in order to ensure the safety of the HTTR-IS system.

29.3 Conclusion and Future Work

Since a massive amount of hydrogen is necessary in the future hydrogen society, nuclear-produced hydrogen is one of the most progressive candidates for hydrogen production. This section describes the HTTR-IS program in Japan which aims to demonstrate nuclear hydrogen production. Program overview was summarized including the achievement of the HTTR project. Also, newly proposed safety philosophy which aims to apply conventional chemical plant standard to hydrogen production plant is described. Furthermore, conceptual design results of the HTTR-IS system is illustrated.

Based on these achievements, it is desired to carry out the further studies to which one can increase a reliability of the design and evaluation methods. Pilot-scale hydrogen production experiments using an apparatus made of industrial materials and driven by helium gas heating are expected to validate a system evaluation method. In addition, operational data of tritium concentration data during the HTTR continuous high-temperature operation is required for tritium movement method validation. Moreover, detail design with a cost evaluation is required to show the potential of nuclear hydrogen demonstration.

References

1. Council for Science and Technology Policy, 2008. Strategy for Innovative Technology.
2. Japan Atomic Energy Commission, 2008. The innovative nuclear technology roadmap for contributing to the mitigation of global warming.
3. Saito, S., Tanaka, T., Sudo, Y., Baba, O., Shindo, M., Shiozawa, S., Mogi, H., et al., 1994. Design of high temperature engineering test reactor (HTTR). Japan Atomic Energy Research Institute, JAERI 1332.
4. Fujikawa, S., Hayashi, H., Nakazawa, T., Kawasaki, K., Iyoku, T., Nakagawa, S., Sakaba, N., 2004. Achievement of reactor-outlet coolant temperature of 950°C in HTTR. *J. Nucl. Sci. Technol.*, 41(12):1245–1254.
5. Kubo, S., Nakajima, H., Kasahara, S., Higashi, S., Masaki, T., Abe, H., Onuki, K., 2004. A demonstration study on a closed-cycle hydrogen production by the thermochemical water-splitting iodine–sulfur process. *Nucl. Eng. Des.*, 233: 347–354.
6. Nakagawa, S., Takamatsu, K., Tachibana, Y., Sakaba, N., Iyoku, T., 2004. Safety demonstration tests using high temperature. *Nucl. Eng. Des.*, 233: 301–308.
7. Kunitomi, K., Yan, X., Nishihara, T., Sakaba, N., Mouri, T., 2007. JAEA'S VHTR for hydrogen and electricity cogeneration: GTHTR300C. *Nucl. Eng. Technol.*, 39(1): 9–20.
8. Nishihara, T., Mouri, T., Kunitomi, K., 2007. Potential of the HTGR hydrogen cogeneration system in Japan, *Proc. 15th Int. Conf. Nucl. Eng. (ICONE15)*, Nagoya, Japan, April 22–26, ICONE-10157.
9. Sakaba, N., Sato, H., Hara, T., Kato, R., Ohashi, K., Nishihara T., Kunitomi, K., 2007. Conceptual design of the HTTR-IS hydrogen production system, JAEA-Research 2007–058: 1–31.
10. Sakaba, N., Sato, H., Ohashi, H., Nishihara, T., Kuhnitomi, K., 2008a. Development scenario of the iodine-sulphur hydrogen production process to be coupled with VHTR system as a conventional chemical plant. *J. Nucl. Sci. Technol.*, 45(9): 962–969.
11. Ohashi, K., Nishihara, T., and Kunitomi, K. 2006. Fundamental philosophy on the safety design of the HTTR-IS hydrogen production system. *Trans. Atom. Energ. Soc. of Japan*, 6(1): 46–57 (in Japanese).

12. Sakaba, N., Kasahara, S., Onuki, K., Kunitomi, K., 2007. Conceptual design of hydrogen production system with thermochemical water-splitting iodine–sulfur process utilizing heat from the high-temperature gas-cooled reactor HTTR. *Int. J. Hydrogen Energy*, 32: 4160–4169.
13. Sakaba, N., Ohashi, H. Sato, H., Hara, T., Kato, R., Kunitomi, K., 2008. Hydrogen production by high-temperature gas-cooled reactor; conceptual design of advanced process heat exchangers of the HTTR-IS hydrogen production system. *Trans. Atom. Energ. Soc. Jpn.*, 7(3): 242–256 (in Japanese).
14. Sakaba, N., Kasahara, S., Ohashi, H., 2007. Hydrogen production by thermochemical water-splitting IS process utilizing heat from high-temperature reactor HTTR, *Proc. 16th World Hydrogen Energy Conference (WHEC16)*, Lyon, France, 13–16 June, S04–043.
15. USNRC, 1995. RELAP5/MOD3 Code Manuals, Idaho National Engineering Laboratory, NUREG/CR-5535.
16. Takamatsu, K., Katanishi, S., Nakagawa, S., Kunitomi, K., 2004. Development of plant dynamics analytical code named conan-GTHTR for the gas turbine high temperature gas-cooled reactor, (I). *Trans. Atom. Energ. Soc. Japan*, 3(1): 76–87 (in Japanese).
17. Sato, H., Kubo, S., Sakaba, N., Ohashi, H., Tachibana, Y., Kunitomi, K., 2009. Development of an evaluation method for the HTTR-IS nuclear hydrogen production system. *Annals Nucl. Energy*, 36: 956–965.
18. Ohashi, H., Sakaba, N., Nishihara, T., Inagaki, Y., Kunitomi, K., 2007. Numerical study on tritium behavior by using isotope exchange reactions in thermochemical water-splitting iodine–sulfur process, *J. Nucl. Sci. Technol.*, 44(11): 1407–1420.
19. Murakami, T., Terada, A., Nishihara, T., Inagaki, Y., Kunitomi, K., 2006. Analysis on characteristic of hydrogen gas dispersion and evaluation method of blast overpressure in VHTR hydrogen production system. *Trans. Atom. Energ. Soc. Japan*, 5(4): 316–324 (in Japanese).
20. Van den Berg, A.C., 1985. The multi-energy method: A framework for vapour cloud explosion blast prediction. *J. Hazard. Mater.*, 12(1): 1–10.
21. CD Adapco Group, 2004. User Guide STAR-CD Ver. 3.22.
22. Murakami, T., Nishihara, T., Kunitomi, K., 2008. Safety assessment of VHTR hydrogen production system against the fire, explosion and acute toxicity. *Trans. Atom. Energ. Soc. Japan*, 7(3): 231–241 (in Japanese).
23. Morgan Jr., D.L., Kansa, E.J., Morris, L.K., 1983. Simulations and parameter variation studies of heavy-gas dispersion using the SLAB model. UCRL-88516 Rev. 1.
24. Ohashi, K., Nishihara, T., Tazawa, Y., Tachibana, Y., Kunitomi, K., 2009. Basic principles on the safety evaluation of the HTGR hydrogen production system. *JAEA-Technology* 2008–093: 1–33 (in Japanese).

30

Nuclear Hydrogen Project in Korea

Won Jae Lee

CONTENTS

30.1 Hydrogen Economy Road Map

Clean alternative energy that can replace fossil fuels is required in order to resolve the problems of dwindling fossil fuels and climate change. Hydrogen is considered one of the promising future energy solutions due to its high energy density and its clean, abundant and storable nature. Though controversy remains over the prospect of a future hydrogen economy, there is an increasing consensus that the hydrogen economy is a practical and inevitable future rather than merely an option to protect against climate change and the exhaustion of fossil fuel supplies. Hydrogen has already been used as chemical feedstock in oil refining, fertilizers and chemicals, and so on. In the future hydrogen economy, hydrogen will additionally be used as the fuel for fuel cells in transportation, distributed electricity generation, and portable electronic devices, as well as for the feedstock in emerging markets such as synthetic fuel, clean iron ore reduction, coal-to-liquid conversion, and so on. World hydrogen demand that was 50 Mtons/year in 2008 is projected to increase to 78 Mtons/year in 2030 and to 166 Mtons/year in 2040. Hydrogen supply in Korea was 1 Mtons/year in 2008 and is expected to increase to 3.6 Mtons/year in 2030 and to 11.8 Mtons/year in 2040. The market demand will increase rapidly when entering the hydrogen economy in the 2030s.

In light of this anticipated increase in reliance on and demand for hydrogen, advanced countries have put forward their road maps to the hydrogen economy and have launched extensive programs in support of those road maps. In Korea, rapid climate changes and heavy reliance on imported fossil fuels have motivated the government to set up a road map to the hydrogen economy (MOCIE, 2005) [1]. As shown in Figure 30.1, the

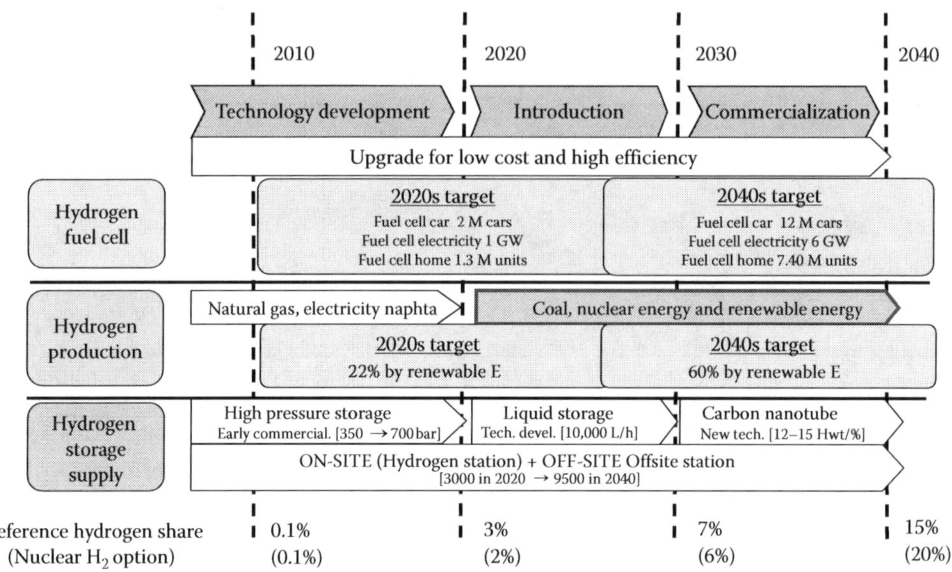

FIGURE 30.1
Road map to Hydrogen Economy in Korea (MOCIE, 2005).

road map consists of four phases: (1) the technology development phase in the 2010s, (2) the introduction phase in the 2020s, (3) the commercialization phase in the 2030s, and (4) the hydrogen economy in the 2040s. For each phase, specific targets are given for hydrogen utilization (fuel cells for cars, central and distributed electricity generation), hydrogen production (fossil energy such as natural gas and coal, nuclear energy, and renewable energy), hydrogen storage (high-pressure gas, liquid and carbon nanotube, and so on), and hydrogen supply (onsite and offsite stations). The reference projection of the hydrogen share in the total energy demand is 7% in the 2030s and 15% in the 2040s. The nuclear hydrogen option adds 5% more to the hydrogen share in the 2040s.

30.2 Nuclear Hydrogen Perspectives

The road map to the hydrogen economy requires the balanced development of hydrogen production, storage/delivery and utilization technologies. One of the major challenges is determining how to produce massive quantities of hydrogen in a clean, safe, and economic way. Among various hydrogen production methods, the massive, safe, and economic production of hydrogen by water splitting using a very high-temperature reactor (VHTR) can provide a successful path to the hydrogen economy. Particularly in Korea, where usable land is limited, the nuclear production of hydrogen is deemed a practical solution due to its high energy density. Another merit of nuclear hydrogen is that the nuclear is a sustainable and technology-led energy not affected by the uncertainties plaguing fossil fuel supplies. Advanced countries like the United States and Japan have launched extensive nuclear hydrogen projects to meet their roadmaps to the hydrogen economy. Korea also launched a nuclear hydrogen project in 2004.

Current hydrogen demand is mainly from oil refinery and chemical industries. Hydrogen is mostly produced by steam reforming using the fossil fuel heat. Against the fossil fuel run-out and climate changes, there is a growing interest in introducing the nuclear hydrogen to the existing hydrogen markets. In Korea, more than 0.6 Mtons/year of hydrogen is produced and consumed at oil refinery industries. Consider that a commercial scale 600 MWt VHTR can produce 0.06 Mtons/year of hydrogen. More than 10 nuclear hydrogen units are required to replace current steam reforming hydrogen production in the oil refinery. In the hydrogen economy in the 2040s, it is projected that the 25% of total hydrogen demand is supplied by the nuclear hydrogen, which is around 3 Mtons/year. For this, it is expected that 50 nuclear hydrogen units will be required.

30.3 Nuclear Hydrogen Programs in Korea

Korean Atomic Energy Commission (AEC) officially approved the long-term development plan for a nuclear hydrogen system in 2008 and the nuclear hydrogen project has now become the national agenda. Nuclear Hydrogen Project in Korea consists of two major programs; the nuclear hydrogen key technologies development program and the nuclear hydrogen development and demonstration (NHDD) program. Figure 30.2 illustrates the implementation plan of the nuclear hydrogen project. The key technologies development program was launched at Korea Atomic Energy Research Institute (KAERI) in 2006 and it focuses on the development and validation of key and challenging

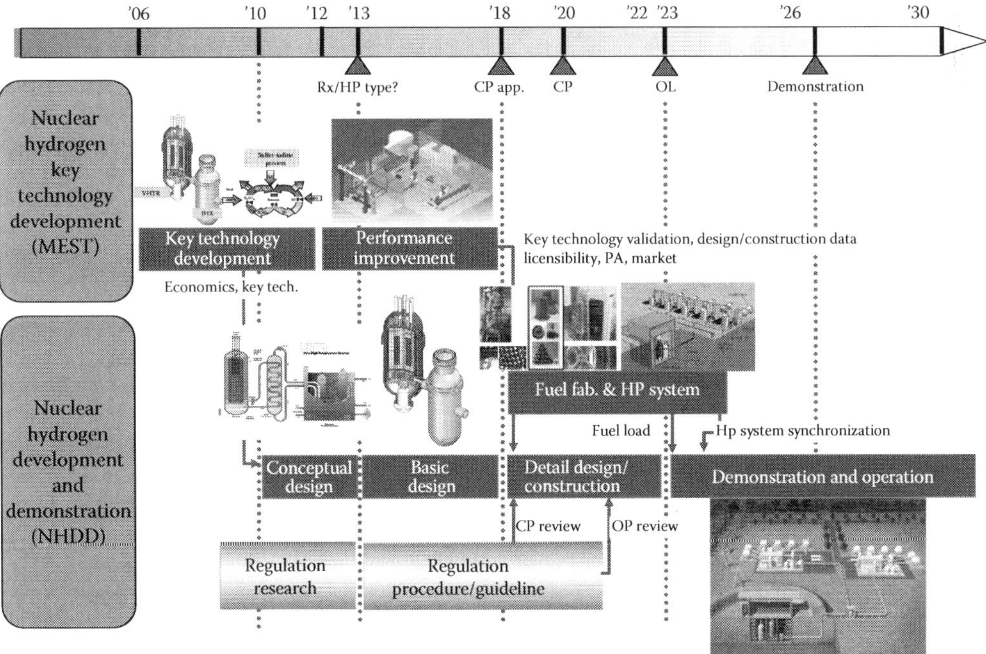

FIGURE 30.2
National Nuclear Hydrogen Project Plan (approved by AEC in December 2008).

technologies required for the realization of the nuclear hydrogen system [2]. The program will run up to 2017 in phase with Gen-IV International Forum projects and the NHDD program. The NHDD program is to design and construct a nuclear hydrogen demonstration system and to demonstrate its hydrogen production and safety. The program is expected to start after 2011 aiming at the completion of construction in 2022 and the demonstration by 2026. Commercial venture for the NHDD program is being discussed. The final goal is to commercialize the nuclear hydrogen technology for 2030s.

Figure 30.3 illustrates a reference nuclear hydrogen demonstration system. It is fully dedicated to the hydrogen production and consists of a VHTR at 950°C, five modules of sulfur–iodine (SI) water-split hydrogen production processes and an intermediate loop that transports the nuclear heat to the hydrogen production process. An underground reactor and indirect loop configuration are adopted for ensuring the safety. Power level of 200 MWt is selected and a cooled-vessel design is adopted [3], which enables the use of a domestic fabricated reactor pressure vessel and ensuring the operation and maintenance cost of the demonstration system by selling the product hydrogen. Both the block and pebble cores are the candidates and the selection is expected in 2012 [4]. And, the ranges of operating parameters are being studied for maximizing the hydrogen production efficiency and considering the integrity, sizing, and manufacturing capability of components [5].

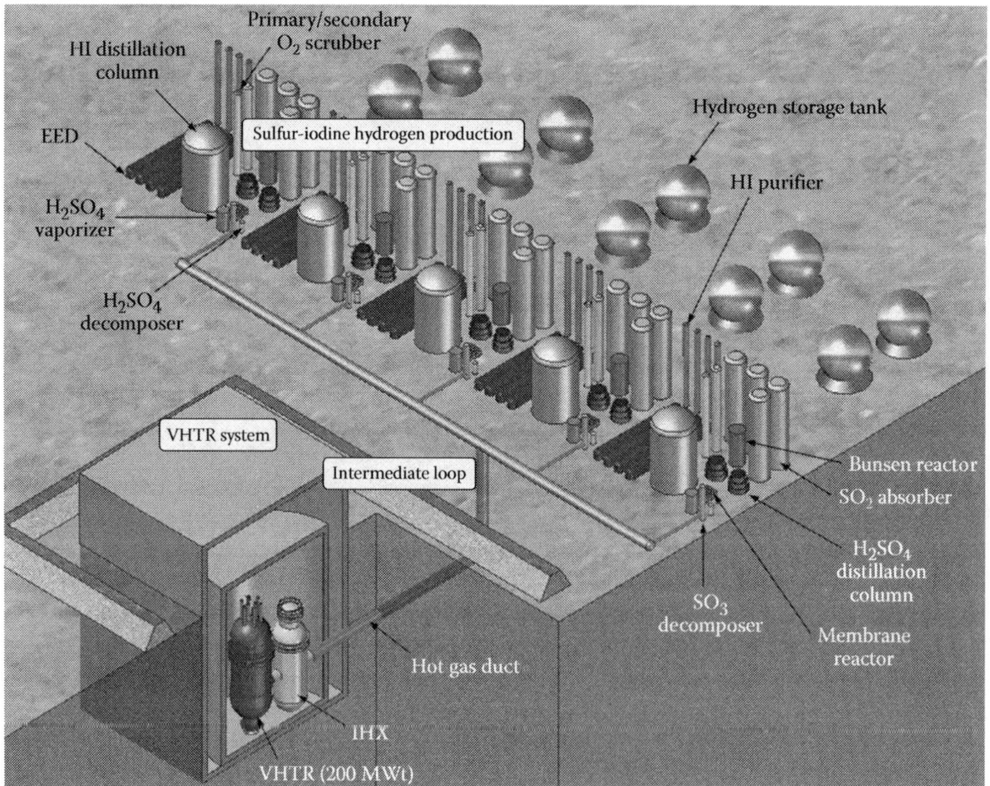

FIGURE 30.3
Nuclear hydrogen demonstration system.

30.4 Nuclear Hydrogen Key Technologies Development Program

Through a preconceptual design study in 2004 and 2005, key technological areas in the VHTR core and system, hydrogen production process, coupling between the reactor and the chemical side, and the coated fuel were identified. Based on this, the nuclear hydrogen key technologies development program was launched at KAERI in 2006 as a national program of Ministry of Education, Science and Technology. It aims at the development and validation of key and challenging technologies that are required for the realization of the nuclear hydrogen system. KAERI takes the leading role of the program and the development of the VHTR technologies. Korea Institute of Energy Research (KIER) and Korea Institute of Science and Technology (KIST) takes the role of developing the SI hydrogen production technologies.

The program consists of two stages; the first stage (2006–2011) for the development of technologies and the second stage (2012–2017) for the performance improvement and validation of technologies. The first-phase study (2006–2008) of the development stage was completed in February 2009 and the second-phase study (2009–2011) had just started. The program is a 12-year program and run in phase with Gen-IV International Forum and the NHDD programs.

30.4.1 Results of the First-Phase Study

From the first-phase study, key basic technologies in the design and computational tools, the high-temperature materials and components, the tristructural-isotropic (TRISO) fuel manufacturing and the SI thermochemical hydrogen production process were developed. The technologies developed can be used for the conceptual design of the NHDD system. Major outcomes of the first-phase study are [6]

1. Design and Computational Tools
   - Design concepts of the 200 MWt block and pebble cores at 950°C satisfying safety and economy
   - The design concept of a cooled-vessel that enables the use of conventional SA508/533 reactor pressure vessel by preventing the direct contact of hot helium with pressure boundary and by additionally cooling of the vessel
   - Computational tools for nuclear design and thermofluid/safety analysis, and so on. Figure 30.4 shows the code systems under development
2. Materials and Components
   - High-temperature material tests in Helium environment
   - Graphite oxidation tests
   - The design concept of a process heat exchanger (PHE: SO_3 decomposer) which is a metal-based and surface-coated hybrid-type heat exchanger as shown in Figure 30.5. A printed circuit-type design is applied to the Helium side and a plate-fin-type design is selected to the SO_3 side to permit replacement of SO_3 decomposition catalyst. In order to ensure the corrosion resistance, the surface of the SO_3 side is coated and mixed with SiC by ion-beam bombardment. Use of metal substrate enhances the manufacturability

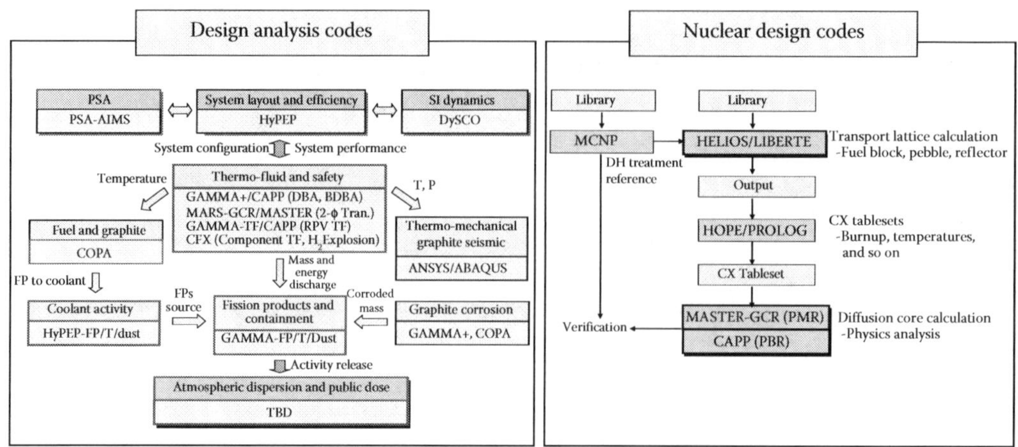

FIGURE 30.4
KAERI computational tools.

- 10 kW high-temperature and -pressure nitrogen/sulfuric acid gas loop to validate the design concept of the PHE and gas loop technologies

3. Fuel
    - Basic TRISO manufacturing and qualification technology at 20 g/batch scale
    - Fuel performance analysis modules
4. SI Hydrogen Production
    - Demonstration of 3.5 L/h of hydrogen production at atmospheric pressure and continuous operation
    - Basic technologies for catalysts, material corrosion, and vapor–liquid equilibrium data
    - Basic technologies of the individual process unit at pressurized conditions

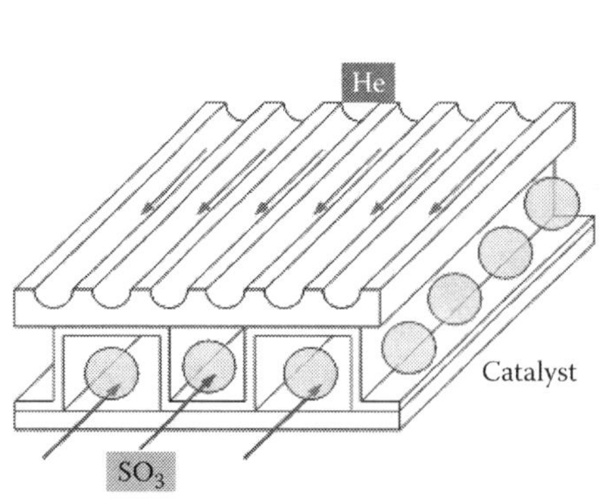

FIGURE 30.5
Process Heat Exchanger (SO_3 decomposer).

30.4.2 Ongoing Activities of the Second-Phase Study

The second-phase study focuses on the improvement and validation of the selected technologies developed in the first-phase study to the level that can be applied to the basic design of the NHDD system. The following study is planned and being performed:

1. Design and Computational Tools
   - Design concepts of the NHDD system configuration and optimization of operating parameters
   - Verification and validation and the documentation of computational tools
2. Materials and Components
   - Continuation of material tests
   - Design validation and manufacturing technology of the PHE
   - Construction of 150 kW high-temperature and pressure helium loop
3. Fuel
   - Optimization of fuel manufacturing and qualification technology at 20 g/batch scale
   - Verification and validation and the documentation of fuel performance code
   - Preparation and initiation of fuel irradiation tests
4. Pressurized SI Hydrogen Production
   - Detail design of Bunsen reaction skid at 50 L/h
   - Construction of HI decomposition skid at 50 L/h
   - Construction of H_2SO_4 decomposition skid at 200 L/h

30.4.3 Longer-Term Plan

In the second-stage (2012–2017) development, the technologies developed in the first-stage development will be improved and refined for application to the detailed design of the NHDD system. Design and safety issues derived from the interaction with licensing authority will be resolved. The PHE design will be commercialized by validating its performance and design through 500 kW helium loop. Fuel manufacturing technology at 2–3 kg/batch scale will be established and the manufactured TRISO particles will go through irradiation tests. For the demonstration of SI performance, it is scheduled to integrate the pressurized SI processes for continuous operation at 50 L/h and then to construct pilot-scale facility.

In order to resolve technical challenges and to improve associated technologies in the VHTR and SI process, various multilateral and bilateral international collaborations are in progress. As multilateral collaborations, KAERI participates in the fuel, hydrogen, materials and computational methods validation and benchmark projects of the Generation-IV International Forum (GIF) VHTR system and in the OECD and IAEA code benchmarks. As bilateral collaborations, I-NERI projects with INL and ANL of the United States are being performed. Nuclear hydrogen joint development center with General Atomics of the United States and the nuclear hydrogen joint research center with INET of China were established for joint research and development of nuclear hydrogen technologies. Information exchange and joint collaborations are in progress with JAEA of Japan, PBMR of South Africa, and CEA of France, respectively. Fuel manufacturing technologies at 2–3 kg/batch scale are being transferred from FZJ of Germany. In order for planning the joint commercial venture of the NHDD program, a domestic council consisting of industries, research organizations, and universities is being established. Such international and domestic collaboration will contribute a lot in ensuring the soundness and integrity of the technologies developed and in mitigating the risk of the NHDD program.

30.5 Two-Step Approach for Process Heat and Hydrogen

By virtue of high temperature feature of the high-temperature gas-cooled reactor (HTGR), HTGR can deliver efficient heat not only to replace the fossil fuel in the industrial process heats, but also to produce hydrogen [7,8]. Figure 30.6 shows the potential markets for nuclear heat applications. They include the district heating and desalination at low temperature ranges, the oil refineries, coal-to-liquid production, high-temperature steam for industrial complexes, steam-methane reforming hydrogen production and high-efficiency electricity generation at medium temperature ranges, as well as coal gasification and water-split hydrogen production at high temperatures. Major interests of the HTGR applications are in the medium and high temperature ranges.

Strategically, a two-step approach is being reviewed in Korea. The first step aims at a mid-term deployment of the advanced HTGR (AHTGR) by the late 2010s to replace the process heats now supplied by fossil fuel. The AHTGR at 850°C is based on a semimature technology with low technical and licensing risks so that the target can be met. Two application areas are being considered. Both produce hydrogen by steam-methane

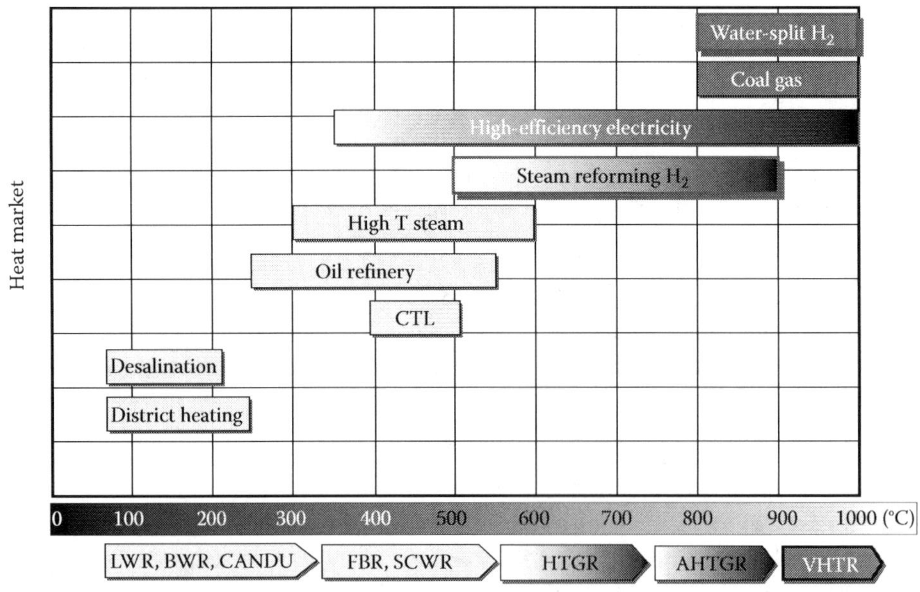

FIGURE 30.6
Potential process heat markets using nuclear energy.

reforming and then the product hydrogen is used (1) to produce the synthetic methanol by recycling captured CO_2 and (2) to reduce iron ore. The second step aims at a long-term deployment of the VHTR for the production of hydrogen. Considering that there are still challenges to be resolved in the VHTR and water-split hydrogen production technologies, target is setup to demonstrate the nuclear hydrogen production by the middle of the 2020s. This two-step approach is deemed practical and achievable by introducing the semimature technologies of the AHTGR first and then extending them to the future technologies of the VHTR.

30.6 Summary and Conclusion

Massive, safe, clean, and efficient production of hydrogen by water splitting using the VHTR can provide a practical path to the future hydrogen economy. Recent government approval of the long-term development plan for a nuclear hydrogen system in Korea has laid a cornerstone for the foundation of the hydrogen economy. A two-step approach that introduces a semimature technologies of the AHTR first and then extending them to the future technologies of the VHTR is deemed a practical and achievable solution for the current issues of dwindling fossil fuels and climate changes and for the future hydrogen economy. In conclusion, the HTGRs and their applications to the industrial process heats and clean hydrogen production will contribute to the low-carbon green growth, clean environment, and energy security of Korea.

Acknowledgment

This work has been supported by the Nuclear Research & Development Program of the Korea Science and Engineering Foundation (KOSEF) grant funded by the Korean government Ministry of Education, Science and Technology (MEST).

References

1. Bu, K.J., National vision for hydrogen economy and its action plan set up, *Energy Economy Research*, **4**(2), 129–147, 2005.
2. Lee, W.J. et al., Status of nuclear hydrogen project in Korea, *ANS Embedded Topical on ST-NH2*, Boston, USA, June 24–28, 2007.
3. Kim, M.H. et al., Computational assessment of the vessel cooling design options for a VHTR, *Proceedings of the Korean Nuclear Society Autumn Meeting*, Pyeongchang, Korea, October 30–31, 2008.
4. Jo, C.K. and Noh, J.M., Preliminary core design analysis of a 200 MWt pebble bed-type VHTR, *Proceedings of the Korean Nuclear Society Spring Meeting*, Jeju, Korea, 2007.
5. Shin, Y., Pre-evaluation of nuclear hydrogen process, KAERI Report NHDD-HI-CA-08-06, 2008.

6. Chang, J. et al., A study of a nuclear hydrogen production demonstration plant, *Nuclear Engineering and Technology*, **39**(2), 111–122, 2007.
7. Kim, Y.W. et al., Gas cooled reactor, its potential applications for process heat, *IAEA TM on Non-electric Applications of Nuclear Energy*, Daejeon, Korea, March 3–6, 2009.
8. Lee, W.J. et al., Perspectives of nuclear heat and hydrogen, *Nuclear Engineering and Technology*, **41**(4), 413–425, 2009.

31

NGNP and NHI Programs of the U.S. Department of Energy

Matt Richards and Robert Buckingham

CONTENTS

31.1 Introduction

The U.S. Congress passed the Energy Policy Act of 2005, which required the Secretary of the U.S. Department of Energy (DOE) to establish the Next Generation Nuclear Plant (NGNP) Project. In accordance with the Energy Policy Act, the NGNP Project consists of the research, development, design, construction, and operation of a prototype plant that (1) includes a nuclear reactor based on the research and development activities supported by the Generation IV Nuclear Energy Systems Initiative, and (2) shall be used to generate electricity, to produce hydrogen, or to co-generate electricity and hydrogen. The NGNP Project supports both the national need to develop safe, clean, and economical nuclear energy and the national hydrogen fuel initiative, which has the goal of establishing greenhouse-gas-free technologies for the production of hydrogen. The DOE has selected the helium-cooled high-temperature gas-cooled reactor (HTGR) as the concept to be used for the NGNP because it is the most advanced Generation IV concept with the capability to provide process heat at sufficiently high temperatures for production of hydrogen with high thermal efficiency. The DOE has also selected the Idaho national laboratory (INL), the DOE's lead national laboratory for nuclear energy research, to lead the development of the NGNP. Concurrently with the NGNP program, the nuclear hydrogen initiative (NHI) was established to develop hydrogen production technologies that are compatible with advanced nuclear systems and do not produce greenhouse gases. As of 2009, the DOE has been considering nearer-term process heat applications for the NGNP and has placed less emphasis on the long-term hydrogen production mission. As discussed below, this decision was based in part on feedback from potential industrial end users of the

NGNP technology. The current DOE schedule for the NGNP Project calls for startup of the NGNP by 2021.

The Energy Policy Act also stipulated the NGNP Project should be undertaken in partnership with the industrial end users of the technology. To that end, a working group was assembled consisting of suppliers of the technology, nuclear plant owners/operators, potential end users, and other supportive technology companies. The objective of the working group is to form an alliance that would provide the private sector perspective and direction for completion of the NGNP in partnership with DOE. The Alliance will support the selection of the specific operating conditions and configuration for the NGNP to ensure it meets private sector expectations, commence management of the project using commercial processes, share the cost of design and construction with the government, and secure a commercial operating company to operate the plant. An outcome of meetings among the working group members was that the NGNP Project should also demonstrate the use of process heat and process steam for industrial applications that could support nearer-term commercialization. These industrial applications include petroleum refining, oil recovery, coal and natural gas derivatives, and petrochemical production. A survey performed by MPR Associates, Inc. for INL [1] indicated near-term industrial applications in the United States could utilize about 78 GW(t) at process temperatures ranging from 250 to 500°C, about 65 GW(t) at process temperatures ranging from 500 to 700°C, and about 13 GW(t) at process temperatures ranging from 700 to 950°C. This thermal energy utilization would require approximately 275 HTGR modules rated at 600 MWt, which represents a significant opportunity for commercialization of this technology. Figure 31.1 shows an artist rendition of the NGNP plant that includes demonstration of industrial applications.

31.2 NGNP Project

As defined in the NGNP Preliminary Project Management Plan [2], the NGNP Project objectives that support the NGNP mission and DOE's vision are as follows:

1. Develop and implement the technologies important to achieving the functional performance and design requirements determined through close collaboration with commercial industry end-users.
2. Demonstrate the basis for commercialization of the nuclear system, the hydrogen production facility, and the power conversion concept. An essential part of the prototype operations will be demonstrating that the requisite reliability and capacity factor can be achieved over an extended period of operation.
3. Establish the basis for licensing the commercial version of the NGNP by the United States Nuclear Regulatory Commission (NRC). This will be achieved in major part through licensing of the prototype by NRC, and by initiating the process for certification of the nuclear system design.
4. Foster rebuilding of the U.S. nuclear industrial infrastructure and contributing to making the U.S. industry self-sufficient for its nuclear energy production needs.

In accordance with the Energy Policy Act stipulations, the NGNP preconceptual design work was focused on concepts for simultaneous electricity generation and hydrogen

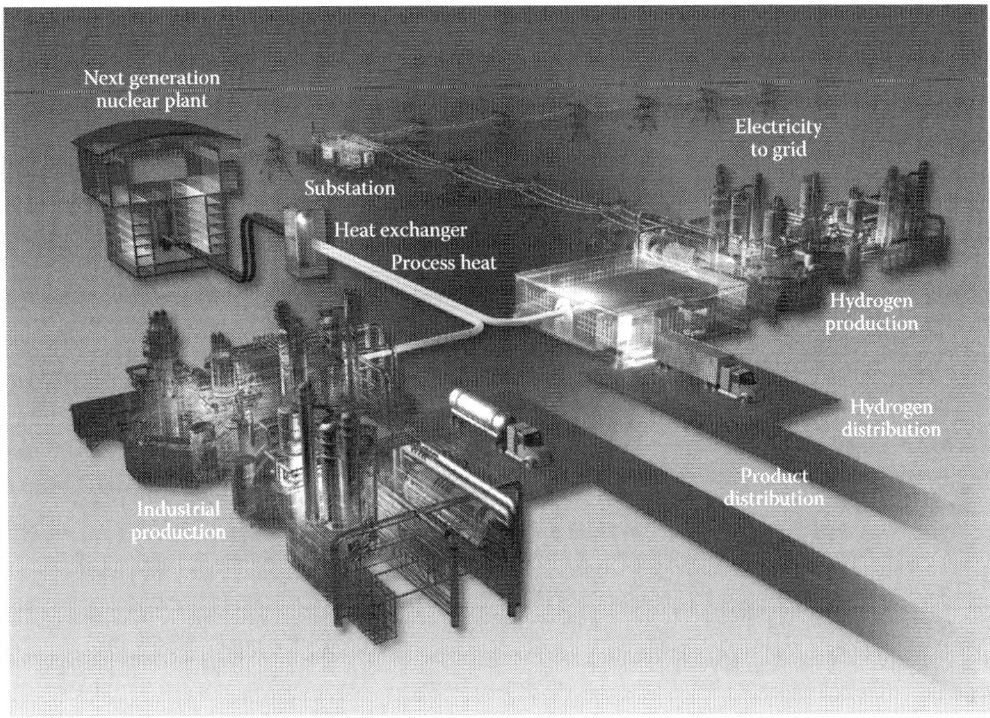

FIGURE 31.1
Artist rendition of the NGNP Plant. (Courtesy of Idaho National Laboratory.)

production. In 2007, NGNP preconceptual design work was completed with the objective of developing a framework in which the design and technology development of the NGNP could progress and to begin to develop bases for selection of the specific design and operational characteristics of the NGNP [3]. This work was completed by three contractor teams with extensive experience in HTGR technology, nuclear power applications, and hydrogen production. These teams were led by Westinghouse Electric Company, LLC; AREVA NP, Inc., and General Atomics (GA). The scope of work included completion of special studies to address key aspects of the NGNP, including reactor type (prismatic block-type core vs. pebble-bed core), power level, power conversion system (PCS) design, heat transport system (HTS) design, licensing, and end-product disposition. These special studies provided the basis for the preconceptual NGNP designs developed by each contractor, which in turn provided the basis for economic assessments for Nth-of-a-kind (NOAK) commercial plants. All three contractors recommended reactor power levels in the range 500–600 MW(t), reactor outlet temperatures in the range 900–950°C, and capability to use approximately 10% of the reactor thermal power for demonstration of hydrogen production at an engineering scale. AREVA and GA recommended prismatic block-type cores, whereas Westinghouse recommended a pebble-bed core.

The GA NGNP preconceptual design is based on the GA Gas Turbine Modular Helium Reactor (GT-MHR) concept shown in Figure 31.2, which utilizes a direct Brayton cycle PCS to produce electricity with a thermal efficiency of 48% [4]. GA has also been developing MHR concepts to support the nearer-term industrial applications discussed above, including a concept that utilizes a steam generator to produce process steam at 585°C.

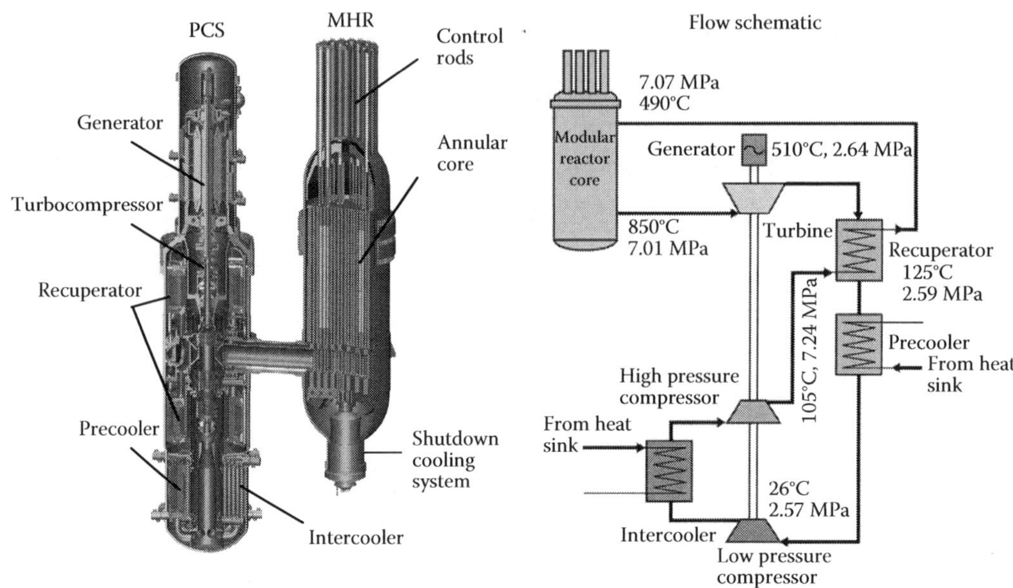

FIGURE 31.2
The gas turbine modular helium reactor.

The NGNP project was funded at US$164.3 million in FY2010. In FY2010 the DOE awarded approximately US$22 million to a team led by GA to complete a conceptual design for the NGNP. In FY2011, the Nuclear Energy Advisory Committee will be asked to review the conceptual design results, the state of NGNP R&D led by INL, and the licensing activities, and make recommendations on whether or not to proceed to Phase 2 of the project including the design selection and development, and construction of NGNP.

31.2.1 HTGR Design and Safety Features

The growing international interest in MHR concepts is the result of MHR design and safety features, which include [5]:

1. *Passive safety, competitive economics, and siting flexibility.* The MHR does not require active safety systems to ensure public and worker safety. The high-energy conversion efficiency of the MHR, combined with the elimination of active safety systems, result in a design that is passively safe and economically competitive with other nonpassively safe reactor concepts. Because of its high efficiency, the MHR rejects less waste heat than other reactor concepts. This design feature, combined with passive safety, allows for more flexible siting options for the MHR.
2. *High-temperature capability and flexible energy outputs.* The MHR is capable of producing process-heat temperatures up to approximately 950°C. This high-temperature capability translates into a high-energy conversion efficiency for a variety of energy outputs, including electricity, hydrogen production, and synthetic fuel production.

3. *Flexible fuel cycles*. The MHR can operate efficiently and economically with several different fuel cycles. MHR designs have been developed utilizing low-enriched uranium (LEU) fuels, high-enriched uranium (HEU) fuels, mixed uranium/thorium and plutonium/thorium fuels, and surplus weapons-grade plutonium fuels. The thermal neutron spectrum of the MHR, combined with robust, ceramic-coated particle fuel, allow for very high burnup in a single pass through the reactor. More recently, an MHR design has been developed to deeply burn plutonium and other transuranic (TRU) actinides recovered from light-water reactor (LWR) spent fuel. As shown in Figure 31.3, the flexible fuel cycle capability of the MHR, combined with its flexible energy output capability, result in a design concept that is very well suited for a wide variety of energy-growth scenarios.

As shown in Figure 31.4, the MHR prismatic block-type fuel element consists of coated particle fuel formed into fuel rods and inserted into graphite fuel elements. The fuel kernels are manufactured using a sol–gel process and are coated with pyrolytic carbon and silicon carbide (SiC). The coating system consists of four layers and three materials; a low-density pyrolytic carbon layer (buffer), an inner high-density pyrolytic carbon layer (IPyC), a SiC layer, and an outer high-density pyrolytic carbon layer (OPyC). The coatings are deposited under conditions to produce isotropic material properties, and the layers are referred to collectively as a TRISO coating, which stands for TRI-material, ISOtropic. The coated particles are consolidated with a carbonaceous matrix into compacts which are loaded into graphite fuel elements. A standard fuel element has 210 blind fuel holes, 108 coolant holes, and contains 3126 fuel compacts. The MHR core design developed by GA consists of 102 fuel columns in three annular rings with 10 fuel blocks per fuel column, for a total of 1020 fuel blocks in the active core.

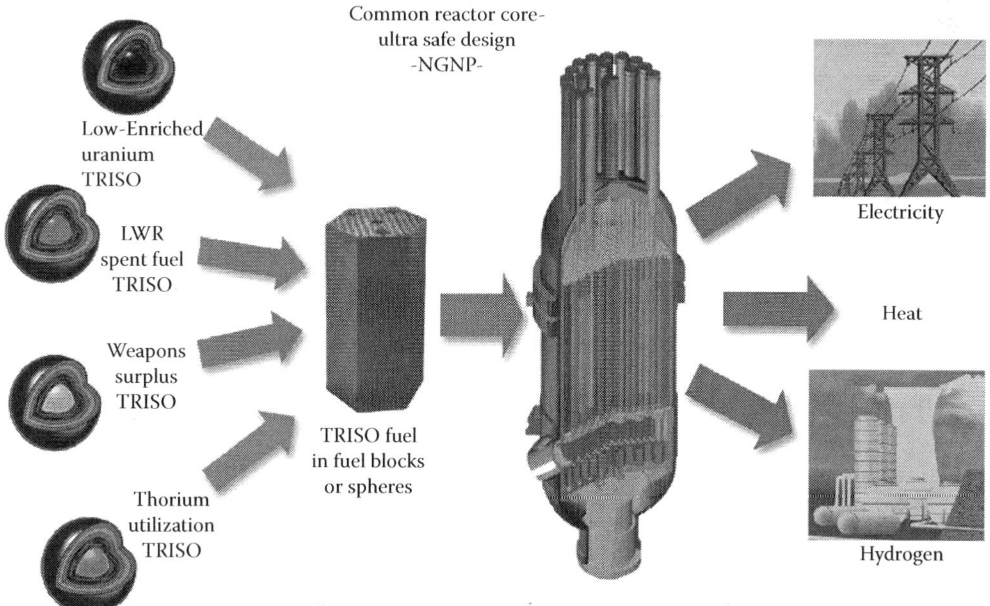

FIGURE 31.3
MHR fuel cycle and energy output options.

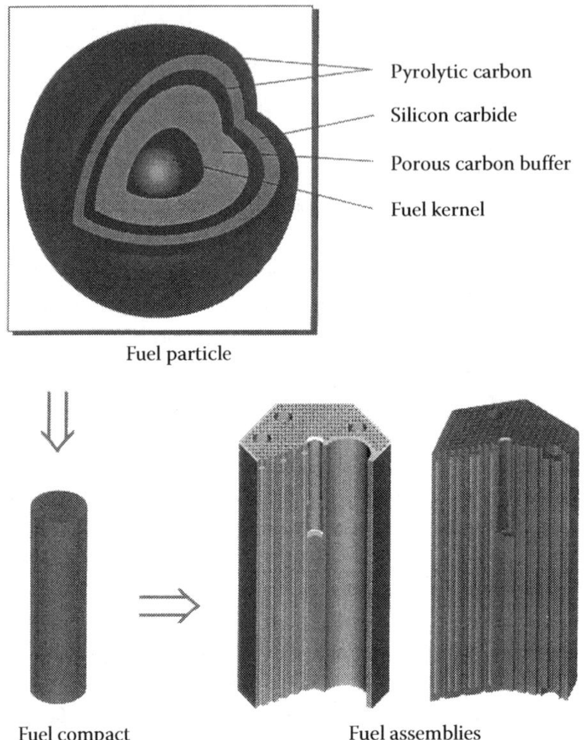

FIGURE 31.4
MHR fuel element components.

Passive safety features of the MHR include (1) ceramic, coated-particle fuel that maintains its integrity at high temperatures during normal operation and loss-of-coolant accidents (LOCAs); (2) an annular graphite core with high heat capacity that limits the temperature rise during a LOCA; (3) a relatively low power density that helps to maintain acceptable temperatures during normal operation and accidents; (4) a negative temperature coefficient of reactivity; and (5) an inert helium coolant, which reduces circulating and plate out activity. During a LOCA, the decay heat is conducted through the graphite to the vessel. The heat is transferred from the vessel by thermal radiation and natural convection to a reactor cavity cooling system (RCCS). As shown in Figure 31.5, the RCCS has no active components and rejects heat using natural convection airflow. Figure 31.6 shows the peak fuel temperature as a function of time during loss of flow accidents and LOCAs. The peak fuel temperature remains below 1600°C during these accidents, which is well within the demonstrated performance capability of coated-particle fuel.

31.2.2 GA NGNP Preconceptual Design

Figure 31.7 shows the NGNP concept developed by the GA-led team [6]. The nuclear heat source for the NGNP consists of a single 600 MW(t) MHR module with two primary coolant loops for transport of the high-temperature helium exiting the reactor core to a direct cycle PCS for electricity generation and to an intermediate heat exchanger (IHX) for hydrogen production. The GA NGNP concept is designed to demonstrate hydrogen production

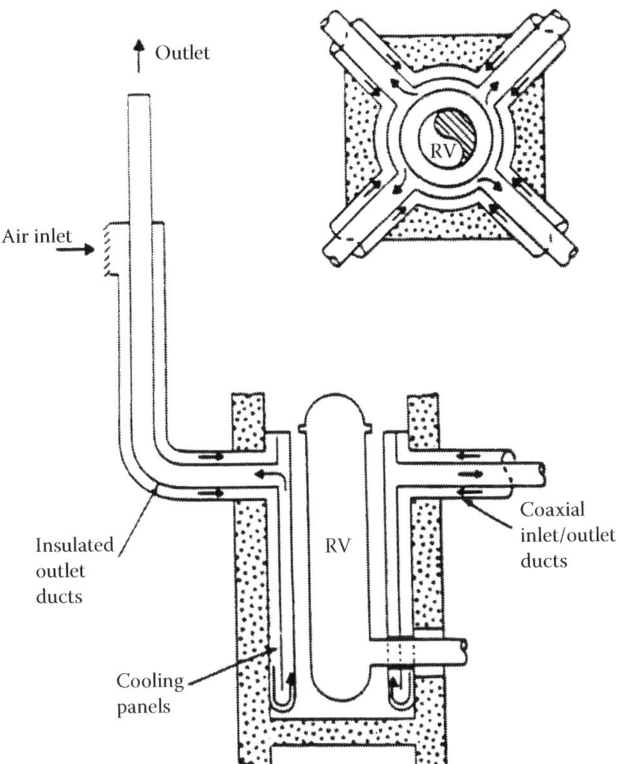

FIGURE 31.5
Passive air-cooled RCCS.

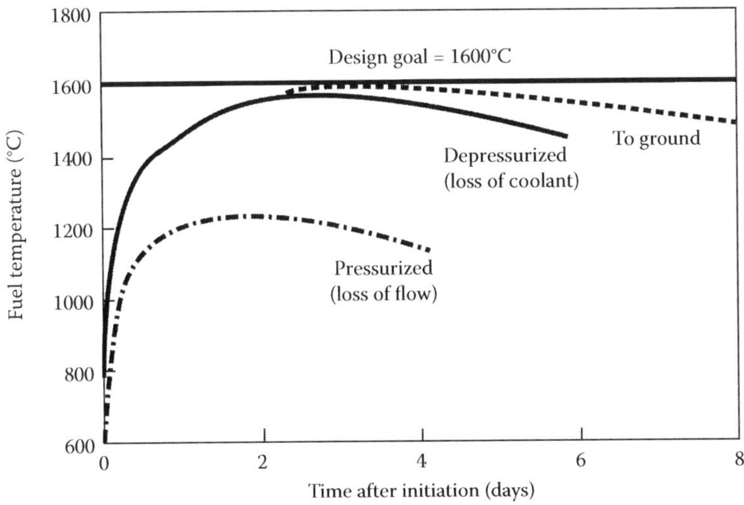

FIGURE 31.6
Peak fuel temperatures during accidents.

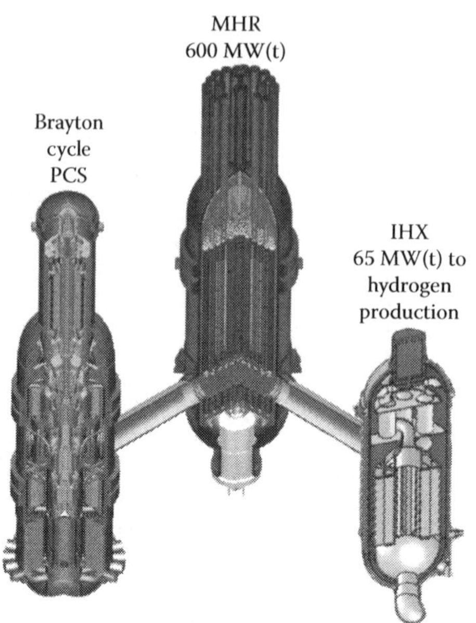

FIGURE 31.7
NGNP concept developed by the GA team.

using both the thermochemical sulfur-iodine (SI) process and high-temperature electrolysis (HTE). The two primary coolant loops can be operated independently or in parallel. The reactor design is essentially the same as that for the GT-MHR, but includes the additional primary coolant loop to transport heat to the IHX and other modifications to allow operation with a reactor outlet helium temperature of 950°C (vs. 850°C for the GT-MHR). The IHX transfers a nominal 65 MW(t) to the secondary heat transport loop that provides the high-temperature heat required by the SI-based and HTE-based hydrogen production facilities. Table 31.1 gives the nominal plant design parameters for operating in both cogeneration and electricity-only modes with a reactor outlet helium temperature of 950°C.

The plant layout consists of the reactor building, the two hydrogen production facilities, and several support buildings and facilities. Systems containing radionuclides and safety-related systems are located in the Nuclear Island, which is separated physically and functionally from the remainder of the plant. A key consideration for safety and licensing of the NGNP (and commercial nuclear hydrogen production plants) is colocation of the MHR module with the hydrogen production facilities. It is proposed to locate the two hydrogen production facilities at a distance of 90 m from the MHR in order to limit the distance over which high-temperature heat is transferred. This separation distance is consistent with an INL engineering evaluation [7] which concluded that separation distances in the range of 60–120 m should be adequate in terms of safety.

31.2.3 Economic Assessments for Commercialization

Two commercial nuclear hydrogen plant variations were evaluated with respect to their hydrogen production costs versus a projection of the future market value of hydrogen:

1. A NOAK nuclear hydrogen production plant consisting of two 600 MW(t) MHR modules providing process heat to a SI-based hydrogen production facility and

TABLE 31.1

NGNP Nominal Plant Design Parameters

	Electricity Only	Cogeneration
MHR System		
Power rating	600 MW(t)	600 MW(t)
Core inlet/outlet temperatures	590/950°C	590/950°C
Peak fuel temperature—normal operation	1250–1350°C	1250–1350°C
Peak fuel temperature—accident conditions	< 1600°C	< 1600°C
Helium mass flow rate	321 kg/s	321 kg/s
MHR System pressure	7.0 MPa	6.4 MPa
Power Conversion System		
Mass flow rate	321 kg/s	286 kg/s
Heat supplied from MHR System	600 MW(t)	535 MW(t)
Turbine inlet/outlet temperatures	948/617°C	948/617°C
Turbine inlet/outlet pressures	7.0/3.0 MPa	6.2/2.6 MPa
Electricity generation efficiency[a]	50.5%	50.5%
Heat Transport System		
Primary helium flow rate	N/A	35 kg/s
Secondary helium flow rate	N/A	35 kg/s
IHX heat duty	N/A	65 MW(t)
IHX primary side inlet/outlet temperatures	N/A	950/590°C
IHX secondary side inlet/outlet temperatures	N/A	925/565°C
HTE-Based Hydrogen Production System		
Peak SOE temperature	N/A	862°C
Peak SOE pressure	N/A	5.0 MPa
Product hydrogen pressure	N/A	4.95 MPa
Hydrogen production rate	N/A	6000 Nm^3/h
Plant hydrogen production efficiency[b]	N/A	~53%
SI-Based Hydrogen Production System		
Peak process temperature	N/A	900°C
Peak process pressure	N/A	7.0 MPa
Product hydrogen pressure	N/A	4.0 MPa
Hydrogen production rate	N/A	9000 Nm^3/h
Plant hydrogen production efficiency[b]	N/A	~45%

[a] Neglects parasitic heat losses from the RCCS and SCS; neglects reduction in efficiency due to turbine blade cooling.

[b] Based on the higher heating value of hydrogen (141.9 MJ/kg).

two 600 MW(t) MHR modules dedicated to electricity production to provide the electric power needed by the SI-based hydrogen production facility and to provide surplus electricity to the grid.

2. A NOAK HTE-based nuclear hydrogen production plant consisting of four 600 MW(t) MHR modules providing both process heat and electricity to the HTE-based hydrogen production facility, which consists of 292 modular hydrogen production units with solid oxide electrolyzer (SOE) cells.

TABLE 31.2

Hydrogen Production Costs for Commercial NOAK Plants

	SI-Based Plant		HTE-Based Plant	
	Credit ($/kg)	H_2 Cost ($/kg)	Credit ($/kg)	H_2 Cost ($/kg)
H_2 Production Cost without Credits	—	3.14	—	2.40
Electricity Credit at 106 mil/kW-h	0.70	—	—	—
H_2 Cost with Electricity Credits	—	2.44	—	2.40
O_2 Credit at $23/Tonne	0.18	—	0.18	—
H_2 Cost with Electricity and O_2 Credits	—	2.26	—	2.22

The SI- and HTE-based hydrogen production plant designs were based on previous work performed by GA as part of a U.S. DOE Nuclear Energy Research Initiative (NERI) [8–10].

The commercial assessment for each of the plant variations involved development of a capital cost estimate and operating cost estimate (including operations and maintenance, fuel, and decommissioning costs) for the plant and an estimate of the amount of hydrogen produced by the plant in order to calculate the unit cost of hydrogen production. This unit cost was then compared against the projected market value of hydrogen that was estimated as part of an NGNP end-products study. Table 31.2 summarizes the result of the commercial assessment. The overall hydrogen production cost for the SI-based plant and the HTE-based plant were estimated to be about $2.26/kg and $2.22/kg, respectively. For both plants the hydrogen production cost is about 10% below the projected market value of hydrogen.

31.3 Nuclear Hydrogen Initiative

31.3.1 Overall Objectives

The objectives of the NHI are:

1. Operate laboratory- and pilot-scale experiments of thermochemical and HTE hydrogen production technologies to demonstrate feasibility and scale-up.
2. Select the hydrogen production technology to be coupled with the NGNP.
3. Demonstrate a commercial-scale hydrogen production system for use with advanced nuclear reactors.

After reviews and assessments of a number of hydrogen production processes were performed, the NHI is presently supporting laboratory-scale research and development of HTE at INL (Chapter 18), and the SI thermochemical cycle at GA, and the hybrid sulfur thermochemical cycle at Savannah River National Laboratory (Chapter 19). These processes were determined to have the highest potential for successful demonstrations that could lead to future commercialization. Since the work done on the other processes is already reported elsewhere in the book, only the work being performed at GA on the SI process is described below.

31.3.2 The SI Process Development

GA proposed the original SI process in 1970s and examined the application of nuclear and solar energy to thc process in the 1980s. Table 31.3 includes the milestones of the research and development works undertaken until now at GA.

The work performed at GA since 2005 involves collaboration among GA, Sandia National Laboratory (SNL), and the French Commissariat L'Energie Atomique (CEA) for integrated laboratory-scale (ILS) experimental demonstrations of the SI process. The SI process is illustrated in Figure 31.8 and involves decomposition of sulfuric acid (H_2SO_4) and hydrogen iodide (HI), and regeneration of these reagents using the Bunsen reaction. Process heat is supplied at temperatures greater than 800°C to concentrate and decompose H_2SO_4. The exothermic Bunsen reaction is performed at temperatures below 120°C and releases waste heat to the environment. Hydrogen is generated during the decomposition of HI, using process heat at temperatures greater than 300°C. Figure 31.9 shows the ILS skids housing CEA's Bunsen reaction unit, SNL's H_2SO_4 decomposition unit, and GA's HI decomposition unit.

The ILS demonstration is performed to achieve incremental technical objectives. First, it performs open loop component reaction experiments that demonstrate the key chemical reactions and verify engineered materials for construction. This is followed by the development of process control elements and interface components such as pumps and buffer tanks to assemble an integrated process experiment. The closed loop process experiment of hydrogen production under at prototypical temperatures and pressures is carried out to develop the technical basis for evaluating the SI nuclear hydrogen production. The last task involves the preparation for a process flow sheet for follow-on pilot scale plants that can be used to predict the process efficiency and preliminary production costs.

TABLE 31.3
The IS Process Research and Development Milestones at GA

Year	Milestone
1972	GA initiates thermodynamic based investigation into possible thermochemical cycles
1974	Begin active investigation into practical thermodynamic cycles
1975	Apply initial patent for SI cycle to make sulfuric acid and hydrogen
1977	Follow with multiple patents covering complete SI cycle
1978	Begin DOE support of SI cycle development
1979	Start construction of glass loop
1981	Demonstrate Bunsen flow reactor at 60 Lph equivalent hydrogen
1980	SI tops DOE evaluation of SI, HyS, and LTE
1981	DOE ends support of nuclear SI cycle
1981	DOE starts support of solar SI
1983	Demonstrate solar decomposition of sulfuric acid on GIT power tower, atm operation, limited operation due to lack of sunshine
1986	Complete high pressure simulated solar sulfuric acid decomposition 5–10 atm operation, Pt and Fe catalysts, 700–900°C outlet temperature
1999	GA begins NERI evaluation of potential nuclear TC processes, jointly with SNL and UL
2000	Identify top two processes: SI top candidate, UT-3 second place
2003	Complete NERI final report on SI flowsheet and begin the 1st phase I-NERI
2005	Start the 2nd phase I-NERI, jointly of SNL, GA CEA
2009	Conclude I-NERI SI ILS Experiment at GA site

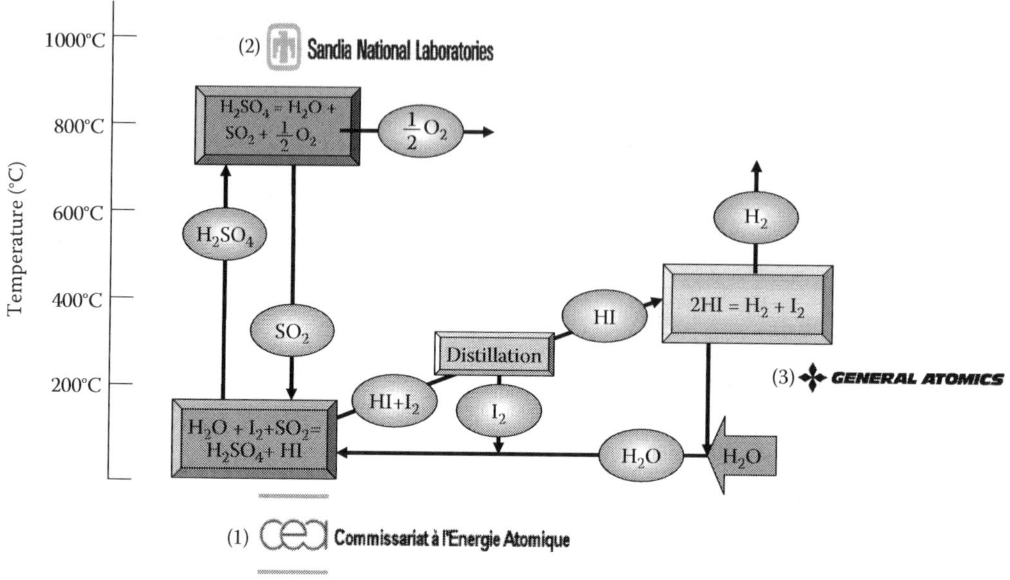

FIGURE 31.8
SI process schematic.

The ILS is designed for production capacity up to 200 L/h hydrogen. All process components are fabricated of engineering materials to the maximum extent and validated for use in future pilot plants. Materials of SiC and glass-lined steels are heavily recommended based on the ILS experience for pilot plant construction in Table 31.4. The three process sections are fabricated into skid-mounted units facilitating modular shipment to and assembly

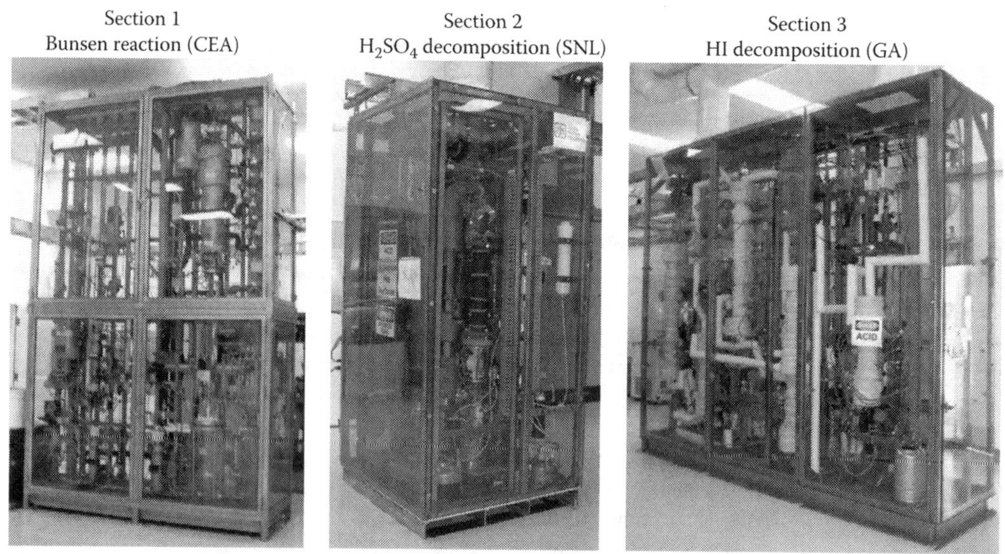

FIGURE 31.9
Skids for laboratory-scale demonstration of the SI process at GA.

TABLE 31.4

The ILS Program Recommended Materials for Pilot Plant Construction

Equipment	Material Options	Material Recommended for Pilot Plant
HI Decomposition Section		
Iodine extraction vessel	Lined steel (glass, Teflon) Ta alloys, Ta-coated steel Nb alloys, Nb-coated steel ceramics	Glass-lined steel
HI distillation vessel	Brick-lined coated steel, coated steel Ta alloys, Ta-coated steel ceramics	Brick-lined coated steel
HI decomposition vessel	Hastelloy B2, B3, C276 Brick lined, coated-steel ceramics	Hastelloy C276 pipe
I_2 and HI recycle vessel	Lined steel Ta alloys, Ta-coated steel Nb alloys, Nb-coated steel ceramics	Glass-lined steel
Balance of system piping, pumps, valves	Lined steel Ta alloys, Ta-coated steel Nb alloys, Nb-coated steel Glass-lined, ceramic-lined	Glass-lined steel
Sulfur Acid Section		
Acid concentration	Lined steel (glass, Teflon), Ceramics	SiC, Teflon
Acid decomposer	Ceramics, SiC, SiO_2 800 H, 617, High Si Irons Lined, or coated steels	SiC, SiO_2
H_2SO_4 piping, tanks	Glass-, Teflon-lined steels, Ta-coated steels, ceramics	SiC, glass and Teflon-lined steels
Bunsen Reaction Section		
Bunsen reactor	Glass-lined steel Ta-coated, lined steel ceramics	Glass-lined steel
SO_2, O_2 separation and compression	Stainless steels (with H_2O removed) glass-lined Ta lined, coated	Stainless steels
H_2SO_4 phase piping, tanks	Glass-lined Ta coated, lined Ta, TaW alloy tubing PTFE tubing	Glass-lined steel
HI phase piping, I_2 buffer tanks	Glass-lined steels, Ta-lined steels Teflon-lined steels ceramics	Glass-lined steel

at GA site. The integrated process is continuously operated with buffer storage of process chemicals between the sections. The control system can operate each section in standalone and integrated operation modes. The design does not require process heat integration.

Because of the relatively small scale of the loop, some difficulties were encountered with pumping and level control of iodine in the Bunsen reaction section, whose process schematic is shown in Figure 31.10. As a result, sustained operations proved difficult. Several improvements to this skid have been effective in dealing with these issues, including adding differential pressure cells to the top of the Bunsen reactor to better determine the iodine interface level, using pressurized nitrogen in combination with a metering pump to regulate the flow of iodine, and shortening the distance between the iodine feed tank and the Bunsen reaction skid to reduce the length of heat-traced tubing. The SO_2 processing was fully tested for interfaced operation with the sulfur acid skid, and the Bunsen reaction was performed with iodine preloaded. The Bunsen reactor in Figure 31.11 has produced an upper phase with H_2SO_4, concentrated to 12.2 wt%, but longer runs are needed to reach equilibrium conditions. Target specifications were achieved for the lower phase, with HI concentrated to 57.8 wt%, and were accepted for feed to the HI skid. Figure 31.12 shows samples taken from the upper and lower phases of the Bunsen reactor.

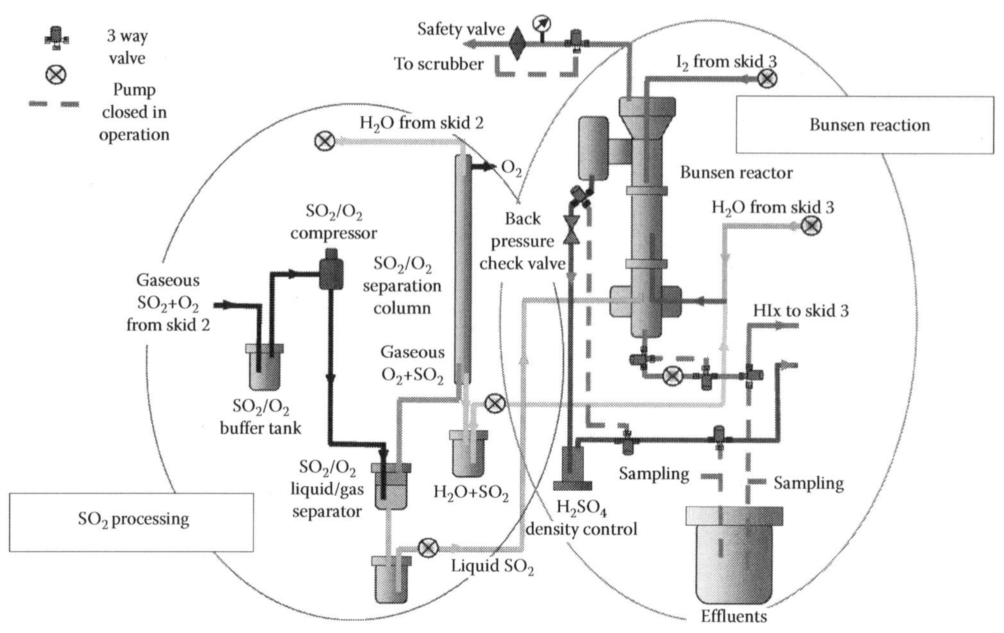

FIGURE 31.10
The ILS's Bunsen section comprises SO_2 processing and Bunsen reaction.

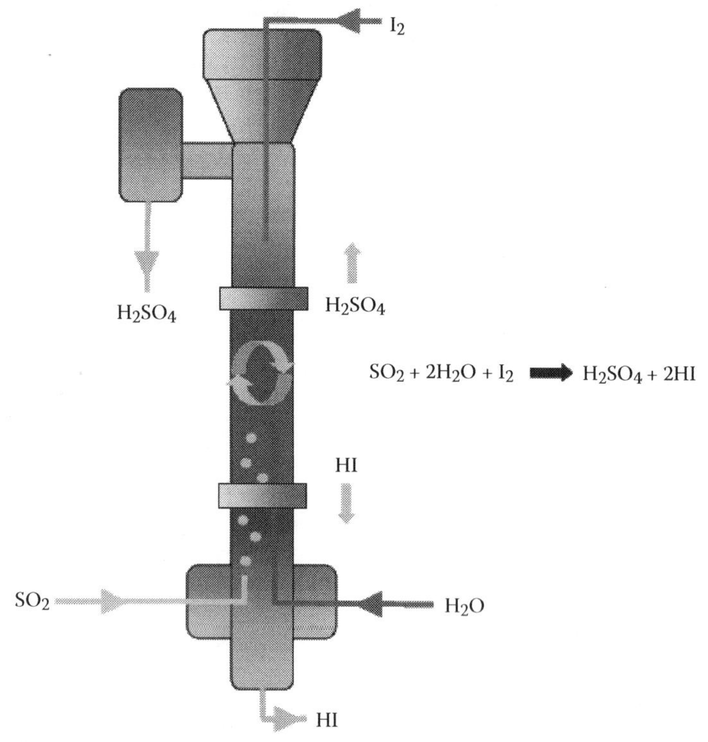

FIGURE 31.11
Reaction flow streams of the Bunsen reactor design.

FIGURE 31.12
Upper and lower phase samples form the Bunsen reaction skid.

The H_2SO_4 decomposition skid is built on the process scheme in Figure 31.13 and has operated successfully utilizing the SiC bayonet-type decomposer illustrated in Figure 31.14. Liquid H_2SO_4 flows into the outer annular flow path and the reaction products flow out of the central tube. With this design, acid boiling, superheating, decomposition, and heat recuperation are combined into a single unit. All connections are at the lower-temperature end of the bayonet. Decomposition was demonstrated with this equipment at temperatures ranging from 700 to 850°C, and the decomposition rate increases two and a half times in over this temperature range. At 850°C, the skid has produced SO_2 at a rate of approximately 300 standard liters per hour, as shown in Figure 31.15. The operation of the

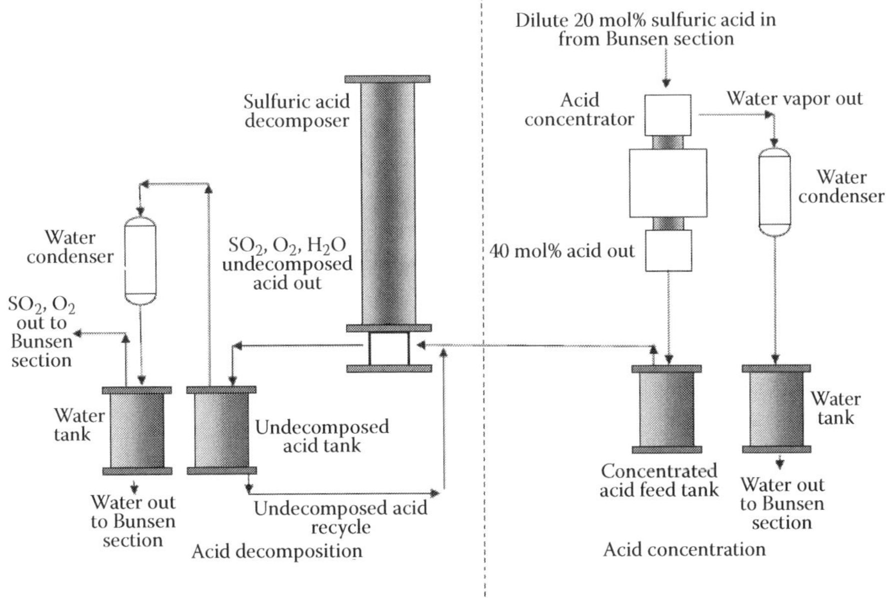

FIGURE 31.13
The ILS's sulfuric acid section comprises two subsections.

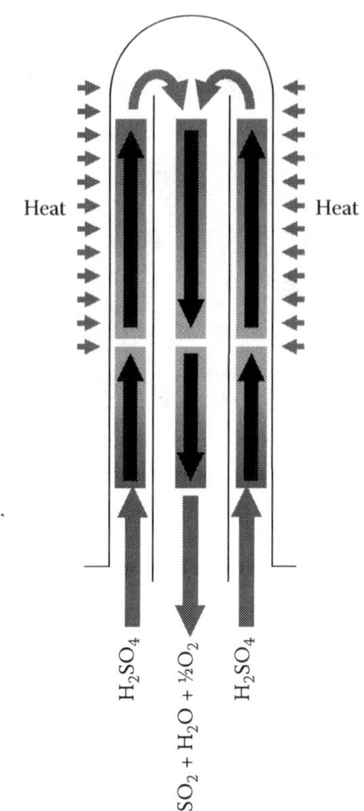

FIGURE 31.14
Bayonet concept of H_2SO_4 decomposer.

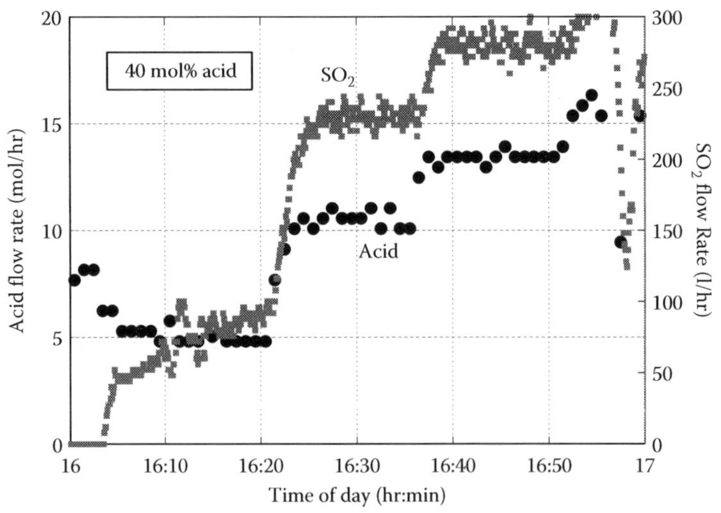

FIGURE 31.15
H_2SO_4 decomposition and SO_2 gas generation data at 850°C process operation.

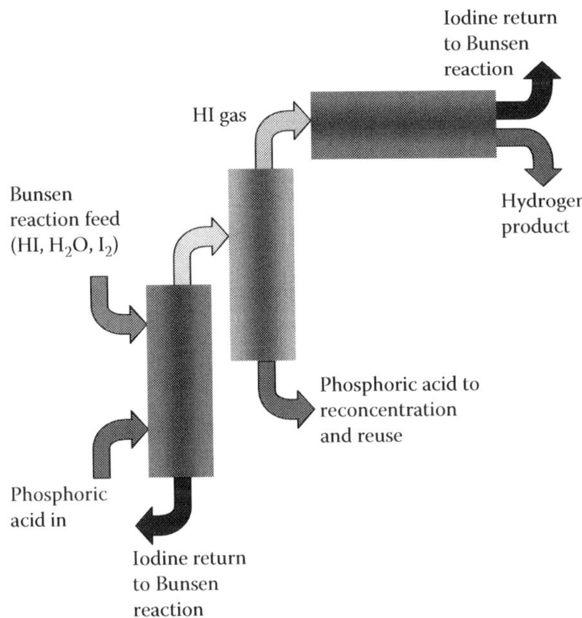

FIGURE 31.16
The ILS's HI decomposition process uses phosphoric acid for extraction of I_2.

sulfur acid decomposer has been reproducible through approximate 20 cycles and SO_2 produced in the process has been fed to the Bunsen section skid. The only problem associated with this section seems to be catalyst durability that requires further evaluation.

For the HI decomposition skid, it was initially intended to use a reactive distillation process to separate I_2 from HI/H_2O. However, testing of this process using realistic feed compositions was not successful, and further investigation is planned. The skid was then incorporated with a previously proven process that utilizes phosphoric acid to extract I_2 from the mixture. The overall process built into the skid is illustrated in Figure 31.16.

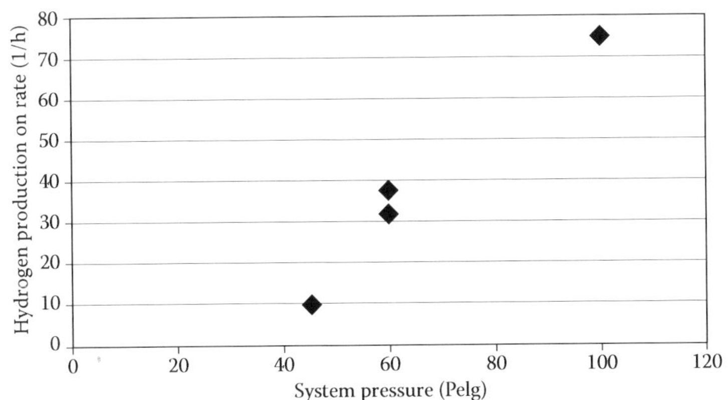

FIGURE 31.17
The HI section hydrogen production rate versus HI reaction pressure.

After the I_2 is extracted, HI is distilled from the phosphoric acid/water mixture, and the HI is then decomposed over a catalyst bed into H_2 and I_2. Unreacted HI is condensed at 0–5 °C and fed back to the reactor. Higher system pressures result in less HI recycle and higher H_2 production rates, as shown in Figure 31.17. This skid has produced hydrogen at up to 75 standard liters per hour, which is below the target rate of 100 to 200 standard liters per hour.

The loop-test program was funded through an international NERI (I-NERI) project which concluded in April 2009. International collaborations are being investigated to support further development of the SI process. Future work will focus on process and component improvements, including development of the reactive distillation process for HI decomposition, materials development and optimization, and scale-up of component designs.

References

1. Survey of HTGR Process Energy Applications, MPR-3181, Rev. 0, MPR Associates, Inc, Alexandria, VA, May 2008.
2. Next Generation Nuclear Plant Project, Project Management Plan, INL/EXT-05-00952, Rev. 1, Idaho National Laboratory, Idaho Falls, ID, March 2006.
3. Next Generation Nuclear Plant Pre-Conceptual Design Report, INL/EXT-07-12967, Rev. 1, Idaho National Laboratory, Idaho Falls, ID, November 2007.
4. Labar, M., Shenoy, A., Simon, W., and Campbell, E., The gas turbine-modular helium reactor, *Nuclear News*, 46(11), 2, 2003.
5. Richards, M., Shenoy, A., Venneri, F., LaBar, M., Schultz, K., and Brown, L., The modular helium reactor for future energy needs, *Proceedings of ICAPP '06*, Reno, NV, June 4–8, 2006, Paper 6154.
6. NGNP and Hydrogen Production Preconceptual Design Studies Report, RGE 911107, Rev. 0, General Atomics, San Diego, CA, July 2007.
7. An Engineering Analysis for Separation Requirements of a Hydrogen Production Plant and High-Temperature Nuclear Reactor, INL/EXT-05-00317, Rev. 0, Idaho National Laboratory, Idaho Falls, ID, March 2005.
8. Richards, M., Shenoy, A., Brown, L., Buckingham, R., Harvego, E., Peddicord, K., Reza, S., and Coupey, J., H2-MHR Pre-Conceptual Design Report: SI-Based Plant, Final Technical Report for the period September 2002 through September 2005, GA-A25401, General Atomics, San Diego, CA, April 2006.
9. Richards, M., Shenoy, A., Harvego, E., McKellar, M.G., Peddicord, K., Reza, S., and Coupey, J., H2-MHR Pre-Conceptual Design Report: HTE-Based Plant, Final Technical Report for the period September 2002 through September 2005, GA-A25402, General Atomics, San Diego, CA, April 2006.
10. Brown, L.C., Besenbruch, G.E., Lentsch, R.D., Schultz, K.R., Funk, J.F., Pickard, P.S., Marshall, A.C., and Showalter, S.K., High efficiency generation of hydrogen fuels using nuclear power, GA-A24285, Rev. 1, *General Atomics*, San Diego, CA, December 2003.

32

International Development of Fusion Energy

Satoshi Konishi

CONTENTS

32.1 The International Thermonuclear Experimental Reactor Project

The largest and most important project of the fusion energy research and development in the world is the International Thermonuclear Experimental Reactor (ITER) being built in France. The ITER machine will be the first experimental fusion device that generates fusion energy in engineering scale, typically 500 MW for several hundreds to thousands of seconds by sustained reaction, aiming at the demonstration of possible commercial energy production from fusion. This project originally started by the four parties, the European Union, Japan, former Soviet Union, and the United States was based on the Summit discussion in Geneva in 1985. Following the Conceptual Design Activity in 1988–1990, Engineering Design Activity conducted from 1992 to 2001 by the four parties generated a detailed design of the plant to be ready for construction, supported by a number of developments and experimental verification of the components that posed various kinds of technological challenges. Despite scientific confidence and expectation, recognition of the difficulty in both technical and financial scale took the next phase a few years to start international cooperation in the implementation of the construction and operation of this large device. In 2003, Peoples Republic of China and the Republic of Korea joined, and India followed in 2005, and the current joint efforts by the seven parties supported by more than one half of the world population.

The ITER agreement was signed in 2006 deciding Cadarache, in Southern France as a construction site of the plant, and ITER Organization that is responsible for construction, operation, and decommissioning of the entire ITER project. ITER organization is staffed with approximately 500 people from the seven parties in 2010, led by the Director General Kaname Ikeda of Japan, and is expected to expand to have 1000 people by 2020. The ITER Council consisting of high level representatives of the governments of the seven member parties meets at least twice a year to decide major important issues such as finance, personnel, schedule, and regulations.

The construction started in 2007 and is planned to take more than 10 years, and the test program lasting approximately 20 years will demonstrate the generation of fusion energy, typically energy multiplication by factor 10 by the reaction of the plasma heated by external energy input. All seven parties share the fabrication of the components of the device by providing "in kind" contribution and the experimental results, and technical information obtained by the project will be equally shared. Total project cost for 30 years including operation is estimated to be approximately 10 billion euros.

ITER is a "Tokamak" machine that has been the most successful plasma confinement device so far. Tokamak was originally invented by the scientists in the Soviet Union in 1960s, using torus shaped plasma that is confined with both external coils and its own plasma current. As the largest tokamak in the world, ITER holds approximately 800 cubic meter of plasma in the torus of 6.2 m major radius. Strong magnetic field generated by 18 superconducting "toroidal field" coils, 6 "poloidal field" coils and a central solenoid surrounding the torus vacuum vessel holds plasma for several hundreds of seconds or longer. Plasma contains hydrogen isotopes of deuterium and tritium, that reacts to yield alpha particles and fast neutrons with large amount of energy at the temperature above 150 million degree C. External heating system of 50 MW composed of neutral deuterium beams and microwaves are used to heat the plasma to this high-temperature. Figure 32.1 [1] shows the tokamak device of ITER.

Concerning the utilization of the energy, however, ITER does not have a plan to extract all the generated fusion energy to be demonstrated for actual use, because 440 blanket modules that surround the burning plasma only absorb the fusion energy for shielding purpose. In the fusion reactor, a blanket is expected to produce tritium fuel from lithium compound with neutron irradiation, and the test blanket module (TBM) program is the

FIGURE 32.1
ITER tokamak device. (Adapted from http://www.iter.org)

unique and important experiment toward the future energy application of fusion in ITER, because as one of the most important mission, testing tritium fuel production and extraction of high-grade heat and electricity production in a future reactor is planned. For this purpose, ITER has 3 test ports for the 6 TBMs that are mock-ups of demonstration breeding blankets to be tested with the fusion neutron in the reactor environment.

Although ITER device and its plasma are shared by the member parties, those blanket modules are developed and fabricated by members, and will not become part of the ITER facilities. Various blanket concepts, such as helium-cooled solid breeder, water-cooled solid breeder, helium-cooled lithium lead breeder, are developed as test modules by the member parties to pursue different concepts of fusion-energy utilization based on their domestic fusion programs, to be applied for fusion power plants in the future. In this context, ITER has a function of user facility to provide testing capability of the TBMs. For the governance of this program, TBM program committee, represented by the governments of the member parties was formed as advisory committee of the ITER council. The energy utilization of the fusion is therefore, studied through this TBM programs for ITER. Extraction of high grade heat, such as hot water of PWR condition or high-temperature helium above 400°C will be demonstrated with ITER TBM when its burning plasma is tested. It is possible to use for the demonstration of electricity generation, but the higher temperature blankets that enable hydrogen production will have to wait for the later program, and are currently being studied in parallel [2].

32.2 The ITER-BA Technology Development Activities IFERC for Demonstration

The "Broader Approach" (BA) projects carry great importance for the advancement of fusion energy and will complement the global efforts on realizing ITER. It is a research activity between the European Atomic Energy Community (EURATOM) and the Japanese government on the development of fusion energy, particularly toward demonstration fusion reactor that is the common target for both parties. Agreement was signed in February 2007 to establish a framework for Japan to conduct research for a period of 10 years, mainly in Aomori, Japan. Although this activity is operated between Europe and Japan, ITER parties can participate. Based on the recognition that various research and development efforts are needed to realize fusion energy other than ITER, the scope of the research of BA is selected to complement research with ITER. While ITER emphasizes burning plasma experiment and system integration of the tokamak plant, the BA is concerned with technical issues related to demonstration technology, material research particularly designing of irradiation facility, and plasma physics that is not covered by ITER. Within the BA, three projects were set into motion: JT-60SA advanced plasma experimentation with a superconducting tokamak, engineering validation and engineering design activity (EVEDA) of the international fusion materials test facility (IFMIF), and International fusion energy research center.

Satellite tokamak JT-60SA is being built in Japan to support and supplement ITER experiments. It is a superconducting tokamak of 3 m in major radius, and expected to make high performance deuterium plasma for 100 seconds to 8 hours. Facility of the JT-60 such as building, plasma heating system, and diagnostics are effectively used. Because it is a joint project with the European Union many of the new components such as toroidal coils, cryogenics, and cryostat are supplied by the European Union. Operation of JT-61SA is scheduled to start around 2016.

IFMIF is a high-flux neutron source based on deuteron beam of 30 MeV hitting flowing liquid lithium metal target. Fast neutron of 14 MeV that simulates fusion demonstration condition is generated at the area of 20 cm × 5 cm and used for the irradiation test of fusion reactor materials such as ferritic steel. In the EVEDA, major components of the facility, such as prototype accelerator with ion source, RFQ and DTL, 1/3 liquid lithium target loop, and test cell for irradiation of miniature material specimens at prescribed temperature, are developed. Engineering design of the facility will be made at the same time and will be supported by the technology R&D.

IFERC is established to conduct and coordinate demonstration design and development. Blanket materials for demonstration reactor, such as reduced activation rerritic/martensitic steel, SiC fiber-SiC composite ceramic, advanced lithium compound tritium breeder, and inter-metallic compound Be neutron multiplier, are jointly studied by Europeans and Japanese. Demo design and tritium fuel handling are also studied. Together with the development activity, concept of the demonstration that will utilize the fusion energy at high efficiency is expected to be established. Other than the demonstration design and R&D, simulation center equipped with peta-flops supercomputer is another major program within the IFERC project. Remote experiment of ITER will also be prepared when ITER will be ready.

32.3 Roadmap from ITER to Fusion Energy Plant

Various fusion power plant designs have been made in the past, but no concrete demonstration fusion plant following ITER as a next step is seriously considered in the world yet. However, many countries including the European Union, Japan, China, and Korea officially state and consider their demonstration plans immediately follow the successful demonstration of the burning plasma experiments in ITER, and have shown rough schedules or "Roadmaps" toward them. Most of them are considered to be used for electricity generation for demonstration in late 2030s to 2040s. It is considered that the development and verification of the materials and blanket will be needed to utilize fusion energy for commercial purpose, and the technical programs in these countries include R&D programs for them in parallel to ITER. Fusion is expected to be introduced into the energy market around 2050 in their schedules.

References

1. http://www.iter.org
2. T. Ihli, T.K. Basu, L.M. Giancarli, S. Konishi, S. Malang, F. Najmabadi, S. Nishio, et al., Review of blanket designs for advanced fusion reactors, *Fusion Eng. Des.* 83(7–9), 2008, 912–919.

Section VI

Appendices

Appendix A: Chemical, Thermodynamic, and Transport Properties of Pure Compounds and Solutions

Seiji Kasahara

This section provides chemical, thermodynamic, and transport properties of the key pure compounds and solutions present in hydrogen production processes and fuels.

TABLE A.1

Melting and Boiling Points of Pure Compounds

	H_2	O_2	H_2O	HI	I_2	H_2SO_4	SO_2	SO_3
Melting point (°C)	−259.198[a]	−218.79	0.00	−50.76	113.7	10.31	−75.5	62.2
Boiling point (°C)	−252.762	−182.953	99.974	−35.55	184.4	337	−10.05	

	CaO	**$CaBr_2$**	**Fe_3O_4**	**$FeBr_2$**	**HBr**	**Br_2**	**Cu**	**CuCl**
Melting point (°C)	2613	742	1597	691	−86.80	−7.2	1084.62	423
Boiling point (°C)	2850[b]	1815			−66.38	58.8	2562	1490

	$CuCl_2$	**HCl**	**C[c]**	**CO**	**CO_2**	**CH_4**
Melting point (°C)	598	−114.17	4489[d]	−205.02	−56.558[a]	−182.47
Boiling point (°C)	993	−85	3825[e]	−191.5	−78.464[e]	−161.48

Source: [a,c–e]Adapted from Lide, D. R. 2006–2007. *CRC Handbook of chemistry and physics (87th edition)*. CRC Press, Boca Raton; [b] Perry, R. H., D. W. Green, and J. O. Maloney. 1997. *Perry's Chemical Engineers' Handbook (7th edition)*. McGraw-Hill, New York.

Note: At pressure 760 mmHg.

[a] Triple point.

[c] Graphite.

[d] Triple point at 10.3 MPa.

[e] Sublimination.

TABLE A.2

Standard Formation Enthalpy $\Delta_f H^\circ$ of Pure Compounds (Unit: kJ/mol)

Temperature (K)	H_2	O_2	HI	I_2	H_2SO_4	SO_2	SO_3	CaO	$CaBr_2$	Fe_3O_4	$FeBr_2$	HBr
250	0	0	26.939	0			−394.937					−30.157
298.15	0	0	26.359	0	−813.989	−296.842	−395.765	−635.089	−683.247	−1120.894	−248.948	−36.443
300	0	0	26.336	0	−813.936	−296.865	−395.794	−635.086	−683.296	−1120.869	−248.986	−36.485
350	0	0	25.665	0			−396.543					−52.077
400	0	0	16.987	0	−812.526	−300.257	−399.412	−634.738	−712.744	−1118.728	−277.955	−52.261
450	0	0	15.710	0			−400.656					−52.449
500	0	0	−5.622	0	−808.704	−302.736	−401.878	−634.242	−711.325	−1115.219	−276.138	−52.636
600	0	0	−5.961	0	−802.804	−304.694	−403.675	−633.787	−709.990	−1110.550	−274.360	−52.996
700	0	0	−6.243	0	−794.924	−306.291	−405.014	−633.449	−708.767	−1104.774	−272.199	−53.320
800	0	0	−6.466	0	−785.146	−307.667	−406.068	−634.112	−708.463	−1098.011	−270.537	−53.599
900	0	0	−6.633	0	−826.632	−362.026	−460.062	−634.064	−707.289	−1090.611	−269.078	−53.830
1000	0	0	−6.754	0	−811.918	−361.940	−459.581	−634.273	−706.160	−1091.694	−217.513	−54.018
1100	0	0	−6.838	0		−361.835	−459.063	−634.749	−674.053	−1095.556	−216.269	−54.168
1200	0	0	−6.897	0		−361.720	−458.521	−643.226	−678.708	−1097.841	−214.482	−54.287
1300	0	0	−6.942	0		−361.601	−457.968	−643.009	−674.712	−1095.254	−211.057	−54.381
Solid → Liquid	—	—	—	386.750	283.456	—	—	—	1015.000	—	650.000[f]	—
Liquid → Gas	—	—	—	457.666	—	—	—	—	—	—	964.000[g]	—
Referred table	a	a	b	c	d	b	b	e	d	e	h	b

Temperature (K)	Br_2	Cu	CuCl	$CuCl_2$	$CuO\cdot CuCl_2$[i]	HCl	C	CO	CO_2	CH_4	H_2O
250		0				−92.219	0			−73.426	
280											−286.410
298.15	0	0	−138.072	−205.853	−350.079	−92.312	0	−110.527	−393.522	−74.873	−285.830
300	0	0	−138.059	−205.828	−350.077	−92.316	0	−110.516	−393.523	−74.929	−285.771
320											−285.137
340											−284.506
350		0			−349.948	−92.443	0			−76.461	
360											−283.874

400	0	0	−136.933	−204.433	−349.673	−92.589	0	−110.102	−393.583	−77.969	−243.009
450		0			−349.286	−92.747	0			−79.422	
500	0	0	−135.437	−202.947	−348.814	−92.913	0	−110.003	−393.666	−80.802	−243.896
600	0	0	−133.803	−201.393	−347.675	−93.251	0	−110.150	−393.803	−83.308	−244.797
700	0	0	−132.097	−199.800		−93.579	0	−110.469	−393.983	−85.452	−245.658
800	0	0	−119.759	−198.181		−93.881	0	−110.905	−394.188	−87.238	−246.461
900	0	0	−117.701	196.550		−94.152	0	−111.418	−394.405	−88.692	−247.198
1000	0	0	−115.709	−194.919		−94.388	0	−111.983	−394.623	−89.849	−247.868
1100	0	0	−113.795	−193.306		−94.591	0	−112.586	−394.838	−90.750	−248.468
1200	0	0	−111.976	−191.732		−94.766	0	−113.217	−395.050	−91.437	−249.004
1300	0	0	−110.296	−190.242		−94.916	0	−113.870	−395.257	−91.945	−249.479
Solid → Liquid	265.900	—	703.000	—	—	—	—	—	—	—	—
Liquid → Gas	332.503	—	—	—	—	—	—	—	—	—	372.780
Referred table	c	c	d	e		b	j	b	b	b	k

Source: [a–h,j,k]Adapted from Chase, M. W. Jr. 1998. *J. Phys. Chem. Ref. Data*, Monograph No. 9; [i] Zamfirescu, C., I. Dincer, and G. F. Naterer. 2010. *Int. J. Hydrogen Energy*, 35:4839–52.

Note: Pressure: 0.1 MPa; Phase is the one at 0.1 MPa; "—" in the columns of "Solid → Liquid" and "Liquid → Gas" shows the transitions are outside of the temperature range in the table.

[a] Reference state—ideal gas.
[b] Ideal gas.
[c] Reference state.
[d] Crystal—liquid.
[e] Crystal.
[f] Transition from crystal I to crystal II.
[g] Transition from crystal II to liquid.
[h] Crystal (I–II)—liquid.
[i] Calculated by an equation in Ref. [4], "Cu_2OCl_2" in the original paper.
[j] Reference—graphite.
[k] Liquid real gas, p = 1 bar.

TABLE A.3

Standard Entropy S° of Pure Compounds (Unit: J/(K·mol))

Temperature (K)	H_2	O_2	HI	I_2	H_2SO_4	SO_2	SO_3	CaO	$CaBr_2$	Fe_3O_4	$FeBr_2$	HBr
250	125.640	199.990	201.456				248.192					193.567
298.15	130.680	205.147	206.589	116.142	156.895	248.212	256.769	38.212	129.704	145.266	140.666	198.699
300	130.858	205.329	206.770	116.479	157.753	248.459	257.083	38.473	130.168	146.179	141.163	198.879
350	135.325	209.880	211.268				265.191					203.373
400	139.216	213.871	215.176	174.062	200.356	260.448	272.674	51.300	152.206	191.952	164.560	207.270
450	142.656	217.445	218.640				279.637					210.717
500	145.737	220.693	221.760	279.920	237.744	270.495	286.152	61.980	169.774	232.441	183.209	213.812
600	151.077	226.451	227.233	286.764	271.844	279.214	298.041	71.052	184.352	269.299	198.853	219.216
700	155.606	231.466	231.964	292.571	303.674	286.924	308.655	78.917	196.836	303.578	213.067	223.861
800	159.548	235.921	236.160	297.618	333.844	293.829	318.217	85.858	207.851	335.969	225.121	227.965
900	163.051	239.931	239.946	302.082	362.745	300.073	326.896	92.072	217.816	366.919	236.015	231.660
1000	166.216	243.578	243.404	306.087	390.646	305.767	334.828	97.700	227.004	388.079	298.490	235.032
1100	169.112	246.922	246.589	309.724		310.995	342.122	102.849	266.061	407.220	308.659	238.140
1200	171.790	250.010	249.544	313.058		315.824	348.866	107.596	275.891	424.695	317.942	241.025
1300	174.288	252.878	252.300	316.144		320.310	355.133	112.005	284.933	440.770	326.482	243.720
Solid → Liquid	—	—	—	386.750	283.456	—	—	—	1015.000	—	650.000[f]	—
Liquid → Gas	—	—	—	457.666	—	—	—	—	—	—	964.000[g]	—
Referred table	a	a	b	c	d	b	b	e	d	e	h	b

Temperature (K)	Br_2	Cu	CuCl	$CuCl_2$	$CuO \cdot CuCl_2$[i]	HCl	C	CO	CO_2	CH_4	H_2O
250		28.915				181.770	4.394			180.113	
280											65.215
298.15	152.206	33.164	87.027	108.085	539.443	186.901	5.740	197.653	213.795	186.251	69.950
300	152.674	33.315	87.328	108.530	540.154	187.081	5.793	197.833	214.025	186.472	70.416
320											75.279
340											79.847
350		37.127			558.239	191.574	7.242			192.131	
360											84.164

400	256.093	40.484	102.675	129.678	574.436	195.467	8.713	206.238	225.314	197.356	198.473
450		43.489			589.081	198.906	10.191			202.291	
500	264.329	46.206	115.711	146.679	602.429	201.989	11.662	212.831	234.901	207.014	206.428
600	271.110	50.982	126.773	160.906	626.007	207.353	14.533	218.319	243.283	215.987	213.003
700	276.873	55.103	136.354	173.142		211.942	17.263	223.066	250.750	224.461	218.712
800	281.883	58.739	159.827	183.881		215.978	19.826	227.277	257.494	232.518	223.809
900	286.316	62.009	167.712	193.453		219.602	22.221	231.074	263.645	240.205	228.448
1000	290.293	64.994	174.765	202.093		222.903	24.457	234.538	269.299	247.549	232.730
1100	293.898	67.763	181.146	209.974		225.944	26.548	237.726	274.528	254.570	236.725
1200	297.198	70.368	186.971	217.222		228.768	28.506	240.679	279.390	261.287	240.481
1300	300.240	72.871	192.329	223.937		231.406	30.346	243.431	283.932	267.714	244.032
Solid → Liquid	265.900	—	703.000	—	—	—	—	—	—	—	
Liquid → Gas	332.503	—	—	—	—	—	—	—	—	—	372.780
Referred table	c	c	d	e		b	j	b	b	b	k

Source: [a–h,j,k]Adapted from Chase, M. W. Jr. 1998. *J. Phys. Chem. Ref. Data*, Monograph No. 9; [i] Zamfirescu, C., I. Dincer, and G. F. Naterer. 2010. *Int. J. Hydrogen Energy*, 35:4839–52.

Note: Pressure: 0.1 Mpa; Phase is the one at 0.1 MPa; "—" in the columns of "Solid → Liquid" and "Liquid → Gas" shows the transitions are outside of the temperature range in the table.

[a] Reference state—ideal gas.
[b] Ideal gas.
[c] Reference state.
[d] Crystal—liquid.
[e] Crystal.
[f] Transition from crystal I to crystal II.
[g] Transition from crystal II to liquid.
[h] Crystal (I–II)—liquid.
[i] Calculated by an equation in Ref. [4], "Cu_2OCl_2" in the original paper.
[j] Reference—graphite.
[k] Liquid real gas, p = 1 bar.

TABLE A.4

Standard Formation Gibbs Energy $\Delta_f G^\circ$ of Pure Compounds (Unit: kJ/mol)

Temperature (K)	H_2	O_2	HI	I_2	H_2SO_4	SO_2	SO_3	CaO	$CaBr_2$	Fe_3O_4	$FeBr_2$	HBr
250	0	0	5.615	0			−374.943					−50.348
298.15	0	0	1.560	0	−689.918	−300.125	−371.016	−603.501	−664.139	−1017.438	−237.362	−53.513
300	0	0	1.406	0	−689.149	−300.145	−370.862	−603.305	−664.020	−1016.797	−237.290	−53.619
350	0	0	−2.698	0			−366.646					−55.615
400	0	0	−6.428	0	−647.968	−300.971	−362.242	−592.755	−651.460	−982.386	−227.334	−56.108
450	0	0	−9.279	0			−357.529					−56.577
500	0	0	−10.088	0	−607.248	−300.871	−352.668	−582.316	−636.304	−948.681	−214.890	−57.026
600	0	0	−10.948	0	−567.476	−300.305	−342.647	−571.975	−621.427	−915.794	−202.807	−57.869
700	0	0	−11.756	0	−528.851	−299.444	−332.365	−561.701	−606.765	−883.776	−191.053	−58.655
800	0	0	−12.528	0	−491.486	−298.370	−321.912	−551.362	−592.168	−852.657	−179.575	−59.398
900	0	0	−13.275	0	−454.377	−296.051	−310.258	−541.024	−577.703	−822.428	−168.295	−60.109
1000	0	0	−14.006	0	−413.788	−288.725	−293.639	−530.677	−563.365	−792.602	−159.039	−60.796
1100	0	0	−14.727	0		−281.409	−277.069	−520.297	−551.654	−762.463	−153.237	−61.466
1200	0	0	−15.441	0		−274.102	−260.548	−509.239	−540.043	−732.139	−147.605	−62.124
1300	0	0	−16.151	0		−266.806	−244.073	−498.082	−528.650	−701.771	−142.172	−62.773
Solid → Liquid	—	—	—	386.750	283.456	—	—	—	1015.000	—	650.000[f]	—
Liquid → Gas	—	—	—	457.666	—	—	—	—	—	—	964.000[g]	—
Referred table	a	a	b	c	d	b	b	e	d	e	h	b

Temperature (K)	Br_2	Cu	CuCl	$CuCl_2$	$CuO\cdot CuCl_2$[i]	HCl	C	CO	CO_2	CH_4	H_2O
250		0				−94.809	0			−54.536	
280											−240.123
298.15	0	0	−120.876	−161.680	−369.7	−95.300	0	−137.163	−394.389	−50.768	−237.141
300	0	0	−120.769	−161.405		−95.318	0	−137.328	−394.394	−50.618	−236.839
320											−233.598

340											−230.396
350		0				−95.809	0			−46.445	
360											−227.231
400	0	0	−115.157	−146.806		−96.280	0	−146.338	−394.675	−42.054	−223.937
450		0				−96.732	0			−37.476	
500	0	0	−109.882	−132.569		−97.166	0	−155.414	−394.939	−32.741	−219.069
600	0	0	−104.924	−118.638		−97.985	0	−164.486	−395.182	−22.887	−214.018
700	0	0	−100.244	−104.971		−98.747	0	−173.518	−395.398	−12.643	−208.819
800	0	0	−97.249	−91.535		−99.465	0	−182.497	−395.586	−2.115	−203.501
900	0	0	−94.560	−78.302		−100.146	0	−191.416	−395.748	8.616	−198.086
1000	0	0	−92.096	−65.251		−100.799	0	−200.075	−395.886	19.492	−192.593
1100	0	0	−89.828	−52.362		−101.430	0	−209.075	−396.001	30.472	−187.035
1200	0	0	−87.731	−39.619		−102.044	0	−217.819	−396.098	41.524	−181.426
1300	0	0	−85.779	−27.004		−102.644	0	−226.509	−396.177	52.626	−175.775
Solid → Liquid	265.900	—	703.000	—	—	—	—	—	—	—	—
Liquid → Gas	332.503	—	—	—	—	—	—	—	—	—	372.780
Referred table	c	c	d	e		b	j	b	b	b	k

Source: [a–h,j,k]Adapted from Chase, M. W. Jr. 1998. *J. Phys. Chem. Ref. Data*, Monograph No. 9; [i] Zamfirescu, C., I. Dincer, and G. F. Naterer. 2010. *Int. J. Hydrogen Energy*, 35:4839–52.

Note: Pressure: 0.1 MPa; Phase is the one at 0.1 MPa. "—" in the columns of "Solid → Liquid" and "Liquid → Gas" shows the transitions are outside of the temperature range in the table.

[a] Reference state—ideal gas.
[b] Ideal gas.
[c] Reference state.
[d] Crystal—liquid.
[e] Crystal.
[f] Transition from crystal I to crystal II.
[g] Transition from crystal II to liquid.
[h] Crystal (I–II)—liquid.
[i] Referred from [4], "Cu_2OCl_2" in the original paper.
[j] Reference—graphite.
[k] Liquid real gas, p = 1 bar.

TABLE A.5

Standard Constant Pressure Heat Capacity C_p° of Pure Compounds (Unit: J/(K·mol))

Temperature	H_2	O_2	HI	I_2	H_2SO_4	SO_2	SO_3	CaO	$CaBr_2$	Fe_3O_4	$FeBr_2$	HBr
250	28.344	29.201	29.134				46.784					29.130
298.15	28.836	29.376	29.156	54.436	138.584	39.878	50.661	42.120	75.040	147.235	80.232	29.141
300	28.849	29.385	29.158	54.506	138.940	39.945	50.802	42.242	75.124	147.695	80.274	29.141
350	29.081	29.694	29.216				54.423					29.167
400	29.181	30.106	29.328	80.669	158.205	43.493	57.672	46.626	77.990	171.126	82.500	29.220
450	29.229	30.584	29.502				60.559					29.313
500	29.260	31.091	29.736	37.464	177.619	46.576	63.100	48.982	79.454	192.380	84.726	29.453
600	29.327	32.090	30.348	37.612	197.033	49.049	67.255	50.480	80.500	212.547	86.952	29.870
700	29.441	32.981	31.064	37.735	216.447	50.961	70.390	51.555	81.630	232.714	89.178	30.427
800	29.624	33.733	31.798	37.847	235.860	52.434	72.761	52.400	83.471	252.881	91.404	31.055
900	29.881	34.355	32.496	37.959	255.274	53.580	74.570	53.112	85.856	273.050	93.630	31.698
1000	30.205	34.870	33.135	38.081	274.688	54.484	75.968	53.735	88.617	200.832	106.692	32.319
1100	30.581	35.300	33.706	38.232		55.204	77.065	54.300	112.968	200.832	106.692	32.897
1200	30.992	35.667	34.211	38.431		55.794	77.937	54.831	112.968	200.832	106.692	33.425
1300	31.423	35.988	34.655	38.699		56.279	78.639	55.329	112.968	200.832	106.692	33.902
Solid → Liquid	—	—	—	386.750	283.456	—	—	—	1015.000	—	650.000[i]	—
Liquid → Gas	—	—	—	457.666	—	—	—	—	—	—	964.000[j]	—
Referred table	a	a	b	c	d	b	b	e	d	e	f	b

Temperature (K)	Br_2	Cu	CuCl	$CuCl_2$	$CuO\cdot CuCl_2$[k]	HCl	C	CO	CO_2	CH_4	H_2O
250		23.782				29.130	6.816			34.216	
280											75.563
298.15	75.674	24.442	48.534	71.881	114.845	29.136	8.517	29.142	37.129	35.639	75.351
300	75.624	24.462	48.771	71.923	115.034	29.137	8.581	29.142	37.221	35.708	75.349
320											75.344
340											75.388
350		24.975			119.528	29.149	10.241			37.874	
360											75.679

400	36.723	25.318	56.902	75.061	122.986	29.175	11.817	29.342	41.325	40.500	35.982
450		25.686			125.634	29.223	13.289			43.374	
500	37.077	25.912	59.831	77.278	127.695	29.304	14.623	29.794	44.627	46.342	35.699
600	37.301	26.481	61.505	78.785	130.960	29.575	16.844	30.443	47.321	52.227	36.521
700	37.460	26.996	62.760	79.956		29.985	18.537	31.171	49.564	57.794	37.595
800	37.586	27.494	66.944	80.885		30.494	19.827	31.899	51.434	62.932	38.780
900	37.692	28.049	66.944	81.655		31.055	20.824	32.577	52.999	67.601	40.023
1000	37.787	28.662	66.944	82.370		31.628	21.610	33.183	54.308	71.795	41.292
1100	37.876	29.479	66.944	83.001		32.186	22.244	33.710	55.409	75.529	42.554
1200	37.962	30.519	66.944	83.607		32.713	22.766	34.175	56.342	78.833	43.78[illegible]
1300	38.049	32.143	66.944	84.190		33.203	23.204	34.572	57.137	81.744	44.954
Solid → Liquid	265.900	—	703.000	—	—	—	—	—	—	—	—
Liquid → Gas	332.503	—	—	—	—	—	—	—	—	—	372.780
Referred table	c	c	d	e		b	g	b	b	b	h

Source: [a–j]Adapted from Chase, M. W. Jr. 1998. *J. Phys. Chem. Ref. Data*, Monograph No. 9; [k] Zamfirescu, C., I. Dincer, and G. F. Naterer. 2010. *Int. J. Hydrogen Energy*, 35:4839–52.

Note: Pressure: 0.1 MPa; Phase is the one at 0.1 MPa. "—" in the columns of "Solid → Liquid" and "Liquid → Gas" shows the transitions are outside of the temperature range in the table.

[a] Reference state—ideal gas.
[b] Ideal gas.
[c] Reference state.
[d] Crystal—liquid.
[e] Crystal.
[f] Crystal (I–II)—liquid.
[g] Reference—graphite.
[h] Liquid real gas, p = 1 bar.
[i] Transition from crystal I to crystal II.
[j] Transition from crystal II to liquid.
[k] Calculated by an equation in Ref. [4], "Cu_2OCl_2" in the original paper.

The values in Tables A.2, A.3, A.4, and A.5 have relations as follows [3]:

$$H^{\circ}(T) = \int_0^T C_p^{\circ}(T)\mathrm{d}T$$

$$S^{\circ}(T) = \int_0^T \frac{C_p^{\circ}(T)}{T}\mathrm{d}T$$

$$\Delta_f H^{\circ}(T) = \Delta_f H^{\circ}(298.15K) + \left[H^{\circ}(T) - H^{\circ}(298.15K)\right]_{\text{compound}} - \sum\left[H^{\circ}(T) - H^{\circ}(298.15K)\right]_{\text{elements}}$$

$$\Delta_f G^{\circ}(T) = \Delta_f H^{\circ}(T) - T\{S^{\circ}(T)_{\text{compound}} - S^{\circ}(T)_{\text{elements}}\}$$

C_p°: Standard constant pressure heat capacity (kJ/(mol · K))
H°: Standard enthalpy (kJ/mol)
S°: Standard entropy (kJ/(mol·K))
T: Temperature (K)
$\Delta_f H^{\circ}$: Standard formation enthalpy (kJ/mol)
$\Delta_f G^{\circ}$: Standard formation Gibbs energy (kJ/mol)
compound: Objective compound
elements: Elemental substances of reference state (the phase in the condition of 298.15 K and 0.1 Mpa) composing the objective compound

TABLE A.6

Enthalpy Change with Phase Shift of Pure Compounds (Unit: kJ/mol)

		H_2	O_2	H_2O	HI	I_2	H_2SO_4	SO_2	SO_3
Fusion	m.p.	0.12[a]	0.44	6.01	2.87	15.52	10.71		8.60[b]
Vaporization	b.p.	0.90	6.82	40.657	19.76	41.57		24.94	40.69
	25°C			43.990	17.36			22.92	43.14
		CaO	**$CaBr_2$**	**Fe_3O_4**	**$FeBr_2$**	**HBr**	**Br_2**	**Cu**	**CuCl**
Fusion	m.p.	80	29.1	138	43	2.41	10.57	13.26	7.08
Vaporization	b.p.				123[c]		29.96		
	25°C					12.69	30.91		
		$CuCl_2$	**HCl**	**C[d]**	**CO**	**CO_2**	**CH_4**		
Fusion	m.p.	15.0	2.00	117.4	0.833	9.02	0.94		
Vaporization	b.p.		16.15		6.04		8.19		
	25°C		9.08						

Source: [a,b,d]Adapted from Chase, M. W. Jr. 1998. *J. Phys. Chem. Ref. Data*, Monograph No. 9; [c] The Chemical Society of Japan (eds.), 2001. *Kagaku Binran Kisohen (5th edition)*. The Chemical Society of Japan, Maruzen, Tokyo.

Note: m.p.—Melting point at 101.325 kPa; b.p.—Boiling point at 101.325 kPa.

[a] Triple point at –259.198°C.

[b] γ–form.

[d] Graphite.

TABLE A.7

Heat of Vaporization of Pure Compounds

	H_2	O_2	H_2O	SO_2	SO_3	HBr
C_1	1.013	9.008	52.053	36.760	73.370	24.850
C_2	0.698	0.4542	0.3199	0.4	0.5647	0.39
C_3	−1.817	−0.4096	−0.212	0	0	0
C_4	1.447	0.3183	0.25795	0	0	0
T_c (K)	33.19	154.58	647.13	430.75	490.85	363.15
T_{min} (K)	13.95	54.36	273.16	197.67	289.95	185.15
T_{max} (K)	33.19	154.58	647.13	430.75	490.85	363.15
	Br_2	HCl	CO	CO_2	CH_4	
C_1	40.000	22.093	8.585	21.730	10.194	
C_2	0.351	0.3466	0.4921	0.382	0.26087	
C_3	0	0	−0.326	−0.4339	−0.14694	
C_4	0	0	0.2231	0.42213	0.22154	
T_c (K)	584.15	324.65	132.92	304.21	190.564	
T_{min} (K)	265.85	158.97	68.13	216.58	90.69	
T_{max} (K)	584.15	324.65	132.5	304.21	190.56	

Source: Adapted from Perry, R. H., D. W. Green, and J. O. Maloney. 1997. *Perry's Chemical Engineers' Handbook (7th edition)*. McGraw-Hill, New York.

$$\Delta H_v = C_1 \cdot (1 - T_r)^{C_2 + C_3 \cdot T_r + C_4 \cdot T_r^2}, \quad T_r = \frac{T}{T_c}$$

T_c: Critical temperature (K)
T_{min}: Minimum temperature of applicable range (K)
T_{max}: Mininim temperature of applicable range (K)
ΔH_v: Heat of vaporization (kJ/mol)

TABLE A.8

Thermodynamical Parameters of Aqueous Solutes

	H_2	O_2	I_2	HI	H_2SO_4	SO_2
State	**ao**	**ao**	**ao**	**ai**	**ai**	**ao**
Standard formation enthalpy ($\Delta_f H^o$, kJ/mol)	−4.2	−11.7	22.6	−55.19	−909.27	−322.980
Standard entropy (S^o, J/(K·mol))	57.7	110.9	137.2	111.3	20.1	161.9
Standard formation Gibbs energy ($\Delta_f G^o$, (kJ/mol))	17.6	16.4	16.40	−51.57	−744.53	−300.676
Standard heat capacity (C_p^o, J/(K·mol))				−142.3	−293	

continued

TABLE A.8 (continued)

Thermodynamical Parameters of Aqueous Solutes

	$CaBr_2$	Br_2	HBr	$CuCl_2$	HCl
State	ai	ao	ai	ao	ai
Standard formation enthalpy, ($\Delta_f H^\circ$, (kJ/mol)	−785.92	−2.59	−121.55		−167.159
Standard entropy (S°, (J/(K mol))	111.7	130.5	82.4		56.5
Standard formation Gibbs energy ($\Delta_f G^\circ$, (kJ/mol))	−761.49	3.93	−103.96	−197.9	−131.228
Standard heat capacity (C_p°, (J/(K·mol))			−141.8		−136.4

Source: Adapted from Wagman, D. D. et al. 1982. *J. Phys. Chem. Ref. Data*, 11, Supplement No. 2.

Note: At 298.15 K and 0.1 MPa; ai: Standard state of ionized substance, the hypothetical ideal solution at unit mean ionic molality, $m_\pm$ = 1 mol/kg–H_2O containing the ions of which it is assumed to be composed at infinite dilution; ao: Standard state of unionized substance, the ideal solution at unit molality, m = 1 mol/kg–H_2O.

$$m_\pm = (m_R^a \cdot m_X^b)^{\frac{1}{a+b}} \quad \text{for species of } R_a X_b$$

m: Molality (mol/kg-H_2O)

R, X: Ionic species

TABLE A.9

Standard Formation Enthalpy $\Delta_f H^\circ$ of Compounds in Aqueous Solutions (Unit: kJ/mol)

n	*m*	HI	H_2SO_4	SO_2	$CaBr_2$	$FeBr_2$	HBr	$CuCl_2$	HCl
1.0	55.555556		−841.791				−72.72		−121.55
1.5	37.037037		−849.888				−85.86		−132.67
2.0	27.777778		−855.440				−93.72		−140.96
2.5	22.222222		−859.611				−99.37		−145.48
3.0	18.518519	−35.82	−862.912				−103.26		−148.49
3.5	15.873016		−865.611						
4.0	13.888889	−42.80	−867.879				−108.621		−152.917
4.5	12.345679	−44.89	−869.808				−110.437		−154.503
5.0	11.111111	−46.40	−871.477				−111.738		−155.774
5.5	10.10101		−872.937						
6.0	9.2592593	−48.367	−874.222				−113.583		−157.682
7.0	7.9365079		−876.372						
8.0	6.9444444	−50.522	−878.075				−115.683		−160.005
9.0	6.1728395		−879.435						
10	5.5555556	−51.610	−880.527				−116.955	−245.60	−161.318
12	4.6296296	−52.258	−882.134				−117.734		−162.180
15	3.7037037	−52.944	−883.623				−118.436	−250.50	−163.025
20	2.7777778	−53.530	−884.916				−119.077	−253.72	−163.845
25	2.2222222	−53.781	−885.585				−119.411	−255.73	−164.339

TABLE A.9 (continued)

Standard Formation Enthalpy $\Delta_f H^o$ of Compounds in Aqueous Solutions (Unit: kJ/mol)

n	m	HI	H_2SO_4	SO_2	$CaBr_2$	$FeBr_2$	HBr	$CuCl_2$	HCl
30	1.8518519	−53.928	−885.983				−119.641	−257.32	−164.670
40	1.3888889	−54.099	−886.460				−119.959		−165.096
50	1.1111111	−54.208	−886.774				−120.160	−261.08	−165.356
75	0.7407407	−54.375	−887.292				−120.416		−165.724
100	0.5555556	−54.451	−887.636	−326.578			−120.562	−264.55	−165.925
115	0.4830918		−887.811						
150	0.3703704		−888.188	−327.298			−120.721		−166.147
200	0.2777778	−54.601	−888.627	−327.837			−120.809	−266.90	−166.272
250	0.2222222			−328.268					
300	0.1851852	−54.664	−889.372	−328.641			−120.918		−166.423
400	0.1388889	−54.702	−889.974	−329.243	−784.04		−120.980	−269.0	−166.506
500	0.1111111	−54.735	−890.493	−329.745	−784.17		−121.026	−269.53	−166.573
600	0.0925926	−54.760	−890.983				−121.064		−166.619
700	0.0793651		−891.359				−121.093		−166.657
750	0.0740741			−330.687					
800	0.0694444	−54.802	−891.728		−784.37		−121.118	−270.41	−166.686
900	0.0617284		−892.050				−121.139		−166.711
1000	0.0555556	−54.836	−892.343	−331.377	−784.475		−121.160	−270.70	−166.732
1500	0.037037		−893.522	−332.465			−121.223		−166.804
1650	0.03367					−336.4			
2000	0.0277778	−54.923	−894.476	−333.222	−784.776		−121.261	−270.79	−166.850
2500	0.0222222			−333.783					
3000	0.0185185	−54.969	−895.941	−334.264			−121.311		−166.908
3500	0.015873			−334.674					
4000	0.0138889	−54.994	−897.112	−335.005			−121.340		−166.933
5000	0.0111111	−55.015	−897.970	−335.594	−785.102		−121.361		−166.963
7000	0.0079365	−55.045	−899.330				−121.390		−166.992
7500	0.0074074			−336.574					
10000	0.0055556	−55.066	−900.752	−337.163	−785.307		−121.415		−167.017
11500	0.0048309					−333.9			
15000	0.0037037		−902.342						
20000	0.0027778	−55.103	−903.326		−785.458		−121.453		−167.055
30000	0.0018519		−904.715						
50000	0.0011111	−55.137	−906.024		−785.617		−121.491		−167.092
70000	0.0007937		−906.698						
100000	0.0005556	−55.149	−907.321		−785.697		−121.508		−167.117
150000	0.0003704		−907.807						
200000	0.0002778		−908.104						
300000	0.0001852		−908.430						
500000	0.0001111		−908.719		−785.818				
700000	7.937E−05		−908.853						
1000000	5.556E−05		−908.957						
2000000	2.778E−05		−909.087						
Infinite		−55.19	−909.27		−785.92		−121.55		−167.159

Source: Adapted from Wagman, D. D. et al. 1982. *J. Phys. Chem. Ref. Data*, 11, Supplement No. 2.
Note: n: Mole number of H_2O per 1 mol of component; m: Molality of component, mol/kg-H_2O.

TABLE A.10

Constant Pressure Heat Capacity of Solutions (Unit: J/(K·g))

Component	Temperature (Temperature Range °C)	Molality (mol/kg–H_2O) 0.3	0.5	0.7	1.0	2.0	3.0	4.0
H_2SO_4	25–45	4.089	4.029	3.969	3.882	3.626	3.424	3.237
$CuCl_2$	19–51	3.988	3.880	3.783	3.655			
HCl	25 (below 1.0 mol/kg–H_2O), 20.5 (above 2.0 mol/kg–H_2O)	4.101	4.051	4.003	3.933	3.700	3.508	3.342

Component	Temperature (Temperature Range °C)	Molality (mol/kg–H_2O) 6.0	10.0	20	40	70	90
H_2SO_4	25–45	2.944	2.553	2.067	1.780	1.769	1.682
$CuCl_2$	19–51						
HCl	25 (below 1.0 mol/kg–H_2O), 20.5 (above 2.0 mol/kg–H_2O)	3.074	2.712				

Source: Adapted from The Chemical Society of Japan (eds.), 1994. *Kagaku Binran Kisohen (4th edition).* The Chemical Society of Japan, Maruzen, Tokyo.

TABLE A.11

Vapor Pressure of Pure Compounds

	H_2	O_2	H_2O	SO_2	SO_3	HBr
C_1	12.69	51.245	73.649	47.365	180.99	29.315
C_2	−94.896	−1200.2	−7258.2	−4084.5	−12060	−2424.5
C_3	1.1125	−6.4361	−7.3037	−3.6469	−22.839	−1.1354
C_4	3.2915E−04	2.8405E−02	4.1653E−06	1.7990E−17	7.2350E−17	2.3806E−18
C_5	2	1	2	6	6	6
T_{min} (K)	13.95	54.36	273.16	197.67	289.95	185.15
T_{max} (K)	33.19	154.58	647.13	430.75	490.85	363.15

	Br_2	HCl	CO	CO_2	CH_4
C_1	108.26	104.27	45.698	140.54	39.205
C_2	−6592	−3731.2	−1076.6	−4735	−1324.4
C_3	−14.16	−15.047	−4.8814	−21.268	−3.4366
C_4	1.6043E−02	3.1340E−02	7.5673E−05	4.0909E−02	3.1019E−05
C_5	1	1	2	1	2
T_{min} (K)	265.85	158.97	68.15	216.58	90.69
T_{max} (K)	584.15	324.65	132.92	304.21	190.56

Source: Adapted from Perry, R. H., D. W. Green, and J. O. Maloney. 1997. *Perry's Chemical Engineers' Handbook (7th edition).* McGraw-Hill, New York.

$$P = \exp\left[C_1 + \frac{C_2}{T} + C_3 \ln T + C_4 T^{C_5}\right]$$

P: Vapor pressure (Pa)
T: Temperature (K)
T_{min}: Minimum temperature of applicable range (K)
T_{max}: Maximum temperature of applicable range (K)

TABLE A.12

Vapor Pressure of Pure HI and I_2

	Parameters for Antoine Equation			Parameters for Harlacher–Braun Equation
		I_2		
	HI	Soild	Liquid	HI
A	5.6089	9.8109	7.0181	33.884
B	416.04	2901.0	1610.9	−3013.08
C	188.09	256.00	205.0	−2.673
D				1.23
T_c (K)	424	819	819	

Source: Adapted from The Chemical Society of Japan (eds.), 2004. *Kagaku Binran Kisohen (5th edition)*. The Chemical Society of Japan, Maruzen, Tokyo.

$$\log P = A - \frac{B}{C + t} \qquad \text{(Antoine equation)}$$

$$\ln P = \exp\left[A + \frac{B}{T} + C \ln T + \frac{DP}{T^2}\right] \qquad \text{(Harlacher–Braun equation for HI)}$$

P: Vapor pressure (Torr) [1 (Torr) = 133.322 (Pa)]
T: Temperature (K)
t: Temperature (°C)

Antoine equation is applicable to the condition of lower temperature than 0.75 T_c and in the range of 1–1500 Torr. T_c is critical temperature and is shown in the table. Harlacher–Braun equation can be applicable to the temperature above the applicable range of Antoine equation.

TABLE A.13

Temperature at Some Vapor Pressures of Pure Compounds (Unit: °C)

Pressure (mmHg)	Cu	CuCl[a]	C[b]
1	1628	546	3586
5	1795	645	3828
10	1879	702	3946
20	1970	766	4069
40	2067	838	4196
60	2127	886	4273
100	2207	960	4373
200	2325	1077	4516
400	2465	1249	4660
760	2595	1490	4827

Source: Adapted from Perry, R. H., D. W. Green, and J. O. Maloney. 1997. *Perry's Chemical Engineers' Handbook (7th edition)*. McGraw-Hill, New York.

[a] CuCl or Cu_2Cl_2.

[b] Solid graphite.

TABLE A.14

Partial Pressures of H_2O and SO_2 over Aqueous Solution of Sulfur Dioxide (Unit: mmHg)

H_2O													
SO_2 Concentration ($g\text{-}SO_2$)/($100g\text{-}H_2O$)	**Temperature (°C)**												
	10	**20**	**30**	**40**	**50**	**60**	**70**	**80**	**90**	**100**	**110**	**120**	**130**
1.0	9.2	17.4	31.7	55.1	92.2	149.0	233	354	524	757	1071	1484	2022
2.0	9.1	17.4	31.6	55.0	91.9	148.6	233	353	523				
3.0	9.1	17.3	31.5	54.7	91.6	148.1	232						
4.0	9.1	17.3	31.4	54.6	91.4								
5.0	9.1	17.2	31.3	54.4									
6.0	9.0	17.2	31.2	54.3									
7.0	9.0	17.1	31.1										
8.0	9.0	17.1	31.0										
9.0	9.0	17.0											
10.0	8.9	17.0											
11.0	8.9	16.9											
12.0	8.9												
13.0	8.8												
14.0	8.8												
15.0	8.8												
16.0	8.8												

SO_2													
SO_2 Concentration ($g\text{-}SO_2$)/($100g\text{-}H_2O$)	**Temperature (°C)**												
	10	**20**	**30**	**40**	**50**	**60**	**70**	**80**	**90**	**100**	**110**	**120**	**130**
1.0	42	59	85	120	164	217	281	356	445	548	661	775	879
2.0	86	123	176	245	333	444	581	746	940				
3.0	130	191	273	378	511	682	897						
4.0	176	264	376	518	698								
5.0	223	338	482	661									
6.0	271	411	588	804									
7.0	320	486	698										
8.0	370	562	806										
9.0	421	638											
10.0	473	714											
11.0	526	789											
12.0	580												
13.0	635												
14.0	689												
15.0	743												
16.0	799												

Source: Adapted from Munemi et al. A. (Eds.), 1977. *Ryu-san Hando Bukku Kaiteibann*. Ryu-san Kyoukai , Tokyo.

TABLE A.15

Total and Partial Pressure of Aqueous Sulfuric Acid Solution

H_2SO_4 in Liquid Phase (wt%)	Temperature (°C)																	
	0	10	20	30	40	50	60	70	80	90	100	110	120	130	140	150	160	170
10	5.82E-03	1.17E-02	2.23E-02	4.04E-02	7.03E-02	0.117	0.189	0.296	0.449	0.664	0.957	1.349	1.863	2.524	3.361	4.404	5 685	7.236
	5.82E-03	1.17E-02	2.23E-02	4.04E-02	7.03E-02	0.117	0.189	0.296	0.449	0.664	0.957	1.349	1.863	2.524	3.361	4.404	5 685	7.236
	6.44E-30	1.49E-28	2.78E-27	4.26E-26	5.49E-25	6.02E-24	5.73E-23	4.77E-22	3.52E-21	2.33E-18	1.39E-19	7.56E-19	3.77E-18	1.74E-17	7.43E-17	2.97E-16	1.11E-15	3.93E-15
	5.76E-22	6.34E-21	5.88E-20	4.68E-19	3.24E-18	1.97E-17	1.07E-16	5.26E-16	2.35E-15	9.60E-15	3.53E-14	1.27E-13	4.18E-13	1.29E-12	3.75E-12	1.03E-11	2.72E-11	6.82E-11
20	5.34E-03	1.07E-02	2.05E-02	3.73E-02	6.49E-02	0.109	0.175	0.275	0.417	0.617	0.891	1.258	1.740	2.361	3.149	4.132	5.342	6.810
	5.34E-03	1.07E-02	2.05E-02	3.73E-02	6.49E-02	0.109	0.175	0.275	0.417	0.617	0.891	1.258	1.740	2.361	3.149	4.132	5.342	6.810
	1.03E-28	2.23E-27	3.94E-26	5.77E-25	7.14E-24	7.57E-23	6.99E-22	5.67E-21	4.10E-20	2.66E-19	1.57E-18	8.44E-18	4.18E-17	1.91E-16	8.15E-16	3.25E-15	1.22E-14	4.30E-14
	8.43E-21	8.74E-20	7.69E-19	5.84E-18	3.89E-17	2.29E-16	1.21E-15	5.81E-15	2.54E-14	1.02E-13	3.81E-13	1.32E-12	4.32E-12	1.32E-11	3.85E-11	1.06E-10	2.79E-10	7.02E-10
30	4.48E-03	9.09E-03	1.74E-02	3.19E-02	5.58E-02	9.39E-02	0.152	0.239	0.365	0.542	0.786	1.113	1.544	2.101	2.810	3.697	4.793	6.127
	4.48E-03	9.09E-03	1.74E-02	3.19E-02	5.58E-02	9.39E-02	0.152	0.239	0.365	0.542	0.786	1.113	1.544	2.101	2.810	3.697	4.793	6.127
	2.05E-27	3.95E-26	6.26E-25	8.32E-24	9.41E-23	9.21E-22	7.89E-21	5.99E-20	4.08E-19	2.50E-18	1.40E-17	7.19E-17	3.40E-16	1.50E-15	6.15E-15	2.37E-14	8.62E-14	2.96E-13
	1.41E-19	1.31E-18	1.04E-17	7.21E-17	4.41E-16	2.41E-15	1.19E-14	5.35E-14	2.21E-13	8.44E-13	3.00E-12	9.97E-12	3.12E-11	9.24E-11	2.59E-10	6.94E-10	1.78E-09	4.36E-09
40	3.26E-03	6.70E-03	1.30E-02	2.41E-02	4.27E-02	7.25E-02	0.119	0.188	0.290	0.434	0.634	0.904	1.264	1.732	2.333	3.090	4.031	5.185
	3.26E-03	6.70E-03	1.30E-02	2.41E-02	4.27E-02	7.25E-02	0.119	0.188	0.290	0.434	0.634	0.904	1.264	1.732	2.333	3.090	4.031	5.185
	6.88E-26	1.13E-24	1.56E-23	1.81E-22	1.81E-21	1.58E-20	1.22E-19	8.43E-19	5.24E-18	2.96E-17	1.53E-16	7.30E-16	3.23E-15	1.33E-14	5.17E-14	1.88E-13	6.49E-13	2.12E-12
	3.44E-18	2.76E-17	1.93E-16	1.19E-15	6.49E-15	3.20E-14	1.44E-13	5.92E-13	2.25E-12	7.98E-12	2.64E-11	8.24E-11	2.43E-10	6.78E-10	1.81E-09	4.60E-09	1.12E-08	2.64E-08
50	1.93E-03	4.05E-03	8.02E-03	1.51E-02	2.72E-02	4.70E-02	7.82E-02	0.126	0.196	0.298	0.441	0.638	0.903	1.253	1.708	2.289	3.021	3.930
	1.93E-03	4.05E-03	8.02E-03	1.51E-02	2.72E-02	4.70E-02	7.82E-02	0.126	0.196	0.298	0.441	0.638	0.903	1.253	1.708	2.289	3.021	3.930
	3.68E-24	5.22E-23	6.21E-22	6.30E-21	5.55E-20	4.29E-19	2.94E-18	1.81E-17	1.01E-16	5.16E-16	2.42E-15	1.05E-14	4.24E-14	1.60E-13	5.69E-13	1.91E-12	6.08E-12	1.84E-11
	1.09E-16	7.69E-16	4.74E-15	2.59E-14	1.27E-13	5.62E-13	2.28E-12	8.51E-12	2.95E-11	9.56E-11	2.91E-10	8.35E-10	2.27E-09	5.89E-09	1.46E-08	3.46E-08	7.89E-08	1.74E-07
60	8.36E-04	1.80E-03	3.67E-03	7.10E-03	1.31E-02	2.32E-02	3.95E-02	6.51E-02	0.104	0.161	0.244	0.360	0.519	0.734	1.020	1.392	1.870	2.475
	8.36E-04	1.80E-03	3.67E-03	7.10E-03	1.31E-02	2.32E-02	3.95E-02	6.51E-02	0.104	0.161	0.244	0.360	0.519	0.734	1.020	1.392	1.870	2.475
	3.41E-22	4.15E-21	4.26E-20	3.76E-19	2.88E-18	1.95E-17	1.18E-16	6.43E-16	3.19E-15	1.45E-14	6.06E-14	2.36E-13	8.58E-13	2.93E-12	9.43E-12	2.87E-11	8.33E-11	2.31E-10
	4.38E-15	2.73E-14	1.49E-13	7.25E-13	3.17E-12	1.26E-11	4.62E-11	1.56E-10	4.92E-10	1.45E-09	4.02E-09	1.06E-08	2.64E-08	6.31E-08	1.44E-07	3.16E-07	6.70E-07	1.37E-06
70	2.07E-04	4.67E-04	9.95E-04	2.01E-03	3.87E-03	7.15E-03	1.27E-02	2.17E-02	3.60E-02	5.78E-02	9.05E-02	0.138	0.206	0.301	0.431	0.605	0.837	1.138
	2.07E-04	4.67E-04	9.95E-04	2.01E-03	3.87E-03	7.15E-03	1.27E-02	2.17E-02	3.60E-02	5.78E-02	9.05E-02	0.138	0.206	0.301	0.481	0.605	0.837	1.138
	7.84E-20	7.96E-19	6.85E-18	5.09E-17	3.31E-16	1.91E-15	9.85E-15	4.61E-14	1.97E-13	7.75E-13	2.83E-12	9.61E-12	3.07E-11	9.22E-11	2.62E-10	7.10E-10	1.83E-09	4.53E-09
	2.49E-13	1.35E-12	6.49E-12	2.78E-11	1.08E-10	3.80E-10	1.24E-09	3.73E-09	1.05E-08	2.79E-08	6.98E-08	1.66E-07	3.75E-07	8.14E-07	1.69E-06	3.40E-06	6.59E-06	1.24E-05

continued

TABLE A.15 (continued)

Total and Partial Pressure of Aqueous Sulfuric Acid Solution

H_2SO_4 in Liquid Phase (wt%)	Temperature (°C)																	
	0	10	20	30	40	50	60	70	80	90	100	110	120	130	140	150	160	170
75	7.47E-05	1.75E-04	3.88E-04	8.11E-04	1.62E-03	3.09E-03	5.65E-03	9.97E-03[a]	1.70E-02	2.81E-02	4.52E-02	7.08E-02	0.108	0.162	0.236	0.339	0.478	0.662
	7.47E-05	1.75E-04	3.88E-04	8.11E-04	1.62E-03	3.09E-03	5.65E-03	9.97E-03	1.70E-02	2.81E-02	4.52E-02	7.08E-02	0.108	0.162	0.236	0.339	0.478	0.662
	1.74E-18	1.58E-17	1.21E-16	8.08E-16	4.73E-15	2.46E-14	1.16E-13	4.92E-13	1.92E-12	6.93E-12	2.32E-11	7.29E-11	2.15E-10	6.01E-10	1.59E-09	4.03E-09	9.74E-09	2.26E-08
	2.00E-12	1.01E-11	4.47E-11	1.78E-10	6.43E-10	2.12E-09	6.46E-09	1.83E-08	4.85E-08	1.21E-07	2.87E-07	6.44E-07	1.38E-06	2.85E-06	5.65E-06	1.08E-05	2.00E-05	3.59E-05
80	1.97E-05	4.90E-05	1.15E-04	2.53E-04	5.31E-04	1.06E-03	2.04E-03	3.76E-03	6.68E-03	1.15E-02	1.92E-02	3.12E-02	4.93E-02	7.60E-02	0.115	0.170	0.246	0.350
	1.97E-05	4.90E-05	1.15E-04[a]	2.53E-04	5.31E-04	1.06E-03	2.04E-03	3.76E-03	6.68E-03	1.15E-02	1.92E-02	3.12E-02	4.93E-02	7.60E-02	0.115	0.170	0.246	0.350
	5.31E-17	4.17E-16	2.80E-15	1.64E-14	8.51E-14	3.95E-13	1.65E-12	6.34E-12	2.23E-11	7.31E-11	2.23E-10	6.41E-10	1.74E-09	4.46E-09	1.09E-08	2.56E-08	5.75E-08	1.25E-07
	1.61E-11	7.43E-11	3.05E-10	1.13E-09	3.79E-09	1.17E-08	3.34E-08	8.88E-08	2.22E-07	5.22E-07	1.17E-06	2.49E-06	5.08E-06	9.95E-06	1.88E-05	3.43E-05	6.08E-05	1.04E-04
85	3.43E-06	9.52E-06	2.45E-05	5.89E-05	1.34E-04	2.86E-04	5.84E-04	1.14E-03	2.13E-03	3.83E-03	6.66E-03	1.12E-02	1.83E-02	2.91E-02	4.51E-02	6.83E-02	0.101	0.147
	3.43E-06	9.52E-06	2.45E-05	5.89E-05	1.33E-04	2.86E-04	5.84E-04	1.14E-03	2.13E-03	3.83E-03	6.66E-03	1.12E-02	1.83E-02	2.91E-02	4.51E-02	6.82E-02	0.101	0.147
	2.29E-15	1.41E-14	7.67E-14	3.71E-13	1.62E-12	6.43E-12	2.34E-11	7.91E-11	2.49E-10	7.34E-10	2.04E-09	5.38E-09	1.35E-08	3.24E-08	7.45E-08	1.65E-07	3.51E-07	7.25E-07
	1.21E-10	4.90E-10	1.79E-09	5.94E-09	1.81E-08	5.13E-08	1.35E-07	3.36E-07	7.86E-07	1.75E-06	3.71E-06	7.52E-06	1.47E-05	2.77E-05	5.03E-05	8.89E-05	1.52E-04	2.55E-04
90	5.18E-07	1.59E-06	4.49E-06	1.17E-05	3.85E-05	6.53E-05	1.41E-04	2.91E-04	5.71E-04	1.07E-03	1.95E-03	3.40E-03	5.75E-03	9.44E-03	1.51E-02	2.35E-02	3.57E-02	5.32E-02
	5.18E-07	1.59E-06	4.48E-06	1.17E-05	2.85E-05	6.52E-05	1.41E-04	2.90E-04	5.69E-04	1.07E-03	1.94E-03	3.38E-03	5.71E-03	9.38E-03	1.50E-02	2.33E-02	3.54E-02	5.26E-02
	6.71E-14	3.45E-13	1.59E-12	6.64E-12	2.54E-11	8.97E-11	2.94E-10	9.04E-10	2.61E-09	7.12E-09	1.84E-08	4.56E-08	1.08E-07	2.44E-07	5.33E-07	1.12E-06	2.29E-06	4.53E-06
	5.34E-10	2.00E-09	6.77E-09	2.11E-08	6.07E-08	1.63E-07	4.11E-07	9.76E-07	2.20E-06	4.73E-06	9.73E-06	1.92E-05	3.66E-05	6.72E-05	1.20E-04	2.07E-04	3.48E-04	5.72E-04
92	2.43E-07	7.65E-07	2.21E-06	5.90E-06	1.47E-05	3.44E-05	7.59E-05	1.59E-04	3.19E-04	6.12E-04	1.13E-03	2.01E-03	3.46E-03	5.78E-03	9.39E-03	1.49E-02	2.30E-02	3.47E-02
	2.42E-07	7.62E-07	2.20E-06	5.87E-06	1.46E-05	3.41E-05	7.54E-05	1.58E-04	3.16E-04	6.06E-04	1.12E-03	1.98E-03	3.41E-03	5.69E-03	9.23E-03	1.46E-02	2.25E-02	3.40E-02
	2.16E-13	1.07E-12	4.75E-12	1.92E-11	7.09E-11	2.42E-10	7.71E-10	2.30E-09	6.43E-09	1.71E-08	4.30E-08	1.03E-07	2.38E-07	5.26E-07	1.12E-06	2.30E-06	4.59E-06	8.86E-06
	8.03E-10	2.96E-09	9.93E-09	3.06E-08	8.70E-08	2.31E-07	5.75E-07	1.35E-06	3.02E-06	6.42E-06	1.31E-05	2.56E-05	4.82E-05	8.79E-05	1.55E-04	2.66E-04	4.44E-04	7.23E-04
94	1.09E-07	3.48E-07	1.02E-06	2.79E-06	7.08E-06	1.69E-05	3.80E-05	8.13E-05	1.66E-04	3.24E-04	6.07E-04	1.10E-03	1.92E-03	3.27E-03	5.39E-03	8.66E-03	1.36E-02	2.08E-02
	1.07E-07	3.44E-07	1.01E-06	2.75E-06	6.96E-06	1.66E-05	3.72E-05	7.95E-05	1.62E-04	3.15E-04	5.90E-04	1.07E-03	1.86E-03	3.15E-03	5.19E-03	8.32E-03	1.30E-02	1.99E-02
	6.77E-13	3.26E-12	1.41E-11	5.57E-11	2.01E-10	6.69E-10	2.07E-09	6.02E-09	1.65E-08	4.26E-08	1.05E-07	2.47E-07	5.55E-07	1.20E-06	2.50E-06	5.04E-06	9.83E-06	1.86E-05
	1.12E-09	4.09E-09	1.36E-08	4.15E-08	1.17E-07	3.09E-07	7.65E-07	1.79E-06	3.96E-06	8.35E-06	1.69E-05	3.28E-05	6.14E-05	1.11E-04	1.95E-04	3.32E-04	5.50E-04	8.89E-04
96	4.16E-08	1.36E-07	4.07E-07	1.13E-06	2.93E-06	7.12E-06	1.64E-05	3.57E-05	7.42E-05	1.48E-04	2.83E-04	5.21E-04	9.29E-04	1.61E-03	2.70E-03	4.41E-03	7.03E-03	1.10E-02
	4.01E-08	1.30E-07	3.90E-07	1.08E-06	2.78E-06	6.72E-06	1.54E-05	3.34E-05	6.91E-05	1.37E-04	2.61E-04	4.79E-04	8.51E-04	1.46E-03	2.45E-03	3.99E-03	6.33E-03	9.83E-03
	2.40E-12	1.14E-11	4.82E-11	1.86E-10	6.55E-10	2.14E-09	6.47E-09	1.84E-08	4.92E-08	1.24E-07	3.00E-07	6.89E-07	1.52E-06	3.21E-06	6.56E-06	1.29E-05	2.47E-05	4.59E-05
	1.48E-09	5.40E-09	1.79E-08	5.43E-08	1.53E-07	4.00E-07	9.85E-07	2.29E-06	5.04E-06	1.06E-05	2.13E-05	4.12E-05	7.67E-05	1.38E-04	2.41E-04	4.08E-04	6.73E-04	1.08E-03
97	2.35E-08	7.74E-08	2.35E-07	6.59E-07	1.73E-06	4.25E-06	9.87E-06	2.18E-05	4.58E-05	9.21E-05	1.78E-04	3.32E-04	5.98E-04	1.04E-03	1.77E-03	2.93E-03	4.71E-03	7.41E-03
	2.18E-08	7.13E-08	2.15E-07	5.98E-07	1.55E-06	3.79E-06	8.75E-06	1.92E-05	4.00E-05	8.01E-05	1.54E-04	2.85E-04	5.11E-04	8.86E-04	1.49E-03	2.45E-03	3.93E-03	6.14E-03

	5.00E-12	2.34E-11	9.86E-11	3.76E-10	1.31E-09	4.24E-09	1.27E-08	3.57E-08	9.46E-08	2.37E-07	5.65E-07	1.28E-06	2.80E-06	5.86E-06	1.18E-05	2.31E-05	4.38E-05	8.06E-05
	1.67E-09	6.09E-09	2.01E-08	6.11E-08	1.71E-07	4.49E-07	1.10E-06	2.56E-06	5.62E-06	1.18E-05	2.37E-05	4.57E-05	8.49E-05	1.53E-04	2.66E-04	4.49E-04	7.40E-04	1.19E-03
98	1.17E-08	3.91E-08	1.21E-07	3.44E-07	9.14E-07	2.28E-06	5.38E-06	1.20E-05	2.57E-05	5.24E-05	1.03E-04	1.94E-04	3.54E-04	6.26E-04	1.07E-03	1.80E-03	2.93E-03	4.66E-03
	9.80E-09	3.23E-08	9.78E-08	2.75E-07	7.20E-07	1.77E-06	4.13E-06	9.12E-06	1.92E-05	3.88E-05	7.52E-05	1.41E-04	2.54E-04	4.45E-04	7.57E-04	1.25E-03	2.02E-03	3.19E-03
	1.24E-11	5.78E-11	2.41E-10	9.11E-10	3.15E-09	1.01E-08	2.99E-08	8.33E-08	2.18E-07	5.41E-07	1.27E-06	2.87E-06	6.19E-06	1.28E-05	2.57E-05	4.97E-05	9.32E-05	1.70E-04
	1.87E-09	6.79E-09	2.24E-08	6.80E-08	1.91E-07	4.98E-07	1.22E-06	2.83E-06	6.22E-06	1.30E-05	2.61E-05	5.03E-05	9.35E-05	1.68E-04	2.92E-04	4.93E-04	8.10E-04	1.30E-03
98.5	7.68E-09	2.61E-08	8.12E-08	2.34E-07	6.30E-07	1.59E-06	3.79E-06	8.56E-06	1.84E-05	3.90E-05	7.51E-05	1.43E-04	2.63E-04	4.70E-04	8.15E-04	1.37E-03	2.26E-03	3.63E-03
	5.69E-09	1.88E-08	5.72E-08	1.61E-07	4.24E-07	1.05E-06	2.45E-06	5.44E-06	1.15E-05	2.34E-05	4.55E-05	8.55E-05	1.55E-04	2.78E-04	4.67E-04	7.76E-04	1.26E-03	1.99E-03
	2.24E-11	1.04E-10	4.33E-10	1.63E-09	5.62E-09	1.79E-08	5.28E-08	1.46E-07	3.81E-07	9.40E-07	2.20E-06	4.94E-06	1.06E-05	2.19E-05	4.35E-05	8.37E-05	1.56E-04	2.83E-04
	1.96E-09	7.14E-09	2.36E-08	7.14E-08	2.00E-07	5.23E-07	1.28E-06	2.97E-06	6.52E-06	1.36E-05	2.74E-05	5.27E-05	9.77E-05	1.75E-04	3.04E-04	5.14E-04	8.44E-04	1.35E-03
99	4.79E-09	1.66E-08	5.28E-08	1.55E-07	4.25E-07	1.09E-06	2.64E-06	6.05E-06	1.32E-05	2.77E-05	5.55E-05	1.07E-04	2.01E-04	3.63E-04	6.39E-04	1.09E-03	1.83E-03	2.99E-03
	2.68E-09	8.88E-09	2.71E-08	7.66E-08	2.02E-07	5.03E-07	1.18E-06	2.63E-06	5.59E-06	1.14E-05	2.23E-05	4.20E-05	7.66E-05	1.35E-04	2.32E-04	3.87E-04	6.29E-04	9.99E-04
	5.02E-11	2.32E-10	9.61E-10	3.60E-09	1.23E-08	3.91E-08	1.15E-07	3.16E-07	8.20E-07	2.01E-06	4.70E-06	1.05E-05	2.24E-05	4.59E-05	9.10E-05	1.74E-04	3.24E-04	5.86E-04
	2.06E-09	7.50E-09	2.47E-08	7.49E-08	2.10E-07	5.48E-07	1.34E-06	3.10E-06	6.81E-06	1.43E-05	2.85E-05	5.49E-05	1.02E-04	1.82E-04	3.16E-04	5.34E-04	8.76E-04	1.40E-03
99.5	3.13E-09	1.13E-08	3.73E-08	1.14E-07	3.23E-07	8.61E-07	2.16E-06	5.14E-06	1.17E-05	2.53E-05	5.27E-05	1.06E-04	2.06E-04	3.87E-04	7.08E-04	1.26E-03	2.19E-03	3.72E-03
	7.75E-10	2.58E-09	7.89E-09	2.24E-08	5.95E-08	1.49E-07	3.50E-07	7.84E-07	1.68E-06	3.43E-06	6.74E-06	1.28E-05	2.33E-05	4.14E-05	7.11E-05	1.19E-04	1.94E-04	3.09E-04
	1.82E-10	8.39E-10	3.46E-09	1.29E-08	4.40E-08	1.39E-07	4.05E-07	1.11E-06	2.86E-06	6.98E-06	1.62E-05	3.59E-05	7.64E-05	1.56E-04	3.08E-04	5.88E-04	1.09E-03	1.96E-03
	2.17E-09	7.88E-09	2.60E-08	7.86E-08	2.20E-07	5.74E-07	1.40E-06	3.25E-06	7.12E-06	1.49E-05	2.98E-05	5.72E-05	1.06E-04	1.90E-04	3.29E-04	5.54E-04	9.09E-04	1.45E-03
100	3.23E-09	1.24E-08	4.35E-08	1.41E-07	4.25E-07	1.20E-06	3.19E-06	8.04E-06	1.93E-05	4.41E-05	9.66E-05	2.04E-04	4.14E-04	3.14E-04	1.55E-03	2.87E-03	5.15E-03	9.05E-03
	1.96E-10	6.55E-10	2.01E-09	5.75E-09	1.53E-08	3.84E-08	9.10E-08	2.05E-07	4.39E-07	9.03E-07	1.78E-06	3.39E-06	6.23E-06	1.11E-05	1.91E-05	3.21E-05	5.26E-05	8.40E-05
	7.55E-10	3.47E-09	1.42E-08	5.28E-08	1.79E-07	5.60E-07	1.63E-06	4.44E-06	1.14E-05	2.76E-05	6.38E-05	1.41E-04	2.98E-04	6.06E-04	1.19E-03	2.26E-03	4.16E-03	7.46E-03
	2.28E-09	8.27E-09	2.73E-08	8.24E-08	2.30E-07	6.00E-07	1.47E-06	3.39E-06	7.43E-06	1.55E-05	3.10E-05	5.95E-05	1.10E-04	1.97E-04	3.41E-04	5.74E-04	9.41E-04	1.50E-03

Source: Adapted from Perry, R. H., D. W. Green, and J. O. Maloney. 1997. *Perry's Chemical Engineers' Handbook (7th edition)*. McGraw-Hill.

Note: Total pressure, H_2O partial pressure, SO_3 partial pressure, H_2SO_4 partial pressure from the top of each column.

[a] A considerable error in original paper is corrected.

TABLE A.16

Density of Pure Liquid Compounds

	H_2	O_2	H_2O			SO_2	SO_3	HBr
C_1	5.414	3.9143	5.459	4.9669	4.3910	2.106	1.4969	2.832
C_2	0.34893	0.28772	0.30542	0.27788	0.24870	0.25842	0.19013	0.2832
C_3	33.19	154.58	647.13	647.13	647.13	430.75	490.85	363.15
C_4	0.2706	0.2924	0.081	0.18740	0.25340	0.2895	0.4359	0.28571
T_{min} (K)	13.95	54.35	273.16	333.15	403.15	197.67	289.95	185.15
T_{max} (K)	33.19	154.58	333.15	403.15	647.15	430.75	490.85	363.15

	Br_2	HCl	CO	CO_2	CH_4
C_1	2.1872	3.342	2.897	2.768	2.9214
C_2	0.29527	0.2729	0.27532	0.26212	0.28976
C_3	584.15	324.65	132.92	304.21	190.56
C_4	0.3295	0.3217	0.2813	0.2908	0.28881
T_{min} (K)	265.85	158.97	68.15	216.58	90.69
T_{max} (K)	584.15	324.65	132.92	304.21	190.56

Source: Adapted from Perry, R. H., D. W. Green, and J. O. Maloney. 1997. *Perry's Chemical Engineers' Handbook (7th edition)*. McGraw-Hill, New York.

$$\rho = \frac{C_1}{C_2^{\left[1+\left(1-\frac{T}{C_3}\right)^{C_4}\right]}}$$

ρ: Density (kmol/m^3)
C_1, C_2, C_3, C_4: Constants
T_{min}: Minimum temperature of applicable range (K)
T_{max}: Maximum temperature of applicable range (K)

TABLE A.17

Density of Molten Liquids of Pure Compounds

	$CaBr_2$	Cu	CuCl
t_m (°C)	742	1084.62	423
ρ_m (g/cm^3)	3.111	8.02	3.692
k (g/(cm^3·°C))	0.0005	0.000609	0.00076
t_{max} (°C)	791	1630	585

Source: Adapted from Lide, D. R. 2006–2007. *CRC Handbook of Chemistry and Physics (87th edition)*. CRC Press, Boca Raton.

$$\rho = \rho_m - k(t - t_m)$$

ρ: Density (g/cm^3)
ρ_m: Density at melting point (g/cm^3)
k: Constant (g/(cm^3·°C))
t: Temperature (°C)
t_m: Melting point (°C)
t_{max}: Maximum temperature of applicable range (°C)

TABLE A.18
Density of Aqueous Solutions

	HI	$CaBr_2$	HBr
α	−1.33539E-08	2.13197E-08	−5.1394E-09
β	1.72047E-06	−1.03920E-06	1.07265E-06
γ	2.25058E-05	9.16716E-05	2.43298E-05
δ	7.2801E-03	8.01701E-03	7.0596E-03
ε	0.99676	0.997614	0.99683
Temperature (°C)	25	25	25
Concentration range (wt.%)	1–45	2–50	1–65

Source: Adapted from 2004. The Chemical Society of Japan (eds.), *Kagaku Binran Kisohen (5th edition)*. The Chemical Society of Japan, Maruzen, Tokyo.

$$\rho = \alpha c^4 + \beta c^3 + \gamma c^2 + \delta c + \varepsilon$$

ρ: Density (g/cm^3)
c: Concentration (wt.%)
α, β, γ, δ, ε: Constants

TABLE A.19
Temperature Dependence of Density of Aqueous Solutions (Unit: g/cm^3)

		H_2SO_4									
Concentration (wt%)		**10**	**20**	**30**	**40**	**50**	**60**	**70**	**80**	**90**	**100**
Temperature (°C)	0	1.0735	1.1510	1.2326	1.3179	1.4110	1.5154	1.6293	1.7482	1.8361	1.8517
	10	1.0700	1.1453	1.2255	1.3103	1.4029	1.5067	1.6198	1.7376	1.8252	1.8409
	20	1.0661	1.1394	1.2185	1.3028	1.3951	1.4983	1.6105	1.7272	1.8144	1.8305
	30	1.0617	1.1335	1.2115	1.2953	1.3872	1.4898	1.6014	1.7170	1.8038	1.8205
	40	1.0570	1.1275	1.2046	1.2880	1.3795	1.4816	1.5925	1.7069	1.7933	1.8017
	50	1.0517	1.1215	1.1977	1.2806	1.3719	1.4735	1.5838	1.6971	1.7829	1.8013
	60	1.0460	1.1153	1.1909	1.2732	1.3644	1.4656	1.5753	1.6873	1.7729	1.7922
	80	1.0338	1.1021	1.1771	1.2589	1.3494	1.4497	1.5582	1.6680	1.7525	
	100	1.0204	1.0885	1.1630	1.2446	1.3348	1.4344	1.5417	1.6493	1.7331	

		CuCl[a]						HCl			
Concentration (wt%)		**1**	**4**	**8**	**12**	**16**	**20**	**10**	**20**	**30**	**40**
Temperature (°C)	0	1.0095	1.0387	1.0788	1.1208	1.1653	1.2121	1.0528	1.1067	1.1613	
	10							1.0504	1.1025	1.1533	
	20	1.0072	1.036	1.0754	1.1165	1.1595	1.2052	1.0474	1.0980	1.1493	1.1980
	40	1.002	1.0305	1.0682	1.107	1.151	1.1953	1.0400	1.0888	1.1376	
	60							1.0310	1.0790	1.1260	
	80							1.0206	1.0685	1.1149	
	100							1.0090	1.0574	1.1030	

Source: Adapted from Perry, R. H., D. W. Green, and J. O. Maloney. 1997. *Perry's Chemical Engineers' Handbook (7th edition)*. McGraw-Hill, New York.

[a] "Cu_2Cl_2" in the original paper.

TABLE A.20

Temperature Dependence of Density of Pure Compounds (Unit: kg/m^3)

Gas Phase							
Temperature (K)	**H_2**	**O_2**	**H_2O**	**SO_2[a]**	**CO[a]**	**CO_2**	**CH_4**
260						2.052	0.7441
273.15	0.08871	1.410				1.951	0.7080
280	0.08654	1.376				1.902	0.6906
300	0.08077	1.284		2.648	1.1382	1.773	0.6442
320	0.07572	1.203		2.473		1.661	0.6038
340	0.07127	1.132		2.321		1.562	0.5681
350				2.252	0.9752		
360	0.06731	1.069		2.187		1.474	0.5364
380	0.06377	1.013	0.5784	2.069		1.396	0.5081
400	0.06058	0.9622	0.5478	1.963	0.8532	1.326	0.4826
420			0.5205	1.868		1.262	0.4596
440			0.4960	1.781		1.205	0.4386
450	0.05385	0.8552		1.741	0.7583		
460			0.4739	1.703		1.152	0.4195
480			0.4537			1.104	0.4020
500	0.04847	0.7696	0.4352		0.6824	1.059	0.3859
550	0.04407		0.3951			0.963	0.3508
600	0.04039	0.6413	0.3619		0.5687	0.882	0.3215
650	0.03729		0.3339			0.814	
700		0.5496	0.3099			0.756	
750			0.2892			0.706	
800		0.4809	0.2711		0.4265	0.662	
850			0.2551			0.623	
900		0.4275	0.2409			0.588	
950			0.2282			0.537	
1000		0.3848	0.2168		0.3412	0.529	
1050			0.2065				
1100						0.481	
1200		0.3206					
Reference	[9]	[9]	[10]	[9]	[9]	[10]	[9]

Liquid Phase			
Temperature (K)	**H_2O**	**I_2[b]**	**Br_2[b]**
273.15	999.83		
280	999.98		3157
290	998.92		3125
300	996.66		3094
320	989.47		3020
340	979.48		2954
350			2921

continued

TABLE A.20 (continued)

Temperature Dependence of Density of Pure Compounds (Unit: kg/m^3)

Liquid Phase			
Temperature (K)	**H_2O**	**I_2[b]**	**Br_2[b]**
360	967.23		2887
386.75		3940	
400		3910	2742
450		3790	2530
500		3660	2284
550		3520	1955
584.2			1180[c]
600		3360	
650		3190	
700		2980	
750		2730	
800		2320	
819.15		1610[c]	
Reference	[10]	[10]	[10]

Solid Phase (At Room Temperature)								
I_2	**CaO**	**$CaBr_2$**	**Fe_3O_4**	**$FeBr_2$**	**Cu**	**CuCl**	**$CuCl_2$**	**C[d]**
4.933E + 03	3.34E + 03	3.38E + 03	5.17E + 03	4.636E + 03	8.96E + 03	4.14E + 03	3.4E + 03	2.2E + 03

Source: Adapted from Lide, D. R. 2006–2007. *CRC Handbook of Chemistry and Physics (87th edition)*. CRC Press, Boca Raton.

Note: At 0.1 MPa except SO_2 and CO.

[a] Gas phase is at 101.325 kPa.

[b] Vapor–liquid equilibrium condition, pressure varies with temperature.

[c] Critical temperature.

[d] Graphite.

$$\rho = \rho_r \cdot \exp\left[a\Delta T + b(\Delta T)^2\right]$$

(Equation for solid phase CuO·$CuCl_2$ in the range of 298–675 K and at 1 atm [4])

a, b: Parameters ($a = -2.660831725 \times 10^{-24}$, $b = -1.221274397 \times 10^{-13}$)

T: Temperature (K)

T_r: Reference temperature (=298.15 (K))

ΔT: $T - T_r$ (K)

ρ: Density (kg/m^3)

ρ_r: Reference density (−4080 (kg/m^3))

TABLE A.21

Viscosity of Pure Compounds (Units (Pa ·s))

			Gas Phase				
Temperature (K)	**H_2**	**O_2**	**H_2O**	**HI**	**I_2**	**SO_2**	**SO_3**
270	8.32E–06	1.896E–05		1.712E–05		1.172E–05	
273.16					1.230E–05		1.20E–05
280	8.53E–06	1.954E–05	8.32E–06	1.775E–05		1.216E–05	
290	8.73E–06	2.011E–05	8.73E–06	1.837E–05		1.260E–05	
310	9.14E–06	2.123E–05	9.54E–06	1.961E–05		1.347E–05	
330	9.54E–06	2.231E–05	1.035E–05	2.084E–05		1.433E–05	
350	9.94E–06	2.337E–05	1.117E–05	2.207E–05		1.518E–05	
370	1.033E–05	2.440E–05	1.198E–05	2.329E–05	1.728E–05	1.603E–05	
373.16							1.68E–05
470	1.221E–05	2.920E–05	1.605E–05	2.927E–05	2.166E–05	2.012E–05	
473.16							2.11E–05
570	1.40E–05	3.35E–05	2.012E–05	3.50E–05	2.60E–05	2.40E–05	
573.16							2.48E–05
650	1.53E–05	3.67E–05	2.338E–05	3.93E–05	2.93E–05	2.66E–05	
670	1.56E–05	3.74E–05	2.419E–05		3.01E–05	2.76E–05	
673.16							2.80E–05
711.15					3.250E–05		
770	1.73E–05	4.11E–05	2.825E–05			3.11E–05	
796.15					3.604E–05		
870	1.88E–05	4.44E–05	3.233E–05			3.44E–05	
970	2.03E–05	4.76E–05	3.640E–05			3.75E–05	
1000	2.07E–05	4.85E–05	3.762E–05			3.84E–05	
1100	2.22E–05	5.14E–05				4.13E–05	
1200	2.36E–05	5.42E–05				4.42E–05	
1250	2.43E–05	5.56E–05				4.54E–05	
1300	2.50E–05	5.69E–05					
Reference	[11]	[11]	[11]	[11]	[11], [12] (at 273.16, 711.15, 796.15 (K))	[11]	[13]

Temperature (K)	**Br_2**	**HCl**	**CO**	**CO_2**	**CH_4**
270		1.312E–05	1.638E–05	1.359E–05	1.021E–05
280	1.46E–05	1.364E–05	1.687E–05	1.406E–05	1.054E–05
290	1.51E–05	1.416E–05	1.734E–05	1.453E–05	1.087E–05
310	1.60E–05	1.518E–05	1.827E–05	1.545E–05	1.152E–05
330	1.69E–05	1.619E–05	1.918E–05	1.636E–05	1.215E–05
350	1.79E–05	1.719E–05	2.005E–05	1.726E–05	1.277E–05
370	1.88E–05	1.817E–05	2.091E–05	1.814E–05	1.337E–05
470	2.36E–05	2.291E–05	2.485E–05	2.230E–05	1.618E–05
570	2.84E–05	2.74E–05	2.84E–05	2.61E–05	1.87E–05
650	3.22E–05	3.06E–05	3.10E–05	2.90E–05	2.06E–05

continued

TABLE A.21 (continued)

Viscosity of Pure Compounds (Units (Pa ·s))

Temperature (K)	Br_2	HCl	CO	CO_2	CH_4
670	3.32E−05		3.16E−05	2.96E−05	2.10E−05
770	3.78E−05		3.45E−05	3.29E−05	2.32E−05
800	3.92E−05		3.54E−05	3.39E−05	2.38E−05
870			3.73E−05	3.60E−05	2.52E−05
970			3.99E−05	3.89E−05	2.70E−05
1000			4.06E−05	3.97E−05	2.76E−05
1100			4.30E−05	4.24E−05	
1200			4.53E−05	4.49E−05	
1300			4.75E−05	4.74E−05	
Reference	[11]	[11]	[11]	[11]	[11]

Liquid Phase

Temperature (K)	H_2O	I_2	H_2SO_4	SO_2	SO_3	Br_2
273.15	1.7919E−03			3.80E−04		1.252E−03
283.15	1.3069E−03		4.52E−02			
293.15	1.0020E−03		2.86E−02	2.97E−04	2.19E−03	
298.15						9.44E−04
303.15	7.973E−04				1.52E−03	
313.15	6.529E−04		1.50E−02	2.37E−04	1.10E−03	
323.15	5.470E−04			2.10E−04	8.05E−04	7.46E−04
328.15	5.042E−04				6.80E−04	
333.15	4.667E−04		9.4E−03			
353.15	3.550E−04		6.0E−03			
373.15	2.822E−04		4.0E−03			
388.15		2.29E−03				
393.15		2.210E−03				
413.15		1.916E−03				
423.15	1.8E−04					
433.15		1.675E−03				
453.15		1.445E−03				
468.15		1.288E−03				
473.15	1.3E−04					
523.15	1.1E−04					
573.15	9.0E−05					
643.15	5.2E−05					
Reference	[5], [12] (between 423.15 and 643.15 (K) inclusive)	[12]	[12]	[12]	[12]	[1]

Note: At 1 atm.

TABLE A.22

Viscosity of Solutions (Unit: Pa·s)

Concentration (wt%)		H_2SO_4							
		10	20	40	60	70	80	90	100
Temperature (°C)	0	2.16E–03	2.71E–03	4.70E–03	1.00E–02	2.00E–02	4.3E–02		
	10	1.56E–03	2.01E–03	3.48E–03	7.50E–03	1.40E–02	3.1E–02	3.9E–02	3.9E–02
	20	1.23E–03	1.55E–03	2.70E–03	5.70E–03	1.020E–02	2.2E–02	2.4E–02	2.7E–02
	30	9.8E–04	1.23E–03	2.16E–03	4.58E–03	7.7E–03	1.54E–02	1.6E–02	1.9E–02
	40	7.9E–04	9.9E–04	1.80E–03	3.71E–03	6.1E–03	1.09E–02	1.3E–02	1.4E–02
	50	6.6E–04	8.3E–04	1.53E–03	3.20E–03	5.1E–03	8.1E–03	9.E–03	1.05E–02
	60	5.6E–04	7.1E–04	1.31E–03	2.80E–03	4.4E–03	6.2E–03	7.E–03	8.E–03
	70	4.9E–04	6.4E–04	1.15E–03	2.48E–03	3.7E–03	5.1E–03	6.E–03	6.5E–03
	80	4.5E–04	5.8E–04	1.03E–03	2.21E–03	3.2E–03	4.5E–03	5.E–03	5.5E–03

Concentration (wt%)		$CuCl_2$			HCl				
		12.01	21.35	33.03	5	10	15	20	30
Temperature (°C)	0				1.84E–03	1.89E–03			
	10				1.38E–03	1.45E–03			
	15	1.56E–03	2.18E–03	3.20E–03					
	20				1.08E–03	1.16E–03	1.24E–03	1.36E–03	1.70E–03
	25	1.21E–03	1.72E–03	2.46E–03					
	35	9.9E–04	1.38E–03	1.93E–03					
	45	8.2E–04	1.13E–03	1.56E–03					

		HBr			
Concentration (mol/dm^3–H_2O)		0.125	0.25	0.5	1.0
Temperature (°C)	25	8.963E–04	8.987E–04	9.048E–04	9.187E–04

Source: Adapted from The Chemical Society of Japan (eds.), 2004. *Kagaku Binran Kisohen (5th edition).* The Chemical Society of Japan, Maruzen, Tokyo.

TABLE A.23

Thermal Conductivity of Pure Compounds (Unit: W/(m·K))

Temperature (K)	Gas Phase						
	H_2	O_2	H_2O	HI	I_2[a]	SO_2	SO_3[b]
250	1.560E−01	2.254E−02	1.40E−02[a]	5.1E−03		7.80E−03[a]	
273.15							8.15E−03
293.15							9.25E−03
300	1.815E−01	2.674E−02	1.81E−02	6.2E−03		9.6E−03	
313.15							1.01E−02
333.15							1.16E−02
350	2.033E−01	3.056E−02	2.22E−02	7.2E−03		1.18E−02	
353.15							1.29E−02
373.15							1.43E−02
386.8					4.26E−03		
400	2.212E−01	3.420E−02	2.64E−02	8.3E−03	4.4E−03	1.43E−02	
450	2.389E−01	3.77E−02	3.07E−02	9.3E−03	4.9E−03	1.70E−02	
500	2.564E−01	4.12E−02	3.57E−02	1.03E−02	5.5E−03	2.00E−02	
600	2.91E−01	4.80E−02	4.64E−02	1.23E−02	6.5E−03	2.56E−02	
700	3.25E−01	5.44E−02	5.72E−02	1.43E−02	7.6E−03	3.05E−02	
785					8.7E−03		
800	3.60E−01	6.03E−02	6.8E−02	1.62E−02		3.52E−02	
900	3.94E−01	6.61E−02	7.8E−02	1.82E−02		4.00E−02	
1000	4.28E−01	7.17E−02		2.00E−02			
1100	4.62E−01	7.71E−02					
1200	4.95E−01	8.21E−02					
1300	5.28E−01	8.71E−02					
Reference	[14]	[14]	[14]	[14]	[15]	[14]	[13]

Temperature (K)	Br_2[b]	HCl	CO	CO_2	CH_4
250		1.19E−02	2.141E−02	1.289E−02	2.77E−02
300	4.7E−03	1.45E−02	2.52E−02	1.662E−02	3.43E−02
350	5.7E−03	1.70E−02	2.88E−02	2.050E−02	4.12E−02
400	6.8E−03	1.95E−02	3.23E−02	2.441E−02	4.84E−02
450	8.0E−03	2.18E−02	3.55E−02	2.834E−02	5.78E−02
500	9.9E−03	2.40E−02	3.86E−02	3.228E−02	6.71E−02
584	2.8E−02[c]				
600		2.81E−02	4.44E−02	4.03E−02	8.58E−02
700		3.21E−02	4.97E−02	4.87E−02	1.041E−01
800			5.49E−02	5.60E−02	1.24E−01
900			5.96E−02	6.21E−02	1.46E−01
1000			6.44E−02	6.80E−02	1.69E−01
1100			6.92E−02	7.33E−02	
1200			7.38E−02	7.80E−02	
1250			7.61E−02		
1300				8.25E−02	
Reference	[14]	[14]	[14]	[14]	[14]

TABLE A.23 (continued)

Thermal Conductivity of Pure Compounds (Unit: W/(m·K))

			Liquid Phase				
Temper-ature (K)	**H_2O[c]**	**HI[b]**	**I_2[b]**	**H_2SO_4**	**SO_2[c,e]**	**SO_3[b]**	**Br_2[c]**
273.15		4.62E−02		3.08E−01		2.73E−01	
280	5.818E−01				2.070E−01		
290	5.918E−01				2.007E−01		
293.15				3.30E−01			
300	6.084E−01				1.945E−01		1.22E−01
310	6.233E−01				1.88E−01		1.20E−01
313.15				3.51E−01			
320	6.367E−01				1.82E−01		1.18E−01
323.15		3.84E−02				2.12E−01	
330	6.485E−01				1.76E−01		1.16E−01
333.15				3.73E−01			
340	6.587E−01				1.70E−01		1.14E−01
350	6.673E−01				1.63E−01		1.11E−01
353.15				3.95E−01			
360	6.743E−01				1.57E−01		1.09E−01
370	6.793E−01				1.51E−01		1.06E−01
373.15		3.02E−02		4.16E−01		1.51E−01	
380	6.836E−01				1.45E−01		1.04E−01
386.8			4.26E−03				
400	6.864E−01		4.40E−03		1.32E−01		9.9E−02
423.15		1.30E−02				1.03E−01	
473.15						6.80E−02	
500	6.348E−01		5.60E−03				7.3E−02
584							2.8E−02[d]
600	4.81E−01		7.40E−03				
700			1.04E−02				
785			2.65E−02				
Reference	[14]	[13]	[13]	[13]	[14]	[13]	[14]

continued

TABLE A.23 (continued)

Thermal Conductivity of Pure Compounds (Unit: W/(m·K))

	Solid Phase						
Temperature (K)	I_2[f]	Fe_3O_4[g]	**Cu**	C[h,i]	C[h,j]	C[k,l]	C[k,m]
250	5.12E–01		4.04E + 02	1.31E + 02[a]	9.7E + 01[a]	2.45E + 03	1.16E + 01
273.2	4.81E–01		4.01E + 02	1.31E + 02[a]	9.8E + 01[a]	2.23E + 03	1.06E + 01
300	4.49E–01		3.98E + 02	1.29E + 02	9.8E + 01	2.00E + 03	9.5E + 00
317.1		4.44E + 00					
335.7		4.35E + 00					
350	4.01E–01		3.94E + 02	1.24E + 02	9.5E + 01	1.69E + 03	8.0E + 00
353.9		4.35E + 00					
385.6		4.31E + 00					
386.8	3.75E–01						
400			3.92E + 02	1.18E + 02	9.0E + 01	1.46E + 03	7.0E + 00
453.2		4.14E + 00					
500			3.88E + 02	1.06E + 02	8.1E + 01	1.13E + 03	5.4E + 00
600			3.83E + 02	9.5E + 01	7.3E + 01	9.3E + 02	4.4E + 00
700			3.77E + 02	8.5E + 01	6.5E + 01	7.9E + 02	3.8E + 00
800			3.71E + 02	7.7E + 01	5.9E + 01	6.8E + 02	3.2E + 00
900			3.64E + 02	7.0E + 01	5.4E + 01	6.0E + 02	2.8E + 00
1000			3.57E + 02	6.4E + 01	4.9E + 01	5.3E + 02	2.5E + 00
1100			3.50E + 01			4.8E + 02	2.3E + 00
1200			3.42E + 02	5.5E + 01	4.3E + 01	4.4E + 02	2.1E + 00
1300			3.34E + 02			4.0E + 02	1.9E + 00
Reference	[15]	[15]	[15]	[15]	[15]	[15]	[15]

At 1 atm except H_2O, SO_2 in the liquid phase, and Br_2 in the vapor and liquid phases.

[a] Estimated value.
[b] Read from a graph.
[c] Coexisting with saturation liquid of the component, pressure varies with temperature.
[d] Critical point.
[e] Extrapolation from an experimental data between 260 and 298 K.
[f] Polycrystalline. Estimated values except at 300 K. Accuracy around room temperature is with in 10%.
[g] Single crystal.
[h] All graphite, the pitch-bonded petroleum–coke–base graphite with typical room-temperautre density 1.73 g/cm^3 and is produced by the Carbon Production Division of Union Carbide Corporation.
[i] Perpendicular to the direction of the molding pressure.
[j] Parallel to the direction of the molding pressure.
[k] Pyrolytic graphite, produced by the deposition of carbon from a gaseous hydrocarbon onto a heated surface at high temperature of the order of 2300 K.
[l] Parallel to layer planes.
[m] Perpendicular to layer planes.
[n] Estimated value.

TABLE A.24

Thermal conductivity of solutions (Unit: W/(m·K))

		H_2SO_4				HCl				
Concentration (wt.%)		**25**	**50**	**75**	**96**	**10**	**30**	**50**	**70**	**90**
Temperature (°C)	0	0.491	0.436	0.381	0.317	0.574	0.618	0.648	0.668	0.68
	10					0.535	0.576	0.611	0.64	0.65
	20	0.531	0.469	0.400	0.326	0.488	0.518	0.548	0.57	0.59
	30					0.442	0.465	0.490		
	35					0.419	0.440	0.460		
	40	0.563	0.493	0.420	0.333					
	60	0.587	0.513	0.438	0.340					
	80	0.609	0.529	0.454	0.347					
	100	0.624	0.538	0.462	0.352					

Source: Adapted from The Japan Society of Mechanical Engineers (eds.), 1983. *JSME Databook: The Thermophysical Properties of Fluids*. The Japan Society of Mechanical Engineers, Tokyo.

TABLE A.25

Properties of Pure Compounds as Fuel

	Temperature (K)	Unit	Hydrogen H_2	Methane CH_4	Ethane C_2H_6	Propane C_3H_8	Ref.
Phase			Gas	Gas	Gas	Gas	
Higher heat value (HHV)	298.15	MJ/m³	−11.7	−36.5	−64.5	−93.3	a
	298.15	MJ/kg	−141.8	−55.5	−51.9	−50.3	a
Lower heat value (LHV)	298.15	MJ/m³	−9.9	−32.9	−59.1	−85.9	a
	298.15	MJ/kg	−119.9	−50.0	−47.5	−46.3	a
Melting point		°C	−259.198[c]	−182.47	−182.79	−187.63	[1]
Boiling point		°C	−252.762	−161.48	−88.6	−42.1	[1]
Heat capacity	298.15	J/(K·mol)	28.84	35.79	52.70	73.51	[5]
Density	298	kg/m³	0.0824	0.657	1.243	1.854	[9]
Viscosity	298	Pa s	8.90E−06	1.110E−05	9.40E−06	8.21E−06	[9]
Thermal conductivity	298	W/(m·K)	1.806E−01	3.37E−02	2.12E−02	1.80E−02	[9]

	Temperature (K)	Unit	Methanol CH_3OH	Ethanol C_2H_5OH	Dimethyl Ether CH_3OCH_3	Ref.
Phase			Liquid	Liquid	Gas	
Higher heat value (HHV)	298.15	MJ/m³	−17836.8	−23293.4	−60.5	a
	298.15	MJ/kg	−22.7	−29.7	−31.7	a
Lower heat value (LHV)	298.15	MJ/m³	−15675.2	−21044.0	−55.0	a
	298.15	MJ/kg	−19.9	−26.8	−28.8	a

continued

TABLE A.25 (continued)

Properties of Pure Compounds as Fuel

	Temperature (K)	Unit	Methanol CH_3OH	Ethanol C_2H_5OH	Dimethyl ether CH_3OCH_3	Ref.
Melting point		°C	–97.53	–114.14	–141.5	[1]
Boiling point		°C	64.6	78.29	–24.8	[1]
Heat capacity	298.15	J/(K mol)	81.6	111.4	65.8[c]	[5]
Density	298	kg/m³	787	785	1.908[d,e]	[9]
Viscosity	298	Pa s	5.55E–04	1. 078E–03	9.16E–03[e,f]	[9]
Thermal conductivity	298	W/(m·K)	2.03E–01	1.67E–01	1.34E–02[e,f]	[9]

Source: [c] Adapted from Perry, R. H., D. W. Green, and J. O. Maloney. 1997. *Perry's Chemical Engineers' Handbook (7th edition)*. McGraw-Hill, New York; [e] Yaws, C. 2001. *Matheson gas data book (7th edition)*. McGraw-Hill, New York.

Pressure: 101 kPa.

[a] Calculated using heat of conbustion data in Ref. [2].

[c] Triple point.

[d] At 70°F (about 21°C).

[f] At 25°C.

TABLE A.26

Properties of Gas Fuels

		Coal Gas	Coke Oven Gas	Off–Gas of Petroleum Refining	Natural Gas (Produced in Japan 1)	Natural Gas (Produced in Japan 2)	Natural Gas (LNG)
Composition (mol%)	H_2	49	10	54	0	0	0
	CO	4	27	0	0	0	0
	CH_4	34	0	23	97	94	90
	C_2H_4	3	0	5	0	0	0
	C_2H_6	1	0	13	0	5	6
	C_3H_6	0	0	1	0	0	0
	C_3H_8	0	0	2	0	1	4
	C_4H_{10}	0	0	0	0	0	0
	N_2	6	56	2	1	0	0
	O_2	1	1	0	0	0	0
	CO_2	2	6	0	2	0	0
Average molar weight		11.71	26.41	11.95	16.72	17.02	18.01
Higher heat value (HHV)	MJ/m³	21.1	4.3	28.4	35.4	38.3	40.1
Lower heat value (LHV)	MJ/m³	18.9	4.1	25.7	31.9	34.7	36.5
Viscosity	Pa·s	1.3E–05	1.9E–05	1.1E–05	1.2E–05	1.1E–05	1.1E–05

TABLE A.26 (continued)

Properties of Gas Fuels

		Liquified Petroleum Gas	City Gas (From LNG)	City Gas (From LPG)	City Gas (From Naphtha etc.)	City Gas (From Coal etc.)
Composition (mol%)	H_2	0	0	0	37	46
	CO	0	0	0	3	5
	CH_4	0	88	0	29	22
	C_2H_4	0	0	0	3	4
	C_2H_6	1	5	0	1	1
	C_3H_6	0	0	0	2	0
	C_3H_8	96	5	0	0	0
	C_4H_{10}	3	2	22	0	0
	N_2	0	0	62	8	10
	O_2	0	0	16	2	2
	CO_2	0	0	0	15	10
Average molar weight		44.38	18.99	35.27	17.71	15.24
Higher heat value (HHV)	MJ/m^3	86.5	41.9	26.0	19.3	17.1
Lower heat value (LHV)	MJ/m^3	85.6	38.2	24.0	17.3	15.3
Viscosity	Pa·S	9.0E−06	1.1E−05	1.5E−05	1.4E−05	1.5E−05

Source: Adapted from The Japan Society of Mechanical Engineers (eds.), 1983. *JSME Databook: The Thermophysical Properties of Fluids*. The Japan Society of Mechanical Engineers, Tokyo.

Note: Temperature: 25°C.

TABLE A.27

Composition of Petroleum Oils (Unit: wt.%)

		Gasoline	Kerosene	Diesel (#2)	No.6 Fuel Oil (Residual Oil)
Alkanes		47	32	41	13
Cycloalkanes		3	52	37	15
Alkenes		10	1	1	
Aromatics		35	14	22	34
Others	Total	0.3			
	Insolubles and Polar materials				44
	Inorganics			0.1	
Total		95.6	100.1	101.4	106.2
Reference		[17]	[18]	[17]	[17]

Note: Each data is an average of many data of different samples. Therefore "Total" is not 100%.

TABLE A.28

Temperature Dependence of Thermodynamic and Transport Properties of Petroleum Oils

	Heat Capacity (kJ/(kg·K))		Density (kg/m³)			Viscosity (Pa·s)			Thermal Conductivity (W/(m·K))		
Temperature (°C)	Gasoline[a]	Kerosene	Gasoline[a]	Kerosene	Diesel Oil	Gasoline[a]	Kerosene	Diesel Oil[b]	Gasoline[a]	Kerosene	Diesel Oil
0						7.35E-04	2.15E-03		0.1204	0.1192	
10						6.43E-04	1.73E-03				
20	2.06	2.00	751	819	878.7	5.29E-04	1.49E-03	7.856E-03			0.1169
30	2.11	2.04	743	814							
40	2.15	2.09	735	808	865.4	4.11E-04	1.08E-03	4.154E-03			0.1146
50	2.20	2.14	721	801					0.1105	0.1114	
60	2.24	2.18	717	795	852.0	3.28E-04	8.32E-04	2.590E-03			0.1122
70	2.30	2.23	708	788							
80	2.35	2.28	699	781	838.5	2.69E-04	6.64E-04	1.794E-03			0.1099
90	2.41	2.33	690	774							
100	2.46	2.38	681	766	825.1	2.25E-04	5.45E-04	1.337E-03	0.1005	0.1042	0.1076
110	2.51	2.43	671	759							
120	2.57	2.48	660	751		1.91E-04	4.57E-04				
130	2.62	2.53	650	744							
140	2.68	2.58	639	736		1.65E-04	3.90E-04				
150	2.74	2.63	628	728					0.0919	0.0965	
160	2.80	2.68	617	720		1.46E-04	3.38E-04				
170	2.86	2.73	605	711							
180	2.92	2.79	594	703		1.26E-04	2.96E-04				
190	2.98	2.84	582	694							
200	3.04	2.89	570	685		1.11E-04	2.62E-04				
210	3.11	2.94		676					0.0800	0.0891	
220		3.00		668		9.9E-05	2.34E-04				
230		3.05		658							
240		3.11		649		8.9E-05	2.11E-04				
250		3.16		638						0.0816	
260		3.21		628		8.1E-05	1.91E-04				
270		3.26		618							
280						7.3E-05	1.74E-04				
290											
300						6.7E-05	1.59E-04			0.0738	

Source: Adapted from Vargaftik, N. B. 1975. *Tables on the Thermophysical Properties of Liquids and Gases (2nd edition)*. Hemisphere, Washington.

At about 0.1 MPa.

[a] Soviet fuel designation B-70.

[b] calculated using kinetic viscosity and density data in Ref. [20].

TABLE A.29

Element composition of Coals (Unit: wt%)

Element	Anthracite	Brown Coal		
		Bright	Lignate	Ordinary
C	94.0	73.0	68.0	70.0
H	2.0	5.0	5.1	5.2
O	2.2	18.0	25.6	22.9
S	0.8	3.0	0.5	1.0
N	1.0	1.0	0.7	0.9

Source: Adapted from Ražnjević, K. 1976. *Handbook of Thermodynamic Tables and Charts*. Hemisphere, Washington.
Note: Moisture and ash-free, average value of samples.

TABLE A.30

Properties of Solid Fuels

	Unit	Coal			Coke	Reference
		Anthracite	Bituminous	Lignite		
Net calorific value[a]	MJ/kg	26.7	25.8[d]	11.9	28.2	[21]
Heat capacity	kJ/(Kg·K)	1.089[b]		1.243[b, f]	0.837[e]	[12]
Density	kg/m^3	1370[c]	1100[e]	965[c, f]	1350[e]	[12]
Thermal conductivity	W/(m·K)	0.238[c]	0.174[e]	0.155[c, f]	0.163[e]	[12]

Ambient temperature, atmospheric pressure when no notation is shown.
[a] Lower heating value (LHV) not including condensation heat of moisture within fuels.
[b] Average of 0–100°C.
[c] At 30°C.
[d] Data of "Other Bitumineous Coal" in original paper.
[e] At 20°C.
[f] 3.4 % water.

References

1. Lide, D. R. 2006–2007. *CRC Handbook of Chemistry and Physics (87th edition)*. CRC Press, Boca Raton.
2. Perry, R. H., D. W. Green, and J. O. Maloney. 1997. *Perry's Chemical Engineers' Handbook (7th edition)*. McGraw-Hill, New York.
3. Chase, M. W. Jr. 1998. NIST-JANAF thermochemical tables 4th edition, *J. Phys. Chem. Ref. Data*, Monograph No. 9.
4. Zamfirescu, C., I. Dincer, and G. F. Naterer. 2010. Thermophysical properties of copper compounds in copper–chlorine thermochemical water splitting cycles, *Int. J. Hydrogen Energy*, 35:4839–52.
5. The Chemical Society of Japan (eds.). 2004. *Kagaku Binran Kisohen (5th edition)*. The Chemical Society of Japan, Maruzen, Tokyo.
6. Wagman, D. D., W. H. Evans, V. B. Parker, R. H. Schumm, I. Halow, S. M. Bailey, K. L. Churney, and R. L. Nuttall. 1982. The NBS tables of chemical thermodynamic properties. *J. Phys. Chem. Ref. Data*, 11, Supplement No. 2.

7. The Chemical Society of Japan (eds.). 1994. *Kagaku Binran Kisohen (4th edition)*. The Chemical Society of Japan, Maruzen, Tokyo.
8. Munemi, A. et al (Eds.), 1977. *Ryu-san Hando Bukku Kaiteibann*. Ryu-san Kyoukai, Tokyo.
9. The Japan Society of Thermophysical Properties (eds.). 2008. *Thermophysical Properties Handbook*, Yokendo, Tokyo.
10. The Japan Society of Mechanical Engineers (eds.). 1983. *JSME Databook: The Thermophysical Properties of Fluids*. The Japan Society of Mechanical Engineers, Tokyo.
11. Touloukian, Y. S., S. C. Saxena, and P. Hestermans. 1975. *Thermophysical Properties of Matter Volume 11: Viscosity*. IFI/Plenum, New York.
12. Ražnjević, K. 1976. *Handbook of Thermodynamic Tables and Charts*. Hemisphere, Washington.
13. Horvath, A. L. 1975. *Physical Properties of Inorganic Compounds*. Edward Arnold, London.
14. Touloukian, Y. S., P. E. Liley, and S. C. Saxena. 1970. *Thermophysical Properties of Matter Volume 3: Thermal Conductivity: Nonmetallic Liquids and Gases*. IFI/Plenum, New York.
15. Touloukian, Y. S., R. W. Powell, C. Y. Ho, and P. G, Klemens. 1970. *Thermophysical Properties of Matter Volume 1: Thermal Conductivity: Metallic Elements and Alloys*. IFI/Plenum, New York.
16. Touloukian, Y. S., R. W. Powell, C. Y. Ho, and P. G, Klemens. 1970. *Thermophysical Properties of Matter Volume 2: Thermal Conductivity: Nonmetallic Solids*. IFI/Plenum, New York.
17. Yaws, C. 2001. *Matheson gas data book (7th edition)*. McGraw-Hill, New York.
18. Potter L. T. and K. E. Simmons. 1998. *Composition of Petroleum Mixtures, Total Petroleum Hydrocarbon Criteria Working Group Series Volume 2*. Amherst Scientific Publishers, Amherst.
19. Armstrong, G. T., L. Fano, R. S. Jessup, S. Marantz, T. W. Mears, and J. A. Walker. 1962. Net heat of combustion and other properties of kerosine and related fuels. *J. Chem. Eng. Data* 7:107–16.
20. Vargaftik, N. B. 1975. *Tables on the Thermophysical Properties of Liquids and Gases (2nd edition)*. Hemisphere, Washington.
21. Eggleston, S., L. Buendia, K. Miwa, T. Ngara, and K. Tanabe (eds.). 2006. *2006 IPCC Guidelines for National Greenhouse Gas Inventories Volume 2: Energy*, The Institute for Global Environmental Strategies, Hayama.

Appendix B: Thermodynamic and Transport Properties of Coolants for Nuclear Reactors Considered for Hydrogen Production

Seiji Kasahara

CONTENTS

B.1 Light Water

B.1.1 Introduction

The thermodynamic and transport properties of light water (as well as heavy water) reactor coolant are well documented by a 6-year IAEA Coordinated Research Project [1]. For light water, the newest property formulation for scientific use from the International Association for the Property of Water and Steam (IAPWS) is IAPWS-95 adopted in 1995 [2], replacing the old formulation, IAPS-84. A minor revision of parameters in IAPWS-95 was made [3]. Simultaneously, a separate formulation IAPWS-IF97 was maintained for use in steam power industry [4], replacing IFC-67 formulation. IAPWS-IF97 formulation is consistent with IAPWS-95. Tables of calculated results of IAPWS-IF97 are summarized (e.g., [5]). A revision of IAPWS-IF97 mainly on the extension of pressure range up to 50 MPa at high temperature (Region 5) was made [6]. Temperature scale of the revision is International Scale of 1990 (ITS-90) [7].

B.1.2 Thermodynamic Properties

Thermodynamic properties by the revision of IAPWS-IF97 [6] are introduced. Thermodynamic properties can be calculated from basic equations. Five different basic equations for the different range of pressure and temperature, which is called "region" here, are used. All properties are calculated from the basic equations. Figure B.1 illustrates the regions. Each region except Region 4 is separated by solid lines. The boundary between Regions 2 and 3 is defined as Equation B.1.

$$\pi = n_1 + n_2\theta + n_3\theta^2, \tag{B.1}$$

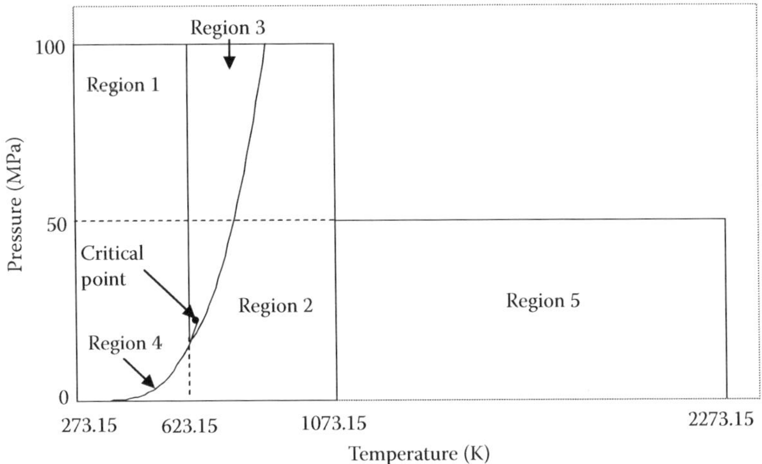

FIGURE B.1
Regions of basic equations for light water.

$$\pi = \frac{p}{p^*}, \quad \theta = \frac{T}{T^*}, \quad p^* = 1\,\text{MPa}, \quad T^* = 1\,\text{K},$$

It is noted that the unit of pressure, p is MPa, not Pa, through Appendix B.1. Temperature range of Equation B.1 is from 623.15 K at 16.5292 MPa to 863.15 K at 100 MPa. Parameters n_i used in Equation B.1 are shown in Table B.1.

Region 4 is the saturation condition as shown in Equation B.2.

$$\frac{p}{p^*} = \left[\frac{2C}{-B + \left(B^2 - 4AC\right)^{0.5}}\right]^4, \tag{B.2}$$

$$A = \vartheta^2 + n_1\vartheta + n_2, \tag{B.3}$$

$$B = n_3\vartheta^2 + n_4\vartheta + n_5, \tag{B.4}$$

$$C = n_6\vartheta^2 + n_7\vartheta + n_8, \tag{B.5}$$

TABLE B.1

Parameters for the Boundary between Regions 2 and 3 of Light Water (Equation B.1)

i	n_i
1	3.4805185628969E + 02
2	−1.1671859879975E + 00
3	1.0192970039326E − 03

TABLE B.2

Parameters for Saturation Condition (Region 4) of Light Water (Equations B.3 through B.6)

i	n_i
1	1.1670521452767E + 03
2	−7.2421316703206E + 05
3	−1.7073846940092E + 01
4	1.2020824702470E + 04
5	−3.2325550322333E + 06
6	1.4915108613530E + 01
7	−4.8232657361591E + 03
8	4.0511340542057E + 05
9	−2.3855557567849E − 01
10	6.5017534844798E + 02

$$\vartheta = \frac{T}{T^*} + \frac{n_9}{(T/T^*) - n_{10}}. \tag{B.6}$$

$$p^* = 1\ \text{MPa}, \quad T^* = 1\ \text{K}$$

Parameters n_i used in Equations B.3 through B.6 are shown in Table B.2. Equation B.2 is applicable along entire vapor–liquid curve from the triple point (273.16 K and 611.657 Pa) to critical point (647.096 K and 22.064 MPa). And the equation can be extrapolated to 273.15 K and 611.213 Pa.

Uncertainty of specific volume, constant pressure heat capacity, and speed of sound in all regions is summarized in the literature [6]. And uncertainty of enthalpy is illustrated in another literature [8].

B.1.2.1 Region 1

This region shows the state of compressed liquid. The temperature is between 273.15 and 623.15 K. The pressure is over the saturation pressure and below 100 MPa. The basic equation is temperature and pressure dependence of Gibbs free energy as in Equation B.7.

$$\frac{G(p,T)}{RT} = \gamma(\pi,\tau) = \sum_{i=1}^{34} n_i \cdot (7.1 - \pi)^{I_i} \cdot (\tau - 1.222)^{J_i}, \tag{B.7}$$

$$\pi = \frac{p}{p^*}, \quad \tau = \frac{T^*}{T}, \quad p^* = 16.53\,\text{MPa}, \quad T^* = 1386\ \text{K},$$

Parameters, n_i, I_i, and J_i are listed in Table B.3.

Thermodynamic and transport properties are calculated as follows. "$\times 10^{-3}$" is added to the equation of specific volume, Equation B.13 and "$\times 10^{3}$" is added to the equation of speed of sound, Equation B.14 in the original literature [6] to fit test values shown in it. The same modification is applied to equations of specific volume and speed of sound in other regions.

TABLE B.3

Parameters for Gibbs Free Energy in Region 1 of Light Water (Equation B.7)

i	I_i	J_i	n_i
1	0	−2	1.4632971213167E − 01
2	0	−1	−8.4548187169114E − 01
3	0	0	−3.7563603672040E + 00
4	0	1	3.3855169168385E + 00
5	0	2	−9.5791963387872E − 01
6	0	3	1.5772038513228E − 01
7	0	4	−1.6616417199501E − 02
8	0	5	8.1214629983568E − 04
9	1	−9	2.8319080123804E − 04
10	1	−7	−6.0706301565874E − 04
11	1	−1	−1.8990068218419E − 02
12	1	0	−3.2529748770505E − 02
13	1	1	−2.1841717175414E − 02
14	1	3	−5.2838357969930E − 05
15	2	−3	−4.7184321073267E − 04
16	2	0	−3.0001780793026E − 04
17	2	1	4.7661393906987E − 05
18	2	3	−4.4141845330846E − 06
19	2	17	−7.2694996297594E − 16
20	3	−4	−3.1679644845054E − 05
21	3	0	−2.8270797985312E − 06
22	3	6	−8.5205128120103E − 10
23	4	−5	−2.2425281908000E − 06
24	4	−2	−6.5171222895601E − 07
25	4	10	−1.4341729937924E − 13
26	5	−8	−4.0516996860117E − 07
27	8	−11	−1.2734301741641E − 09
28	8	−6	−1.7424871230634E − 10
29	21	−29	−6.8762131295531E − 19
30	23	−31	1.4478307828521E − 20
31	29	−38	2.6335781662795E − 23
32	30	−39	−1.1947622640071E − 23
33	31	−40	1.8228094581404E − 24
34	32	−41	−9.3537087292458E − 26

It should be noted that unit of the gas constant, R in Appendix B.1 is kJ/(kg · K), not kJ/(mol K).

$$\textit{Internal energy:} \quad \frac{U(\pi,\tau)}{RT} = \tau\gamma_\tau - \pi\gamma_\pi \tag{B.8}$$

$$\textit{Enthalpy:} \quad \frac{H(\pi,\tau)}{RT} = \tau\gamma_\tau \tag{B.9}$$

$$\textit{Entropy:}\quad \frac{S(\pi,\tau)}{R} = \tau\gamma_\tau - \gamma \tag{B.10}$$

$$\textit{Constant pressure heat capacity:}\quad \frac{C_p(\pi,\tau)}{R} = -\tau^2\gamma_{\tau\tau} \tag{B.11}$$

$$\textit{Constant volume heat capacity:}\quad \frac{C_v(\pi,\tau)}{R} = -\tau^2\gamma_{\tau\tau} + \frac{(\gamma_\pi - \tau\gamma_{\pi\tau})^2}{\gamma_{\pi\pi}} \tag{B.12}$$

$$\textit{Specific volume:}\quad v(\pi,\tau)\cdot\frac{p}{RT} = (\pi\gamma_\pi)\times 10^{-3} \tag{B.13}$$

$$\textit{Speed of sound:}\quad \frac{w^2(\pi,\tau)}{RT} = \left(\frac{\gamma_\pi^2}{\dfrac{(\gamma_\pi - \tau\gamma_{\pi\tau})^2}{\tau^2\gamma_{\tau\tau}} - \gamma_{\pi\pi}}\right)\times 10^3 \tag{B.14}$$

$$\gamma_\pi = \left(\frac{\partial\gamma}{\partial\pi}\right)_\tau,\quad \gamma_{\pi\pi} = \left(\frac{\partial^2\gamma}{\partial\pi^2}\right)_\tau,\quad \gamma_\tau = \left(\frac{\partial\gamma}{\partial\tau}\right)_\pi,\quad \gamma_{\tau\tau} = \left(\frac{\partial^2\gamma}{\partial\tau^2}\right)_\pi,\quad \gamma_{\pi\tau} = \left(\frac{\partial^2\gamma}{\partial\pi\partial\tau}\right).$$

B.1.2.2 Region 2

This region shows the superheated steam. Maximum pressure of the region is the saturation pressure obtained in Equation B.2 in the range from 273.15 to 623.15 K, the boundary defined in Equation B.1 from 623.15 to 863.15 K and 100 MPa from 863.15 to 1073.15 K, respectively. The basic equation is the temperature and pressure dependence of Gibbs free energy as in Equation B.15.

$$\frac{G(p,T)}{RT} = \gamma(\pi,\tau) = \gamma_0(\pi,\tau) + \gamma_r(\pi,\tau), \tag{B.15}$$

$$\gamma_0 = \ln\pi + \sum_{i=1}^{9} n_{0i}\tau^{J_{0i}}, \tag{B.16}$$

$$\gamma_r = \sum_{i=1}^{43} n_i\pi^{I_i}\cdot(\tau - 0.5)^{J_i}, \tag{B.17}$$

$$\pi = \frac{p}{p^*},\quad \tau = \frac{T^*}{T},\quad p^* = 1\,\text{MPa},\quad T^* = 540\,\text{K},$$

Parameters n_{0i} and J_{0i} in Equation B.16 are in Table B.4. And parameters n_i, I_i, and J_i in Equation B.17 are summarized in Table B.5.

TABLE B.4

Parameters for γ_0 in Region 2 of Light Water (Equation B.16)

I	J_{0i}	n_{0i}
1	0	−9.6927686500217E + 00
2	1	1.0086655968018E + 01
3	−5	−5.6087911283020E − 03
4	−4	7.1452738081455E − 02
5	−3	−4.0710498223928E − 01
6	−2	1.4240819171444E + 00
7	−1	−4.3839511319450E + 00
8	2	−2.8408632460772E − 01
9	3	2.1268463753307E − 02

TABLE B.5

Parameters for γ_r in Region 2 of Light Water (Equation B.17)

i	I_i	J_i	n_i
1	1	0	−1.7731742473213E − 03
2	1	1	−1.7834862292358E − 02
3	1	2	−4.5996013696365E − 02
4	1	3	−5.7581259083432E − 02
5	1	6	−5.0325278727930E − 02
6	2	1	−3.3032641670203E − 05
7	2	2	−1.8948987516315E − 04
8	2	4	−3.9392777243355E − 03
9	2	7	−4.3797295650573E − 02
10	2	36	−2.6674547914087E − 05
11	3	0	2.0481737692309E − 08
12	3	1	4.3870667284435E − 07
13	3	3	−3.2277677238570E − 05
14	3	6	−1.5033924542148E − 03
15	3	35	−4.0668253562649E − 02
16	4	1	−7.8847309559367E − 10
17	4	2	1.2790717852285E − 08
18	4	3	4.8225372718507E − 07
19	5	7	2.2922076337661E − 06
20	6	3	−1.6714766451061E − 11
21	6	16	−2.1171472321355E − 03
22	6	35	−2.3895741934104E + 01
23	7	0	−5.9059564324270E − 18
24	7	11	−1.2621808899101E − 06
25	7	25	−3.8946842435739E − 02
26	8	8	1.1256211360459E − 11
27	8	36	−8.2311340897998E + 00
28	9	13	1.9809712802088E − 08
29	10	4	1.0406965210174E − 19
30	10	10	−1.0234747095929E − 13

continued

TABLE B.5 (continued)

Parameters for γ_r in Region 2 of Light Water (Equation B.17)

i	I_i	J_i	n_i
31	10	14	−1.0018179379511E − 09
32	16	29	−8.0882908646985E − 11
33	16	50	1.0693031879409E − 01
34	18	57	−3.3662250574171E − 01
35	20	20	8.9185845355421E − 25
36	20	35	3.0629316876232E − 13
37	20	48	−4.2002467698208E − 06
38	21	21	−5.9056029685639E − 26
39	22	53	3.7826947613457E − 06
40	23	39	−1.2768608934681E − 15
41	24	26	7.3087610595061E − 29
42	24	40	5.5414715350778E − 17
43	24	58	−9.4369707241210E − 07

Thermodynamic and transport properties are calculated as follows:

$$\textit{Internal energy: } \frac{U(\pi,\tau)}{RT} = \tau\cdot(\gamma_{0\tau}+\gamma_{r\tau}) - \pi\cdot(\gamma_{0\pi}+\gamma_{r\pi}) \tag{B.18}$$

$$\textit{Enthalpy: } \frac{H(\pi,\tau)}{RT} = \tau(\gamma_{0\tau}+\gamma_{r\tau}) \tag{B.19}$$

$$\textit{Entropy: } \frac{S(\pi,\tau)}{R} = \tau\cdot(\gamma_{0\tau}+\gamma_{r\tau}) - (\gamma_0+\gamma_r) \tag{B.20}$$

$$\textit{Constant pressure heat capacity: } \frac{C_p(\pi,\tau)}{R} = -\tau^2\cdot(\gamma_{0\tau\tau}+\gamma_{r\tau\tau}) \tag{B.21}$$

$$\textit{Constant volume heat capacity: } \frac{C_v(\pi,\tau)}{R} = -\tau^2\cdot(\gamma_{0\tau\tau}+\gamma_{r\tau\tau}) - \frac{(1+\pi\gamma_{r\pi}-\tau\pi\gamma_{r\pi\tau})^2}{1-\pi^2\gamma_{r\pi\pi}} \tag{B.22}$$

$$\textit{Specific volume: } v(\pi,\tau)\cdot\frac{p}{RT} = \left[\pi\cdot(\gamma_{0\pi}+\gamma_{r\pi})\right]\times 10^{-3} \tag{B.23}$$

$$\textit{Speed of sound: } \frac{w^2(\pi,\tau)}{RT} = \left[\frac{1+2\pi\gamma_{r\pi}+\pi^2\gamma_{r\pi}^2}{(1-\pi^2\gamma_{r\pi\pi}) + \dfrac{(1+\pi\gamma_{r\pi}-\tau\pi\gamma_{r\pi\tau})^2}{\tau^2\cdot(\gamma_{0\tau\tau}+\gamma_{r\tau\tau})}}\right]\times 10^3 \tag{B.24}$$

$$\gamma_{0\pi} = \left(\frac{\partial \gamma_0}{\partial \pi}\right)_\tau, \quad \gamma_{0\pi\pi} = \left(\frac{\partial^2 \gamma_0}{\partial \pi^2}\right)_\tau, \quad \gamma_{0\tau} = \left(\frac{\partial \gamma_0}{\partial \tau}\right)_\pi, \quad \gamma_{0\tau\tau} = \left(\frac{\partial^2 \gamma_0}{\partial \tau^2}\right)_\pi, \quad \gamma_{0\pi\tau} = \left(\frac{\partial^2 \gamma_0}{\partial \pi \partial \tau}\right),$$

$$\gamma_{r\pi} = \left(\frac{\partial \gamma_r}{\partial \pi}\right)_\tau, \quad \gamma_{r\pi\pi} = \left(\frac{\partial^2 \gamma_r}{\partial \pi^2}\right)_\tau, \quad \gamma_{r\tau} = \left(\frac{\partial \gamma_r}{\partial \tau}\right)_\pi, \quad \gamma_{r\tau\tau} = \left(\frac{\partial^2 \gamma_r}{\partial \tau^2}\right)_\pi, \quad \gamma_{r\pi\tau} = \left(\frac{\partial^2 \gamma_r}{\partial \pi \partial \tau}\right).$$

B.1.2.3 Region 3

This region shows the critical and overcritical water. Temperature range is between 623.15 K and 843.15 K. Pressure range is from the boundary line defined in Equation B.1 to 100 MPa. It is noted that the basic equation is of Helmholtz free energy and density instead of pressure is used as a parameter of the equation only in this region. The equation is shown in Equation B.25.

$$\frac{F(\rho,T)}{RT} = \phi(\delta,\tau) = n_1 \ln \delta + \sum_{i=2}^{40} n_i \delta^{I_i} \tau^{J_i}, \tag{B.25}$$

$$\delta = \frac{\rho}{\rho^*}, \quad \tau = \frac{T^*}{T}, \quad \rho^* = \rho_c = 322\,\text{kg/m}^3, \quad T^* = T_c = 647.096\,\text{K}$$

Parameters used in Equation B.25, n_i, I_i, and J_i are in Table B.6.

Thermodynamic and transport properties are calculated as follows. "$\times 10^{-3}$" is added to the equation of pressure (Equation B.31) in the original literature [6] to fit test values shown in it.

$$\textit{Internal energy:} \quad \frac{U(\delta,\tau)}{RT} = \tau\phi_\tau \tag{B.26}$$

$$\textit{Enthalpy:} \quad \frac{H(\delta,\tau)}{RT} = \tau\phi_\tau + \delta\phi_\delta \tag{B.27}$$

$$\textit{Entropy:} \quad \frac{S(\delta,\tau)}{R} = \tau\phi_\tau - \phi \tag{B.28}$$

$$\textit{Constant pressure heat capacity:} \quad \frac{C_p(\delta,\tau)}{R} = -\tau^2\phi_{\tau\tau} + \frac{(\delta\phi_\delta - \delta\tau\phi_{\delta\tau})^2}{2\delta\phi_\delta + \delta^2\phi_{\delta\delta}} \tag{B.29}$$

$$\textit{Constant volume heat capacity:} \quad \frac{C_v(\delta,\tau)}{R} = -\tau^2\phi_{\tau\tau} \tag{B.30}$$

$$\textit{Pressure:} \quad \frac{p(\delta,\tau)}{\rho RT} = (\delta\phi_\delta) \times 10^{-3} \tag{B.31}$$

TABLE B.6

Parameters for Helmholtz Free Energy in Region 3 of Light Water (Equation B.25)

i	I_i	J_i	n_i
1	—	—	1.0658070028513E + 00
2	0	0	−1.5732845290239E + 01
3	0	1	2.0944396974307E + 01
4	0	2	−7.6867707878716E + 00
5	0	7	2.6185947787954E + 00
6	0	10	−2.8080781148620E + 00
7	0	12	1.2053369696517E + 00
8	0	23	−8.4566812812502E − 03
9	1	2	−1.2654315477714E + 00
10	1	6	−1.1524407806681E + 00
11	1	15	8.8521043984318E − 01
12	1	17	−6.4207765181607E − 01
13	2	0	3.8493460186671E − 01
14	2	2	−8.5214708824206E − 01
15	2	6	4.8972281541877E + 00
16	2	7	−3.0502617256965E + 00
17	2	22	3.9420536879154E − 02
18	2	26	1.2558408424308E − 01
19	3	0	−2.7999329698710E − 01
20	3	2	1.3899799569460E + 00
21	3	4	−2.0189915023570E + 00
22	3	16	−8.2147637173963E − 03
23	3	26	−4.7596035734923E − 01
24	4	0	4.3984074473500E − 02
25	4	2	−4.4476435428739E − 01
26	4	4	9.0572070719733E − 01
27	4	26	7.0522450087967E − 01
28	5	1	1.0770512626332E − 01
29	5	3	−3.2913623258954E − 01
30	5	26	−5.0871062041158E − 01
31	6	0	−2.2175400873096E − 02
32	6	2	9.4260751665092E − 02
33	6	26	1.6436278447961E − 01
34	7	2	−1.3503372241348E − 02
35	8	26	−1.4834345352472E − 02
36	9	2	5.7922953628084E − 04
37	9	26	3.2308904703711E − 03
38	10	0	8.0964802996215E − 05
39	10	1	−1.6557679795037E − 04
40	11	26	−4.4923899061815E − 05

$$\text{Speed of sound: } \frac{w^2(\delta,\tau)}{RT} = \left[2\delta\phi_\delta + \delta^2\phi_{\delta\delta} - \frac{(\delta\phi_\delta - \delta\tau\phi_{\delta\tau})^2}{\tau^2\phi_{\tau\tau}}\right] \times 10^3 \tag{B.32}$$

$$\phi_\delta = \left(\frac{\partial\phi}{\partial\delta}\right)_\tau, \quad \phi_{\delta\delta} = \left(\frac{\partial^2\phi}{\partial\delta^2}\right)_\tau, \quad \phi_\tau = \left(\frac{\partial\phi}{\partial\tau}\right)_\delta, \quad \phi_{\tau\tau} = \left(\frac{\partial^2\phi}{\partial\tau^2}\right)_\delta, \quad \phi_{\delta\tau} = \left(\frac{\partial^2\phi}{\partial\delta\partial\tau}\right).$$

B.1.2.4 Region 4

This region shows the saturation pressure equation line. The range is already explained in the beginning of this section. Basic equation and properties calculation of both Region 1 and Region 2 can be used. The average differences of the value of Gibbs energy between regions are 0.006 and 0.001 kJ/kg below 623.15 and above 623.15 K, respectively.

B.1.2.5 Region 5

This region shows the high-temperature water. Temperature and pressure are from 1073.15 to 2273.15 K and from 0 to 50 MPa, respectively. The basic equation is Equation B.33, which is pressure and temperature dependence of Gibbs free energy.

$$\frac{G(p,T)}{RT} = \gamma(\pi,\tau) = \gamma_0(\pi,\tau) + \gamma_r(\pi,\tau), \tag{B.33}$$

$$\gamma_0 = \ln\pi + \sum_{i=1}^{6} n_{0i}\tau^{J_{0i}}, \tag{B.34}$$

$$\gamma_r = \sum_{i=1}^{6} n_i\pi^{I_i}\tau^{J_i}, \tag{B.35}$$

$$\pi = \frac{p}{p^*}, \quad \tau = \frac{T^*}{T}, \quad p^* = 1\,\text{MPa}, \quad T^* = 1000\,\text{K},$$

List of parameters n_{0i} and J_{0i} in Equation B.34 is Table B.7. Table B.8 shows the parameters n_i, I_i, and J_i for Equation B.35.

TABLE B.7

Parameters for γ_0 in Region 5 of Light Water (Equation B.34)

i	J_{0i}	n_{0i}
1	0	−1.3179983674201E + 01
2	1	6.8540841634434E + 00
3	−3	−2.4805148933466E − 02
4	−2	3.6901534980333E − 01
5	−1	−3.1161318213925E + 00
6	2	−3.2961626538917E − 01

TABLE B.8

Parameters for γ_r in Region 5 of Light Water (Equation B.35)

i	I_i	J_i	n_i
1	1	1	1.5736404855259E – 03
2	1	2	9.0153761673944E – 04
3	1	3	–5.0270077677648E – 03
4	2	3	2.2440037409485E – 06
5	2	9	–4.1163275453471E – 06
6	3	7	3.7919454822955E – 08

The equations to calculate other thermodynamic and transport properties are just the same as for Region 2. That is, Equations B.18 through B.24 are used (see Appendix B.1.2.2).

B.1.2.6 Metastable States

For Regions 1 and 3, basic Equations B.7 and B.25 are applicable to metastable superheated liquid states close to the saturated liquid line, Equation B.2. Though basic Equation B.15 can be used for the metastable vapor state in Region 2, the parameters are not valid for below 10 MPa. In the pressure range below 10 MPa, different parameters n_{01} and n_{02}, have to be used in Equation B.16. n_{0i} and J_{0i} used in the lower pressure range are listed in Table B.9. And Equation B.36 for γ_r has to be used instead of Equation B.17.

$$\gamma_r = \sum_{i=1}^{13} n_i \pi^{I_i} \cdot (\tau - 0.5)^{J_i}, \tag{B.36}$$

$$\pi = \frac{p}{p^*}, \quad \tau = \frac{T^*}{T}, \quad p^* = 1\,\text{MPa}, \quad T^* = 540\,\text{K}.$$

Parameters for Equation B.36 are shown in Table B.10. Equations to calculate other properties are the same as for Region 2, that is, Equations B.18 through B.24 (see Appendix B.1.2.2).

TABLE B.9

Parameters for γ_0 in Metastable Vapor State of Lower Pressure than 10 MPa in Region 2 of Light Water (Equation B.16)

I	J_{0i}	n_{0i}
1	0	–9.6937268393049E + 00
2	1	1.0087275970006E + 01
3	–5	–5.6087911283020E – 03
4	–4	7.1452738081455E – 02
5	–3	–4.0710498223928E – 01
6	–2	1.4240819171444E + 00
7	–1	–4.3839511319450E + 00
8	2	–2.8408632460772E – 01
9	3	2.1268463753307E – 02

TABLE B.10

Parameters for γ_r in Metastable Vapor State of Lower Pressure than 10 MPa in Region 2 of Light Water (Equation B.36)

I	I_i	J_i	n_i
1	1	0	−7.3362260186506E − 03
2	1	2	−8.8223831943146E − 02
3	1	5	−7.2334555213245E − 02
4	1	11	−4.0813178534455E − 03
5	2	1	2.0097803380207E − 03
6	2	7	−5.3045921898642E − 02
7	2	16	−7.6190409086970E − 03
8	3	4	−6.3498037657313E − 03
9	3	16	−8.6043093028588E − 02
10	4	7	7.5321581522770E − 03
11	4	10	−7.9238375446139E − 03
12	5	9	−2.2888160778447E − 04
13	5	10	−2.6456501482810E − 03

Nomenclature of B.1.2

Parts of dimensionless basic equations and derivations of them are explained in the text in order to avoid this list too long; p^* and T^* have different values in different subsections.

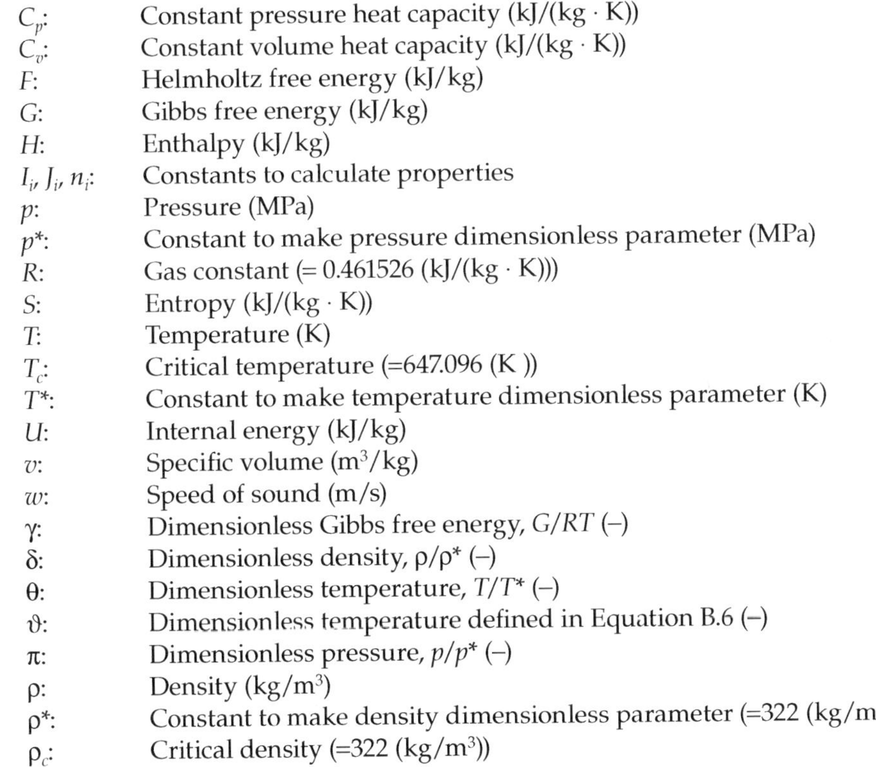

A, B, C: Constants defined in Equations B.3, B.4, and B.5, respectively
C_p: Constant pressure heat capacity (kJ/(kg · K))
C_v: Constant volume heat capacity (kJ/(kg · K))
F: Helmholtz free energy (kJ/kg)
G: Gibbs free energy (kJ/kg)
H: Enthalpy (kJ/kg)
I_i, J_i, n_i: Constants to calculate properties
p: Pressure (MPa)
p^*: Constant to make pressure dimensionless parameter (MPa)
R: Gas constant (= 0.461526 (kJ/(kg · K)))
S: Entropy (kJ/(kg · K))
T: Temperature (K)
T_c: Critical temperature (=647.096 (K))
T^*: Constant to make temperature dimensionless parameter (K)
U: Internal energy (kJ/kg)
v: Specific volume (m^3/kg)
w: Speed of sound (m/s)
γ: Dimensionless Gibbs free energy, G/RT (–)
δ: Dimensionless density, ρ/ρ^* (–)
θ: Dimensionless temperature, T/T^* (–)
ϑ: Dimensionless temperature defined in Equation B.6 (–)
π: Dimensionless pressure, p/p^* (–)
ρ: Density (kg/m^3)
ρ^*: Constant to make density dimensionless parameter (=322 (kg/m^3))
ρ_c: Critical density (=322 (kg/m^3))
τ: Dimensionless temperature, T^*/T (–)
ϕ: Dimensionless Helmholtz free energy, F/RT (–)

B.1.3 Transport Properties

Viscosity and thermal conductivity are calculated by other methods than basic equations shown in the previous section.

B.1.3.1 Viscosity

The newest release of formulation on viscosity by IAPWS [9] is a revision of the IAPS Formulation 1985 [10]. Temperature scale of ITS-90 [7] is used. Density value from IAPWS-IF97 (see Section B.1.1) is recommended for Equation B.39.

$$\bar{\mu} = \bar{\mu}_0(\theta)\cdot\bar{\mu}_1(\theta,\delta)\cdot\bar{\mu}_2(\theta,\delta), \tag{B.37}$$

$$\bar{\mu}_0(\theta) = \frac{100\sqrt{\theta}}{\sum_{i=0}^{3} n_i/\theta^i}, \tag{B.38}$$

$$\bar{\mu}_1(\theta,\delta) = \exp\left[\delta\cdot\sum_{i=0}^{5}\left\{\left(\frac{1}{\theta}-1\right)^i\cdot\sum_{j=0}^{6} n_{ij}\cdot(\delta-1)^j\right\}\right], \tag{B.39}$$

$$\bar{\mu} = \frac{\mu}{\mu^*},\quad \delta = \frac{\rho}{\rho^*},\quad \theta = \frac{T}{T^*}\quad \mu^* = 1.00\times10^{-6}\,\text{Pa}\cdot\text{s},\ \rho^* = 322.0\,\text{kg/m}^3,\ T^* = 647.096\,\text{K},$$

Parameters n_i, n_{ij} are shown in Tables B.11 and B.12, respectively. $\bar{\mu}_2$ is called critical enhancement. This term means modification around the critical point, 645.91 K < T < 650.77 K and 245.8 kg/m^3 < ρ < 405.3 kg/m^3. The effect of this term is more than 2% only within the range. Influence of this critical enhancement outside of the region is less than the uncertainty in the formulation. Therefore, when computing speed is required, this term can be considered as 1 in all temperature and density range. Detailed calculation of this term is not shown here because of its complicacy. You can see the detail in the original paper [9].

The range of pressure and temperature where the equation is applicable is

$273.16\text{ K} \le T \le 1173.15\text{ K},\quad \text{for } 0 < p \le p_t$

$T_m(p) \le T \le 1173.15\text{ K},\quad \text{for } p_t < p \le 300\text{ MPa}$

$T_m(p) \le T \le 873.15\text{ K},\quad \text{for } 300\text{ MPa} < p \le 350\text{ MPa}$

$T_m(p) \le T \le 433.15\text{ K},\quad \text{for } 350\text{ MPa} < p \le 500\text{ MPa}$

$T_m(p) \le T \le 373.15\text{ K},\quad \text{for } 500\text{ MPa} < p \le 1000\text{ MPa}$

TABLE B.11

Parameter for μ_0 in Viscosity of Light Water (Equation B.38)

i	n_i
0	1.67752
1	2.20462
2	0.6366564
3	−0.241605

TABLE B.12

Parameter for μ_1 in Viscosity of Light Water (Equation B.39)

i	j	n_{ij}
0	0	5.20094E − 01
1	0	8.50895E − 02
2	0	−1.08374E + 00
3	0	−2.89555E − 01
0	1	2.22531E − 01
1	1	9.99115E − 01
2	1	1.88797E + 00
3	1	1.26613E + 00
5	1	1.20573E − 01
0	2	−2.81378E − 01
1	2	−9.06851E − 01
2	2	−7.72479E − 01
3	2	−4.89837E − 01
4	2	−2.57040E − 01
0	3	1.61913E − 01
1	3	2.57399E − 01
0	4	−3.25372E − 02
3	4	6.98452E − 02
4	5	8.72102E − 03
3	6	−4.35673E − 03
5	6	−5.93264E − 04

n_{ij} of the set i and j not shown in the table is 0.

T_m is melting temperature and p_t (6.11657×10^{-4} MPa) is the triple point pressure in literature [11].

Estimated uncertainty is illustrated in the literature [9].

B.1.3.2 Thermal Conductivity

The newest release of formulation on thermal conductivity by IAPWS [12] is a revision of the IAPS Formulation 1985 [10]. It is made to conform to the IAPWS-IF97 for thermodynamic properties (see Section B.1.1) and ITS-90 temperature scale [7].

Thermal conductivity is represented as in Equations B.40 through B.47. The range of pressure and temperature where the equations are applicable is

$$p \le 100\ \text{MPa}, \quad \text{for } 0°\text{C} \le T \le 500°\text{C}$$
$$p \le 70\ \text{MPa}, \quad \text{for } 500°\text{C} \le T \le 650°\text{C}$$
$$p \le 40\ \text{MPa}, \quad \text{for } 650°\text{C} \le T \le 800°\text{C}$$

$$\bar{\lambda} = \bar{\lambda}_0(\theta) + \bar{\lambda}_1(\delta) + \bar{\lambda}_2(\theta,\delta), \tag{B.40}$$

$$\bar{\lambda}_0(\theta) = \sqrt{\theta} \cdot \sum_{i=0}^{3} a_i \theta^i, \tag{B.41}$$

$$\bar{\lambda}_1(\delta) = b_0 + b_1\delta + b_2\exp\left[B_1 \cdot (\delta + B_2)^2\right], \tag{B.42}$$

$$\bar{\lambda}_2(\theta,\delta) = \left(\frac{d_1}{\theta^{10}} + d_2\right)\delta^{\frac{9}{5}} \exp\left[C_1 \cdot \left(1 - \delta^{\frac{14}{5}}\right)\right] + d_3 S\delta^Q \exp\left[\left(\frac{Q}{1+Q}\right) \cdot \left(1 - \delta^{1+Q}\right)\right]$$
$$+ d_4 \exp\left(C_2\theta^{\frac{3}{2}} + \frac{C_3}{\delta^5}\right), \tag{B.43}$$

$$Q = 2 + \frac{C_5}{(\Delta\theta)^{\frac{3}{5}}}, \tag{B.44}$$

$$S = \frac{1}{\Delta\theta}, \quad \text{for } \theta \geq 1 \tag{B.45}$$

$$S = \frac{C_6}{(\Delta\theta)^{\frac{3}{5}}}, \quad \text{for } \theta < 1 \tag{B.46}$$

$$\Delta\theta = |\theta - 1| + C_4. \tag{B.47}$$

Here,

$$\bar{\lambda} = \frac{\lambda}{\lambda^*}, \quad \delta = \frac{\rho}{\rho^*}, \quad \theta = \frac{T}{T^*}, \quad \lambda^* = 1\,\text{W}/(\text{m}\cdot\text{K}), \ \rho^* = 317.7\,\text{kg/m}^3, T^* = 647.26\,\text{K}.$$

Parameters for Equations B.41 through B.47 are shown in Table B.13. Uncertainty of the equation is not shown in the literature. Calculation results are summarized in tables in the literature [12].

TABLE B.13

Parameters for Thermal Conductivity of Light Water (Equations B.41 through B.43)

i	a_i	b_i	B_i	C_i	d_i
0	0.0102811	−0.397070			
1	0.0299621	0.400302	−0.171587	0.642857	0.0701309
2	0.0156146	1.060000	2.392190	−4.11717	0.0118520
3	−0.00422464			−6.17937	0.00169937
4				0.00308976	−1.0200
5				0.0822994	
6				10.0932	

Nomenclature of B.1.3

T^* and ρ^* have different values in different subsections.

a_i, b_i, B_i, C_i, d_i:	Constants to calculate thermal conductivity
i, j, n_i, n_{ij}:	Constants to calculate properties
p:	Pressure (MPa)
p_t:	Triple point pressure (=6.11657 × 10^{-4} (MPa))
p^*:	Constant to make pressure dimensionless parameter (MPa)
Q, S:	Constants to calculate thermal conductivity defined in Equations B.44, B.45, and B.46, respectively
T:	Temperature (K)
T_m:	Melting temperature (K)
T^*:	Constant to make temperature dimensionless parameter (K)
δ:	Dimensionless density, ρ/ρ^* (–)
$\Delta\theta$:	Constant to calculate thermal conductivity defined in Equation B.47
θ:	Dimensionless temperature, T/T^* (–)
λ:	Thermal conductivity (W/(m · K))
λ^*:	Constant to make thermal conductivity dimensionless parameter (=1 (W/(m · K)))
$\bar{\lambda}$:	Dimensionless thermal conductivity, λ/λ^* (–)
$\bar{\lambda}_i$:	Term of dimensionless thermal conductivity (–)
μ:	Viscosity (Pa · s)
μ^*:	Constant to make viscosity dimensionless parameter (=1.00 × 10^{-6} (Pa · s))
$\bar{\mu}$:	Dimensionless viscosity, μ/μ^* (–)
$\bar{\mu}_i$:	Term of dimensionless viscosity (–)
ρ:	Density (kg/m^3)
ρ^*:	Constant to make density dimensionless parameter (kg/m^3)

B.2 Heavy Water

B.2.1 Introduction

IAPWS has summarized properties of heavy water. The minor revision of IAPS Formulation 1984 is the newest version of thermodynamic property formulation at present [13]. The original IAPS Formulation 1984 was adopted by International Association for the Properties of Steam (IAPS), former name of IAPWS [14]. The original formulation was the dimensionless version of the formulation [15]. The minor revision of the IAPS Formulation 1984 includes the change of temperature scale from the Practical Temperature Scale of 1968 (IPTS-68) to the International Temperature Scale of 1990 (ITS-90) [7]. Here, the revised version of IAPS Formulation 1984 is roughly introduced. Viscosity and thermal conductivity were obtained from a different formulation [16].

B.2.2 Thermodynamic Properties

Thermodynamic properties are shown based on equations in the revised version of IAPS Formulation 1984 [13]. The formulation uses the ITS-90 temperature scale [7]. Symbols are changed from the original literature to fit the ones in Appendix B.1. The basic equation is the (dimensionless) Helmholtz free energy. Applicable temperature and pressure range are 276.95 (triple point temperature) –800 K and 0 MPa–100 MPa. However, this method cannot be used in the region around critical point as shown below.

$$|T - T^*| \leq 10, \quad |\delta - 1| \leq 0.3$$

The basic equation is as follows:

$$\frac{\rho^* \cdot F(\rho, T)}{p^*} = \phi(\delta, \theta) = \phi_0 + \phi_1, \tag{B.48}$$

$$\phi_0 = (n_{00} + n_{01}\theta) \cdot \ln\theta + \sum_{j=2}^{7} n_{0j}\theta^{j-2} + n_{08}\theta \ln\delta, \tag{B.49}$$

$$\phi_1 = \theta\delta \cdot \left(\frac{1}{\theta} - \frac{1}{\theta_1}\right) \cdot \sum_{i=1}^{7}\left[\left(\frac{1}{\theta} - \frac{1}{\theta_i}\right)^{i-2} \cdot X_i(\delta)\right], \tag{B.50}$$

$$X_i(\delta) = \sum_{j=1}^{8} n_{ij} \cdot (\delta - \delta_i)^{j-1} + e^{-z_0\delta} \cdot (n_{i9} + n_{i10}\delta), \tag{B.51}$$

$$\theta = \frac{T}{T^*}, \quad \delta = \frac{\rho}{\rho^*}, \quad T^* = 643.847\,\text{K}, \quad \rho^* = 358\,\text{kg/m}^3, \quad p^* = 21.671\,\text{MPa},$$

Values of T^*, ρ^*, and p^* are similar to critical values, but not the same. It is noted that the unit of pressure is MPa. Table B.14 shows parameters n_{ij}. And θ_i and δ_i are listed in Table B.15.

Other thermodynamic and transport properties are calculated from the basic equation as follows:

$$\textit{Internal energy: } U = \left(\frac{p^*}{\rho^*}\right) \cdot (\phi - \theta\phi_\theta) \tag{B.52}$$

$$\textit{Enthalpy: } H = \left(\frac{p^*}{\rho^*}\right) \cdot (\phi - \theta\phi_\theta + \delta\phi_\delta) \tag{B.53}$$

$$\textit{Entropy: } S = \left(-\frac{p^*}{\rho^* T^*}\right) \cdot \phi_\theta \tag{B.54}$$

$$\textit{Gibbs energy: } G = \left(\frac{p^*}{\rho^*}\right) \cdot (\phi + \delta\phi_\delta) \tag{B.55}$$

$$\textit{Constant pressure heat capacity: } C_p = C_v + \left(\frac{p^*}{\rho^* T^*}\right) \cdot \left(\frac{\theta\delta\phi_{\delta\theta}^2}{2\phi_\delta + \delta\phi_{\delta\delta}}\right) \tag{B.56}$$

TABLE B.14

Parameters for ϕ_0 and X_i in Thermodynamic Properties of Heavy Water (Equations B.49 and B.51)

i	j	n_{ij}
0	0	5.399322597E – 03
0	1	–1.288399716E + 01
0	2	3.087155964E + 01
0	3	–3.827264031E + 01
0	4	4.424799189E – 01
0	5	–1.256336874E + 00
0	6	2.843343470E – 01
0	7	–2.401555088E – 02
0	8	4.415884023E + 00
1	1	1.15623643567E + 02
1	2	–1.61413392951E + 02
1	3	1.08543003981E + 02
1	4	–4.7134202123 8E + 01
1	5	1.49218685173E + 01
1	6	–3.60628259650E + 00
1	7	6.86743026455E – 01
1	8	–9.51913721401E – 02
1	9	–1.57513472656E + 03
1	10	–4.33677787466E + 02
2	1	6.07446060304E + 01
2	2	–9.27952190464E + 01
2	3	6.32086750422E + 01
2	4	–2.64943219184E + 01
2	5	9.05675051855E + 00
2	6	–5.78949005123E – 01
2	7	6.65590447621E – 01
2	8	–5.25687146109E – 02
2	9	–3.41048601697E + 03
2	10	–1.46971631028E + 03
3	1	4.44139703648E + 01
3	2	–5.80410482641E + 01
3	3	3.54090438940E + 01
3	4	–1.44432210128E + 01
3	9	–1.02135518748E + 03
3	10	–1.36324396122E + 03
4	1	1.57859762687E + 01
4	2	–1.94973173813E + 01
4	3	1.14841391216E + 01
4	4	–1.96956103010E + 00
4	9	–2.77379051954E + 02
4	10	–4.81991835255E + 02
5	1	–6.19344658242E + 01
5	2	7.91406411518E + 01

continued

TABLE B.14 (continued)

Parameters for ϕ_0 and X_i in Thermodynamic Properties of Heavy Water (Equations B.49 and B.51)

i	j	n_{ij}
5	3	−4.84238027539E + 01
5	4	1.91546335463E + 01
5	9	1.28039793871E + 03
5	10	1.86367898973E + 03
6	1	−7.49615505949E + 01
6	2	9.47388734799E + 01
6	3	−5.75266970986E + 01
6	4	1.73229892427E + 01
6	9	1.37572687525E + 03
6	10	2.31749018693E + 03
7	1	−2.60841561347E + 01
7	2	3.28640711440E + 01
7	3	−1.86464444026E + 01
7	4	4.84262639275E + 00
7	9	4.30179479063E + 02
7	10	8.22507844138E + 02

n_{ij} of the set i and j not shown in the table is 0.

$$\textit{Constant volume heat capacity: } C_v = \left(-\frac{p^*}{\rho^* T^*}\right)\cdot \theta\phi_{\theta\theta} \tag{B.57}$$

$$\textit{Specific volume: } v = \frac{1}{\rho^*\delta} \tag{B.58}$$

$$\textit{Pressure: } p = p^*\delta^2\phi_\delta \tag{B.59}$$

TABLE B.15

Parameters for ϕ_1 and X_i in Thermodynamic Properties of Heavy Water (Equations B.50 and B.51)

i	θ_i	δ_i
1	1.000038832E + 00	1.955307263E + 00
2	6.138578282E − 01	3.072625698E + 00
3	6.138578282E − 01	3.072625698E + 00
4	6.138578282E − 01	3.072625698E + 00
5	6.138578282E − 01	3.072625698E + 00
6	6.138578282E − 01	3.072625698E + 00
7	6.138578282E − 01	3.072625698E + 00

$$\text{Speed of sound: } w = \left[\left(\frac{p^*}{\rho^*}\right)\cdot\left(\frac{C_p}{C_v}\right)\cdot\left(2\delta\phi_\delta + \delta^2\phi_{\delta\delta}\right)\right]^{1/2} \tag{B.60}$$

Here, $\phi_\theta = \phi_{0\theta} + \phi_{1\theta}$, and so on. And

$$\phi_{0\delta} = \left(\frac{\partial\phi_0}{\partial\delta}\right)_\theta,\quad \phi_{0\delta\delta} = \left(\frac{\partial^2\phi_0}{\partial\delta^2}\right)_\theta,\quad \phi_{0\theta} = \left(\frac{\partial\phi_0}{\partial\theta}\right)_\delta,\quad \phi_{0\theta\theta} = \left(\frac{\partial^2\phi_0}{\partial\theta^2}\right)_\delta,\quad \phi_{0\delta\theta} = \left(\frac{\partial^2\phi_0}{\partial\delta\partial\theta}\right),$$

$$\phi_{1\delta} = \left(\frac{\partial\phi_1}{\partial\delta}\right)_\theta,\quad \phi_{1\delta\delta} = \left(\frac{\partial^2\phi_1}{\partial\delta^2}\right)_\theta,\quad \phi_{1\theta} = \left(\frac{\partial\phi_1}{\partial\theta}\right)_\delta,\quad \phi_{1\theta\theta} = \left(\frac{\partial^2\phi_1}{\partial\theta^2}\right)_\delta,\quad \phi_{1\delta\theta} = \left(\frac{\partial^2\phi_1}{\partial\delta\partial\theta}\right).$$

Calculation results of Helmholtz free energy, pressure, and constant volume heat capacity and uncertainty of specific volume are shown as a Table [13].

Nomenclature of B.2.2

Parts of dimensionless Helmholtz free energy and derivations of them are explained in the text in order to avoid this list too long.

- C_p: Constant pressure heat capacity (kJ/(kg · K))
- C_v: Constant volume heat capacity (kJ/(kg · K))
- F: Helmholtz free energy (kJ/kg)
- G: Gibbs free energy (kJ/kg)
- H: Enthalpy (kJ/kg)
- i, j, n_{ij}: Constants to calculate properties
- p^*: Constant to make pressure dimensionless parameter (=21.671 (MPa))
- S: Entropy (kJ/(kg · K))
- T: Temperature (K)
- T^*: Constant to make temperature dimensionless parameter (= 643.847 (K))
- U: Internal energy (kJ/kg)
- v: Specific volume (m³/kg)
- w: Speed of sound (m/s)
- X_i: Constant defined in Equation B.51 (–)
- z_0: Constant (= 1.5394) (–)
- δ: Dimensionless density, ρ/ρ^* (–)
- θ: Dimensionless temperature, T/T^* (–)
- ρ: Density (kg/m³)
- ρ^*: Constant to make density dimensionless parameter (= 358 (kg/m³))
- ϕ: Dimensionless Helmholtz free energy, $\rho^* F/p^*$ (–)

B.2.3 Transport Properties

The equations from IAPWS are introduced [16]. This data set is a minor revision of the release of IAPS in 1983. The formulation consistent with the revised release of IAPS Formulation 1984 [13] and the ITS-90 temperature scale [7].

B.2.3.1 Viscosity

Viscosity is calculated by Equations B.61 through B.63. Applicable temperature and pressure range of the equation is from 277 K (melting point) to 775 K, and 0 MPa to 100 MPa, respectively.

$$\bar{\mu} = \bar{\mu}_0(\theta) \cdot \bar{\mu}_1(\theta,\delta), \tag{B.61}$$

$$\bar{\mu}_0(\theta) = \frac{\theta^{1/2}}{\sum_{i=0}^{3} A_i/\theta^i}, \tag{B.62}$$

$$\bar{\mu}_1(\theta,\delta) = \exp\left\{\delta \cdot \sum_{i=0}^{5}\left[\left(\frac{1}{\theta}-1\right)^i \cdot \sum_{j=0}^{6} B_{ij} \cdot (\delta-1)^j\right]\right\}, \tag{B.63}$$

$$\theta = \frac{T}{T^*}, \quad \delta = \frac{\rho}{\rho^*}, \quad \bar{\mu} = \frac{\mu}{\mu^*} \quad T^* = 643.847\,\mathrm{K}, \ \rho^* = 358\,\mathrm{kg/m^3}, \mu^* = 55.2651\,\mu\mathrm{Pa \cdot s},$$

Parameters, A_i and B_{ij} are shown in Tables B.16 and B.17, respectively. Calculation result is in the literature [16]. Uncertainty is illustrated in the same literature.

B.2.3.2 Thermal Conductivity

Thermal conductivity is calculated by Equations B.64 through B.73. Temperature and pressure range is from 277 K (melting point) to 825 K, and 0 MPa to 100 MPa, respectively.

$$\bar{\lambda} = \lambda_0 + \Delta\lambda + \Delta\lambda_c + \Delta\lambda_L, \tag{B.64}$$

$$\lambda_0 = \sum_{i=0}^{5} A_i \theta^i, \tag{B.65}$$

$$\Delta\lambda = B_0 \cdot \left[1 - \exp(B_e \delta)\right] + \sum_{j=1}^{4} B_j \delta^j, \tag{B.66}$$

TABLE B.16

Parameter for μ_0 in Viscosity of Heavy Water (Equation B.62)

i	A_i
0	1.000000
1	0.940695
2	0.578377
3	−0.202044

TABLE B.17

Parameter for μ_1 in Viscosity of Heavy Water (Equation B.63)

i	j	B_{ij}
0	0	0.4864192
1	0	–0.2448372
2	0	–0.8702035
3	0	0.8716056
4	0	–1.051126
5	0	0.3458395
0	1	0.3509007
1	1	1.315436
2	1	1.297752
3	1	1.353448
0	2	–0.2847572
1	2	–1.037026
2	2	–1.287846
5	2	–0.02148229
0	3	0.07013759
1	3	0.4660127
2	3	0.2292075
3	3	–0.4857462
0	4	0.01641220
1	4	–0.02884911
3	4	0.1607171
5	4	–0.009603846
0	5	–0.01163815
1	5	–0.008239587
5	5	0.004559914
3	6	–0.003886659

B_{ij} of the set i and j not shown in the table is 0.

$$\Delta\lambda_c = C_1 \cdot f_1(\theta) \cdot f_2(\delta) \cdot \left(1 + \left[f_2(\delta)\right]^2 \cdot \left\{\frac{C_2 \cdot \left[f_1(\theta)\right]^4}{f_3(\theta)} + \frac{3.5 f_2(\delta)}{f_4(\theta)}\right\}\right), \tag{B.67}$$

$$\Delta\lambda_L = D_1 \cdot \left[f_1(\theta)\right]^{1.2} \cdot \left\{1 - \exp\left[-\left(\frac{\delta}{2.5}\right)^{10}\right]\right\}, \tag{B.68}$$

$$f_1(\theta) = \exp\left(C_{T1}\theta + C_{T2}\theta^2\right), \tag{B.69}$$

$$f_2(\delta) = \exp\left[C_{R1} \cdot (\delta - 1)^2\right] + C_{R2} \exp\left[C_{R3} \cdot (\delta - \delta_{r1})^2\right], \tag{B.70}$$

$$f_3(\theta) = 1 + \exp[60 \cdot (\tau - 1) + 20] \tag{B.71}$$

TABLE B.18

Parameters for Thermal Conductivity of Heavy Water (Equations B.65 through B.70)

i	A_i	B_i	C_i	C_{Ti}	C_{Ri}	δ_{ri}	D_i
0	1.00000	−167.310					
1	37.3223	483.656	35429.6	0.144847	−2.80000	0.125698	−741.112
2	22.5485	−191.039	5.0E + 09	−5.64493	−0.080738543		
3	13.0465	73.0358			−17.9430		
4	0.0	−7.57467					
5	−2.60735						

Be −2.50600

$$f_4(\theta) = 1 + \exp[100 \cdot (\tau - 1) + 15], \tag{B.72}$$

$$\tau = \frac{\theta}{|\theta - 1.1| + 1.1}, \tag{B.73}$$

$$\theta = \frac{T}{T^*}, \quad \delta = \frac{\rho}{\rho^*}, \quad \bar{\lambda} = \frac{\lambda}{\lambda^*} \quad T^* = 643.847\,\text{K},\ \rho^* = 358\,\text{kg/m}^3,\ \lambda^* = 0.742128\,\text{mW/(m} \cdot \text{K)}.$$

Parameters are summarized in Table B.18. Calculation result is in the literature [16]. Uncertainty is illustrated in the same literature.

Nomenclature of B.2.3

$A_i, B_i, B_e, C_i, C_{Ti}, C_{Ri}, D_i, \delta_{ri}$:	Parameters for calculation of thermal conductivity
B_{ij}:	Parameters for calculation of viscosity
f_i:	Term of $\Delta\lambda_c$, $\Delta\lambda_L$
T:	Temperature (K)
T^*:	Constant to make temperature dimensionless parameter (=643.847 (K))
$\Delta\lambda, \Delta\lambda_c, \Delta\lambda_L$:	Terms of dimensionless thermal conductivity (–)
δ:	Dimensionless density, ρ/ρ^* (–)
θ:	Dimensionless temperature, T/T^* (–)
λ:	Thermal conductivity (mW/(m · K))
λ^*:	Constant to make thermal conductivity (= 0.742128 (mW/(m · K)))
$\bar{\lambda}_0$:	Term of dimensionless thermal conductivity (–)
$\bar{\lambda}$:	Dimensionless thermal conductivity, λ/λ^* (–)
μ:	Viscosity (μPa · s)
μ^*:	Constant to make viscosity dimensionless parameter (=55.2651 (μPa · s))
$\bar{\mu}$:	Dimensionless viscosity, μ/μ^* (–)
$\bar{\mu}_i$:	Term of dimensionless viscosity (–)
ρ:	Density (kg/m^3)
ρ^*:	Constant to make density dimensionless parameter (=358 (kg/m^3))
τ:	Parameter defined in Equation B.73 (–)

B.3 Helium

B.3.1 Thermodynamic Properties

Helium is used as a coolant for HTGRs (see Chapter 10). Thermodynamic properties of helium are calculated based on the equation of state [17]. Though this equation is for helium 4, it can be used for natural helium because other isotopes are minute. The equation of state is written as Equation B.74. It is noted that the unit of p is atm, ρ is mol/L, and T is K. Gas constant R is 0.0820558 atm/(mol · K). This equation is applicable in the range from 15 K to 1500 K and from 0 atm to up to 1000 atm.

$$p = \rho RT \cdot \left[1 + B(b_i, T) \cdot \rho\right] + \sum_{j=1}^{8}\left[n_{1j}\rho^3 T^{\left(1.5-\frac{j}{2}\right)}\right] + \sum_{j=1}^{4}\left[n_{2j}\rho^4 T^{(1.5-j)}\right] + \sum_{j=1}^{6}\left[n_{3j}\rho^5 T^{\left(0.75-\frac{j}{4}\right)}\right]$$
$$+ \sum_{j=1}^{3}\left[n_{4j}\rho^3 e^{\gamma\rho^2} T^{(1.0-j)}\right] + \sum_{j=1}^{3}\left[n_{5j}\rho^5 e^{\gamma\rho^2} T^{(1.0-j)}\right] + \sum_{j=1}^{2}\left[n_{6j}\rho^6 T^{(1-j)}\right], \quad \text{(B.74)}$$

$$B(b_i, T) = \sum_{i=1}^{9} b_i T^{\left(1.5-\frac{i}{2}\right)}. \quad \text{(B.75)}$$

n_{ij} and b_i are shown in Tables B.19 and B.20, respectively. And $\gamma = -5.00 \times 10^{-4}$.

Thermodynamic and transport properties are calculated as follows. Constants p^* and ρ^* are added to the equations in the original literature [17] to converse units.

Enthalpy:

$$H = H_{T_0}^0 - \int_0^\rho \left[\frac{p}{\rho^2} - \frac{T}{\rho^2} \cdot \left(\frac{\partial p}{\partial T}\right)_\rho\right] \cdot \left(\frac{p^*}{\rho^*}\right) d\rho + \left(\frac{p}{\rho}\right) \cdot \left(\frac{p^*}{\rho^*}\right) - RT \cdot \left(\frac{p^*}{\rho^*}\right) + \int_{T_0}^{T} C_p^0 \, dT. \quad \text{(B.76)}$$

Entropy:

$$S = S_{T_0}^0 - R \cdot \left(\frac{p^*}{\rho^*}\right) \cdot \ln\left(\frac{\rho RT}{p_0}\right) + \int_0^\rho \left[\frac{R}{\rho} - \frac{1}{\rho^2} \cdot \left(\frac{\partial p}{\partial T}\right)_\rho\right] \cdot \left(\frac{p^*}{\rho^*}\right) d\rho + \int_{T_0}^{T} \frac{C_p^0}{T} \, dT. \quad \text{(B.77)}$$

Constant pressure heat capacity:

$$C_p = C_v + \frac{T \cdot \left[(\partial p/\partial T)_\rho \cdot p^*\right]^2}{(\partial p/\partial \rho)_T \cdot (p^*/\rho^*)} \cdot \left(\frac{1}{\rho^2}\right). \quad \text{(B.78)}$$

Constant volume heat capacity:

$$C_v = C_v^0 - \int_0^\rho \frac{T}{\rho^2} \cdot \left(\frac{\partial^2 p}{\partial T^2}\right)_\rho \cdot \left(\frac{p^*}{\rho^*}\right) d\rho. \quad \text{(B.79)}$$

TABLE B.19

Parameter for Pressure of Helium (Equation B.74)

i	j	n_{ij}
1	1	−3.6027735292E − 05
1	2	1.6079946555E − 03
1	3	−2.7441763615E − 02
1	4	1.4739506957E − 01
1	5	−4.3559344838E − 01
1	6	1.3447956078E + 00
1	7	−1.7040375125E + 00
1	8	9.0262674040E − 01
2	1	1.9661380688E − 06
2	2	1.7122932666E − 04
2	3	2.3051000563E − 04
2	4	−9.6564739100E − 04
3	1	−2.3326553271E − 07
3	2	4.0855110880E − 07
3	3	1.0900567964E − 05
3	4	−5.0060952775E − 05
3	5	1.1312765043E − 04
3	6	−1.2539843287E − 04
4	1	5.6875644111E − 03
4	2	−1.4438146625E − 01
4	3	3.3768874851E − 03
5	1	1.0754201218E − 06
5	2	−4.5264622308E − 05
5	3	3.8597388864E − 05
6	1	−1.4802195348E − 08
6	2	4.1721791119E − 07

TABLE B.20

Parameter for B_i in Pressure of Helium (Equation B.75)

i	b_i
1	−5.0815710041E − 07
2	−1.1168680862E − 04
3	1.1652480354E − 02
4	7.4474587998E − 02
5	−5.3143174768E − 01
6	−9.5759219306E − 01
7	3.9374414843E + 00
8	−5.1370239224E + 00
9	2.0804456338E + 00

T is considered as a constant in terms of integration by ρ in Equations. B.76, B.77, and B.79.

Speed of sound:

$$w = \left[\left(\frac{C_p}{C_v} \right) \cdot \left(\frac{\partial p}{\partial \rho} \right)_T \cdot \left(\frac{p^*}{\rho^*} \right) \right]^{1/2}. \quad \text{(B.80)}$$

Calculation results of thermodynamic properties are summarized in detail in a table of the literature [17]. Error of these properties should not be less than 1% except C_p in the transposed critical region [17].

Nomenclature of B.3.1

b_i: Constant for the equation of state
$B(b_i,T)$: Second Virial coefficient defined in Equation B.75 (L/mol)
C_p: Constant pressure heat capacity (J/(kg · K))
C_p^0: Constant pressure heat capacity of ideal gas (=(5R/2) · (p^*/ρ^*), 20.7858 (J/(kg · K)))
C_v: Constant volume heat capacity (J/(kg · K))
C_v^0: Constant volume heat capacity of ideal gas (=(3R/2) · (p^*/ρ^*), 12.4715 (J/(kg · K)))
H: Enthalpy (J/kg)
$H_{T_0}^0$: Reference enthalpy (= 2.1823×10^4 (J/kg))
n_{ij}: Constant for the equation of state
p: Pressure (atm)
p_0: Reference pressure (= 1 (atm))
p^*: Constant for pressure unit conversion (= 1.01325×10^5 (Pa/atm))
R: Gas constant (= 0.0820558 ((L · atm)/(mol · K)))
S: Entropy (J/(kg · K))
$S_{T_0}^0$: Reference entropy (= 9.3717×10^3 (J/(kg · K)))
T: Temperature (K)
T_0: Reference temperature (at normal boiling point, 4.22 (K))
w: Speed of sound (m/s)
γ: Constant for the pressure (= -5.00×10^{-4})
ρ: Density (mol/L)
ρ*: Constant for density unit conversion (= 4.0026 ((kg/m^3)/(mol/L)))

B.3.2 Transport Properties

Transport properties, viscosity, and thermal conductivity at low density are calculated based on kinetic theory of gas. Chapman–Enskog theory using Lennard–Jones potential was applied [18]. The calculated result is summarized in a table of the literature.

B.3.2.1 Viscosity

Viscosity is calculated as in Equations B.81 through B.91.

$$\mu(T) = \frac{5}{16} \cdot \left(\frac{mkT}{\pi} \right)^{1/2} \cdot \frac{f_\mu}{\sigma^2 \Omega^{(2,2)*}}, \quad \text{(B.81)}$$

$$f_\mu = 1 + \frac{3}{196} \cdot \left(8E^* - 7\right)^2, \tag{B.82}$$

$$E^* = 1 + \frac{T^*}{4} \cdot \frac{\mathrm{d}\left(\ln \Omega^{(2,2)*}\right)}{\mathrm{d}T^*}, \tag{B.83}$$

$$\Omega^{(2,2)*} = \left(\rho^*\right)^2 \alpha^2 \cdot \left[\sum_{i=0}^{4} a_i \left(\ln T^*\right)^{-i}\right], \tag{B.84}$$

$$\alpha = \ln V_0^* - \ln T^*, \tag{B.85}$$

$$a_0 = 1.04, \tag{B.86}$$

$$a_1 = 0, \tag{B.87}$$

$$a_2 = -33.0838 + \left(\alpha_{10}\rho^*\right)^{-2} \cdot \left[20.0862 + \left(\frac{72.1059}{\alpha_{10}}\right) + \left(\frac{8.27648}{\alpha_{10}}\right)^2\right], \tag{B.88}$$

$$a_3 = 101.571 - \left(\alpha_{10}\rho^*\right)^{-2} \cdot \left[56.4472 + \left(\frac{286.393}{\alpha_{10}}\right) + \left(\frac{17.7610}{\alpha_{10}}\right)^2\right], \tag{B.89}$$

$$a_4 = -87.7036 + \left(\alpha_{10}\rho^*\right)^{-2} \cdot \left[46.3130 + \left(\frac{277.146}{\alpha_{10}}\right) + \left(\frac{19.0573}{\alpha_{10}}\right)^2\right], \tag{B.90}$$

$$\alpha_{10} = \ln V_0^* - \ln 10. \tag{B.91}$$

Equation B.84 is applicable in the range of T^* above 10, that is, above 104.0 K. Accuracy of the viscosity in the range from 50 to 1000 K is 0.3% [18].

B.3.2.2 Thermal Conductivity

Thermal conductivity is obtained by the way similar to viscosity calculation.

$$\lambda(T) = \frac{75}{64} \cdot \left(\frac{k^3 T}{\pi m}\right)^{\frac{1}{2}} \cdot \frac{f_\lambda}{\sigma^2 \Omega^{(2,2)*}}, \tag{B.92}$$

$$f_\lambda = 1 + \frac{1}{42} \cdot \left(8E^* - 7\right)^2. \tag{B.93}$$

E^* and $\Omega^{(2,2)*}$ are the same value as those for viscosity.
Accuracy of the thermal conductivity in the range from 50 to 1000 K is 0.7% [18].

Nomenclature of 3.2

a_i: Parameter to calculate $\Omega^{(2,2)*}$
E^*: Dimensionless collision integral ratio (–)
f_λ: Higher order correction factor for thermal conductivity (–)
f_μ: Higher order correction factor for viscosity (–)
k: Boltzmann constant (= 1.380662×10^{-23} (J/K))
m: Mass of a helium atom (average isotopic composition) (=6.6466×10^{-27} (kg))
T: Temperature (K)
T^*: Standardized temperature, kT/ε (–)
V_0^*: High-temperature scaling parameter (= 8.50×10^5 (–))
α, α_{10}: Parameters defined in Equations B.85 and B.91, respectively
ε: Scaling parameter ((ε/k) = 10.40 (K))
λ: Thermal conductivity (W/(m · K))
μ: Viscosity (Pa · s)
π: Circle ratio (ca. 3.14)
ρ^*: High-temperature scaling parameter (= 0.0797 (–))
σ: Length scaling parameter (= 2.610×10^{-10} (m))
$\Omega^{(2,2)*}$: Reduced collision integral (–)

B.4 Sodium

The properties of liquid sodium for Sodium Fast Reactors (see Chapter 11) summarized by Argonne National Laboratory (ANL) [19] in 1995 are introduced here. Calculated results are summarized in tables of the literature. Uncertainty of properties except entropy is shown in Table 21 [19].

B.4.1 Thermodynamic Properties

B.4.1.1 Enthalpy

CODATA value along the saturation curve is recommended below 2000 K.

$$H_l(T) - H_s(298.15K) = -365.77 + 1.6582T - 4.2395 \times 10^{-4}T^2 + 1.4847 \times 10^{-7}T^3 + 2992.6T^{-1}$$

$$(371\ \text{K} \le T \le 2000\ \text{K}) \tag{B.94}$$

Average of liquid and vapor enthalpies minus one half of the enthalpy of vaporization is considered as the liquid enthalpy above 2000 K to the critical temperature, 2503.7 K. The equation is

$$H_l(T) - H_s(298.15\,K) = (2128.4 + 0.86496T) - \frac{1}{2}\cdot\left[393.37\cdot\left(1-\frac{T}{T_c}\right) + 4398.6\cdot\left(1-\frac{T}{T_c}\right)^{0.29302}\right]$$

$$(2000\ \text{K} < T < 2503.7\ \text{K}) \tag{B.95}$$

TABLE B.21

Uncertainty of Thermodynamic and Transport Properties of Liquid Sodium

Enthalpy

Temperature (K)	Uncertainty (%) $(\delta H)/H$
$370.98 \le T \le 1000$	1
$1000 \le T \le 1600$	$0.17 + 8.3 \times 10^{-4}T$
$1600 \le T \le 2000$	$-0.5 + 1.25 \times 10^{-3}T$
$2000 \le T \le 2400$	10
$2400 \le T \le 2500$	$-38 + 0.02T$

Heat Capacity

	Uncertainty (%)	
Temperature (K)	$(\delta C_p)/C_p$	$(\delta C_v)/C_v$
$370.98 \le T \le 1000$	2	5
$1000 \le T \le 1600$	3	10
$1600 \le T \le 2000$	20	40
$2000 \le T \le 2200$	30	65
$2200 \le T \le 2400$	35	80
$2400 \le T \le 2503$	50	90

Density

Temperature (K)	Uncertainty (%) $(\delta\rho)/\rho$
$370.98 \le T \le 700$	0.3
$700 \le T \le 1400$	0.4
$1500 \le T \le 2503$	$-32.22 + 0.0233T$

Speed of Sound

Temperature (K)	Uncertainty (%) $\delta w)/w$
$370.98 \le T \le 1700$ [(a)]	1
1700 [(a)] $\le T \le 2503$	$-48 + 0.029T$

[(a)] "1600" in original paper is considered as an error.

Viscosity

Temperature (K)	Uncertainty (%) $(\delta\mu)/\mu$
$370.98 \le T \le 1500$	$2.3 + 0.0018T$
$1500 \le T \le 2500$	$-10 + 0.01T$

Thermal Conductivity

Temperature (K)	Uncertainty (%) $(\delta\lambda)/\lambda$
$370.98 \le T \le 700$	5
$700 \le T \le 1100$	$-7.25 + 0.0175T$
$1100 \le T \le 1500$	$3.75 + 0.0075T$
$1500 \le T \le 2500$	15

B.4.1.2 Entropy

Entropy is obtained by the equation below [19]:

$$S_l(T) - S_l(T_m) = \int_{T_m}^{T} \frac{C_\sigma}{T} \mathrm{d}T \tag{B.96}$$

C_σ is calculated as Equation B.99 in Appendix B.4.1.3. Entropy calculation by a simpler equation, Equation B.97 [20], is also shown because Equation B.96 is very complicated.

$$S_l = -5.90356 + 1.51103 \ln T - 5.73462 \times 10^{-4} T + 1.57165 \times 10^{-7} T^2 - 3425.81 T^{-2}$$

$$(370.98 \text{ K (melting point)} \le T \le 1644.26 \text{ K}) \tag{B.97}$$

It is noted that the equation is based on other data sets than Appendices B.4.1.1 and B.4.1.3. Consistency with other properties is not guaranteed. Estimated error is 0.5%.

B.4.1.3 Heat Capacity

Constant pressure heat capacity is calculated as in these equations. "× 10³" is added to Equations B.98, B.99, and B.102, which are taken from corresponding equations in the original literature [19] in order to convert units.

$$C_p = C_\sigma + \frac{T\alpha_p\gamma_\sigma}{\rho} \times 10^3, \quad \text{(B.98)}$$

$$C_\sigma = \left(\frac{dH}{dT}\right) - \left(\frac{\gamma_\sigma}{\rho}\right) \times 10^3, \quad \text{(B.99)}$$

$$\alpha_p = \alpha_\sigma + \beta_T\gamma_\sigma, \quad \text{(B.100)}$$

$$\alpha_\sigma = -\left(\frac{1}{\rho}\right)\cdot\left(\frac{d\rho}{dT}\right), \quad \text{(B.101)}$$

$$\beta_T = \left[\frac{\beta_s C_\sigma + \left(\frac{T}{\rho}\right)\cdot\alpha_\sigma\cdot(\alpha_\sigma + \beta_s\gamma_\sigma)\cdot 10^3}{C_\sigma - \left(\frac{T}{\rho}\right)\cdot\gamma_\sigma\cdot(\alpha_\sigma + \beta_s\gamma_\sigma)\cdot 10^3}\right], \quad \text{(B.102)}$$

$$\beta_s = \beta_{s,m}\cdot\left(\frac{1+\theta/b_\beta}{1-\theta}\right), \quad \text{(B.103)}$$

$$\theta = \frac{T - T_m}{T_c - T_m}, \quad \text{(B.104)}$$

$$\gamma_\sigma = \frac{dp}{dT}, \quad \text{(B.105)}$$

$$\ln p = a + \frac{b}{T} + c\ln T. \quad \text{(B.106)}$$

Density ρ is calculated as shown in Appendix B.4.1.4.

Constants used in these equations are as follows. The constants have to be substituted to Equations B.98 through B.106 without converting unit.

$a = 11.9463$, $b = -12633.7$, $c = -0.4672$, $f = 275.32$ kg/m³, $g = 511.58$ kg/m³, $h = 0.5$, $b_\beta = 3.2682$, and $\beta_{s,m} = 1.717 \times 10^{-4}$ 1/(MPa).

Constant volume heat capacity is calculated as follows:

$$C_v = C_p\cdot\left(\frac{\beta_s}{\beta_T}\right). \quad \text{(B.107)}$$

β_s and β_T are obtained by the same equation as for constant pressure heat capacity.

B.4.1.4 Density

Fitting analysis of data from the melting point to 2201 K to the form of Equation B.108 is recommended in the literature [19].

$$\rho = \rho_c + f \cdot \left(1 - \frac{T}{T_c}\right) + g \cdot \left(1 - \frac{T}{T_c}\right)^h \quad (370.98\,\mathrm{K} \le T \le 2503.7\,\mathrm{K}) \tag{B.108}$$

$$f = 275.32\ \mathrm{kg/m^3}, \quad g = 511.58\ \mathrm{kg/m^3}, \quad \text{and } h = 0.5.$$

B.4.1.5 Speed of Sound

A fitting of data from literatures to a quadratic equation is recommended below 1773 K. "×10⁴" is added to the Equation B.110 taken from the original literature [19] to convert unit.

$$w = 2660.7 - 0.37667T - 9.0356 \times 10^{-5}T^2 \ (370.98\ \mathrm{K} \le T \le 1773\ \mathrm{K}) \tag{B.109}$$

$$w = \frac{1}{\sqrt{\rho\beta_s}} \times 10^4 \ (1773\ \mathrm{K} < T \le 2503.7\ \mathrm{K}) \tag{B.110}$$

β_s and ρ are calculated in Equations B.103 and B.108, respectively.

B.4.2 Transport Properties

The recommended equations for transport properties, viscosity, and thermal conductivity are as shown below.

Viscosity:

$$\ln\mu = -6.4406 - 0.3958\ln T + \frac{556.835}{T}. \ (370.98\,\mathrm{K} \le T \le 2500\,\mathrm{K}) \tag{B.111}$$

Thermal conductivity:

$$\lambda = 124.67 - 0.11381T + 5.5226 \times 10^{-5}T^2 - 1.1842 \times 10^{-8}T^3$$
$$(370.98\ \mathrm{K} \le T \le 2500\ \mathrm{K}) \tag{B.112}$$

Nomenclature of B.4

a, b, c: Parameters for pressure (Equation B.106) ($a = 11.9463$, $b = -12633.7$, and $c = -0.4672$)
b_β: Constant for adiabatic compressibility (= 3.2682)
C_p: Constant pressure heat capacity (kJ/(kg · K))
C_v: Constant volume heat capacity (kJ/(kg · K))
C_σ: Heat capacity along the saturation curve (kJ/(kg · K))
f, g, h: Parameters for density calculation (Equation B.108) ($f = 275.32$ (kg/m³), $g = 511.58$ (kg/m³), and $h = 0.5$)

H: Enthalpy (kJ/kg)
p: Pressure (MPa)
S: Entropy (kJ/(kg · K))
T: Temperature (K)
T_c: Critical temperature (= 2503.7 (K))
T_m: Melting point (= 370.98 (K))
w: Speed of sound (m/s)
α_p: Thermal expansion coefficient (1/K)
α_σ: Thermal expansion coefficient along the saturation curve (1/K)
β_s: Adiabatic compressibility (1/(MPa))
$\beta_{s,m}$: Constant for adiabatic compressibility (=1.717×10^{-4} (1/(MPa)))
β_T: isothermal compressibility (1/(MPa))
γ_σ: Temperature derivative of the pressure with respect to temperature along the saturation curve ((MPa)/K)
$(\delta C_p)/C_p$, $(\delta C_v)/C_v$, $(\delta H)/H$, $(\delta w)/w$, $(\delta\lambda)/\lambda$, $(\delta\rho)/\rho$, $(\delta\mu)/\mu$: Uncertainty of properties calculation (%)
θ: Dimensionless temperature defined in Equation B.104 (–)
λ: Thermal conductivity (W/(m · K))
μ: Viscosity (Pa · s)
ρ: Density of liquid (kg/m^3)
ρ_c: Critical density (=219 (kg/m^3))

Subscripts

l: Liquid
s: Solid

B.5 Liquid Lead

Liquid lead is utilized as coolant in the reactor for Secure Transportable Autonomous for Hydrogen production (STAR-H2) (see Chapter 14). Not all properties are shown here.

B.5.1 Thermodynamic Properties

B.5.1.1 Enthalpy and Constant Pressure Heat Capacity

Empirical equations of enthalpy and heat capacity from a literature of measurement data [21] are shown. "$\times 10^{-3}$" is added to corresponding equations in the original literature to convert units.

Enthalpy of solid phase (face centered cubic) is

$$h_s(T) - h_s(298.15K) = (-7850.973 + 24.489T + 4.477 \times 10^{-3}T^2 + 4.520 \times 10^4 T^{-1}) \times 10^{-3}$$

$$(298.15\text{ K} \le T \le 600.65\text{ K (melting point)}) \qquad \text{(B.113)}$$

Uncertainty is less than 1%.

Enthalpy of liquid phase is

$$h_l(T) - h_s(298.15K) = \begin{pmatrix} -7389.27 + 36.287T - 5.140 \times 10^{-3}T^2 + 3.158 \times 10^5 T^{-1} \\ + 1.371 \times 10^{-6}T^3 - 1.0875 \times 10^{-10}T^4 \end{pmatrix} \times 10^{-3}$$

$$(600.65\text{ K (melting point)} \le T \le 3600\text{ K}) \quad \text{(B.114)}$$

Uncertainties for enthalpy are 0.4 kJ/mol (1.6%) at 1000 K and 2.5 kJ/mol (6%) at 1500 K. The enthalpy of melting at the melting point is

$$\Delta h_{melt} = 4.812 \pm 0.040 \quad \text{(B.115)}$$

Constant pressure heat capacity of solid (face centered cubic) is

$$c_p = (24.489 + 8.954 \times 10^{-3}T - 4.52 \times 10^4 T^{-2}) \times 10^{-3}$$

$$(298.15\text{ K} \le T \le 600.65\text{ K (melting point)}) \quad \text{(B.116)}$$

Uncertainty is less than 1%.
Constant pressure heat capacity of liquid is

$$c_p = (36.287 - 1.0280 \times 10^{-2}T - 3.158 \times 10^5 T^{-2} + 4.113 \times 10^{-6}T^2 - 4.35 \times 10^{-10}T^3) \times 10^{-3}$$

$$(600.65\text{ K (melting point)} \le T \le 1300\text{ K}) \quad \text{(B.117)}$$

Difference of the heat capacity calculation with other data is 0.1% at melting point and 3% at 2000 K.

Calculation results of enthalpy and heat capacity are shown in figures and tables of the literature [21].

B.5.1.2 Other Thermodynamic Properties

Other thermodynamic property in literature [22] is summarized. The data in this literature are the compilation of several literatures.

Density:

$$\rho = 11367 - 1.1944T \ (600.65\text{ K (melting point)} \le T \le 2000\text{ K}) \quad \text{(B.118)}$$

Difference of the value by Equation B.118 from the data to make the equation is less than 0.7%.

Speed of sound:

$$w = 1951.75 - 0.3423T + 7.635 \times 10^{-5}T^2$$

$$(600.65\text{ K (melting point)} \le T \le 2000\text{ K}) \quad \text{(B.119)}$$

Estimated error is estimated as ca. 0.05%.

B.5.2 Transport Properties

Transport properties of liquid lead in the same literature as Appendix B.5.1.2 [22] are shown.

Viscosity:

$$\mu = 4.55 \times 10^{-4} \cdot \exp\left(\frac{1069}{T}\right) \quad (600.65\,\text{K (melting point)} \le T \le 1470\,\text{K}) \tag{B.120}$$

Estimated error is 4%.

Thermal conductivity:

$$\lambda = 9.2 + 0.011T \; (600.65\text{ K (melting point)} \le T \le 1300\text{ K}) \tag{B.121}$$

Estimated error is estimated as ca. 10%.

Nomenclature of B.5

h: Enthalpy (kJ/mol)
c_p: Constant pressure heat capacity (kJ/(mol · K))
T: Temperature (K)
w: Speed of sound (m/s)
Δh_{melt}: Melting enthalpy (kJ/mol)
λ: Thermal conductivity (W/(m · K))
μ: Viscosity (Pa · s)
ρ: Density (kg/m³)

Subscripts

l: Liquid
s: Solid

B.6 Molten Salts

B.6.1 Introduction

2LiF–BeF_2 molten salt is a candidate coolant of molten salt reactor, and so is NaF–ZrF_4 (see Chapter 13). Because data of thermodynamic and transport properties of the molten salt mixtures are not sufficiently known, consistency of several different properties is not necessarily kept.

B.6.2 LiF–BeF_2 System

B.6.2.1 Thermodynamic Properties

Enthalpy, entropy, and constant pressure heat capacity of pure LiF and BeF_2 liquids are shown as in equations below [23].

$$h(T) = h_{ref} + \int_{298.15}^{T} c_p \, dT, \tag{B.122}$$

TABLE B.22

Parameters for Thermodynamic Properties of Pure LiF and BeF_2 Liquids (Equations B.122 through B.124)

	$\Delta_f h°$(298.15 K) (kJ/mol)	s°(298.15 K) (kJ/(mol K))	Parameters for c_p			
			a	*b*	*c*	*d*
LiF(l)	−599.931	4.3085E − 02	6.4219E − 02	0	0	0
BeF_2(l)	−1021.658	6.0495E − 02	4.0984E − 02	4.4936E − 05	0	0

$$s(T) = s^{\circ}(298.15K) + \int_{298.15}^{T} \frac{c_p}{T} dT, \qquad \text{(B.123)}$$

$$c_p = a + bT + cT^{-2} + dT^2 \qquad \text{(B.124)}$$

Reference state enthalpy h_{ref}, has the value as the standard formation enthalpy at 298.15 K and 1 bar, $\Delta_f h°$(298.15 K). $\Delta_f h°$(298.15 K), s°(298.15 K), and parameters for c_p are shown in Table B.22.

For the mixture of LiF–BeF_2 liquid, constant pressure heat capacity is estimated as 2.39×10^{-3} kJ/(g · K) for the mixture of LiF:BeF_2 = 66:34 mol% [24]. Temperature dependence of the heat capacity is not clear. An empirical method of Dulong and Petit (Equation B.125), which assumes that each atom in a mixture contributes 8 cal/(mol · K) (ca. 3.3472×10^{-2} kJ/(mol · K)) to heat capacity, is accurate to 10% for the mixture of alkali halides and BeF_2 [25].

$$c_p = \sum_i x_i N_i \times \left(3.3472 \times 10^{-2}\right). \qquad \text{(B.125)}$$

Temperature- and composition-dependent property of Gibbs energy is only known. Gibbs energy is calculated as the sum of that of pure components, ideal mixture Gibbs energy, and excess Gibbs energy [26].

$$g_{\text{LiF-BeF}_2}(T) = x_{\text{LiF}} g_{\text{LiF}}(T) + x_{\text{BeF}_2} g_{\text{BeF}_2}(T) + x_{\text{LiF}} \cdot \left(R \times 10^{-3}\right) \times T \ln x_{\text{LiF}} + x_{\text{BeF}_2} \cdot \left(R \times 10^{-3}\right) \cdot T \ln x_{\text{BeF}_2} + {}_{xs}g, \qquad \text{(B.126)}$$

$$g_i(T) = h_i(T) - Ts_i(T). \qquad \text{(B.127)}$$

Redlich–Kister polynomials, Equation B.128 [23], are applied for excess Gibbs energy [26].

$$_{xs}g = x_{\text{LiF}} x_{\text{BeF}_2} \sum_{k=0}^{N} L_{\text{LiF-BeF}_2,k} \cdot \left(x_{\text{LiF}} - x_{\text{BeF}_2}\right)^k, \qquad \text{(B.128)}$$

$$L_{\text{LiF-BeF}_2,k} = p_{\text{LiF-BeF}_2,k} + q_{\text{LiF-BeF}_2,k} T, \qquad \text{(B.129)}$$

$p_{\text{LiF-BeF}_2,k}$ and $q_{\text{LiF-BeF}_2,k}$ are in Table B.23.

TABLE B.23
Parameters for L_k in Excess Gibbs Energy of LiF–BeF_2 Mixture Liquid (Equation B.129)

K	$p_{LiF-BeF_{2,k}}$ (kJ/mol)	$q_{LiF-BeF_{2,k}}$ (kJ/(molK))
0	−15.58	−1.1645E − 02
1	71.32	−6.3487E − 02
2	4.8058	0

Density of pure LiF, BeF_2, and LiF–BeF_2 mixtures of some specific compositions are in Equation B.130 [27].

$$\rho = A_{\rho BeF_2-LiF} - B_{\rho BeF_2-LiF}T. \tag{B.130}$$

Parameters A and B are shown in Table B.24. Uncertainty of data of 100 mol% BeF_2 and 100 mol% LiF is 0.5%. For the mixture of composition not shown in Table B.24, approximation of molar average of single component molar volume is applicable with accuracy better than 5% [25].

$$\rho = \left(\frac{\sum_i x_i M_i}{\sum_i x_i v_i} \right) \times 10^{-3}. \tag{B.131}$$

B.6.2.2 Transport Properties

Viscosity of pure LiF, BeF_2, and mixture of LiF–BeF_2 of some specific compositions are in Equation B.132 [27]

$$\mu = A_{\mu BeF_2-LiF} \cdot \exp\left(\frac{B_{\mu BeF_2-LiF}}{RT} + \frac{D_{\mu BeF_2-LiF}}{T^2} \right). \tag{B.132}$$

Parameters A, B, and D are summarized in Table B.25. Uncertainty of data of pure LiF and BeF_2 is 10%.

Thermal conductivity is not investigated enough. A review of literatures recommends temperature-independent value of thermal conductivity of LiF–BeF_2 mixtures of

TABLE B.24
Parameter for Density of BeF_2–LiF Mixture Liquid (Equation B.130)

Composition (mol%)		Parameter		
BeF_2	LiF	$A_{\rho BeF_2-LiF}$	$B_{\rho BeF_2-LiF}$	Temperature Range (K)
100	0	1.972E + 03	1.45E − 02	$1073 \leq T \leq 1123$
89.2	10.8	2.075E + 03	1.48E − 01	$1020 \leq T \leq 1130$
74.9	25.1	2.158E + 03	2.39E − 01	$860 \leq T \leq 1130$
50.2	49.8	2.349E + 03	4.24E − 01	$930 \leq T \leq 1130$
34	66	2.413E + 03	4.88E − 01	$800 \leq T \leq 1080$
0	100	2.3581E + 03	4.902E − 01	$1149 \leq T \leq 1320$

TABLE B.25

Parameters for Viscosity of BeF_2–LiF Mixture Liquid (Equation B.132)

Composition (mol%)		Parameter			Temperature Range (K)
BeF_2	LiF	$A_{\mu BeF_2\text{–}LiF}$	$B_{\mu BeF_2\text{–}LiF}$	$D_{\mu BeF_2\text{–}LiF}$	
0	100	1.8359E − 04	2.183201701E + 04	0	$1125 \le T \le 1317$
36.00	64.00	5.94E − 05	3.82842405E + 04	0	$740 \le T \le 860$
45.00	55.00	2.07E − 05	4.95812295E + 04	0	$700 \le T \le 820$
50.00	50.00	8.45E − 06	5.87025021E + 04	0	$660 \le T \le 840$
55.01	44.99	6.27E − 06	6.46857222E + 04	0	$680 \le T \le 840$
60.00	40.00	4.21E − 06	7.24680924E + 04	0	$720 \le T \le 840$
65.00	35.00	3.11E − 06	8.03759847E + 04	0	$740 \le T \le 980$
70.00	30.00	2.02E − 06	8.9873823 6E + 04	0	$760 \le T \le 940$
75.00	25.00	9.2E − 07	1.026352371E + 05	0	$780 \le T \le 960$
79.99	20.01	5.98E − 07	1.145598366E + 05	0	$840 \le T \le 980$
85.00	15.00	4.57E − 07	1.254384186E + 05	0	$820 \le T \le 1000$
90.02	9.98	1.71E − 07	1.441412115E + 05	0	$880 \le T \le 1120$
91.02	8.98	1.99E − 07	1.462750872E + 05	0	$820 \le T \le 1100$
93.01	6.99	1.05E − 07	1.569444657E + 05	0	$860 \le T \le 1100$
94.91	5.09	7.69E − 08	1.647268359E + 05	0	$840 \le T \le 1100$
96.01	3.99	6.05E − 08	1.709192595E + 05	0	$880 \le T \le 1080$
97.00	3.00	2.31E − 08	1.835551509E + 05	0	$880 \le T \le 1180$
98.01	1.99	6.62E − 09	1.98324918E + 05	0	$900 \le T \le 1240$
99.01	0.99	1.53E − 09	2.174461179E + 05	0	$980 \le T \le 1240$
100	0	7.603E − 10	2.200402413E + 05	1.471000E + 06	$847 \le T \le 1252$

LiF:BeF_2 = 66:34 mol% is 1.1 W/(K · m) [24]. And an empirical temperature- and composition-dependent equation, Equation B.133, is proposed [28].

$$\lambda = -0.34 + 0.5 \times 10^{-3} T + 32.0 \cdot \left(\frac{\sum_i x_i}{\sum_i x_i M_i} \right). \tag{B.133}$$

B.6.3 NaF–ZrF_4 System

B.6.3.1 Thermodynamic Properties

Mixing enthalpy of liquid NaF and ZrF_4, Δh_{mix}, and enthalpy of mixture of them, $\Delta h_{mixture}$, at 900°C are experimentally measured [29]. The relationship of these two properties is shown in the equation below.

$$\Delta h_{mixture,NaF\text{-}ZrF_4}(900^\circ C) = \left[h_{NaF,S}(900^\circ C) - h_{NaF,S}(25^\circ C) \right] + \left[h_{ZrF_4,S}(900^\circ C) - h_{ZrF_4,S}(25^\circ C) \right]$$
$$+ \Delta h_{m.,NaF}(900^\circ C) + \Delta h_{m,ZrF_4}(900^\circ C) + \Delta h_{mix,NaF\text{-}ZrF_4}(900^\circ C)$$

TABLE B.26

Enthalpies of Pure NaF and ZrF_4 (Equation B.134)

	$h_S(900°C) - h_S(25°C)$, (kJ/mol)	Melting Enthalpy at 900°C, Δh_m (kJ/mol)
NaF	49.09	32.84
ZrF_4	110.27	65.24

(B.134)

The values used in Equation B.134 are shown in Tables B.26 and B.27. Temperature dependence of mixing enthalpy is not shown because heat capacities data is not known enough.

Constant pressure heat capacity of NaF–ZrF_4 mixture of NaF:ZrF_4 = 57:43 mol% is estimated 1.066×10^{-3} kJ/(g · K) by a Molecular Dynamics simulation in the range between 850 K and 1100 K [30]. For other composition, the empirical Dulong and Petit method (Equation B.125) (see Appendix B.6.2.1) is applicable to the value of alkali halides and ZrF_4 with accuracy of 10% [25].

Density of NaF–ZrF_4 mixture at 1323 K is as shown in Equation B.135 [27].

$$\rho = 1.91 \times 10^3 + 2.307 \times 10 C_{ZrF_4} + 1.019 C_{ZrF_4}^2 - 8.043 \times 10^{-2} C_{ZrF_4}^3 + 1.364 \times 10^{-3} C_{ZrF_4}^4 \quad (B.135)$$

Composition range of ZrF_4 is 0–25 mol%.

Temperature dependence of density of composition of NaF:ZrF_4 = 57:43 mol% is as in Equation B.136 [25].

$$\rho = 3.650 \times 10^3 - 8.8 \times 10^{-1}\, T \quad (B.136)$$

B.6.3.2 Transport Properties

Viscosity of the mixture is shown below [27].

$$\mu = A_{\mu \text{NaF-ZrF}_4} \cdot \exp\left(\frac{B_{\mu \text{NaF-ZrF}_4}}{RT}\right). \quad (B.137)$$

Parameters A and B are shown in Table B.28. Uncertainty of data of pure NaF is 1%.

TABLE B.27

Mixing Enthalpy of Liquid NaF and ZrF_4 and Enthalpy of NaF–ZrF_4 Mixture Liquid (Equation B.134)

Composition (mol%)		Mixing Enthalpy at 900°C	Enthalpy of Mixture at 900°C
NaF	ZrF_4	$\Delta h_{mix,NaF-ZrF_4}$ (kJ/mol)	$\Delta h_{mixture,NaF-ZrF_4}$ (kJ/mol)
29	71	−10.81 ± 1.72	137.54 ± 1.83
50	50	−23.85 ± 1.64	104.83 ± 1.58
61	39	−31.79 ± 1.45	86.85 ± 1.33
67	33	−33.30 ± 1.34	79.73 ± 1.30
73	27	−33.56 ± 1.99	73.65 ± 1.46
80	20	−25.15 ± 2.06	75.21 ± 2.03

TABLE B.28

Parameters for Viscosity of NaF–ZrF_4 Mixture Liquid (Equation B.137)

Composition (mol%)		Parameter		Temperature Range (K)
NaF	ZrF_4	$A_{\mu NaF-ZrF_4}$	$B_{\mu NaF-ZrF_4}$	
100	0	1.197E – 04	2.646842682E + 04	$1273 \le T \le 1373$
53	47	7.666E – 05	3.323406801E + 04	$873 \le T \le 1073$
50	50	2.993E – 05	3.243532905E + 04	$873 \le T \le 1073$

Thermal conductivity data of the mixture is not known. The recommended empirical equation, Equation B.133 (see Appendix B.6.2.2) has to be used at present.

Nomenclature of B.6

A, B, D: Parameters for calculation of density and viscosity
C: Composition of a component (mol%)
c_p: Constant pressure heat capacity (kJ/(K · mol))
g: Gibbs energy (kJ/mol)
h: Enthalpy (kJ/mol)
h_{ref}: Reference state enthalpy (kJ/mol)
$L_{i,k}$: Intermolecular parameter defined in Equation B.129 (kJ/mol)
M: Molar weight (g/mol)
N: Number of atoms of a component (2 for alkali halides, 3 for BeF_2, and 5 for ZrF_4) (–)
R: Gas constant (= 8.31441 (J/(K · mol)))
T: Temperature (K)
v: Molar volume (m^3/mol)
x: Molar composition (– (mole fraction))
Δh_m: Melting enthalpy (kJ/mol)
Δh_{mix}: Mixing enthalpy of two liquids (kJ/mol)
$\Delta h_{mixture}$: Enthalpy of mixture compound compared with the standard enthalpy of pure components (kJ/mol)
λ: Thermal conductivity (W/(K · m))
μ: Viscosity (Pa · s)
ρ: Density (kg/m^3)

Subscripts

$_i$: Component
$_m$: Melting
$_L$: Liquid
$_S$: Solid
$_\rho$: Density
$_\mu$: Viscosity

References

1. Cognet, G., A. Efanov, V. Fortov, J. K. Fink, K. Froment, G. Gromov, I. S. Hwang, et al. Thermophysical properties of light and heavy water, In *Thermophysical Properties Database of Materials for Light Rater Reactors and Heavy Water Reactors. IAEA-TECDOC-1496*, 345–375, 2006.
2. Wagner, W. and A. Pruss. The IAPWS formulation 1995 for the thermodynamic properties of ordinary water substance for general and scientific use. *J. Phys. Chem. Ref. Data*, 31, 387–535, 2002.
3. International Association for the Properties of Water and Steam (eds.). *Revised Release on the IAPWS Formulation 1995 for the Thermodynamic Properties of Ordinary Water Substance for General and Scientific Use*, The International Association for the Properties of Water and Steam, 2009.
4. Wagner, W., J. R. Cooper, A. Dittmann, J. Kijima, H.-J. Kretzschmar, A. Kruse, R. Mareš, et al. The IAPWS industrial formulation 1997 for the thermodynamic properties of water and steam. *J. Eng. Gas Turbines Power*, 122, 150–182, 2000.
5. The Japan Society of Mechanical Engineers (eds.). *1999 JSME Steam Tables*, 5th edition, The Japan Society of Mechanical Engineers, 1999, The Japan Society of Mechanical Engineers.
6. International Association for the Properties of Water and Steam (eds.). *Revised Release on the IAPWS Industrial Formulation 1997 for the Thermodynamic Properties of Water and Steam*, The International Association for the Properties of Water and Steam, 2007.
7. Preston-Thomas, H. The international temperature scale of 1990 (ITS-90), *Metrologia*, 27, 3–10, 1990.
8. International Association for the Properties of Water and Steam (eds.). *Uncertainties in Enthalpy for the IAPWS Formulation 1995 for the Thermodynamic Properties of Ordinary Water Substance for General and Scientific Use (IAPWS-95) and the IAPWS Industrial Formulation 1997 for the Thermodynamic Properties of Water and Steam (IAPWS-IF97)*, The International Association for the Properties of Water and Steam, 2003.
9. International Association for the Properties of Water and Steam (eds.). *Release on the IAPWS Formulation 2008 for the Viscosity of Ordinary Water Substance*, The International Association for the Properties of Water and Steam, 2008.
10. Sengers, J. V. and J. T. R. Watson. Improved international formulations for the viscosity and thermal conductivity of water substance. *J. Phys. Chem. Ref. Data*, 15, 1291–1314, 1986.
11. International Association for the Properties of Water and Steam (eds.). *Revised Release on the Pressure along the Melting and Sublimation Curves of Ordinary Water Substance*, The International Association for the Properties of Water and Steam, 2008.
12. International Association for the Properties of Water and Steam (eds.). *Revised Release on the IAPS Formulation 1985 for the Thermal Conductivity of Ordinary Water Substance*, The International Association for the Properties of Water and Steam, 2008.
13. International Association for the Properties of Water and Steam (eds.). *Revised Release on the IAPS Formulation 1984 for the Thermodynamic Properties of Heavy Water Substance*, The International Association for the Properties of Water and Steam, 2005.
14. Kestin, J. and J. V. Sengers. New international formulations for the thermodynamic properties of light and heavy water. *J. Phys. Chem. Ref. Data*, 15, 305–320, 1986.
15. Hill, P. G., R. D. Chris MacMillan, and V. Lee. A fundamental equation of state for heavy water. *J. Phys. Ref. Data*, 11, 1–14, 1982.
16. International Association for the Properties of Water and Steam (eds.). *Revised Release on Viscosity and Thermal Conductivity of Heavy Water Substance*. The International Association for the Properties of Water and Steam, 2007.
17. McCarty, R. D. Thermodynamics properties of helium 4 from 2 to 1500 K at pressures to 10^8 Pa. *J. Phys. Chem. Ref. Data*, 2, 923–1041, 1973.

18. Kestin, J., K. Knierim, E. A. Mason, B. Najafi, S. T. Ro, and M. Waldman. Equilibrium and transport properties of the noble gases and their mixtures at low density, *J. Phys. Chem. Ref. Data*, 13, 229–303, 1984.
19. Fink, J. K. and L. Leibowitz. Thermodynamic and transport properties of sodium liquid and vapor. *ANL/RE-95/2*, Argonne National Laboratory, 1995.
20. Fink, J. K. and L. Leibowitz. Calculation of thermophysical properties of sodium, *Conf-8106164-5*, p. 8, Argonne National Laboratory, 1981.
21. Gurvich, L. V. and I. V. Veytes. Lead and its compounds. In C. B. Alcock (ed.) *Thermodynamic Properties of Individual Substances: Elements C, Si, Ge, Sn, Pb, and Their Compounds*, Vol. 2, Parts 1 and 2, 4th edn, pp. 399–451, Hemisphere, Washington, 1990.
22. Sobolev, V. and G. Benamati. Thermophysical and electric properties. In *Handbook on Lead-Bismuth Eutectic Alloy and Lead Properties, Materials Compatibility, Thermal Hydraulics and Technologies*, 2007 edn., NEA-06195, pp. 25–99, Nuclear Energy Agency, 2007.
23. van der Meer, J. P. M., R. J. M. Konings, M. H. G. Jacobs, and H. A. J. Oonk. A miscibility gap in LiF–BeF_2 and LiF–BeF_2–ThF_4. *J. Nucl. Mater.*, 344, 94–99, 2005.
24. Beneš, O. and R. J. M. Konings, Thermodynamic properties and phase diagrams of fluoride salts for nuclear applications, *J. Fluorine Chem.*, 130, 22–29, 2009.
25. Williams, D. F., L. M. Toth, and K. T. Clarno. Assessment of candidate molten salt coolants for the advanced high-temperature reactor (AHTR) , pp. 18 and 20. *ORNL/TM-2006/12*, Oak Ridge National Laboratory, 2006.
26. Beneš, O. and R. J. M. Konings. Thermodynamic study of LiF–BeF_2–ZrF_4–UF_4 system. *J. Alloys Compd.*, 452, 110–115, 2008.
27. Janz, G. J. Thermodynamic and transport properties for molten salts: Correlation equations for critically evaluated density, surface tension, electrical conductance, and viscosity data. *J. Phys. Chem. Ref. Data*, 17, pp. 21, 65, 273 and 291, Supplement 2, 1988.
28. Khokhlov, V., V. Ignatiev, and V. Afonichkin. Evaluating physical properties of molten salt reactor fluoride mixtures. *J. Fluorine Chem.*, 130, 30–37, 2009.
29. Lin, I-Ching, A. Navrotsky, J. Ballato, and R. E. Riman. High-temperature calorimetric study of glass-forming fluorozirconates. *J. Non-Crystalline Solids*, 215, 113–124, 1997.
30. Salanne, M., C. Simon, P. Turq, and P. A. Madden. Heat-transport properties of molten fluorides: Determination from first-principles. *J. Fluorine Chem.*, 130, 38–44, 2009.

Index

B

D

E

I

M

N

U

V

W

X

Y

Z